W9-AEZ-885

THE MINERAL RESOURCES OF AFRICA

THE MINERAL RESOURCES OF AFRICA

by

NICOLAS DE KUN
Columbia University in the City of New York, N.Y.

ELSEVIER PUBLISHING COMPANY
AMSTERDAM LONDON NEW YORK
1965

ELSEVIER PUBLISHING COMPANY
335 JAN VAN GALENSTRAAT, P.O. BOX 211, AMSTERDAM

AMERICAN ELSEVIER PUBLISHING COMPANY, INC.
52 VANDERBILT AVENUE, NEW YORK, N.Y. 10017

ELSEVIER PUBLISHING COMPANY LIMITED
RIPPLESIDE COMMERCIAL ESTATE
BARKING, ESSEX

LIBRARY OF CONGRESS CATALOG CARD NUMBER 64-14180

WITH 136 ILLUSTRATIONS AND 172 TABLES

PRINTED IN THE NETHERLANDS

A smelter rises on the heights of Kalabi
A large smelter inherited from our father Lupadila
A smelter in which copper flows and undulates.
XVIII–XIXth century song of African miners

In the day-time you felt that you had got high up, near to the sun, but the early mornings and evenings were limpid and restful and the nights were cold.
The geographical position, and the height of the land combined to create a landscape that had not its like in all the world.
Isak Dinesen (Karen Blixen): Out of Africa

Preface

More than one and a half million years ago ancestors of man excavated stone in Africa to make weapons and tools. Preceding most other continents, metals appeared in Africa before several millenia, and more than a thousand years ago the gold trade flourished on the coasts. Ever since mining played a special role in the lifes of the Africans.

Africa has again come of age, and the main base of the new national economies, at least in the wealthier countries, are minerals and metals. They are their greatest material contribution to the rest of the world. For the first time this book endeavours to describe in detail this mineral wealth and these sources of energy, their development and geology.

In the elaboration of this volume, it is a pleasure to acknowledge helpful information received from directors and members of geological surveys and mines departments, and notably of H. Besairie, C. Boocock, R. Causse, C. Chatue Kamga, J. Colonna Cimera, O. Dumaz, B. Fabre, G. Faglin, J. Giri, J. Grez, Ch. Kanyamaranga, H. Martin, Ph. Morin, R. K. Puri, J. G. Richards, M. Schnell, A. A. C. da Silva Pinto, S. Srivastavu, W. H. Swift, J. Zafimahova and others. Furthermore, the writer would like to thank H. Ancelle, Th. C. Denton, H. de Michaux, H. Farber, M. A. Geith, W. Hance, S. H. Haughton, Th. G. Murdock, J. Nichols, A. S. Rogers, M. Rush, M. Thomas, W. J. van Biljoen, F. A. Williams and others for information they so kindly provided. Indeed, the list of references shows the work accomplished in this difficult field and used in the writing of this book. In addition to this documentation, mainly in the English and French languages, the writer had access to numerous unpublished reports and he could draw on 15 years' experience in Africa. Mrs. E. Baer helped in the compilation of data and in the correction of copies. The research work has been carried out under the auspices of the Henry Krumb Fund of Columbia University.

It is hoped that it appears at the right moment. Indeed, the production of the traditional African commodities, gold, copper and diamonds, continues to increase, but a change is perceptible. Following the large-scale development of iron ore and bauxite deposits, the oil fields, in 1965, claim first rank as mineral producers. Moreover, the importance of fertilizers, water and hydropower resources on the one hand, and of mineral based manufacturing industries on the other hand, is increasingly recognized.

When reading the names of places, hills and rivers, we should remember that north of the Sahara these are transliterated in various forms from the Arabic, and that south of the desert we find originally only spoken languages. Consequently, there is no consistency.

The various forms of geographic names illustrate the spellings in which the reader may find them in the literature. Finally, unless otherwise specified, distances are given in direct lines or 'air miles'.

This book is dedicated to the development of these resources and of the countries of Africa.

Contents

CONTENTS

Part II

Economic Geology

Appendices

List of figures

List of tables

PART I

Industrial development and mineral economics

SECTION I

Africa's share of world resources and production

Africa's metal resources are surpassed only by those of North America, and they dwarf the ore reserves of South America, western Europe, southern Asia and Oceania. Its overall resources rival those of the communist world. Its poverty in coal hampers the development of heavy industries, but this will be counter-balanced by its vast hydroelectric potential.

Africa, better than any other area, can specifically provide the substances that are scarce in North America; productions of the two continents are, in fact, complementary. Africa holds an important share of the world's bauxite reserves and also vast iron resources, which are scheduled to become one of its principal exports. The continent is the free world's principal producer of manganese.

Africa dominates the world markets of the strategic minerals: cobalt, chromium, diamonds, lithium, beryllium, tantalum and germanium. The continent is also the world's principal producer of gold, radium, scandium, caesium, corundum, and sheet graphite; however, it contains hardly any mercury or molybdenum.

In the following discussion, the latest available data on resources are only tentatively presented (US BUREAU OF MINES, 1962). Reserves are essentially an economic concept, depending on fluctuations of the market and supply, and evaluations will vary according to criteria chosen for each deposit. The maintenance of given or projected levels of production through long periods of time is considered the most accurate basis of comparison. Even so, statistics differ with the years and within the same year. Consequently, all figures and percentages have been rounded.

STEEL INDUSTRY METALS

Estimates of *iron ore* resources vary greatly. World resources have increased from an estimated 30,000 million long tons[1] in 1958 to 250,000 million long tons in 1962. Africa contains little of the measured reserves because of a lag in exploration. However, the iron ore potential of Africa attains another 33,000 million long tons. Its 4% share in total world iron ore production of 500 million long tons[2] will soon be multiplied at a rate far more rapid than that of expansion of total world supply. *Pig iron* and *steel* output are expanding from their present rate of 1% of the world supply.

[1] Unless otherwise specified all tons in this text are short. See Appendix, p. 637.

[2] Unless otherwise specified all production totals cited are for annual periods.

Africa contains approximately 90% of the world reserves of *chromite* and produces 30% of the world output of 5 million tons; both the Soviet Union and Asia participate in similar proportions. *Ferrochromium* production started in Africa only a couple of years ago, but by 1965 the continent will become one of the world's major producers of both low-carbon and high-carbon ferrochromium.

Ghana boasts a monopoly of battery and type B grade *manganese* reserves. Although the Soviet group produces more than half of the world's 15 million tons of manganese ore, the African share of more than 20% surpasses all the remaining countries. However, *ferromanganese* production is still in its infancy on the continent.

In South Africa we can also report a small output of *ferro-silicon*.

The Soviet bloc and tropical areas enjoy an overwhelming share of the world's *nickel* resources. Mainly low grade resources of the non-communist world and of Cuba are in excess of 70 million tons. This compares with an African potential of a million tons. With 250,000 tons per year, Canada dominates the nickel markets, followed by the Soviet Union. African production is in the range of 3,000 tons.

China dominates the world's *tungsten* resources of more than 170 million tons of concentrates and also the world's production of 70,000 tons; Africa's share ranges from 1 to 2% of the output.

Similarly, African *molybdenum* resources are insignificant; North America excels in this metal. World molybdenum production is in the range of 90,000 tons.

The case of *cobalt* is quite different; 90% of the world resources are located in the Congo which contributes 70% of the global production of 15,000 tons.

While world *vanadium* production almost doubled to more than 13,000 tons during the last few years, African output more than trebled. The supply of the non-communist world is 60% in North America, 30% in Africa and less than 10% in Europe. Vast, low grade resources are available in Africa.

BASE METALS

Chile is the only country which holds more of the world's 170 million ton *copper* reserve than Zambia (Northern Rhodesia) and the Congo, where a majority of the high grade ores are located. But the United States still produces more than a third of the world's copper output of 5 million tons. African capacity is expected to expand within the decade. The continent also ranks among the world's principal producers of blister and electrolytic copper.

With respect to other base metals, Africa plays a minor role. The continent holds, for instance, only 5 million tons of the estimated world *zinc* reserve of more than 80 million tons. The African share is less than 10% of the mine production of 4 million tons, compared to North America's one-third. However, most of the concentrates are smelted locally.

North America also contains much of the world's 50 million ton *lead* reserve; Africa attains only 3.5 million tons. World production has not yet passed the mark of 3 million tons, to which Africa contributes less than 10%.

Malaysia dominates *tin* resources and production; Africa's share of the output is slightly over 10% of the world total of 200,000 long tons.

LIGHT METALS

The world's *bauxite* reserves containing 60% alumina are in excess of 3,000 million long tons; the African share attains 800 million long tons. Bauxite resources containing 40% alumina reach the mark of 8,000 million long tons, of which 2,400 million are found in Guinea. African output trebled, while world production of bauxite, which reached approximately 30 million long tons, only doubled. African production is scheduled to treble again later in this decade. Its present share of world output is 6%; North America and the Caribbean produce one-third of the world's bauxite. With her vast bauxite and hydropower resources, West Africa is rapidly becoming an important producer of *alumina* and aluminium.

African share of world *magnesite* production of 8 million tons is insignificant; the Soviet Union dominates. Africa produces 10% of the world's *titanium* mineral output of more than 2 million tons. However, the production of the continent is expanding five times as rapidly as that of the rest of the world.

STRATEGIC METALS

African resources are outstanding in the field of metals used in strategic industries.

North America contains 60% and South Africa almost 35% of the world *uranium* resources, excluding communist countries. Africa contributes more than 10% to the non-communist world's 30,000 ton uranium production, and North America provides 80%. The share of Africa in thorium resources and output is also very large. *Zirconium* is the only metal originating mainly in Australia, which provides almost 70% of the 170,000 ton world production; North America contributes 20% and Africa 10%. African output fluctuated, while world output doubled. African resources are abundant. Most of the non-communist world's *hafnium*, zirconium's twin metal, has been extracted from Nigerian zircons.

Africa still dominates the world's *lithium* ore production, contributing two thirds, versus North America's one-third of the presumed 100,000 ton output, which excludes production of communist countries. Africa contains most of the reserves.

Africa produces less than half, and South America a quarter of the 8,000 tons to more than 10,000 tons of *beryllium* ore produced annually. The output of other continents is still small.

Africa has dominated the world's 2,000–6,000 ton *columbium (niobium)* ore output since the beginning of production. Brazil will contribute substantially to the expanding market. The Congo and Brazil share most of the world resources, exceeding 10 million tons.

Africa is distinguished by 80% of the world resources of *tantalum* ore which are in the range of 15,000 tons. It contributes the majority of the current production of 200–600 tons.

Data on *rare earth* outputs and resources are secret and consequently unavailable, although Africa is known to contribute a very important share. Reserves of Egypt and South Africa are in excess of 300,000 tons. Thorium outlook is similar.

PRECIOUS MINERALS

Few realistic data are available on *gold* reserves, which attain 500–1,000 million troy ounces (17,000–34,000 tons), of which the South African share is predominant. Africa produces more than half of the world output of 50 million troy ounces (1,700 tons) of gold, the Soviet bloc contributes 25%, and North America 10%. While the gold output of the rest of the world has not expanded, South African production has doubled during the last 10 years. A few million ounces of gold find their way to markets annually without being reported at customs or in statistics.

Silver resources have been estimated to attain 5,000 million troy ounces (170,000 tons); the African share is in the range of 100 million troy ounces. North America produces half of the world's annual supply of 250 million troy ounces (8,600 tons); Africa contributes only 4%. Both world and African production are subject to fluctuations.

World resources of the *platinum* group metals have been estimated to exceed 25 million troy ounces (860 tons). South Africa boasts a 40% share of the world resources of 15 million troy ounces of platinum, 7.5 million troy ounces of palladium and 2.5 million troy ounces of iridium, osmium, rhodium, and ruthenium found combined in the world. North America, the Soviet Union and South Africa respectively provide 40%, 30% and 25–30% of annual sales of 1.2 million troy ounces (42 tons). Production is believed to be larger than sales.

Africa provides 93% of the world *gem diamond* production, which attains 6–7 million carats (1.4 tons), and contains an equal percentage of the resources. A considerable quantity of diamonds is not reported in statistics, being exported without the benefit of customs control.

Africa is distinguished by a quasi monopoly of (natural) *industrial diamond* production, and attains 98% of the 27 million carat (6 ton) production. A 7 square mile area at Bakwanga, Congo, contains 95% of the world's resources, which are estimated to attain 500 million carats (110 tons). Moreover, Africa boasts an artificial diamond producing capability.

Africa has an important share in the production of *semi-precious stones* and abrasives.

METALS USED IN THE ELECTRICAL INDUSTRY

Cadmium reserves have been estimated to attain 400,000 tons; the African contribution is small. North America provides almost 50%, Europe 30%, and Africa 8% of the slowly expanding production, which now attains 25,000 tons. African output has increased tenfold, while world production progressed by 50%.

World resources and production of *mercury* have attained 6 million flasks (230,000 tons) and 250,000 flasks (9,000 tons). Mediterranean Europe produces 40% of this amount. Only occasionally does Mediterranean Africa contribute a few tons.

Brazil dominates the world production of high grade *quartz* of 2,000 tons per annum, of which Africa has but a small share.

North America provides two-thirds of the *selenium* production of more than 1,000 tons. With the Congo, Zambia (Northern Rhodesia) contributes 3–7%. African production has doubled, while world production expanded by 30%.

METALS USED IN THE CHEMICAL INDUSTRY

Africa is the principal producer of the transistor metal, *germanium*, and contributes two-thirds of the world's 80–100 ton production. The Congo and Southwest Africa hold most of the high grade reserves of the metal. Large, low grade, reserves exist in coal deposits.

METALS USED IN THE CHEMICAL INDUSTRY

Africa contributes only 3% to the world's *pyrite* and *sulphur* supply of 8 million long tons (contained sulphur). Western Europe, the principal producer, provides 40% of the total.

China produces a third and Africa a fifth of the world *antimony* output of 60,000 tons. African resources are adequate, but production regressive.

The United States boasts most of the world's *baryte* resources which exceed 100 million tons. North America and western Europe each contribute 40% of the 3 million ton world production, Africa provides only 3–4%.

The situation relating to *fluorspar* is similar; total output attains 2.5 million tons, of which North America and western Europe share one-third of the production, and Africa produces 4%. African output has expanded more quickly than the world supply. Large reserves are known to exist in Southwest Africa.

Western Europe and North America share most of the world's 1.5 million ton *diatomite* production, leaving only 2% to Africa, where production is stagnant.

Western Europe attains 50% of the 60,000 ton world *arsenic* production. African reserves are low and most output has ceased.

Africa contributes less than 1% of the *bismuth* supply, which attains 7,000 tons per annum.

The United States supplies one-third of the world's *sodium* carbonate; Africa provides 1%. World capacity attains 20 million tons per year.

MINERALS USED IN AGRICULTURE AND FOOD

North America and Europe each produce roughly one-third, and Africa only 2% of 100 million tons of *salt* recovered annually. However, African resources are abundant.

Morocco and its neighbors include 25,000 million long tons of the world's *phosphate* reserves, which are in excess of 50,000 million long tons. North America produces more than 40% of the world production of 40–50 million long tons, and Africa contributes 25%.

The world supplies of *nitrogen* and *potash* compounds attain respectively 15 million and 12 million tons. Western Europe and North America together supply more than half of this output. Africa contributes 2% of the nitrogen–ammonia production and potash sales will soon start. Little is known about potash resources.

BUILDING AND CERAMIC MATERIALS

Western Europe and North America are the world's principal producers of *cement*. In 1963, global production was in excess of 2,000 million barrels (340 million long tons), but out of this African output only attained 2.5%. Limestone and other cement raw materials are abundant, but fuels frequently have to be imported.

Western Europe and North America also provide 60% of the world supply of *gypsum* of 50 million tons per annum. Out of vast resources, Africa contributes only 2% of this output. However, African output doubled while world production gained only 50%.

Canada and the Soviet Union each offer 40% and Africa less than 15% of the *asbestos* production of 3 million tons. Most of the high grade fibrous asbestos of Africa is produced in Southern Rhodesia.

If we consider all grades, North America produces almost 60% and Africa 2% of the total world supply of *mica*, which attains 200,000 tons. Madagascar is the world's only source of phlogopite.

North America produces 70% and Africa almost 30% of the non-communist world's 300,000 ton output of *vermiculite*. The largest reserves are in South Africa.

Conversely, Africa contributes less than 0.1% of the world supply of 3 million tons of *talc* minerals per annum. Asia and North America respectively provide 40% and 30% of this output. Similarly, Africa contributes only 0.1% to the *pumice* market of 14 million tons, but African output increases fairly rapidly. Mediterranean Europe provides 75% of this light material.

Paradoxically, Madagascar contributes only 3% to the global production of half a million tons of *graphite* of all grades. Western Europe and Asia now dominate this market, but the Malagasy Republic boasts vast resources of high grade flake graphite.

SCARCE SUBSTANCES

The Congo monopolizes the world's *radium* supply of roughly 50 g per year. The radium is extracted in Belgium. African *caesium* reserves equal those of Canada. Befanamo in Madagascar is the only natural source of *scandium*. On the other hand Tanganyika is the leading producer of sepiolite or *meerschaum* (10 tons per year). From this substance, pipes are made in Nairobi, Kenya. Guinea is believed to contain much of the world's *gallium* reserves, in view of its huge bauxite resources. The highest grade gallium mineral, germanite, occurs only at Tsumeb, Southwest Africa. No tellurium and indium are being produced in Africa, although these metals exist on the continent. Rhenium is not being produced in Africa.

ENERGY SOURCES

African coal is low grade. Reserves, largely concentrated in South Africa, are large by local standards, but they represent only 1% of the world total. Communist countries produce more than half of the world's *anthracite* production of almost 200 million tons. The combined share of South Africa and Morocco is less than 1%. The Soviet Union and the United States each contribute 20% to the bituminous *coal* production of 2,000 million tons, the supply of South Africa and of the rest of the continent attains 2.5%. Hardly any *lignite* is produced in Africa. Reserves are in the range of a few hundred million tons.

World reserves of *bituminous shales* are not accurately known, but vast low grade resources are available in Africa, south of the Equator, mainly in the Congo and in Madagascar.

Crude *oil* reserves of the Arabian peninsula and its periphery (all integral parts of Africa from the point of view of the earth sciences) represent almost two-thirds of the world total of 300,000 million barrels or 40,000 million long tons. The rapidly expanding share of Africa proper constitutes 5% of the world total. North America contributes almost 40%, the Near East only 25% and the Sahara 5% to the world production of roughly 9,000 million barrels (1,100 million long tons). The African rate of expansion is tens of times as high as the increase of world production.

African natural *gas* resources are surpassed only by those of North America, which markets 75% of the world's 20,000,000 million cubic ft. supply. The supply from Hassi er Rmel (Sahara), one of the world's largest gas fields, is scheduled to expand rapidly.

The scarcity of conventional fuels is compensated by Africa's vast *hydroelectric* potential, which represents 40% of the world total of 205 million kW. Resources of the Inga site, on the Congo downstream from Leopoldville, are in excess of the total capacity of North America. Africa's share of present production attains only 1% of the total, but will be considerably multiplied during the next years.

SECTION II

The value of African production

Mining is the only sector of economic activity in Africa entirely in the monetary economy of which it has been and still is one of the decisive initiators in many parts of the continent (UN, 1959, p. 61).

SUBSTANCES

Income from the mineral industry is a more important part of the gross national product of Africa than it is on any other continent. Moreover, most of the investment in Africa is in the mining industry. An important, predominant part of the income of African governments (in well-endowed countries) comes from export duties, taxation and royalties assessed on mineral products and to a lesser extent from import duties on equipment and material ordered by mining companies.

In 4 years, the total value of African mineral output expanded from $[1]2,000 to $2,500 in 1961. Since then, we have witnessed accelerated progress and by early 1964 the rate of production attained $4,000 million per year. In other words, proceeds from minerals and oil have doubled in seven years!

Gold represents one-third of the value of African production. The value of gold output has increased from $600 to $800 million (in 1961) in 4 years. In 1964, its share will top $1,000 million. However, by 1965, the percentage share of gold is scheduled to decline to one-quarter of the continent's total mineral proceeds as a result of accelerated expansion of oil, gas, iron and bauxite production. Gold will loose its first place to oil within a few decades or years.

The value of *crude oil* sales has increased in 4 years from $30 million to $150 million, and in 1963 it is believed to have passed the mark of $750 million, even if estimated at the supposed sales price and not at the higher posted price. As soon as the Libyan and Algerian pipe lines are completed, it will rival gold. To oil, we have to add *natural gas*. The marketing of this fuel has just begun.

Copper comes third. The share of copper proceeds represents now less than 15% of the total, and it will continue to decline for the reasons noted above. However, the absolute value of copper production increased slightly to $550 million in 1962.

Diamonds hold fourth place. Expanding production provides more than $200 million per year, in addition to uncontrolled sales. The value of *uranium* production increased

[1] All figures are expressed in US $ at the present rate of conversion even if they refer to the past. Present rates are: $1.50 equals an Egyptian £; $2.80 equals a £ (Commonwealth countries and Libya); $1.80 equals a Tunisian dinar; $2 equals a Sudanese £; $1.40 equals a South African rand; the American dollar is the currency of Liberia.

One dollar equals 2.5 Ethiopian dollars, 4.9 French francs, 5 Moroccan dirhams, 7.15 East African shillings or dinars, 28.6 Portuguese escudo, 50 Belgian francs, 245 CFA Malagasy and Mali francs.

from $100 to $150 million, then declined in 1962 to $110 million. Productive capability is much larger, but for some time the uranium glut and market conditions in the United States and in the United Kingdom will limit further expansion of the world's least expensive uranium production in South Africa.

Cement materials proceeds are difficult to assess. Processed cement provides a gross income in the range of $150 million. *Phosphate* rock produces almost as much. Based on this, a phosphate processing industry is expanding rapidly in North Africa. The value of the *coal* output attains $90 million per year. *Iron ore* has already reached a similar amount. But iron ore production is scheduled to attain $300 million in a few years' time and it will become, by the end of the century, one of the continent's principal industries. At the same time, the steel and ferroalloy industries are expected to grow.

Manganese ore, with a 1962 proceeds of $60 million, also shows a growth trend. Asbestos, zinc, lead, tin and cobalt each provide $40–60 million yearly. *Bauxite* is still valued at $30 million, but bauxite, alumina and aluminium output are scheduled to outstrip $100 million soon.

The share of other products attains less than 1% of the total, that is approximately: *platinum* group metals and *chromite*, $20–40 million each; columbium (niobium) with tantalum, up to $10 million; antimony, $4 million; salt, soda, silver, titanium, germanium and tungsten ore, $3 million each; gypsum, beryl, lithium ore and diatomite, $2 million each; and zirconium concentrates, $1 million.

COUNTRIES

Both production and growth are unequally distributed (Fig. 1). Indeed, an overwhelming part of the expansion of the last years can be attributed to three countries: Algeria, Libya and South Africa. With an estimated total of more than $1,400 million in 1963, *South Africa*[2] still supplies a third of the continent's mineral proceeds. The growth rate of 15% in 4 years has been lately accelerated and she is expected to keep the first place as long as conditions permit normal production. In 1963, *Algeria*, *Libya*, *Zambia* (Northern Rhodesia) and the *Congo* came next with proceeds of $300–370 million each. In 1964, Libya and Algeria already hold second place. The production of these countries doubled in a couple of years. With $150 million per year, *Moroccan* output is more stagnant. These six countries contribute 85% of the continent's production!

Six other countries, namely Southwest Africa, the United Arab Republic, Southern Rhodesia, Nigeria, Ghana and Sierra Leone are provided by a mineral income of $60–80 million per year. However, some of the Egyptian production comes from Asiatic Sinai. Nigerian output is expected to expand.

The *Liberian* output is anticipated to attain $150 million in a few years and Guinean minerals proceeds, more than $100 million, probably somewhat later. The expansion of Guinean production is dependent on the investment climate.

The mineral income of Angola, Gabon and Tanganyika ranges from $20 million to $30 million per annum. *Angolan* and *Gabonese* production are expected to expand rapidly. Similarly, Mauritania will obtain a $50 million iron ore income in a couple of years.

[2] The reader is referred to the chapters concerning each country for further details.

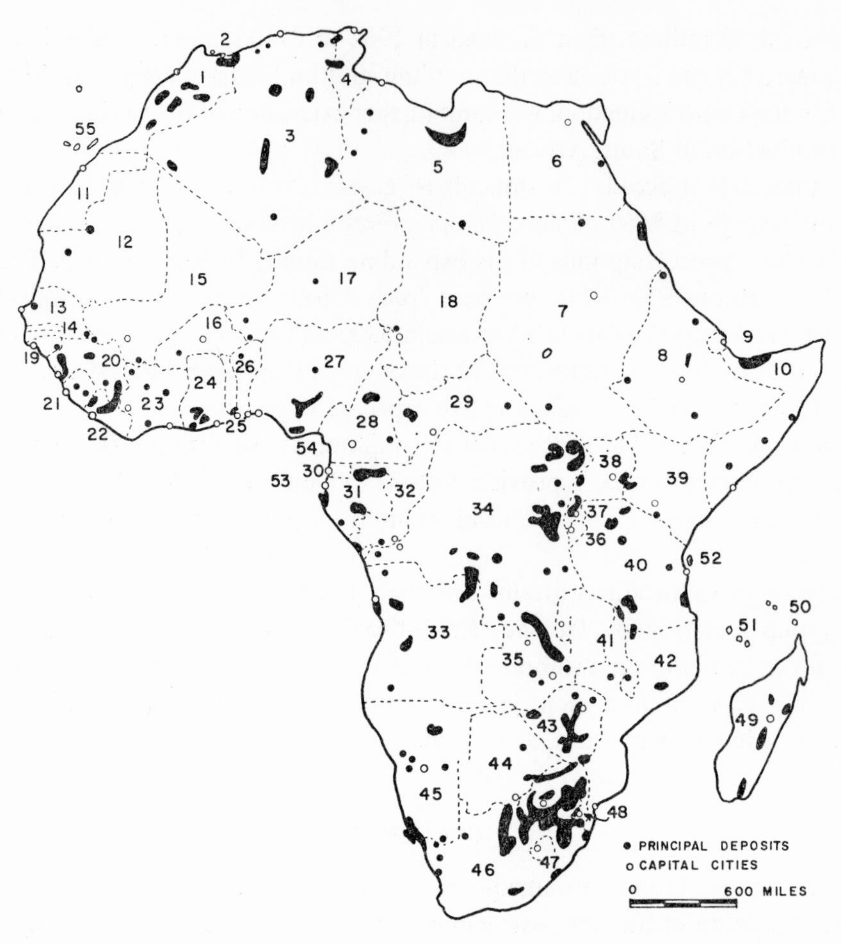

Fig. 1. Map of Africa, showing locations of principal mineral deposits. 1 = Morocco, 2 = Melilla, 3 = Algeria, 4 = Tunisia, 5 = Libya;
6 = United Arab Republic, 7 = Sudan, 8 = Ethiopia, 9 = French Somali Coast, 10 = Somali Republic;
11 = Spanish Sahara, 12 = Mauritania, 13 = Senegal, 14 = Gambia, 15 = Mali, 16 = Upper Volta, 17 = Niger Republic, 18 = Chad;
19 = Portuguese Guinea, 20 = Guinea, 21 = Sierra Leone, 22 = Liberia, 23 = Ivory Coast, 24 = Ghana, 25 = Togo, 26 = Dahomey, 27 = Nigeria;
28 = Cameroons, 29 = Central African Republic, 30 = Rio Muni, 31 = Gabon, 32 = Congo Republic, 33 = Angola, 34 = Congo, 35 = Zambia;
36 = Burundi, 37 = Rwanda, 38 = Uganda, 39 = Kenya, 40 = Tanganyika (Republic of Tanganyika and Zanzibar), 41 = Malawi, 42 = Mozambique;
43 = Southern Rhodesia, 44 = Bechuanaland, 45 = Southwest Africa, 46 = South Africa, 47 = Basutoland, 48 = Swaziland;
49 = Malagasy Republic, 50 = Seychelles Islands, 51 = Comoro Islands, 52 = Zanzibar (Republic of Tanganyika and Zanzibar), 53 = São Thomé, 54 = Fernando Póo, 55 = Canary Islands.

Tunisia benefits from $15 million mining proceeds. Uganda, Kenya, Rwanda and Swaziland have about $10 million of mineral income, but *Swazi* production will soon expand.

The output of other countries adds up to less than 1% of the African total, they contribute less than $30 million per year. Cement, alloys, steel, processed materials and power produced from falling water, although of mineral origin, have not been included in this review.

Concentration of mineral income to specially restricted areas is a typical African phenomenon. A zone of 35,000 square miles, extending from Kimberley to north of Johannesburg, supplies 35% of the continent's mineral proceeds. An area spreading 10,000 square miles from Roan Antelope in Zambia (Northern Rhodesia) to Kolwezi in the Congo provides another 15–20% of these. Furthermore, an area of 40,000 square miles in the bay of Syrte, Libya, produces 10%, but the remaining 99% of the continent's surface contributes only 40% of the mineral proceeds.

Similarly, the *per capita* mineral income is also unevenly distributed regionally and locally. This indicator attains its peak in Libya, where it already attained $250 in 1963. It will expand by at least 50% in a couple of years. Following Libya, Zambia (Northern Rhodesia) and Southwest Africa boast $130 per capita. South Africa trails with $80. Within a few years, mineral income per inhabitant will fluctuate around $100 in Mauritania and it will exceed $50 in Liberia. Gabon will attain a similar figure. But the continent's average has expanded from $10 to $20 per capita in 1964. This figure should be compared with $100 in the United States.

SECTION III

The distribution of African output and resources

In the first chapter, substances have been arranged according to their contribution to various transforming industries. In the following review, minerals will be grouped according to source and origin. Only major deposits are considered here, but other occurrences are included in the second part of the book.

IRON–BAUXITE GROUP

Iron resources are abundant in the western Sahara, on the Guinea coast, in the Congo and in South Africa. Manganese reserves are large in Gabon, Katanga and in the Kalahari. Bauxite resources are concentrated on the Guinea coast (Fig. 2).

Iron. Profitable extraction of iron ore is a function of economic and geographic conditions. In general, the availability and coakability of coal, the phosphorus, sulphur and arsenic content of the ore, and distances and topography are considered in addition to grade in defining a deposit's workability. For Saharan resources, the availability of water also remains an important factor.

Ten years ago, only northern African deposits, in the sphere of Mediterranean industry, and South African reserves located near the coal seams were considered as resources. The depletion of high grade iron deposits of the (American) Great Lakes and the expansion of industries of the Common Market and of Japan have thrust long-known, but previously unworked, deposits into the commercial category. Medium grade deposits located as far as 400 miles from the coast are now considered workable. Lower grade and more distant occurrences will become extractable in a few decades.

Recent evaluations of deposits indicate existence of the following high grade resources (±55% iron), located relatively near the coast or outlets: 800 million long tons in Gabon; 500 million long tons each in Guinea and Liberia; 200 million in Sierra Leone; 150 million in Mauritania; 100 million each in South Africa, Swaziland and Algeria; and 70 million each in Morocco and Southern Rhodesia. More distant, but high grade resources are located in Guinea, South Africa and the Congo, each in excess of 1,000 million long tons. The order of magnitude of the principal resources containing more than 40% iron has been estimated to attain 4,500 million long tons of ore in the Algerian Sahara, 5,000 million long tons in Mauritania, 5,200 million in Guinea, 5,000 million in the Congo, 7,300 million in South Africa and unknown amounts in the southern Sudan and the Central

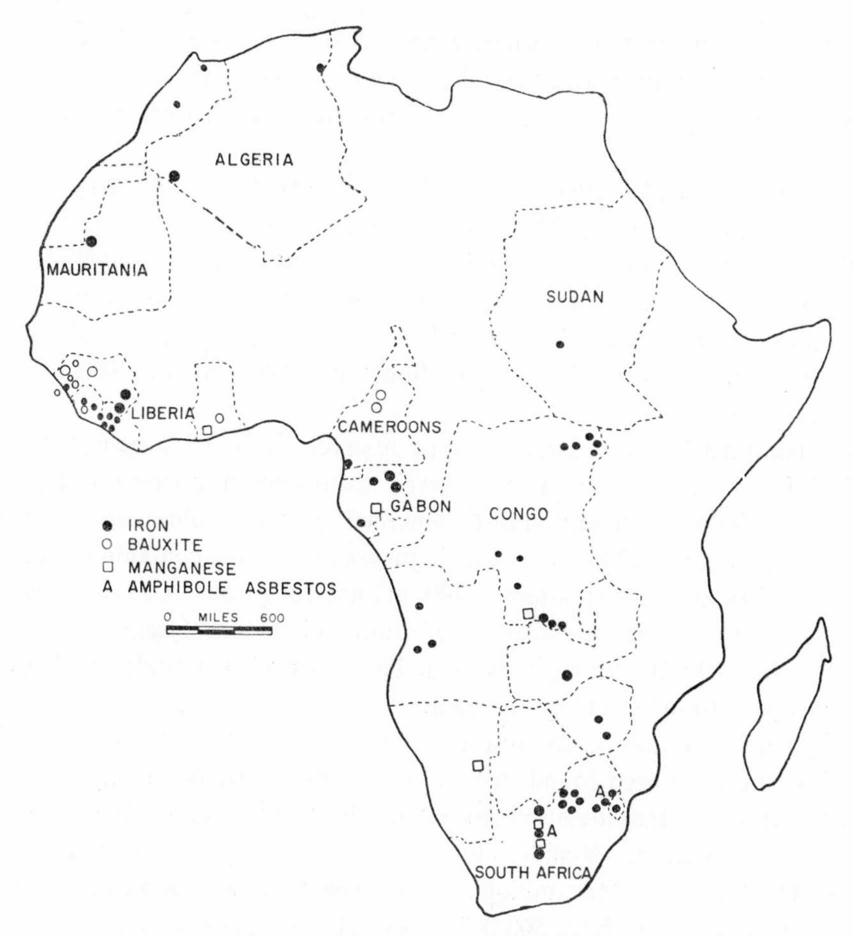

Fig. 2. The principal deposits of iron, bauxite, manganese and amphibole asbestos.

African Republic. Resources of the Senegal, Mali, the Niger and Congo Republics, Zambia, Tanganyika, Somalia, Ethiopia, Egypt and Libya may possibly total 2,000 million long tons.

Africa produces 18 million long tons of iron ore and will produce 25 million long tons. At the present, Algeria, Liberia, Sierra Leone and South Africa each contribute 2–4 million long tons, Morocco supplies one million. The bulk of the remainder comes from Tunisia, Guinea and Angola; some also originates in Southern Rhodesia and Egypt. Liberia is projected to ultimately produce 15 million long tons; Mauritania, 4 million; Sierra Leone, 3 million; and Angola and Gabon, a yet undisclosed amount. Swaziland and Southern Rhodesia are scheduled to begin large-scale production soon.

South Africa has a *steel* producing capacity of 4 million tons, but Southern Rhodesian and Egyptian production is much smaller. Algeria has produced steel; Morocco, Ghana and Senegal plan to do so. Because of the size of economically feasible steel mills, the problem of steel production is envisaged on a regional basis. In West Africa, for instance, a steel producing capability of 400,000 tons is recommended, with a possible expansion to one million tons by 1970.

Manganese. South African resources have been estimated to attain more than 100 million tons of good chemical grade ore plus a few hundred million tons of less accessible ore, dwarfing other African reserves. Gabonese reserves are calculated at 70 million tons. As previously mentioned, Ghana contains all of the non-communist world's battery grade manganese reserves. Manganese nodules are also scattered on the Atlantic Ocean floor west of Africa.

Production has exceeded 3 million tons and has increased six-fold in 20 years. By South Africa 47% is provided and 12–20% each by Ghana, Morocco and the Congo. Egypt contributes 6%, and the Ivory Coast, Zambia and Southwest Africa[1] contributed 2% each. Angola and Bechuanaland are smaller producers, while Ethiopia and Southern Rhodesia have insignificant productions. Angola and Ivory Coast production will expand, with Gabon being scheduled to supply half a million tons. South Africa also produces *ferro-manganese*, an alloy.

Bauxite. Like other low-priced ores, bauxite reserves also have to be near the sea or in the vicinity of inexpensive hydroelectric power. Low silica and iron content contribute to aluminium grades in the definition of commercial reserves. Guinea supplies 600 million long tons, and Ghana gives 230 million, to high grade reserves containing 60% alumina. Potential resources of Guinea attain 2,400 million long tons. Cameroonese reserves, located 400 miles from the sea, are in excess of a million long tons. Sierra Leone and Malawi contain more than 50 million long tons each. The order of magnitude of Mali and Ivory Coast reserves is 5–10 million long tons each.

The Congo also has low grade resources in the vicinity of hydroelectric potential. Indices of bauxite have been found in Zambia, Southern Rhodesia, Madagascar, South Africa, and other areas; they are also believed to exist in Ethiopia. African bauxite production has passed the mark of 1.8 million long tons, most of it coming from Guinea. Ghana supplies 200,000 long tons. Mozambique is a smaller producer. Sierra Leone started production in 1963. Guinea produces 500,000 tons of alumina per year and Cameroon manufactures 70,000 tons of aluminium, which will also soon be produced in both Guinea and Ghana.

BASE METAL GROUP

Of reserves 90% are concentrated in the metal belt of Katanga and Zambia (Northern Rhodesia), with an additional 7% provided by Morocco and Southwest Africa (Fig. 3).

Copper. Twenty-five million tons of copper have been estimated to lie on the Zambian side and 15–20 million tons on the Congo side of the frontier between these two areas. Large reserves are found at the two extremities; the greatest concentration is at Kolwezi, Katanga, and there are somewhat smaller deposits at Nchanga and Mufulira in Zambia. Southwest African reserves represent half a million tons. South African reserves are similar, while low grade resources contain 1.5 million tons of copper. Mauritania, Uganda, Southern Rhodesia and possibly the Sudan each contain 200,000 tons. Reserves of other countries are rather low.

The Copperbelt produces 90% of the continent's annual production of more than one million tons of copper, of which almost 60% originates in Zambia and 30% in the Congo.

[1] In 1962 exports ceased.

Fig. 3. The principal deposits of the base metals.

Copper output has been trebled in 20 years. Expanding South Africa now has a production representing 5% of the remainder, Southwest Africa 2%, Uganda and Southern Rhodesia 1.5% each. Morocco, Angola and Tanganyika are minor producers.

Zinc. African zinc reserves, estimated to attain 4–5 million long tons, are distributed between the Kipushi mine, Congo, Zambia and Nigerian, Moroccan and western Algerian deposits. Also, small reserves are available in east and southern Africa, Egypt, and Mali.

African output increased tenfold in 20 years but it has not yet reached 300,000 tons. Of this, the Congo contributes almost 40%; Zambia, Algeria and Morocco 10–20% each. Southwest Africa and Tunisia share the remainder.

Lead. Morocco contains most of the continent's reserves of 3.5 million tons. Southwest African resources attain one million tons. The continent's total production is below the mark of 250,000 tons, of which Morocco provides 45%; Tsumeb in Southwest Africa, 35%; Tunisia and Zambia, 7% each; and Algeria, 4%. The Moanda mine in Tanganyika stopped production in 1960. The two Congos, Nigeria and Egypt are minor producers.

Cobalt. The ores of Kolwezi (Congo) represent 90% of African reserves, and provide 80% of the 13,000 ton production. Zambia and Morocco respectively contribute two-thirds and one-third of the remainder. Ugandese reserves attain 80,000 tons. Nodules of the Atlantic sea floor contain 0.5–1% cobalt.

Cadmium. African output is approximately 2,000 tons, all of the cadmium coming from Kipushi, Congo, and Tsumeb, Southwest Africa, except for 1% produced in Zambia.

Vanadium. Titan-magnetite ores of South Africa contain as much as one million tons of vanadium, although high grade reserves are comparatively small in other parts of southern Africa. Northern African phosphate deposits constitute low grade reserves. South and Southwest Africa share the 4,500 ton output of the continent. The Congo, Angola and Zambia have also produced vanadium.

Germanium. African resources can be estimated to attain a few thousand tons. The zinc-copper ores of Kipushi, Congo, and Tsumeb, Southwest Africa share the 70 ton production.

Selenium. African production, in excess of 50 tons, comes from the Congo and Zambia. Resources are large.

Silver. African resources are in the range of 100 million troy ounces (3,400 tons). Half of this amount is contributed by gold deposits, especially in South Africa, and the other half is provided by the zinc–copper–lead ores of Kipushi (Congo), by the continent's principal silver deposit, Tsumeb (Southwest Africa), and by various Moroccan deposits. Until 1961, African production ranged from 10 to 12 million ounces and the output of Kipushi attained 4 million ounces. In 1962, African production decreased to 7 million ounces, but South African output expanded to 2.5 million ounces. At Tsumeb, recovery ranges from 1 to 2 million ounces and Moroccan production is believed to fluctuate from 700,000 to 900,000 ounces per year. Estimates for Algeria stand at 300,000 ounces per year. Minor producers include Southern Rhodesia, Kenya, Tanganyika and Tunisia.

Barium. Few data are available on barytes reserves, but large quantities are known to exist in North African deposits and in eastern African carbonatites. Morocco provides 60–80% of the continent's 120,000–150,000 ton yearly output, and Algeria supplies most of the remainder. The United Arab Republic and South Africa are minor producers. Barytes also occur in Tunisia.

Strontium. Strontium minerals are found in the same countries as barytes and in other countries of eastern Africa. The continent's only producer, Morocco occasionally supplies roughly 400 tons per annum. Production is limited by demand.

Fluorspar. Fluorite is widespread in Africa. Large reserves are believed to exist in South and Southwest Africa. South Africa is distinguished by some of the best optical fluorspar. This country boasts an annual production of 110,000 tons. Morocco, Southwest Africa and Rhodesia add another 1,000 tons to this amount.

Mercury. Some mercury has been found with zinc ores in North Africa and in Swaziland. In Tunisia, 2–7 tons per year of mercury are occasionally produced. Algerian and South African output have ceased.

GOLD GROUP

Gold occurs in most of the countries of Africa, along with traces of silver, discussed in the preceding chapter. Uranium is a satellite of South African gold ores and of other deposits,

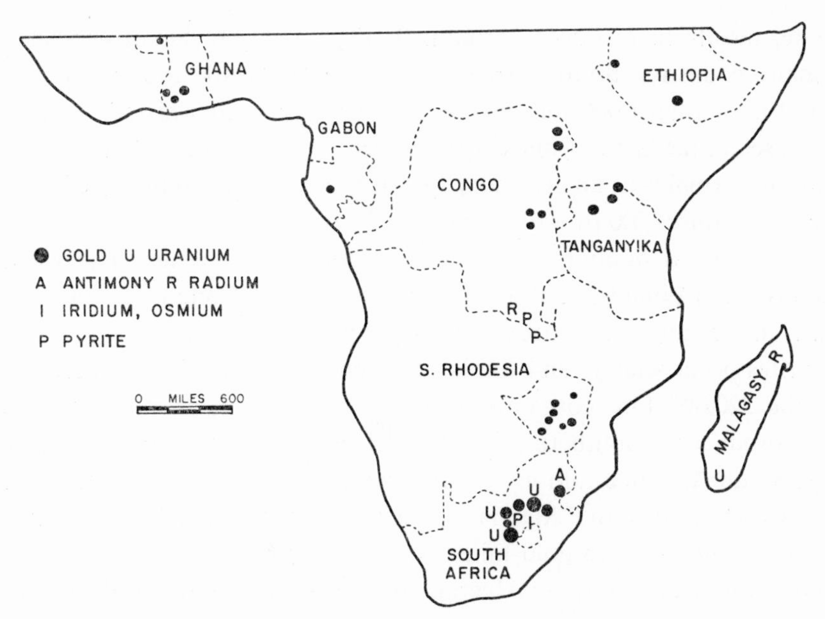

Fig. 4. The principal deposits of gold, uranium, antimony, radium, iridium, osmium and pyrite.

as are osmium, iridium and ruthenium. Antimony, arsenic and pyrites accompany gold, but the latter is also associated with base metals. Radium is extracted from uranium concentrates (Fig. 4).

Gold. South African gold reserves, 92% of the continent's, have been evaluated to suffice for 10 years. As exploration progresses, they can be expected, however, to last till the end of the century. The Congo, Southern Rhodesia and Ghana each contain approximately 2% of the resources, Tanganyika, 1%. Ghanean reserves have decreased.

South Africa produces more than 28 million troy ounces of gold (900 tons), an output that has been expanding at an average rate of 5% per year. The rest of Africa produces 2 million troy ounces (70 tons), an output that has remained stagnant. Other important producers include Ghana, Southern Rhodesia and the Congo, which each contribute 1–4%. Congolese production has been temporarily affected by political conditions. Tanganyika contributes 0.3% of the total, Ethiopia and Upper Volta combined 0.3%, Gabon less than 0.1%. The combined share of other countries attains less than 30,000 troy ounces, i.e., 0.1% of the total. Kenya, Zambia, the Congo Republic and Rwanda produce more than 3,000 troy ounces each. Swaziland, Sudan, Burundi, Liberia and Egypt (United Arab Republic) each contribute more than 1,000 troy ounces. The production of Nigeria, Uganda, Cameroon, the Central African Republic, Mozambique and Bechuanaland ranges between 200 and 600 troy ounces each, while Morocco, Angola and Madagascar provide 100 troy ounces each.

Considerable amounts of gold are not included in statistics, as they escape the control of government agencies.

Uranium. South Africa contains most of the continent's 380,000 ton reserve of U_3O_8; 8,000 tons remain in the Congo and 4,000 tons have been discovered in Gabon. Madagascar has a small reserve of uranium–thorium ore, as have several other countries. Phosphates consistently carry uranium. Small Nigerian phosphate deposits are distinguished by high uranium values, as are certain granites too hard to work; 25,000 million long tons of phosphate, mostly in Morocco, contain 0.013% of U_3O_8. Small deposits have been found in many countries. South Africa produces 4,000–6,400 tons per year. Congolese production of more than 1,000 tons has ceased, having played a historic role in providing the uranium used at the birth of nuclear industries. Deposits of Zambia have been worked out, but uranium remains as an accessory metal. Madagascar produces 500 tons of uranothorite. Gabon produces 200 tons of uranium.

Radium. Commercial radium is scarcer than gold or diamonds. The world stock of radium attains seven pounds. Radium is a by-product of Congolese uranium ores. The production is in the range of 1.5 troy ounces per year.

Pyrite. Pyrites occur widely, and the largest deposits are in the Copperbelt and South Africa. Of the 220,000 ton (contained sulphur) output 75% is a by-product of South African gold production. Southern Rhodesia, Algeria and Morocco supply the rest.

Antimony. South African antimony reserves have been estimated to attain 250,000 tons, and resources of Morocco and Algeria are in the range of 100,000 tons. Antimony is a co-product of a South African gold mine, which contributes more than 90% of the continent's 12,000 ton yearly output. Algeria, Morocco and Southern Rhodesia share the remainder.

Arsenic. Arsenic is widely scattered in Africa. Reserves are low; however, they exist in some degree in Morocco, Algeria, South Africa and other areas. Southern Rhodesian arsenic production, a by-product of gold mining, has oscillated between 200 and 1,000 tons. Algeria is scheduled to reenter production.

Iridium, osmium. The platinum group metals fall into two categories. A predominant part of the world reserves of the first group are included as accessories of South African gold ores. Half a million troy ounces (17 tons) are distributed as 35% osmium, 33% iridium and 15% ruthenium. Iridium and osmium production attains 2,500 troy ounces for each metal.

ULTRABASIC GROUP

The chromium-platinum group follows the north–south axis of eastern Africa. Deposits acquire importance only in southern Africa. Unlike other continents, some diamonds occur in most of Africa south of the Sahara. Important concentrations are however, restricted to the Congo basin, the Guinea coast and southern Africa (Fig. 5).

Platinum, palladium, ruthenium, rhodium. Ten million troy ounces of platinoid metals are concentrated in South Africa, consisting of more than 70% platinum and 25% palladium, 2% rhodium and 1% ruthenium. Reserves of other African countries are low. South African production capacity attains as much as 600,000 troy ounces per year; sales have ranged lately from 300,000 to 400,000 troy ounces (11 tons) per annum. Ethiopian output has fallen to 200 troy ounces and Congolese platinum production has ceased. Some platinum occurs in Southern Rhodesia, in the Algerian Sahara and Sierra Leone.

Chromium. Low grade ores of South Africa containing more than 45% chromite (chrome oxide) represent a reserve of 200 million tons; lower grades may reach as high as 2,000 million tons. The chromium/iron ratio is 1.6/1. Reserves of Southern Rhodesia have been estimated at 300 million tons of easily accessible high grade ores and vast resources of

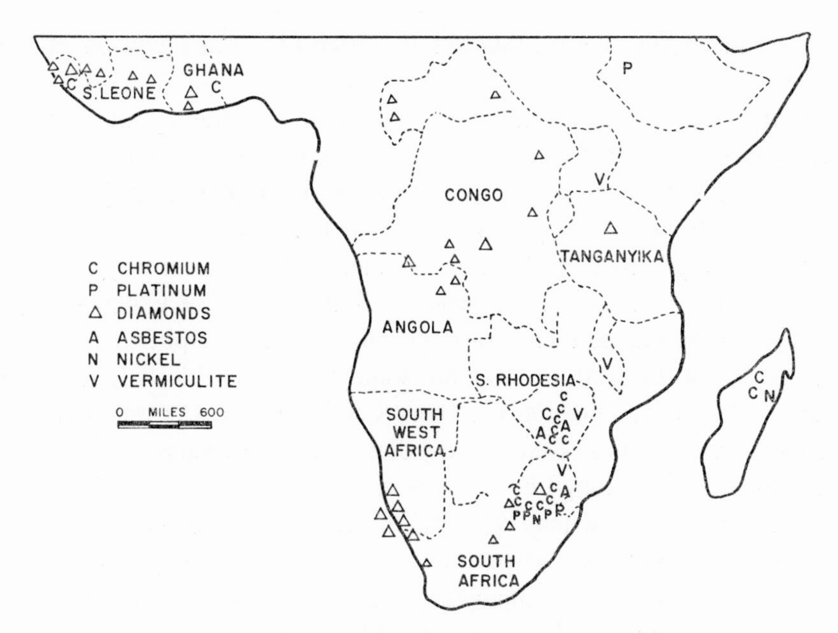

Fig. 5. The principal deposits of the ultrabasic group.

less attractive ore. Togolese resources represent 50,000 tons. South Africa now produces one million tons, and Southern Rhodesia 0.5 million tons. Out of 7 million tons of reserves, Madagascar produces 20,000 tons. Sierra Leonian output is low. Some chromium has been found in Egypt, the Sudan, Tanganyika, Dahomey, Guinea and Ghana. Nodules of the Atlantic Ocean floor also contain chromium. *Ferrochromium* is manufactured in South Africa and Southern Rhodesia. Madagascar may join them later.

Asbestos. While data on reserves lack precision, it is known that resources sufficient for 30 years are available. A majority of the world's high grade reserves are located in Southern Rhodesia and South Africa. South Africa contributes one half, Southern Rhodesia 35% and Swaziland 8% of the 400,000 ton production. Bechuanaland, Egypt, Kenya and Mozambique share the remainder.

Asbestos also occurs in the Congo, the Sahara, Ethiopia and Tanganyika.

Nickel. South African deposits occur in the same environments as platinum and chromium. Low grade reserves attain 100,000 tons. Madagascar resources are in the range of a million tons. Nodules of the ocean floor contain 0.5–1.5% of nickel.

South Africa contributes 90% of the continent's 3,000 ton production, and the rest originates in Morocco and Southern Rhodesia. Egyptian production has ceased.

Magnesium. Magnesite reserves are concentrated in southeastern Africa. Low grade magnesium resources are widely disseminated. South Africa furnishes 90% of the continent's 115,000 ton magnesite supply, and Southern Rhodesia contributes most of the rest. Tanganyika and Kenya are small producers.

Talc. Both talc and pyrophyllite are widespread in the eastern third of Africa. Commercial production of more than 20,000 tons is derived from Egypt and South Africa, and Swaziland supplies the remainder.

Vermiculite. South Africa is distinguished by the world's largest reserves, and contributes to the continent's more than 90,000 ton production all but 200 tons provided by Kenya, the Sudan and Tanganyika. Vermiculite has also been found in Egypt, in Malawi and in the Rhodesias.

Gem diamonds. Resources are generally supposed to provide an adequate supply for many years. Southwest African reserves are considered to be particularly large. The two principal producers, South Africa and Southwest Africa, contribute 50% of the total value of more than 200 million dollars, and Sierra Leone provides another 15%. Expressed in weight, South Africa provides 30%; Sierra Leone, approximately 20%; and Southwest Africa, Ghana and the Tshikapa–Lunda field, on both sides of the Congo–Angola border, 13% each of the 6.5 million carat (1.4 ton) officially recorded output. Guinean production has increased to 6% of the total till 1960, but some diamond fields of Upper Guinea were closed in 1961. Tanganyika, Liberia, the Ivory Coast and the Central African republics share the remainder. Estimates of unreported diamond supply vary.

Industrial diamonds. The Congo contributes two-thirds of the 27 million carat (6 ton) production and contains more than 95% of the estimated 500 million carat (110 ton) reserve of natural diamonds. Minor producers include Ghana, 10%; South Africa, 9%; Sierra Leone, 7%; Angola with Tanganyika, 3%; and Guinea, the Ivory Coast, Liberia, Southwest Africa and the Central African Republic share the rest. South Africa can also produce artificial diamonds.

TIN GROUP

Tin, columbium and tantalum are concentrated in eastern Congo and northern Nigeria. Beryllium, lithium, mica and titanium are abundant in southern Africa (Fig. 6).

Tin. The Congo and Nigeria evenly share most of the continent's reserves. Low grade resources are available in South Africa and on a smaller scale in the Congo. Reserves of other producing countries are low.

The Congo with Nigeria supplies 80%, Rwanda 7%, South Africa 7%, Southern Rhodesia 3% and Southwest Africa 1% of the continent's production formerly limited to 20,000 long tons by international agreement. The remaining 2% originates in Tanganyika, Cameroon, the Niger and Congo republics and in Uganda. Morocco and Swaziland show an insignificant production.

Tungsten. The continent's tungsten ore reserves are in the range of 10,000 tons, located mostly in Rwanda and Congo. The Congo contributed 50% of the production of roughly 1,000 tons, Rwanda 20%, Southwest Africa 10%, Uganda 10%, with South Africa and Southern Rhodesia sharing the rest. Activity at the principal Congolese tungsten mine was temporarily interrupted in 1961 and Rwanda became the most important producer.

Molybdenum. Some molybdenum occurs in the South African tin districts, in Moroccan copper-tungsten ores and in Sierra Leone lead veins. Production has ceased.

Columbium (Niobium). The Congo contains two-thirds of the continent's resources, estimated to attain 3–5 million tons. Other reserves include approximately half a million tons each in Uganda, Kenya and Tanganyika, and 100,000 tons each in Malawi and Nigeria. Zambia, Angola and the Niger Republic also have resources.

Nigeria still produces almost 90% of the columbium ore production of more than 2,000 tons, which reached its peak of 4,000 tons during the Korean war. The Congo with

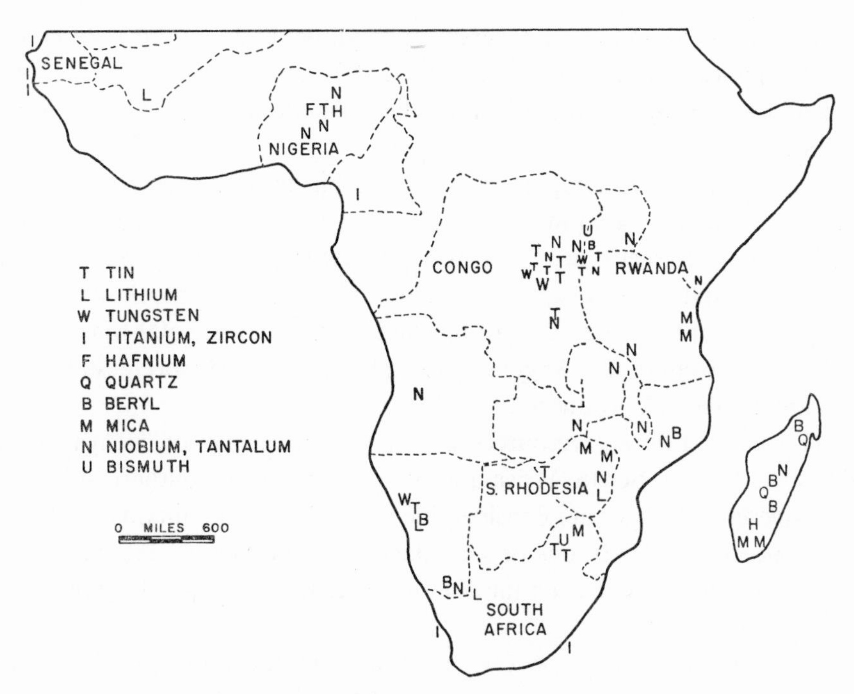

Fig. 6. The principal deposits of the tin group.

Mozambique contributes 8%, and Rwanda and Madagascar the remainder. Southwest African, Southern Rhodesian, and Ugandese productions are insignificant.

Tantalum. 80% of the continent's resources of approximately 13,000 tons of ore are concentrated in the Congo; Mozambique and Rwanda are the second and third largest producers, respectively.

Supply (200–600 tons) covers the demand with difficulty; the Congo contributes 30%, Southern Rhodesia and Mozambique more than 20% each, and Rwanda, Uganda, Nigeria, South and Southwest Africa the remainder. Madagascar is a minor producer.

Lithium. Resources are hard to assess. Low grade Congolese resources attain several million tons. High grade Southern Rhodesian resources are high. Southern Rhodesia provided 90% of total output of 50,000–90,000 tons of ore, Southwest Africa and South Africa combined contributed 4,000 tons. Occasionally some lithium is worked in Mozambique and Rwanda.

Caesium and rubidium. Caesium is a satellite and by-product of lithium; 100,000 tons of caesium ore are believed to exist in Southern Rhodesia and 50,000 tons in Southwest Africa. Production is expanding beyond 100 tons. No rubidium has yet been produced in Africa. Reserves, contained in lithium ores, are large.

Beryllium. Reserves are hard to evaluate owing to inadequate exploration. The 4,000 ton production of beryl fluctuates with demand. The principal producers: Uganda, Madagascar, Southern Rhodesia, Mozambique, the Congo, Rwanda, South Africa (and occasionally Southwest Africa) each contribute 10–25% of the output. Most of the jewelry grade beryl comes from Madagascar (Malagasy Republic).

Scandium. A small Madagascan deposit contains almost all of the world's high grade resources.

Bismuth. Bismuth occurs in the same areas as beryllium and tantalum. Reserves are low. Mozambique supplies 90% and Uganda almost 10% of a production which exceeds 20 tons. South Africa, Southwest Africa and Southern Rhodesia are small producers. Output has ceased in the Congo.

Mica. Large resources of scrap mica are available in southern Africa. Reserves of Tanganyikan sheet mica remain low. Madagascar and Tanganyika dominate the production of sheet and block mica, each providing roughly 100 tons of phlogopite and muscovite. Southern Rhodesia and Angola supply an additional 100 tons of block muscovite. Madagascar also produces 1,000 tons of splittings and South Africa 2,500 tons of scrap. Production in Kenya, Mozambique, Angola and Zambia is small; Morocco, Sudan and Southwest Africa are occasional producers.

Rare earth metals. Primary resources of the cerium group of Kenya, Tanganyika, Zambia, Malawi, Angola, South Africa and other countries of southern Africa are large. Deposits of Nigeria and Niger are considerable. The Nile delta contains 200,000 tons and South African beaches 100,000 tons of secondary ores. Production data are published incompletely, but South Africa has an important share. Resources of the yttrium group are small.

Thorium. The distribution of the cerium group and thorium is similar, though not identical, as thorium occurrences are more widely spread. Madagascar is one of the continent's principal producers. Market conditions limit production.

Titanium. Reserves in coastal beaches are difficult to assess. The Nile delta contains more than 20 million tons of concentrates. Ilmenite production is in excess of 200,000 tons, and rutile output attains 5,000 tons. South Africa produces 4,000 and Egypt 1,000 tons of rutile. South Africa and Egypt contribute 90% of the ilmenite; Senegal production has declined. Mozambique and Madagascar are small producers. Gambian production was interrupted. Titanium production is scheduled to start in South Africa.

Zirconium. Resources are large, especially on the beaches, in the Congo, Uganda and Nigeria. Other resources are available in eastern and southern Africa and in the Niger Republic. At 7,000 tons per annum, South African output is stable but the production of Senegal ranges from 12,000 to 2,000 tons. Nigeria, the United Arab Republic and Malagasy are small producers.

Hafnium. Most of the world's hafnium is separated from Nigerian zircons. Hafnium production is in the range of 10 tons, and resources are approximately 30,000 tons.

CARBON-FUEL GROUP

Graphite and coal have been largely limited to southern Africa; oil and gas are generally restricted to northern Africa (Fig. 7).

Graphite. Madagascar flake graphite reserves are very large, and attain several million tons. Potential resources are huge. Graphite also occurs in Kenya, Katanga, Zambia, Mozambique, South Africa, Nigeria, Sudan, etc. Production exceeds 15,000 tons, of which Madagascar contributes most; South Africa provides generally 1,000 tons. Kenya is an occasional producer.

Anthracite. Anthracite is scarce in Africa. South Africa and Morocco share, in a propor-

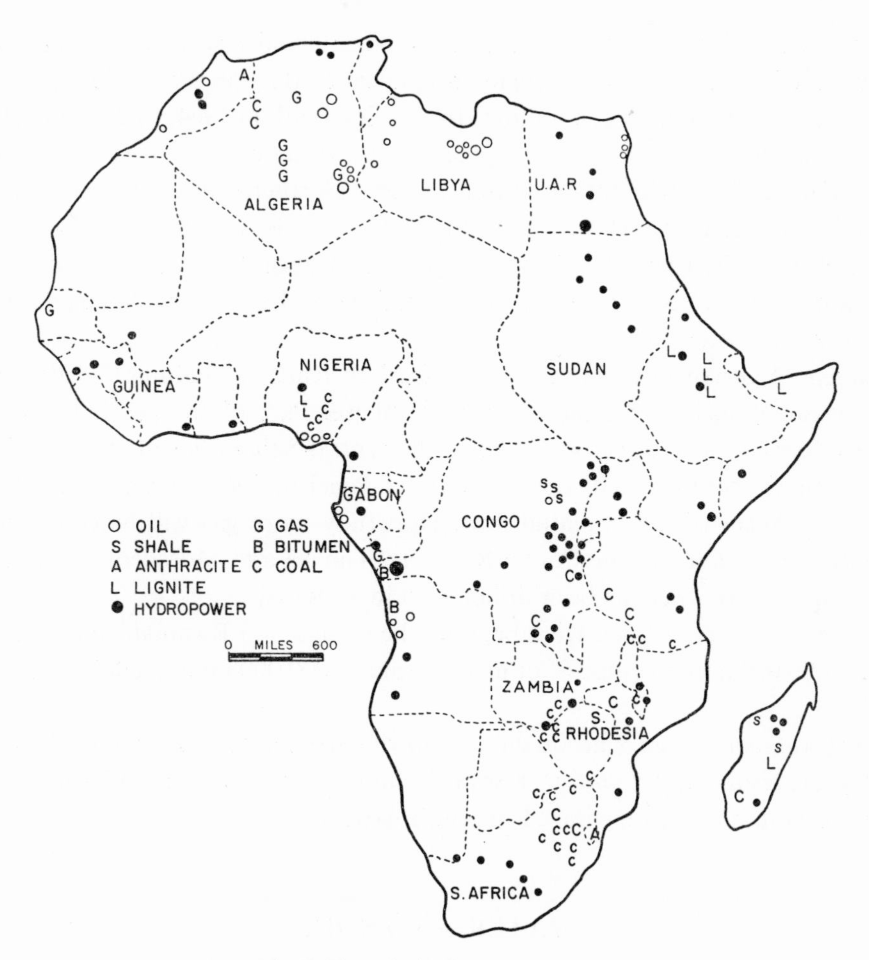

Fig. 7. The principal energy resources.

tion of 3/1, a production in excess of 1.5 million tons. Algerian production has ceased, but Swaziland may become a producer.

Bituminous coal. Resources have been estimated to attain roughly 75,000 million tons of which South Africa contributes 92%; Southern Rhodesia, 7%; Madagascar with Tanganyika, 0.4–0.5%; and Mozambique, the Congo, Malawi, Nigeria, Algeria and Angola the rest. South Africa represents 90% of an output of 50 million tons; Southern Rhodesia provides 6%. Nigeria, Mozambique, the Congo, and Algeria share the remainder. Swaziland and Tanganyika have insignificant productions.

Lignite occurs in Madagascar, Ethiopia, Sudan, Egypt and many other countries. Kerogene sediments occur in South Africa. Nigeria boasts a 200 million ton lignite reserve.

Bituminous shales. Bituminous sediments occur widely on the periphery of the oil fields, mainly in the coastal basins (e.g., Angola, Congo). The largest deposits are in the Congo Basin; they cover 4,000 square miles and extend far below the cover. Similar horizons are undoubtedly interbedded in sediments of southern Africa and Madagascar. Malagasy resources of contained oils and tars have been estimated at 9,000 million barrels.

Crude oil. Rapidly expanding reserves appear to be in excess of 15,000 million barrels (2,000 million tons); most are located in the Libyan and Algerian Sahara. Exploration is being actively carried out in Africa, and new discoveries are anticipated with certainty.

The share of the Atlantic coastal basins, mainly Nigeria and Egypt, does not attain 10%, and shows in Rift Valley sediments remain insignificant. Algerian production has expanded from 10 million to 60 million, 120 million, 160 million, and, in 1964, to 200 million barrels in 5 years. New finds become active as pipelines are being completed. The case of Libya is similar; yield attained 320 million barrels in 1964. Further expansion is foreseen with a considerably increased throughput before the end of the decade. In 1964, Egypt reported an output of 40 million barrels, Nigeria 50 million, Gabon 8 million, Angola 9 million, Morocco a million barrels and the Congo Republic less. Nigerian and Angolan production are also expected to increase.

Natural gas. Algerian reserves are large. Hassi er Rmel, containing $7 \cdot 10^{13}$ cubic ft. (60% recoverable), has entered the commercial phase. Part of the gas is being liquified (condensed) for shipment. Other gas fields in the central Sahara have not yet achieved the productive phase, because of distances. Hassi er Rmel in 1962 marketed 13,000 million cubic ft., a fragment of the potential flow. Hopefully more gas will be commercialized. Total production of Gabon, the Congo Republic, Tunisia and Morocco achieve less than 1,000 million cubic ft. Gas has been drilled also in Somalia.

Methane gas. Waters of Lake Kivu between the Congo and Rwanda, contain a concentration of 2,000,000 million cubic ft. of methane and 7,000,000 million cubic ft. of carbonic anhydride.

Carbon dioxide and helium. Carbon dioxide flowing from many points of the East African Rift Valley has been actively used in Kenya. Helium could be tapped in Tanganyika. The utilization of both has been envisaged in South Africa.

WATER AND SOIL

Water and soil are the most important mineral resources. Water is used for the generation of electricity, and is indispensable for most industries as well. The role of water (H_2O) in the human metabolism is as vital as the contribution of soil in food growing. Soil, an alteration product of minerals, is partly biogene, as is some iron ore, phosphate, copper, uranium, etc. African water resources exhibit the antithesis of an abundant hydroelectric potential and inadequate agricultural and alimentary water supply. Water and soil formation is scarce in the deserts and savannahs.

Hydroelectricity. The Congo boasts two-thirds of the continent's 205 million kW hydropower reserves (in mean annual flow). The Sudan and the Rhodesias each contain approximately 7% of these resources. Important contributors include Malawi, Egypt, Ethiopia, the Congo Republic, Cameroon and Ghana (Fig. 7).

Installed capacity, unfortunately, represents only a fraction of the potential. The Congo, Zambia and Southern Rhodesia are each equipped with generating capacity exceeding 500,000 kW; Cameroon and Uganda have acquired a capacity of 150,000 kW. The installed hydropower of the rest of Africa is 150,000 kW. It is, however, expanding rapidly. The largest project, Inga, near Léopoldville, represents a capacity of 25,000,000 kW.

Water. Industrial water is in sufficient supply in approximately half of Africa. The lack of water hampers the development of industries in the deserts. Water for alimentation and drinking is inadequate or unsuitable if untreated in most of the continent. More personnel is engaged in the search for water than in the exploration of most other substances combined. Surficial water supply is inadequate for the formation of soil and agriculture in more than 30% of the continent. Geothermal steam escapes from rocks in Kenya, Tanganyika and other areas.

Soil. Soils are in adequate supply in 60% of the continent. Their quality depends on the parent rock and human care. The ratio of acid to basic rocks is particularly high in Africa, and the quality of soil is less than average. The destruction of vegetation gradually caused the disappearance of soils, which enhanced the process of lateritization.

SEDIMENTS

Large phosphate and smaller salt, gypsum, diatomite, nitrate and clay deposits occur in northern Africa. South Africa is the principal exploiter of cement-rock and limestone (Fig. 8).

Phosphates. Moroccan phosphate reserves have been estimated to attain 20,000 million long tons. Algerian and Tunisian reserves are each in the range of 2,000 million long tons.

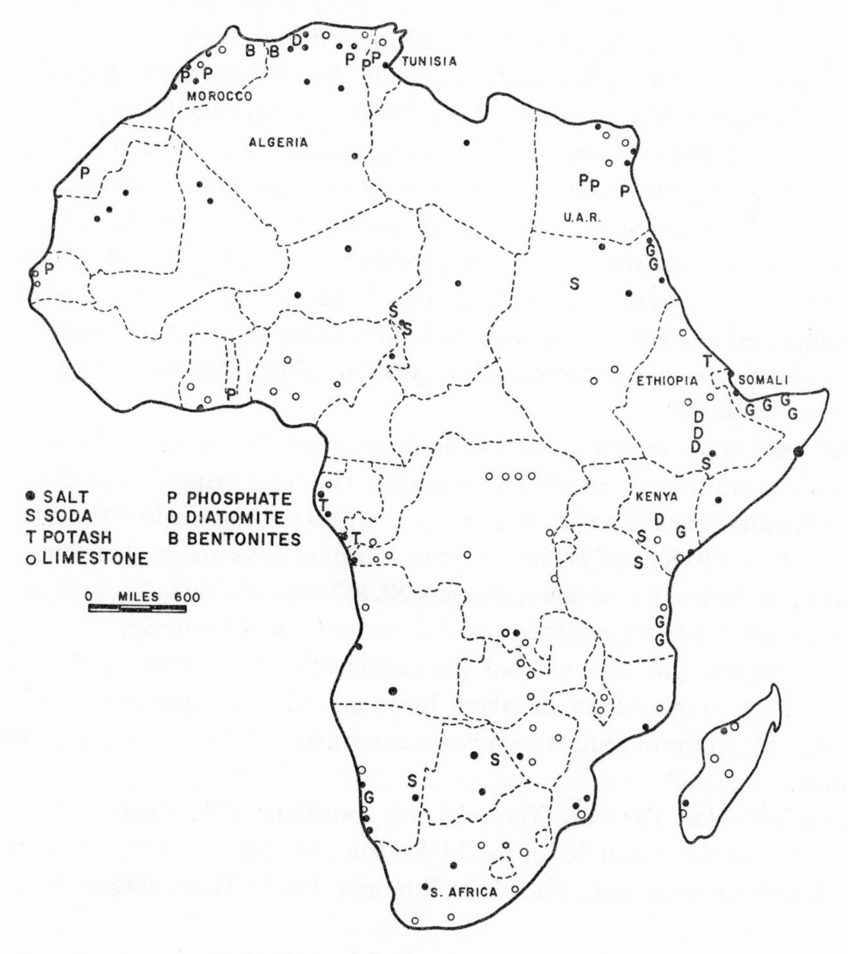

Fig. 8. The principal resources of sediments.

Low grade Egyptian resources are large. Reserves of the Spanish Sahara attain 500 million tons. Senegalese and Togolese reserves attain 50 and 100 million tons, respectively. Nigerian reserves are low. Non-sedimentary reserves in eastern and southern Africa are in the range of 500 million tons.

Morocco produces 65% of the continent's 13 million long ton output, Tunisia 16%, Algeria and Egypt each 5%, South Africa and Togo each 2% and Senegal 4%. Some phosphate is being recovered in Madagascar, in the Seychelles (Astove), in Uganda, Southern Rhodesia and Southwest Africa. The production of Senegal and Togo is expanding. While Algerian output has declined lately, the opening up of the new Djebel Onk deposit will again boost production to its previous level. With the increased demand for fertilizers, the manufacture of phosphate products is rapidly expanding.

Limestone. Resources are adequate in most of Africa. While small-scale quarrying is widespread, large industrial output tends to be concentrated in the more industrialized areas, such as South Africa, northern Africa and the United Arab Republic. *Marble* quarrying is conditioned by the needs of developed building industries, like those of South Africa and Morocco. The purest form of *calcite* is iceland spar. Labourers collect this mineral in Southwest Africa and South Africa.

Cement plants provide the largest market for low-magnesia limestone and for clays and marls. All these raw materials are in adequate supply, although not always in the most populated areas. By 1962, Africa supplied 50 million barrels (8.7 million long tons) of cement per year. The principal producers include: South Africa, 30%; the United Arab Republic, 25%; Algeria, Morocco and Tunisia, 20–25% combined; Nigeria, 5%; Kenya, 4%; the Congo and Southern Rhodesia, 3% each; Angola, Mozambique and Senegal, 2% each; and Zambia and the Sudan, 1%. Because of political and economic conditions, the productions of Algeria and the Congo have fluctuated lately. On the other hand, the Nigerian output increased sharply. Smaller producers include Uganda, the Canary Islands, Ethiopia, Rwanda, the Malagasy Republic and Cameroon. A large quantity of cement is still being imported into Africa. But with the high transport/cost ratio of cement there is an incentive for local production. Among others, Gabon, Ghana, Madagascar and Tanganyika plan new cement plants.

Clays are used in the cement industry and in the manufacture of bricks. In addition to these Africans build houses out of clay or adobe. The total extraction of clays cannot be assessed. A country with the population of Egypt is said to produce 30,000 tons of common clay and more than 10 tons of kaolin per year. Ceramic clays are produced in many other countries. South Africa, for instance, extracts 30,000 tons of china clay each year. Algeria and Morocco are important producers of *detergent clay* and *bentonite.*

Diatomite. Algeria provides 85% of the continent's 30,000 ton supply, with Kenya contributing 10%. South Africa, Southern Rhodesia, Mozambique and the United Arab Republic total 5% of the output. Other resources are available in Nigeria, Mali, Tanganyika and Ethiopia.

Pumice and pozzolan. The Cape Verde Islands contribute 50%, Egypt 30%, the Canary Islands and Kenya the remainder of the 14,000 ton per year production of these volcanic materials. Large resources are available in Ethiopia, less in Mozambique, Swaziland and Morocco.

Gypsum. Extensive beds of gypsum and anhydrite have been discovered above the oil fields of the Sahara and of the Atlantic coast. Similarly, gypsum is widespread on the eastern coast of Africa, between the Sudan and Mozambique. Somali resources represent a potential of 200,000 million tons. Sudanese reserves are also vast as are those of Tanganyika. As to production, the United Arab Republic ranks first with half of the million ton total. South Africa and Algeria each contribute another 20%. Furthermore, small outputs have been reported in Kenya, Morocco, Tunisia, Angola, Tanganyika and occasionally in Madagascar, but the Somali deposits are not yet worked.

Salt is widely spread in the deserts and in the oil fields which lie under the desert. Similarly, the coastal belts contain thick layers of salt, where most of the salt is evaporated from sea water. Most countries produce some salt, amassing a total of 2 million tons. The United Arab Republic contributes 30% of the output, South Africa 14%, Algeria, Tunisia and Ethiopia 10% each, and Southwest Africa near 4%, Angola and the Sudan 3% each and Tanganyika 2%. The production of Morocco, Senegal, the Cape Verde Islands and Mozambique ranges from 1% to 2% each. Finally, Mauritania, Ghana, Libya, Kenya and Madagascar tend to contribute 1% each. Smaller outputs were recorded in the Canary Islands, Chad, Cameroon, the Congo, Uganda, Somalia and Mauritius.

Potash salts accompany common salt in the salt horizons of some oil fields. Potash layers have been explored in Libya, Morocco, Gabon and in the Congo Republic. Reserves of northern Ethiopia have been estimated at 6 million tons. Both these and the Congo deposits are being developed.

Sodium carbonate. Lake Magadi, Kenya, the continent's largest, steady, producer provides 100,000–160,000 tons yearly. Soda has been found in other parts of the Rift Valley and in areas of southern and northern Africa.

Nitrogen (ammonia). The 130,000 ton per year nitrogen compound production of Egypt is expanding. South Africa produces 40,000 tons. Ammonia capacity has expanded to 160,000 tons. Similar production is also anticipated in Southern Rhodesia and Morocco. Algerian plans call for a production of 60,000 tons of nitrogen per year. Salpetre occurs in the sub-desertic areas of northern and southern Africa.

MISCELLANEOUS

Feldspar. Commercial production is limited to South Africa (30,000 long tons) and to a symbolic output in Erythrea. Extraction has been suspended in Southern Rhodesia and Madagascar.

Sillimanite. South Africa is distinguished by larger reserves of sillimanite than Southwest Africa. Southern Rhodesia kyanite reserves attain 300,000 tons. Other kyanite resources have been found in eastern Congo, Uganda, Kenya, Tanganyika and Malawi. Andalusite occurs in eastern Congo and in Ghana. South Africa produces 60,000 tons and Southwest Africa 2,000 tons, largely sillimanite.

Corundum. Southern Rhodesia produces almost 4,000 tons, South Africa 100 tons. Mozambique production has ceased. Evaluation of a Tanganyika deposit has begun.

Garnet. Industrial garnet production has reached 100 tons in Madagascar and also continues from Tanganyika. Large reserves of garnet are available in Africa.

Semi-precious stones. Madagascar production of citrine, labradorite, ornamental garnet, amazonite, quartz, tourmaline, jasper, etc., excluding jewelry beryl, is in excess of 10 tons. Southwest Africa produces emeralds, garnet, jade, topaz, amazonite, aquamarine, etc. South African production of tourmaline, amethyst and tiger eye is in excess of 5 tons. Emerald production reaches the markets unreported. Nephrite jade and emeralds have been found in South Africa, and ruby and sapphire in Tanganyika.

Meerschaum. Tanganyika production of sepiolite has diminished to just over 10 tons. South African production has ceased. The mineral also occurs in Kenya, Somalia and in northern Africa.

Quartz. Madagascar has produced 4 tons of piezo-electric and 15 tons foundry quality quartz. Electrical grade quartz also occurs in Tanganyika.

Scarce metals. Bauxites and zinc ores contain 0.004% of *gallium.* African sources are undoubtedly considerable. The highest concentrations of gallium occur in germanite, found only in Southwest Africa; 0.0024% *indium* occurs in zinc ores. *Tellurium* is a by-product of gold, lead, copper (and oil) deposits; 0.00007% rhenium accompanying molybdenum is scarce in Africa.

Building and roadbuilding materials are abundant in most of Africa. Their supply is limited by demand. Production of these materials, and also of sand and gravel is greater in the more developed parts of the continent, i.e., northern and southern Africa.

SECTION IV

The history and development of mining

THE BEGINNINGS OF MINING

The search for water and salt is as old as the human race. At the dawn of civilization, metal mining was started in Africa. Indeed, when underdeveloped nomads still roamed in Europe and North America, Africans already extensively used metals. By the Vth millennium BC *copper* made its appearance in Egypt. Under the Pharaohs, slaves of the Egyptian state worked the *gold* mines of the Southeastern Desert and Berenice; and until lately, forced labour continued to work Ethiopian gold mines. In the VIIIth century BC, Ethiopian invasions forced the Ptolemeans to abandon the mines of the Wadi Olaki. Here, Agatharcides and Diodoros described sophisticated mining and smelting methods, including sluice boxes and cupellation which are methods that are still used in our times. To the west, the Berbers and then the Romans worked lead, limestone and salt and used bitumen. Another northern-African, Hannon, the Carthaginian navigator, visited the coast and bought gold from the Bambutos. Although the Romans returned there, it was not until 1714 that a French explorer saw the placers of the Upper Falémé in Senegal. Long before this, in the Xth century AD, Persians established their gold trading posts in Monomatopa, south of the Zambezi, and Arab geographers described the metal mining of western Africa.

The fame of the golden throne of the King of Ghana reached Leon the African, in the XIth century AD; El-Bekri by this time had already visited the *iron* mines of Fort Gouraud which, rediscovered by pilots, recommenced operations in 1963. In 1354, Ibn Batuta, another great Arab explorer, reached Azelick and In Gall on the caravan trails of the Niger and found that copper mining had gone on there for many years. Indeed, already in the IVth century, *iron* appeared in Egyptian tombs and western African artifacts. In central Africa, the age of iron appears to have come only a few centuries ago, and small scale smelting continues even now. Gold mining continued in the Red Sea Hills and in Ghana until 1873, when British soldiers looted Ashanti. Obuasi reopened as an industrial operation in 1899. However, geologists have evaluated that the total pre-industrial production was less than a recent year's output.

FROM THE XVIth TO THE XXth CENTURY

After the establishment of the Lunda Empire, the *copper* of Katanga reached Europe.

Indeed, the XVIth century archives of the Oud West Indische Compagnie of The Hague record that copper transported from the interior to the estuary of the Congo River was regularly shipped to Europe. In the hinterland of the coast in the XVIIth century Catholic Kingdom of the Congo, specialists already worked the deposits of the Mavoio–Niari area. Similarly, in the XVIIIth century, each caravan travelling the Bihé trail, a route the Benguela railway now follows, or the Pungo–Andongo path to Ambriz or the Zaire transported, along with ivory, 5–7 tons of copper. The slaves spent 3–8 months en route before they reached the coastal strip controlled by the Portuguese. Ladislas Magyar, a naval officer who married the daughter of the King of Bihé, vividly described the horrors of caravans. In 1853, the Hungarian explorer also visited another area where Africans smelted copper, namely, Tsumeb in today's Southwest Africa. In 1688, south of the Orange River, Dutch explorers reported on the copper of O'okiep and by 1837 a British company planned the exploitation of these. But to return to Katanga, in the second half of the XIXth century under the control of the Bayeke warriors, the exploitation of copper took place each season after the sorgho harvest. However, experts believe that during that period not more than 1,000 tons of copper were smelted on both sides of the Copperbelt. Then, at the turn of the century, Africans led the British and Belgians to the ancient workings, and when the railway reached the Copperbelt in 1911, the world's largest copper complex commenced operations.

Let us now turn southward to the Cape. Unlike gold, Africans did not appreciate *diamonds* until after their discovery at Kimberley in 1867. By 1887, 2 years before one of the great recessions of our economic system, production already was in excess of 3 million carats! (Compare with the present 4.4 million carats). But miners soon turned their attention again to *gold*. While for centuries Africans washed gold, not until 1872 did Australian and American prospectors find gold in South Africa. The Witwatersrand was not discovered until 1884, and the goldfield of the Orange Free State had to wait for its exploitation until after World War II. This activity triggered the mining of *coal*, a fuel Africans already used. In 1888, the Portuguese granted a concession at Tete. In South Africa, from its beginnings in 1859, its output was raised to 70,000 tons in 1891 and to a million tons in 1894! It was not until 1904, when the railway reached Wankie and the bridge on the Zambezi, that this colliery could sell coal. The Malagasy already had noticed another form of carbon, and for centuries used graphite as a pigment. However, graphite was not sold for export until 1907. Meanwhile, in Southern Rhodesia, industrial gold mining commenced in 1896, in northeastern Congo in 1905 and south of Lake Victoria in 1921.

If we now turn our attention to another focus of industrial development, northern Africa, we witness the discovery of the pyrites of Guelma (in Algeria) in 1848, the beginnings of 90 years of mercury mining at Ras el Ma and Taghi in 1860, and the opening of the Aïn Mokra iron mine in 1865. In 1873, Philippe Thomas, a geologist, discovered the first *phosphates* of Algeria and, in 1886, those of Gafsa in Tunisia. In 1917, one of his associates, Major Bursaux, discovered Khouribga and, also in 1917, General Lyautey founded the Office Chérifien des Phosphates, which was to 'outproduce' Tunisia. Meanwhile, in 1910 already, R. Chudau explored the phosphates of the Thiès Plateau, but not until after World War II did production start in Senegal. As to coal, lagging behind South Africa, the deposits of Colomb-Béchar (Algeria) and Djerada were not discovered until the first decade of this century. Their exploitation started respectively in 1918 and 1931, and they are now already nearing their closure.

THE INDUSTRIAL REVOLUTION

After the railway reached Selukwe, Southern Rhodesia exported 9,000 tons of *chromite* in 1906. Although chromite had been discovered in 1868, the first shipment did not leave South Africa until 1929. Union Minière commenced the exploitation of cobalt in the same year. Platinum, which neighbours chromite in the Bushveld, was first recovered in 1924. Although Czarist Russians also discovered this noble metal in Ethiopia, output always remained modest. However, Katangan palladium–platinum production reached the mark of 12,000 ounces in 1936. The British worked *tin* in northern Nigeria in 1905, and Manono in the Congo followed in 1920. However, with the added production of Maniema in 1942, Congo production took the lead. From the same areas, the export of columbite–tantalite commenced in 1933. At Broken Hill, miners extracted *zinc* in 1913, and at Kipushi, in the Congo, production started in 1936. In 1913, mica mining commenced in Madagascar, and was followed in 1920 by South Africa and Southern Rhodesia.

As to *diamonds*, following their 1909 discovery in Southwest Africa, geologists found the gems in the southern Congo in 1913, in Angola in 1917, in Ghana in 1919, in Sierra Leone in 1930, and off the coast of South Africa in 1960. Following the early beginnings of *manganese* mining at Nsuta in Ghana (1916), and after World War I in South Africa and Morocco, operations started in the Congo in 1956 and in the Gabon in 1962. The second Iron Age started slowly in South Africa in 1934, in Liberia in 1951 and in Sierra Leone in 1953, but with our decade it has gathered impetus in Liberia and Mauritania, and other deposits are awaiting the capital investments permitting their exploitation.

Let us now take a look at the remaining 'late comers'. Pyrites production in South Africa did not start until 1938, and the manufacture of sulphuric acid in Katanga also started late. Conversely, uranium production in Katanga lasted from 1922 until 1961. The gold mines in South Africa initiated uranium recovery in 1951.

Production of oil at Gemsa did not start until 1911, although previous to this people saw *oil* seep from many areas and already used Egyptian bitumen. Although drills occasionally tested the Sahara as early as 1871 and the Gabon Basin in 1931, the exploration of oil needed another stimulus. In 1948, Conrad Kilian, the Alsatian geologist, deposited a sealed document with the French Academy of Sciences in which, to those who could read it after his death, he revealed that the Sahara must contain oil. Another geologist, Nicolas Menchikoff, shared this view and the Suez crisis gave impetus to accelerated exploration. Following the discovery of the small Djebel Berga gas field in 1954, the year 1956 saw the finds of Hassi Messaoud, Edjeleh and the gas of Hassi er Rmel. In 1957, Shell found oil in Nigeria. In 1959, Esso located the Zelten field; the latest find of the Oasis group, Gialo, only came in 1962.

THE 'UNFOLDING' OF THE INDUSTRY

Prior to the industrial revolution, simple, labour-intensive operations, and primitive communications and marketing systems effectively limited output. On the other hand, the building of railroads and river navigation systems, i.e., the opening up of Africa, was generally based on mineral prospects. However, two forces combined to obtain these results; these forces were the rapid capital formation in Europe at the turn of the century, and the dynamics of nascent industries, liberating capital and talent which individuals, and later companies, invested in Africa. With technocratic management, wages rose rapidly from their previous subsistence levels to considerably surpass the wages prevailing in other African industries or in many other underdeveloped areas; however, compared to the western world, they still remained low. Thus, not only could Africa provide metals and precious stones which were scarce elsewhere, but the companies could guarantee an incentive to investors and also start autofinancing of major projects. At the same time, the agglomeration and structuring of the big corporations provided living standards that were hitherto unknown.

As we have seen, the mining revolution in Africa started from three centres: southern Africa, northern Africa and the Copperbelt. But it spread rapidly. The period following the Great Depression was spent in regaining the lost ground, but then rapid expansion started and is still gathering impetus.

Except for industrial agriculture and trade, in which the Lever Brothers diarchy is undoubtedly the leader, and for developing communications, textile, cement and similar industries of the first manufacturing phase, mining still dominates the scene. What is then the structure of the African mining world? With shares issued to holders, frequently small investors, mainly in the United Kingdom, South Africa, Belgium, the United States and France, the control of groups is more important than ownership proper. Bearing this in mind, we immediately find a dozen financial groups and states controlling important segments of the industry.

THE STRUCTURE OF THE SOUTHERN AFRICAN GROUPS

The Anglo-American Corporation manages a quarter of the continent's mining industry. Although assets of the group may reach $4,500 million, the Corporation's own assets, having lost 20% of their value after the Sharpeville shootings, are only in the range of $300 million. The Corporation has five 'prongs'. It controls a third of the continent's gold, copper and coal production but somewhat less of the uranium output. Marketing even Soviet (but not Ghana) diamonds in a near monopoly, the Central Selling Organization of De Beers disposes of $250 million worth of gemstones each year. Although only two of its members sit on these boards, the Oppenheimer family is believed to hold a substantial part of the shares of the group.

Being the strongest and largest of the mining houses, the Corporation also finances projects in other groups. With assets of $300 million, the expanded Rand Selection Corporation now carries out a considerable part of these activities. Indeed, this financial corporation was in a position to raise $100 million on the international markets. To facilitate both-ways investment in southern Africa and the United Kingdom and North America, the base of Rand Selection was enlarged in 1962 through acquisition of the capital of De Beers Investment Trust (a subsidiary of De Beers Consolidated Mines) and of substantial investments of the Anglo-American Corporation, De Beers, British South Africa Company subsidiaries, Central Mining, Johannesburg Consolidated (JCI), Engelhard Hanovia, Federale Mijnbouw Beperk and INCO (the International Nickel Company of

Canada). Many of these acquisitions have been achieved through exchanges of shares. Consequently, we find (minority) cross holdings and common members on boards. Among these, the British South Africa Company (BSA, which owns the royalty rights in the Rhodesias.) stands out because the Anglo-American Corporation appears to be the leading holder of its stock.

The Gold Fields group and the Union Corporation have considerably smaller interests in BSA. A recent exchange of stock between JCI and Rand Selection introduced Anglo-American to the platinum industry, and simultaneously facilitated the financial position of Johannesburg Consolidated. As to the General Mining and Finance Corporation (gold mines), observers believe that Anglo-American has controlling interests in this group. In the Rand Mines and in Anglo-American, the Engelhard interests represent the relatively largest American participation in South Africa. Indeed, the American–South African Investment Company has assets in the range of $30 million. Engelhard Hanovia and the Oppenheimer group have a 10% interest in each other, with Engelhard industries controlling 72% of the stock of Engelhard Hanovia. (While the latter owns 20% of the stock of Minerals and Chemicals Philipp, the directors of this company have acquired an 8% interest in Hanovia).

On the other hand, the name of the Anglo-American Corporation does not reflect United States ownership. To return to the Central Mining, minority interests of the French branch of the Rothschild are also represented in this group, as well as in De Beers, the Franco-Algerian mixed state oil companies and the Mauritanian iron mines.

However, traditionally with South Africans, Britons are the most heavily represented on the boards of the various groups. Among the leading shareholders of the General Mining and Finance group, we should point out the Scott family, and in the Gold Fields and Anglo-Transvaal groups, the Drayton family (who are also active in western Africa). The connection between the Union Corporation and Hambro's Bank (in London) is self-explanatory.

THE STRUCTURE OF OTHER GROUPS

Controlling less than 10% of African production, the Société Générale de Belgique ranks second in its influence which spreads from the Congo to Angola and Rwanda. For further details concerning this complex, we refer the reader to the Congo chapter. Recently the big oil companies, especially Esso Standard which was founded by John D. Rockefeller, the Oasis triad and Royal Dutch Shell developed large stakes in Africa. Through their holdings of Tanganyika concessions shares, the Rockefeller Brothers have also strong but indirect interests in Union Minière du Haut Katanga. The Selection Trust group (not to be confused with Rand Selection) cooperates with De Beers in Ghana and Sierra Leone and with the Newmont Corporation in Southwest Africa. Its portfolio includes Central Mining shares and it also operates in Bechuanaland and Southern Rhodesia; however, the majority of its interests are concentrated in Zambia.

Whereas such companies as Rio Tinto, Newmont, Union Carbide, Peñarroya and the Vanadium Corporation have a long tradition of African investments, the interests of Frobisher's, Bethlehem Steel, United States Steel, Republic Steel and Swedish companies, especially in the new iron and manganese industry, and of Olin–Mathieson and Péchiney in the development of bauxites, is more recent.

As to the states, undoubtedly France has the greatest stake. Although Shell has a majority interest in CREPS, French mixed state companies manage both of the big Algerian oil corporations. Through other state corporations and especially the Bureau de Recherches Géologiques et Minières (BRGM), France has invested heavily in the exploration of sixteen

TABLE I

PARTICIPATION OF MINING GROUPS IN SOUTHERN AFRICAN MINING

Groups	*Share* %	*Value* *$ Million*
Gold Fields	14	180
Central Mining	12	150
Union Corporation	8	100
General Mining	8	100
Anglo-American:		750
South Africa	30	380
Zambia	65	250
Southwest Africa	60	50
Tanganyika	80	15
Southern Rhodesia	10	10
Swaziland	40	10
Congo–Angola		(5)

TABLE II

PARTICIPATION OF OTHER ORGANIZATIONS IN AFRICAN MINERAL INDUSTRIES

Organization	*Country*	*Share %*	*$ Million*
Société Générale	Congo	85	340
	Angola	30	8
	Rwanda	30	2
French state	Algeria	65	>300
	Mauritania	34	10
Shell	Algeria	20	>100
Shell-BP	Nigeria	80	100
Oasis	Libya	50	>200
Esso	Libya	45	200
Selection Trust	Zambia	35	130
	Southwest Africa	10	10
	Sierra Leone	30	15
	Ghana	30	15
Moroccan state	Morocco	60	100
United Arab Republic state	Egypt	100	100

republics. Consequently, BRGM has accepted a minority interest in Mauritanian, Senegalese and Gabonese mining companies. As to African states, the United Arab Republic controls Egyptian industry and the Chérifian state corporations dominate the Moroccan scene. To illustrate this, we have summarized in Tables I and II the order of magnitude of *(a)* the value of the output controlled by the major groups during one of the years of the early sixties (or, for oil, the potential value) and *(b)* the share of these within the mining industry. Similar income from profit sharing and taxes is derived by the states. Undoubtedly, with its 70% share of the revenues of mining companies, the Congolese state ranks first in this respect. The historic cause of this is that in the Congo, the state or state participations, instead of the British chartered companies, received the royalty or conceding rights which were capitalized in the mining corporations. This contrasts with a taxation averaging 30% of the profits in southern Africa. But in a more modern development, the '50–50' profit sharing agreements are now standard features of new mining ventures. This principle has been applied in the Libyan oil industry, in the Liberian, Mauritanian and Gabonese iron ore industries and elsewhere. Of course, the largest profits are to accrue to Libya. Indeed that country will become one of the wealthiest in Africa.

MINERAL DEVELOPMENT

We have already indicated the future: further expansion of the gold, copper, asbestos and other industries; and the accelerated development of a new generation of minerals: oil, gas, iron ores, manganese and bauxite. This entails the industrialization of hitherto undeveloped areas of Algeria, Liberia, Mauritania, Guinea and Gabon. The second industrial revolution demands local processing of the continent's wealth, the development of building and manufacturing industries, and a higher consumption. Unlike previous expansions financed by private means, foreign and international organizations will take an ever increasing role in the financing of these projects. Africa received almost 40% of the aid given to all developing areas. Table III gives the proportions of foreign aid to Africa in 1960; of this foreign aid only a minority is devoted to mining.

During the same year, the Soviet Union committed $220 million to the United Arab Republic and $40 million to Ghana. The International Bank for Reconstruction and Development is the first among various agencies, such as the Common Market, the United Nations and others, whose aid if compounded represented 40% of the total. Out of this aid, hydropower, water and educational projects have received a considerable share.

TABLE III

FOREIGN AID TO AFRICA IN 1960

Source of aid	*$ Million*
France	730
United States	230
United Kingdom	140
International Bank	130
Belgium	90

In 1962 United States aid to Africa reached the mark of $700 million, but $220 million of this amount was concentrated in the United Arab Republic and $130 million in Ghana; these are neutralist countries in which the United States competes with the USSR. On the other hand, IBRD loans totaled $230 million, and with the signing of the Douala treaty between the European Economic Community and the 18 associated (overwhelmingly French speaking) countries, the $130 million aid of the European Development Fund (FED) is expected to be expanded considerably and to complement abundant aid provided to these countries by the French Fund of Aid and Cooperation (FAC, the successor of FIDES). Conversely, the scope of British aid and loan facilities to English speaking Africa is strictly limited. United Nations aid is, however, increasing within its modest limits. Out of a 1946–1962 total of $46 million the UN Special Fund has provided $23 million in 1962, with the UN Technical Assistance contributing $10 million that year. Prior to 1962 the International Development Association and the International Finance Corporation supplied $10 million. Indeed, a new phenomenon, multilateral aid, is rapidly progressing and coordinated plans for the development of the continent and each country have become imperative.

We will now discuss the resources of each country. To do this, we have divided the continent, in addition to the islands, into five areas. Political frontiers correspond neither to geographic nor to ethnic boundaries. Consequently, these divisions are arbitrary. We believe, however, that they best serve our purposes. The smallest of the areas, southern Africa, will remain for some years the leading producer, with both middle (that is, central and eastern) Africa and northern Africa outranking western Africa. Although it is as large as the other areas, northeastern Africa trails the list. Although the following chapters give the first comprehensive account of these industries, the reader will find additional data in the second part of this book, especially concerning smaller occurrences of ordinary minerals and oil, gas, hydropower, quarry products and water.

SECTION V

Northern Africa

The Maghreb of the Arabs, Morocco, Algeria, Tunisia and Libya, is distinguished by the continent's largest oil, gas, phosphate and lead resources and by a considerable degree of industrial development.

MOROCCO

The kingdom of Morocco, ruled by a descendant of a long line of sultans, is inhabited by 12 million Moroccans. Within its area of 150,000 square miles, the kingdom holds the world's largest phosphate resources, which provide more than half of all minerals sold by Morocco. In 1961, the total value of the mineral product reached $160 million, a 30% increase in a year. But plans call for a 50% increase of phosphate production by 1965. The value of the output would then automatically reach $200 million.

Mines are arranged in four main areas. Eighty miles south of Rabat and Fès, the geographic centre (C) of the country contributes the phosphates, lead and zinc and some iron. Its northeastern corner supplies more of these metals and all the coal, antimony and clays (Fig. 10). Smaller centres include the southeastern corner, the area of Marrakech and others (Fig. 56). In 1960, seventeen deposits contributed 93% of the total value of minerals as can be seen in Table IV. How then did this production develop? A glance at Fig. 9 and Table V gives an idea of its expansion.

TABLE IV

DISTRIBUTION OF PRODUCTION OF MINERAL RESOURCES IN MOROCCO

Mineral resource	*Mining site*	*% of total prod.*
Phosphates	Khouribga, Youssoufia (C)	100
Lead }	Bou Beker, Touissit (NE)	98
Zinc }	Aouli, Mibladen (C), jbel Aouam	76
Manganese	Imini, Bou Arfa	80
Iron	Ouichane, Setolazar (NE), Aït Amar	95
Anthracite	Djerada (NE)	100
Oil	Sidi Kacem	100
Cobalt	Bou Azzer, Aghbar	100

TABLE V

DEVELOPMENT OF PRODUCTION (1,000 TONS) OF COMMODITIES IN MOROCCO

Mineral resource	*1922*	*1933*	*1942*	*1951*	*1962*	*% of value*	
						1960	*1962*
Phosphates	90	200	800	5,100	9,000	54	62
Iron ore			3	600	1,200	9	7
Manganese ore	1	5	50	400	520	12	12
Lead	3		7	70	100	12	8
Zinc			1	20	40	4.5	3
Anthracite		30	130	420	410	4	3
Oil (1,000 barrels)		7	30	500	970	1.5	1.5
Cobalt		0.1	0.3	0.8	1.6	1.5	1.5
Copper					2.3		
Antimony				0.9	0.5		
Fluorite				1.1	0.5		
Salt				50	30		
Barytes				3	100		
Fuller's earth				13	40		

What then is the structure of Moroccan industry? The state owns the principal sources of production, the Office Chérifien des Phosphates (OCP) and the Société Chérifienne des Pétroles; with the investments of the Bureau de Recherches et de Participations Minières (BRPM), it dominates the Charbonnages Nord-Africains, i.e., all the phosphate, coal and oil production; with the Société Anonyme Chérifienne d'Études Minières, it controls half of the manganese output, as it does five other mixed companies in which BRPM is associated with private capital. Nevertheless, all these are managed as private corporations as are six others in which this type of capital exclusively prevails. Phosphates represent more

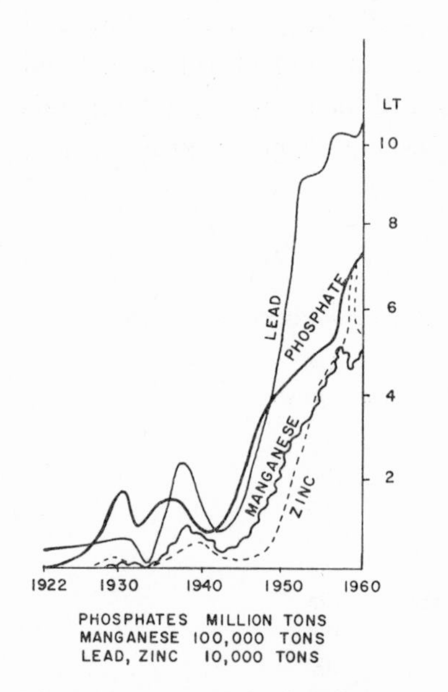

Fig. 9. Moroccan production of phosphates, lead, manganese and zinc.

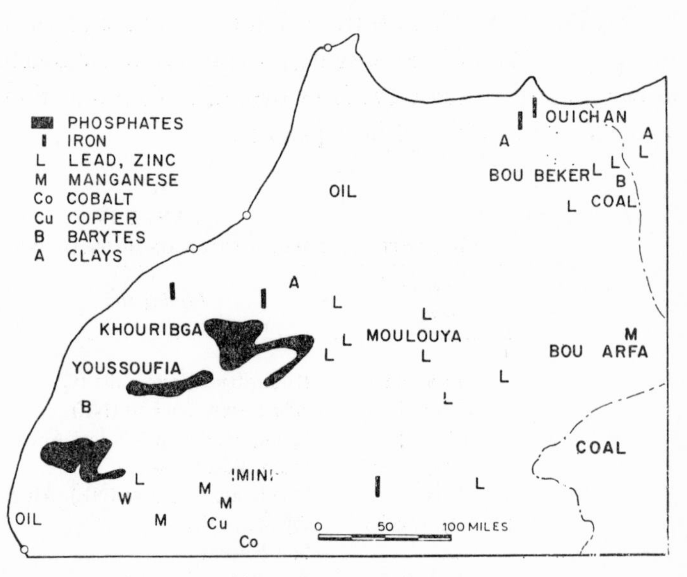

Fig. 10. The mineral resources of Morocco.

than 20% of all Moroccan exports, with other minerals contributing another 20%. France is the outstanding importer of Moroccan products, followed by other Common Market countries and the United Kingdom (phosphates and iron ore) and the United States (manganese).

PHOSPHATES

Reserves of 21,000 million long tons, mainly at Khouribga but also at Youssoufia, place Morocco at the head of the world's phosphate resources. Indeed, these represent half of the world total. The dynamism of a private corporation distinguishes the OCP, which is the largest corporation of Morocco. In 1960, it paid $22 million in wages (excluding social benefits) to its 14,000 employees and it has consistently contributed 8.5% of the Moroccan budget; its dividends represent $20 million each year and various taxes $29 million. To emphasize their importance, we should say that phosphates represent more than 40% of all the foreign exchange earned by the Moroccan treasury, 40% of the transports of the national railways, 80% of the traffic of Casablanca Harbour and 87% of shipping of Safi Harbour where the ores of Youssoufia are routed. In 1962, Youssoufia, located 50 miles from Safi, produced 2.2 million long tons; Khouribga, 90 miles southeast of Casablanca, contributed 6 million long tons. OCP markets four grades (Table VI).

To develop the highest grades of phosphate from low grade ores, OCP and Dorr–Oliver plan new methods of chemical washing.

In 1926, OCP established the superphosphate plant of the Société Chérifienne d'Engrais

TABLE VI

COMPOSITION OF FOUR MOROCCAN PHOSPHATE GRADES

Substance	*80/82*	*Calcined*	*75/77*	*70/72*
P_2O_5	37.4	35.8	34.5	32.5
CaO	53.9	54.7	52.4	51.8
F	4.6	3.3	4.5	4.4
CO_2	2.3	0.6	4.1	5.7
SiO_2	1.3	2.4	2.4	2.7
Moisture, Organics	1.1	0.6	1.5	1.9
SO_3	0.6	1.7	1.6	1.4

TABLE VII

SALES DISTRIBUTION OF MOROCCAN PHOSPHATE EXPORTS IN 1962

Country	%	*Country*	%
France	15	Scandinavia	12
United Kingdom	11	Poland	6
Belgium	10	The Netherlands	5
Western Germany	9	Italy	4
Spain	8	South Africa	4

et de Produits Chimiques at Casablanca. Since 1941, OCP has produced fine grained Kourifous at Khouribga and Calcofos at Youssoufia. Since 1944, the Société Chérifienne des Engrais Pulvérisés has produced hyperphosphate at Berrechia and Mohammedia, and the Société des Hyperphosphates Reno has built a similar plant at Safi. Also, a new chemical complex at Safi will produce super triple phosphate and ammonium phosphate with sulphuric acid processed from the pyrrhotites of Kettara (near Marrakech). With inadequate soil resources, the use of phosphates in Morocco is imperative.

Where do Moroccan exports go? The Common Market takes 40% of the production. In recent years, while exports to Italy, South Africa and China decreased, Poland, Ireland and Japan became important importers (Table VII).

PHOSPHATE MINING

Deposits extend from Tadla to Lake Zima

Middle Atlas Bekrit–Timhadit district
- Oulad Abdoun
  - Khouribga–Oued Zem (worked)
  - El-Borouj
  - Kasba–Tadla
  - Northern Atlas border
- Ganntour (worked) and Adrar
- Meskala (near Essaouira)
- Chichaoua–Imi-n-Tanout

South of the Atlas
  - Oued-Ergita (Agadir to Tichka)
- Khela-Ouarzazate

Of these deposits, only two are worked. In 1921, OCP started operations at Khouribga and in 1931, at Youssoufia (EYSSAUTIER, 1952). On the Oulad Abdoun plateau, Khouribga covers an area 50 by 30 miles (Fig. 132). In the beginning, miners worked underground in the richest bed which contained 75% tricalcic phosphate. Surveyors divided the deposit into numerous sectors containing 10–20 million tons of ore. Opened in 1929, Sector I (or Bou Lannouar) has now been worked out, as has Sector III (Bou Jniba), which was

TABLE VIII
MOROCCAN PHOSPHATE PRODUCTION

Sector	*Long tons per day*
II	8,000
VI	4,500
Grouni	4,500
Opencast	3,000

opened in 1931. During the Great Recession, the working of Sectors II and IV (or André Delpit in the east) were suspended. The present pattern of production is shown in Table VIII.

At the end of the incline shafts, miners drive 300–660 ft. long crosscuts every 12–15 ft. Since the hanging wall is not very strong, miners can only extract 15 ft. wide trenches; the pillars are then worked out through retreat drives. Miners use picks, hammers to dig the ore and the phosphate reaches the surface on conveyor belts. At Sidi Daoui, seven inch holes are blasted at 12 ft. intervals, shovels remove up to 45 ft. of overburden (overburden/ore ratio = 4/1). Trucks and, at Sector II, conveyor belts move the ore to the plant.

Here, wet screening eliminates 10% of the gangue (larger than three-fifths of an inch). Then horizontal rotary ovens, each rated at 60 tons per hour, reduce moisture content from 15 to 1.5%. From 100,000 ton bins, the Khouribga 75/77 phosphate reaches railcars. From part of the stock, dry screens eliminate 4% of the impurities and then tilt rotary furnaces decarbonate the phosphate at 950° C. After elimination of lime and organic matter, the calcined phosphate is ready for processing. Since 1961, at Sidi Daoui, experts have deschlammed or podzuolized phosphates in situ by trickling water. Then, after being oversized and screened, this ore is moved through logwashers, cyclones and Escher-Wyss centrifugators, and is finally dried. The plant can produce 750,000 long tons of 80–82% phosphate each year.

At Youssoufia, which has an even weaker roof than Khouribga, drifts are only 7 ft. wide and crosscuts are driven at intervals of 34 ft. On the other hand, the thickness of each layer now attains 20 ft. Since this relatively low grade (70–72%) ore is dryer, it requires less heating.

IRON ORE

Although iron ores are widely scattered, resources and output are concentrated in the hinterland of Melilla. Recently, the harbour of Nador drained the ores of Ouichan (Uixàn) and Setolazar which combined represent 50 million tons of ore. Other estimates specify a total of 55 million tons containing 50–60% iron. Based on these resources, the government plans to set up a steel mill. While the ore of Ouichan easily finds markets in the United Kingdom, Germany and France, the products of Setolazar contain more sulphur and the ores of Aït-Amar have to compete with high grade ores. In 1960, the United Kingdom bought 46% of all exports, Germany 27%, with France, The Netherlands, Spain and

TABLE IX

DISTRIBUTION AND COMPOSITION OF MOROCCAN IRON DEPOSITS

Deposit	*Area*	*Million tons*	*Fe* %	*SiO_2* %	*P* %	*Others* %
Kettara	NW of Marrakech	100(?)	51	13.5	0.03	S 0.5
Tidsi	E of Agadir	100(?)	42	7.5	0.1	CaO 12
Jebel Ougnat	SE corner	50	52.5	7.5	0.8	
Khenifra	NE of Tiznit	30	43	9	0.3	BaO 15
Oulad Saïd	SW of Casablanca	20	35	23	0.7	
Aït-Amar	Khouribga	10–25	45.5	14	0.6	Mn 0.3
Ouarzemine	S of Agadir	10	42.5	13.5	0.6	

Czechoslovakia following in that order. The Uixàn mine contributed 77% of the national total (two-thirds of this is no. 1 ore) with the production of Setolazar representing 18% of the total. Recently, the Compañia Española de Minas de Rif switched from the room and pillar method to gloryholes. The management also prepared the exploitation of the Sidi-Brahim sector and miners rebuilt several kilns for desulphuration. The plant of Ataloyan now comprises screening, heavy media, Humphrey spiral and spiral classifier sections. In the Setolazar sector, while geologists explored the Chérif and Iberkanen Zones, miners developed various levels of Gibraltar, Chérif and Bocoya. Contractors work the small deposits bordering the Masin and Uixàn Rivers.

In addition to those of Nador, the resources available in 1959 are shown in Table IX. In 1961 the output of Aït-Amar expanded to 300,000 tons, but the mine was closed in 1962. Of these low grade ores three-quarters were stripped opencast, the rest was mined underground. The Société Marocaine des Mines et Produits Chimiques occasionally exploited the mine of Tidsi in the Sous Valley. The contribution of all other deposits only attains 7% of the total, which has declined in 1962 to a million long tons.

MANGANESE

Increasing exports from Gabon reduce the demand for manganese. Consequently, the output is rapidly decreasing. Traditionally, the Société Anonyme Chérifienne d'Études Minières (SACEM) produces at Imini all the chemical (high) grade ore containing 84% MnO_2 and almost half of the metallurgical grades (containing 48% Mn). Each year the Société Anonyme des Mines de Bou Arfa ships 100,000 tons of ore from its deposit located north of Colomb-Béchar (Algeria) and 180 miles south of Oujda. In 1963, Tiouine (south of Imini) ceased production. To counteract this, Tieratine expanded its production to 25,000 tons, with Tisgui–Illane contributing. Furthermore, Idikel, southeast of Agadir and M'koussa between Marrakech and Khouribga, contributes a few thousand tons. We could also mention Tifernine and Fint. However, with 80 workings in the area of Ouarzazate, and some others around Oujda, miners collect some manganese in many localities.

Beyond the ridges of the High Atlas, there is the problem of communications. Indeed, although known since 1918, mining only started at Imini in the thirties. No recent data are available but in the early fifties, geologists estimated 10 million tons of ore in the sectors of Sainte Barbe, Tazoult, Bou Azzer, Tighermit and Assaoud. While SACEM ships high grade ores directly to the Sidi Marouf plant near Casablanca, lower grades are first treated on the pneumatic tables of Bou Azoult. After adding 10% fine anthracite and 15% water, the company exports sinter containing more than 56% manganese. In its Aïn Beïda sector, SMBA follows a similar flowsheet. The sinter is transported by rail from Bou Arfa to Nemours harbour and thence to France. Similarly, on the national scale, France takes 80% of the metallurgical and 20% of the chemical grades of manganese, with the United States importing 70% of the chemical and 10% of the metallurgical grades.

THE BASE METALS OF OUJDA

The lead–zinc deposits of Bou Beker and adjacent areas in Algeria have brought fame to the Société des Mines de Zellidja, and the Company made the Lacaze family famous.

Mr. Lacaze chairs a fourteen-man board on which Peñarroya, Newmont and OCP hold vice-chairmanships, with Moroccan State organizations — the Bureau d'Études et Participations Industrielles (BEPI) and the Bureau de Recherches et Participations Minières (BRPM) — holding two other memberships and the Newmont Corporation and Peñarroya each occupying one more seat. Zellidja, with a capital of $5 million, has invested $27 million.

To achieve this, the company contracted loans, which it repaid between 1951 and 1963, from the Newmont Corporation and from another United States company, St. Joseph Lead, at the flat rate of 1% of its output. Consequently, although Newmont only owns 2.5% of its equity, the company, as we have seen, is strongly represented on the board of Zellidja. Other common interests include the Société Nord-Africaine du Plomb (NAP) and ALZI. Zellidja controls 51% of adjacent NAP, with Newmont sharing 32% and St. Joseph Lead 17%. Zellidja leased a 40 ton per day mill to this operation. Algerian ores represent 30% of the throughput of this plant, but the reserves of NAP are being depleted. Indeed, in 1951, with loans of the United States Administration of Economic Cooperation, the group built its Cheraga plant a mile east of Bou Beker.

During the Algerian war, output varied with local conditions, as did dividend payments to Newmont which declined from a maximum of $500,000 per year. Furthermore, with the BRPM, Zellidja has explored the Tazzeka Mountains. In 1961, Zellidja invested $600,000, mainly in exploration. However, with low prices, the profits of the Company declined to $600,000.

While for all practical purposes, Bou Beker remains the country's only major producer of zinc, at Touissit, the lead output of the Compagnie Royale Asturienne des Mines now equals its neighbour. Combined, the two companies now supply almost half of Morocco's lead; however, if we consider the total production since the beginning of operations (after World War I), Bou Beker produced twice as much lead as Touissit. With a capital of $7 million, the Compagnie Royale Asturienne des Mines boasts $12 million of investments, not only at Touissit but also in Tunisia and in European smelters. The Laloux family and the Nagelmackers Bank, founded as early as 1747 in Liège (Belgium), control Asturienne des Mines with the Banque de Paris et des Pays-Bas and Palilux (a Luxemburg-based holding company of Liège investors). In 1961, this company made a profit of $3 million on its various operations.

Within their area of 2.5 square miles, in 1951 the twin deposits encompassed 30 million tons of ore containing a million tons of lead and an equal amount of zinc. Whereas the ratio of lead oxides and sulphides is highly variable, approximately two-thirds of the zinc occurs in sulphides. Both mines now work the room and pillar method with drives in the ore at Bou Beker and in the footwall at Touissit. In 1961 at Bou Beker, miners excavated a million tons of rock, mainly from Sector VI; in 1960 at Touissit, lead output increased by 50% and in 1961 the new shaft system no. 5 of Chebket was inaugurated. Since miners now work lower levels, the zinc output of Touissit has increased. As to Zellidja, in 1960 zinc concentrates consisted of 50% sulphides, 23% oxides and 27% carbonates, but in 1961 the proportion of sulphides rose to 60%.

Since 1951, the Oued el Heimer smelter of Zellidja and the Société Minière et Métallurgique de Peñarroya, 15 miles from the mine, processes all the concentrates of Bou Beker and others; in 1961, it recovered 27,000 tons of lead, a million ounces of silver (a majority of the Moroccan production) and in 1960, 300 tons of sodium antimonate.

OTHER METAL MINES

Producing 30% of the national total in the Midelt 100 miles south of Fès, Mines d'Aouli is the leading producer of lead. The boards of Peñarroya and Aouli share the same chairman. In addition to at least two other members sitting for Peñarroya, the BRPM and the Achellah Company are also represented on the board. With a capital of $4 million, the company has invested $5 million. The company lost $500,000 in an unfavourable lead market in 1961. The Aouli Mine boasts a somewhat larger output than Mibladen, located 4 miles north-east. In 1950, geologists estimated 4.5 million tons of ore containing 4% lead in the Henri Reef. This vein contributes half the ore of Aouli, and the Edmond-Engil Reef supplies one-third. Since 1926, miners have worked underground here, but at Mibladen the ore is predominantly stripped, with 80% coming from the A, B2 quarry. Undoubtedly, the principal discovery of late years was made by the BRPM nearby in the Midelt. This lead deposit also nears its productive phase.

There are numerous other occurrences of lead in the eastern High Atlas and its western end, in the eastern Atlas and elsewhere. Among the smaller mines, we should remember jbel Aouam (between Mibladen, Meknès and Khouribga, 8,000 tons), Ksar Moghal (70 miles east of Mibladen, 10,000 tons), the area of Tafilalt (Tafilalet) and Bou Dahar, Aït Labbès (south of Mibladen) and Erdouz (southwest of Marrakech). The last two and Touissit and Bou-el-Baroud produce some zinc also. The Coopérative d'Achat et de Développement de la Région Minière de Tafilalet, i.e., the state, buys the ore of smaller operators.

Traditionally, France imports 80% of Moroccan lead (followed by the United States and Italy) and 90% of its zinc (with Germany as a poor second). In the Bou-Azzer Graara zone of the Anti-Atlas, the *cobalt* resources of Aghbar (or Arhbar) now dominate the reserves of Bou-Azzer and possibly those of Ightem. Since 1959, cobalt arsenide exports (half of them to France) have remained fairly stable at 1,500 tons contained metal. Nickel also shows in the district.

Though widespread in the Anti-Atlas, *copper* resources are hardly adequate. Followed by the Jebel Klakh and Azegour, the principal producer remains Bou Skour, 170 miles southeast of Marrakech. Germany and Belgium share the imports. In addition to the Anti-Atlas, tin shows in the area of Oulmès (southeast of Rabat), which is distinguished by the 15 ton per year production of El-Karit.

While *antimony* is widely distributed, mainly in the Mrirt southeast of Rabat, only 50 tons per year now come from Tarmilet; the readily accessible Beni Mesala deposit of the Rif dominates the market. Between 1926 and 1947, virtually alone in Africa, Azegour exported 1,700 tons of molybdenum concentrates, but the tungsten production of this mine was never significant. From Tiouit in the Anti-Atlas, a company occasionally exports concentrates containing 100 ounces of gold, silver and 35% copper.

FUELS AND ENERGY

While Morocco is the continent's principal producer of anthracite, her fuel position is far from excellent. Indeed, in addition to all the coking coal requirements of her cement plants (and of steel mills if these are to be established) the country imports more *oil* products than the Société Chérifienne des Pétroles can produce at Sidi Kacem, east of Mohammedia. During the last decade, both oil and anthracite production have decreased by 10%; oil because of the depletion of the northern fields, anthracite because of falling demand. The Sidi Kacem refinery imports 900,000 barrels per year of crude oil from the Soviet Union. The Société Marocaine Italienne de Raffinage (SAMIR), a company shared by ANIC, (a subsidiary of the Italian Ente Nazionale Idrocarburi) and by the state, inaugurated its 9 million barrel per year refinery at Mohammedia in 1952. In 1962, crude production increased to almost a million barrels, but natural gas declined to 260 million cubic ft. At Essaouira on the southern coast, experts of the SCP discovered a 20,000 million cubic ft. gas field, complementing the find of Sidi-Rhalem. SOMIP, the Société Marocaine Italienne des Pétroles continues its operations in the province of Tarfaya. Southwest of the Kechoula gas field, the Sidi-Rhalem oil field of Société Chérifienne des Pétroles is interesting.

Morocco uses half of the *anthracite* the Charbonnages Nord-Africains produces at Djerada south of Oujda, and exports 200,000 tons (mainly to France, Benelux, Algeria and Tunisia); this is the same amount as the coking coal it is obliged to import. Charbonnages Nord-Africains has a capital of $5 million and an investment of $6 million, but the state, which has a controlling interest, is obliged to contribute $120,000 to meet yearly losses. Indeed, the state is obliged to compete with its own hydropower plants centred on the Oued el'Arib and Oum er Rebia. Between 1950 and 1954 power production increased from 500 million to 800 million kWh, reaching 1,000 million by 1960.

INDUSTRIAL MINERALS

Lately, the rate of increase of *barytes* production has been impressive. With 60,000 tons, the Compagnie Minière et Industrielle ranks first. The company extracts barytes on the northeastern and southern limbs of the Jebel Ighoud southeast of Safi Harbour. Further south are the underground workings of the Société Marocaine Minière et de Produits Chimiques. The Isserder, Almezi and Aït-Sourn sectors of Tessaout contribute 30,000 tons. The stripping of the Tnine vein continues.

Having reached its peak of almost 70,000 tons in 1958, the recovery of *salt* has decreased each year, standing now at 30,000 tons. At Lake Zima in the southwest, the Société Chérifienne des Sels is the principal producer, followed by the Compagnie Salinière du Maroc at Sidi-Brahim. Conversely, in the Beni-Oukil area (in the hinterland of Nador), SERAMA expanded the *Fuller's earth* production of Rio Masin and Kasmeoun to 35,000 tons in 1960; the bentonite output of Providencia (also in the northeast) is not far from 10,000 tons. In addition to these, considerable resources of lead have been worked at Gara Ziad near Moulouya. In the Upper Moulouya Valley, Tamdafelt supplies 3,000 tons of detergent clays per year. Each year, the mine of Kettara near Marrakech (see "Iron" chapter) can provide 15,000 tons of pyrrhotite for the manufacture of *sulphuric acid* and 1,500 tons of iron oxides (pigment). The acid requirements of the Safi chemical complex are based on several million tons of pyrrhotite resources.

From its 1952 peak of 3,500 tons, *fluorite* production plummeted to zero by 1957. In 1961, there was again a small output. The best known deposits of this flux are perhaps jbel Tirremi, Bergamou and Gouaïda. Similarly, from its 1948 peak of 800 tons, *asbestos* output (mainly from Bou Offroh near Bou-Azzer) decreased to zero by 1958. The *graphite* exports of Frag el Ma ceased earlier. After the war, workers recovered a total of 550 tons of *beryl* and 300 tons of *mica* in the Tazenakht of southern Morocco.

In 1961, Morocco was the continent's fourth cement producer with 3.7 million barrels. Indeed, limestone is most abundant, notably in the vicinity of the cement plants of Casablanca, Meknès and Saïs. Production is increasing.

As we have seen, Morocco has ample and varied mineral resources. However, the whole country is sub-arid and its southeastern third is almost a desert. Consequently, the further development of water resources is of the greatest importance.

CEUTA AND MELILLA

With a population of 160,000 (half of whom are military), the Spanish presidios cover 12 square miles on the Morocco coast. While there are no mineral resources other than building materials and stone, Melilla used to be the outlet of the iron ores of Ouichan (Uixàn). There is no production in the Ifni enclave.

ALGERIA

While northern Algeria heals the scars of the war, oil continues to flow from the Sahara. Indeed, combined with gas, oil dwarfs all other sectors of Algerian industry, making the republic the continent's fourth mineral producer. This country of 900,000 square miles and 11 million people also holds other important resources (Fig. 11). While the value of yearly iron ore production is in the range of $20 million, zinc and lead fetch up to $6 million, phosphates $3.5 million, with diatomite, bentonite, pyrite and barytes adding more than a million dollars. With peace, these industries look forward to expansion and before the end of this decade we believe that the yearly oil (and gas) output may range around $700 million.

In 1922, Conrad Kilian started mapping the Sahara but few shared his 'dreams' of oil in the desert. Then, after the war, exploration started. In 1956, SN REPAL struck oil at Hassi Messaoud and gas at Hassi er Rmel, and CREPS struck oil at Edjeleh. Before 1965, reserves and yearly output are expected to attain respectively 8,000 and 200 million barrels. These discoveries are the result of intense geological and geophysical exploration for which the

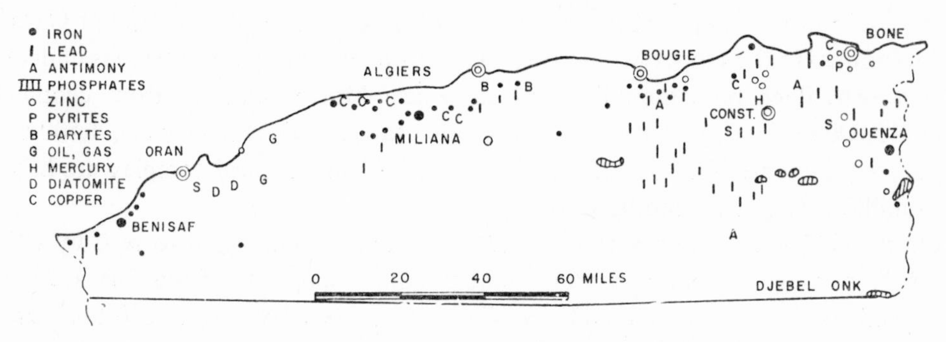

Fig. 11. The mineral resources of northern Algeria.

TABLE X

CAPITAL STRUCTURE OF ALGERIAN OIL INDUSTRY[1]

Share holders		*SN REPAL* %	*CFP(A)* %	*CEP* %	*CREPS* %	*CPA* %
State: France	BRP	48.45		57.4	4.5	4.5
State: France	RAP				51	24
State: Algeria		50				
CFP			86			
Shell					35	65
Cofirep		1.55	7.5	8	3	3
Finarep			7.5	10	16.3	3.5
BP				2.1		

[1] For an explanation of the abbreviations, see the list on p. 638.

French Government provided capital under various forms, with smaller participation from oil companies (Table X).

We can distinguish four groups of basins (Fig. 58):

The gas field of Hassi er Rmel, 200 miles south of Algiers, controlled by SN REPAL (50.9%) and CFP (49.1%).

The oil field of Hassi Messaoud, 300 miles south of Constantine, owned by SN REPAL.

The Fort Polignac Basin, 400 miles south of Gabès (or 250 miles southeast of Hassi Messaoud) owned by CREPS.

The In Salah gas basins, 300 miles south of Hassi er Rmel, with their northern part controlled by CPA and their southern half owned by CREPS.

THE NORTHERN SAHARA

SN REPAL, a company founded in 1946, obtained its first concession in 1952. Following a campaign of seismic refraction, geophysicists, in early 1956, selected a drilling site 70 miles south of the oasis of Ouargla, near the waterhole of cameleers that has since become famous, (Fig. 125). On June 15th of the same year, drillers struck oil at a depth of 11,000 ft. With its northern half belonging to CFP(A)[1], the deposit covers 620 square miles; its reservoir is 330 ft. thick. Experts believe that the deposit will yield 2,800 million barrels under natural pressure, i.e., 16% of its total resources. In 1962, production reached the mark of 70 million barrels; in early 1963, the field operated at a daily rate of 210,000 barrels. Since the oil is light (density 0.8), the company utilizes it after centrifugation and degasification in its local plant. Conversely, Saharan crude (apart from being expensive) hardly accords with the shift towards heavier fuel oils. Consequently, the French Government assures the markets of UGP by decree. Experts also estimate that the light oil will require the injection of less gas to increase recovery. Indeed, with each unit of oil, 220 times as much gas escapes. In other words, gas resources can be estimated to reach the mark of several million million cubic ft. At present, the syndicate utilizes daily more than 3

[1] CFP participates mainly in the international oil companies of Iraq, Iran and Abu Dhabi. The French government owns 35% of the shares of the company.

million cubic ft. of gas to run the 26,000 kW pumping plant of Haoud el Hamra. From here, each day 300,000 barrels of oil flow to Bougie Harbour through the 22 and 24 inch pipe line of SOPEG.

In the district of Hassi Messaoud there are other deposits; southeast of this centre, El Gassi Touil contains less gas than Erg el Agreb of the Société Nationale des Pétroles d'Aquitaine. This zone already produced 10 million barrels in 1963. Indeed the Compagnie des Pétroles France-Afrique, (COPEFA) discovered oil and gas condensate at Gassi Touil. Blowouts and fire accompanied these finds. COPEFA owns 40% of this concession, Omnium de Recherches et d'Exploitations Pétrolières, (OMNIREX) controls 30%, the Phillips Petroleum Company 25% and CFP(A) 5%. South of Gassi Touil, at Rhourde Nouss, the Société de Participations Pétrolières (PETROPAR), El Paso France and Compagnie Franco-Africaine de Pétroles (FRANCAREP) association identified two levels of gas at Rhourde Nouss.

East of Hassi Messaoud and north of Gassi Touil, the Sinclair Mediterranean Petroleum Company, which manages the concession of Sinclair, Newmont, and Société Anonyme Française de Recherches et d'Exploitation de Pétroles (SAFREP), located a major oil field at Rhourde el Baguel in 1962. The order of magnitude of reserves reaches the mark of 300 million barrels. However, northwest of Hassi Messaoud and 25 miles north of the oasis-town of Ouargla, the Oulougga oil field of PREPA–FRANCAREP–AFROPEC does not appear to be commercial. A new pipe line is to connect these fields with Arzew.

In 1952, crews of SN REPAL started drilling at Beriane in the northern M'Zab. Using first the seismic refraction method and then reflection, geophysicists located, 30 miles northwest of the oasis of Ghardaïa, the structure that was to become Hassi er Rmel (Fig. 131). It was not until November, 1956, that drillers struck gas at a depth of 7,100 ft. Out of 250 ft., 157 are productive; according to the latest estimates, they contain 70 million million cubic ft. of gas and dissolved in this are 2,100 million barrels of oil. According to other reports of the same company, resources attain 40 million million cubic ft. In 1961, the government decreed the transfer of part of the concession to SEHR, the Société d'Exploitation de Hassi er Rmel. SEHR works the deposit. At the present, each day 140 million cubic ft. of gas, collected by 25 miles of pipes, can be degasolined. In 1961, total output reached the mark of 21 million million cubic ft., but in 1962, throughput attained only 12 million million cubic ft.; in early 1963 the field was operating at a daily rate of 34 million cubic ft. Algerian condensate production, mainly from Hassi er Rmel, rose from one million barrels in 1961 to 1.7 million in 1962.

The gas pipe of SOTHRA (Société de Transport de Gaz Naturel d'Hassi er Rmel à Arzew), a 20 and 24 inch line of 300 miles, transfers gas compressed under its own pressure. Three stations of compression between Hassi er Rmel and Relizane will double throughput. COMES, the Société Commerciale du Méthane Saharien, sells gas at Relizane, Damesne, Tiaret, Mostaganem and on the new tieline which reaches Algiers. From their 1962 rate of 15,000 million cubic ft., sales are expected to expand. Furthermore, a 200 mile long pipe line transfers the oil condensate to Haoud el Hamra (Hassi Messaoud).

With headquarters at Hydra near Algiers, SN REPAL has a capital of only $9 million. Since its investments reached $14 million by the end of 1961, the company has contracted loans worth $12 million. In 1961, for the first time, the company made a profit of some importance: $1 million.

In 1961, the SN REPAL group had interest in many other companies (see Table XI).

TABLE XI

SN REPAL AFFILIATES

Name	*Function*	*Capital $ million*	*REPAL share %*
SEHR	Operation of Hassi er Rmel	1.3	50.98
CAREP	Exploration and operation	0.4	36.08
SOPEG	Haoud el Hamra–Bougie pipe line	2.5	49.98
SOTHRA	Hassi er Rmel–Arzew pipe line	1.2	32.49
CAMEL	Liquefaction of gas	0.25	15.00
SRA	Algiers refinery	0.8	10.00
UGP	Oil processing and sales	3.0	33.33

In addition to SN REPAL, CFP(A) has a 49% share in SEHR, and SNPA, CEP and COFEPA also utilize the pipe line of SOPEG. As to UGP, RAP and GEP have an equal share. Of the equity of CAMEL, Conch International Methane Ltd. owns 50%, with CFP and BRP sharing the remainder. CAMEL has contracted to deliver 35,000 million cubic ft. of gas each year to the British Gas Board and, beginning in late 1964, to supply 17,000 million cubic ft. per year to Gaz de France.

SN REPAL owns 15% of the equity of SOMOS (the Société des Monomères de Synthèse), SNPA and other French groups own additional 35%, and El Paso Oil owns the rest. The company will process butadiene.

Finally, southwest of Hassi er Rmel, oil shows at Belketeief.

THE SOUTHERN SAHARA

Resembling the distribution of fields in neighbouring Libya, but not the Hassi Messaoud district, the Fort Polignac-Tinrhert Basin is complex; until 1963, this basin was the largest oil producer in Africa. It consists, in fact, of a field of the first magnitude (by African standards), namely Zarzaïtine, of Tiguentourine, Edjeleh and a number of others. The distribution of fields is visualized in Fig. 125 and below.

Ohanet

Tin Fouyé — Reculée — Zarzaïtine

Edjeleh

Tiguentourine

El Adeb — Ouan Taredert

Larache — Dôme à Collenia

Assekaïfaf

Production of the basin has risen from 50 million barrels in 1961 to 80 million barrels in 1962, with new discoveries coming in.

Zarzaïtine boasts various oil levels; in the Devonian level alone, geologists have estimated 1,800 million barrels of oil. The production of Zarzaïtine has expanded rapidly from 10 million barrels in 1960 to 56 million barrels in 1962. In 1961, engineers replaced the vertical separators of four centres by horizontal separators of greater capacity, and also installed a third stage of atmospheric separation and degasification at centre 4, which serves the whole field. Consequently, the yield reached the rate of 175,000 barrels per day in 1961 and 210,000 in 1962. The first major find was Edjeleh, and for a couple of years people designated the field under this name. Geologists estimated the reserves of the 1,000 ft. level at 1,000 million barrels. Between 1960 and 1962, output quadrupled to reach the mark of 14 million barrels per year and 45,000 barrels per day. Indeed, by then gas lifts extended to 55 wells and the total number of producing wells expanded to 230.

At Tiguentourine and El Adeb Larache, CREPS started production in 1962 at a combined rate of 17,000 barrels per day or 5 million barrels in that year. With CEP controlling part of Ohanet, CREPS has reached agreements with EURAFREP to join the production of this field; EURAFREP delivered 4 million barrels in 1962 and CREPS 1.5 million barrels. This field may attain a total rate of 20,000 barrels per day. The small field of Tan Emellel South is in the same category. Furthermore, CPA shares Tin-Fouyé, a field that started production in mid 1963 at the rate of 20,000 barrels of crude per day. In 1962, CEP – Mobil Oil initiated production at Askarene, Tamadenet and Guelta, respectively with 220,000, 35,000 and 20,000 barrels. But CREPS has already charted its next step with the exploitation of Asse-kaïfaf North and, in the south, of Erg Bouhanet, Ouan Taredert and Dôme à Collenias.

With the help of Shell, the marketing position of CREPS is relatively easy, but to deliver the additional flow of oil, TRAPSA (a subsidiary of CREPS) has lately built two additional pumping stations between the terminals of In Amenas (in the desert), La Skhirra (on the sea) and Station no. 3. In 1961, the small field of Taouratine delivered 900 million cubic ft. of gas to the power plant of In Amenas, a modern city of a few thousand inhabitants. With a capital of $23 million and an investment of $115 million, TRAPSA made a profit of $900,000 in 1961. Furthermore, TRAPSA built a tieline which connects with the Ohanet-Haoud el Hamra pipe line of TRAPES. In 1961, this pipe evacuated an overflow of 5 million barrels via Bougie. In 1962, experts rated the line at 160,000 barrels per day.

By 1961, CREPS boasted a capital of $60 million, an investment of $320 million and a maiden profit of $12 million. As to markets, Royal Dutch Shell takes its 35% share, with RAP distributing its 65% share to Shell, Mobil Oil Française SACFP, Antar Pétroles de l'Atlantique, La Mure, UGP and others.

TABLE XII

THE SOUTHWESTERN SAHARA GAS FIELDS

Name	*Distance from In Salah miles*	*Reserves million cubic ft.*
In Salah		450,000
Zini	40 SE	110,000
Djebel Berga	50 SSW	450,000
Oued Djaret I	50	140,000
Bahar el Hammar	80 S	10,000
En Bazzene	70 SSW	10,000
Mahbes Guenatir	70	7,500
Thara 101	70	1,000
Meredoua	150 S	50,000

Unfortunately, (Table XII), the resources of the gas fields that form a long chain trending south of In Salah cannot yet be economically evacuated.

Other gas fields cluster around Bahar el Hammar, Mahbes Guenatir and the Gara Azel Matti, 140 miles southwest of In Salah; the exploration of the western Reggane and Tindouf Basins has not yielded major results.

OTHER RESOURCES OF THE SAHARA

In addition to gypsum, salt, iron, platinum, nickel, tin and tungsten, we find phosphates and coal on the northern edge of the desert. In the basins, which also include the oil and gas fields, several hundred ft. thick layers of *gypsum–anhydrite* and also of *salt* are abundant. Distances do not permit yet their exploitation. Similarly, the radioactive (probably uraniferous) sediments which lie at 7,000–10,000 ft. of depth in the southern basins of the Sahara now only have a curiosity value. Distances also hinder development of the resources of the Hoggar Mountains in the centre of the Sahara (Fig. 100); platinum and chromium at Tibeghim and G'oudène west of Tamanrasset, cross fibre asbestos at Tin Rallès near Tibeghim and Imzaden in the south, tin at In Tounine and Djilouet with tungsten at Laouni in the Adrar Renaissance, 140 miles south of Tamanrasset (west of this town a few diamonds were also collected).

We hope, however, that the fate of the *Gara Djebilet* will not be similar. In 1952, P. Gevin, a geologist working for the Bureau de Recherches Minières d'Algérie (BRMA), discovered in the district of Tindouf — the westernmost corner of Algeria, 400 miles from the sea — a series of hills or Garas containing iron ore (Table XIII). Unfortunately, the ore also contains 0.6% phosphorus, 0.17% titanium, 0.08% vanadium and some sulphur. Nevertheless, SERMI, a subsidiary of the Bureau Industriel d'Algérie (BIA), has studied its exploitation. Similar deposits extend on the other side of the Tindouf Basin towards the frontiers of Morocco. If we then move northeastward, we find, 90 miles south of Colomb-Béchar, the Djebel Guettara deposit with a million tons of ore containing 44% *manganese*, 1–2% iron, 10–15% silica and 0.45 arsenic. At a distance of 40 miles from this deposit, the Compagnie de Mokta el-Hadid has explored the Bou Khdeissar occurrences.

Moving northward, we reach the railhead of Colomb-Béchar. There, in 1907, G.B.M. Flamand discovered the Kenadza *coal* basin. Between 1928 and 1938 annual output oscillated at the costly collieries of Béchar Djedid and Ksi-Ksou from 15,000 to 25,000 tons; their output reached 220,000 tons in 1946 and has since decreased continually. In 1963, the government closed its Houillères du Sud-Oranais. Nevertheless, more than 50 million

TABLE XIII
IRON ORE OF GARAS IN DISTRICT OF TINDOUF

Tons ore	*Iron %*
800	55.5
1,600	53
500	48
1,300	39
200	30

tons of coal remain at Ksi-Ksou, with additional resources in other parts of this basin of Guir and in the Mézarif field, 40 miles east of the town.

IRON ORE

Since the times of the Phoenicians and Romans, the metal industry has flourished.

While small occurrences are widespread, important deposits are concentrated in the east around Constantine and on the Moroccan frontier, west of Oran (Fig. 56). Between 1865 and 1961, three deposits contributed three-quarters of the 110 million tons of iron ore Algeria produced. Between 1865 and 1904, the Compagnie de Mokta el Hadid produced 7 million long tons of iron ore from the area of Bône (in the east) and prior to 1927, the same company extracted 17 million long tons from Beni Saf near Oran. But now miners have to work 120 ft. below the sea and by 1965 the 150,000 ton per year output is expected to cease. Following the Great Recession, yearly production averaged 2 million tons, with exports to the United Kingdom, Germany, Italy and the Benelux countries passing the mark of 3 million long tons per year in 1938, in the early fifties and sixties.

By far the largest deposit, *Ouenza* boasts above-the-water-table reserves of 80–120 million long tons of high grade ore plus 20 million long tons of brown ore containing 45% iron. Out of the latter, the projected steel mill of Bône will use a million tons each year. Meanwhile, with its production capacity of 3 million tons per year, the Société d'Ouenza has to compete with the ores of western Africa for its export markets. Deriving up to $2–3 million of profits per year from this company, the state controls half of its capital. Further south, this company also operates, with draglines and trucks, the *Bou Khadra* deposit containing 20 million tons of high grade ore and non-evaluated barytic resources. A 15 mile tieline connects the mines to Oued Kebrit station, situated 100 miles south of Bône. The Compagnie des Phosphates de Constantine, associated with the Compagnie Mokta el Hadid, can extract 200,000 long tons of ore containing 50.9% iron each year from Khanguet (140 miles from Bône). In 1960, resources stood at 4 million long tons.

We now move from Bône westwards to Bougie, and thence beyond Algiers (see Table XIV).

During the Algerian war, Sidi Marouf and adjacent Tissimiran suspended operations, but Zaccar continued to produce 200,000–400,000 long tons per year. In addition to Beni Amrane and Chabet Ballout, there are numerous small occurrences in the hinterland of Bougie (the Kabylia of the Babors).

TABLE XIV

COMPOSITION AND DISTRIBUTION OF ALGERIAN IRON ORE DEPOSITS

Mine	*Location*		*Group of companies*	*1960 Reserves million long tons*	*Grade iron %*
	Miles				
Filfila	15	W of Philippeville	Miliana	0.1	57
Sidi Marouf	15	E of Djidjelli	Zaccar	1	51
Timezrit	25	S of Bougie	Zaccar	1	56.5
Zaccar	70	W of Algiers	Zaccar	4	51.7
Breira	30	W of Cherchel	Miliana	0.2	52
Beni Aquil	30	W of Cherchel	Miliana	0.5	40.8
Rouina	110	W of Algiers	Miliana	0.2	49.5

NON-FERROUS METALS IN NORTHERN ALGERIA

While the Berbers and Romans worked *zinc* and lead in the Ghar Rouban (Fig. 57), only the beginning of operations in the frontier district adjacent to Zellidja gave impetus to exports (Fig. 56). The Zellidja Company controls the Société Algérienne du Zinc, or ALZI, operating at El-Abed, and the Société des Mines d'Aïn-Arko. The Newmont Corporation has a 31.8% interest in ALZI, and the St. Joseph Lead Company holds 17.15% of its shares. Both companies have their headquarters at El-Abed, near Sidi-Djilalli in the county of Sebdou. In 1961, the Zellidja Mill in Morocco treated 300,000 tons of concentrates from ALZI and El-Abed. Between 1959 and 1960 production evolved as shown in Table XV.

Although profits declined in 1961, Zellidja provides 80–90% of the zinc production of Algeria which, by 1956, had expanded from 15,000 tons in 1929 and 1940 to more than 30,000 tons per year. Rising against general market trends, output in 1961 topped 40,000 tons. As to other mines, prior to 1930, Hammam n'Bails (near Guelma, Constantine area) exported 180,000 tons of ore. This mine also produced 800 tons of *antimony* in 1960, and 1,200 tons in 1961. During the years, antimony production varied considerably as did the output of mercury. But to return to zinc, smaller deposits are situated in the Ouarsenis Mountains at Sidi Kamber, Kef Semmah, Mesloula, Guergour and Djebel Felten. While the Société de la Vieille Montagne (controlled by the Société Générale de Belgique, the French Hottinguer, Lazard and Rothschild groups) owns the Ouarsenis mines, the Peñarroya group manages Sidi Kamber and Montmins.

As to *lead*, output averaged a few hundred tons per year in the XIXth century, and expanded to more than 10,000 tons per year by 1956. However, in 1961, Meslaoua, Bou Kamia and the Djebel Gustar were obliged to cease operations. Although a byproduct of lead, silver recovery has increased from 230,000 ounces per year to 400,000 ounces since 1959. Until a method is found to process Cavallo ores near Bougie, only Aïn Barbar and Boudoukha produce nine hundred tons of copper. To summarize, among 80 base metal mines (including also those of Collo, Bou Thaleb and Clairefontaine) only 10 are operating.

In 1961, non-ferrous metals respectively fetched: zinc, $4.6 million; lead, $1.8 million; pyrites, $400,000; antimony, $80,000; and copper $40,000.

At El Halia near Constantine, the production of *pyrite* containing 47.8% sulphur passed from 15,000 tons in 1927 to 45,000 tons in 1938, and levelled out between 20,000 and 30,000 until 1960, when the company started a programme of expansion. Indeed, with an output of 90,000 tons per year, operations would become profitable. In this case, France would

TABLE XV

NORTHERN ALGERIAN ZINC AND LEAD PRODUCTION

	El-Abed				*Oued Zounder*			
	1,000 tons		*Grade %*		*1,000 tons*		*Grade %*	
	1959	*1960*	*1959*	*1960*	*1959*	*1960*	*1959*	*1960*
Ore	115	140			85	120		
Concentrates zinc	45	45	62.4	62.3	22	23	61	60.6
lead	5	7	73.1	70.9	2.5	2.5	75.7	71.5

have to import the quantities not taken by the Algerian phosphate industry. Finally, in 1961, tungsten mining started in the small arsenic deposit of Béléliéta near Constantine.

INDUSTRIAL MINERALS

In 1957, the siliceous phosphate deposit of M'Zaïta ceased operations and in 1965–1966, after 46 years of operations, the Compagnie des Phosphates de Constantine expects to close its low grade (65–68%) underground mine at the Kouif. To replace the Kouif, the Compagnie du Djebel Onk and the government prepared a deposit (Fig. 134) L. Joleaud discovered in 1907 on the frontiers of Tunisia. The company's $600,000 capital consists of:

Compagnie des Phosphates de Constantine	40 %
Bureau d'Investissement d'Algérie	18 %
Caisse d'Équipement d'Algérie	16 %
Compagnie Financière pour le Développement Économique de l'Algérie	8.5%
Société Algérienne de Développement et d'Expansion	8.5%
Société Algérienne des Produits Chimiques, Banque Nationale pour le Commerce, Compagnie Algérienne Schiaffino	9 %

Lately the Office Chérifien des Phosphates acquired a minority interest.

Since an investment of $30 million is required, France has decided to take direct and indirect responsibility in the equity, and to build a 70 mile long railway line from Bir el Ater to Tébessa connecting with the 150 mile long Bône railway. There are other problems: to upgrade to 75% 600 million tons of ore forming thick layers averaging only 57% tricalcic phosphate, the company is obliged to desilicify, calcinate and then wash the material. In the future, plants may upgrade the ore to 80% tricalcic phosphate. But to concentrate 900,000 tons each year, the plant requires 1,000 barrels of oil per day and 8–12 million gallons of water. Fortunately, just north of the phosphates, geologists of a company shared by SN REPAL (65%) and Djebel Onk (35%) have located oil; unfortunately, however, the company is obliged to import water from Chéria, a distance of 40 miles.

Salt is abundant in the sea, in the salt lakes and depressions, in the sebkhet and chotts and in the salt mountains of the Saharan Atlas, such as Djelfa and El Outaya. Disregarding small scale operations, experts estimate that, since 1958, industrial production has been reaching the mark of 150,000 tons per year, i.e., 40,000 tons less than requirements. At the end of the war, *cement* requirements stood at 160,000 tons per year; these were covered by the plants of Oran and Algiers. In 1961, cement production reached the mark of 6.3 million barrels. Limestone resources are adequate for expansion. As to clay manufacture, 500,000 tons of bricks reached markets in 1961 and the manufacture of tiles increased by 8% in one year. Only after World War II did the large *bentonite* deposit of Marnia (on the western border) enter its productive phase. Each year, 100,000 tons of this expanding clay fetch $400,000, half of this amount from abroad.

Similarly, *gypsum* resources are abundant in several areas, including the Chélif Basin west of Oran. Reaching the mark of 190,000 tons in 1958, production doubled in 2 years. Also, in the same area, there are numerous quarries of *diatomite* around Saint Denis du Sig and Bosquet (Fig. 136), and a small output at Ouillis (near Mostaganem). Each year, the El-Ksar quarry (at Saint Denis) produces 25,000–30,000 tons of this material

TABLE XVI
KABYLIA BARYTES RESOURCES

Keddara area	*tons*	*Other areas*	*tons*
Vein 3	145,000	Affensou vein	170,000
Veins 5 and 6	80,000	Mouzaïa les Mines	45,000
Vein 4	40,000	Gouraya	45,000
Veins 0, 1 and 2	35,000	Koudiat el Madène	35,000
Tala Roumi	80,000	Narrows of Palestro	15,000
Oaks (Chênes)	5,000	Bou Mahni	13,000
Total Zouggara	385,000	Hamiz Dam	12,000
Aïn-Sultane	15,000		
Saddle	45,000		
Total	445,000	Total	335,000

(half of it for export). Its value reaches half a million dollars. Since 1959, vermiculite working recommenced and reached 400 tons by the following year.

East of Algiers, in Small Kabylia (Fig. 93), especially around Palestro, *barytes* resources are abundant. Table XVI gives the resources, estimated by the BRGM in 1962. Half of the Keddara ore is of white grade barytes. Production increased from 3,000 tons in 1938 to 25,000 tons of rock barytes or 15,000 tons powder by 1950. But in 1961, output decreased from its 1960 peak of 60,000 tons of rock or 40,000 tons of powder. The value of the production varies from $300,000 to 600,000. Of course, output is dependent on drilling. In addition to Palestro, there are quarries at Bou-Zega, Rivet, Saint Pierre and Saint Paul (near Algiers). At Bou Mani in Great Kabylia, operations have been interrupted. Small Kabylia (Oued Agrioun etc.) is also distinguished by an installed hydropower capacity of 75,000 kW, however, in most of the country, water resources are inadequate. For a further discussion of this problem, we refer the reader to the chapter on water and soil. On the other hand, some oil still remains in the Oued Guétérini.

TUNISIA

On the eastern edge of the Maghreb, Tunisia is less well endowed than its neighbours. Nevertheless, the Néo-Destour Government, which rules 4 million Tunisians, insures a favourable investment climate. Within the 40,000 square miles of Tunisia, phosphates (the principal export commodity) are concentrated inland of the Bay of Gabès, with iron ore and base metals in the northwestern triangle of the country. During the last decade, mineral production has slightly decreased. Through the harbour of La Skirra, south of Gabès, Tunisia controls the outlet of the pipe lines of the Edjeleh-Zarzaïtine field (Algeria), i.e., an important fraction of the African oil potential. Moreover, in 1963, a 23,500 barrels per day refinery commenced operations at Bizerta.

In 1886, Philippe Thomas, a French geologist, discovered the *phosphates* of Gafsa (Fig. 134). The Compagnie des Phosphates et du Chemin de Fer de Gafsa commenced the exploitation of Metlaoui in 1899, and, by 1900, attained 100,000 tons output. In 1927, the working of Redeyef started with output reaching its peak of 3 million tons. Later agreements reserved a larger share of the market for Morocco; nevertheless, out of reserves attaining 2,000 million tons which are second only to those of Morocco, Tunisia contributed for many years somewhat more than 2 million tons of phosphate per annum, mostly of the 65% variety which fetches low prices. Consequently, the Gafsa Company conducted tests at Metlaoui for the calcination and upgrading of a million tons of phosphate per year to a grade of 75%. Productive capacity attains 1,750,000 tons at Gafsa, 450,000 tons at M'Dilla and 300,000 tons at Kalaa Djerada. Near the latter is the extensive low grade deposit of Gouraya. While the phosphate zones extend from the Algerian frontier and the vicinity of the Djebel Ongg deposit 100 miles east, the princiapl mines are centred on Gafsa, 80 miles from the harbour of Gabès. The small occurrences of Meheri-Zebbeus, Djebel Rechaib, Chaketma and Skour extend along a belt connecting the district of Gabès with Djérissa in the northwest. While hyperphosphate production decreased to 60,000 tons in 1960, the output of super-triple phosphate increased to reach 100,000 tons. Union Phosphatière Africaine markets the products of Tunisia, Algeria and Togo.

As to *iron* ore, the second most important product, a capacity of a million tons per year has been maintained, with Djérissa contributing practically all the output. From Djérissa near the Algerian frontier, the ore is shipped by rail northeast to Tunis. Since only Tamera is active in the Douaria area, the output of this district has now fallen to 10,000 tons per year. A belt of minor occurrences of iron ore extends from Djérissa to Bizerta in the north-northeast; these are, however, inactive.

We now turn to base metals. As we have already stated, all the deposits have been concentrated in the triangle of the Algerian frontier, the line connecting Djérissa with Bizerta and the sea (Fig. 57). Although averaging 25,000 tons per year in the last decade, the output of *lead* dropped to 20,000 tons in 1959, essentially because of unfavourable market conditions. Consequently, Kangouet and Kef Fout were obliged to curtail production and Djebel Azered and Fedj el-Adoum await better days to recommence operations. Lately, Djebel Hallouf, Sidi Amor and Sidi Youssef provided all the output. Recovery of *silver*, a byproduct of lead, reached a peak of 135,000 ounces in 1958; it fell to 40,000 ounces by 1959, where it remains. The country also produces a few hundred tons of antimonium lead concentrates each year and, alone in Africa, some *mercury*. At Oued Maden on the frontier 30 miles from the sea, both of these accompany lead and some zinc. Rising from 20 flasks in 1956, none in 1957 and 40 in 1958, mercury exports now range between 150 and 200 flasks. Like lead, *zinc* is affected by market conditions. Even El Akhouat, the mine that produces practically all the 3,500 tons per year of zinc, is obliged to restrict output.

TABLE XVII

THE MINERALS CONTAINED IN THE TUNISIAN 'TRIANGLE OF METALS'

Mineral resource	*Occurrence*
Manganese	Thuburnic
Arsenic	Fedj Assene
Copper	Aïn el Bey, Chouïchia
Fluorite	Bou Jaber
Barytes, strontium ore	Bechater, Bazina, Dogra, Zag et Tin, Sidi Amor, Bou Jaber
Vanadium ore	Djebba

Fedj el-Adoum operates occasionally, but the country's productive capacity reaches the mark of 5,000 tons per year. Table XVII shows the distribution of ores in the 'Triangle of Metals' near the Algerian frontier (Fig. 57).

The Compagnie Royale Asturienne des Mines (see the "Morocco" chapter) manages a mixed state–private company and is the largest producer.

Less fortunate than its neighbours, the Cap Bon field near Tunis only provides 250,000 cubic ft. of gas. Lately, at the southernmost tip of Tunisia at Makrerouza, drillers struck some oil at a depth of 6,200 ft. and gas at Bir Ali Ben Khalifa. On the other hand, hydropower resources are largely restricted to the Medjerda Basin in the northwest. *Salt* recovery from the sea fits into the pattern: from 150,000 tons in the early fifties and up to 190,000 tons in 1958, output has decreased in 1962 to 180,000 tons. Naturally, salt is also abundant in the Chotts of southern Tunisia. Conversely, the output of gypsum at 17,000 tons per year is fairly stable. With adequate limestone resources in the north, Tunisia, like the other more industrialized countries is a major producer of cement, achieving 2 million barrels per year. Resembling the rest of northern Africa, water and soil resources are generally insufficient and in the south particularly inadequate.

LIBYA

OIL PRODUCTION

By now we associate the name of Libya with oil (Fig. 57), but we did not always do so: just after the war, an international mission decided that the 680,000 square miles of this country were almost barren. Then, prompted by discoveries in neighbouring Algeria and the promulgation of a modern oil law in 1955, 17 companies started exploration and invested more than $800 million by the beginning of 1963. The law requires companies to relinquish one quarter of their concession every 5 years. Such redistribution took place in 1961 and the next date is 1966. The efforts of the Oasis group — which is the association of Amerada Petroleum Corporation, the Marathon International Oil Company (a subsidiary of the former Ohio Oil Company) and of the Continental Oil Company (a wholly owned subsidiary of Standard Oil of New Jersey) — and of Esso Sirte were particularly successful. Amerada operates Oasis but markets its one-third share through Continental Oil.

Experts believe that the Dahra, Waha and Gialo fields of Oasis and the Zelten and Raguba fields of Esso can provide, in the first 5 years of production, an income of $500 million to the royal government of Libya which shares 50% of the profits. Since 70% of this amount has already been earmarked for development (mainly agricultural), the country's population of 1.2 million will derive important benefits. Indeed, to provide additional water and soil resources, considerable investment is necessary.

Since the value of the yearly output is expected to rise to $500–900 million, Libya will rank among the leading mineral producers of Africa. Indeed, by 1965, throughput has reached the mark of 1,000,000 barrels per day (a 100% increase in a year) from which $80 million profit has accrued to the government; the 88 mile long, 30 inch Es Sider pipe line of Oasis alone has a capacity of 310,000 barrels per day. However, the first producer was

Esso. On August 8th, 1961, its 100 mile pipe line reached Port Brega (Marsa el Brega) and, for the first time, on October 25th of the same year, a cargo of 200,000 barrels loaded Libyan crude. The loading device, a 140 ft. high cylindrical tanker, is anchored 4,000 ft. offshore. A 56 mile long, 20 inch spurline connects Raguba with Zelten in the east and, in 1963, Oasis completed a 150 mile long, 32 inch tieline connecting Waha with Dahra in the west. The market structure of Libyan oil is still evolving: the United States, United Kingdom, Panama and Germany are important customers, and in Italy, a 10,000 barrel per day refinery was built for Libyan crude. In addition to these, in 1963, Esso initiated refining at its 8,000 barrel per day plant of Brega. The company produces gasolines, kerosene, diesel fuel and fuel oils.

OIL FIELDS

But where are the oil fields? Oriented east-southeast, i.e., parallel to Es Sider, the Bay of Syrte (Fig. 123), the fields dot a 300 mile long and up to 20 mile wide band which can be visualized thus:

Bahi						
Mabruk						
Dahra	100 miles		20 miles		80 miles	
Zahaligh	←——→	Raguba	←——→		←——→	Amal
Hofra				Zelten		
		Beda				Gialo
		Samah		Waha		
				Defa		

In its second well, Oasis struck oil at Dahra; experts now evaluate the productive capacity of that field at 120,000 barrels per day. Mabruck, an Esso field drilled by the American Libyan Oil Company, could only produce through secondary recovery. Similarly, the productivity of Hofra (27,000 barrels per day, Mobil Oil) and of Zahaligh (20 miles south of Mabruck), with its yield of 2 million cubic ft. of gas, is restricted. On the other hand, Esso estimates the rate of Raguba at 40,000–60,000 barrels per day. As to Zelten, with an investment of $50 million, the company is building a 110 mile long, 36 inch pipe line to pump 500,000 barrels of salt water from the Mediterranean. Its injection is expected to increase the field's present 240,000 barrels per day throughput.

We now return to Oasis. The Waha Pool (100,000 barrels per day) can be circumscribed within a perimeter of 4 by 2.5 miles. Though smaller, the group also believes that Defa is commercial. By far, the largest appears to be the latest find, Gialo, where drillers struck oil in August, 1961, at a depth of 6,300 ft., with additional reservoirs at only 2,900 ft. In one of the wells of the 8 by 5 mile area, the productive zone is 200 ft. thick! In the western part of Concession 59, the L and V area finds define an area of 35 square miles. In 1962, Oasis baptized this zone the Samah field. The group obtained these results with an investment of $120 million in 1960 and 1961. The discovery of the L–G field in Bloc 71, 30 miles south of Defa, shows that the Syrte Basin bulges southward. Moreover, the EE and BB gas fields south of Gialo and the P oil field 25 miles southeast of Gialo indicate the eastern extensions of this trend, which has been pursued already as far as the B–L field of Compagnia Ricerche Idrocarburi (CORI) 80 miles east of Gialo and of the C6 and C7–65 fields of BP-Hunt 100 miles southeast of Gialo which lie on the same latitude as Edjeleh in Algeria. Other companies such as Gulf Oil, Mobil Oil (who controlled Hofra and Amal), NB-Hunt and Atlantic Refining and Phillips (offshore) also continue their operations. In 1962, they were joined by the National Oil Company (NOCOL); Libyans own 51% of its equity, Kewanee Overseas Oil and other American companies 10.1%, a Swedish marketing group 19%, with the rest attributed to other European and American interests.

Oil and gas also show in a broad zone of northwestern Libya extending from west of Tripoli to Edjeleh. This belt includes the Oued Chebbi gas field in the northwestern corner of Tripolitania and Bir el-Rhezeil, yielding more than 2 million cubic ft. of gas. The Bir Tlacsin (Esso), Bir el-Rhezeil, Emgayet and Atshan (Esso) oil fields — located respectively 120, 200, 300 and 500 miles from Tripoli — had an initial yield of 1,200, 1,500, 1,500 and 900 barrels per day. Both Atshan and El-Haghe are dwarfed by their western neighbour, Tiguentourine (Algeria). Zuara, west of Tripoli is the nearest port to Emgayet (Gulf Oil) and to the CPT-fields. As of now, the potential of the western belt appears to be smaller, but in the east, geologists have evaluated reserves of 3,500 million barrels by 1961 and

TABLE XVIII

LIBYAN OIL WELLS

Concession and productive wells	*Bloc*	*Initial potential barrels per day*	*Depth or interval ft.*
Oasis			
Gialo	59	1,200	6,300
F 1	59	1,200	7,650
E 1	59	1,200	6,300
E 2	59	1,000	2,700
L 1	59	2,000	6,300
M 1	59	1,900	7,000
Q 1	59	1,700	7,400
D 1	26	1,000	4,300
Esso Syrte			
Raguba	20	2,250	5,500
BP-Hunt			
C 1	65	3,910	8,700
Gulf			
O 1	66	1,100	6,620
Z 1	66	1,140	9,720
CPT(L)			
E 1	23	1,930	11,180
Mobil			
Hofra	11	1,500	3,100
D 1	12	1,070	9,500

2,500 million barrels in 1962, which with exploration continuing might even rise to more than 10,000 million barrels. For further details on yields and depths we refer the reader to the "Oil" chapter of the second part. Productive wells of 1960 – mid 1962 are listed in Table XVIII.

OTHER PRODUCTS

In addition to abundant oil and limited amounts of gas, Libya has long produced *salt*, 15,000–20,000 tons per year according to statistics. Libyans recover salt from sebkhet, dry salt lakes and the sea, but considerably larger salt, gypsum-anhydrite resources are available in the oil fields and the surrounding sediments. Similarly, in the Syrte, Italians discovered the *potash* deposits of Marada long ago, and now, with fuels available, a processing plant appears to be feasible. In the same area there is also soda. Each year, Libyans manufacture more than 10 million bricks from coastal clays. At Shekshuk and Nalut, lignite seams could not compete with the liquid fuels. In the northern Fezzan, AFRISCO, an Italian concern, explores the shallow deposits of Brakk which are believed to consist of 200–300 million tons of ore containing more than 44% *iron*. In 1963 20,000 tons of lime were slaked and a cement plant was planned at Tripoli.

SECTION VI

Northeastern Africa

Undoubtedly the poorest of the five regions of the continent, the Nile Valley and the Horn of Africa hold resources of iron, hydropower, phosphates, gypsum, salt and potash.

UNITED ARAB REPUBLIC

Egypt is a gift of the Nile. However, the limited mineral resources of her 460,000 square miles are concentrated in the oases, the Sinai and in the Red Sea (Fig. 12). Nevertheless,

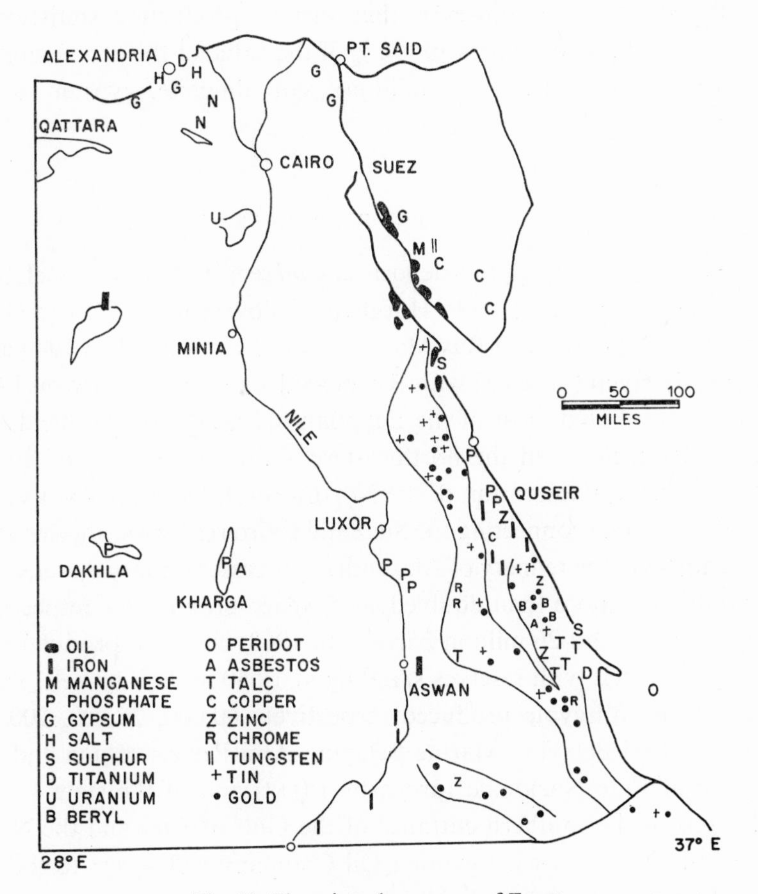

Fig. 12. The mineral resources of Egypt.

TABLE XIX

PRODUCTION OF EGYPTIAN MINERAL RESOURCES

Mineral resource	*1962*	*Percentage increase over 1956*	*Mineral resource*	*1962 tons*	*Percentage increase over 1956*
Oil	32[1]	270	Sulphur	11,000	380
Phosphates	620,000[2]	90	Talc	6,500	85
Iron ore	450,000[2]	330	Pumice	4,300	250
			Diatomite	300	30
Manganese ore	320,000 tons	6,400	Barytes	3,000	–[3]
Salt	570,000 tons	100	Asbestos	300	–[3]
Gypsum	440,000 tons	200	Zircon	400	300
Ilmenite } titanium	120,000 tons	270	Lead	2	1.5
Rutile } titanium	1,000 tons	–[3]	Gold, ounces	1,000	7

[1] Million barrels.
[2] Long tons.
[3] No production in 1956.

the Nile offers water and hydropower which are essential for Egypt's 30 million inhabitants. Alone in Africa, the state owns all these resources but the government of the United Arab Republic encourages their exploitation. Indeed, since 1956, production did increase (Table XIX). We should note, however, that various production statistics differ. With approximately $100 million, oil represents 90% of the value of the production, followed by phosphates ($4 million), iron ore ($3 million), salt, ilmenite, gypsum and manganese (combined $3 million).

PRINCIPAL RESOURCES

In 1868, miners of the Gemsa sulphur deposit saw *oil* seeping from a tunnel, but the Gemsa oil field was not discovered until 1909. Hurghada followed in 1913, Abu Darba in 1918, Ras Gharib in 1938, Sudr, Asl and Ras Matarma and Feiran between 1949 and 1953, and finally the largest, Belayim (in Sinai) with Rudeis-Sidri and Bakr between 1955 and 1959. While Hurghada contributed most of the Egyptian oil, many of these fields have by now been depleted, both in Sinai on the northwestern shore of the Gulf of Suez and on its southern shore in Africa. Exploration of the Northwestern Desert has not yet met success. In addition to the pipe line connecting the Suez and Cairo refineries, another pipe line binds Port Saïd to Ismailia and the refinery of Alexandria processes imported crude. Nevertheless, the national production more than doubled in 5 years progressing respectively in 1959, 1960 and 1961 to 22, 24 and 26 million barrels. In 1963, crude oil production reached the mark of 40 million barrels with reserves standing at 1,560 million barrels. The Compagnie Orientale des Pétroles d'Égypte produced respectively 60,000, 25,000, 200, 360 and 600 barrels per day at Land Belayim, Marine Belayim, Abu Rudeis, Ekma and Feiran. COPE — an association of Ente Nazionale Idrocarburi (ENI) and of the state — also explores the Zeit-Kalig area, at the southern entrance of the Gulf of Suez and the Nile delta. Half of the profits of the International Egyptian Oil Company will revert to the United Arab Republic with the remainder being divided between COPE and ENI.

The Ras Gharib and Hurghada fields of El-Nasr Oil Fields produced in 1962 at a rate of 14,000 and 550 barrels per day respectively. Furthermore, the yield of the Bakr field of the General Petroleum Company has attained the mark of 5,300 barrels per day and each day the Asl and Sudr fields of El-Nasr Oil Fields and Mobil Oil Egypt produced 2,500 barrels each. The development of the Kareim field has been completed.

While small by Northwestern African standards, *phosphate* deposits play a considerable role in the mineral economy of the United Arab Republic. They are found in distant areas, but until now miners have concentrated their efforts on the most accessible deposits: at Quseir and Safaga on the Red Sea. The deposits situated west of Hamrawein Bay contain only 2 million tons of phosphate. Others include Sebaiya, Mahamid and Basaliya, south of Luxor at the confluence of the Wadi Quena with the Nile, and the oases of Kharga and Dakhla, respectively 150 and 250 miles west of Luxor. In 1961, according to certain statistics, output was normal; according to others, production dropped from its habitual half a million tons to 300,000 tons. Traditionally, half of this amount was exported, with India and Japan each taking 30%, and Ceylon 20%. The importance of phosphates for the farmers of the Nile Valley cannot be underestimated. Miners both strip and exploit underground the El Naser Mine, and the Société Financière et Industrielle d'Égypte expanded the Kafr ez-Zaiyat superphosphate plant.

Let us note here that the manufacture of another fertilizer, nitrogen, increased from 35,000 tons in 1956 to 200,000 tons in 1960; however, it decreased to 120,000 tons in 1961.

Since the government plans to build steel mills, the 45 million tons of *iron* ore in the Gebel Ghorabi of the Baharia Oasis (150 miles southwest of Cairo) acquire special importance. While low grade (47.9%) by international standards, they can nevertheless be used in the mill that SENTAB, the Swedish engineering firm, is building for the state at Minia (Minya) on the Nile. In the Ghorabi, 110 million tons of even poorer ores, containing only 28.2% iron, are also available. For some time, the Egyptian Iron and Steel Company each year has extracted more than 300,000 tons of minette from Hawan, east of Aswan. This ore contains 44% iron, with considerable amounts of silica. In the Nile Valley, south of Aswan and as far as the Sudanese Wadi Halfa deposit, other occurrences include Kalabsha, Garf Husein, Kurusko and Abu Simbil (viz. Temple of Abu Simbel). According to plans, these deposits will provide ore for steel mills to be built at Aswan. The cumulative effect of these projects will be the expansion of output to 2 million tons per year. Until now, the iron-*manganese* deposits of the Um Bogma district (in the southwestern Sinai, near the Gulf) have been the most productive. Since 1959, exports have doubled to reach the mark of 200,000–250,000 tons of low grade and 20,000 tons of high grade ore containing 21% manganese and 37% iron. With new mines opening at Beda Thora (east of Bogma), production is expected to rise. The government also plans the building of a ferro-manganese plant of 10,000 tons per year at Bogma. In addition to these, in the Elba region of the Eastern Desert and in the southeastern corner of the country (adjacent to the Sudanese deposits), miners also win some low grade manganese ore. Also in the Eastern Desert, some iron shows south of Quseir at Wadi Karim, Um Shaddad, Um Nar, Um Ghamis, Siwiqat, Um Lassaf, Dabbah and Um Hagalig.

ENERGY

The establishment of a steel industry is hampered by the lack of coking coal and solid fuels in general. However, lately, in the Oyun Mussa of the Sinai, teams have discovered non-coking seams extending 9 miles at a depth of 1,500 ft. The Jebel Maraga could supply 150,000 long tons of coking coal per year (ash content 3%).

In contrast to fuels, the Nile can provide abundant *hydropower*. Political controversy has made Sadd el-Aali, the High Dam of Aswan, famous. In 1956, Sir Alexander Gibbs and his partners developed the plans of I.C. Steele, Lorenz G. Straub and Karl Terzaghi of the United States, Max Pruess of Germany and André Coyne of France, but the planners of the USSR Ministry of Power Stations Construction moved the dam 4,500 ft. upstream in the reservoir of the old or low Aswan Dam, thus eliminating the upstream banket. The Soviet Government has already assigned $270 million to the scheme and costs are expected to rise to $400 million during the decade the Sadd el-Aali Executive Organization will require to complete the project.

While the dam is smaller in many respects than others, it will impound a record 104 million acre ft. of water, which is three times as much as the Hoover Dam. The dam is also technically remarkable since notable portions of the structure will consist of dune sand deposited in more than 100 ft. of water, and because it will rest on 650 ft. of silt, sand and gravel. Other important characteristics include:

Dam:	length, crest	11,500 ft.
	height above bed	365 ft.
Reservoir:	length	300 miles
	surface	1,500 square miles
	flood storage	24 million acre ft.
	annual loss	8 million acre ft.
Diversion channel:	length	6,000 ft.
	maximum discharge	385,000 cubic ft. per sec.
Power:	total installed	2,100,000 kW.

The High Dam will also permit utilization of the Low Dam power plant in all seasons; however, it will flood the temples of Abu Simbel, which would cost UNESCO $70 million to preserve.

INDUSTRIAL MINERALS

With a yearly output of more than half a million tons, Egypt remains the continent's principal producer of *salt*, which is recovered at Alexandria, Mersa Matruh, Rosetta and Idku on the Mediterranean and in a number of other pans. Similarly, with 13 million barrels in 1962, Egypt produced a quarter of the continent's cement, ranking second only to South Africa.

The greatest production of *gypsum* comes from the 350 ft. thick layers of Ras Malaab (north of the Um Bogma manganese deposit) in the Sinai. However, gypsum is also deposited in the shallow lakes of Maryut (near Alexandria), Manzala and El Ballah; and miners quarry gypsum at El Hammam near Alexandria. The largest resources lie at some depth on the coast of the Red Sea and the Gulf of Suez. Various data estimate output at 400,000,

450,000 and 600,000 tons per year. Furthermore, in 1960, workers recovered 300,000 tons of common clay in the Nile Valley and 10,000 tons of kaolin, mainly east of Aswan.

For a long time, native sulphur was known to occur with gypsum and aragonite at Gemsa, at the entrance of the Gulf of Suez and at Ranga on the Red Sea, due east of Aswan.

Natron precipitates at Wadi el-Natrun, northwest of Cairo and in the area of Baharia (see "Iron"). Aluminium and magnesium sulphates (alum and epsomite, the drug) impregnate part of the Khargla and Dakhla Oases (see "Phosphates").

Some uranium shows in the Gebel Qatrani, north of the famed Coptic Oasis of Fayum and at Elquosen, around Quseir on the coast (also see "Phosphates"). Another source of radioactive materials is the Nile delta; experts have evaluated it to contain several hundred thousand tons of black sands containing 64% ilmenite (titanium ore), 5% zircon and 0.02% of monazite. From this source, production has ranged yearly around 17,000 tons of ilmenite, 1,000 tons of rutile and about 3,000 tons of zircon.

In the southeast, another source of titanium, the Abu Ghalaqa (or Chalaga), entered its production phase in 1961. Generally this type of hard rock deposit is uneconomical, but with Egypt's low labour, housing and similar welfare costs, each worker carves 8 tons of rock per day, of which 50% is waste and 50% ore. From the jetties of the Abu Gushun (Gussum), the state can ship each year 100,000 tons of ilmenite (40–50% to Italy).

OTHER PRODUCTS

There is small scale production of a number of substances, mainly in the Southeastern Desert (Fig. 58). Workers recover talc and steatite (soapstone) in the southeast, at Wadi Gulan el-Atshan, Derhib, Makbi, Bir Disi and Wadi Kharit; some low grade asbestos and formerly vermiculite at Hafafit (northeast of Aswan); marble at Wadi Dagbag (north of Wadi Mia) and Gebel Rokam; and also pumice and diatomite.

Since the days of the Pharaohs, gold production has not failed to decrease. It now stands at 1,000 ounces per year; it is found with traces of other minerals in a 350 mile long belt which follows the coast 50–100 miles inland in the Eastern and Southeastern Desert. Southeastern resources are schematized below.

Gold: Um Rus, Hanglia, Um Udd, Gebel Nugrus, Abu Dabbab and many others.

Lead, copper and zinc: Samiuki, Atshan, Derhib, Abu Swayel, El Atawi (only copper, now abandoned), Sarabit el Khadim (copper).

Beryl: Zabara, Sikait, Nugrus, Um Kabu.

Tin and tungsten: Igla, Nuweibi, Abu Dabbab, El Mueilha.

Tungsten: El Dob, Abu Marwa, Abu Kharif, Um Bissila, Fatiri, Gebel Maghrabiya, Hammad.

Molybdenum: Gattar, Gebel Um Harba, Wadi Dib, Abu Marwa, Gebel Um Disi, Gebel el-Dob, Gebel el Shayeb.

Baryte: east of Aswan, Siwiqat el-Soda (with lead).

Low grade chromite: Barramia, Ras Shait, Abu Dahr, Um Kabu.

Graphitic schists: Wadi Bent, Abu Geraiya, Wadi Sitra.

In addition to these, copper shows in the Sinai at Regeita, Abu el-Nimran and Samra, and nickel is found on Saint John's, an island of the Red Sea famous for its peridots.

SUDAN

Upstream of Egypt, 12 million Sudanese inhabit almost a million square miles. In the Sudan, the Nile accumulates more power than in Egypt. The government fosters the exploitation (Fig. 13) of the country's scattered mineral resources:

Salt, 60,000 tons yearly since the fifties.

Gypsum, 5,000 tons in early fifties; 1956–1959: 2,000 tons per year.

Iron ore, 1962: 20,000 tons.

Manganese, 1956–1958: 7,000–9,000 tons per year.

Gold, 1951–1955: 1,700 ounces per year; 1956–1958: 6,000 ounces total; 1959–1960: 2,500 ounces per year.

Mica, 1957: 6 tons; 1958: 190 tons.

Unlike Egypt, *iron* resources are extensive. They are broadly distributed from the Congolese border to the northern frontier. There, at Wadi Halfa on the Nile, oölithic ironstones lie at shallow depth. On the other hand, the iron ores of Fodikwan are of the magnetite type. The Jebel Abu Tulu rises in Kordofan (in the south–central Sudan), 21 miles from Rigl el-Fula (a railway station located 590 miles from Khartoum or 1,100 miles from Port Sudan). The deposit offers, along with 7 % silica, 39 million tons of high grade (61.1 %) ore; its quarrying started in 1960. In the southwestern Sudan, deposits resembling those of the adjacent Congo are believed to extend. Indeed, iron ores outcrop at Mvolo, Rumbek, on the Sobat, as well as at Um Semeima, Jebel Haraza and Nahud in Kordofan, and in the eastern corner at Tokar.

Following the Congo, the Sudan boasts one of the greatest hydropower potentials of Africa: 1.5–3 million kW. The latest dam at Sennar will also serve irrigation. Indeed,

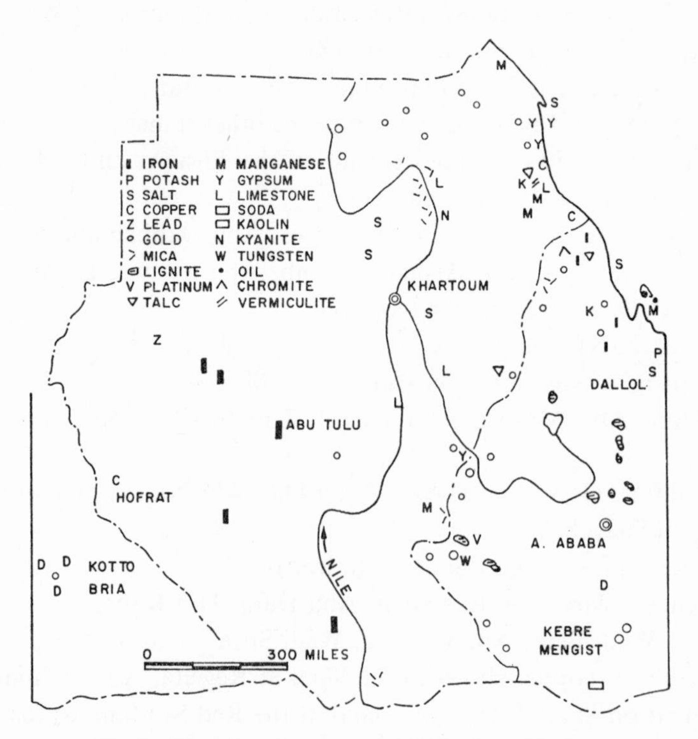

Fig. 13. The mineral resources of Ethiopia and Sudan.

emulating the famous Gesira Scheme, much remains to be done to provide adequate water facilities. By 1965, the Port Sudan refinery of Shell and BP (Sudan) will deliver 20,000 barrels of oil per day. Each year, Sudanese industry manufactures 500,000–700,000 barrels of *cement*. An important deposit of *limestone* is at Shereik. Not far from Kosti (on the Nile, south of Khartoum), Nyefr Regaig is distinguished by 50 million tons of limestone (0.3–3.3% MgO). Farther southeast on the railway line at Bagger el Hadari are another one and a half million tons of similar grade, with limestones of good quality also occurring at Sagadi, Mashata and the Jebel Dud. A large cement plant is to rise at Rank.

Near the frontier of the Central African Republic, Hofrat en Nahas, with a million tons of ore containing 500,000 tons of copper metal, after the ancients has attracted the attention of the Pera Trading Company of Livorno (Italy) and its United States affiliate, African Mining Company. These expect to invest more than $17 million in the development of the deposits, located 180 miles southwest of Nyala (a station situated 300 miles west of Rigl el-Fula). The government plans to extend the rail to Wau, south of Nyala. Gold also shows at Hofrat, as does *lead* in the Jebel Kutum (Kutub) north of Kobe. Also in the south, *graphite* has been reported to occur near the Bongo River and on the Yambio–Meridi road in Bahr el-Ghazal.

Some *fluorspar* is found between Kosti and Rigl el-Fula in the Jebel Sermeih. Opposite Shereik in the Rubatab belt, workers collect ruby *mica* of fine stained quality. Although mica was discovered as early as 1918, mining did not start until the fifties and in 1960, operations were restricted to the Wadi El-Koro and to the area comprised between the Gebels Absol and Qarn Dam El-Tor. Close at hand on both banks of the Atbara bend of the Nile, *kyanite* is disseminated in rocks. Mica also shows in the south in the Khor Buddini (Fung district) and near Longairo and Lufira (Mongella district).

In arid Sudan, *salt* is widespread in beds of the Selima Oasis, in the sands of Butana (east of the Blue Nile, between Rufâa and Khartoum) and on the lagoons of the Red Sea. There, at Ras Roweiya, salt has been recovered on a large scale. In the same area are abundant resources of *gypsum*. Along the coast, as far as 40 miles from Port Sudan, extend 30 ft. thick beds; they also cover Makawa Island. Harbours in Khor Donganab could serve those deposits. Conversely, workers recover soda inland, mainly near the Jebel Kashaf in the Wadi Natrun. In the Northern Province, stringers and lenses of low grade coal and lignite outcrop. Since Pharaonic times, *gold* has brought fame to the Sudan (Fig. 58). Ancient workings are scattered in the northeast between the frontier and Abu Hamed (i.e., the northernmost point of the Atbara bend), from Dereheib southward towards Port Sudan, in the Red Sea Hills, and south of Wadi Halfa in Dongola. Between the two World Wars, Om Nabardi (in the first of these areas) was the only source of primary gold, with Fazogli (where the Blue Nile enters the Sudan) supplying placer gold. Incidentally, in the centre of the country (east of the Abu Tulu iron deposit) placer gold also shows around Tira Mandi. However, at the present only the Doishat and Bir Kateib mines attract interest.

From southeastern Egypt and Ankalidot, *manganese* deposits extend in the Red Sea Hills to Matateib. There are three other manganiferous areas: Sinkat, southwest of Port Sudan; Abu Samr-Allokaleib, south of Sinkat; and in the southeast, east of Malakal, Paloich-Wabuit. In the hinterland of Port Sudan, at Dirbat Well, 250,000 tons of wollastonite (mineral wool) are available; prospectors report tungsten ore from Khor Abent, low grade kaolin from the Khors Odrus and Eishaff, and vermiculite from 16 miles south of Tohamiyam. There is activity at the last locality, but the copper deposits of Tokar (the hinterland of the coast, north of Erythrea) remain idle.

During the last years, another district, Qala en Nahl, attracted attention northwest of Lake Tana. Here, resources of talc-magnesite schists reach the mark of 200 million tons and at Umm Sagata, geologists measured in each foot of rock one-half to two inches of low grade asbestos with fibre lengths ranging from one-eighth to three inches. Furthermore, chromite and gold show.

ETHIOPIA

The Negus, a descendant of the legendary Queen of Sheba, rules 15–22 million Ethiopians. The empire's 460,000 square miles boast varied but scattered resources (Fig. 13), among which gold has been prominent since the time of the legends. But today, the exploitation of potash and exploration for oil are in the foreground. A detailed assessment of the mineral potential of Ethiopia is overdue.

NON-METALS

For a number of years, the Sinclair and Elwerath companies explored for oil in Ogaden, the eastern triangle of Ethiopia. In 1962, drilling commenced south of Scillave, but we should not forget that oil seeps from the Dahlac (Dallak) Islands in the Red Sea. While the search for oil continues, Italians have already evaluated the *coal* and *lignite* resources of the Chelga Basin, of Waldia, Debra Brehan, Mush and of the Alaltu-Didessa area in the west. Seams are thin and their calorific value ranges from 5,000–9,000 BTU per lb. However, downstream of the natural reservoir of Lake Tana, the Blue Nile boasts ample hydropower resources. To complement the Awash power plant, the United States Agency for International Development has planned their development. On the other hand, in 1966, a 10,000 barrel per day refinery costing $13 million is expected to commence operations at Assab on the Red Sea; this plant is offered by the USSR.

On the coast, the Società Saline di Massaua and the Société des Salines d'Assab each recover more than 200,000 tons of *salt* from the sea each year. Therefore, Ethiopia is one of the continent's leading exporters of salt. However, the United States Bureau of Mines statistics record productions increasing from 180,000 tons in 1951 to 190,000 tons by 1962. In addition to these, in Dankalia the Ras (Chief or Prince) Seyoum markets 20,000 tons of rock salt each year; there is little activity in the south, in the Gestro Valley. Also, in the Danakil Depression is the principal industrial project, the *potash* mine of Dalol (Fig. 58). As early as 1912, the emperor authorized Adriano Pastori to explore the potash beds and sources of Dalol (Dallol). The Pastori brothers, the Compagnia Mineraria Coloniale after 1929 and Mr. Orighetti after 1933 produced some potash. Lately, in 1958 exploration, teams recovered 450 tons. Finally in 1961, the imperial government granted a 2,000 square miles concession to the Ralph M. Parsons Company of Los Angeles. The company is investing $13–16 million in the shallow open pit mine, processing plant, power and water supply, in a 65 mile road to the harbour of Ras Andargas and in the building of new jetties;

it has already invested $200,000 in exploration. Each year, starting in 1965, the company plans to export 300,000 tons of potash, mainly to India and Japan. The government will share 50% of the profits. Between Djibouti and Dire Dawa, labourers recover sulphur from the crater of Mount Dofane.

In 1936, the Società Anonima de Cementi Africa Orientale built a 40,000 ton per year plant at Gurgussum near Massawa (Erythrea). In the preceding year, the Ethiopian Cement Corporation started operations at Dire Dawa near the French Somali Coast.

In 1959, 1960 and 1962, the state-owned corporation produced respectively 28,000, 35,000 and 36,000 tons (or 240,000 barrels) of cement and its new plant at Addis Ababa will process 70,000 tons of cement per year from the limestones of the Mugher Canyon, 30 miles north of the capital. Limestones also outcrop at Jijiga, Gelemso on the upper Tacaze, Akaki Haik, Waldia (travertines) on the road to Ginda, Enda Eish and other localities. Each year seven plants produce 10 million bricks. *Pumice* and pozzuolana are abundant on Mount Dula, at Auash, on Lake Tana and further south in the Rift Valley. Gypsum is restricted to the northeast: Aisha on the frontier of the French Somali Coast, Dankalia, Massawa (and the shores of Lake Rudolf). On the Asmara Plateau (at Zolot, Zanaf, Gorbati and Bet Meka), there is a steady production of kaolin; however, the diatomite resources of the Webi Shebeli and the Lake Tabo Valley remain unexploited.

Another mineralized area surrounds Harar (south of Djibouti). We have already mentioned Dire Dawa; asbestos, mica and garnet are known to occur east of this locality as far as Jijiga. Whereas garnets are also found in Gedem, lapidaries collect peridot, a semi-precious stone, on Kod Ali and other islands of the Red Sea.

METALS

For centuries Ethiopians worked gold. Between 1951 and 1957 each year, operators reported an output of 30,000 ounces. According to the United States Bureau of Mines, almost 60,000 ounces in 1959 and 45,000 ounces in 1962 were declared; however, the French Embassy reports 25,000 ounces in 1960 and 30,000 ounces in 1961. Out of this amount, Erythrea generally contributes 5,000 ounces except for 1959, when her output reached the mark of 17,000 ounces. Deposits are widespread at Seroa, Ad Teclesan, Asmara, Goala and Obel in Erythrea, and in the westernmost part of the country between the Blue Nile and the Acobo, in Beni Shangul and Wallega Province where workers recovered one mass of 45,000 ounces at Tullu Kapi and the Acobo. However, only Kebre Mengist near Adola is of some importance (Fig. 52).

This field is located at an altitude of 4,500 ft., 350 miles south of Addis Ababa. The Gudji inhabitants discovered the gold and Italians worked it first. In 1940, production attained 12,000 ounces. Workers built rudimentary dams and, in 1956, a dragline dredge was installed on the Beda Kesa. However, labourers recover most of the gold by wooden batea panning and some ground sluicing, without the benefit of appliances. Each month, the 3,500 man labour force collects 0.75–1 ounce per man. A $1.4 million plan calls for the exploitation of the adjacent Bore Valley. This project would double the output.

In 1904, Kurmakoff, a Russian engineer, discovered *platinum* at Yubdo in the Wallega gold province (USONI, 1952). From 1926, production increased until 1932 when it attained 6,500 ounces. By 1935, workers panned or washed 2,000 ounces per year; but output decreased slightly since 1950 to its present level of 180 ounces per year. In addition to platinum the alloy contains 3–5% rhodium, iridium, osmium and palladium (combined). New reserves of a few million tons of ore warrant the building of a dam on the Birbir River and an output of 10,000 ounces per year.

Since 1959, a company exports 1,500 tons of *manganese* yearly from Maglalla (20 miles southwest of Dalol) and the Krupp group might barter 300,000 tons per year of ore containing 60% *iron* and manganese at Agametta, 20 miles south of Massawa. Here, in 1921, geologists estimated 2.5 million tons of ore containing 59.6% iron and 0.34% manganese. While the resources of Gedem at Massawa do not appear to be significant, in 12 occurrences of the Hamasien Plateau near Massawa, 25 million tons of ore containing 40–56% iron are available. Iron also shows on Mount Tulului in northern Erythrea. However, the limonitic iron ores of northern and central Ethiopia are not as well known. Nevertheless, a company has built a 10,000 tons per year iron foundry near Addis Ababa.

Let us now consider the other metals. In the northernmost part of Erythrea, between Mount Tulului and the sources of the Anseba, copper (Raba) and nickel (Barca) show with chromium, asbestos and magnesite at the Shamege. In the placers of Kebre Mengist, Baro and Wallega, some wolfram occurs. While molybdenite shows at Kiltu in Wallega, vanadium ores are believed to show in the placers of Bilu (Kebre Mengist).

FRENCH SOMALI COAST

Surrounding Djibouti Harbour (the head of the Ethiopian railway), the 8,500 miles of this arid enclave offer little to its 70,000 Dankali and Somali inhabitant. As we have seen, layers of gypsum extend to its borders, but *salt* was the only mineral product. And even the production of salt, which averaged 56,000 tons between 1951 and 1956, decreased to 8,000 tons in 1956 and 2,000 tons in 1957, when recovery ceased.

SOMALIA

The Somali Republic is located on the eastern and southeastern coast of the Horn of Africa. Two million Somalis inhabit this sub-desertic area of 250,000 square miles. The capital, Mogadishu[1], is situated on the southern coast. Berbera and Zeila are the harbours

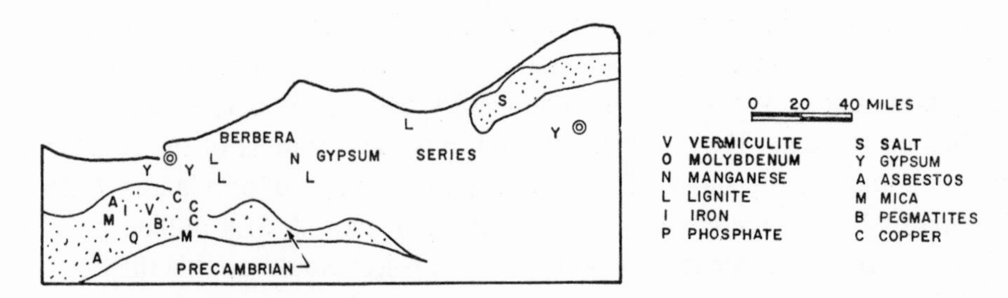

Fig. 14. The mineral resources of northern Somalia. (After PALLISTER, 1958).

[1] Also spelt Mogadiscio.

in the Gulf of Aden. The government accepts aid from Italy, Great Britain, the United States, the United Arab Republic, the Soviet Union and other sources.

Anhydrite and *gypsum*, accompanied by some limestone (Fig. 14) and shale, constitute a 2,000 ft. thick column covering 14,000 square miles of former Somaliland. Gypsum outcrops in a triangle extending from Berbera to Erigavo 160 miles in the east and to Las Anod 150 miles south of Erigavo. Ten miles from Berbera, the Suria Malableh deposit is the most favourably located. The 400 ft. thick column includes the following reserves:
over 95% anhydrite, less than 1% carbonate, 390,000 metric tons;
over 95% gypsum, less than 2% carbonate, 1,980,000 metric tons;
over 90% gypsum, less than 3% carbonate, 6,500,000 metric tons;
over 85% gypsum, less than 4% carbonate, 8,150,000 metric tons.

In 1958, the Geological Survey drilled three holes nine and a half miles from Berbera. Anhydrite is abundant in the upper layers. Partings of shale and marl in the gypsum are lenticular. Two horizons of 'marbled' gypsum have been distinguished. Their sulphate content is variable: in one hole at 257 ft. of depth, shale and dolomite form the drill cores, while in other holes, the layer of gypsum has not yet been pierced at 240 and 300 ft.

Bituminous shales are intercalated in gypsum. A seam of shale has also been found at Las Anod at a depth of 40 ft. Exploration for oil continues.

Other resources are of little present value. Geologists have found some coal and lignite in the Hedhed Tug near Onkhot, Daban and Biyo Gora. Some sulphur shows west of Berbera and workers recover salt near Zeila. Rock salt shows at Heis and probably also at Daga Der and Darraboh Hills, 25 miles from Berbera. In 1935–1936, an Italian company evaporated 260,000 tons of salt at Hordio on the Bay of Hafun (in the Horn) and, before the British dismantled the plant, experts rated its capacity at 440,000 tons per year. Indeed, the evaporating basins covered 2,200 hectares. Saline Somalia SA planned to replace a ropeway by a 9 mile long narrow gauge railway to produce 150,000 tons of salt per year, but this would necessitate an investment of $450,000. Although salt has also been stockpiled at Dante, in 1962 the nation's output only attained 2,000 tons.

At Manja Yihin or Magaian, at the border of the two former Somalias, veins average 1% *tin*. Reserves are moderate and World War II interrupted exploration. The mineralization extends west to Dagan Kuled-Asileh. Washed downstream of Manja Yihin south of Bender Ziada, tin shows in gravels. In 1956, workers collected 16 tons of beryl, some columbite-tantalite and monazite in the Henweina Valley, 30 miles south of Berbera. In the same district, mica and asbestos have been found at Dala'as near Lafarug. Vermiculite occurs in the same area and molybdenite shows near Dobo on the Ethiopian border. While graphite is reported to occur at Ala 'Ule Tug' near Baroma, talc shows near Mora Mountain 30 miles northwest of Hargeisa. Barytes are accompanied by fluorspar in the Bihendula Range south of Berbera, by lead at Unkah, Kul and Arargob and by iron ores at Galan Galo in the Horn. Galena also shows between Lalis and Dananjiehh near Bihendula. Furthermore, copper shows near the same locality, east of Hudiso and west of Lalis. Corundum occurs northwest of Hargeisa.

Between Adad and Salawel and near the Bihen Gaha Pass, geologists have identified low grade manganese. In the south, currents deposit black sands between Gobuin and the old mouth of the Juba River; a sample of which contains 15% TiO_2 and traces of rare earths. HOLMES (1954) discovered that *meerschaum* or sepiolite covers larger areas of El Bur than in most parts of the world. The substance contains 16% MgO, 12% CaO, 1% Al_2O_3 and 42% SiO_2. From Mait Island near Cape Humbeis in the north, 150 tons of guano have been shipped per year.

West of Mogadishu, experts estimate the *iron* ore resources of the Bur Range at a few tens of millions of tons (Fig. 52): between Bur Mun, Bur Safarnolai and Bur Galangol, geologists observed magnetite outcrops within an area of 25 by 20 miles. Samples average a grade of only 35% iron with a maximum of 43%. In addition to the Burs, iron shows at Galan Galo, at Tomassoh west of Hudiso and near Sheikh in the northwest.

SECTION VII

West Africa

The larger Saharan and sub-desertic part of western Africa contains significant mineral concentrations only in Mauritania and Senegal, relatively near the coast. This belt consists of eight countries. The ten countries of the southern tier are smaller but generally richer. Among these, the most favoured states are Guinea, Sierra Leone, Liberia, Ghana and Nigeria.

SPANISH SAHARA

The Canary Islands lie off the coast of 100,000 square miles of desert inhabited by 70,000 Maures. The principal province of the Spanish Sahara is Rio de Oro. Vila Cisneros harbour, its capital, is located west of Fort Gouraud. As the Mauritanian iron ore deposits are so near the frontier, there is a mineral potential. Oil companies had important investments in the exploration of the coastal basins. Furthermore, according to reports, phosphate resources are considerable.

MAURITANIA

Covering 420,000 square miles, the Islamic Republic of Mauritania occupies the southwestern portion of the Sahara. Largely nomads, 600,000 Maures are estimated to inhabit the oases. The coast extends from Port Etienne to the Senegal River; the capital, Nouakchot, is located 150 miles north of Saint Louis du Senegal. The country's industry and the revenue of the francophile government is based, aside from fishing, on the iron deposits of Fort Gouraud near the southeastern corner of the Spanish Sahara (Fig. 55). The Koediat d'Idjil (Fig. 61) consists of the F'Derik, Rouessa and Tazadit Hills, respectively containing 43, 14 and 87 million tons of iron ore (64% Fe). Twenty million tons of F'Derik ore necessitate underground mining but the rest can be worked open cast. Additional resources containing 45% iron attain several thousand million tons. Output is programmed to rise from 4 million tons in 1963–1964 to 6 million tons per annum by 1967. Water has been discovered 40 miles from the deposit and a 400 mile long railway has been built from Pointe de Chacal at Port Etienne to Idjil. Furthermore, the harbour will be expanded to handle ships of 60,000 tons.

Total investment is planned to attain $170 million: 34% by the Bureau de Recherches

Géologiques et Minières (BRGM); 27% by Compagnie du Chemin de Fer du Nord, Compagnie Financière pour l'Outremer, Denain-Anzin, and Union Sidérurgique du Nord de la France combined; 20% by BISC (Ore) Ltd., British Ore Investment Corporation Ltd. and British Steel Corporation Ltd. combined; 15.7% by Società Finanziaria Siderurgica and Società Mineraria Siderurgica combined; and 3.3% by German steel groups. A 15 year loan of $66 million has been provided by the World Bank (IBRD), a $10 million loan by the Caisse centrale de Coopération Économique, and another $20 million loan has been guaranteed by the French Republic. Mauritania will receive a 6–9% export duty on the ore and 50% of the profits of MIFERMA (Société Anonyme des Mines de Fer de Mauritanie).

Northeast of Fort Gouraud, the iron belt continues towards Sfariat. Further north, MIFERMA and BRGM have also explored the beryl and lithium occurrences of the Bir Oum Greine district.

Near Akjoujt, halfway between Nouakchott and Port Etienne, the Guelb Moghrein is located 150 miles inland of Cape Timiris. The deposit contains *(a)* 8 million tons of iron ore (68% Fe), *(b)* 450,000 tons of copper (grade 1.5%) and 14 million ounces of gold (grade 2/3 dwt. per ton) in the sulphide zone and *(c)* 200,000 tons of copper (2.9% grade) and 11 million ounces of gold in the silicate-oxide zone. Drillers have struck water 30 and 80 miles from the deposit. Experts have worked out three projects necessitating, respectively, an investment of $60, 30 and 15 million. The latest of these would provide 25,000 tons of copper and 40,000 ounces of gold per year to be trucked to Nouatchott, where a new wharf would be built. The Société des Mines de Cuivre de Mauritanie (MICUMA) has an initial capital of $1.6 million. A new corporation to be formed by the Homestake (Mining Company) with the Bureau de Recherches Géologiques et Minières and the Mauritanian government plans to invest $36 million in the latest project.

In the Legleitat el Khader, 15 million tons of ore contain 52–55% iron with copper showing in the wadi Jennin and on the Akjoujt-Atar trail. Ilmenite and zircon are disseminated in coastal sands at Gendert, Lemsid, El Mahara and on Arguin Island. South of Cape Timiris, 6 million tons of sand contain 2.5–5% titania (TiO_2). For centuries, labourers produced salt at Trarza and Idjil. Furthermore, *oil* exploration has started on favourable structures of the interior Taoudéni basin. However, inadequate water resources pose a major problem for soil formation, stock breeding and mineral recovery. Finally, phosphate occurs at Civé.

SENEGAL

Like Mauritania, the Senegal Republic has been more fortunate than the other sub-Saharan countries. A population of 3 million inhabits its 76,000 square mile area, supplying a mineral production (Fig. 15) valued at $6 million per year. The principal product is phosphate and the government welcomes foreign investment. The centuries-old ties of the Senegalese and French, high standards of learning and the first French speaking university in Africa provide the infrastructure of development. Remaining an active metropolis, Dakar, the capital, has been the centre of the West-African Federation.

Northeast of Dakar, *phosphates* cover the plateau extending from Tivaouane to Louga. Production continues to expand; in 1961, the Péchiney Company exported from Pallo

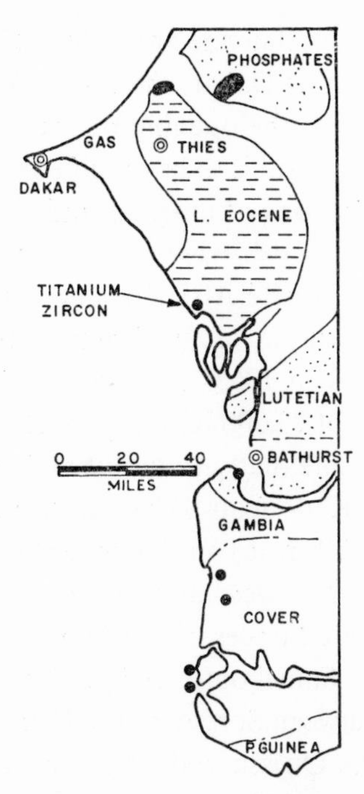

Fig. 15. The mineral resources of the Senegal coast.

140,000 tons of aluminium phosphate, 20,000 tons of the dehydrated variety and 7,000 tons of phosphal; 400 tons of baylifos, a fertilizer, is used directly after thermal treatment. (The lower ore contains 0.03% uranium.) For 1964, the company plans a 200,000 tons output. The production of Taïba lime phosphate is now in excess of 400,000 tons, and short range plans call for an output of 600,000 tons; the potential capacity of the deposit is a million tons per year. Reserves attain 39 million tons (Keur Mor Fall, 13 million tons, 58% BPL; N'Domour-Diop, 13 million tons, 61.5% BPL; intermediate zone, 8 million tons, 57% BPL). An additional 60 million tons of probable reserves are available. Compagnie des Phosphates de Taïba, founded by the Constantine and Oceania (Makatea) phosphate companies, Péchiney, Pierrefitte, COFIMER, Auxon, and the Bureau de Recherches Géologiques et Minières plan to invest $20 million. However, International Minerals and Chemicals of Skokie, Illinois now have a majority interest. A million ton output would necessitate a $55 million investment. Other deposits are available at Pire-Goureye southeast of Mékhé and at Guiers.

From 4,400 tons in 1949, SOCOCIM expanded *cement* production based on Rufisque marls and limestones to 160,000 tons in 1957 and 200,000 tons in 1960 (or a million barrels in 1962).

Black sands concentrated in the Djiffère plant contribute $750,000 yearly to the Senegalese economy. From 40,000 tons in 1957, the Gaziello Company reduced *ilmenite* exports to 25,000 tons in 1960. With grades diminishing, production has further decreased by 50%. However, in five years *zircon* production has tripled to reach 11,000 tons in 1960, declining to 3,000 tons in 1963. At M'Bour-Joal, a million tons of black sands containing 650,000 tons of recoverable ilmenite are available. Furthermore, the Casamance (southern Senegal, south of Gambia) dunes contain 500,000 tons of concentrates carrying 375,000 tons of ilmenite. Grades attain 10%. At Cayar Lampoul, in the dunes extending from Dakar north to Longo, 4 million tons of ilmenite are included in 6 million tons of heavy sands, but grades attain only 26%. South of Dakar, the Fata (Djiffère) deposit, distinguished by grades of 25%, has been exhausted; the opening of the high grade Niaming-Joal deposit has born fruit in 1964. The Djiffère plant is to be modified and a new plant erected at Dakar, thus increasing capacity to 60,000 tons of ilmenite and 12,000 tons of zircon. Investment will exceed $4 million.

At Cap Vert near Dakar, SAP (Société Africaine des Pétroles) has already proved 1,200 million cubic ft. of gas, but the output of the Diam Niade oil field has decreased from 17,000 barrels in 1961 and ceased in 1963. The Société Africaine de Raffinage did start operations in 1963–1964 at a rate of 23,000 barrels per day. Generally representing 70,000–80,000 tons per year, experts estimate that salt output of Senegal and Mali now attains only 55,000 tons. In the southeastern Kedougou-Saraya district, 40 million tons of ore contain 63% iron, with 45 million tons (53% iron) on the upper Gambia, 500 miles from the coast. Meanwhile, at Dakar, a 30,000 ton per year mill will use imported iron ore and scrap. Also at Saraya, geologists have discovered traces of uranium, tin and lithium. Columbite-tantalite and tin show at Diambaloye, Samékonto, Bafoundou and in the Falémé. Also in the Falémé River, wolfram occurs between Sorćto and Saboussiré. Molybdenum, copper, lead and gold indices are found in the Ouessa and Tenkoto districts, and alluvial gold in the Gambia and Falémé at Kossanto and Tenkoto. At Tomborokoto, geologists have identified traces of copper. Finally, graphite occurs at Patassi.

GAMBIA

This country of 300,000 people consists of a 200 mile long and 25 mile wide strip of land following the Gambia River. The Senegal Republic and its southern province, Casamance, surround Gambia. Of course, the black sand deposits continue between Joal and the Casamance delta (Fig. 15). Between 1957 and 1959 they have produced 60,000 tons of *ilmenite* but working has been suspended. Around Bathurst, the capital, oil exploration has not yet met with success.

MALI

Covering 470,000 square miles, landlocked Mali spreads from the bend of the Niger into the Sahara. A majority of the 3.7 million people live in the south. The government pursues a positive neutralist policy accepting aid from France and the USSR. Complementing the Senegal railway, freight is being trucked from the railhead in the Ivory Coast to Bamako, the capital. Unfortunately, the mineral potential (Fig. 54) of the republic is compromised by the inadequate infrastructure of transport.

The *manganese* deposit of Ansongo on the Niger border includes 1.5 million tons of siliceous ore (43–46% Mn). Low grade *phosphates* (27% P_2O_5) outcrop at Tamaguilel 80 miles north of Gao. The government plans to distribute this fertilizer locally. In 1962 1,000 tons of phosphates were produced at Tilemsi. Capacity ranges up to 10,000 tons per year. Geologists have found low grade copper ore at In Darset, and 150,000–200,000 tons of zinc (grade 7–10%) at Tessalit in the north. In the Balia area, at Kéniéba on the Guinea border and west of Bamako, resources of *bauxitic* laterite attain 550 million tons (39–44% Al_2O_3). Plans call for the construction of a dam at Guina on the Senegal River and of a harbour at Kayes.

Though interesting, 30 million tons of *iron* ore in southwestern Mali are not competitive in the present market. In a similar category, the Sinsinkourou pegmatite represents 100,000 tons of spodumene (6.7% lithia), with three others containing 60,000 tons near Bougouni in the southwest. Prospectors have also found cassiterite and columbite–tantalite in these dykes and in the Adrar of the Iforas in the east. Gold occurs in the placers of the Falémé, and in the same area geologists have discovered a few *diamonds* at Kéniéba. Indices of molybdenum, lead, copper, chromium, nickel and of the other metals show in various districts, including the Iforas, Kayes, Sikasso and the Mandingue tableland. However, the only active production remains the recovery of *salt* in the empty quarter of Taoudenni in the northwest. Production could be increased from 2,500 tons in 1962 to up to 20,000 tons. Unfortunately, water supply is inadequate and soils are generally poor. Cement rock and marble are found at Bafoulabé; limestone occurs at Goudam.

UPPER VOLTA

The Voltaic Republic lying north of the Ivory Coast and Ghana communicates with the sea through the Abidjan-Bobo Dioulasso-Ouagadougou railroad. A population of 3.7 million, many of which are Mossi and Bobo, inhabit 106,000 square miles. In matters of exploration (Fig. 54) the government cooperates with the Bureau de Recherches Géologiques et Minières (BRGM). Located near the northwestern corner of Ghana, the Poura vein contains at least 4.5 million ounces of *gold* (grade 11 dwt. per ton). Engineers plan a plant for a throughput of 1,000 tons of ore per day and an output of 50,000 ounces per year. Furthermore, workers have also recovered some gold from tailings and placers. Gold also shows at Dossi, Bagassi, Gaoua and south of Banfora.

Eighty miles northeast of Bobo-Dioulasso, the Tiéré deposit represents 300,000 tons of ore with grades varying between 30% and 45% of manganese. At Tambo, 5 million tons contain 52% manganese. Furthermore, manganese shows at Bonéré, Kalgo and Duo. In 1940, 300 tons of copper were produced at Gongondy near Gaoua in the southern reentrant. Nearby, copper also shows at Dienemera. Titaniferous iron ore is disseminated between Dori and Djibo, principally at Hoka and Tin-Edia with the latter, unfortunately, containing grades of 12% TiO_2. Traces of other metals are known to occur in the rest of the republic but water resources are inadequate.

NIGER

Inhabited by 2.5 million people, the Niger Republic spreads north of Nigeria from Mali to Chad over 460,000 square miles. In exploration, the government cooperates closely with BRGM. The country's mineral resources (Fig. 49) are located in the Aïr Mountains (north–centre). An official production of 50–60 tons of *tin* ($100,000) has been maintained each year from the Tarrouadji Mountains in the southern Aïr, 40 miles east of Agadès. Observers believe, however, that effective output is higher. Unfortunately, the paucity of water and labour hampers recovery. In the Adrars Elmeki and Guissat, 70 miles north of Agadès, tin, *tungsten* and columbite–tantalite are associated. Tungsten minerals are also disseminated 15 miles northeast of Agadès. *Copper* and *uranium* show along a 100 mile western rim of the Aïr passing west of Agadès through Assouas. Furthermore, traces of copper occur in the wadi Enneg. Rare earths, thorium and lead have been reported from the Tatouss-Dabaga Mountains. Lithium, molybdenum minerals and manganese ore (39% Mn) show north of Tera (southwestern Niger).

Traces of chromite are found at Makalondi and titaniferous iron ore at Saoua. The low grade iron ore plateaus of northern Togo and Dahomey cross diagonally into neighbouring Niger. The principal deposits, Doguel Kaina Say and Dyabon, are located south of Niamey (Fig. 24). The Kainji Dam project (Nigeria) would facilitate navigation on the Niger and access to these deposits. Meanwhile, water supply is inadequate in the greater part of the country, even for soil formation. The sebkhet (salt pans) of Teggidda N'Tesemt, Seguedine, Bilma and Fachi in the area of Kaouar, can provide 3,000 tons of salt each summer. Other salt pans include Birini N'Kouni, Filugue and Gouré. Hopefully, oil exploration is centred on the Djado and Tarak basins in the northeast.

CHAD

Between Niger and Sudan, most of Chad (République du Tchad) is part of the Sahara. However, at Fort Archambaud, equatorial forest covers the country's southern edge. In Chad, we find 2.6 million people inhabiting 500,000 square miles. At Mundu on the Logone south of Fort Lamy, the capital, geologists have discovered 10 million tons of bauxite. Salt extraction, averaging 6,000 tons yearly in the Ennedi, Borku and Tibesti between 1955 and 1958, dipped in 1959 to 2,000 tons. From the Yédri Mountains in the Tibesti (northern Chad), 22 tons of wolframite have been airlifted. Unfortunately, the occurrence is too remote and the climate too arid for recovery.

In 1939, the Compagnie Minière de l'Oubangui Oriental (CMOO) produced 6,000 ounces of gold at Kamboké in the southwest, but by 1940 output decreased to 60 ounces.

In Lake Chad *soda* precipitates. North of the Lake (Fig. 24), e.g., at Kalia, workers cut 50 pound slabs of soda, selling each for 50 cents. Camels carry the slabs 10 miles south to Baga Sola, whence they are shipped to Fort Lamy. Authorities believe that the output reaches the mark of 7,000 tons each year. Out of this, 1,000–2,000 tons are sold in Chad, 700 tons in the Central African Republic, roughly 2,000 tons may leave the country without the benefit of customs and, in 1960, a maximum of 4,400 tons worth $140,000 were exported to Nigeria. Around Lake Chad, diatomaceous earths occur, but Chad's water resources are insufficient. The vogue of wildcat drilling is now reaching the Republic; SNPA manages exploration for PETROPAR in the Erdis concession. Finally limestones occur at Mayo-Kebbi.

We now turn to the southern tier of countries.

PORTUGUESE GUINEA

Little is known about the resources of this country of 14,000 square miles and 600,000 people. Black sands are believed to have continued to sediment south of Casamance as far as Bissau, the capital. Oil companies have also explored the vicinity of Bissau. Laterites containing 40% iron are known to occur along several roads. The Guinean alignment of bauxite deposits extends from Boké into the northeastern part of the country.

GUINEA

Guinea fronts the sea, extending inland north of Sierra Leone and Liberia. A population of 3 million inhabits the country's 100,000 square miles. Guinea contains the world's largest bauxite resources and vast resources of iron ore (Fig. 54). Consequently, in the next 20 years, the country could become one of the continent's major mining areas. The government of the Democratic Party of Guinea pursues a policy of positive neutralism, accepting aid from the USSR and the US. In 1961, Canadian and French owned properties of Bauxites du Midi have been expropriated, but reorientation of policies has improved the investment climate.

BAUXITE

Tablelands or bové of the Upper Cogon include 420 million tons of bauxite containing 55% alumina, including a large share of higher grades (Fig. 72). Sangarédi, the planned railhead, is located between Portuguese Guinea and Conakry, the capital. An 80 mile long railway bridging the Rio Nuñez and Kataba was to be built to Kansar on the ocean. A production of at least 1,500,000 tons has been planned, of which a million tons were to be exported and 500,000 tons were to be reduced to 220,000 tons of alumina at the Boké plant. The Société des Bauxites du Midi, a subsidiary of ALCAN, has invested $20 million and planned to invest a total of $80 million. In 1964 the Compagnie des Bauxites de Guinée was formed by the government and American investors to produce at least a million tons of bauxite. The Harvey Aluminium Company studies a new plan of exploitation of Boké.

Other deposits of bauxite cover the plateaus of the middle Fatala halfway between Boké and Fria. Further south, the Fria deposit, one among several on the confluence of the Konkouré and Badi, contains 250 millions tons of bauxite with grades surpassing 40% alumina. The reduction of 1.5 million tons of bauxite to 500,000 tons of alumina was achieved in 1962. A 90 mile long railway has been built to Conakry with roads connecting Fria to Ovassou and Soubétibé. Conakry harbour has been doubled with wharfs now extending to 600 and 450 ft. Hydropower resources of the Konkouré attain almost a million kW. The share of Souapiti, 18 miles from Fria, is 350,000 kW with Amaréa boasting 250,000 kW. The earth dam and hydroelectric plant were scheduled to cost approximately $160 million; the alumina plant, $120 million; the railway and roads built by TRANSFRIA, $1.5 million; and the extension of the harbour, $10 million. The Olin–Mathieson Chemical Corporation owns 48.5% of the shares of the Fria Company; Péchiney-Ugine, 26.5%; the British Aluminium Company Ltd. and Aluminium Industrie Aktiengesellschaft, 10% each;

and Vereinigte Aluminium Werke AG, 5%. Output is divided according to shares, votes of the board being weighted thus: Péchiney, 55%; Olin, 34%; British Aluminium and Aluminium Industrie AG, 4%; and VAW, 2%. Péchiney manages Fria. Exports are divided between France, $10 million; rest of western Europe, $7 million; Cameroon (Edea), $7 million; North America, $800,000. Alumina output is to expand to 1,200,000 tons.

Resources of Kindia, 90 miles east of Conakry on the Niger railway, attain 100 million tons of bauxite (>40% alumina). Minor deposits are found in Bassari country north of the Cogon. The Tougué deposits, east of the Fouta Djalon Mountains and 300 miles from Conakry, represent several hundreds of million tons of high grade bauxite, but only 10 million tons are left in the Los Islands off Conakry. From 1952 to 1961, the year of nationalization, 3.7 million tons were exported to Canada and 60,000 to Germany. Production now continues at a slower rate under the supervision of Polish and Czech experts. Let us add that total resources of Guinea have been estimated to exceed 2,400 million tons!

OTHER RESOURCES

Bordering the Ivory Coast and located north of Lower Buchanan Harbour in Liberia, the Guinean or northern sector of the Nimba Range (Fig. 63) contains 225 million tons of *iron ore* (Fig. 18). Resources of the Simandou Range (Fig. 62) are in excess of 700 million tons and these may attain 2,200 million tons. COFRAM (Compagnie Franco-Américaine des Métaux) manages CONSAFRIQUE, the Consortium Européen pour le Développement des Ressources Naturelles de l'Afrique, which consists of the Belgian holding companies BRUFINA and AUXIFIX, the Banque de l'Indochine, the Deutsche Bank, Hambro's Bank, London, and the Algemene Bank Nederland. The seven largest Japanese steel mills contribute 20% of the capital. North of Marampa, Sierra Leone, the Lola deposit lies inactive, but since 1953, the deposit of the Kaloum peninsula and Mount Kakoulima near Conakry has been worked. This chromiferous iron ore is easily reductible. Reserves of soft ore attain 1,000 million tons (55% iron) and of hard upper ore 200 million tons (50% iron). Production and investment attain, respectively, 700,000 long tons per year and $12 million. However, in 1961, production decreased to 540,000 tons ($1.6 million). Of the exports were sent 60% to Poland and 40% to the United Kingdom. To facilitate exports, experts plan the enlargement of Conakry harbour.

In 1936, the production of *diamonds* started. In 1960, recorded output attained 50,000 carats in the Macenta district (north of Liberia, Fig. 18). Unofficial production may have been ten times as great, exporting 140,000 carats of gems and 210,000 tons of industrials. SOGUINEX, a subsidiary of the Sierra Leone Selection Trust, is the largest producer. A buying office has been opened at Kankan, but in 1960 the Bounoudou workings were closed. In 1963, the government reserved the area extending from Balue to Bimbaka and Tougbarako for private enterprise. *Gold* output of eastern Guinea has also been largely unofficial, with statistics representing only the declared production. Exploration continues in the Siguiri district. Finally, graphite also occurs at Lola.

SIERRA LEONE

The value of the mineral production of this small (30,000 square miles) but rich country (Fig. 17) has expanded from a million dollars in 1934 to 10 million in 1952, $36 million in 1959 and $60 million in 1961 (Fig. 16). Thus the per capita mineral income of 2.5 million Sierra Leonians attains $20. A moderate government provides a favourable investment climate.

DIAMONDS AND IRON ORE

Diamonds, discovered in 1930, constitute the principal exports and 90% of the mineral output. The diamond scheme has succeeded in reverting part of the clandestine exports to Freetown, the capital and principal harbour. Licensed mining carried out in a 9,000 square mile area of eastern Sierra Leone, including the Kono district, the Sewa drainage and the land between Tongo and the Moa, produces 1,040,000 carats. The administration assists and polices private miners. Sierra Leone Selection Trust (SLST) has provided 720,000 carats from Yengema and Tongo. The company holds mining rights in a 210 square mile area until 1985. SLST has also started dredging the Bafo River on a large scale. Additional kimberlite veins have been discovered at Monkey Hill, the first deposits of this type having already been located in 1948 at Koidu. The production of SLST is now also sold through the governmental diamond office. Indeed, although De Beers has considerable interests in it, the company has terminated its contract with the Central Selling Organization. The diamond company operating in Ghana, the Consolidated African Selection Trust (CAST, an associate of the Rhodesian Selection Trust), owns the $4.5 million capital of SLST. However, the State receives 50–60% of the profits of SLST and a majority of the profits of CAST come from SLST. The proportion of industrials to gems is higher than in any other field with the value of the carat averaging almost $20. In 1960 and also in 1961, a million carats of diamonds changed hands in Monrovia. A considerable part of these is believed to originate in Sierra Leone. Consequently, the total value of production may have been in the range of $90 million.

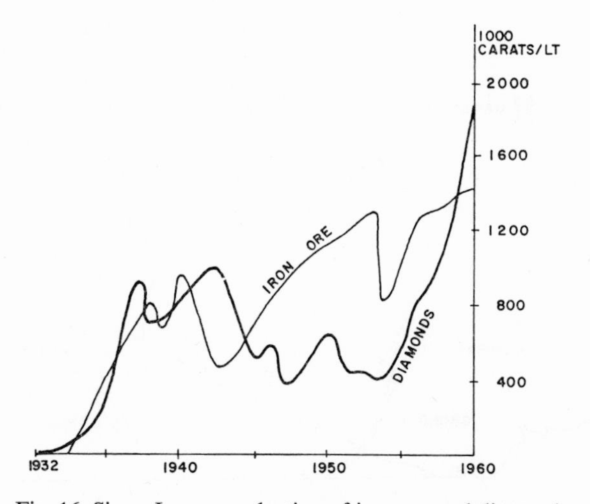

Fig. 16. Sierra Leone production of iron ore and diamonds.

In central Sierra Leone, ore reserves of Marampa containing 65% *iron* are in excess of 30 million tons. The mills of the Sierra Leone Development Corporation (SLDC) work at capacity. The Northern Mercantile and Investment Corporation, William Baird and Company, and the United Africa Company (Levers') share the $17 million capital of SLDC. In 1961, the company installed a 500,000 ton per year concentrator costing a million dollars. Output attained 2.1 million tons; $18 million worth of ore has been shipped through Pepel to Western Germany, the United Kingdom and The Netherlands, France being a minor importer. A new entrance channel has been inaugurated and the new quays will facilitate servicing of 35,000 ships. Sixty miles northeast of Marampa, the low grade ores of Tonkolili (170 million ton reserve, grade 56% iron) have to be dressed. An additional $90 million is needed for the financing of this project. Ironstones have also been found in the Bagha Hills on the Liberian border.

OTHER PRODUCTS

The Geological Survey has discovered an 18 mile long band of *bauxite* in the Mokanji Hills extending beyond the Jong in the Moyamba district, at Kpaka near Pujehun and earlier in the Gbonge Hills. In 1962, Sierra Leone Ore and Metals Ltd., a subsidiary of Aluminium Industrie AG, started extraction, exporting 20,000 tons in 1963. Far away in the Sula Mountains, 250,000 tons of bauxite have been estimated. The adverse *titanium* market has hampered rutile production at Hangha. Consequently, in 1960, the mill was closed. Consolidated Zinc Corporation and Southern Chemical Corporation Inc. have explored a 2,500 mile area of the Bonthe, Moyamba, Port Loko and Kambia districts between Freetown and the Guinean frontier. Ilmenite sands are known to spread between Tokeh and

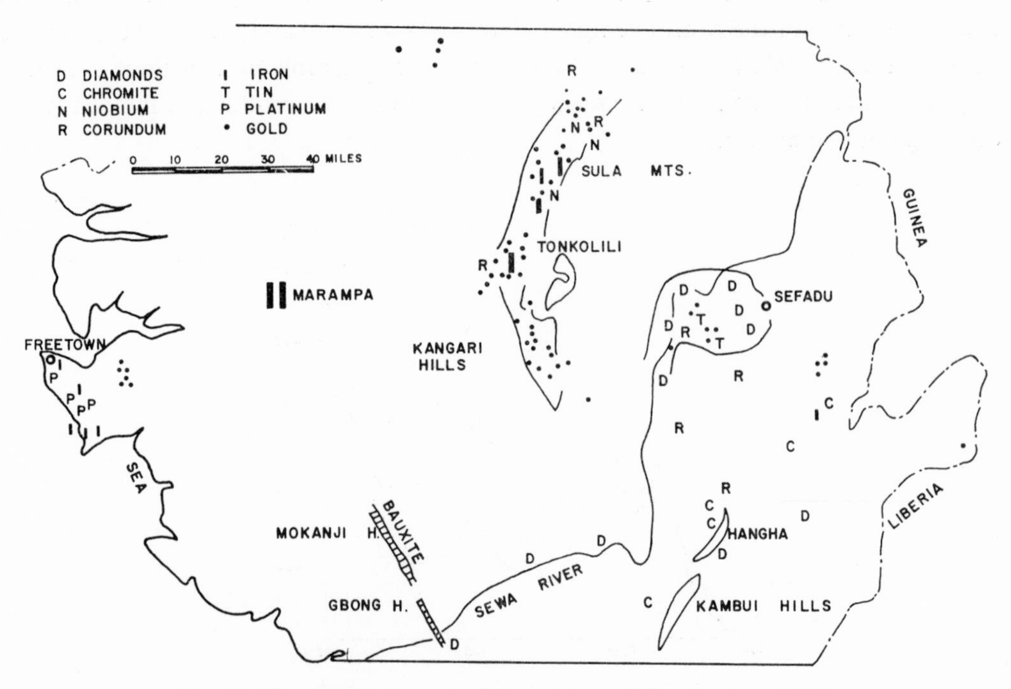

Fig. 17. The mineral resources of Sierra Leone. (After POLLETT, 1952).

York and in the estuary of the Whale River. At Sherbro a company of Pittsburgh Plate and Glass will produce 200,000 tons of rutile out of a reserve of 30 million tons.

Although market conditions have limited the underground production of *chromite* at Hangha in southeastern Sierra Leone, an output of 20,000 tons per annum was maintained between 1952 and 1956. Attaining only 6,000 tons in 1960, production of lump ore has been suspended, but in 1963 9000 tons were exported. Reserves are limited. Between 1929 and 1949 when output ceased, miners extracted 5,200 ounces of platinum from the Colony Hills near Freetown. Between 1930 and 1956, when industrial production was interrupted, 340,000 ounces of *gold* were won from placers of Sula, Kangari, Gori, Loko, the Nimini and Kambui Hills. Output reached its peak of 40,000 ounces in 1936. But vein gold is being explored at Dalakuru, Yirisen and Baomahun, situated in the Diang, Kafe and Valonia chiefdoms. Workers recover salt mainly in the deltas of the Samu chiefdom.

Minor concentrations of columbite and niobian rutile are found in the Sula-Kangari zone in the Valonia chiefdom. The Sende Stream carries the more tantaliferous variety. As usual, the pyrochlore of the ceramic grade Bagbe nepheline syenites is uneconomical. Corundum is disseminated in the placers of the Kono and Koinadugu districts. Molybdenite is associated with copper, lead and zinc minerals at Kalakuru in the Koinadugu district. Graphite is disseminated in gneisses of the Gbangbama Hills, e.g., at Moselelo. Finally, in the Koya chiefdom, 3 million tons of lignite lie below the sea level and overburden.

LIBERIA

Because of the development of its high grade iron ore resources (Fig. 18), the rubber growing and agricultural economy of Liberia has undergone intense structural changes. An estimated population of 1–2 million inhabit the country's 40,000 square miles. The mineral production

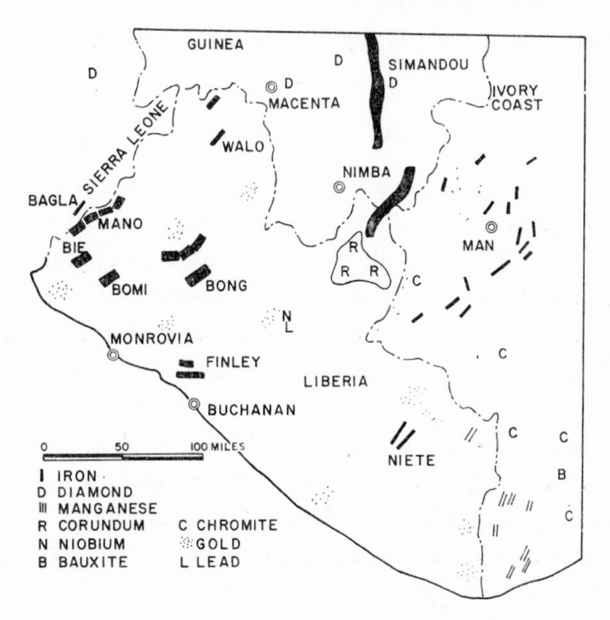

Fig. 18. The mineral resources of Liberia, Guinea and the Ivory Coast.

is scheduled to attain a value of $150–300 million in a few years' time, providing one of the continent's highest incomes. Furthermore, a pro-American government provides a favourable investment climate.

Iron ore resources are widely distributed in the northwestern third of the country and, by the end of the decade, Liberia is scheduled to become the world's third exporter of iron ore.

NIMBA

On the Guinea–Ivory Coast border, the southern sector of the Nimba Range contains explored reserves of 250 million tons averaging 65% iron, probable reserves of 20–50 million tons (60% iron) and 50 million tons of possible ore (62% iron). A 165 mile railway has been built to Buchanan. This harbour, constructed at a cost of $35 million, will be distinguished by a 6,200 ft. breakwater, with wharfs receiving ships of 45,000 tons maximum (compare with the projected 60,000 tons at Port Étienne). Investment in the ore complex has attained $200 million, with the Export–Import Bank providing $30 million. Whereas the Liberian government owns half the equity of LAMCO, the Liberian American Swedish Minerals Company holds five of the seats including the vice-presidency of the eleven-man board. Bethlehem Steel Corporation owns 25% of the shares and six Swedish companies, IAAC (International African American Corporation), Canadian–Liberian Iron Ltd. (60% owned by the Swedish consortium and 40% by IAAC), also providing the presidency, share the rest.

Trafik AB Grängesberg–Oxelösund are the managers of the complex, and the selling organization. Swedish SENTAB controlled construction. Raymond International built the railway and harbour, and the Viviani Company constructed the road. By 1965, production is scheduled to attain 6 million tons and eventually 15 million tons with Bethlehem Steel absorbing 2.5 million tons. German steel companies have guaranteed a $52 million loan of the Kreditanstalt für Wiederaufbau, will receive ore as will French steel corporations. East of Monrovia, LAMCO is also exploring the Finley Mountains.

OTHER IRON MINES

In the *Bong* Range, 60 miles northwest of Monrovia by road, reserves of the Zaweah Hills attain 220 million tons certain, 40 million tons probable and 20 million tons possible ore averaging 37.4% iron. The ore will be concentrated to contain 65% iron. A yearly production of 1.5–3 million tons will be expanded considerably. DELIMCO, the German-Liberian Mining Company, a subsidiary of Gewerkschaft Exploration of Düsseldorf, is half owned by the Thyssen group of the Ruhr and half by the Liberian government. To attain a 3 million ton output, DELIMCO will invest $60 million and an American group $20 million. DELIMCO is also exploring the Kpo and Wologisi Hills. *Bomi* means heavy rock or iron ore in the Gola language. The Bomi Hills deposit belonging to the Liberia Mining Company is controlled by Republic Steel. Situated 40 miles north of Monrovia, the mine contains 90 million tons of ore. A quarter of this is high grade (66% iron) lump ore, the rest containing 43–45% iron. Sulphur content varies between 0.05 and 1%, phosphorus averaging 1%. The output of this deposit is now in excess of 3 million tons, fetching

$35 million. Enlargement of Monrovia harbour will eliminate a bottleneck of production for Bong, Bomi and Mano.

North of Bomi, reserves of the *Bie* Mountains owned by Mine Management Associates (MMA) are believed to be large, averaging a grade of 54.4% iron. MMA consists of the Liberian government (50%), Liberian investors (35%, four-fifths of this amount having been lent by Mr. Christie), of Mr. Christie and other foreign shareholders (15%). Furthermore, MMA manages the concession of the National Iron Ore Company at *Mano* River on the Sierra Leone border. From here, more than 2 million tons of ore are being exported yearly via Bomi and Monrovia and by 1965, plans call for a 4.5–5 million ton production. Estimates of reserves vary between 50 and 200 million tons of ore containing more than 54% iron. In addition to these, probable and possible resources attain 200 million tons containing 40–50% iron. Reserves of the *Bassa* deposit owned by the Liberia Development Company (LDC) near Robertsfield airport and 30 miles from Buchanan have reached the mark of 50 million tons of ore containing 51% iron. Finally, the *Mount Jedeh* deposit is located in eastern Liberia not far from the Ivory Coast border, and the Putu Range is situated 100 miles from Harper City and Cape Palmas.

OTHER RESOURCES

Bauxite is also believed to exist in eastern Liberia, as is manganese, near Webo and east of Monrovia. Although most of the *diamonds* exported from the country (1 million carats) have found their way to Monrovia from Sierra Leone, Guinea and the Ivory Coast, observers believe that a few stones may also originate in the country, especially in the Lofu Basin extending from south of Bomi to Weasu in western Liberia. Furthermore, diamonds have been reported from south of Nimba and from Tapeta in the east. Similarly, gold production has not entirely been controlled. The principal deposit has been Boken Jede, east of one of the Firestone plantations in southern Liberia. Workers also have washed gold *(a)* at Pleebo and Webo in the southeastern corner at Tchien, *(b)* south of the Cavally River at Kokoya, *(c)* about halfway between Nimba and Buchanan, *(d)* northeast of Monrovia at Gola Kenneh and Wenju, *(e)* north of Robertsfield at Tawalata and Yangaya, *(f)* Belle Yella near the Bong Mountains and *(g)* at Neepu and Rivercess near the Guinean border. Other heavy minerals of these placers include columbite–tantalite, galena (lead), and pyrite at Tawalata and Kokoya. Finally, corundum shows in the diamond placers and at Golella north of Tawalata.

IVORY COAST

A population of 3 million inhabits the 125,000 square miles of the Ivory Coast, which is one of the most developed of the western African republics. The favourable position of Abidjan harbour, the opening of a new channel in the reef, have attracted the products of Upper Volta and partially reoriented the traffic of Mali towards the Ivory Coast. Consequently, the city of Abidjan is expanding rapidly. In line with this policy, a considerable effort is being made: the Development and Resources Corporation of New York has drawn up an $8 million program of geophysical, geochemical and geological exploration for the new state prospecting company.

DIAMONDS AND MANGANESE

The value of the mineral production ($15 million) is principally provided by *diamonds* (Fig. 19). Between 1948 and 1960, 1.5 million carats have been officially reported. Uncontrolled production has been considerable, and a large part of the diamond production of the Upper Sassandra and Seguéla has been attracted by buyers in Monrovia (Liberia) and Kankan (Guinea). Buying offices operate now and the output of small scale workings of the Sassandra oscillates between 50,000 and 390,000 carats. Large sections of the Bandama and Comoé drainages also carry diamonds.

Twenty miles north of Seguéla, the variable production of the Société Diamantifère de la Côte d'Ivoire or SODIAMCI (500–11,000 carats) is hampered by the paucity of water. Hopefully, SODIAMCI plans to expand its output to 20,000 carats per year. However, exploitation of the Tortiya placers discovered in 1952 between Bouaké and Korhogo has been better organized by the Société Anonyme de Recherches et d'Exploitation Minière de la Côte d'Ivoire or SAREMCI. Indeed, increased production now attains 170,000 carats. Here the ratio of industrials to gems is only 1.5. A new plant has been built and intensive exploration continues. New reserves have been evaluated north of Seguéla. The Watson Company also explores diamonds.

In 1960, the production of metallurgical grade *manganese* started at Grand Lahou, 90 miles west of Abidjan. In 1961, a 100,000 tons per year output of lump ore and 25,000 tons per year of fines has been achieved, representing $3 million (Fig. 19). The first of these categories contains 47% manganese, the second 41%. With investments reaching $4 million, reserves attain only 1 million tons of ore (grade 47.6%) or 1.5 million tons at 44%. However, the Ziémougoula deposit 25 miles north of Odiénné in the northwestern corner of the country is somewhat larger. This ore averages 46% manganese. Other occurrences include: Boundiali west of Seguéla (4 million tons), Mbouésso near Yaouré in the central Ivory Coast (3 million tons, 25% manganese), Dabakala near Grand Lahou (1 million tons, 45% manganese) and the Boundoukou zone in the east including Assoufouma-Broumba. Plans call for further exploration in the favourable zones of Féttékro-Mokta, Aboisso (east), Yaouré-Soubré and possibly Boundiali-Seguéla.

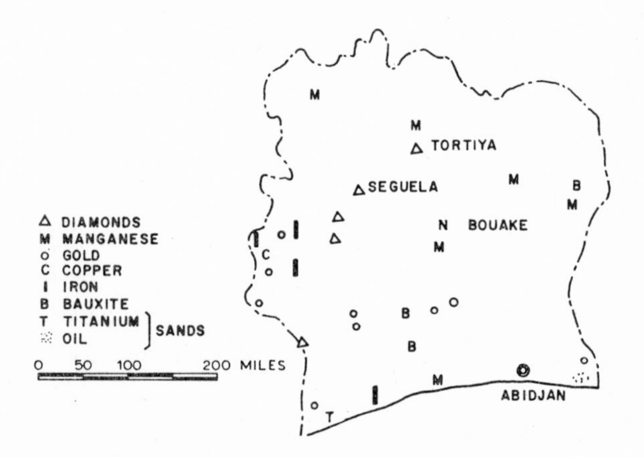

Fig. 19. The mineral resources of the Ivory Coast.

OTHER RESOURCES

The *bauxite* line extending from Guinea to Ghana crosses the country. But until now, only traces were discovered at Lakota, Divo, Oumé-Toumodi and Boundoukou.

The French Office of Geological and Mining Exploration (BRGM) acting in association with the European Coal and Steel Community has localized large reserves of low grade (40% to 50%) *iron ore* in the Ivorian part of the Nimba chain (Fig. 18). Low grade deposits containing more than 40% iron cover 2 square miles, but high grade deposits located on crests are small. Banded ironstones of the Baoublé Range between Man and Douékoué include several hundred million tons of ore containing 30–40% iron. Northwest of Guiglo, the Mount Gao syncline comprises 150 tons of magnetic ore containing 42% iron, but north-northwest of Man, the deposits of Sipilou, Mount Douan and Mount Kpao are smaller. West of the Sassandra, geologists have located 200 million tons of low grade oölithic ore; they have evaluated 140 million tons of medium grade ore. While eliminating 40% of this ore in the slimes, granules measuring >3 mm can be concentrated to a grade of 40–50%. Further east, between Grand Lahou and Dabou, large reserves of low grade ore contain 35–40% iron and 12–15% alumina. Nickel shows southwest of the Marabadiassa.

In the coastal basin, $10 million have been invested in the exploration of *oil*. At Groguida, 4 miles west of Grand Lahou, oil and gas seep at depths of 5,000–5,500 and 6,670–8,670 ft. Furthermore, at Beoou, asphalt shows at a depth of 12,000 ft. Bituminous materials are also found at Eboinda. Consequently, exploration continues on the western coast. The Abidjan refinery will produce 15,000 barrels per day in 1965.

In the western Ivory Coast, workers have washed gold in the placers of Toulepleu, Tabou and Soubré. Panning now continues at Seguéla. Between 1895 and 1900, 30,000 ounces of gold were produced at Afema. The peak production of 3,000 ounces was attained in 1949, dropping by 1958 to 150 ounces. The Ity veins near Toulepleu contain 180,000 ounces of gold, with the grade averaging 15 dwt. per ton. On high grade ores, experiments of chlorination have been successful and recovery attains 85%. At the centre of Fluoto, geologists distinguish a 100 ft. deep and sub-surficial mineralization. Minor occurrences of gold include the Hiré veins near Divo, the Bia, N'zi, lower Comoé and Aboisso Basins, Kokoumbo, and Asupirri near Afema in the southeast. In addition to these, copper shows at Bianfla near Bouaflé, at Toulepleu-Tai and in the northeast.

At Grand Lahou and Addah, geologists have discovered 400,000 tons of ilmenite containing 50% titanium. Grades attain 18% ilmenite, 0.2% rutile and 0.5% zircon. A production of 20,000 tons per year is planned. Ilmenite also shows at Tabou on the Ghana border, 2,000 tons of rutile having been discovered in the Man placers. In fact, low grade black sand deposits extend 300 miles from Fresco to the Ghana border. Since 1951, SAREMCI won one to a few tons of columbite–tantalite yearly near Bouaké. Furthermore, columbite occurs north of Tabou.

Monazite shows in the Touba-Odienne area with uraninite occurring in the lower Cavally and at Tabou. At Adzopé north of Abidjan, geologists have discovered some spodumene. Finally, lithium minerals and beryl might also exist in the northwest at Bouaké and northeast of Abidjan.

GHANA

This country is smaller but more densely populated than the Ivory Coast; 7 million Ghanese inhabit 90,000 square miles. After years of stagnation, the value of mineral exports has now attained $80 million (Fig. 20). However, the per capita mineral income is only $16. The principal products are gold and diamonds. The production of aluminium will, however, revolutionize the country's cocoa-based economy. The dominating People's Convention Party pursues a positive neutralist policy obtaining, however, (in addition to the Soviet Union) ample credits from the United States. The government now owns a large section of the mining industry. The principal deposits are located in a 150 mile long zone (Fig. 21) extending from the coast to the capital of Ashanti, at Obuasi and Konongo in Ashanti at Prestea, Tarkwa and Bogosu near the southwestern coast and at Bibiani.

GOLD

Ghana is the continent's second producer of gold which has given its name to the Gold

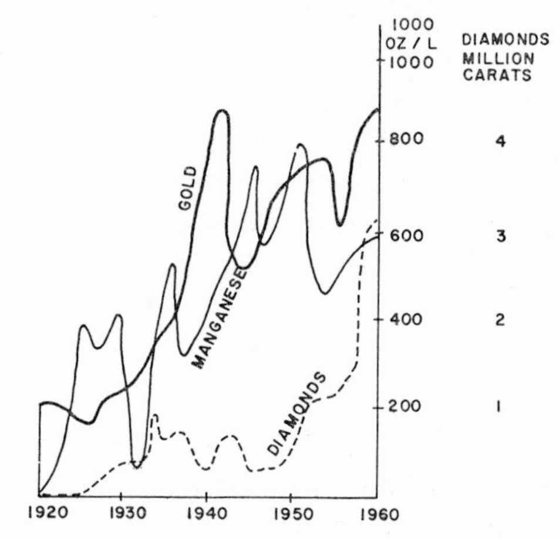

Fig. 20. Ghana production of gold, manganese and diamonds.

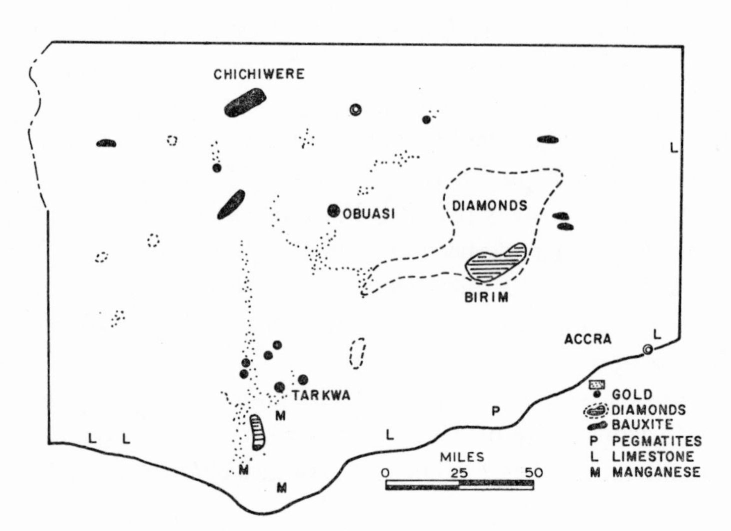

Fig. 21. The mineral resources of Ghana.

Coast. Gold has been exported for at least 7 centuries and used at the court of Asantehene the King of the Ashanti. Industrial output started at the turn of the century. Between the war and 1951, production varied between 500,000 and 700,000 ounces. Except for the poor year of 1955, output oscillated between 700,000 and 800,000 ounces from 1952 to 1957. Increasing further, a 900,000 ounce production was attained in 1962 and a record 1,100,000 ounces in 1961. Similarly, the output of alluvial gold has increased from 20,000 in 1945 to 60,000 in 1960, the rest of the production being lode gold. A thousand cubic yards have to be washed or 2.5 tons of ore crushed to recover an ounce of gold.

Although the number of employees of the gold industry has decreased from 30,000 to 20,000 in the sixties, the value of yearly production now attains $40 million. The principal producer is the government-owned Ghana State Mining Corporation, followed closely by Ashanti Goldfields Corporation Ltd. This company, distinguished by a capital of $7 million, tends to make a yearly profit of $3 million. The Drayton group, which also has South African investments, is one of its shareholders. Operations of the corporation are concentrated at the continent's principal non-banket gold mine Obuasi. Here mining reaches water levels and the shaft is being deepened. Out of reserves of 2 million tons of ore, recovered mill grades attain almost 14 dwt. per ton. Mines of the Ashanti Corporation include Côte d'Or, the Justice's, and Ayeinm to which the old workings of Old Chief, Tom Collins, Ashanti and Kokortaswia might be added. The Corporation has also explored the Mamkwadi area in south–central Ghana. At Ahyenase north of Obuasi and east of Kumasi (capital of Ashanti), a small company, Konongo Gold Mines Ltd., operates a new reef found in 1961. The mine includes the Boabedroo and Zongo sections. Output is in excess of 40,000 ounces, grade 8 dwt. per ton.

Operations of the State Mining Corporation are more scattered. In 1961 in the Bogosu-Prestea-Tarkwa district, the Ariston mine produced 140,000 ounces (grade 6 dwt. per ton). The shaft has been sunk to 530 ft. The Prestea field contributed 50,000 ounces of gold (grade 6 dwt. per ton). The Tuapim ore body is being developed. The Tarkwa banket field includes Tarquah, Abbontiakoon, Abosso, Tamsoo and the Bippos. This field produces 130,000 ounces (grade 4 dwt. per ton). The Akoon incline has been deepened. Between Kumasi and the frontier, production of the Bibiani field has decreased to 80,000 ounces (grade 4 dwt. per ton). Evaluated reserves of the Corporation are relatively low with grades averaging 6 dwt. per ton. The principal placers are in the Offin, Ankobra and Tano Rivers. Three dredges operate in the Offin, at Subin Hill and Dunkwa, another in the Jimi River. However, the Ntronang field in western Akim and the Nangodi district in the northeastern corner of the country are inactive (Fig. 54).

Silver and arsenic are byproducts of gold. From each 100 ounces of Ghana gold, 2–5 ounces of silver can be recovered. Production has oscillated between 15,000 and 45,000 ounces yearly, averaging lately 25,000 ounces.

DIAMONDS AND MANGANESE

In 1919, N. R. Junner discovered the Birim *diamond* field and the small Bonsa field in 1922. The Birim field includes the Atiankama, Subinsa, Subin, Supong, Abansa and Esuboni drainages, the Bonsa field comprising also the Pintotum and the Nimasin Rivers. Production reached the million carat mark by 1932, 2 million in 1951 and 3 million in 1957.

Ghana diamonds are small, often coloured and the ratio of gems to industrials is unfavourable. Although the mean value of a carat does not attain $9, the value of the production has increased from $1.6 million in 1942 to $3 million in 1951 when smallscale African mining started. During the next decade, the value of this type of product was in excess of the company's output. This situation was reversed in 1961. The total value of the diamonds has decreased from $17 million in 1951 until 1957. Production has decreased by 200,000 carats in 1961, but the value has increased to $24 million. More than 300 persons hold digging licenses, employing 7,000 people. Their output has declined to 1.3 million carats.

In 1960, the Consolidated African Selection Trust (CAST, a sister company of the Rhodesian Selection Trust) terminated its contract with another of its associates, the Central Selling Organization (De Beers'), selling its 1.4 million carat output through dealers of the Accra Government Diamond Market. Boasting a capital of $14 million, CAST owns the Sierra Leone Selection Trust and makes yearly profits in the range of $5 million. In 1961, the washed grade was in excess of 2 carats per cubic yard. To complement these operations, the Selection Trust is completing the Anincheche project. Smaller companies include the Holland Syndicate at Apryensua (output 270,000 carats), Cayco Ltd. exploiting the Anyinatsu and Abusukutu placers (30,000 carats), and Akim Concessions Ltd. at Nomasua and Sentonkwakwa (output 16,000 carats, grade only 0.5 carat per cubic yard). Clandestine mining affects operations. Isolated stones have also been found in the Pra River in southeastern and northern Ghana.

With a capital of $40 million, the African Manganese Company (an affiliate of the Union Carbide Corporation) has produced most of the western world's battery grade *manganese*. In 1916, production started at Insuta (Nsuta) near Takoradi Harbour, increasing from 400,000 tons in 1925 to almost 900,000 tons in 1952 and decreasing thence to 500,000 tons. Reserves, though, are decreasing. The value of the production attained $3.6 million in 1942, $23 million in 1952 and $16 million in 1961. Other occurrences include Yakau in the southwest, Odumase, Birrinsim in Ashanti, Kalimbi, the Siro and Pongo areas in the north.

OTHER RESOURCES

Water supply is inadequate in the north. Hydropower resources will, however, be exploited actively with the *bauxite* deposits. The principal of these, the Chichiwere Tinte Hills near Yenahin (in Ashanti) include 180 million tons of bauxite containing 44% Al_2O_3, 15–20% Fe_2O_3, 2–4% TiO_2 and 0.2–1.5% SiO_2. Furthermore, southwest of Yenahin, deposits of the Sefwi Hills contain 32 million tons of ore. Smaller deposits include Kwamtama in the southwest, Atiwa, Ejuanema and Atiwedru in the southeast, and Impuesso. At Awaso, the British Aluminium Company has increased its output from 190,000 tons in 1942 to 300,000 tons in 1962.

The Volta Aluminium Company (VALCO), a consortium of Kaiser and Reynold Aluminum, will smelt 90,000 tons of aluminium at Tema Harbour. Moreover, according to plans output will expand to double this amount. This will necessitate an initial bauxite production of approximately half a million tons. The company is investing $42 million itself, and has received a loan of $126 million from the United States government. Total investment in the smelter is scheduled to attain $280 million. VALCO will consume 300,000 kW of the generated 770,000 kW capacity of the Akosombro site on the Volta River 80 miles from its estuary. The dam will span 2,100 ft. and rise to a height of 370 ft. Its waters will inundate 250 miles of the Volta. The hydroelectric project of the Volta River Authority alone will cost $196 million. To finance this scheme, the International Bank for Reconstruction and Development is loaning $47 million, the United States Agency for International Development, $27 million; the United States Export Import Bank, $10 million; and the British government, $14 million. In 1961, Impresit–Girola–Lodigiani and F. Recchi have begun the construction of the $43 million dam. In addition to this, a secondary dam has been planned 12 miles downstream of Akosombo at Kpong and at Bui on the Black Volta.

Graphite occurs within and close to the manganese and gold deposits. In northeastern Ghana on the Togolese frontier, the low grade deposit of the Shiene area includes 100 million tons of ore containing 45% iron (to be concentrated to 56.7%), 0.04% phosphorus and 0.018% sulphur. The level of the artificial lake of the Volta River will rise to a distance of 30 miles from the deposit. Moreover, lateritic iron ores cover other areas. But the Tema smelter will utilize each year 30,000 tons of scrap iron. *Salt* is evaporated along the coast between Elmina (castle) and Keta and recovered at Daboiya in the north. Here low grade barytes also occur. However, the extension of the adjacent Togolese deposits has not been located yet. Between Cape Coast and Winneba, beryl is found in numerous pegmatites with mica occurring in the same district and further north. Similarly, at Akim Oda and in Makoba lagoon in the same district, geologists have located columbite–tantalite. Conversely, the lithium pegmatites of Ejisu are situated in Ashanti. In the same province, andalusite is abundant at Abodun near Bekwai. Rare earth, thorium and titanium minerals are found in the Kpong River north of Accra. However, ilmenite also occurs in the Hô district. Garnets concentrate in the Adaklu-Ho districts in the east and in the Mankwadzi-Abrekum area on the southern coast. On the other hand, corundum occurs at Neibwale in the north. Asbestos, chromium and talc show at Anum and Akuse in southeastern Ghana, but oil exploration has not yet met success. An Italian company operates the 25,000 barrel per day refinery of Tema.

TOGO

Extending 300 miles inland between Ghana and Dahomey, the Togolese Republic has a coastline of 50 miles. A million people inhabit its 22,000 square miles. The government of Lomé pursues a favourable investment policy.

When *phosphate* production (Fig. 135) reaches its ultimate goal of a million tons per year, it will represent 40% of the government revenues. In 1961, output started with 300,000 tons. Present capacity attains 750,000 tons. The government owns 20% of the stock of Compagnie Togolaise des Mines de Bénin, in which it is represented by the vice-chairman of the board. Other stockholders include the Compagnie des Phosphates de Constantine, Peñarroya, COFIMER (Compagnie Financière pour l'Outre-mer), Compagnie Internationale d'Armement Maritime, Pierrefitte and the Banque de Paris et des Pays-Bas. Incorporated with a capital of $5 million, the company has investments attaining $24 million. In the deposit extending 15 miles inland north of Lake Togo, reserves are in excess of 100 million tons. Five times more overburden than ore has to be excavated. Working started at the

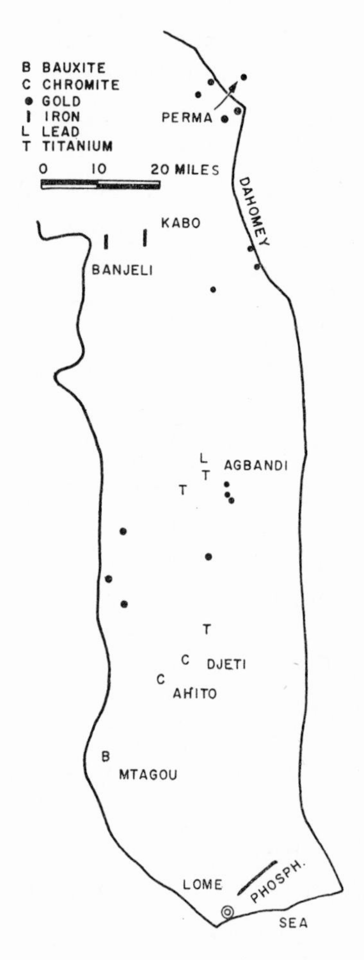

Fig. 22. The mineral resources of Togo.

northern end (Hahotoé). Phosphates are transported by rail to the Kpémé refinery near Porto Seguro. Conveyor belts carry the end product, ore of 80% tricalcic phosphate grade (and some of 76%) to a 4,000 ft. long wharf for loading in a 40 ft. deep sea. The company exports phosphate to France, Japan, the United States, Denmark, with Germany and Spain being minor importers.

Unfortunately, the search for oil and titanium has not yet met success. Near Palimé station, 70 miles from the sea, the Mount Agou deposit contains only one million tons of bauxite (Fig. 22). On Mount Djolé at Bangeli near Bassari station (300 miles from the sea), geologists evaluate a small reserve of 50 million tons of open cast *iron ore* containing 44–55% iron. This zone extends towards the Kabré Mountains. Only 50 miles from Lomé at Ahito, the Péchiney company has evaluated 50,000 tons of low grade chromite. The ultrabasic belt includes Djedi, continuing in the north-northeast in Dahomey. Furthermore, the iron and manganese belt of northern Togo, Dahomey and southern Niger includes Sansanné-Mango and Dapango near Oti.

DAHOMEY

Extending further north than neighbouring Togo, Dahomey borders Nigeria. A population of 2 million people inhabits the 40,000 square mile republic. The government has shifted the capital from Porto Novo to adjacent Cotonou. Standards of education are high in the country, but mineral resources are limited. At Athiémé and in Nigeria, only traces of the Togolese phosphate belt remain. Also on the coast, the Société des Pétroles (SAP) has undertaken exploration for oil. During World War II, workers recovered 8,000 ounces of *gold* monthly from the marginal Perma placers (Fig. 22) near Natitingou in northern Dahomey. Similarly, the alluvia of the Konadogo and M'Binah Streams contain some gold. Here the Office of Geological and Mining Research (BRGM) located lately the Camera and Patrick veins containing 5,000 ounces of gold. Further north and 500 miles from the coast the same organization has evaluated, at Loumbou–Loumbou near Kandi, 200 million tons of ore containing 38–53% iron. Similarly, near Tanguiéta, the Bontomo *chromite* occurrence is not commercial, nor are the copper showings northeast of the town (Fig. 24).

NIGERIA

Nigeria is distinguished by having the continent's largest population (45 million). Consisting of the large northern and several southern regions and of the capital Lagos, the Federation covers 340,000 square miles. In coalition with the eastern Ibo, representatives of the northern Muslim emirates control the government. An educated middle class, a favourable position along the coast, oil, gas and water power potential, low grade iron ore and coal resources (Fig. 24, 53) are many other assets for the country's development. In addition to palm oil, peanuts and cotton, the country's principal contribution to the world economy remain oil, tin and columbite (Fig. 23). Undoubtedly, crude oil is the main mineral product. In 1962, the value of oil produced reached the mark of $50 million.

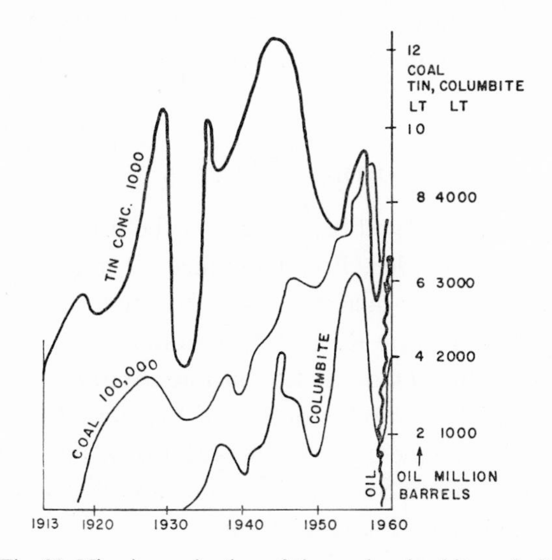

Fig. 23. Nigeria production of tin, coal, columbite and oil.

TIN AND COLUMBITE

In the north on the Jos Plateau (Fig. 109) geologists have estimated 100,000 tons of *tin* concentrates, mainly in placers. Fluctuations of the tin market and the International Tin Agreement control production. Capacity is in excess of 8,000 tons or $15–20 million, and the new smelter of Jos has a capacity of 10,000 tons per year. Nigeria is the world's leading producer of columbite. Indeed, we believe that approximately 80% of all niobium (columbium) recovered in the world prior to 1965 was a byproduct of the plateau tin fields. In placers and altered rock, geologists estimate resources in the range of 40,000 tons. During the Korean War, production reached its peak of 4,000 tons, and attained 2,500 tons in 1962. Future production will be influenced by the Brazilian pyrochlore output. Deposits extend from Rayfield to Bukuru and Jantar, with others concentrating 20 miles south around Ropp and Buka Bakwai.

Amalgamated Tin Mines of Nigeria, ATMN, produces 40% of the tin output. Another company of the London Tin group, Anglo-Oriental of Nigeria Ltd., manages the properties of ATMN. This group is also one of the leading tin producers of Malaya, and, consequently, of the world. In 1960, on a capital of $4.5 million, ATMN made a $600,000 profit. ATMN boasts reserves of 40,000 tons of tin concentrates in grades averaging 0.7 lb. per cubic yard. South of Bukuru, Bisichi Tin makes a $300,000 yearly profit on a capital of $2.8 million. On a capital of $900,000, its associate, the Jantar Nigeria Company, tends to make a profit of $70,000 per year. Each year, output of Bisichi ranges from 400 to 600 tons of tin and 300 to 500 tons of columbite, with Jantar producing 200–300 tons of tin and 200–350 tons of columbite. Bisichi also produces a few tons of wolfram. ATMN now supplies only 300–400 tons of columbite and there are numerous other small producers.

Like thorite, Nigerian pyrochlores containing niobium (columbium), uranium and thorium in hard rocks are uneconomical. But these columbites contain 5% of the rare metal tantalum which could become recoverable by solvent extraction. Furthermore, each year a few tons of tantalite–columbite are produced off the Plateau, mainly in the Wamba–Jemaa and Egbe fields. Zircon, an abundant byproduct of tin mining, is only occasionally exported (800 tons in 1961). However, most of the world's hafnium, a reactor metal, (about 20 tons) has been extracted from these. Some mica and beryl are found in Kabba Province, with low grade beryllium minerals occurring also in the Plateau granites.

OTHER PRODUCTS

Twenty-five miles from the sea, the oil basin includes the principal field Bomu located east of Port Harcourt (Fig. 24). In 1963, Bomu supplied 19 million barrels of crude oil (or 53,000 barrels per day), i.e., three-quarters of the Nigerian production. With 3.6 million barrels (10,000 barrels per day), Imo River was the only other major producer. Whereas medium yield fields like Afam, Oliobiri, Ebubu and Apara averaged respectively 2,000, 3,000, 1,500 and 2,000 barrels per day, the output of Elelenwa, Oza, Korokoro and Bonny was less important. The development of Oza, Korokoro, Kokori, Apara, Ughelli, Bonny, Nun River, Krakama and Eriemu continued in 1962, and Bodo West and Opukushi were discovered that year, followed by Soku and Ughielli. In 1962 reserves stood at 700 million barrels.

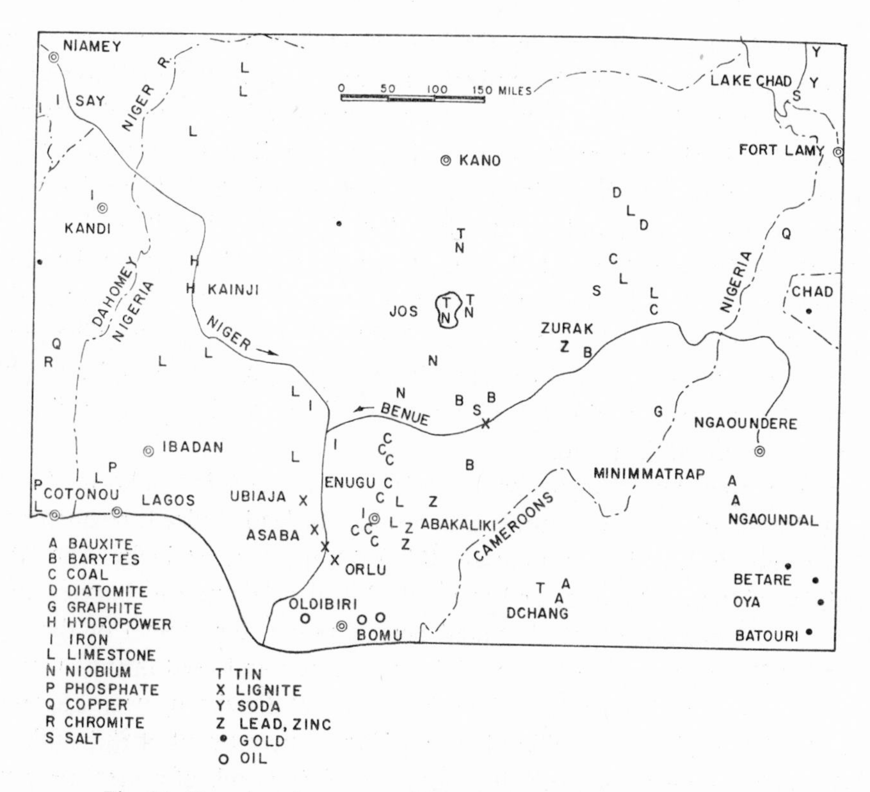

Fig. 24. The mineral resources of Nigeria and neighbouring countries.

Since 1961, the total production of Shell-BP Petroleum Development Company of Nigeria has increased by 47% to reach the mark of 26 million barrels. For 1964, the company forecasts a production of 50 million barrels, with 80 million barrels projected for 1965 and double this amount for 1970. Indeed, by the end of the decade, total investment will attain almost $600 million. Furthermore, Shell and BP cover the interest of a $13 million loan contracted by the Nigerian port authority for the dredging of the Port Bonny deep water channel downstream of Port Harcourt to a depth of 35 ft. which will permit the pumping of the scheduled outputs from Bonny terminal. Finally, offshore exploration continues. The refinery of Alese is designed for a throughput of 32,000 barrels per day.

As to natural gas, Shell-BP has contracted to supply gas to the Trans-Amadi Industrial Complex through the Eastern Nigeria Development Corporation.

Unfortunately, water is in insufficient supply in large areas of Nigeria, especially in the northern region and also in Onitsha, Oweri and Benin Province in the south. Indeed, hydrogeology remains one of the vital functions of the Geological Survey. However, considerable *hydropower* resources are concentrated in the north. With an investment of $190 million, the Kainji Dam project would generate 860,000 kW. An artificial lake would flood 100 miles of the Niger, including the Forge Island Basin. The northern Niger would be opened up for navigation, facilitating access to the Republic of Niger. Moreover, at Jebba downstream of Kainji and at Shiroro on the Kaduna, supplementary dams could be built.

Sub-bituminous *coal* resources of the eastern province attain (Fig. 122): Inyi, 10 million tons; Enugu, 40 million tons; and Nsukka-Ezimo, 25 million tons. Those in adjoining areas of the northern province attain: Orukpa, 40 million tons; Okaba, 60 million tons; and Ogbagaga, 70 million tons. Five collieries are operating and these include reserves of: Hayes, 22 million tons; Ekula, 10 million tons; and Iva, 4 million tons; Olwetty and Sitting Pretty are smaller. Each year, production attains half a million tons or more ($1 million). On the border of the central province, *lignite* reserves of Ogwashi-Asaba represent 60 million tons with those of Obomkpa attaining 10 million tons. Lignite and coal also show in other localities (Ibadan, Lagos, Sokoto). At Enugu, 50 million tons of laterite covered by a 50-ft. overburden can be screened to an *iron* content of 43%. Moreover, laterites are abundant in other areas of the Federation. In Kabba Province near the Niger, the Agbaja deposit includes 30 million tons of ore containing 51% iron, 1% phosphorus and 0.08% sulphur. Possible reserves attain 100 million tons. Plans call for a steel mill using local coal, scrap and imported iron ore.

Through Abakaliki near Enugu, the *lead–zinc* belt extends 350 miles to Zurak off the Plateau. At Zurak, *silver* could become a byproduct of lead. Indeed, resources are large. Though moderate, production could attain 100,000 tons/year. At Lefin, Aba-Gbandi and other localities of Benue Province and at Dumgel in the Adamawas, barytes are also by-products of lead–zinc. Some gold is found near Kaduna, the capital of the north. Ten million tons of *limestone* are easily accessible at Nkalagu and used in the manufacture of *cement*, as is Enugu coal. Another plant is located at Ewekoro, near Lagos. In 1962, cement production attained 2.8 million barrels. With an investment of $8.6 million the new plant of Kalaimbana, near Sokoto, will produce 600,000 barrels of cement per year. Limestone reserves attain here 100 million tons and 40 million tons at Jakura. Limestone and marble occur at Lokoja and several other localities. Near Lake Chad, medium grade diatomite outcrops in the northeast at Abakire and Bularaba. Unfortunately, the Ososhun phosphate deposits (near Lagos) are of low grade. The principal low grade graphite occurrence, is Birnin Gwari near Kaduna. Some asbestos is found at Shemi. On the other hand, ceramic *clays* are known in many localities, for instance at Enugu where glass sands also occur. However, the salt output remains inadequate.

CAMEROON

The Federal Republic of Cameroon consists of the former large French speaking Camerounese Republic and of the English speaking southern Cameroons (capital Bouea). While Yaoundé, the capital of the Federation, is in the inner highlands, the principal city is Douala, the main harbour. Four million people inhabit this country of 170,000 square miles. The government welcomes foreign investment.

ALUMINIUM AND IRON

Bauxites constitute the republic's principal resources (Fig. 24, 53). The Bureau de Recherches Géologiques et Minières, acting for the Syndicat des Bauxites du Caméroun in which the Péchiney and Ugine companies participate also, has explored the Minim-Matrap deposit which is one of the largest of the world. Geologists estimate reserves in excess of 1,000 million tons of *bauxite* containing 43% alumina and 3.4% silica. The deposit is located southwest of Ngaoundéré, 300 miles from the sea. Furthermore, south of Minim-Matrap, Ngaoundal represents 90 million tons of bauxite containing 41.7% alumina. This deposit is situated on the projected Douala–Chad railway. In Bamileke country, other smaller deposits are located nearer to the sea and near the border of the two Cameroons and to Nkongsamba, head of the Chemin de Fer du Nord. At Fongo-Tongo near Dchang, 34 million tons containing 45% alumina and 4% silica can be concentrated from 46 million tons of bauxite. An extension of the Bamboutos deposit contains 4 million tons of ore (46% alumina and 4.3% silica). Deposits located further east between Bafany and Bafoussam have not yet been evaluated.

East of Douala, the Edea *aluminium* plant has been designed in connection with the Sanaga irrigation project. The hydropower plant boasts a capacity of 120,000 kW. The Compagnie Camérounaise de l'Aluminium Péchiney-Ugine (ALUCAM) has been incorporated with a capital of $20 million. The plant and dwellings costing $30 million have been partly covered by a loan of the Caisse Centrale de Coopération Économique. Production of aluminium started in 1957 and the rate exceeded 43,000 tons per year. Because of the partial stoppage of the electrolysis plant during the dry season which affects the Sanaga, the projected rate of 65,000 tons cannot be attained. Alumina is imported from Gardanne (France) and Fria (Guinea). After processing, half of the aluminium is re-exported to France, the rest to the United States, Belgium, The Netherlands and the United Kingdom. In 1962, the manufacture of aluminium sheet for roofs started.

Between Kribi and Campo on the border of Spanish Guinea, preliminary estimates of the reserves of the Mamelle Range have been reported to attain 150 million tons of ore averaging only 35% *iron*. Conversely, the silica content is high. Fortunately, the deposit is located 10 miles from the sea. The Syndicat de Recherches de Fer et de Manganèse du Caméroun, which is formed by the French Office of Geological and Mining Research (BRGM) and the High Authority of the European Coal and Steel Community (Luxembourg), has carried out further studies of concentration. Experts distinguish three types of ore. Unfortunately, the two top categories containing, respectively, 52% and 68% iron are scarce.

OTHER RESOURCES

In the Douala Basin SOREPCA, the Société de Recherches et d'Exploitation des Pétroles du Caméroun, incorporated in 1951 with a capital of $20 million, has evaluated 100,000 barrels of *oil* at Suellaba, 11,000 million cubic ft. of *gas* at Logbaba and 900 million cubic ft. at Bomono. Further tests are carried out in the southern part of the basin and in southern Cameroons.

In 1933, prospectors discovered *gold* in the Lomkader region, in the Batouri and Bétare districts in the east. In 1942, output reached its peak of 25,000 ounces. Out of a total production of 280,000 ounces, 120,000 ounces were produced since 1956. By now, however, output has decreased to 500 ounces, supplied largely by labourers selling gold to the Compagnie Minière du Caméroun at Bétare-Oya. In 1959, at Faro northwest of Ngaoundéré in the north, 50 ounces were recovered. In 1928, *tin* was discovered at Mayo Darlé and Taparé on the frontier of Cameroons. Output started in 1932, and reached its peak of 320 tons in 1939. The last full year of operations (1960) of Société des Étains du Caméroun (SEC) produced 70 tons. In 1960, extraction of the eluvial placers stopped, but SOFIMEC (Société de Fibre et Mécanique) of Douala plans to work the tailings. Some tin also shows in southern Cameroons. Tungsten associated with lead and zinc is found at Goutchoumi and with bismuth at Ribao. A small deposit of molybdenum and uranium is known to exist at Ekomedion. There are other uranium indices at Poli.

As early as 1908, rutile, the high grade *titanium* ore, was discovered in the region of Edea, and at Banyo east of Kribi. Extraction started in 1935, and reached its war peak of 3,300 tons in 1944. But with prices declining, output (exported to Great Britain) slid to 40 tons in 1957. Although considerable reserves remain, no activity has taken place since. Between Edea and Pouma, the Bureau de Recherches Géologiques et Minières plans to exploit the *kyanite* placers of the Niyba. In 1958, the pilot operation supplied 5 tons out of a reserve of 50,000 tons. In the north, in the Tiffel-Goutchoumi area northwest of Garoua, geologists have explored indices of *copper*. The search for diamonds has met with little success. Since the northern part of country lacks water, a considerable effort of hydrogeology is made. At Garoua, in the north, a company dominated by the Società Cimentiere del Tirreno studies the building of a 180,000 barrel per year cement plant.

SECTION VIII

Middle Africa

Middle Africa is comprised of the 14 countries between latitudes 2° N and 16° S. We have added the Central African Republic to these as it is also in the Congo drainage. Since the majority of the mineral resources of Mozambique are located north of latitude 16° S, we will also include this country. The area can be divided into three parts: *(a)* Atlantic coast; Spanish Guinea, Gabon, Congo Republic, Angola, *(b)* central Africa; Central-African Republic, Congo, Zambia, Burundi, Rwanda, *(c)* eastern Africa; Uganda, Kenya, Tanganyika, Malawi and Mozambique. The principal resources are located in the Congo, Zambia and Gabon.

RIO MUNI

A population of 160,000 inhabits this country of 10,000 square miles. Bata, the capital of Rio Muni, is on the sea north of Libreville. Oil companies have explored the coast. Iron ores are known to occur in the province as well as north of it, in Cameroon and southeast in Gabon.

GABON

The richest of the republics of equatorial Africa covers 110,000 square miles inhabited by only 460,000 people. In addition to large iron resources, oil, uranium and manganese are the mainstays of the rapidly expanding Gabonese mineral economy (Fig. 25, 53). The stability of the government offers a favourable climate for investment. Moreover, ties with France are traditionally close.

Each year, SPAFE, the Société des Pétroles d'Afrique Équatoriale, extracts 8 million barrels of crude *oil* from the Port Gentil Basin; output has increased in 1964. Results are particularly encouraging in the Tchengué, Pointe Clairette, M'Béga, Cap Lopez and Batanga fields (Fig. 129) which yielded, in 1963, respectively, 1.5, 1.3, 1, 1.1 and 0.7 million barrels of crude. Other producing fields include Ozouri, Animba, Rembo Kotto, Simany, Anguille (offshore), Allewana, and Illigoué, with productions of, respectively, 240,000, 110,000, 10,000, 60,000, 110,000, 30,000 and 130,000 barrels. In 1962, Anguille the field operated only 5 months and the Illigoué field only one month. The Pointe Weze field,

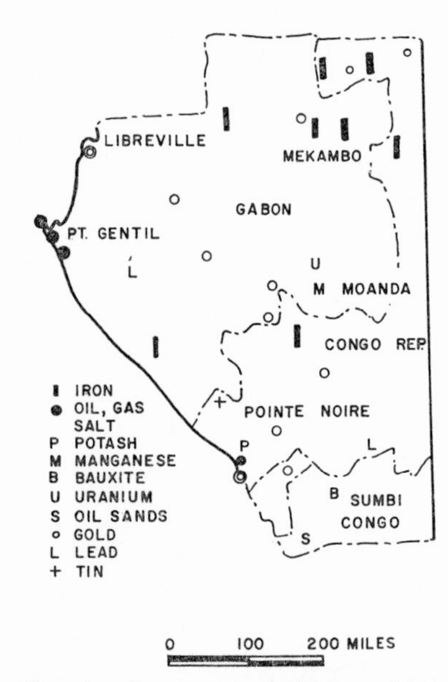

Fig. 25. The mineral resources of Gabon and lower Congo.

between Cap Lopez and Libreville, and the N'Tombenyoni fields did not produce. On the other hand, oil shows south of Anguille at Dorade and offshore Espadon. Favourable structures were located at Olende and M'Polunié in the Delta, the Tchengué Océan field was identified on the west coast and exploration started in the Libreville Basin. Reserves reached 150 million barrels. Almost 200 million cubic ft. of gas are delivered annually to the Société d'Énergie Électrique de Port Gentil, and projects of a steel industry based on iron and possibly gas resources have been discussed. However, much more gas escapes; indeed, for each barrel of oil more than 1,000 cubic ft. of gas were reported at M'Bega and Batanga in 1962, with 800 cubic ft. in the Illigoué, 600 in the Cap Lopez and 500 in the Anguille fields.

At Moanda on the Congolese border, the mining of *manganese* will provide a revenue of $7 million each year (Fig. 68). The tax structure has been guaranteed for 5–25 years. Investing $90 million, the Compagnie Minière de l'Ogoué (COMILOG) has a capital of $10 million: the United States Steel Corporation contributes 49%; Bureau de Recherches Géologiques et Minières, 22%; Compagnie de Mokta el Hadid, 14%; and Société Auxiliaire de Franceville, 15%. One half of the initial 500,000 ton yearly production will be absorbed by United States Steel; thus Gabon provides for a third of the American manganese consumption. France and other members of the Common Market import the balance. A 50 mile long ropeway connects Moanda with Mbinda. A 170 mile long, narrow gauge railway ties in with the Congo-Ocean Line. The ore is shipped from Pointe Noire Harbour, Congo Republic. Manganese also shows at Okondja, in the southeast on the Lower Mandjibe, at Ikoy east of Port Gentil and at N'Gounié east of Lambaréné.

In 1961, the Mounana *uranium* mine belonging to Compagnie Minière d'Uranium de Franceville started production (Fig. 80). In nine months the company produced concentrates containing 400 tons of uranium. Ores are worked opencast and concentrated to a grade of 30%. Magnesium uranate is shipped by the Mbinda manganese railway. Reserves attain 5,000 tons of uranium (grade 0.5%). The Commissariat à l'Énergie Atomique and the Compagnie de Mokta el Hadid have invested $10 million, and annual value of the output will be similar. Uranium also shows at Kaya-Kaya on the Congolese border. At Makokou-Mékambo in northeastern Gabon, geologists have evaluated 620 million tons of ore containing 62.2% iron at Belinga Hill, 210 million tons containing more than 60% iron on Boka-Boka Hill and 110 million tons on Batouala Hill; other hills have not yet been evaluated (Fig. 64). Plans call for the building of a 400 mile long railway and of a new harbour opposite Port Gentil. The Bethlehem Steel Corporation owns 50% of the shares of the Société des Mines de Fer de Mékambo (SOMIFER); Bureau de Recherches Géologiques et Minières, the exploring company, 12%; German steel corporations, 10%; French and Belgian groups, plus 2%; the Banque de Paris et des Pays-Bas and Compagnie Financière de Suez, 5% each; and FIAT and COFIMER, 3% each.

Northeast of Kango, the ferrugineous quartzites of Mela contain 45% iron, with each ft. of depth representing a reserve of 300,000 tons. East of Mitzic near the southeastern corner of Rio Muni, quartzites of Mebaga Hill contain 50–55% iron. Other banded ironstones are found at Ngama. Finally, 25 miles from the southwestern coast, Milingui Hill, the principal deposit of the Tchibanga area, includes 80 million tons of siliceous ore containing 43.5% iron.

Reaching 16,000 ounces in 1962, the *gold* production of the Eteke district in southern-central Gabon continued to slide. In the M'Bigou, Koula-Moutou and Lastourville districts, small scale production has expanded to 1,800 ounces. Although the alluvial diamond placers of Makongonio in the southern Gabon have been exhausted, geophysical exploration for pipes continues. A diamond has been found near Kango. At Kroussou and Offoubou south of Lambaréné, lead, zinc, accessory copper and silver show. In the Abamie Basin, columbite–tantalite traces have been explored. Geologists have found tin on the periphery of the Etéke gold field on the Ogoué River. More important, the Achouka *limestones* might provide the rock for a cement plant at Port Gentil. Finally, salt and *potash* occur above the coastal oil–gas fields, e.g., in the area of Lake Azingo.

CENTRAL AFRICAN REPUBLIC

A million people inhabit the landlocked 190,000 square miles of former Ubangi-Shari, where the late Barthélémy Boganda founded the republic. The succeeding governments guarantee foreign investment. From Bangui, the capital and principal river port on the Congo River, boats sail to Brazzaville where freight is loaded on the railway. The mining areas are located on the Cameroonese border around Berbérati (Fig. 53) and in the northeastern corner of the country in the Kotto district (Fig. 13). During the last 25 years, 2 million carats of *diamonds* and 420,000 ounces of gold were produced. In 1914, the first diamond was discovered at Ippy. Between 1948 and 1958, an output of more than 100,000 carats per year was achieved. However, since the greatest alluvial concentrations have been depleted, production has decreased from its peak of 156,000 carats in 1952 to 70,000 carats. In addition to these 200,000 carats of Congo diamonds were exported through Bangui in 1962. Up until 1960, a total of 1,800,000 carats have been produced in the western sector and 250,000 in the eastern. Although lately 60% of production originated in the west, the east supplies an evergrowing share. In the order of their importance, the operating companies include:

(a) in the west:

Compagnie Minière de l'Oubangui Oriental (CMOO),
the Société Anonyme de Recherches et Exploitations Minières Centre Oubangui (SAREMCO),
at Ouadda, the Société Minière Intercoloniale (SMI),
at Berbérati, Société Minière de Carnot (SOMICA), and Sangha Mine at N'Dem;

(b) in the east:

Société Minière Est Oubangui (SMEO) at Yalinga,
Société Auxiliaire de Mines (SAM) at Ouandjia,
Société Minière du Zambe (SMZ) at Bria, and CDDC at Boungou.

Gems represent roughly 40% of the output. The largest diamond weighed 150 carats and was found by SMEO at Yalinga. While in the eastern district Yuba Consolidated tests the dredging of larger rivers, the BRGM explores the large flats of the Bia district. In the west, the government carries out the search for primary deposits.

Since 1929, CEM has produced a total of 160,000 ounces of *gold* at Pouloubou and Roandji in the eastern district and CMOO has contributed 100,000 ounces from Sosso Polipo and Bouar-Baboua in the west. Other deposits include Lobaye, Mobaka, the Upper Sangha and Iraésé. Unfortunately, both have been worked out. Already production has decreased, plunging during the last years from 900 to 500 and 100 ounces. At the present, the diamond workings of *(a)* Kotto Dar el-Kouti contribute 40–60% of the gold, *(b)* small operations of Bouar-Baboua in the western sector contribute a similar amount and *(c)* the western diamond district of the Upper Sangha supplies a couple of per cent. If extended to the western sector, small scale production could increase to 1,500 ounces. At Bambari, gold is associated with *tin*. Tin also occurs at Yaloké-Bossangoa and Bocaranga.

Although *iron ore* deposits are known to exist at Toropro, Pili, Kaga Yama and Kaga Bagongolo (mainly in the Bangui district), geologists have not evaluated these because distances are still prohibitive for shipment. While chromium shows at Bossangoa and mica southeast of Ippy, there are traces of nickel, cobalt and platinum in the centre of the country.

CONGO REPUBLIC

Between the Congo and Gabon, extending northward from the sea, the Congo Republic (République du Congo, République Congolaise or in United Nations parlance Congo Brazzaville) covers 130,000 square miles. This Congo has 800,000 inhabitants, and the government actively encourages mineral investment.

Through the discovery of *oil* at Pointe Indienne, these received a great impetus (Fig. 25). In 1962, its third year of production, SPAFE (the Société des Pétroles d'Afrique Équatoriale) sold 900,000 barrels for $2 million. Since then, wells yield each day 1,700 barrels of 37.5° API oil (viscosity 3.6, sulphur content 0.12%) and experts believe that the ultimate capacity may have been reached at 3,000 barrels per day. A few miles north of Pointe Noire, the principal harbour, at Rivière Rouge (the Red River), SPAFE built tanks stocking 170,000 barrels of crude. From these, pumps can transfer oil through an underwater pipe line to ships stationed beyond the reef. While oil exploration continues, the company has already found *gas* at Pointe Indienne and bituminous and glass sands at Pointe Noire. In the oil

TABLE XX
ESTIMATED BASE METAL RESOURCES IN THE CONGO REPUBLIC

Occurrence	*Ore tons*	*Lead %*	*Zinc %*
M'Passa	55,000	21	35?
Palabanda	40,000	13	
Yanga-Koubenza	15,000	14	

basin at Hollé, rigs drilled 10–40 ft. thick layers of potash and considerably thicker salt ridges. With an investment of $47 million, the Société d'Exploitation des Potasses de Hollé plans to exploit each year 2 million tons of ore and to sell 400,000 tons contained K_2O. The French state potash mine of Alsace, the Congo government, BRGM and American capital participate in the company, which has obtained an IBRD loan. Nearby at Tchivoula, sediments contain 45–55% tricalcic phosphate (Fig. 53).

In 1900, copper was discovered at Mindouli between the coast and Brazzaville, however, between 1910 and 1935, the Compagnie Minière du Congo Français (CMC) sold only 10,000 tons of copper. Close at hand, the company extracted 80,000 tons of lead and 11,000 tons of zinc from M'Fouati and adjacent Hapilo between 1938 and 1961. Whereas these deposits are now dormant, the CMC and the BRGM have evaluated a couple of other occurrences in the vicinity (Table XX).

While M'Passa takes the place of M'Fouati, gold production plummeted from its 1943 peak of 28,000 ounces to 3,000 ounces in 1961. The Mountains of Mayombe supply two-thirds of this output with the remainder coming from the placers of Fort Rousset in the northwestern corner of the country. Near the gold fields of Mayombe, workers recovered 30 tons of tin, 10 tons of columbite–tantalite and 5 tons of wolframite from the placers of Mayoko in 1961. Another occurrence lies nearby on the road leading to the Kouilou dam site. The Kouilou project could produce up to 800,000 kW of power. Between 1944 and 1947, the same area produced 200 tons of corundum. At some distance at Dolisie, workers could quarry limestone.

Diamonds show in the flats of Lali-Bouenza, in an island of the Congo River, in the far north, but mostly stones originating in the Congo (Leopoldville) are sold. Indeed, these represented a half of all the exports of the Congo (Brazzaville) in 1962. A bauxite deposit reportedly containing 60 million tons of alumina was found 20 miles from the Pointe Noire – Mbinda railway. However, according to reports, 40 miles from Zanaga and at the same distance from the Gabon border, there extends a 4 mile long, 3,000 ft. wide and 60–90 ft. thick body of ore containing 65% iron. The exploitation of 400 million tons (?) of ore necessitates the construction of a 110 mile long railway and an investment of $60 million.

ANGOLA

The mouth of the Congo, the sources of the Kasai and Zambezi, the lower Cubango and Cunene define a quadrangle of 480,000 square miles which we call Angola. Within these limits, administrators have estimated a population of 5 million.

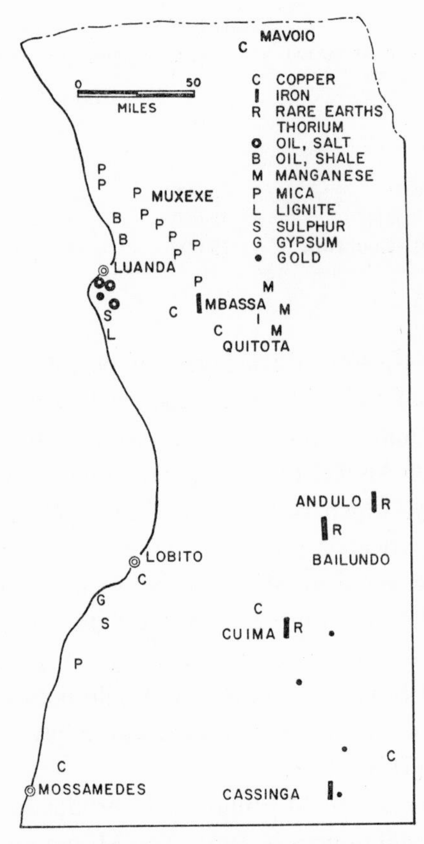

Fig. 26. The mineral resources of western Angola.

Traditionally, diamonds of the northeast were the mainstay of the mineral economy, but the discovery of oil near Luanda will shift the emphasis (Fig. 26). In addition to these centres, miners are active in the hinterland of Luanda and the highlands of Nova Lisboa (Bihé). Evolution of Angolan outputs is shown in Table XXI.

TABLE XXI

ANGOLA MINERAL PRODUCTION

Mineral resource	*Output*			*Value 1962 $ 1,000*
	Units	*1956*	*1962*	
Oil	barrels	50,000	3.400,000	11,000
Diamonds	carats		1,080,000	22,000
Iron ore	tons		830,000	4,000
Manganese ore	tons	30,000	14,000	200
Salt	tons	90,000	60,000	
Oil shale	tons	?	40,000	160
Copper	tons	1,600	1,100	1.700
Mica	tons	400		
Ferro-manganese	tons	?	1,600	12
Gold	ounces	30	80	2

THE COAST

After many years of exploration, PETRANGOL (the Sociedade dos Petroleos de Angola, a subsidiary of the Belgian oil company) discovered *oil* in the Cuanza Basin near Luanda. PETROFINA owns 33.5% of the equity of PETRANGOL, and another affiliate, Companhia de Combustiveís di Lobito, owns 11%. On the other hand, the Compagnie d'Anvers (a business associate of the Société Générale de Belgique) controls PETROFINA, and the Banco Brunay (an affiliate of the Société Générale) has interests in Companhia de Combustiveís. It was common knowledge for decades that oil seeped from coastal sediments from the Cabinda enclave (north of the estuary of the Congo) to as far south as Novo Redondo. Indeed, for many years at Libongos north of Luanda, the Companhia dos Asfaltos de Angola exploited sediments containing 18–22% bitumen, as did the Companhia dos Betominesos de Angola at Husso Norte. In addition to the bituminous limestones of Caxito and Libongos, there is asphaltite (libollite) at Calucala and Quilungo. The deposits form, in fact, the northeastern edge of the Luanda Basin with Luali and Dondo located in Cabinda. In 1963, Libongos and Husso were active; however, adjacent Glória, Terra Nova Undui and Iembe remain inactive as does Musserra further north (Fig. 26).

The *oil fields*, on the other hand, are situated in the southwestern part of the basin. In 1962, the principal field, Tobias, produced at the rate of 9,000 barrels per day and, in 1964, a 25,000 barrel per day production was scheduled by PETROFINA. Angolan output increased spectacularly from 700,000 barrels in 1961 to 8.7 million barrels in 1964, out of which Tobias supplied 8,000,000 barrels; Benfica, 700,000; Luanda, 60,000; and Galinda, only 150 barrels. (PETRANGOL operates Benfica and Luanda). Offshore drilling is programmed and the throughput of the Luanda refinery has already been increased to 12,000 barrels per day. Finally, in 1962, the production of gas reached the mark of 700,000 million cubic ft. (Fig. 130).

With other fuels, the country is less lucky. For many years, at Bom Jesus, peat was known to occur as was lignite at Cabo Ledo further south. Also, in the coastal strip north of Cabo Ledo at Dombe Grande, sulphur occurs in gypsum, a material that has been exported from Ouche near Benguela. Low grade phosphates show in the coastal strip of Cabinda and the northwest and workers recover salt from the sea. At some distance of the bituminous shales lie the *mica* mines: Cassalengues (Companhia Mineira do Lobito), Quizambilo (União de Micas Ltd.), Muxexe (Sociedad Angolana de Minas), Cambaça, Ambriz, Dui, Chesce and Fandongo. Production was always small and decreased between 1960 and 1961 from 13 tons sheet mica to 2 tons, from 8 tons of semiprocessed mica to 1 ton, and from 430 tons waste to 20 tons. No output is reported in 1962.

DIAMONDS

Until 1961, diamonds represented 15% of all exports of Angola. Following the discovery of diamonds in the southern Congo, the Belgian FORMINIÈRE Company expanded its activity to the northeastern corner of Angola (Fig. 104). In 1917, for the first time, the Companhia de Diamantes de Angola (DIAMANG) exported 4,100 carats of diamonds from the Lunda district. By 1930, production reached the mark of 300,000 carats, doubling by 1951 and then again by 1961. More than half of the stones are gems, each averaging 8 carats, but they are of good quality. DIAMANG has its headquarters in Tshokwe country near the Congolese border. Through that frontier used to be the easiest access by road fromTshikapa, a now closed district of FORMINIÈRE. Dundu also has an airstrip. However, supplies now follow the long route by rail from Lobito to Vila Luso, then by truck up north to Dundu.

DIAMANG employs 25,000 workers in its four sectors: Cassanguidi, Andrada, Maludi and Camissombo. Southwest of these, geologists have discovered kimberlites (the host rock of diamonds). Each year shovels move 4 million cubic yards of gravel. Hydropower moves excavators and concentrators. With a capital of $10 million, the company makes yearly profits of $8 million, half of which is the share of the Angola government. In addition to this, the company also provided a $3.5 million loan to the authorities in 1955. Like the Congo production, diamonds are marketed through the Diamond Corporation Ltd. (De Beers). The Société Générale de Belgique and its affiliate, FORMINIÈRE (INTERFOR), respectively hold 11.5 and 7% of the shares of DIAMANG in addition to the Banco Brunay (in which the same group and the Banque de l'Union Parisienne are interested). But the state enjoys a dominating position on the board of DIAMANG on which the De Beers group, the Guggenheim Brothers, and the Ryan and Tuck families are also represented. West of Dundu, diamonds also show on the upper Cuango (Kwango).

OTHER PRODUCTS

Since its inception in 1957, the production of *iron* ore steadily increased, but reached a plateau in the early sixties. In the hinterland of Luanda, the Companhia Mineira de Mombassa controls the M'Bassa mines (the Saia occurrences are further east); however, the principal deposits (the mines of Cuima, the Bailundo and Andulo deposits of Companhia Mineira do Lobito) are located on the plateau of Nova Lisboa. In 1964 this company plans to export a million tons of ore. Cassinga, where the Companhia Mineira de Lombige has estimated 1,000 million tons of ore containing 45–60% iron, is situated in the southwest. The Krupp Company plans to invest $45 million in this mine which will be connected through a 60 mile long spur with the Mossamedes railway. In the same area, between the Cubango and the Cunene, there are also titaniferous iron ores (and sillimanite deposits). Germany imports as much ore containing 60–66% iron as France, Israel, Czechoslovakia and Japan combined.

The exploitation of *manganese*, west of Malange 200 miles east of Luanda, commenced in 1943 and reached its peak of 1953. The Companhia do Manganés de Angola controls the principal deposit Quitota as well as Quichuinhe, Quiaponte, Gungungo, Quicolo and Serra Bé; the Sociedad Mineira de Malanje owns Quiluco II. At Quitota, there is also a modest production of ferro-manganese ore.

The sovereigns of the XVIIth century Kingdom of the Congo exploited the *copper* deposits of the northwest. However, only in 1950 did the yearly production of Empresa do Cobre de Angola at Mavoio reach the mark of 1,000 tons. Between 1956 and 1959, a total of 25 tons of vanadium ore was recovered at adjacent Lueca. However, the lead–zinc traces strewn between Mavoio and the Zombo, and the copper occurrence of Brutué east of Mossamedes, remain small.

For centuries, Africans have washed *gold* but production has never exceeded its 1942 peak of 5,000 ounces. At the present, in the Cabinda enclave, workers pan gold only at

Macende, with Gunda, Malumba, Quissoque lying idle. Similarly, south of Nova Lisboa, only Cuenguè operates, with Chiriva, Canjanja, Chombo and Samboto remaining closed (as do Ponte and Lombige in the hinterland of Luanda).

While central and southern Angola are sub-arid, hydropower is available near the coast. Near Luanda, the Mabubas plant can produce 56 million kWh each year. Also, on the Cuanza River, the capacity of the Cabubas project can be expanded from 700 million to 4,000 million kWh yearly. Respectively near Lobito and Mossamedes, the Biopio and Matala plants produce 40 million and 70 million kWh each year.

THE CONGO

In the Congo, industry mainly means mining. During the crises the country has undergone, the mining of copper (the principal product) and its byproducts has been less frequently interrupted by military operations than in the eastern Congo, where the tin and gold mines had to face great difficulties, as did also the diamond mine of Bakwanga. Nevertheless, the Congo's industrial base remains intact. Consequently, when the communications and exchange systems will be reestablished with security, the mining industry will again generate the resources which always have been the mainstay of the country's economy. What then does the Congo produce? (Table XXII). To the official data (Fig. 27), we should add clandestine exports which, in 1961, according to certain observers, represented a half of the diamonds and a third of the gold produced (and most gold recovered in 1964).

Everyone now knows that in the heart of Africa 5 million Congolese inhabit 900.000 sq. miles. In the mines, 100,000–120,000 people find their livelihood. The capital, Leopoldville, is situated 200 miles from the Atlantic, i.e., on the opposite side of the mining areas, which are in the east (Fig. 29). A glance at the preceding table shows that whereas in Katanga (the southeast), copper output increased, in Maniema–Kivu (east) and Kibali–Ituri (northeast), gold and tin production declined during the troubled period. However, the

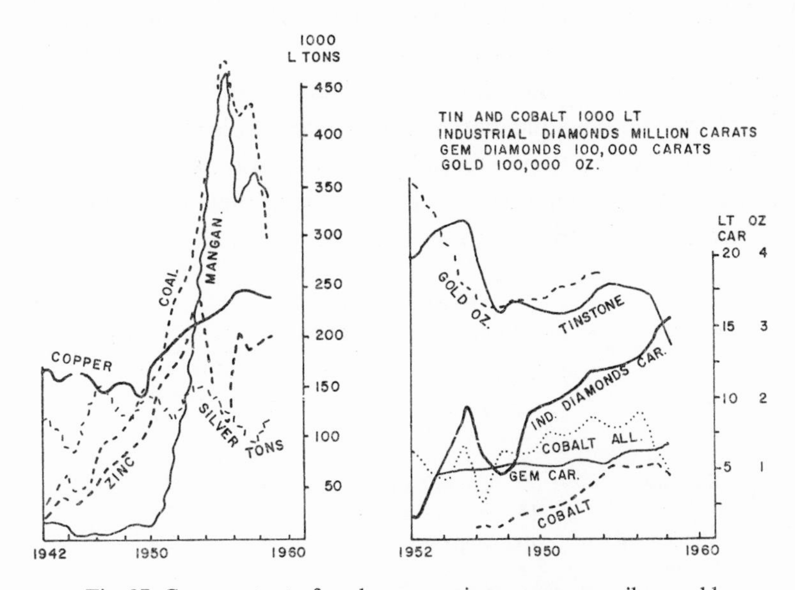

Fig. 27. Congo output of coal, copper, zinc, manganese, silver, gold, tin, cobalt, gem and industrial diamonds.

TABLE XXII

CONGOLESE MINERAL PRODUCTION

Mineral resource	Units	1962 Weight	1958 Weight	1958 Value $ 1,000	Percentage of 1958 output over 1948
Copper	tons	325,000	260,000	125,000	150
Uranium oxide	tons		2,300	±50,000	0
Industrial diamonds	carats	17,700,000	16,000,000	35,000	300
Gem diamonds	carats	460,000	670,000	5,000	120
Cobalt	tons	10,700	9,300	24,000	200
Tin	tons	7,900	10,600	20,000	70
Manganese ore	tons	340,000	370,000	19,000	2,700
Zinc	tons	105,000	125,000	12,000	250
Gold	ounces	210,000	350,000	12,000	110
Silver	ounces	1,600,000	3,800,000	3,000	100
Germanium	tons	9	20	3,000	[1]
Coal	tons	80,000	320,000	2,300	250
Tungsten ore	tons	600	900	650	350
Tantalum ore	tons	80	300 (Tantalum and Niobium ore)	600 (Tantalum and Niobium ore)	170 (Tantalum and Niobium ore)
Niobium ore	tons	50			
Cadmium	tons	200	150	500	700
Beryl	tons	300	1,000	400	[1]
Iron ore	tons	6,000			
Salt	tons	1,000	500	15	40
Bituminous sand	tons		4,000	15	[1]
Palladium	ounces	small	120	2	130 (Palladium and Platinum)
Platinum	ounces	small	30	2	
Radium	grams	27	75	1	[1]
Lead	tons		5	1	100

[1] No production in 1948.

closing of the Shinkolobwe uranium mine and the weakness of the zinc market caused a greater reduction of revenues. Resembling its unequal geographic distribution, the capital control of the industry is also concentrated. (See Table XXIII).

The Société Générale de Belgique diarchy not only controls most of the mining, but also other sectors of Congolese industry, especially transportation and manufacture. However, the state receives an overwhelming part of the revenue of these corporations, considerably more than in other parts of Africa or in most of the world.

TABLE XXIII

GEOGRAPHIC DISTRIBUTION OF CAPITAL CONTROL OF CONGOLESE MINES

Value of production				State share of revenues
Area	%	Group control	%	%
Katanga	72	Union Minière	67	70
Kasai	15	Sibéka–Interfor	23	70
Maniema-Kivu	10	Empain	4	25
Kibali	3	State	3	90
		Brufina	2	40

THE SOCIÉTÉ GÉNÉRALE DE BELGIQUE

In 1822, William, king of The Netherlands, founded the Société Générale, with the Capouillet, the Counts Baillet and Meeus, the Barons Van Zuylen van Nyvelt and also the Rittweger, Schumacher, Tiberghien and Van der Elst families holding a minority of the shares (CRISP, 1961). Its own subsidiaries retain a majority of the seats on the board of directors of Société Générale. Indeed, on the twelve man board we find *(a)* three representatives of Assurance Générale Vie et Incendie who simultaneously represent their personal interests, namely, the Viscount Jonghe d'Ardoye, Count Lippens and the Baron Hankar (Solvay group) with one board member of an associated group, *(b)* two representatives of another subsidiary in the insurance field, the Royale Belge, *(c)* two representatives of the Saxe-Coburgs, the royal family of Belgium, *(d)* a representative each of a subsidiary in the electricity field, of the Marquis de Boëssière-Thiennes and the Barons Goffinet.

Through the initiative of Léopold II and of Captain Thys (his aide-de-camp), the Société Générale, with a minority participation of De Groof and Company, founded CCCI (the Compagnie du Congo pour le Commerce et l'Industrie); in 1891, the King initiated a subsidiary of CCCI, the Compagnie du Katanga, to explore all and to own one-third of the lands located south of the zone then controlled by the Arab slave traders. Compagnie du Katanga has two types of shares: priority shares which get 5% of profits with taxes already acquitted (CCCI owns 8,500 out of 12,000 of these shares); and common stock which receives 95% of the remaining profits with 5% reserved for directors and executives, that is, largely the controlling group. Out of 1.2 million common shares, CCCI owns less than 160,000. On the eleven member board of the Compagnie du Katanga, we also find the Baron Lambert who controls a bank associated with the French branch of the Rothschild family, and the Prince Napoléon who is related to the Saxe-Coburg family ruling in Belgium. Minority interests include: the Anglo-American Corporation and Tanganyika Concessions Ltd. in which, along with the Rockefeller family(?), Anglo-American and three United States groups are also interested. We can illustrate thus the holdings of these groups in percentages:

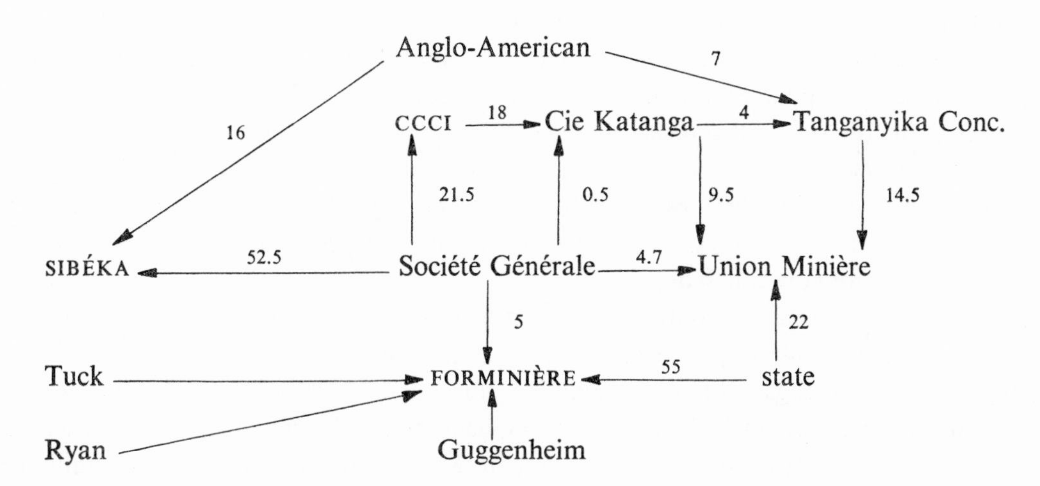

Furthermore, SIBÉKA owns 2% of the equity of Union Minière and 1% of the shares of TANKS.

Since shares are widely distributed, mainly in Belgium and France, no external group ever attempted to take control. Consequently, the corporation became auto-cephalous with its hierarchy renewing itself from its own ranks. Combined with these technocratic tendencies, the Société Générale has always tended to pursue an expansionist policy, creating numerous enterprises which it controls frequently through a relay system of intermediate companies and through a system of preferential shares which multiplies the effect of its (minority) holdings. Indeed, being the founder of companies in which the public was eager to invest, the corporation was in a position of attributing special rights to its founding shares.

In 1906, Léopold II, who also controlled the Congo Free State, simultaneously founded Union Minière du Haut Katanga, the Société Internationale Forestière et Minière (FORMINIÈRE) and the Société des Chemins de Fer du Bas Congo au Katanga (the parent company of SIBÉKA). Traditionally, the governor (i.e., chairman) of the Société Générale was chairman of the board of these corporations in which the state has had commanding interests since their founding but which have always been managed as dynamic private enterprises. On the other hand, executives who have risen from the European and African corporations of the Société Générale have followed each other in the governorship.

HISTORY OF KATANGA MINING

Since the XVIIth century, the Lunda Empire expanded eastwards to cover an area as big as France. The Kazembe who governed in the name of the Mwatayamvo, the emperor (a hierarchy with which the Tshombe family is related), what we today call Upper Katanga controlled mining and smelting. Similarly, under the auspices of the Lunda State, traders organized caravans with the copper regularly reaching its European markets. Fifty years after the visit of the Brazilian De Lacerda in 1850, Ladislas Magyar, the Hungarian explorer, lived in Katanga. Later, coming from south of Lake Victoria, the Bayeke conquered the mining area. Their king, M'siri, has not failed to exploit it.

Following Captain Le Marinel, the Compagnie du Katanga took over the Delcommune expedition of the CCCI, but these failed to conclude an agreement with M'siri, as did the expeditions sent by Cecil Rhodes. A year later, in 1891, Captain Bodson, a member of the expedition led by Captain Stairs killed M'siri (an ancestor of Godefroid Munongo, former Minister of the Interior of Katanga). By the time the expedition of Captain Bia and Lieutenant Francqui (who was to become governor of the Société Générale de Belgique and president of Union Minière) reached Bunkeya, the capital of Garanzane (Katanga), famine reigned. In 1892, Jules Cornet, the geologist of the expedition, visited the Kambove deposit at what we now call Jadotville.

In 1899, coming from the south, George Grey (a member of the famous Whig family) rediscovered Kipushi after N'kana. In 1900, the state with two-thirds of the shares and the Compagnie du Katanga with the remaining third combined to establish the Comité Spécial du Katanga (CSK) which, among others, held the royalty rights on the southeastern corner of the Congo. (In 1961, the Compagnie du Katanga made a profit of $3.5 million.) The next year, Robert Williams, an associate of Cecil Rhodes, obtained exploration permits from the CSK and formed Tanganyika Concessions Ltd. (TANKS or TCL). As in Zambia, 'discovery' meant African guides leading explorers to ancient workings; thus, a delegate of Chief Panda showed Michael Holland the old drifts and pits of Kambove.

The first industrial mines were hardly more sophisticated than these workings and it took the oxcarts of Major Boyd-Cunningham seven months to do the trek from Benguela to Ruwe with some equipment. Finally, in 1906, the state, TCL and the Compagnie du Katanga pooled their interests and Union Minière du Haut Katanga was created to exploit the area until 1990. Jean Jadot, an engineer, became the first executive of the corporation. The railway reached Elisabethville in 1910 and industrial production started, as shown in Table XXIV.

TABLE XXIV

FIRST YEAR OF EXTRACTION OF DIFFERENT PRODUCTS IN THE CONGO

Copper	1911	Cobalt	1926	Lead ore	1937
Tin	1918	Zinc ore	1936	Cadmium	1941
Uranium	1922	Manganese ore	1937	Zinc metal	1953
Radium	1922			Germanium	1953

With a few exceptions, the production of Union Minière has not failed to expand. Thus, from 2,700 tons in 1912, copper output increased to 20,000 tons by 1920, to 45,000 tons in 1922 and to 150,000 tons in 1929, with cobalt production attaining 600 tons. During the First World War, the formation of African skilled personnel started which, in contrast to most of the continent, was to remain a distinctive trait of Congolese industry. In the late twenties, the corporation initiated the policy of housing the families and stabilizing workers, whose numbers were to reach 25,000. During the same period, the Prince Léopold Mine started operations and radium sales and coal mining commenced. The Francqui hydropower plant commenced deliveries during the Great Depression. In 1937, the Kolwezi and Musonoi mines started operations in the west with important developments in smelting: the sulphides of Kipushi, the production of sulphuric acid and the recovery of cobalt from the solutions of Shituru.

During the Second World War, the mines of the Elisabethville district declined and emphasis shifted further to the west and Kipushi. After the war, the quarrying rate increased rapidly and in 1956, before its fiftieth anniversary, the corporation sold more than 5 million tons of copper, 93,000 tons of cobalt, and almost a million tons of zinc. Trends indicate that around 1970, total copper production will reach the mark of 10 million tons. By 1953, the extraction rate at Kipushi topped 100,000 tons per month. With the trend towards hydrometallurgy, the hydropower plants Bia, Delcommune and Le Marinel were inaugurated in close succession in 1950, 1952 and 1956 and experts now believe that yearly productive capacity has already reached the mark of 370,000 tons of copper.

UNION MINIÈRE

A majority of fifteen members of the board of Union Minière rose from its own ranks with three each making their career in the Société Générale and Tanganyika Concessions (TANKS) and two with the former Comité Spécial du Katanga. Rhodesian Anglo-American Ltd. probably holds 7% of the stock of TANKS. Furthermore, another member of the Anglo-

American group, Rand Selection Trust Corporation holds additional shares of TANKS. In addition to the Compagnie du Katanga, the Banque Lambert (Brussels) and Lazard Frères (of Paris) have interests in this holding which has historic contacts with the other 'royalty holder', the British South Africa Company. Departing from earlier practice, a different person occupies the chair of the Société Générale and Union Minière.

Union Minière now has a capital of $160 million. Since the late fifties, yearly sales have ranged around $200 million and although earnings have decreased from their mark of $100 million and profits from $70 million since 1959, the corporation is basically healthy. Out of its total profits, the state takes, under various forms, about 70% (i.e., double of the amount paid in neighbouring Zambia, in South Africa and more than the '50%' of oil and iron ventures). How are then profits distributed? The company pays a 10% royalty on distributed profits exceeding $1.8 million. As mentioned, two-thirds of these go to the state with one-third being attributed to the Compagnie du Katanga. Furthermore, according to the 1963 Leopoldville proposal, half of the Congo's share would revert to the states with the central government receiving the remainder. For the time being, Union Minière has authority to keep profits and foreign exchange which it uses for operations, turning over the rest to the central government. In 1964, taxes have increased (Table XXV).

In 1962, the government received all of the profits of $12 million. However, in the first half of 1963, the company was authorized to utilize $40 million out of $84 million total receipts.

TABLE XXV

UNION MINIÈRE PROFIT AND TAX STRUCTURE

Taxes and profits	*1961* *$ million*	*1960* *$ million*
Export duty	24	28
Profits tax	6	5
Dividends tax	4	7
Other taxes	7	8
Two-thirds of royalty	1	3
Profits state	5	8
Total state	47	59
Profits others	25	38
Other	660,000	

TABLE XXVI

UNION MINIÈRE CAPITAL STRUCTURE

Share holders	*Shares*	*Voting right*
State	210,000	120,000
Tanganyika Concessions	180,000	130,000
Compagnie du Katanga	120,000	80,000
Société Générale	60,000	30,000
Congométaux–Vielmont	7,000	
Empain group	3,000	
Total	580,000	410,000
Other	660,000	

TABLE XXVII

UNION MINIÈRE PRODUCTION

Year	*Copper* 1,000 tons	*Cobalt* 1,000 tons	*Zinc* 1,000 tons	*Germanium* tons	*Silver* 1,000 ounces	*Cadmium* tons
1940	150	2,5	12		3,700	
1950	260	9,4	75		4,000	130
1960	333	9	120	15	4,000	230
1962	325	10,7	105	9	1,600	100

Excluding the concession, the corporation values its assets at $190 million; during the last decade, it has invested $300 million (including $50 million in education, housing and welfare) and plans to continue to do so as we will see below. In 1961, investments attained $24 million, and $30 million in 1962. The corporation receives $2–5 million each year from its subsidiaries. The development of the corporation's production is shown in Table XXVII.

In addition to this, the corporation produces hydropower, sulphuric acid, radium, metallurgical cement, lime, iron ore, manganese, some gold, and it formerly produced uranium, vanadium, platinum, palladium, nickel and tin.

UNION MINIÈRE MINES AND PLANTS

The deposits are strewn along a 200 mile long stretch (which is rarely wider than 15 miles) especially in three areas (Fig. 28):

Western district: Dikuluwe, Mupine (Co), Kamoto (Co), Musonoï (Co), Kolwezi, Ruwe.

Central district: Tenke (Co), Fungurume (Co), Kakanda, Sesa, Kambove (Co), Kisanga (iron), Kanunka (iron), Kakontwe (limestone), Kamatanda, Shinkolobwe (uranium).

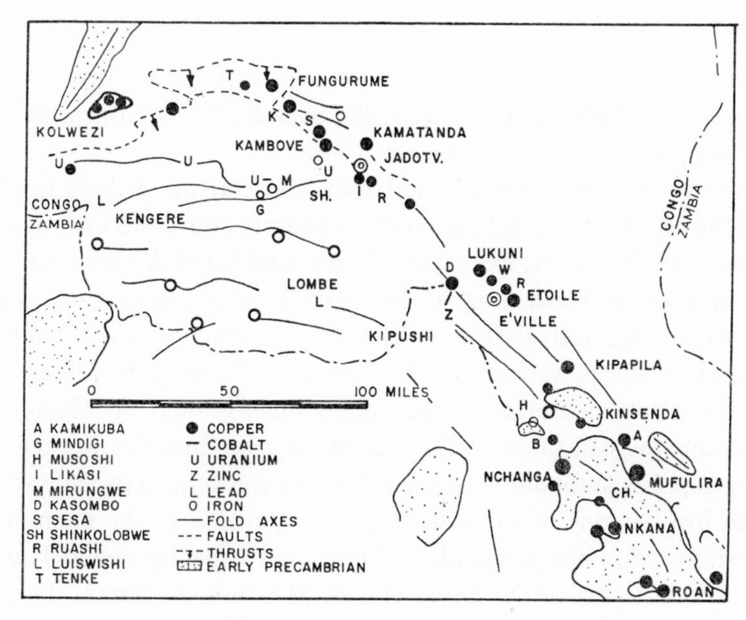

Fig. 28. The mineral resources of the Copperbelt.

Southern district: Lukuni, Luiswishi (Co), Ruashi (Co), Étoile du Congo (Co), Karavia, Kipushi (zinc etc.).
Far South: Kipapila, Musoshi, Kasumbalesa (iron), Kinsenda, Kamikuba.

Unmarked place names refer to copper and in those followed by (Co), i.e., cobalt, copper is also the dominating metal. In addition to these, lead shows at Kengere 40 miles south of Kolwezi, copper appears alone at Mirungwe and cobalt appears at adjacent Mindigi 40 miles south of Fungurume. Let us mention here that copper also shows at Kapulo near Lake Mweru, at Lubi and with lead and zinc in the Lower Congo. There is some lead at Lombe, 60 miles south of Shinkolobwe. However, only a few of the mines are now operating, with Kamoto in the west contributing 45% of the ore and 60% of the copper. Indeed, these adjacent quarries are among the world's largest copper deposits (Table XXVIII).

TABLE XXVIII
UNION MINIÈRE ORE EXTRACTION

Region	Ore million tons	
	1962	1959
West:		
Kamoto	3.0	2.4
Ruwe	2.7	1.2
Musonoï	0.8	1.9
Kolwezi		
South:		
Kipushi	1.0	1.0
Centre:		
Kambove	2.0	0.2
Shinkolobwe	2 0	0.2
Total	10	6.9

In addition to these, in 1961, 17 million cubic yards of overburden were removed (23 million cubic yards in 1960). Thus, by now, the overall grade of opencast mines has decreased to 3% copper and 0.1% cobalt.

While miners still use churn drills in soft rock, most of the drilling is done by heavy equipment. In the big quarries, electric powered shovels dump ore on wagons and heavy duty trucks. From 1960 to 1961, extraction almost doubled at *Kamoto*, and, by now, this mine of siliceous oxides provides most of the ore of Kolwezi concentrator. The excavation of a trench across the Kamoto East orebody facilitates the working of the lower levels of Kamoto Main where shafts were sunk below the 1,200 ft. level. In 1963, pilot extraction started and by 1965 this large mine will be entirely underground. At Musonoï, equipment strips and then tips the overburden of the southern zone into the worked out northern zone. The mine supplies sulphuretted-oxidized ore to *Kolwezi*. Between 75% and 90% of the time of the five sections of the concentrator is devoted to the palm-oil flotation of the Kamoto oxides, with the remainder of time and capacity devoted to the surficial sulphurization and flotation of the mixed ores of Musonoï. At Ruwe, the overburden of the lower levels of two southern extensions are removed. The washing plant preconcentrates

TABLE XXIX

UNION MINIÈRE PRODUCTION OF COBALT AND COPPER CONCENTRATORS IN 1962

Area	*Source*	*Concentrates 1,000 tons*	*Copper %*	*Cobalt %*	*Contained 1,000 tons*	
					copper	*cobalt*
Kolwezi	oxides	760	25.6	1.78	190	14
	mixed	80	1.2	3.19	28	2
Ruwe	breccia	99	22.3		22	
		110	5.1		5	
Kambove	mixed	140	31.7	1.91	44	3

the breccia ore. Close at hand at the Kolwezi quarry, a new skip operates. The production of concentrators is given in Table XXIX.

At Fungurume, pilot operations commenced in 1961. The evaluation of the Tenke-Kakanda zone was completed, and the 800,000 ton Kakanda concentrator was commissioned in 1962. Following the depletion of *Kambove*-Main, Kambove-West now provides the mixed sulphurettes-oxidized ores of the new concentrator which commenced operations in 1961, and already attains all of its ultimate 1.1 million tons per year capacity. In the mine, three shafts will reach a depth of 600 ft. At the same time, the Shinkolobwe mine which produced the uranium used in World War II will be dismantled. Indeed, the decreased resources of this mine of uranium, cobalt, gold, platinum, palladium and molybdenum cannot compete with the large uranium deposits now operating.

At Kakontwe, Union Minière quarried 210,000 tons of (metallurgical) limestone and at Kisanga, 6,000 tons of iron ore. The Kamatanda mine (in the centre), Ruashi and Karavia (west of Elisabethville) operate occasionally to combat unemployment. In 1961 at Ruashi, ores containing 2,000 tons of copper were washed. In 1960 operations were suspended at Étoile du Congo and Lukuni, but the Fungurume mine started operations in 1962.

The Prince Léopold Mine at *Kipushi* is the continent's principal zinc, silver, germanium and cadmium deposit; it strongly resembles Tsumeb. An interior shaft has now reached the depth of 2,300 ft. and a sub-vertical shaft is being sunk to 2,800 ft. where grades remain promising. In 1960, the winding shaft connecting levels 1,300 and 2,300 ft. was commissioned, and in 1961, miners prepared the exploitation of the 2,000 ft. level by top-slicing with metal supporting props. Only in the low grade friable ores do shrinkage and sub-level caving prevail. That year, because of the Katanga War, the Kolwezi concentrator could not operate at its full capacity of 1.1 million tons per year (See also Table XXX).

While complex sulphuretted ores undergo differential concentration, magnetic separators recover reniérite, the germanium sulphide found in specific zones of the deposit.

TABLE XXX

KIPUSHI ZINC, COPPER AND LEAD PRODUCTION

Year	*Concentrate*		*Concentrate*		*Contained metal, 1,000 tons*		
	1,000 tons	*zinc %*	*1,000 tons*	*copper %*	*zinc*	*copper*	*lead*
1962	156	57.3	150	26.1	90	40	
1960	200	56.6	235	25.1	110	60	3

UNION MINIÈRE SMELTERS AND PRODUCTS

The corporation operates four smelters. Near Elisabethville, the Lubumbashi water jacket furnaces process mainly the complex concentrates of Kipushi into matte. With added concentrates, the converter section treating the matte has a capacity of approximately 140,000 tons of blister copper, but in 1962, military operations reduced output to 100,000 tons. In 1960 at *Kolwezi*, the $50 million automated electrolysis plant, one of the most modern in the world, commenced operations. The copper section, boasting a capacity of 110,000 tons, leaches the oxidized concentrates of Kolwezi producing mainly cathodes with a couple of thousand tons of mother sheets. The Jadotville smelter refines a majority of these cathodes. Furthermore, the cobalt section has a capacity of 3,800 tons of cobalt per year.

Near *Jadotville*, the Shituru electrolytic plant lixiviates the oxidized, sulphuretted and carbonated concentrates of Kolwezi and Kambove, and recovered 135,000 tons of copper cathodes and 17,000 tons of mother sheets in 1962. These are used in the sections of marketable electrolytic copper at Shituru and Luilu and for packing cathodes. In addition to these, the plant produces 6,700 tons of cobalt cathodes. The copper refinery further processes the cathodes of Shituru and Luilu as well as the raw copper of Lubumbashi and of the Jadotville electric smelter; it produced 147,000 tons of electrolytic copper and 43,000 tons of soluble anodes in 1961. The cobalt refinery produces granules from a majority of the Shituru and from all the Luilu cathodes. (The remainder of the Shituru cathodes is refined in Belgium.) The three-phase furnaces of the electric smelter can produce *(a)* 5,000–7,000 tons of copper-cobalt alloy for export, and *(b)* 12,000–15,000 tons of blister for the manufacture of soluble anodes for mother sheets. From germanium concentrates of Kipushi, a single phase furnace can produce 10 tons of germanium dusts by volatilizing fusion. The 1962 production can be read from Table XXXI.

TABLE XXXI

UNION MINIÈRE PRODUCTION METHODS AND PRODUCTION OF COBALT AND COPPER IN 1962

Copper	*Tons*	*Cobalt*	*Tons*
Electrolytic, ingots	148,000	Electrolytic, granules	8,500
Blister	61,000	In copper alloy	1,500
Cathodes	114,000	Cathodes	300
With cobalt and zinc	2,000	Other	100
Total copper	325,000	Total cobalt	10,700

While the corporation exports a majority of the copper to France, Belgium and Germany, i.e., the Common Market, the United States is an important importer of cobalt.

From 120,000 to 150,000 tons of *zinc* concentrates of Kipushi, SOGECHIM, a subsidiary, roasts 100,000–130,000 tons of sintered concentrates, recovering the same amount of *sulphuric acid.* METALKAT, another subsidiary, buys 100,000–120,000 tons of these concentrates, resells almost all of the raw and roasted material to the Belgian industry and produces 60,000 tons of electrolytic zinc with 2,000 tons of copper. The company also plans to recover zinc from the slags of Lubumbashi.

The Jadotville smelter treats the *germanium* concentrates of the Kipushi separator, but METALKAT, at Kolwezi, processes the flue-dusts retained in the filter bags of the Lubumbashi furnaces, with the rate of recovery increasing substantially. From these, the Société Générale Métallurgique de Hoboken, another subsidiary, extracts germanium oxide and high purity germanium, equivalent to 10–30 tons metal, with 40% coming from the magnetic concentrates and 60% from the dusts. Each year, the same company at Oolen recovers, from residual slimes of the products of Kipushi, 1.8–5 million ounces of *silver* which the Hoboken electrolytic smelter purifies and refines.

To return to METALKAT, from the dusts of zinc concentrate roasting and *cadmium*-bearing cements, this subsidiary recovers 100–220 tons of electrolytic cadmium bars per year in its flue-dust plant at Kolwezi. Although Shinkolobwe has been closed down, by cooperating with Euratom (one of the European Common Market authorities) the Hoboken plant still sells minor amounts of uranium. On the other hand, *radium*, of which the Congo is the sole commercial source in the world, enjoys a stable market. Following their record of 100 g, sales levelled out at 27 g/year. The refineries of Hoboken also recover 1,000 ounces of gold per year which comes mainly from Ruwe, and some palladium and platinum, which originates in this mine and Kolwezi.

Some vanadium and molybdenum have been recovered, respectively, at Ruwe and Shinkolobwe. Talc is particularly abundant at Kolwezi, as is fluorite at Kakontwe, barytes (also occurring in the Lower Congo) and amphibole asbestos. Graphite is found in several localities. In 1962, Union Minière quarried 230,000 tons of limestone at Kakontwe and, with a capacity of 110,000 tons, the corporation burns 90% of the Congo's lime. Finally, the principal salt works are at Guba, with other salines at Kalamoto, Mwashya and Nganza.

UNION MINIÈRE SUBSIDIARIES

Although other members of the Société Générale group have considerable shares in these subsidiaries, they delegate, in effect, their control to Union Minière. We have already mentioned two subsidiaries. SOGECHIM, the Société Générale Industrielle et Chimique de Jadotville, is the almost wholly-owned Congolese subsidiary of the Société de Gestion et d'Exploitation d'Industries Chimiques. Union Minière owns $1 million of the $7.2 million equity of this company which, in addition to sulphuric acid, produces 1,600–2,000 tons of sodium chlorate for AFRIDEX (the explosives company), 140–180 tons of glycerine, and the drinking water of Jadotville. METALKAT, the Société Métallurgique Katangaise (the Congolese branch of the Société Métallurgique du Katanga), produces zinc, germanium and copper. Union Minière owns one-third of its $12 million capital. At Hoboken (Belgium), the Société Générale Métallurgique refines blister and cobalt alloys, and its electrolytic smelter at Oolen produces germanium, cadmium and radium, as well as products coming from other parts of the world. With 150,000 of its shares, Union Minière controls a third of its equity.

SOGEFOR, the Société Générale des Forces Hydroélectriques du Katanga, operates the power plants of Union Minière, and produced 2,200 million kWh in 1962, out of which 400,000 kWh were exported to the Zambian Copperbelt. With 100,000 shares, Union Minière controls SOGEFOR.

With minor support from the Compagnie du Katanga, Union Minière formed SUD-KAT, the Société de Recherche Minière du Sud-Katanga, to exploit 10,000–30,000 tons of *manganese* ore at Kasekelesa and the now dormant *lead* and zinc properties of Kengere and Lombe. North of Kolwezi, the production of the Charbonnages de la Luena fell from 260,000 tons in 1959 to 80,000 tons in 1962 and the capacity of the adjacent Kisulu *colliery* now stands at 35,000 tons per year. Union Minière holds 49,000 of the shares of this company and Compagnie du Katanga retains 15,000 of these as well as 5% of the equity of the Wankie Colliery in Southern Rhodesia. In northern Katanga, there were two other collieries at Albertville and Greinerville, the latter controlled by GÉOMINES. To return to Jadotville, Ciments d'Afrique Centrale (25% owned by Union Minière) operates at 12,000 tons per year below its capacity. In addition to these, the corporation's subsidiaries produce flour and explosives, they build houses and distribute electricity.

DIAMONDS

In 1903, experts identified a stone of a tributary of the Lualaba in Katanga as a diamond, but the kimberlite pipes of central Katanga remained barren. In 1909, N. Janot, P. Lancswert and M. K. Shaler accidentally found a gem in the Tshiminina (Fig. 104), an affluent of the Kasaï in the southern Congo, but only in 1918, when George Young discovered Bakwanga, did the world's largest diamond enterprise start. By 1920, H. de Rauw concluded that the diamondiferous gravels were related to Karroo sediments and in 1946, I. de Magnée discovered the kimberlite breccia of *Bakwanga* (Fig. 105). Containing with its 4 square mile cover of sediments an overwhelming majority of the world's diamonds, this mixed rock has already produced 200 million carats! While 97–98% of the Bakwanga diamonds are industrials and 25–35% crushing boart, those of the Tshikapa area (southeast of Luluabourg) are gems. A glance at Table XXXII shows how production progressed.

As to SIBÉKA (the Société d'Entreprise et d'Investissement du BÉCÉKA), MIBA (the Société Minière de Bakwanga) manages its assets of $100 million (including the mining rights valued at $44 million). In 1961, out of $15 million working revenue and $8 million of investments in other companies, SIBÉKA paid 70% of its revenues to the state (Table XXXIII).

In 1961 at Bakwanga, machines excavated 4 million cubic yards of gravel and overburden, i.e., an overall grade of 4.5 carats per cubic yard! Tournapuls strip and excavators and rotary shovels remove waste. Shovels load the ore on dumpers. The material moves first through logwashers and pans and, in the fully automated plant, experts achieve preconcentration by heavy-media separation. Closed circuit television controls the equipment. The Young hydropower plant provides electricity.

TABLE XXXII

BAKWANGA DIAMOND PRODUCTION

Year	*Million carats*
1920	0.2
1930	2
1940	9
1950	10
1960	13
1961	18
1962	18

TABLE XXXIII

SIBÉKA'S DUTIES, TAXES AND PROFITS IN 1961

Revenue distribution	*1961* *$ million*	*1951* *$ million*
Export duties	5	1.2
Taxes	2.5	2.2
Profits state	6	2.4
Profits private	7	2.0

On the other hand, SIBÉKA suspended operations in the Luebo west of Bakwanga. Indeed, from 120,000 carats in 1960, production fell to 70,000 carats in 1961.

INTERFOR (the Société Internationale Commerciale et Financière de la FORMINIÈRE) has a capital of $10 million, dominated by the state; the Ryan family, who control the Royal McBee typewriter business machine corporation in the United States; and Kaysam, a latex producer, controlled by the American Guggenheim and Tuck (associates of the Solvays) families (and De Beers'). FORMINIÈRE (the Société Internationale Forestière et Minière du Congo) managed Bakwanga from 1929 to 1960 and also its sisters, the Sociétés Minières of Kasaï, Lueta and Luebo, while simultaneously operating its own Tshikapa field (and controlling its subsidiary, the Société Minière de la Télé, a gold producer east of Stanleyville).

In the drainages of the Kasaï River and its tributaries, the Luebo and Lulua, this diamond field extends 150 miles north from the Angola border. Until now, FORMINIÈRE has recovered 16 million carats from its own concessions. From 360,000 carats in 1950, although reserves were declining, output rose to a peak of 440,000 carats in 1958; however, in 1960, tribal troubles caused operations to cease. All the mines of the Tshipaka district were placers. Gravel excavated by hydraulicking or other means was washed in pans, thence moving through trommels and jigs or sink and float concentrators.

SIBÉKA and FORMINIÈRE-INTERFOR also have holdings (Table XXXIV).

Paralleling the South-African Ultra-High Pressures unit, SIBÉKA and the De Beers group are establishing a plant for the production of industrial diamonds at Shannon in Eire.

Diamonds also show in Kwango southeast of Leopoldville, in several districts of Maniema, notably at Mutandulwe and in the Uélé gold field.

TABLE XXXIV

FORMINIÈRE-INTERFOR AND SIBÉKA HOLDINGS

Company	*Shares $*	
	Sibéka	*Forminière*
Bécéka-Manganèse	337,000	10,000
Diamang		95,000
S.A. Diamant Boart	119,000	20,000
Industrial Distributors Ltd.	735,000	40,000
Diamond Trading Company Ltd.	30,000	15,000
Diamond Purchasing and Trading	75,000	37,000
Bauxicongo		12,000

IRON AND MANGANESE

In 1945, when Polinard explored in the southwestern corner of Katanga, the Kisenge manganese deposit (POLINARD, 1946), SUD-KAT, had been producing manganese for many years. But these deposits, like those of Zambia, were small; their production rarely exceeded 15,000 tons per year and only in 1952 did BÉCÉKA-Manganèse (a subsidiary of the diamond company) start operations. Since then the company has invested $11 million in its quarries and concentrators. Consequently, since 1954, the output of the Congo expanded, averaging 350,000–420,000 tons of ore containing 50% metallurgical grade manganese. In 1962, with a personnel of only 700, the subsidiary company Société Minière de Kisenge produced 340,000 tons of ore. With a capital of $4 million, the company obtained a revenue of $3.6 million in 1959 (Table XXXV).

At Kisenge, draglines load the ore on Euclid trucks. While high grade ore is simply washed in a 100 ton per hour plant, a sink and float plant has treated ore containing 30% manganese since 1958. A 200 mile long, 110,000 Volt line connecting with the hydropower network of Kolwezi provides power for these installations.

Out of the $1.6 million capital of SEDEMA (the Société Européenne des Dérivés du Manganèse, which produces metallic and alloy manganese at Tertre in Belgium), BÉCÉKA-Manganèse owns $220,000, SIBÉKA owns $200,000, sister companies and the Manganese Chemical Corporation controls the rest.

Although high grade iron ore is abundant in the northeastern and southeastern Congo, prohibitive transportation costs still do not permit its export. Estimates of these resources range from 3 to 10 million tons of ore containing 43–63% iron, but no systematic evaluation has yet been carried out. In the northeast where the largest deposits are located, at least two districts can be distinguished: Uélé east of Stanleyville, and the Kilo-Moto area with its surroundings. The iron mountains of Duru, Tina, Ibi, Lai, Kai, Bahengo, Asonga, Kuma, Ami, Zani-Kodo and others range in length from 1 to 6 miles, with their width varying from one-third to 1 mile. The area of Duru alone would hold 2,000 million tons of ore. In the northwestern corner of the Congo south of Bangui, other iron mountains rise.

Along the frontier of Zambia, a belt of iron deposits extends from Mutuntu 70 miles east to Musombwishi. Furthermore, isolated deposits are found along the frontier at Musaka and Kasumbalesa, and in the Copperbelt between Kolwezi and Kisanga (Jadotville). From the latter and Kasumbalesa, Union Minière used to extract up to 20,000 tons of iron for use as a flux in the water jacket furnaces. In addition to these, unworked deposits of iron show in western Katanga and southern Kasaï and in the Mayombe (Lower Congo).

TABLE XXXV

SOCIÉTÉ MINIÈRE DE KISENGE DISTRIBUTION OF THE 1959 REVENUE

Revenue distribution	*$ million*
Export duties	6
Taxes	4
Royalties	4
Total state	14
Shareholders	6

TIN

In 1903, Robyns discovered tinstone at the confluence of the Lualaba and Lufupa in Katanga. North of Kolwezi, Union Minière operated the small Busanga tin mine as well as Mwanza and Shienzi, but only in 1910 did prospectors discover the big primary deposit of Manono in northwestern Katanga in the corner of the confluence of the Lualaba (Congo) and the Lukuga. At the beginning of the century, Léopold II invited Baron Edouard Empain (the builder of the métro, the Paris underground (subway), and of Héliopolis in Egypt) to build a railway connecting the Lualaba to Lake Tanganyika. As an inducement, CFL (the Compagnie des Chemins de Fer du Congo Supérieur aux Grands Lacs Africains) received the mining rights in an area extending from the Congo to the vicinity of Lake Kivu, which later proved to be the principal tin zone. But the first missions found little gold and not until 1923 did Lallemand discover tinstone in the western part of Maniema. Following the prospecting campaign of the early thirties, SYMÉTAIN and the COBELMIN group established themselves in this virgin area. The total tin production of the Congo developed as shown in Fig. 5 and Table XXXVI.

TABLE XXXVI
TIN PRODUCTION IN THE CONGO

Year	*Tons*	*Year*	*Tons*
1913	14	1945	19,000
1930	1,000	1950	15,000
1940	13,000	1962	10,000

Vieille Montagne, the world's leading zinc smelter controlled by the Société Générale de Belgique and by the French Hottinguer, Rothschild and Lazard groups, has important interests in GÉOMINES (the Société Géologique et Minière des Ingénieurs et Industriels belges); but succeeding to the Comité Spécial du Katanga (CSK), the state owns 18% of the equity and the Compagnie du Katanga holds 9%. On its $14 million capital and $26 million investment, GÉOMINES has also borrowed $900,000 from the CSK to exploit hard rock deposits. Indeed, at Manono west of Albertville, the cost problem of crushing large reserves of hard ore was aggravated by the Katanga War. The Manono tin smelter (the largest in Africa) boasts a capacity of 10,000 tons, but even in the year 1959–1960, the company only produced 2,500 tons of tin and 140 tons of tantalite–columbite.

While the Banque de Bruxelles was founded in 1871, only in 1935 did the Counts de Launoit amalgamate the companies they control in Europe and Africa. BRUFINA, the Société de Bruxelles pour la Finance et l'Industrie, owns 38.5% of the shares of the Compagnie Africaine Foncière which controls SYMAF (the Syndicat Minier Africain and its subsidiary SYMÉTAIN) who operates the Kalima and Punia districts in western Maniema. Philipp Brothers of New York and the Swiss Société Générale pour l'Industrie also have interests in SYMÉTAIN. Alluvial and eluvial placers extend from Kalima 30 miles south to Amikupi, north to Baselele and 30 miles east to Messaraba. The largest deposit of the Punia-Kasese district (north of Kalima) is Tshamaka. North of Punia, the Malimba and Masaba deposits produced 1,000 ounces of gold and 1,000 carats of diamonds. With an investment of $18 million, SYMÉTAIN had a productive capacity of 4,000 tons per year,

TABLE XXXVIIa

SYMÉTAIN DATA I

Location	*Tin tons*	1959 *Wolframite tons*	1959 *Hydropower million kWh*	1962 *Tin tons*	1962 *Hydropower million kWh*
Kalima	2,000	1,400 (Kalima and Punia together)	20	1,000	11
Punia	1,300		4	650	3

TABLE XXXVIIb

SYMÉTAIN DATA II

Revenue distribution	*1959 $ 1,000*	*1962 $ 1,000*
Taxes	500	120
CFL royalty	300	100
Net profit	500	220

but tin production was first curtailed by the International Tin Council and then by events in Maniema (see Table XXXVII).

Administered by FORMINIÈRE, COBELMIN, a subsidiary of the Empain and Société Générale groups, manages six mining districts of four companies which had, in the late fifties, the productive capacities given in Table XXXVIII.

With one hydropower plant, each of these districts includes 6–12 mines. At Kampene, the smallest, workers only extract primary ores; at Kaïlo, both primary ore and eluvial placers; at Moga, eluvials and alluvials; but in all other tin fields, mostly river gravels. The Lulingu district holds the largest resources, followed by Moga. With taxes and profits at a quarter of their normal rate, all have been strongly affected by adverse conditions in Maniema.

With a capital of $2 million MINERGA (the Compagnie Minière de l'Urega, a wholly owned subsidiary of MGL controlled by the Empain group) produced, in 1962 at Lulingu, 500 tons of tin with some columbite–tantalite and made a profit of $200,000. In the same year MILUBA (the Compagnie Minière du Lualaba, a combination of the Empain group and

TABLE XXXVIII

COBELMIN PRODUCTION CAPACITY (1959)

District	*Location*	*Company*	*Tin tons*	*Columbite tons*	*Wolframite tons*	*Gold ounces*
Lulingu	E of Kalima	Minerga (Miluba)	900	100		
Kima	Kasese	Miluba, Belgikamines	800	50		
Kampene	S of Kalima	Belgikamines	400			
Kaïlo	E of Kindu	Kinorétain	500		100	
Moga	near Kalima	Kinorétain	500		50	
Namoya	SW of Bukavu	Kinorétain				30,000

BELGIKAMINES boasting a capital of $1 million) produced 400 tons of tin, paid $100,000 in taxes and made $160,000 in profits. With the Wielemans Breweries, the Banque Lambert (a group related to the Rothschilds) controls BELGIKAMINES. The Banque de Paris et des Pays-Bas, has an important interest in KINORÉTAIN (Mines d'Or et d'Etain de Kindu, originally founded by J. de Mathelin de Papigny). Until 1960, yearly production of KINORÉTAIN was in excess of $3 million.

From Maniema, the tin fields extend into the mountains bordering Lake Kivu where KIVUMINES, the Congolese branch of SOBAKI and Minière des Grands Lacs (see 'gold' section) operate.

When CFL relinquished its mining rights on western Kivu, AUXILACS (the holding of the Empain group) received a third of the equity of the predecessor of SOBAKI (the Société Belgo-Africaine du Kivu); but other interests, including the Belgian State, alimenta (Swiss Nestlé), CCCI, the Société Générale and the Saxe-Coburgs (reigning in Belgium) are believed to control this company which operates, with a capital of $2 million, the districts of Kigulube (east of Lulingu), Ona (Kasese) and Utu (northwest of Bukavu). Each year, labourers can produce 800 tons of tin with some columbite–tantalite and 10,000 ounces of gold. North of Kigulube, sharing with Philipp Brothers of New York, SOBAKI owns the mixed ore concession of Kabili and Katulu. In 1961, paying taxes equivalent to $200,000, the company made a profit of $600,000.

Since the depletion of its primary Nzombe deposit southwest of Bukavu, the placer tin production of the Kamituga district of Minière des Grands Lacs decreased from 600 tons to 200 tons. Northwest of Bukavu, the Kabunga district ceased its activities. Prior to the troubles, the mines of Maniema and Kivu had employed 30,000 people.

TUNGSTEN, TANTALUM, BERYL

We will now consider the ores usually found with tin. By 1956, the output of *tungsten* ore reached a peak of 1,300 tons, and has decreased ever since. In the largest deposits, Mount Mokama in the Kaïlo district and Mount Misoke in the Moga district, the tin/tungsten ratio ranges from 1/9 in veins to 5/1 in eluvial placers. West of Lake Edward, in the Etaetu deposit of Minière des Grands Lacs, tungsten predominates. Out of its capacity of 500 tons per year, workers still recovered 250 tons of wolframite here in 1962. In addition to smaller occurrences such as Bakwame and Sinsibi in the Kasese district of SYMÉTAIN, some wolframite shows in various tin concentrates of Kalima. Topaz accompanies wolframite, especially at Mokama and Kibila.

Maniema and Manono are the world's leading producers of *tantalum*. Indeed, in these columbite–tantalites, the niobium/tantalum ratio reaches the mark of 5/3. By 1956, the production of MINERGA and GÉOMINES, with minor amounts contributed by MILUBA, Minière des Grands Lacs and the predecessor of SOBAKI, attained 400 tons. While at Manono tin concentrates average 5% columbite, and at Lulingu 10%, there is pure columbite–tantalite south of Kasese in the Idiba district of MILUBA. This district boasts a majority of the world's tantalum resources. As to *niobium* (columbium), southwest of Lake Edward the primary Lueshe deposit of SOMILU (a company dominated by Union Carbide with interests of Minière des Grands Lacs and SOBAKI) boasts a few million tons of this metal. Further north at Bingu, geologists have located similar deposits. While

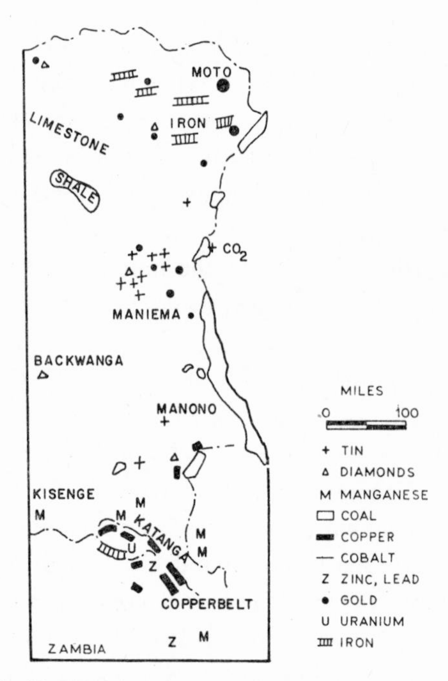

Fig. 29. The mineral resources of the eastern Congo.

zircon and cerium–thorium minerals are abundant in these complexes, yttrium minerals are associated with garnet in the columbite placers (Fig. 29).

Also, during the Korean War, *beryl* production reached its peak of 1,900 tons. Although beryl shows in most of the pegmatite fields of Kivu and eastern Maniema, the Kobokobo mine near Kamituga remains the only producer. Since reserves are depleted, the output of Minière des Grands Lacs diminished to 300 tons by 1962. Manono holds the world's largest resources of spodumene, a *lithium* ore. However, until now distances precluded the commercial working of this low grade material. On the other hand, Minière des Grands Lacs occasionally produced some amblygonite, another lithium mineral. Similarly, at Ngussa south of Kamituga, workers frequently washed for bismuth. Amethyst, tourmaline, feldspar and kaolin occur in the same rocks as beryl.

GOLD

In 1903, Hanan and O'Brien (two prospectors of the state) found placer gold near Kilo west of Lake Albert, and the discovery of Moto followed in 1906. Although Katanga already produced gold and in the thirties the Kivu-Maniema was to enter its productive phase, the Société des Mines d'Or de Kilo-Moto has contributed, since the beginning, two-thirds of the total output which evolved, as shown in Table XXXIX.

In the northeastern corner of the Congo lies Kilo-Moto. Seventy miles south of the Sudanese border, Agbarabo and Gorumbwa, near the town of Watsa (the Moto district), boast extensive resources. While the Durba mill concentrates these ores, smaller mines deliver their ore to the Rambi, Bulabula and Egbeu plants, respectively 40 miles east, 15 miles southwest and 25 miles west of Watsa, and to Beverendi. Furthermore, a dredge is operating 70 miles downstream of Durba on the Kibali River. The Zani deposits are located

TABLE XXXIX

CONGO GOLD PRODUCTION

Year	Congo ounces	Kilo-Moto ounces	Corporation	Ounces	
				1950	1960
1910	25,000	25,000	MGL	64,000	62,000
1920	100,000	100,000	Télé	16,000	
1930	180,000	130,000	COBELMIN	12,000	26,000
1940	560,000	250,000	Aruwimi	6,000	
1950	340,000	200,000	CNKI	6,000	11,000
1960	290,000	190,000	Union Minière		1,000

between Rambi and the northwestern tip of Uganda. Forty miles west of Lake Albert, within its 800 square miles, the Kilo district traditionally used to be the principal gold producer of the Congo. But the resources of Old Kilo, Nizi and others have been depleted and now the mines surrounding the Yedi plant 50 miles west of Kilomines, with Senzere and Nzebi near Kanga save the district from closure. There are also smaller mills such as Alosi, Akwe and Aboa. Recent production of the mines is shown in Table XL.

From the washing and dredging of placers, operations progressively shifted to quarrying (by 1950, more ore was crushed than washed) and then, at the turn of the decade, underground. Consequently, the hydropower consumption of the mines and mills has risen to 35 million kWh per year and the number of employees decreased to 30,000. Milling practice includes crushing, ball-mills, amalgamation and, finally, cyanadation of residues. The state owns an overwhelming part of the shares of the Société des Mines d'Or de Kilo-Moto, a company distinguished by a capital of $4 million, investments of $30 million and, in normal years, by a profit of $2 million and an employment of up to 30,000 people.

West of Watsa and north of Stanleyville, the Bondo district has been worked out. Similarly, even before the troubles of 1960, around Babeyru east of Stanleyville, the Société Minière de la Télé maintained only its Adumbi mill and tributed the other mines to workers. In like manner, west of Lake Edward, the gold mines of the Butembo district of the Compagnie Minière des Grands Lacs Africains (MGL) and west of these, the mines of the Angumu district of Somiba, have been depleted. However, in the Kamituga district southwest of Lake Kivu, in addition to smaller placers at Kilunga and Mapale, the gravels of

TABLE XL

KILO-MOTO PRODUCTION

	Units	1958	Site	1961
Moto: Durba	ounces	100,000	Durba	90,000
Moto: others	ounces	22,000	Zani	9,000
Kilo: Kanga	ounces	36,000	Kanga	14,000
Kilo: others	ounces	40,000	Yedi	22,000
Placers	ounces	20,000		3,000
Total ore milled	tons	730,000		380,000
Grade	dwt. per ton	10		20

TABLE XLI
GOLD PRODUCTION OF LUGUSHWA AND MOUNT MOBALE

Gold	*1959 ounces*	*1962 ounces*
Placer	44,000	42,000
Vein	21,000	15,000

Lugushwa hold extensive, and the veins of Mount Mobale have adequate resources (see Table XLI).

The company employs 4,000 people; it has a hydropower plant at Zalya. AUXILACS, i.e., the Empain group, directly controls 38% of the shares of MGL. The company has a total investment of $10 million and a capital of $3 million. In 1959, MGL paid $110,000 in taxes and made $500,000 in profits, but in 1962, pre-tax profits plummeted to $160,000.

Southwest of Lugushwa, the Namoya district of KINORÉTAIN was closed during the troubles of early 1961, but the Mwendamboko vein deposit holds underground and opencast resources containing 7 dwt. per ton gold for at least a decade's operation. The hydropower and cyanidation plants lie idle. In the forties, other districts (Kampene, Shabunda and Kima) managed by COBELMIN contributed 50,000–100,000 ounces of gold per year, but only low grade reserves remain. East of Lulingu, KIVUMINES continues to operate the Muta placers, and north of Punia, SYMÉTAIN occasionally recovers some gold. North of Albertville, this group has worked out the Kiyimbi mine and in the mountains bordering Lake Tanganyika, SOREKAT closed down Mutotolwa. We have already discussed the gold production of Union Minière at Ruwe. In addition to these, traces of gold show in the southern Kasaï and in the Lower Congo.

OTHER PRODUCTS

As we have already mentioned, limestone is widespread in Katanga, in the Lower Congo, and particularly north of Stanleyville, Bumba and Lisala where, however, the demand is limited. Three *cement* plants operating in Katanga (Lubudi, Albertville), in the Lower Congo and near Bukavu in Kivu have a productive capacity of 3 million barrels per year, (or of $15 million). But insecurity reduced sales to 1 million barrels. In the Lower Congo, the Lukala quarries can also produce marble. In addition to the lime burning of Katanga, in the Central Congo and Kivu, respectively, 8,000 and 4,000 tons of lime found markets in 1958. In the same year, i.e., at the height of the expansion triggered by the Ten Year Plan, we have the breakdown of stone quarries in the six former provinces (Table XLII). In other words, Leopoldville province contributed a half and Katanga gave a quarter of these building and road building stones with the other four provinces sharing one quarter. The inadequate communications system failed to contribute to the country's coherence. Unfortunately, the Ten Year Plan only achieved a fraction of its highway building objectives (but all the projected expenses). Consequently, as soon as the economy again reaches its point of 'take off', the quarrying industry will be involved in another expansive phase.

At Sumbi, near the border of the Cabinda enclave, BAMOCO, the predecessor of BAUXI-

TABLE XLII

CONGO QUARRY PRODUCTS IN 1,000 TONS (1958)

Province	*Sandstone Quartzite*	*Granite*	*Limestone*	*Gravel and sand*	*Ballast*	*Large stone*	*Crushed*	*Gravel*	*Sand*
Léopoldville	1,200	150	100	310					
Equator	3,000			70,000					
Katanga					160	60	240	50	130
Kasaï					50	80	3	50	60
Kivu						50	15	3	15
Oriental						25	15	25	7

CONGO, has explored and planned the development of 50 million tons of ferrugineous *bauxites* in conjunction with the Inga hydropower project. FORMINIÈRE owns half of the $500,000 equity of BAUXICONGO (the Société de Recherches et d'Exploitation des Bauxites du Congo) with other members of the Société Générale diarchy sharing most of the remainder. When exploitation starts the state will receive 20% of the shares. In addition to bauxite, we can enumerate other low grade resources of the Central Congo province and North Kivu and Kibali-Ituri (Table XLIII).

Although the solid fuel position of the Congo is weak, the country is distinguished by having the only hydrocarbon resources of Africa that are not confined to the coastal strips or the Sahara. Hydropower is both plentiful, well distributed in the vicinity of industrial complexes and inexpensive per unit cost. As we have already mentioned, the principal colliery is at Luena, the coal is of low grade and reserves are low. In northern Katanga, Greinerville Colliery, controlled by GÉOMINES, sells coal to the cement plant of Albertville where there are also other coal seams. Nevertheless, coal production decreased from 460,000 tons in 1956 to 80,000 tons in 1962. According to reports, there is some lignite near Thysville and peat in the Kivu Highlands. Here, gases escape from hot springs. In the same area, south of Lake Albert, some oil seeps. In the late fifties, the Union Chimique Belge evaluated 10 million cubic miles of methane and carbon dioxide, which are easily released from the depths of Lake Kivu. Experts believe that this cheap and renewable source can provide the energy requirements of the nascent industries of that area. At the other end of the country, SOBIASCO (a Société Générale subisidiary) quarried 800 tons of bituminous sands in 1958. In addition to the bitumen rights of Maniangu and Mavuma, a FORMINIÈRE-PETROBELGE venture explores oil in the Lower Congo. But the largest hydrocarbon resources are those of Stanleyville; sediments extend more than 6,000 square miles and average 30 gallons of oil products per ton. But the 12,000 barrel per day refinery of SOCIR, a subsidiary of ENI, will use imported crude at Banana.

TABLE XLIII

OCCURRENCE OF LOW GRADE RESOURCES IN THE CONGO

Potash	Lower Congo	Great Lakes area
		Virunga lavas (Lake Kivu)
Asbestos	Matadi area	Mahagi (Lake Albert)
Phosphate	Coastal strip	Mount Homa
Gypsum	Lovo, Bangu	

TABLE XLIV

CONGO HYDROPOWER

Potential		Installed capacity			
Location	*Million kW*	*Mining industry*	*kW*	*Non-mining ind.*	*kW*
Lower Congo	85	Union Minière	500,000	Lower Congo	60,000
Greater Inga	25	GÉOMINES	35,000	Stanleyville	20,000
Smaller Inga	3	SYMÉTAIN }		Bukavu	20,000
Rest of Congo	45	COBELMIN }	25,000	Albertville	20,000
		MGL }			
		Kilo-Moto	18,000		
		Bakwanga	16,000		
Congo total	130	Total	594,000	Total	120,000

However, representing one quarter of the world total, the *hydropower* resources of the Congo are too extensive (Table XLIV).

Traditionally in the Congo, hydropower generation has been the 'hand maiden' of mining. During the last decade, installed capacity doubled and we hope that, with its mineral wealth intact, the country will soon join again the road of development.

ZAMBIA (NORTHERN RHODESIA)

Encompassing the second largest copper resources in an area of 80 by 30 miles. Zambia ranks third as a producer. In this country of 290,000 square miles and 2.5 million inhabitants, 45,000 people work in the copper mines which contribute 90% of the country's foreign exchange. There is also cobalt (a byproduct of copper), zinc, lead, manganese and iron. Since the government encourages their exploitation, its establishment in 1962 produced an influx of capital.

THE COPPERBELT

At the southeastern end of a 300 mile long band extending from Kolwezi, the Rhodesian Copperbelt consists of two belts. The Mufulira-*Bwana-Mkubwa* belt in the northeast borders the Katanga Pedicle (i.e., the Congo frontier). Running parallel 20 miles southwest, the other belt comprises *Bancroft*, *Nchanga*, Chambishi, *Nkana*, Chibuluma, Baluba and Roan Antelope. There is another division: whereas the Anglo-American group manages those in italics, the Rhodesian Selection Trust controls the others. These groups followed, in 1925, in the steps of the Emperors of the Lunda and Bemba Chiefs who for centuries maintained an active trade in copper. In 1902, Africans led W. C. Collier and J. J. Donohue to the ancient workings of Roan Antelope and Bwana Mkubwa. These prospectors recognized the old workings of Chambishi in 1903. While already in 1906, an unknown explorer had pegged a claim on Nkana, only in 1923 and 1924 did prospectors of the Rhodesia Congo Border Concession discover Bancroft and Nchanga. Also, in 1924, His Majesty's Government took over the administration of the present Zambia from the British South Africa Company (BSA) which was founded by Cecil Rhodes. In the following years, R. J. Parker, A. Gray, R. Brooks, J. A. Bancroft, T. D. Guernsey and other geologists directed

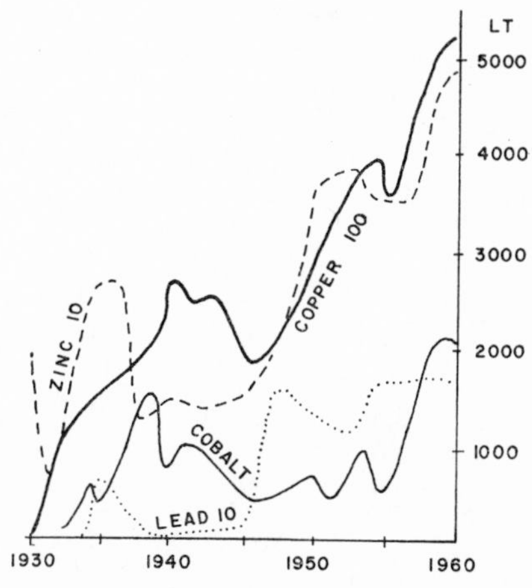

Fig. 30. Northern Rhodesia production of cobalt, copper, lead and zinc.

the evaluation of the deposits, the last of which, Chibuluma, was only drilled in 1939. Meanwhile, in 1913, Bwana Mkubwa commenced operations (to be closed in 1931); followed by Roan Antelope and N'kana in 1931; by Mufulira in 1933; by Nchanga in 1936; and then, after two decades, by Chibuluma in 1955 and by Bancroft in 1957; Baluba still awaites exploitation (Fig. 86). Conversely, productive capacity increased to 700,000 tons.

The country's output (Fig. 30) developed thus: 240,000 tons in 1940; 300,000 tons in 1950; 400,000 tons in 1955; 630,000 tons in 1960 (1961). At this rate of expansion, present reserves will suffice for only one generation. However, with evaluation continuing, reserves decrease only slightly.

The 1959–1960 results of the copper mines can be assessed from Table XLV. Whereas, both in 1959 and in 1960, Rhodesian Anglo-American Ltd. (RHOANGLO) derived a profit of $20 million from its mining companies, the Rhodesian Selection Trust obtained a revenue

TABLE XLV

ZAMBIA COPPER MINES

Mines	*Capital $ million*	*Copper reserves produced*			*1962–1963 results*				
		%	*million tons*		*ore mill.tons*	*copper 1,000 tons*	*$ million*		
							value	*profit*	*tax*
Rhoanglo	211		17.8	4.4	11.8	348	234	86	25
Nchanga	78	4.4	10.2	1.8	4.3	194	150	50	18
Bancroft	38	3.7	3.9	0.2	1.9	47	26	3	
Nkana	75	2.9	3.6	2.4	5.6	107	58	33	7
Bwana Mkubwa		4.0	0.1	0.03					
Selection Trust	167[1]		13.3	4.8	12.0	210	140	35	14[1]
Mufulira	52	3.3	5.9	2.3	5.3	100	80	25	10
Roan Antelope	62	2.9	2.7	2.3	6.1	90	50	8	2
Baluba		2.4	2.7						
Chambishi		3.4	1.2						
Chibuluma	3	4.6	0.5	0.2	0.6	20	10	2	
Total	378		31.1	9.2	23.8	558	374	121	39

[1] Includes corporation items.

TABLE XLVI

CAPITAL STRUCTURE IN % OF ZAMBIA COPPER COMPANY SUBSIDIARIES

Subsidiaries	*Rhoanglo*	*Rhokana*	*BSA*	*RST*
Nchanga	55	34		
Bancroft	19	43		
Rhokana	52.5	100		
Mufulira	3.6	26.6	4	64.7
Roan Antelope				100
Chambishi	3.6	26.7	4	64.7
Baluba	3.6	26.7	4	64.7
Chibuluma	3.6	26.7	4	65

of $7 million, mainly from Mufulira. By 1965, the productive capacity of this group will attain 300,000 tons. To understand the capital structure of the Copperbelt, we should further add that the Anglo-American group has a considerable minority share in most Rhodesian Selection Trust subsidiaries (Table XLVI).

Furthermore, the Anglo-American group is one of the principal shareholders of the British South Africa Company. Anglo-American also manages Chartered Explorations Ltd., in which its 40% interest is outnumbered by the 55% interest of the BSA; the Gold Fields group also have a 5% interest. The British South Africa Company has held the royalty rights in the Rhodesias and Bechuanaland until 1964. Until lately, a flat 10% remittance by the Copperbelt companies on the price of copper mined represented $30 million per year, i.e., an overwhelming part of its receipts. Out of this royalty, the company since 1949 transfered only 20% to the Zambia government, which now negotiates the re-purchase of these rights.

RHOANGLO and RST share the Rhodesia Congo Border Power Corporation. In 1961, a quarter of the power this company distributed still came from the Congo, but the share of the Kariba Dam is expected to increase further.

In 1963, refined shapes represented 76% of the copper output, blister 23% and others 5%. On the other hand, exports are widely distributed (including the Soviet Union), with the United Kingdom taking 30%, the Common Market 15%, Japan and India 5% each.

RHOANGLO GROUP

At Chingola 20 miles west of the centre of Kitwe, *Nchanga* is one of the world's largest copper deposits. While properties containing less than 1% copper were being developed in the Americas between 1955 and 1964, the high grade reserves of the three ore bodies of Nchanga have not failed to increase (Table XLVII).

In 1963, reserves increased further to 232 million channel tons. Table XLVIII shows that, similarly, output expanded spectacularly.

The 1963 output is similar to that of 1960, but profits after taxation decreased to $36 million in 1961 and to $31 million in 1963. The opening of the Nchanga Open Pit in 1957 and, to a smaller extent, the beginning of operations at the Chingola Open Pit in 1958 permitted this expansion. Indeed, the capacity of these installations is, respectively, in the range of

TABLE XLVII
NCHANGA COPPER RESERVES

Year	*Nchanga, Nchanga West million tons*	*% Cu*	*Chingola million tons*	*% Cu*	*River Lode million tons*	*% Cu*	*Total million tons*	*% Cu*
1956	146	4.74	2	7.00	2	4.21	150	4.76
1958	149	4.75	11	4.81	2	4.21	162	4.75
1963	214	4.36	15	4.89	4	3.86	232	4.39

TABLE XLVIII
NCHANGA OUTPUT

Year	*Ore milled million tons*	*Copper, %*		*Sales, 1,000 tons*		*$ million*	
		sulphide	*oxide*	*electrolytic*	*blister*	*sales*	*profits*
1956	3.2	2.88	1.96	95	27	100	70
1958	3.5	2.42	2.45	110	23	65	20
1962–1963	4.3	3.08	2.3	176	18	150	50

1,000,000 and 300,000 tons of ore per year. With this additional output, the RHOANGLO group was in a position to meet the shortfall and interruption of the production of Bancroft. At Nchanga Open Pit, a one ton bucket wheel excavator strips the overburden at a cost of 15 cents (US) per cubic yard. With one-third of the ore to be mined opencast, operations are planned to a depth of 800 ft. On the other hand, Nchanga West is successfully worked by the continuous longwall method. Not content with these results, the corporation has invested $10 million in a new low grade oxide leach plant and a roasting plant which will improve extraction, further increase output by 3,500 tons per year and recover 7,000 tons of copper per year from concentrates that were previously smelted in N'kana. After grinding and flotation, the sulphides are transported to N'kana and Roan Antelope, while oxides are leached at Nchanga. At Kitwe, reserves of *N'kana* evolved as shown in Table XLIX.

During the same period of time, sales increased 20%; in 1963, they stood at almost 96,000 tons of electrolytic copper and 13,000 tons of blister. Following the lean years of 1957 to 1959 (low copper prices, Bancroft), profits are again increasing. Miners drill long holes in the sub-level stoping method. In 1960, a new shaft costing $350 per ft. was commissioned at Mindola which is by far the larger of the two mines. The smelter receives its

TABLE XLIX
N'KANA RESERVES

Year	*Mindola*		*N'kana North*		*N'kana South*	
	million tons	*% Cu*	*million tons*	*% Cu*	*million tons*	*% Cu*
1956	68	3.37	29	3.02	26	2.64
1958	82	3.25	25	3.15	20	2.65
1960	81	3.15	24	3.06	15	2.62
1963	83	2.94	23	2.99	18	2.74

TABLE L
BANCROFT COPPER RESERVES (1963)

Location	Million tons	Copper %
Kirila Bomwe South	43	4.30
Kirila Bomwe North	22	4.10
Konkola	32	2.48

concentrates from the 475,000 ton per month crushing, grinding and flotations plant and from Nchanga and Bancroft. Each month, five reverbatory furnaces and the converter produce 9,000 tons of blister and 18,000 tons anode copper. On the other hand, with its 1,664 electrolytic cells, the refinery produces tough pitch high-conductivity copper. The RHOKANA and Nchanga companies each own $800,000 of the equity of Rhodesia Copper Refineries Ltd.

Bancroft, halfway between the border and Nchanga, has attracted considerable attention. Operations commenced in 1957; they were suspended between 1958 and 1959, the mine was flooded, and 1960 can be considered as the first full year of operations, and of maiden profits. In 1961, profits reached $5 million. Table L shows the reserves. To contend with changing dips, widths and difficult ground conditions, the management adopted flexible mining methods. Since the ore is wet and, at both walls, sandy, the washing plant operates underground and 30% of the material that is less than three-sixteenths of an inch in diameter is pumped to the surface. This arrangement increases the relative capacity of shaft system No.1 of Kirila Bomwe South and ultimately will permit the expansion of the mill from its monthly 150,000 ton capacity to 300,000 tons. While the sulphide/oxide ratio still attained 1 in 1960, the development of Kirila Bomwe North will provide harder ore. At the present, the whole production is blister.

RHODESIAN SELECTION TRUST GROUP

Sir Chester Beatty, a mining engineer of American origin organized the Selection Trust Ltd., which registered the Rhodesian Selection Trust in 1928 in London. In 1930, an exchange of holdings took place with the predecessor of American Metal Climax Ltd. (AMAX). Consequently, Seltrust Investments Ltd. and the Selection Trust Ltd. own 12.3% of the shares of AMAX, with the Hochschild family holding 10%. On the other hand, with 43.5% of outstanding shares, AMAX is the principal shareholder of RST.

Twenty miles north of Kitwe, *Mufulira* is the second largest deposit of the Copperbelt. While output and costs per ton each increased by 20% between 1955 and 1960, copper prices lost 10%. Consequently, in ten years, costs doubled. However, costs were reduced in 1962. In 1963, 5.9 million tons of ore were milled (compared to 4.1 million tons in 1959), but with the sinking of five new shafts and development of Mufulira West, output is expected to further increase by 50%. Depending on their thickness, miners block-cave or open-stope three ore bodies. The caving of the hangingwall is controlled and stations can pump up to 20 million gallons of water per day. Ore crushed underground to less than six inches is transferred to Symons crushers, ball mills and froth flotation cells. In response to the development of Mufulira West, the concentrator has been expanded from 400,000 to 600,000 tons per month. From reverberatories and converters, blister reaches two anode furnaces whence it is moved to 976 purifying tanks. In addition to refined shapes, the company also produced 34,000 tons of other and unrefined copper in 1963.

The Mufulira and Roan Antelope companies have a substantial interest in Kadola, Luapula, Mwinilungwa and Chisangwa Mines Ltd., and each has a 42% interest in Rhodesian Selection Trust Exploration Ltd.

Roan Antelope is situated at Luanshya, 20 miles southeast of Kitwe and somewhat nearer to the airport of Ndola (Bwana Mkubwa) in the north. Since 1953, output has remained similar but smelting costs have doubled. Here, mining covered five miles and the new MacLaren shaft is almost a mile west of the Irwin shaft which, in turn, is located less than three miles from the Storke shaft. Miners develop the main level mainly in the footwall and electrical trolleys haul the ore. The ground ore moves to Dorr classifiers, hydrocyclones and then to Agitair and Mineral Agitation flotation.

In 1959, the plant switched from blister to anodes which are treated at Ndola. The 1960 production consisted of 67,000 tons refined shapes, 20,000 tons blister and 5,000 tons cathodes (1961: 90,000 tons anode copper). RST owns two-thirds of the equity of the Ndola refinery with British Insulated Callender's Cables holding the remainder. On a 90,000 ton per year throughput, the refinery makes a $1 million profit. Consequently its capacity has been doubled.

In addition to the 7.4 million tons (5.03% copper) of *Chibuluma*, west of Nkana, at Kalulushi town, 2.4 million tons (4.47%) were available 9,000 ft. west. With its copper and cobalt production, the company repaid the $14 million loan it received from the United States General Services Administration before starting operations in 1957. Miners use back stoping, bench and trail stoping and cut and fill reclamation of rib pillars. The Mufulira plant recovers blister from the Chibuluma concentrates. Table LI gives the level of costs of the group in 1963.

The development of Chambishi commenced in 1962. The ultimate cost is expected to attain $20 million out of which $3 million have been spent on evaluation. Commencing partial production in 1965, the mine will produce 28,000 tons of copper each year.

TABLE LI

SELECTION TRUST COSTS, IN DOLLARS PER TON IN 1963

	Electrolytic		*Fine refinable*	*Rise of costs in % 1953–1963 (1953=100%)*	
	Roan	*Mufulira*	*Chibuluma*	*Roan*	*Mufulira*
Mining	270	210	210	140	130
Milling	50	50	70	120	110
Smelting				220	150
Metallurgy	42	20	29		
Royalty	60	60	60		
Railway	42	43	43		
Shipping	13	13	13		
Grand total[1]	500	410	450		

[1] Includes overhead.

MINOR METALS

Although not on the same scale as those of the Congo, the *cobalt* resources of the Copperbelt are considerable, and each year the production fetches almost $3 million (Table LII).

TABLE LII

PRODUCTION AND RESERVES OF COBALT IN THE COPPERBELT

Mine	*Reserves 1962–1963*		*Output in tons*	
	% Co	*tons Co*	*prior to 1960*	*1963*
N'kana	0.18	220,000	25,000	1,400
Baluba	0.16	180,000		
Chibuluma	0.15	15,000	2,000	300
Total		415,000	27,000	1,700

In addition to these, cobalt also shows at Bancroft. Since 1933, each year *N'kana*, the oldest producer, has extracted 1,100–1,300 tons from 0.11% grade ore. In the plant, concentrates containing 3% cobalt are roasted, leached, evaporated and purified. Then by rising the pH iron, copper and finally cobalt precipitates, and are recovered by electrolysis. The rated capacity of the plants is in excess of 2,000 tons per year.

Baluba, as yet undeveloped, is located northwest of Roan Antelope. *Chibuluma* commenced cobalt production in 1959 to meet what were then considered to be the requirements of the United States Strategic Stockpile. The Chibuluma orebody boasts 0.21% cobalt, but Chibuluma West only contains 0.07%. From the flotation cells, concentrates containing 3—3.5% cobalt move to cleaners and filters whence they are transported to the Ndola refinery (owned by Roan Antelope and Callender's). The matte is refined in Belgium. This brings the total cost of a pound of cobalt delivered in the United States to $1.70.

Between 1957 and 1959, N'kana also produced 110 tons of uranium oxide and the *selenium* production (recovered in Belgium) increased from 12 tons each in 1957 and 1958 to 16 tons in 1959 and 31 tons in 1963. In addition to these, abundant resources of *pyrite* are available both in the upper beds and above.

A broad band of 300 copper occurrences extends west and south of the Copperbelt from Mwinilungwa on the border of the Congo and Angola beyond Lusaka. In 1957, the only producer of some importance, Kansanshi south of Jadotville, was flooded. Since 1955, Mtuga northeast of Broken Hill has produced 700 tons of copper in certain years. Anglo-American owns 43% of the equity of the Kansanshi company, Tanganyika Concessions and an associate 42%, the Selection Trust 10% and the British South Africa Company 5%. Northwest of Lusaka in many of these deposits, such as the Silver King, Matala, King Edward and others, gold and silver are associated with copper. Since 1957, the country's gold production has increased from 3,000 to 5,000 ounces. Between 1956 and 1958, *silver* output oscillated around 600,000 ounces, and reached the mark of 950,000 ounces in 1959 and 880,000 ounces in 1963. However, most of this amount is recovered from refinery slimes and not from small mines. Indeed, between 1958 and 1960, the silver production of Broken Hill ranged from 50,000 to 260,000 and 140,000 ounces.

BROKEN HILL

In the geographic centre of the country, the Broken Hill deposits were discovered in 1902. The Rhodesia Broken Hill Development Company established a *lead* furnace in 1915 and a plant to recover electrolytic *zinc* in 1928. With a capital of $10 million, the company lately sold a similar value of metals each year, making a profit of $3 million. However, prices are affected by the base metal markets. Cadmium and sulphuric acid production was 130 and 57,000 tons respectively in the period 1957–1958, and reached the level of 20 and 18,000 tons respectively in 1961. The zinc and lead position can be gauged from Table LIII.

The company is a subsidiary of the Anglo-American Corporation. Consequently, it was fortunate enough to be assured of its outlets in southern Africa. As a result, $11 million was invested in Imperial Smelting furnaces which will double the capacity. After its 1943 peak of 800 tons, vanadium production ceased in 1953. On the other hand, at Star Zinc the company has proved 100,000 tons of ore containing 20% zinc (but no lead).

OTHER PRODUCTS

There are three other products whose yearly value is in excess of $1 million: hydropower, manganese and cement. Without the coal of Wankie and power imported from Katanga, the Copperbelt could hardly have operated. The coal basins of Kandabwe, Gwembe and Luano could not compete with Wankie whence 700,000 tons of coal were imported in 1961. The $300 million hydropower plant of Kariba started operations in 1960. The first power house is located in Southern Rhodesia, but this country uses little of the supply. With the increase of the installed capacity the cost of power will fall. In 1961, a 260,000 kW-capacity could supply the Copperbelt's maximum demand. Out of the total generation of 1.8 million kWh, the Kariba plant produced 1 million, 460,000 came from Katanga, and the remainder was supplied by local plants.

Manganese exports have oscillated from 40,000 tons in 1956 to 50,000 tons in 1958 and 35,000 tons in 1963 (representing in 1963 a value of $1 million). With 30,000 tons, Gypsum Industries of South Africa is the principal producer in the Broken Hill area. North of Lake Bangwelu near Fort Rosebery, the Rhodesian Vanadium Corporation (a subsidiary of the American Vanadium Corporation) was the greatest producer (Table LIV).

The Premier Portland Cement Company dominates the *limestone* industry. In 1963,

TABLE LIII

BROKEN HILL PRODUCTION OF ZINC AND LEAD

	Reserves 1961			*Production*		
	million tons	*grades %*		*1955*	*1960*	*1963, tons*
Proved	3.7		Ore milled tons	130,000	180,000	
Indicated	2.4		Zinc	27.9%	29.8%	49,000
Total	6.1		Lead	16 %	14.4%	19,000
Oxide	2.7					
Sulphide	3.4					
Zinc		27.9				
Lead		14.4				

TABLE LIV

ZAMBIA MANGANESE PRODUCTION IN 1959

	Tons	% Mn		Tons	% Mn
Mashimba	16,000	52.9	Bahati	3,000	47.2
Kampumba	18,000	50.0	Lubemba	6,000	17.2
Chiwefwe	9,000	45.0	Areius		
Luano	4,000	47.0	Kemparembe	2,000	
MMNR	4,000	52.2	Fanie's		

out of a total of half a million tons, 250,000 tons were used in the manufacture of cement (a lower figure because of the completion of the Kariba Dam), 200,000 tons as a flux in the copper mills and 30,000 tons for lime burning. At Chilanga south of Lusaka, the largest producer boasts reserves of 23 million tons of limestone; the Rio Tinto group has brought its Ndola reserves to the Northern Rhodesia Lime Company. Consequently, the company invested $1 million in expansion.

Although not as extensive as those of Katanga, *iron* resources are considerable. Out of 350 million tons of ore, 200 million are concentrated west of Lusaka at Nambala (Table LV).

In the southeast in the Nchoma district, a few tons of tin are produced. Tin also shows in the Chambeshi drainage in the north. The niobium–rare earth–thorium deposits of Nkumbwa in the northeast and of the Feia district (where the Zambezi leaves the country) are extensive but not competitive. However, apatite is available in these complexes (see phosphate chapter). West of Feia, lapidaries recover emeralds and on Lake Mweru, workers win salt. Beryl is found in many localities, but mainly at Lundazi, roughly, halfway from the eastern border line. North of there, on the Lunzi, there are barytes, and southwest, in the Petauke district, there is graphite. In 1963, 17 tons of amethyst were recovered.

TABLE LV

COMPOSITION OF THE ZAMBIA IRON ORES

Deposit	Approximate reserves million tons	Iron %	Impurities %
Nambala	200	65	7 silica
Kampumba	56	61	1–24 silica
King Edward Mount	10	55	0.1 sulphur
Sanje	7	67	0.2 phosphorus 0.7 sulphur
Chinda	5	70	0.2 phosphorus
Shimuyoka	5	69	6 silica
Cheta	5	65	0.1 sulphur
Namantombwa	4	70	0.3 phosphorus
Nabutali	3	68	0.2 phosphorus
Venter's Ridge	0.5	67	0.02 sulphur
Strauss' Ridge	small	65	0.1 sulphur
Nagaibwa	30		
Pamba	12		
Chongwe	10		
Chisamba	0.5		

BURUNDI

At the southernmost source of the Nile, and at the northern tip of Lake Tanganyika, the king or Mwami rules Burundi, a country of 10,000 square miles and 3 million inhabitants. In Bujumbura, the capital, the government invites investments. Compared to the coffee crop, the country's mineral production is limited, mainly to *gold*. On the Congo/Nile divide (Fig. 51) (east of the frontier of the Congo), the annual gold output of MINÉTAIN, the Société des Mines d'Étain du Ruanda et de l'Urundi (with headquarters in Kigali, Rwanda, q.vid.), and of its lessees has decreased from 3,000 ounces in the fifties ($100,000) to 1,500 ounces in 1960 and 1,000 ounces in 1961 and 30 ounces in 1962 because of clandestine operations. SOMUKI, the Société Minière de Muhinga et de Kigali, is a much smaller producer. Some gold is also found east and south of Bujumbura (where some oil also seeps). Furthermore, every ounce of gold bullion contains 10% silver. East of and between the Congo/Nile gold belts, in the districts of Ngozi and Bubanza, are scattered tin and tantalite–columbite occurrences, the most famous of which are Mogare and Ndora. In this area, MIRUDI, the Société Minière du Ruanda-Urundi, a subsidiary of Compagnie Minière des Grands Lacs Africains (MGL, Empain group), has operated. There is also some tin east of Bujumbura in the district of Muramvia, near Muhinga and southwest of the Mbuye deposits of Rwanda, but production only attains 30–50 tons per year.

At Karonge southeast of Bujumbura, SOMUKI used to produce 300 tons of bastnaesite each year. Although reserves are adequate, the mine cannot compete with *rare earths* recovered from beach sands and at Mountain Pass in California. Consequently, large scale production ceased in 1957. Though limestone is available north of Bujumbura and in the Maragarasi basin (southeast), builders import cement from the Congo. Similarly, although abundant hydropower can be tapped in the Maragarasi Valley and in the streams falling from the highlands, Bujumbura uses the plant of the Ruzizi, near Bukavu, Congo.

RWANDA

The Republic of Rwanda differs from its southern neighbour, Burundi, in many respects. From Kigali, the new capital, the government administers 10,000 square miles of high mountains with a population of 3 million. Rwanda produces ten times as many minerals as Burundi, which are trucked to Kampala in Uganda. From Bujumbura, communications flow towards the Tanganyika Railway and Dar es Salaam harbour. The government gives ample encouragement to the struggling mining industry which is divided into relatively small companies operating scattered deposits with a limited capital.

By 1926, the exploration team headed by the Abbé Salée had identified tin and rare metals, which are still the mainstay of the industry. In 1941, when the Malayan tin mines were not available to the Allies, Rwanda mineral production reached its first peak and declined thence until 1944. It reached its absolute maximum by 1955 (the Korean War); but its value has declined ever since (from $6 million in 1956 to $5 million and $4 million in 1957 and 1958). Furthermore, during the struggle of the Bahutu and the tall Watutsi royalists, the country suffered considerably. Consequently, output plummeted. But we will now describe their situation during the last year of production towards which we hope

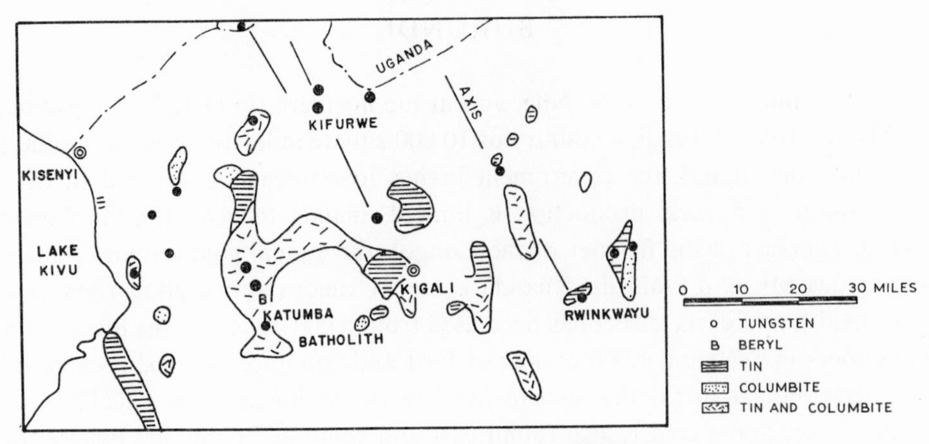

Fig. 31. The mineral resources of Rwanda.

they will soon return. Central and eastern Rwanda boast significant resources of tin and rare metals (Fig. 31) which are in excess of those of South Africa and which, in 1960, produced an income in excess of $3 million. While tinstone normally represents 80–90% of this value, during the war years high tungsten (and columbium–tantalum) prices accounted for sales of $1 million. The 1962 results compare thus to the 1951–1955 average: tin, tungsten ore and tantalite, respectively, decreased from 2,200 to 1,600, from 660 to 160, and from 100 to 50 tons. Gold decreased from 800 to 300 ounces, but beryl and lithium ore *increased* from 70 to 400 and from 400 to 500 tons. We can distinguish four principal mining areas: Katumba (or Gatumba, between Kigali and Lake Kivu); Kigali; Rwinkwavu (Kibungu) further east; and Ruhengeri, on the Uganda frontier. Furthermore, gold shows in the extension of the Congo/Nile divide of Burundi, as it does east of the tungsten mines of Kifurwe. East of Lake Kivu, some tin, tungsten and gold also occur (Fig. 51).

From the placers of the Nyabarongo, Lake Muhazi and Rwinkwavu, workers recover *tinstone*; at the last locality, and at Rutongo and Katumba, they also exploit veins and pegmatites. In the Kigali district, wolframite is associated with tin. At Kifurwe, the continent's largest deposit, and at Bugambira, *tungsten* ores are found in veins; on Lake Bulera (in the north), in veins and placers; and on Lake Rweru and Giciye, mainly in placers.

TABLE LVI

MINING IN RWANDA IN 1958

Miners	*Area*	*Tin conc. tons*	*Tantalite tons*	*Wolframite tons*	*Number of employees*
Minétain	Katumba, Kibungu	700	66	50	4,000
Somuki	Kigali	650			1,600
Géoruanda	Rwinkwavu	470			1,000
Corem	Kigali	170	0.1		400
Marchal	Kifurwe	50		70	600
Stinclhamber	Bugarama			90	300
Mirudi	Katumba	30	3		300
Others	North	10	0.1	20	100
Total		2,080	70	230	8,300
Value $		3,100,000	180,000	180,000	

TABLE LVII

PRODUCTION OF MINÉTAIN IN TONS

Year	*Tin*	*Columbite*	*Wolframite*	*Beryl*	*Lithium ore*
1960	480	50	40	230	2,600
1961	510	50	100	480	2,000
1962	550	50	70	400	350

Miners have extracted most of the columbite–tantalite from the placers of Katumba, Bijojo, Mutaho and Nyungwe, but the primary deposits have rarely yielded much of these ores (Mbuye). Although some of them were found at Katumba, the search for *beryl* and lithium ore (amblygonite) has not yet been as successful as in neighbouring Uganda. Nevertheless, Rwanda is an important producer of beryl; the demand is steady, and the production is increasing and could be increased further. Finally, some bismuth shows in the Katumba district as do traces of uranium at Kanyaru.

MINÉTAIN, the Société des Mines d'Étain du Ruanda et de l'Urundi, dominates the scene. Others include SOMUKI (the Société Minière de Muhinga et de Kigali), GÉORUANDA (the Société géologique et minière du Ruanda, which is a subsidiary of the GÉOMINES company which operates the Manono deposit), COREM (the Compagnie de recherches minières), MIRUDI (the Société minière du Ruanda et de l'Urundi, Empain group, which has suspended operations), and private miners like Marchal and Stinclhamber (Table LVI).

In 1959, MIRUDI and MINÉTAIN respectively produced 340 and 120 tons of beryl valued at more than $1 million, and, since the beginning of operations, MINÉTAIN has recovered 10,000 tons of amblygonite, a lithium ore. Furthermore, GÉORUANDA and MINÉTAIN produce some gold.

In 1929, MINÉTAIN, the largest company, was founded with a capital of $1 million. After numerous changes, the capital now stands at $3 million (at the official rate of exchange). The Société Métallurgique of Hoboken (Belgium) owns 14,000 shares out of 190,000 issued, and MINÉTAIN holds many of its own shares. On the other hand, Union Minière du Haut Katanga is in possession of one-third of the shares of Société Métallurgique and, in effect, the Société Générale de Belgique controls MINÉTAIN (Table LVII). The 1961 output sold for $900,000; consequently, the deficit was reduced from $140,000 to 12,000. By 1961, the hydropower line of the Taruka plant extended from Musha to the N'tunga mine, but market conditions limit tungsten production. From the tailings of the tin–columbite mines of Katumba and from Nyarigamba, workers recover more and more beryl, but the amblygonite deposit of Rongi is at the end of its useful life.

UGANDA

Almost seven million people inhabit the 94,000 square miles of Uganda, which is a member of the Commonwealth and of the East African Organization; it gives ample guarantees to the investor. The capitals, Kampala and Entebbe, are situated near Lake Victoria; however, most of the mining activity is carried out (and expanding) in the western provinces of Toro, Kigezi and Ankole. Indeed, by 1960, the value of the annual output reached the level of $8.5 million (compared to a government revenue of $19 million).

THE WESTERN PROVINCES

The yearly production of the Kilembe copper mine northeast of Lake Edward has been expanded to 17,000 tons of blister copper and its value has reached the mark of $10 million; however, all other mineral products (including cement) provide an income of only $2.5 million (Fig. 51). This compares with a national income of $50 million. In 1926, Tanganyika Concessions Ltd. (one of the major shareholders of the giant Union Minière du Haut-Katanga) started the exploration of the Kilembe prospect. It was brought to the stage of export production in 1956 by Kilembe Mines Ltd., a company owned by Kilembe Copper Cobalt Ltd. (of which the Falconbridge Nickel owns 70%. The Uganda Development Corporation and Colonial Development Corporation, that is the state, own 30%). The Company recently published an ore reserve of 14.6 million tons of which 8.2 million tons average a grade of 2.3% copper (i.e., 190,000 tons of metal) and 0.18% cobalt. Pyrite grades are very much in excess of traces of nickel.

In 1937, export production of *tungsten* ores started in the southwest and, in 1947, the yearly output passed the mark of 100 tons. Between 1937 and 1961, private wolfram miners generated an income of $4 million. However, the well-known fluctuations of the tungsten market and small scale operations control the output to which Nyamulilo contributes 70%, Kirwa gives 20%, and Bahati, Ruhizha, Mutolere and others (all near the Rwandese border) contribute the rest. Bismuth, a rare metal in Africa, occurs near and in the tungsten districts. Small operators have exported 50 tons from the area of Kayonza and Kitawulira since 1948, but the production of the Rwanzu mine near the Congolese border has been reduced. In 1962 output dipped to 12 tons.

Ugandan hopes are based on *beryl*. Prior to 1940, some beryl reached the export markets; by 1955, they reached the mark of 140 tons and then in 1960 (after a decline), they reached a peak of 430 tons or more than $1 million. Production expanded again in 1961, but 1,100 tons fetched only $400,000. More than 40 mines (including Bugangari) near the border of Rwanda contribute the beryl containing 10–12% BeO. However, some beryl is also found north of Lake Victoria. In 1962 output fetched only $300,000.

Between 1927 and 1960, 10,500 tons of tin concentrates were exported from north of the borderland of Rwanda and Tanganyika of which 3,300 tons came from Mwirasandu. However, since 1962, sales have increased to 120 tons per year producing foreign revenue of about $170,000. Lithium comes from the same pegmatite fields, but production has dwindled from its 1950 peak of 300 tons to just a few tons. The exploitation of the feldspars of Bulema, Nyabakweri and other areas is restricted by limited demand.

Between 1936 and 1960, private miners exported 200 tons of *columbite*–tantalite from Rwentali, Rwenkanga, Kashozo and Rwabuchacha and other small concessions of the southwest; however, production is now only 10 tons. Tantalum concentrates are also recovered at Bulema.

TORORO AREA

In the soils of the Sukulu complex north of Lake Victoria (Fig. 33) near the border of Kenya, geologists have estimated 300,000 tons of niobium oxide in 0.2% grade ore with additional but poorer resources. In the same soils are found 10 million tons of *phosphate* with, as in the preceding case, other resources. The Uganda Development Corporation plans to extract 125,000 tons of apatite each year and to process it into 25,000 tons of superphosphate with an additional 500 tons of niobium (columbium) oxide. Final plans call for a 250,000 ton per year programme that is ten times as much as at the neighbouring Busumbu Ridge of the Bukusu complex, the reserves of which are also proportionately smaller. The Busumbu Mining Company upgrades the ore from 25% P_2O_5 to a maximum of 35% P_2O_5 (and 5% iron) and exports it to Magadi, Kenya, where it is converted to soda phosphate (Fig. 52).

Furthermore, the Sukulu and adjacent Tororo complexes can be considered as deposits of magnesian *limestone*, baryte and vermiculite. The Uganda Lime Company produced 100,000 tons of cement in 1958 and 60,000 tons in 1961; lime production reached a peak of 17,000 tons in 1962. Kilembe Mines use the limestone of Hima River; other deposits of Lake George include Muhokya and Dura. In 1962, the value of cement and lime production was nearing the mark of $2 million. The soils of the North Valley of Bukusu contain 3.7% *baryte*, but grades of other valleys are smaller. Similarly, reserves of Mugatuzi Hill in the southwest are small. Geologists have evaluated a small segment of Namakera Hill (part of the Bukusu complex) to a depth of only 50 ft. and have estimated 130,000 tons of good quality *vermiculite* with an additional 250,000 tons of poorer grade. At Sukulu, 220 million tons of soil contain 0.25% zirconium oxide.

OTHER AREAS

Salt has been extracted since time immemorial. By 1958, a yearly output of 10,000 tons was reached at Katwe on Lake George (an annex of Lake Edward), but recovery was later reduced to 5,000 tons. Furthermore, salt is won from Lake Bunyampaka and from the hot springs of Kibiro (also in the west). While bromine is found in the lake, only little iodine shows there.

Like other shores of Lake Victoria, the beaches and islands of the north are also covered by high grade glass sands (98–99.9% silica!), e.g., Kome Island at Entebbe, and Kyaggwe. While pottery clays tend to be restricted to Buku and Mukono, Ugandans manufacture bricks and tiles from clays of many other localities such as Nansana, Kisubi and Bugungu.

According to statistics, *gold* was first exported in 1931 from the Buhweju field east of Lake Edward. By 1938, production had reached its peak of 22,000 ounces and was never to reach it again. Indeed, since 1949, output averages only 500 ounces, coming mainly from Amonikakine and Tira Busia in Bukedi; thus the 140,000 ounce output of the 1931–1961 period now belongs to the domain of the past. From the beds of the swamps of the Buhweju, where a diamond has also been found Africans collect piezo-electric quartz of low grade. The gold/silver ratio averages 10. Silver production has remained below its 2,000 ounces peak of 1938. In 1959, small scale production of lead and accessory zinc ceased at adjacent Kitaka and Kampono.

In 1944, 12 tons of low grade mica were exported from the remote Karamoja–Acholi area of the northeast, but production has ceased since. Prospectors have also discovered some mica around Kampala and Mahagi, north of Lake Albert. From the same areas, 50 tons of asbestos were recovered in 1950 and 10 tons in both 1954 and 1956. The chromite deposits of Moroto in Karamoja are unfortunately small as are widespread talc occurrences. Chemists have determined platinum in the chromite of Nakiloro.

Geologists' estimates of iron resources of southwestern Uganda are in excess of 30 million tons; at Sukulu and Bukusu, titaniferous iron ore reserves are of the same order of magnitude. South of Bukusu, thorite is found, and other thorium–rare earth–uranium minerals are scattered in placers and in pegmatites of the southwest.

Diatomite is restricted to Panyango in the northwestern corner of Uganda (where geologists have estimated a reserve of 75,000–100,000 tons of material under a cover of 15 ft.), and to Atar and Alui, north of Lake Albert. Not far from the northern borders, graphite is widely scattered; the mineral is more concentrated at Mobuku, near Kilembe, with gypsum showing at Mobuku and Lake George at a little distance. Kyanite is widespread; in the northwest, Erusi and Azi Hill respectively carry 10 and 80% of that mineral. The lavas of the borderland of the Congo and Rwanda include 5% potash. A little oil seeps from beds of the Rift Valley, especially south of Lake Edward. Further north at Buranga, geothermal steam can be drilled but only in depth.

The hydropower resources of the Victoria Nile are large; the capacity of Owens Falls attains 150,000 kW, some of which reach Kilembe and Kenya. Downstream, the Bujagali Falls could provide additional capacity. Ground-water resources are not adequate but drilling gives good results.

KENYA

Nine million people inhabit the 220,000 square mile area of Kenya. Population centres are located in the southern part of the country where the mineral industry developed

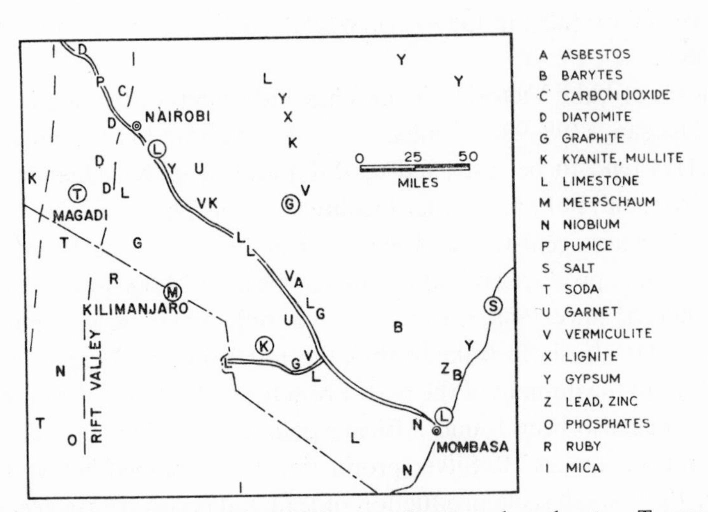

Fig. 32. The mineral resources of southeastern Kenya and northeastern Tanganyika.

(Fig. 32). Mining does not play as important a role in the largely agricultural economy of Kenya as it does in the former Rhodesias. Nevertheless, between 1960 and 1961, the value of the mineral product doubled, and now oscillates around $15 million.

INDUSTRIAL MINERALS

Out of this, the share of *soda* is $4 million. Since 1914, the Magadi Soda Company, affiliated with the Imperial Chemical Corporation, has worked the trona of Lake Magadi southwest of Nairobi, the capital. Production expanded from 2,500 tons in 1916–1917 to 50,000 tons in 1922, to 60,000 tons in 1941, to 100,000 tons in 1946–1947, and to 150,000 tons in 1961. Buckets dredge the trona, then on the pontoons, crushers and electromagnetic separators concentrate the material which is pumped to the ash plant on the shore. There, the liquor goes through trommels, settling cones and centrifuges, and finally undergoes calcination and pulverization. The ash contains 97% sodium carbonate; sodium bicarbonate (bath salt base) is also locally abundant. The company has also produced washing soda by recovering sodium fluoride from lake salts. Trona has also accumulated in the Elmenteita and Nakuru Lakes of the Rift Valley; unfortunately, these are often covered by water. Magadi is also the principal producer of *salt*, which is evaporated on the Fundi Isa peninsula 80 miles north of Mombasa. The recovery of salt generally oscillates around 20,000 tons per year, which corresponds to a value of $500,000.

In 1962, *cement* production reached almost 330,000 tons, i.e., a value of more than $5 million. At Bamburi, near Mombasa, the British Standard Portland Cement Company has a capacity of 200,000 tons and at Athi River, Nairobi, the East African Portland Cement Factory has already installed a capacity of 120,000 tons per year. At Bamburi, coral limestone, shale and gypsum from Roka (50 miles north) are used. On the other hand, the Athi factory uses limestone from Sultan Hamud on the Mombasa Road, volcanic ash from the Rift Valley and gypsum from the Tana River in the east. Workers also quarry limestone at Mombasa, Koru and Turoka, near Kajiado.

The development of extensive resources in the Tana River Basin, eastern Kenya, is unfortunately hampered by inadequate transport facilities. Gypsum from Tula Valley, near Garissa on the Tana, is used locally and transported to the coastal cement plants; it is also used for agricultural needs. Other deposits in the coast include Wajir, el-Wak, and Mandeca; these are all near to Somalia and Konza. Some gypsum is also recovered by evaporation at Gougoni, 80 miles north of Mombasa. The cement plant also uses gypsum from the neighbouring Mida Creek; other occurrences include Homa Mountain and Mombasa shales. The production of gypsum has steadily expanded from 80 tons in 1951 to more than 20,000 tons at the present and its value exceeds $200,000.

Extensive deposits of diatomite are located in depressions of the Rift Valley, at Kariandus and Gicheru, near Gilgil, at Koora, near Magadi, on the Soysambu Estates and other localities near Elmenteita, and at Nderit, Eburru and Subukia. The Kariandus deposit, the largest of those explored (Fig. 52), carries 84% silica. Locally, the product is used in soaps and it is also exported as a filtrate and insecticide. Output has increased from 2,000 tons in 1940 to almost 5,000 tons in 1958, attaining now 3,000–4,000 tons ($250,000). In 1946, *carbon dioxide* was tapped at Esegeri on the western rim of the Rift Valley (pressure attained 80 lb. per square inch), and in 1957, at Kerita on its eastern rim (under 35 lb. per square inch pressure). In the Kedong Valley and near Lake Magadi, other emanations are known to exist. Supply has increased from 300 tons in 1951 to 700 tons yearly since 1956, and its value reaches almost $150,000.

MINOR INDUSTRIAL PRODUCTS

The principal *kyanite* area is located between Tsavo and Taveta, east of Mount Kilimanjaro; it is accompanied by corundum. Since 1951, kyanite has been converted to mullite. In 1958, large scale operations started at Taveta. Until late 1958, this site produced 70,000 tons of kyanite and 30,000 tons of mullite, most of which was exported. Other deposits are known to exist nearby at Kevas and Loosoito, as well as in the Sultan Hamud area, south of Nairobi.

Graphite resources of Kenya are not as large as those of Madagascar; however, a steady production has been maintained since 1951, which reached its peak of 1,000 tons in 1957. Market conditions limit its further expansion to 2,000 tons per year. A company quarried a mine in southern Kitui, and other deposits are known in the Tsavo-Mtito Andei, Machakos, Voi, Namanga, Loperot and Merti areas. There were no exports in 1962.

For years, ceramic and glass quality *quartz* was quarried on Kinyiki Hill, near Sultan Hamud south of Nairobi, but piezo-electric quartz is scarce. At Kinyiki, workers have also recovered gem *sapphires*. Numerous occurrences of *kaolin* include those of Fort Hall, Machakos, Ndi, Eburru and Voi. At Kisii, east of Lake Victoria, some soapstone is sold. *Meerschaum* from the Amboseli Basin near the Kilimanjaro is of good quality, but the Nairobi factory manufactures most of the pipes from Tanganyika meerschaum. At Naivasha, Longonot, Menengai and in the rest of the Rift Valley, pumice is abundant with *pozzuolanic* materials occurring in the Highlands. From Kinyiki, workers have extracted magnesite, which also exists at Kipiponi, Magongo and other localities. At Baragoi, West Suk, Sultan Hamud, Tsavo, Taita, etc., miners have found and sporadically won mica and asbestos; vermiculite and talc come from the same areas. Titanium grades of the coastal sands are low. Natural *steam* has been used, e.g., at Eburru. Kenya, like most countries of Africa, lacks *water*, except for the southwest. Hydropower resources are only of the order of 130,000 kW. Frobisher Ltd. drills for oil in the northeast (Marararii) as does the Shell Development Corporation on the east coast. Shell and British Petroleum operate the Mombasa refinery (38,000 barrels per day).

METALS

Copper supplied by the Macalder mine, east of Lake Victoria near the Tanganyika border, has become the second mineral product of the country. In 1955, geologists evaluated reserves in excess of 30,000 tons of contained copper (grade 2%), 50,000 tons of zinc (grade – 3.5%), 140,000 ounces of gold. (1.6–2.7 dwt. per ton) and 100,000 ounces of silver (grade 1–1.8 dwt. per ton). Production of zinc and copper ores started in 1951; cement copper, gold and silver have been produced since 1956. By 1961, copper output neared 3,000 tons and $1.5 million. Almost a thousand tons of zinc were produced in 1951–1952. Copper also occurs at Kitere near Macalder; lead, zinc and baryte accompany copper at Vitengeni, north of Mombasa. Since 1926, official records have tabulated an output of gold with a cumulative value attaining $20 million. The principal deposits surround Lake Victoria with minor occurrences at Kibigori in the Rift Valley and at Kisii. Reserves have decreased, consequently output has now diminished to 9,000 ounces per year. On the other hand, silver extraction, also associated with lead in the copper mine, has witnessed an expansion varying between 50,000–20,000 ounces per year.

TABLE LVIII

TANGANYIKA OUTPUT IN $ 1,000

Minerals	1926 $	1940 $	1950 $	1961 $
Diamonds	100	40	2,100	14,000
Gold	80	3,400	2,300	3,600
Salt	50	150	320	900
Tin	20	180	210	600
Others	80	30	1,070	900
Total	330	3,800	6,000	20,000

Homa Mountain near Lake Victoria includes 10 million tons of iron ore per 100 ft. of depth. At Mrima, south of Mombasa, geologists have evaluated only 100,000 tons of manganese and 400,000 tons of niobium (columbium), much of it unrecoverable. Finally, at Machakos, workers have recovered some beryl, a mineral which also shows at Baragoi, Sebit and in the far north.

TANGANYIKA

Tanganyika covers 360,000 square miles and the republic has ten million inhabitants. The principal city is Dar es Salaam, the capital and main harbour. The government of TANU (the Tanganyika National Union) provided a highly favourable investment climate. Africans have smelted iron ore for centuries and, between 1884 and 1918, $1.4 million worth of gold and mica were recovered under German direction. But industrial production only started in 1923 and a glance at Table LVIII gives an idea of its growth.

Mining thus represents one quarter of the revenues of the Republic. Since the Mpanda mine was closed in 1960, copper, lead and silver production valued at $2.3 million each year has now ceased. While cement, phosphates and gypsum can supplement the mineral income, the exploitation of distant coal resources would be of major benefit to the country.

DIAMONDS

It is said that the first diamond was found in Tanganyika in 1913 near what became the Mabuki mine in 1925 (Fig. 33). The gravels of Mabuki have contributed 95,000 carats; the neighbouring Kizumbi, Usongo and Uduhe occurrences reportedly have produced only 17,000 carats (Fig. 106). The kimberlite underlying these fields has not proved to be payable. In 1940, J. T. Williamson, the Canadian geologist, a pathetic and lonely figure, discovered a large deposit at Mwadui in the same area south of Lake Victoria. The production of Mwadui passed the mark of 100,000 carats in 1945, 300,000 carats in 1952, 500,000 carats in 1958 and 670,000 carats in 1962. However, between 1952 and 1959, the average value of the carat decreased from $34 to $23, and the average weight of stones simultaneously decreased from 0.48 carat to half of this amount. Since mill capacity was increased to 7,000 tons per day in 1956, a continuous high output can be expected, at least until grades start decreasing in the pipe (Fig. 107). Development has reached the 1,200 ft. level.

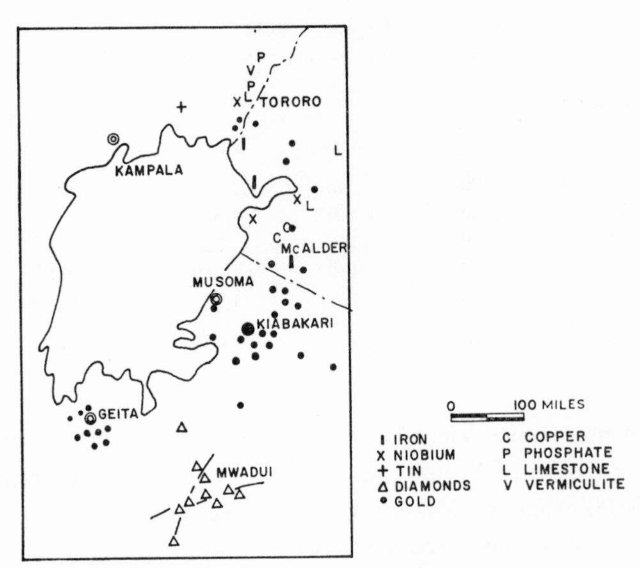

Fig. 33. The mineral resources of the Lake Victoria area.

After the owner's death, the De Beers group purchased the 1,200 shares of Williamson Diamonds Ltd. and then transferred 320 shares to the Tanganyika government to pay the estate duty and inheritance tax levied on Dr. Williamson's shares. Furthermore, De Beers sold, for $3.5 million, 250 additional shares to the government on a loan repayable with 6% interest out of the dividends of the state's 600 shares. With a $3 million loan by De Beers Prospection Ltd., the company is actively exploring southwestern Tanganyika.

Adjacent to Mwadui, the new plant of Alamasi cannot cover costs and the company has obtained tax concessions. Production is in excess of 20,000 carats.

OTHER IMPORTANT MINERALS

This category includes gold and silver, salt and soda, and tin with its accessory, tungsten.

Gold is found in five areas: Musoma and Geita (respectively east and south of Lake Victoria), Iramba (farther southeast), Lupa (northwest of Lake Nyasa), and Mpanda (east of Lake Tanganyika) (see Fig. 52 and Table LIX). In 1962 output attained 100,000 ounces.

As already mentioned, Mpanda ceased operations after having produced 2.5 million ounces of silver. During the last 20 years, the annual output of the Iramba-Sekenke field has been only twice in excess of 1,000 ounces of gold; it has now reached the low of 40

TABLE LIX

TANGANYIKA GOLD AND SILVER OUTPUT IN OUNCES

Year	Geita		Musoma		Lupa	
	gold	*silver*	*gold*	*silver*	*gold*	*silver*
1939	40,000		40,000		48,000	
1949	35,000	8,000	10,000	1,000	21,000	17,000
1959	44,000	17,000	38,000	4,000	3,000	600

ounces. The Kiabakari mine, east of Lake Victoria, holds the largest reserves. Estimated reserves of Geita, the second producer, are however decreasing, as are its profits. The Geita Gold Mining Company is indebted by $1.3 million (compared with its authorized capital of $2 million). In 1958, Kentan Gold Areas sold 78.63% of the Geita shares to Zambesia Exploration Company, a wholly owned subsidiary of Tanganyika Concessions Ltd.

Ever since its inception, *salt* production has steadily increased and has more than doubled in each decade; however, the output still does not cover consumption and the export market of the Congo and Zambia. Nyanza Salt Mines Ltd. extract, from the brines of Uvinza near Lake Tanganyika, more than half of the Republic's 35,000 ton yearly production and ship 7,000 tons to the Congo. While some salt is extracted at Ivuna north of Lake Nyasa, a majority of the remainder is extracted from the sea by the saltworks of Tanga, and from eight operations at Bagamoyo, Dar es Salaam and Kunduchi and from three pans at Lindi and Mtwara in the south.

As at Lake Magadi in adjacent Kenya, soda salts also accumulate in the Tanganyika section of the Rift Valley (150 million tons at Lake Natron and 500,000 tons at Lake Balangida); however, the costly fractional crystallization process the mixed salts necessitate is not yet capable of competing with the Kenya products. As to potash, somewhat more of this substance shows in the saline springs of the Western Rift Valley than in the Central Rift.

In 1924, prospectors discovered *tin* at Kyerwa, not far from the larger Rwanda tin fields. Production reached its peak of 400 tons in 1938, and averaged 60 tons per annum during the last decade. With the inauguration, in 1960, of a new plant capable of recovering 240 tons of the low grade ores of Kyerwa, production has increased again. Operations on the other sites (Kaitambuzi, Katera, the Ilama and Murongo areas) are on a smaller scale than those of Kyerwa and the total wolframite production of Karugu and Chamuyana adds up to only 260 tons.

Mica mining has a tradition dating back to the year 1902 of the former German East Africa. Small companies and private miners still operate with small capital and inadequate investments, and this is reflected in the production data. Since the Second World War, output of sheet mica has averaged 80 tons per annum; its value has been influenced by market conditions and it decreased from a peak of $260,000 to as low as $150,000. During the same period, the miners also sold 1,000 tons of low-priced waste mica. During the last years, the companies transferred production into the hands of Africans who now can sell most of their output to the graders, F. F. Christien and Company and New African Mica Company at Morogoro. Near that centre, northwest of Dar es Salaam, are situated the principal fields Uluguru and Mikese; Kisitwe, Mangalisa and Rubeho lie somewhat farther away.

Between 1957 and 1959, some vermiculite of Mikese was exfoliated at Dar es Salaam; other occurrences of this mineral are found at Kwekivu and Lufusi. Furthermore, a few tons of beryl with mica have been recovered from Makanjiro and Sibweza. Similarly, feldspar is available in these rocks and near Dodoma.

LARGELY UNTAPPED RESOURCES

Tanganyika boasts coal resources which are large but unfavourably situated even by African standards. In the Ruhuhu field east of Lake Nyasa, the Colonial Development

Corporation has estimated 100 million long tons at Mbalawala, 15 million long tons at North Ngaka and 200 million long tons at Mchuchuma with other seams awaiting evaluation. Ruhuhu coals average 15% ash and a calorific value of 12,800 B.T.U. per lb.; in the Songwe-Kiwira field north of the Lake, 20 million long tons of fuel cannot boast as high a calorific value. Reserves of the Galula and Ufipa fields in the southwest and of the Mhukuru, Mbamba Bay and Njuga basins, situated south of Ruhuhu, appear to be smaller.

Like coal, the iron ore resources of the Republic are large, remote and low grade. Indeed, the ores of Liganga east of Lake Nyasa contain 13% titania and 0.7% vanadium oxide; these amounts would be considered as prohibitive but Swedish engineers have demonstrated, at Smalands-Tåberg, that adequate iron can be produced from these ores by the Krupp-Renn process. Certain geologists estimate resources of several hundred million tons of this material with minor reserves at Hundusi, northwest of Dar es Salaam, at Mbabale, and at Manyoro, east of Lake Tanganyika.

Similarly, the hydropower resources of the Republic have hardly been tapped; the lower Pangani could generate 68,500 kW which is twice all the electricity now generated. The plans of TANESCO call for a storage dam at Nyumba ya Mungu (church in Swahili) capable of impounding a flow of 450 cubic ft. per second and for tunnels which would absorb $ 5.5 million of the $14 million cost of the new Hale station. Other power resources of the east comprise the Rufiji and Ruvu Rivers. Meanwhile Italian ANIC plans to build a 12,000 barrel per day refinery at Dar es Salaam.

Gypsum resources of the southern coast are extensive. At Pindiro, each foot of depth represents 17,600 tons of high grade material; furthermore, at neighbouring Mbaru and Mkomore, the deposit extends to a depth of several hundred ft. The Mkomazi, Msagali and Itigi occurrences are, however, smaller. In 1962 only 2,500 tons were produced.

Since Tanganyika's limestone resources are adequate, the writer hopes that plans for the establishment of a cement industry will soon reach the stage of implementation. The site of Tanga, with its 10 million-ton reserve of limestone and a water and power supply, must compete with the existing cement plant of Mombasa in neighbouring Kenya. Near Dar es Salaam, Wazo boasts reserves equivalent to 20 million tons; Mjimwena could also sustain a yearly output of 100,000 tons. Farther from the capital, Lugoba and Tarawanda each contain 6–10 million tons of limestone. Other sites include the Songwe River in the southwest. With an investment of $3,5 million a factory producing 130,000 tons of cement per year is planned at Wazo Hill.

In like manner, the country's *phosphate* resources should also soon be exploited. In 1956, experts discovered phosphate at Minjingu east of Lake Manyara; Consolidated Gold Fields Ltd. have since proved a reserve of 10 million tons averaging 20% P_2O_5. The deposits of Chali and Chamoto are smaller, and Mbeya and Zizi are of the less attractive apatite type.

OTHER PRODUCTS

In 1907, Germans initiated the production of precious *garnet* from the Luisenfeld in the southeast. In 1955, the claim was resurrected to a third life and rebaptized Namaputa. The mine can now produce more than 300 tons of garnet per year. At some distance in the Pare Mountains, geologists have estimated in the cover of Lolubokoi, Buiko and Kichaa, respectively, 25,000, 800 and 100 tons of garnet with 10,000 tons in harder rock at Sangarama. In addition, at Nyarumba in the centre of the country, 10 million tons of overburden contain 15% garnet with underlying hard rock boasting tenors of 30–40%. In 1961, $40,000 worth of rubies and corundum reached markets and in 1962, workers recovered rubies and sapphires from the Umba River in the Lushoto district. A silicate rock from Merkerstein in the north is marketed as Tanganyika artstone and the 'anyosite' of Longido fetches a price of $500 per ton. Chrysoprase and amethyst show near Itiso and Kilosa. However, at Mleha and Ilende, reserves of low grade corundum are larger with production totalling 30 tons. Tanganyika Crystals Ltd. have explored the zircons of Mashowa.

Between 1950 and 1960, Uruwira Minerals Ltd. produced, at Mukwamba near Mpanda, $1,100,000 worth of lead, $700,000 worth of copper, $470,000 worth of silver and the equivalent of $360,000 in gold. At some distance, between Sikito and Kapapa, a subsidiary of the Union Corporation Ltd. (of South Africa) has also proved the existence of the same metals. Copper also shows near Kigoma, Kigugwe, Kidete and elsewhere. Geologists have explored the nickel prospects of Kapalugulu, Kabalwanyele, Twamba, Mwahanza and Ngasamo, but reserves are limited.

Meerschaum was discovered on the shore of Lake Amboseli, near the Kilimanjaro (Fig. 31), in 1953 and later was exported to the pipe factory of Nairobi. Production at Sinya decreased from 20 tons to 1 ton in 1962, but reserves are available. At Gelai near Lake Natron, Industrial Minerals Ltd. have inferred 115,000 tons of magnesite, accompanied by bentonite. In addition, magnesite is available at Haneti, Itiso, Chambongo and Merkerstein. Prior to 1940, private miners exploited low grade soapstone at Hedaru in the north; more than 130,000 tons of this material are available around Kilombo.

While graphite is widespread (e.g., around Nachingwea, in the Uluguru Mountains, Idete and Daluni), workings are hampered by their remoteness and limited by market conditions, as are also the kyanite prospects of Chankuku and Uguruwa. The asbestos showings of Ikorongo and Mbembe are small. The diatomites of the Kagera (one of the sources of the Nile) contain 76% silica; a million tons of the best glass sands lie close at hand, at Bukoba on the shores of Lake Victoria. In the Pugu Hills, near Dar es Salaam, experts have estimated a kaolin reserve of 2,000 million tons, with subsidiary glass sands; the kaolin deposits of Matamba and Malangali are more remote. The nomadic Masai paint themselves with ochre won at Monduli.

Although large by earlier standards, the 800,000 ton niobium reserve of the Mbeya complex in the southwest is not competitive now, nor are those of Oldoinyo Dilli, Galappo, Hanang, Wigu, Ngualla and Musensi; however, the rare earth–thorium reserves of Wigu are considerable. Grading and markets limit the collecting of piezo-electric quartz from Rubeho and Idibo. Between Tukuyu and Mbeya in the southwest, pumice (a volcanic glass) covers the hills; similar ashes appear east of Kilimanjaro. Sulphur fills a crater of that peak, and also rises with the springs of Wingayongo. In a like manner, helium ascends with springs and gases (7 cubic ft. per hour at Manyeghi, 3.6 cubic ft. per hour at Maji Moto, east of Lake Victoria and from the Rift Valley).

MALAWI

Malawi, a country of 37,000 square miles surrounding Lake Nyasa, is wedged into Mozambique and inhabited by a population of 3 million. The government pursues a favourable investment policy but the country's mineral resources are limited (Fig. 34).

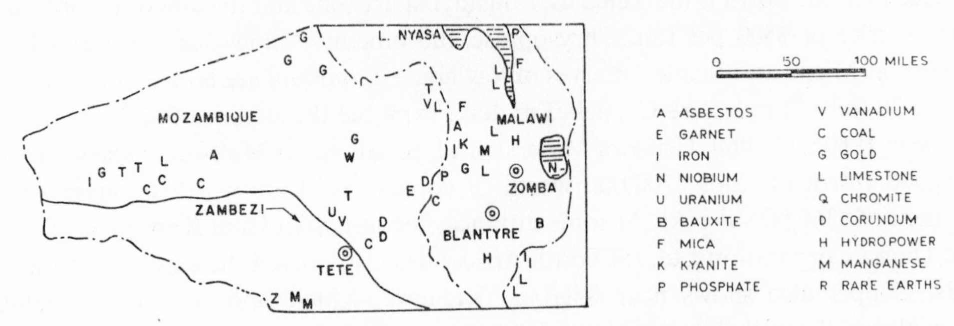

Fig. 34. The mineral resources of western Mozambique and Malawi.

The Changalumi *cement* plant near Zomba produces 90,000 tons of clinker per year. Limestone resources of several million tons remain available, but only with difficulty can lime be burned from this material and from the magnesian limestones of Matope and the Middle Shire. Downstream of Lake Nyasa, natural cement clays and some diatomite cover the floor of Lake Malombe. In the area of Blantyre, brick clay resources are adequate, but the kaolin deposits of Rivi Rivi, Chinteche and Kumbanchenga and the gypsum marls of Chiromoare are of low grade. From Mozambique, *coal* seams extend to Sumbu and Nkombedzi; another field of high ash coal appears at Chiromo. While the southern basins would have to compete with the Mozambique colliery, the northern fields of N'kana and Livingstonia are unfortunately near to the larger resources of Tanganyika.

In 1960, geologists estimated 60 million tons of siliceous *bauxite* on Mount Mlanje with minor occurrences on the Zomba and Nyika Plateaus. Five square miles of the same Mauze complex contain 30% fusible nepheline, and 90 square miles of Zomba include 50% nepheline with other resources at Port Herald, Nyika and Nathace. Close at hand on Tundulu Hill, the Anglo-American group and the Geological Survey recently evaluated, to a depth of 100 ft., 170,000 tons of hard *phosphate* in grades of 20%, 100,000 tons in grades of 10% and 25,000 tons contained in rock with tenors of 5%. The phosphate resources of the soils of Mlinde and of the guano islands of Lake Nyasa are, however, smaller. With relatively low grades, niobium reserves of Chilwa Island, south of Lake Nyasa, do not appear to attain 100,000 tons, but at Ilomba in the northwest, resources of uraniferous pyrochlore are believed to be more abundant. Most of these deposits include more zirconium than niobium and the complex of Chilwa also comprises 300,000 tons of manganese and 2,000 tons of fluorite (Fig. 52).

Laporte Titanium Ltd. owns the prospecting licenses of the Port Herald titanium sands which contain 4,000 tons of abrasive garnet; Rhodesia Chrome Mines Ltd. has explored the black sands of the eastern shores of Lake Nyasa. In addition to these, Rhodesian Selection Trust has proved reserves of thorium and rare earths in the Monkey Bay on the south shore. The Geological Survey has identified these metals on the Palombe plain and in the area of Fort Johnstone; the Survey has also evaluated, in the Kangankunde complex, a reserve of 26,000 tons of rare earth monazite in grades averaging 4.7% with more strontium and inferred even greater resources of these. The extent of areas containing 6.5% monazite is, however, smaller and at Kangankunde, Nathace and Bangala, veins carrying 50% baryte are unworkable.

In the Kapirikamodzi deposit south of the Lake, the Geological Survey estimates that only 80,000 tons of vermiculite are available to a depth of 50 ft. out of which only 5% exfoliate sufficiently to command markets.

The rocks of Lilongwe and Dowa contain up to 10% graphite; those of Wamkuramadzi, Port Herald and the Kirk Range contain even smaller grades, but all of these would have to compete with the extensive Malagasy deposits. In 1952 and 1954, at Kapiridimba near Ncheu, General Refractories Ltd. produced 3,000 tons of kyanite out of more than 300,000 tons available in rock averaging 15% grades. Geologists of the Survey have located other deposits at Chingoma and Chakulambela, near Neno and in the Kirk Range where amphibole asbestos has also been found. Furthermore, between 1942 and 1946, 1,000 tons of corundum were exported from Tambani in the south.

Gold shows in the Lisungwa and Dwangwa Valleys, lead near Dowa, copper and nickel at Cholo and Blantyre. Experts have evaluated only 500,000 tons of iron ore on the Chilwa Island, 50,000 tons at Dzonze and 80,000 tons at Mindale; but, around Mzimba and Nyika, Africans have smelted iron for a long period of time. The long list of minor finds of mica consists of the Port Herald Hills, Tambani, Cholo, the Senzani area, the Kirk Range, the Fort Johnstone area, Kasho Hill, Mount Hora and the Lombadzi Valley. In addition to these, low-priced soapstone is also abundant in the northern Kirk Range and near Ntonda, with porcellaneous magnesite occurring in the Middle Shire and pumice at Karonge.

MOZAMBIQUE

With the harbours of Lourenço Marques and Beira, Portuguese speaking Mozambique controls vital export routes of South Africa, Zambia, Southern Rhodesia and Malawi. More than 6 million people inhabit the 300,000 square miles of this moderately endowed country. With the First World War, mineral production passed the mark of $400,000 and remained at this level, except for the lean years of the Great Recession. In 1950, the opening of the colliery and of the Alto Ligonha pegmatite field initiated a rapid expansion, but in 1960 the value of mineral products still represented only 17% of the national revenue as can be inferred from Table LX.

TABLE LX
MOZAMBIQUE OUTPUT IN $1,000

Year	*Coal*	*Columbite*	*Beryl*	*Gold*	*Asbestos*	*Bauxite*	*Total*[1]
1900				170			170
1920				250			320
1940	20			350		7	400
1950	300	6	80	30		13	450
1960	1,800	850	490	20	20	70	3,300

[1] Including other minerals.

PRINCIPAL PRODUCTS

In 1836, the Governor of Rios de Sena discovered the *coal* field of Tete on the Zambezi; production started in 1932 at Moatize and since 1957, the yearly output of Companhia Carbonífera de Moçambique has risen to 330,000 tons, thus totalling more than 3 million tons. Other coal seams are located 150 miles upstream on the banks of the Zambezi, east of Lake Nyasa in Cahora Bassa district, and on the Rio Mepotepote on the border of Southern Rhodesia (Fig. 34). The Chiasa–Pangura field holds 150 million tons of coal.

TABLE LXI

ALTO LIGONHA OUTPUT

Mineral	*Production in tons*		*Value $*
	1962	*1937–1958*	*1937–1958*
Beryl	440	7,300	2,200,000
Columbite–tantalite	110	570	1,900,000
Tourmaline	?	1.4	230,000
Lithium ore	200	10,300	300,000
Other micas	1	150	
Samarskite	?	20	10,000
Bismutite	15	20	
Monazite	?	15	
Kaolin	?	650	

The pegmatite field of Alto Ligonha is located nearer to Mozambique City than to the estuary of the Zambezi (Fig. 115). Empresa Mineira do Alto Ligonha is one of the world's major producers of beryllium and tantalum and also contributes other minerals (Table LXI).

Samarskite is a rare radioactive mineral of niobium, tantalum and the rare earths which also coexist with thorium in monazite. There are also other uraniferous rare minerals. Most of the tourmaline was sold in 1942, but miners still occasionally recover this semi-precious stone as they do kaolin at Boã Esperança. The carbonatite of Muambe is another source of the rare earths, thorium and niobium.

Prior to the arrival of the Portuguese, Africans extracted gold at Manica on the western, Rhodesian border; but the Missale and Chifumbázi deposits (north of Tete and southwest of Lake Nyasa) of the Companhia da Zambézia do not appear to be as ancient. By 1914, production had reached its peak of 15,000 ounces; by 1930 output decreased to a few ounces, but reached a secondary peak in 1940, whence it declined to 100 ounces in 1962. Silver production started in 1906 and reached its peak of 3,030 ounces per year before gold output culminated; it has decreased since 1913. Until now, 4,000 ounces of silver have been produced, with the gold/silver ratio averaging 7.

Although *salt* has been recovered since time immemorial, industrial production still does not cover all the country's needs; between 1946 and 1956, 45,000 tons of salt were imported from Portugal, Angola, Egypt, South Africa and Goa. During the same period, five Mozambiquan districts supplied salt by evaporation (Table LXII).

While this salt contains 0.1–2.7% impurities and 1.6–5.2% other salts, the brines of Chibuto south of the Limpopo contain 0.2% impurities and 10% other salts; between 1946 and 1956 they provided 12 tons of salt. In 1961, output stood at 44,000 tons.

In 1930, 1943–1946 and 1948–1951, experts explored the *asbestos* deposits of the Serra Mangota near Manica on the border; the short quarter inch chrysotile fibre has to compete with the products of Southern Rhodesia. Production started in 1951 and, following an interval, oscillated between 200 and 300 tons yearly from 1954–1958; in 1962, output plummeted to 110 tons. While the low grade fibres of Tete belong to the first type, the tremolite of the Munhinga near Manica does not fiberize. In 1914, The Companhia do Moçambique discovered bauxite at Mount Snuta near Manica. Small scale production started in 1935, and since 1938, output has steadily expanded from 400 tons to the present

TABLE LXII

MOZAMBIQUE SALT PRODUCTION

District	*1,000 tons*
Mozambique	60
Zambezia	30
Lourenço Marques	20
Cabo Delgado	12
Manica	3

yearly rate of 5,000 tons. Wankie Collieries, the owners, export the refractory bauxite, which contains 59.2–64.3% alumina and 5.9–14.1% silica, to their collieries in Rhodesia and to South Africa, where it is processed into aluminium sulphate (Fig. 52).

OTHER RESOURCES

In 1934, the Société Géologique et Minière du Zambèze explored the copper occurrences of Cacanga near Tete; the deposits of M'Panda-Unkua (Pandemakua) and of Cónua of the Companhia da Zambézia were already known by then. Edmundian and Copper Mine provided a small production after 1902, but during the years of activity, 1903–1910 and 1915–1920, most of the 2,600 tons of copper came from Tete. Copper also shows at Massamba and Chichie, tin near Ichope. Between 1919 and 1952, the Mines Division of the Companhia de Moçambique exported 230 tons of ton concentrates from the area of Mount Doeroi near Manica. Wolframite shows there and also in the placers of the Machinga and Catoa Rivers in Tete.

The chromite occurrences of Mount Nhantreze in the Tete district and of Mount Munuca in Zambezia are of low grade. In 1955, nine tons of *corundum* were exported from Revuè on the Rhodesian border. Other showings include Cachissene in the Niasa district, Canchoira and Zóbuè; and only 30 miles south of Villa Pery, private miners, with an investment of $7,000, can recover at least 100 tons of corundum each month. Diatomite including 87% silica is fairly abundant south of the Save River near the Goba railway. At the present stage of exploration, iron occurrences appear to be numerous but limited in extent; they include the Moatize Valley near Tete, the Serra Manga near the Zambezi, Gogói and Lucite near the Rhodesian border and numerous other outcrops which have been worked for centuries by Africans. The Messica and Lundi mines reportedly have a capacity of 500,000 tons of iron ore and the Macheduda deposit includes 200 million tons of ore with 17% titania. Kyanite shows in the placers of the Muaguide River in the north and at Zóbuè close to the Rhodesian border. Limestone is widely scattered in the country; at Mossurize in the Manica district, at Maputo in the south (a district in which they are abundant), near Marávia, in Tete, around Naguema in the district of Mozambique and near Vila Cabral in the district of Niassa.

Fluorite only shines near Macossa in the southwest, but garnet shows at Báruè, Ancuabe and Zóbuè. The Companhia da Zambézia owns the Angonia graphite deposit near Tete; private miners have explored the Metochéria and Nacoto deposits and Grafites de Moçambique has prospected Otaco-Ancane. The Sociedade Mineira do Itotone produced 2,000 tons (with 70–74% carbon) and, in 1951, the Evate mine contributed 300 tons. Explorers report magnesite and talc in the Manica district, respectively from the road of Vila Pery and the Muza Valley; close at hand. The Alto Chimezi includes 40% of manganese. Near Vilanculos and the Save River, birds have covered caves with 3 ft. of guano representing a volume of 20,000 cubic yards. With depth, the phosphate (P_2O_5) content increases from 8.5 to 10.6%, concomitantly nitrogen grades augment from 5.7–2%, but potash content remains fairly constant at 0.8%. The Imamputo plant, 35 miles from Lourenço Marques, can produce each year 7,500 tons of bentonite, valued at $175,000.

Expanding perlite is abundant in the Libombos Mountains in the hinterland of Lourenço Marques. The black sands of Pebane on the coast of Zambezia average 30–60% ilmenite, 0.5–1.5% rutile (both ores of titanium), a few per cents of zircon and traces of rare earth phosphates. But, in the south, the black sands of Incomáti are perhaps more favourably located. Vermiculite, an exfoliating mica, is reported to occur at Tundumula and Matema in the Tete district. Unfortunately, on the coast, gas has not found markets yet, but the refinery of Lourenço Marques provides 12,000 barrels of oil per day. Finally, the largest hydropower site, Cahora Bassa, on the Zambezi, can produce more than 700,000 kW.

SECTION IX

Southern Africa

Southern Rhodesia, Bechuanaland, Southwest Africa, South Africa, Basutoland, and Swaziland form this area.

SOUTHERN RHODESIA

Between the Zambezi and the Limpopo, Southern Rhodesia covers 150,000 square miles and is inhabited by a population of 3 million. The country's mining industry is not as closely knit as are the great corporations of South Africa, Zambia or the Congo, and it faces a number of political and economic problems: In recent years, the chrome industry has had to face Soviet competition; the asbestos companies compete with Canadian sales; the Wankie Colliery is obliged to compete with the cheap hydropower of the Kariba Dam. The small gold producers are also vulnerable (Fig. 36).

Prior to 1964, Southern Rhodesia produced a total of $1,600 million worth of minerals (which was the order of magnitude of the 1964 annual output of South Africa). This sum can be divided into six headings (Fig. 35 and Table LXIII).

ASBESTOS

Following Canada and the Soviet Union, Southern Rhodesia contributes 5% of the world production of asbestos. By 1949, the yearly output was in excess of 80,000 tons and $10

TABLE LXIII

SOUTHERN RHODESIA PRODUCTION

Mineral resource	*Beginning to 1960 $ million*	*1963 $ million*
Asbestos	310	17
Gold	690	20
Coal	130	8
Chromium	130	5
Copper	30	9
Others	40	7
Total	1,330	66

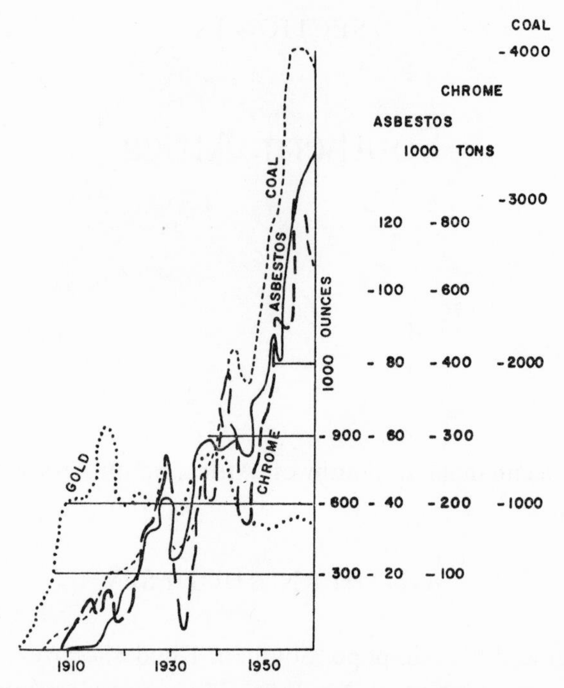

Fig. 35. Southern Rhodesia production of asbestos, coal, chrome and gold.

million and increased rapidly until 1957; since then, production has oscillated between 140,000 and 160,000 tons per year. While the Soviet Union penetrates western markets, the Southern Rhodesian industry finds increasing outlets in the communist bloc (e.g., in Czechoslovakia) in addition to its traditional markets in the United Kingdom (1/3), western Europe (1/6) and India (1/6). As to grades, in 1959, no. 4 constituted 40%, no. 5, 30%, no. 3 and no. 6, more than 10% each. Processed asbestos and higher grades can support shipping costs better than low grade material. Responding to the expansion of the building trades which use lower grades, in 1960 Vanguard Asbestos Mines expanded its

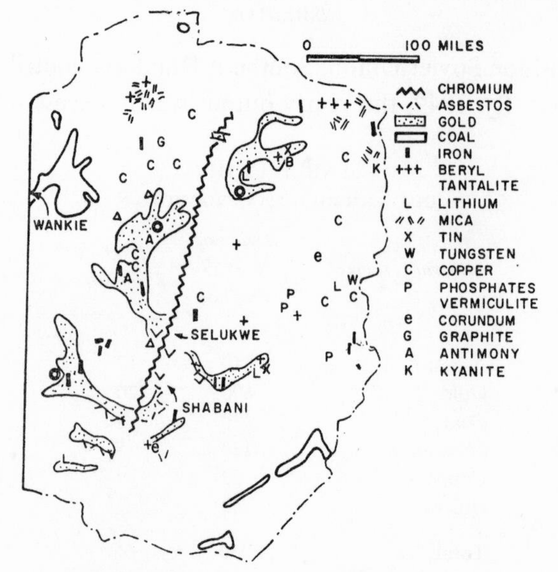

Fig. 36. The mineral resources of Southern Rhodesia.

Belingwe short fibres plant which now produces material for the plastics industry. The Government has waived royalties on grades 6 and 7 and assists this reorientation; consequently, production is expanding at a greater rate than its value. In 1963, the unit price of a ton decreased to $120.

Exploration started in 1907 and Mr. Kerr pegged the Shabanie claims in 1915; African Asbestos Mining Company took over the principal mines in 1919 and these are now controlled by Turner and Newall Ltd. This Manchester-based, but world-wide asbestos, cement and plastics manufacturer operates with an authorized capital of $165 million, and now produces three quarters of the Rhodesian output. Since 1954, dividends of Turner and Newall have increased steadily and reached $9 million in 1962. In the south, the Rhodesian and General Asbestos Corporation, a subsidiary of the group, exploits the Shabani (Fig. 101), Mashaba and Filabusi deposits (employing 4,400) and another affiliate, African Asbestos Mining Company, works Nil Desperandum. Small scale producers include Southern Rhodesian Chrysotile Corporation (Cullinan group) at Serpentine Hill, Gurumba Tumba and Chikwanda, Rhodesia Montelo Asbestos Ltd. (Charter Holdings group), the Mashaba Rhodesian Asbestos Company Ltd., and the Ethel Asbestos Mines in the northern part of the Great Dyke.

Operators prefer block caving to methods based on shrinkage. The flow sheet of the African Associated Mines group comprises drying, screening, and separation through aspiration, followed by the opening of fibres and grading. For each ton of ore, 12 kW of power and five tons of air are used.

GOLD

Africans have worked gold for centuries and certainly prior to 1505 when the Portuguese established their trading settlement of Sofala on the Mozambique coast. At what Rhodesians now call the Prestwick and Farvic mines, African workings reached depths of 150 ft. Consequently, they were not 'primitive'. However, this type of industry disappeared by 1880 when King Lobengula of the Matabele granted Cecil Rhodes the concessions which became the nucleus of the British South Africa Company, a chartered company which still retains royalty rights in three countries. From 1933 (the recession year) to 1940, gold output increased; it fell from its peak of 810,000 ounces to 490,000 ounces in 1950, then slowly increased to 570,000 ounces by 1963. In 1962, 300 producers (half of the 1949 number) milled 2.0 million tons of ore, recovering 5.7 dwt. gold per ton.

In the centre of the country, the Hartley district (Fig. 76) contributes more than 44% of the production; the adjacent Gwelo district provides 17%; the Bulawayo area in the south, 21%; and the Salisbury district in the north, 17%. A few hundred ounces are also extracted in the Fort Victoria field. The industry still employs 17,000 people but, with scattered reserves and small scale operations, costs continue to rise and profits to decline. In 1962, the twelve largest producers included those shown in Table LXIV.

In 1960, the Rio Tinto group acquired 90% of the shares of the Cam and Motor Gold Mining Company, which showed a pre-tax profit of $1.5 million in 1959; but production decreased in 1962. Profits of the Globe and Phoenix Company are similar, but the Connemara Mine of Frobisher Ltd. is close to the end of operations. At Cam and Motor, work has now reached a depth of 6,500 ft. (and a temperature of 115°F), but the small mines

TABLE LXIV
PRINCIPAL GOLD MINES

Mine	*Group*	*Locality*	*Ounces*
Cam and Motor	Rio Tinto	Eiffel Flats	110,000
Dalny	Falcon Mines	Chakari	60,000
Globe and Phoenix	Globe and Phoenix	Que Que	30,000
Arcturus	Arcturus	Arcturus	30,000
Muriel	Homestake	Banket	16,000
Kanyemba	Kanyemba	Umsweswe	17,000
Barberton	New Consolidated Goldfields	Queens	20,000
Long John	Forbes and Thompson	Gwanda	19,000
Pickstone	Rio Tinto	Eiffel Flats	17,000
Connaught	Mazoe Consolidated	Jumbo	13,000
Bell and Riverlea	Globe and Phoenix	Que Que	10,000
Patchway	Rio Tinto	Golden Valley	19,000

cannot afford this type of operation. The usual flowsheet includes crushers, stamp mills, tables, Dorr classifiers, flotation, roasting and precipitation.

CHROMIUM

When the railway reached Selukwe in 1906, chromite was first exported. In 1908, Rhodesia Chrome Mines Ltd. took over the claims of J. L. Popham. Ever since, Rhodesia has been a major producer of chromium, and production has increased steadily with peaks in 1957 and 1960 of 670,000 tons. In 1961, the United States imported more metallurgical chromite from Rhodesia than from Turkey, but Soviet exports reduced production in 1963 to 40,000 tons. In 1962, the United States imported 300,000 tons, South Africa 25,000, Norway, Japan and the United Kingdom combined 50,000 and Sweden 10,000 tons of chromite from Rhodesia. Exports include 55% metallurgical, 30% chemical and 15% refractory grade ore. In 1962, out of 19, the principal mines included those shown in Table LXV.

High grade reserves of Selukwe (in the centre of the country, Fig. 94) are considerably larger than those of Mashaba, but around Belingwe (Rhonda, Mlimo, Spinel, Eureka) similar lenticular ores contain much iron. Aer is another small mine. Nevertheless, the resources of the Great Dyke which crosses the country from north to south are much larger; they include, to an incline depth of 500 ft.: 190 million tons of metallurgical ore

TABLE LXV
CHROMITE EXPORTS

Rhodesia Chrome Mines	*1,000 tons*	*Others*	*1,000 tons*
Selukwe Peak	190	African Chrome Mines (Mtoroshanga)	50
Railway Block	110	Rhodesian Vanadium (Umvukwes)	60
Prince Mine (Mashaba)	8	Rhodesian Mining (Mtoroshanga)	40

(with 48% Cr_2O_3, Cr/Fe = 2.8/1), and 350 million tons of chemical/refractory ore (with 49% Cr_2O_3, Cr/Fe = 2.3/1). These metallurgical grade resources are thus the largest of the non-communist world. In the north Rhodesian Mining Enterprises exploit the large eluvial (i.e., fine grained) reserves of the Great Dyke which are believed to reach the mark of 60 million tons of ore. However, with demand for chemical grade chromite decreasing, the Jester and O'Meath plants are not operating at capacity.

At Selukwe, Gardner Denver drifters handle the stoping. The funnel-like form of ore bodies demands the establishment of numerous faces. At Vanad on the Great Dyke (Fig. 95), stripping has now been abandoned, and inclined shafts reach a depth of more than 3,000 ft. From 25 ft. below, crosscuts are driven to the seam every 300 ft. The chrome industry employs 8,000 people.

With the development of secondary industries at Que Que, Windsor Ferro Alloys produce 11,000 tons of ferro-chromium yearly, and at Gwelo, Rhodesian Alloys Ltd. produces 15,000 tons of this material.

COAL

Africans described the burning stones of the Victoria Falls area to A. Giese, and in 1893, the German explorer pegged the claims of what later became *Wankie*, the continent's biggest colliery. The Wankie Colliery Company (managed since 1953 by the Anglo-American Corporation) has an authorized capital of $15 million and yearly profits of about $4 million; it operates at about 60% of its annual capacity of 5.2 million tons of coal. Between 1949 and 1957, coal production increased from 2 million tons to 4 million tons per year, and has slowly decreased ever since to less than 3 million tons.

In Zambia (Northern Rhodesia) and Katanga, the significant role of Wankie coal is now counterbalanced by the Le Marinel and Kariba hydropower plants. The closing of no. 1 Colliery leaves 18 to 24 thick seams lying at a depth of 150–350 ft. in no. 2 and no. 3

TABLE LXVI

PRODUCTS AS PERCENT DRY BASIS

%, Dry basis	*Fixed carbon*	*Volatile*	*Ash*	*Sulphur*	*BTU tons*
Unwashed coal	60	26	14	2.5	13,000
Washed coal	63	27	10	1.5	13,500
Coke	84	1	15	1.5	

TABLE LXVII

WANKIE DISTRICT, COAL RESERVES IN MILLION TONS (ASH CONTENT IN %)

Coal	*Wankie*		*West*			*East*
	Wankie	*Entuba*	*Lubimbi*	*Sengwe*	*Sebungu*	*Marowa*
Good coking	820(13)	70(13)	20(13)			
Idem, possible	300(13)					
Poor	260(25)	70(30)	50(19)	45(32)	90(23)	16
Idem, possible	110(25)					

Collieries. Miners extract only the lower half of the seam during development and the remainder is won on the retreat. Electrical coal cutters excavate the fuel which is then transported by belt conveyors or track. After washing, coal yields good coke and, on carbonation, it yields tars, naphtalene, creosote, benzols, other hydrocarbons and ammonia liquor (Table LXVI).

From Wankie, coal measures extend towards Sebungwe (Fig. 121) in the east. The untapped low grade resources of the *Sabi-Limpopo* field in the southeastern corner of the country are considerably larger. Indeed, the Makushwe, Malilolongwe *A* and *B* seams, respectively, include reserves of 4,730, 96, and 170 million tons of coal containing 35, 37 and 32% ash (Table LXVII).

In the east, a majority of the reserves of Sebungwe remain to be proved. Finally, the colliery has to compete with the 20,000 barrels per day refinery of Umtali.

BASE AND RARE METALS

Africans smelted the red metal, *copper*, long before the arrival of Europeans who initiated the small mines of Umvuma, west of Sinoia, Copper King and Copper Queen. In 1954, however, the Messina group started larger production which has reached the level of 15,000 tons of contained copper (Fig. 51). Furthermore, in 1961, the Alaska (Sinoia) smelter of the Messina Rhodesia Smelting and Refining Company was completed. In 1962, MTD Ltd. (Mangula) estimated the sulphide reserves of its largest mine, Molly (in the northwest), at 330,000 tons of copper contained in 1.27% ores. In the year 1959, the company, which has a capital of $14 million, paid its first dividends. The group equipped the Alaska Mine to produce at a rate of 500 tons per day. Reserves of the Copper Queen were given at only 5 million tons of ore containing 2.85% zinc, 1.4% copper and 1.4% lead. In another area, the Bradley deposit of the Sabi Valley (southeast) provided 2,000 tons of copper, and the Muriel Mine (Banket district) belonging to another group contributed, in addition to gold, 5,000 tons of copper. In 1963 the country produced 18,000 tons of copper and 83,000 ounces of silver.

Silver is a byproduct of copper and gold mining, and in 1959, the northern Salisbury, Hartley, Gwelo and Bulawayo districts, respectively, produced 26,000, 3,000, 1,300 and 1,000 ounces of this precious metal. As to zinc, the Elbus Mine, near Wankie, produced a few tons. Near Wankie, the Billiton Maatschappij now owns Kamativi Tin Mines Ltd,, of which it has increased the capital to $8.5 million and the production to 1,200 tons of concentrates. The company also smelts the tin of smaller operators including Kapata, and San and Gwaai.

Bikita Minerals Ltd., a company largely owned by the Rhodesian Selection Trust, is a leading producer of *lithium* minerals and *beryl*. Each year between 1957 and 1963, 50,000–100,000 tons of lithium minerals (priced at more than $1 million) were produced in addition to a few tons of pollucite, the mineral of caesium, of which large reserves are also available. However, market conditions curtail the lithium ore production of the Enterprise belt near Salisbury and of the even smaller producers of Odzi, Mtoko, Filabusi and Gwanda. At Bikita, beryl is a byproduct of lithium mining, but small workings of this mineral are widespread, especially east and north of Salisbury (God's Gift, Pope, Mauve, Augustus, etc.), at Benson and Matake. From its inception in 1949, beryl production rapidly expanded to 1,700 tons in 1953, and then decreased until 1956, whence output continues at the yearly rate of 300–600 tons. Each year, the Benson Mine in the northeast, plus Makanga, Bindura, Kamativi and Bikita produce 40–80 tons of tantalum and niobium ore. Furthermore at Bikita, large, unused resources of rubidium are available.

OTHER PRODUCTS

Both Rhodesian Cement Ltd. and Premier Cement Company Ltd. quarry *limestone* at Colleen Bawn near Gwanda, and each makes yearly profits in excess of $1 million. Portland Cement Company exploits the Sternblick Mine. Lime is burned at Shamva (Snow White), Redcliff and Lalapanzi; at Sambawisa, near Wankie, Luveve Stone Crushing produces metallurgical calcite, and high grade limestone comes from the Early Worm. The value of these products is in excess of $1,000,000.

For high grade *iron ore*, the outlook is favourable; rapidly increasing production will soon pass the mark of 800 thousand tons per year. Nevertheless, with a reserve of 70 million tons, the largest deposit, Buhwa (in the southeast), still lies idle. At Mwanesi, 27 miles from Umvuma the ore averages 40% iron with concentrations of 60%. This deposit could produce many millions of tons per year. After 1961, the Rhodesian Iron and Steel Company doubled its steel output of 2,000 tons of pig iron per week, which is extracted from a yearly output of 150,000 tons of ore. But its Que Que deposit, which has reserves of 6 million tons, is small by African standards. South of Que Que, mines will export 300,000 tons of iron ore in three years. The steel mills of Kobe in Japan also showed considerable interest in the Beacon Tor (Que Que), Block Mamba, Yank, B1 (Gatooma), and Nyuni (Fort Victoria) claims. From Norie (near Belingwe), out of a reserve of 2 million tons, 130,000 tons are being sent to Japan via Lourenço Marques each year. From Glendale (Mazoe), the Iron Duke Mining Company exports 70,000 tons of pyrite containing 41.5% *sulphur* to the Copperbelt yearly.

The *mica* fields of Southern Rhodesia are located in the north. The principal mines of the Miami field (Nzoe, Grand Parade, Gil Gil, Locust and Lynx) and Ubique, near Wankie, produce 15 tons of mica yearly, in addition to scrap, the demand for which is decreasing. F. F. Chrestien and Company are active in the mica trade. In 1952, *tungsten* concentrate production reached its peak of 500 tons, but production at Sequel (Wankie), Essexvale and Karoi, and at Bindwa diminished. Each year, Cullinan Refractories Ltd. recover 6,000 tons of corundum at O'Brien's Prospect; lapidaries collect emeralds at Zeus, aquamarine and heliodor at Karoi. At Somabula and in the southwest, diamond mining has ceased but the production of sapphire, amethyst (9 tons in 1963) and agate (2 tons, 1961) continues.

The Anglo-American Corporation estimates the *phosphate* reserves of Dorowa at 37 million tons with a P_2O_5 content of 8%, and a subsidiary of African Explosives and Chemical Industries Ltd. produces 6,000 tons of concentrates each month. Clays are won at Wankie, Firdon, Barry and Tipton, and Rhodesian Alloys Ltd. produce quartz at Gwelo. Antimony is a by-product of the Sebakwe gold mine, and nearby at Que Que, the state recovers arsenic in its roasting plant. Each year, only 400 tons of diatomite reach the market. Some nickel is found at Noel, Empress and Trojan, with platinum showing in the Great Dyke at Makwiro and Wedza. Other products with their sources include uranium at Cripmore, quartzite and fluorspar at Wankie, manganese at Que Que, baryte at Que Que River,

Shamva, and Dodge, vermiculite at Mtoko, graphite at Graphite King and other localities, magnesite in the north near Gatooma and also with talc as a by-product of asbestos mining, and sillimanite at Miami.

BECHUANALAND

The Kalahari covers most of the country; on its eastern edge, where the desert recedes, there is a long tradition of mining. Indeed, in 1867, Mauch, the German geologist, had already described the Tati gold fields. In the northeast, gold production of the Francistown area has already decreased from more than 1,000 ounces in 1954 to 140 ounces in 1963. Similarly, during the same period, the recovery of silver declined from 300 to 20 ounces, and the output of copper from the Bushman Mine in the Phudulooga area ceased long ago. The search for nickel, a rare metal in Africa, has reactivated the area: Rhodesian Selection Trust, in association with International Nickel and Minerals Separation Company (Mond), has evaluated a deposit containing 0.4–0.7% nickel at Magogaphate, 65 miles southeast of Francistown near the meeting point of the borders of South Africa and Southern Rhodesia. West of this area, workers of the Consolidated Selection Trust extracted three diamonds from the Macloutsie River; prospectors of De Beers operate in Ghanzi land.

Interest has shifted to other areas of Bechuanaland's 280,000 square miles, which are inhabited by about 300,000 people. As in Mauritania (where Nouakchott now replaces Saint Louis de Sénégal), the capital is concomitantly shifting from Mafeking (abroad in South Africa) to Lobatsi. North of Lobatsi at Mamabule, the railway conveniently crosses a *coal* field where geologists already have proved a 250 million ton reserve of coal lying under shallow cover. The coal seams of the Morapule basin (150 million tons reserve) are more erratic (Fig. 50).

Even the desert will contribute to the country's wealth: investing $6 million, the Rhodesian Selection Trust plans to build a *soda* plant at Nata, on the Sua, which is a pan of the Makarikari swamp. Following solar evaporation, the chloride/carbonate ratio of the liquor will be upgraded to 3/1, and then the pulp will flow through a 10 inch pipe line to a processing plant located 10 miles south of Francistown (Fig. 46). Since the soda market is favourable, production is scheduled to attain 180,000 tons of sodium chloride and 60,000 tons of sodium carbonate.

Until collieries and soda plants are built, asbestos remains the principal product: the 2,300 ton production of the Marlime Chrysotile Corporation at Moshaneng near Kanye is valued at $420,000. Asbestos also shows at Topsi. Also near Kanye, at Kwakgwe, manganese output will increase, it is hoped, to 25,000 tons yearly. Similarly, it is hoped that Bamalete Manganese Ltd. will reopen the Ootsie manganese workings 12 miles north of Lobatsi. In addition to these, other manganese and asbestos prospects have attracted attention in Bangwaketseland.

SOUTHWEST AFRICA

A triangle on the Atlantic coast, Southwest Africa covers 320,000 square miles, but only about half a million people inhabit this arid country. At Windhoek, the Government

TABLE LXVIII
SOUTHWEST AFRICAN PRODUCTION

Year	Value in $ million		Net profits in $ million	
	Diamonds	*Other*	*De Beers*	*Tsumeb*
1956	49	48	31	22
1957	47	39	28	20
1958	38	27	25	12
1959	43	30	31	12
1960	44	32	33	14
1962	32	26	43	12

derives a majority of its revenues from De Beers' diamonds and the Tsumeb Corporation's base metal operations. Their production is shown in Table LXVIII and Fig. 37. Taxation represents approximately 30% of the profits.

DIAMONDS

In 1908, a worker from Kimberley picked up the first gem in the Namib or 'Diamond desert' of Pomona east of Lüderitz Bay. Production started and reached 560,000 carats by 1910 and a peak of 1,600,000 carats by 1914. The deposits, as H. Merensky proved in 1928, extend, in fact, along the whole coast, concentrating mainly between Pomona and the Orange River and within this stretch, along a 30 mile long strip north of Oranjemund (Fig. 108). Consolidated Diamond Mines of Southwest Africa, a subsidiary of De Beers' (Anglo-American group), with assets of $90 million and a capital of $15 million, produces the triple of this amount in profits each year! Following the recession of 1933, production reached another peak of 980,000 carats in 1951. Although, in 1962, the throughput of 940,000 carats was smaller, profits were practically identical. The company's concession lapses in 1990 but it has other investments in the De Beers group, especially in South Africa.

In 1962, machines trenched and stripped 14 million cubic yards of coastal sands. A narrow gauge railway transports this material to the heavy media-grease table plant. The quality of the Southwest African diamonds is particularly high for this type of deposit. Indeed, 80% are gems!

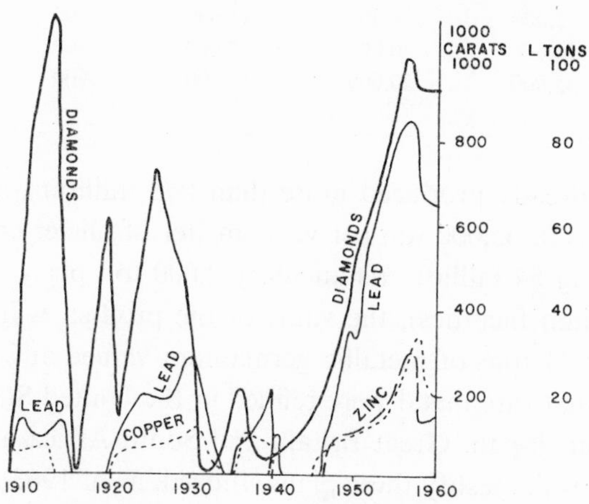

Fig. 37. Southwest Africa production of diamonds, copper, lead and zinc.

Furthermore, 150 miles north of Oranjemund, pumps suck diamondiferous sands from the ocean floor, and concentrate the salt-covered stones on barges. Indeed, experts believe that diamonds are scattered on the ocean floor at least 200 miles north and south of Oranjemund; and that, in the Chameis Reef alone (100 miles north of the estuary of the Orange), at least 14 million carats are available with grades currently attaining five carats per ton! In early 1964, Marine Diamonds Ltd. recovered 21,000 carats per day with suction equipment. Of the shares of this venture, Sea Diamonds Ltd. (a company of the Texan oilman, Sam Collins) holds 44%; the General Mining and Finance group, 25%; the Anglo-Vaal group, 16% and its subsidiary, Middle Witwatersrand Ltd., 7.5%. In 1963, Sea Diamonds agreed to invest an additional $1.5 million and General Mining and Anglo-Vaal will contribute, together, an additional $1.5 million. On the other hand, De Beers probably provided $3 million, and received an option on 25% of Sea Diamond and the right of first refusal on the Collins holdings in Sea Diamonds Ltd., which are believed to represent 80% of the company's equity.

TSUMEB

Prior to 1850, Africans worked copper in the northern areas, and in 1910, the Otavi Minen und Eisenbahngesellschaft initiated production. However, during the Second World War, the South African Government confiscated the mine and in 1948, the Tsumeb Mining Corporation – consisting of the United States companies, American Metal Climax Company, 29.13%; Newmont Mining Corporation (the managers), 29.15%; O'okiep Copper Company (a subsidiary of the preceding); and of the Rhodesian Selection Trust – acquired the mine for less than $3 million, which is only one fourth of a present year's profit. Production developed as shown in Table LXIX.

TABLE LXIX
TSUMEB OUTPUT

Year	*Lead tons*	*Copper tons*	*Zinc tons*	*Cadmium tons*	*Silver ounces*
1908	5,000	3,000	—	—	200
1920	8,000	5,000	—	—	—
1940	3,000	1,000	11,000	300	460,000
1960	50,000	30,000	15,000	900	1,700,000
1962	65,000	30,000	15,000	900	1,300,000

Thus the mine has already produced more than two million tons of metals, and with new metallurgical plants (a 35,000 ton per year smelter of blister copper, a 90,000 ton per year lead smelter costing $4 million, its ancillary 9,000 ton per year sulphuric acid plant and expanded germanium facilities), the value of the product will increase considerably. In 1962, in addition to 31 tons of metallic germanium valued at $1,800,000, concentrates containing 12 tons of that rare metal were refined in the United States and with other ores at Hoboken in Belgium. Japan, Great Britain and South Africa also import products of Tsumeb. Exploration results justify these plans. Indeed, as of 1963 and down to the 3,150-ft. level, reserves attained 6.8 million tons of ore containing 13.2% (15.1%) Pb, 4.8%

(5.9%) Cu, and 3.9% (7.0%) Zn. The figures in parentheses show grades recovered prior to 1960, thus demonstrating that zinc tends to disappear in the deeper oxides whereas other grades remain fairly constant. Similarly, 3 million tons of probable ore contain 12.5% Pb, 4.6% Cu and 3.7% Zn. The stope and pillar method of extraction is successful and plans call for eventually sinking the De Wet Shaft to a depth of 4,150 ft.

The Corporation is also building a new mill 40 miles south of Tsumeb at the Kombat Mine, a 3.6 million ton deposit containing 3% Cu and 3% Pb (Fig. 91).

OTHER MINES

By 1959 exports of the Otjosundu manganese mine (halfway between Windhoek and Tsumeb) of the South African Minerals Corporation increased to 100,000 tons, and thence declined in 1961 to 50,000 tons ($1,500,000). Most of the ore, containing 48% manganese, was shipped through Swakopmund to the United States. Output ceased in 1962.

In addition to Tsumeb, the *Otavi Mountains* are distinguished by numerous smaller deposits of lead, zinc, copper and vanadium, a metal of which Southwest Africa is one of the leading producers (Table LXX).

Following its 1951 low point of 200 tons, the vanadium production of Southwest Africa has steadily increased; it reached 800 tons by 1960, and 2,000 tons in 1962, but mines and reserves are small. In 1961/62, the Southwest Africa Company produced 11,000 tons of lead-vanadate concentrates and 100 tons of zinc at Berg Aukas. At that date, reserves stood near 880,000 tons containing 3% V_2O_5, 6.2% Pb and 35% Zn. In 1957, the Gold Fields group acquired control with the Anglo-American Corporation and the British South Africa Company who also participate in Tsumeb Exploration Ltd.

Small prospects are periodically worked and then closed. At the time of writing, the

TABLE LXX

THE ORES OF OTAVI

	Vanadium	*Lead*	*Zinc*	*Copper*
Abenab (+ West)	+	+	+	
Alt Bobos, Friesenberg	+			+
Harasis, Uitsab	+	+	+	
Berg Aukas	+		+	
Baltika	+			
Kombat, Guchab		+		+
Rietfontein		+	+	

following copper mines appeared to be of some significance: Kopermyn, halfway between Tsumeb and the sea; Onguati, northeast of Walvis Bay; Onganja, north of Windhoek; and Kobos, south of the capital.

Small operators also mine lead, silver and occasionally copper, south of Usakos, at Hohewarte, near Windhoek, Namib Siding near Swakopmund, and other localities. Lavas of Goboseb, Awahab and Rehoboth can also be considered as low grade copper ores (Fig. 38).

In 1962, following a rapid decline, 940 tons of *tin* were sold for almost 700,000 dollars

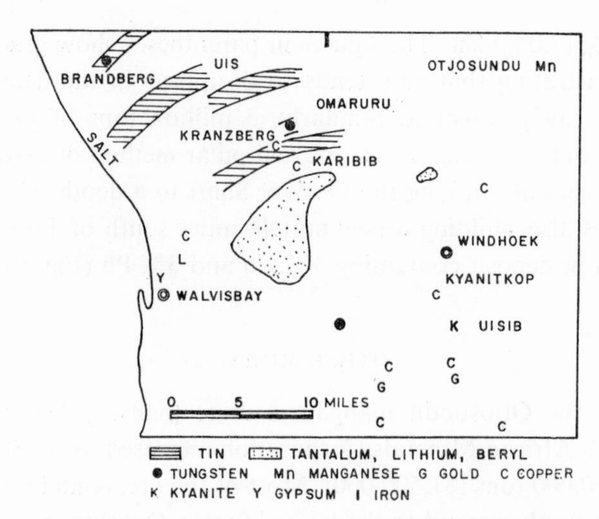

Fig. 38. The mineral resources of central Southwest Africa.

with a few tons of tungsten. Eluvial deposits have been worked out, therefore, small primary deposits north of Swakopmund are costly to operate. At Brandberg West, the Southwest Africa Company has estimated 7,000 tons of tin with tungsten; at Uis, only tin ore and at Kranzberg and at Natas, southwest of Windhoek, only tungsten ores were produced. On the southern border, tungsten also shows in the pegmatite fields of the Warmbad district. As to other pegmatitic minerals, their production depends on foreign markets and freight costs (Table LXXI).

For a number of years, miners worked more petalite than all other lithium minerals combined. Large reserves of caesium and rubidium associated with lithium also await markets. The principal pegmatite deposits include Helikon, Rubicon and Aurora, in the Karibib field (northeast of Swakopmund) and Kinderzitt, Tantalite Valley and Dassiefontein in the far south. Lapidaries occasionally extract a few pounds of semi-precious tourmaline (0.1 tons in 1963), heliodor, aquamarine, morganite (e.g., at Spitzkopje and Rössing) and a few tons of cat's eye, amethyst, rose quartz, but less chalcedony. Bismuth is found near Erongo, and rutile at Giftkuppe, and garnets occur both in the pegmatite fields and at Swakopmund. The most attractive marbles are those of Karibib. In Otaviland, limestone is also abundant.

Salt resources of the coast are vast. Between Swakopmund and Cape Cross, workers recover salt from pans but, with gypsum and anhydrite, rock salt also forms thick layers Production averages 60,000–90,000 tons ($350,000). Also along the coast, workers recover 1,000 tons of guano yearly.

At Kyanitkop and Uisib, south of Windhoek, kyanite and sillimanite production reached

TABLE LXXI

PRODUCTION OF THE PEGMATITE FIELDS IN TONS

Minerals	*1954*	*1957*	*1960*	*1963*
Lithium ores	7,000	6,000	5,000	2,800
Beryl	550	400	400	150
Columbite–tantalite	10	10	10	6

its peak of almost 4,000 tons in 1957. A steady flow of exports continues, mainly to the United States. Indeed, resources are large.

At Okarusu, north of Otjiwarongo, abundant resources of metallurgical fluorspar are available. Following its 1954 peak of 3,000 tons and the cessation of production in 1957, there is again a limited output. The Eisenberg iron deposit has occasionally contributed flux to Tsumeb. Other resources include those of Otjiwarongo and Windhoek and the conglomerates of Otavi. In 1952, graphite production from Keetmanshoop and Bethanie ceased. Among other resources, we should list barytes in the Khomas Highlands, near Windhoek and in the Otavi Highlands; asbestos near Swakopmund and Keetmanshoop; iceland spar at Rehoboth, Gibeon and Swakopmund; and soda on the fringes of the Kalahari.

SOUTH AFRICA

With its industrialized economy based on vast mineral resources, South Africa is a class in itself on the continent. Excluding oil and gas, the value of South African mineral output ranks ninth in the world. Furthermore, South Africa holds a number of 'firsts': the largest gold, chromium, platinum group, manganese, vanadium, asbestos and vermiculite resources of the world; the largest iron, coal, uranium, nickel, antimony, titanium, zirconium and fluorspar resources of Africa; and the largest production of numerous precious and industrial minerals in Africa (Fig. 46).

With its output, South Africa plays the same role in the African mineral industry as does the United States in the Americas. Conversely, there are flaws; these are, mainly, the lack or scarcity of oil, coking coal, aluminium, zinc and lead. These and the necessity of exporting her products and importing capital and labour continue to oblige South Africa to rely heavily on the rest of the world. South African mining is oriented towards the precious; since the beginnings, gold has represented three-quarters of the total value produced, diamonds 8% and coal 6%, and during the last decade gold still had a two-

TABLE LXXII

RESULTS OF SOUTH AFRICA IN 1963

Mineral	*Production*[1]	*$ thousand*	*Percentage of 1963 over 1951*
Gold (ounces)	27.400,000	960,000	240
Uranium	4,500	93,000	—[2]
Coal	47,000,000	95,000	270
Diamonds (carats)	4,400,000	51,000	170
Asbestos	205,000	30,000	200
Copper	60,000	30,000	150
Platinum group (ounces)	350,000	±32,000	330
Manganese ore	1,500,000	18,000	150
Iron ore	4,900,000	17,000	770
Limestone, lime	8,200,000	13,000	250
Chrome ore	870,000	6,800	150
Vanadium oxide	2,500	4,200	—[2]
Nickel	2,900	4,700	380

TABLE LXXII (continued)

Mineral	*Production*[1]	*$ thousand*	*Percentage of 1963 over 1951*
Sillimanite	62,000	2,500	—[2]
Pyrite	460,000	3,800	2,200
Tin conc.	3,800	3,800	160
Antimony	20,000	4,100	47
Phosphates, crude	500,000	5,600	5,600
Salt	220,000	2,800	180
Silver (ounces)	2,700,000	3,500	350
Granite	?	3,400	770
Fluorspar	58,000	1,300	840
Flint clay	150,000	1,800	—[3]
Vermiculite	99,000	1,800	500
Titanium conc.	32,000	400	—[2]
Gypsum	210,000	760	180
Fire clay	250,000	460	?
Sandstone		1,100	250
Silica	300,000	950	330
Osmiridium (ounces)	6,000	400	60
Slate		600	130
Magnesite	108,000	720	700
Monazite	2,300	450	—[2]
Feldspar	46,000	430	190
Andalusite	11,000	200	290
Zircon	2,600	130	—[2]
Marble	5,000[4]	90	100
Lead conc.	25	12	4
Kaolin	38,000	270	400
Emeralds	0.4	410	—[2]
Mica	2,300	200	720
Cement shale	190,000	110	110
Pigments	4,400	100	50
Wonderstone	2,000	180	630
Silcrete	16,000	52	40
Bentonite	8,600	120	—[2]
Magnetite	580,000	28	—[2]
Graphite	670	28	480
Beryl	420	100	40
Baryte	2,770	27	130
Talc	7,600	35	100
Corundum	80	4	1
Tantalite	30	62	1,600
Fuller's earth	500	2	40
Lithium ore	420	14	—[2]
Diatomite	220	3	230
Tiger's eye	130	55	290
Total value		1,415,000	250

[1] Production is shown in tons unless stated otherwise.
[2] Not sold in 1951.
[3] Not listed in 1951.
[4] Including others.

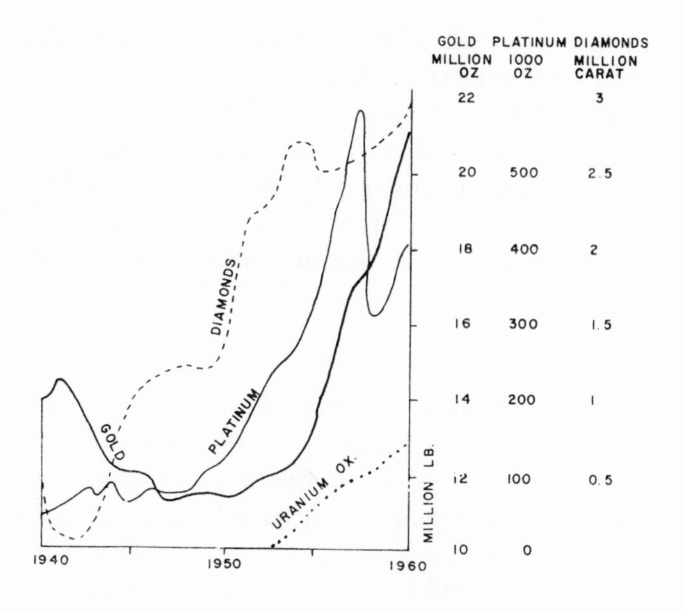

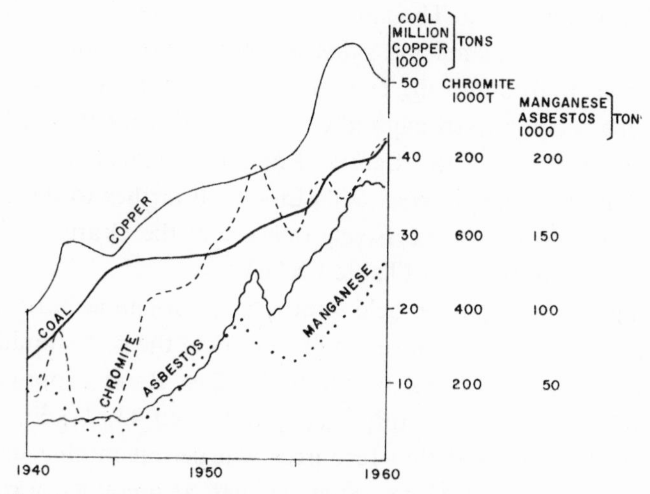

Fig. 39. South Africa output of coal, asbestos, chromite, copper, gold, manganese, platinum, uranium and diamonds.

TABLE LXXIII

SOUTH AFRICAN GROUPS

Group	*1963 product $ million*	*Share, %*			*Others*
		Gold	*Uranium*	*Coal*	
Anglo-American	430	34	14	33	diamonds 90%, titanium 100%
Gold Fields	190	17	10	3	tin 40%
Central Mining	170	15	16	10	
General Mining	120	9	32	4	
Union Corporation	130	14			chromium
Johannesburg	40	2	10	5	antimony 100%, manganese
Anglo-Transvaal	110	7	19	5	platinum group, nickel 60%

thirds share. The dynamics of South African mining can be best judged by the doubling of its value within seven years' time (Fig. 39, Table LXXII).

What then is the outlook? With its broad industrial base, the potential for expansion and diversification of the steel and manufacture oriented industries is large. These developments will have to complement the lapsing of uranium contracts in 1973, and the possible depletion of gold resources by the end of the century. Meanwhile, their sheer size will help the seven groups which control the industry to face what the future holds (Table LXXIII).

GOLD

Africans have worked gold for centuries, but not until 1886, did prospectors find the noble metal in the Witwatersrand. But from that year to 1964, South Africa produced almost 750 million ounces (about 23 thousand tons) of gold, which realized more than £6,200 million or, at the present rate of exchange, $17,200 million. Nevertheless, we should not forget that, beginning in 1933, the value of an ounce of gold changed from £4.3 to £8, whence it rose in 1950 to £12.4 ($35). In other words, in our century as before, currencies have lost their value, but not gold. The mines of the Rand are a living tribute to the technical, organizing and financial talents and skills of the 450,000 people engaged in this enterprise. Not only did geologists find the extensions of the Witwatersrand by reasoning and drilling, but also, engineers devised methods to reach and work these (the world's deepest mines). South of Johannesburg, the Central Rand first extended east and west, thence into the detached Far West Rand (West Wits Line) and farther to the Klerksdorp Fields. And then in 1946, experts drilled excessively rich ore in the Orange Free State which, in 1960, became the principal producer (Table LXXIV).

With the opening of the Evander field southeast of Johannesburg, the gold industry lately again showed its vitality. But the great producers of the past should not be forgotten. Indeed, prior to 1964, Crown Mines, Randfontein, East Rand Proprietary, Government GM Areas and New Modderfontein, respectively produced 43, 21, 32, 27 and 19 million ounces of gold, and it is believed that resources will permit mining until the end of the century. Indeed, the Chamber of Mines estimates that, at present working costs ($6.30 per ton) recoverable reserves attain less than 700 million ounces, or 27 years of activity. But, costs have doubled since 1949 and since 1959, the average rate of increase attains $0.30 per ton. A 25% increase of costs would render 8% of the reserves unexploitable. The

TABLE LXXIV

OUTPUT OF THE GOLD FIELDS

Field	*Tons milled % of total*	*Sold ounces recovered % of total*	*Working profit % of total*
Evander	2	3	2
East Rand	29	18	11
Central Rand	12	7	1 (Central Rand and West Rand combined)
West Rand	7	2	
Far West Rand	14	20	26
Klerksdorp	12	16	17
OFS	24	34	43

TABLE LXXV

RESULTS OF GOLD MINES IN SOUTH AFRICA IN 1963

Mines	*Gold 1,000 oz.*	*Ore 1,000 tons*	*Grade dwt per ton*	*Profit*[2] *$1,000*
Anglo-American				
Western Holdings	1,640	2,280	14.4	40,000
President Brand	1,480	2,030	14.6	37,000
Free State Geduld	1,540	1,490	20.6	38,000
Vaal Reefs	840	1,810	9.4	16,000
Daggafontein	540	2,780	3.9	10,000
Western Reefs[1]	610	1,870	6.5	8,900
President Steyn	700	1,990	7.1	9,700
Welkom	580	1,730	6.7	7,500
South African Lands	300	1,400	4.3	1,900
East Daggafontein	250	1,310	3.8	1,600
Brakpan	150	1,630	1.8	400
Western Deep Levels	800	2,280	7.1	11,000
Gold Fields				
West Driefontein	1,800	2,550	14.1	39,000
Doornfontein	670	1,500	8.9	11,000
Venterspost	530	1,480	7.2	5,000
Libanon	440	1,440	6.1	4,400
Vlakfontein	250	650	7.7	3,300
Vogelstruisbult[1]	190	900	4.3	1,800
Free State Saaiplaas	260	990	5.3	700
Sub Nigel	180	790	4.7	400
Luipaardsvlei[1]	180	1,260	2.8	4,800
Simmer and Jack	70	330	4.2	30[3]
Robinson Deep	120	640	3.7	20
Spaarwater	40	130	6.5	40
Rietfontein	30	100	5.2	40
Dominion Reefs[1]	3	20		4,500
Rand Mines				
Blyvooruitzicht[1]	1,110	1,760	12.6	26,000
Harmony[1]	970	2,440	8.0	15,000
East Rand Proprietary	760	2,970	5.1	3,600
Durban Deep	440	2,410	3.7	1,700
Crown Mines	290	1,650	3.5	400[3]
City Deep	290	1,390	4.2	300[3]
Consolidated Main Reef	80	690	2.2	200[3]
Rose Deep	50	290	3.2	100[3]
Union Corporation				
St. Helena	1,020	2,410	8.5	22,000
Grootvlei	550	2,740	4.0	6,900
East Geduld	380	1,410	5.4	5,400
Bracken	470	1,130	8.4	8,900
Leslie	450	1,440	6.2	6,200
Winkelhaak	440	1,270	6.8	6,200
Marievale	280	1,180	4.8	3,800
Geduld	130	930	2.9	300
Van Dyk	70	380	3.7	400

TABLE LXXV (continued)

Mines	*Gold* *1,000 oz.*	*Ore* *1,000 tons*	*Grade* *dwt per ton*	*Profit*[2] *$1,000*
General Mining				
Stilfontein	1,110	2,420	9.1	17,000
Buffelsfontein[1]	950	2,200	8.6	19,000
West Rand[1]	270	2,470	2.2	2,000
Ellaton[1]	10	20	8.3	200
Anglo-Transvaal				
Hartebeestfontein[1]	750	1,620	9.2	13,000
Virginia[1]	480	1,810	5.3	9,500
Loraine	410	1,030	7.9	4,900
Rand Leases	190	1,010	3.8	40
Johannesburg				
Freddies Consolidated	140	660	4.2	1,200
Randfontein[1]	100	9,300	2.1	7,700
Western Areas	340	1,360	5.0	1,900
East Champ d'Or	5	110	0.9	200
New Kleinfontein	90	640	2.8	140[3]
Witwatersrand Nigel	50	260	4.4	100
Total with others	27,430	78,420	6.9	438,200

[1] Uranium producers.
[2] Profits include uranium and acid.
[3] Losses.

great producers would be less effected: in 1962, the 24 mines that have opened since World War II contributed 90% of the industry's profits and 73% of the gold production. These mines have experimented with concentrated mining. In order to cut unit costs, stopes are concentrated to a limited area in which ventilation, equipment and supervision can be intensified. The worked out faces can be abandoned and rapid progress resumed. With a fixed price of gold and rising wages and costs, the mines fight in effect for time. The old mines cannot compete in this race. Twenty of these are expected to cease operations before 1970. Consequently, while output continued to rise at a rate of 10% in 1963 and the industry acquired more than $300 million of stores and equipment in 1964, the government was obliged to follow the example of other gold producing countries, such as Canada and Australia, and offer subsidies to the old mines of the Central and Eastern Rand. As a first step, marginal producers received a pumping cost subsidy and loans to cover losses.

During half a century of progress, exploration and valuation techniques evolved which could guarantee investments of up to $50 million. Modern shafts boasting a diameter of 27 ft. are sunk with record speeds (1,000 ft. per month) to record depths (7,000 ft. and more). At present, longwall stoping is used more frequently than any other method. Miners now build piles or mats of slabbed timber as pillars or use hydraulic props. Locomotives haul up to 20 ton-cars to skips holding 10 tons of ore, which double drum, electrical machines or Koepe installations hoist at speeds of 40 miles per hour. Ventilation averages 100 cubic ft. of air each minute per miner working. In a number of West

Wits and Free State mines it is necessary to pump as much as 15 million gallons of water each day. In the mills, machines crush and grind the ore; the gold is recovered by gravity concentration and cyanidation, which now achieves a recovery of 97.6%. Over the years, efficiency and costs have increased. As of 1963, working costs stood at $5.3 per ton milled or $21 per ounce recovered. According to various formulae in taxation, the Government takes less than 30% of the gold profit or 20% if uranium and other byproducts are included.

ANGLO-AMERICAN GOLD MINES

Orange Free State and Klerksdorp

The corporation's holdings comprise five of the 13 mines in the *Orange Free State* (Fig. 40). While working profits of Western Holdings Ltd. increased from $13 million in 1958 to $27 million in 1960 and $40 million in 1963, reserves augmented from 4.3 to 5.9 million tons, and recovery improved by 10%. In 1964, capital expenditure requirements were planned to reach $4.2 million, mainly for mill extension. As a result of faulting, the Basal Reef is found at shallow depth near its sub-outcrop, but values tend to decrease.

The profits of President Brand, located south of Western Holdings, have also been increasing rapidly, with ore reserves increasing slightly. A third shaft has been added to the two preceding ones. The President Steyn Company repaid a $5.6 million loan to the Anglo-American Corporation in 1962. In 1964, capital expenditure on shaft sinking reached the mark of $4 million. With the commencement of stoping in the no. 3 shaft area, the ore tonnage is increasing. In the Free State Geduld shaft, the Basal Reef lies at a depth of 4,540 ft. The $8 million cost of sinking and equipping the no. 4 shaft is financed from profits which increased from $12 million in 1958 to $38 million in 1963. Exploratory drilling

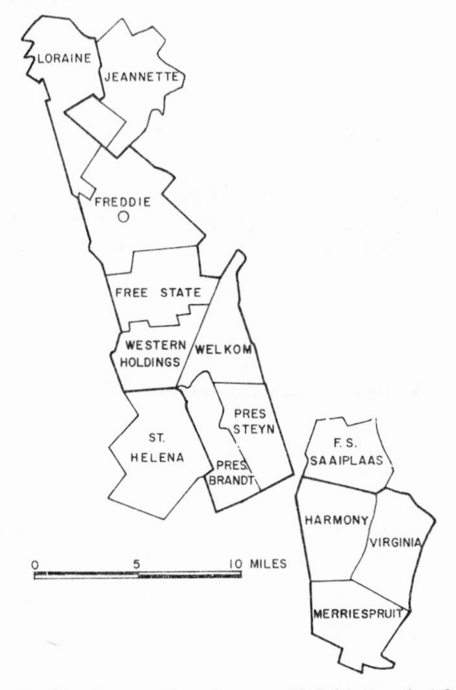

Fig. 40. The Orange Free State gold field, South Africa.

is continuing actively. Similarly, the outlook of the Welkom Company has improved; since 1954, revenues have increased constantly but so have working costs per ton. Nevertheless, capital expenditure is now low.

The Anglo-American group controls the two southwestern mines of the *Klerksdorp* field. Between 1958 and 1961, profits, reserves, gold and uranium grades increased steadily at Vaal Reefs where, in 1961, a new shaft system costing $15 million was commissioned. Similarly, profits and production of adjacent Western Reefs, a mine opened in 1940, continues to expand (Fig. 42).

Witwatersrand

Western Deep Levels, in the *Far West Rand* south (Fig. 41a) of Blyvooruitzicht, is the largest individual gold mining project ever undertaken. The Central Mining Finance and Gold Fields group are also included in this loan syndicate, but after the exhaustion of the $43 million capital (mainly on shaft sinking) only the Anglo-American Corporation provided loan facilities. Expenditures of 1961 have expanded to reach $27 million. The mineralized Ventersdorp Contact Reef has been intersected by two shafts, respectively at 5,600 and 6,400 ft. of depth. Further sums will be needed before the Carbon Leader is reached between 7,500 ft. and 12,500 ft. of depth. Since water is particularly abundant underground, special sinking and pumping techniques are used. In March of 1963, this deep mine are forwarded to the mill 200,000 tons of ore.

The Anglo-American Corporation controls five of the *East Rand* mines. At Daggafontein, ore has been mainly drawn from haulage pillars. The milling rate, which during the last decade has averaged 2.7 million tons yearly, is expected to decline as will profits and reserves. At East Daggafontein, decline of the milling rate was reversed in 1951; production comes mainly from the Kimberley Reef of the western sector. Rising costs reflected in-

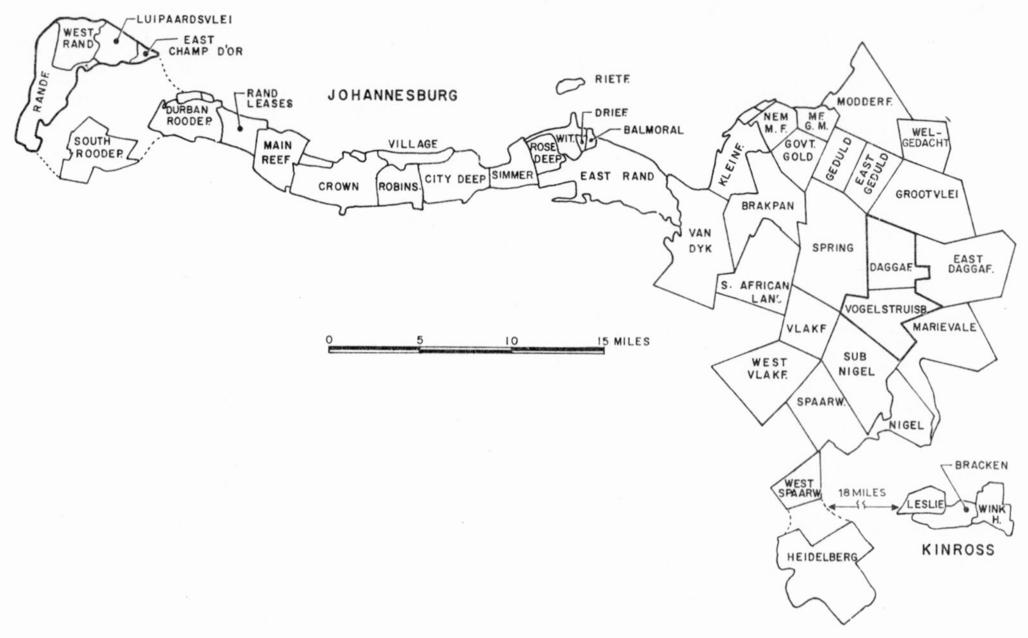

Fig. 41a. The Witwatersrand.

creased distances of haulage to shaft no. 1 of South African Land and Exploration Company, where the cost of the new shaft system was estimated to amount to $7.5 million. Between 1911 and 1963, Brakpan produced a revenue of about $300 million at contemporary prices. Reserves are decreasing, and at Springs they have been rapidly depleted.

GOLD FIELDS GROUP

The *Consolidated Gold Fields of South Africa*, the historic successor of the original company initiated by Cecil Rhodes, now draws a majority of its revenue from West Driefontein, the second largest profit maker of the Rand. The company also exploits uranium in the Free State Saaiplaas shaft; it has substantial interests in platinum mining at Rustenburg, both on a local level and through Johnson and Mattheys.

The Gold Fields group controls a majority of the mines of the *Far West Rand* (Fig. 41a). Between 1960 and 1964, the quantity of recovered gold increased by 33% at West Driefontein. In 1963, $1.5 million were spent on capital expenditure of which mill building required $12 million. Until 1965, an investment of $37 million will be covered from profits. The Carbon Leader Reef continued to provide a majority of the ore (130,000 tons per month) with the Ventersdorp Contact Reef giving encouraging results. The milling rate from this Reef is expected to reach 75,000 tons per month and cumulative capacity will reach the mark of 230,000 tons per month by 1965. In the Venterspost Mine, the sinking of shaft no. 3 and incline 2B permitted the opening of the southern and deep sectors of the claim. During these operations, capital expenditure varied between $1 and 1.5 million, partly necessitated by water flowing from fissures. A steady output is maintained at Doornfontein where sinking of the no. 2 shaft is planned to cost $7 million. The Carbon Leader dips from a depth of 4,150 ft. to an estimated 8,500 ft. and the capacity of the reduction works has been increased to 125,000 tons per month. At Libanon, the Harvey

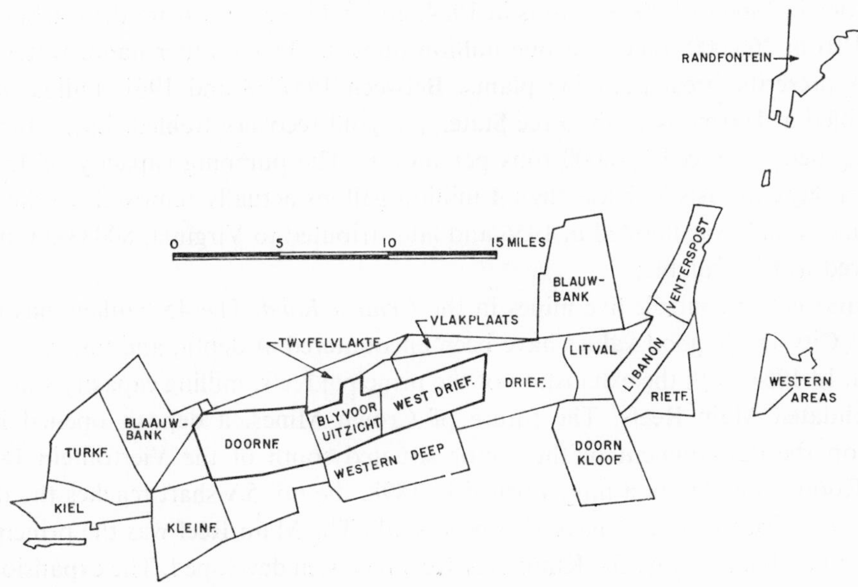

Fig. 41b. The ar West Witwatersrand.

Watts Shaft has pierced the Carbon Leader Reef at 6,170 ft. of depth; capital expenditures attained $1.2 million. However, the Main Reef is the principal productive horizon.

In the *West Rand* at Luipaards Vlei, ore reserves of the Bird Reef and Main Reef are relatively stable; a little tonnage is expected to be added, however, beyond the Witpoortjie Fault. Activity tended to concentrate on the uranium ores of the Bird Reef where grades remain at 2 lb. per ton. Three mines are located in the *Central Rand* near Johannesburg: operations will soon cease at Simmer and Jack (a mine which has operated since 1888); at Robinson Deep, milling rates have been reduced, partly because of pressure bursts; and profits are decreasing at the old Rietfontein mine.

The group is active in the *East Rand* where it controls four mines (Fig. 41a). At Vlakfontein, gold occurs in shoots at 6,600 ft. with 16 million tons of coal lying at less than 320 ft. of depth. At Vogelstruisbult, reserves are decreasing but uranium provided in 1964 revenue. At Sub Nigel, both yields and profits have progressively declined; the mine has acquired the tribute of the deep lying northeastern sector of the Spaarwater claim. Similarly, at Spaarwater, efforts have been made to increase reserves of the western payshoots. In the Orange F.S. field at Free State Saaiplaas, development has been hindered by water-bearing fissures. The capacity of the plant attains 120,000 tons per month; $7 million of new capital have been subscribed, thus bringing the total to $40 million.

RAND AND UNION CORPORATION GOLD MINES

Rand Mines Ltd.

Financed by Central Mining, the group controls Blyvooruitzicht and Harmony (two mines that provide a majority of its profit), eight other gold mines and an asbestos mine in Southern Rhodesia. Results of Blyvooruitzicht in the West Rand (Fig. 41a) are favourable. The completion of the B series of sub-incline shafts caused mill tonnage to increase from 400,000 tons in 1960 to 1,700,000 tons in 1964, and gold recovery from the Carbon Leader increased from 260,000 ounces to one million ounces. On the other hand, water-bearing dolomites prescribe great pumping plants. Between 1957/58 and 1961, milled ore more than doubled at Harmony in the Free State, and gold recovery trebled. Later, the milling rate is planned to exceed 200,000 tons per month. The pumping capacity of 13 million gallons per day contrasts with less than 4 million gallons actually removed. In the Merriespruit Mine, which was flooded in 1956 and later tributed to Virginia, 800,000 tons of ore are believed to be available.

The Rand group controls five mines in the *Central Rand.* The 45 haulage has been re-opened at City Deep, good values have been encountered at depth, and losses have lately decreased. In 1964, with the exhaustion of the principal reefs, milling capacity was reduced at Consolidated Main Reefs. The future of Crown Mines, a deposit opened in 1897, depends on the development of the sector situated south of the Vierfontein Dyke. At Durban Roodepoort Deep, a mine opened in 1898, the no. 5A shaft reaches the depth of 9,146 ft.; no. 1 East and no. 8 have also been sunk. The Main Reef was the principal productive banket, but recently, the Kimberley Reef has been developed. The expansion of the united plant to a capacity of 200,000 tons per month has cost $2 million. Rose Deep, which dates from 1897, now nears the stage of depletion; also, notice has been given to the government of the final closure of Modderfontein East in the *East Rand*. East Rand Proprietary, a mine which started operations in 1908, is being transformed to operate at a depth of 12,000 ft., for which capital expenditures have been planned at $60 million.

Union Corporation Ltd.

This group produces in excess of $150 million worth of gold and base metals yearly. The group entirely controls the *Evander Field* east of the Witwatersrand, where its investments are expected to reach the mark of $80 million by the time the Bracken and Leslie Mines have completed the stage of full production. By 1970, Evander–Brendan township will have a population of 10,000, and the number of inhabitants of Kinross will increase to almost 20,000. The corporation owns the South African Pulp and Paper industries of Tugela, Natal, whose earnings have reached the mark of $25 million.

At St. Helena, the first of the *Orange Free State* mines, the sinking of no. 7 shaft to a depth of 5,340 ft. and the opening of deeper levels in no. 2 shaft were completed in 1961; no. 7 shaft does facilitate access to the upfaulted eastern zone of the lease and the elimination of 2 million cubic ft. of air per minute. The amount of water pumped is in excess of 1,000 million gallons. Indeed, part of the no. 2 shaft area is heavily faulted. The reduction plant is capable of milling 200,000 tons of ore per month. The Corporation controls five mines in the *East Rand*. At Grootvlei, working revenue averages more than $19 million. The exploitation of the Main Reef has priority over the Kimberley Reef, but at Marievale more than 25% of the ore is already drawn from the latter. Since the bulk of the remaining ore of East Geduld is contained in drive pillars, costs are expected to rise. Van Dyk is expected to make small profits for only a few more years, and Van Dyk Consolidated will be converted into an investment company.

In the *Evander Field* (Fig. 41a), 1959 was the first full year of production of Winkelhaak. The tonnage extracted from the Kimberley Reef is slowly increasing but profits have risen rapidly. No. 2 shaft was planned to reach a depth of 4,300 ft. and the capacity of the reduction plant is to be increased from 100,000 to 150,000 tons per month. Before it started production in 1963, the investment in Leslie Mines was expected to total $28 million; financing was facilitated by the drawing of $3 million loan from the National Finance Corporation. No. 1 shaft reaches a depth of 3,170 ft. The capacity of the mill has been expanded from 75,000 to 100,000 and later to 150,000 tons per month. The investment and loan requirements of Bracken are similar. Here, the Kimberley Reef lies at a depth of only 700 ft. Trial milling was started in 1961, which was earlier than at Leslie, because no. 1 and 1A shafts have reached their depths of more than 2,600 ft. North of Bracken, 4 years and $38 million will be needed to bring the Kinross mine to a 100,000 ton per month production.

OTHER GOLD MINES

General Mining and Finance Corporation

This group controls three mines of the Klerksdorp (Fig. 42) field (one of which, Ellaton, is at the end of its useful life) and the West Rand mine. Working profits of Buffelsfontein

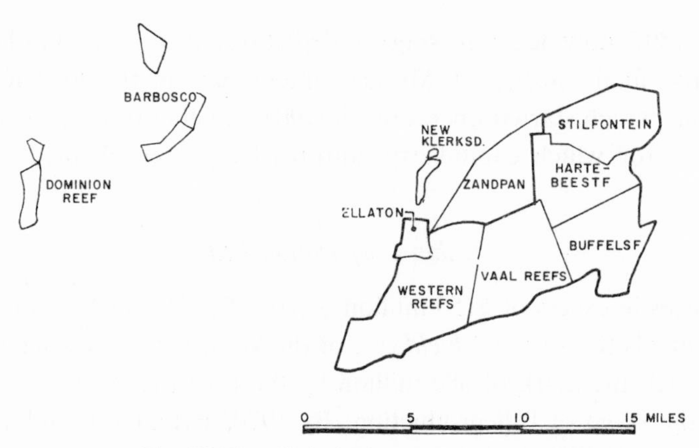

Fig. 42. The Klerksdorp area, South Africa.

are generally maintained above the level of $17 million. However, the cessation of uranium production will tend to reduce profits. Three subvertical shafts will permit access to the 5,000–8,000 ft. levels of the Vaal Reef. Their sinking will be staggered between 1958 and 1970; and during the next seven years, the required capital expenditure will be $8 million per annum. At Stilfontein, the quantity of milled ore has steadily risen during the last year; the milling capacity reached 160,000 tons per month. The mine has been handicapped by an increased flow of water and the low grades of the Margaret Shaft. The sinking of the Toni Shaft, the extension of the pumping facilities and of the milling capacity to 200,000 tons have required an investment of $7 million per year. At *West Rand*, a large uranium producer, a ferrochrome plant will be erected. Finally, in 1961, 22,000 ounces of gold were produced at South Roodepoort Main Reef.

Anglo-Transvaal Consolidated Investment Company

The group controls the new Hartebeestfontein mine and three others; as well as the Murchison gold-antimony mines; the small Agnes, New Consort and Sheba mines; the Associated Manganese Mines at Postmasburg; the Ermelo refinery; the Union Lime Company; and, through Anglo-Transvaal Industries Ltd., at least part of fish, glass, iron equipment, engineering, plastics and other industries.

Hartebeestfontein in the Klerksdorp field (Fig. 42) already draws half of its ore from deeper levels. Since a fault traverses the mine, it is planned to increase the number of shaft systems from four to five. On the other hand, since reef widths attain less than 40 inches, one third of the broken ore has been sorted out, thus permitting increase of the milling rate to not more than 130,000 tons per month. Since capital expenditures have progressively decreased from the yearly $7 million rate, a $2.5 million loan has been repaid to the Anglo-American Corporation. At Zandpan near Hartebeestfontein, shaft stations have been cut at vertical intervals of 150 ft., between 6,500 and 7,250 ft. of depth. The vicinity of the first shaft is heavily faulted, but the company is assured of loans totalling $20 million, in addition to its capital of $15 million. At Virginia, the shaft system has been expanded to facilitate mining areas tributed from Harmony.

In the Orange Free State (Fig. 40), the Loraine mine is relying more and more on ore from the Elsburg Series Reefs which are also exploited in the Riebeck area. Capital expenditure is decreasing more rapidly than at Hartebeestfontein, but hoisting capacity is now limited by ventilation facilities to 150,000 tons per month. A majority of the ore of Rand Leases (Vogelstruisfontein) in the Central Rand is of marginal payability. The milling rate has been maintained at 130,000 tons per month until 1964, but capacity equals 206,000 tons.

The Anglo-Transvaal group, through Eastern Transvaal Consolidated Mines, controls the New Consort, Sheba, Agnes and Mamre mines (which produce respectively 30,000, 30,000, 15,000 and 1,000 ounces) and the Frantzina's Rust and Mount Morgan properties.

The Johannesburg Consolidated Investment Company

The group controls four mines, and brought the Western Areas gold mine into production; it has a 37½% share in the exploration scheme adjacent to this area and a 44% participation in the syndicate exploring the area north of Kinross, near Klerksdorp, and at Kroonstad. The company also controls the Rustenburg Platinum and the African Asbestos Cement Corporations.

The profits of Randfontein Estates in the *West Rand* (Fig. 41a) have lately been maintained around $8 million. The rate of milling and capital expenditures will be reduced from the $300,000 yearly level. At Western Areas, in the Far West Rand, shaft sinking started in 1960; the main shaft has been designed to hoist more than 100,000 tons of ore per month and milling operations have started on Elsburg or Ventersdorp Contact Reef ore. Reserves are estimated at 68 million tons of ore averaging 9.3 dwt. per ton or approximately 1,000 tons of gold! At Government Gold Mining Areas (Modderfontein), the quantity of milled ore was reduced in 1962, but pyrite production continues. In 1961, Freddies Consolidated in the Free State made a small profit for the first time in many years. Nevertheless, the future of the mine depends on finding reserves in the Elsburg and *B* Reef horizons in addition to the decreasing resources of the Basal Reef.

Smaller groups

The Henderson Group controls the Tweefontein and Witbank collieries in Transvaal the Arcturus and Muriel mines, and Witwatersrand Nigel, and are now concentrating on the shoots of the Poortje section. The Northern African Group controls New Kleinfontein and, through Central South African Lands and Mines, the Afrikander Lease, a uranium producer; it plans to maintain a moderate rate of activity at Kleinfontein as long as possible.

URANIUM AND OTHER BYPRODUCTS

Providing additional profits, uranium production had an enormous impact on the gold industry. In 1923, geologists identified uranium in the gold fields; however, only when the direct leaching process proved to be applicable (five years after the evaluation of 1945) did systematic recovery commence. In the Atomic Energy Board, the South African Government vested the right to own and purchase uranium. The Board passed various contracts with gold producers, the last of which provides for the production of 28,000 tons of uranium oxide between 1961 and 1973 at a price of $11.20 per pound of U_3O_8 which is a profit of $3 per pound. By 1973, contractual output will have declined from 4,800 to 1,300 tons per year but resources of 220,000 tons of uranium metal in grades averaging 0.5 lb. U_3O_8 per ton will still be available. During the same period of time, the number of participating mines will decline from 29 to 6, with most mines stockpiling uranium dumps. The effect of this evolution can best be gauged from the distribution of working profits according to products (Table LXXVI).

TABLE LXXVI

GOLD AND BYPRODUCT PROFITS

Minerals	*1960* *$ thousand*	*1963* *$ thousand*
Gold	275,000	388,000
Uranium	77,000	55,000
Sulphuric acid	4,000	4,000
Silver	300	400
Osmiridium	30	30

TABLE LXXVII

URANIUM OUTPUT THROUGH 1963

Group	*Tons U_3O_8*
Anglo-American Corporation	8,900
General Mining and Finance	11,000
Johannesburg Consolidated	8,200
Gold Fields of South Africa	6,900
Anglo Transvaal Consolidated	6,500
Rand Mines	6,200
North African Mining	900
Total	48,600

TABLE LXXVIII

OUTPUT OF PRINCIPAL URANIUM MINES IN 1963

Group	*Mine*	*U_3O_8 tons*	*Ore treated 1,000 tons*	*U_3O_8 lb./ton of ore*
General Mining	West Rand	770	840	1.59
	Buffelsfontein	375	1,450	0.52
Gold Fields	Luipaardsvlei	510	410	2.51
	Dominion Reefs	25	20	2.28
	Vogelstruisbult	45	110	0.92
Rand Mines	Harmony	405	2,060	0.40
	Blyvooruitzicht	340	1,760	0.39
Anglo-American	Western Reefs	700	1,850	0.75
Johannesburg	Randfontein	445	540	1.65
Anglo-Transvaal	Hartebeestfontein	485	1,620	0.61
	Virginia	455	1,805	0.50

Uranium production is more evenly distributed between groups than gold output. The output of the groups from the beginning through 1963 is shown in Table LXXVII.

Of this total, Randfontein and adjacent West Rand Consolidated (Fig. 41a) contributed almost third; the 1963 results are included in Table LXXVIII.

By 1970, six mines will produce 18,000 tons or 60% of the projected output. The 1963 reserves of these mines are shown in Table LXXIX.

TABLE LXXIX
1963 RESERVES

Mine	*Field*	*1963 Reserve*	
		thousand tons of ore	*U_3O_8 lb./ton of ore*
West Rand	West Rand	3,400	1.3
Buffelsfontein	Klerksdorp	5,700	0.7
Western Reefs	Klerksdorp	2,500	0.8
Hartebeestfontein	Klerksdorp	4,700	0.7
Blyvooruitzicht	Far West Rand	4,800	0.6
Harmony	Free State	5,100	0.6

At West Rand Consolidated, reserves of the Black Reef boast a grade of 20 ounces per ton; thus, this mine is expected to easily continue production after 1973. Fairly late in 1961, in the East Rand at Daggafontein (Anglo-American group), the unit cost of uranium has been reduced. The principal uranium producer of the Johannesburg Consolidated group is Randfontein, but the reserves of East Champ d'Or are decreasing. Hartebeestfontein, a low cost producer, has a plant equipped for the ferric leach process. In the Free State (Fig. 40), the concentrates of Vaal Reefs are processed at Western Reefs, but Harmony has not been authorized to process all its slimes. In addition to the plants of Welkom and President Steyn, Anglo-American is expected to continue mining uranium at Western Holdings. What methods of concentration do they use? From the cyanide residues of tailings dumps of 27 gold mines, 19 uranium plants have recovered uranium. In the usual flowsheet, sulphuric acid leaches the filtrate to which manganese dioxide is added before secondary filtration. Then nitric acid and ammonia exchange the ions of the pregnant solution and produce ammonia diuranate. Finally, Calcined Products Ltd. dry and pack this precipitate which is sold to the American and British atomic energy commissions. By now, metallurgists have improved the recovery to 85%.

To concentrate one ton, the mills require 44 lb. of sulphuric acid and companies have built plants to convert half a million tons of *pyrite* into acid each year. Reserves are abundant. In 1963, Virginia (Anglo-Vaal Group, Orange Free State) made a $1 million profit on these operations; Buffelsfontein, $700,000; Harmony, $500,000; Blyvooruitzicht, $400,000; and three others, a total of $1,100,000. In addition to these, the O'okiep copper and the Rooiberg tin companies also produce pyrite, but sulphur production will concomitantly decrease with uranium output.

Gold bullion contains *silver* and the concentrates include osmiridium; these are twin metals related to platinum. On the other hand, osmiridium averages 16% of another rare metal, ruthenium. Whereas the gold/silver ratio is fairly constant at 9/1, *osmiridium* grades vary considerably. The Main and Kimberley reefs of the East Rand are distinguished by the highest grades, with others in the Black Reef of Modderfontein; values in other gold fields are insignificant. Since 1925, the mean gold/osmiridium ratio increased from 1,500/1 to 3,500/1. While silver prices have risen, osmiridium prices have rapidly declined from their 1952 peak. Output and value evolved as shown in Table LXXX.

TABLE LXXX

PRODUCTION OF SILVER AND OSMIRIDIUM

Year	Silver		Osmiridium	
	Ounces	*$*	*Ounces*	*$*
1910	820,000	260,000	—	—
1930	1,000,000	260,000	5,600	200,000
1960	2,200,000	2,100,000	6,300	200,000

In addition to these, methane gas containing 8% helium escapes from the Orange Free State gold fields.

COAL

Africans have used coal for centuries, but industrial production only started in 1859. Subsequently, to satisfy the needs of the diamond and later of the gold mines, coal output expanded rapidly to 13 million tons in 1927, 19 million tons in 1940 and 47 million tons

TABLE LXXXI

SOUTH AFRICAN COAL RESOURCES

Field	Reserves million ton		BTU
	proved	*probable*	
Transvaal			
Soutpansberg	—	5,000	11,700
Waterburg	17,500	—	11,700
Springbok Flats	—	5,000	9,700
Witbank	4,100	—	11,700
Middelburg-Belfast	1,000	—	11,400
Bethal	—	7,000	10,700
Breyten-Ermelo	6,200	11,000	11,700
Springs	8,200	—	9,700
Vereeniging	200	—	9,700
Natal			
Kliprivier	400	1,700	12,200
Vryheid	200	—	12,700
Utrecht	500	—	12,200
Orange Free State			
Vierfontein	200	—	10,200
Odendaalsrus	—	1,000	9,700
Vereeniging	2,200	—	9,700

in 1963. While coal underlies vast areas of Transvaal and of the Orange Free State, the fuel, at depths that can be economically reached, lies only along the rim of this basin. The chain of fields arches from north of Durban towards Johannesburg and beyond; however, with the exception of 300 million tons in Natal, all this coal is non-coking. Also in Natal, and neighbouring Swaziland, some anthracite is found. However, in one of the fields of northern Transvaal, geologists have evaluated 6,200 million tons of coking coal. Total reserves of 80,000 million tons of coal averaging only 10,600 BTU are distributed as shown in Table LXXXI.

From 10 ft. thick horizontal seams averaging only 200 ft. of depth, 50 collieries extract the world's cheapest coal. Indeed, the pithead price averages only $1.90, but with distances, the freight to Lourenço Marques (in Mozambique) quadruples the price. Consequently, exports fell from 4 million tons in World War II to the present 1.6 million tons. By the pillar and stall method, twin borers and rail-mounted arc wall cutters cut and load coal in mechanized collieries; these methods tend to replace manual operations.

The Anglo-American Corporation controls one third of the South African coal production. Moreover, in 1964 Anglo-American contributed General Mining and Finance stock to Meinstraat Beleggings formed by the Drydens and Federale Mynbou. The company will finance the opening of the Camden field. Its production will increase from 5 million tons to 9 million tons in 1970. The South African and General Investment Trust own the Clydesdale Collieries which produce 10% of the country's output. In 1962, other groups had a smaller output (Table LXXXII).

TABLE LXXXII

OTHER COLLIERY GROUPS

Group	*Million long tons*
Central Mining and Delagoa	4.0
SASOL (state)	2.2
Johannesburg Consolidated	2.0
Henderson	1.3
Anglo-Transvaal	1.1
Gold Fields	1.0

Anglo-American Corporation collieries

Through Amalgamated Collieries of South Africa, the corporation controls the Cornelia, Schoongezicht and Springfield (Fig. 43) mines. The profit of Amalgamated Collieries has risen to almost $3 million, but authorized capital only attains $10 million. The shafts of *Cornelia*, the continent's second largest colliery, are situated 40 miles south of Johannesburg on both sides of the Transvaal/Orange Free State border. The no.2 shaft has been established to provide additional fuel to the Vaal Power Station, and the colliery's output has steadily risen from 3.5 million long tons in 1956 to 4.1 million long tons in 1960, with yearly profits averaging $2.5 million. The *Springfield* colliery, located north of Cornelia, normally supplies 2 million tons of coal to the Klip Power Station; its profits average $600,000.

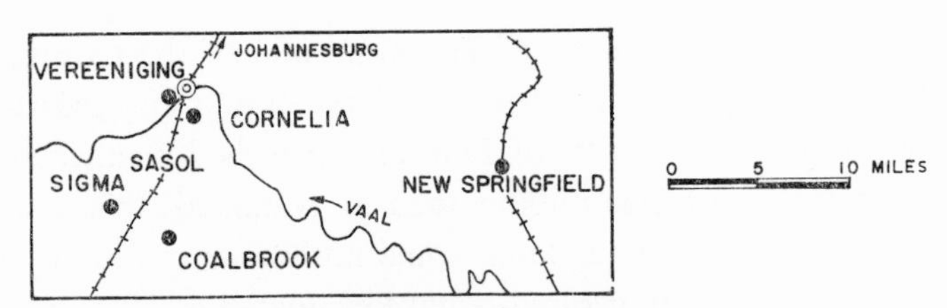

Fig. 43. The coal fields south of Johannesburg.

The Vierfontein colliery, situated in the Klerksdorp gold field in the Orange Free State, produces up to 1.5 million long tons yearly with a profit of $700,000.

In the *Witbank* basin (Fig. 44) east of Johannesburg, South African Coal Estates owns the Navigation and Landau collieries, the combined production of which attains 1.8 million long tons and rising profits of $1.4 million. ISCOR, the steel company, buys the output of Navigation. The New Largo mine situated west of Witbank has supplied, since 1953, 1.2 million long tons of coal for the power requirements of the Rand; another colliery is also ready for production. North of Witbank, the Coronation produces more than 1 million tons of coal yearly at a profit of $800,000. Three Anglo-American collieries are located southeast of Witbank. New Schoongezicht started operations in 1952 with its output of 900,000 long tons being disposed of through the Transvaal Coal Owners' Association; its profits range around $500,000. Since 1960, adjacent Blesbok has supplied 600,000 long

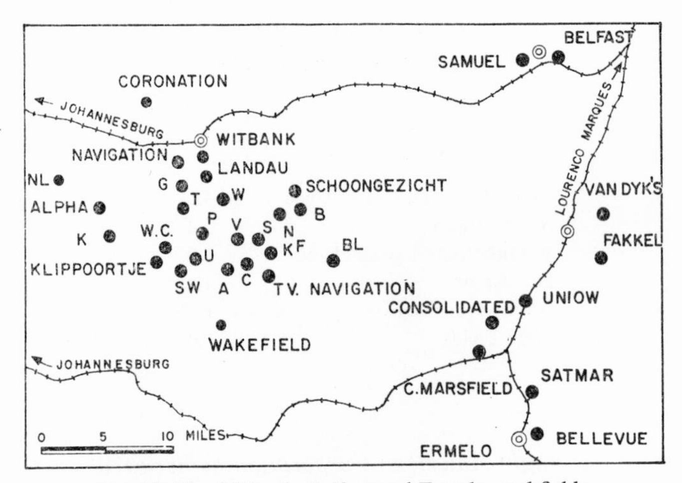

Fig. 44. The Witbank, Belfast and Ermelo coal fields.

tons of blend coking coal to ISCOR and its profits range around $300,000. The Springbok company (distinguished by a capital of $3 million) has invested $1.5 million in the development of the blend coking coal of seam no.5, which will supply 750,000 long tons per year to ISCOR with 250,000 long tons being absorbed by the coal trade.

Anglo-American has only two collieries in Natal. Natal Coal Exploration, situated at Ballengeich south of Newcastle, has operated electrically since its opening in 1951 and extracts 700,000 long tons yearly. Vryheid Coronation in northeastern Natal, with a capital of $7 million, produces 500,000 tons of coke yearly.

Transvaal and Free State

Six collieries of the Anglo-American Corporation and one mine of the South African General Investment Trust are located in the *Witbank* field; many have been discussed above. The basin can be divided into western, central, southern and southeastern sectors.

East of Johannesburg, the production of the Vlakfontein colliery of the Meinstraat Beleggings has steadily increased to more than 300,000 tons, but with profits declining, it is planned to transfer operation to Klipportjie. At some distance the western sector of Witbank includes the New Largo, Alpha, Acme, Kendal and Coniston collieries.

In the central sector, the Witbank mines, controlled by Rand Mines (the Central Mining group), produce 1.7 million long tons yearly and profits now exceed $1 million. The Phoenix colliery of the Johannesburg Consolidated group sells 1.5 million long tons of coal yearly and also fire clay; they formerly sold oil shale also. Other mines of the central sector include Navigation, Blackhill, Landau and Greenside. The Tweefontein mine of the Henderson Corporation produces up to 1.4 million long tons yearly with a profit of $600,000. Belonging to the same group, the collieries of Witbank Consolidated are located near Oogies in the southern sector; their capacity now attains almost half a million long tons yearly.

The Klipportjie colliery of Meinstraat Beleggings will increase its capacity from 450,000 long tons to 700,000 long tons yearly. The South Witbank (Anglo-African Trust) and Tavistock and Uitspan (Marlin Trust) collieries each sell 700,000 long tons of coal yearly.

Along the southeastern sector of the basin, collieries extend from Raleigh to Wakefield. At Douglas, the Transvaal and Delagoa Bay Company (Rand Mines) has financed, from profits, mechanization costing $1 million; their yearly output attains 1.5 million long tons. At Koornfontein, a mechanized mine of Meinstraat Beleggings, only 650,000 tons are produced yearly; however, its capacity is as much as one million long tons. Nearby, Van Dyks Drift colliery of Transvaal Consolidated Land and Exploration Company (Rand Mines) and the Transvaal Navigation mine at Vlaklaagte (Meinstraat Beleggings Edms., Beperk) respectively produce 600,000 and 700,000 long tons yearly.

Clydesdale (Transvaal) Collieries are distinguished by an issued capital of $7 million and by the trebling of production and profit between 1955 and 1959. The 1 million long ton output of New Clydesdale, south of Witbank, has remained stationary but the production of Coalbrook in the *Orange Free State* has increased from 1 million long tons to 2.7 million long tons in 1962 (Fig. 43). At North Coalbrook, the disaster of January 1960 demanded the lives of 437 men.

With an investment of $2.8 million and a yearly coal output of 2.2 million long tons, the state-owned South African, Coal, Oil and Gas Corporation (SASOL) mechanically operates the 600 million ton Sigma colliery near Coalbrook. Since the Sasolburg plant is at a distance of only 50 miles from Johannesburg but 400 miles from the coastal refineries, the gasoline distilled from coal costs only 60 cents (US) at the pithead and enjoys a 4 cents (US) price advantage over its competitors. The Lurgi gasification plants produce hydrogen and carbon monoxide (with tars and ammonium sulphate); then, the material is synthesized into 4,000 barrels of C–C_4 hydrocarbons and gasoline each day by the Kellog–Synthol process; then into 1,500 barrels of diesel oil and waxes by the Lurgi–Rurchemie method. However, the Government has decided to double the annual gasoline output to the rate of 66 million gallons by 1967. Indeed, apart from considerations of autarchy, the plant makes a $3 profit on each ton of coal; the state company does not pay taxes.

The Natal coal fields

Two basins (Fig. 45) and two isolated collieries can be distinguished in Natal; one of these extends from Newcastle (Star Colliery) 40 miles south, beyond Dundee. Natal Navigation Collieries (a sister company of Transvaal Navigation collieries) and its subsidiaries are the largest producers of the field, with yearly profits totalling $1.2 million (compared with a capital of $2.2 million). The Kilbarchan produces 1.4 million long tons yearly with

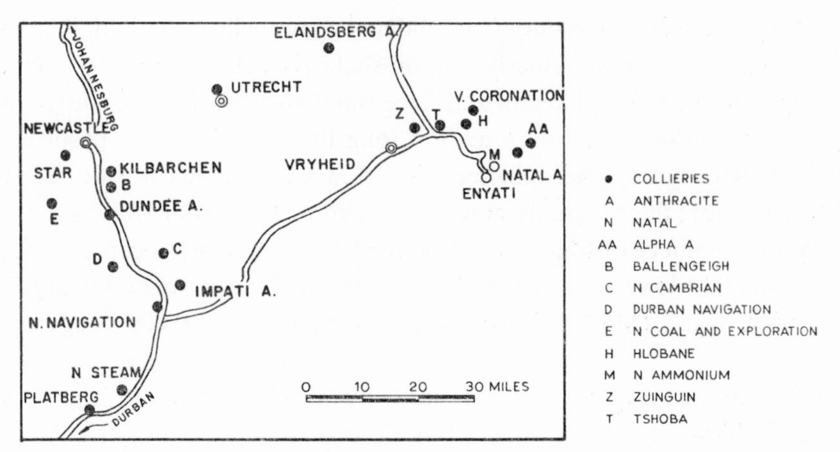

Fig. 45. The Natal coal fields.

the Northern Natal Navigation Mine supplying 200,000 long tons of coke. The Northfield property, near Dundee; Hlobane, near Vryheid; and Elandsberg, an isolated colliery, also belong to the group. The Natal Cambrian Collieries, a subsidiary of Johannesburg Consolidated Investment, has a capital of $1 million; its Ballengeich and Cambrian mines produce respectively 500,000 and 40,000 long tons yearly. Durban Navigation sells 470,000 long tons yearly to its parent company ISCOR. On the other hand, Dundee Anthracite now sells 100,000 long tons yearly, but Dundee Coal processes tar only. Finally, the isolated Utrecht (Welgedacht) Company has supplied coal since 1910, and now produces more than 200,000 long tons, and small profits.

In another field extending east of Vryheid, Zinguin sells more than 220,000 tons each year for a profit of $140,000. Natal Ammonium was obliged to dump 40,000 long tons of its 180,000 production for lack of transport; nevertheless, the company made a profit of $300,000. Natal Anthracite has a similar performance and reserves of 18 million long tons. Also belonging to the Stafford Mayer Company, Enyati has a capacity of ½ million long tons, but Alpha Anthracite is at the end of its useful life.

DIAMONDS

Children found the first diamond on the Orange River in 1867, but the successive find of an African Griqua, the 82 carat 'Star of South Africa', gave the initial impetus to the economy of the country we now call South Africa. By 1870, diggers had switched from river beds to the yellow and blue grounds or soils of what later proved to be the pipes of Kimberley. By 1888, they were persuaded to amalgamate their workings of a few square yards in De Beers Consolidated Mines, which grew into the giant of the international diamond industry. At the turn of the century, the largest single diamond pipe, the Premier near Pretoria, was discovered. In 1926, alluvial diggings extended from the Vaal–Harts Valleys and Bloemhof to the Lichtenburg–Ventersdorp field of the western Transvaal (Fig. 50). After a collector bought the 'Star of South Africa' for $70,000, production expanded until 1926; it rose again after the Great Depression and the slump of the Second World War. In 1949, the Jagersfontein mine and in 1950, the largest producer, the Premier pipe recommenced production. Consequently, ouputs increased (Table LXXXIII). In 1965 the De Beers mine is in full operation.

TABLE LXXXIII

INCREASE OF DIAMOND OUTPUT

Year	*Million carats*	*$ Million*
1940	0.5	4
1950	1.6	25
1963	4.4	5.1

How do the big companies extract the precious stones 90 years after the primitive beginnings? We should first remember that, in each 10 tons of rock, there are only 0.6–3.5 carats or an average of 0.3–2 stones. Whereas the Cullinan weighed 3,000 carats (1.4 lb.), the smallest diamond is not more than powder. Furthermore, grades decrease with depths which reach several thousand feet. Since 1958, the mines have switched from traditional chambering to block caving, and thus have reduced costs to 70 cents (US) per ton. To recover one stone from 10–200 million other particles, rotary pans wash the ore, from which the screw conveyors rake the slurry to vibrating screens. While the undersize of these screens adheres to grease tables, the oversize is treated in heavy media liquids which operators of the Premier mine also use for primary concentration.

The organization of the industry is now as simple as it was confused in its beginnings; De Beers produces more than 90% of the diamonds. This associate of the Anglo-American Corporation boasts cash assets of $120 million; it has two producing units: De Beers, who operates Bultfontein, Wesselton, Dutoitspan and Jagersfontein in the Kimberley district with a capital of $18 million, and the Premier Diamond Mining Company with a capital of $330,000 (!). The Government takes 60% of the profits of the Premier mine (about $1.5 million) in addition to taxes which, for the two companies, represented only $30 million out of a combined profit of $70 million. In addition, there is a 10% export tax on uncut diamonds. In 1958, the company built a new treatment plant at Kimberley for $4 million and, because of the reorientation of Ghana and Sierra Leone, it decided to double the capacity of the Premier mine which has already produced half of the South African diamonds. Statistics show these results (see Table LXXXIV).

Dutoitspan produces yellow diamonds, but Koffiefontein remains closed. While the

TABLE LXXXIV

DIAMOND OUTPUT

	1962		1962, 1,000 carats		
	1,000 carats	*carat per 100 ton*	*gems*	*industrials*	
Premier	1,700	9	430	1,260	Premier
Wesselton	900	32	880	880	De Beers
Bultfontein	500	33	30	80	other pipes
Dutoitspan	300	17	290	190	alluvials
Jagersfontein	100	5			
Others	900				
Total	3,900		1,630	2,280	

company recovers an additional 100,000 carats from other sources, the State Alluvial Diamond Diggings operate in Alexander Bay near Oranjemund. As in adjacent Southwest Africa, these deposits are believed to continue off the coast. Furthermore, at Nooitgedacht, Mallin diamonds recover 100,000 carats; the production of Star Diamonds, Treasure Trove and Industrial Diamonds add up to even less.

The Diamond Producer's Association, a cartel consisting of De Beers and the Government comprises De Beers Consolidated, the Premier Company, the Consolidated Diamond Mines of Southwest Africa, the Diamond Corporation, and the South African and Southwest African governments. The Association sells gems through the Diamond Trading and Purchasing Company, a De Beers subsidiary authorized by the Central Selling Organisation (the international diamond cartel); Industrial Diamonds Ltd. only markets boart and drilling grit. A subsidiary of the latter, Ultra High Pressure Units Ltd., manages a plant capable of producing (in competition with the artificial diamonds and similar borazon of the General Electric Corporation) a large caratage of boart at Springs near Johannesburg. However, as long as the Congo continues to supply industrial diamonds, De Beers will restrict the activities of these Units.

ASBESTOS

Capable of providing all three varieties of asbestos and with amosite and crocidolite resources of several square miles, South Africa occupies a unique position. What are these three types of asbestos? (Table LXXXV).

Since it is purer, crocidolite, originating in Griqualand (called Cape blue), also fiberizes

TABLE LXXXV

TYPES OF ASBESTOS

Trade name	*Uses*	*Qualities*	*Source*
Crocidolite	spinning, non spinning	chemical resistance	Griqualand, Transvaal
Amosite	felting, insulation	acid resistant	Transvaal
Chrysotile	spins best	tensile strength	Barberton district

better than Transvaal blue, a variety with which some amosite is always mixed. For obvious reasons the shape is most important in the grading of asbestos:

(1) The trade grades crocidolite according to lengths from one eighth to more than one and a half inches. Spinnable fibre of more than three quarters of an inch rarely represents more than 10% of the ore.

(2) Experts classify amosite (of + or > ½ inch) according to colour (they favour white) which indicates its quality and strength.

(3) The Bauer McNett machine or the Canadian Test Box screens chrysotile (from ⅛ to 1½ inches) into eight grades.

While the Cape blue belt extends 250 miles through Prieska and Kuruman, between Chunies Poort and the Steelpoort River, the eastern Transvaal blue and amosite belt extends 60 miles with lower grades occurring in the east. However, South Africa is only a minor producer of chrysotile which is recovered in the southeastern Transvaal in the De Kaap Valley which includes the New Amianthus mine near Barberton, and in 20 miles of the Carolina belt which comprises Kalkkloof and Stoltzburg. The evolution of the asbestos industry can be gauged from Table LXXXVI.

TABLE LXXXVI

ASBESTOS PRODUCTION IN THOUSANDS OF TONS AND $

Year	*Cape blue*		*Transvaal blue*		*Amosite*		*Chrysotile*	
	tons	*$*	*tons*	*$*	*tons*	*$*	*tons*	*$*
1930	7	360	—	—	3	70	15	450
1940	7	320	3	170	17	730	0,4	30
1950	16	2,200	15	2,100	43	3,600	13	1,800
1960	68	11,500	13	2,000	72	6,100	2,4	4,000
1963	84	15,800	11	14,200	76	9,700	3	2,600

Since the fibre has to be protected from damage, in the resuing method, the stope face advances as a stope cut in the footwall. However, the most delicate operation is sorting, during which 20–30% of the fibre is lost, much of it in the mills.

Like chromium, asbestos is not a highly organized industry. Nevertheless, a few companies are pre-eminent (Table LXXXVII).

TABLE LXXXVII

ASBESTOS GROUPS

	Capital $ thousand	*1961 Profit $ thousand*	*Mines*
Cape Asbestos Co.	12,000	3,200	Koegas (15,000 ton per year), Pomfret and Penge (Egnap Co.) (25,000 ton per year), Warrendale, Mimosa and Riries (10,000 ton per year).
Griqualand Exploration and Finance Co.	600	300	Whiterock, Mt. Vera, Arcadia, Asbes, Bretby
Kuruman Cape Blue Co.			Whitebank, Newstead, Bestwell, Kuruman
Msauli Asbestos Co.	600	600	Komati

In 1962, markets were widely distributed (Table LXXXVIII).

TABLE LXXXVIII
1963 ASBESTOS EXPORTS

Importing territories	*Cape blue* %	*Amosite* %	*Chrysotile* %	*Transvaal blue* %
Common Market	30	15	33	20
United States	10	23	—	—
United Kingdom	0	25	3	25
Japan	10	5	6	13
Australia	—	10	10	—

BASE AND OTHER METALS

When Phalaborwa starts operations, the country will boast a *copper* output of 150,000 tons. Meanwhile, the production of the older O'okiep and Messina deposits is stagnating. Although, in 1688, copper was already known to occur at O'okiep, operations did not start until 1863. The O'okiep Copper Company (a subsidiary of the Newmont Corporation and of American Metal Climax) with a capital of $3.5 million, mills 1.8 million tons of ore each year and produces 40,000 tons of copper valued at $23 million, half of which is profit. The headquarters, Nababeep, are located 80 miles from Port Nolloth on the Atlantic. In 1963, reserves attained 550,000 tons of metallic copper in grades of 2.01%.

With a capital of $7 million, the Messina Development Company near the frontier of Southern Rhodesia produces 14,000 tons of copper ingots each year with a profit of $3 million. But in 1962, reserves stood at only 100,000 tons of copper in grades of 1.6%. Furthermore, the Rustenburg Platinum Mines can produce matte containing 3,000 tons of copper.

By far the largest resources are located at Phalaborwa near the Kruger Park; they include, for each 100 ft. of depth, 72,000 tons of copper in grades of 1.26% and 130,000 tons of copper at a grade of 0.5%, from which the Palabora Mining Company expects to extract 2.1 million tons of copper metal in grades averaging 0.7%. State-owned FOSKOR and the Transvaal Ore Company (Merensky Trust) hold the mining rights; the Rio Tinto group formed with the latter Palabora Holdings who are expected to control 51% of the Palabora Mining Company. Rio Tinto-Zinc will directly and indirectly hold 39% of the equity. In addition to 'holdings', the Newmont Corporation has a 30% interest; a subsidiary of Rio Tinto, Rio Algom Mines (a Canadian uranium producer), and American Metal Climax (10%) also participate. To mill 12 million tons of ore containing 20% magnetite and to extract 80,000 tons of copper yearly, financing will attain $105 million. The Kreditanstalt für Wiederaufbau offered $25 million fixed interest finance at a rate of 7%, and 50,000 tons per year are ear-marked for Germany.

In the area of Gravelotte in the eastern Transvaal (Fig. 81), Consolidated Murchison Goldfields (a company of the Anglo-Transvaal group) produces 25,000 tons of hand-spalled ore and flotation concentrates containing 60% *antimony*. Thus, with sales of $3 million yearly, South Africa is second only to China. Each year, the Company makes a profit of $1,000,000, which is the double of its issued capital. Reserves of antimony and gold are decreasing in the Murchison Range. Out of resources of several million tons (with 2 million tons certain), Umgababa Minerals Ltd., a subsidiary of the Anglo-American Corporation, can produce 110,000 tons of ilmenite containing 50% *titanium* oxide, 7,500 tons of high grade rutile and 9,000–11,000 tons of zircon each year. With a loan of $2.6 million, the company installed new monitors, cyclones, Humphrey spirals, electromagnetic and electrostatic separators. Most of the recovered minerals are shipped to Britain. Furthermore, around Rietfontein north of Cape Town, at least 300,000 tons of ilmenite and 100,000 tons of zircon are available. Finally at Umbogintwini a plant produces titanium oxide and the Umgababa plant has been moved to Morgan Bay.

In the Transvaal (mainly in the Marico district), both lead and zinc production have oscillated between 100 and 1,000 tons per year since World War II. Minor resources, mainly lead, are also available in the tin fields. Recently, the Argent Lead and Zinc Company of New Consolidated Gold Fields built a 6,000 ton per month plant to float lead and zinc. Only 10 tons of zinc were produced in 1962.

IRON AND MANGANESE

For centuries, Africans have smelted some *iron*. Indeed, South African resources are vast but they are handicapped by low grades, inadequate communications and by the paucity of coking coal (Table LXXXIX).

TABLE LXXXIX

PRODUCTION AND RESERVES OF IRON ORE

Area	*Reserves million tons*	*Grade %*	*Production million tons total*	*1960*
Postmasburg	1,300	55	8	2
Transvaal	6,000	40–55	28	2
Bushveld	2,200	titaniferous	—	—
Griqualand and Transvaal	2,000,000	25–40	—	—

Low grade reserves extend in belts of several hundred miles; since 1931, the opencast Thabazimbi mine (in the northwestern Transvaal) has provided the ore of the Pretoria steel mill and the Sishen mines (in Griqualand) have supplied the ore of the Vanderbijlpark steel mill (south of Johannesburg). With a capital of $70 million and yearly profits of $30 million, state owned ISCOR (the South African Iron and Steel Industrial Corporation) produces cheap steel and dominates the industry. In 10 years, its steel output has doubled and a $680 million plan calls for its expansion to 4.5 million ingot tons per year. In addition to the government, the Prestwick mill of the African Metals Corporation receives 600,000 tons from Manganore in Griqualand each year.

In 1922, *manganese* was discovered, and in 1930, the deposits were opened up. Indeed, resources are vast but, unlike iron, manganese is largely restricted to eastern Griqualand (Table XC).

TABLE XC

MANGANESE RESOURCES

Area	*Grade %*	*Million tons*
(South) Postmasburg	>43	5
	35–42	7
	25–34	8
(North) Kalahari east	high	70
west	high	200

Resources of the western sector, unfortunately, lie under several hundred feet of sand. On the other hand, the manganese/iron ratio of low grade ores tends to increase favourably to 10/1. With ore occurring in irregular pockets, miners resort to hand-sorting but, in the western quarries, companies mechanized waste removal. Manganese mining continues to develop rapidly (Table XCI).

TABLE XCI

DEVELOPMENT OF MANGANESE MINING

Year	*Tons*	*$ Million*	*1963 Sales*		*1963 Grades*	
			Area	*%*	*% Mn*	*% of total*
1940	450,000	2	Common Market	35	<40	40
1950	900,000	9	South Africa	25	40–45	40
1963	1,600,000	20	United States	10	>45	20

In addition to several smaller operators and minor deposits, two corporations dominate the industry in the Transvaal: South African Manganese Ltd. operates its Manganore deposits near Lohathla with a capital of $900,000, and makes a $2 million profit each year. In 1959, the company established a new mine at Hotazel. Both South African Manganese and its competitor, Associated Manganese Mines, supply ore for the new Cato Ridge plant. With a capital of $5 million, Associated Manganese (a subsidiary of Anglovaal) transports more than 500,000 tons of ore each year from Kuruman and Hay. The company also holds leases at Devon, Adams and Blackrock, and has a considerable interest in Ferro Alloys Ltd.

METALS OF THE BUSHVELD

With the world's largest platinum, palladium, chromium and vanadium resources, South Africa is rightly famous. All these, and nickel and tin are found in the Bushveld north of Johannesburg. Following the work of two decades, Dr. Hans Merensky and A. F. Lombaard identified platinum in the Lydenburg district in 1924, and, in 1942, discovered what was to be called the Merensky Reef near Rustenburg. In the same area, another German geologist noted chromite in 1868 and, by 1910, nickel was known to occur at Vlakfontein, northwest of Rustenburg.

Platinum group

The Merensky Reef is now known to lie in numerous stretches of 1.5–4 miles over a distance of 180 miles at Rustenburg (Fig. 99), 100 miles in the Lydenburg district and 40 miles near Potgietersrus in the north. While grades otherwise average 2 dwt. per ton, at Rustenburg, tenors of 6 dwt. per ton persist over stoping widths of 2.5 ft. and for several hundred ft. along the dip. Whereas the platinum/palladium ratio ranges from 66–78% Pt to 12–24% Pd at Rustenburg and Lydenburg, palladium content increases to 44–62% at Lydenburg. In addition to these, concentrates contain 2% rhodium, 1.5% ruthenium and 0.3% iridium. Furthermore, the Norite Reef extends 100 miles across Rustenburg with a grade of 5 dwt. per ton over a width of 4 ft., and a number of platiniferous pipes occur at Lydenburg. At Rustenburg, the stope advances in a straight line and the rest of the operations, sorting and tramming, is also standardized with the utmost precision. Capacity attains 2 million tons per year with each ton costing $1.40. After crushing and corduroy tabling in the Klipfontein and Waterval mills, the pulp is floated, whence the concentrate moves to blast furnaces and converters. The daily capacity of the plants attains 2,500 ounces of platinum group metals, 40,000 lb. nickel and 20,000 lb. copper. Finally, at the Brimsdown plant in Britain, Messrs. Johnson, Matthey and Company (who also own half of the matte smelter) process and market the product. On the other hand, the Gold Fields group has an interest both in Johnson, Matthey and in the Rustenburg company.

Production started in 1926, and rose to 55,000 ounces by 1930, when prices dropped to one third. From 1948 on, prices and production rose again, and reached the mark of 600,000 ounces in 1957; because of oversupply, prices plummeted again, but increased to $120 per ounce in 1964. Meanwhile, the sole producer, Rustenburg Platinum Mines (a subsidiary of Johannesburg Consolidated Investment Company) has increased sales to the American oil companies which use platinum as a catalyzer of fuel. The price of palladium, a metal used in the electrical industry, fluctuated around $30 per ounce, but the company achieved a net profit of more than $7 million because sales had again increased from 300,000 to 400,000 ounces. As to nickel, another byproduct, the company lately produced matte containing 2,000–3,000 tons which sold for less than $5 million; the deposits of Vlakfontein and of Insizwa in Natal are not active, but a new mine starts operations at the Pilanesberg

Others

To pass to *chromite*, regular production started in 1924 and continued to expand with good, though variable, fortunes (Table XCII).

TABLE XCII
CHROMITE MINING RESULTS

Year	*Tons*	*$ million*	*Grades 1963*		*1963 Exports*	
			% Cr	*% of total*	*Area*	*%*
1940	180,000	0.5	<44	15	United States	70
1950	540,000	1.4	44–48	83	Western Germany	10
1962	1,000,000	8.3	>48	2	United Kingdom	10

Averaging a Cr/Fe ratio of less than 2, most South African chrome ore is of the low grade chemical variety, but users mix most of it with other ores to obtain the metallurgical grade. However, reserves are inexhaustible and production now reaches a million tons per year.

Up to fifteen superposed seams extend 50 miles in the Lydenburg district and 70 miles in the Rustenburg district (Fig. 99). Around Lydenburg, miners work 3–5 ft. of seam no. 5 of the Lower Group, but the Intermediate Group contains the best grades. The large resources of the Thabazimbi area are idle except for the Zwartkop, Groothoek and Tweefontein operations where Chrome Mines of South Africa (Union Corporation Ltd.) stocked 270,000 tons of low grade chromite in 1959, which represent a profit of $600,000. In the Pilanesberg, high grade (50%) seams extend only 7 miles. Far from the railhead, the Ruighoek, S.A. Minerals, Consolidated Chrome and the Palmiet companies operate here. In the Rustenburg sector proper, Henry Gould Ltd. mines Elandsdrift; Buffelsfontein recovers 15,000 tons per year; and at Bultfontein, a plant rises on a 30 million ton deposit. The eastern district is much nearer to Lourenço Marques Harbour and many of the 17 companies stope the seams here. The sorting or separation of contaminations presents a special problem as do inadequate communications. In addition to the Bushveld, chromite is available at Isitilo in Natal.

The Bushveld holds vast *vanadium* resources. While values vary from trace to 2.4%, ore of 1.6% vanadium (i.e., 0.6% more than the limit of payability) is available at Stoffberg, near Middelburg. In the Witbank plant, the vanadium content of the titanomagnetite is leached out as an ammonium vanadate and then oxidized. Production started late in 1957 and expanded rapidly towards its goal of 2,500 tons. The metal finds its principal markets in the United Kingdom, the Common Market and Austria. However, cobalt reserves of the Kruisrivier area are small.

South Africa is a minor producer of tin; the ore is found in the centre of the Bushveld. The output of the four primary producers is rarely in excess of 3,000 tons (Table XCIII).

TABLE XCIII

TIN PRODUCERS

Group	*Company*	*Mine*	*1962 Tons tin*	*Grade %*	*Profits $*
	Zaaiplaats	Zaaiplaats	900	1.3	240,000
Gold Fields	Rooiberg	Rooiberg	700	0.5	500,000
Gold Fields	Union Tin	Naboomspruit	200	0.6	14,000
	Vellefontein	Rooiberg	150	0.9	20,000

In South Africa, molybdenum is scarce. There are occurrences in the Bushveld tin fields; the largest, Groenvley, contains 50 tons of oxide. Some arsenic is also associated with tin; the New Consort mine, near Barberton, produced a few tons prior to 1946.

BUILDING AND CERAMIC MATERIALS

With its industrialized economy, South Africa developed a sector of mining which is in its infancy in most of Africa. The most important products are given in Table XCIV.

TABLE XCIV

QUARRY OUTPUT

Product	*1962 Value $ million*	*Product*	*1962 Value $ million*
Limestone	12	gypsum, sandstone, silica	2.5
Stone	several	slate, magnesite, feldspar	1.2
Sillimanite	2	andalusite, marble, kaolin	0.5
Clays	2	mica, shale, pigment, wonderstone, silcrete, bentonite, magnetite, talc	0.6
Granite	3		
Fluorspar	1.5		
Vermiculite	1.4		

The limestone products used by the industry in 1961 are given in Table XCV, and large plants are shown in Table XCVI.

TABLE XCV

LIMESTONE USED BY THE INDUSTRY IN 1962

Limestone product	*Tons*	*$*
Lumps of burnt lime	500,000	3,900,000
Slaked lime	160,000	1,800,000
Cement limestone	3,900,000	2,500,000
Metallurgical limestone	1,200,000	1,500,000
Agricultural limestone	300,000	750,000

TABLE XCVI

PRINCIPAL GROUPS

Company	*Quarries, plants*	*1961 profits, $*
White's SA Portland Cement	Lichtenburg, White's	2,700,000
Anglo-Alpha Cement	Dudfield, Fouries, Schaapvlakte	2,000,000
Pretoria Portland Cement	Rietvallei, Kalkheuvel	4,000,000
Northern Lime	Taung, Limeacres	1,000,000
Cape Portland Cement	De Hoek, Riebeck	800,000
National Portland Cement	Cape Flats, Holrivier	440,000
Natal Portland Cement	Umzimkulu-Pt. Shepstone	300,000

Both Anglo-Alpha and Natal Cement are Anglovaal group companies. Both Northern Lime and White's boast a productive capacity of 700,000 tons.

Naturally, the *building stone* industry is more widely distributed. Companies quarry sandstone (mainly Ecca), granite (between Johannesburg and Pretoria, and at Cape Town), various road metals, quartzite (in the Witwatersrand), and also slate (in the Transvaal). Marble Lime and Associates, the principal producers of marble, branched out of Marble Hall into the asbestos and chrome industries. We should not forget the wonderstone of Ottosdal in the western Transvaal, which is an ornamental stone that finds ready markets in the United States.

As to *clays* (expressed in percentages of value), stock bricks represent 35%; refractory and facing brick, 20% each; pipes, 10%; tiles, 7%; and pottery, 2% of the total. Furthermore, two of the largest brick manufacturers (not only of the continent but of the Southern Hemisphere) operate in South Africa. With a capital of $5 million, the Vereeniging Brick and Tile Company produces refractories and other clays at Vereeniging, Rietfontein and in the Wolvekraal district, and makes a yearly profit of $1.5 million. Its activities are expanding. Cullinan Refractories, with a capital of $2.7 million, produces large profits at Olifantsfontein. In addition to the Hume Pipe Company, several hundred establishments exploit adequate resources.

In refractories, South Africa has a dominating position: the *sillimanite* resources of distant Pela Mission reach the mark of several hundred thousand tons and there are additional reserves in northwestern Cape Province at Gamsberg near Pofadder. Japan, the Common Market and the United Kingdom receive most of the exports. On the other hand, sillimanite's sister mineral, andalusite, is found in and around the Little Marico River. As to silcrete, the quartzites of Mossel Bay and Riversdale are the raw material of excellent silica bricks.

In 7 years' time, sales of *fluorspar* trebled! This is a multipurpose mineral used as a flux, metallurgical, ceramic and optical material. Traditionally, Japan imports half of the 100,000 ton output, with the United States, South Africa itself and Canada absorbing 40% of the remainder. Since South African steel production will double, local markets are expected to expand. Indeed, the Marico–Ottoshoop area near Mafeking, where most of the fluorspar is mined, is not inconveniently located. In this district, pure fluorspar used in lens making brought fame to Oog van Malmanie. Large reserves are also available in the central Bushveld; Buffalo and Ruitgepoort represent one and a half million tons of ore. In addition to these, explorers found fluorspar at Hlabisa, in Natal and in the lower Orange Valley.

With minor exceptions, the United States and South Africa share the market of *vermiculite*, an exfoliating (expanding) insulator. With a small investment, the Palabora Mining Company (related to Hans Merensky Services) produced one third of the world production of 30,000 tons of vermiculite at its large Loole deposit (near the Kruger Park) in 1963. Out of this, the Common Market imported 40%, the United Kingdom 30% and the United States 15%. Though not exfoliating, mica is similar to vermiculite. Between Mica Siding and the Olifants River, South Africa holds large deposits of mica, but the ratio of sheet (half ton) to waste (2,500 tons) mica is unfavourable. At present only a few mines, Impala, Lily, Lady Godiva, Union and Maury, are operating.

In contrast to vermiculite and to reserve of the rest of Africa, *gypsum* resources are small. In the richest area around Kimberley, experts have evaluated only 7.5 million tons of gypsum, but the distant deposits of the Vanrhynsdorp district north of the Cape are larger; to complement these, gypsum is available at Goedverblyf near the Cape, at Ngobevu in Natal, and elsewhere.

Silica imports have ceased since the war, but sand production is not expanding. Large resources are available at Philippi near Cape Town (12 million tons), in the 50 mile long Moot Valley and at Pienaaspoort near Pretoria. Although production has doubled in the last 5 years, magnesite resources at Kaapmuiden, in the Barberton district, in the Lydenburg district, in the Southerland district and in the northern Transvaal are still in excess of demand. Also in the Barberton district are the principal talc quarries, but similar material is available at Hennops near Pretoria, and at Umhloti in Natal.

With its expanding sales, *kaolin* leads us to other ceramic materials. Seven companies quarry kaolin at Platnek, Rooibosch haak, Rietfontein, Bultfontein, and Kookfontein in the Transvaal, in the Riverdale, Albertinia and Van Rhynsdorp districts of the Cape and at Benvie and Inanda in Natal. Kaolin is derived from feldspar, a material which is also used in glass and earthenware. The Selati mine near Leydsdorp dominates the feldspar scene. Whereas kaolin produces white glazes, mineral *pigments* colour. There are three of these; oxides, ochres and umber. The annual oxide production of Dundee, Hopevale, Hazeldene (5,000 tons reserve), Novembersdrift (10,000 tons reserve), and Bethal reaches the mark of 1,500 tons or $25,000. On the other hand, at Riverdale, 300 tons of ochre fetch $2,000; the market of umber produced at Krugersdorp, Brits and Sabie is limited by demand. At Riverdale, there is also a small production of Fuller's earth, a detergent which is also available at Dover, near Parys in the Free State.

OTHER PRODUCTS

With its diversified economy, South Africa produced $5 million worth of phosphates, $2 million worth of salt, and a number of minor products in 1963; its waterfalls produced 180,000 kW.

In addition to copper, vermiculite and rare earths, Phalaborwa (near the Kruger Park) includes *phosphates* (Table XCVII).

TABLE XCVII
PHOSPHATE RESOURCES

	Million tons ore	P_2O_5%
To the valley level	11	10
For each 100 ft. below	10	10
For each 10 ft.	40	6

By North African standards, these reserves are small and low grade; however, the state-owned Fosfaat-Ontginningskorporasie (FOSKOR) could expand its output to half a million tons of phosphate in response to local demand. As a byproduct, the company recovers 310,000 tons of magnetite (iron ore). At Langebaan (Cape Province), the African Metals Corporation can produce 130,000 tons of langfos fertilizer each year. Other low grade resources are available at Zoetendalsvley, the Saldan hapeninsula and Wondergat. Superphosphates are manufactured at Phalaborwa, Sasolburg and Modderfontein. The plants of Umbogintwini and Somerset West will have a capacity of 300,000 tons per year.

In contrast to most other materials, South African salt resources are inadequate and, to complement the winning of salt from the pans of Kimberley, Calvinia, Vryburg, Hammandskraal and Soutpansberg, recovery from sea water has to expand. Responding to this demand, SALNOVA has established a large refinery at Coegas and another affiliate of

Cerebos Ltd., the National Salt Corporation, exploits the brines of Varsch Vlei. The soda outlook is not favourable and output from the Pretoria pan cannot meet requirements. Similarly, salpetre incrustations, although widespread between Griquatown and Prieska, cannot be considered as adequate.

Some *corundum* is associated with kyanite, and the largest deposits of the abrasive extend from Louis Trichardt to Pietersburg, from Messina to Tatchankop, and from Leydsdorp to the Olifant. In 1952, the value of corundum sales reached its peak of $260,000, but (more rapidly than Rhodesian output) they slid from the world record of 5,000 tons to a hundred tons.

Unfortunately, the diatomaceous occurrences of Bankplaats north of Ermelo, of Kannikwa (80,000 tons) and of Witdraai (20,000 tons) are either low grade, remote, or both. Barytes are widespread but only Schoonoord near Barberton, with its 250,000 tons reserve, has produced significant amounts lately. However, at Gams in western Cape Province, geologists estimate 3 million tons of barytes for each 100 ft. of depth. Similarly, small occurrences of graphite are widely scattered, but only Twyfelaar and Uitkomst in the eastern Transvaal and Groot Spelonken and the banks of the Mutali have been productive.

In northwestern Cape Province (and in the eastern Transvaal), workers collect a number of minerals (Table XCVIII).

TABLE XCVIII

OCCURRENCE OF MINERALS

Beryl	*Lithium ore*	*Tantalite–columbite*	*Bismuth*
Noumaas			Henkries
Straussheim		Oeraroep	Durabies
Steinkopf	Steinkopf	Steinkopf	Steinkopf
Pala Kop	Pala Kop	Letaba	
		Duiwelskloof	

Beryl production reached its peak of 900 tons in 1950 and output of the other metals is also decreasing. In addition to the tin fields and to O'okiep, tungsten shows at McTaggart's Camp, Boksputs and Van Rooi's Vley in the Gordonia district and at Klein Kliphoog near Steinkopf. However, since its peak of 300 tons in 1956, production has not failed to decrease. The position of thorium and rare earths is different: the Anglo-American Corporation owns 75% of the $1.1 million capital of Monazite and Mineral Ventures and, in 1962, mining restarted at its interesting Steenkampskraal deposit in the northwestern Cape.

Also in the same areas, in Namaqualand on both sides of the lower Orange River, in Griqualand (northern Cape Province) and in the eastern and northern Transvaal, lapidaries collect a number of semi-precious stones. From near Gravelotte, 600 lbs. or $300,000 worth of emeralds are exported to Switzerland and the United States each year. Similarly, the United States imports most of the ornamental wonderstone production of Gestoptefontein near Mafeking. Since 1928, miners have recovered some gem beryl in the Zoutpansberg district. In addition to spinels, the area of Bergview near Messina boasts a reserve of 100,000 tons of garnet of occasionally more than ½ inch. While jade garnet occurs in the Rustenburg and Pretoria districts, Tiger's or Cat's eye is the silicified variety of the crocidolite asbestos of Griqualand. In the Barberton district, another serpentine is marketed under the name of verdite.

In addition to the oil-from-coal plant of Sasolburg and to the refineries of Durban (100,000 barrels per day), in 1965 the Cape Town plant of Caltex will supply 30,000 barrels of oil per day. Torbanite, a carbonaceous shale is available at Kromhoek and Mooifontein, but with the exhaustion of the original deposits, the refinery will switch to imported crude. Since coal is abundant, South Africans have neither attempted to exploit the asphaltite (pseudocoal) of the southern Karroo nor to work lignite and peat. However, in the fifties, the Bongwan Gas Springs Company tapped 9 million cubic ft. of gas per year west of Port Shepstone (Natal).

However, the rivers can provide cheaper energy. Investing a total $630 million, the South Africans plan to dam the sources of the Orange in the Ruitgevallei and at Van der Kloof, where each of these will impound 1.3 million acre ft. A 50 mile long tunnel will divert part of the dammed waters towards the south to the drainage of the Fish and Sundays Rivers. Multipurpose dams, tunnels and canals costing $420 million will mainly serve the needs of irrigation. In addition, the hydro-electric power network costing $126 million is planned to include three plants on the upper Orange which will ultimately generate 76,000 kW; seven plants downstream which will supply 20,000 kW; and 65,000 kW will be obtained on the Fish River and 13,000 kW on the Sondags River.

BASUTOLAND

Basutoland, the 'roof' of South Africa, is an enclave of 12,000 square miles surrounded by the republic. Out of a population of 700,000, many go to work on the Rand where Basutos have gained fame because they specialize in such difficult tasks as shaft sinking. It is difficult, indeed, to live off the land; the mineral resources of Basutoland are insignificant. In 1959, not far from Maseru (the capital), at Butha Buthe and Sekameng, workers recovered 8,700 carats of *diamonds* valued at $150,000. Kimberlite pipes are also found at Ngopetvou, Tsepinare Berea, Molibi Berea and Thabantvonyana (Fig. 46). Also, near Maseru, Butha Buthe and Mohale's Hoek, thin seams of coal (calorific value 11,000 BTU) are found. There is little limestone at Mohale's Hoek, but some ochre at Sehonghong, a few quartz crystals, perhaps some nitre, abundant building and roadbuilding stone but insufficient groundwater. Only the *agates*, which Basutos collect in the Drakenberg Mountains, find ready markets. And thus, until new discoveries are made, Basuto miners will remain Basutoland's contribution to the African mining industry.

SWAZILAND

In contrast to Basutoland, another small country, Swaziland (wedged between the southern tip of Mozambique and South Africa) illustrates how mineral development can change the national economy. For the last 20 years, asbestos has been the government's main source of revenue and the 300,000 people who inhabit the 6,700 square miles of Swaziland

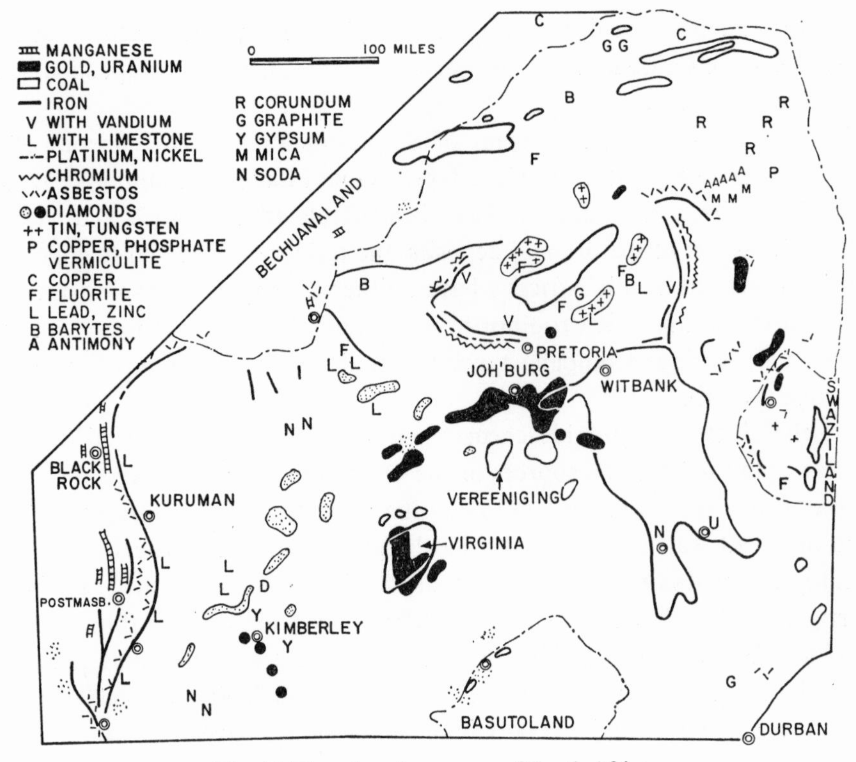

Fig. 46. The mineral resources of South Africa.

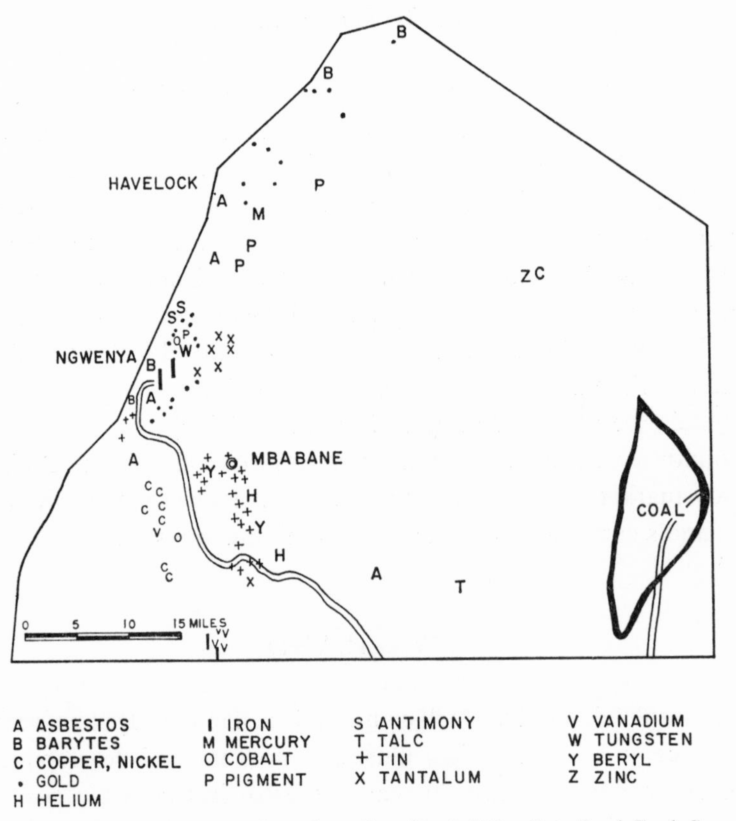

Fig. 47. The mineral resources of northern Swaziland. (After Swaziland Geol. Survey).

will soon benefit from the exploitation of the Ngwenya (Bomvu Ridge) deposit near Darkton. West of the capital (Mbabane) in these mountains, the Swaziland Iron Ore Development Company (a subsidiary of the Anglo-American Corporation) has proved 47 million tons of ore containing 62% *iron*, to which an additional 15 million tons of probable ore might be added. During its 21 year lease, which includes a clause for another 21 year extension, the company will pay only a mineral tax of 2.5% of the value of the ore mined and a yearly rental of $700 for the acreage they use! However, the Anglo-American Corporation with its associate, De Beers Consolidated Mines, has lent the Government $11–14 million for the building of the Swaziland Railway, which the company will construct. Since this will be the country's first rail link with Lourenço Marques, it will stimulate the Swazi economy.

In addition to these ores, the Geological Survey has estimated 280 million tons of low grade ore; north of Ngwenya and near the Havelock asbestos mine, the ores of Iron Hill contain only 48.7% iron, but 22.3% silica. Worse, the occurrences of the south (Gege) only include 31% iron.

But in eastern Swaziland on the Stegi–Manzini road in the Hlatikulu district, the Geological Survey, the Johannesburg Consolidated Investment Company and Central Mining Ltd. estimated 250 million tons of *coal*, including *anthracite* (a rare fuel in Africa) and semi-bituminous coal; their calorific values range from 13,200 to 12,300 BTU. The working of iron and coal will treble the value of the mineral production which will consequently reach the mark of $20–25 million.

Swaziland has a long history of mineral production. Indeed, Mbandzeni, King of the Swazis, granted a concession to exploit the Forbes gold reef as early as 1882. Subsequently, tin and asbestos were found. Table IC illustrates the rise and fall of outputs and their value (with pounds converted to dollars at the rate of 2.8).

TABLE IC

SWAZI OUTPUT

Year	*Asbestos tons*	*Tin tons*	*Gold ounces*	*Asbestos $ 1000*	*Tin $ 1000*	*Gold $ 1000*	*Total $ 1000*
1910	—	450	14,000	—	120	160	280
1930	—	180	—	—	70	—	70
1950	33,000	40	1,800	4,670	70	60	4,800
1963	33,000	5	2,000	5,600	13	60	5,700
1907–1963				108,000	5,000	2,000	116,000

In 1929, H. H. D. Castle found *asbestos* at Havelock in the high ridges of northwestern Swaziland, a couple of miles from the South African frontier; production started in 1939. New Amianthus Mines, an affiliate of Turner and Newall, the great asbestos processors of Manchester and Rhodesia produces six grades of fibre which represent 90% of the Swazi production. By 1908, tin production from the placers of Ezulwini and Mbabane had already reached its peak of only 526 tons. Reserves are insignificant. Swaziland was the world's only producer of a few tons of yttrotantalite, a byproduct of tinstone. In the *gold* belt of Swaziland (east of Bomvu Ridge), Forbes Reef, Piggs Peak and Horo each contributed 30% of the total output. But near Forbes Reef, new resources of 570,000 of gold (tenor 3.1–5.7 dwt per ton) are available, with smaller reserves at Wyldsdale.

From 1942, the beginning of their sales, to 1963, the value of all other minerals produced was hardly more than $200,000! Swaziland is the continent's only exporter of ceramic diaspore which goes to Western Germany. The Mankaiana district contributes 60–800 tons of material each year; it is similar to bauxite but, unfortunately, mixed with pyrophyllite. Nevertheless, 3,000 tons of this soapstone have also found their market in South Africa; in the northwest, numerous other occurrences of talc are found. The Geological Survey estimates reserves of diaspore, pyrophyllite and associated andalusite, respectively, at 26,000, 240,000 and 28,000 tons. Each year, a company exports a few hundred tons of barytes from the one million ton Londosi deposit near Mbabane. In the Mankaiana district, 300,000 tons of kaolin are available. However, the nickel resources of the Forbes Reef area, 140,000 tons metal, are of too low grade (0.11%). Other resources are given in Table C.

TABLE C

OTHER SWAZI RESOURCES

Mineral	*Locality*	*Reserves, tons*
Fluorspar	Hlatikulu district	1,600
Corundum	Hlatikulu district	500
Columbite	Forbes Reef	150
Silica	Pigg's Peak, Mankaiana	250,000
Limestone	Nsalitshe	10,000

In addition to these, some beryl, monazite and scheelite (rare earth–thorium and tungsten ores) have also been recovered.

SECTION X

The islands and the sea

MALAGASY REPUBLIC

East of the Mozambique Strait, Madagascar, the graphite island, rises from the sea. More than five million Malagasy inhabit the Republic where the government is responsible for maintaining a favourable investment climate. The country's mineral economy is diversified and distinguished by rather exotic products but its annual value of more than $4 million is

TABLE CI

MALAGASY ORE PRODUCTION AND RESERVES

Ore	*Production*		*Reserves 1,000 tons productive*	*ore*	*Reserves 1,000 tons*
	$ 1,000	*tons*			
Graphite	1,680	21,000	inexhaustible	Copper	6
Mica	7,400	1,000		Nickel	70
Uranium	1,040	500	low	Baryte	500
Salt	470	18,000	sea water	Bauxite	10,000
Beryl	110	400		Iron	20,000
Chromite	150	12,000	5,000	Bitumen	1,000,000
Monazite	140	700	100	Oil shale	20,000
Columbite	40	20	low	Lignite	20,000
Quartz	90	30		Gypsum	40
Mineral samples	10	lb. 5			
Semi-precious stones	30	1.5			
Coal	16	2,000	60,000		
Garnet	40	5			
Limestone	40	60,000	large		
Gold	30	oz. 9	low		
Ilmenite	3	4,000	4,000		
Zircon	1	430	130		
Total	4,680				

still of moderate size. The production of graphite is regressive; consequently, since 1958 and 1959, new products, uranium ore and mica, rival its value. Chromite production will add further impetus to the mineral economy but low grade coal and oil shale resources remain unexploited (Fig. 48). As of 1963, the situation is shown in Table CI.

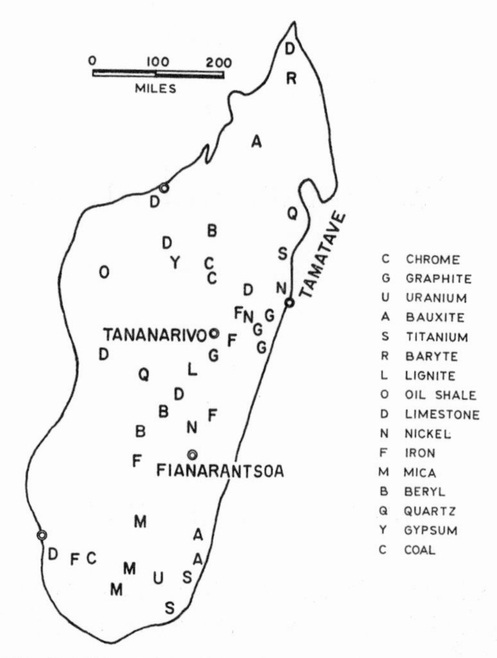

Fig. 48. The mineral resources of Malagasy Republic.

GRAPHITE

For centuries, the Malagasy used the black pigment *graphite*; industrial exploitation started in 1907, and reached its peak of 40,000 tons by 1917. The grades were low and only since 1927, when De Robillard designed his first flotation cell, have miners produced graphite containing 85% carbon. Before that date and later, there were many fluctuations caused by overproduction, lack of labour and small scale operations. By 1947, miners had invested, mechanized and concentrated their operations near Tamatave Harbour on the east coast and thus reduced the cost of exploitation and freight. By 1964, out of more than 30, only eight operators remained in business. Nevertheless, Madagascar, which until recently held the monopoly of flake graphite used in the manufacture of crucibles, must now compete with the Korean production. Meanwhile, resources remain vast, not only west and south of Tamatave but also along a long belt extending in the Eastern Highlands, several hundred miles south of Tananarivo and near Ampanihy at the southern end of the island. Indeed, groups of 10 ft. thick layers containing 5–6% graphite extend for miles, but miners can only quarry weathered, soft rock. These include, in addition to the Société des Mines de Sahanavo which operates in the Sahanavo Valley, the smaller Compagnie Lyonnaise de Madagascar at Ambatomitamba, and the Marovintsy workings, Rostaing at Sahamamy, Gallois at Antsirakambo, Micouin and Pochard at Ampangadiatany and Izouard and Louys at Faliarano–Périnet.

How do they exploit this mineral? Water carries the ore, excavated by bulldozers, towards the plant where it is washed, crushed, floated, screened and finally dried. For each ton of ore containing 3–6% graphite, miners need ten times as much water, and they install for each ton per month of graphite, 1–2 kW of power. Finally, workers load the product on barges or trucks which carry it to the Périnet–Tamatave railway (or, in the case of Marovintsy, by sea). The United States and the United Kingdom, followed by the

Common Market, are the principal importers of flake; the Common Market and the United States, followed by the United Kingdom, share the expanding market of graphite powder.

MICA AND URANIUM

In contrast to widespread graphite, mica is restricted to a triangular area of southern Madagascar and within this district, uranium–thorium is only found in a narrow band of 40 miles. Madagascar dominates the world market of *phlogopite*, a mica distinguished by having greater heat resistance than muscovite. Demand for phlogopite splittings was greatest in the United States, and the combined imports of Japan and France equal this amount; there is less interest in Western Germany and the United Kingdom. Rossi discovered the first phlogopite deposit in 1912, but only the founding in 1926 of the Société des Minerais de la Grande Ile (SMGI, Homburg group), of the Union des Micas (A. Seyrig), and an investment of $800,000 gave impetus to the industry. But with its 5,000 miners and graders, mica is a product requiring intensive labour. Therefore, the companies which provide 60% of the production, as well as the ever decreasing number of private miners, have to compete with the producers of cheaper muscovite. The fluctuations of the phlogopite market are a consequence of this situation. At Benato, the largest deposit, SMGI produces STO and lower grade BSA mica; Union des Micas can export VSP, DSP and FP grades from Ampandrandava. Other producers include the Société de Sahanavo at Mafilefy, the Société des Mines de Sakasoa, Rollet, at Morarano Felli–Defon at Vohitramboa, and Jehl at Marovala.

In contrast to phlogopite, the muscovite mica and uranium resources of the beryl fields are insignificant. But in 1953, geologists discovered 0.3–0.4% *uranium–thorium* ore in phlogopite rocks of the Mandrare Valley and the French Atomic Energy Commission pays up to $40 for each pound of high grade ore (i.e., with a favourable uranium/thorium ratio). Belafa is more attractive than the extensive Sakoatelo deposit which, unfortunately, contains much thorium and little uranium. Ambindandrakemba, Kotovelo, and other occurrences are even smaller. Whereas numerous other leases are available, the end of low cost (quarriable) reserves is in sight.

RARE MINERALS

Madagascar is famous for its 20 ft. long beryl prisms. In addition to this, the pegmatite fields aligned along the axis of the island have contributed columbite-tantalite, piezo-electric quartz and semi-precious stones. But since 1947, out of fifteen fields only four have been of major importance; within these, only five pegmatite lenses, A4 and F3 in the Malakialina district south of Tananarivo, Antsakoa and Analila in the Berere field, and Ambondro in the Analava field, supplied more than half of the *beryl* production. However, this type of deposit is being rapidly exhausted and, with modern beryllometers, the search for new lenses goes on. Berere in the north and Vorondolo southeast of Tananarivo each provided more columbite–tantalite than Malakialina, Vohombohitra and Ampandramaika. Near Berere, the Befanama lens provided 60 lb. of high grade *scandium* ore of which it is the principal source in the world. Furthermore, traces of lithium and bismuth show, and occasionally some fusible feldspar finds markets.

On the other hand, since *piezo-electric quartz* has to face the competition of the artificial product, output fell from its 1955 peak of 20 tons to 10 tons in 1963, with ornamental and foundry quartz plummeting similarly. Workers collect quartz in ten areas which are arranged in a long arc. From the dormant mine of Ramartina (or Cristallina, southwest of Tananarivo), the Société du Quartz has extracted 30 tons of piezo-electric quartz out of a reserve of 2,000 tons. At Antambolehibe (southeast of Cristallina), the Compagnie Générale de Madagascar has already recovered 20 tons opencast and now operates underground. The Bay of Antongil, north of Tamatave, Horombe and Tsivory in the south can provide important quantities, but the potential of Ambatondrazaka, Ankazobe, Ambatofinandrahana (Itremo) and Farafangana is smaller.

Strangely enough, 5 lb. of rare mineral samples fetched prices as high as piezo-electric quartz.

In the centre and the south of the Great Island, a great variety of *semi-precious stones* were found by lapidaries in 1963 (Table CII).

TABLE CII

MALAGASY GEMS

	lb.		*lb.*
Jasper	2,500	Agate	400
Garnet	3,300	Corundum	80
Amazonite	3,200	Moonstone	200
Amethyst	50	Diopside	1
Citrine	140	Cymophane	3
Gem garnet	130	Sapphire	0.2
Labradorite	200		
Tourmaline	40		
Beryl	50		

Morganite and aquamarine are among the most famous gem beryls of Madagascar, and to these triphane, topaz, rhodonite and rare danburite, scapolite, rhodizite and kornerupine should be added. For the distribution of these occurrences, we refer the reader to the 'Miscellaneous' chapter of the second part. To pass to industrial *garnet*, Besosa, Ianatrata and Ampandramaika (near Ambatofinandrahana) hold the largest garnet resources, and Beforona and Ejeda boast reserves of another abrasive, corundum.

SOTRASUM, a subsidiary of the Péchiney company and of the French Atomic Energy Commission, recovers *titanium*, rare earth–thorium ores and zircon from the dune sands that extend on the southeastern coast from Antete through Fort Dauphin to Ialanamainty; however, near Tamatave, low grade deposits remain idle.

OTHER RESOURCES

While Ugine must await the construction of a road to the rail terminal to exploit the 5 million ton *chromite* deposit of Andriamena (110 miles north of Tananarivo), in 1961 the company started operations at the 250,000 ton occurrences located only 25 miles from

Tamatave Harbour. Thus, to the undersaturated chrome market of the franc zone, Ugine contributes each year 20,000 tons of ore averaging only 45% Cr_2O_3 and a Cr/Fe ratio of only 1.5/1; at Bemanevika, a 90,000 ton per year output of high grade chromite (52% Cr_2O_3 and a Cr/Fe ratio of 2.2/1) necessitates an investment of $8 million. In addition to these, other resources are available near Andriamena, particularly at Maevatanàna. To complement chromium, at Valozoro in the southeast, 70,000 tons of its sister ferro-alloy, *nickel* (in grades of 1.75%), can be recovered with an investment of $11 million. However, another nickel occurrence, Ambalotarabe near Lake Alaotra, contains only 3,000 tons of this metal. The Ambatovy-Anamalay iron ores contain a half to a million tons of nickel.

The Malagasy *fuel* situation epitomizes the contrasts of underdevelopment. Indeed, at Bemolanga, in the northwest, vast sediments contain 5–15% bitumen; the low grade coals of Sakoa (containing 17% ash) lie in the southwest far from the lines of communication; and finally, south of Tananarivo near Sambaina, there are lignite and oil shale. We should not forget, however, that right in its centre, mainly in the Ikopa Valley, the Great Island is endowed with 2.7 million kW of hydropower resources.

Since widely distributed iron deposits at Ambatovy (in the northeast, 50 million tons), Betioky (in the south, 10 million tons), Fasintsara (the southeast 70 million tons) and others are either small or of low grade; there is now little hope for their exploitation. Unfortunately, in the north, conditions of the aluminous laterites of Marangoka and Analavory are similar. The manganese position is not better, but the cement materials (limestone and clay) are abundant at the Amboanio plant near Majunga, at Soalara in the southwest, Antsirabe, and at Diego Suarez in the north. In the two latter localities, another construction material, pozzuolana, is also found; however, the only gypsum deposit, Ankay, lies in the northwest far from the lines of communication. Similarly, the baryte deposit of Andavokoera (much larger than Ampandrana) is located in the far north. Also at Andavokoera, gold production had reached its peak by 1909, and totalled that year (with numerous smaller mines) 120,000 ounces. It has declined ever since.

In addition to these, lead and zinc show at Besakay, and copper at Vohibory and Vohémar; kyanite is found at Mananjary, sillimanite and talc in numerous rocks, wollastonite (mineral wool) in the south, kaolin and alunite (the aluminium sulphate) at Ampanihy, and guano on Juan de Nova and other islands of the Mozambique Straits.

SMALLER ISLANDS

Small islands rise from the seas that surround Africa. Since most of them are not more than volcanic cones ringed by reefs, their resources, apart from pumice (volcanic ash), coral limestone, salt and guano, are much smaller than those of the surrounding seas (Table CIII).

The Canary Islands produce each year 400,000 barrels of cement and 5,000 tons of sulphur.

In 1956 and since 1959, the Canary Islands (Fig. 55) have exported 1,600 tons of pumice each year. The pozzuolana resources of Santo Antão and Santago in the Cape Verde archipelago (southwest of Dakar) are larger; in 1959, exports commenced with 10,000 tons, but dropped the next year to 7,000 tons. While *phosphate* is available on all islands

TABLE CIII

ISLANDS RESOURCES

Islands	*Square miles*	*Population*	*Resources (excluding limestone)*
Mediterranean			
Alhucemas, Chafarinas, Peñon de Velez	70	190,000	
Atlantic			
Canary	3,000	900,000	*pumice*, salt, *cement*, *sulphur*
Cape Verde	1,600	200,000	*pumice*, salt
Fernando Póo	800	40,000	
São Tomé and Principe	400	70,000	hydropower, (guano: I. Cabras)
Ascension, St. Helena, Tristan da Cunha	100	5,000	manganese, phosphate, (guano: Gough I).
Indian Ocean			
Réunion	1,000	340,000	hydropower, (gold)
Mauritius	800	700,000	salt, gold, (guano: Rodriguez)
Comoro	800	200,000	*pumice*
Zanzibar and Pemba	1,000	300,000	(gold)
Seychelles group	160	40,000	*guano*
Socotra	1,400	10,000	(guano: Kal Farun)

south of the Guinea Coasts and Arabia, only the Seychelles, Amirantes and other archipelagos east of Tanganyika are consistently distinguisheid by resources (see the 'phosphate' chapter) and within these only *Astove* Island north of Madagascar boasts a production of 8,800 tons per year (4,500 tons in 1956, 19,000 tons in 1958). As to salt, the Cape Verdians win the greatest amount from the sea as is shown in Table CIV.

TABLE CIV

SALT PRODUCTION IN TONS

Year	*Cape Verde*	*Canary*	*Mauritius*
1956	24,000	20,000	4,000
1962	26,000	15,000	4,000

THE SEA

Naturally, the sea also offers other resources such as magnesium. Indeed, its waters dissolve most of the elements and from them, manganese, chromium, nickel and other metals precipitate to its bottom. Unfortunately, we can only infer the following hypothetical pattern (Dr. Bruce Heezen, personal communication, 1963): Since north of Walvis Bay (Southwest Africa) the temperature of the bottom water is relatively warm and because the Walvis Ridge obstructs circulation, little *manganese* collects in nodules even below 4,000 ft. of depth—in contrast to the situation prevailing in the West Atlantic where manganese is found on rocky slopes even above the isobath of 4,000 ft. However, west of Walvis Bay, south of the Cape, and in the Indian Ocean as far as Cape Gardafui of Somalia, conditions are different. In the Malagasy depth (south of Madagascar), the Mascarene depth (between the Mauritius–Seychelles Mount and Madagascar) and the basins lying off the shelf of Kenya and Somalia, a few hundred miles from the coast, considerable deposition of nodules can be expected below 4,000 ft. or on slopes above this depth. And even in the Mozambique Channel, currents would scour sediments away, but not manganiferous nodules, consequently increasing their grade. Further north on the rocky ridge of the Gulf of Aden, nodules might also accumulate, but the Red Sea and the Mediterranean do not appear to be favourable. Similarly, between Mauritius and the Antarctic, towards the Kerguelen Archipelago, the high rate of diatome sedimentation precludes the concentration of manganese.

TABLE CV

METAL CONTENT OF NODULES

Metal	%	*Metal*	%
Iron	17	Copper	0.2
Manganese	16	Lead	0.1
Titanium	0.8	Molybdenum	0.03
Nickel	0.4	Yttrium	0.02
Cobalt	0.3	Chromium	0.002

As to the quantities involved, we should bear in mind that in the Pacific Ocean 60 million tons of nodules are believed to form each year (MERO, 1962). In that ocean, manganiferous provinces lie farther from the coast than iron rich areas, nickel–copper concentrates lie even farther from the continent and islands, and cobalt is restricted to topographic heights. Furthermore, in the Atlantic, four nodule samples average metals shown in Table CV.

PART II

Economic geology

In Part I, we have seen mining develop into the principal industry of Africa; we now turn to the base of these resources, economic geology. To describe the deposits, we have established nine groups of substances. Undoubtedly, this classification, like so many others, is arbitrary. Nevertheless we favour it because substances of similar or common origin can be discussed in the same group. Since a majority of minerals is polygenetic, we have attributed each of these to the group distinguished by the principal deposits (e.g., phosphates among the sediments, pyrites in the gold group, and so forth). But to understand their distribution, we will first endeavour to fit the deposits into a pattern:

SECTION I

The mineralogenetic provinces of Africa

The African continent is distinguished by extensive mineralized, Precambrian orogenies, divided by younger basins and tectonic zones. The resulting pattern is more distinctive than the structure of other shield areas. It is more complex, however, than the pattern of the younger Andean and Alpine orogenies. Indeed, in Africa between 3,300 and 450 million years ago, at least five groups of cycles were superimposed (Fig. 49). In the following pages we will describe the mineral provinces of Africa; for the general and structural

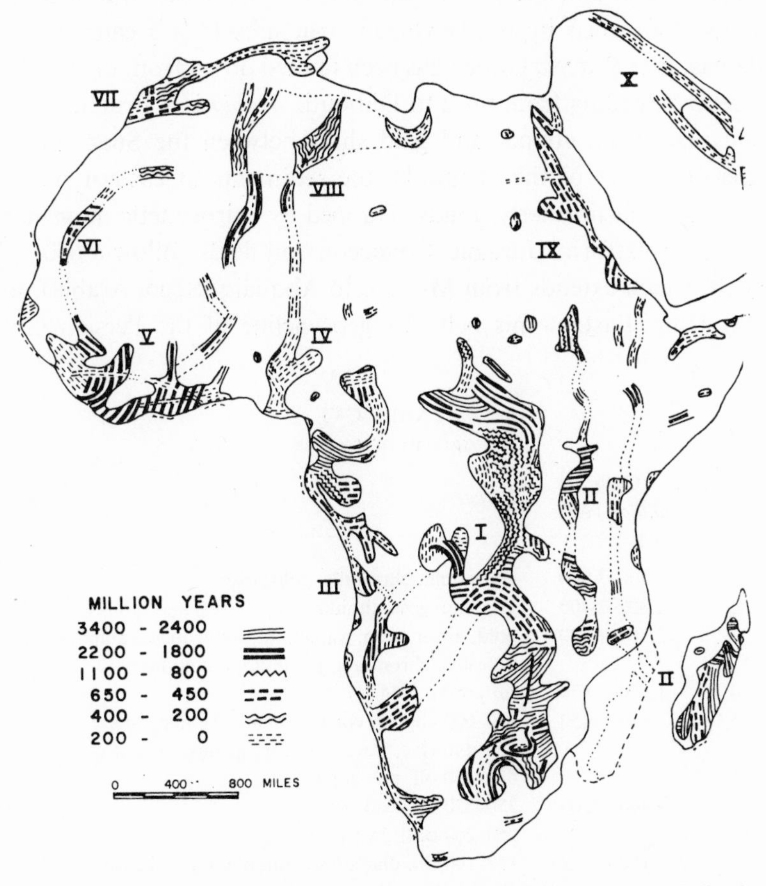

Fig. 49. The mineral belts of Africa. I = High Africa; II = Eastern Africa; III = The Atlantic Rim; IV = The Nigerian Arc; V = The Guinean Shield; VI = The Mauritanian Arch; VII = The Atlases; VIII = Sahara; IX = Red Sea Rift; X = Arabia.

geology of the continent we refer the reader to the map of Africa published by the Association of African Geological Surveys.

BELTS AND CYCLES

While west of a line connecting Mount Cameroun with Aswan, the post-Cambrian Sahara surrounds a shield, south of this line, the orogenic–metallogenetic belts of Precambrian Africa surround interior basins. Sedimentation, orogenic evolution and six alignments control ten mineralogenetic belts. Within a 2,000 mile long and 500 mile wide complex of orogenies extending from the Orange Free State to the Sudan, *High Africa* rises between the Congo, Angola and Kalahari basins and the East African folds. This belt contains the majority of the continent's resources. We can distinguish several minor provinces in *Eastern Africa* between Ethiopia and Mozambique, which continue in Madagascar. Rising between the basins and the sea, the *Atlantic Rim* has been mineralized, between the Orange River and Cameroon, symmetrically to, but less intensely than High Africa. The young *Nigerian Arc* divides the continent, and the *Guinean Shield* forms the backbone of western Africa. Its synclinal troughs have been mineralized between the coast and the Saharan Hoggar. The *Mauritanian Arch* marks the western rim of Africa. In the north, antedating the Alpine orogenies and mineralizations, the *Atlases* border the continent. The oil basins of the *Sahara* are controlled by pre-Devonian structures to a greater extent than other oil fields. In the east, the *Red Sea* graben has been faulted down from the poorly mineralized Precambrian trend extending from Suez to Djibouti. *Arabia* is a structural part of Africa. Here oil, manganese, iron, copper and gold show between the Suez and Red Sea Rift Valleys. On the other end of the peninsula, the sediments of eastern Arabia have been warped following ancient northerly trends. Trapped by epirogenetic movements along the eastern edge of this platform, Jurassic–Cretaceous oil fields follow a 650 mile long and 30 mile wide belt, which extends from Murban to Abquaiq (Saudi Arabia), Burghan (Kuwait) and Nahr Umr. East of this belt, the geosyncline of the Persian Gulf developed

TABLE CVI

GROUPS OF CYCLES

	Cycles million years	*Parageneses*
1	3,400–3,000	chromium, chrysotile, gold, iron
2	2,800–2,400	graphite, gold, lithium
3	2,300–1,900	gold, silver, iron, vanadium, manganese, amphibole asbestos, chromium, platinum, diamonds
4	1,050– 850	tin group, gold
5	650– 450	650–600 copper, cobalt, vanadium, manganese
		550–480 zinc, lead, silver, tin group
		480–450 oil, phlogopite
6	400– 200	370–320 oil, lead, zinc
		250–200 coal, hydrocarbons
7	150– 0	150–100 tin, diamonds, columbium, hydrocarbons
		80– 50 phosphates, hydrocarbons
		– 0 bauxite, salt, titanium, zirconium

parallel to the Red Sea. Along the eastern limb of the geosyncline, oil accumulated in Oligo-Miocene folds between *(a)* Garzan (Turkey) and Naft-i-Shah (Iraq), and *(b)* Lali and Kuh-i-Mund (Iran).

All Precambrian belts strike in northerly directions. However, within these, internal structures are controlled by five cyclically rejuvenated arcs: the Birrimian, Nyasan, Somalian, Libyan and Atlasic alignments (N20°E, N, N10°W; N45°E; N75°E, N75°W).

After the consolidation of the shields, metallization gathered impetus until it reached its peak 2,100 million years ago, and attained a secondary maximum 650 million years ago. Finally, around 200 million years ago, the accumulation of oil and coal represented a ternary peak. Table CVI gives the cycles and their paragenesis.

Of course, we should consider the geochronological data as tentative but useful orders of magnitude. Furthermore, as we shall see, many mineralizations are polycyclic and polygenetic.

HIGH AFRICA

High Africa, the world's principal Precambrian orogenic–metallogenetic belt, consists of cyclically and spatially alternating trends. Arcs of 2,200 million years old Transvaal sediments are oriented northwards; another branch of these and the Bushveld complex trend essentially east. However, *(a)* the 3,000 million years old arcs of the Rhodesian Shield trend northeastwards, *(b)* the 2,000 million years old north-northeast striking Great 'Dyke' graben intersects these trends and farther, the 650 million years old, east-northeast oriented Copperbelt divides High Africa into two parts. North of this geosyncline, older belts bend from the north-northeast, *(c)*, to the north-northwest. Then the 2,200 million years old, earlier Ituri gold belts revert again to the easterly trend. Furthermore, transversal Permian valleys cut the mega-belt of High Africa at *a*, *b* and *c*. We distinguish five groups of provinces:

A. South Africa: iron, manganese, asbestos, gold, uranium, chromium, platinoids, coal, diamonds;

B. Rhodesian Shield: gold, antimony, chromium, chrysotile, rare metals;

C. Katangian Geosyncline: base metals, cobalt, silver, vanadium;

D. Great Lakes: manganese, tin, gold, diamonds;

E. Ituri: gold, iron.

South Africa

In South Africa 2,000 million years ago, ferruginous sediments of the Transvaal system were folded into the Griqualand geosyncline. The branches of this north trending structure are distinguished by large deposits of iron ore (1)[1], amphibole asbestos (2), and manganese (3). From Bechuanaland, another wing of this system trends east across the Transvaal. Similarly, these sediments boast iron deposits, e.g., at Thabazimbi (4), and amphibole asbestos at Penge (5). South of this wing in the Witwatersrand brachy-syncline, pyrite, gold (6), uranium, carbon and osmiridium sedimented cyclically. Although uranium dates are spread between 3,000 and 1,300 million years ago, mineralization culminated between 2,400 and 2,200 million years ago.

[1] Numbers designating the provinces refer to the figures.

Farther north, the ultrabasics of the Bushveld were injected 2,000 million years ago across the Transvaal sediments. In this vast ellipse, chromium (7), nickel, copper, platinum, palladium (8), vanadium, titanium and iron (9) bearing sheets lie upon each other. However, in the granites of the centre, tungsten, molybdenum and bismuth accompany tin. Following 1,500 million years of erosion, the Permian Karroo transgression flooded the tip of Africa. In this basin, thick measures of low grade coal extend 300 miles (10). Similarly, in the central depression of the Bushveld lies the Warmbad coal field (11). During the Jurassic, 250 kimberlite pipes pierced the western edge of the carbonaceous Karroo. Some of these contain diamonds (12). From others, erosion transferred the gems into placers (13).

Rhodesian Shield

If we turn north of the Limpopo, a poly-metamorphic and granitized shield of 700 by 250 miles extends as far as the Mid-Zambezi. Attached to this, a 300 mile long pedicle projects through the eastern Transvaal to Swaziland. In that country, the ores of Ngwenya (14) belong to the oldest iron cycle, the Swaziland system. Close at hand at Havelock and at Barberton (in South Africa), chrysotile seams grew, while 3,300 million year old gold was disseminated along east-northeast oriented stresses (15). Farther north, the mica pegmatites of the Olifants River run parallel (16). They point towards the copper–iron–phosphate–vermiculite–thorium carbonatite of Phalaborwa (17), which is with an age of at least 2,000 million years, one of the oldest of the world. Distinguished by antimony (18), gold and mercury, the Murchison Range also follows the east-northeast trend.

Similarly, Permian coal fields have been aligned in east-northeast oriented basins which extend 300 miles from Mamabule (Bechuanaland, 19) to the Waterberg, Soutpansberg (South Africa, 20) and Sabi-Limpopo fields (Southern Rhodesia, 21).

The schistbelts which float on the Rhodesian granites bend from the west towards the northeast (Somali line). In these, asbestos and chromite deposits are arranged along arcs which strike from Filabusi to Belingwe east-southeast and then bend northeast to Shabanie and Mashaba (22). South of the preceding belt, at West Nicholson, minor deposits are oriented east-southeast (23). Whereas the principal chromium and asbestos deposits, Selukwe (24) (northeast of Bulawayo) and Shabanie, are believed to be older than 3,000 million years, the others belong possibly to the 2,750 million year old cycle. Itabirites (25) are widespread, e.g., at Buhwa (Belingwe) and also at Que Que. The 2,500–2,700 million years old lithium–beryllium–tantalum paragenesis (26) of Bikita distinguishes the wing extending to Vila de Manica. In the northern part of the shield, minor pegmatite-placer fields include Kamativi, Choma, Miami and Mtoko. Galena from the Kamativi tin field and Miami mica are respectively 1,200 and 450 million years old.

In the centre of the shield, schistbelts tend to bend northward. Here the age of galenas accompanying gold, antimony and arsenic culminates around 2,750 million years. North of the Bulawayo belt (27) the Gwelo belt connects with the principal Hartley gold district (28). The less important Salisbury belt lies further north (29). However, the copper mineralization (30) of Mangula in Rhodesia, Francistown in Bechuanaland and Messina in the Transvaal is more sporadic. In the 330 mile long rift of the 'Great Dyke' (31), chromiferous seams have settled in the same cycle as in the Bushveld. Nevertheless, nickel minerals and asbestos occur only sparingly. Much later, Permian sediments flooded most of the Zambezi Valley. In these, seams of low caloric coal (32) are interbedded *(a)* between Wankie (Victoria Falls) and Mafungabusi (western Rhodesia), *(b)* between Gwembe and Kandabwe (Zambia) and *(c)* between Caringe and Chicoa in Mozambique. Finally, an arc of Jurassic carbonatites extends from Feia, on the border of the former Rhodesias and Mozambique to the phosphate complex of Dorowa in the south-southwest (33).

North of the Zambezi, early Precambrian remnants outcrop, distinguished by a 400 mile long copper zone (34) which extends from Kansonso through Kalulu (Zambia) to Cacanga in the western outlier of Mozambique. This hydrothermal belt is oriented west-northwest, parallel to the neighbouring but younger Katangian belt.

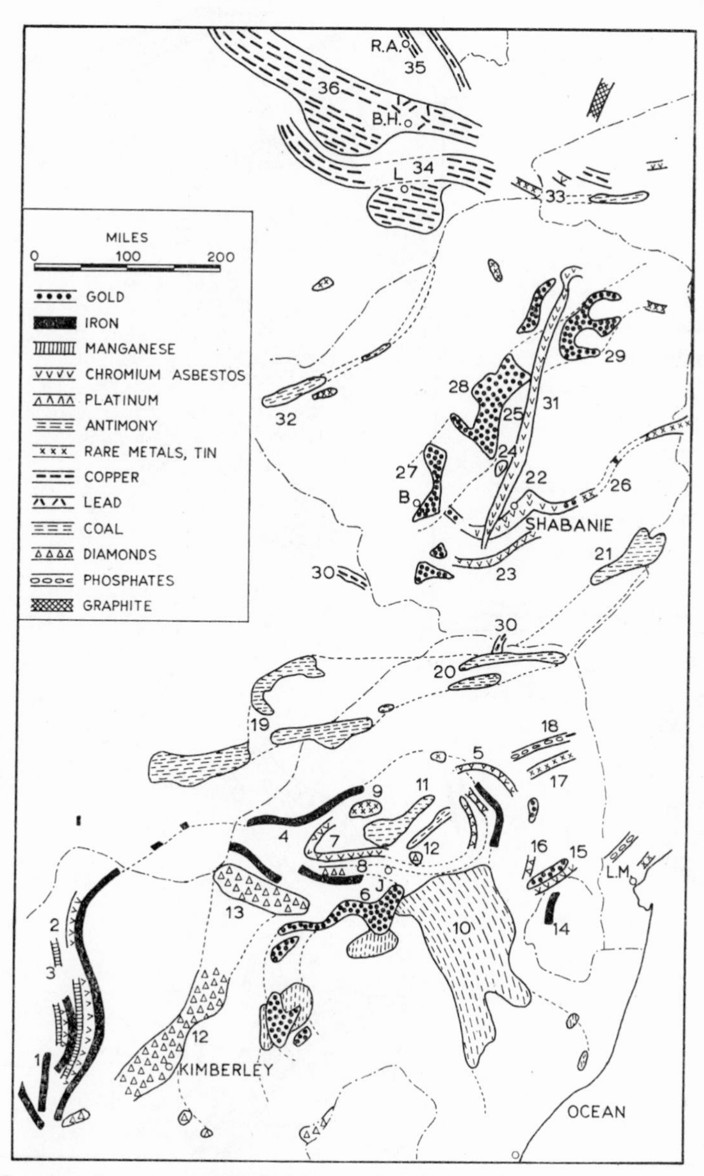

Fig. 50. The mineral resources of High Africa, south (for explanation of the figures, see text).

The Katangian Geosyncline

Between more than 620 and 550 million years ago, the Congo Basin was surrounded and partly filled by sediments similar to those of the 2,200 million year old Transvaal system (Fig. 50). Remnants of this rim are particularly distinguished by a paragenesis of base metals in the Copperbelt (35). The early Katangian synclines have been bent into a 600 mile long arc extending from eastern Angola, through Jadotville, beyond Broken Hill. In two parallel but arcuate trends, the deposits extend from Kolwezi to Bwana Mkubwa and Roan Antelope (Fig. 51).

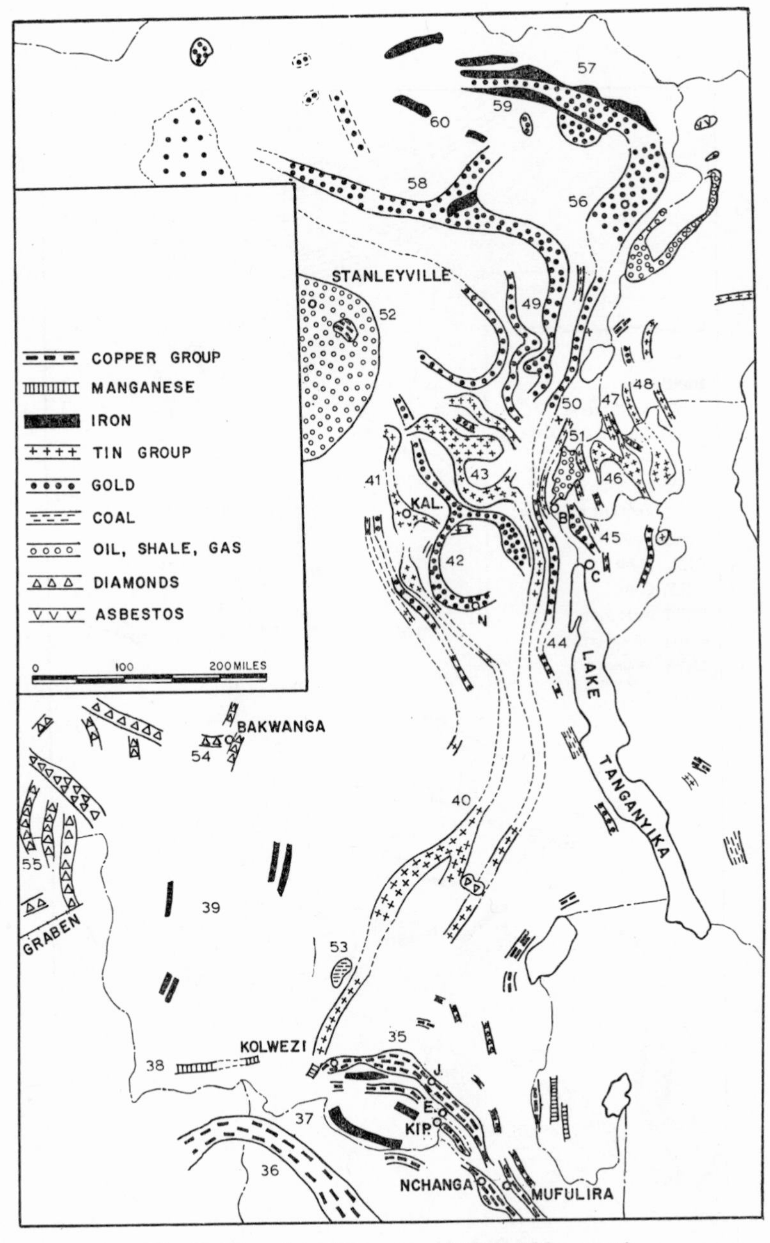

Fig. 51. The mineral resources of High Africa, north.

South of the Copperbelt, a broad cupriferous zone extends from Calunda (Angola) to Chifumpa, Sable Antelope and Broken Hill and covers 500 miles (36). Copper also shows *(a)* between Lusaka and the Mid-Zambezi, *(b)* in the northern part of Southern Rhodesia, *(c)* in the Mokambo trough, north of Mufulira and *(d)* at Kapulo in central Katanga.

Pyrite is omnipresent, but cobalt values decrease radically south of Elisabethville. The gold–platinum–palladium–vanadium horizon of Ruwe, near Kolwezi, is located above the copper beds. The same metals plus nickel, molybdenum and cobalt are associated with epigenetic but stratiform uranium deposits controlled by tectonics. Above the copper beds, hydrothermal lead and zinc replace crushzones. At Kipushi (the continent's principal zinc deposit), germanium, silver and cadmium are also included in the paragenesis. On the other hand, vanadium occurs in the vicinity of the Broken Hill lead deposits. Overlying the copper horizons, iron ores are widespread. In addition to Nambala, the principal iron belt extends from Mutuntu to Musombwishi along the Congo–Zambia border (37). Furthermore, a broad zone of iron deposits connects Kolwezi with Fort Roseberry, west of Lake Bangwelu. Manganese occurs at the two extremities of this zone.

Great Lakes

Beyond the Katangian plateau, middle and late Precambrian orogenies extend 900 miles north-northeast from Kolwezi (Fig. 51) to Lakes Edward and Victoria (Birrimian line) and their width attains 300 miles. They are distinguished by 1,000 million years old tin belts. Along Lake Tanganyika, a pedicle virgates southeastwards (Nyasa line).

At Kisenge (38) in the southwestern corner of Katanga, manganese ores are interbedded with east-west trending middle Precambrian rocks. Beyond the northeastern corner of Angola, iron ores (39) are found in the Congo in a broad area between Kapanga and Charlesville. To the northeast, the principal Kibara tin belt extends from Kolwezi 200 miles north-northeast to Manono (40). A parallel 1,000 million years old belt extends from Mitwaba to Muika.

Between the Congo–Lualaba and the Great Lakes, as well as in southwestern Uganda, Rwanda and Burundi, we distinguish several tin and gold belts. While tin and columbite–tantalite are related to batholiths, tungsten favours smaller stocks and, in Rwanda and Uganda, a sedimentary belt. However, gold occurs in the synclinoria between the lines of granitic stocks. Beyond the tin–tungsten–copper–zinc deposits of Kailo, the western tin belt (41) extends from Punia through Kalima–Moga south to Kampene. The western gold belt (42) connects Kima with Shabunda (130 miles) where it bifurcates; both the Namoya and Kamituga branches are 950 million years old. The eastern tin belt (43) follows the exocontact of the Kasese batholith. On the roof of this batholith, most of the world's tantalum is located. Less important gold and tin–tungsten belts alternate *(a)* east of the batholith and *(b)* west of the northern basin of Lake Tanganyika (44). On the other side of the lake at Mpanda, silver and lead point towards the Lupa field (68, Fig. 52). A 100 mile long tin–tantalum belt spans the islands of Lake Kivu, the Rwanda shore and the Congo/Nile divide in Burundi. Gold occurs along the virgating trend which follows the southern frontier of Burundi. Between 870 and 970 million years ago, tin, beryl, tantalum and columbium (46) concentrated in Rwanda around Gatumba as far as Kigali. In the north-northwest, a 60 mile long tungsten belt (47) connects Nyakabingu and Kifurwe with

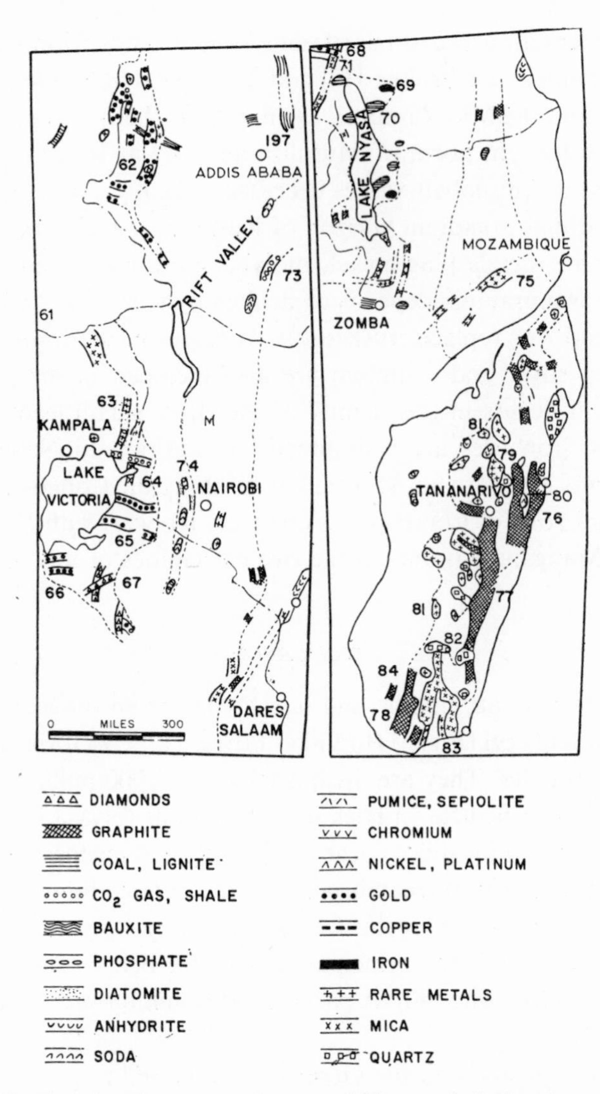

Fig. 52. The mineral resources of eastern Africa, north (left) and south (right).

Kirwa (Uganda). The Ankole (48) beryl, tantalum and niobium field is located further east. Close at hand, tin shows in Karagwe (Tanganyika).

West of Lake Edward, three gold belts undulate along the north-northeast striking folds and stocks of North Kivu (49), and the Mumirwa field spreads southeast of the lake. Platinum occurs in the Lenda alluvia and tungsten in the Etaetu eluvia, where tectonic lines turn west towards the Uele gold belt. On the eastern slopes of the Ruwenzori at Kilembe, copper and cobalt are disseminated in plagioclasites, but west of the lakes, a series of niobium carbonatites (50) has been emplaced. Finally, oil seeps from Pleistocene sediments of the Albertine Rift and methane (51) accumulates in the waters of Lake Kivu.

Karroo seas flooded the valleys and the western edge of the Great Lakes orogenies. Near Stanleyville, Triassic bituminous shales cover 100 miles of the bend of the Congo (52). Farther south, isolated basins of Permian coal (53) include Luena (north of Kolwezi), Greinerville and Albertville (west of Lake Tanganyika). The Cretaceous diamond rim of Kasai–Angola, Katanga, Maniema, Uele, Central African Republic, Gabon and Kwango

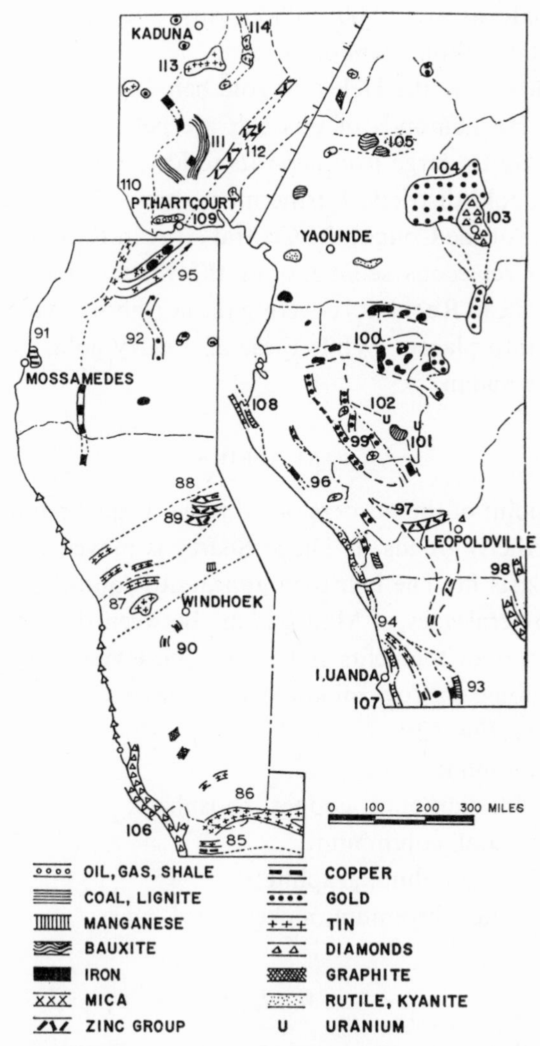

Fig. 53. The mineral resources of the Atlantic Rim, north(right) and south(left).

surrounds the Congo Basin. As in South Africa, kimberlites are intrusive along the western edges of the Karroo in Congo–Angola. Whereas most of the world's industrial diamonds occur at Bakwanga (54), gems are found at Dundu (Angola) and Tshikapa (Kasai). South of these, kimberlites have extruded along the edge of a Rift pointing towards the Angolan carbonatites (95, Fig. 53).

Ituri

The east-southeast oriented Ituri belt follows the 'Libyan' line, i.e., the northern frontier of the Congo, for 600 miles. Resembling other belts of this epoch, Ituri is distinguished by a mineralization of gold (55) and iron. In the highlands west of Lake Albert (in the continuation of the mineral axes of the Great Lakes) mineralization follows the orientation of roof screens. We can distinguish three east-northeast trending alignments in the southern or Kilo district (56), but gold is disseminated in the Moto district (57) (northeastern corner of Congo, Fig. 51). Galenas are 1,900 million years old, but 400 miles west, the Kule Matundu veins are 3,300 million years old.

Between Babeyru and Yindi, the Uele (58) gold belt follows the southern contact of the Kibali batholith. Alluvial diamonds of this belt are believed to be Precambrian. In the roof screens of the batholith, large iron deposits follow east-west trends in Uele (60) and in the Kilo-Moto (59) gold districts. Furthermore, banded ironstones outcrop as far as the northwestern corner of the Congo, the Central African Republic and the southwestern Sudan. Finally, from Cretaceous sediments of the Kotto–Bria district (61, the eastern corner of the Central African Republic) covering the northern rim of the Ituri belt, diamonds have been transferred into placers. While these also carry gold, copper occurs nearby at Hofrat en Nahas in the Sudan.

EAST AFRICA

As far as the meeting point of the borders of Ethiopia, Uganda and Kenya, the Mozambiquian belt follows northerly trends. If Diego Suarez is moved to coincide with Mozambique City, the two coasts fit. The belt then appears to continue along the 900 mile long south-southwesterly mineral axes of Madagascar. Between the Mozambiquian and High Africa, various Precambrian segments have been preserved. From western Ethiopia to Malawi, a 1,800 mile long, narrow mineralogenetic axis traverses these. We distinguish five groups of provinces (Fig. 52):

A. Wallega: gold, platinum;
B. Lake Victoria: gold, diamonds, copper, phosphate;
C. Lake Nyasa: iron, coal, columbium;
D. Mozambique: beryl, tantalum, graphite;
E. Madagascar: graphite, chromium, nickel, beryl, mica.

Wallega

The 400 mile long Wallega gold province (62) connects Roseires on the Blue Nile in the Sudan with the tributaries of Lake Rudolf in western Ethiopia. In these Precambrian windows rising from the Abyssinian lava flows, we distinguish six belts of deposits. In the centre of this province are the platiniferous laterites of Jubdo.

Lake Victoria

Along Lake Victoria, a 500 mile long belt extends from Mount Elgon to central Tanganyika. From the Sudan to Ruri-bay of Lake Victoria, a double line of Miocene carbonatites (63) follows the Uganda–Kenya border southward. Respectively northeast, east and south of the Lake, the Kakamega–Migori (64, Kenya), Musoma (65, Tanganyika) and Geita (66) gold belts are perpendicular to these phosphate–niobium trends. Indeed, the Musoma belt is oriented east-southeast, the others strike east-northeast. The hostrocks antedate the mineralized granites of Lunyo (1,800 million years), and possibly attain an age of 2,800 million years. South of Lake Victoria, kimberlite pipes follow an arc extending 60 miles south-southeastwards from Mwadui (67). Furthermore, the arc of the Iramba plateau is located southeast of Mwadui.

Lake Nyasa

A variety of deposits surrounds the northern shores of the Lake. The Precambrian Lupa gold field (68) trends northwest along the Rukwa Rift; at its intersection with the Luangwa Rift, the niobium carbonatites of Mbeya (71), Musensi and Nkumbwa were extruded during the Jurassic. West of Lake Nyasa at Petauke (Zambia) and Dowa (Malawi) graphite shows. However, northeast of the lake at Njombe (69) titaniferous magnetite settled in ultrabasics. At Kiwira Songwe, Ruhuhu and Nkana faulted coal fields (70) face each other on the lake. Furthermore, coal measures outcrop at Maniambe on the eastern shore of the Lake. The Tete, Nkombedzi and Chiromo coal fields appear in the Karroo valley, connecting the Zambezi with the Shire south of Lake Nyasa. The monazites of the Monkey Bay and the Tambane betafites are 600 and 550 million years old. Finally, three 140 million year old alignments of carbonatites follow Birrimian lines.

Mozambiquian

The principal gold placers of Ethiopia are located at the northern end of this belt, at Kebre Mengist (73). Early Precambrian graphite is disseminated in the gneisses of Tsavo and Voy in southeastern Kenya. The mica pegmatites of the Uluguru Mountains, Tanganyika, are 600 million years old, but the Mrima rare earth carbonatite (near Mombasa) was extruded in the Jurassic. The soda and diatomite deposits of Lake Magadi in the Kenya Rift Valley are Tertiary (74). In addition to the guano cover of islands of Lake Manyara (Tanganyika), sepiolite (meerschaum) precipitated at the foot of the Kilimanjaro. Parallel to the northern Mozambique coast (N 50° E), 430–570 million years old pegmatites extending from Alto Ligonha (75) to Mocuba contain beryl, lepidolite and columbite–tantalite. Pegmatites of the same cycle occur along a westerly line extending to the Mid-Zambezi (in Southern Rhodesia).

Madagascar

The lithologic and mineralogenetic character of Madagascar is distinct. The more than 2,000 or 2,400 million years old Graphite System forms the backbone of the island. Since no trend can be distinguished in the north, all significant graphite deposits are located in three belts. The 150 miles long Tamatave belt comprises a coastal and western branch, both striking south-southwest (76). The Central Belt extends 300 miles south from Tananarivo to Betsiramy (77). Very far, the southern belt strikes 120 miles south-southwest towards Ampanihy (78).

Chromepicotite (79) occurs at Andriamena and along the axis of the northern half of the island, but nickel (80) shows between Lake Alaotra and Valozoro, 200 miles in the south-southwest. Pegmatite fields are also aligned along the axis of the island. However, only the fields of the Central Highlands, extending 300 miles from Berere to Ampandramaika, contain significant amounts of beryl and columbite (81). As in Mozambique, a majority of the ores belongs to a 500 million years old cycle. Eight districts of piezo-electric quartz (82) follow a long arc. In the Benato–Ampandandrava and Ambindandrakemba–Fort Dauphin branches of southern Madagascar, phlogopite (83) has grown in ultrabasics. West of the second branch, uranothorianite was disseminated in these 480 million years ago.

If Madagascar is shifted to its 'Gondawanian' position, the southern Sakoa coal field (84) moves to the latitude of the South African Carolina coal basin. The Malagasy Ridge projects north-northeastward. In these archipelagos, guano accumulated on Astove Island.

THE ATLANTIC RIM

For 2,500 miles, from Mount Cameroon to the Cape, the Atlantic coast follows fold belts. The width of the orogenies decreases from 500 miles in the north to 150 miles in the south. While a narrow strip of younger marine sediments covered the coast, at the end of the Precambrian, sedimentation started in the basins, dividing the Atlantic Rim from High Africa. Early Precambrian orogenies cover *(a)* part of the Cameroon and Gabon, *(b)* western Angola and *(c)* Namaqualand. Late Precambrian geosynclines *(d)* connect *a* and *b*; *a*, *b* and *d* strike north-northeast and north-northwest. The late Precambrian folds of Southwest Africa however, strike northeast. Finally, *c* trends east. We can distinguish seven groups of provinces (Fig. 53):

A. Namaqualand: copper, beryl, lithium;
B. Otavi: zinc, vanadium, copper, tin, lithium;
C. Angola: iron, manganese;
D. Lower Congo: zinc, lead, copper;
E. Gabon: iron, manganese, gold;
F. Cameroon–Ubangi: bauxite, diamonds, gold;
G. Coast: diamonds, oil, gas, potash.

Namaqualand

The Namaqualand bridge of early Precambrian rocks connects the Transvaal orogenies with the coast. In these, copper (and tungsten) are disseminated at O'okiep (85, South Africa), with sillimanite occurring at Pela, southeast. Within these ancient rocks, early Precambrian thorium and rare earths combine at Steenkampskraal. Farther north, the contemporaneous Orange River belt extends from Jackalswater through Warmbad (Southwest Africa) 200 miles east to Upington. This beryl–tantalum–tungsten paragenesis is 950 million years old (86).

Otavi

In Otavi land, warped sediments extend 300 miles from Walvis Bay to Tsumeb and point northeast towards Elisabethville. Along similar trends, acid and alkaline ring-complexes are aligned and between the coast and Omaruru (87), four tin belts strike N 60° E. However, 510 million years old lithium, caesium, beryllium and tantalum also occur south of Karibib (87). In addition to the zinc–lead–copper–silver–germanium ores of Tsumeb (88), 560–780 million years old vanadium shows in the Abenab belt (89). While copper is also disseminated in Karroo lavas (90), at Otjosundu, manganese occurs 160 miles south of Abenab.

Angola

In the aureole of the Huila batholith in southwestern Angola, titaniferous magnetite and limonite deposits extend from Ongaba in Southwest Africa 250 miles north-northwest to Picolo in the Mossamedes area. Copper shows in the Pedra Grande district (Mossamedes, 91) and a gold belt extends from Cassinga to (92) Gandavira, near Nova Lisboa. Copper shows between Cassinga and Serpa Pinto in the east.

The Jurassic alkaline–carbonatite belts of Capuia and Longonjo (95) are distinguished by a paragenesis of iron, thorium, rare earths and niobium. Their N 20°E and N 60°E oriented arcs point towards contemporaneous kimberlites in the northeast (Somali line, 54).

A north-northwest striking horizon of manganese lenses (93) extends from Quitota to Galungo Alto, near Vila Salazar. Mica pegmatites are aligned along a belt extending from Mussaca (94) to Ambriz on the coast.

In the narrow, Precambrian orogenies extending to the Gabon in the north-northwest, bauxites cap the ultrabasics of Tshela (Congo). Gold *(a)* occurs in the Mayombe belt between Chivolo Grande (Cabinda) and the upper Niari, *(b)* tin, niobium and tungsten, further north in the Congo Republic, with itabirites *(c)* outcropping at Tchibanga in southwestern Gabon (96).

Lower Congo

The Lower Congo geosyncline, a narrow trough extending from northwestern Angola 900 miles north-northwest to Gabon, was filled during the last stage of the Precambrian by sediments of the Katanga type. Zinc and lead show at its nose near Kroussou and Offobou (east of Port Gentil) and in a transverse fault zone extending from Boko Songo beyond Mindouli, near Brazzaville (97). The mineralization is 680 million years old. Through Mavoio (Angola), the copper belt continues 200 miles south-southeast to Bembe. Here traces of gold and vanadium also show. East of the syncline, diamondiferous Cretaceous beds outcrop in the Kwango district in Congo and Angola (98). Similarly, diamonds are also scattered on the Precambrian of southern Gabon (Makongonio).

Gabon

In Gabon, roof pendants float on the 1,200–1,700 million years old granites of central Gabon. Iron occurs at the southern end of this zone at Zanaga (Congo Republic), but gold shows exclusively in the screens of a 250 miles long belt extending from Ndjole through Eteke into the Congo Republic (99). Scarce tin and niobium are related to the intrusives.

In northern Gabon, itabirites and iron ores are wedged into troughs surrounded by 2,500 million years old granites (100). Deposits occur in two 150 mile long belts extending from the southeastern corner of Rio Muni beyond Mékambo into the Congo Republic in the east. Similar deposits constitute the Kribi or Mamelles Range in the southwestern corner of the Cameroon. Traces of gold are also scattered in northern Gabon, but hardly warped late Precambrian sediments occupy the southeastern corner of the country. These include, at Moanda (101, near Franceville), manganese, and uranium and vanadium precipitate nearby at Mounana (102).

Cameroon–Ubangi

Various minerals have accumulated in the sediments of this 600 mile long zone extending from the Central African Republic northwestwards across the Cameroon. In addition to the placer diamonds of Berbérati and Carnot (103, western part of the Central African Republic) derived from Cretaceous conglomerates, alluvial gold is found in a larger area comprising the Bouar district as well as the Batouri and Bétaré-Oya zone in the Cameroon (104). Indeed, gold shows as far north as the Logon (Chad). Extensive mantles of bauxite, at Minim-Matrap and Ngaoundal (105, central Cameroon), cover the eastern slopes of the Adamawa Chain on the same latitude as the Ghana bauxites. Minor deposits are located in the southwest, e.g., at Fongo-Tongo. Finally, tin shows in the placers in neighbouring Mayo Darlé.

Coast

Diamonds occur along the 1,000 mile long coastal strip of Southwest and South Africa. The principal zone extends from Pomona through Oranjemund to Port Nolloth and Kleinzee (106). Furthermore, coastal and submarine sands also contain diamonds. Their rhythmic redistribution resembles the recycling of the Congo Basin deposits.

During the Cretaceous, marine sediments started accumulating between the Gulf of Guinea and southern Angola. At Biopio near Lobito, copper shows in these. The Precambrian relief and younger tectonics control the diapyrism of salt domes and hydrocarbon reservoirs. Near Luanda (Angola), Lower Cretaceous reservoirs are aligned (107), and as far north as Cabinda, sediments are saturated by bitumen. In the Congo Republic, oil and gas accumulates at Pointe Indienne, with potash at Hollé. Farther north, the Port Gentil oil and gas basin (108, Gabon) follows a 25 mile long line above Middle Cretaceous to Miocene salt diapyrs.

NIGERIA

Between the Precambrian orogenies of middle–southern Africa and the Guinea–Hoggar Shield, young sediments covered the gap. This is the 'hinge' of Africa, extending from Saint Helena north-northeast to São Tomé, Mount Cameroon, the Tibesti Mountains, the Bay of Syrte and the Dinarides. In the Tibesti, tungsten is related to late granites. On the other hand, guano and manganese accumulated on the tuffs of Saint Helena. But in Nigeria, mineralization attained its climax 2,000 million years later than in the neighbouring belts. A Mesozoic mineralogenetic arc extends from the delta of the Niger, 1,100 miles north to the Saharan Aïr (Fig. 49, 53).

On the coast, oil reservoirs connect Bomu (near Port Harcourt) with Oliobiri (109). Farther west, phosphate beds outcrop between Ososhun (near Lagos), Toffo (Dahomey) and Lake Togo (110, Fig. 54).

In a 60 mile long zone connecting Orlu with Ujaba through Ogwashi, Triassic seams of lignite outcrop, but Cretaceous coal fields are aligned between Enugu and Ogboyaga, 60 miles north-northeast (111). On both sides of the confluence of the Niger and Benue, ironstones developed on Cretaceous surfaces. Farther east, lead, zinc and barytes occur in a 350 miles long Cretaceous belt (112) connecting Ishiagu with Abakaliki (near Enugu) and Zurak in the northeast. Finally, diatomite and natron occurrences surround the Chad Basin.

Metamorphic and granitic rocks occupy the western half of Nigeria, Dahomey and Togo. The 'old' tantalite pegmatites of the Egbe and Wamba–Jemaa fields (113) appear to be 500 million years old. However, most of the granitic ring complexes were emplaced 160 million years ago, i.e., during the Jurassic cycle of diastrophism. The 600 mile long alignment of these tin, niobium, hafnium (and tungsten) deposits extends from Afo on the Benue northwards though Jos (114) to the Adrars Elmeki and Guissot in the Aïr (Niger Republic). West of the Main Mineral Axis, gold shows, e.g., at Birnin Gwari, Kontangora (Nigeria) and in Dahomey.

THE GUINEAN SHIELD

The Guinean Shield connects Conakry with Accra (900 miles) and averages a width of 400 miles. Furthermore, a 200 miles wide branch of the shield projects north-northeast through Upper Volta. Between Gao and Niamey, the Niger graben separates this branch from its extension, the Hoggar in the Central Sahara. The shield consists of gneisses, granites and middle Precambrian metasediments. All Precambrian axes follow the north-north-

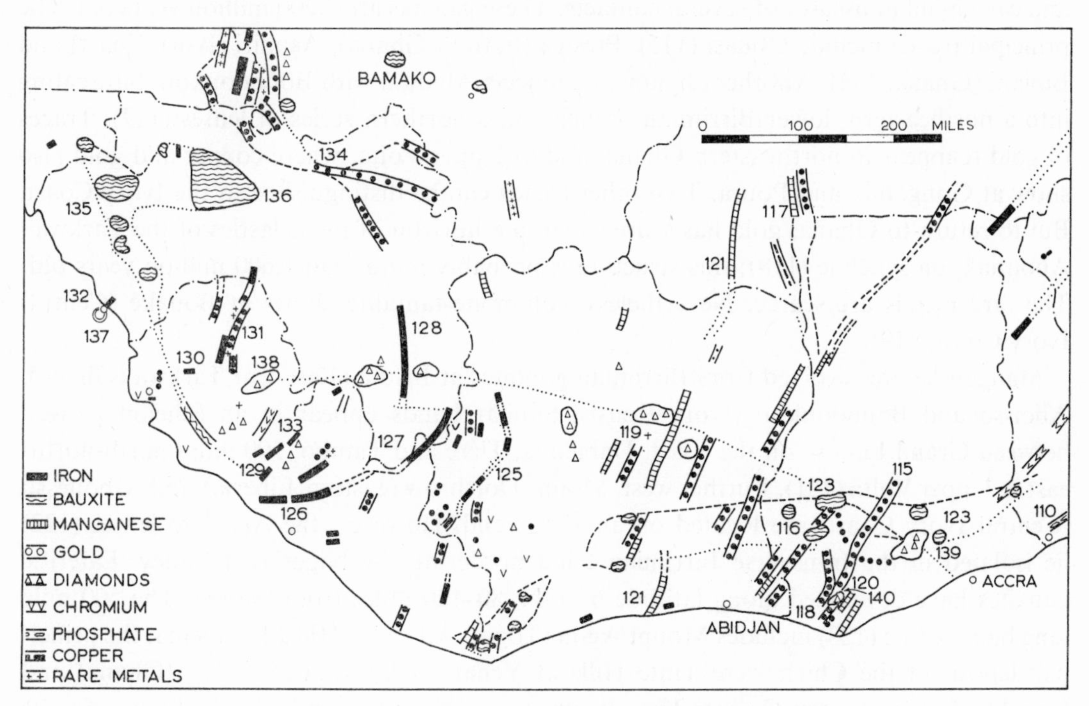

Fig. 54. The mineral resources of the Guinean Shield.

easterly trend, distinguished by iron, manganese and gold deposits. However, the mineralization is significant only in a 200 mile wide band, limited by the coast and the Diamond Arc connecting Freetown with Accra. Finally, bauxites of the inland scarp represent the last phase of mineralogenesis. We distinguish four groups of provinces (Fig. 54):

A. Togo: chromium, iron;
B. Ghanaian: gold, manganese, bauxite;
C. Liberian: iron, bauxite, gold;
D. The Diamond Arc: diamonds.

Togo

In Togo, a separate belt of orogenies extends from Accra north-northeast to the graben of the Niger. In this band, an alignment of chromiferous serpentinites comprises Aouam (southeastern Ghana), Mount Ahito (Togo) and Bontomo (Dahomey). In northern Togo and Dahomey, an iron ore zone includes Mount Djolé and Loumbou-Loumbou. While gold shows nearby at Perma, oolithic iron ores cover the Dapongo and Doguel Kaina Say plateaus (northern Togo, western Niger). Finally, pisolitic bauxite appears at Gaya (southwestern Nigeria).

Ghanaian

Arenaceous metasediments of the lower Birrimian occupy most of the 250 miles long stretch between Grand Lahou and Takoradi (Ghana). Their compact folds extend 100 miles inland in Ghana, and as far as 250 miles inland in the Ivory Coast. Interlayered metasediments and volcanics constitute the narrow bands of the upper Birrimian, but these screens are scarcer north of the Diamond Arc and west of Grand Lahou. Gold, sulphides and carbon fill in fissures of several contacts. These galenas are 2,200 million years old. The principal trends include Obuasi (115), Prestea (both in Ghana), Asupiri (Ivory Coast) and Bibiani (Ghana, 116). Another alignment connects Abidjan with Boundoukou, bifurcating into a northeastern, lower Birrimian branch and a northern series of lenses (117). Traces of gold reappear in northeastern Ghana, and in Upper Volta, where copper and gold also show at Gongondy and Poura. Two other trends can be distinguished in the Ivory Coast. But to return to Ghana, gold has sedimented in a horizon of metaclastics of the Tarkwa–Abontiakoon syncline (118); this structure is probably more than 1,600 million years old. The rare metals are scarce. Nevertheless, columbite-tantalite shows at Bouaké (Central Ivory Coast, 119).

Manganese ores evolved from Birrimian gondites at Insuta (120) near Tarkwa (Ghana), Aboisso and Boundoukou (Ivory Coast). Similar bands appear in an échelon pattern between Grand Lahou, on the coast, Dabakala, Tiéré and Tambo, 500 miles north-northeast in Upper Volta (121). Further west, Mount Dorthrow (eastern Liberia) and Mbouésso (Central Ivory Coast) are located on another trend. However, the Ansongo lenses (122) lie isolated in the transverse Birrimian band bordering the Niger Rift Valley. Lateritic bauxites have developed along latitude 6°30′N, 80–120 miles from the sea. The 500 mile long bauxite line (123) includes Mount Agou (Togo), Atiwedra Hills, Ejuanema, the principal deposit of the Chichewere Tinte Hills at Yenahin, Affo and Nsisreso (Ghana) and Dagolila–Lakota (Ivory Coast). The alignment is believed to extend into Liberia in the direction of the Guinean bauxite belt. East of Grand Lahou, laterites extend along the coast with oolithic iron ores accumulating west of Sassandra (Ivory Coast, 125).

Liberian

Arcs of more or less magnetic quartzites are wedged into the gneisses of Guinea, Liberia and Sierra Leone. Intense metamorphism and compressive forces have affected these discordant, tightly folded synclines. The sequence is believed to be older than 2,200 million years in Guinea and older than 3,000 million years at Kambui (Sierra Leone). Along the 100 mile long line of the Mount Gao syncline and Mount Jedeh, in the western Ivory Coast and eastern Liberia (125), itabirites rise already. However, in the 300 mile long arc extending from Robertsville eastwards and bending northward in central Liberia, are arranged en échelon the four principal iron ore ranges: Bassa, Bomi (126), Nimba (127) and Simandou (128).

Along the Liberia–Sierra Leone border, the Bagha Hill, Mano River and Bié Mountain belt (129) strikes north-northeast. Farther west, in a 90 miles long schist belt extending from Central Sierra Leone north-northwestward into Guinea, occur the Yamboeili, Marampa (130) and Laia deposits. Similarly, in the Kambui schist belt, we find Tonkolili (131). As to ultrabasics, iron laterites cover the dunites of the Kaloum peninsula (132, Conakry, Guinea). At Freetown (Sierra Leone), titanomagnetite and platinum occur in the ultrabasic Colony complex. Chromite lenses developed in the serpentinites of Hangha (133, eastern Sierra Leone), and also show in the southwestern Ivory Coast. On the other hand, gold placers include the Sula, Kangari and Kambui belts in Sierra Leone, and Boken Jede, Kokoya, Tawalata and the Cavally in Liberia. Other forms of gold occur at Kato (Guinea) and Mount Flotouo (near Ity in the western Ivory Coast). Resembling Ghana, metals of the acid group remain rare: molybdenum accompanies the base metals at Koinadugu (Sierra Leone) and traces of columbite occur at Sula, Sende (Sierra Leone), Kokoya and Tawalata (Liberia).

Birrimian enclaves reappear along the edge of the shield, covering northern Guinea. The Nzérékoré manganese deposit (northwestern Ivory Coast) is comprised in one of these lenses. South of Bamako, lithium concentrated in the Sinsinkourou pegmatites. Extending from the Upper Niger to the Falémé, a 250 mile long gold belt crosses the largest enclave (134) in a northwesterly direction, thus connecting with the tip of the Mauritanian Arch.

The world's principal bauxite deposits (135) form a 300 mile long morphologic belt extending from the Corubal drainage, Portuguese Guinea, southeast to Sierra Leone. The deposits of Bové, Badi–Konkouré and Kindia emerge in Guinea 90–30 miles from the sea, but in southern Sierra Leone, the Mokanji and Nbong Hills rise higher. On the other side of the Fouta Djalon Mountains 200 miles from the sea, bauxitic laterites cover a 100 by 80 mile area between Dabola, Tougué and Siffray (136). Bauxite also shows *(a)* in Bassari country, north of the Fouta Djalon, *(b)* at Telikan, at the sources of the Niger, *(c)* at Koulouba (Bamako), *(d)* at Mpébougou (Ségou) in Mali and *(e)* in the Loos Archipelago, off Conakry (137).

The Diamond Arc

Diamond deposists are aligned along a distinctive 700 mile long arc, connecting Sefadu (Sierra Leone) with the Macenta field, Bounoudou, Bassam (Guinea), Seguéla, Tortiya (Ivory Coast) and the Birim field (Ghana). This arc is the perpendicular pair of the prevailing Precambrian trend. We distinguish three cycles: *(a)* the principal field, Sefadu (Sierra Leone, 138), and those of Guinea are Mesozoic, *(b)* the small Bonsa field (Ghana, 140) is probably Lower Tarkwaian and, finally, *(c)* the Tortiya (Ivory Coast) and Birim (Ghana,

139) districts appear to be Lower Birrimian. Finally, diamonds were carried as far as 100 miles in the drainages of the upper Niger, and south in the Sewa, Cavally, Sassandra and Pra.

THE MAURITANIAN ARCH

The Mauritanian Arch forms the western rim of Precambrian Africa; crossing Mauritania, a 700 mile long orogenic belt extends northwards along the Guinea–Senegal–Mali border. Then along the frontier of the Spanish Sahara, the arch turns east-northeastward. This wider part, 700 miles long, consists of granites and metamorphics. Thus the arch, the Guinean Shield and the Hoggar constitute the framework of the western basins of the Sahara. After the Cretaceous, a 150 mile wide sedimentary strip was deposited between the Arch and the Atlantic.

Phosphatogenesis started at Civé, on the Sénégal River. However, the principal phosphate plateau developed during the Lower Eocene (Lutetian) at Taiba, northeast of Dakar (141), and aluminium phosphates formed at Pallo. Near Dakar, gas and oil accumulated in Lower Cretaceous shales (142). In southwestern Mali, along the southern edge of the Arch, diamonds are scarce at Kéniéba as is bauxite at Fallo Kéréwane and Témbéré. Nearby, the iron ores of Koudekourou and Sicasso occur in southern Sénégal and Mali In the same district, the North-Guinea gold belt continues in the Falémé Valley of southeastern Sénégal where rare metals also show.

A 600 mile long, N 30°E striking line connects Akjoujt with Tindouf. Northeast of Nouakchott, in the Guelb Moghrein (Akjoujt), gold, cobalt and graphite accompany copper and iron (143). Beyond similar occurrences, the iron ores of the Kediat d'Idjil (144) are located at Fort Gouraud, Mauritania, near the southeastern border of the Spanish Sahara (Fig. 55). Sfariat prolongs this zone northeastwards. North of Fort Gouraud, beryl has been disseminated around Bir Mogreine. Farther north as far as the Anti-Atlas,

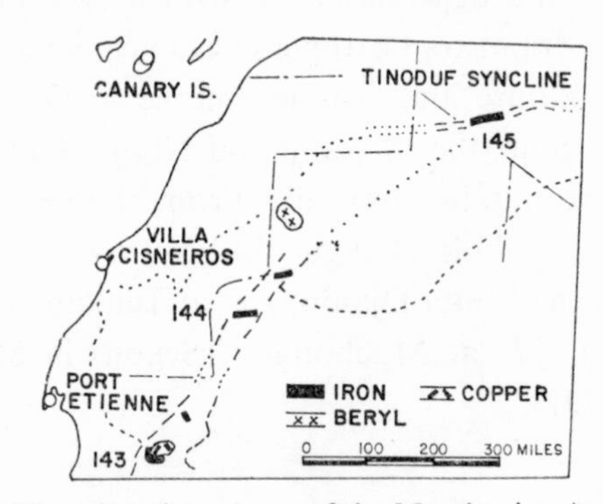

Fig. 55. The mineral resources of the Mauritanian Arch, north.

Mesozoic transgressions flooded the Tindouf basin with oolithic iron ores coagulating in the Devonian Garas (Hills) including the Djebilet (Algeria, 145).

THE ATLAS

The Atlasic orogenies extend from the southern tip of Morocco, 1,300 miles east-northeast

to the Bay of Gabès in Tunisia. This trend first appears in the Precambrian backbone of the Anti-Atlas and in the remnants of the eastern Algerian coast. We distinguish two directions: N 50°, 80°, and again 50°E. At the Oujda and Constantine 'hinges' connecting these, important ore deposits are located. The tectonic line connecting Agadir with Sfax divides *(a)* the Paleozoic Anti-Atlas of Morocco from *(b)* the Mesozoic orogenies of the High and Pre-Saharan Atlas and *(c)* the Middle- and Tell Atlas *(d)*. The pre-Triassic windows of the Moroccan Meseta lie north of these *(e)*. Finally, Tertiary sedimentation, volcanism, the Alpine cycle and tectonic style have affected the East Coast and the Rif.

The belt is distinguished by: *(a)* a mineralization of phosphates, *(b)* the migration of Paleozoic parageneses into three younger carbonate horizons and *(c)* the coexistence of replacement and vein deposits of lead, zinc and iron. We can distinguish five groups of provinces:

A. Anti-Atlas: cobalt, molybdenum, copper, coal, iron;
B. Meseta–High Atlas: phosphate, zinc, lead, antimony, iron, manganese;
C. Rif–Tell Atlas: zinc, lead, iron, coal, barytes, diatomite;
D. Eastern Atlas: zinc, lead, antimony, iron, mercury;
E. Gafsa basin: phosphate, iron, oil.

Anti-Atlas

The Anti-Atlas extends 300 miles in an east-northeasterly direction along the southern border of Morocco (Fig. 56). The northern part of the range has been mineralized in the vicinity of Precambrian windows. In the Tirghrarghine pegmatites, 1,700 million years old beryl shows, and in the lower Precambrian and earliest Cambrian copper is scattered, with cobalt, nickel and copper (146) in the ultrabasics of Bou Azzer. Furthermore, gold shows at Tiouit and Cambrian molybdenum and tungsten occur at Azegour (147) in a northern outlier of this belt. In addition to late Precambrian manganese ores (Tiouine),

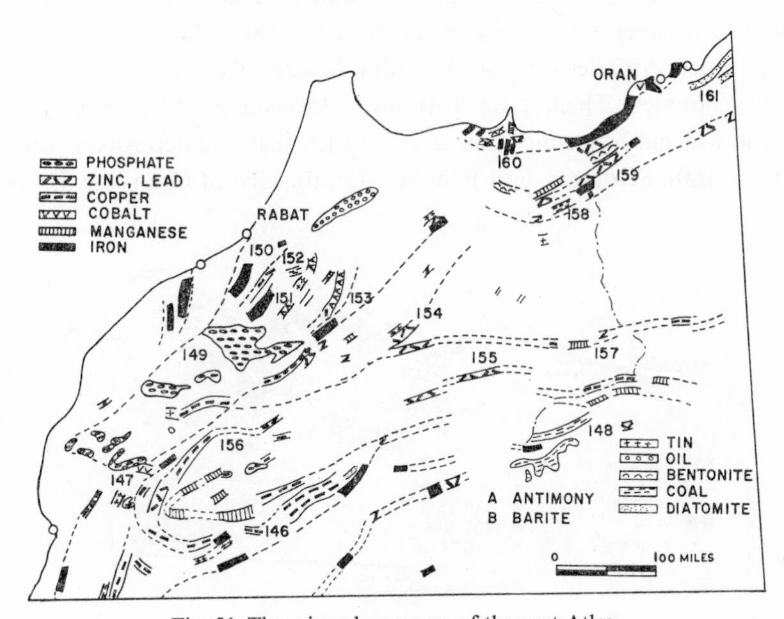

Fig. 56. The mineral resources of the west Atlas.

Silurian, oolithic iron ores outcrop *(a)* in the west in the Ouarzamine syncline (Mogador) and *(b)* in the east at Imi n'Tourza. East of this deposit, in the Colomb-Béchar window of Algeria, the Paleozoic reappears with the coal measures (148). Finally, west of the Anti-Atlas, oil shows in the Tarfaya and Sous basins.

Meseta–High Atlas

Phosphates sedimented on the shelf separating the Anti- from the High Atlas. During the early Eocene, phosphatization reached its peak on the large Khouribga plateau (149) and at Youssoufia. The Silurian iron ores include Oulad Said, south-southeast of Casablanca and Aït Amar, but the Khenifra deposit is Carboniferous (153). The N 25 °E oriented deposits of the Moroccan Meseta comprise antimony and tin at Mrirt and Oulmès (151), with a lead–zinc–baryte paragenesis at the Jebel Aouam (152).

Covered by Lower Jurassic limestones, the Middle-Atlas virgates N 50 °E. We distinguish two east-northeast oriented zones in the Mibladen–Moulouya lead–zinc district. Here, 320–520 million years old galenas antedate their hostrocks. In the Jurassic limestone of the High Atlas, the Bou Dahar zone (155) resembles Mibladen. In the Paleozoic window (Azegour, 147) connecting the western High Atlas with the Anti-Atlas, the lead–zinc veins (e.g., Goundafa) are 400–520 million years old. At the western and eastern ends of the High Atlas, we find Cretaceous manganese at Imini (156) and Bou Arfa (157) and also in the Saharan Atlas of Algeria. Finally, north of the Meseta, oil accumulated in the Paleo–Caenozoic Rharb basin.

Rif–Tell

Between Melilla and Bougie in the east, the Rif, the Oujda horst and the Tell Atlas cover 500 miles (Fig. 56, 57). The two former include the principal ore deposits of the Atlases (Ouichan and Bou Beker), but the intensity of mineralization decreases eastwards. Coal measures outcrop between Christian (on the Meseta), Djerada (158) and Oran. Between Bou Beker and Oued Zounder (Oujda) 300 million years old zinc and lead (159) migrated into Jurassic carbonates. Then, Late Tertiary volcanism produced the Camp Bertaux, Providencia and Marnia bentonite beds as well as the iron ore deposits of Melilla, Nador, Ouichan (Uixán, 160), Setolazar, and Beni Saf (Oran). East of Oran, the diatomites of the

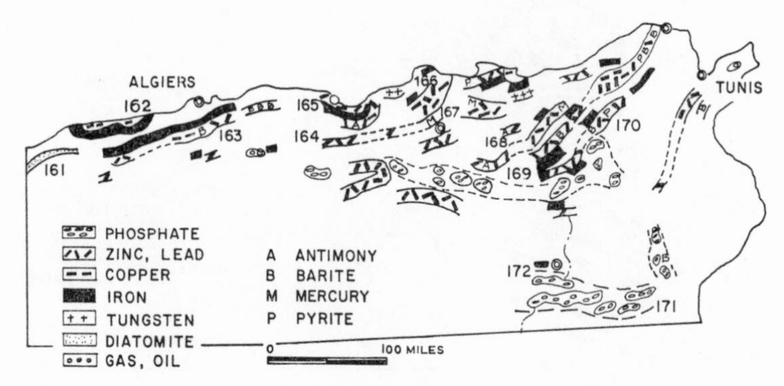

Fig. 57. The mineral resources of the east Atlas.

Chélif syncline rest upon Upper Miocene gypsum (161). Also, during the Miocene at Zaccar (162) west of Algiers, pyrite was rearranged. However, the lead–zinc paragenesis of the Ouarsenis Mountains (163), the Jebel Gustar (164) and the iron ores of Timezrit (165) are Jurassic–Cretaceous. At Palestro, baryte veins cut limestones and at Cavallo (Bougie), copper is related to Miocene andesites.

Eastern Atlas

The Saharan and Tell Atlas coalesce in eastern Algeria. North of Constantine, the Flysch (166) bends northeastward around the Collo stock. This area includes pyrite (El Halia), antimony (Hamman n'Bails) deposits, and mercury (Oued Maden, 167). However, the principal belts extend in the Cretaceous nappe from Ouenza, 150 miles northeast, to Bizerta, attaining an aggregate width of 60 miles. Among others, we find here lead, zinc and iron (Table CVII).

TABLE CVII

LEAD-ZINC AND IRON DEPOSITS IN THE EASTERN ATLAS

	Lead–zinc (170)	*Iron (169)*
Miocene		Mesloula (Tunisia)
Eocene	Djebel Ressas	
Lower Cretaceous	Sakiet Sidi Youssef	Ouenza
	Sidi Amor	Djérissa
	Mesloula (Algeria)	

Finally, gas accumulated at Cap Bon near Tunis.

Gafsa Basin

The early Eocene phosphate sea flooded the Gabès–Chott-el-Djerid basin along latitude 34°N. The belt (171) extends 200 miles from Négrine (Algeria) through Redeyef–Metlaoui to the Djebels Berda and Chemsi (Tunisia). Containing phosphates, oil and iron, the Djebel Ongg (172) warp is located in Algeria 30 miles north of Négrine. Another branch of the sea flooded the lake region between the two Atlases, extending between M'zaita, the Kouif deposit (Algeria) and the Djebel Ressas (Central Tunisia).

SAHARA

The Precambrian mass of the Hoggar constitutes the anchor of the Sahara. During the Paleozoic, a 150 mile wide zone of sediments was first deposited around this 'island'. The Thetys flooded *(a)* the edge of the Devonian, *(b)* western Libya and *(c)* the Tademaït Arch, and thus extended the northerly (Birrimian) trends from the Hoggar towards Algiers. Post-Jurassic formations cover the two Ergs on both sides of this ridge. Finally, Eocene sediments gained the Syrte basin from the Mediterranean. The Saharan hydrocarbon provinces overlap these formations in a 1,500 mile long belt, expanding from a width of

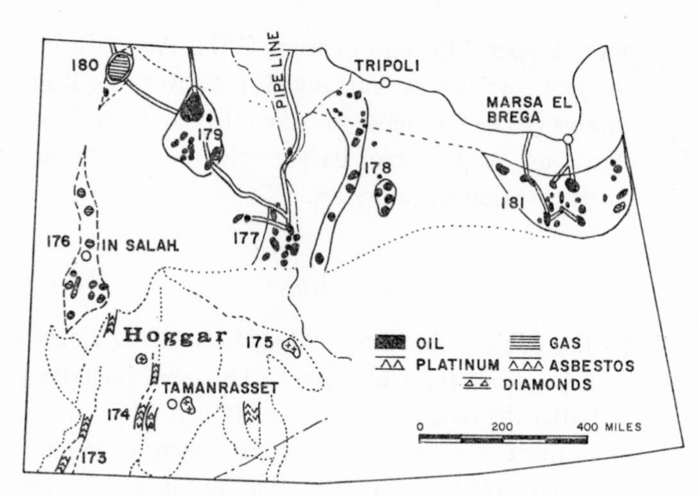

Fig. 58. The mineral resources of the Sahara.

100 miles in the Syrte to 500 miles on the meridian of Algiers. The age of oil reservoirs increases in the same direction, but pre-Silurian structures control oil accumulation. We distinguish the following groups of provinces (Fig. 58):

A. Hoggar: platinum, asbestos, tin, zinc;
B. Paleozoic (South):
C. Triassic (North): } oil, gas, anhydrite, salts.
D. Eocene (East):

While exploration of the basins continues south of the Hoggar, oolithic iron ore, manganese, coal and phosphates occur on the periphery of the Sahara.

Hoggar

Six belts, belonging to two Precambrian orogenies, alternate in the 500 by 500 mile area of the Hoggar; they all strike north. During both epochs, platinum was disseminated in the Goudène belt (southwest, 173), between Tibeghim and Tin Dahar (174), and in the enclave of longitude 7°30′E. Diamonds show along the eastern edge of a graben, and asbestos is more widely scattered. This ultrabasic mineralization compares with the chrome parageneses of the Guinean Shield. In the youngest granitic ring-complexes, tin and tungsten occur at In Tounin (near Tamanrasset), at Elbema (north of Tin Dahar) and at Djilouet (175), north-northeast of the Aïr (the southeastern wing of the Hoggar) and the northern end of the Nigerian Arc (114). In the southwestern wing of the Hoggar, zinc shows at Tessalit and also in the Adrar of the Iforas. Finally, south of the Iforas, phosphates cover the Tilemsi Valley.

Paleozoic (South)

Gas started to accumulate with some oil during the Cambrian along a 250 mile long trend extending south-southwest from El Goléa to In Salah and Azzal Matti. The principal reservoirs are Lower Devonian (176). However, in the Reggane basin further west, oil shows in the Carboniferous. In the Fort Polignac field 500 miles south of Gabès, oil accumulation

(177) culminated in the Lower Devonian. We can distinguish two north-northeast oriented series of reservoirs: Edjeleh–Zarzaïtine and Tiguentourine. Underlying these, radioactive Ordovician graptholite shales are contemporaneous with the second Katangian uranium phase. From Atshan (east of Edjeleh), the western Libyan oil arc extends 400 miles north-northeast towards Bir Tlacsin and Tripoli (178).

Triassic (North)

Oil reservoirs developed after the Permian transgression. Superimposed on Precambrian structures, the Hercynian epeirogeny controls them. This province includes the principal oil and gas field of the Sahara. In the north-northeast oriented Hassi Messaoud–El Agreb trend (179) 400 miles south of Constantine, oil was trapped in the Triassic and Cambrian. The Hassi er Rmel gas field (180) is on the same trend as the older In Salah fields. Oil also shows at Makrerouza (Tunisia) and Ouled Chebhi (Tripolitania). As in other oil fields, salt and gypsum are abundant.

Eocene (East)

The Atlasic trends continue to coastal Cyrenaica. But during the Eocene, this orogeny was cut by a trough connecting Calabria with Syrte. Here the oil basins of Mabruck, Zelten, Gialo and Sarir occur in a 400 mile long 'integral' oriented parallel to the Bay of Syrte (181). Upper Cretaceous sediments and Paleocene–Eocene limestones form the best traps (with Infra Cambrian–Ordovician limestones at Amal). In addition to the potash of Marada, we find iron ores at Brak.

RED SEA

A poorly mineralized, 1,000 mile long orogenic belt rises along the coast of the Red Sea; it reappears at and bends around the Djibouti 'hinge' (Fig. 59). Sedimentary deposits and evaporites favour the coastal strip: Miocene oil fields extend along the Gulf of Suez (182) between Rahmi and Hurghamra. But some reservoir rocks are Carboniferous and Cretaceous. In addition to the Miocene sulphur and anhydrite of Gemsa (Egypt), anhydrite covers the coast between Dunganab and Marsa Salak (north of Port Sudan) and Mukawwar Island (183). Furthermore, in the Dahlac Islands petroleum and at Dallol (184, northeastern Ethiopia) potash shows.

Latest Cretaceous phosphates (185) occur along latitude 25°N at Quseir (on the Red Sea), Idfu (near Luxor) and in the Oases. At Um Gheig, lead and zinc migrated into Miocene sediments (186). East of Aswan (187) and near Wadi Halfa (Sudan, 188), Upper Cretaceous iron ores cover the western rim of the belt.

A 450 million year old and 40 mile wide mineral belt extends from Fawakhir (near Luxor) 500 miles south-southeast to Port Sudan. Between Baramaia, Hafafit, Abu Dahr and Atshan, this belt comprises an ultrabasic chromite, talc (189), asbestos and vermiculite zone. Some beryl, mica, tin and tungsten show, e.g., at Igla (United Arab Republic). The gold fields include Fawakhir (190), Baramaia–Abu Had (189), Eigat Deraheib (191) and the Red Sea Hill (Gebeit, 192) zones. At the Sudanese border, the orogenic belt expands to a width of 300 miles and on the inner, Nubian side of this belt, gold occurs in a broad zone connecting Wadi Halfa with the Atbara bend of the Nile (193), and reappears in the window extending from Wadi Halfa south to Dongola (194).

In the Tokar district (Sudan), copper shows; on the Anseba in Erythrea, nickel and chromium occur. From the sources of the Anseba, gold belts extend to Asmara and from the southwestern corner of Ethiopia, to Agordat (195). Similarly, iron appears on the Anseba, on Mount Tulului, south of Asmara (196), and with manganese at Agametta, near Massawa. The Erythrean trends are south-southwesterly.

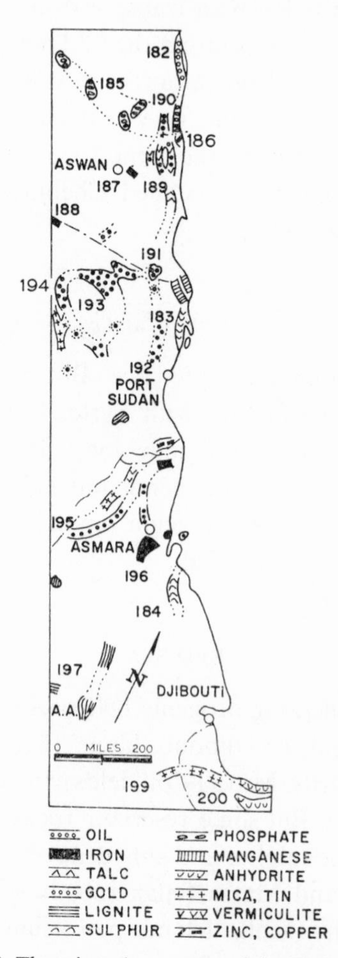

Fig. 59. The mineral resources of the Red Sea belt.

Lignites evolved on the scarp of the Ethiopian plateau, between Waldia and Debra Brihan (197), and at Chilga, north of Lake Tana. As in Kenya, pumice and diatomite occur in the Rift Valley. In southern Ethiopia and Somalia, metamorphic rocks reappear south-southeast of the Red Sea belt. West of Mogadishu, the Somali coast is oriented parallel to the itabirites (198) of the Bur window; north of the city, meerschaum (sepiolite) layers spread around El Bur.

Precambrian remnants rise between Harar, eastern Ethiopia, and Socotra Island, 800 miles in the east-northeast (Atlasic line). Along a parallel zone connecting Jijiga (Harar) with Warran Weis, Theb and Humbeleh (Berbera), mica, beryl and columbite (199) pegmatites outcrop. However, the ultrabasic mineralization of this province is related to ring complexes. In the Horn of Africa, tin reappears at Dayah Kul-Magiaian. Finally, the Eocene Anhydrite Series covers 15,000 square miles of northern Somalia, attaining its maximum thickness near Berbera (200).

THE MINERAL 'PATTERN'

What laws govern the distribution of these metallogenetic provinces and the relative concentration of ores? Of course there is no simple answer. Indeed, according to the value our economy attributes to the substance, we arbitrarily select grades which define a mineralization. These grades may mean one useful part in seven million for diamonds, one in 200,000 for gold, one in 1,000 for tin, two in 100 for copper and 95 in 100 for oil. Consequently, metallogenetic units do not correspond to simple mineral provinces. Nevertheless, we can deduce a few rules.

Although we lack information on the early history of the sedimentary basins and the Precambrian rocks they cover, from the little we know and subject to new data, we might infer that in addition to the ancillary nucleus of Guinea, the continent grew from 'High Africa'. At the northern and southern ends of this megabelt, we find rocks more than 3,000 million years old and if we travel northward from South Africa, we see fold belts more than 2,000 million years old. Similar ages appear around Lake Victoria, on the Malagasy Plateau, in southern Cameroon, and in the Guinean Shield where, incidentally, a few rocks more than 3,000 million years old were also dated. These areas are characterized by an overwhelming majority of the continent's resources of: iron (Griqualand, Transvaal, Mekambo, Kilo-Moto, Bomi–Nimba); gold (Witwatersrand, Southern Rhodesia, Kilo-Moto, Ghana); uranium (Witwatersrand); chromium–platinum (Bushveld, Great Dyke). Why did these metals appear so early? No doubt ultrabasics should crystallize first. Indeed, before 3,000 million years ago, small but high grade bodies of *chromium* and asbestos appear in the Rhodesian Shield and its pedicle. However, north of the Zambezi this mineralization is both weak and rare, and even south of the River these metals, now accompanied by the platinum group, reach their climax later and in another type of deposits. Indeed, 2,000 million years ago, enormous ultrabasics sheets poured out through narrow channels. Was further consolidation of the crust necessary before pressures enhancing violent extrusions and the differentiation of metals could build up? We do not know, but in the interval of the two ultrabasic cycles, most of the world's *gold* appears in placers, within a relatively limited period of early 'time', and a variety of epigenetic deposits play a minor role.

And conversely, in one identical type of deposit, narrow synclinal belts, an overwhelming majority of the continent's *iron* prot-ores sedimented. These iron troughs are located both near the edges and in the middle of the fold belts. We can only infer that the iron content of the seas changed rather suddenly or, alternatively, the seasons permitted the alternation of deposition of ions. Anyhow, traces of gold are frequently associated with these ferruginous belts, but the cycle of gold is more widely spread both in alluvial and hydrothermal deposits. Some gold even reappears in the following 1,000 million year cycle, disappearing then as did chromium, platinum and asbestos 1,000 million years earlier.

As we have seen, uranium accompanies gold, but this metal, among others, reappears in the latest Precambrian of the Congo and Gabon. The behaviour of *graphite* is different. Before probably 2,500 million years, this mineral appeared 'illogically' as the first form of carbon in the apparently sedimentary syngenetic deposits. There are also somewhat younger and much smaller deposits of graphite, but even these are generally restricted to the same East African belt. Traces of other forms of carbon, hydrocarbons and diamonds, first appear at the 'same time' in the Witwatersrand. But in a complete reversal of what might be expected, the 'lower' forms of carbon culminate 2,000 million years later!

Although geochronological data show an orogenic cycle in the midst of it, from 1,850 million years ago until 1,100 million years ago, we witness a virtual gap of metallogenesis. Then, projecting from earlier fold-masses, new orogenies appear in the Great Lakes and Orange River belts where tin and rare metals crystallized in the aureole of acid intrusives. We should not forget, however, that the rare metals had already appeared in Southern Rhodesia 2,700 million years and that other pegmatitic cycles oscillated around 500 million years ago, frequently in older rocks in Mozambique, Madagascar, Southwest Africa and central Nigeria. Conversely, in the Nigerian Arc, a different type of tin deposits, granitic ring-complexes, are related to the Jurassic cyle of diastrophism which is distinguished by another (a carbonatic) type of niobium–rare earth–phosphate ring complexes and by the sudden extrusion of *all* major diamond pipes and veins. We can collect diamonds in several Precambrian zones but, proportionately, these are insignificant. Can we again infer that only under an even thicker and older crust could sufficient pressures accumulate to trigger the extrusion of kimberlites?

Meanwhile, the interior basins dividing the orogenies of High Africa and the Atlantic Rim were filled in several cycles. During the latest Precambrian, the first of these cycles is distinguished by the deposition of a majority of the world's copper, cobalt and germanium and most of Africa's zinc, silver and cadmium. Although sediments of this 600 million year old cycle cover the broad edges of earlier folds, an overwhelming majority of the ores is confined to the narrow gap dividing the Rhodesias from the Great Lakes belt and within these, the earliest furrows carry most of the metals! Within the same belt, the principal metal, copper, appears to be sedimentary, but all the others tend to be hydrothermal! We can mention manganese here. At least in the grades we today consider economical, this metal has many peaks. Although the principal coincide with the middle and latest Precambrian, there are others. In fact, with smaller quantities of time becoming more meaningful, we can distinguish two parts of this cycle. In the second 500 million year old sub-cycle in Southwest Africa, the base and the rare metals crystallized 'simultaneously' but under very different conditions. The base metals, especially lead with its capacity of migration, reappeared in the Atlas around 350 million years ago and even later, frequently in newer sediments in which their deposits were rejuvenated.

Although appearing in the Cambrian, hydrocarbons culminate in the Devonian, Triassic and Paleocene of the Sahara. Indeed, after the Paleozoic, the northern half of the continent received its sedimentary cover with great leaps. Conversely, the empty basins of southern Africa were suddenly flooded in one epoch, the Permian, when coal measures appeared. Interestingly enough, a majority of the diamond kimberlites pierce the edge of these carbonaceous sediments, i.e., possibly a zone of weakness (?). Then, soon following this diastrophic cycle which also put phosphates on the maps of southeastern Africa, an entirely different sedimentary type of phosphate deposit suddenly appears in a distinctive belt of northern Africa. At the same time, the land surfaces on which the world's principal bauxite deposits still evolve started to develop in another belt, on the Guinea Coast.

What conclusions can we draw from this metallogenetic history? First, the continent, at least as far as we can ascertain, appears to have grown or at least thickened from several nuclei, most of which coalesced by the late Precambrian into 'High Africa'. This large and early belt acquired most of the metals, but in the east, adjacent and contemporaneous belts remained poor! Secondly, for more than 2,500 million years, sediments played a most important role in this growth and an even greater role in the metallogenesis. Indeed the principal ores of Africa, although frequently Precambrian, are at least partly syngenetic. Thirdly, within megabelts, internal trends tend to form an alternating or grid pattern. Similarly, in time many metals oscillate, but there is a distinctive distribution pattern initiated in southern Africa (and Guinea) and progressing towards the Mediterranean, or an 'age pattern' starting with gold and iron and finishing with bauxite.

SECTION II

Iron-bauxite group

This group includes iron, manganese and bauxite, three of the principal ores of Africa, and amphibole asbestos, a mineral closely related to banded ironstones. Alteration processes play important roles in the genesis of the three ores. The principal iron and manganese deposits can be described as sediments in which metals were concentrated by metamorphism and leaching. While their genesis has been controlled by paleo-climates, the development of the lateritic bauxites, iron and manganese ores is conditioned by more recent climates. Minor iron occurrences are widespread, but larger deposits are *(1)* restricted to West Africa, where bauxite and manganese are also found, and *(2)* with manganese to the periphery of the Congo basin and to South Africa, where *(3)* amphibole asbestos appears.

IRON

Iron occurrences are as widespread as those of gold, but the principal deposits can be classified among the itabirites or banded ironstones, magnetitic quartzites or jaspylites. In the Precambrian seas iron oxide and silica precipitated seasonally, their layers alternating with the pH. The sediments crystallized under the weight of the column and under dynamo-metamorphic pressure, which also tended to stack the magnetic layers, thus eliminating silica. Martitization bears witness to several cycles of leaching that controlled the development of the ore bodies themselves. Such deposits tend to be found in curved troughs of the Guineo-Liberian shield, Gabon, Congo, South Africa (and in the rest of Precambrian Africa), with the age of the original sediments varying from 2,500 to 2,000 million years. However, in Swaziland and the eastern Transvaal similar sediments were deposited earlier. In South Africa and the northern Congo carbonatic sediments are prominent in the sequence but they are absent in other areas. The late Precambrian deposits of Katanga replace dolomites, as do Paleozoic to Tertiary occurrences of the Atlas. Finally, iron ores related to ultrabasics (e.g., in the Bushveld) are excessively titaniferous, but their lateritic cover is commercially exploited in Guinea (Fig. 2).

TUNISIA

Djérissa, the principal deposit, is 140 miles from La Goulette Harbour, 75 miles from the coast, and 18 miles from the Algerian border. A fault cuts a dome structure, and near this feature replacement affected both the lower and recifal (Lower Cretaceous) Aptian horizons. Goethite and stilpnosiderite are accessory minerals of the wedgelike hematite lenses. Iron content attains 56%, manganese 2%. The small occurrences of Slata (copper and lead sulphides) and of the Nebeur group also belong to the Cretaceous domes of the Saharan Atlas. At *Tamera* and *Ganara*, in Central Tunisia, goethite and stilpnomelane, accompanied at lower levels by hematite, chalcophane, pyrolusite and pyrite, are underlain by Upper Eocene (Priabonian) clays and continental sediments. The iron ore beds are three to 50 ft. thick. Conglomerates outcrop locally.

The *Douaria* district is 10 miles from the coast and 25 miles from the Algerian border. The Flysch hematite, stilpnosiderite and goethite beds lie in discordance upon the Oligocene, and a conglomerate horizon divides the two ferruginous beds of diagenised marls. The principal deposit of the district is El Harrech. The ore (except for Bourchiba in the Nefza group) contains 0.4% arsenic. Gheriffa is a similar deposit.

The *Djebel Ank* deposit (see "Phosphate" chapter) is situated 18 miles from Gafsa and 130 miles from Sfax Harbour. Eocene beds of ooliths are underlain by phosphates and covered by Mio–Pliocene clays. The beds extend over three miles in the northern part of the syncline, where their mean width attains ten ft. The ooliths are embedded in clay consisting of goethite, stilpnosiderite and wavellite. Other deposits are located at Chouïchia, Hemeïma, and at Djebel Oust, Thuburnic, Ras Rajel and Harraba.

ALGERIA

There are numerous small deposits in the Atlas. The second largest deposit, Ouenza, is in the neighbourhood of Tébessa on the Tunisian border; the third largest deposit is at the Djebel Bou Khadra, located south-southwest of Ouenza. The smaller Khangouet deposit is in the same area. Timezrit, a small deposit, lies 25 miles southwest of the petroleum harbour Bougie (Fig. 11); Zaccar (near Miliana) is situated 70 miles southwest of Algiers; Sidi Marouf is 60 miles from Djidjelli Harbour, and Beni Saf is near the coast between Oran and the Moroccan frontier. Finally, Gara Djebilet, the largest deposit, is located in the Tindouf district of the western Sahara.

Ouenza and Bou Khadra

The marls of Ouenza consist of *(1)* Vraconian and (Middle Cretaceous) Lower Cenomanian fibrous green-gray marls, *(2)* Upper Albian and Vraconian black slaty marls, *(3)* dark Middle Albian limestones and marls, *(4)* Lower Albian argillaceous marls, *(5)* Clansayensian yellow and gray marls divided by a sandstone reef, and *(6)* lenticular and sandy Lower Cretaceous Aptian marls. The northeast striking Pyrenean (Eocene) anticline consists of an outer zone of Middle Albian marls and an inner Albian–Aptian zone, affected in the northeast by the Miocene Béni Barbar downthrust. The southeastern compartment of the fold has been thrust down by the main fault which is parallel to the anticline. The main fault has been complicated by later faulting and the principal diapyr which is believed to continue under the northwestern border of the belt of pyrometasomatic deposits. The principal diapyr fills in the core of the anticline, joining two marginal diapyrs (north of the mine) placed on both sides of the fold. The Triassic diapyrs displace and occasionally overturn the Cretaceous beds.

Lower Cretaceous Aptian limestones have been replaced by siderite and pyrite, particularly around the open fissures. Above the hydrostatic level, that is, at a depth of 300 ft., the ore essentially consists of powdery black goethite and limonite, but the brown ore contains more silica. Siderite and pyrite crystals have been found in geodes and as relicts in compact ore. In the southwestern Sainte Barbe sector of the deposit, the Aptian limestones are homogeneous, with ore lenses showing irregular shapes. At the other end, in the northeastern Hallatif zone, ore contacts are predetermined by the limits of the limestone lenses and facies changes. The iron content is 54%, while manganese attains 1.9%, silica 3.3%, and phosphorus and sulphur 0.025% and 0.036% respectively.

The *Bou Khadra* anticline has a northeasterly trend. In the southwest a fault of the Tébessa–Morsott rift cuts the fold, thus facilitating the emplacement of Triassic diapyrs. North of the fault compact lenses of Aptian limestones outcrop, with (Middle Cretaceous) Turonian–Senonian strata in the south. In the north replacement has affected (Lower Cretaceous) Aptian beds. In addition to low grade barytic iron, Bou Khadra ores contain only 46.5% iron and 1.5% manganese, but silica attains 7.6% while phosphorus and sulphur grades attain 0.024% and 0.1%. Khangouet hematites contain 47% iron, 2% manganese and 4% silica.

Other deposits

The complicated Timezrit structure rises from Cretaceous strata as a faulted syncline. Three subvertical (Lower Jurassic) Liassic beds have been wedged into Neocomian layers, and the border of the Lower Lias has been mineralized along a north-northwest trending fault.

The complex *Zaccar* deposit illustrates the multi-stage evolution of these typically non-African occurrences of the Mediterranean orogeny: prior to the Miocene, fumarolic dissemination of pyrite was followed by large scale impregnation and zonal replacement processes. During the next period of folding and open fissurization, iron was recycled in a telethermal or diagenic phase, from which the principal hematite veins originate. The subsequent extrusion of rhyolite had a variable effect on the deposits and created a contact zone of sulphides, azurite and hematite. Finally, minor dislocations enriched the lower layer of volcanogene ores. Siderite exists under the hydrostatic level, while hematitization apparently had progressed per ascensum (GLANGEAUD, 1952).

At *Breira*, in the same district, the upper level of a ferruginous limestone lens has been replaced by siderite. Near Breira, the Middle Miocene Beni Aquil vein traverses all strata and consists of baryte and siderite, accompanied at the upper levels by hematite. The Sidi-Marouf hematite ores contain 51% iron, 3% silica and 2.5% manganese. The large *Beni-Saf* deposit (now almost exhausted) appears to be a product of hydrothermal replacement phenomena, observed at the contact of limestones and shales, and essentially related to distant basic intrusives. Here most of the ore minerals are concentrated in Lower Jurassic Lias and Upper Jurassic limestones.

Arseniferous ores have been mined at Rovina, 110 miles west of Algiers, and low manganese hematites at Filfila near Philippeville Harbour. Other small occurrences include Bou Amranen, Gueldaman and Gouraya.

Gara Djebilet

The Gara Djebilet is located 80 miles southeast of Tindouf and 300 air miles from Agadir Harbour in Morocco (Fig. 55). In the western Sahara, the deposit is included in the southern limb of the Devonian Tindouf syncline. These sediments have been only slightly warped, striking almost east and dipping 1° north. The layers consist of:

(1) Coarse Cambrian–Ordovician sandstones, increasing in thickness from 15 ft. in the valleys to a maximum of 300 ft. in the hills;

(2) Silurian shales and sandstones, transgressive towards the east, forming a mile wide band;

(3) Lower Devonian Gedinian shales, sandstones including an oolithic horizon, and sandy shales and coarse yellow sandstones bulging to a maximum width of three miles;

(4) Siegenian clays and shales of the footwall;

(5) Emsian layer of *(a)* oolith, hangingwall ore incorporating argillaceous cylinders and coarsening southwards, *(b)* fine sandstones and ferruginous quartzites terminating the ore formation laterally, *(c)* sandstones and sandy limestones, carrying *Spirifer pollicoï*, constituting the hangingwall in the western Gara, in the central basin and in the eastern wing, and *(d)* a thin layer of Crinoid limestone, although it is lacking in the central Gara;

(6) Middle Devonian Eifelian *(a)* pink sandstones carrying *Spirifer cultrijugatus*, including inliers of ferruginous quartzites, followed by *(b) Spirifer speciosus* clays and marls and concluded by *(c)* a thin bed of polyp limestones.

The conglomeratic and transgressive facies emphasize littoral changes. Emersion of shelves controlled the differential oxidization. Finally, Quaternary eluvia followed Tertiary Hammada. The iron is believed to have been eroded and leached from the Yedri stock and deposited in littoral–lagoonal conditions in the flat trough. However, under unfavourable pH conditions and emergence, silica and lime took the place of iron. Three hills or Gara extending 50 miles constitute the deposit. The western Gara outcrops over 14 square miles, and the central or Djebilet occupies an oval of 50 square miles. The dip being only 1°, the ooliths continue northwards. Thus, the average two mile width of the Gara Djebilet is widened to three miles, but beneath the cover the deposit also extends laterally.

The ooliths, consisting of siderite and chlorite, have been partially oxidized to Fe_2O_3 and Fe_3O_4. Oxidization and martitization have progressed downwards. Two types of ore have been distinguished in the upper layers: *(1)* violet powdery ooliths and *(2)* fine grained compact ore. Magnetite and maghemite have been intermingled with some silica and late calcite, with iron phosphates unfortunately contributing the 0.6% P grade. A representative analysis gives 79.4% Fe_2O_3 (55.6% Fe), 6.2% SiO_2, 6.4% H_2O, 0.51% TiO_2, 0.08% S, 0.08% V_2O_5, and 0.004% Cr_2O_3. Grades have been distributed in lenses: lower values surround intermediate values, which again enclose high grades located in the eastern corner of the central and western Gara. In the Gara Djebilet, a 20 ft. thick layer of high grade ore of 58.75% iron is surrounded by a 4 by 3 mile oval containing 53.5% iron. Combined they represent 400 million tons of ore containing 56.7% Fe. Values of 53.5% occupying the eastern half, and an inlier in the south, represent 1,200 million tons of ore. Furthermore, the western half and the northern rim include 500 and 950 million tons, respectively containing 48% and 38% iron. Nevertheless, the overburden of the high grade zone increases from 10 ft. in the south to 100 ft. in the north, averaging 25 ft. The high grade zone of the western Gara covers a trapezium 10 miles square in its northeastern corner, containing 400 million tons of 54.5% ore. Finally, other values are distributed thus: 400 million tons (51.5%), 300 million tons (41.8%), and 200 million tons (30%) plus an additional 600 million tons of ore.

MOROCCO

The principal deposits are situated at Uixán (Ouichan)–Setolazar in the Beni-Ifrour Mountains, 18 miles south of Melilla, Imi n'Tourza (Djebel Ougnat) 60 miles west of Colomb Béchar (capital of the Algerian Sahara), the Khenifra district 25 miles south of Meknès, and the Oulad Saïd district in the southern vicinity of Casablanca and Ouarzemine, 5 miles from the coast south of Agadir. Minor deposits include Aït Amar, 25 miles southeast of Casablanca, Taouz, east of the Djebel Ougnat, and a number of smaller occurrences (Fig. 10, 59).

Beni Bou Ifrur district

The deposits cover an area of 3 by 7 miles, 10 miles inland from Nador. Jurassic limestones, including some Lias, lie in contact with andesites, trachytes and dacites. The deposits can be divided into three groups: *(1)* Ouichan, Achara, Setolazar, *(2)* Iberkanen Cherif, Gibraltar and La Tardina, and *(3)* Afra and Alicantina (JEANETTE, 1961; Fig. 60).

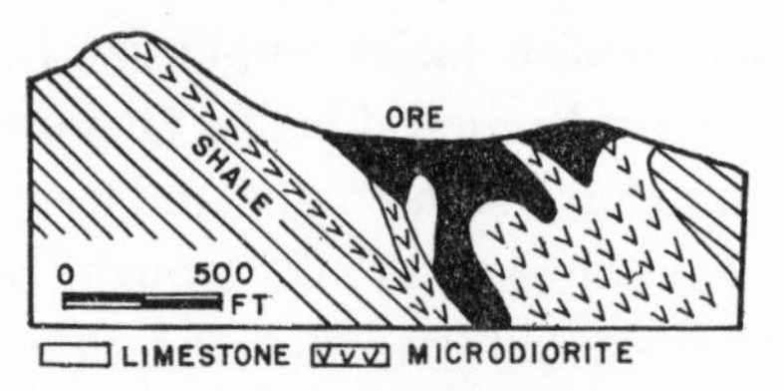

Fig. 60. Section of Ouichan.

While the *Ouichan* limestones, which attain a thickness of 830 ft., are interbedded in green argillites and rarer quartzites, andesites, cinerites and sands cover the northern part of the structure. An andesine–hornblende–quartz–diorite, probably of the Jurassic age, represents the principal facies of the intrusive, but some of the dykes are more acidic. The principal Alpine fold plunges northeast. However, faults strike east–west, dipping 30° north; subvertical joints are oriented north-northeast. The Ouichan fault dips 60° northwest with its thrust attaining 1,000–1,500 ft. The central diorite has been intruded 150 ft. above the limestone–argillite contact (RHODEN, 1961). In an area covering 3 square miles microdiorites surround three sides of the sediments, while also being injected along the main fault. Limestones outcrop only in the southwestern quarter of this structure, but in the andesite xenoliths of argillites have been preserved.

Mineralization followed the contacts of limestone and argillite and of the central microdiorite body. The ore bodies, extending 1,000–2,000 ft., thus exhibit most irregular and digitated shapes. While 80% of the ore was formed as a replacement, 20% crystallized in stockworks adjacent to dykes or impregnating the Upper intrusive and the Sidi Brahim mylonite contact. Indeed, no ore is found more than 300 ft. from the diorite. Pennine developed at the dioritic contacts of the ore, which includes 75% magnetite (diameter 1 mm), 10% pyrite, quartz, calcite and ankerite. Martite and hematite also belong to the paragenesis; chalcopyrite rims pyrite, but pyrrhotite is scarce. The sequence was: silicates, magnetite–hematite, pyrite, carbonates. The calcite of the ore may have originated in depth! Late veins of galena, sphalerite and baryte cut the complex, with phosphorus content averaging 0.05%. Contacts between the ore and the limestone, believed to have been controlled by a thermal gradient, are sharp. The limestone was recrystallized in a 600 ft. aureole, and later oxidized to a depth of 300 ft. In this zone all magnetite has been martitized, the content of sulphur being reduced. Finally, collapse breccias are caused by excessive leaching.

Acharà and *Setolazar* are located on the eastern side of Acharà Mountain, where late tangential faults form écailles. The mineralization extends $^{3}/_{4}$ mile between Chérif and Gibraltar, attaining a width of 400 ft. The thickness of sediments is smaller than in the west, but the beds alternate more frequently. Quartzites are rare, and diorites are injected in sills. However, the main intrusive has not yet been found. The ore bodies or sheets, intersected by repeated faults dipping east, have completely replaced the limestone. Relict carbonates have been found at Acharà 2, and the limestone argillite cover has been transformed into albite–epidote–garnet hornfelses and into garnetites. The Setolazar ore contains 50% Fe, 7.15% SiO_2, 4.2% Al_2O_3, 0.06% P and 4% CO_2. Magnetite and calcite are interlocked in the ore. Ouichan ores attain higher grades.

Other impregnations of the eastern Rif include the Tres Forcas group north of Melilla, the Tistutín Mountains, and Beni Saïd and Beni Ulichek in the Afrau group, 80 miles from Melilla.

Other deposits

At *Imi n'Tourza*, a 90 ft. wide oolithic bed is interbedded in Middle Llandeilo (Middle Ordovician), *Acidaspis buchi* carrying sandstones. The ore contains 55% iron, 5% silica and 0.3% phosphorus. The following levels can be distinguished (upwards) in the iron layer: *(1)* fine grained ooliths in the sandstones; *(2)* large randomly arranged ooliths, pebbles and shell fragments; *(3)* two feet of conglomerate, including oolith pebbles; *(4)* ooliths, with the lower levels containing magnetite.

Llandeilo strata also include oolithic beds in other countries. Typical Bracchiopodes (Lingulides) as well as Lamellibranches *(Aviculopecten)* are found in the ooliths. In most deposits, the quartz core of the oval ooliths is surrounded by iron chlorites, and the grains are cemented by fine grained quartz and chlorite; their diameter is 1 mm. At the surface the ore has been weathered to limonite.

Oulad Saïd district. On the right bank of the Oum er Rbia River, between Keradid and the sea, Llandeilo quartzites and sandstones were folded during the Hercynian cycle. They rise from Pliocene–Holocene sediments in a north-northeast trending range of hills. The principal outcrops are Sokhrat Meknassi, Rhimline and Keradid. The 7–12 ft. thick horizon of oöliths has been interbedded in schists, 300 ft. above a thick quartzite layer. The lower layer of laminated ore consists of quartz cores surrounded by chlorites and embedded in an iron carbonate, but the upper layer contains only 42% iron, 24% silica and 0.7% phosphorus. Here again the ore has been weathered to limonite at the surface.

At *Ouarzemine–Tachilla*, east of Tiznit, a north-northeast trending, 20 mile long syncline of quartzite beds has been divided by schist horizons. The up-faulted Middle Llandeilo column rises from Quaternary sediments. The thickest oolith bed is underlain by quartzites, but other lenses are interbedded with sandstones. At the surface most of the chlorite has been limonitized. The total thickness of the layers is 30 ft., the best ores occurring in a three to ten ft. wide zone; values are variable: 35–45% iron, 12–18% silica, and 0.5% phosphorus. The Tazagmouat deposit outcrops in the extension of this trend, 60 miles south-southeast of Ouarzemine.

Khenifra district. Lower Carboniferous Viséan shales, conglomerates and limestones, including iron ore, were folded into the South Khenifra syncline. Whilst the Djebels Bou Ousel and Bou Guergour constitute the southeastern limb of the fold, Djebel Hadid lies opposite to them. At Bou Ousel, the 150 ft. thick limestone bed has been completely replaced along the conglomerate contact. However, at Bou Guergour metasomatism only affected the lower levels, and it is believed that the mineralizing solutions travelled along the discordant contact, opened during post-Viséan orogenic movements. Siderite has been altered to limonite, and baryte occurs among the alteration products and in veinlets. The ore contains 43% iron, 15% baryte and 10% silica.

At *Aït Amar*, a Middle Llandeilo shaly oolith lens is interbedded in the slates of a larger anticlinorium. The lens is two thirds to one mile long and 45–55 ft. thick, with the width increasing along the dip. Thick and narrow beds of shales emphasize the foliation of the ooliths. The outer zone of the grains consists of chlorite and magnetite, constituting 30% of the ore which contains 50% iron, 14% silica, 0.4% phosphorus and almost 1% sulphur. Other deposits of the district include Sidi Abd en Nour, Haït el Hamar and Bled ech Cheikh. Whereas oolithic deposits characterize Boulhaut and the Tiflet district, the Taza district is metasomatic. Among many minor occurrences Khenifra, Jebel Tekzim, Sokhrat el Jaja, Aït Ahmane, Goulits, Tanefert and numerous vein deposits may be noted.

MAURITANIA

At Fort Gouraud, near the southeastern border of the Spanish Sahara, the *Kediat d'Idjil* rises 1,000–2,000 ft. from surrounding regs, or stone deserts. The range trends (Fig. 55, 61)

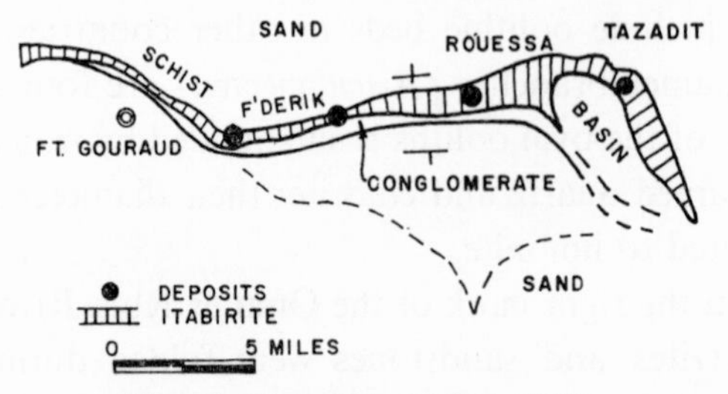

Fig. 61. Map of the Kediat d'Idjil.

east, attaining a length of 11 miles in the north and seven miles in the south, its width increasing from 1,000 ft. in the west to more than a mile in the east. Itabirites form the contact of a triangular syncline of brecciated ironstones which dip gently in the south under Cambrian conglomerates, grits and sandstones. The itabirites dip 60°–80° south, that is, towards the breccia. Furthermore, narrow strips of banded ironstones extend seven miles west of F'dérik to Toumba Atomai, also bending northward into the surrounding orthogneisses and schists. Whereas dolerite dykes have been injected into the latter, quartz–microcline pegmatites traverse the western branch of the Kedia[1]. The F'dérik–Toumba branch dips north, but beyond Tazadit the eastern branch dips southwest, thus surrounding the internal anticlinal fold of Achoui. While the ridges consist of schists, quartzite and hematite itabirite, the area south of the Kedia consists of brecciated or conglomeratic itabirite recemented by hematite. The hematite has recrystallized under intense dynamo-metamorphic pressure into a banded mosaic of polyhedral grains, with iron oxides occasionally migrating to rim silica grains or to form crystallites in the latter. There is some orientation of quartz grains, but no strain polarization. Apatite and argillaceous matter are present, with relics of martite, ankerite and siderite appearing locally.

Five ore bodies have been distinguished: F'dérik east of Fort Gouraud, Azouazil, Rouessa, El Mamariat and Tazadit, 3, 11, 13 and 16 miles from F'dérik respectively. The ore bodies are surrounded by hematitic quartzite.

Four types of ore have been distinguished (LETHBRIDGE and PERCIVAL, 1954). The hard, compact variety consists of fine grained massive hematite, traces of magnetite and 1–2% silica, with the ore containing 68% iron. The blue-gray colour alters to black on long exposure, and to red when powdered. This type of ore reports at the two ends of the range, especially at F'dérik, but it rarely occurs in the breccia south of it. Compact ore laterally changes to hematite itabirite. The laminated ore of Rouessa and Tazadit, containing 66–68% iron, is more porous. Slump or biscuit ore develops through the leaching of silica from the itabirite. Since its plaquettes are frequently covered by float derived from it, winds blow desert sands on its surface. On the southern flanks of Rouessa and Tazadit, samples of this type contain, respectively, 59.6% and 65.1% iron, 12.4% and 2.8% silica, and 3.8% and 5.1% alumina. Powdery ore containing 67% iron has also been leached from itabirite. A majority of the Rouessa powder passes the 200 mesh screen, with hard nodules of one to two inches remaining on the sieve. On the other hand, thin lateritic canga covers some of the breccia.

The itabirites contain up to 40–47% iron. However, the hematite bodies average 65–68% iron and 2% silica. They developed in several phases: The main zone of F'dérik is believed to have formed during the Precambrian era by sedimentation or leaching of sub-horizontal itabirites. Little ore has developed by stacking here (as at Nimba or Simandou) or by weathering and leaching at the surface (GROSS and STRANGWAY, 1961), but the main ore zone of Rouessa is believed to have concentrated by that process.

The curving body of F'dérik extends 1,000 ft. Compact hematite remains concordant to the bedding to a depth of 750 ft. The ore body plunges north and attains a width of 150 ft. At the surface the ore body is ramifying and unconformable. At Rouessa West several elongated lenses of deep compact hematite strike east; the two longest bodies extend 600 and 800 ft. The eastern lens has a width of up to 100 ft. South of the western body five narrow bands strike east. North of the eastern ore body a fatter small lens is located. This complex is surrounded by a mass of plaquette ore 1,700 ft. long, 500–1,000 ft. wide, and 250 ft. deep. A ravine divides this type of ore in two. In the midst of the plaquette ore, hard, surficial hematite covers two areas. Finally, at the surface the bodies occupy areas of 500 by 400, and 300 by 150 ft. respectively.

[1] Also written Koediat.

The hematite bodies of *Legleitat el Kader* contain 52–55% iron and 1% phosphorus, and in the Guelb Moghrein near Akjoujt copper minerals crystallize in hematite bodies of an iron ore formation (see 'Copper' chapter).

SIERRA LEONE

The principal deposits are located at Marampa 50 miles east of Freetown, and in the Tonkolili district of north-central Sierra Leone (Fig. 17, 54). Minor occurrences include the banded ironstones of the Gori and Kambui Hills, and the layered titaniferous iron ores of the ultrabasic Colony Complex which dip into the sea near Freetown. Furthermore, lateritic iron ores are abundant in the western, non-granitic part of the country.

Tonkolili

The belt of Kambui Schists wedged in the Basement strikes north for 80 miles, with its width attaining nine miles. The series consists of quartz, talc and chlorite and occasionally also of mica and magnetite schists, metavolcanics, quartzites and conglomerate. Greenstone constitutes the indicator horizon. The monocline appears to dip steeply westwards. In the *Sula-Kangari* Range conglomerates outcrop at the southern end of the deposit, with amphibolites and hornblende schists appearing along the Pampana River. Five bands of magnetite–quartz–hornblende schists, associated with quartzites, occur in three belts divided by thicker zones of greenstone. Thinner relics of the greenstone are interlayered with the banded ironstone. The ironstone belts appear to thicken at secondary folds. The *Numbara* concentration is distinguished by dips changing from 15°–45°. The laminated hematite bodies are concordant with the ironstones. The principal hematite bed extends more than 12 miles, including two ore bodies on Simbili Hill; the Pampana River body outcrops in the prolongation of this alignment. In addition to the important Waka ore body, other bodies include Kegbema, Sangbaya and Sokoya. Hematite, accompanied by some limonite, appears to be a product of silica leaching and magnetite oxidization of the ironstones, identified at a depth of 100 ft. The ores contain 58–62% iron, 3% silica, 1–2% alumina, and traces of phosphorus (POLLETT, 1952).

Marampa

From Guinea, the 4 miles wide belt of Marampa Schists extends 90 miles into Sierra Leone along a largely south-southwest striking line, attaining its greatest width at Marampa. Between the older Kasila Series and the Rokell Series, the Marampa Series generally dips steeply eastwards. Marampa Schists contain quartz, sericite, biotite, hematite and garnet; quartzites and hornblendites also occur with the latter grading into the Kasila Series. The hematite schist bed extends 12 miles; at Lunsar its foliation changes toward the northeast, and a strike fold located in the central portion of the Ghafal and Masoboin ore bodies has produced the deposit. Indeed, normally the hematite schist shows a thickness of only 45–100 ft., but at the bends the enriched beds expand to 600 ft. Both the ore bed and the bordering muscovite schists have been overfolded and puckered. Dips are subvertical, diminishing, however, south of Masoboin to 30°–60°. The interfoliated specularite, containing 65% iron and the hematite–quartz–sericite schist, have been leached to hard red–brown hematite to a depth of 30 ft.

GUINEA

The Republic of Guinea extends from the coast inland and eastwards as far as the Ivory Coast. Thus, Sierra Leone and Liberia are situated between the Ivory Coast and Guinea (Fig. 18, 54).

Three major types of deposits can be distinguished:

(1) Laterites are located mainly in western or lower Guinea, at Kaloum, Koumbia, Cape Verga and Doumbiaghi. They cover ultrabasic and basic rocks.

(2) Hematite lenses developed at Yamboieli, Akoundiat and Kafou, near and similar to Sierra Leone deposits.

(3) The itabirite ranges of Simandou and Nimba in eastern or upper Guinea extend into Liberia.

Kaloum

Conakry, the capital of Guinea, has been built on the Kaloum peninsula. There, Precambrian schists and gneisses have been intruded by the dunite stock of Mount Kakoulima and by acid rocks, and finally covered by Silurian schists and Ordovician sandstones:

(1) The olivine and magnetite of the dunite has been partially altered to serpentine, chlorite, carbonates and hematite; stringers of chalcedony have been redeposited.

(2) Above this thin zone is soft yellow ore. A majority of the iron minerals have been weathered to goethite and stilpnosiderite, 70% of which is finer grained than 10 microns. Pyroxene relics can still be recognized in this zone, overlain by:

(3) The transitional zone of porous, dusty red ore, containing nodules of:

(4) The upper, dark and hard fissure-crossed laterite of porous or compact ore, characterized by auto-agglomeration or hardening when exposed to air.

(5) The final and unalterable stage consists of dense granules of hematitic aspect.

The chemical composition of the various stages is similar. Between the last stages, absorbed (hygroscopic) water appears to coagulate in larger colloids. Ores of stage *3* have to be agglomerated artificially. The most frequently found width of the hard upper crust is 25–35 ft. but only the upper 20 ft. have been worked. The limit between the soft and hard

TABLE CVIII

LATERITIZATION OF DUNITE

	Laterite %	*Laterite/dunite ratio of contents*
Fe	51	5: 1
Fe_2O_3	10	5: 1
H_2O, absorbed	12	5: 1
H_2O, absorbed	11	11: 0
SiO_2	2.5	1: 15
MgO	0.3	1:1,200
CaO	0.05	1: 45

stage appears to be the low level of the hydrostatic table. Since most of the silica, calcium and magnesium have been eliminated through lateritization, iron, aluminium and water are relatively concentrated five fold (Table CVIII).

Most of the iron is oxidized (70–77%). The highest iron values, indicated by elevation of the surface, are just below the crust. The combined aluminium and iron content increases moderately as a function of iron content, but chromium values appear to be an inverse function of iron values. The most probable chromium and phosphorus values were recorded as 1.25% and 0.06%. Finally, the free water content increases from 11% in the durcicrust to 33% in the soft ore.

Yamboieli

In the Forecariah district, near the Sierra Leone border, hematite bodies are wedged along lamination surfaces into the Middle Precambrian Marampa Series of itabirites and gritty and argillaceous lenses. The stratigraphic column of the syncline includes: itabirite; overlain by the low grade, lower hematite level (35–40% Fe); quartzite; the upper horizon of powdery hematite (50–55% Fe), and again itabirite. When underlain by high grade ore, supergene canga is layered.

Hematite also occurs at Akoundiat and Kafou.

Simandou

The Simandou chain, 60 miles north of Nimba in eastern Guinea, strikes south for 50 miles. The 2,500–3,000 ft. thick discordant Atacorian (Middle Precambrian) monocline is wedged into alternating Dahomeyan amphibolites and gneisses, intruded by post-Birrimian dolerites and granites (GODFRIAUX et al., 1957). Near *Fon* peak, the upper eastern slopes of the chain consist of 900 ft. of light coloured amphibole quartzites, with incipient mylonitization, interbedded with infrequent grey schists. This column includes amphibolites, amphibolitic quartzites, quartzites, black schists, and, in the north, more micaceous metasediments. Most of the western slopes consist of 900 ft. of pyrite or garnet schists, and interbedded violet quartzites including rare beds of magnetic quartzite 30 ft. thick.

The upper reaches of Simandou consist of 1,200 ft. of banded ironstone. On the eastern crest, 750 ft. of magnetite–hematite quartzite are followed by 100 ft. of graphitic schists and ferruginous schists, and finally by 150–300 ft. of magnetite–hematite quartzites (Fig. 62).

In its linear trough, dynamo-metamorphism has affected the rigid chain. Itabirite beds are believed to have been sheared, gliding then along layers of schist, which they compressed. The strike of faults varies; most, however, are transverse. Thus the deposits were created by shear overfolding and faulting, which thickened the ironstone series.

At Lamadougou Pass, itabirite layers have glided on each other, thus eliminating the interbedded schists. In the middle of the 1,500 ft. thick series, a tectonic breccia divides the banded ironstones into two parts. East of *Foko* signal the schists have also been 'extruded', but in the westernmost part they are still extant. At the centre of Foko Mountain the magnetitic quartzites of the crest, interbedded with allochtonous amphibole quartzite, attain a thickness of 3,000 ft. At *Fon* peak the thickness of the ironstone series increases to 3,500 ft., doubled by complicated thrust faulting and the gliding of identical layers into parallel positions. Here the top series of schists is also affected by shear. Again at *Diodio*, 30 miles north, the itabirites expand.

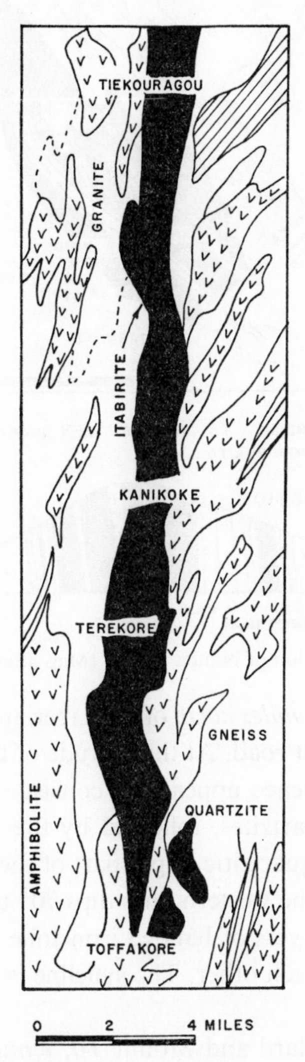

Fig. 62. Map of Simandou, south. (After OBERMULLER, 1941).

On the sides of the chain, e.g., at Bétoubé, Sorayo, between Simandou and Nimba, and at many passes as well, the ratio of itabirites to schists is lower. Consequently, dynamometamorphic phenomena are essential criteria of the genesis of this type of deposit.

Nimba (North)

From the Liberian border, the Nimba range trends 15 miles towards the northeast. The average width of the Middle Precambrian (Atacorian) chain proper is 4 miles (Fig. 63). The western slopes of the range consist essentially of quartzites, amphibolites and schists, but in the east itabirites predominate. The surrounding calc-alkaline Basement granites intrude or migmatize gneisses, schists and amphibolites.

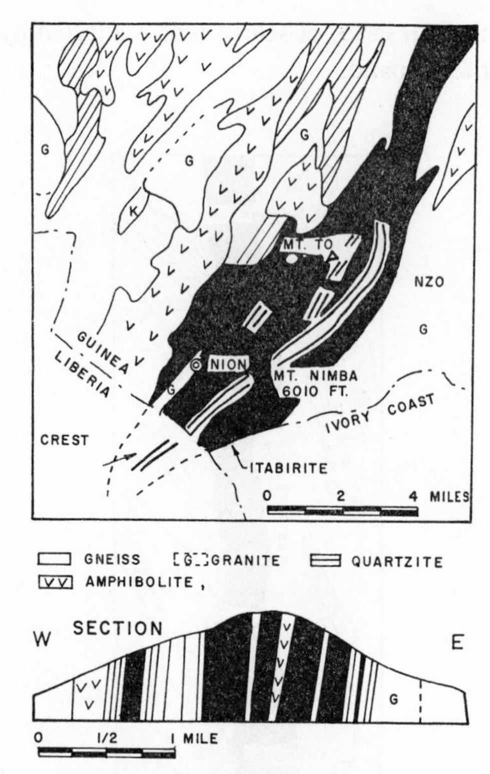

Fig. 63. Map and section of Nimba Range. (Map after OBERMULLER, 1941).

Banded ironstones outcrop *13 miles north* of the Liberian border at an altitude of 1,800 ft., south of the Guinea–Ivory Coast road. At the latitude of the rock peak Richaud, itabirites start their expansion. The sequence appears to consist of 1 mile of magnetic quartzite, and of white and feldspathic quartzites, followed by the northern extremity of the main itabirite ridge. Outcrops of the quartzitic upper part of the eastern limb and of the western wing of the syncline are poor. The western limb dips 70° towards the west, but the eastern plunges 80° east. The various layers of banded ironstone appear successively, like a stack of cards which have glided on each other. The syncline is wedged into granites in the east and into gneisses in the west.

Between Mount Richard–Molard and Mount Tô, *7 miles north* of the Liberian border, both the chain and the itabirite layers widen. At the Goué River, the width of the itabirites doubles to 1.3 miles, expanding further in the gliding 'shoe-string' structure of *Mount Tô*. The magnetite (–hematite) itabirites are overlain by graphitic slates, schists and a bed of magnetite quartzite, followed by amphibolite. Norite (corresponding to the southern dolerite) is injected between the itabiritic eastern, and the quartzitic western limb. In the east, biotite gneisses constitute the country rocks. At Mount Leclerc the syncline turns north-northwest, with the itabirites expanding to a width of almost two miles.

At the height of Nion, *4 miles north* of the Liberian border, the Atacorian syncline contracts to a width of three miles. Towards the east, in the Ivory Coast, the series commences with amphibolites, followed by quartzites, micaschists and itabirites. Magnetite grades increase towards the crest, the iron content attaining at least 60% Fe_2O_3, but ore concentrations are partly related to surficial leaching of silica. The linear beds dip 70° towards the northwest, i.e., inward. At a width of 3,600 ft., along the Ya River, a thin layer of biotite-graphite schists is interbedded. West of this layer, itabirites extend for several hundred ft. with more and more quartzites and schists intercalating, and 3,500 ft. west of the schist interbed the last magnetite-quartzite horizons disappear. Dolerites are followed by a synclinal facies of schists and amphibolites that are distinguished by garnet and actinolite. The quartzites and micaschists of the western limb dip 85° east. Finally, pyritiferous micaschists terminate the series.

Structure. Two limbs of the dissymetric syncline are divided by sub-concordant basic rocks, consisting of dolerites, peridotites and amphibolites. In the north, the itabirite layers glided on each other, *décrochements* shearing softer beds. However, in the central part of the Guinean Nimba the longitudinal component of pressure did not prevail, and thin quartzite and schist laminae have simply been pressed out from the banded ironstone. Farther towards the Liberian border, the two limbs of the syncline regain their lithologic and stratigraphic balance. The Middle Precambrian rift has apparently opened along a zone of weakness of the Basement. Later, when the eastern lip of this trough was thrust, carrying along the deposited ferro-siliceous sediments, the western lip remained stable. Between Nzo and Ziéla, the gliding northern limb collided into a rigid transverse trend, thus deflecting the range's northern tip. South of this obstacle the itabirites piled up to their largest width, three miles.

The durcicrusts of the crests recrystallized from circulating solutions and on the slopes below an altitude of 4,000 ft. erosion and weathering formed limonitic levels, consisting of layers of ferruginous clay, durcicrust and conglomerate ore breccia.

LIBERIA

In the northwestern third of the country, on the Guinea–Ivory Coast border, iron ore deposits occur extensively throughout the extension of the Simandou–Nimba trend. Major deposits include the Bomi Hills, Bie Mountains, Kpo Range, Nimba Range, Putu Hills, Mount Jedeh in Eastern Province, Bong Range, Wologisi Mountains, Wanigisi Mountain, Kpandemai Mountains, and low lying hills along the Mano River on the Sierra Leone border. Other deposits are found in the Niete Mountains of eastern Liberia (Fig. 18).

Nimba (South)

The Liberian Nimba iron ore deposits are located in the Sanekole District, one mile south of the Guinean border (Fig. 63), 200 miles northeast of Monrovia, and 160 miles from Lower Buchanan Harbour. Isoclinally and tightly folded quartzites and itabirites strike southwest, dipping 60°–80° northwest. Numerous symmetric and asymmetric secondary folds have been wedged into Precambrian granites and partly migmatic gneisses. Banded ironstones occur along the main ridge of the range, frequently as cliffs and in many places interbedded with schists; they grade into quartzites and micaschists. Itabirites consist of alternating layers of fine grained quartz, magnetite and hematite grains. In the central part of the main ridge their total width attains 3,000–5,000 ft. Itabirites are mostly finely laminated, and generally the lamination has not been obliterated, but in many places the itabirite has been strongly contorted (SRIVASTAVU, 1961).

Phyllites, consisting of a very finely grained mixture of chlorite, sericite and quartz, and quartzites accompany the itabirites. Quartz veins have frequently been injected into the Nimba Range. As in Guinea, there are a few outcrops and many boulders of a dark to medium green or black compact and slightly weathered diabase. These dykes or sills appear to be a late feature.

Whilst OBERMULLER (1941) and BOLGARSKY (1950) consider the formation as Lower Birrimian, Terpstra (in SRIVASTAVU, 1961) attempted to correlate the Nimba and ancient Kambui Schist belts. Standard itabirites appear to be of the same type as those of Guinea and of the Bomi Hills; they grade into nine ore bodies, attaining a length of 1 mile to a few hundred yards, and a thickness of a few tens to hundreds of yards. The ore bodies are located on or near the crest of lateral ridges of the Main Range. Most of the Precambrian, high grade blue ore has been formed through the leaching of the itabirites by supergene solutions, redepositing and recrystallizing the iron at greater depths and partially replacing quartz, the only gangue mineral. Martitization of hematite progressed per descensum. The brown ore of the shallow ore bodies and crust ores clearly indicates formation by a large scale lateritization process, probably ranging from the Cretaceous period to recent times.

The Main, 'N 74', Tail and Nick's ore bodies are the largest. The Main Ore body is the most remarkable, consisting entirely of high grade hematite and reaching to a depth of more than 1,900 ft. below the crest of the ridge. The other ore bodies lie in the upper layers, and reach a depth of 160–330 ft. The high grade ore shoots in them follow the crest of the highest itabirite chain.

Three types of ore have been distinguished and subdivided into:

(1) Blue ore (deep ore bodies). Dark blue, generally laminated, fine grained, predominantly hematite (martite) ore. It makes up the bulk of the Main Ore body. *(a)* Blue medium hard ore: of peak *2* type; from the crest to more than 660 ft. depth. *(b)* Blue soft ore: powdery to arenaceous in consistency; through all levels, mostly mixed with platy blue 'biscuit' ore. *(c)* Blue recemented ore: penetrating to a certain depth, encountered at more than 1,600 ft. below the crest level.

TABLE CIX

COMPOSITION OF ORE (%)

	Blue ore peak no. 0.2	*Itabirite (tunnel entrance)*	*Soft brown ore tail ore body*	*Lateritic ironstone*
Fe_2O_3	91.21	43.29	83.62	82.05
FeO	0.73	18.40	0.24	0.43
P_2O_5	0.031	0.08	0.18	0.29
SiO_2	0.18	36.11	8.33	1.32
MnO	0.05	0.04	0.15	0.03
Al_2O_3	0.48	0.58	2.06	5.95
TiO_2	0.03	−0.01	0.01	0.17
S	0.004	0.008	0.009	0.043
CaO	—	0.03	0.04	—
MgO	traces	0.02	0.09	traces
loss on ignition	0.54	0.35	5.18	9.12
Total Fe	69.3	45.92	58.65	58.10

(2) Brown ore (shallow ore bodies). Limonite coated hematite (martite), mostly brown to dull black. Rarely met at a greater depth than 100–330 ft. below the surface. This hematite ore constitutes all hitherto known shoots outside the Main Ore body, as well as marginal parts of the latter. *(a)* Brown soft ore: largest part of the shallow ore bodies, mostly mixed with brown 'biscuit' ore. *(b)* Recemented 'biscuit' ore: occurring at a depth of 30–120 ft., transitional to laminated crest ore (laminated lateritic ironstones type *3a*). *(c)* Hard dense ore of the N 74 type: irregular lumps and fissure-fillings of steel-gray, extremely hard (approaching type *3a*).

(3) Lateritic ironstones. Less than 660 ft. deep, hematite cemented by limonite, frequently including large pores, exhibiting varieties of gel structures. *(a)* Lateritic ironstone: *(i)* laminated; *(ii)* unlaminated. *(b)* Canga: breccia to fanglomarate of ore and/or itabirite boulders and pebbles, cemented by a hematite–limonite matrix. *(c)* Orefloat: loose boulders and pebbles of all types of ore in screenfans, as cover on slopes and in creek and river terraces.

Zaweah–Bong Range

The Bong iron deposit is located 60 air miles from Monrovia, southeast of the Saint Paul River in the Kakata–Salala District. The western portion of this range is accessible from the Dibli Island Road. The Bong Range can be considered as the continuation of Bomi Hills, separated by a graben in which the Saint Paul River flows. The Bong Range also strikes east and dips north, but more steeply, averaging 75°–80°. Towards Sanoye the strike curves towards the northeast in the direction of the Nimba Range.

The itabirites are assumed to belong to the same Middle Precambrian horizon and alignment as those of Bomi and Nimba. They extend 9 miles between metamorphic rocks. Chlorite schists have been interbedded locally with the ironstones. Amphibole is abundant in the eastern sector of the range, concordant pegmatites occurring frequently in the west where itabirites have been tightly folded in a syncline. Also in the west, banded ironstones

TABLE CX

THE CONCENTRATION OF IRON

	Fe %	*SiO₂* %	*P* %
Hard itabirite	38.2	41.6	0.03
Slightly altered itabirite	37.4	40.4	—
Altered itabirite	44	32	0.05
Lateritic ironstone	42	24	0.06
Boulders	46	25	0.06

expand to a thickness of 100 ft. in Zaweah I, and to 200 ft. in Zaweah II. The itabirites consist of hematite, martite and quartz, accompanied by some hornblende, but soft itabirites grade into less ferruginous quartzites in depth. Five categories of ore have been distinguished as shown in Table CX.

Compact itabirites, drilled at 90 and 830 ft. depth, contain 38.2% Fe, 41.6% SiO_2, 1.3% MgO, 0.9% CaO, 0.75% Al_2O_3, 0.14% Mn, 0.08% TiO_2, 0.03% P, 0.03% S. The depth of silica leaching averages 130 ft., reaching 230 ft. locally. Lateritic ironstones have been restricted to the ridges, also filling in fissures in itabirite. The diameter of boulders covering the slopes varies between 1 inch and 12 ft. The evolution of the ore can be illustrated as in Table CX.

Bomi Hills

The east–west trending chain consisting of Bomi, Jupi and West Hills is located 45 miles north-northwest of Monrovia. The 9,000 ft. long chain rises 300–500 ft. above the surrounding country. The slightly curving *Bomi Hill* ridge is 1½ mile long and 500 ft. high, expanding to a width of 3,000 ft. in the west and tapering towards the east. The extremities of the ridges bend northward. *Jupi Hill* is the western continuation of Bomi Hill, from which it is separated by a low saddle; this is a straight, east trending ridge, 2,000 ft. long and 250 ft. high. *West Hill*, separated from Jupi Hill by a 500 ft. wide swamp, consists of a main ridge, 2,500 ft. long and 300 ft. high.

Bands of itabirite, 120–180 ft. thick, occur mainly on the crests. The itabirite appears to be a polymorphic roof pendant, grading into and interlayered with medium to fine grained quartz–garnet–chlorite–mica schists, including a quartzite facies. The schists grade into pink gneissic microcline granite and gray granitic gneisses. Massive or schistose chloritic rocks, grading into coarse grained hornblende–garnet–pyroxene spinel gneisses, are also in contact with the ore bodies and skarns form part of the footwall. The itabirites

TABLE CXI

COMPOSITION OF ORE

	High grade ore %	*Siliceous magnetite ore* %	*Itabirite* %
Fe	67.98	63.63	43.77
SiO_2	1.95	8.11	36.58
P	0.032	0.034	0.071
S	0.015	0.012	0.052
Mn	0.07	0.07	0.06
Al_2O_3	0.47	0.82	0.39
TiO_2	0.07	0.07	0.22
CaO	—	0.13	0.14
MgO	—	1.89	0.16

strike N 70 °E and dip 50 °N–90 °N, averaging 75 °N. The Bomi Hills form a shallow, basin-like compression fold, bounded on the north by an anticline, the northern part of which is overfolded. The steeply dipping and pinching lenses of iron ore and banded ironstone of Jupi are in line with the northern limb of this anticline. Another synclinal fold forms the eastern end of Bomi and West hills. Irregular ore lenses develop in fracture zones and folds, also branching out into the itabirite along favourable zones. The ore bodies occur in chlorite and cummingtonite schists adjacent to or in itabirites. While these include magnetite-amphibole lenses, itabirite relics float in the schists. Two types of iron ore have been distinguished:

(1) Massive to coarsely banded high grade ore, consisting almost entirely of magnetite, and grading in depth into schistose magnetic quartzite and, at higher levels, into hematite ore. Martitization progresses downwards, alteration reaching a depth of 120 ft.

(2) Thinly foliated hematite or magnetite and quartz layers. The two facies are generally distinct, although they intergrade occasionally. While massive ore includes 66% Fe, the altered itabirite contains 43–54% Fe, and the bulk of the itabirite contains only 32–46% (averaging 37%). In crush and shear zones the phosphorus grade of 0.1% increases from 1.5–2%. Sulphur grades included in pyrite average 0.05–1%, but these increase at the footwall (Table CXI).

The high grade magnetite ores are believed to have been formed by metasomatic or hydrothermal replacement of the sedimentary iron formation, the end product of the process being the massive ore. The mineralizing solutions appear to have entered and replaced the rocks along closely spaced fractures, subparallel to the layering of the iron formation and the gneiss. Finally, lower temperatures have been attributed to the hematite facies.

Mano River and Bie Mountains

The *Mano* deposit is located 100 miles northwest of Monrovia on the eastern bank of the River dividing Liberia from Sierra Leone. There amphibole–chlorite itabirites, chlorite schists, and quartzites, dipping 30°–45°, are bordered at the footwall by gneisses and migmatites. The itabirites grade into amphibolites and chlorite schists injected by basic rocks. The silica content increases upwards! Silica has been leached from the overlying soft, aluminium rich ores. Unlike most other Liberian deposits, limonitization played a major role in the formation of the ore bodies. As at Tonkolili (Sierra Leone) the top layer has been lateritized to a depth of 3 ft., and to a maximum depth of 20 ft. Consequently a lateritic ground mass embays martite and hematite. In depth the hard ore of the surface becomes argillaceous, and, further downwards, sandy. Its composition at 18 ft. of depth varies between 55% and 61% Fe, 1.2% and 4.7% SiO_2, 0.7% and 7% Al_2O_3, 0.06% and 0.1% S, 0.01% and 0.06% P, and 6.5% and 12.5% H_2O.

On the Sierra Leone border, the *Bie Mountains* rise 20 air miles southeast of the Mano River mine, and 20 miles northwest of Bomi Hills. The ridge of hills extends 25 miles, generally trending east. Parallel to the main range, several smaller ridges rise with a few minor outlying hills. Into the oldest rocks—granites, gneisses and schists—quartzose gneisses, quartzites, itabirites and mica schists have been folded. Paralleling the main ridge, the formation strikes N 75°–85°E. Most dips are quite steep, but local differences in the attitudes do exist. Gray dioritic dykes apparently have been injected into the formation. The layered iron formation has been folded, possibly into a tight isoclinal fold, with the ends of the two limbs forming the top of the main ridge and topographic heights.

The itabirite consists of fine to medium grained foliated magnetite and hematite alternating with layers of silica. Through the leaching out of the silica some of the banded ironstones have been superficially enriched, resulting in the formation of hematite, limonite and goethite. Explored ore averages 54.4% iron.

Bassa and Mount Jedeh

In Grand Bassa County, east of Robertsfield airport, the low lying Bassa Hills are accessible by road from Harbel through Owensgrove. The Hills strike east–west. Similar to Bomi Hills and Nimba, the ore bodies are also of the supergene enrichment type.

Mount Jedeh, the highest peak of the Putu Range, is the prominent landmark in the Tehien–Putu area of the Eastern Province. Its summit is 2,250 ft. above sea level and 1,500–1,600 ft. above the surrounding country. Mount Jedeh proper is $^2/_3$ mile long, standing 800–1,000 ft. above the rest of an irregular ridge which is 8–10 miles long, striking N 30 °E. The summit of Mount Jedeh is approximately 3½ miles northwest of Putu Mission.

Precambrian granitic gneisses of the lowlands weather to light coloured lateritic soils, yielding white quartz-sand after rains. Biotite or amphibole gneisses alter to red laterite and canga. Scarce bedrock exposures show granite–gneiss, quartz–mica schist, at least two varieties of amphibolite, diabase, itabirite and iron ore. Higher parts of the ridge, extending from the southern end of Mount Jedeh proper to the northeast (½ mile north of the Pahnwroll–Matown trail crossing), are composed of itabirite. At the northern end of the ridge a $^1/_4$ mile long itabirite lens also outcrops, but its remainder consists of diabase, black amphibolite and layered green actinolite–talc schists. Southeast of Mount Jedeh, exposures in a creek reveal that light coloured feldspathic gneisses and schists constitute the bulk of the ridge. Finally, at its northernmost end, medium grained diabase occurs.

The layered rocks of Mount Jedeh strike parallel to the ridge, that is, N 80 °E, dipping 60 °–80 °SE. But most of the itabirite appears to pitch 70 °–80 °E. Notwithstanding small contortions, its major structure appears to be a monoclinal lens. Itabirite forms cliffs 100–200 ft. high on the crests and lower ridges and along the slopes. The rock is fine grained, thinly layered or laminated, grading locally into accompanying fine grained gray quartzite. The itabirite is magnetic, averaging 35% Fe. High grade massive magnetite ore, a few inches thick, is occasionally interlayered with itabirite, with magnetite float also appearing on the southern slopes.

THE REST OF WEST AFRICA

In southeastern *Senegal*, at Diaamou, Fadoula, Makoudana, Tengoma, Talary, Tanbacoum, Bafara and Bagouko, hematite beds and lenses are interbedded in Ordovician sandstones of the Kayes district. Laterites occur in the same district at Aité. In the southeastern corner of Senegal, the Kénioba deposits on the Falémé appear to be a differentiation and impregnation products of micro-diorite which intrude Upper Birrimian rocks. The three stocks of Koudekouru, Karakaène and Kour ou diako represent a reserve of at least 100 million tons of canga, containing 53–66% Fe in soft and hard ore. In the Siracoro area of southwestern *Mali*, at Makadougou, Narafilia, Diamino-Lahaly and Ganganhan, metasomatic magnetite deposits follow the contact of Ordovician dolomites and calcareous sandstones with dolerites (BLONDEL, 1952, Fig. 54).

In the *Ivory Coast* itabirites of the eastern crests of the Nimba range, containing 40% Fe, occupy 2 square miles, but hematite bodies containing 60% Fe ore are of only limited extent. East of Nimba, the itabirites of the Mount Gao syncline contain (northwest of Guiglo, Fig. 18) 42% Fe. Other magnetic quartzites of the Man district include Mounts Douan and Kpao. In *Ghana* lateritic replacement ores occur at Shiene on the Togolese border near latitude 9 °N. At Pudo, near Navrongo, at longitude 2°30′E on the Voltaic border, titanmagnetite lodes are found.

Three north-northeast trending iron belts traverse *(1)* southern Togo and central Dahomey, *(2)* northern Togo and Dahomey, and *(3)* the same area and southern Niger. The Benguéli (Banjeli) deposit is located in *Togo*, 250 miles from the coast, on the Ghanaian border. The principal deposit, Mount Djolé, consists of compact hematite in slightly folded quartzites and jasperoids of the Precambrian Buem series. Similar hematite deposits are found in other parts of Togo, with itabirites occurring at Kouniangou and Schinga. While in *northern Dahomey*, in the Niger drainage, hematite and lateritized ooliths occur with Mesozoic sandstones and continue into Niger, magnetite shows near Carnotville.

In the *Niger Republic*, interbedded hematites occur at Tacananhalt in the Tahoua area, and titaniferous magnetite was found at Saoua, Firu and Mardaga (Fig. 24). Furthermore, south of Niamey, oolithic iron ores cover the Doguel Kaina Say and Dyabou plateaus. In *Nigeria* the principal iron deposits are located on both banks of the Niger near the confluence of the Benue, between Lokoja and Baro. On Mount Patti near Lokoja, and on the Agbaja plateau, the cellular laterite crust is underlain by pisolitic ore.

GABON

The northern zone trends east into the Congo Republic 40 miles north of the Gabon border; its western member, the Ivindo deposit, overlaps the frontier. The southern zone is located at Gabon's northeastern border with the Congo, north of the Makokou–Mékambo road; this zone also tends to trend east. Its western member is the elongated mass of Mount Djaddie-Djouah; Bamba is situated nearby; Batouala is located 20 miles and Boka-Boka 40 miles east (Fig. 53, 64).

At *Batouala* rise two ferruginous crests in a shape resembling that of Fort Gouraud. A narrow, 4 mile long ridge strikes N 10°W. Another 7 mile long and broader crest trends N 60°W, connecting with the southern part of the preceding ridge. The habitual five hematite facies can be distinguished (AUBAGUE, 1957): itabirite, plaquette ore, hard ore, surface ore, breccia.

A 5 mile wide inlier of itabirites strikes northwest in the Precambrian granites and gneisses. The thickness of foliation of the banded ironstone is most variable and frequently indistinct. In it, only remnants of magnetite have been preserved, the rest having been martitized. Plaquette ore occupies 7 miles of ridges, averaging a width of 165 ft. The hematite plaquettes are 1–3 mm thick and 20–40 mm (1–2 inches) long (plaquette ore becomes powdery under little pressure). Thin bands of silica remain between the concordant plaquettes. The ore bodies pinch out at the end of crests and in depth. A body of hard ore conforms to the trend of itabirites; it reaches a length of 2,000 ft., with its width varying from 500 to 820 ft. The ore is generally compact, with ghosts of plaquettes, micro- and crushed breccia textures, imperfectly hematitized itabirite and disseminated quartz seldom remaining. Interlocked hematite grains of various sizes tend to be small and to contain some quartz. Narrow laminae of limonite are aligned along the foliation. On plaquette ore and on desilicated itabirite a 3–70 ft. thick crust of plaquette, hard or cracked texture, develops with clay. Its texture recalls primary ores. Covering a width of ½–1 mile, two facies of breccia have been distinguished. On the ridges fine grained, homogeneous hematite breccia developed from ore, with poorer heterogeneous breccia accumulating on the slopes.

In the surrounding gneisses and granites, the *Djaddie-Djouah* lens of itabirites strikes

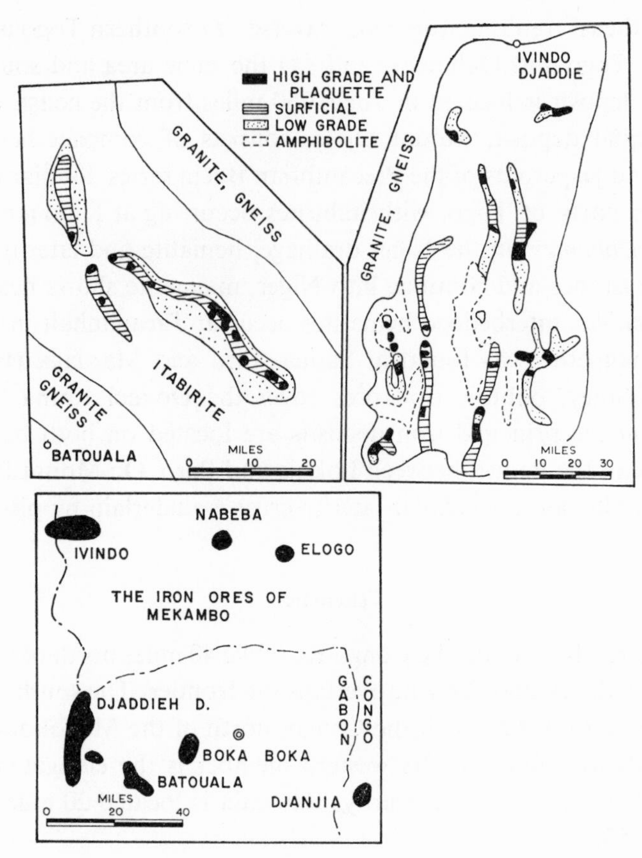

Fig. 64. Iron ores of Gabon. (After AUBAGUE, 1957).

north, attaining more than 30 miles of length and up to 12 miles of width. The itabirites consist of thin (1 millimeter) laminae of hematite, accompanied by magnetite and quartz. Chlorite and sericite schists outcrop sparingly, mainly between the ridges of itabirite; ferruginous breccia covers the slopes of the ridges. At the surface, beds of deep blue or brecciated and cracked ore occur along four parallel crests of a total length of 15 miles. However, the soft ore forming plaquettes or a powder rarely outcrops. Unlike Batouala, no hard ore is found here.

In the Ivindo body, extending 20 miles, surface ore occurs mainly in the west, but the eastern part is more heterogeneous.

THE REST OF WEST CENTRAL AFRICA

The principal iron deposits of the *Cameroon* are the banded quartzitic ironstones of the Mamelles (Nipple) chain and of Mewongo Hill in the Kribi area, 17 miles from the coast and near the Rio Muni border. Roughly 100 million tons of ore contain 50–55% Fe in laterites, and 35–40% Fe in hard rock. The iron belts of the northeastern Congo extend into the *Central African Republic* and Sudan. East of the diamond deposits of Ouadda, itabirites enclose lenses of hematite. In the M'Bari drainage between Rafaï and Yalinga, also in the eastern part of the republic, hematite bodies are found. However, itabirites rising near Roandji, 30 miles north of Bambari in the centre of the country, contain both hematite and magnetite. Similarly, lenses of both minerals developed in ironstones near Bogoin, northwest of Damora and north of Bangui, the capital.

Congo Republic. Three distinct zones can be distinguished: In the northwestern corner of the republic, the Ivindo-Elogo belt strikes east, parallel to the border of Gabon. Mount Nabeba is situated 40 miles, and Elogo 60 miles, from Ivindo, but the Makokou–Mékambo belt of Gabon can be traced further east into the Congo Republic. A deposit occurs somewhat east of Lébango and 60 miles east of Boka-Boka (the nearest occurrence in Gabon). The Abolo deposit is situated 40 miles south of Lébango. The mineralization recalls that of Gabon. Other deposits are located in the south at Zanaga and in the Mayombe belt (Fig. 25).

ANGOLA

No systematic exploration or valuation has been carried out on the deposits occurring in the Basement complex, in the Precambrian Oendelungu metasediments and in the carbonatite belts (Fig. 26, 53).

At *M'Bassa* on the Lucala River, 17 miles southeast of Zenza do Itombe (east of Luanda), schlieren of magnetite segregate in hornblendite, replacing schists and limestones. The titaniferous magnetite contains 55% Fe, with silica grades reaching a maximum of 7.3%. In the manganese belt at *Saia*, also on the Lucala, 12 miles south of Quizenga railway station, specularite replaces Oendelungu sandstones. Near Andulo, at Chilesso (Sambundi), Saraquete, Essale, Changa, Bailundo or Teixeira de Sousa (Lupel, Hnia and Hanga) 50 miles north of Nova Lisboa, and at Cuima, 100 miles south of that town, hematite and magnetite bodies are included in the carbonatite ultrabasic belts of west-central Angola. These complexes are intrusive into early Precambrian gneisses and schists. Ultrabasic alkaline rocks occupy most of their area. However, the veniform, lenticular and irregular specularite–magnetite bodies are associated with sövite and alvikite. The composition of the samples is shown in Table CXII.

Titaniferous magnetite lenses, associated with elongated anorthosite and alkaline syenite bodies, occur in a south-southeast oriented triangle at Picolo, Otchinjao, Chibia and Chitado on the Cunene River, the southern frontier. The ores of Chitado contain 48.3% Fe, 15.3% Ti, 1.3% SiO_2 or silica, and traces of P. At Dongo, 180 miles east of Sà da Bandeira (east of Mossamedes), folded, subvertical, banded hematite quartzites include fine grained hematite lenses. Furthermore, in Cuanhama country in the southwest, iron pans developed 1–2 ft. beneath the Kalahari Desert from ferruginous solutions drawn up by capillarity. One of these, the Mupo pan, 150 miles east of Otchinjao, consists of limonite, hematite and quartz, but its one foot layer contains only 20–25% Fe.

TABLE CXII

COMPOSITION OF COMPLEXES

	Fe %	*Si* %	*Ti* %	*P* %
Sambundi	60.5	0.6	1.4	0.34
Saraquete	51.5	0.6	1.9	0.08
Essale	58.7	0.4	3.5	1.1
Lupel	60.1	0.6	1.1	0.29

The Katanga iron belt extends into the eastern entrant of Angola, e.g., in the drainages of the Niela and Manhinga Rivers.

SOUTHWEST AFRICA

The two principal low grade deposits, Kaokoveld and Gaseneirab, are both related to late Precambrian, Otavi dolomites (Fig. 53). Extending the southwest Angolan belt, the Central *Kaokoveld* deposits outcrop on hundreds of miles between Ohopoho in the north and Sesfontein in the south. The iron horizon is part of broadly folded Middle Otavi Dolomites, including conglomerates, sands and glacial mud, and some ferruginous dolomite. Axes strike south; dips are subhorizontal. The character of the ore is variable; it may be compact, showing sedimentary relic structures, or else occur in repetitive beds among other layers. Selected bodies of magnetite, hematite partly weathered to limonite, and ochre contain 37% Fe and 20% SiO_2 or silica. However, the rest contains only 30% Fe.

In the Cauas Okawo-Gagarus area of the southwestern Kaokoveld the Choabendes Series, correlated to Witwatersrand quartzite, includes a hematite itabirite horizon. The beds are vertical and hard, forming a characteristic relief; one of them is 2 miles long and 7–20 ft. thick. Highest grades attain 58% Fe, but ferruginous quartzite carries only 21% Fe.

Between Outjo and Fransfontein, the Gaseneirab ore zone consists of limonite and carbonate rock bedded in grey dolomite. The ore is self-fluxing.

Southeast of Windhoek, the capital, two hematite bearing zones strike northeast. Lenses of specular hematite and quartz, associated with some magnetite, are interbedded with steeply dipping, siliceous Damara metasediments. On the best outcrops, on Tsatsachas and Elisenhöhe, stringers and veins of megacrystalline quartz traverse the lenses. The ore contains 0.2 % P and 0.1 % S.

At *Otjosundu*, the ratio of iron, manganese and quartz (see 'Manganese' chapter) is variable in each lens. Iron minerals consist of hematite, magnetite, braunite and ferruginous jacobsite. The interbedded lenses included in isoclinal folds have been truncated at the surface and affected by granitization. Steeply dipping itabirite lenses, similar to those of Windhoek, outcrop 9 miles inland from Walvis Bay Harbour. Quartz garnet granulite constitutes their wall rock. With each ton of ore 3.5 dwt. gold and 28 dwt. silver are associated, and phosphorus and sulphur content is low.

In the Keetmanshoop district, a 1–4 ft. thick black sand bed of the Fish River Series, Nama system, consists of ilmenite, chromite, zircon sphene and, locally, magnetite.

At Okorusu, 33 miles north of Otjiwarongo (titanmagnetite altered to hematite and limonite) and at Kalkfeld (lime rich limonite) iron deposits have been identified in alkaline carbonatite complexes and at the contact of Damara–Otavi dolomites and alkaline intrusives.

EGYPT AND LIBYA

In *Libya*, the surface ores of Brakk, in the northern Fezzan, contain 44% Fe.

In the *United Arab Republic* four broad types can be distinguished: *(1)* the ooliths of the Sahara, *(2)* the laterites of the Nile, *(3)* metamorphic and *(4)* replacement deposits of the southeastern desert (Fig. 12, 58).

Southwest of Cairo, the *Baharia Oasis* sinks below the level of the sea. In the northeastern part of the oasis, the Gebel Ghorabi is composed of *(1)* Middle Cretaceous, Cenomanian sandstones and shales, overlain by *(2)* thin basal conglomerates and quartzites and *(3)* covered by *Nummulites atacicus* carrying Eocene goethite, pisolitic and pseudopisolitic yellow ore, ochres, limonite, siderite and hematite. Barite and salt fill in cavities and fissures, with iron percolating into the underlying strata. The faulted, subhorizontal beds cover a square mile, and average 13 ft. in thickness. The ores are believed to have precipitated in shallow waters, partially replacing sediments (EL SHAZLY, 1959).

East of *Aswan* two oolithic horizons have been distinguished in the Senonian (Upper Cretaceous) Nubian sandstones. The discontinuous lower horizon of hematite averages only 1–$1^1/_4$ ft. in thickness, and reaches a maximum of 8 ft. In the upper horizons, two or three bands of ooliths and ferruginous sandstone are divided by sandstone and clay. They total 1–5 ft. of thickness, attaining a maximum of 12 ft. The ore averages 46.85% Fe and 14.14% Si, with traces of sulphur reaching the mark of 0.3%. At Karim, Um Hagalig, Siwiqat Um Lasaf, Um Shaddad and Um Ghamis, south of Quseir on the Red Sea, lenses of banded iron ores are concordant with the metasediments. At Um Shaddad and Um Ghamis the ores contain equal amounts of hematite and magnetite, 44–49% Fe, and 24–27% SiO_2 or silica. Finally, at Halaib and Wadi Malik, in the southeast, manganiferous iron ores fill fissures.

SUDAN

At Wadi Halfa, on both banks of the Nile, oolithic iron ores are interbedded in Nubian sandstones. Manganiferous iron ore horizons occur in the Umm Ruwaba Series at Paloic and Wabuit in Upper Nile Province. Lateritic iron is found in the Nubian series and in the Basement complex in Darfur, Kordofan, and especially in Bahr el-Ghazal. Jebel Barberi hill in Darfur consists of hematite. Specularite occurs near Tohamiyan on the Red Sea, and at Khor Hamrik in Tokar (Fig. 13, 58).

As in Egypt, probably Mid-Tertiary laterites of variable iron content cover vast areas at various altitudes. They rise 3,000 ft. near Rejaf, Ninule, Juba and Xaj, and 2,200 ft. at Yanbio, but only 1,300 ft. at Wan. The durcicrust reappears on the Blue Nile, at Singa and Sennar, at Torit, north of Bahr el 'Arab, in the Nubian Mountains, and in Darfur and Kordofan. Furthermore, red loams cover large areas (ANDREW, 1952).

The Jebel Abu Tulu deposit is located in the south-central Sudan on latitude 12° North, 1,100 miles by rail from Port Sudan. Two hills rise from the Precambrian plain, covered by recent clay. Iron ore lenses occur at the crest of the round main hill in a northeast trending band, 2,000 ft. long and 3,000 ft. wide. Other ore bodies are located east and southeast of the preceding band, and a smaller hill rises northeast of Main Hill. Ore lenses are aligned there along a 500 ft. long and 150 ft. wide zone trending N 30°W. The ore bodies are surrounded by quartz–sericite–chlorite schists, with meta-andesite occurring in the southeast. East trending faults traverse Main Hill, brecciating the fault planes. In addition to these, two sets of joints have been distinguished. In its present position, the foliation of the iron ore (when apparent) is also easterly. The hard, compact ore becomes streaky at the contacts; magnetite is largely martitized and pyrite and quartz are scarce. The ore averages 61.1% Fe and 6.5% and 5.7% SiO_2 (or silica) in the main and annex hills, respectively. Other components include 1.7% Al_2O_3, less than 0.4% S, 0.5% MgO, 0.4% CaO, and 0.06% MnO, but no P.

At *Fodikwan*, a massive magnetite lens is generally concordant with the metasediments and volcanics of the Oyo Series, but it participated in later tectonic movements and shearing. Ramifying veins and stringers of magnetite traverse the country rock in which magnetite is also disseminated. At *Wadi Halfa*, the upper band of ooliths is divided from the lower, generally finer grained, horizon by 40 ft. of Nubian sandstone. While the upper band outcrops north of Wadi Halfa, the lower band has been measured southeast of the town, where it has a thickness of 7–20 inches.

ETHIOPIA AND SOMALIA

The principal deposits of *Erythrea* appear to be Ad Cushet, Shikketti and Lamza. Smaller occurrences include Abarda, Ad Haushia, Zeban Dibta and Ad Taclai. At Ualet Shek, and on Mount Tulului in the Falcat and Ghinn Valleys of northern Erythrea, hematite lenses appear to be related to Precambrian limestones, schists and granodioritic intrusives. Banded ironstones, including hematite, magnetite, amphibole and calcite (?), occur at Agametta, 18 miles south of Massawa Harbour. Rather similar deposits were found on the sea at Del Ghedem. Furthermore, pisolitic and cavernal ironstones occur at Hamasien, in the area of Asmara (Fig. 58).

The principal resources of lateritic limonites appear to occur in the Adigrat Series in the Edaga Sunni Valley at Enticho and Adi Abo, in the Barachit zone, and in many other localities of northern *Ethiopia*, e.g., in Tigra province. Laterites have been found in western Ethiopia in the Goggiam region, at Finfinni (with gold), at Tullu Aira, near Jubdo also in Galla province, and in Chercale. Limonite of the Mai Gudo 'iron mountain' (Dombova area, in Galla) contains 54% Fe. Finally, magnetite and hematite occur in the Borana and Matacapersa areas.

Somalia. Iron segregates in the gabbro of Sheikh in the northwest, and magnetite is associated with apatite in the amphibole schist of Hudiso. At Galan Galo in northern Somalia, a crush zone traversing Eocene limestone is filled in by hematite, limonite and baryte. The Bur Range trends N 70°E (Fig. 52), 80 miles west of Mogadishu. Outcrops of itabirites and granite form the backbone of a 45 miles wide window of Precambrian schists there. The following areas of outcrops can be distinguished: south, between Bur Dividen and Bur Geluai (25 miles); north, between Bur Mun and Bur Safannolai (25 miles); centre, around Bur Galan and Bur Dur, Bur Eile and Jacoe, 50 miles east-northeast of Bur Dur. At Bur Galan, subvertical itabirite, marble and granite outcrop, and at Bur Galangadl near Geluai boulders of itabirite can be followed for a mile (R. HOLMES, 1954). In these, magnetite, black specularite grains of 1–5 mm, and quartz laminae are accompanied by some pyroxene and sphene. Samples of Bur Galan, Kuli-Kuli and Mun contain from 37.8% to 42.8% Si, and from 38.9% to 35.6% Fe. At Jessoma and Jirta Gagno, in southern Somalia, Cretaceous sandstones show grades of more than 50% Fe.

CONGO

The largest deposits of the republic are in the northeast, Katanga and Kasaï (Fig. 29, 51). Systematic valuations have not been completed.

The Northeast

Most of the itabirites are part of the Middle Precambrian Kibali system, remnants floating on the granites which occupy most of the northeastern Congo. Graphitic quartz–chlorite–amphibole schists, albitites, amphibolites and carbonatic rocks build up the column. The banded ironstones consist mainly of magnetite and hematite with accompanying siderite, ankerite, chlorite, pyrite and arsenopyrite. Both silica and iron content are variable; see, e.g., Table CXIII.

TABLE CXIII

IRON AND SILICA CONTENT OF BANDED IRONSTONES

	Fe %	*SiO_2* %
Gaima	69	5
Likamva	50	25
Maie	46	30

DUHOUX (1950b) feels that the banded ironstones were formed by lateral secretion from the ferro-magnesian minerals of the schists or from carbonatic rocks, which seem to have been abundant in the Precambrian of the northeastern Congo. Silicification of carbonatic layers would be a concomitant phenomenon of itabirization.

Resources are vast and cover Uélé province east of Stanleyville. Deposits are less frequent but abundant in the Kibali–Ituri province, west of the Albertine Rift. Several itabirite crests rise from the relief as 'iron mountains'. Itabirites of the Duru area alone contain 2,000 million tons of iron, according to DE DORLODOT and MATHIEU (1929), excluding lateritic deposits. One of the elevations, Mount Tina, is 2½ miles long, 1 mile wide, and rises 1,500 ft. above the level of the plateau. Other 'iron mountains' include Ibi, Laï, Kaï and Edu. Furthermore, in the area of Paulis, Uélé, there are several large banded ironstone deposits such as the Mont de Fer and Mount Bahengo (between Pawa and Zino). Other vast itabirite deposits occur at Mount Asonga and down the Nepoko River, 30 miles west of Wemba, in Uélé. Dimensions of banded ironstone hills of Kibali–Ituri province are shown in Table CXIV.

TABLE CXIV

IRON 'MOUNTAINS'

Mountain	*Length* *miles*	*Width* *ft.*
Kuma	6	1,500
Ami	3	500
Go	2	250
Gobu	1¼	150
Zani Kodo	⅔	1,200

Many of the banded iron ore bodies include areas of more than 60% Fe, with only traces of sulphur, chromium and nickel. Since they emerge from the relief they can be quarried. The belt extends west between the Bili and Uélé Rivers to the northwestern Congo, and reappears at Bosobolo and Zongo. Finally, lateritic–limonitic crusts have not been prospected at all, but are known to occur both as alteration products of bedded ironstones and of basic rocks at Libenge, east of Buta (Uélé) and near Watsa (Kibali-Ituri).

Katanga

In northeastern Katanga itabirites are known to be part of the Muhila Series. In southwestern Katanga, banded ironstones interbedded with quartzite form a chain of hills, 20 miles long, in the valley of the Lupwezi River, a tributary of the Lubudi. At some distance, Mount Konongo Kielu, or Songe Munonga, is 50 miles from the Bukama railroad. Other deposits have been found in the vicinity of Mutombo-Mukulu, Kanonge village, and between the Mujyay and Kibondobondo signals. These mountains are more than a mile long, hundreds of ft. wide and 100–300 ft. high. Itabirites consist of alternating beds of $^{1}/_{4}$–1 inch thick quartzite, chert, schists or siliceous dolomite, and of $^{1}/_{3}$–4 inches thick magnetite–hematite layers, including thicker magnetite bodies. Iron values vary between 35% and 65%. These deposits extend into Kasaï (see below).

West of Lake Tanganyika a number of replacement deposits have been discovered on the plateau of Marungu. While the bedded Kabue Malue ores are related to basic intrusives, the Mount Kalolo–Fela Peak deposit of magnetite, martite and hematite (covering 2 square miles) developed at the contact of hornblendite and granites intruding limestones. Similarly, further east at Kondoka on the lower Tabo River magnetite lenses also occur in limestones.

Southwest of Elisabethville and part of the Copperbelt lies the iron zone of Upper Katanga at the sources of Lualaba and the Mwemashi basins, on the divide of the Zambezi and the Congo. The belt overlaps into Zambia. More than 30 major replacement deposits are situated in this zone. Irregular magnetite, hematite and martite lenses lie in the Katangian, i.e., latest Precambrian, Kakontwe limestones of the Kundelungu and Mwashya groups, generally adjacent to gabbros and amphibolites. Both the host rock and the gangue is affected by scapolitization, and lead, zinc, copper and silver precipitated between the iron and carbonate horizons. In the Kengere Valley near Lombe, corundum and colourless tourmaline are associated with hematite. North of this main iron zone the Kisanga deposit near Kambove (Jadotville) is already in the Copperbelt. In the Sakabinda area three deposits are related to pink limestones of the upper Kundelungu, and to Mwashya calcschists.

Deposits of the lower Katangian Roan group can be divided into two categories: interbedded and massive. Between Elisabethville and Lukafu, Moa–Mululu Hill is part of a northwest trending chain of iron occurrences. Some quartz is associated with the hematite bed, wedged between layers of quartzite and siliceous ooliths. Kanunka, between Luambo and Mulungwishi, as well as Kasumbalesa in the southeastern corner and Kisanga near Jadotville, are exploited.

Other provinces

Replacement deposits of magnetite and hematite occur in the Precambrian Urundi system, e.g., in the southern part of Maniema province, southeast of Kindu on the Lualaba–Congo at Mwanankusu, Kabotshome, Mukukutshi etc.

The West Katanga banded ironstones extend across the border into Kasaï province, and underlie Karroo sandstones west of Charlesville in an area, of 20 by 7–12 miles, between the Kasaï, Luebo and Lulua Rivers. Some of them rise from softer sediments for as long as three-quarters of a mile. Other deposits are found further south between Luisa and Lueta (Mulundu Mountains), and at the confluence of the Tshiumbe and Lubembe Rivers. In addition to these, lateritic durcicrusts outcrop, e.g., at Bakwa–Nianga in the northeast.

In the *Lower Congo*, magnetite, hematite and pyrite have been partially altered to limonite at Sinyinya, Sali Mountain, and near Chambanze, 15 miles west of the Tshela bauxite deposit. Furthermore, large areas are covered by lateritic crusts, e.g., on the plateau between Matadi Harbour and Seke-Banza.

SOUTHERN RHODESIA – ZAMBIA

In *southern Zambia*, the middle stage of the lower Katangian Lusaka Series of conglomerates, dolomites, quartzites and shales is the host of most deposits, but the Kampumba iron-manganese deposit in the north occurs in Basements schists (Fig. 51). In the west, the biggest replacement deposit, *Nambala*—a 4½ mile long ridge—is located in upper Katangian Kundelungu breccias and also in shales. Other replacement deposits of the western region include *(1)* magnetite with some pyrite at Shimwyoka, *(2)* the Namantombwa magnetite body with accompanying hematite, and *(3)* the Chinda and Nabutaly magnetite occurrences. Replacement deposits of the western area are of two categories: Sanje, Venter's and Strauss' Ridge consist mainly of hematite, but Cheta and King Edward mine contain magnetite–hematite bodies. Similarly, most of the sedimentary iron ore beds contain both hematite and magnetite. Pamba and Chongwe also contain quartz, and at Nagaibwa magnetite, quartz and hematite beds alternate.

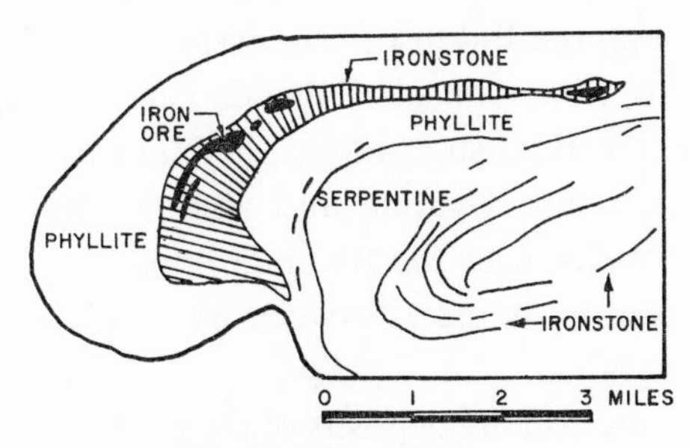

Fig. 65. Map of Buhwa.

In *Southern Rhodesia*, banded ironstones occur in a series of lenses in the schist belts. The North Hill deposit of RISCO, near Que Que, is exploited. In the southeastern part of the country, 90 miles southeast of Fort Victoria large lenses of hematite, magnetite and itabirite, containing 43–51% Fe, are interbedded in granulites and paragneisses between the Lundi and Sabi Rivers. The largest lenses form Manyoka and Mongula Hill. The Shawa, Dorowa and Chishanya carbonatites also include magnetite deposits. Metasomatic occurrences have been reported in the Precambrian Beitbridge gneisses and metamorphic carbonates. Early Precambrian and Umkondo schists of the Lusite River, Melsetter district (near the Mozambique border), also contain iron.

The *Buhwa* deposit is located in an east-northeast trending (Fig. 65) schist belt east of the southern end of the Great Dyke. The 1–2 mile wide belt of serpentine, actinolite schists and quartzites expands to a synclinal structure up to 5 miles wide that includes from west to east, quartzite, phyllite, ironstone, phyllite, a greenstone lens and the ultrabasic schists. The principal kidney-shaped ironstone lens extends from south of Buhwa, 6 miles to the east, to Hwikwi (WORST, 1962). Concordant with the hematite–itabirite into which they grade are six isolated hematite bodies. While itabirite relics remain in the ore bodies, phyllite lenses are scarcer. Dimensions of the principal bodies are shown in Table CXV.

TABLE CXV
BUHWA ORE BODIES

	Length maximum ft.	*Width average ft.*	*Depth minimum ft.*
Summit	1,600	60	—
West	5,700	140	580
Middle	1,200	290	750
Hwikwi	1,900	220	960

The Summit body consists of 64% Fe, but the others average 62.9–64.1% Fe, 3–4.5% SiO_2 3.4–4.3% Al_2O_3, 0.1–0.8% MnO_2, and 0.06–0.13% P. The hard ore fingers out into soft ore and itabirite in depth.

EASTERN AFRICA

In Kenya the principal banded ironstones and lateritic pyrite deposits occur in Nyanza province. In addition to these, the Bukura pyrite impregnation and the Homa Bay (Lake Victoria) carbonatite contain iron. The east Uganda carbonatite belt also includes minor iron reserves, especially in the Bukusu complex, at Nangalwa, and at Sukulu. Laterites continue from southern Sudan into Uganda and also develop from Precambrian hematitic quartzites in Kigezi province near the Rwanda border.

In Burundi banded ironstones outcrop between Makambo and Rutana, as they do in and near Muranda in Rwanda.

Tanganyika. The titaniferous iron ore deposits of the Ukingwa and Upangwa chiefdomes, Njombe district, are situated near the northern tip of Lake Nyasa (STOCKLEY, 1948). Partly granitized metasediments are overlain by metasediments injected by ultrabasics. The titanmagnetite bodies of the Basement are concordant with *(1)* banded charnockite in northern Upangwa and with *(2)* banded anorthositic granulites in Ukinga (Fig. 52). The ore is associated with thrust zones at Ukinga, and with permeation products and carbonatites at Ukisi on Lake Nyasa. Finally, metasomatic, hematite bodies are partially altered to limonite. The largest deposits of the area are related to the anorthosite-gabbro complex injected in several stages and mainly along a thrust as sheets, laccoliths and veins. However, differentiation was largely completed *before* injection. Magnetite bodies (11–

22% TiO_2) are most frequently found in the 2,000–3,000 ft. thick Low 'coarse' zone, followed by the Meso-zone of 1,000 ft. Relic screens of country rock separate the jointed megacrystalline mafic seams of ferromagnesian minerals and feldspar. While thin concordant magnetite seams appear to be a primary syngenetic feature, larger and younger discordant magnetite bodies appear to have crystallized, mostly before the gabbro. At *Liganga*, the largest deposit, magnetite seams seem to have been introduced late along joints; contacts are sharp. In the protoclastic magnetite bodies, chloritized mafic relics and fluidal swirls remain suspended.

In *Malawi* ironstone horizons tend to be thin. On Mindale Mountain near Blantyre, rather steeply dipping, banded hematite–magnetite is covered by float and rubble of iron ore, while on Dzonze Mountain, south of Ncheu, specularite has crystallized in lenticular bodies. At Mzimba hematite is associated with veins of vitreous quartz, while on the edge of the Nyika Plateau quartz–hematite schists outcrop.

Mozambique. The titanmagnetite bodies located 13 miles east of Tete contain 50% Fe and 19% TiO_2. At Chiunda, between the Ruo River and the Manga Mountains in Zambezia, magnetite lenses trend N 25°E, and occasionally N 45°–70°E, in amphibole gneisses. Hematite shows on the western, Rhodesian border at the Lucite River and along the Gogói–Metove road. Finally, in the Nyasa district, segregations are related to ultrabasics.

SOUTH AFRICA

As already stated, South Africa contains a considerable part of the world's iron ore resources. The principal districts include metasediments of Griqualand and Transvaal and the Bushveld complex. Minor occurrences are spread over the country (Fig. 46, 50).

Griqualand

In a north trending arc the iron ore formations extend 300 miles from Postmasburg to the Kalahari, forming the Doornbergen Chain and the Asbestos and Kuruman Hills. The Gamagara quartzites, belonging to the lower Griquatown Stage of the Pretoria Series, have glided on the Dolomite Series. Other Gamagara rocks contain red and diaspore shales, conglomerates and breccias (WAGNER, 1928). The folded and overfolded banded ironstones are largely concordant. However, the underlying dolomites have frequently been crushed and silicified; because of the slumping of the carbonate and its lower resistance, the ironstone sheets have been broken up into domes, basins and troughs. A typical section at Lucan Dam in Griqualand West shows (Fig. 66):

700 ft. of jasper, including magnetite layers,
50 ft. of quartzite,
500 ft. of banded ironstone,
300 ft. of quartzite and grit,
300 ft. of ferruginous rocks,
500 ft. of quartzite,
800 ft. of thinly banded ironstone and quartzite.

The upper Griquatown ironstones attain a thickness of 2,500 ft., largely covered by Kalahari sand. Three facies of iron ore have been distinguished: *(1)* Gamagara shales, *(2)*

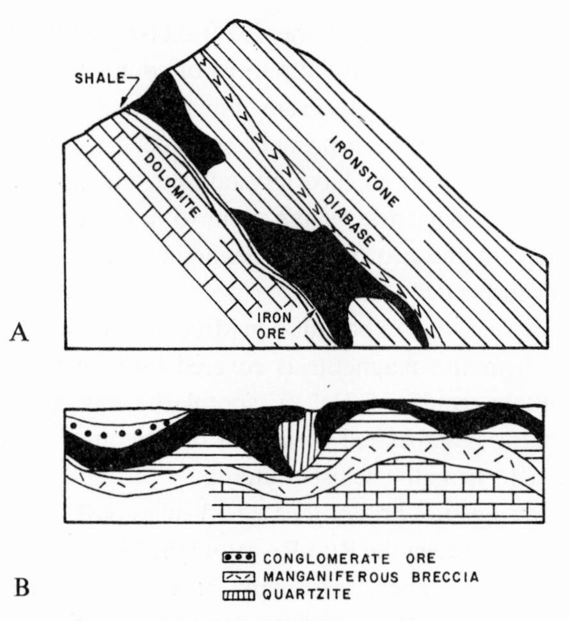

Fig. 66. A. Section of Thabazimbi. B. Section of Griqualand.

conglomerates, and *(3)* the underlying Blinkklip. The folded and brecciated tectonic ores of the Blinkklip are the most heterogeneous. The conglomerate consists of hematite pebbles embayed by fine grained, largely specular hematite. The layer is rather irregular, with an average thickness of 10–30 ft. (maximum 120 ft.). The intermediate bed of laminated, high grade blue hematite derived from the shale averages a thickness of 60 ft. (maximum 200 ft.).

The principal deposits are found in the northern area of *Sishen* around Bruce, King, Lylyveld and Mokanning, but important ironstone bands are also believed to occur with the manganese ores of Black Rock.

Pretoria Series

The Pretoria Series of the Transvaal System surrounds the Bushveld complex in two largely west–east and north–south bending arcs, each extending 100 miles. The principal iron formations of the east include the magnetitic quartzite of the Time Ball Hill and Daspoort horizons. At Time Ball Hill, 12 ft. of banded oolithic ironstones are interbedded in a column of shales and some quartzite. The average diameter of the ooliths is 0.3 mm. The original quartz nuclei, partly replaced by calcite, are embayed in alternating zones of magnetite and chamosite, surrounded by magnetite. Frequently, two nuclei are enclosed in the same shell and small subhedral quartz crystals and ooliths are partly replaced by magnetite. Finally, the groundmass of hematite, magnetite, chamosite and calcite is injected by calcite stringers.

Around *Thabazimbi* high grade ores occur in irregular, 60–80 ft. thick sheets. The ore bodies include lenses of banded ironstones which finger out in depth (Fig. 66).

Pisolitic ironstone and the Clayband also contain oolithic ore. The lower bed of the *Daspoort* horizon is 2½–5 ft. thick; it consists of hematite, limonite and pisolites, one inch in diameter, embayed in fine grained sand. Concentric pisolites are built of hematite, limonite and chamosite, and surrounded by a sandy ferruginous groundmass. The upper bed consists of foliated oolithic ironstone and ferruginous shale. Since the ores have recrystallized in the aureole of the Bushveld complex, muscovite is accompanied by the vanadium mica, roscoelite.

At Syferfontein, Cyferfontein, Buffelsfontein and Zandfontein, the thickness of the ore horizon varies between 5 and 15 ft. out of a column of 150 ft. The iron content oscillates around 44%; sulphur content is low, but phosphorus attains 0.01–0.1%, and the average grade of this 100 billion ton reserve is roughly 30% Fe. The beds can be traced for 50 miles.

Southeast of the Bushveld the Daspoort horizon is absent. The Time Ball Hill Stage extends for 180 miles. The upper bed attains a thickness of 27 ft. at Airlie Station, but the thinner, lower bed of 5 ft. contains only 30% Fe. Overall reserves of ore, containing 35–40% Fe, reach the mark of 25,000 million tons.

Southwest of the Bushveld, the Daspoort horizon consists of shaly sandstone and white quartzite. Here an outer zone of hematite surrounds the composite core of the ooliths. An estimated 1,000 million tons of ore contain an average of 47% ferric iron.

In the area of the Crocodile River, the Transvaal facies contains 10,000 million tons of iron ore, with 30% Fe.

Small but high grade deposits have been found in the Bronkhorstspruit, Potchefstroom, Wolmaransstad, Ermelo, Middleburg and other districts.

Bushveld

The iron horizon is located approximately 6 miles inward from the Merensky Platinum Reef, that is, 2 miles above the latter in the layering of the complex. Allotriomorphic labradorite and diallage norite enclose the labradorite–bytownite anorthosite. Bands and pipes of magnetite, accompanied by ilmenite, occur in the latter. The minerals surround anhedral feldspar, having developed hypersthene coronas. The lower layer of iron ore attains a thickness of almost 4 ft., but the upper band is thinner. Interlocked maghemite, magnetite and ilmenite are accompanied by some chromite, corundum and nepheline. The chromitite phases of the Bushveld underlying the Merensky Reef contain 25% Fe.

Titaniferous iron ores have also been found on Insimbe (iron) Hill in the Tugela Valley, Natal and Piet Retief in the Transvaal.

Other occurrences

Kromdraai, north of Pretoria, is contact metamorphic, while Phalaborwa is carbonatitic. The pre-Transvaal, Swaziland and Witwatersrand systems also contain banded ironstones, bedded ores being included in the Natal Karroo. Finally, primary and secondary hematite are interfoliated in the high grade replacement products of the Crocodile River.

SWAZILAND

Iron occurs in several horizons of the early Precambrian Swaziland System, but only deposits of the Fig Tree shales, located approximately in the middle of the column, are of economic significance. The schist belt extending along the northwestern border of Swaziland attains a width of 5 miles, consisting, from east to west, of *(1)* basic schists, *(2)* shales, graywackes and cherts, *(3)* the iron horizon, *(4)* a much wider zone of the upper or arenaceous series, and *(5)* basic schists. As the shales have been ferruginized with a variable intensity, the ironstone horizon is, in fact, discontinuous. It extends 5 miles in Swaziland, from Clarry's Camp beyond Bomvu Ridge, and 12 miles in Transvaal, reentering Swaziland at its northern tip, the Iron Hill deposit. The ironstone band expands at its southern extremity of 1 mile to a width of 2,600 ft. In the west it is bordered by a narrow band of chert, and in the east by quartzites. Somewhat younger tremolite schists surround its southern end, also overlying it in the narrow band extending towards the north. Samples of this band or crest contain 41.4% Fe and 36.9% SiO_2 or silica. The structure can be interpreted as two synclinal folds separated by an anticline. In the expanded southern zone, more than three hematite ore bodies up to 50 ft. wide are separated by low grade shales, and the bedding of the shales can still be distinguished in the compact hematite. Specularite is often found in vugs of both rocks. At the southern end samples of Clarry's Camp contain 63.7% Fe and 6.11% SiO_2.

The *Ngwenya* (Ingwenya) or Bomvu Ridge zone extends 6,500 ft., attaining a maximum width of 320 ft. One can distinguish five types of ore: *(1)* compact hard blue, *(2)* foliated blue, *(3)* hard, fat reflection, *(4)* friable soaplike touch, and *(5)* biscuit coloured ore in the southern portion of the ore body. The ore body dips 40°–60°. Its northern, central and southern sectors respectively contain 61.0%, 55.9% and 64.6% Fe, and 7.4%, 17.3% and 3.5% SiO_2 or silica.

The *Iron Hill* deposit averages 48.9% Fe, 22.3% SiO_2, 4.1% Al_2O_3, 0.4% Mn, 0.1% TiO_2, 0.07% S and 0.03% P. Iron shows in the lower series in the south, with limonite occurring in quartzites and conglomerates, especially at Silotwane Peak. Hematite accompanies jaspers of Masali Peak and north of Lyly Valley, and the shales of Mahlanganpeppa and Ngwenya Beacon.

In southwestern Swaziland, Middle Precambrian (upper Pongola System) shales and quartzites of the Hlatikulu and Mankaiana districts contain up to 40% Fe, 8% Mn and 0.15% V (WAY, 1962). At Cascade in the Piet Retief district iron content ranges from 25% to 29% and attains 31% at Gege.

MADAGASCAR

Lateritic crusts form the largest deposits. The principal among these is located at *Ambatovy* and *Analamay*, not far from the Tananarivo–Tamatave railroad. The durcicrust covers soft laterite, probably resting on early Precambrian ultrabasics. This crust is up to 10 ft. thick, and averages 3–5 ft. Both the hard and soft ore contain approximately 50% Fe (see Table CXVI).

At Tsiandava, in western Madagascar, quartz gravels have been recemented by hematite, limonite, and stilpnosiderite derived from laterite. In the southwest at Betioky, a similar but finer layer of laterite sandstone extends 30 miles, with its width averaging 1 ft.

Beds of magnetite quartzite, containing 10%–50% Fe, extend 70 miles inland from the east coast *(1)* at Lake Alaotra, *(2)* south of Tananarivo, *(3)* between Ambositra and Vangaindrano, *(4)* at Betroka in the south, and *(5)* at Maevatanàna in the northwest. At Maevatanàna seven 3 ft. thick beds, extending $^2/_3$ mile, contain 20–45% Fe; the Ambatoa-

TABLE CXVI

COMPOSITION OF THE MORAMANGA ORES

	Durcicrust %	*Soft ore* %	*Crust* %
Fe	52.2	50	50
SiO_2	1.3	1.1	0.8
Cr_2O_3	1.3	0.4	0.9
NiO	0.2	0.1	—
TiO_2	2.2	1.1	0.7
P_2O_5	0.09		
S	0.6		

laona beds (near Tananarivo) average 25% Fe with peaks of 40%; and in south-central Madagascar, at Bekisopa, lenses of compact magnetite contain up to 67% Fe. Finally, west of Tananarivo, titaniferous magnetite segregates in the gabbros of Vangoa. Perhaps the largest deposit, Fasintsara, northeast of Fianarantsoa, is a band of pyroxene itabirites. Grades decrease with the depth from 40% to 30% iron.

AMPHIBOLE ASBESTOS

Several amphiboles have fibrous habits. Tremolite and actinolite fibres occur sparingly in shears and calcomagnesian rocks. Antophyllite, a more abundant form, is related to serpentinized ultrabasics, but the two principal asbestiferous amphiboles are not endogene. Fibrous riebeckite, cummingtonite or crocidolite, and amosite occur exclusively in sedimentary ferruginous quartzite and dolomite, their composition being similar to that of this sequence (Cilliers in KEEP, 1961). Banded ironstones deposited in shallow seas precipitate during alternating seasons, either directly or from colloids. Between two orogenic cycles asbestiform crocidolite needles developed on regular layers of already crystallized fine grained magnetite, but, when magnetite layering failed to provide the base for oriented growth, only randomly radiating crocidolite and grains of common riebeckite formed. The load of sediments in Griqualand and the aureole of the Bushveld in the Transvaal are believed to have contributed to the dynamo-metamorphic pressure needed for the opening of the fibre channels (Fig. 2).

SOUTH AFRICA

Two asbestos areas can be distinguished: Griqualand in northern Cape Province, and the eastern Transvaal. Crocidolite is found in both districts, amosite occurring only in the Transvaal. Both forms of asbestos crystallize in the banded ironstones of the 2,200 million year old Transvaal System (Fig. 50).

Griqualand

The asbestos deposits are included in the same belt as the iron and manganese occurrences described in the preceding chapters. The principal arc extends 280 miles, from 28 miles south of the Bechuanaland border to Lovedale, 12 miles south of Prieska and the Orange River. Lovedale is situated 130 miles southwest of Kimberley. A pair of shorter arcs extends from the Orange River through Postmasburg to Black Rock, with a small outlier fanning out towards the northwest. The principal deposits are concentrated in the Asbestos Range and Kuruman Hills of the main belt.

The thickness of the Lower Griquatown Stage of the Transvaal System diminishes from 6,000 ft. at Koegas to 3,000 ft. at Kuruman and 1,000 ft. at Pomfret on the Bechuanaland border. The lowest asbestos horizon is found in banded ironstones, 50 ft. above the underlying Dolomite Series. Except for Koegas, all crocidolite is won from this level. At Koegas, an intermediate horizon is found at 1,100 ft., the Prieskite horizon at 2,000 ft., the Westerberg at 2,500 ft., and the Upper horizon at 3,400 ft. Asbestos levels are more ferruginous than the rest of the series, which consists of quartz and hematite–limonite chert, interlayered with sodic horizons of non-oriented fibrous riebeckite or crocidolite. However, carbonatic and conglomeratic cycles are rare in the shallow water sediments. The strata have been warped into two broad synclines and overfolded at Koegas. The subhorizontal thrust faults of the Postmasburg area contrast with the transverse faults of Doornberg. At Koegas, dolerite and diabase dykes intersect the lower horizon, thus transforming crocidolite into riebeckite. A 300 ft. thick diabase sill occurs 400 ft. below the Westerberg horizon, a thinner sill overlying the seam by 35 ft. As a result of surface silification, silica pseudomorphs develop after crocidolite (see 'Semi-precious stone' chapter). Nevertheless, amphiboles resist weathering better than the country rock, but they are finally transformed into a limonitic dust. Asbestos horizons of standard size occupy 750,000 square ft., but they are exceeded by the 20 million square ft. existing at Koegas (Fig. 67).

In the *southern sector*, groups of cross fibre seams are interbedded in favoured zones

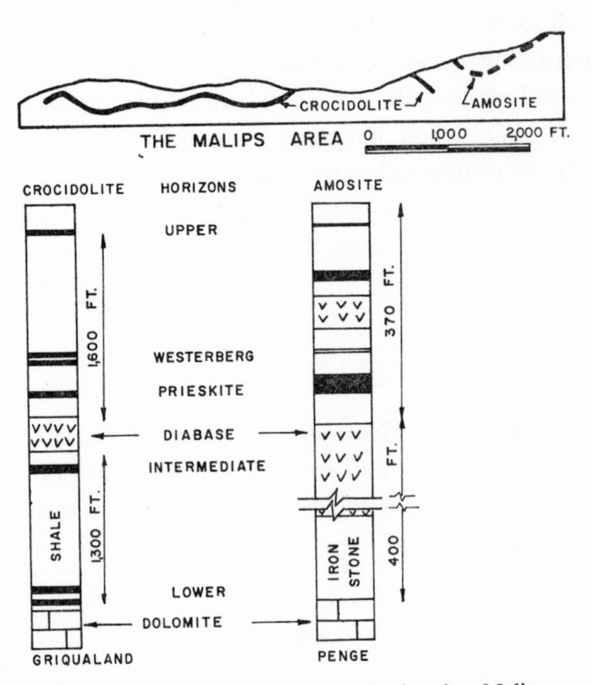

Fig. 67. Asbestos zones in Griqualand and at Malips.

with remarkable regularity. Seams may thin out after several to 30 ft., but new seams appear frequently in the same channel.

Walls are distinct, but sometimes undulated (A. L. HALL, 1930), with the seams bulging in ichtyolites. Along the dip, the highest points of these undulations may form a ridge or a transverse wave pattern. The slopes of the waves are most variable. In one of the sides of the seams, cone structures exhibit brusque pyramidal protuberances, frequently filled in by carbonates. On the other hand, some crocidolite protrudes into the cones. Lines of parting, habitual in chrysotile seams, are scarce, as are impurities. The fibrosity is regular, with the angle of the prisms generally falling 10°–15° short of the perpendicular. The normal blue colour occasionally becomes paler, signalling a silky, fleecy texture. If in contact with darker shaly ironstones, darker fibres represent the best grade. The length of fibres generally varies between ½–1 inch, extending locally to 5 inches at Stilverlaats!

When in contact with underlying limestones, seams of the *northern sector*, interbedded in sub-horizontal jaspylite and itabirite, are distructed. Complicated structures are most frequent in the lower 50 ft. of the ironstone. Totalling 5–6 inches, with $^{3}/_{4}$ inch fibre in a column of 7 ft., the grouping and spacing of individual seams is more erratic than in the South. Fibre lengths are comparable here with those prevailing in the southern sector. However, the crocidolite is generally paler and fleecier. The asbestos is lavender blue, different colours appearing in adjacent seams.

Crocidolite deposits

The mines commence at the high grade Pomfret deposit 125 miles west of Mafeking. Heuningvlei is situated 40 miles south of Pomfret. Riries, Whitebank and Asbes, 90 miles south of Heuningvlei (in the area of Kuruman), include considerable asbestos bodies. Adjacent deposits include Depression, Mount Vera, Langley, Mansfield, Newstead, Bretly, Hurley, Crawley, Grasmere, Klipvlei, Schietfontein, Mimosa, and Warrendale, which is an important concentration. Blackridge, Lelykstaat and *Koegas*, the largest mining areas, are located respectively 50, 80 and 88 miles south of these. In the Koegas, Prieska, Carn Brae stretch, commercial crocidolite is found at Brimitsa, Geduld, Orange View, Kliphuis, Klein Naawte, Erfrust, Buisolei, Glen Allan and Asbestos Reef.

In the old Westerberg deposit, two concordant series of seams form a broad arc, each group consisting of three to seven seams over a cumulative thickness of 9–15 inches. The length of fibre increases with folding. Indeed, seams are particularly well developed in overfolded ferruginous shales.

At *Koegas*, on the Orange River, a horseshoe-shaped syncline pitches 13°. The width of nine folded seams (exceeding fibre lengths of $^{1}/_{8}$ inch) varies between 3 inches and 7 ft., and averages 3 ft. The total thickness of the horizon attains 110 ft. and the width of channels totals 18 ft., varying individually from 15 to 56 ft. The cumulative length of fibres has been recorded to be from 1 to 7 inches, i.e., 0.07–0.5% of the channel. However, only in the three upper seams and in the third seam from the bottom do fibres average a length of more than ½ inch. The Main Reef may contain 2 inch fibres. At Lelykstaat, the jaspylite host rock dips 40°–90°. Two groups of three to five seams are interbedded in the horseshoe structure of Keikaamspoort.

In the Kuruman district, the asbestos horizons immediately overlying the Dolomite

Series are worked. At Hopefield, 15 miles north of Griquatown, secondary structures of a bracchyanticline are hosts of the seams. A group of thirteen seams in a layer of 2 ft. occurs in overfolds of the foliated jaspylites of Blackridge, with another group of six seams in a column of 8 ft. At Pannetjie, crocidolite is found especially high in the Transvaal System. The *Warrendale* deposit near Postmasburg extends 1½ miles. Seven major channels, attaining a maximum width of 200 ft., have been distinguished in a column of 250 ft. Other deposits are rather small. Individual seams that pinch out over relatively short distances have been observed to form groups of three seams. Well-developed parallel fibres are tightly packed. Movements affecting the folded ironstones caused their orientation to oscillate from the perpendicular. Fibres tend to extend somewhat beyond the line of parting, bending on this plane.

The lower seam of Pomfret covers more than 600,000 square ft., dipping 4°NE at a depth of 80 ft. The channel of the lower asbestos body is 30 ft. wide.

Eastern Transvaal

The Pietersburg–Lydenburg field extends in a 60 mile long east trending arc from Chuniespoort, at the northern end of the eastern branch of the Bushveld ultrabasics, to the confluence of the Olifants and Steelport Rivers. The arc of the Dolomite Series of the Transvaal System surrounds the Bushveld, bending farther southward. After a gap of 100 miles, chrysotile reappears in the Series at Carolina, 30 miles west of the Swaziland border. Amphibole asbestos occurs between dolomite layers within the 800 ft. thick Banded Ironstone Stage, most probably in the same sequences as in Griqualand. When unfolded, the beds dip at increasing angles towards the centre of the Bushveld, downwarped by the outpouring of the ultrabasics. The sediments have been compressed into several folds between the early Precambrian granites on which they rest and the gabbros. The limit of the Banded Ironstone and lower Dolomite Stage was particularly affected. Secondary folds and faults (Wonderkop) complicate the structure. Furthermore, lateral variations of facies have been frequently observed in the area, with ironstone developing in dolomitic horizons along the strike. The Griqualand and Transvaal itabirites resemble each other: bedding is close spaced and minor folding is distinct. Nontronite and magnetite develop especially in the fibrous horizons. However, red jaspylites and agglomerated riebeckite do not appear, with patches and bands of prismatic cummingtonite (amosite slates) distinguishing this formation.

Crocidolite is restricted to the western half of the field, both amphiboles tending to occur at specific levels above the dolomite. The lower zone includes two to five packs of asbestos, with amosite predominating, 90–150 ft. above the indicator level. Between the 230 and 340 ft. levels the Main Zone contains two to six packs, mainly crocidolite. The Short Fibre Zone, at 440 ft., consists of a crocidolite seam. In the Limited Zone, at 510 ft., some amosite has developed. The last mixed seam appears in the Upper Zone at the 770 ft. level. Duplex seams, however, are of no commercial value, and in economic deposits crocidolite seams are separated from the overlying amosite seams by 200–500 ft. of ironstone. The Main Zone is particularly well-developed between Kromdraai and Lot 124. The Short Fibre Zone appears in this sector, and, after a gap, between Hooggenoeg and Horn Gate. Major seams tend to be accompanied by a group of minor stringers. Transvaal crocidolite is harder than the Griqualand variety because of disseminated magnetite crystals, and it is also not as fibrous because of the admixturing amosite. However, its light colour has been also attributed to the presence of amosite, a lower iron and sodium content, and less silica.

Amosite

Although amosite occurs in many horizons of the Pretoria and Dolomite Series of the eastern Transvaal, commercial concentrations favour zones composed of several seams. The lateral extensions of the layers remain variable, but the seams thicken in the fold arches. Generally, only one of the two hornblendes has developed in each deposit. However, they grade into each other along the strike. Exceptionally, the two silicates are intimately associated or else only divided by a colour contact in the seam. A thin band or film of either amphibole may separate the other fibre from the wall. Fibres are particularly long in the eastern Lydenburg district, where they attain lengths of up to 3 ft. The quality of amosite appears to improve with depth, its fibrosity and tint ameliorating concomitantly. Montasite of Montana, near Malipsdrift, is a softer, high silica amosite.

Even in the Lydenburg sector, which covers 200 square miles, amosite is generally more abundant than crocidolite. Although amosite occurs at many horizons, commercial fibre is restricted to zones. Highest grades are found in the Lower and Limited Zones, and seams persist along the strike for hundreds of ft., i.e., longer than in Griqualand. Parallel chains of fibre lenses tend to occupy larger zones than crocidolite seams. At lower levels seams may coalesce to a total width of 20 inches, divided only by slaty partings. The ratio of fibre to ironstone is very favourable. As in Griqualand, cones containing ironstone or magnetite cross fibres protrude into the seams. The fibres bend rather frequently at their extremities. Magnetite pockets and stringers occur mainly in the upper horizons, but amosite continues along the dip for more than 1,000 ft. Commercial fibre is found far beneath the hydrostatic level, becoming springier as the thickness of the overburden increases.

The amosite seams are spread over a 25 mile long line, extending from Chuniespoort East through Malipsdrift to plot 36. (Explored crocidolite horizons begin at plot 19, 10 miles east of Chuniespoort, extending 35 miles to Dublin.) Amosite seams are known to occur between plots 14 and 17, halfway between Malipsdrift and Dublin, and also at the latter locality. Beyond a gap of 10 miles, the amosite horizon outcrops for 10 miles in the eastern portion of Streatham, at Penge, Weltevreden and part of Kromellenborg. Between Penge and Streatham fibres are longer, averaging 1½ inches. The best fibres of the western sector concentrate in its eastern subdivision at the Malips River, but the quality decreases on both sides.

At *Penge*, the principal deposit is *(1)* the Main Zone, which attains a thickness of 40 ft., lying 450 ft. above the Dolomite. *(2)* A 5 ft. thick zone is located at the 540 ft. level, with *(3)* the limited Higher Zone of 14 ft. located 690 ft. above. A 200 ft. thick diabase sill has been injected into the basal part of the Dolomite Series. In the Main Zone three packs of seams are spread over a column of 40 ft. These seams may attain a height of 2 ft., the average height being 3–6 inches. Extensions of fibre zones along anticlines are frequent (Fig. 67).

At Weltevreden, 6 miles south of Penge, amosite seams are narrower. At Kromellenborg, beyond Weltex, the grade is generally low. At *Malipsdrift*, 40 miles from Penge, the younger crocidolite band is part of a synclinal structure; it is situated 5,000 ft. north of amosite seams occupying 5 million square ft. The principal amosite seam extends more than 5,000 ft. along the strike, with fibre lengths varying from 1 to 6 inches. Two other seams are of low grade. In several seams three to five folded horizons, each containing a total of an inch of crocidolite occupy 7 million square ft. These faulted horizons are more extensive than the amosite channel.

Other areas

At Rhenosterfontein (in the Marico district) west of the Bushveld, pink amosite occurs in shales of the Dolomite Series, slip amosite covering early Precambrian (cf. Sebakwian) ironstones 40 miles west of Messina. In the same region brittle crocidolite is found in the ironstones of Thabazimbi and the Warmbad districts. Fibrous anthophyllite, surrounded by magnesite rock, is included in altered ultrabasics, floating in the early Precambrian granites of *Corea* in the Rhodesian frontier district of Soutpansberg. Similar asbestos occurs in the Murchison Range. Finally, slip tremolite relics have remained in the pale schists of Klip River, near Pomeroy in Zululand.

THE REST OF AFRICA

The Griqualand belt extends north of the Molopo River into Bechuanaland; most of the fibre, however, has been silicified. Some crocidolite is found at Nakala Phala. In the banded ironstones of Rehoboth and Omaruru, Southwest Africa, some crocidolite is associated with tremolite and antophyllite.

In the Serra Mangota, north of Vila de Manica, Mozambique, tremolite fibres $^2/_3$ inch long occur in serpentine, while in the Munhinga Valley of the Tseterra and Revuè streams, long semi-oriented tremolite developed in talcschists. In Madagascar, amphibole asbestos occurs *(1)* at Belasiray, 3 miles from Antsiafabositra in the north, *(2)* at Sahatany near Antsirabe, and *(3)* at Lohanifotsy near Tamatave. In Malawi, blue asbestos and antophyllite are found in the serpentinized ultrabasics of the Kirk Range in the Ncheu district and near Uzumara Hill.

In *Tanganyika*, 70 miles east of Lake Victoria, stringers of slip fibre tremolite represent 1% of the gabbros and syenites of Ikorongo. At Mbembe, northwest of Dar es Salaam, antophyllite and tremolite–actinolite fill in a pod in migmatites. In addition, antophyllite shows at Rubeha and Haneti. In *Kenya*, antophyllite occurs at Kinyiki near Mtito Andei, in southern Kitui, West Suk, Baragoi, and in the Taita Hills, and crocidolite is found at Sultan Hamud, southern Machakos. Amphibole asbestos occurs in the Acholi, Karamoja and West Nile provinces of Uganda. Near Mufulira, northern Zambia, a lens of $^1/_8$–½ inch long amosite–crocidolite occurs in quartzite, while large deposits of crocidolite occur near the Katanga Copperbelt (HUGE and EGOROFF, 1947).

At Hafafit in Egypt, feldspar–vermiculite veins injected into sheared serpentinite are rimmed by antophyllite.

In the Kangari Hills near Masati, Sierra Leone, antophyllite occurs in ultrabasics of the epidote–amphibole facies. However, shortly after its crystallization it altered to antigorite and silica (MARMO, 1951).

MANGANESE

The largest, polygenetic, deposits are situated in South Africa and Gabon. The former were derived from leached manganiferous dolomites, but the latter are essentially eluvial. Similarly, alteration processes play an important role in gondite, carbonatite and vein deposits. Manganese ores may develop on or from rocks of any age. However, in South Africa and West Africa, the principal mother rocks are attributed to the 2,200–2,400 million year old Precambrian cycle. In the Congo they are attributed to the 1,700 million year old cycle, in Gabon to the Francevillian of probably late Precambrian age, and in Morocco to the Cretaceous (Fig. 2).

The largest gondite deposits are those of Katanga, but similar occurrences are widespread in West Africa. The history of their formation appears to be clearest at Nsuta and may be summarized as follows:

(1) Syngenetic sedimentation of manganese in clays, sandstones, and grits.

(2) Volcanic activity of the synclinal phase, marked by generally interbedded tuffs and basic lavas.

(3) Epimetamorphism transforms the silica rich sediments into manganese oxide bearing quartzites, and the aluminous rocks into gondite, ie., roughly kaolin into spessartite.

(4) Isoclinal or overfolded folding turns the sea shore subvertical, including the manganese horizon (during the Birrimian/Tarkwaian interval of sedimentation).

(5) Milky quartz veins cut the Precambrian strata prior to Caenozoic weathering.

(6) Lateritization occurs between the eroded surface and the hydrostatic level. The upper part of the lenses is epigenetized by descending water, creating the mammillary ores. Quartz and minerals of the kaolin–halloysite group are leached out; silica is then dissolved from spessartite and finally the aluminium of the garnet enters the clay minerals.

Algae of the Chaetophoraceae *(Gongorosira)* type, and the iron bacteria *Chlamydothrix*, are found on the ores and may also have catalyzed the deposition of manganese. Early hydrothermal activity and lateritization are not believed to have played a major role in the formation of these deposits.

MOROCCO

More than half of the manganese produced in the Moroccan Kingdom has been extracted from Imini, about one third from Bou Arfa and the remainder elsewhere (BOULADON and JOURAVSKY, 1952). The remainder comes from small sedimentary and vein deposits located *(a)* mostly along the Precambrian front of the Anti-Atlas, *(b)* in the Middle Cretaceous north of this orogeny, and *(c)* in the Lias (Lower Jurassic) of the Oujda district, 10 miles from the coast near the Algerian border (Fig. 10, 59).

Imini

Imini is 20 miles south of Marrakech. Middle Cretaceous sandstones are underlain by Precambrian and Permo-Triassic rocks. The manganese bearing Cenomanian–Turonian layers are at the contact of these sandstones and of the overlying fine grained dolomites. The lowest dolomite beds carry rolled quartz, some feldspar and muscovite along with fossil *Exogyra africana*. A thin manganiferous bed is included in the upper layers of the dolomite that is overlain by sub-discordant Senonian (Lower Cretaceous) sandstones, gypsum, clays and coral limestone. The occurrences trend 15 miles west-southwest from Tazoult to Bou Aggioun. The principal deposits of central Bou Tazoult-Bou Azzer and Sainte Barbe are, respectively, 2 miles and 1 ½ miles long. The overall width of the two manganese bearing beds is 3–8 ft. At *Bou Tazoult* in the west, the ore bed overlies the dolomite, but it underlies the dolomite in the east. A thin sandstone horizon divides the manganese horizons from each other in the west, and a dolomite bed separates them in the east. The lower contact is sharp. A red clay and sandstone level precedes the dolomitic upper contact. The northern limit of the manganese bodies is sharp, the southern border being represented by a progressive thinning of the reef, the average thickness of which is 1,200 ft.

The veins can be followed for 25 miles towards the east-southeast through Amzerko to *Aoufour* and *Tasdremt*. In these 10 mile long deposits, the two lower ore layers are very thin. Only the 2–5 ft. thick top layer of lead, barium bearing psilomelane, and coronadite is being worked.

The principal ore mineral is polianite associated with goethite and psilomelane stringers. The unworkable upper layer consists of coronadite. Whilst the polianite shows copper traces, psilomelane and coronadite contain 1.5% BaO, 1% PbO and 0.25% K_2O.

The deposits are supposed to be syngenetic, shore or lagoon sediments, slightly affected by supergene alteration. Other hypotheses attribute their origin to Precambrian lavas, to waters percolating through the overlying sandstones as at Franceville, and to fossil lateritization.

Bou Arfa

Bou Arfa is located south of the Casablanca latitude, near the Algerian border. A heterogeneous, lower detrital series is underlain by typical Precambrian rocks. The upper layers of this series consist of arkoses, clays, polianite bearing dolomites, and thin manganiferous beds. The following dolomite–limestone series is composed of siliceous dolomites, dolomitic limestones including the Hamaraouet manganese lenses, detrital beds, and the upper limestone–dolomite–marl beds which include the Aïn Beïda ores. An unconformable upper detrital series terminates the Lower Jurassic Lias.

The *Hamaraouet* level is best developed in the west. As at Postmasburg, the ore is found in dolinae of this fossil karst. Polianite and baritic psilomelane, associated with hausmannite, fill in 50 ft. wide cavities continuing in broad, open fissures. Manganite nodules are found in pryolusite, with goethite zones forming part of concretional polianite growths The upper *Aïn Beïda* level is thickest in the east. This body is traversed by a fault on its southern contact. Above the bottom shears, conforming lenticular ore leads are more than 500 ft. long and more than 50 ft. wide, and average a width of 5 ft. Polianite is associated with secondary dolomite and calcite.

The deposits are believed to be syngenetic, the manganese having been derived from Precambrian rocks and veins or from younger basalts.

Tiouine

Tiouine is located southeast of Imini in the eastern part of a 10 mile long belt, 30 air miles from Marrakech. The late Precambrian sedimentary-volcanic column consists of 600 ft. of lavas overlain by a detrital stage of agglomeratic breccias. Red tuff layers include the manganese bearing beds and conglomerates. The upper detrital and dolomitic series is earliest Cambrian, Georgian. The deposit is ½ mile long and has an easterly pitch. Of the five to fourteen ore beds that have been found (averaging seven), the thickness varies from 2 to 3 ft., locally attaining more than 10 ft.

Red tuff levels have been identified in and above the ore beds, into which they grade laterally. Braunite and criptomelane are accompanied by baritic psilomelane, polianite and coronadite, and by quartz and lava breccia attacked by the manganese minerals. The ores contain 2–13% BaO, 0.5–2% PbO, and traces to 5% K_2O.

Minor occurrences

Syngenetic sedimentary deposits occur in the late Precambrian *(1)* of the Anzi-Tafraout basin (Idikel) southeast of Agadir, *(2)* at Alib en Nam, in the Oujda district, and *(3)* at Migouden and Offremt. Manganese has also been found in *(4)* Permo-Triassic sandstones at Narguechom, Jebel Mahsseur and other localities of the Oujda district, and *(5)* in Cretaceous sandstones at Garet el Asset and Ben Zireg. *(6)* Deposits in Lias dolomites include M'Koussa, east of Marrakech and south of Rabat, Tizi n'Rechou and Boulbab, near Meknès.

Manganese bearing, fissure-filling swarms of lenses traverse late Precambrian lavas and conglomerates of almost 2,000 square miles of the Ouarzazate district. The veins are accompanied by dolomite at Taourat, Tachgagalt and Tamegra, and by baryte at Tisguililane, Bachkoun, Tikirt, Tinzaline, and in the Siroua and Manabha districts. At Bourdine, the veins intersect Paleozoic granodiorites; at Zekhara they occur in Silurian schists.

ALGERIA AND TUNISIA

The *Djebel Guettara* is located 110 miles southwest of Colomb-Béchar. Arseniferous manganese ores are related to Precambrian tuffs and rhyolites. Syngenetic manganese has been reconcentrated by replacement processes. Other occurrences have also been found in the Ougarta and Bou Kaïs Mountains (Lucas, 1956). The Oujda belt of Morocco extends as far as Marnia, near Oran.

Karstic, Cretaceous and Tertiary deposits of western Algeria include the Kahal de Brézina, Oued (wadi) Zelmon, Djebel Keracha and Marnia. Vein occurrences in Tunisia, like Djebel Batoum, Thuburnic, and Djebel el Aziza are generally Cenomanian–Turonian, i.e., Middle Cretaceous.

Vein and replacement deposits like Laghouat, Kef el Aguel and Aïn Kerda have little economic importance.

LIBERIA

Manganese ore is known to occur in Kingsville, Montserrado County, 30 miles east of Monrovia and on Mount Dorthrow, Eastern Province. The small Kingsville deposit has been found to be of low grade, containing about 23% of manganese.

Mount Dorthrow is located approximately 500 ft. north of Dwarbo and about 28 miles northeast of Zia town, Konobo Chiefdom. The Cavalla River flows 1,600 ft. to the southeast. The mountain rises at a distance of 90 miles from the coast and 20 miles from the Ivory Coast border. The deposit is accessible by a 60 mile long trail from Nyaake, a town on the Webbo motor road. Mount Dorthrow may be considered as a Birrimian anticlinal ridge striking approximately east–west, about 3,000 ft. in length and rising 150 ft. above the surrounding country.

The manganese ore occurs in several scattered, dome-shaped outcrops on the top of the ridge. These outcrops average about 25 ft. in length and 11 ft. in width, striking N 55 °W–N 75 °E; with an average dip of 67° S. Three small outcrops of the southwestern part of the ridge are grouped to constitute ore body no.1. Ore body no.2 of the northeastern section, which is considerably larger than no.1, also consists of a series of isolated outcrops. Both appear to be lenticular and bedded in structure.

Manganese is also found as float and in boulders throughout the ridge, especially on the slopes (SRIVASTAVU, 1961). Surface chip and pit samples have shown an average of 31 % Mn.

It is reported that there exists another manganese prospect between Wautiken and Fronwodoe, 40 miles north of Harper Harbour, west of the Ivorian border. The prospect is said to lie at a distance of about 18 miles from the motor road, passing through Nyaake, Webbo Chiefdom, and could be considered important because of its proximity to the coast. Samples from the area appear to resemble Mount Dorthrow ore macroscopically, and would seem to be better in quality and grade.

IVORY COAST

Manganese occurs at more than 40 localities of the Ivory Coast Republic (Fig. 19).

Boundoukou

In the northeastern district of *Boundoukou* near the Ghana border, manganese ore is found in *(1)* a north-northeast trending Birrimian horizon at Motiambo (20 % MnO), *(2)* a north-northeast trending range at Niniango (10 miles west of Boundoukou), *(3)* Djeré, Sapis, Koïssi, Ndawa, Assoufouma-Broumba (300,000 tons of tuffaceous schists and quartzites), and Sapli with *(4)* minor indices at Zéré, Sorobo, Siminimi and Koubo.

Manganese also occurs *(1)* northwest of Agboville, *(2)* north of Bouaflé and *(3)* in the southwestern corner of the country, forming the north-northeast trending belt of the *Tabou* district between Divo and Dabakala. A continuous zone extends between *Youkou*, *Grabo* and *Niépa*. Other occurrences are found near to Nekaouiné on the Tabou road, at *Noubaki* in red tuffaceous schists, and at *Baourobly*, on the Man-Duékoué road, in lenses in quartzites. At Nero, 17 miles from the coast, gondites are included in a quartzite and schist column, intruded by small basic stocks.

Grand Lahou

The *Grand Lahou* deposit, which is the principal producer, lies on the lagoons west of

Abidjan. The Upper Birrimian includes metabasic rocks and the manganese horizons of parametamorphic, grey-black schists and gondites, injected by greenstones (PECCIA-GALLETTO, 1960). The Lower Birrimian consists of sericite–chlorite schists and gneisses. The manganese zone exhibits a northeasterly trend and a steep dip.

Interstratified gondite lenses contain a maximum of 22% MnO. Pyrolusite stringers traverse the psilomelane above the water table. First silica and then garnets and diallage have been leached out. In the semi-altered schists, graphite, pyrite and rutile occur. Layered lenses of psilomelane and spessartite are connected by pyrolusite. The ore consists of scales of thin layers of psilomelane or manganite, and of argillaceous relics of bedded shales which disappear near the surface. Adjacent to the main manganese horizon, the lateritic shale contains more iron, with psilomelane encrusting quartzite and goethite nodules.

Korhogo area

The overall length of the manganese horizon, in the north-northeast trending Birrimian schists, is 90 miles. The northern branch of minor deposits begins at Diaouala on the Upper Volta frontier and continues for 25 miles. The southern branch begins at Pinion, west of Korhogo, and is interrupted along its total length of 30 miles only by one fault between Lopin and Lagnokaha. Indices are found between the two branches *(1)* at Nangolo (6 miles east-southeast of Diaouala), *(2)* northwest and southwest of *Kapérogo*, *(3)* southeast of Korokaha, at Zambo–Kalaha, and *(4)* east and south of Sindia, between Sédékaha and le Bou. A small deposit is situated at Lopin. The Birrimian schists may be divided into an eastern zone of leptynites and schistose orthoamphibolites and into a western zone of less metamorphic arkoses and greenstones with late intrusives.

The *Dassoumblé* deposit, 18 miles west-southwest of Korhogo, contains 3,300,000 tons of manganese in gondites and manganese oxide bearing garnetites. The ore consists of quartz veins and aggregates, partly oxidized garnet, and pyrolusite with some psilomelane and wad. Both MnO and SiO_2 content varies from 20% to 40%.

The Lagnokaha deposit is situated 20 miles southwest of Korhogo. The country rocks are underlain in the east by the brecciated leptynites of Mount Kafiplé, and in the west by the gabbro of Mount Sonloubakaha. Little gondite is seen, while quartz veinlets and pyrolusite–psilomelane stringers traverse the ore. One million tons of high silica manganese (30% MnO) is available here.

Ziémougoula

In the northwestern corner of the Ivory Coast, 30 miles north-northeast of Odienné, a group of manganese beds extends 17 miles, with their width exceeding 1 mile. The thickness of each bed ranges from 3 to 60 ft. Manganese grades vary from 58.05% to 16.09%, averaging 46%, with the silica content varying from 0.5% to 20%. Here geologists estimate a reserve of 3,000,000 tons.

GHANA

Nsuta, the world's principal producer of high quality manganese, is situated near the gold mining centre of Tarkwa. Manganese also occurs in southwestern Ghana, west of Takoradi Harbour and of the Butre River, south of Anibon, near the port of Axim. In Ashanti Province, manganese is found at the bauxite deposit of Yenahin and at Konongo in the Juaso district (Fig. 21, 54).

The Upper Birrimian of Nsuta includes carbonatic–andesitic greenschists, porphyritic basic lavas, altered tuffaceous sediments (above the manganese horizon), porphyroblastic biotite greenstones in a hornfelsic breccia, and quartz–sericite–graphite phyllites. Phyllites include only rare gondite lenses. The horizon of manganese phyllites and gondites is 1,500–2,000 ft. below the top of the Birrimian and in the upper part of the phyllite group. The manganese zone is 375–400 ft. thick.

The beds are isoclinally folded along the north-northwest trending, Upper Birrimian axis; dips are subvertical. Deposits occur in two folded beds: *(1) Asikuma* and Nyankumasi I and II, and *(2) Nsuta*. Between the two axes, a major syncline has a generally southerly pitch. Gondite and manganese phyllite are concordant in the folds, consisting of garnet, quartz and manganese oxides in various degrees of alteration. Bedded slabby gondites are more mineralized than the common crumbly variety. Rough bedding and strong banding are apparent.

Lenticular ore bodies and pipe veins are 1,000 ft. long and 100 ft. thick. Pyrolusite, psilomelane and manganite are believed to be the principal ore minerals; wad, gibbsite and variscite are rare. Manganese oxides show pseudomorphs after spessartite. Leached oxides have apparently coagulated in colloidal form in groundwater of variable composition, precipitating and migrating in several phases. Concentrations can be divided into three types: *(1)* elimination of impurities by leaching, *(2)* replacement of gondite, muscovite gondite and muscovite schists, and *(3)* fissure fillings. All three types consist of gamma MnO_2 and cryptomelane, often forming colloformous textures. Pyrolusite is abundant in fissures but rare in replacement ores. Traces of lithiophorite and goethite occur in both. Impurities include garnet, zircon, muscovite, quartz and amphibole. The central part of the lenses consists of crumbly black oxides. The rest is black and dull. The veins contain 50–55 % MnO, whereas the top 30 ft. of detrital ore contains only 40 % MnO. In the upper part of the pipe veins, the ore is hard stalactitic or mammillary.

Some quartz veins are spotted and mixed with manganese minerals. Manganese oxides also intersect clayey bands. Halloysite is disseminated in the ore bodies and concentrated on its borders or at the contact of quartz stringers. Shattered milky quartz was deposited later. Silica was removed during lateritization. Detrital ores of nodular-concretional rounded boulders are of little commercial value. Under these blocs, no workable ore is found.

UPPER VOLTA AND MALI

At *Tiéré* in Upper Volta, 80 miles northeast of Bobo Dioulasso, north trending phyllites are bordered in the east by sheared andesites. The principal subvertical gondite lens forms a virgate, and strikes east 1 ½ miles southeast of Tiéré, with a minor occurrence appearing in the southeast. Diagonal crush zones terminate the 1,600 ft. long lens, which attains its maximum width of 160 ft. in the east. The gondite contains less than 30 % Mn, but two narrow lenticular bodies include 40–45 % Mn. They attain lengths of 120 and 80 ft., respectively.

Three types of ore are found: *(1)* black ore, with an amorphous outer zone surrounding a metallic interior; *(2)* compact ore, with quartz stringers perpendicular to the direction of the manganese vein; *(3)* gritty ore near the wall.

The mineralized alignment extends to Bouéré (Houndé). Other occurrences include Sokoura (Ouo) and Kalye–Kadémé.

Ansongo, on the left bank of the Niger in Mali, is a halt of the Sahara trail linking Oran to Lagos. The principal deposit, *Tikanasité* Hill, includes a reserve of 7,000,000 tons, and is located 23 miles southeast of Ansongo (SERVANT, 1956). The two other deposits, Agaoula and Tondibi Hill, 16 miles south of the halt, represent 500,000 and 300,000 tons of manganese.

The conformous manganese horizon trends parallel to the Precambrian schists for more than 20 miles. Most of the ore is compact; however, concretional structures and inclusions of quartz are also found. Spessartite appears in the quartzites on the edges of probably syngenetic ores similar to those of Nsuta (Ghana). Barite is associated with psilomelane and pyrolusite in the compact zone, followed by a concretional and finally by a layered horizon.

NORTHEASTERN AFRICA

Egypt. The principal deposits are situated in the southwestern part of the Sinai peninsula, which is out of the scope of this book. They are located along a west-northwest trending arc roughly parallel to the Red Sea Graben. The pyrolusite, psilomelane, hematite and goethite ores of the largest and northernmost district, Abu Zenima, are tied to the replacement horizon dividing Lower Carboniferous limestones from sandstones. In the Gebel Mûsa district, manganese is found in various dykes and veins. On the Sherm el-Sheikh cliffs, breccia filling conglomerates contain some manganese.

The small Wadi Malik deposit is located near the Red Sea on latitude 26° N. A group of ramified, northwest striking, fracture filling veins traverses Miocene conglomerates and sandstones underlain by probably Precambrian diorite. Pyrolusite, psilomelane and hematite are believed to be primary. They have been altered and oxidized near the surface. The Gbel Alda vein is north of the Quena bend of the Nile, downstream from Aswan.

Sudan. Four manganese districts have been distinguished (Fig. 13):

(1) Halaib, in the northeastern corner of Sudan.

(2) Sinkat, 80 miles southwest of Port Sudan.

(3) Abu Samr-Allaikaleib, 100 miles south of Sinkat.

(4) Paloich-Wabuit, between the Blue and White Nile.

While the first three districts are located in the Red Sea Hills and the coastal strip, the fourth area is situated in the southeastern Sudan. The occurrences of *Halaib* are aligned in a 40 mile long belt trending southwest, 3–30 miles inland. Here, 3–6 ft. wide manganese and calcite veins strike north and east in sandstones, limestones and gypsum. Samples contain 23–53% Mn at Amar, 52% at Ankalidot and 29–33% at Adargab. Other occurrences include El Hobal, Eishumhai, Takamanyai and Metateib.

Near the Agwampt, 40 miles from *Sinkat*, a 3–12 ft. thick vein of manganese oxides fills in fractures of hornblende and chlorite schists. Manganese content varies from 47% to 55%. Another group of occurrences trends from Abu Samr east to Warreiba. At Abu

Samr, veins and lenses occur in garnet quartzites and gneisses intruded by granite. Similarly, at Tolik, the manganese lode, changing along the strike into a magnetite–spessartite–quartz vein, traverses magnetite gondite. At Allaikaleib, rhodonite, rhodocrosite and spessartite are found along the contacts of the vein, and at Wurreiba stringers of manganese oxide intersect rhodonite. Samples taken show 26–38% Mn at Abu Samr, 8–35% at Tolik, 16–43% at Allaikaleib, 10% at Boshikwan, and 33% at Wurreiba (KABESH and AFIA, 1961).

At Paloich and Wabuit several horizons of manganiferous iron ore are interbedded with Pliocene–Pleistocene clays of the Umm Ruwaba Series. The 260–272 ft. level of Paloich contains 18–25% Mn, but at Wabuit grades are lower.

In Ethiopia, manganese is abundant in the iron ores of the eastern sector of Mount Gedem, near Massawa. Since the lenses occur in travertines, they are believed to have been deposited from springs. Finally, between Adad and Salawel in northern Somalia, rhodonite rich lenses occur in magnetite gondite, while in the Bihen Gaha Pass wad is deposited from hot springs.

GABON

The *Moanda* deposits are among the biggest of the non-communist world. Other occurrences in the Gabonese Republic and in the Congo (Brazzaville) are similarly related to latest Precambrian, unfolded, Francevillian grits and shales. Such indices are found at *Okondja* (60 miles northeast of Moanda), 10 miles from this locality on the Okondja-Kedé road, and 18 miles and 25 miles from Okondja on the Ekalla road (Fig. 53). A similar deposit of manganese, dissolved and redeposited from Francevillian sandstones and shales, is found at the *Kanda* stream, a sub-tributary of the Ogoué. Deposits in the Lower Precambrian Ogoué System occur in Central Gabon (BAUD, 1956), east of *Kimongo*, in fissure fillings of jasperoids, and at Alémbé in graphitic schists. Manganiferous veins cut Precambrian rocks at the Ikoy, Offoubou and Ezalé streams.

Moanda

The Francevillian System, underlain by the Basement Complex, consists of: *(1)* the lower feldspathic sandstones (sometimes dolomitic), *(2)* the lower middle stage of shales, *(3)* the upper middle stage of feldspathic sandstones and jasperoids, and *(4)* the thick, upper conglomeratic sandstones. The Francevillian column is 1,300–2,000 ft. thick. The manganese horizon covers five plateaus (Fig. 68).

Middle Francevillian shales and graphitic schists underlie the principal, *Bangombé* plateau. Most of these rocks are covered by manganese bearing nodular pisolites. This plateau and Okouma were probably upfaulted from the surrounding lowlands. The ores form a kind of 'manganicrust' divided into:

(1) Surface soils with occasional pisolites; width ½–1 ft.

(2) Powdery soil including manganese bearing pisolites of 1–2 inches; width 10–12 ft. The outer zone of the pisolites is stilpnosiderite, with the core consisting of psilomelane and/or goethite.

(3) A transitional zone of richer pisolites, boulders, manganiferous nodules and slabs included in clays; average width 3–5 ft.

(4) The mineralized horizon of 15 ft. subdivided into: *(a)* manganiferous slabs and nodules in clay, *(b)* subhorizontal slabs and semi-bedded clays, and *(c)* a cementation zone of hard polianite (2 inches–1 ft.).

Values attain:

30–45% Mn
3–4% SiO_2
7% Al_2O_3
0.04–0.13% P
0.03–0.09% S

The ore may consist of layered slabs, concretion slabs, cellular or concretional boulders of manganese oxides and quartz, semi-lateritic agglomerates and nodules. Concretional growths are rich, and fine grained layered slabs are frequent. Psilomelane is the principal

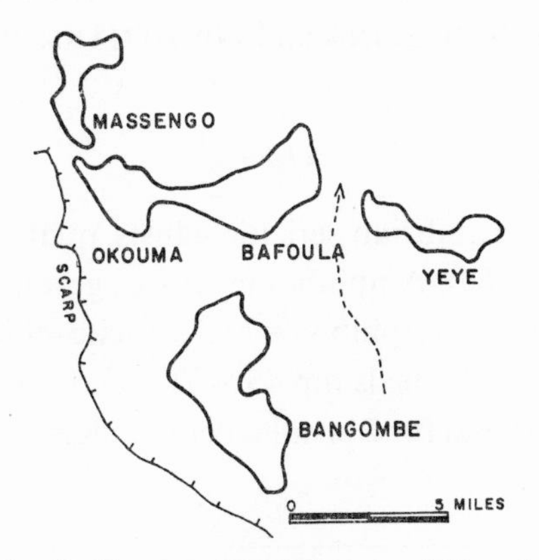

Fig. 68. The manganese plateaus of Moanda.

mineral of pisolites, slabs consisting mostly of manganite. In the concretional growths, psilomelane and manganite are associated with polianite. The lowest subdivision consists of polianite accompanied by diallage, but the bed-rock is barren. The mineralized horizon is concordant, i.e., horizontal as the underlying Francevillian, but its upper level undulates as does the present land surface. The 20 square mile deposit appears to be eluvial. The bed-rock has been partially transformed into layered manganese ores. The concretional growths were undoubtedly deposited in nodular aggregates or concentric shells. The Middle and Upper Francevillian schists and sandstones contain only 0.3%–0.01% of Mn. The ores appear to have been derived from eroded strata or from circulating solutions.

The *Okouma* plateau is located 3 miles north of Bangombé. Of the 10 square mile surface half is workable. Okouma appears to have been upfaulted in comparison with its eastern extension, the Bafoula plateau. The eluvia of the Okouma are similar to the Bangombé deposit. The third level of Okouma is 3 ft. thick and sometimes workable. An intermediate, fourth horizontal layer of slabs and vacuolar ore, followed by slates, overlies the zone of cementation polianite. The sandstones of the Bafoula plateau grade into the ore zone of relic quartz and layered manganese oxides.

Only the southern ½ square mile of the northern *Massengo* plateau, another compartment of the block faulting, is mineralized. The ores are more siliceous here and they are underlain by graphite bearing schists. The occurrences of the rest of the plateau tend to prove that the manganese might have been reconcentrated from the sandstone. Manganese and iron bearing boulders have been eroded from the sandstones of the *Yéyé–Maila* plateau (east of Bafoula); they appear to have been formed by percolating waters.

CONGO

Indices of manganese are frequently found in the whole country, particularly in the Lower Congo. The principal deposits of Katanga and northern Zambia, however, are situated in the larger aureole of the Copperbelt, extending from the sources of the Zambezi to Lake Bangwelu. *Kisenge*, the principal deposit, is located in southern Katanga near the Angola border (Fig. 27, 51). Kasekelesa is 60 miles west of the copper centre of Kolwezi; Kiale is near by. The Mwaba Fita, Kipupa and Buyofwe cluster of occurrences is located 90 miles north of Jadotville, on the same latitude as Fort Rosebery. Other small occurrences are also known in southern Katanga (SCHUILING and GROSEMANS, 1956).

Kisenge

The general trend of early Precambrian sericitic schists, quartzites and the included manganese deposits is east-west (Fig. 69). Amphibolite schists, granites and gabbros outcrop in the south, while amphibole schists, amphiboles and sericite–chlorite schists outcrop in the north (MARCHANDISE, 1958). The beds dip 45°–70° south. A series of manganese ore and gondite lenses trend east-northeast for 2½ miles between Kisenge, Kamata and Kapolo. The

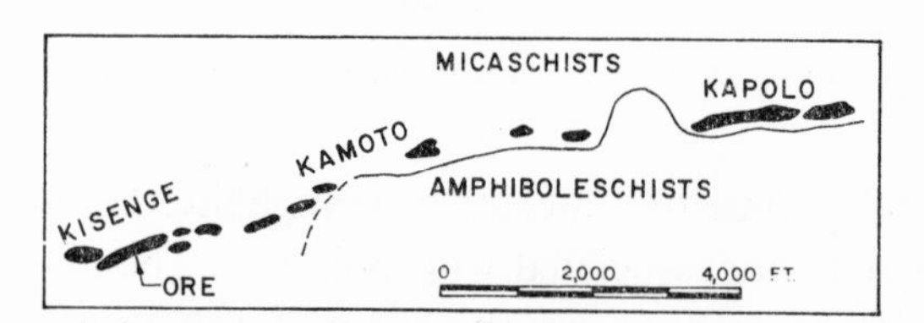

Fig. 69. The manganese deposit of Kisenge.

width of the lenses rarely exceeds 300 ft. Some of them form swarms, with the distance between clusters increasing towards the east. On the crest of Kisenge Hill in the south, a large bed of rich manganese ore is followed by alternating beds of poor and rich ore and barren schists, completed by a thin lens of rich ore. Lithological variations are essentially functions of the proportions of garnet, graphite and mica in the schists surrounding the ore bodies. Muscovite–biotite–tremolite and garnet quartzites constitute the footwall, changing progressively into an andalusite–sillimanite–kyanite–tremolite–garnet schist. Manganese carbonates have been discovered in depth and may have constituted part of the original ore.

The original garnet is a calcareous spessartite containing subsidiary portions of the almandite molecule. These portions contain 35% Mn instead of the theoretical 43%, and 2% calcium. Some titanium substitutes for iron. Garnet crystals have been fractured into sectors. Always widening cracks and planes of twinning have been filled in by black manganese oxides. In the final stage, siliceous screens bear witness to the original spessartite in oxide filled polyhedra. Manganese grades are an inverse function of the silica content of poorer ores. Manganese is believed to have been liberated by laterite-type tropical weathering which eliminated silica. The remaining oxides have dissociated, and coagulating manganese concretions have been compacted by diagenetic changes. Subsurface concentration of manganese is followed by a gradual decrease of values.

The principal ore mineral is cryptomelane, accompanied by pyrolusite and lithiophorite. Polianite is believed to occur in the upper reaches. Much of the bluish grey cryptomelane ore exhibits zonal structures, and is divided by thin layers of impurities from which radiate ore nodules. Pyrolusite layers show a more metallic lustre. Textures can be subdivided into *(1)* honeycombs of polyhedral pseudomorphs embayed by cryptomelane infillings or veinlets, and *(2)* colloform or distinctly zonal ore. Average Mn content attains 50%, Fe 4.5%, SiO_2 plus Al_2O_3 9.5%, and P 0.18%.

Predominant cryptomelane contains 56% Mn ($MnO_2/MnO = 12$), 3% FeO, 2% SiO_2, 2% Al_2O_3, 1% SiO_2, 1% K_2O and 0.4% BaO. Potash values are subnormal. Most of the cryptocrystalline cryptomelane consists of acicules, embayed in an isotropic groundmass, which float or aggregate in bunches or stringers. Pyrolusite constitutes only 5–10% of the ore. The mineral contains 92% MnO_2, 1% MnO, 0.4% $MnSiO_3$, and 1.5% Fe_2O_3. Anisotropic pyrolusite crystals terminate in rounded extremities. Some form aureoles or thin layers around or in cryptomelane. The third mineral, lithiophorite (?), does not contain lithium, just as in the Brazilian Sierra do Navio deposit. Prismatic crystals are lighter in colour, less anisotropic, and softer than pyrolusite.

Kasekelesa

Kibarian (Middle Precambrian) light coloured and white quartzites alternate with closely bedded rubefied schists. The large, late Precambrian Kundelungu conglomerate covers the northern sector as well as a spot further south. Post-Kundelungu faults trend north-south, as does the mineralized zone which measures 2 by ½ miles. Veins, soft and lateritic eluvia and manganiferous durcicrusts constitute the deposit. In the large conglomerate, vertical and horizontal veins of manganese ore extend to a depth of 60 ft. The shape of lenses is more irregular. Pyrolusite is the principal ore mineral. Ore grades vary between 50% and 55%, with 0.6%–2.5% Pb appearing in some parts of the deposit. Vanadinite traces are also present. The pyrolusite contains 71% MnO_2, 9% MnO, 2.5%–6% SiO_2, 1.8% BaO, 1%–2% Fe_2O_3, and 0.05% P_2O_5.

ZAMBIA

The principal deposits of high metallurgical grade manganese are located along the north-east trending alignment of the Luapula (Congo) Rift Valley, in the *Fort Rosebery* district, between the Bangwelu marshes and the Congolese border. The deposits exhibit a northerly strike and a subvertical dip. The average width of ore bodies is 2–3 ft. (FOCKEMA and AUSTIN, 1956). Ore is massive at Lukwina and in southern Kitakwe. The rest of the Kitakwe ore body consists of boulders with wad covering breccia. The Bahati and Mulozi bodies also contain large detriti. The host rocks are Precambrian, pink granite and intrusive quartz porphyries. The ore bodies consist of psilomelane, hausmannite and manganite relics, terminating in hematite lenses. The deposits fill in fissures caused by the peripheral effects of Rift faulting, with the grade diminishing downwards. The small deposits of Mashimba, Mwanasassa and Cirpili appear to be similar.

The small Chiwefwe deposit is in the extension of the Copperbelt. Sheet-like psilomelane ore rubble, with some pyrolusite and manganite, overlie early Precambrian and Muva System schists and quartzites. Manganese has been leached out of the country rocks and redeposited by percolating water. The Musofu boulder bed is situated 30 miles northwest of Chiwefwe. The Kampumba deposits, southeast of Broken Hill, may be divided into compact and rubble ores. They have been derived from manganiferous dolomite, the iron ore appearing to be syngenetic. The small deposit of Chowa is located north of Broken Hill.

THE REST OF MIDDLE AFRICA

Surface indices are found along the Fort Archambaud–Fort Lamy Road (in the Chad) and at Ta-Yoro, northwest of Goutekeri in the diamond district of Berberati (Central African Republic). At Bolo near Bouar, also in former Ubangi–Shari, manganese occurs at the contact of Precambrian rocks and intrusives. In the folded Precambrian rocks of the Mayombe, manganese oxides precipitate in boulders and fissures in the Bamba Mountains, at Romano near Kimongo, and at Mémé near Londela–Kayes. Furthermore, manganiferous laterites occur 11 miles from Kimongo.

The manganese belt of *Angola* extends 60 miles from Galungo Alto to Pungo Andongo in the southeast. The district is located 200 miles east of Luanda. At Quicuinhe, a pinching and swelling body is sub-concordant with folded Precambrian or Paleozoic felspathic grits. There the subvertical vein strikes west-northwest, extending more than 3,000 ft. Near the ore body the country rocks have been crushed and veined by rhodonite. Offshoots and stringers of psilomelane penetrate into the wall rocks. The ore contains up to 40% silica. Similarly, the ore shoots of Quitota strike west-northwest, with manganese also impregnating granite. Other ore bodies occur along the strike between the two mines west of Quicuinhe and near Galungo Alto.

Carbonatite complexes, e.g., Mrima (600,000 tons of ore at 20% Mn) and Kiwara (Kenya), Chilwa Island and Tundulu (Malawi), include manganese carbonate zones (see 'Columbium' chapter). Recent hot spring impregnations are found near Gilgil, north of Nairobi, and residual oxides at Chonyi, north of Mombassa.

In Tanganyika, manganese ores are interbedded with volcanic ash at Matamba; they concentrate at the surface of the ironstones of Manyoro.

Low grade manganese ores have been found on the Alto Chimezi, near Manica, and at Chundize and Benga of the Tete district in Mozambique.

The gondites of the manganiferous zone of southern Madagascar occur in the Precambrian 'Graphite System'. The principal occurrences include Bepeha, Latona, Soakibany, Begorago, Befamata, Ankara and Ambatomainty, Besosa and Befaraha. At Masokoamena, north of Tananarivo, veins traverse laterite. Similar deposits are found at Mahamavo and Antanandava. At Saronara, manganese oxides fill in fissures.

SOUTH AFRICA

The principal deposit is the so-called Kalahari field at *Black Rock*, on the Bechuanaland border. The *Postmasburg* deposit is also situated in Griqualand, the northern part of Cape Province. Minor deposits are found in the Transvaal dolomites, in various quartzites, and in other small occurrences (Fig. 46, 50).

'*Kalahari*' *Field*

The *Black Rock–Smarlt* deposits constitute one of the major manganese reserves of the non-communist world. While the eastern zone contains more than 70,000,000 tons of ore, the resources of the sand-covered western zone are measured in hundreds of million tons. Between *Mamatwan* and *Lehating* and beyond, under the sand and the pre-Karroo surface, ironstone beds strike north-northwest and dip 25°W. A similar or thrusted eastern zone of banded ironstones has been discovered at Kipling, Langdon, and other localities. In the main belt, quartzites have been observed above the ironstone horizon. In the eastern zone, lavas have been found under this level. As in West Africa, pre-Karroo, acid and mafic intrusives are wedged in the metasediments. The thickness of the overlying, post-glacial Karroo sediments, and of the top Kalahari sands, ranges from 12 to more than 400 ft. The structural relationship between the overthrusted Ongeluk lava, the Gamagara quartzite, and the upper or lower Griquatown manganese-ironstones is not yet certain (Fig. 70).

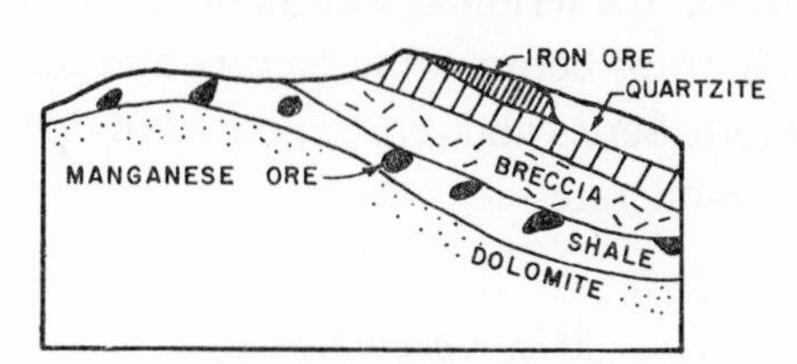

Fig. 70. Occurrences of ore in Griqualand.

The prospected length of two to three manganese horizons in the banded ironstones is over 40 miles. The beds are 30–40 ft. thick. Ore bodies are concordant and co-laminated with the banded ironstones. In the lenses, layers of manganese oxides and calcium plus manganese carbonate alternate.

According to BOARDMAN (1961), the manganese horizon was derived during periods of erosion and epigenetic concentration (pre-Waterberg) from Upper Griquatown manganiferous limestones and dolomites containing 10–30% Mn. According to DE VILLIERS (1956), asbestos was derived from ironstones after the Lower Griquatown stage. Folding and thrusting were followed by metasomatic processes during this replacement period. Manganiferous solutions derived from the allochthonous Dolomite Series have enriched the asbestos. Ore minerals include braunite, ferrian braunite and some psilomelane, jacobsite and hausmannite. Bixbyite is abundant in the central section of the western belt. The manganese content of 50% diminishes westwards to 42% but desirable calcite is added. Phosphorus content is low and silica moderate; iron content is 10–25%.

Replacement and hydrothermal manganese is found 18 miles west of the Black Reef, in the Korannaberge at Groenwater, Blauwkrantz and Neiskop.

Postmasburg

More is known about the geology of the smaller deposit located 100 miles south of the Kalahari field. The stratigraphic column consists of the limestones and dolomites of the Dolomite Series and lavas, banded ironstones, cherts, jaspers and quartzites of the three Griquatown Stages of the Pretoria Series, Transvaal System. The basal conglomerate of the Loskop System was deposited in depressions of the folded Transvaal rocks. Quartzites and aluminous shales constitute the Loskop System in the east (Gamagara formation) and in the west (Lower Matsap). Following the Waterberg epoch, the strata were overthrusted and brecciated. The banded ironstones of the Lower Griquatown Stage form the Blinkklip breccia, whereas blocks of chert were derived from the Dolomite Series. Carbonatic rocks of this Series are shattered, forming tectonic 'dolinas' and fissures, and facilitating the concentration of manganese.

The *Klipfontein* Hills form the eastern belt between Postmasburg and Sishen. The western *Gamagararand* is 40 miles long. Minor occurrences are located south and west of Postmasburg at Rooinekke and Aucampsrust. Ferruginous ores of the western belt between Mokanning and Lace's Goat consist of bixbyite with some manganese diaspore and ephesite in *(1)* 'dolina' accumulations of the Dolomite Series, and *(2)* replacement bodies in the Gamagara shale.

In the eastern belt, and also in the terminal sectors of the western belt, ore occurs in gulfs of the siliceous breccia and in traps between the breccia and the dolomite. Manganese has been derived from the Dolomite Series and reprecipitated in open spaces. Phosphorus content is low, and reserves are moderate.

Minor occurrences

Transvaal dolomites contain 3% MnO. Psilomelane-type ores precipitate mainly in dry soils of large dolinae. Some deposits are 20 ft. thick. The principal ones are situated between Krugersdorp and Ventersdorp. Other occurrences reach into Bechuanaland and into the area between the Thabazimbi iron deposit and Middelwit in northern Transvaal. Manganese content is high and the ore is used by the uranium industry. Late pyrolusite-type ore is associated with an iron ore vein cutting dolomite and chert at Wolvekrans, Krugersdorp district. Wad occurs frequently (e.g. at Graskop).

Fissure veins of the Loskop and Waterberg Systems contain manganese at Warmbad, Nylstrom, Soutpansberg and De Wagendrift in the northern Transvaal. Manganese occurs in the quartzites of Bronkhorstfontein and Weenen. In the Table Mountain System of Cape Province, veins and impregnations contain pyrolusite and psilomelane in Cedarberge, Citrustal and Clanwilliam, as well as at Hout and Kogel Bay and in the Franschhoek Mountains. In Namaqualand, manganese–iron ores similar to those of West Africa are found in the Kaaien Series at Gams Portion 1 farm. Lenses covering a distance of 5 miles contain 30% MnO. Small deposits occur on Derdepoort, Rooisloot (Transvaal), Caledon south of Amsterdam, Beechwood, Buffelsdrift and elsewhere.

THE REST OF SOUTHERN AFRICA

The banded ironstone of the Dan 2 Mine, near Que Que in Southern Rhodesia, averages

29% Mn and 25% Fe. In Bangwaketsee country, e.g., at Kgwakgwe in southeastern *Bechuanaland*, mineralization is believed to be related to chert breccias of overlying shales of the Black Reef Series. On Ootsi Mountain in Bamalete country pods and bands of botryoidal manganese oxides occur in shales of the same horizon. South of Ramoutsa, manganese impregnates and replaces sandstone beds in quartzites of the Pretoria Series, although only to a depth of 65 ft.

Otjosundu is located 100 miles north-northeast of Windhoek, the capital of *Southwest Africa*. The Precambrian schists and gneisses of the Damara System have been folded into part of a larger synclinorium and intruded by the Salem granite. The two manganese horizons are divided by hematite itabirites and biotite schists which include manganese lenses located on drag folds. The lenses may be as long as 5,000 ft., and average a width of 15 ft. In them braunite, jacobsite, hausmannite, bixbyite and hollandite are layered. Garnet, quartz and barium feldspar occur at the contact and in the lenses. The core, and sometimes the hanging wall of the ore bodies as well, shows higher values. Conversely, iron values decrease upwards.

SAINT HELENA

In Prosperous Bay (in the northeastern part of the island) manganese concentrates along the perimeter of the Lower Complex tuffs, altered to argillite. Coalescing veins traverse the tuff and have been redeposited in boulders of wad and in 2 ft. wide seams in sand, but little manganese is found in adjacent lavas.

BAUXITE

Lateritic soils cover large areas of Africa: a belt extending 200 miles inland from the Guinea coast, a 300 mile long strip in the Cameroon, a broad zone of the Gabon and of the southern Cameroon, other districts of the Congo Republic, the northeastern Congo, and parts of Kenya and Madagascar. However, the belt of lateritic bauxites is more restricted. Indeed, tropical bauxites have developed only between latitudes 6° and 11° north in *Guinea*, Ghana, the Cameroon, Guyana, the Caribbean, Malaya and other countries. The principal belt extends 1,200 miles from Portuguese Guinea to Togo, 80–200 miles from the sea. However, no equivalent deposits are known south of the equator, the Congo and Malawi occurrences being of only minor importance (Fig. 2).

The genesis of bauxites

Given sufficient time, proper temperature, hygroscopic grade, salinity of water and paucity of vegetation, aluminous laterites form from any pelitic or sub-psammitic rock on flat lying, morphologically conditioned surfaces. The lateritic capping then protects the erosion surfaces, but the changing of laterites to bauxite is a secondary phenomenon: because of the differential solubility of aluminium and iron hydroxides, regressive erosion dissolves the ferruginous rim of lateritic basins. Lateritization of bauxite is controlled by:

(1) the type of the parent rock; *(2)* the facies of meteoric water: pH, relative concentra-

tion of ions; *(3)* their dynamics: drainage, suspension; *(4)* climate: temperature, variations and extremes of seasons.

Since subcolloïdal bauxites are more plastic than limonite, they weld better into the morphology of the undulating land surface of Guinea. Indeed, ferruginous laterites tend to dehydrate through the surface by insolation, with the cracks forming polygons characteristic for each area. Facilitated by isostatic movements or faulting, erosion subsequently destroys these forms. Hard and soft layers of laterite represent various stages of desication and hydration. When percolation intensifies, iron migrates with the water. Since iron is in the soft layer in the dry season, the iron content of the hydrostatic table decreases at its end. However, the hydrostatic table does not considerably affect the behaviour of iron on the Konkouré dolerites. The resilification of surface layers is a late phenomenon.

The structural types of bauxite include:

(1) Massive, occurring rarely in the Bassari Hills and in the Bové of the Cogon.

(2) Nodular or pseudo-conglomeratic ore, a frequent type in laterites and bauxites which includes the dimensional sub-facies: *(a)* meganodular (> 2 inches); *(b)* nodular (½–2 inches); *(c)* pisolitic (½–1 inch); *(d)* oolithic (< 2 mm).

(3) Brecciated, developing in rearranged surface bauxites.

Vacuolar, schistoid and arenaceous laterites contain little alumina. Colour and specific gravity are also functions of the gibbsite/goethite ratio.

Most of the laterite minerals are crystalline, but they have colloidal dimensions. During the earliest alteration of the feldspar, when alumina is still insoluble, gibbsite begins to crystallize by hydrolysis in a pH of 8–9, although at Souapiti gibbsite also crystallizes at a pH of 5–6. Its solution and crystallization have been attributed to the system of two seasons (BONIFAS, 1959). Gibbsite also develops from kaolinite. Boehmite has been observed in surface crusts, e.g., at Koulouba in Mali, and with diaspore in Ghana where zircon, rutile, ilmenite and gold also occur occasionally.

GUINEA

The three principal Guinean bauxite deposits form a structural-morphologic alignment trending south of southwest from Portuguese Guinea to Bové (north of Boké), to the Kon-

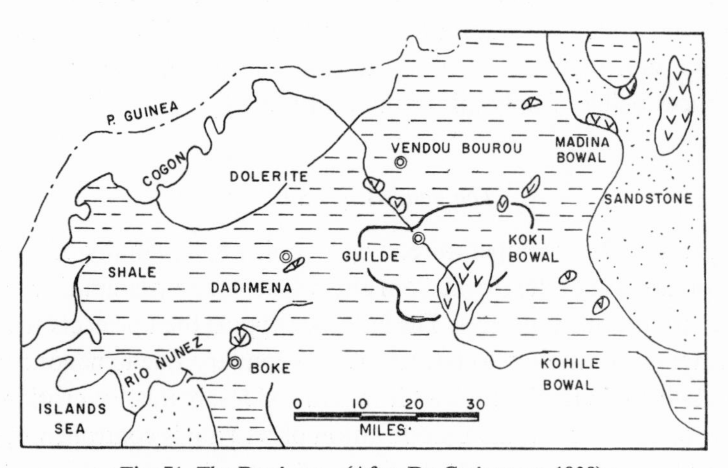

Fig. 71. The Bové area. (After DE CHÉTELAT, 1938).

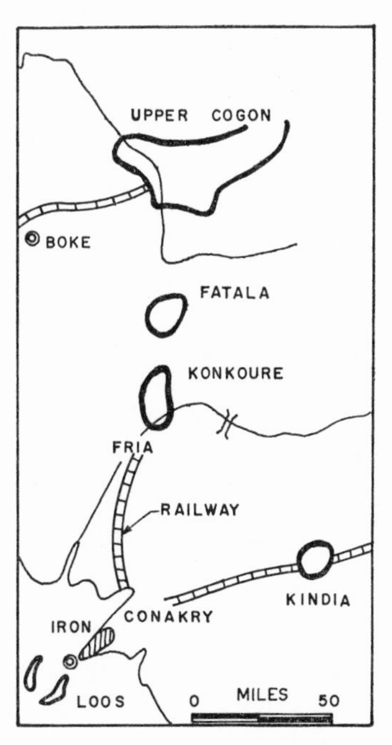

Fig. 72. The bauxite deposits of western Guinea.

kouré near Fria, and to Kindia (Fig. 71, 72). The Tougué–Dabola group of deposits is located on the eastern rim of the Fouta Djalon Mountains, 300 miles from the capital. Finally, the Loos Islands complex lies off the coast of Conakry.

Morphology of the deposits

In large areas of Western Guinea, horizontal Silurian shales, micaschists and subordinate quartzites cover folded Precambrian metamorphics. In a 30 miles wide band, siliceous late Silurian sandstones and quartzites outcrop along the coast and east of the longitude of Conakry. On the border of Senegal and Portuguese Guinea a triangle is occupied by early Silurian red sandstones and conglomerates. Both smaller and larger basic bodies are scattered largely along a north-northeasterly trend. Complementing this, another important tectonic trend strikes N 30°W, with the conjugated systems forming an oblique checkerboard. Cyclical isostatic movements appear to have uplifted the hinterland along a line connecting the southern bank of the Cogon River with tributaries of the Fatala and Konkouré. Subsequently, accelerated erosion attacked the lateritic mantle. A majority of lateritic bauxites form on sub-arid raised depressions which are called bowal by the Peulh or Fulani people (plural: *bové*). Slopes of more than 8° are, in fact, detrimental to the formation of West African bauxites. Several surfaces have been distinguished:

(1) The isolated high plateaus surrounding the Fouta Djalon chain are only drained by evaporation.

(2) The 200 ft. thick bauxitic and ferruginous laterites of the Bassari country in the north have been derived from folded Precambrian schists and quartzites. The bové are intensely drained, even developing caves. Indeed, the destruction of this level has advanced considerably.

(3) The formation of bauxite reached its maximum in the undulating intermediate surface that covers tabular shales. Small lakes form depressions, but late valleys are V shaped and vegetation is scarce. The bové of the upper Cogon, Féfiné and Santa contain the largest deposits.

(4) The limonitic lower horizon of *(1)* the coast and *(2)* lying between the Fouta Djalon and Bassari Hills has been derived from folded rocks; it has been covered by sand.

Three types of crust have been distinguished: lateritic bauxites, laterites and limonitic laterites.

(1) Lateritic bauxites are generally pink–white, massive, rarely oolithic or pisolitic, and contain less than 20% Fe. They are typically surrounded by other laterites, e.g., on the Cogon and Bassari.

(2) Polychrome bauxitic laterites contain more than 20% Fe, exhibiting nodular, rarely pisolitic or brecciated textures. They mainly cover the bové of the Cogon–Féfiné drainage. Both varieties, which contain more than 50% Al, have been derived from argillaceous schists and shales.

(3) Sub-conglomeratic, ferruginous bauxites include light coloured bauxite nodules in a lateritic ground mass, or vice-versa. They cover large areas between Bové, and Badiar and in Bassari country; the Fe/Al ratio oscillates around one. This bauxite has been derived from sandy shales, sandstones and even placers.

Gibbsite and amorphous alumogels occur in most Guinean bauxites, but the silica of halloysite and kaolinite has largely been leached.

Bové

The centre of the district is located 55 miles northeast of *Boké* and 80 miles from the coastline. The plateaus rise 900–1,000 ft., i.e., 600–750 ft. above the valley of the upper Cogon (HARDER, 1952). The laterite is 40–50 ft. thick. Diabase dykes and sills which intrude the horizontal sandstones, siltstones and shales are particularly abundant in the bauxite area. The fine grained diabase contains andesine–labradorite and augite. The distribution of iron and alumina is irregular. A typical column consists of ferruginous durcicrust, bauxite, ferruginous laterite, and from several inches to several ft. of kaolinic clay or lithomarge, which contains 27–37% Al. When the clay is thick, iron content decreases (to 9%), and silica content increases (to 45%) with depth, but alumina content varies less. The bauxite mainly consists of gibbsite, hematite and goethite. By insolation, up to 5% of boehmite develops near the surface. Finally, exoquartz is present in the upper kaolinite in the lower layers.

A sample taken from the left bank of the Cogon River, northeast of Boké, shows subschistose fragments completely replaced by fine grained gibbsite, accompanied by some iron and rare fragments of zircon. Second generation gibbsite also occurs in the ferruginous groundmass. This formation is believed to have been lateritized from eluvia lying on horizontal shales. The sample contains 58.2% Al_2O_3, 4.2% SiO_2, 7.2% Fe_2O_3, 2.5% TiO_2, and 28% H_2O. Another sample from Sinthiouiou, near Bové, carries 58.3% Al_2O_3, 7% SiO_2, 5.4% Fe_2O_3, 0.6% TiO_2, and 28% H_2O. Considerable areas contain more than 50% Al_2O_3.

Southern Guinea

The *Fatala* deposits, halfway between Fria and Bové, cover a 40 mile long area, and have a width varying from 7 to 17 miles. The deposit contains 300 million tons of bauxite averaging more than 45% Al_2O_3, although large sectors contain over 50% Al_2O_3. Lateritic bauxites outcrop north-northeast of Boffa (north of Conakry) near Bokoro, Barenja, Bassaya etc. (DE CHÉTELAT, 1938). Two textural types have been distinguished: *(1)* massive subporous bauxites, forming boulders at Bassaya, and *(2)* banded bauxites derived from schistose rocks.

Fine grained gibbsite fills rare vugs in amorphous compact samples from Bassoya and Farenja. Relic bed joints, covered by ferruginous solutions and gibbsite, include small cavities filled in by microscopic gibbsite. This bauxite appears to have been derived from the grey argillaceous schists of Toromélé. An analysis gives 62.3% Al_2O_3, 5.1% SiO_2, 1.1% Fe_2O_3 and 30% water. Bauxites have been formed in situ at Filu-Foré on the Fatala River.

The *Kindia* deposit includes 100 million tons of bauxite containing 40% Al_2O_3.

Among several deposits of the area of the Badi-*Konkouré* confluence, Fria contains 250 million tons of bauxite with more than 40% Al_2O_3. At *Souapiti*, on the southwestern end of the Fria deposit, horizontal Silurian sandstones have been injected by dolerite sills ranging in grain size from gabbro to basalt. Amphibole quartzites and hornfelses have developed at the contacts. The alternating sedimentary sub-volcanic column has been intensely lateritized, affecting dolerites and shales the most. More rain falls in the two seasons than in the Bové, totalling 80 inches. Evaporation averages 40 inches per annum, with day temperatures varying between 22°C and 32°C. The alteration of dolerite to kaolinite is rapid, and gibbsite concentrates in certain cases near the surface.

Upper Guinea

At the sources of the Gambia, south of the Senegalese border, lateritic bauxites are surrounded by ferruginous laterites in the area of the *Bassaris*. The bauxites have been derived from Precambrian schists. Generally the rock is compact, nodular and breccialike. In a sample from the Guingan bowal, gibbsite replaces the schist fragments. Similar bauxites occur at Akoul, and contain 70.1% Al_2O_3, 1.4% SiO_2, 2% TiO_2, 3.4% Fe_2O_3, and 23% H_2O.

In Central Guinea, bauxites and laterites cover the eastern slopes of the *Fouta Djalon* which extend from the Senegal-Mali border 100 miles south towards the railway at *Dabola*, and then spread 80 miles in an east–westerly direction towards the Bafing River. A line connecting the towns of Mali and Labé represents the western limit. Horizontal Cambrian–Silurian sandstones cover Precambrian granites, which include relics of schist and gneiss. Dolerite sheets and dykes have been injected into this sequence. The peneplains rise 600–1,000 ft. from the valleys. Lateritization is most intense on the plateaus (high bové), which are covered by grass in the rainy season and then barren following burning at the beginning of the dry period. Walls of these flat hillocks attain 30–50 ft. in height. Boulders of this formation are included in the basin laterites of larger valleys which, however, generally contain less alumina. The parent rock is considered to have been dolerite, while shales and psammites occurred less frequently.

The laterites are heterogeneous with intermingled bauxitic and limonitic patches. Samples collected *(1)* between the Sitaouma and Koya, and the Missira and Sobori, contain 58–59% Al_2O_3, 12–13% Fe_2O_3 and 2% SiO_2, *(2)* between the Diavoya and Loufa 58–63% Al_2O_3, 2–6% Fe_2O_3 and 4% SiO_2, *(3)* 2½ miles north of Kouné 55–67% Al_2O_3, 1–14% Fe_2O_3 and 2–9% SiO_2. Grades reported from samples originating in the Horé bowal between Sobori and Ndiré Yangueya, Lallabara, between Lallabara and Lagui, and from the bowal northeast of Tougoué, oscillate between 63 and 71% Al_2O_3, 1 and 3.5% Fe_2O_3, and 2.5–4% SiO_2.

The principal deposits of dolerite derived bauxites of the *Dabola* area include Dabola, Siffray and Bissikrima.

In northeastern Guinea, between *Siguiri*, Fatoya and Bougourou, the basin of the Niger is underlain by micaschists intersected by veins of auriferous quartz (LACROIX, 1914). Their lateritization is more irregular, progressing both along and across planes of anisotropy. The silicates are kaolinized and finally kaolin is oxidized. Banded and pisolitic samples of Fatoya and pisolites of Siguiri contain: 40, 46 and 51% Al_2O_3; 4.4, 1.2, and 0.7% SiO_2, plus 13, 6 and 2% exoquartz; 58, 32 and 18% Fe_2O_3; 0.2, 0.7 and1% TiO_2, and 21–25% water.

In the Bô–Oulé, Tamakaya and Kambarégna Mountains, and on the western slopes of Tétikan Kourou, remnants of the high, lateritic peneplain of the *Upper Niger* have been preserved on the western edge of the Niandan Banié chain north of Korossa (Seguri district, CHERMETTE, 1946). Télikan is located 20 miles, and Kambarégna 40 miles, from Kouroussa, which is situated 370 miles from Conakry by rail. The laterite includes a considerable amount of pisolitic bauxite. Samples from Télikan and Kambarégna contain 65.1 and 71.8% Al_2O_3, 3.1 and 0.9% SiO_2, 4.3 and 0.4% Fe_2O_3, 4 and 3.2% TiO_2, and 23.6 and 23.5% water.

Los archipelago

Kassa and *Tamara* Island, connected in the south by islets, lie off the coast at Conakry. They are in the direct extension of the 30 mile long, southwest oriented, Kakoulima–Ka-

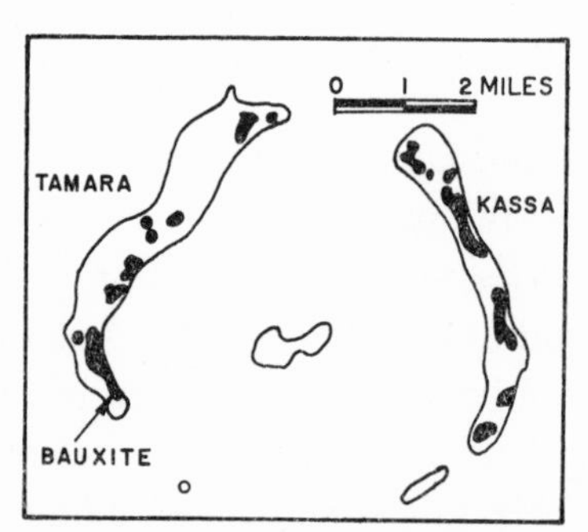

Fig. 73. Map of Los Archipelago.

loum peridotite–gabbro sill which is intrusive into horizontal Silurian sandstones (MILLOT and DARS, 1959). The syenite ring of Los (Loos) has a diameter of five miles, with Roume Island, in the centre, consisting of låvenite syenite. Windward and coastal slopes have not been lateritized, nor have the islets been swept by marine breezes. However, the inner coast has been, especially on Tamara. Lateritization attains its maximum in sub-horizontal, plant-covered depressions. The main areas (Fig. 73) include *(1)* a patch inland of Pointe Lajeune and Plate, on Kassa, *(2)* the whole crest of this island extending 2½ miles from Mangue to Kouromandia, *(3)* smaller patches in the south, and *(4)* at Pointe de l'Aréthuse on Tamara, *(5)* in the middle of this island, and *(6)* its southern rim as far as Topsail. Variations of composition do not affect the process in any way (LACROIX, 1914). One can distinguish three zones: *(a)* 2–4 inches of porous semi-altered rock, *(b)* 2–30 ft. of concretional bauxite, and *(c)* ferruginous bauxite crusts of 3–10 ft. The average thickness of bauxite varies from 30 to 45 ft.

The lateritic layer, resting on the alkaline rocks, is porous because of the alteration of the feldspathoids. It averages 11 ft. of thickness. While gibbsitization is non-oriented, 'centripetal', ferruginous solutions travel along joints and anisotropies too. Aluminium and iron exhibit a different behaviour in the early stages of alteration, but later their migration coincides (BONIFAS, 1959), with the exception of surface areas. The upper layers are massive and polychrome because of the banded distribution of iron (the latter also concentrates at the surface). At Kassa, the changes of composition between the alkaline syenite and bauxite can be illustrated thus: Al_2O_3 22–51%, Fe oxides 4–7%, SiO_2 57–0.4%, TiO_2 0.3–0.9%, alkalies and magnesia 16–1.6%, and water 1–33%. However, the proportion of alumina in the groundmass is considerably greater. The bauxite proper consists of cavernous aggregates and stringers of gibbsite, accompanied by hematite and scarce anatase. The crust also carries kaolinite and goethite. The composition of porous rock, bauxite, and durcicrust compares as follows: Al_2O_3: 56, 58 and 44%; Fe oxides: 8.1, 6.9, and 23%; SiO_2: 3.3, 0.4 and 5.4%; TiO_2: 1.7, 2.1 and 1.4%; alkali and magnesia: 2, 2.5 and 1.5%; water: 30, 30 and 25%. The average composition of the laterites is 53% Al_2O_3, 11% Fe oxides, 6% SiO_2, and 26% H_2O, but only ores containing more than 50% Al_2O_3 and less than 10% SiO_2 are worked.

SIERRA LEONE

The *Mokanji Hills* deposit in southwestern Sierra Leone sheds the waters of the Banta and Dassa chiefdoms (Fig. 17, 54). Here residual laterite develops on a band of hypersthene gneiss of the Kasila Series, which extends more than 18 miles, with its width varying between 100 and 700 yards. The band is locally broken by lateral faults and injected by dolerite dyke (J. D. Pollett, personal communication, 1962). Covering a square mile, the band extends from three miles south of the Muyambu–Matru road, south-southeast to Gbowala, Kekerihun, Ngiewoma and beyond. Alteration and bauxitization attain a depth of 60 ft. and probably average 30 ft., but the banding of the gneiss can still be recognized in the spongy bauxite. Because of the leaching of iron oxides, the ore is alveolar. A majority of the bauxite contains 62% Al and only 2.5% Si. The cream, fawn or pinkish brown material consists of hard gibbsite, some limonite, and minor amounts of clay silicates of the kaolin group.

Along the strike of the same hypersthene gneiss band, the Gbonge deposit has developed 25 miles southeast of Mokanji. Gbonge Hill is located 12 miles from the tide water mark. On the western slope of the Hill, the zone of lateritic bauxite extends 2 miles. Best grades occur in a 3,600 by 1,800 ft. area, but because of its slope the bauxite body is irregular in shape and section. A third of the samples contains 44% Al_2O_3 and 4% SiO_2, on the other hand, clay stringers increase the silica content.

In addition, lateritic bauxites are found near Hastings. Lateritic bauxite occurs in a 30 square mile area of the northern *Sula* Mountains (south and east of Worowaia) in central Sierra Leone. The 6–38 ft. thick durcicrust, which averages 14 ft., rises above 2,000 ft. in altitude (WILSON and MARMO, 1958). Banded structure has been preserved only above laminated rocks, and not in laterites derived from detritals and soils. A 1½–30 ft. thick soft brown laterite underlies the crust and averages 18 ft. Another generation of laterite emerges locally under the deeper clay (5–20 ft.). Several cycles of lateritization and fossilization have been distinguished.

The durcicrust is hardest where there is no vegetation. In the upper layers vugs and cavities, covered by limonite and occasionally by gibbsite, are most abundant. But at a depth of a few inches the durcicrust softens to less ferruginous laterite. The underlying rocks generally consist of chlorite, talc schists and amphibolites. From 45% at the surface of the durcicrust, maximum alumina content increases downwards, attaining 52% below 20 ft. in soft laterite. Because of percolation and abstraction of silica, alumina content augments towards the top of the clay and decreases to 42% below 30 ft. Maximum ferric oxide content increases in the durcicrust from 55 to 60%, but diminishes rapidly in the soft laterite to 42%. Vegetal decay, fire and plants attract iron towards the surface. Silica is leached out in the durcicrust, and during the dry season insolation hardens the upper clay layer. Once formed, the durcicrust thickens downwards until erosion destroys the laterite. An estimated 250,000 tons of bauxite contain 40% Al_2O_3 and 7.5% SiO_2.

GHANA

The principal deposits are derived from Birrimian inliers of sediments and volcanics, folded and sheared along north-northeast trending axes. The small Ejuanema deposit has been altered from sub-horizontal Voltaian shales. Arenaceous, ferrugineous and siliceous Tarkwaian sediments diminish the grade of bauxites, as do ferro-magnesian basic rocks. Deposits are located on parallel 6° 30′ along a 5–40 mile wide belt 100 miles inland. This belt actually represents an east-southeast trending tectonized zone. The best ores are found 1,600–2,600 ft. above sea level, on slopes of from less than 5° to 12°. No significant deposits have developed below 1,000 ft., or farther than 140 miles from the sea. The principal deposit, the Chichiwere Tente Hills near Yenahin, are situated 130 miles from Takoradi Harbour as the crow flies, and 35 miles west of Kumasi, capital of Ashanti. Abundant bauxite was formed from acid lavas, tuffs and amphibolites. The Affo group deposits (Sefwi Bakwai) are 30 miles south of Yenahin; Asafo, Nsisreso and Asampanaiye (in the vicinity of the Ivory Coast border) are 100 miles from the sea. The Ejuanema deposit is 90 miles north-northwest of Accra, while the *Atiwedru* Hills, near Kibi, are situated 55 miles from the capital (Fig. 74).

Minor occurrences of southeastern Ghana include Nkwanta Bepo, Odumparara, and the Atewa Range (Oba scarp) as well as Mount Kawkawti and Sumanchichi in the southwest. Asempanaiye is located west of Sefwi and Yenahin, not far from the Eburnean border. Other deposits are found west of Sui, near Asafu and Asunsu. Smaller deposits of

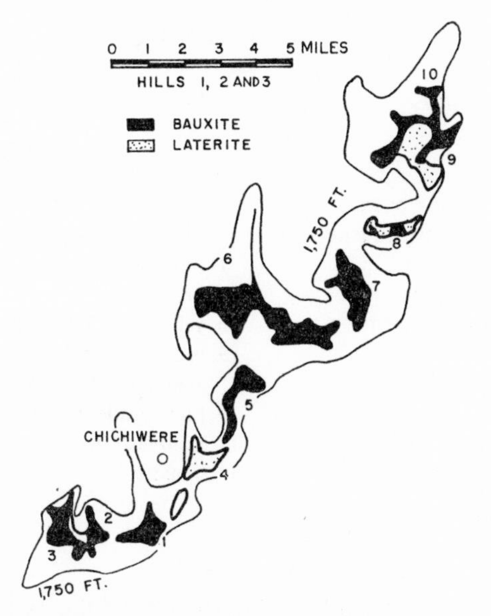

Fig. 74. The Chichiwere Tente Hills bauxite deposit.

Ashanti include: Kwamissa and Bosumkessi, Mount Mia near Impuessu and Kensere Hills, and the Sansu and Oboum ranges of the Oba scarp.

Yenahin

The Chichiwere Tente Hills extend 20 miles north-northeast. The hills consist of sub-vertical, upper Birrimian acid lavas, tuffs, phyllites, and dykes bordered on the west by quartz–mica schists and on the east by slates and phyllites. The plateau rises to an altitude of from 2,000 to 2,400 ft., and the ridge sheds the waters of the Tano and Ofin Rivers. Their tributaries, the Boni, Agogo, Subin, Dissiri and Apemperame, dissect the crest into the hills numbered 1–9, from south to north. The lower layers of vermicular bauxite are soft, but upper levels are harder. Whereas the overburden attains a maximum thickness of 6 ft., the laterite is from 20 to 50 ft. thick. Whatever the surface grades, they increase at 5 ft. of depth, but thence values decrease slightly to a depth of 20 ft., increasing again to the average depth of 33 ft. Bauxite may have a clear brownish tint. Since sources drain the iron of the laterites, they increase the relative alumina content. The iron–alumina relationship can be expressed in the formula: $Al_2O_3/70 = 1 - Fe_2O_3/85$. Because no streams have worked themselves up to it, much of the central sector of Hill 9 is too ferruginous. Conversely, bauxite of spurs drained by headwaters attains the highest grades.

In the Asuakwo swamps near Nsiseram, Hill 6, humic acids have drained iron and concentrated alumina so that it comprises 64% of the bulk. Sectors 1 and 2–3 cover the crest of the southern hill. Sector 1 has a lozenge shape; its longest dimensions attaint 5,000 and 4,000 ft. Sectors 2 and 3 form a 1,400 ft. long and 3,000 ft. wide arc with protruding spurs. The tints of bauxite which have been distinguished are shown in Table CXVII.

Sector 1 bauxite contains 46–50% Al_2O_3, but Sectors 2 and 3 include 55 and 49%. Iron content oscillates between 13 and 15%.

Hill 4, sloping towards the west, is ferruginous. Hill 5 forms a 7,000 ft. long crest bulging

TABLE CXVII
BAUXITE TINTS

ft.	*Hill 1*	*ft.*	*Hill 3*
3	red soil	1/2	loose bleached bauxite and sand
5	boulders of pink bauxite	5	orange and yellow brick bauxite
6	cream and mauve bauxite	7	orange and yellow brick bauxite
14	purple and cream vermicular bauxite	1	red and yellow bauxite including breccia
1	purple, powdery cream bauxite	6	yellow brick bauxite and breccia
1	pink bauxite clay	3	reddish ferruginous brick bauxite
		10	high grade yellow brick bauxite
		17	yellowish brick bauxite
24	bauxite	49	bauxite

towards the northeast. The ore consists of 20 ft. of yellow, flaggy breccia, 8 ft. of red-stained ore, 2 ft. of argillaceous bauxite, and 1 ft. of lithomarge containing bauxite. Hill 6 is lenticular, bulging in the northwest. The ridge extends 4 miles, the width of the crest proper attaining 1,500–2,500 ft. The bulge averages 7,000 by 5,000 ft. The thickness of red, orange and yellow ferruginous, locally gritty, bauxite attains 42 ft. Alumina content varies between 42 and 63%, silica between 0.3 and 0.8%, iron oxide between 18 and 26%, and titanium oxide between 4 and 5%. East of Hill 6, Hill 7 spreads over 8,000 by 3,500 ft. across the headwaters of the Boni. Hill 8 is small. Further north, Hill 9 contains high grades only on its periphery, and Hill 10 constitutes one of its small extensions.

The reserves of Hill 6 attain 86 tons, with respectively 11, 9, 7, 13, 23, 15 and 4 tons having been evaluated in Hills 1, 2, 3, 5, 7, 8 and 9.

Affo group

Afumba-Nfatahun, Boka Khirri, Supirri Bepo, Ichiniso and the Kanainyeribo Hills form a 5 mile long, east-northeast trending arc, while Angwinyare Boka and Suman Chichi lie 10

TABLE CXVIII
BAUXITE COMPOSITION

	Al_2O_3 %	Fe_2O_3 %	SiO_2 %	TiO_2 %	H_2O %	*Reserves million tons*
Afumba	50.2	20.2	0.4	1.5	26.4	3
Nfatahun	52.1	16.3	0.5	1.3	27.6	
Boka Khirri	50.6	19.6	0.6	1.3	27.0	4
Supirri	56.1	9.2	0.8	1.2	30.0	3
Ichiniso	48.7	21.2	1.7	1.1	26.2	11
Kanaiyeribo	48.5	21.6	0.2	1.3	26.5	10
Angwinyare Boka	50.2	18.8	1.1	2.0	26.6	}
Suman Chichi	53.2	15.5	0.6	0.9	28.4	}

miles northeast. The parent rock, sub-vertical Birrimian schists, also trends N 40°–55° E. Covered by 20–60 ft. thick tables of lateritic bauxite, the hills rise 1,000–1,800 ft. above the sea. Lithomarge has developed between the laterite and the schists, and grades have increased by erosion at the ends at Afumba, Angwinyare and Suman Chichi. Fragments of schists, resembling the brecciated bauxites of Boké, are found at Supirri and elsewhere; laterite and gibbsite replace them either largely or completely. When substitution is completed, relic banding remains (cf. Mokanji Hills, Sierra Leone). For the composition of typical samples see Table CXVIII.

Gas escaping from the bauxite contains 76% methane, 12% nitrogen and argon, 10% CO_2, and 1% hydrocarbon. Bauxites at the Affoh deposit contain up to one dwt./ton of gold and 17 dwt./ton silver.

Ejuanema and Nsisreso

Mount Ejuanema rises to an altitude of 2,500 ft., 1,700 ft. above the valley. The summital deposit lies on late Precambrian (Voltaian) sandstones and shales which have been intensely block faulted. Favourable conditions for bauxitization only occur at Ejuanema. The sediments typically consist of 2–5 ft. of red soil and bauxite granules, 10–20 ft. of massive bauxite, 5–8 ft. of rubbly bauxite, 2–7 ft. of clay with vestigial bauxite, and 2 ft. of clay resting on a bed of shale. The shale was originally interbedded with largely eroded sandstone, and exoquartz derived from it occurs in the soil. Tints vary considerably, as do values which average 60.5% Al_2O_3, 9.7% Fe_2O_3, 1.4% SiO_2, 2.2% TiO_2 and 26% H_2O. Reserves attain 4 million tons.

The undulating *Nsisreso* deposit extends 3 miles in a northeasterly direction. The crest sheds tributaries of the Bia and Sayere Rivers. Its width varies between 1,500 and 2,000 ft. (bauxite lenses and laterite are also found further upstream on the crest). The bauxite crust is 25–38 ft. thick; 90% of the ore, representing 25 million tons, occurs to a depth of 20 ft. It contains 48.7% Al_2O_3, 21.5% Fe_2O_3, 1% SiO_2, 3.8% TiO_2, and 23% H_2O. Moreover, 20 million tons contain 50% and 15 million tons contain 55% Al_2O_3. A meadow, covering one-fifth of the area, occupies the north-central part of the deposit without any overburden. Other occurrences probably exist in the area.

CAMEROON

The eastern slopes of the Adamawa chain in the Ngaoundéré district of the north-central Cameroon are covered by the continent's second largest bauxite deposits (Fig. 24, 53). The Adamawa bauxites average 44% Al_2O_3, 3% SiO_2 (2% being soluble), 23% Fe_2O_3, 4.5% TiO_2, 25% H_2O, 0.5% MgO, less than 0.5% K_2O, 0.2% Cr_2O_3, 0.01–0.02% V, 0.03–0.07% Mn and 0.05–0.15% P_2O_5 (gibbsitic, ferruginous and aluminous layers have been distinguished in them). About 300 miles directly northeast of Douala Harbour, the lateritic plateaus cover 800 square miles. The plateaus adjacent to *Minim* and *Matrap*, 70 miles southwest of Ngaoundéré, cover 0.6 square miles, attaining 24–27 ft. in thickness at an altitude of 4,000 ft. They include more than 1,000 million tons of bauxite averaging 43% Al_2O_3 and 3.7% total SiO_2. The *Ngaoundal* deposit, southeast of Minim and Matrap, includes 90 million tons of bauxite containing 41.7% Al_2O_3 and 1.4% SiO_2. The thickness of the bauxite averages 37 ft.

Northwest of Dschang, the *Fongo–Tongo* deposits are located near the southern end of the Adamawa range, 100 air miles from Douala, near the railhead Nkongsamba and the limit of the two Cameroons. The crusts and boulders attain 20–22 ft. in thickness, and include 46 million tons of bauxite. When screened, the product contains 34 million tons of ore of 45% Al_2O_3 and 4% SiO_2. Furthermore, in the Bangam sector east of Dschang, voluminous but heterogeneous deposits exist between Bafang and Bafoussam, and northeast of Fongo–Tongo (in the southern part of the Bambouto area) thinner layers contain 4 million tons of ore of 46% Al_2O_3 and 4.3% SiO_2.

THE REST OF WEST AFRICA

Bauxitic laterites extend from the Cogon Valley into eastern *Portuguese Guinea*. Bauxite occurs in *Mali* at *(1)* Sangalan and Fontofa (45% Al_2O_3) in Satadougou country, extending from the Fello Kéréwane plateau and Tembéré to the Guinea–Senegal border, *(2)* at Koulouba, near the capital, Bamako, and *(3)* at Mpébougou in the Ségou district further east. The M'Pébougou bauxites have been lateritized from argillaceous sandstones and detrital material. Samples contain 1.5–2.5% SiO_2 and up to 50–60% Al_2O_3. In addition to this, ferruginous laterites are also present. Pisolitic bauxite boulders are found below Gaya, Niger Republic, on the Dahomeyan border, and white and pink bauxite occurs at Kaya, Upper Volta, also showing at Sikasso and Bobo Dioulasso. Bauxite is also believed to outcrop in the eastern province of Liberia.

The principal deposit of the *Ivory Coast* is situated in the area of Lakota, 130 miles northwest of Abidjan, the capital. Near Lakota and Gagnoa, at Oumé, pisolitic bauxite boulders derived from durcicrust cover dolerite. Ferruginous pisolitic bauxite also occurs at Boussèra near Gaoua, northeastern Ivory Coast. In the western Ivory Coast, boulders are found at Tombokro on the left bank of the White Bandama, with bauxite also showing at Didizo, near Lahou (on the coast). South of Lakota near Sassandra, the *Dagolila* ridge is a remnant of a vast lateritic crust that was derived from Birrimian schists. Rising at an altitude of 930 ft. and resting on red clays, 2 ft. of pisolitic bauxite are covered by 1 ft. of yellow clay, including bauxite nodules, and by a thin layer of sandy ferruginous and aluminous soil. Furthermore, boulders and nodules are found on the Babokou road 3, 4 and 5 miles from Lakota, on the Gagnoa road near Niakpalilé at mile 10 (km 16), and on the Divo road 3–4 miles from Lakota.

Rising at the same altitude, other ridges of the area include pockets of bauxite, e.g., 13 miles northeast near Grodilé(?). Bauxites have been seen on the slopes of the ridges situated on the Tiassalé road, 7 miles east of Divo, and 30 miles east near Eramankono.

In a small zone of Mount Agou, *Togo* (AICARD, 1957) albite–garnet pyroxenites have been bauxitized at an altitude of 3,000 ft. In this occurrence, which is located near Palimé (70 miles northwest of the capital, Lomé), Al_2O_3 values attain 46–53%, with Fe_2O_3 varying between 30 and 18%, SiO_2 1–1.5%, TiO_2 1.2 and 1.3%, and H_2O 23–28%.

In *Chad* bauxite shows in the Logon basin and at Laï near the southwestern corner of the country. On the Koro Plateau (near Laï) an average of 17 ft. of bauxite is covered by a ferruginous durcicrust, and underlain by 30 ft. of kaolin and a 300 ft. thick column of alternating sand and kaolin. Geologists attribute the sequence to the Tertiary Paleo-Chadian and believe that the bauxites were derived from kaolin by lateritization, i.e., desilicification and gibbsitization. They average 57.3% Al_2O_3, 8.2% Fe_2O_3, 3.8% TiO_2 and 3.7% SiO_2. The plateau covers 85 acres, including 3.5 million tons of bauxite, while only 500,000 tons of bauxite are available in the Koutou Kouma Mountains west of Koro.

CONGO

From Isangila on the lower Congo, near the Inga Cataracts, 3 mile wide lenses of basic lava extend in a northwesterly direction towards Sumbi, and then along the border of the Congo Republic to Kai M'Baku and the Cabinda frontier (Fig. 53; GROSEMANS, 1959). On a mean altitude of 1,800 ft., Sumbi is situated on latitude 5° South and has a rainfall averaging 300–325 inches per year. Since the area is bordered by higher quartzite ridges (STAS, 1959), the temperature is lower than in the vicinity. Microdolerites, basalts and andesites overlying tillites form the core of a north pitching anticline, with basalt outcropping mainly along the anticlinal axis. The Ndongi, Lukula, Nicki and Tsamvi streams have cut their valleys into this elongated tableland. It forms an erosion surface of 1,800–2,000 ft. which is believed to be Mid-Tertiary. With numerous interruptions, the most favourable area extends from Kipunda, through Makwanzi to Kimongo–Vangui in the east, and from Buende–Sundi through Londelu–Kai and Kimbaku beyond Longo–Nbengo. The length of this zone attains 15 miles. These two topographic bands are separated by 3 miles of irregular terrain. Surrounded by bauxitic laterites, the best bauxites consist of gibbsite, kaolinite and goethite containing 42.6% Al_2O_3, 36.9% Fe_2O_3 and 0.5% SiO_2. Gibbsite values appear to be greatest between 15 and 30 ft. of depth.

Pisolitic aluminous clays are known to exist, for example in the area of Stanleyville at Basoko and Elisabethville.

SOUTH EASTERN AFRICA

Following the bauxite plateaus of the Cameroon and Chad, the belt extends into the Sudan and Ethiopia, but geologists have not yet explored these areas systematically. In Tanganyika, aluminous clays of Mombo and the Uluguru Mountains contain respectively 26–34% Al_2O_3 and 39–44% SiO_2, but bauxitic nodules in the soils of Amani include 58% Al_2O_3.

In *Malawi*, thin bauxitic laterites cover parts of the Nyika Plateau, with some siliceous bauxite capping part of Mount Zomba. However, the principal deposit developed from the rings of various syenites of Mount Mlanje. The peneplained surface of the *Lichenya* plateau rises at an altitude of 6,500 ft., i.e., 3,500 ft. below the peak. Here the undulating cover of bauxites averages 20 ft. of thickness, while the maximum attains 50 ft. The ore averages 42.7% Al_2O_3, 13.93% Fe_2O_3, 1.57% TiO_2, 2.22% silicates, and 15.65% exoquartz. Much of the exoquartz has been derived from pegmatites injected into the syenites. We should emphasize that the Malawi deposits are located on the same latitude as the Malagasy bauxites, i.e., 15° south (see below).

On the Beira railroad, the deposits of the Moriangane Range near Vila de Manica, *Mozambique*, straddle the Rhodesian frontier. The bauxites rest on hornblende–orthoclase syenites and granites which include incompletely assimilated metamorphic xenoliths and, at Mount Snuta, massive or veniform dolerites. Although certain samples contain 59–64% Al_2O_3, 5.9–14.1% SiO_2 and only 0.7–2.7% Fe_2O_3, the sediments average, in fact, more than

15% SiO_2. In consequence, they are used only as a refractory and for the production of aluminium sulphate.

In Concession 50 of the Mankaiana district in *Swaziland*, fine grained diaspore of ceramic grade covers a hill. On its southwestern slopes outcrop dark, spotted schists containing hydromica, sericite and remnants of leucoxene. The diaspore is associated with pyrophyllite and subsidiary andalusite of the Insuzi Series, and the product contains 51.7% alumina, 35.6% silica, 1.8% alkalies, 0.9% titania and 0.5% ferric oxide.

Among durcicrusts of the high plateau (tampoketsa in *Malagasy*) lateritic bauxites are believed to exist. Indeed, in small depressions of the Ankazobe tampoketsa, geologists have identified compact lenses attaining 20 inches of thickness. They contain 50% Al_2O_3, 12–18% SiO_2 (mostly as exoquartz), 6–13% Fe_2O_3, and 2% TiO_2. Bauxite also shows at Manjakatombo and Mandrosohasina, south of Ankazobe, but another deposit is located in northern Madagascar. Rising at an altitude of 6,000 ft., bauxitic crusts cover the basalt flows of *Marangaka*. The ore contains 2% SiO_2, but only 38% Al_2O_3. Finally, at Analavory (25 miles south of Marangaka), a bauxitic crust, attacked by erosion, rises at an altitude of 5,300 ft. on granite. The ore contains 54% Al_2O_3 and 3.5% SiO_2.

Between Esama and Farafangana, 3–10 miles from the southeastern coast, peneplains developed on basalts and leptytes. This part of the Great Island enjoys a tropical climate with an evenly distributed rainfall of more than 100 inches per year. In the southern sector of this belt, leptytes of the high peneplain, originally already containing up to 20% Al_2O_3 have been bauxitized. A majority of the bauxitic crusts, which attain a thickness of 2–7 ft., rises at an altitude of 180 ft. and is surrounded by eluvial sands or, less frequently, by lateritic clays. Limonitized garnet, quartz and sillimanite accompany gibbsitized feldspars and possibly some boehmite. These bauxites contain 51% Al_2O_3, 11% SiO_2 (in screened samples) and 10% Fe-oxides. On these peneplains the 'islands' of bauxite extend from Esama to *Manantenina*, thence to the Analalava forest and Marovony, a total of 15 miles. Within this belt siliceous bauxites are believed to cover a few square miles.

North of this sector traps cover the coastal scarp extending 45 miles from Vangaindrano to Farafangana. On this intermediate land surface lateritic crusts locally attain a thickness of up to 18 ft. Even at the end of the dry season the water table can be found a ft. below the top of the crust. Of course these laterites are ferruginous, and, as in Ghana, only at head waters of streams do we find aluminous zones. Thus the restricted surfaces of Ankarana, Enato, Marovary and Ampefivato average respectively 27, 37, 37 and 40% Al_2O_3. However, boulders embedded in the clay underlying these laterites contain up to 40% Al_2O_3. Finally, the laterites of the upper land surface or crests contain only 24% Al_2O_3 (HOTTIN and MOINE, 1963).

SECTION III

Gold group

While gold is one of the most widely spread metals of Africa (Fig. 4), uranium, antimony, arsenic and the pyrites, normally its satellites, tend to concentrate in southern Africa, where most of the gold also accumulates. The principal deposits of gold, uranium and pyrites are bankets, but antimony favours hydrothermal veins, while arsenic also frequently forms impregnations. In addition to their major deposits in the gold paragenesis, pyrite is disseminated in other sediments, uranium occurs in hydrothermal associations from which gold is missing, and antimonite crystallizes in Morocco in veins traversing the aureole of stocks.

GOLD

Gold has been disseminated all over Precambrian Africa. Minor deposits, scattered over the continent, consist essentially of hydrothermal (pneumatolytic ?) veins and stockworks, and of the placers eroded from them. Impregnations and disseminations in shear zones, metabasites or metalavas and albitites appear in Southern Rhodesia, the northeastern Congo, Tanganyika and Ghana; auriferous quartz veins similarly favour the horizons mentioned above, and occasionally include itabirites. Pyrite and high temperature sulphides are common accessory minerals of these ores, with arsenopyrite and antimonite appearing mainly in southern Africa. The dating of accessory galena indicates that mineralization reached its climax in Ghana, the Congo and Rhodesia 2,200 million years ago, preceded by secondary maxima of 3,300 million years, in the Barberton, Uele (Congo) districts, and 2,700 million years in Rhodesia and Tanganyika. The 1,000 million year old cycle of the Great Lakes (Congo) concludes and terminates gold mineralization, if the 500 million year old mica ages are not attributed to the auriferous quartz veins of Egypt and Sudan.

The largest deposits, however, are the bankets or metamorphic placers of South Africa. The age distribution of accessory uraninite strongly resembles the pattern of hydrothermal veins, reaching its maximum before 2,000 million years, but the Ghana bankets are only 1,800 million years old. Pyrite is included in both parageneses, and carbon acquires a greater importance in the bankets than in the veins. Scarce diamonds show in both banket areas; ultrabasic accessories, however, are limited to the paleo-alluvia of South Africa, which represent the world's largest concentration of gold.

WEST AND NORTH AFRICA

The only *Moroccan* deposits of significance are located in a ½ by 2½ mile zone along the contact of the Precambrian Tiouit stock, 250 miles east of Marrakech by road. The low-lying quartz veins of the east are distinguished by a sulphide-hematite paragenesis: *Jemaa N'ougoulzi*, at the western end, also includes sphalerite, galena and stannite, but copper veins are sub-vertical. The silver/gold ratio decreases from 10/1 at Tj'ouit, to 4/1 at Jemaa N'ougoulzi. In addition, gold is associated with the cobalt ores of Bou Azzer (see 'Cobalt' chapter).

In *Mauritania*, the Fort Gouraud hematite ores contain 0.06 ounces/ton of gold, and the Akjoujt copper–hematite paragenesis includes 0.03 ounces/ton. Alluvia of the Gambia River (Mako and Bantakokoula in Senegal) and Falémé carry gold as do veins of Banora and the Blue Lode of Kaho and Kéniéro, on the frontiers of *Guinea*, *Senegal* and *Mali*. All these are located in Lower Birrimian rocks, as are the placers of the Upper Niger and Tinkisso, and of Fitaba and Kaba Houré in Mamou and Dabola county, central Guinea. The proliferous stockworks of the Fatoya district of eastern Guinea have been protected from erosion by laterites. While the Nankoba stringers follow a 3 mile long and 600 ft. wide crush zone, the arsenopyrite–gold sequence of Soubako is related to an intraformational Birrimian contact. In addition, traces of gold show in the Simandou banded ironstone chain, north of Nimba. The *Sierra Leone* concentrates have been derived from the contact and exocontact of Kambui schists and muscovite granite, injected by pegmatites, and to a lesser degree from the Marampa schists of the Sula Mountains, Kangari, Gori, Nimi and Loko Hills (see 'Iron' chapter). Most of the gold is very pure, but the impregnations of Bomahun in quartz and arsenopyrite are distinguished by a gold/silver ratio of 3/1. Finally, gold and cassiterite are associated in the Nimi, Lake Sonfon and Sanden river placers.

Placer gold occurs throughout *Liberia* (Fig. 18), and a few grains of gold were even found in concentrates of the Soni stream flowing across *Monrovia* (SRIVASTAVU, 1961). The gold fields of Liberia are located in the Tchien district, Eastern Province; Boken Jede, Sinoe County; Yangaya, Belle–Yella, and Tawalata, Suehn–Bopolu district; Jeblun, Todee district; Pleebo and Webbo, Maryland county; Kokoya, Gbarnga district; Gola Konneh and Wenju, Grand Cape Mount county; Neepu, Rivercess, Ziahma–Buleama and Gisima chiefdoms, Voinjama–Kolahun district. The principal *Boken Jede* placers include the hydrographic systems of the Ni Jeday and Tadobo, and also include the Yanefor, a subsidiary of the Tadobo, a tributary of the Dugbay which enters the sea near Nanakru. The source of the placers is believed to be east trending diorite, impregnated by gold and intersected by a series of auriferous quartz veins. The deposit is residual rather than transported, and grades of 'Kumasi', the surface, red-brown, alteration product of diorite, diminish with depth. The lower gravels of creeks crossing this formation bear gold above the bed-rock.

The principal placers of the *Ivory Coast* follow the triple frontier with Guinea and Liberia, along the Cavally (Fig. 54) and its tributary, the Dibo. Other alluvia are situated in the Soubré. On the other hand, primary deposits of the Ivory Coast (Ghana and Upper Volta) are arranged along the north-northwest trending Birrimian axes; the Asupiri veins, for example, are at the southern end of the Kumasi alignment, with similar stockworks outcropping at Aféma and Hiré. However, at Nuon, near Toulépléu, gold shows in the aureole of granite and diorite intrusive into Birrimian rocks, and at Lao gold and silver bearing veins intersect Upper Birrimian amphibolite adjacent to a granodiorite contact.

The principal gold district of *Upper Volta* is located 50 miles north of the northwestern corner of Ghana, in a 15 mile wide bulge of a Birrimian band. The overall length of the north-northwest trending swarm of quartz lenses attains 15 miles; they cut sericite and chlorite schists mainly at the intersection of transverse faults. Ophiolites and old dolerites are the preferred host-rocks of veins, with amphibolites occurring east of *Poura*. One can distinguish five sectors: *(1)* Fara-Diansi, in the south (4 miles), *(2)* Poura (4 miles), *(3)* three ft. thick lenses between Poura and Soumbo, *(4)* Soumbo (3 miles), ending in a dyke north of Koussa, and *(5)* Baporo. The Baporo swarm is oriented west-northwest, and the Filampou veins east of this occurrence strike parallel. The Ton veins are located east of Diansi. The principal swarm has been faulted at the Black Volta. The sub-vertical Poura vein proper has a width of 8–15 ft. and grades of 10–18 dwt./ton; the principal ore body extends 600–1,200 ft. and to a depth of more than 580 ft. The paragenesis includes pyrite, chalcopyrite and stibnite; gold forms patches, disseminations and bands, but grades decrease with depth. The auriferous Bavillers quartzite is in the aureole of the Zamo stock. The *Lay* veins are near the Rio Réga intrusions, and the Songo and Zergouré veins extend the Birrimian gold trend of northeastern Ghana. Auriferous veins also outcrop at Ligmaré, Loubiou, Ourbera and Malba, with placers occurring at Sissé, Iridiaka, Djikando and Midebo.

We now pass from the Ghana deposits, to which a separate chapter is devoted, to the Kéran, Bungba and Momoin gold fields surrounding Mount *Togo*, and to the Agbandi vein system in Atakpamé county of south-central Togo. The upper Perma placers, near Natitingou, in northeastern *Dahomey*, have been derived from quartz veins and from lenses, and attain a maximum length of 1,000 ft. in transverse fissures, cutting an anticline of schists, quartzites and amphibolites. From this pyrite, galena and arsenopyrite paragenesis, erosion has transferred gold into the placers of an incipient canyon. Sulphide veins contain gold near Niamey, capital of the Niger Republic, and in the Aïr Mountains gold shows at Tunan, Tilisdak and in the Oued Hgayague. In *Chad*, gold shows in the alluvia of the Mayo N'Dala stream, near Kamboké, Pala county.

In northern *Nigeria*, gold has been concentrated from veins and stringers of Precambrian roof-pendants between Gwaria, Minna and Maradi, west of the Jos plateau. The Kontangora belt, 100 miles to the west, extends from Boussa on the Niger to the limit of the Tertiary sediments. The principal, southeastern *Cameroon* placers flow on graphitic talc and amphibolite schists between Meigonga and *Betare Oya*, overlapping into the Baboua–Bouar field, Central African Republic, where argentiferous gold forms large grains. The *Sosso* drainage flows on roof pendants. Gold has been carried downstream in the hydrographic system of the Sanaga to Ouesso, and east of Fort Rousset into the northern part of the Congo Republic.

THE REST OF AFRICA

The *Egyptian* gold mineralization follows a 700 mile long orogeny, in which the majority of the auriferous quartz veins has been injected in the aureole of small stocks. A series of veins follows the southern contact of an elongated batholith, east of the Luxor bend of the

Nile; these include Hadrabia, Abu Gerida, and the wadis Esch, Hamima and Jemna. The Um Rus–Baramia group of gold–pyrite–quartz veins outcrops in an area of minor stocks or domes of the granite relief, and at Wadi Zeidum the basal, Nubian conglomerate is auriferous. The belt branches southeast of Aswan. The paragenesis of Um Haimur–Um Garaiart–Murra, the northernmost district of the western branch, also includes graphite and sulphides (Fig. 12, 58).

The eastern branch reappears at the frontier of the *Sudan*, in the fold belt running parallel to the Red Sea. Two districts can be distinguished there: *(a)* Eigat–Daraheib, (Onib, Ceiga and Giafferie), covering an area of 80 by 40 miles, and *(b)* the Red Sea Hills belt, extending nearer the coast, 100 miles from Gebeit to southwest of Port Sudan (Akilebleigh, Rajarkindie, Tearomeur). Swarms of quartz veins cut metamorphics, granites and basic rocks. The western branch reappears southeast of Wadi Halfa, and gold occurs in a 150 by 120 mile area of the Nubian Desert between the frontier and the Atbara bend of the Nile where mineralization was intensive. Extensive biotite and hornblende granite stocks are intrusive into schists, quartzites, limestone and marble, and gabbro, diorite, younger basalts and felsites also occur. Numerous auriferous swarms and veins are closely related to the intrusives and especially to porphyritic felsite, but placers also carry gold. At Um Nabardi, zones of up to 1,500 ft. in length, cutting mica and chlorite schists, include veins. The belt re-emerges east of the Atbara bend of the Nile and of the Rubatab mica field, and further north the Dalgo–Dongola belt, extending south of Wadi Halfa, outcrops in a window, piercing Nubian sandstone.

In the southern Sudan, (khors) placers and veins of the Blue Nile system extend from Roseires and Qala en Nahl to the Wallega district of Ethiopia. In addition alluvia have been mineralized in the Tira Manti window south of El Obeid. Gold accompanies copper at Hofrat en Nahas, in the vicinity of the Central African Republic. There the *Roandji* placers have been derived from a banded ironstone horizon, and the *Pouloubou* field in eastern Ubangi extends the Moto trend of the northeastern Congo.

Some gold shows in northern *Uganda*. The principal gold belt of that country is, however, located south-southeast of Lake George. Most of the placers of Buhweju flow on sandstones and micaschists, but veins are scarce. South of Buhweju, the Mashonga gravels carry gold, as do alluvia and veins of the rare metal fields of southwestern Uganda. Gold is also associated with sulphides in the Busia field, an extension of the Kenya gold belts, north of Lake Victoria. The gold district of northern *Rwanda* is wedged between the tin–columbite zones of Gatumba and Kigali. On Kalenda Hill, near Myovi, quartz and gold have impregnated a 3–10 ft. wide, sub-vertical crushzone in a fold-shear which favours the contact of sandstones and schists. On the Nyongwe, east of Lake Kivu (in southern Rwanda), a late Precambrian quartzite–sandstone horizon carries disseminated gold and quartz stringers. The principal *Burundi* gold zone trends 200 miles S30°W, overlapping into southeastern Rwanda near Muhinga. Gold has also been found on the Nile–Lake Tanganyika divide, south of Bujumbura, and in the Lugogo stream (a tributary of the Ruzizi, connecting Lakes Kivu and Tanganyika) alternating hard quartzites and soft schists have trapped alluvial gold. Special chapters are devoted to the Congo, Tanganyika, Kenya and Gabon deposits.

The principal deposits of the *Congo Republic* are the placers of the Mayombe belt between the bend of the Niari River and Dolisie, west of Brazzaville on the frontier of

Cabinda. Gold also shows at Mayoko and in the southern extension of the Gabon gold fields.

The Mayumbe gold belt (as it is spelt by the Portuguese) continues in the tributaries of the Luali, especially the Chivolo Grando in the Cabinda enclave of *Angola.* However, young placers also developed in the Lombije and Calumbo Rivers 150 miles east of Luanda, the capital, and in the Colui drainage at Vila da Ponte and Cassinga, in southwestern Angola. Paleo-alluvia were deposited between the latter and Nova Lisboa after the Cretaceous, e.g., at Chipindo and Gandavira. The wash is finer there than the gravel of modern placers, and the thickness of the cover may be as much as 30 ft., recalling the old tin placers of Nigeria. Furthermore, gold has been eroded from the Jurassic–Cretaceous carbonatites of the Nova Lisboa plateau, particularly from Coola. As in Katanga, the bedded copper ores of Bembe and Mavoio, in northern Angola, contain 3–10 dwt. gold per ton of concentrate, and gold also shows in other copper occurrences of Cuma, Sa da Bandeira and Mossamedes.

The Monarch, Vukwe and Old Tati fields of the *Francistown* belt of *Bechuanaland* strike 60 miles north–south between the Shashi and Ramaquabane Rivers, on the Rhodesian border. Vein and funnel-shaped deposits can be distinguished there. In *Southwest Africa,* pegmatites and (feldspar–) quartz veins, mostly of the exocontact, contain gold in association with sulphides and scheelite (and calcite), but basal conglomerates of several epochs only show traces. Gold has been transferred into eluvia from the quartz lodes cutting early Precambrian schists at Rehoboth, south of Windhoek. This gold field includes Neuras, Noois, Kobos and Dymoeb, and the paragenesis includes copper, bismuth, antimony and arsenic. In the Ondundu Otjiwapa field quartz stringers occur in metamorphics.

The principal gold field of *southern Zambia* is located at Matala, 80 miles west of the capital Lusaka. Pinching and swelling quartz–tourmaline veins penetrate into a hematitized shear at the apex of a granite filled anticline there. Between the Chongwe and Luangwa Rivers, gold is associated with copper minerals and bismuth carbonates. Deposits of the Broken Hill area include Velocity and Katie. At Sasare and Lutembwe, in the country's eastern corner, stringers and pyrite bearing veins contain gold. The deposits of Southern Rhodesia are described separately.

In *Malawi,* the Dwangwa (Fuliwa) and Lisungwe placer systems contain gold, as do the silicified faults which limit the Karroo of the Shire Valley at Mwanza and Namalambo. In *Mozambique,* the Luenha placers at Missale, and at Chifumbázi in the Tete district (jutting into the Rhodesias), carry gold in an area underlain by the Basement. However, in the Alto Ligonha pegmatite district (see 'Beryl–Tantalite' chapter), gold is associated with bismuth, and also occurs at Mount Muiane and in the Namírruè placers. At Manica, an extension of the Rhodesian Umtali schistbelt, gold has precipitated in veins, on both sides of the contact, in iron bearing quartzites and later in alluvia. The banded ironstones, conglomerates and schists of the *Swaziland* gold belt stretch 40 miles south-southwest, in the extension of the Barberton district of South Africa. The deposits follow two alignments in the narrow schistbelt: the Forbes Main Reef follows the granite contact, and along the Oban Art Union alignment veins containing arsenopyrite are injected on the contact of calc- and talcschists, near a quartzite horizon. Pockets of gold and wad are associated in the brecciated and contorted banded ironstones of the Devil's Reef Mine; however, a majority of Swaziland gold occurs in the zone of oxidization.

To pass to the islands, *Lithothamnium* limestones of Weti, *Zanzibar*, show traces. The alluvial gold of *Madagascar* has been eroded from pegmatites, quartz lenses, and itabirites outcropping in schist pendants of the intrusives of the Ambositra district, southeast of Tananarivo. In the Maevatanana district, northwest of Tananarivo quartz–carbonate–sulphide gold veins and funnels are arranged along the contact of schists and quartzites. Finally, on *Mauritius* traces show in the veins of Fressanges and Midlands.

GHANA

In southwestern Ghana two major placer belts have been distinguished. The Ankobra River (with its placers) flows 90 miles from the north to the sea at Axim, in the south. The primary districts, including the Tarkwa field, lie largely east of the Ankobra. In its mineralized section the Ofin River meanders towards the southeast (distances measured along the river are naturally longer). Upstream, the heads of the two placer systems are only 20 miles apart. The Obuasi field lies northeast of the Ofin. The Tano placers are located 25 miles west of the Ankobra. The two belts coalesce, in an auriferous background which also extends north and west. Other deposits include Bibiani, north of the Ankobra and Ofin placers, the Sewumo, Ajumodium and east of Bondokou areas (near the Ivory Coast border), and the Nangodi field on the Upper Volta frontier.

The primary gold alignments exhibit the N30°E trend, favouring (Fig. 21, 54): *(1)* the intraformational contact of Lower Birrimian schists and Upper Birrimian metalavas, *(2)* the Birrimian/Tarkwaian contact, *(3)* the Upper Birrimian greenstone horizon bordering gondite (manganese), or *(4)* Dixcove granites, *(5)* the aureole of younger granites in the Lower Birrimian, and *(6)* the Tarkwaian banket.

The Birrimian and Tarkwaian are older than 2,200 and 1,800 million years. Synclines of the isoclinally folded strata have locally been uplifted, sheared and faulted by overthrusting, thus forming the gold horizons. Ore shoots generally pitch in the plane of the lenticular fissure fillings, with values diminishing downwards. Gold is generally associated with sulphide bearing (grey) quartz, fractured quartz, arsenopyrite and country rock xenoliths, but galena, pyrite and tourmaline rarely form indicators.

Tarkwaian synclines include conglomerates, the auriferous bankets and phyllites; little gold is therefore left in the following Akwapimian System. The principal trend extends from Konongo through Obuasi to Akropong (100 miles). Arranged in échelon, further south, the Prestea belt is shorter, as is the parallel Tarkwa trend.

Konongo and Oboum

The Upper Birrimian belt is 12 miles long and 1 ½ miles wide. The principal Akyenase reefs are subconcordant in and with isoclinally folded, carbonaceous, chlorite, biotite, garnet schists and hornfelses 500 ft. from the Tarkwaian contact. Horizons of carbonaceous material have been found on strike faults between the Awere and Odumase reefs. Subvertical faults strike east and east-southeast. The Akyenase lenses are 1–3 ft. thick, and exhibit a pinch and swell structure; the Odumase reef follows a tectonized line, while the Awere lens is contorted. Arsenopyrite is the best indicator; apophyses have been injected into crush zones, but the fissure fillings have also been affected by post-mineralization shears and transverse faulting. The Boabedroo reef is a mile north of and in the same channel as Odumase, while the *Zongo* fissure is parallel. The reef shows the 'streaky bacon' texture, due to the inclusion of graphite. Blue quartz is the indicator of gold. At Obenemasi, arsenopyrite, pyrite and pyrrhotite have impregnated a carbonatized schist at the contact of a calc-chlorite metalava.

To pass to the *Oboum* belt (between Konongo and Obuasi), veins and impregnations largely follow the limit of the two Birrimian stages. At Oboum, e.g., phyllites and amphibolites or metalavas are bordering the veins. Other deposits of the belt include Obuoso, Adakwai, Begoso and Wiawso.

The Obuasi belt

The reefs follow a 5 mile long alignment, 500–2,000 ft. west of a curved bed of metabasalt and albitite; they are included in a sequence of carbonaceous schists and phyllites, interlayered with graywacke, dipping 70–85°. The lenses are located in arcuate shear zones, striking between north and northeast and dipping sub-vertically. At the Ashanti mine, the shear runs along the strike. Close at hand at Obuasi, the 50 ft. wide shear zone is largely filled by quartz and terminates in a graphitic material. The mile long lens dips 65–70° northwest and pitches northeast in the fissure. However, movement did not cease with the emplacement of the quartz, which was subsequently shattered, and the wall has been drag-folded. The graphitic Ashanti quartz apophyse attains a length of 1,000 ft. and a width of 40 ft. The reef dips east and also pitches northeast. Nearer to the metabasalt, the Sansu, Kan Su, Akapoli Su and the Côte d'Or-New Make reefs fill in a well defined overthrust fault which dips 45° northwest. The Insintsiam ore shoot averages a thickness of 6 ft.; its dip is variable. However, the 'New Make' cuts across the bedding locally.

The Obuasi reef widens from 20 to 40 ft. at the branching point of apophyses. It contains thin laminae of graphitic country rock in a banded structure which is particularly prominent in the New Make. The Ashanti and Obuasi veins coalesce locally in depth, but they are also separated by relics of brecciated carbonate schist.

The quartz reefs carry ankerite, pyrite, arsenopyrite, sericite, galena, pyrrhotite, sphalerite, sericite, carbonaceous substance, albite and gold. The ore shoots have been localized by coalescing and the intersection of reefs and faulting, and the auriferous liquids appear to have followed oblique chimneys. The grapitic shear material, probably formed between the Birrimian and Tarkwaian, is typomorphic for the deposits; indeed even sulphide grades vary with it.

The Justice's deposit consists of a series of lenses of pyrite and arsenopyrite (cut by stringers) impregnating schists. Following oxidization, a limonitic capping developed. The Ayeinm vein is a straight fissure filling while Kan Su resembles the Côte d'Or, but Sansu and Korkortaswia are lenticular.

The Akropong and Prestea belts

Granites and porphyries outcrop in the vicinity of the sericite schists of the *Akropong* belt; metabasites lie in the southeast. Another less mineralized horizon runs parallel to the main belt in the northwest. The thickness of the Bokitsi reefs varies between 2 and 20 ft. The

Twoofoo and Doumake Etika reefs are lenticular. The Atasi reef strikes east-northeast, dipping 55° south. The mile long Ettadoom vein traverses phyllites, greywacke and grit. Xenoliths of graphite schist are suspended in this vein which contains some ankerite arsenopyrite, pyrite and galena.

The tectonic style of the *Prestea belt* is more intense: One or several horizons of auriferous veins and crush zones mainly follow the Birrimian/Tarkwaian contact; however, they lie farther from this limit in the north. The Dixcove granite and basic dykes have intruded the thrust faulted and sheared metasediments and -lavas. The main series of *Prestea* (Ariston) lenses extends for a mile in sub-vertical carbonaceous sericite schists and greywackes and attains a thickness of 35 ft., while the principal sub-horizontal ore shoot attains a length of two-thirds of a mile. The sulphide paragenesis also includes chalcopyrite, bournonite and pyrrhotite, as well as ankerite, calcite and rutile. Furthermore, at Broomasie laminated quartz lenses, carrying gold, ankerite and sulphides, were injected into a graphitic breccia. Other deposits of the belt include Anfargah, Ekotokroo, Tuappim and Ampassam.

The southern and western districts

Birrimian deposits of the Tarkwa–Axim district occur in phyllite, shales and amphibolites. At Akanko, manganese bearing schists have been impregnated, as have carbonate-chlorite schists at Akoko. In the Sefwi–Sunyani district, gold has precipitated in amphibolite in the aureole of Dixcove granite, but in the *Akim–Birim* district most of the quartz has been eroded into placers. In addition, a belt of numerous, rather poorly mineralized veins extends parallel to the main trend for 100 miles from Kumasi, capital of Ashanti, through Mansu Kwanta to Mape and thence to the Ivory Coast border. The country rocks are less tectonized there, and the gold is coarser and more patchy. It occurs along lines vaguely indicating old granitic stocks.

Bibiani is located near the Yenahin bauxite deposit. The gold field follows the eastern limb of a 3 mile wide isoclinally folded, Birrimian screen in gneissic biotite granite, but hornblende granites emerge from the syncline, and porphyries have been intruded during the first phases of folding. The Bibiani reefs outcrop and undulate intensely in phyllites and schists of the earlier division of the Lower Birrimian close to their contact with greenstones. Minor occurrences of the south, Sariehu, Mamnao and Bepaiyasi, are also located in the greenstones. The contact of the roof pendants and granite bands is sub-concordant with the regional trend, as is jointing. Frequently graphite shears cut through the bedding, and graphitic seams often follow the contacts of the reefs. Four generations of quartz contain 10% carbonate, sericite, graphite, pyrite, arsenopyrite and anhydrite. Finally, the auriferous quartz has been sheared or crushed. The folded, Bibiani North ore bodies consist of quartz and impregnations of schists; nearby the combined maximum thickness of the East reefs attains 300 ft., while other veins narrow to 20 ft.

Tarkwaian gold

The alternating lavas and sediments of the Upper Birrimian are followed by the arenaceous Tarkwaian. Its two lower series have a rather similar composition of sandstones, quartzites, breccias and conglomerates. The second member, the Banket Series, includes the auriferous sediments in its upper third, phyllites. Sandstones and quartzites conclude the sequence.

Relics of the 6,000 ft. thick column have been conserved in a 10 mile wide, tightly folded megasyncline, complicated by overthrusting at its rims. The Anantanfro–Kotraverchy anticline, pitching south-southeast, emerges in the centre of this structure; it is bordered in the west by an overthrust fault. Elongated batholiths have been intruded along the main axes. North of Akontansi a short, subsidiary anticline defines the Tarkwa syncline. The principal deposits follow the eastern limb of this syncline for 10 miles from Damang to Cinnamon Bippo, and thence for 12 miles through Aboso, Abbontiakoon and Tarkwa to Effuenta. The nose of the structure is affected by two parallel, transverse faults, and the banket is bent for 5 miles around this nose, resuming the regional northwesterly trend between Tebarébe and Pepe. In this, the richest part of the field, the two limbs are 1 ½ mile apart. Further west the banket follows the undulating extremities of overfolded parallel structures and finally pinches out.

Sericite and cross bedded specularite are typomorphic components of the white-gray Banket Series. In the gold field the series forms the Banket Range, but elsewhere it outcrops only in narrow inliers among older rocks. The auriferous conglomerates and breccias are interbedded in sedimentary streaks or lenses with the quartzites of the banket zone. The thickness of the conglomerate varies from less than an inch to more than 20 ft. Pay shoots or channels are distinguished by glassy quartz pebbles better sized and rounded than the rest on the contacts or in the midst of the banket. Hematite is the indicator mineral.

The banket attains almost 2 ft. in thickness at the Cinnamon Bippo where it dips west. In the north, however, faults cut the pay shoots. At *Adjo Bippo* the thickness of the banket column varies between 1 and 7 ft., but the dip oscillates too. The banket has been overfolded at *Abosso*. Most of the gold of *Abbontiakoon* is found in the bottom foot of the banket, which consists of larger pebbles than the western reef. At *Abosso* the banket has been overfolded, and at Tarquah the deposit has been mainly folded into synclines with the dip averaging 25°W. Greatest concentrations are found here in the top couple of ft. of several hundred ft. long pay shoots.

The banket reemerges, though poorly mineralized, between Banka and Bintempi, southeast of Konongo, at the northern extremity of the Tarkwaian megasyncline.

Placers

The most important alluvial placers occur in major rivers like the *Ofin*, from its intersection at the Kumasi–Ivory Coast belt to its crossing at the Tarkwaian line, and the *Ankobra* (with its tributary the Fura), from the Prestea belt to the sea. Farther west, the Tano placers derived from the Kumasi-Ivory and Bibiani belts, are 1 mile wide and 20 ft. thick. Gold values and the grain size increase downwards in the alluvia, which also contain ilmenite, rutile, zircon and staurolite. Finally, other alluvia are mineralized in the Birim, its tributary, the Pusu–Pusu, the Pra, Jimi and Nwi.

GABON

The only gold field of significance in the expanse separating Ghana from the Congo covers a 30 by 15 mile area of the Ikoye and Ogoulu basin in the Ogoué drainage, central Gabon (COSSON, 1959; Fig. 21, 53). North-northeast trending screens and lenses of polymetamorphic and migmatic amphibole gneisses with or without gneisses, mica-microcline gneisses, quartzites and micaschists are intermingled there with granite. The complex is bordered in the east by the Chaillu batholith and in the southwest by the sub-horizontal, late Precambrian (cf. Katangian?) sediments of the peripheric basin of the Congo. The principal faults appear to strike north-northwest, but dolerite dykes have been injected along a northerly trend near the contact of homogeneous granite and the migmatic zone.

In the gold field proper, amphibolites predominate in the *Etéké* roof pendant. However, 20 miles west, the southern sector of the *Pounga* band consists of biotite gneisses, with or without amphibolite, with a varied sequence of the kyanite–staurolite–mica facies building its northern part. The northeastern zone of the *Upper Ikoye* or Ovala is similarly varied. Only streams having flown on the metamorphic roof pendants, or on outliers of the Basement, carry gold, and values persist downstream to 1 or a few miles from the granitic front. However, rivers crossing granite remain barren. Similarly, in the Chaillu batholith, farther south on the Congo Republic/Gabon border, the small placers of Mayoko, Malanga, Magnima and Bitolo have eroded their gold from minor metamorphic enclaves. Concordant veinlets of the Etéké field carry gold, as do banded ironstones, quartzites, soapstones and micaschists containing altered, auriferous pyrite. Apparently the placers have concentrated gold from a mineralized background situated farther from the old granite than tin.

The lower Ikoye field, 70 miles northwest of Etéké, is supposed to be similar to the main district. The Mitendi placers, 70 miles north of Etéké, have eroded and reprecipitated gold from quartz lenses containing auriferous arsenopyrite, some free gold, pyrite, hematite and galena. These veins occur in north-northwest trending, isoclinal folds of sandstones and schists of the Ogoué System surrounded by Middle Precambrian rocks. However, 80 miles west of Mitindi, in the *Ndjolé* field, alluvia have been eroded from gold–quartz lenses which only rarely contain arsenopyrite.

ETHIOPIA

Kebre Mengist or Adola is in the far south, 250 air miles or 350 road miles south of Adis Ababa, and east of Lake Abaya of the Rift Valley and of the plateau lavas. In the Ghido Hadadi Mountains 15 miles south-southwest of Kebre Mengist (Fig. 52), only streams which have flown on basic rocks appear to carry gold (USONI, 1952). The principal (10 mile long) zone of placers trends north, and connects the bends of the Awata and its tributary the Marmora. And, indeed, at the sources of these rivers a north striking and steeply dipping series of amphibolites, pyroxenites and schists is cut by quartz veins. The streams flow eastward, and mineralization starts at their upper course. Chakiso Camp is located between the Shanka and Kalacha streams; Laga Gora and Reggi are situated on the divide separating the Awata and Marmora. Placers are included in the Baddacessa, Chakisso (1 ½ mile by 750 ft.) and Kalacha Rivers, and in the Dembi or Regi and its tributary, the Wallena. However, gold also shows in the *Marmora* and Dembi, 15 miles downstream of the main field. West of the divide the Iddi and Iddidima, tributaries of the Marmora, also carry gold. The Awata Valley reaches a depth of from 250 to 300 ft., and the Marmora canyon attains 800–900 ft. Under an overburden of 3–40 ft., both stream beds and terraces contain gold. The gold is coarse, and nuggets of 5–10 ounces are not uncommon; however, 40 ounces is the maximum. Gold tends to concentrate in axial runs, especially above the bedrock.

North of Adola the Bilu stream, and the western part of the Ababa Valley in the Doria drainage, carry gold, and 50 miles south-southwest of Adola the Dogga and Uggiuma torrents of the Dawa drainage include placers.

A gold province is situated in the bulge of the border, west of longitude 36° (that is, the meridian of Lake Rudolf). The placer belt of *Gubba*, trending north, 40 miles east of the Sudanese border, extends 100 miles from the Albiu to the lower Durra Abido, an affluent of the Blue Nile. Basalts cover only part of the Basement, the Beni Shangul belt extends from the Blue Nile on the Sudanese border, 150 miles south. Micaschists in the north and chlorite schists in the south are underlain by gneiss and amphibole granite, with traps extruding more frequently in the south.

The easternmost field, *Ondonoc*, is located 30 miles southwest of the bend of the Blue Nile. Veins strike in two directions there. There are two north-northwest alignments, a mile apart, and west-northwesterly trends are longer and later, but less important. Siderite, tourmaline and pyrite are associated with gold. Placer gold occurs in the Sirecole, 10 miles from the border, and at a number of other localities.

In the Wallega district, which is situated 100 miles east of the border, north striking quartz lenses traverse amphibole, sericite and chlorite schists and granites, bordered by leucite basalt flows. The swarm covers an area of 15 by 15 miles. Pyrite is the principal accompanying mineral, chlorite and calcite being scarcer. In the Emyo Valley, breccias have been impregnated. Eluvial and alluvial placers are also abundant. On the border between the Acobo and its tributary, the Cari, 100 miles north of Lake Rudolf, quartz veins of the exocontact contain macroscopic gold.

Five gold fields have been distinguished in Erythrea: Seroa, Ad Teclesan, Asmara, Gaala, and the Obel Valley plus Addi Hezza in neighbouring Tigrai province; most of the concordant lenticular veins traverse schists in well-defined alignments.

TANGANYIKA AND KENYA

Gold belts surround the southern and eastern shores of Lake Victoria, including the Geita and the more important Musoma district, and the Nyanza field beyond the border (Fig. 33, 52), in Kenya. In the *Singida* field, halfway between Lake Victoria and Dar es Salaam and 200 miles southwest of the Kilimanjaro, only scattered occurrences are left. The Lupa field is situated farther south between Lakes Nyasa and Tanganyika. Gold is also disseminated at many localities in the Basement.

At *Kakaméga*, northeast of Kisumu on Lake Victoria (Kenya), an enclave of subvertical tuffs and lavas strikes east-northeast between later Precambrian conglomerates, grits and granites. The contorted lenses of the Rosterman mine are best developed in diorite rather than in the finer grained volcanics. The Kimingini veins are also in a shear zone, but the Muchang ore body occurs in grits and sandstones.

The *Migori* gold field is located in old dolerites and shales of a 3 mile wide and 50 mile long belt. In Kenya's principal gold deposit, the *Macalder* copper mine, situated 10 miles east of Lake Victoria, ramified quartz lenses, veins attaining a width of 20 ft., and apophyses follow the contact and fissures of graywackes and metabasalts. Thin banded ironstones constitute the footwall of the southern vein. Electrum is associated with galena at the end of the paragenetic sequence of pyrite, arsenopyrite, pyrrhotite and chalcopyrite. The Blackhall reef follows a shear zone, and the Kahancha veins occur in crushed graphitic schists. Finally, the Lolgorien lenses are found in slates and graywackes on a faulted contact.

The stratigraphic sequence of the *Musoma* district includes amphibolites and gneissose granites overlain by quartzites, graphitic schists, metalavas and banded ironstones which are favoured by the gold quartz reefs. Arenaceous metasediments conclude the sequence. Faulting has taken place prior to and after the effusion of the lavas. The deposits have spread over 50 miles from Victoria Nyanza along an east-southeast trend. Most of the gold clusters in a triangle of 15 by 15 miles between Kiabakari, Muzangumbe and Simba. Tanganyika's principal gold deposit, *Kiabakari*, is located 30 miles south of Musoma Harbour by road, in a 300 ft. wide band of itabirites, sericite and graphite schists, surrounded by dolerite altered to quartz-amphibolite. Auriferous aplite dykes have been injected along the same alignment into the diorite, which is cut by quartz and dacite prophyry veins. The amphibolites have been intensely altered at the contact of the aplites, silicified, and then impregnated by pyrite and arsenopyrite. The granite lies 1,000–2,000 ft. south. The formation of quartz veins and stringers coincided with propylitization and with the dissemination of auriferous pyrite, arsenopyrite, fluorite, apatite, chlorite, zoisite and garnet! (STOCKLEY, 1935). Banded ironstones are believed to have been derived from partly felsitic tuff; they have been impregnated at Kiori. At Maji ya Moto (Hot Water), gold has been disseminated in altered amphibolite, the favoured host-rock. Veins also occur at Nyamongo, Kitarahota and Simba.

At *Geita*, on the southern shores of Victoria Nyanza, banded ironstones, interbedded with partly extruded plastic tuffs, dolerites and metalavas have been isoclinally folded into an elongated roof pendant. This series generally dips north, but the fold pitches northwest and the ironstone formation strikes east-northeast. The deposit is located south of a major syncline at the intersection of east striking shears and crossfolds of buckling. In the ironstones, the sulphide ore bodies follow the nose of a fold-faulted anticline at the contact of albitites. These albitites were derived from tuffs and volcanics by albitization; they have also been impregnated. Finally, following the cycle of mineralization, sub-concordant shears and connecting faults cut this structure.

In the *Lupa* field a series of west-northwest striking, pinching and swelling quartz lenses is aligned en échelon. Laminated quartz is the indicator of mineralization. The Saza veins extend 2 miles in granites and gneisses; they contain free gold, auriferous pyrite, baryte and tourmaline.

CONGO

The Congo belts extend for 700 miles from the Sudanese border into northern Katanga. The belts trend west-southwest in the northeastern Congo, bend south-southwest at Lake Albert, and then continue in the more extensive Kivu–Maniema belt. The principal deposits are located in the Kilo and Moto districts, precisely at this bend (Fig. 29, 51). There have been distinguished two major cycles: 2,200 million years in the northeast and 950 million years in Kivu–Maniema. Minor deposits, located 300 miles west of Moto, are 3,300 million years old, but gold also occurs in the Katangian.

Northeastern Congo

Amphibolites, albitites, banded ironstones, diorite and schists containing variable amounts of carbonates, albite, talc, chlorite and mica float on the batholith which occupies the northeastern corner of the Congo. The mineralization is related to these Middle Precambrian (Kibalian) roof pendants or screens. Basic dyke rocks, mainly of the Rift cycle, cut

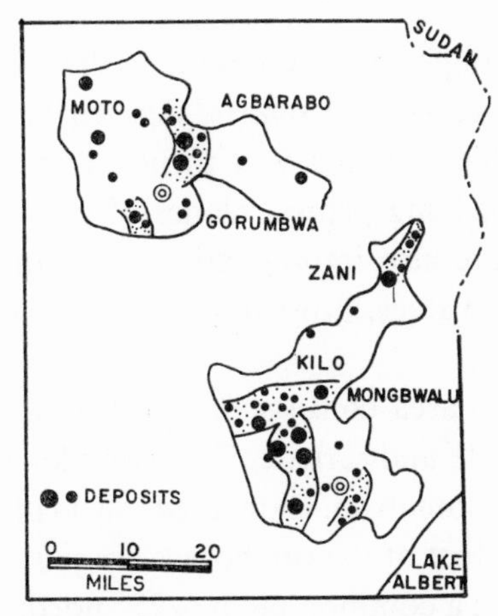

Fig. 75. The Kilo-Moto gold field.

the complex, but most of the deposits follow the orientation of the schists. Two gold fields can be distinguished (Fig. 75): Kilo, near Lake Albert, and Moto or Watsa further northwest. In the Kilo area three west-southwest trending alignments have been distinguished (DUHOUX, 1950a): *(1)* Tsi – Nizi – Montjeri – Luma – Tsele – Labo; *(2)* Sindani – Old Kilo – Dindo – Alosi; *(3)* Pokwo – Yemoliani – Agoye – Galaya – Yedi.

The orientation of low dipping reefs and sills of Kanga and Mongbwalu (Senzere, Creek, Issuru, Kopatele, Guellay), which is related to shear zones in the batholith, is more variable. The structural control of the northern Moto, or Watsa, district is less apparent, however.

Locally, the floating sub-horizontal schists have been thrust over the amphibolites, with carbonate-talc and actinolite schists marking the gliding plane. The complex has been further affected by two or three cycles of young, post-Karroo faulting (WOODTLI, 1961).

The Andissa veins occur in a silicified hornblende–chlorite albitite (previously considered as granite), surrounded by metasediments, near the contact of sericite schists. The country rock has been foliated near the reef contacts, either following planes of lamination or a sub-perpendicular orientation. Contacts are distinct, but xenoliths of both rocks are also immersed beyond these, along planes of anisotropy. Quartz also follows the foliation of albitite, intersected by joints, although it overlaps into the schists. On *Mount Tsi*, a stockwork of quartz lenses (typical thickness three-quarters of an inch) and stringers of less than one-tenth of an inch mineralized the upper reaches of a sericitized albitite, previously considered as a granite. The albitite exhibits an apparently intrusive relationship to the surrounding amphibolites and chlorite schists. On *Mount Nzi* sub-parallel veins form a dense, $^2/_3$ mile long lode system of elongated lenses, exhibiting pinch and swell structures, of up to 3 ft. in thickness. The veins are related to a sericitized–chloritized quartz albitite, similar in composition to quartz diorite, emplaced in amphibolites and metalavas.

The Yedi veins partly follow an en échelon pattern, defined by the plane of anisotropy of quartz–biotite amphibolite, and by sub-perpendicular fractures. At Nzebi, gold bearing, concordant mylonitic ankerites lie in partly mineralized albitites interlayered in diorites; amphibolites cover the outcrop. Here, a first cycle of albitization and silicification of the ankerites and schists was followed by shearing and lamination, and later by the impregnation of gold.

The Senzere deposit near Mongbwalu is confined to the axis and secondary anticline of a synclinorium of talc–chlorite–carbonate schists, derived from dolomite and interbedded with microalbitite. Finally, this complex was thrust over amphibolite. In the large but dissected lens quartz covers bedding planes, planes of two successive foliations, microfissures and microfolds. There are also impregnations, and quartz sometimes embays fragments of biotite schists. Gold values, however, are independent of pyrite and pyrrhotite concentrations.

The substratum of the *Moto* area consists of ankerite–albite–quartz schists of variable composition, foliated by chlorite and sericite. This complex has largely been derived from carbonate rocks by sodic metasomatism. Elongated inliers of itabirite, graphitic and pyrite bearing schists also outcrop. Most of the ore bodies are parallel to the strike (south-southwest), although their dip varies; a common pitch is 25° northeast. The impregnated funnels occur more frequently in areas of inliers, but their composition and structure cannot be distinguished from the barren halo. The *Agbarabo* deposit is identical in composition to the albitized and sericitized ankerites in which it developed; it is bordered, however, by a barrier of itabirite, siderite and quartzite. The ore body, an ellipsoidal, 60 ft. wide funnel, exhibits concentric zones of impregnation which decrease in intensity outwards. However, quartz, sulphide and magnetite values bear no relationship to the grade of gold, which attains 400 dwt./ton. There have been distinguished two generations of quartz. The *Gorumbwa* ore sheet, measuring 200–400 ft. by 10–100 ft., also merges into the country rock. The sheet of values (maximum 200 dwt./ton) pitches 25° north-northeast, but the sulphides rarely contain more gold than the gangue. Gold containing 20–25% Ag (and 3% Cu) is, indeed, the penultimate product of crystallization, followed by galena.

Perhaps half of the gold of the Kilo–Moto district has been eroded into eluvial and alluvial placers. Although concentrations are in the area of primary deposits, gold has also been transported 100 miles down the Aruwimi–Ituri, towards the Congo River. In contrast to the mineralization of the Kilo and Moto roof pendants, the Uéle gold belt, east of Stanleyville, follows the southern (east–west trending) contact of the Kibali batholith and the surrounding schists. In the past, veins and stockworks constituted the principal deposits like Durba, Yindi and Babeyru. Placers and minor vein deposits are largely aligned along north–south trending axes west of the Lake Albert and Edward Rift Valley (Biakatu, Mobissio, Tabili, Bela, Lubena, Mununzi, Luhule, Bilati, Lutunguru and others), but minor placers follow a hoof or a fold shaped arc southeast of Stanleyville. One of these branches is sub-parallel to the Uele belt and the other already follows the southerly trend.

Kivu–Maniema

A belt of minor deposits follows the tightly folded ranges bordering Lake Kivu and the northern basin of Lake Tanganyika. They extend west of the Lake to both sides of the Middle Precambrian (Urundian) Itombwe syncline, wedged in older strata. Minor deposits occur on Idjwi Island in Lake Kivu and in the Twere and Shampumu belt west of the Lake. At Lubongola, 70 miles west of Bukavu, hydrothermal gold and silver are related to the differentiation of uralitized gabbro. Further south at Kabilombo, Mayamoto, Tshikunga, Kiandjo and Miki, gold, cassiterite and occasionally bismuth and arsenopyrite are associated. Further south, at the sources of the Luiko, auriferous quartz lenses which contain hematite outcrop in narrow folds of chlorite achists, floating in granite. The deposits of 'parallel 5°' in the Kiyimbi area, dissected by radial rift faulting, extend this trend. Finally the belt terminates in the Mutotolwa deposit, near Baudouinville, in the uplifted chain bordering Lake Tanganyika.

The position of the major gold belts of Maniema (Fig. 110), between the Congo-Lualaba and the Great Lakes, is controlled by the Kasese batholith and by a series of relatively smaller stocks further west. Gold belts follow the synclinoria dividing them. The principal gold zone starts with the Kima–Binga–Katshungu vein, stockwork and placer districts, extending for 150 miles to *Shabunda*, and branching there towards Kamituga in the east and Namoya in the south. While minor deposits of the 120 mile long Kamituga branch include Nyawaza, Tshakindo and Muta, Wampongo, Dunga, Imonga, Kama and Mukukutshi are aligned along the longer Namoya branch. West of this belt and of a series of granitic stocks, but east of the Lualaba, another belt extends from Songwe to Baseme and Kabotshome (DE KUN, 1957).

At *Kamituga*, 80 road miles southwest of Bukavu on Lake Kivu, an auriferous aureole surrounds the tourmalinized dome of Mount Kibukira. The worlds' sixth largest mass of gold, weighing 50 lb., was found south of this mountain downstream from basalt flows. Nevertheless, the neighbouring quartz veins are barren. Lava (related to the rift cycle) filled valleys protecting them from erosion, and gravels of the narrow, unconsolidated placers, emerging now on crests, contain gold, as does the present inverted hydrographic system. However, the principal deposits are located in the biotite schists and biotitic quartzites of the Smaller (Petite) Mobale. There schists are impregnated by pyrite, arsenopyrite and chalcopyrite adjacent to auriferous quartz veins, but sulphides contain only moderate values of gold. The flat lying A and B lodes traverse the schists and extend for 1 mile, attaining a thickness of 2–4 ft. Apophyses, veins and stringers related to them are abundant in a 1 square mile area. Furthermore a large shattered amphibolite lens was impregnated by gold, as were the schists; spodumene pegmatites subsequently developed. The extensive placers of *Lugushwa*, with subsidiary quartz veins, occur 50 miles southwest of Kamituga.

The *Namoya* (Saramabila) district is situated on the raised rim of the Kama graben, 100 miles west of the Bay of Burton in Lake Tanganyika.

Tightly folded sericite–chlorite schists and quartzites trend west-northwest, as does the gold zone. The last folds drop abruptly along an intraformational conglomerate, downfaulted by the parallel Kama graben, one of the radial outliers of the Rift Valley. Karroo sediments have invaded this channel. In the Namoya (Saramabila) area proper the local trend veers towards the north; the deposits of Mounts Mwendamboko, Kakula and Namoya are arranged there in an échelon structure. A faulted quartz vein of Mount Namoya contains galena and coarse gold. Auriferous quartz veins are also abundant on the two other parallel hills. The *Mwendamboko* deposit is located at the intersection of a shear zone, forming an acute angle with the foliation; the position of the shorter Kakula deposit is similar. Albite and chlorite schists are prominent rocks of the thightly layered mineral area. Arsenopyrite and pyrite impregnate several beds as far as 300 ft. from the undulating elliptical deposit. Immediately below the zone of alteration, the mineralization extends for almost 1,000 ft. with the widest axis attaining 250 ft.; the central depth of the funnel reaches 1,000 ft. or more. In this chimney three major and several minor vein zones have been distinguished, their direction oscillating between the planes of schistosity and shear. However, the orientation of stringers varies considerably in each of these elongated lenticular zones. At different levels and locations various lenses gain importance, but the major lenses continue to great depth, pitching from each other. Furthermore, minor concentrations tend to be spread over a larger section. The bottom of the deposit has a keel-like shape. Alteration to talc, and silicification, are more intensive at the surface, and tourmalinization continues where the quartz bodies coalesce to a 500 by 15 ft. capping. The outcrop covers only a few square ft.

Microscopic gold (200 mesh) occurs both in the quartz veinlets and bodies and in their tectonized aureoles. Arsenopyrite and pyrite contain more gold than their hosts. Scarce chalcopyrite, pyrrhotite and cassiterite traces are also included in the paragenesis.

At *Musefu*, in Kasai province (southern Congo) near the Angola border, quartz feldspar lenses of the quartz-diorite (plagioclasite) phase of the Luisa norite body contain gold. In the Lower Congo gold shows with galena and baryte at the *Isanghila* waterfalls, and in the *Sansikwa* veins.

In the supergene alteration zone of the Katangian, *Ruwe* copper–cobalt deposit (near Kolwezi) gold, palladium, platinum and copper vanadates are associated. While the mineralization of Kambove is similar, gold is part of the copper–uranium–nickel paragenesis of Shinkolobwe.

SOUTHERN RHODESIA

The schist belts of Southern Rhodesia are extensions of those of the northern Transvaal. In the granitized Basement they follow a 300 mile long and 100 mile wide, north-northeast trending alignment (Fig. 36, 50). The major districts include, from north to south: Salisbury, Hartley-Gwelo, Bulawayo-Selukwe and Gwanda. Furthermore, minor schist belts branch off from the south in a north-oriented alignment. They include the Umtali and Victoria gold fields.

Gold has precipitated in quartz veins, crush lodes and impregnations in early Precambrian (Bulawayan) greenstones and metadolerites, interbedded with schists and quartzites. A majority of the deposits favours the metamorphics rather than the granite. However, almost all follow the schistosity of the country rock, the granite being banded or containing oriented roof pendants or screens. The principal deposits include quartz–pyrite–gold fissure veins and impregnations, but the veins of late quartz are generally narrow, with gold occurring in hetero- or homogeneous irregular shoots and lenses. Pyrite may be accompanied by sphalerite, galena, scheelite and other minerals. In rather wide crush zones, impregnated by pyrite (and quartz), gold occurs in irregular shoots of precise boundaries. This paragenetic association may consist of pyrite, arsenopyrite, chalcopyrite, pyrrhotite, sphalerite and galena. Finally, auriferous stibnite is associated with gold, both in veins and in impregnations.

Salisbury belt

The Arcturus replacement bodies or 'reefs', 25 miles east of Salisbury, are part of the Enterprise schist alignment. The Arcturus, Slate and Planet outcrops exhibit variable orientations and pitches in homogeneous epidiorite, injected by acid rocks of variable texture. Mineralization is heterogeneous, arsenical sulphides being disseminated both in the felsite and epidiorite, with the latter also containing pyrite. The Prince of Wales reefs, south of Bindura, have been injected along a mile long east–west trending line, which forms an acute angle with the undulating contact of diorite and grits. Granite outcrops 1,000 ft. to the north, and there is a barrier of greenstone at the same distance to the south. There have been distinguished three ore channels, with parallel fissures playing an important role. Concentrations are related to inclusions and frequency of sulphides, rather than to quartz. On both sides of the contact of the Bindura cupola, veins and stockworks (e.g., Phoenix Prince) are noted for their paragenesis of gold, pyrite, pyrrhotite, (cobaltiferous) arsenopyrite, molybdenite and other sulphides. North of Salisbury, in the aureole of the small Jumbo stock, gold and pyrite bearing veins, including the Flowing Bowl, Commonwealth and Jumbo, are arranged in a concentric pattern. Finally at Shamva, east of Bindura, arkoses are impregnated.

Hartley belt

The principal gold province forms the Hartley Gwelo belt, west of the Great Dyke in the centre of the country (Fig. 76). The irregular belt of suspended metamorphics is 120 miles long and 10–30 miles wide. The deposits tend to concentrate in the 10 mile wide eastern exocontact, but a number of deposits also occur in the batholith in the northeast and on both sides of the contact of the western granitic stocks. The *Cam and Motor* (90 miles southwest of Salisbury, Fig. 77) is the most important of several hundred commercial deposits. There are two perpendicular systems of lodes that traverse a tectonized arkose lens, interbedded in tightly folded and faulted but altered Bulawayan greenstones, which are derived from metalavas and intruded by dolerite. Bed faults have developed into an east–west trending fracture system in their upper horizon. It is cut by its pair, including Cam. Nearby the sub-concordant Motor zone divides greenstones from schists. As quartz veins and stringers are intimately mixed in the lodes with thin or thick sheets of country rock, the wall can be determined only by economic considerations of grades. The lode zone tends to subdivide in depth, first into two and then into three major, and several minor lodes or apophyses partly following the walls, and disappearing and branching off in depth. In fact the zone consists of a series of lenses in a modified pinch and swell structure, welded together in low pressure areas. Pyrite and arsenopyrite impregnate mainly schistose fragments, and arsenopyrite also favours greenstone. Conversely, gold grains as large as

1 mm are found in quartz. Similarly, stibnite occurs in quartz and in thin or lenticular bodies, particularly in greenstone.

At *Dalny*, 20 miles northwest of Cam and Motor, zones of quartz stringers, schist, and occasionally larger quartz lenses fill a major shear zone which trends northeast and dips 65° north in carbonatized greenschist. A dolerite dyke, forming an acute angle with the shear, intersects the main ore body.

In the Golden Valley, 12 miles northwest of Gatooma, a mile long, grey-white quartz lens strikes generally north-northeast, and dips 40° west. The faulted lode, containing felsite and greenschist xenoliths, is cut by dolerite. Gold shoots are widely separated, with scheelite patches accompanying widespread pyrite. At some distance around Chakari, quartz and schists are mineralized by gold, pyrite, arsenopyrite and pyrrhotite, and along a line extending 20 miles east of Gatooma quartz reefs traverse pre-Bulawayan granite.

At Pickstone deposit, 20 miles east-northeast of Gatooma, bulging and branching quartz lenses are injected into a shattered greenstone. Other deposits include Thistle Etna, Inez and Shepherd's Reef (SWIFT, 1961).

The complexly faulted and tightly folded anticline of Kanyemba (2 miles south of Umsweswe) pitches 30° to the southwest. A 9 inch –4 ft. thick reef, containing pyrite and arsenopyrite associated with gold, traverses chloritized greenstone there.

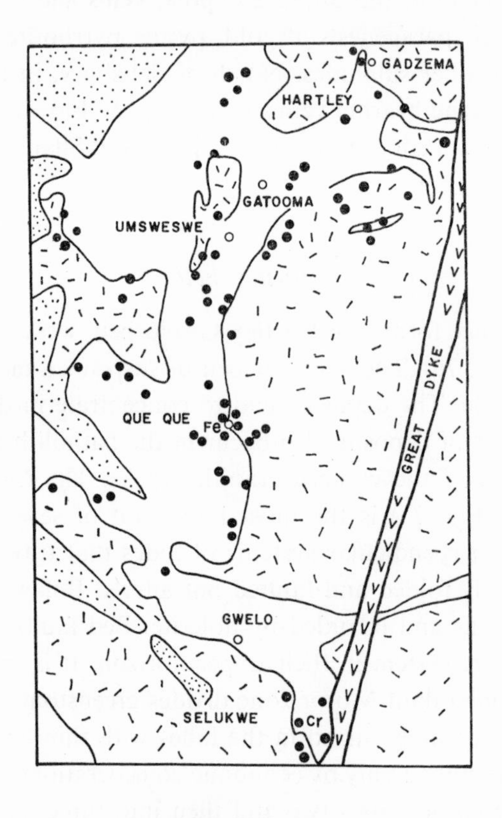

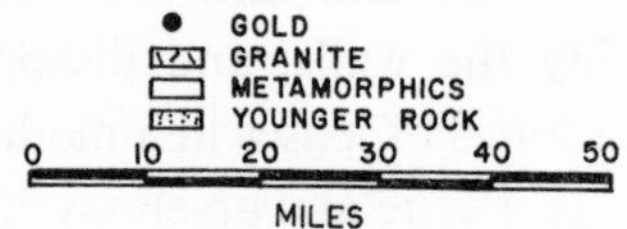

Fig. 76. The Hartley and Gwelo area.

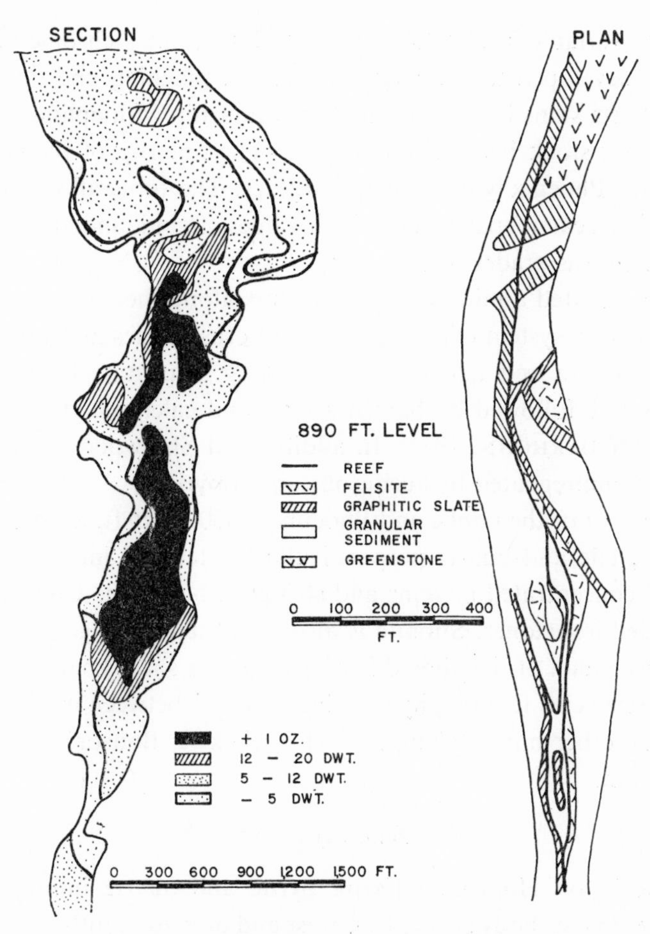

Fig. 77. Section and plan of the Motor Lode, central.

At Gadzema, north of Hartley, metamorphics surrounded by granite have been mineralized, and in the Giant mine a layer of quartzite, ironstone and chlorite schist has been impregnated.

Gwelo district

In the Que Que belt, arsenical impregnations are distinguished from younger and more important auriferous quartz veins containing stibnite or pyrite.

The *Phoenix* mine is at the Que Que steel centre, while the *Globe* lies 500 yards to the northeast. The lode zone strikes along a granitic contact, but pitches differ. This zone of weakness has been rejuvenated in several cycles. Bulawayan dolerite dykes were injected into the earlier (Sebakwian) metabasites under pressure from the east, and early barren quartz reefs penetrated into new fissures formed in the dolerite. Subsequently, a second injection of dolerite and a lamprophyre was followed by the precipitation of quartz and gold, and by a late phase of auriferous stibnite bodies. Accompanying minerals include tetrahedrite, pyrite, scheelite, sphalerite and galena. At *Phoenix*, unlike the Globe, fissure fillings and their numerous apophyses are distinct. The fractures have been reopened several times, however. The compact, banded quartz contains macroscopic gold, particularly concentrated in the wall zone. The transformed Sebakwian metabasites have been interlaminated in variable proportions of magnesite, talc, chrome, mica and quartz. Fissures are well developed in the magnesite rocks, pinching and degrading in the talc schists, but the quartz veins have an intimate relationship with aplites and porphyries, the border of the parent stock being microgranitic. The outcrop of the Globe is in granite, with the trace of the Phoenix lying in metamorphics. The footwall dolomite body, associated with keratophyrs, carries particularly high values of gold, and at lower levels a similar, elongated carbonate lens grades into quartz.

The Gaika lode, located a mile south of Phoenix along the contact of the same narrow schist belt, consists of a system of veins, shear and crush zones and various impregnations in alternating magnesite and talc schists, including serpentine lenses. The injection of porphyritic dykes was followed by barren quartz and calcite, and finally by quartz gold and stibnite (SCHNEIDERHÖHN, 1931). In addition, at Sherwood Starr a pipe of banded ironstone has been impregnated by auriferous arsenopyrite.

At Bell, 9 miles west of the Globe and Phoenix, a 350 by 25 ft. lens of crush breccia consisting of chlorite, talc and quartz perpendicularly intersects impure quartzites. Gold is disseminated and concentrated in veins and stringers, but in the Connemara deposit gold impregnates banded ironstones. Stibnite is also associated with gold at Riverlea. Finally, along the western contact of the altered Selukwe stock a shear zone has been mineralized. At Wanderer, quartz veins favour phyllites interbedded between ironstones and grits, but the majority of the other mines (Camperdown, Tebekwe, Bonsor, Blue Bird and Yankee Doodle) are dormant.

Bulawayo belt

In the Turk deposit, on Huntsman Farm, pyrite and quartz impregnate carbonatized greenschists; the lenticular body strikes east-west and dips 70° south. In the Sabiwa deposit, which includes *Long John* between Bulawayo and Nicholson, a series of lenses, oriented north–south and dipping 70° west, impregnates banded ironstone. Lenticular gold and sulphide shoots are spread along the strike. The paragenesis includes pyrite, arsenopyrite and pyrrhotite.

While the fissure filling *Fred* vein near Filabusi (asbestos) strikes east and dips south, one of the ore shoots exhibits a westerly pitch. Xenoliths of sheared greenstone are cemented by quartz in the fracture; the ore contains 1.3% arsenopyrite. In the Slope and True Blue deposits, near Filabusi, phyllites have been mineralized between a barrier of serpentine and granite. Other active deposits include Marvel and Teutonic. The Germania reef, near Belingwe, follows the contact of the granite, and in Essexvale (south of Bulawayo) a greenstone belt and granite are traversed by faults. The Bushtick, Eveline and Wollvinder shoots fill in fissures. Scheelite accompanies gold in the stock. At Gwanda (south of Bulawayo), a 40 mile long, northwest striking shear zone is distinguished by arsenical impregnations and quartz sulphide veins. At Antelope, 60 miles south of Bulawayo, gold, magnetite, pyrrhotite, pyrite and garnet have crystallized in a fissure.

Auriferous quartz veins, such as those of the Lonely and Queen mines, occur in the Bembesi Valley north of Bulawayo. At Sunace the reefs developed on thrust planes, and in the Turk mine shattered and carbonatized metabasites include lenticles of quartz and pyrite. In the brecciated green schists of the B. and S. mine gold is associated with arsenopyrite, and grades increase at the intersection of ironstones.

Umtali

In the Penhalonga Valley, near the Mozambique border, e.g., in the Rezende mine, a varied sulphide paragenesis is found in veins traversing hornblendic quartz-diorite, but only rarely ironstones. Stockworks include Old West, Iona West and Howat's Luck.

SOUTH AFRICA

While minor occurrences are widespread in the Transvaal, e.g., at Barberton and Pilgrim's Rest, the world's principal gold province follows the northwestern rim of the Witwatersrand bracchy-syncline (Fig. 78), a 200 mile long and 100 mile wide structure surrounding

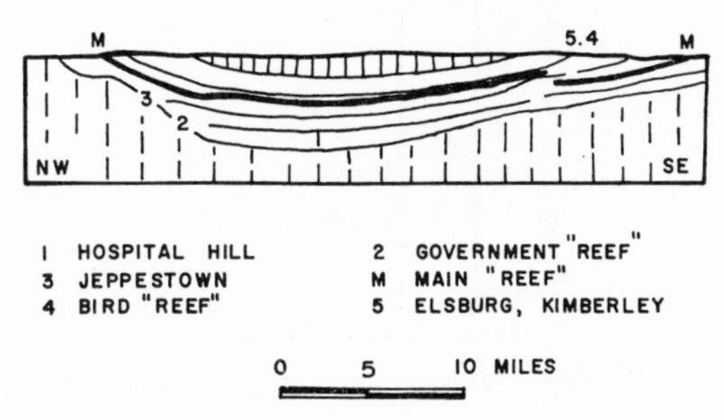

Fig. 78. Section of the central Witwatersrand.

the Vredefort dome. Four districts are distinguished: the Orange Free State, Klerksdorp, the Witwatersrand, and Heidelberg–Evander (Fig. 46, 50).

Orange Free State

In the Odendaalsrus–Virginia gold field, 150 miles south-southwest of Johannesburg, bankets are buried at a depth of from 3,000 to more than 10,000 ft. The western graben of the V-shaped field extends for 40 miles in a north-northwesterly direction, and the eastern syncline, which is even more dissected, trends 12 miles towards the northeast. The average width is 7 miles. The greatest concentrations are located around Welkom, in the middle of the western branch (Fig. 40).

The gold field has been intensely faulted. The (inferred) Border fault limits the field in the west, with the dextral De Bron fracture zone constituting the eastern boundary of the graben (COETZEE, 1960). The triangular, central horst divides the two branches of the field. The Ararat fault, the eastern limit of the Kameldoorns–Welkom block, intersects the graben in an acute angle, and west of Virginia north-northwest striking faults border the Rift Valley. The Homestead fault dragged the western limb of the syncline upwards. The old and long strike faults appear to have a downthrow towards the east; in the north, however, the Erfdeel fault, one of the younger transverse features, thrusts strata down. Faulting appears to have started in Witwatersrand times, but during the following lower Ventersdorp volcanic cycle some of the tectonic movements were reversed.

Intraformational folding along the north-northwest axis has been inferred from the transgression of local synclines at Riebeck and Free State Geduld Mine. Some folding also took place after the Witwatersrand. Along the western edge the sediments are tightly folded. Indeed, in the western branch most of the dips are easterly, with the exception of the south-

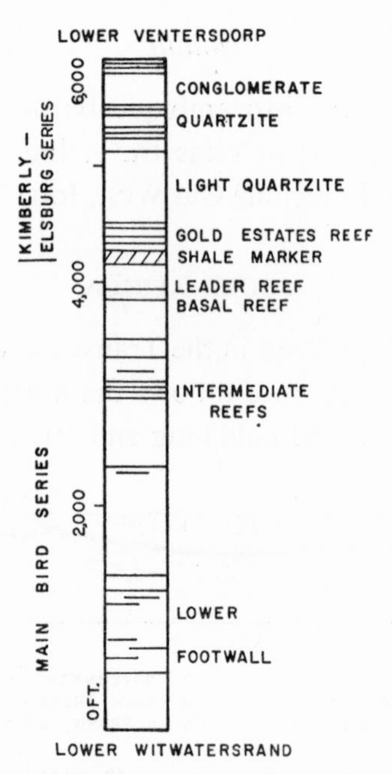

Fig. 79. The upper Witwatersrand division in the Free State.

ern and northern extremities which exhibit northern and western dips, respectively. Finally, in the shorter western branch the bankets or reefs trend north and west.

The average thickness of the earliest, main stage of the upper Witwatersrand Division is 3,000 ft. (Fig. 79). During this stage, banded gray and light gray quartzites were deposited, with some grit; in the following Bird Stage, grits and conglomerates were covered by speckled quartzite. The intermittent khaki shale divides the Basal and Leader reefs, with the big conglomerate separating the A and B reefs in the overlying Kimberley Stage. The Elsburg Stage starts with a conglomerate horizon, announcing alternating clastic sediments and quartzites. Finally, andesitic Ventersdorp lavas conclude the sequence. However, 4 miles east of Odendaalsrus, little sorted, highly pyritic, quartz and chert conglomerates of the young Van den Hevers Rust or Rainbow reefs form a 5 mile long band of high values. Gold was probably eroded from pre-Elsburg reefs which were tectonized in the west. The Gold Estate Leader is also young, but its extension appears to be smaller.

The thickness of the group composing the reefs attains 850 ft. in the Lorraine-Jeannette area and 230 ft. elsewhere.

The oligomictic reef conglomerates consist largely of sorted, closely packed, rounded quartz pebbles of about an inch in a vitreous groundmass of quartz, with pyrite forming transgressive lenses. In the principal Basal Reef, underlying most of the basin, facies changes are common: At Odendaalsrus and north of Welkom the reef forms a narrow bed, becoming coarser in the south of the President Brand Mine and in the Harmony Lease; again, in the Virginia and Merriespruit mines, the thickness of the reef varies between 0 and 10 ft. Where well developed, conglomerate and quartzite beds alternate; frequent pyrite stringers are also mineralized. At the hangingwall, which grades into the quartzite, the number of inclusions diminishes. In the St. Helena mine the Basal Reef, which attains a thickness of 18 inches, exhibits large pebbles in a dark matrix. The greatest channel width reaches 4 ft., and best values of gold are associated with carbon which accumulates at the base of the bed. The St. Helena area is also distinguished by facies changes: mainly the increase of grain size of pebbles and pyrite.

The slightly mineralized Leader Reef is located at an average of 50 ft. above the Basal Reef, and the B and A reefs are 380 and 440 ft. higher. Finally, the Van der Hevers Rust reefs overlie the base by 800–1,700 ft.

Because of the discontinuities, correlations are rather difficult; however, radiometric logging helps: The Basal and Leader reefs have been correlated to the Bird Reef group of the Rand, the A and B reefs to the Kimberley Reef group, and the Van der Hevers Rust reefs to the Elsburg reefs.

Since intraformational unconformities play an important role in the mineralization, reefs frequently occur above them. Erosion appears to have prevailed along the fossil shores of the sedimentary basins. The Basal Reef is frequently subconcordant in the median and eastern part of the V formed by the district and the 'Leader' is transgressive on the Basal Reef, but the reef is cut off by B. Finally in the west of the graben, adjacent to a pre-Ventersdorp fault, the top of the Elsburg Stage is transgressive on the folded Upper Witwatersrand Series. It is believed that the basin receded periodically because of local movements, thus enhancing erosion or a hiatus of sedimentation followed by younger strata. The resulting pattern is an alternating succession of concordant edges and step-like unconformities.

Values of the Basal Reef appear to average 500 inch-pennyweights (in.-dwt.) in 50% of the cases (8% exceeding 1,000 in.-dwt.), 20% attaining 90 in.-dwt. and 10% 0–50 in.-dwt. Ore shoots or bands divided by low grade strips are oriented northwest, and the broader ore zones of the western branch appear to consist of a series of thinner bands. In the Virginia mine, the Leader has eroded the Basal Reef over large areas of high grades, and this banket is also believed to exist in the Merriespruit and Harmony mines. On the other hand, the widespread but low grade B Reef appears to contain commercial concentrations only at Odendaalsrus. Lenses of the Middle Reef, including more than 1,000 in.-dwt. gold, occur in various layers between the Basal and Leader reefs.

Klerksdorp

At Klerksdorp, 100 miles west-southwest of Johannesburg, several basins, characterized by irregular or streaky gold values, trend in parallel directions for 25 miles (Fig. 42).

Great concentrations have been formed in the ancient and uraniferous Dominion Reefs, separated only by a band of coarse quartzite from the underlying, early Precambrian granite; this sequence is best developed south of Hartebeestfontein. Lenses of the upper quartzites of the Government Reef Series have shown good values at the Afrikander and Barbosco mines (in the west). Conglomerate lenses of the Lower Jeppestown quartzite (Lower Witwatersrand Series) are also mineralized there and at Buffelsdoorn.

The *Vaal Reef*, correlated to the Main-Bird Series, is distinguished by 50% of the district's concentrations in the Stilfontein, Hartebeestfontein and Buffelsfontein mines, near Strathmore (in the south). The thickness of this small-sized conglomerate decreases from 80 inches at Buffelsfontein to 3–4 inches at the other localities, but gold accumulates mainly in the thin basal carbon horizon. At Klerksdorp Townlands, two to three conglomeratic bands of the Commonage Reef in the early Main-Bird Series are spottily mineralized, and in the overlying 20–200 ft. thick, discordant Gold Estates Reef, commercial concentrations are even more restricted.

The Elsburg reefs consist of *(1)* four bands in the upper group, separated by 80 ft. of quartzite from the *(2)* lower group of bands which are spread over a thick layer of quartzite. The Elsburg reefs have been partly eroded and rearranged below the cover of Ventersdorp lava. To conclude, the only commercial concentrations of the younger Black Reef series have been exploited at Nooitgedacht. Most of the Vaal Reef lies in the five mile wide graben, bordered by the Buffelsdoorn and Kroomdraai fault. Rift and block faulting have been accompanied by the effusion of the Ventersdorp lavas there. The dip of the Vaal Reef varies between sub-horizontal and 25°, and its strike oscillates according to transverse and normal faults. Furthermore, the eastern lip of the graben is complicated by younger parallel faults producing minor horsts.

East of Klerksdorp (and southwest of Johannesburg) several conglomeratic inliers of quartzites surround the Vredefort–Parys stock in a long arc; these bands are poorly mineralized.

Witwatersrand

The northern rim of the brachy-syncline, the Witwatersrand, can be divided into the Far West, West, Central and East Rands. East of Boksburg, the Karroo covers Witwatersrand sediments; further east, at Springs, Transvaal metasediments also appear. The granites of an early Precambrian anticlinal batholith extend from Heidelberg to Delmas in the north, while Witwatersrand sediments reappear near Trichardt (Fig. 41). Beyond Randfontein, the Western Rand divided by the Bank fault trends west-southwest. Concentrations follow the northwestern rim of the trough and extend its bends. In the median part of the syncline the dip of the beds diminishes from 60° to 35°, decreasing farther to the subhorizontal. Transverse folding has taken place in the east. Strike faults, similar to those of the Orange Free State field, mainly affect the Lower Witwatersrand Division. They include the Witpoortje and Roodepoort in the Western Rand, and the Rietfontein in the centre. Faulting is sub-contemporaneous with the extrusion of the lower Ventersdorp lavas, and reached its greatest intensity before the Transvaal cycle. Minor, complicated normal, reverse and pivotal faults also cut the reefs.

Mineralization is widespread among the conglomeratic members of the Upper Witwatersrand Division. Commercial values are restricted to the *Main Reef* and, to a lesser degree, to the overlying clastic beds. The thickness of 600 ft. of quartzite and grit, overlain by the 3–12 ft. thick Main Reef, diminishes from Florida on the West Rand outwards. In the West Rand this reef appears 300–400 ft. above the lower Witwatersrand group, which is distinguished by the greatest concentrations. Further west, and 130–200 ft. below the Carbon Leader, is the principal gold–uranium carrier. Lenses of the Black Bar may divide the Main Reef from its Leader, intraformational channels being filled by detriti. The thickness of the most important ore horizon of the centre, the larger grained Main Reef Leader, varies between a few inches and 10 ft., and west of Crown Mines towards the Consolidated Main Reef Mine, the Leader thins out from 8 to 1 ft. and becomes irregular lenticular.

In the centre 70–100 ft. of quartzite are overlain by the South Reef, but at the Simmer and Jack mine this banket borders the Leader. A similar banket of alternating reef and quartzite also occurs in the West Rand. The Upper Leaders of the West consist of several ore shoots and horizons. Furthermore, the Johnstone and Livingstone reefs are younger local features of the Far West and the West Rand. On the other hand, in the Far East Rand high grade series of lenses are embayed in low grade layers.

The 300 ft. thick Bird Reef Group, so well developed in the Orange Free State, lies in the West Rand between Randfontein and Consolidated Main Reef Mines, 1,600 feet above the Main Reef Group. However, the gold bearing White Reef lies 100 ft. beneath the uraniferous Monarch Reef.

On the West Rand, and also on part of the East Rand, bands of conglomeratic pebble of the auriferous Kimberley Reef Group are interbedded in 600 ft. of quartzite, 1,500–3,500 ft. above the preceding group. In the eastern part of the West Rand gold values increase as a function of pyrite content. The unconformous May Reef (belonging to this group in the East Rand) has eroded its gold and uranium from earlier bankets. The large pebbled but diffuse Elsburg Group is included in quartzite and grit at the top of the sequence; widespread but medium grade values are found in these reefs southeast of Westonaria.

Finally, in the western part of the Witwatersrand, the terminal Ventersdorp Contact Reef has eroded gold when in contact with underlying mineralized reefs.

Heidelberg and Evander

The thickness of Witwatersrand sediments, reemerging between Heidelberg and Greylingstadt 25 miles southeast of Johannesburg, is considerably diminished compared to the type locality. The heterogeneous and lenticular banket is also thinner. The earliest Witwatersrand Government Reef Series, at Edenkop, the Main Reef Group at Molyneux mine, and the late Kimberley Reef of Roodepoort contribute commercial concentrations. At Evander (70 miles east-southeast of Johannesburg, Fig. 41) as at Heidelberg, the Main Reef Group is insufficiently mineralized, but the overlying sedimentary column is thin. At Winkelhaak, the Kimberley Reef overlying the Kimberley shale and quartzite unconformably, dips 25–30° north; the banket carries 300 in.-dwt. of gold.

Genesis

Crystals and grains of variable size form 2% of the Rand banket; in an Orange Free State sample, the heavy mineral concentrates contain 92% pyrite, 3% zircon, 2% chromite and 2% gold. The size of one half to two thirds of the Rand gold is between 10 and 70 μ. Angular and jagged grains of gold, 1–100 μ in diameter, containing only 10% Ag, occur in pyritized patches of the groundmass, but traces are also found in veins traversing the pebbles. Although gold is not genetically related to pyrite, it precipitated on and around pyrite, and in the Orange Free State macroscopic gold covers cavities of quartz and sulphides. Pyrrhotite sometimes associates with gold and chalcopyrite occurs in traces, e.g., in the Virginia area. The indicator tucholite, a hydrocarbon irradiated by uraninite, forms concordant seams consisting of grains perpendicular to the bedding, particularly at the base of the Carbon Leader and Main Reef in the Free State. Hydrocarbon itself, however, is scarcer than neogene quartz. Conversely, grains of mostly detrital uraninite (see 'Uranium' chapter), which attain 80 μ in diameter, are more abundant than are iridosmine, ilmenite, leucoxene, small green diamonds, chlorite, galena and sphalerite.

The banket differs from non-mineralized sediments: *(1)* pebbles consist almost exclusively of quartz; *(2)* they are well sorted; *(3)* a direct relationship exists in several areas between pebble size and greater concentrations of gold. According to Reinecke, the following rule would prevail:

$$\text{grade (g/ton)} = \text{largest pebble size (mm)} - 7$$

In parts of the Rand and the Free State, gold often occurs in pay streaks oriented northwest. The length of these scoured channels is generally ten times as great as their width, and in the West Rand their length reaches almost a mile. However, in the East Rand it is only a third of this. The streaks are surrounded by low grade areas; they may strike parallel, as in the Central Rand, or fan out as in the East. In low grade zones the streaks become lenticular, with the conglomeratic horizons contracting. Since earlier bankets have contributed to each conglomerate after each gap of sedimentation and resorting, unconformities play a considerable role in the accumulation of gold, especially between the Vaal and Basal reefs. A cyclical pattern is evident. Each cycle has its own rhythm of deposition, terminating rather abruptly.

The conglomerates are believed to have been deposited in a generally subsiding inland sea, perhaps in a sub-arid climate affecting the barren Precambrian landscape. During the long sedimentary history, cycles of more intensive erosion produced the mineralized oligomictic banket. Transportable and soluble gold may have been derived from far lying deposits. Pyrite is atypomorphic, hardly justifying the hydrothermal hypothesis. The discovery of rounded pyrite by P. RAMDOHR (1958) indicates, indeed, its alluvial origin. Nevertheless, 2,000 million years of subsequent sedimentation have strongly affected the deposits, recementing the conglomerates and recrystallizing the gold.

Barberton

Barberton is situated on the latitude of Pretoria, near the Swaziland border. The deposits are located on both sides of the contact of the ancient Swaziland rocks and of the intrusive hornblende granite phase, with many occurrences on the margin of a later biotite granite. The latter Nelspruit granite may have contributed much of the gold, which is also found disseminated with sulphides in cherts and shales. Mineralization is related to minor folding, shearing and faulting. Indeed, ore lenses are found in strike or fracture shear zones, but the mineralization of transverse fractures is variable. Ladder reefs occur in tension fractures. Nelspruit pegmatites at the Consort mine traverse a mineralized contact zone; impermeable cherts determine the location of ores which, incidentally, fade into the country rock.

In the shear breccia, silver, pyrite, arsenopyrite, and pyrrhotite (associated locally with chalcopyrite, galena, quartz, carbonates and graphite) are disseminated along with gold. Ore types are characterized by pyrrhotite, pyrite, lead and antimony minerals, while submicroscopic gold is enclosed in pyrite and arsenopyrite. Surface oxidization was intense.

In the Consort mine arsenopyrite and gold are disseminated along a silicified contact. In the Sheba mine replacement gold occurs in a ramified fracture system, and the greatest concentrations of the Golden Quarry are located in shears radiating from a zone of mylonitization. Other deposits include Fairview, Three Sisters, the banded ironstones of Vesuvius and Makonjwa, French Bobs, and Barbrook.

Pilgrim's Rest

Varied deposits are situated on the Drakensberg, 70 miles north of Barberton and east of the Bushveld. Poorly mineralized large fracture fillings or sheets extend for 30 miles at Bokwa. Sub-horizontal contact sheets or sills, distributed over a column of 8,000 ft., mainly impregnate favoured zones of weakness; they exhibit bed slip and mylonitization. The greatest concentrations of this type are in the area of Vaalhoek, Graskop and Pilgrim's Rest, principally in the Dolomite and Pretoria Series of the post-Witwatersrand, Transvaal System, but sediments correlated to the Witwatersrand have also been discovered. Offshoots of the sub-horizontal sheets are intensely mineralized on Mount Anderson, in the Finsbury Valley. Their point of departure shows particularly high values, and ore bodies arched on one side project from the Sandstone reef and Klein Sabie. At Lindenau, gold and bismuth ochre are found in a polyp-shaped irregular body and at Elandsdrift in a quartz blow.

Pyrite, arsenopyrite, bismuthinite, sphalerite and galena are associated with gold in the horizontal and vertical quartz carbonate sheet fillings. Gold values grow with auriferous sulphide content of coarse sediments, e.g., the Black Reef of the Transvaal System (which is associated with iron and manganese in Griqualand). Numerous deposits were also altered and then transported in eluvia and alluvia.

Minor occurrences

In the Schweizer–Reineke district, 70 miles west of Klerksdorp, auriferous veins are injected into Swaziland banded ironstones, and in northeastern Transvaal, in and along the Sutherland Range of the Klein Letaba district, reefs and mineralized contact zones are found in a Swaziland syncline, intruded by granite. At Eersteling and Mount Maré, north of the Bushveld, the veins are injected into early Precambrian granite, and in Natal and Zululand the quartz veins are also of Swaziland and lower Dominion Reef age. Other small occurrences are scattered over the Transvaal and the neighbouring area of Mafeking.

OSMIUM, IRIDIUM AND RUTHENIUM

Most of the world's resources of these three platinum group metals are associated with gold

TABLE CXIX

PLATINOIDS

	Witwatersrand gold %	*Rustenburg platinum* %
Osmium	34—35	0.1
Iridium	30—33	0.1
Ruthenium	13—16	1.0
Platinum	11	
Rhodium	0.7	
Gold	0.6	

in the *Witwatersrand* and Orange Free State bankets. In them, values range from 3 to 0.3 mg osmium–iridium per ton. The ratio of the heavier metals to gold varies between 500 and 10,000/1, and averages 4,000/1. Furthermore, some ruthenium also occurs in the Bushveld platinum ores (Table CXIX).

Iridosmine exhibits a sympathetic relationship with gold, especially with coarser grains, with uraninite, and with rare green diamonds (RAMDOHR, 1958). Rounded grains of the alloy tend to occur in the lower portions of the reefs or in carbon seams. Moreover, sulphides including skutterudite cover some of the osmiridium and platinum grains. In the East Rand, iridosmine concentrates in the Main, Leader and the Kimberley reefs. However, at Modderfontein the last auriferous banket, the Black Reef, also contains osmiridium. On the other hand, the fields extending from the West Rand to the Orange Free State are poorer, and the Central Rand only shows traces.

In Southern Rhodesia, traces of iridium are associated with the Antenor gold deposit at Gwanda.

URANIUM

In Africa uranium is widespread in a variety of parageneses. Sedimentary deposits of the Katanga Copperbelt contributed the fissionable material used in the Second World War. In Gabon, the Moanda fissure fillings represent another hydrothermal variety. Uranothorianite occurs in carbonatites and Malagasy pyroxenites. Similarly, uranium is an accessory element of the rare metals in Malagasy and Mozambique pegmatites, and of Congolese columbite placers. Furthermore, the granitic ring complexes of Nigeria and the Niger Republic contain uranium, as do north and west African sedimentary phosphates (Fig. 4).

However, in the largest deposits of the Witwatersrand System uraninite and tucholite are associated with gold (Fig. 50). These minerals also occur in the underlying ancient Dominion Reef System, at the base of the overlying bankets of the young Ventersdorp Contact Reef and Black Reef, and at the base of the Transvaal System (see 'Gold' chapter). The pyritic material of the young reefs strongly resembles the Witwatersrand banket from which it was eroded.

DOMINION REEF SYSTEM

North and northeast of *Hartebeestfontein* in the Klerksdorp basin two reefs contain uranium accompanied by gold. In fact uranium values are greater than in the Rand, but gold grades are inferior. While most of the strata consist of lava flows, at the beginning of the cycle sandstones and quartzites (including lenses of grit and conglomerate) were eroded from the Basement granite. In the two reefs rounded quartz pebbles are embayed in a groundmass of quartz and sericite. The thick lower reef, which contains disseminated uraninite, fills low points of the granitic relief. On the other hand the upper Reef, consisting of fine pebble and grit, covers most of the area with a 4 ft. – 1 inch thick banket. Since uranium values are an inverse function of its thickness, 1 inch thick strips contain 1% U_3O_8. Containing 2.5% pyrite, some cassiterite and monazite, most of the banket was probably eroded and reworked from the underlying conglomerate.

WITWATERSRAND SYSTEM

Except for the Afrikaner and Inner Basin Reef of the Government Reef Group at Klerksdorp, the fine grained, lower Witwatersrand sediments contain little uranium. In the upper stage facies changes, signalled by large grained clastic material, enhance the transport and precipitation of commercial concentrations of uranium. Nevertheless, the best oligomictic conglomerates represent only 4% of the column. With gold and uranium values exhibiting a sympathetic (though irregular) relationship, most of the well known gold reefs, especially the Carbon Leader in the Main Reef Group, and the May and related reefs in the Kimberley Group of the East Rand, contain uranium. In the West Rand, the White, Monarch and other reefs of the Bird Reef Group are uraniferous, as is the Klerksdorp equivalent, the Vaal Reef. In the Orange Free State the Basal and Leader reefs are also mineralized. In the Commonage Reef (Main Reef Group) and certain Elsburg bankets, including the Van der Hever's Rust, and reefs at Klerksdorp, uranium constitutes a by-product. Furthermore, uranium shows in the A, B, and Rainbow reefs of the Kimberley Elsburg Series of the Orange Free State.

Uranium tends to concentrate above manifestations of sedimentary interruptions, erosion, and channelling distinguished by partial recycling of deposited material. Unlike syngenetic copper deposits, the mineralization does not transgress sedimentary horizons. Indeed, uranium values are homogeneous on a regional scale but variable on a local scale, where greater values accumulate in pay streak channels more often at the base of bankets than at the top or middle. Fingering out in lenses or in a seam of tucholite, the thin, uraniferous, oligomictic conglomerates generally jut into the basin farther than barren quartzites In the Footwall or Hybrid Reefs of the East Rand minerals have been deposited parallel to the banks of meandering, deep scour channels. Values decrease as a function of the distance from the banket. On the other hand, islands of reef conglomerate are also found in the midst of barren quartzites.

The cycle of mineralization follows a distinctive pattern: a single peak anomaly in the Basal Reef contains the best values, two to three maxima are in the Leader Reef with gold values at the base, a single peak occurs in the B reef and treble maxima in the BPC and A reef. Grades attain 0.30 –0.7 pounds U_3O_8 per ton (150–350 g/t). Most of the mines contain commercial concentrations.

Although finely disseminated in the quartz sericite groundmass and in the hydrocarbon, uranium granules tend to concentrate even more at the base than other heavy minerals. Transported rounded grains have occasionally been replaced by hydrocarbon. Moreover, secondary pitchblende occurs in specks and inclusions, occasionally associated with the original uraninite.

URANINITE

Uraninite is more abundant than tucholite in all three systems. The Dominion Reef, the bottom layer, contains little uraniferous hydrocarbon. Conversely, at the top of the column some uraninite has accumulated in distinct leads of Black Reef quartzites. However, grades are low in them and tucholite predominates. Alteration of uranium minerals is, indeed, most intensive in the Black Reef, but hardly perceptible in the Basal Reef.

Widely but heterogeneously disseminated in the conglomerates, *detrital uraninite* is believed to be derived from a high temperature granitic environment. The average diameter of uraninite is 75 μ in Witwatersrand bankets and 100 μ in the Dominion and Ventersdorp Contact Reef. Clusters of grains are embayed in phyllosilicates of the groundmass, but oval grains occur sparingly in pyritic quartzites.

In the Dominion Reef, galena and gold are frequently included in uraninite and galena also fills cracks of the latter. At Hartebeestfontein a lens of the Vaal Reef shows uraninite concentrations. On the other hand, in the Ventersdorp Contact Reef a greater proportion of uraninite occurs in quartz and associated with shulpides. Uraninite appears to invariably contain 1–2% of rare earths, 1.5–2.5% thorium, and 1–1.5% zirconium oxides.

Covering locally detrital uraninite, zircon and chromite, and partly replacing some pyrite grains, *secondary uraninite* fills in patches and veinlets mainly in the groundmass of the banket and also in the quartz. Veinlets attain a thickness of 70 μ and a length of 400 μ, but patches cover as much as one square mm. They are particularly large in the ancient Dominion Reef. At West Rand Consolidated, East Champ d'Or, Luipaardsvlei, Randfontein Estates, Stilfontein and Ellaton in the West, Far West Rand and Klerksdorp districts; secondary uraninite is the principal radioactive mineral of the Bird Reefs Group. In the Ventersdorp Contact Reef, the morphology of the mineral is similar. However, in the pyritic quartzites of the Witwatersrand System and in the Black Reef the mineral has mainly replaced quartz.

TUCHOLITE

Tucholite, an amorphous uranium bearing hydrocarbon, contains 30–70% C and 1.5–9% U_3O_8. With its granules accumulating at the base of each banket, this substance is found in every epoch, but in the old Dominion Reef only its traces show and its values increase upwards in the stratigraphic column. On the Far West Rand, at Blyvooruitzicht, Doornfontein and West Driefontein, tucholite concentrates in parting planes and thinned out facies of the banket, such as the Carbon Leader. With its concentrations decreasing upwards in the Carbon Leader of Blyvooruitzicht, columnar laminae of tucholite stand on the footwall. Moreover, in the Western Reefs Mine, tucholite accumulates in the Ventersdorp Contact Reef.

The dimension of grains varies betweeen 0.2 and 0.45 mm, but in the quartzites the grain size decreases to 70–300 μ. Hydrocarbon tends to replace uraninite in tucholite, in which they were deposited together. However, uraninite crystallites create reaction rims in the surrounding substance. The hydrocarbon also includes infrequent sulphides, mainly galena and quartz. Occurring in microfissures, gold also replaces these sulphides. Uranium minerals, especially detrital uraninite, tend to be 'tucholitized'. Although non-uraniferous hydrocarbon or bitumen is rare, gases escape from the reefs (see 'Carbon dioxide' chapter).

CARBONATITES

In the Transvaal five thoriferous complexes have been identified:

(1) Phalaborwa, located near the Kruger Park; *(2)* Glenover in the northwest near the Bechuanaland border; *(3)* the belt of Goudini; *(4)* Tweerivier, and *(5)* Kruidfontein in the western Bushveld.

At Phalaborwa (see 'Vermiculite', 'Phosphate' and 'Copper' chapters) euhedral uranothorianite, which averages a diameter of 0.3 mm mainly crystallizes in the inner, magnesian sövite. Especially with magnetite, uranothorianite is also disseminated in the outer sövite and in other rings. Expressed in uranium oxide units, the grade reaches the mark of 0.16 lb./ton. Since the thorium/uranium ratio averages 3/1, this proportion is, in effect, lower than in the Malagasy pyroxenites. Furthermore, uraniferous columbite and alteration products occur in the Glenover complex.

Surrounded by breccia, the large Goudini carbonatite (6 by 4 miles) has been extruded through the ultrabasics of the Bushveld.

CONGO

Uranium is especially widespread in the Copperbelt in the Congo, but its deposits are separate from the copper–pyrite beds. *Shinkolobwe–Kasolo* is located on the southern edge of the Jadotville section where the Katanga geosyncline bends west (Fig. 28). Accompanied by numerous secondary faults, the overthrust Tantara fold traverses an overturned, double syncline there.

Following the magnesium enrichment of Roan dolomites, the ascending uraniferous solutions penetrated along a zone of inclined dolomitic shales, and expanded sub-horizontally under a nappe. Selenides, monazite and molybdenite were introduced during a later phase of chloritization, followed by the precipitation of the cobalt–nickel sulphides. The sulphides were then crushed and copper was deposited. Whereas molybdenite is widespread, copper remains scarce. Finally, during supergene alteration, uraninite was oxidized (DERRICKS and VAES, 1956). We can distinguish the following phases:

(1) quartz–tourmaline–apatite–chlorite–talc; *(2)* uraninite; *(3)* Fe–Ni–Co–Cu–Mo sulphides; *(4)* carbon.

Near the surface, uranium veins have penetrated into two spurs of shaly dolomite along the foliation and fissures. At the sub-surface level uraninite is associated with cobalt and nickel, with subsidiary gold and palladium. Occurring also in hydrothermal chlorite of the lower Série des Mines, gold coats uraninite. At lower levels uraninite, accompanied by vaesite (NiS_2) and molybdenite, occurs in fine veins and stockworks. Copper, cobalt and nickel sulphides are included in this uraninite. At the deepest level, uranium is no longer localized and nickel disappears progressively. More than 25 new minerals have been discovered at Kasolo; sulphides include siegenite (nickel linnaeite) with or without selenium and cattierite, a mineral exhibiting an antipathic relationship to uraninite. Vaesite forms veins and disseminations, but torbernite, kasolite and sklodowskite only cover vugs. Finally uraninite has been oxidized in the zone of alteration, which extends to a depth of 200 ft.

At *Kalongwe* (in the southern part of the Katangese Copperbelt) uraninite impregnates a fault zone intersecting the Série des Mines at the beginning of a three phase cycle in which cobalt, then cobalt–copper, and finally copper sulphides followed uranium. The age of this uraninite is 600 million years, i.e., 15 million years less than at Shinkolobwe. At *Swambo*, (west of Shinkolobwe) uraninite is disseminated in a transverse crush zone which cuts an écaille of anticlinal remnants that extends from Shinkolobwe to Kalongwe. Closely resembling Kasolo, uraninite is followed by chlorite, cobalt–nickel sulphides and, after faulting, by chalcopyrite. Occurrences of the Jadotville and Elisabethville districts include Ruashi, Luiswishi, Luishia and Kambove. To them we may add Kamoto in the Kolwezi district.

GABON

On the northwestern rim of the Congo basin a horizon of unfolded (Upper Precambrian) Francevillian sediments shows extensive traces of uranium, but the *Mounana* deposit, situated (Fig. 25, 80) in southeastern Gabon, belongs to a lower horizon. Here the synclinal sequence consists of coarse felspathic sandstones and interbedded conglomerates, sandstones, schists and shales overlain by jaspers. Two paired fault systems (BERNAZEAUD, 1959), north, east-northeast, north-northeast and northwest, have produced a pattern of

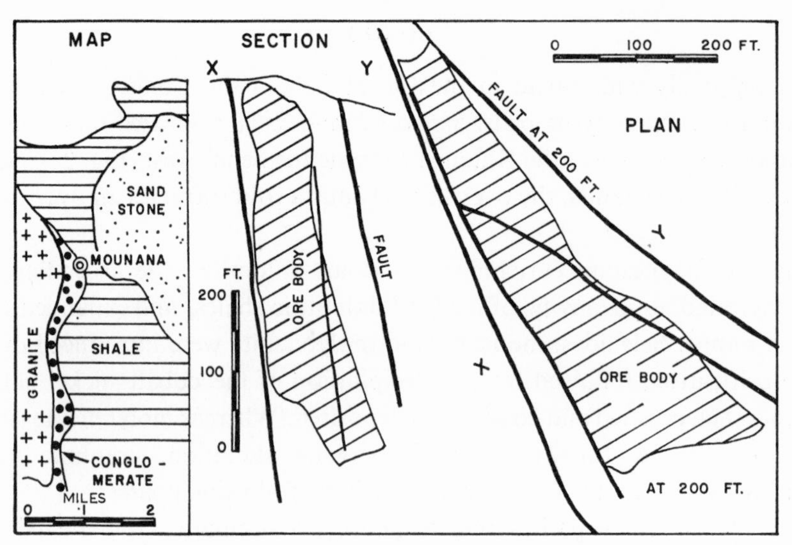

Fig. 80. The Mounana uranium deposit.

elongated intersected blocks. Striking north-northwest, the sub-vertical Mounana fault borders a horst of Basement gneisses and granites on the west, and coarse sandstones exhibiting cavernous, conglomeratic and schistose facies in the east. Penetrating between quartz grains into fissures and joints, francevillite, the barium–lead uranovanadate, impregnates the porous sandstones. Extending to a depth of 400 ft., the principal oreshoot of the hoodzone covers an area of 500 by 60 ft. Moreover, the lens has been enriched in ferghanite, carnotite, chalcocite, uranopilite and bassetite, but the vanadium mineralization of vanadinite, melanovanadinite, corvusite, and roscoelite does not exhibit a sympathetic relationship with uranium. The paragenesis also includes pyrite, galena, chalcopyrite, sphalerite, bornite, baryte, calcite and wulfenite.

MADAGASCAR

Consisting of gneisses, quartzites, plagioclase and diopside pyroxenites, the zoned ultrabasics of *Mandrare* cover 20,000 square miles of southeastern Madagascar. There are two types of granite that are intrusive in troughs and anticlines of the complex. Thorianite has been disseminated in the calco-magnesian *Tranomaro* horizon: at Ambindandrakemba in pyroxenite, at Kotovelo in wernerite, at Amboanemba in phlogopitite and phlogopite pyroxenite, at Ankerotsy in cipolinos, and in both at Belafa. Showing scapolite remnants, thorogummite is present in pyroxenite (MOREAU, 1959). Controlled by folds, impregnated lenses are concordant to schistosity, but they are less foliated than barren strata. Furthermore, lenticular bodies of acid intrusives are adjacent to concentrations. Calcite occurs occasionally, and phlogopite is controlled by faulting. The thorium/uranium ratio varies between 2/1 and 12/1. Uranothorianite is believed to have precipitated from marls or crystallized during metamorphic, pegmatitic–hydrothermal phases or granitization.

The largest deposit, *Belafa*, is located in the north. Here a sub-horizontal mineral zone is included in pyroxenites lying in a syncline. Uranium content of the thorianite varies from 13 to 26%, with the thorium content attaining 32%. The neighbouring Sakoatelo ore body is included in spinel pyroxenites. Located in the centre of the belt, the Ambindandrakemba pyroxenite lens attains a length of 1,600 ft. and a maximum width of 230 ft. Here the concordant thorite zone is 1,000 ft. long and up to 35 ft. thick. Uranium content is 22%. Situated southwest of the preceding lens, the concordant Kotovelo body is 650 ft. long and 220 ft. wide; these ores average 0.4% thorianite.

In addition to this, uranium and thorium show in pegmatites at Itasy and Vakinarkaratra in betafite, euxenite, and monazite (and biotite), and at Ambositra and Fianarantsoa in columbite–tantalite, monazite, beryl, and uraninite (and muscovite). On the other hand, autunite and uranocircite, derived from the Basement by late volcanic activity, reprecipitate in marly clays and sandstones of the Quaternary Vinaninkarena Lake.

THE REST OF AFRICA

At Azegour in *Morocco*, uranium accompanies molybdenum and tungsten. In *Algeria* uraninite is associated with pyrite, molybdenum and magnetite in the pyrometasomatic Aïn Sedma deposit (Kabylia) and at Icherridène. In the Saharan Hoggar, silicified breccia zones of the Tiguelamine granite and the later Precambrian Traourirt granites of Elbema, Tesnou and Aït Oklan, contain pitchblende, pyrite and fluorite. The granitic ring complexes of Maïdenia, Aïr (Niger Republic), the Kaffo Valley (Northern Nigeria), Arayé, Tibesti (Chad) and Poli (Cameroon) carry uranium, as do crush zone and fissures of the younger granites. The late Precambrian zinc–copper deposits of the lower Congo geosyncline show traces of uranium.

In *Rwanda*, biotite granites of Gitarama and certain metamorphics of Kibirira, near the Gatumba stock, are highly radioactive. Microlite of the Buranga, Shore, Nyarigamba and Mutato pegmatites contains 1% uranium. At Buranga and Shore uranium is partly altered into autunite. Pitchblende stringers are included in the Karago pegmatite. Similarly, near the Rwandese border in *Uganda*, the Bulema pegmatite carries tantalite containing 0.04% U_3O_8 and microlite (0.4–1.4% U_3O_8), as do other pegmatites of the Kigezi–Ankole district. For instance, the limit of the shattered quartz core of the Nanseke pegmatite includes euxenite. Quaternary tuffs of Ndale contain betafite. Furthermore, uranium and thorium have been disseminated in the Lunyo granite, north of Lake Victoria.

Malawi. Sodalite syenites of the Ilomba Hill basic alkaline complex contain betafite, but pyroxenites contain pyrochlore (12% U_3O_8). Biotite feldspar aplites of Tambani contain betafite, columbite–tantalite, hematite and zircon.

In the *northern Zambian* Copperbelt at Mindola, pitchblende and coffinite are disseminated with copper minerals in a quartz–feldspar mosaic. This lens lies near the copper beds of Nkana. In *Southern Rhodesia* some uraninite is disseminated in the copper arkoses of the Molly deposit in the northern part of the country.

URANIUM IN PHOSPHATES

In apatite and collophane, uranium (radius 0.97 Å) substitutes for calcium (radius 0.99 Å). Uranium has undoubtedly percolated into sediments from the seas, which today contain 4 g of uranium per square mile (DAVIDSON and ATKIN, 1952). African phosphates generally contain more uranium than others. Uranium is distributed homogeneously in soft rock in them, concentrating on the edges of compact phosphate minerals. Mean uranium content of Moroccan and Algerian phosphates is 0.012–0.014%. Carnotite covers joints of a uraniferous horizon of the Djebel Onk (Algeria, Fig. 134). Ferghanite and tyujamunite have been recognized in Morocco (Fig. 133). Tunisian phosphates contain only 0.004–0.007% U_3O_8 at Gafsa, and 0.007–0.009% at M'dilla. The Safaga phosphates in Egypt, however, include 0.007–0.012% uranium oxide, and the Qoseir ores contain 0.008–0.01%, but hardly any thorium.

Wavellite of Thiès, Senegal, contains 0.033% U_3O_8. The aluminium phosphates of Abeokuta Province, Nigeria, include little uranium; granular and nodular rock from Ososhun contain 0.004 and 0.007% U_3O_8 respectively. Vesicular phosphates from Shoyinka, Ifo, include 0.006–0.007% U_3O_8, and Apon nodules contain 0.008%.

In the Saharan Aïr area (Niger Republic) near the Tibesti Mountains, a complex of bone beds, phosphates and marly sandstone shows traces, as do the Cretaceous phosphates of Hollé (Pointe Noire), Congo Republic.

RADIUM

A majority of the world's stock of radium bromide and salts has been recovered from the uranium slimes of Shinkolobwe–Kasolo in the Congo (see preceding chapter). In uraninite and the complex uranium oxides of this deposit, the uranium/radium ratio averages 2,900/1. Selected ore contains 0.02% radium.

ANTIMONY

Antimony is considerably scarcer in Africa than arsenic. The metal is associated with gold in the only major deposit, the South African Murchison Range. Stibnite tends to impregnate shear zones and carbonates. In Morocco antimonite crystallizes with sulphides in quartz–calcite–baryte veins related to microgranite. However, in North Africa antimony minerals have frequently been oxidized. In addition, stibnite occurs in a quartz vein at Zuarungu, northern Ghana.

THE MURCHISON RANGE

Structure

The *Murchison* Range is part of a 40 mile long, 4–1 mile wide syncline. Striking east-northeast in the roof of ancient granites, this fold extends east of Leydsdorp in the eastern Transvaal. The stratigraphic column consists of (Fig. 81):

(1) schists, amphibolites, sheared quartzites and metalavas;

(2) quartzites, grits, quartz–carbonate rock and schists with inliers of sheared conglomerate; and

(3) carbonatized slates and metalavas.

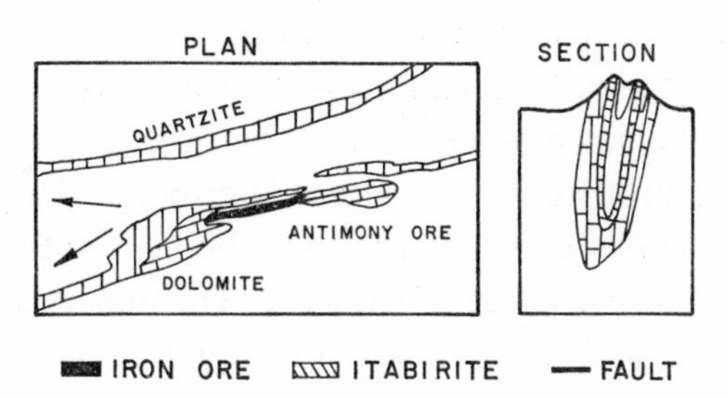

Fig. 81. Plan and section of the Murchison Range.

Along most of its length in the north the schist belt has been intruded by the wedge-like Rooiwater complex. In the west a narrow band of Rooiwater hornblende granites, followed by partly schistose gabbroic rocks in the north, is in contact with the schist belt.

At Plessis Kop and Pike's Kop, and further east between Pioneer and Mashawa Kop, the Murchison syncline appears to have been cross folded. Further east the structure branches with granites wedged in its midst. The syncline has a variable pitch. In the west, the core of the tightly folded structure consists of carbonatized and chlorite–hornblende schists, followed by one of the hard quartzite bands. Incidentally, another quartzite band is interbedded in chlorite–mica schists. The northern branch of the Murchison syncline consists of sub-vertical, siliceous sediments interbedded with carbonates.

The Antimony Bar

The Antimony range is subdivided into two crests of quartzite: the Antimony Bar including the Gravelotte, Maid of Athens, Monarch, United Jack and Free State Hills, and the Chloritoid Bar. At Castle Hill and elsewhere, dolerite dykes fracture and traverse the Antimony Bar.

Gold and sulphides impregnate fractured quartzite beds *(1)* with or *(2)* without quartz, *(3)* in alignments of discontinuous indistinctly defined lenses mainly containing antimony, and *(4)* in well-defined quartz reefs. Lenses of type *(3)* consist of coarse ankerite accompanied by chlorite schists, quartz–mica schist and crush conglomerate, sometimes invaded by quartz stockworks. Stibnite and chalcostibnite crystallize in irregular bodies, and veins and veinlets include rounded quartz and gold. Tetrahedrite, berthiérite and cinnabar occur in the same type of ore bodies, but antimony oxides form through supergene alteration.

Excluding traces, stibium occurs only in the 32 mile long and 400 yard wide *Antimony Line* extending from beyond the railway to Caledonian Camp. North of the quartzites of the Antimony Bar mineralization is widespread but scattered.

In fact a shear zone, the Line, is located north of the talc–chlorite zone and 400–950 ft.

south of the Antimony Bar quartzites. The principal mines — Gravelotte, Jack West, United Jack, Mulati, Weigel, B. Ironstone, Free State, Monarch and Monarch–Cinnabar — are all included in a 7.6 mile long span (SAHLI, 1961). Antimonite concentrates at Gravelotte and United Jack in coarse ankerite, and at Weigel, Free State and Monarch in chlorite schists. Whereas north of the quartzites of the Antimony Bar mineralization is widespread but scattered, arsenopyrite, pyrite and pyrrhotite lenses lie closer to the Bar.

The northern part of the southern limb of the tight synclinal wedge moved downwards in the granite along a bar of quartzite. Only within this horizon have high temperature sulphides been followed by stibnite precipitation. During a period of carbonatization, antimony minerals were disseminated along younger vertical dragfolds and dip slip or transverse zones, located near protrusions of the granitic relief. The principal dragfolds have reduced the distance of the two quartzite bars. Ore bodies are located there along fissures representing the diastrophic end phase of shearing. The mineralization favours tectonized dolomites enfolded in noses and embayments of the quartzites.

Gravelotte

Located 60 miles northeast of Leydsdorp, the principal deposit of the belt is separated from the rest of the Antimony Line, which extends further east. The syncline, defined by the quartz–mica schists of the Chloritoid and Antimony Bars, is 1,500–2,000 ft. wide here, and Karroo dolerites occupy the depression. Irregular auriferous sulphide lenses, found in banded ironstone, hug the western sector of the Chloritoid Bar. Similarly, south of and parallel to the Antimony Bar the pyrrhotite reef extends 100 ft., in banded ironstones. A dyke of sheared granite has been injected between the quartz–mica schists of the reef and the carbonate schists. Finally, 350 ft. south of the Bar the stibnite reefs extend in an integral shaped, east-northeast striking arc.

At the contact of massive carbonatite and talc schists, the North Reef fills in a fissure in the nose of a fold in the Antimony Bar. Fine grained stibnite, intergrown with berthiérite and accompanied by calcite and quartz, fills in a fissure. Its apophyses follow bulging carbonate–chlorite schists. In the east, where carbonatization was more intense, coarse grained replacement stibnite crystallizes up to 25 ft. from the reef. In a typical section, the quartz chlorite schists are followed southward by a pre-ore basic dyke. In the South Reef, carbonatite is replaced by coarse stibnite at the contact of talc schists bordering quartz–chlorite schists. Connecting through narrow channels with the surface, the main ore body forms a flat, sub-vertical bulge of 800 by 800 ft., while its width varies from a few inches to 8 ft. Above the 400 ft. level, replacement veins split in a horsetail structure, thus quadrupling their strike length. Values increase between the 350 and 1,100 ft. levels. Here, at the 1,100 ft. level, the eastern boundary of the ore body retracts with a chimney of high grade ore which remains beneath this level. However, at its extremities the vein swings south, splitting and becoming barren.

OTHER SOUTH AFRICAN DEPOSITS

The United Jack fissure extends 350 ft. at the contact of quartzite and schists, filled in, in the upper reaches, by coarse grained stibnite. Grain size diminishes in depth, but grades increase. Quartz, pyrite and gold are also represented in this paragenesis. In the Mulati Winze quartz lenses develop at the contact of carbonate and carbonate–chlorite schist; stibnite precipitates in them and near faulted carbonate bodies. In the *Weigel* mine the gold, arsenopyrite and stibnite veins, divided by quartz–chlorite schists, unite locally. Grades increase here between the 450 ft. and 800 ft. levels. The *Monarch* veins fill in the sheared contact of talc and hornblende schists. The stibnite lenses bulge where the shear bends. In addition, a stibnite funnel has developed in carbonate. At the end of the belt stibnite also occurs at Castle Koppies, County Down and Caledonia. At Tsama gold, antimonite, arsenopyrite and pyrite are associated.

In the *Barberton* district, subconcordant antimonite lenses impregnate the rocks of Soodoorst and Nooitgezien near Steynsdorp. In the Morning Mist deposit near Ango, ramifying lenses of antimonite exhibit a sub-vertical dip.

MOROCCO

Most of the antimony deposits are related to the Hercynian microgranites of Central Morocco, the area located south of Meknès and east of Casablanca. Granitic aureoles of the same area are distinguished by tin–tungsten–copper mineralization. Other considerable occurrences are at *Beni–Mesala* in the Ceutà Rif in the north and around the Tazzeka Stock.

In the Bled Zaïan and Azarar as far as the Oued Beth, a broad belt of low-volatile microgranite dykes and sills trends northeast for 50 miles. The belt consists of:

(1) the Bled Zaïan zone with the Djebel Timerhdoudine, Mouhajbat, Sidi Embarek and Aberki veins;

(2) central Mrirt including the Koudiat Tibtahine, Jebel Achek west of the Tirza veins in Paleozoic metasediments, Tirza and the Djebels Izougeuerz, Asfah, Bou Iskra and Mguedn in microgranites;

(3) northern Mrirt deposits, the most important in the country: Tourtit, Tafgout, Ich ou Mellal and Masser Amane; and finally

(4) near the Beth stream: Talet Tametot, Ighoud, Igouda, Bou Ichilfane.

Few microgranites outcrop in the Smala area west of Bled Zaïan; this group of occurrences extends from Smala to Enta through Kheneg el Brak and El Hadda. However, the Tafoudelt area is located on the lower Ouadi Beth. Finally, at Aguerd'oufoullous and Mechtoit antimony is associated with lead sulphides.

Antimonite veins 1 ft. or more thick frequently traverse the microgranite perpendicularly to the contact of schists. Impregnating the wall rock, antimonite fills in apophyses. In the Paleozoic metasediments antimony has crystallized in pinching and swelling lenses aligned in fissures, particularly at the contact of rocks of variable hardness. The paragenesis includes rare pyrite, antimonite and scarce zinkenite, jamesonite, arsenopyrite and chalcopyrite, sphalerite, quartz and baryte. Calcite is found in limestones with native antimony occurring at Aberki (AGARD et al., 1952). Nevertheless, veins which traverse microgranite rarely show gangue minerals. Jamesonite and antimonite are associated without zoneographic differentiation, but galena veins cut the antimony mineralization. Remaining intimately related to fine grained acid intrusives, antimony can replace both quartz and microgranite. At the surface, antimonite weathers to antimony ochres.

THE REST OF AFRICA

In *Tunisia*, antimony is associated with the Oued Maden lead deposit and with the Upper Cretaceous (Campanian) replacement paragenesis of copper at Chouichia, near Souk el Arba. The principal *Algerian* deposit, Hammam-n'Baïls, is located in the Nador area near Guelma. Here irregular bodies of ferruginous smithsonite have been impregnated by flageolite and surface nadorite. At Kheneg, near Aïn Kerma, sub-concordant Lower Cretaceous (Aptian) antimonite bodies have been altered to cervantite. On the other hand, in the Taya Mountains antimonite which also covers fissures has been disseminated along the contact of Upper Cretaceous (Senonian) marls and Cretaceous limestones.

At Zuarungu, in northern Ghana, stibnite occurs in a quartz vein.

In *Southern Rhodesia*, stibnite accompanies gold in the early Precambrian ore bodies of Cam and Motor, in their vicinity, and at Sherwood Starr. Antimonite has occupied part of the lower levels of the H reef of the Globe mine, and also occurs in other parts of this deposit in the Phoenix and Bell sectors. In other veins of the Que Que-Gatooma schist belt, narrow quartz lenses containing 40% stibnite may expand to a width of 2 ft. The centre of these fissures is filled in by quartz, surrounded by a zone of crush breccia. Whereas stibnite has been broadly disseminated, gold has concentrated in schists bottoming out in depth. The principal antimonite bearing quartz veins of the Gwelo schist-gold belt are located at Gothic and Pagamesa on the endocontact of moderately sheared granite.

ARSENIC

Arsenopyrite and its alteration product, scorodite, are widespread in Africa, especially in and adjacent to gold impregnations and veins. All disseminations cannot be described. However, arsenopyrite is not as abundant as pyrite. Major commercial deposits are unfortunately rare, but some are found in Algeria and Southern Rhodesia.

North Africa

In *Tunisia*, the Tabett ben Ksouri sandstone is impregnated by realgar, and traces of arsenic also occur in limonites. Several lead–zinc deposits carry realgar and mimetesite. Djebel Tabett ben Ksouri is located between Oued Maden and Fedj Assene. In the Medjerda trough two sandstone levels, separated by Triassic sheets or 'blades', have been impregnated by auripigment and realgar in a chain-like pinch and swell pattern. Values attain 7–8% arsenic.

In eastern *Algeria* at Aïn–Achour, 5 miles north of the Nador railway station, a mimetesite–cerussite body of 30,000 tons replaces Lower Cretaceous (Lias) limestones. The ores contain 15–25% arsenic and 12–23% lead. In the Djebel Debar, isolated patches and indistinct bodies of scorodite–pharmacosiderite and ferruginous–arsenical sandstones fill in karstic basins and depressions of the early Cretaceous limestone.

Morocco. The cobalt–nickel–silver ores of Bou Azzer are all arsenides, hardly including sulphides (see 'Cobalt' chapter). In the quartz veins of Koudiat Cheïba and Jebel Aouam arsenopyrite is associated with scheelite. In the quartz veins of Tikkadarine, Oukilal–Agourzi, Jebel Tisguine near Azegour, and Bou Ijda arsenopyrite is accompanied by some gold. All are related to granitic stocks.

West and Middle Africa

Arsenic minerals occur in the Akjoujt copper deposit, Mauritania, and in the gold district of Poura, Upper Volta.

Ghana. Arsenopyrite occursin the auriferous quartz veins and lenses, e.g., at Beposo, Ashanti and Bibiani North, near the Ivory Coast border. Arsenopyrite is abundant at Nanwa.

Congo. Arsenopyrite shows in the tin fields of Maniema, where it occurs separately from the acid paragenesis. Arsenopyrite is found in the Manono pegmatite. Loellingite impregnates schists at Namoya, averages 7 dwt. gold/ton. Arsenic is included in the hydrothermal and uranium paragenesis of the Copperbelt.

Tanganyika. Arsenopyrite shows in the Karagwe tin field. The mineral is more abundant in the Musoma gold field, and also occurs at Geita, Mwanza and in the Kavirondo gold field, Kenya.

Southern Africa

At Tsumeb, in Southwest Africa, enargite and tennantite are abundant.

In *Southern Rhodesia*, arsenopyrite is abundant in the Gwanda and Umtali districts. Arsenic is recovered in the Que Que roasting plant. Arsenopyrite is widespread in the Hartley district. It is the main carrier of gold in the Pickstone deposit, earlier pyrite being poorer. The sulphides impregnate banded lodes of fine grained carbonates, especially in greenstones. At Glenmore arsenopyrite is associated with stibnite and pyrite in quartz. The felsites of the Road Mine are traversed by stringers of arsenopyrite, pyrite, quartz and calcite. Other occurrences include Eiffel Blue, Concession Hill and Giant (WILES, 1957).

In *South Africa*, the New Consort mine contains the greatest concentrations (1.8%) of the Barberton gold district. Arsenopyrite is associated with the Stavoren–Mutue Fides cassiterite pipes, and also occurs in the monomineral Gaasterland–Roodewal pipe, and in the auriferous lens of Rhenosterhoekspruit in the Bushveld granite. At Geweerfontein, 70 miles northeast of Pretoria, arsenopyrite and bismuth are associated in a fracture zone. At Nooitgedacht in the Lydenburg district arsenic accompanies gold.

PYRITES (SULPHIDES)

Sulphides are the principal source of sulphur in Africa. Although widespread in impregnations and independent or complex sulphide veins and sediments of all African countries, the large sedimentary pyrite deposits of the Witwatersrand and Copperbelt are on a considerably larger scale. Indeed, in these parageneses gold and base metals are the accessory minerals of pyrite. However, even in the Copperbelt, the complex zinc (and copper) sulphides of Kipushi form the base of a sulphur industry which rivals South Africa. Similarly, at Broken Hill in central Zambia sulphuric acid is manufactured from mixed sulphides. For the description of these deposits we refer the reader to the Base Metal Group. Finally, in Morocco pyrrhotite is considered to be a source of sulphur.

South Africa

In the gold districts and beyond, pyrite is the principal heavy mineral of numerous conglomerate horizons of the Witwatersrand (Fig. 91) and adjacent systems (see 'Gold' chapter), occurring as high in the sequence as the basal Black Reef of the Transvaal System. Pyrite constitutes 90–95% of the concentrates, with values ranging from 0.5 to 5%. The size of pyrite grains or nodules of fibroradiant structure varies between a few microns and an inch. Some of the pyrite coats and substitutes iron oxides and even pebbles, and also cover chloritoids. Pyrites may be disseminated in crystals and aggregates; they also impregnate quartz. The present form of the pyrites is considered to be reprecipitated or epigenetic. Transported granules have been identified by P. RAMDOHR (1958). The restrictive concentrations of pyrite in the basal part of the conglomerates and in streaks appear to be less distinct than in the case of gold or uranium. The quartzites also contain pyrite streaks, but values are lower.

Pyrite concentrations were exploited at Krugersdorp, Daggafontein, Buffelsfontein, Vogelstruisbult, Virginia, Government Gold Mining Areas, Transvaal Gold Mining Estates and West Rand Consolidated Mines.

In the Pilgrim's Rest district, pyrite is recovered from the Sandstone Reef, at the top of the Black Reef Series (of Transvaal age). Quartzites of Barberton contain mono-mineralic pyrite. In other deposits of this district and of the Murchison Range, arsenic and antimony reduce the pyrite's suitability.

North Africa

Morocco. The Jebel Kettara deposit is located 20 miles northwest of Marrakech. A subvertical fissure zone which traverses Paleozoic rocks has been filled in by massive pyrrhotite, including some chalcopyrite. The vein outcrops over ⅔ mile, and attains a width of 15–30 ft. Three alteration zones have been distinguished. A 20 ft. high porous pyrite zone is capped by 150 ft. of limonite, followed by the vein below a transitional zone. The upper half of the pyrite zone contains 8% Cu.

Algeria. Small deposits include El Halia, Filfila in the Philippeville Mountains, Aïn Sedma near Collo, and Azouar adjacent to Mansouriah. Pyrite is considered as an impurity in the lower levels of the iron deposits.

Tunisia. In Tunisia, pyrite and melnikovite are found at Aïn Grich and Djebel Kebbouch. The Aïn Grich lens near Bejaoua fills in a cavity of an Upper Cretaceous (Senonian) anticline. At Djebel Kebbouch, near Sakiet Sidi Youssef, an anticline of Middle Cretaceous (Cenomanian) marls and limestones was cut by a Triassic sheet or blade. It was replaced by pyrite, sphalerite and galena in a 450 ft. long and 5 ft. wide are body.

Congo and Zambia

In the column of Katangian sediments, pyrite is widespread. At Roan Antelope, Chibuluma and Mufulira, e.g., copper disseminations grade laterally into pyrite and impregnate whole ore horizons (Fig. 28). Lower Roan ore beds average 2% pyrite. Less pyrite has been disseminated in the various horizons of the Zambian Upper Roan. In the Katangian Upper Roan, pyrite is followed by chalcopyrite, maxima being reached in one horizon of the early silicified, zoned dolomites and in four horizons of the late micaceous dolomitic shales of the Mines Series. Half of them are also distinguished by the abundance of carbon, which may have acted as a reducing catalyst. Along the axis of the Kolwezi *écaille*, pyrite grades are particularly high since an inlier of sandstone is missing there. Pyrite is rare in the copper–*cobalt* parageneses, but it is fairly abundant in the uranium parageneses. Above the Mines Series proper, pyrite reappears in a 230 ft. thick layer of impure dolomites belonging to the Dipeta Stage. At Mufulira pyritic quartzites occur at the base of the overlying carbonaceous Mwashya shales. Pyrite locally represents 10–50% of the 400–1,500 ft. thick Mwashya shales.

The basal tillite of the upper Katangian commonly carries fine grained pyrite and pyrrhotite, as does the small conglomerate of the Middle Kundelungu.

The rest of Africa

East Africa. Commercial quantities of pyrite occur in the Kilembe copper mine in Uganda, in the Lake Victoria gold fields in Kenya, and in the Geita gold deposit in Tanganyika where concentrates contain 10% pyrite.

Southern Rhodesia. In the banded ironstone sequence of Iron Duke Mine near Mazoe, massive auriferous pyrite forms a 30 ft. wide sheet, which dips 70° north along the contact of epidiorite and quartzite and impregnates the latter. Concordant quartz lenses are interbedded in the pyrite body.

SECTION IV

Base metal group

The largest deposits of the principal metal, copper, are at least partly syngenetic; cobalt is a characteristic accompanying metal in these bedded ores (Fig. 3). In hydrothermal replacement deposits copper is associated with zinc, lead, silver, cadmium, germanium and selenium; however, in younger vein deposits lead, accompanied by silver, predominates. Telethermal mercury precipitates in crush- and shear-zones in similar environments but this metal is scarce in Africa. Vanadium tends to occur in the old copper–lead–zinc distritcs, but it is partly diagenetic. Epithermal barium favours the young lead districts; strontium, a satellite of barium, occurs also in carbonatites (like copper) and in sediments. Fluorine crystallizes under broad *pt* conditions, and also exhibits some affinity with the rarer hydrothermal tin mineralization.

Base metal concentrations reached a distinct peak 550–650 million years ago, but secondary maxima are dated between 450–300 million years.

COPPER

Bedded copper ores occur in latest Precambrian geosynclines in Northern Zambia–Katanga, the Lower Congo–Angola and in sediments of Southern Rhodesia. However, stratiform copper ores of the Algerian, Moroccan and Niger Sahara are Mesozoic. Copper sulphides are precipitated in a hot, arid climate, in lagoons or shallow seas during marine transgressions following long cycles of continental sedimentation. Psammites are impregnated, but layered dolomites are replaced by the sulphides. Vertical, mineral zoning is only locally distinct, lateral variations being more frequent. Since metals of generally hydrothermal origin, such as uranium and cobalt, are associated with copper, the deposits are considered to be polygenetic. Replacement deposits of copper, zinc and lead occur in the same sequence in the Congo and Southwest Africa, but small hydrothermal occurrences belong to other cycles.

The Kilembe (Uganda) and O'okiep (South Africa) deposits are related to distinct horizons of plagioclasites and metabasites. Copper also shows in Karroo lavas, and in the Phalaborwa carbonatite (South Africa), copper sulphides impregnate mainly sövite.

CONGO–ZAMBIA

Copper occurs in Katangian sediments: *(1)* in the Copperbelt; *(2)* near Lake Mweru;

(3) in a belt extending from Mwinilungwa in the northwestern corner of Zambia to Kansanshi to Kasempa, Chifumpa, beyond Broken Hill 300 miles east-southeast; *(4)* in the area of Sable Antelope, west of Lusaka; *(5)* around Lusaka, the capital of Zambia, in late Precambrian sediments; *(6)* near Stanleyville; *(7)* in the Lower Congo, in earlier Precambrian rocks; *(8)* at Lubi, in the southern Congo; *(9)* at Mtuga, south of Bwana Mkubwa; *(10)* in a 200 mile long belt extending from Kansonso to north of Lusaka and Kalulu.

The world's largest copper and cobalt deposits form a long, composite unit in Katanga – northern Zambia and (Fig. 28) in this description they will be treated as a unit. Copper has been disseminated along the whole belt; the greatest concentrations occur in the Kolwezi district, in the Katanga and at Nchanga, Nkana and Mufulira, in the northwestern part of the Zambian sector, where one quarter of the belt and 55% of the resources are located. Parageneses include:

Fe (pyrite), Cu, Cu–Co, Cu–Co–U, U–Co–Ni–Cu–(Pd), (Cu)–Au–Pt–V, Cu–Zn–Ag–Pb–Cd–Ge, Fe (hematite), Mn.

Cobalt values are highly variable and unrelated to the distribution of copper; concentrations tend to decrease from the west towards the east. Disseminated uranium is part of the early paragenesis and accumulates in localized veins, lenses and impregnations at Shinkolobwe, Kambove, Kamoto, Luishia, Luiswishi, and Ruashi in Katanga and at Mindola (N'kana) in Zambia. Gold, platinum, palladium and vanadium are associated with coppet at Musonoi only in the low-lying green (foliated) horizon. These metals occur at Ruwe out of the copper zone and at Shinkolobwe, with uranium, nickel and molybdenum. Several cycles can be distinguished in the Cu–Co–U suite. Some zinc has been disseminated in limestones lying 4,000 ft. above the Lower Roan, during which hydrothermal replacement minerals of the Cu–Zn–Pb–Ag–Cd cycle cyrstallized at Kipushi, Kengere and Lombe; germanium appears only at Kipushi.

Arguments in the syngenetic/epigenetic controversy have been presented with considerable detail. In 1913, the syngenetic hypothesis was first enunciated by the German mining engineer Guilmain: 'The Katanga deposits are syngenetic, stratified deposits of sedimentary origin influenced by metamorphic transformation'. During the 15th International Geological Congress, 1929, SCHNEIDERHÖHN (1931) brilliantly supported this thesis, and extended it to cover Rhodesia. The majority of Zambian geologists consider that country's deposits as syngenetic; indeed, isotopic analysis of sulphur shows that a majority, though not all, of sulphide is biogene. However, most Katangese geologists are undecided; at least half of them consider that area's deposits as epigenetic. The latter have achieved their present form by epigenetic and diagenetic processes. The early Shinkolobwe uranium and the late Kipushi zinc deposits are hydrothermal. The mineralization, at least of uranium, spans 50 million years, that is, as long as the Tertiary, and the ores are undoubtedly polygenetic. They have been disseminated mostly in arenaceous–carbonaceous sediments of five different horizons, in a 500 mile long trough, subsequently folded by compressive forces originating in the south.

STRATIGRAPHY

Around the rim of the Congo basin, a thick, psammitic column of tabular or warped sediments covers the Basement in places. Remnants of these latest Precambrian strata have been found west of Lake Victoria, east of Stanleyville, in the Gabon, the Lower Congos and Angola, and in Katanga–Zambia. Similar formations have been drilled beneath the Karroo in the centre of the Congo basin. Unfolded and unmineralized Upper Katangian strata cover the vast Kundelungu Plateau north of Elisabethville, and smaller areas west of the Zambian Copperbelt. Lower Katangian sediments outcrop mainly in the geosyncline. Lateral variations of facies are common. The nomenclature has undergone many uncorrelated changes on both sides of the border, but a tendency for agreement has appeared during recent years. There have been distinguished two divisions in the Katangian: the upper or Kundelungu, and the lower member, called Roan or Shale Dolomite (Système Schisto-Dolomitique) in the French and Portuguese speaking areas. In Katanga, the middle subdivision of the Roan is called the Mines Series (Série des Mines), but in Zambia, the Mine Series has been expanded to designate the entire Lower Katangian.

Lower Roan

Southeast of Elisabethville, mainly in Zambia, the well-developed Lower Roan attains a thickness of 800–2,000 ft., however, west of Elisabethville, the earliest Katangian has not been drilled and the Basement has not been reached.

In *Zambia*, the Lower Roan rests on Basement schists and granites, divided into a Footwall, Ore and Hangingwall Formation.

(1) The lowest member, a conglomerate or locally a pebbly quartzite, fills hollows of the ancient surface (except for Baluba–Muliashi) and averages 100 ft. in thickness.

(2) The overlying eolian oligoclase-quartzites are grey to dark grey; they attain a thickness of 100–300 ft. At Nchanga and Mufulira, grit and arkose take the place of the quartzite.

(3) The 50–200 ft. thick Rhodesian Ore Formation generally remains in one horizon, 400–700 ft. above the Basement. The Formation consists of micaceous dolomite and of bedded argillite (the earliest shale in the sequence), but in the smaller northern or Mufulira branch, the Ore Formation consists of graywackes or quartzites, dolomite and argillite; mineralization descends in the sequence towards the Pedicle of Katanga. Copper accumulates above and beneath the argillite of the main (southern) branch; two-thirds of the copper accumulation concentrates in the more or less carbonaceous shale. Although continuous, the Ore Formation does not always contain commercial concentrations.

(4) The Hangingwall Formation consists of 200–800 ft. of argillites and cross-bedded quartzites, including the late pebbly arkose, which is mineralized at Nchanga West and Mufulira.

In *Katanga*, the sporadically developed Lower Roan is composed of siliceous dolomites, sandstones and conglomerates, overlain by shales and feldspathic dolomitic sandstones.

Upper Roan

The Upper Roan facies vary in the northern and southern parts of the Copperbelt. Argillites and dolomites alternate in the 1,200–2,200 ft. thick sequence southeast of Elisabethville; layers of shale appear at Nchanga; and breccias and quartzites at Mufulira more closely resemble the Congolese sequence, which is generally attributed to the Upper Roan. The *Katangan* Mines Series has been divided into five layers:

(1) C. M. N. 150–1,300 ft. Upper, more or less siliceous, grey-blue or dark grey, carbonaceous dolomites, characterized by *black ore* (M. N.). Subsidiary chlorite bearing shale and sandstone, *stromatolites*, and dolomitic breccia.

(2) S. D. 100–400 ft. Layered, micaceous grey, black or pale green shales or psammites, including carbon-bearing horizons. At their base, massive mineralized stromatolite dolomite of the *main*, *black ore zone* (B.O.M.Z.). Subsidiary felspathic dolomitic sandstone and dolomite.

(3) R. S. C. 0–100 ft. Silicified stromatolite dolomites, of cellular texture when leached. Subsidiary dolomitic psammite.

(4) D. S. I. 20–40 ft. Lower, layered or foliated siliceous, micaceous and chlorite bearing dolomite, comprising the lower foliated rocks (R.S.F.), the stratified dolomites (D.S.), subsidiary chlorite bearing dolomitic shales and locally layered *green ore* (M.V.).

(5) R. A. T. 1–25 ft. Lower *Collenia* dolomite (D.I.), altered to grey chlorite and *ore* bearing, dolomitic shales and sandstone, locally floating on its mylonitized alteration product, the R.A.T. breccia.

Between Jadotville and Kolwezi, 900 ft. of Dipeta sandstones and psammites and alternating Mofya Shales, dolomites and sandstones overlie the Roan proper.

Post Roan

Dolomites, oolithic quartzites, chloritic, graphitic and banded shales of Mwashya (Mwashia) form the latest member of the Lower Katangian, attaining 1,500–6,000 ft. of thickness in Zambia and 150–900 ft. in Katanga. In the Upper Katangian, two or three divisions have been distinguished in the 20,000 ft. thick Kundelungu. The lower or large, locally glacial conglomerate is overlain by the Kakontwe carbonates and calcschists; this series outcrops in the Copperbelt but generally not in the tabular areas of central Katanga. The younger or smaller tillite (petit conglomérat) forms the base of shales, dolomites and sandstones spreading between the Copperbelt and Lake Mweru. Shales and terminal sandstones conclude the Kundelungu.

STRUCTURE

The Middle Precambrian Kibara Range borders the elongated Katanga plateau in the northwest; the Irumide (Muva) Chain limits it in the south. These ancient trends have been rejuvenated by the Tertiary Upemba and Luangwa Rift Valleys. The 600 mile long, late Precambrian, Katangian geosyncline was developed in the centre of the plateau and the trough was bent southwestward before the earlier Kibara orogeny, thus extending this trend 200 miles into Alto Zambeze, Angola. During the late Katangian, the Upper Kundelungu sea spilled over the north. Between Kolwezi and Fungurume, and Kengere–Shinkolobwe (50 miles) the trend is easterly, turning E20°S between Fungurume, Jadotville and Kamwali (60 miles) and thence through Elisabethville and the Zambian Copperbelt due southeast (170 miles). Since pressure has been exerted from the south, south of Elisabethville the belt has been affected less as it forms an angle of 45° with this vector. The pelitic–

dolomitic Roan forms a major or central, two minor, and several short digitated folds. The two southern bands surround the Kafwe anticline in Zambia and the Luina dome, in Katanga, reseparating further northwest. A poorly mineralized northern trough (Mokambo) lies north of these. West of Elisabethville the axes are discontinuous; they have been arbitrarily divided into belts and sectors (see Table CXX).

TABLE CXX

SUBDIVISIONS OF THE COPPERBELT

Belt	*Sector*	*West*	*North*	*East*	*Southeast*
	miles	15	110	80	80
northern	name	Kolwezi	Jadotville	Elisabethville	Mufulira
	from—to	Dikuluwe—Ruwe	Fungurume—Shandwe	Lukuni—Étoile	Kipapila—Bwana Mkubwa
	miles	20	25	30	60
southern	name		Tantara	Kipushi	Nchanga
	from—to	Kalongwe—Kengere	Swambo—Shinkolobwe	Kasombo—M'baya	Musoshi—Nkana

In Katanga, Kundelungu strata tend to occupy synclines or limbs of anticlines, with the Mines Series rising in the core of anticlines, but in Zambia the Lower Roan appears on the limbs of synclines filled in by Kundelungu sediments.

The Mufulira belt is a simple synclinal fold following the anticlinal Kafwe batholith in the northeast; the Musoshi–Nkana belt is, in fact, a trend of irregular Lower Roan ribbons draped around the oval Konkola, Chibuluma and Baluba depressions. Nchanga is situated in gulfs of the middle Precambrian relief. Secondary, drag, pod, recumbent and cross folds have affected the Rhodesian sequence but only minor faults have appeared. This trend disappears in Katanga, northwest of Musoshi, the Mbaya Kasombo belt lying 15 miles northwest. The zinc–copper deposit of Kipushi is located south of Kasombo.

Near the Mufulira axis, a fold reappears at Étoile du Congo, near Elisabethville. Strike faults follow the contact of Mwashya or Kundelungu strata, thus cutting off one of the mineralized limbs of the Étoile anticline. Large synclines are divided by narrow anticlines with their northern limb removed by faulting. Overthrusting and overfolding distinguish the Jadotville district; the roots of the folds, however, lie still nearby. Oblique and strike shears affect the mineralized horizons of R.A.T., D.S.I. and S.D. more frequently. Tangential forces have developed the Congolese equivalent of the Flysch, the mylonitized R.A.T. dolomite; relics of dolomite and psammite have been preserved in this rock among the replacement products, which are mainly chlorite and talc.

From south to north, isoclinal and farther, diapyric folds are thus first affected by strike faults, then overthrust and finally sheared from their roots and thrust northward until they are halted by the rigid Kibara Chain. Since the Katangan Mines Series is more coherent than the rest of the Roan, it forms larger *écailles*, floating in thrust breccias. In this conglomerate, chlorite bearing dolomitic sandstones of the lower stage of the Upper Roan are embayed by a recrystallized dolomitic groundmass.

The following structural types of deposits can be distinguished: *(1)* complete syncline (Roan Antelope); *(2)* one limb of syncline (Mufulira, Nchanga); *(3)* refolded limb of syncline (Nkana); *(4)* complete overfolded anticline (Kasonta); *(5)* parallel, faulted folds (Musonoi, Luiswishi, Lupoto); *(6)* one limb of faulted anticline (Étoile du Congo, Ruashi, Kambove West); *(7)* small écaille (Sasa); *(8)* card stack écaille (Luishia); *(9)* overfolded and overthrust écaille (Kambove); *(10)* polystructural fragments of nappe (Kolwezi); *(11)* diapyric windows (Shinkolobwe); *(12)* replacement funnels in crush zone (Kipushi).

Stratigraphy essentially controls the localization of the southern deposits. Structural control acquires importance in the north where minerals tend to concentrate along the upper faulted zones of anticlines, along the shear planes of écailles at transverse faults, and finally in complex tectonized or crush zones where the remobilized hydrothermal ores concentrate. However, even in the south normally mineralized horizons become barren if they are not in the metallogenetic alignment. Disseminated sulphides have been altered and concentrated in fissures of the same or neighbouring horizons, or in favourable host rocks.

MINERALIZATION

Deposits occur in:

Lower Katangian

(1) Lower Roan argillites between Nkana and Musoshi.

(2) Lower Roan quartzites between Bwana Mkubwa and Kipapila.

(3) Upper Roan dolomites north of Elisabethville and at Sable Antelope.

(4) Latest Roan, Dipeta–Mofya Stage at Fungurume.

(5) Post Roan Mwashya Series at Shituru, Kimbwe, Kansuki and Tilwilembe.

Upper Katangian

(6) Early Kundelungu, Kakontwe Limestone at Kipushi, Lombe, Kengere, Lukila, Tantara, Kirundu and Tenke.

(7) Kundelungu beds at Tufi, Milulu and Lukafu.

(8) Later Kundelungu sandstones intruded by granite on Lake Mweru, at Kapulo, Sampwe and other localities.

While in Katanga sulphides concentrate only in (psammitic) dolomites, a variety of sediments has been mineralized in Zambia. There are dolomitic shales at N'kana, shales at Roan Antelope, sandstones at Bancroft, and quartzite at Nchanga and Mufulira. In Zambia, biotite indicates more intense metamorphism than chlorite does in Katanga.

Zambia

This sector includes the Zambian Copperbelt and Katanga southeast of Elisabethville. About 50 ft. of ore shale constitute the earliest, well-bedded pelite in an arenaceous sequence overlain by dolomite. Abundant anhydrite indicates lagoonal conditions of evaporation. The original hornfelsic argillite alters to shale which is carbonaceous in parts of several ore bodies; only the presence of sulphides distinguishes them from barren shale, though the bedding of the former may be finer. The Ore Formation follows the indentations of the western edge of the Kafwe anticline; this horizon is, however, not mineralized west of the Copperbelt. In the Belt itself copper bearing sectors and bodies are connected by pyritic sediments. Commercial ore bodies represent only a few percent of the ore shale and the impure dolomite in which they are locally found. Between Nkana and Bancroft, footwall quartzites contain low grade ore. Conglomerates underlie the ore bodies. Pyrite and traces of copper occur in most of the column, especially when the sequence is repeated. The mineralized sericitic quartzites of the Mufulira syncline, which exhibit a facies of graywacke or carbonate locally, are in the same position as the ore shale.

Stratigraphic and mineral zoning of pyrite, chalcopyrite, bornite and chalcocite prevails southwest of and parallel to the Kafwe anticline. Chalcopyrite represents half of the copper ores; the order of abundance of sulphides is identical to the paragenetic sequence. Dolomite and quartzite contain larger grains than shale.

Katanga

Most of the primary ore is concentrated in distinctly layered aluminous or sandy carbonates of the Mines Series. Arenaceous sediments are favourable host rocks in the Lower Roan of Zambia, but in Katanga neither they nor massive dolomites are. The same horizons have been mineralized over a stretch of several hundred miles. Pyrite started to precipitate in the deep seas and during the epicontinental phase; at the beginning of the lagoonal phase, pyrite was joined by copper, cobalt (and nickel) sulphides, with less of these precipitating during the later stages of the lagoonal phase. Amorphous carbon ('graphite') is an indicator of pyrite and chalcopyrite, and its absence signals other sulphides.

The most extensive mineralized horizon is the so-called black ore main zone (B.O.M.Z.), near the base of the dolomitic shales. The limit of the mineralized, altered and barren layers is distinct. Another extensive copper bed is the low-lying green ore (M.V.), affecting in the Elisabethville district and the Fungurume fragment the base of the foliated, silicified dolomite (D.S.I.), and in the Kolwezi district the early dolomitic sandstone (R.A.T.). In this area a distance of 30 ft. measured from the barren silicified dolomite determines the accumulation of the ore, irrespective of the stratigraphy. Copper and iron grades have been found to vary in drill holes piercing the Mines Series (see Table CXXI).

TABLE CXXI

COPPER GRADES

	Thickness (ft.)	*Kamoto Cu(Fe) (%)*	*Kilamusembu Cu(Fe) (%)*
C.M.N. upper dolomite	130	— (0.7)	
carbon bearing dolomite	200	tr (0.9)	
coarse dolomitic sandstone	70	0.1 (0.9)	
S.D. dolomitic shales micaceous	12—13	2.8 (2.6)	1.1 (3.9)
grain size 0.3 mm	9—17	0.7 (0.7)	0.2 (1.4)
dolomitic	12—10	1.9 (0.5)	— (1.6)
foliated	23—34	7.7 (0.8)	0.1 (3.0)
R.S.C. massive dolomite		1.1 (0.1)	
D.S.I. lower dolomite layered	15—23	5.3 (0.7)	3.7 (1.3)
fine-grained	13—12	4.8 (0.5)	1.7 (1.1)
R.A.T. dolomitic sandstone		5.8 (0.8)	

Both the host rocks and the sulphides have been affected by silicification. Chloritization is a late replacement process, tourmaline being secondary as well.

One can distinguish five parageneses: *(1)* pyrite in several extensive horizons; *(2)* copper and cobalt in a few extensive horizons; *(3)* uranium, copper and cobalt in lenses of a horizon; *(4)* uranium, (copper), cobalt and nickel in lenses of a horizon; *(5)* copper, zinc and other metals in hydrothermal replacement.

Pyrite and copper–iron sulphides generally *replace* the carbonates. Their precipitation has rarely been affected by the tectonic movements. While simple sulphides of copper are not structurally controlled, copper–cobalt, cobalt–nickel and nickel sulphides are. Although parageneses, including uranium, are restricted to one horizon of the Mines Series, they are strictly epigenetic. The composite paragenetic sequence is unique: *(1)* uranium, *(2)* iron, *(3)* nickel, *(4)* cobalt, *(5)* copper, *(6)* zinc, *(7)* lead. The sequences that can be distinguished are shown in Table CXXII.

TABLE CXXII

PARAGENETIC SEQUENCE

	Uranium		*Copper*		*Zinc*
(1)	uraninite	uraninite	—	—	—
(2)	Fe and Ni-Co sulphides	Fe and Co sulphides	Co-Cu sulphide	pyrite	pyrite
(3)	chalcopyrite	Cu-Fe sulphides	bornite (chalcopyrite)	chalco-pyrite	chalcopyrite
(4)	—	Cu sulphides	Cu sulphides	—	Cu and Zn sulphides
(5)					Pb and Ag sulphides

The grain size of the sulphides appears to be a function of the grain size of the host rock. Pyrite, carrollite and linnaeite crystals and aggregates are idiomorphous, but bornite, chalcopyrite and chalcocite are xenomorphous. The presence of bornite excludes pyrite and an abundance of chalcopyrite. Carrollite precedes bornite in its horizons, but linnaeite crystallizes sparingly in chalcopyrite zones. Pyrite precedes chalcopyrite; however, in ore bodies combining both of these sulphides, only chalcopyrite is found in the stringers. In the zone of cementation, chalcocites mainly affect bornite. In hydrothermal copper–uranium deposits, e.g., at Kalongwe in the Kolwezi district, all sulphides may occur in the same horizon.

The zone of alteration

In Katanga, a majority of the sulphides has undergone supergene alteration to carbonates, oxides and silicates. At the same time, host dolomites have altered, resulting in a decarbonatized, siliceous-aluminous shale. General oxidation reaches a depth of 50–1000 ft. in Katanga, and of 150–200 ft. in Zambia, where oxides can be found as deep as 2,500 ft. Bornite and chalcocite, accompanied by covellite, are characteristic minerals of the zone of reconcentration. Under the present hydrostatic level oxidation is selective, and near the indistinct limit of each zone the various facies may be interlayered. While arenaceous sediments have been impregnated, carbonates have been replaced. The silicified dolomites originally were barren, but alteration products may percolate into them; indeed, at Ruwe, Kambove and Luishia, the mineralization of the basal mylonite appears to be secondary. Heterogenite appears along the contact of the upper dolomite and in the Elisabethville district, the black oxides accumulating in pockets and podholes of the latest dolomitic limestone (C.M.N.). This layer, however, is barren in depth. Fissures, cavities and vugs have also been filled in fractured, mainly silicified rocks. However, sandstones can be completely imbibed by malachite, precipitate also in granules, and filling fissures of alteration. In shales, laminae of the host rock alternate with copper carbonates. Cavities of alteration and leaching are covered by re-arranged, spherulitic or fibroradiant malachite, exhibiting bands of various greens. A large concretion weighs almost 4 tons. Fine, nodular or acicular malachite also fans out, and stalactites of alternating, concentric malachite and chrysocolla reach lengths of up to 3 ft. Cuproferous solutions still continue to deposit malachite. Chrysocolla occurs in impregnations and in laminae, dioptase crystallizing at Tantara.

Azurite, aurichalcite, cuprite, tenorite, trieuite, chalcotrycite, heterogenite, melanochalcite, mindigite and cornuite are rarer, as are the phosphates and arsenates shattuckite, plancheite, katangite, libethenite, pseudo-malachite; the sulphates connellite and chalcantite; and the vanadates calcio-volborthite and cuprodescloizite. Nitrates and uranates like buttgenbachite, gerhardite, kipushite, torbernite and psittacinite have been found occasionally. Native copper is also an alteration product, the greatest mass weighing almost 2 tons.

KOLWEZI DISTRICT

The Kolwezi *écaille* is surrounded by sub-horizontal (Fig. 82) Kundelungu layers 100 miles east of the triple border of the Congo, Angola and Zambia. The Middle Precambrian Kibara chain lies 3 miles northwest of it. The flat basin of the Mines Series trends east-northeast for 15 miles, attaining a width of up to 7 miles. The altered, mylonitized dolomite (R.A.T.) occupies most of the *écaille*, with younger layers of the Lower Roan floating parallel to the general trend in this breccia. The principal alignment of Kamoto–Musonoi–Kolwezi extends for 7 miles in the middle of the *écaille*. Another alignment runs parallel

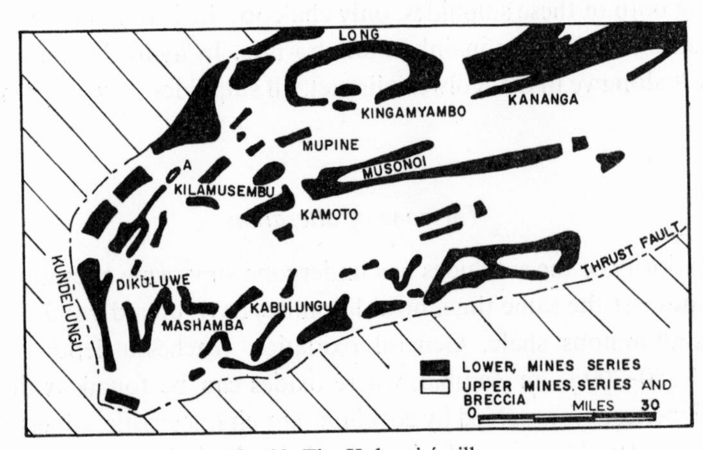

Fig. 82. The Kolwezi écaille.

2 miles north, from Mupine through Kingamyambo and Kananga B, towards Ruwe. At Dikuluwe and further north, narrow bands of the Mines Series also follow the northwestern edge of the *écaille*; the Dikuluwe band of dolomitic shales extends 2 miles, bordered on both sides by silicified, foliated dolomite. East of Dikuluwe, scattered fragments rise between Mashamba and Kabulungu.

The 1,000 ft. wide central alignment essentially consists of a fractured and overthrust band of dolomitic shales, bordered by a narrow band of silicified stromatolite dolomite and foliated silicified dolomite. This formation lies south of the dolomitic shales in the fragments of Kamoto and west of Musonoi, but north of the shales of Musonoi and further east, where the overthrust sequence has been repeated. A ribbon of dark dolomite (C.M.N.) borders the long eastern sector, the same rock also covering large irregular bodies northwest of Musonoi and north of Kamoto. The following layers have been distinguished in the Mines Series:

C. M. N.(1)

Micropsammitic, chloritic, dolomitic sandstone and shale; dolomite, stromatolite dolomite; chloritic and carbonaceous dolomitic shale and dolomite.

Carbonaceous dolomite, with micaceous bedding joints and psammitic inliers in the lower levels.

Crossbedded felspathic, dolomitic sandstone.

S. D. (3)

Carbonaceous dolomitic psammite.

Dolomitic psammite.

S. D. (2)

Carbonaceous dolomitic psammite.

Dolomitic psammite.

Micaceous stromatolite dolomite.

Carbonaceous dolomitic psammite, in the north with dolomitic sandstone.

S. D. (1)

Carbonaceous dolomite.

Dolomitic psammite, with inliers of micaceous dolomite increasing upwards.

R. S. C.

Stromatolite dolomite with inliers of dolomitic psammite.

D. S. I. (2)

Dolomite with micaceous joints, the thickness of the beds thinning upwards.

D. S. I. (1)

More or less chloritic dolomite and dolomitic shale.

R. A. T.

Dolomitic, chloritic sandstone becoming micropsammitic downwards.

The first member, R.A.T., in fact belongs to the preceding stage of the Upper Roan; it is followed by a void of sedimentation and regression, characterized by climatic oscillations, and the accumulation of carbon and sulphides.

At *Kamoto*, five peaks of copper mineralization (Fig. 83) have been recognized. The most

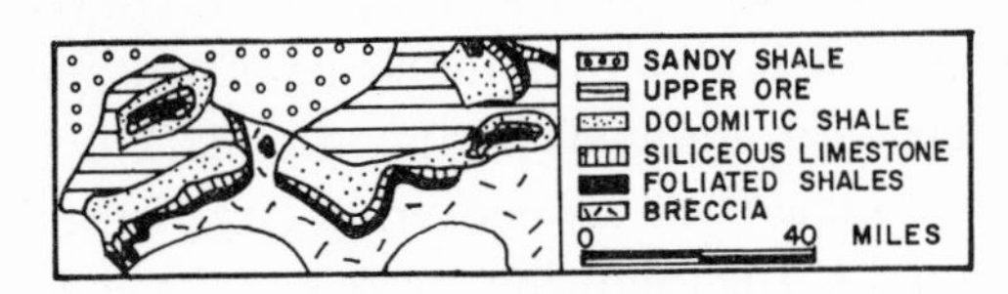

Fig. 83. The Kamoto area.

important is distinguished by bornite and located above and below the stromatolite layer R.S.C. The number of maxima decreases northward: only two are left at Kilamusembu in the D.S.I. dolomite and in the last dolomitic psammite of S.D. (2). The ore horizons pinch out at Kananga. At Kamoto, pyrite reaches its peak in the first and last stage of S.D. (2); its maximum descends northward into lower layers. North and south of the axis of the *écaille* and in every horizon but those immediately overlying or underlying the stromatolite layer, chalcopyrite and pyrite occur instead of pyrite. Only the upper, dolomitic psammites of S.D. (2) and S.D. (3) are distinguished by high chalcopyrite grades. The copper/cobalt ratio averages 10/1.

Carbonaceous, but not very dolomitic beds are indicators of mineralization. With carbon content increasing, pyrite precipitates instead of copper. Chloritic and micaceous horizons (R.A.T. and D.S.I 2) are favourable, but non-micaceous–siliceous beds are barren. As to texture, fine grain size, thin bedding, and the repetition of the sequence characterize the

TABLE CXXIII

MUSONOI SEQUENCE

Layer	*ft.*
Dolomitic limestone	
Upper, red dolomitic shale	330
Black ore main zone (B.O.M.Z.)	15
Dolomitic shale (S.D.)	20
Silicified stromatolite dolomite	50
Foliated dolomite (R.S.C.)	20
Silicified inlier	
Sandy, foliated dolomite (R.A.T.)	20
Breccia of dolomitic shale	

ore formations. At Kamoto, the dolomitic psammites of the lower bed of S.D.I. and the last bed of S.D.(2) contain 9.4 and 2.8% Cu, 0.06 and 0.2% Co, and 1.0 and 2.6% Fe. Table CXXIII shows the sequence at *Musonoi*.

The low lying R.A.T. has been correlated to the green ore horizon (M.V.) of the Elisabethville district. The plastic breccia has invaded the horst of the South Ridge, pushing part of the faulted anticline southward; the other, faulted limb rests on a diapyr. Further north the syncline has been thrown down. The R.A.T. breccia is directly overlain by fragments of any of the later stages of the Mines Series. Crushed zones and voids are also occupied by sediments comprising R.A.T., R.S.C. and S.D. that push asunder smaller fragments or scales. On a large scale the floor of these fragments is horizontal, but it is convex on a small scale.

Ores accumulate 30 ft. above and 30 ft. below the cellular stromatolite dolomites (R.S.C.), i.e., in the dolomitic shales (S.D.) and in the silicified dolomite and shale (R.S.C. and R.A.T.), and concentrate mainly at the contact of R.S.C. and S.D. In a given horizon, the mineralization is lenticular, forming a three-dimensional pinch and swell pattern. Copper oxides occupy large patches in the middle of the cobalt minerals replacing S.D. shales, but the copper–cobalt–vanadium–uranium–gold–palladium–platinum paragenesis descends into the foliated shales (R.S.F. in S.D.) which are barren in the rest of the Copperbelt.

At *Kolwezi* proper, transverse faults have broken up an anticline. Dolomitic breccia (R.A.T.) invades the overlying strata from a thrust fault, and the nose of the fold through an axial fault. The *Ruwe* écaille, occupying a double trough in the Kundelungian shales, has been thrust over the breccia. A faulted anticline divides the twin troughs; the R.A.T. breccia invades the anticline and other voids, coming into anomalous contact with various strata. The 2,000 ft. wide southern trough has been intersected by a transverse fault, its northwestern wing being further subdivided by reverse and thrust faults. The eastern member of this substructure even pierces the overlying dolomitic shales (S.D.) with the central *écaille* thus depressing the axial strip of the *écaille* into the breccia, which has been mineralized by percolating water.

JADOTVILLE DISTRICT

The belt extends from Kamwali 30 miles N60°W to Jadotville, continuing 40 miles to Fungurume. The middle zone comprises: *(1)* Kamwali, Shandwe, Luishia, Kansongwe in the east; *(2)* Shituru, Likasi and Kamatanda, near Jadotville, *(3)* Kalabi 20 miles to the north; *(4)* the Kambove, Kamoya, Kabolela, Sesa, Shangulowe Group, west-northwest of the town; *(5)* Kakanda; *(6)* the Fungurume écaille; *(7)* Shinkolobwe–Kasolo (see 'Uranium' chapter) is located 12 miles due west of Jadotville, *(8)* Tantara being farther on the same line, ending at *(9)* Mirungwe and Mindigi 40 miles west of Jadotville.

This trend points towards the Kengere lead deposit south of Kolwezi. The district is distinguished by its copper mineralization, cobalt accompanying it only at Kamwali, Luishia, Kambove and Fungurume.

The *Fungurume* syncline, bordered by faulted anticlines, pitches south among the east striking Upper Roan *écailles*. The sequence is complete up to the earlier parts of the Mwashya Series, with the structure attaining a width of 10 miles. Faulted contacts, the Kundelungu System and the Large Tillite border both sides of this fold belt. The bordering faults tend to meet and intersect each other in depth and in the west, thus forming a wedge.

There are several ore horizons: the main body consists of 3–7 ft. of dolomite, overlain by 1–2 ft. of chloritic–dolomitic shale, 2–7 ft. of dolomite, and chloritic–dolomitic shale, constituting the altered dolomite of R.A.T. The upper part of the ore body consists of 3–12 ft. of siliceous dolomite overlain by 5–11 ft. of dolomitic and micaceous shale, constituting the layered R.F.S. shales. The second ore body lies in the conglomeratic R.A.T. breccia of poorly stratified shales, sandstones and dolomite. Several phases of dolomitic recrystallization, chloritization and silicification have been distinguished (OOSTERBOSCH, 1951). Complete oxidization reaches a depth of 300 ft., with mixed oxides and sulphides appearing as deep as 750 ft. In this zone oxidized and non-oxidized beds alternate. Bornite–chalcocite and chalcocite are the main ores; bornite dominates at 1,200 ft. of depth. The grain size of the ore and host exhibit a sympathetic relationship, and the upper sulphide zones contain the highest grades (7% Cu). Mineralization appears to be a function of microfissures, in excess of normal porosity; the distribution of cobalt, however, is unrelated to the copper mineralization.

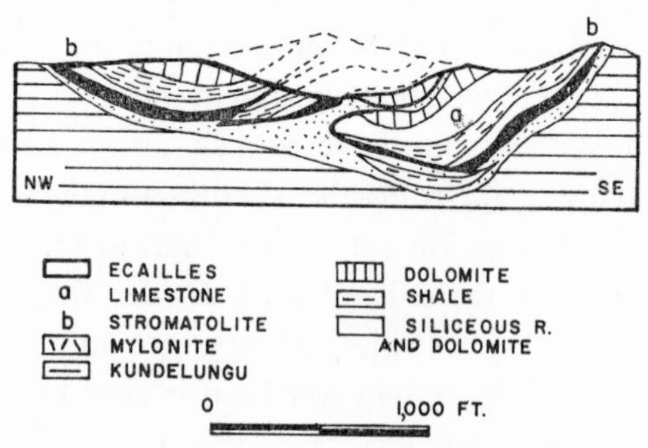

Fig. 84. Section of central Kambove.

The *Kambove* deposits are located on the southern (Fig. 84) limb of the Elisabethville anticline, the northern limb having been sheared off. Complex fault zones divide the core of the Mines Series from the Upper Kundelungu in the northeast, and the Mwashya ooliths from the jaspers in the southwest. At Kambove proper, the low-lying plastic R.A.T. consist of violet, ferruginous and sandy talc–chlorite dolomites. The overthrust *écailles* float in a synclinal trough of the Kundelungian strata filled in by the mylonitized shales; the syncline pitches west. There can be distinguished four *écailles*, all curving downwards. Some of the *écailles* are separated by only a thin layer of breccia. The thickest attain a depth of 300 ft. but the thinner units, having glided from the southeast, are only 50 ft. thick. In the stromatolite layer, gel-malachite accumulates in the voids of cylindrical and oval dolomite cups. Under one of the écailles, percolating solutions impregnate the younger, Kundelungu strata with malachite.

At Kambove West, the mylonites consist essentially of chlorite shale (without dolomitic inliers) embaying over-folded larger and smaller fragments of the rest of the Mines Series. The latter are mainly composed of dolomitic limestones and thinner, faulted beds of the intermediate members. Copper–cobalt oxides and copper silicates are associated to malachite.

The *Likasi* impregnations consist principally of copper silicates, with neogene quartz stringers and patches forming in the breccia.

The two *Shituru* ore horizons are located in the Mwashya Series. The underlying Mines Series, however, has disappeared by faulting, as did part of the Mwashya dolomite from the northern limb of the faulted anticline. The basal, Kundelungu tillite has been conserved. The Mwashya Series consists of carbon bearing shales, pseudo-ooliths, dolomites and breccia.

Chrysocolla accompanied by malachite, which covered bedding planes, has replaced the dolomite. At *Tantara*, chrysocolla, planchéite shattuckite, brochantite, strontianite and cobalt-calcite accumulate in cavities of the Kakontwe limestones.

Malachite is the principal ore of the faulted *Luishia* saddles. Selective alteration results in dolomitic layers containing disseminated sulphides and softer, earthy, copper–cobalt oxides.

ELISABETHVILLE DISTRICT

This district includes Kipushi, the continent's principal zinc and silver deposit 20 miles west of Elisabethville, and a group of occurrences of great historic interest extending from Étoile (Star) du Congo, to Ruashi, Luiswishi and Lukuni. The Kipapila copper deposit is located 30 miles southeast of Elisabethville, in the same trend. The southeast striking Elisabethville anticline comprises a Mines Series core, covered by Mwashya and Kundelungu sediments. The northeastern limb of the anticline is intersected by a southwest dipping strike fault, and only in the southwestern limb has the Mines Series been conserved between faulted abnormal contacts. Transverse faults also intersect the formation.

The Elisabethville anticline widens in the northwest. The *Lukuni–Luiswishi* compartments, floating in the breccia, lie between abnormal contacts of Kundelungu sediments in the northeast, and Mwashya strata in the southwest. Two minor *écailles*, Kisushi and Lukuni, have been overthrust in the north; their silicified stromatolite layer attains only a width of 3–6 ft. Transverse faults separate Lukuni from Luiswishi where several repetitive horizons of the Mines Series are wedged into a trough. At *Ruashi*, an overturned secondary syncline affects the main faulted anticline. Malachite and earthy copper–cobalt oxide constitute the main ores. The *Étoile* (Star) *du Congo* écaille is bordered by the northern fault zone. A 10 ft. thick breccia separates the schistose dolomites from the dolomitic limestones, which thin out to a thickness of 100 ft. Silicates predominate in the mineralization.

SOUTHEASTERN KATANGA

Like the Zambian ore bodies, the Musoshi, Kintenda, Lubembe and Mokambo deposits of the Katanga Pedicle are located in the Lower Roan. At *Kinsenda* and *Lubembe* quartz and quartz–calcite–tourmaline veins crisscross the deposit. The primary paragenesis of these occurrences consists of linnaeite, pyrite, chalcopyrite and bornite; chalcocite and covellite, however, belong largely to the zone of alteration.

North of Bancroft, at *Musoshi*, the Lower Roan column consists of:

Dolomite.
200–300 ft. of shale and sandstone with inliers of arkose.
300–380 ft. of more or less dolomitic shale and sandstone.
Dolomite inlier.
130–270 ft. of arkose, shale and sandstone.
70–130 ft. of dolomite with dolomitic shale and limestone.
Some pyrite and traces of chalcopyrite.
250–500 ft. of arkose, felspathic sandstone and shale.
40– 90 ft. of micropsammitic shales, with pyrite, chalcopyrite and bornite (Ore Formation).
Arkosic conglomerate.

TABLE CXXIV

GRADES OF ORE FORMATION SHALES

	Cu (%)	*Co* (%)	*Fe* (%)
Hangingwall	0.06	0.03	3.2
Ore	4.0	0	3.0
Footwall	0	0	3.9

In the Ore Formation, 1 inch – 10 ft. thick layers of light coloured felspathic sandstone, sandy micaceous green shale and micropsammitic green shale alternate repeatedly. Only an intermediate subgroup of this sequence has been mineralized and the upper member of the subgroup is poorer (Table CXXIV).

Stages of alteration are believed to comprise the crystallization of *(1)* microcline and tourmaline, *(2)* quartz, *(3)* hematite, *(4)* albite, *(5)* apatite and monazite. The mineralized and replacement zones appear to be identical; they are also distinguished by secondary ilmenite, rutile, topaz, zircon and scapolite. Idiomorphous pyrite occurs in isolated grains; chalcopyrite may also surround or penetrate into pyrite, but the presence of pyrite excludes bornite. Chalcopyrite may crystallize either with of later than bornite, and cobalt sulphides are included in both. Blue chalcocite (digenite) is contemporaneous with or older than white chalcocite and covellite. All three are supergene, that is, they replace bornite and chalcopyrite.

BANCROFT AND KONKOLA

At *Konkola*, south of Musoshi, 1½ mile wide Lower Roan siltstones and sandstones embay a stock; the dome slopes towards the southeast, and a fault traverses the ore body in the south. The ore horizon, consisting of dark grey, siliceous siltstone, thickens eastward. Copper sulphides are indistinctly zoned, concentrating at the top and bottom levels.

Bancroft is located 8 miles southeast of Konkola (Fig. 85). A wide band of up to 2 miles of the Lower Roan has been draped around an outlier of older gneisses and granites. Their wedge trends east for 7 miles, bending at the Kirila fault southward, to Bomwe and Kakosa. The Bancroft anticline strikes southeast, but its axis has been deflected and deformed. Sandstones, quartzites and conglomerates constitute most of the footwall. The Ore Formation is transgressive on the conglomerate, and sandstone and dolomite conclude the Lower Roan sequence (SCHWELLNUS, 1961).

The *Kirila–Bomwe* ore body consists of five units: *(1)* calcareous sandstone and siltstone 1.4% Cu; *(2)* banded siltstone, 5.6% Cu; *(3)* alternating pink sandstone and grey siltstone 5.6% Cu; *(4)* grey siltstone, 2.4% Cu; *(5)* sandy siltstone, 1.3% Cu.

Copper values are most constant in the second unit; grades of unit *(3)* vary, however, with the frequence of calcareous sandstone bands. The fourth unit contains oxides and carbonates. The zonal distribution of chalcopyrite, bornite, chalcocite, malachite and chrysocolla reflects the largely supergene sequence, terminating with cuprite, azurite and native copper. The bedded minerals were apparently reconcentrated along the basal cleavage, and later in dissolution cavities.

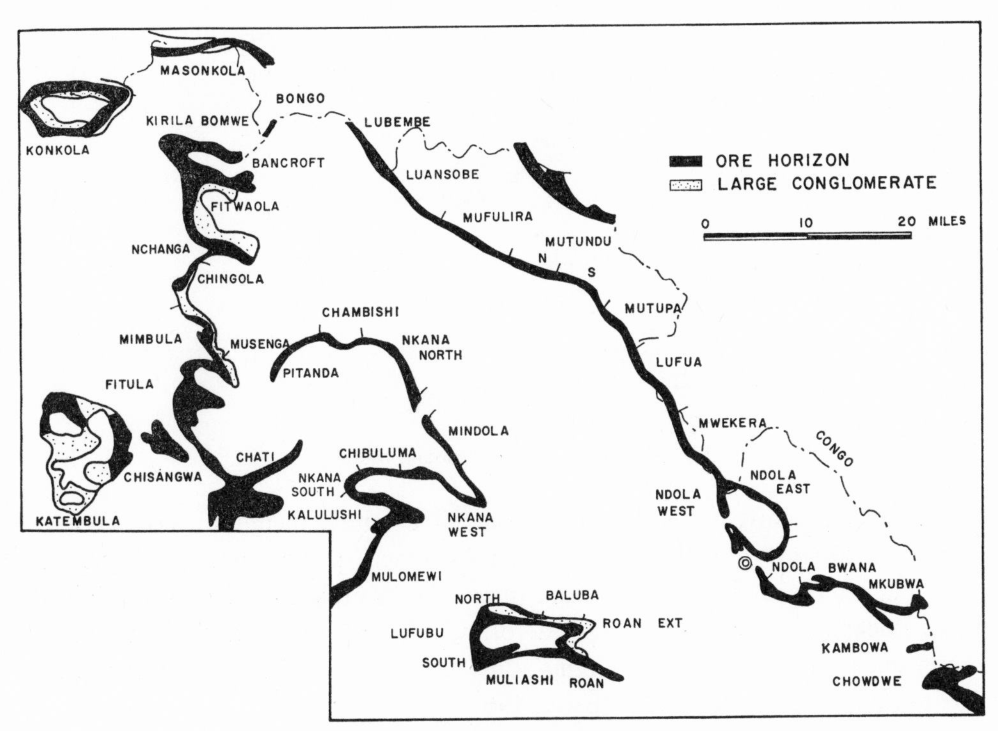

Fig. 85. The Zambian Copperbelt.

NCHANGA

Nchanga is located in an inlier branching out towards the east between Kakosa and Chingola, where a secondary 7 mile long syncline pitches westwards parallel to the schistosity of older Precambrian rocks. While its southern limb dips 20–30°N, the northern limb plunges subvertically. A northwest striking fault, filled in by 200 ft. of breccia, cuts the Luano ore body in two. The *Chingola* ore body, the western extension of Nchanga, rests in one of the recumbent folds.

Copper carbonates occur in the arenaceous Early Roan (McKinnon and Smit, 1961). In the lower, quartz–sericite shale (the principal Ore Formation of Nkana and Roan Antelope), copper minerals are sparsely disseminated. Overlying 120–240 ft. of sediments, the 50–120 ft. thick, mineralized orthoclase–microline quartzite is distinguished by cross bedding, ripples and mud cracks. In the upper laminated shale, dolomite content increases upwards, and the ore body lies 10 ft. above the contact. Upper Roan dolomites are overlain by basic sills, limestones and sandstones.

While ore minerals occur sparingly in much of the 300 ft. thick column, the two main bodies are localized (Fig. 86) above troughs of the old relief; for about 2 miles both trend generally east, with their width attaining less than a mile. Although much of the ore occurs in lenses and beds, mineral zoning overlaps bedding, and fissure filling veins are also found. Chalcopyrite does not descend into the lower ore body. The depth of oxidization attains 2,000 ft. and leaching 1,000 ft., and greatest concentrations correspond to changes of felspathic and siliceous facies. While malachite is omnipresent, chalcocite is more abundant than azurite and chrysocolla; copper and cuprite, uraninite and torbernite occur mainly in the lower layers, but cupriferous vermiculite is found in the upper dolomite.

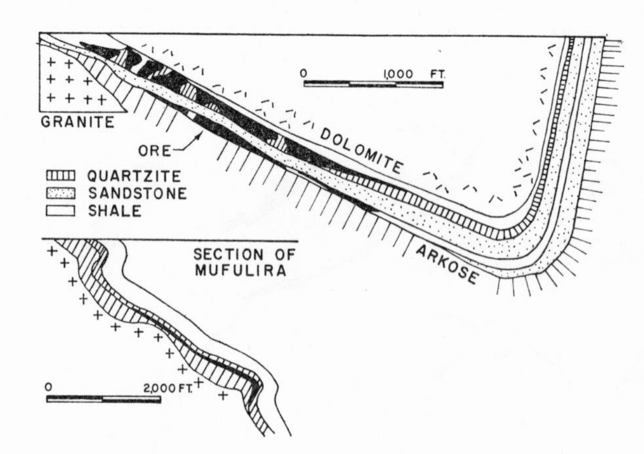

Fig. 86. Sections of Nchanga and Mufulira.

Mimbula, like Bancroft and Nchanga, is located in an embayment that projects eastward. The syncline, situated 6 miles southwest of Nchanga, pitches northwest. Sandstone, quartzite and interbedded siltstone include three subconcordant ore lenses, attaining 300–400 ft. of thickness, at an average altitude of 10, 120 and 180 ft. above their base. The maximum thickness of these lenses attains 30 ft. They consist of bornite, chalcopyrite and covellite accompanied by chalcocite (SMIT, 1961).

NKANA BASIN

South of Mimbula, the Lower Roan beds leave their southwest trending arc. *Chambishi* is located 15 miles southeast of Nchanga (GARLICK, 1961) on the northwestern edge of the Nkana basin, an ellipse of 17 by 10 miles. The southeast pitching syncline of Chambishi is part of an open monoclinal fold but secondary folds prevail in the ore shale, underlain by up to 300 ft. of conglomerate and quartzite. The thickness of the mineralized biotite–quartz argillite varies between 60 and 100 ft.; dolomite–anhydrite–quartz lenses only carry copper in its upper reaches. The shale grades into sandstone, quartzites, and argillites which terminate the Lower Roan. The Upper Roan consists of dolomite and schists injected by an amphibolite sill. The main ore body may attain a thickness of 70 ft,, and a length of at least 4,000 ft. Concentrations vary with lamination and lensing, and bornite values increase in the vicinity of a granite cliff, chalcopyrite accumulating higher. A pyrite aureole surrounds much of the ore body.

The *Nkana* syncline is on the southeastern edge of the basin; its beds dip northwest. The axis of the syncline has been warped, moving harder beds on softer layers. The Ore Formation is underlain by arenaceous sediments and quartzite, and overlain by argillite and sandstone. Kersantite traverses the Lower Roan. The Upper Roan is twice as thick as the Lower Roan, but Mwashya sediments in the basin encircle the Kundelungu. Dolomite, quartz and biotite in laminae of varying thickness constitute two ore types (JORDAAN, 1961). Chalcopyrite and carrollite occur in the micaceous shales of the north ore body, and in quartz and dolomite veins traversing the south ore body. Copper has been disseminated in the whole Ore Formation, particularly at Nkana South; the variation of grades, however, is not controlled by bedding. While the upper, feldspar–quartz chert generally shows the greatest values in northern Mindola, grades are more evenly spread in southern Mindola and Nkana

North, where mineral zoning cuts through bedding and a transitional zone separates the upper bornite from the lower chalcopyrite horizons. At Mindola, zones are not stratiform; bornite even surrounds a smaller oval of chalcopyrite; at Nkana South the two are associated. Veins cutting the sequence show the highest values when traversing ore formations. The usual minerals of the oxide zone, plus tenorite, extend 100–400 ft. in depth. Uranium occurs at Mindola.

Chibuluma, 8 miles west of Nkana, is located on the southern limb of the basin. Conglomerate, quartzites overlie granite–gneiss. The Ore Formation is 30 ft. below the quartzites and conglomerates concluding the Lower Roan syncline. The Upper Roan consists of chlorite schists, dolomite and brecciated shales. Most of the sulphides concentrate in grits of the schistose, sericite quartzite (WINFIELD, 1961). Values vary according to bedding; chalcopyrite, pyrite and barren layers distinguish the zoneography, but bornite remains scarce.

ROAN ANTELOPE – BALUBA

Beyond Chibuluma, Lower Roan beds bend sinuously towards the south. The Muliashi basin, measuring 10 by 4 miles, trends south-southeast in a depression of the older relief.

Baluba is located 20 miles southeast of Nkana, in the northern gulf of the basin (LEE-POTTER, 1961). The syncline pitches west towards the main basin. Its dominant feature is en échelon folding. Mineralized, spotted calcschists lie over 500 ft. of conglomerate, quartzites, grits, argillites and a conglomerate. The thickness of the Ore Formation may attain 180 ft.; argillite and quartzite alternate in the hangingwall. Scapolitic argillite (also containing biotite), arkoses and quartzite terminate the Lower Roan, but the Upper Roan consists of sandy–argillaceous sediments and dolomites.

Chalcopyrite, bornite, chalcocite, carrollite, and some cuprite are most abundant in calcschists and dolomite, both containing tremolite–actinolite, a shear mineral. The Formation measures 3 miles along the folds, attaining a thickness of 27 ft. Low grade chalcocite reoccurs 100 ft. above the main horizon. It includes cobalt sulphides and bornite. In the overlying dolomitic argillite chalcopyrite values are related to the carbonaceous layering of the lower horizons. Roan Extension is located 3 miles south of Baluba. The Roan syncline, a west-northwest pitching satellite of the Muliashi basin, is wedged into older rocks; its limbs have been complicated by secondary or pod folds. The folding is tightest near narrows of the old relief, and at *Roan Antelope* some chalcopyrite accompanies pyrite in pre-Katangan schists adjacent to granitic intrusives (MENDELSOHN, 1961). Dolomites and dolomitic argillite of the Ore Formation rest on the footwall conglomerate. This column attains a thickness of 50–170 ft. Arkoses, argillites and dolomites conclude the Lower Roan, but the sequence is repeated in the Upper Roan which is partly covered by Mwashya shale and Kundelungu tillite.

The ore body averages 25 ft. in thickness, its strike length attaining 8 miles. Pyrite occurs in most of the ore horizon. In the southeastern extremity of the main basin copper minerals have accumulated in low lying tremolite–biotite schists. However, in the Roan basin commercial grades are confined to the median part of the biotite–quartz–feldspar argillite, although some chalcopyrite can be found above the hangingwall at the nose of the fold. On the other hand, dolomite and scapolite have precipitated in the western part of the sub-basin. Mineral zoning overlaps bedding by as much as 1 ft.; minerals concentrate along bedding planes however, especially in argillite. Sulphide granulometry is a function of gangue grain dimensions. Near the main basin, pyrite, chalcopyrite, bornite and chalcocite impregnate higher horizons than at the eastern tip, but chalcocite only appears at eastern Roan. The depth of oxidization generally reaches 200–300 ft.: tenorite, (covellite) and cupreous wad accompany malachite and chrysocolla.

MUFULIRA BELT

The second largest deposit of the Copperbelt is located on the northern fold of the Kafue batholith and anticline, that is, on the southern limb of the Mufulira syncline, with the small Mokambo deposit being situated on its northeastern limb in Katanga. The syncline and partly overturned pod folds pitch northwestward. The Mokambo limb is vertical but the Mufulira limb dips 45° towards the northeast. At its narrowest point the syncline is 8 miles wide. Poorly mineralized granite underlies the Lower Roan in the south. Lower Roan beds extend 16 miles from Katanga to Luansobe, Mufulira and Mutundu South; Kundelungu sediments underlain by the Mwashya Series, however, occupy most of the syncline. The 5,000–6,000 ft. thick Upper Roan consists of two talcose dolomite layers with subsidiary shales and quartzite.

Luansobe, the northwestern extension of Mufulira, is located on the Congolese border (GROEN, 1961). The Ore Formation consists of five quartzite beds interrupted by dolomitic shales. Chalcocite occurs alone in two later quartzites and with chalcopyrite and bornite in the median quartzite.

At *Mufulira*, the Lower Roan conglomeratic grit is overlain by footwall quartzites, the thickness of which may attain 500 ft. (Fig. 86).

The 215 ft. thick Ore Formation is overlain by 180 ft. of argillite and argillaceous quartzite, interrupted by dolomites and the 'marker' grit. The hangingwall formation of 225 ft. repeats the ore sequence, values, however, being uneconomical.

The lower, *c* ore body extends for 3½ miles to all the basins. While the intermediate ore body *b* somewhat overlaps into the central basin, the top *a* ore body is restricted to the eastern basin. However, it excels in its thickness (40, 60, 30 ft.). All three ore bodies consist of grey sericitic quartzite. The footwall of the lower ore body is mineralized in the western basin; between the western and central basin, however, pyrite takes the place of the copper minerals of *c* ore body. In the eastern basin, gritty lenses are abundant, and bedding planes of the lowest felspathic quartzite are mineralized. The first interbed comprises the 'mudseam' of dolomitic siltstone, and the median ore body includes dolomite lenses.

The principal ore mineral of the lower ore body is chalcopyrite with the average copper grade attaining 3%. While bornite predominates in the median oreshoot, chalcocite prevails in the top ore body with grades of up to 10% Cu in lenses and in the footwall. Malachite, cuprite and copper descend to a depth of 200 ft. The pyrite–chalcopyrite paragenesis is characteristic for low grade ore bodies, and the bornite–chalcocite association distinguishes the great copper concentrations. Stratigraphic zoning of ore minerals is indistinct.

The Ore Formation has largely been eroded from adjacent, earlier Precambrian rocks, possibly from the vicinity of the deposit. Indeed, sediments are largely clastic, except for some dolomite and anhydrite.

At *Mutundu North*, 2–8 miles southeast of Mufulira, malachite, bornite and chalcocite occur in dolomitic quartzite at the top of the Ore Formation. The ore quartzite extends southeast to Lufua, Mwekara, and Ndola (where it is believed to lie at a depth of 10,000 ft.), to Bwana Mkubwa, Kambowa and Chondwe. At *Bwana Mkubwa*, 45 miles from Mufulira and 8 miles from the border, a 250 ft. thick Ore Formation outcrops in the southern limb of a rather overturned secondary syncline. The Shovel Pit has been abandoned, but it was not worked out in 1928 (PIENAAR, 1961). Chalcocite and malachite and other minerals are disseminated in the feldspar quartzite of the footwall; the same minerals, as well as chrysocolla and residual bornite, chalcopyrite, and tenorite mainly fill in a fracture in the mineralized, ferruginous sandstones, siltstones and shales, with interbedded dolomites carrying zircon, rutile and tourmaline. Part of the hanging wall has been mineralized, also containing libetenite.

OTHER DEPOSITS

At Bamba Kilenda, near Madimba in the Lower Congo, two parageneses have been found in a zone of fractures. Chalcopyrite is independent from the sphalerite–galena–pyrite association. On the Lubi River (in Kasai), a horizon of dolomites has been impregnated by chalcocite, malachite and chrysocolla.

Around Lake Mweru, copper has been disseminated in the top layer of the Kundelungu (Upper Katangian). While shales and sandstones also contain chalcopyrite, at Kapulo dolomite only carries bornite and chalcocite. However, stromatolite limestones and carbonaceous shales contain pyrite. A stock of granite is located nearby.

At Kansanshi, in northern Zambia south of Jadotville, chalcopyrite, pyrite and gold have crystallized in veins, at the contact of limestone, and in carbonaceous shale. At Chifumpa chalcopyrite and pyrite have been disseminated in dolomitic shale. At Sable

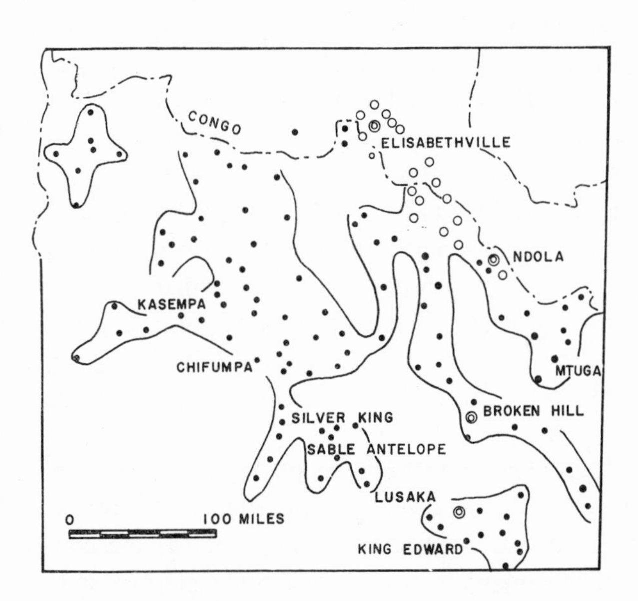

○ COPPER BELT DEPOSITS • MINOR OCCURRENCES

Fig. 87. Copper deposits in Zambia. (After O'BRIEN, 1958).

Antelope (Fig. 87), and at Silver King, copper minerals, including silver, replace a sideritic–dolomitic breccia of the *Upper* Roan. In the King Edward Mine, sulphides have been disseminated in Upper Roan sediments. Other occurrences of the Lusaka area include Chongwe East, Chalimbana and Kachobya.

At Mtuga, chalcopyrite and pyrite impregnate lenses of sheared granite; at Kalulu tightly folded micaschists, and at Kansonso fissure filling gossans, are mineralized. Other occurrences of the area include Hippo, Lewis and Chanobie.

NORTH AFRIÇA

Algeria. Permian and Triassic sandstones or red-beds contain copper at Nédroma, Sidi Yaya, Sidi Driss, the Jbels Kahar, Maïz, and Mellaha. On the other hand, Cretaceous vein deposits are found near Boudjoudoun, at the Jbel Adrar. Copper-silver-antimony-arsenic parageneses prevail at Mouzaïa, in the Mitijda Atlas. Replacement traces are widespread in eastern Algeria, and at Ain Barbar, north of Edough, 20 quartz veins injected into Numidian marls contain chalcopyrite, sphalerite, and pyrite accompanied by galena.

In the Saharan Atlas, northeast of Colomb Béchar, clay horizons and their hangingwall, interbedded in thick columns of Cretaceous (Albian–Aptian) sandstone, carry chalcocite. The largest occurrences cover 700 square miles at Aïn Sefra, Tiloula and Sfissifa. Mineralization is related to vegetal remnants. Sandstones have been mineralized at Ben Hendjir and Tiourtelt, with carbonatic sandstones containing the copper ores at Garet Deba, Garet el Kreil and Dir el Hareich. The *Moroccan* (Fig. 10, 59) deposits are concentrated in the south , in the Precambrian–Paleozoic chains of the Anti-Atlas. Azegour, the only occurrence of some significance, is discussed in the 'Molybdenum' chapter. Most of the chalcopyrite impregnates meta-amphibolites of the tactite–cipolino contact there. The Middle-Precambrian hydrothermal veins include Bou Skour and Azzerballou, but the Sarrho and Oumjerane areas are Paleozoic. Impregnation deposits in Precambrian lavas and other sediments of the Anti-Atlas occur a Issougri, Djebel Gueliz and Tizi Moudou. Hydrothermal veins have also been found in the High Atlas, north of the Anti-Atlas, at Taourirt, Ourika, Tizi n'Israken, Djebel Krakh and in the Jebilet.

Bedded copper deposits are widespread in southern Morocco, in *(1)* the 40 mile long, south-southwest striking, Triassic Argana depression, which forms the western border of the High Atlas, 20 miles east of Agadir; and in four pre-Silurian belts: *(2)* a 25 mile long, N30°E east trending belt, connecting 80 miles east of Agadir, Tataout, Assif Imider, Oued Arrhen and Tanguerfa; *(3)* Iminirfi and Talaat n'Ouamane; *(4)* a 40 mile long, south-southwest striking belt, bordering a Precambrian area, southeast of Agadir and connecting Aït Ourhaïne, Tazalaght, Tasserirt and Agoujgal; *(5)* a 70 mile long belt following the southern border of Morocco, from Tailezza to Jebel Ouansimi, Adrar ou Anas, Taourirt ou Anas and Idm ou Anas.

The three mineralized arenaceous horizons of Argana carry carbonates and three pre-Silurian horizons have been mineralized: *(a)* the contact of most of the Precambrian and the latest Precambrian–earliest Cambrian (Georgian); *(b)* the contact between the first and second stage of the Georgian; *(c)* sandstone–shale interbeds in the Georgian.

While only the basal shales carry copper at Talaat n'Oumane, various arenaceous and dolomitic levels have been impregnated at Agoujgal.

MAURITANIA

Akjoujt is located 120 air miles from the coast and from the Spanish border (Fig. 55). In the syncline, ferruginous and magnetitic quartzites of the centre are successively followed outwards by gneisses, mica and chlorite schists (BLANCHOT, 1958). The Akjoujt Series lies high and uncomformably upon poly-metamorphic gneisses of the Amsaga Series. A syntectonic, calc-alkaline biotite granite stock, surrounded by mylonites, is intrusive into the centre of the syncline. Carbonatite sills and bodies (ferruginous rauhaugite) and dolerite dykes are also injected into the Series. Recent sediments cover all but the eastern extremity of the syncline.

The *Guelb* (or horn) *Moghrein* covers an area of 1,000 by 600 ft. and rises to a height of 250 ft., 3 miles west of Akjoujt. In the west emerges a smaller and somewhat shorter hill. Both consist of chlorite schists including large hematite bodies altering to malachite. Sulphides are sparsely disseminated in late quartz veins traversing the surrounding amphibolites (BLANCHOT, 1955). The copper deposit is associated with one of the carbonate sills overlying amphibolite, intercalated between schists and micaschists. H. RAMDOHR (1957) believes this to be similar to the Tunaberg skarns. The principal carbonate mineral of the pistomesite type and rarer dolomite have been shattered, with magnetite precipitating along the cracks. Chalcopyrite, cubanite and cobaltiferous pyrrhotite have been introduced at a late stage in veinlets cutting the magnetite. The paragenesis also includes arsenopyrite, nickel cobaltite, amphiboles and chlorite, accompanied by gold, graphite, vallériite and skutterudite (VINCIENNE, 1957). Copper minerals cover cleavages and joints in hematite and schists, also forming acicular geodes in the hematite, but melnikovite appears later. Oxidization reaches a depth of 70 ft. distinguished by atacamite, chrysocolla, and malachite accompanied by azurite. Magnetite content attains 45% in the oxide zone and 35% in the sulphide zone; copper grades average 1.8%, and the tenor of gold 16 dwt./ton.

The post-Akjoujt carbonatic phase shows copper in a larger area. At *Tabrinkout*, scheelite and gold bearing bismuthine accompany copper minerals. On the flank of the *Irarchen*, along the wadi *Jennin*, cuprite and malachite are interstitial in hematite lenses of silicified rauhaugite or dolomite. At *Legleitat* the cupriferous carbonatites are associated to albitite and orthoclasite.

THE REST OF WEST AFRICA

Mali. At Nioro, near Sirakoro and near *Lambatara*, malachite and dioptase impregnations in calcareous, Cambrian sandstones, schists and shales are found. A 250 ft. thick ore body with low grade copper occurs at *In Darset*. In the Niger Republic, traces of copper have also been reported from the western border of the Aïr: Cretaceous, crossbedded sandstones of variable grain size include arkose, embayed by shale. Chrysocolla, malachite, cuprite and copper favour argillaceous fluvio-lacustrine beds, distinguished by vegetal remnants. Traces of cobalt, nickel, vanadium, barium and arsenic are independent of the copper mineralization, but uranium is related to the latter.

Upper Volta. At Gongondy, 24 miles east of Gaoua, in the southwest, barren, north striking quartz veins traverse sheared ophiolites, dolerites and brecciated andesites, but the younger, east striking pair of this direction includes veins, pockets and concretions of auriferous chalcocite, chalcopyrite, malachite, some azurite, cuprite and chrysocolla. The width of the outcrop attains only 12 ft. but the lateritized country rocks also contain gold. Values of the vein have attained 33–67% Cu, 8–20 dwt./ton Ag and 3–10 dwt./ton Au; they decrease, however, with depth. Adjacent occurrences include Meïra, Diénamara, Bossora and Koulindiora.

In the Ivory Coast copper shows at Danané–Toulepleu.

CONGO REPUBLIC AND ANGOLA

Base metals are less abundant in the Lower Congo geosyncline than in the contemporaneous Copperbelt. In the *Congo Republic*, copper is associated to lead and zinc in the Niari–Nyanga basin (Fig. 25, 53), west of Brazzaville. Copper and lead–zinc–copper horizons alternate at the contact of the shale–limestone and shale–sandstone series. Copper accompanied by pyrite, tourmaline and albite minerals favour calcareous rocks. Most homogeneous is one of the horizons of the Luila, but the only deposit of some significance is *Mindouli*. Cupriferous soils are not directly related to this mineralization.

In *Angola*, copper predominates in the 20,000 square mile district south of the Congo River (Fig. 26). At *Mavoio*, galena only rarely accompanies chalcopyrite, chalcocite and bornite, as well as cobalt traces with a gangue of siderite, calcite and ankerite in the sub-concordant limestones and dolomites of the shale–dolomite system (CORTESÃO, 1958). Warping is moderate, but the mineralization appears to be related to fold axes and to the 100 mile long 'Hematite' fault system. The oxidized zone is the richest, at *Tetelo*, east of Mavoio, where 8% Cu and 1.2% Co impregnate fissures and breccias of Inkisi sandstones considerably higher in the stratigraphic column. At *Bembe* and *Baua*, copper minerals accompany pyrite and hematite of limestone horizons. Minor occurrences include Lucossa, Dongo, N'Zauevua, Uonde, Muonde, Sanza, Pelo, Caluca, the Serra do Sangue, Xinga, Lucunga and Quifuana.

The arc of the Copperbelt ends in eastern Angola, at the sources of the Zambezi; copper shows at Calunda, southwest of Mwinilungwa and Kolwezi.

At Zenza do Itombe, 100 miles east of Luanda, malachite and black copper oxides impregnate 15 ft. thick conglomeratic arkoses interbedded between Cretaceous limestones and sandstones. The warped bed contains 1.8% Cu. On the Catumbela River and at the abandoned English Mine near Benguela Harbour, lenticles of malachite and iron oxides impregnate Cretaceous sandstones. At Cassenha Hill, near Cuma, 40 miles west of Nova Lisboa, a north-northwest striking, 150 ft. wide fracture zone and adjacent, Precambrian grits have been impregnated by malachite, azurite, torbernite and wad. Copper shows in a 150 mile long trend extending in south-central Angola from Vila da Ponte, east to Cuchi and Menongue near Serpa Pinto. At Menongue tennantites have been identified in quartz veins traversing Precambrian metasediments; malachite, however, is the principal ore mineral, as it is in the stringers of Vimpongos, Pedra Grande and Cambongue, occurring in the aureole of the granites of Mossamedes (T. C. F. HALL, 1949).

SOUTHWEST AFRICA

Of the country's copper 97% occurs in the late Precambrian lead–zinc–silver ores of *Tsumeb*, described in the 'Lead' chapter. However, at Kopermyn and Nosib mine (Fig. 91), in the northwestern part of the country, copper minerals occur in conglomerates, below the Otavi dolomite. The Khan, Ida, Henderson, Hope, Matchless, Gorob and Otjosonjati deposits are associated with silicate–carbonate rocks or veins, but copper also occurs in post-Cambrian veins at Hiebis Ost, Strausenheim and Schweickhardtsbronn.

At *Alt Bobos*, along the southern limb of the Tsumeb syncline, sub-vertical shears and dragfolds are the best hosts of calcitized flow breccia, where sulphides of mainly copper cluster in calcite lenses. In the Otavi Valley, the Kupferberg breccia pipe has long been worked out, as was the calcitized dip slip breccia along the contacts of dolomite and phyllite at Asis, Asis Ost and Gross Otavi. More than 300 Precambrian occurrences are spread over Southwest Africa, especially in the Windhoek area and southeast of Lüderitz Bay. Most of these belong to the Lower Precambrian Ababis Formation (SÖHNGE, 1958). At the Lorelei, chalcopyrite occurs in Ababis granodiorite. At Groendraai and Kalkdraai, near Rehoboth, it occurs in amphibole schists. It also occurs in the Warmbad district, north of the Orange River, in quartz veins and at Outjo in pegmatites.

Copper has been disseminated in Stormberg volcanics of the Karroo. At Goboseb Hills, west of Brandberg (northeast of Walvis Bay), malachite, chrysocolla and cuprite have been concentrated in amygdales and epidotized fracture zones of the lower levels of the lava; the flow averages 25 ft. in thickness (K. Linning, personal communication, 1962). Between Awahab and Khuab, 12.5 miles northwest of the Doros Crater in the Outjo district, malachite, copper and cuprite are associated with prehnite and delessite, crystallizing in melaphyre and in 1–6 inch thick sandstone dykelets in earlier flows. Around Schlip 472, in the Rehoboth district in the south, copper shows in basalts, and at Sinclair and Ginas 20, near Lüderitz, quartz veins traverse lava flows.

NORTHEASTERN AFRICA

Chalcocite fills in fissures at Regeita, Abu El Nimran and Samra in *Egyptian* Sinai. At Samiuki, Atshan (and El Atawi) in the Southeastern Desert, lenses of chalcopyrite and sphalerite replace talc and carbonate in northwest striking shear zones. At Abu Swayel copper is associated with nickel. The Hofrat-en-Nahas deposit is located near the western border of *Sudan*, 1 mile west of the Umbelasher river, a tributary of the Bahr-el-Arab. Silicates outcrop; carbonates and oxides occur to a depth of 80 ft. and sulphides have been identified at a depth of between 390 and more than 420 ft. At Khor Arba'at near Port Sudan, malachite, some chalcopyrite and cuprite occur in a quartz vein. At Tohamiyan in the southeast, malachite covers chlorite schists.

In *Ethiopia*, east of the Raba River (northern Erythrea), cupriferous quartz–epidote veins strike north in diabase. At Semait, copper bearing aplite dykes strike north in epidosite surrounded by schists. At Rassi, south of Asmara, diorite has been impregnated. At Shersher, in Harar Province, lenticules of malachite, azurite and erythrite occur.

In *Somalia*, near the Marodile River northwest of Hargeisa, malachite impregnates joints of quartz–epidosite; the same mineral shows at Elan Gubado.

UGANDA

The Kilembe copper–cobalt mine, on the eastern slopes (Fig. 51) of Mount Ruwenzori, is the only deposit of importance, but south of Kilembe chalcopyrite is also disseminated in gneisses on the Mpanga River. The sulphide accompanies galena at Kitaka, east of Kilembe (BARNES and BROWN, 1958).

The middle Precambrian sequence of *Kilembe* includes:

Gneiss.
50–400 ft. grit, arkose, quartzite.
150–500 ft. biotite–chlorite or amphibole–quartz–epidote.
Albitite.
3–100 ft. fine grained oligoclase ore granulite, with hornblende or biotite.
100–300 ft. foliated amphibole-thulite (quartz) albitite.
2,000 ft. banded epidote garnet granulite.
Gneiss.

The albitites are locally termed amphibolite. Both these and the granulite are essentially albite–oligoclase plagioclasites with accessory silicates. The metabasites and metasediments are partly granitized. Pegmatitic alaskites are injected into the twice folded syncline and into the southern limb of the adjacent, overturned anticline. Younger faults traverse the syncline, causing repetition of the ore horizon in the north. While pyrite has been disseminated between the lower gneiss and the grit, chalcopyrite and pyrrhotite show in the ore granulite and the lower albitite, with only pyrrhotite occuring in the upper albitite. Magnetite is found in the lower granulite.

The ore horizon, as defined by the limit of 2% contained copper, is 20 ft. thick, except for the thicker, northern ore shoots. The sulphides occur in the foliation, in crests and troughs of secondary (but not major) folds, in stringers, impregnations of microfissures, dense stockworks and irregular disseminations. Indistinct zoning transgresses bedding. Copper values peter out in the host granulite, except where it is injected by late alaskite pegmatites. Ore bodies average 2.5% Cu and 0.8% Co; substandard grades of copper are associated with pyrite lenticles in the lower albitite. The upper or principal ore body is situated in granulite. The lower ore body, however, extends into the underlying albitite; a sharp contact divides the first of these bodies from the overlying amphibolite. In the north, the ore body tends to split, a lens of albitite separating it. The indicator of mineralization is amphibole, especially a coarse recrystallized variety, and not biotite. The chalcopyrite/pyrite/pyrrhotite ratio is favourable in granulite, with the iron sulphides concentrating in the upper, though not necessarily top, layers of the albitite. The paragenetic sequence is believed to be: magnetite, ilmentite, cubic pyrite, (molybdenite) disseminations; sphene; octahedral pyrite; massive pyrite; chalcopyrite, magnetite, pyrrhotite, (linnaeite–siegenite), (gold); bird's eye pyrite–marcasite; bornite, chalcocite; azurite, tenorite.

Chalcopyrite contains 32.9% Cu and traces of sphalerite, but in the northern ore body, where grades attain 6% Cu, chalcopyrite has been largely transformed into bornite and chalcocite. Gold is included in linnaeite and massive pyrite.

THE REST OF EASTERN AFRICA

Chalcopyrite occurs in basic lavas of Kitale, *Kenya* near the Uganda border, in the Nyanga–Macalder gold–galena veins on Lake Victoria (see 'Gold' chapter), and in Precambrian schists and gneisses of the *Voi-Tsavo* district, west of Mombasa (PULFREY, 1958).

In *Tanganyika*, chalcopyrite is associated with galena (see 'Lead' chapter) in the Mpanda area, and with gold at Ikoma, Kilimafeza, and Mawa Meru south of Lake Victoria. It also occurs in the Kirundatal and McCallum's mine in the diamond district (WHITTINGHAM, 1958), at Saza, Rukwa, Chunya, Ntumbi, Maher's and Menzies in the Lupa gold field, and in sediments of *Kigugwe*, all north of Lake Nyasa. Copper also occurs in Precambrian gneisses, granulites and pegmatites of the Morogoro–Dodoma district and south of the Kilimanjaro (Usangi, Usambare Mountains, Kidete area, Lufusi, Mgwagwa).

At Cocanga, in the Tete district in western *Mozambique*, pyrite and copper sulphides are disseminated in siliceous–carbonatic lenses of micaschists and gneisses. However, at M'pande–Unkua and Conua, malachite impregnates bedded limestones and marbles. Other occurrences include Boroma and Chidue; indeed this trend appears to extend the Hippo–Kalulu belt of southern Zambia. In the Edmundian and Copper Mines, near Manica on the border of Southern Rhodesia, and at the confluence of the Save and Lundi Rivers, chalcopyrite impregnates a fissure.

In *Malawi*, at Mpemba Hill copper and nickel are associated with pyroxenites, with chalcopyrite occurring in the hornblende gneisses of the Chombe River.

In *Madagascar*, copper shows in numerous small occurrences: In the area of Vohibory (in the south-southwest) both Precambrian and Karroo lenses have been impregnated (Besakoa). Copper, cuprite and dioptase show in the Cretaceous basalt flows of Antanimena in the west, and at Ambatovararahina a cupriferous vein traverses cipolinos adjacent to granite. Other occurrences include Antanivakivaky, Ianapera, Sakalava, Bevalaha, Besatrana, Ankazoabo in the Vohibory area, Matsaborivato and Ambohijanahari, near Vohémar in the north.

SOUTHERN RHODESIA AND BECHUANALAND

Southern Rhodesia

The principal copper districts of *Southern Rhodesia* follow the northeast trending Urungwe–Lomagundi schists belt in the north and the Jurassic(?) Beitbridge–Sabi belt in the south (TYNDALE–BISCOE and STAGMAN, 1958).

The Molly mine is located at *Mangula* in pink, lower Lomagundian meta-arkoses consisting of microline, acid plagioclase and quartz. Calcite, apatite, magnetite and sphene are abundant in the mineralized horizons, of which a large grained granitic arkose is the indicator. Lamination of magnetite, hematite replaced by bornite, and of chlorite distinguish probable bedding. The principal ore body, Molly North (Fig. 36), is, in fact, a one and a half mile long swelling of the copper arkose; Molly South is similar but smaller. Bornite and chalcopyrite ore penetrate into the compact arkose from quartz, chlorite, carbonate, pegmatite or quartz–feldspar veins, especially in the chloritic western zone where both veins and copper are more abundant. Pyrite and chalcopyrite replace bornite; malachite, azurite and chalcocite accompanied by covellite are, however, the principal ore minerals. Uraninite has been altered to uranophane and metatorbernite.

Other deposits. Impregnation deposits include *(1)* the Mercury mine in the northwest, distinguished by malachite and azurite carrying carbonate lenses in quartzite, *(2)* supergene copper minerals in the slates of Skipper mine near Que Que and in the green schists of the Laviron mine, near Bulawayo and *(3)* the bornite shales at Umkondo.

In a shear zone of the *Copper Emperor* deposit, 50 miles northwest of Que Que, chalcopyrite clusters accompany pyrite and sphalerite. At the *Alaska* (Lomagundi district) mine in the north, tabular ore shoots of chalcocite replace pyrite in a crush zone of folded carbonate schists. Commercial ores have been found in anticlinal structures, divided by low grade mineralization, and also in a cone; while synorogenic bornite occurs at depth, enriched malachite and azurite are supergene. In the *Copper Queen* and *-King* at Sanyati, pyrrhotite and chalcopyrite accompany sphalerite and galena (see 'Lead–Zinc' chapter).

In the Falcon veins of the Umvuma schist belt, chalcopyrite is largely altered to copper carbonates and oxides. Malachite and cuprite accompany gold in the Piriwiri belt, Lomagundi district, and chalcopyrite and pyrrhotite are associated with gold in the Muriel mine. At Inyati, malachite precipitates in quartz lenses of a shear zone, and in gneissic granite injected by dolerite. In a similar mine, Sirius, auriferous chalcopyrite and pyrite have not been altered. At Silverside, copper minerals crystallize in dolomite, schists and quartz lenses.

Bechuanaland

The vertical *Bushman* fissure lenses are located in brecciated dolomitic limestones of a 14 mile long shear-zone; their length attains more than 600 ft. with the width ranging between 50 and 150 ft. Chalcopyrite and chalcocite associated to galena, iron minerals and gold–silver traces, have been altered. At *Malokojwe* in the south, high and low grade lenses alternate. Geology of the Phudulooga-Sinti area, further south, is similar, but the Matungu veins are injected into mylonites (BOOCOCK, 1958). Copper carbonates and silicates are scattered in Bulawayan rocks at *Magogaphate* in the northeast, and some copper occurs in auriferous quartz veins at Tati, Mamakobu and in the Rainbow and Monarch mines. At *Tshukutswane*, near Lobatsi, copper and lead are associated with volcanites.

SOUTH AFRICA

The principal deposits are in the O'okiep area in Namaqualand (GROENEWALD, 1958), Phalaborwa and Messina in the northern Transvaal. Small occurrences are scattered over the country (Fig. 46, 50).

Messina

South of the Limpopo, early Precambrian granites have been intruded into folded metasediments and basic rocks. The principal tectonic feature, the Dow–Tokwe fault, branches into satellite faults; the open fissure is believed to have been brecciated and mineralized during or after the Karroo. Largest concentrations are located between the mega-breccia, but brecciated rocks, fissure veins originating in the breccia, and contacts of quartzite and basic dykes, hornblendite, quartzite and granite are also mineralized. Ore bodies are discontinuous, irregular lenses measuring up to 300 by 100 ft. The principal deposits are Campbell, representing almost half of the reserve, and Harper; minor deposits occur at Messina and Artonvilla. The average grade attains only 1.8%. Chalcocite, bornite and chalcopyrite have been oxidized near the surface, and gangue minerals include epidote, calcite, adularia and prehnite.

Phalaborwa

At *Loolekop*, the country's principal deposit, 70 miles west of the Mozambique border (see the 'Vermiculite', 'Phosphate' and 'Uranium' chapters), a 1,000 by 400 yard ellipse of carbonate has been intruded in at least two phases. Outer zones are essentially ultrabasic: phoscorite or magnetite–apatite olivinite, pyroxenite, pegmatitic olivine pyroxenite, shonkinite and alkaline syenites. Late dolerite dykes traverse all zones, and calcite–pyrite veins are also late features. The carbonatite body appears to consist of large grained magnetitic, magnesian sövite with some (dolomite) rauhaugite. Titanoan magnetite forms large clusters; ilmenite lamellae disappear and titanium content descreases with depth. Apatite and chondrodite altering to serpentine are widely disseminated, but olivine, phlogopite, biotite and fluorite remain scarcer. Chalcopyrite spots and striae are well distributed, accompanied by bornite, chalcocite, valleriite, pyrrhotite, cubanite, pentlandite, baddeleyite pyrite, spinel, uranothorianite and chlorite. Bornite is associated either with chalcocite or chalcopyrite; cubanite accompanies the latter or pyrrhotite and valleriite (sometimes with pennine) forms veinlets. Pyrochlore and rare earth minerals are absent, however. The inner plug contains 1.3% Cu, with the carbonatite averaging 0.65% Cu. Values do not diminish at a depth of 1,900 ft. The low copper content of the outer carbonatite zone has been attributed to late sövite veins and alvikite dykes.

O'okiep-Namabeep

Copper shows on both banks of the Orange River, but the main district, covering 600 square miles is located 60 miles from the coast and from the Southwest African border (Fig. 53). The Precambrian dome trends largely east–west. Most of the deposits are situated in the 6 mile wide zone of Nababeep gneisses which pinch out in the east. A narrower band of Concordia gneiss borders the Nababeep north and south. A 6 by 4 mile stock of granite, surrounded by an aureole of granulite, quartzite, and schist, outcrops in the midst of the gneisses. While a fault traverses the granite, another thrust intersects the whole series in the west. The Nababeep-West, -Kloof, O'okiep, Narrap deposits are located along a 6 mile long alignment at the northern exocontact of the granulite. Nababeep Flat, Tweefontein, Wheal Julia, Homeep East, and No. 6 Mine are scattered in and along the northern band of Concordia gneiss; Kopperberg and Carolusberg are located in the thin southern portion of the Nababeep gneiss zone. Copper is disseminated in a series of longish or pipelike lenses and in veins traversing them. The lenses display pinch and swell structures in three dimensions, each covering 100–300,000 square ft. Their contacts are sub-intrusive, but their swarms follow the general structural trend.

Generally limited to the basic suit are 28 ore bodies. The O'okiep ore body is in micadiorite, one of the most frequent of the ultrabasic host rocks. While Nababeep and Carolusberg occur in hyperstenite, the host rock of Tweefontein and Jubilee Mine is a norite. The paragenesis consists of chalcopyrite (with exsolutions of cubanite)–bornite– chalcocite and of chalcopyrite–pyrite–pyrrhotite, the latter showing pentlandite and millerite inclusions. Accessory minerals consist of covellite, siegenite, sphalerite, galena, molybdenite, valleriite, melanite and titano-magnetite.

Other occurrences

At Insizwa, in eastern Griqualand west of the Pilanesberg, western Transvaal, in the Merensky Reef and in Groblersdal, the copper–nickel paragenesis is related to basic–ultrabasic rocks (see 'Nickel' chapter). In the Soutpansberg, chalcocite impregnates a sandstone interbedded in lava. In the Letaba district, copper minerals accompany sphalerite. At Pilgrim's Rest they accompany gold, in the Bronkhorstspruit district siderite, in the Rooiberg–Leeuwpoort area limonite, and in the Sibasa district hematite. The Dania, Santa Barbara, Subeni and Magdalena mines work out the Precambrian deposits of Zululand and southern Natal, which include the *Nkandla* area and the Umhlatuzi, Umfalozi, Insuzi and Buffalo Valleys. In the Vanrhynsdorp district, Cape Province, chalcopyrite, chalcocite and covellite accompany monazite (see 'Thorium' chapter). Other deposits comprise of Gordonia, the Areachap and Maitland mines.

COBALT

A majority of the world's cobalt is found in the unique environment of the Katangan Copperbelt. Cobalt sulphides crystallize before copper in psammitic–pelitic dolomites of the Upper Roan; deposits are stratiform, but not controlled by sedimentation. Most of the sulphides, however, have been altered to complex earthy oxides, and traces have been reconcentrated during this process. In the hydrothermal, Katangian uranium deposits, cobalt accompanying either copper or nickel is also stratiform. It is absent, however, from bedded copper deposits in Katanga outside the Copperbelt and in the rest of Africa. Cobalt is scarcer in the earlier Ore Formation of northern Zambia, which has not been intensively altered. Cobalt/nickel ratios of the bedded sulphides might indicate (though not conclusively) a hydrothermal origin. The Moroccan cobalt–nickel deposits are directly related to epigenized ultrabasics.

CONGO

Cobalt (and occasionally nickel) show in the copper, copper–uranium and uranium deposits of the Copperbelt. Concentrations tend to decrease from the west towards the east; they vary, however, in deposits of the same district and within each deposit. The principal copper–cobalt occurrences include: Kamoto, Musonoï and Kingamyambo in the Kolwezi district, Fungurume, Kabolela, Kambove and Luishia in the Jadotville district, Luiswishi, Lukuni, Ruashi, Étoile du Congo and Kasombo in the Elisabethville district (Chambishi, N'kana and Baluba in the Zambian district).

At Kasombo–Lupoto only minor amounts of copper are found, but cobalt predominates at Luiswishi.

In the *Kolwezi* district (Fig. 82), several peaks of cobalt values have been distinguished; greatest concentrations are achieved in various layers of each *écaille* or deposit. The copper/cobalt ratio averages 10/1, attaining up to 5/1. These variations are not controlled by the stratigraphy or lithology. The distribution of cobalt sulphides and their combinations with copper and nickel has, however, been affected by tectonics. At Kamoto (Fig. 83) the evolution of cobalt values contained in the sulphides of successive layers is given in Table CXXV (OOSTERBOSCH, 1962).

TABLE CXXV

COBALT VALUES IN THE KAMOTO SEQUENCE

	Ft.	%
C.M.N. (upper dolomite)	400	0
S.D. (dolomitic shale)	70	0
	120	0.04
	12	0.2
	24	0.04
	12	0.2
	10	0.06
	8	0.2
	33	0.4
R.S.C. (stromatolite dolomite)	33	0.07
D.S.I. (lower dolomite)	15	0.4
R.A.T. (chloritic shale)		0.04

The mylonitized R.A.T. shale, embaying the *écaille*, also carries cobalt and nickel. At Kilamusembu the following results have been obtained: S.D. (dolomitic shale), lowest layer (13 ft.) 0.2%; D.S.I. (lower dolomite), (12 ft.) 0.3%, (11 ft.) 0.1%.

In the Fungurume *écaille* the distribution of cobalt is similarly unrelated to that of copper. At Tantara, southwest of Jadotville, pink cobalt calcite containing 2% Co occurs with strontianite in the Kakontwe limestone of early Kundelungu age. In the Elisabethville district, cobalt appears in the basal shale (R.A.T.) and in the dolomitic shale (S.D.). At Luiswishi copper–cobalt–uranium and vanadium are associated. In southeastern Katanga cobalt shows in the copper bearing biotite shales and sandstones; in a drillcore of Musoshi, north of Bancroft, the hanging wall and footwall shales of the Lower Roan contain 0.03 and 0.02% Co respectively.

Idiomorphous carrollite crystallizes before bornite and the chalcocites; indeed, bornite crystallites rim carrollite, penetrating also into its fractures. However, no replacement has been observed. Little cobalt is associated to the pyrite, chalcopyrite sequence, occurring in such cases as linnaeite. A majority of the cobalt, however, is leached and transferred into the hydrated earthy, black cobaltiferous oxides ('black ore') accompanying manganese oxides. Asbolane, heterogenite, mindigite, stainierite and trieuite represent variations of the formula: $x\,(CuO)\,.y\,(Co_2O_3)\,.w\,(H_2O)$. At Musonoï, massive hydroxides of cobalt have replaced several horizons of the layered dolomitic shales (S.D.), impregnating the gangue, bedding planes and perpendicular microfissures; cobalt also shows in a complex, Cu–U–Pd–Au–V paragenesis. At Kambove (Fig. 84), and Ruashi, copper–cobalt oxides accompany malachite. The basal shale carries black oxides in these deposits and at Luishia, but in the Elisabethville district pockets and potholes of the upper dolomite (C.M.N.) have been filled in by them. Ferromagnesian spherocobaltite also forms in the zone of oxidation.

Cobalt is associated with bedded copper ores and fissure filling uraninite, e.g., at Kalongwe; south of Kolwezi, it is associated with bedded nickel ores and epigenetic uranium, e.g., at Shinkolobwe–Kasolo and Mindigi in the southern branch of the Jadotville district. Nickel predominates over cobalt, both being generally restricted to the lower dolomite D.S.I. and dolomitic shale. Cobalt has also been found at Shamitumba, south of Shinkolobwe. The paragenetic sequence of these deposits is given in Table CXXVI.

TABLE CXXVI

COBALT SEQUENCE

Co—Cu—U	*Ni—U—Co*
uraninite	uraninite
pyrite	pyrite
carrollite	vaesite
	siegenite
bornite	
chalcopyrite	chalcopyrite
covellite	
digenite	

ZAMBIA

Cobalt is scarcer in the Zambian sector of the Copperbelt than in Katanga. In the principal deposits, N'kana and Baluba, cobalt grades are fairly high and identical. In Zambia, carrollite accompanies (and is included in) chalcopyrite, and not bornite as in the Congo. Pyrite also contains cobalt.

At *Bancroft* cobalt occurs mainly in the alternating shale and sandstone of the 2–4 ft. thick lower *(A)* ore body and in the siltstones and calcareous sandstones of the 2–6 ft. thick fourth *(D)* ore body. Cobalt grades are occasionally in excess of 0.2%, but copper grades of these ore bodies are relatively low (1.4 and 2.4%) and erratic. Cobalt also shows in the footwall conglomerate. At Nchanga, linnaeite is associated with chalcopyrite only in the upper ore body. At *N'kana*, carrollite accompanies chalcopyrite in the northern ore shales, occurring also with cattierite in the quartz and dolomite veins of the southern ore shale. Cobalt values average 0.18% for the deposit; they increase with depth, diminishing to less than 0.1% in two bornite zones. Carrollite occurs in the mixed zone underlying the oxide zone. At *Chibuluma* cobalt distribution is often a reverse function of copper values, with cobalt averaging a grade of 0.18%. Linnaeite occurs somewhat more frequently in the lower horizons, where the mineral accompanies chalcopyrite, cobaltiferous pyrite and pyrite, but cobalt and selenium bearing pyrite is also found in the footwall quartzite.

At *Baluba*, carrollite is most abundant in calcschists, and carrollite and linnaeite are included in chalcocite in the zone of alteration. The copper/cobalt ratio is the lowest in this deposit, that is 15/1, with cobalt values averaging 0.16%. At *Roan Antelope*, both older Precambrian and Katangian pyrite contain cobalt: A sample of older Precambrian pyrite from Roan Extensions contains 2,700 p.p.m. cobalt and 490 p.p.m. nickel. In the Ore Formation, the cobalt content of pyrite decreases from 1,520 p.p.m. in the east to 1,340 p.p.m. in the west, with the nickel content increasing from 110 p.p.m. to 270 p.p.m. Since the intensity of thermal metamorphism increases in the same direction, it is inferred that the distribution of cobalt may be related to this feature. The cobalt/nickel ratio suggests a hydrothermal origin. Some carrollite is associated with chalcopyrite, but not with scarce bornite.

MOROCCO

Although cobalt and nickel occur *(1)* along a north trending line at Azegour (south of Marrakech), *(2)* at Ifergane and Tazalaght and *(3)* also near Mefis (west of Colomb–Béchar), the only district of any importance is *Bou Azzer* (Fig. 10, 59), 100 miles southeast of Marrakech (AGARD et al., 1952). Occurrences of the district are strewn over a 30 mile long east striking alignment, starting at Bou Offroh and continuing through Bou Azzer, Arhbar, Oumlil and Tamdrost, Ahmbed, Irhtem and Aït Ahmane. The alignment is wedged between Upper Precambrian detrital sediments and lavas in the south, and transgressive earliest Cambrian (Georgian) sediments in the north. At Bou Azzer, rocks of the 2 mile wide 'wedge' consist of lower Precambrian granite-gneisses, middle Precambrian, syntectonic quartz-diorites, serpentine, green rocks and post-tectonic quartz diorite. While in the centre a fairly thin serpentine zone divides sediments of the two epochs, at Aït Ahmane a 5 mile wide belt of middle Precambrian basic intrusives, serpentine, chloritic sandstones and upper Precambrian sediments separate the two formations. Following complex east-southeasterly folding and northeasterly faulting, east-southeast striking fissures open up and admit green rocks and serpentine between the Middle and Upper Precambrian rocks.

Deposits are never distant from serpentine contacts and they occur along the northeast trending system of faults. The original gabbros and diorites have been crushed and recrystallized as amphibolites and chlorite schists. Serpentine bodies and dykes consist of antigorite, calcite, garnet, stichtite, chromite and magnetite; they alter along their contact (frequently adjacent to ore shoots) to a pale-green talc–dolomite. Occurrences may be divided into three categories: ore shoots in serpentine, contact veins, and fissure fillings in the country rock. Ore lenses, 50–120 ft. thick, form columnar shoots or wedge-like keels; the shoots include xenoliths of serpentine, talc–dolomite, amphibolite and garnetite.

The principal occurrence, the *Arhbar* ore body, embays the serpentine mass separating it from late Precambrian lavas, but fractures have been mineralized only near the ore body. Sharp contacts are most frequent in the vicinity of lavas which have dammed up the hydrothermal solutions. At Ahmbed 1 and 3, the ore lens continues as a vein; most of the sulphides are concentrated in oreshoots or veinlets in the border zone of ore bodies. Several phases of hydrothermal alteration can be distinguished, namely *(1)* the formation of serpentine, *(2)* its crushing and dolomitization, and *(3)* nickel–cobalt precipitation.

Two paragenetic associations have been distinguished: *(1)* between Arhbar and Bou Azzer including Tizi: skutterudite, cobalt löllingite, chalcopyrite, safflorite accompanied by dolomite, calcite, talc, chlorite and quartz. Some of the Aït Ahmane veins show a similar combination. Gold is associated with these minerals in veins 7 and 5 of Bou Azzer; *(2)* east of Arhbar nickel is found (see 'Nickel' chapter). The depth of thorough oxidation reaches 220 ft. Typical minerals of supergene alteration include copper and nickeliferous cobalt oxides, erythrite, annabergite and garnierite. While erythrite was found at a depth of 600 ft., unaltered loellingite also outcrops.

THE REST OF AFRICA

In the Cunni Valley of the Chercher Mountains, eastern Ethiopia, cobalt shows in schists and quartzites. Traces of cobalt are found in the copper–gold ores of the Macalder mine in Kenya.

Uganda. In the hangingwall albitite of Kilembe, early, cubic pyrite attributed to regional metamorphism shows only traces of cobalt. Iron sulphides of the ore granulite consistently contain 1.7% Co, in the order of crystallization, as Table CXXVII indicates.

TABLE CXXVII

CO- AND NI-CONTENTS OF UGANDAN IRON SULPHIDES

	Co (%)	*Ni (%)*
octahedral pyrite	1.74	1.07
massive pyrite	1.65	0.95
pyrrhotite	1.70	2.08
chalcopyrite	0.03—0.17	0.07—0.033

Octahedral pyrite is occasionally veined by linnaeite, small grains of the latter accompanying chalcopyrite (and massive pyrite), but chalcopyrite contains little cobalt. Linnaeite contains gold and 17% Ni, belonging thus to the siegenite group.

Tanganyika. In eastern Tanganyika, sulphides of the pyroxenite bodies of Chenzema and Mtenga Hill contain 0.5% Co. The metal also shows with manganese at Ukinga and Mangalisa.

South Africa. In the eastern Bushveld the cobalt mineralization is related to basic intrusives. At Kruisrivier, Roodewal, Mineral Range, Winterhoek and Laagersdrift, safflorite, niccolite and gold have been disseminated in stringers traversing quartzite xenoliths and norite. However, at Eenzaamheid, cobalt minerals crystallized in a lode, related to gabbro.

ZINC

Zinc and lead are generally associated. The occurrence of zinc, however, is more restricted, namely to the 500 million year old Kipushi deposit in the Congolese Copperbelt and to the 300 million year old Bou Beker–Oued Zounder zone on the frontier of Morocco and Algeria. Kipushi is the continent's principal deposit of zinc, silver, cadmium and germanium. Other zinc deposits are in the Algerian Atlas and at Broken Hill (Fig. 3) near the Copperbelt. A majority of these are carbonate and crush zone replacements. In other occurrences lead predominates over zinc. The zinc/lead ratio of the principal deposits attains: Kipushi 25/1; Oued Zounder–El Abed 9/1; Broken Hill 3/1; Bou Beker 1/1; Tsumeb 1/5.

Congolese and Zambian occurrences will be discussed in this chapter. The reader is referred to the 'Lead' chapter for details on other deposits.

KIPUSHI, CONGO

Kipushi is part of a discontinuous series of replacement deposits in the Kakontwe limestone

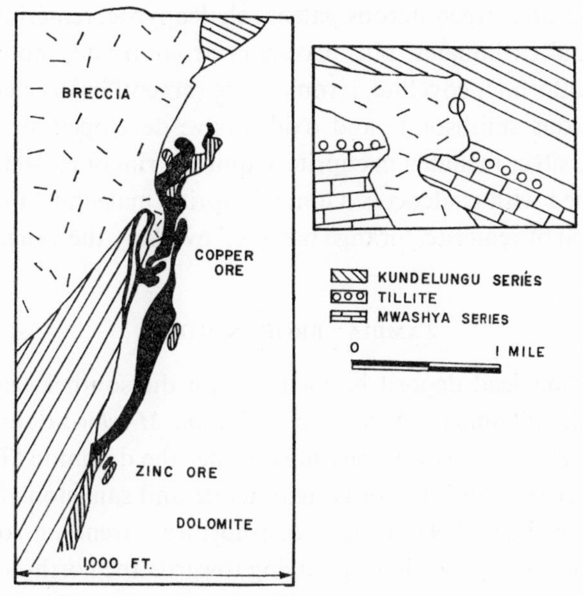

Fig. 88. Section and plan of Kipushi.

overlying the Série des Mines, between the Copper and Iron belts of Katanga. The zinc–lead occurrence of Lombe and the lead–zinc–silver–copper paragenesis of Kengere, south of *Kolwezi*, belong to the same series.

The asymmetric *Kipushi* anticline rises a few hundred yards (Fig. 88) from the Zambian border, pitching east and west. The northern limb dips 60°N, the southern flank lying flat. The fold consists of *(1)* graphitic schists including sandstone inliers, ferruginous zones and oolith, chlorite schist layers of the Mwashya Series; *(2)* 300 ft. of the large, basal tillite; *(3)* the Lower Kundelungu Series, including the Kakontwe carbonates, interlayered schists, calcschists, and sandstones; *(4)* the Upper Kundelungu small conglomerate. The Série des Mines has not been recognized! The summit of the anticline is covered by a sub-vertical hood breccia, dipping northwest, thrust partly over the Kakontwe dolomite. A wedge of calcschists underlies most of the hood. The replacement body developed along the intersection of the hood breccia with a secondary fault zone. The propagation of the mineralizing solutions has been stopped by the less permeable calcschists. The ore funnel strongly resembling the shape of Tsumeb, is located in the Kakontwe dolomite 0–150 ft. from the contact. The thickness of the main vein varies between 30 and 100 ft. An irregular body of copper sulphides is located in the dolomites and at their contact with the breccia. Furthermore, a body of zinc sulphides connects with the eastern part of the upper reaches of the main body. The funnel first dips 45° north, pitching stronger at lower levels. Dolomite relics are included in the chalcopyrite, bornite, sphalerite body, accompanied by some chalcocite. The upper funnel extends to a depth of 350 ft., connecting with a 120–150 ft. thick body, located in the dolomites. A smaller, independent, wedge-like sphalerite body follows the upper edge of the calcschist. The central copper funnel attains 500 ft. in length, finally arching back to the contact. South and below this area, zinc predominates in the paragenesis. While sulphides have also been disseminated in the country rock, pyrite, galena and tennantite are important accompanying minerals. Kipushi is the continent's principal silver deposit, silver being contained in bornite and chalcocite.

The paragenesis includes argentiferous galena, chalcopyrite, reniérite, cadmiferous sphalerite, quartz, mica and carbonates, plus accessory arsenopyrite, molybdenite, vanadinite and cuprodescloizite. Bornite, covellite, stromeyerite (argentiferous chalcocite), chalcocite, native silver, cupriferous smithsonite and oxides have developed as secondary minerals. Cobalt is rare. Native silver included in cuprite acquires a rim of malachite, chrysocolla and chalcocite. If associated with chalcocite, a film of cuprite, malachite and chrysocolla develops. The crystallization of reniérite, 'orange bornite,' overlaps the range of early tennanite.

ZAMBIA: BROKEN HILL

The only important zinc–lead deposit is not far from the southeastern extension of the Katangian Copperbelt alignment. A tongue of *Bwana Mkubwa* dolomite juts into older sediments at *Broken Hill.* Shales and phyllites border the dolomite (TAYLOR, 1954), also forming a narrow interbed. An inlier of conglomerate and sandstone lies north of the ore bodies. These are aligned in a 2,500 ft. long east-northeast trending zone, that is sub-perpendicularly to the prevailing syncline, plunging towards the north-northwest. The dolomites can be divided into *(1)* massive, recrystallized ore bearing; *(2)* schistose; *(3)* compact dark varieties. Ore bodies can be grouped into three or four divisions. They form long, oblique and narrow funnels or veins. Most appear to be related to a transverse structure, possibly a fracture system. While ore bodies follow the contact of dolomite and shale, the intersection of fractures may also have conditioned their shape. Dimensions of the ore bodies vary greatly. There is no primary zoning. Oxidization attains a depth of 1,000 ft. and generally compact sulphides of funnel-cores are surrounded by oxides. Ore body No. 2 has been completely oxidized, containing hardly any lead, and only 0.75% V. Ore body No. 1 is richer in zinc and lead than the other, and exhibits a lower zinc/lead ratio. The paragenetic sequence is: pyrite, chalcopyrite, sphalerite, some bornite and galena. Cadmium, germanium and selenium only show in traces. Some galena enrichment has been recorded during oxidization, beginning with smithsonite and cerussites, and followed by willemite and some hemimorphite. Oxidization coincides with a phase of silicification. Covellite is supergene, but pyromorphite seems to be of ultimately sedimentary origin.

Willemite and native silver form pockets at the nose of folds in the dolomites of Excelsior Zinc, and also occur at Star Zinc, near Lusaka (Fig. 89).

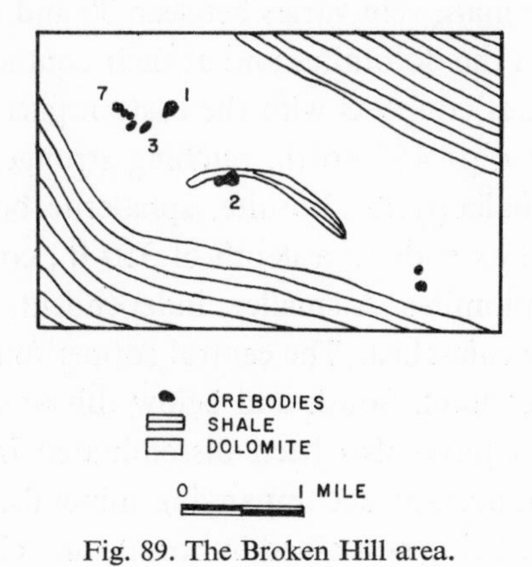

Fig. 89. The Broken Hill area.

NORTH AFRICA

Tunisia

Sphalerite has been reprecipitated at Fedj el Adoum–Djebel El Akhouat, 40 miles east of Touireuf. The deposit should be considered secondary hydrothermal. Galena and pyrite accompany scaly sphalerite. Similarly, zinc predominates over galena in the small deposits of Bechater, the Djebels–Sekarna, and Azered and Aïn Nouba.

Algeria

The biggest deposits are located along the extension of Touissit – Bou Beker (see 'Moroccan' section). Minor occurrences are scattered over the coastal Atlas, mainly in the Triassic and particularly in the zone of anhydrite diapyrs. The occurrences can be divided into the groups of the Ouarsenis, Guergour (Sihb), Bougie, Kabylia of Collo, Bou Thaleb mountains and Clairefontaine.

The principal deposits are *El Abed* and *Oued Zounder* (Fig. 59). Sidi Kamber is of minor importance, as are Mesloula, Djebel Gustar, Bou Kiama and Aïn Barbar. Other occurrences of the Oran province include the lead–copper–silver veins of Gar-Rouban and the sphalerite–galena–smithsonite deposits of Masser and Maaziz. In the *Ouarsenis*, south of Algiers, Lower Jurassic (Lias) windows rising from among the Lower Cretaceous sediments contain irregular bodies of sphalerite, partly altered to smithsonite. The principal vein deposits of this area included Sakamody and Guerrouma. Minor occurrences of the Tablat mountains consist of Rarbou, Nador Chair and Tizi N'zaga.

In the *Collo Kabylia* and Edough chains of eastern Algeria (Fig. 11, 56), the metallogenetic axes trend north and northeast. While Dar Debbagh is distinguished by galena, Aïn–Kechera and Bou Doucka show a Zn/Pb proportion of 4.

In the *Guergour* district, Biban chain, the Cretaceous replacement bodies of Djebel Anini, Kef, Semah and Aïn Roua carry smithsonite. Smithsonite also occurs at the Djebels Gustar, z'Dim and Sfa in the highlands of Sétif. Fissures of the Djebel Soubella, Bou Thaleb, cutting Liassic limestones are capped by smithsonite and galena. Finally, the Djebel Ouasta fissures have been filled and upper Cretaceous (Senonian) limestones replaced by smithsonite.

Other small occurrences include the areas of Bathna, Chebka, Sellaoua and the Aurès.

Morocco

In the largest deposit, Bou Beker, the Zn/Pb ratio almost attains 1. Most of the sphalerite is associated with galena in the fine grained Lower Lias (Lower Jurassic) horizon. Sphalerite is abundant in the intermediate layer appearing in sector III of Bou Beker, in the Goundafa and other areas of the central stock of the High Atlas, in the Central Moroccan tableland in Rehamna, and in the Taouz and Akka regions (Fig. 10, 59). Irregular veins containing galena and impregnated dolinae have been found at Afra, Melilla area, near the contact of diorite and limestone (see 'Iron' chapter). In the southeast 7 miles from Aït Labbès, smithsonite bodies fill in a fracture, but hardly any lead occurs there. At Ouicheddène, in the Marrakech district, a fissure includes ore bodies of lead, at the surface of zinc, and in depth of copper. At Toundout (Igourzane) and Assif el Mall sphalerite crystallized in breccia veins, but the Erdouz veins occur in Cambrian limestones. Sidi Makhlouf is situated north of Marrakech.

At Sidi bou Othmane, a 3–5 ft. thick vein traverses schists injected by pegmatites. Here stringers of quartz cut sphalerite shoots.

THE REST OF AFRICA

At Tessalit and In Darset, in the Adrar of the Iforas, the *Mali* extension of the Hoggar, small, oxidized veins contain 7–10% Zn. Furthermore sphalerite occurs in the Falémé district, at Kayes, in the Air, Niger Republic, and at Chichiwere, Bombiri and Mambali, Ghana.

The zinc belt of southeastern *Nigeria* extends from Ishiagu, located 120 miles from the coast, to Gwona in the northeast, a total distance of 350 miles. The belt crosses the Benue River at an acute angle. Ameka, Aberi and Nyeba, around *Abakaliki*, near Enugu, include veins penetrating to several hundred ft. in depth. Other deposits include Arufu, on the Benue, and Zurak, east of the Plateau. Sphalerite tends to be associated with quartz, but siderite is also abundant.

Galena and sphalerite accompany pyrrhotite and chalcopyrite in the Shanyati deposits of the *Copper Queen* and *-King* in *Southern Rhodesia*. The impregnated dolomites are interbedded in synclines of pre-Lomagundi schists. A small granitic stock lies in the vicinity. Sulphide grades attain 4.6% Zn, 1.6% Pb and 1.8% Cu. Sphalerite accompanies some of the breccia pipes of the Marico district ore bodies, as well as copper in the Murchison range and tin in the Potgietersrus field. At Brakfontein sphalerite is associated with galena in a quartz–siderite vein. Cadmiferous sphalerite crystallized in the Vlakfontein ore body, and sphalerite also occurs in the Dania deposit near Toggekry (South Africa).

LEAD

Lead has been more widely distributed than zinc, veins being somewhat more frequent. However, dolomite and crush zone replacement deposits remain the most important. The association of lead with copper (Tsumeb) is rarer than in the case of zinc. The principal lead deposits include Touissit, Bou Beker, Mibladen and Aouli in Morocco, and Tsumeb in Southwest Africa. Although Algerian, Tunisian, and Broken Hill deposits are minor, small occurrences are fairly widespread. The mineralization favors 700, 400 and 300 million years old carbonates.

TUNISIA AND ALGERIA

More than 20 vein- and impregnation deposits of moderate and minor importance are scattered over the northwestern, Atlasic triangle of Tunisia.

The principal producers of the past were Djebel Ressas, Khanguet Kef Tout (Pb/Zn=1), Sidi Amor ben Salem, Djebel Hallouf, Sidi bou Aouane, Djebel Trozza and El Grefa (lead); Garn Alfaya–Koudiat el Hamra (Pb/Zn=6), and Sidi Ahmed (Pb/Zn=2/5).

Other deposits include Bazina (lead), Bou Jaber (Pb/Zn=3), Fedj el Adoum (Pb/Zn=1); Djebel ben Amara, Zaghouan Djebel Lorbens (zinc), Aïn Allega (Pb/Zn=3/2), Ressas, Touireuf, Oued Maden, Sidi et Taïa (lead), Djebba (Pb/Zn=3) and Djebel Touila (Pb/Zn =1/2). Lead predominates over zinc at Kef Chambi, El Haouaria, Koucha, Djebel Chara, Kebbouch, Sidi Driss and Djebel Semene.

The *Touireuf* chain is one of the Atlas écailles 40 miles south of the coastal border. Quaternary sediments cover a Cretaceous–Eocene monocline, but a number of faults and abnormal contacts complicate the structure. The Ressas East and West, and Saint Jean deposits are related to Eocene limestones, while St. Louis, St. Felix, Jobard and Balloute favor fissures of Upper Cretaceous (Campanian) limestones. In the upper horizons of the intensely tectonized *Ressas-East* deposits galena, accompanied by hydrozincite, precipitated in the upper reaches of veins, impregnating wall rock. Zinc predominates at lower levels. Cerussite appears in the St. Jean lumachelles and in the St. Louis veins, where it is accompanied by smithsonite. At Jobard, calamine and cerussite impregnate a calcite fissure filling.

The Djebel Slata overlaps the border 80 miles south of the sea, near the Djerissa iron deposit. Open faults of Sidi Amor ben Salem, on the western limb of an axially refolded Lower Cretaceous (Aptian) dome, have been mineralized by galena accompanied by berthonite, bleiniérite, cerussite, anglesite, sphalerite, baryte, cesarolite, etc.

The transverse veins and impregnations of Sakiet Sidi Youssef and Koucha are located between the two preceding districts in breccias and Middle Cretaceous (Turonian–Campanian) limestones.

The *Djebel Hallouf* and *Sidi bou Aouque* are further northeast. Here fine grained galena, accompanied by some sphalerite, impregnates Upper Cretaceous (Senonian) –Eocene fractures and faults.

The complex paragenesis of Cavallo, near Bône, *Algeria* (Fig. 11, 56) shows lead, zinc, copper, antimony, silver and gold. The Kef Oum Theboul veins, east of Calle, strike west-northwest. In these galena, chalcopyrite and pyrite are limited to funnels, although veins also impregnate the marls. The argentiferous pyrites may contain gold and arsenic. At *Sidi Kamber* the Pb/Zn ratio is increasing. Furthermore, at Mesloula galena was disseminated in Aptian limestone. Other occurrences include Gar Rouban, the Djebel Tchmoul and Aïn Achour, where arsenic also occurs.

Hardly any zinc and lead have been found between the Atlas and the Equator. However, traces occur in the *Algerian* Sahara at the *Chebka el Haman* and *Charef*, near *Ben Zired* at *Hassi bou el Adam* and *Hassi Chamba*, and on the *lower Daoura.*

MOROCCO

The country's principal lead deposit, Bou Beker–Touissit, is located in the northeastern corner, in the Oujda district. Aouli and Mibladen, in the Upper Moulouya, are situated 30 miles south of Fes, as the crow flies. The Djebels Aouam, Bou Dahar, and also the Tafilalet play a minor role, as do several dozen smaller occurrences (Fig. 10, 59).

Oujda

Bou Beker–Touissit. The mineralized horsts of the Oujda (Fig. 90) district strike east-west for 30 miles (AGARD et al., 1952). Bou Beker and Touissit occupy the eastern 6 miles. Early Carboniferous (Visean) shales are overlain by Lias limestones, dolomites, sandy and pyrite bearing limestones, and iron ooliths, covered by Middle–Upper Jurassic (Callovian–Oxfordian) clays. Dolomite layers, divided by thin argillaceous beds, are less than 3 ft. thick. Disseminated sulphides impregnate the upper 50 ft. of the dolomite and occasionally the overlying bed. Bou Bekker was affected by radial faulting, whilst complex rift, transverse and en échelon faulting distinguish Touissit. The upper layer contains galena and recrystallized dolomite, with the lower layer carrying fine grained galena and sphalerite. The two layers coalesce to a thickness of 30–50 ft. in the southern part of the horst of Bou Beker, and at level 87 of Touissit.

Due to lateral movements dolomite layers were sheared and thrust over each other, tightly warped, and then jointed beneath the parallel ridges. Hydrothermal solutions substituted a late, large grained pink dolomite for the original fine grained carbonate. Galena, containing 0.03% Ag, was introduced during the recrystallization. It is accompanied by some sphalerite and chalcopyrite.

More ore, however, is concentrated in larger irregular bodies of complete replacement measuring 50 by 250 ft., or pseudo-pipes with a diameter of up to 300 ft. Xenoliths of the roof have fallen into the dissolved younger rocks of these caverns or subterranean dolinas, mineralized on their contours. Light coloured, partly zoned sphalerite is accompanied by

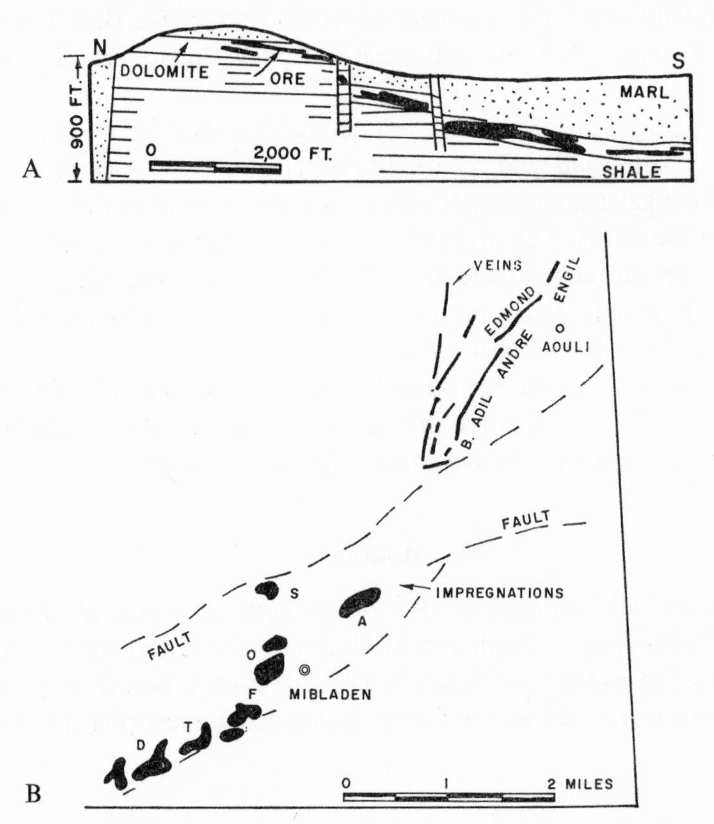

Fig. 90. A. Section of Bou Beker. B. The lead deposits of Aouli and Mibladen.

galena containing 0.05% Ag, spherulitic pyrite, quartz and some chalcopyrite. Rare pink dolomite, associated with galena, may fill late veins which cut the ore.

Gypsum, sulphur, wulfenite, magnesite, smithsonite, anglesite, cerussite, smithsonite and goslarite occur in the zone of oxidation. Furthermore, the Sidi Lahcen lead–silver vein of Oujda highlands is injected into Paleozoic strata, and wulfenite was found at Djebel Mahsseur.

Moulouya

Mibladen–Aouli. The Hercynian fold of Aouli has been faulted into compartments by post-Cretaceous tectonism (Fig. 90).

The structure essentially consists of a Precambrian relic of schists (including granitic stocks) embayed in successive zones of Permo-Triassic lavas and clays, Lower Jurassic (Lias and Toarcian) marls and limestone, and bordered by Cretaceous basal conglomerates. The vein zone forms an intermittent, 12 mile long arc between *Outat Sidi Saïd* and Aouli. However, only the *Henri, Bou Adil,* (Électrique, Frédéric–Edmond and Virgil) veins radiating around Aouli are significant. Replacement deposits strike south-southwest from Bou Almaden beyond Mibladen, mainly along the southern border of the Lias.

The Bou Adil fissure vein is 4–5 ft. wide, a veinlet of large grained galena being 2 inches thick. Quartz and galena seams alternate on its contact. Baryte stringers associated with galena cut quartz veins, particularly in the upper reaches. The 50–100 ft. wide Henri fissure includes several mineralized veins divided by schists. Fine grained galena, accompanied by pyrite and occasional chalcopyrite, occurs both in the fractured quartz and in baryte, but hardly any sphalerite is present. Silver and bismuth values attain an average of 0.04% of the mineral concentrates in both veins.

At *Mibladen*, baryte and galena expanded from veins, impregnating two to five beds of fine grained marly dolomites, marls and limestones in a 10 mile long zone. Nevertheless only a total of 6 ft. in a column of 1,000 ft. contains economic concentrations. There are twelve ore shoots striking east, which form a large angle with the faults. Baryte is abundant in the upper reaches, where ½ inch large galena crystals are largely altered to cerussite. Pyrite is not typomorphic, whilst vanadinite and chalcopyrite are rare. Silver and bismuth values are lower than in the ore veins. Geodes consisting of galena, baryte and cerussite also constitute alignments. The same minerals impregnate joints. The grade attains 1% Pb, but there is less silver and bismuth than at Aouli.

In the northern Moulouya, beds of arkose resting on granite are impregnated by white cerussite, galena and barytes. The ore bodies form thick oval runs several hundred yards long and several to 20 ft. wide, striking north-northeast. Grades attain 3% Pb. The runs are underlain by shear zones traversing the granite. These shear zones contain limonite, tyujamunite, carnotite, baryte and galena.

Jbel Aouam

The northeast striking Aouam veins cut into Paleozoic quartzites and limestones south of the antimony districts of the Moroccan Tableland, 15 miles south of Meknès. The mineralization is believed to be Hercynian. Tungsten is related to later biotite granite and microgranite. The paragenesis consists of galena, sphalerite, chalcopyrite and baryte, with subordinate arsenopyrite. In a quartz vein of the contact arsenopyrite is associated with pyrrhotite and pyrite. The principal, fissure filling Signal reef follows a quartzite contact, its thickness varying between 2 and 12 ft. Zones of galena, sphalerite, baryte, siderite and opal, accompanied by pyrite, chalcopyrite and mimetite, succeed each other around breccia fragments. Oxidized lead ores contain 0.015% Ag.

Ksar es Souk District

This district is situated in southeastern Morocco. The Bou Dahar veins extend along the contact of Lower Jurassic, Domerian limestones and Liassic sediments. While fissure veins occur between Ksar Moghal and Toutia, swarms of veins occur between Ksar es Souk and Taouz. At Mefis, near Taouz, hematite fills in a 30–60 ft. wide zone of fissure. Hematite has been altered to siderite in the upper reaches, and galena impregnates the iron ores along parallel bands.

Other occurrences

In the district of Marrakech, lead accompanies zinc. At Ouled Hassine, 7% Pb is associated with 6% pyrite, and 1.2% Zn. Furthermore, at Boumalne, galena replaces a Jurassic limestone layer. Other occurrences include Demnat, Ksiba, the Beni Tadjit district, and the Tazzeka mountains in the eastern High Atlas.

LOWER CONGO

Lead and zinc predominate in the mineralization of the Katangian geosyncline in the *Congo Republic*, showing also at its nose at Kroussou and Offobou, Gabon. The principal deposits follow the Renéville–Boko Songho alignment between Brazzaville and Loudima. Deposits may form irregular bodies or veins in favorable beds (NICOLINI, 1959). Lead–zinc silicates and carbonates have been worked at M'Fouati on the Lutété. Smithsonite contains 40% Zn. Sulphides of the Hapilo veins, near M'Fouati, include sphalerite and galena. Similarly lead and zinc traces prevail west of the *Mindouli* copper occurrence (Fig. 25, 53).

Most of the mineralization in the Republic of the *Congo*, west of Léopoldville, is localized in the vicinity of faults as fissure fillings or impregnations. At Bamba–Kilenda, near Madimba, chalcopyrite, bornite and chalcocite prevail in the limestones of the shale–limestone (schisto–calcaire) system, below the contact of the shale–sandstone (schisto-grèseux) system. Sphalerite–galena and pyrite, including traces of silver and vanadium, concentrate in felspathic quartzite of the younger shale–sandstone system. The paragenesis also includes tennantite, argyrose and covellite.

Some lead and zinc accompany the predominantly cupriferous mineralization of northwestern Angola, e.g., at *Baua*.

SOUTHWEST AFRICA

Many base metal occurrences have been found in the country (Fig. 38, 53, 91; see 'Copper'

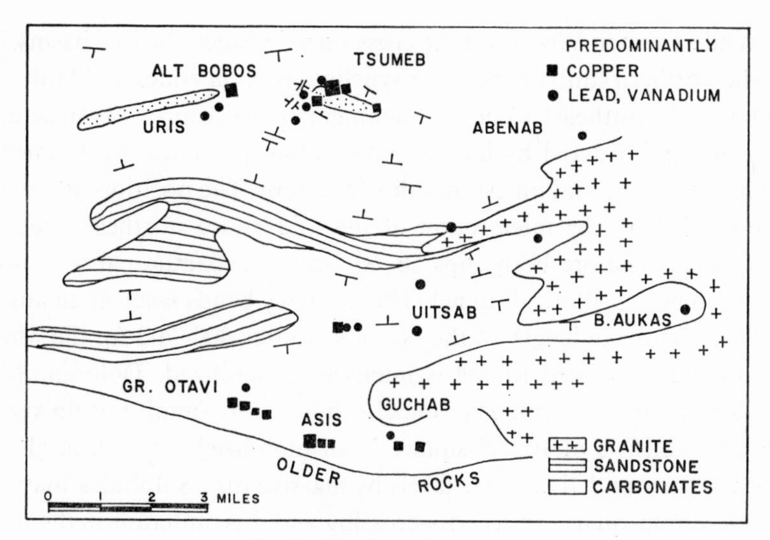

Fig. 91. The Otavi Highlands.

chapter). The only deposit of importance is *Tsumeb*, (Fig. 92) 300 air miles northeast of Swakopmund harbor. Unmetamorphosed, discordant sediments were deposited on the Precambrian formations in a broad arc, beginning on the Kunene river on the border and extending to Tsumeb. This late Otavi–Nama column is composed of the Abenab dolomite, some shale and limestone in the early and late stages, a tillite becoming ferruginous west in the Kaokoveld (see 'Iron' chapter), the Tsumeb dolomite, Mulden felspathic quartzite, arkose, and greywacke. In the Tsumeb area, the Mulden quartzite forms the bottom of an east-northeast trending syncline. Light and dark dolomite, first with subsidiary chert and then with limestone and shale form the limbs (SÖHNGE, 1958). Another trough is situated

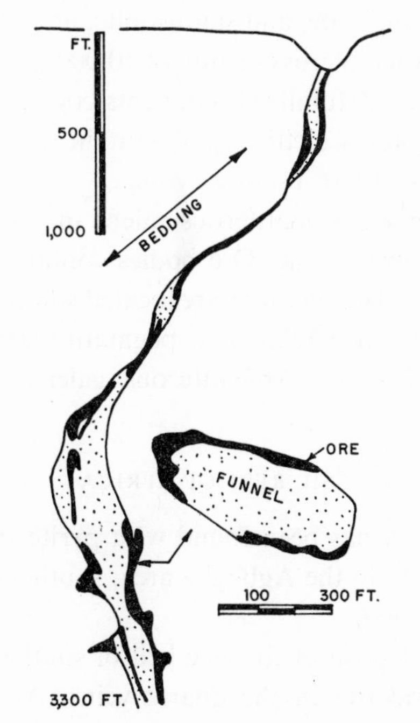

Fig. 92. Section and 2,600 ft. plan at Tsumeb.

further west. While Tsumeb West is on the cross warp between the two basins, *Tsumeb Mine* is located on the northern limb of the main syncline. Breccia zones and faults are arranged along a largely south-southeast trending alignment of the transverse structure. The total width of the syncline, bordered by late dolomite, attains 3 miles. At Tsumeb, dragfolded and dip-slipping dolomites and limestones dip 45° south. The position of the hydrothermal ore pipe is controlled by the intersection of the fold and a northeast trending regional alignment of fissures. The ore body dips 50–60° south, expanding at a depth of 1,800 ft. into a transverse trending elliptical funnel. This chimney bends *north* at an angle of 75° to a depth of 4,000 ft. The strike length of the ore pipe attains 300–700 ft., the width 10–250 ft. The dolomite of the brecciated funnel is intensely mineralized. Dolomite (crystals) may contain 3–9% Zn and 0.5–5% Pb. The funnel exhibits horizontal, but no vertical, zoning. The excentric core consists of 10,000 square ft. of mobilized, carbonatized–silicified feldspathic sediments, surrounded at lower levels by massive ore. Sulphides may replace completely calcite–dolomite–quartz veins criss-crossing and disseminating in the pipe. Calcitization of dolomite, accumulation of chert and graphite announce ore concentrations. Furthermore, ore lenses have also been injected along a bedding fault cutting the funnel. On the other hand, kersantite dykes and sills cut the channel, Karroo dolerites being injected into later dolomites north of Tsumeb.

The sequence of crystallization is: dolomite, quartz, pyrite, bornite, germanite, digenite, sphalerite, sericite, calcite, luzonite, enargite, tennantite, galena, chalcocite (SCHNEIDERHÖHN, 1931). The principal ore minerals include galena, sphalerite and tennantite. Lead is the principal metal. Copper accumulates in the ore near the surface and below 2,700 ft. of depth, zinc being scarce in the same horizons. Sulphides prevail between 1,000 and 2,600 ft. of depth, but minerals have been oxidized both above and *below* these levels. The two germanium minerals tend to accompany galena and tennantite at lower levels.

In the zone of oxidization 50 minerals crystallize, the most important of these being: malachite, azurite, cerussite, anglesite, and smithsonite, and also arsenates and carbonates.

The *Tsumeb West* ore body has a cross section of 10,000 ft. However, at a depth of 300 ft. it peters out in horstail veinlets. Mobilized sediments constitute this lens and the groundmass of projecting breccia-dykes. Calcitized dolomite in the northern section of the outer zone is the copper indicator, as at Tsumeb Mine.

At *Hohewarte*, near Windhoek, argentiferous galena impregnates tremolite marbles in a 1,800 ft. long and 10–100 ft. wide zone. Ore bodies contain 4.7% Pb, some copper, and 30 dwt. silver per ton. Similar disseminations are located south of Usakos. At Namib, near Swakopmund harbour, two gossans related to pegmatite stringers include galena, up to 140 oz./ton of silver, and fluorite. Argentiferous galena also occurs at Aias in the Orange River belt.

THE REST OF AFRICA

Coarse and fine-grained galena has been found with pyrite in the Tawalata and Kokoya placers, Liberia. Galena occurs in the Agbogba area, southeastern Ghana, and with sphalerite in the Agbandi reefs, Togo.

Wase is the principal lead deposit of the zinc belt of southeastern *Nigeria*, where galena tends to be associated with siderite in the quartz veins. At Arena and Akwana, Benue province, argentiferous galena is associated with fluorite.

The telethermal paragenesis of Um Gheig, *Egypt*, found in Middle Miocene lumegrite and other sediments on the Red Sea coast, includes sphalerite (20–25% Zn) and galena (3–5% Pb), altered above the water table to cerussite, smithsonite and wulfenite (El Shazly, 1959). Some lead and zinc has also occurred at Gebel Russas. At Samiuki and Atshan, smithsonite, sphalerite, chalcopyrite and galena replace Precambrian talc (see 'Talc' chapter).

Galena occurs in the Jebel Kutub, Darfur, Western Sudan, also impregnating the limestones of Arargob near Candala, northern Somalia. Galena shows at Goe Ebed and Hulul, near Berbera in a shear zone and at Unkah and Kul, near Erigavo.

At Kitaka in western *Uganda*, east of the Kilembe copper mine, a quartz vein has been injected into a shear cutting a dolerite body in late Precambrian granite–gneisses. The paragenesis includes galena, pyrrhotite, chalcopyrite, coarse gold, reinite (W) and some sphalerite (Barnes and Brown, 1958).

At Vitengeni, 40 miles north of Mombasa on the Kenya coast, some chalcopyrite and quartz is associated with galena and baryte. The mineralization is believed to be related to the carbonatite–alkaline district of Mrima (see 'Rare Earth' chapter).

The Kengere paragenesis, south of Jadotville, *Congo*, includes galena, pyrite and sphalerite, copper and silver. The Lombe deposit contains lead and zinc, as does Kipushi.

Galena occurs in several reefs of the Lupa gold field, *Tanganyika*, especially around Chunya (Saza). The only lead (-copper) deposit of some importance used to be *Mpanda*, not far from the Rukwa rift, 100 miles east of Lake Tanganyika. Shears, several miles long and 25–100 ft. wide, cut Precambrian granite–gneisses, amphibolites and metasediments (Fawley, 1958). The shear zones bend, pinch, branch and split. The distribution of irregularly shaped ore bodies appears to have been controlled by fractured, brecciated or permeable sectors of the shear. Galena, chalcopyrite, pyrite, silver, gold and sphalerite have precipitated with a gangue of quartz, siderite ±baryte. The ore grades into the surrounding shear. Some of the mineralized shear zones show a gradation of oxidized, mixed, enriched and sulphide ore, the dividing lines being at the 100 and 250 ft. levels. In the Mukwamba mine, values attain 2.5% Pb, 0.5% Cu, 0.01% Ag and 1 dwt./ton of gold.

On the Chombe river in southern Zambia, galena impregnates dolomites. At Broken Hill galena accompanies sphalerite. Galena is associated with quartz and calcite in the veins of Mount Licore, near Chioco, western Mozambique, and it also occurs on the Baixo (lower) Mazoe.

At Besakay, near Tananarivo, *Madagascar*, nine quartz lenses are aligned along a 1 mile stretch, their width varying from 1 to 10 ft. The ore contains 7% Pb and some silver. The Besakay vein carries lead, zinc, gold and silver. Other occurrences include Ampandrana and the quartz–baryte swarms of Ankitokazo and Andavokoera.

Galena is an accessory mineral of Southern Rhodesian veins, occurring, e.g., at Elbus, near Wankie.

The long lead belt of the *Marico* district, western Transvaal (*South Africa*), extends from Klaarstroom to Bokkraal. It follows the Upper Dolomite Series and the contact of the largely impervious Pretoria Series, which localized galena in breccia pipes and bodies at *Bornhoek*, *Rhenosterhoek*, *Bokkraal*, *Genadendal*, *and Leeuwbosch* and sphalerite at *Witkop*, *Buffelshoek* and *Kaalplas*. Silver–lead lodes include *Brakfontein*, *Dwarsfontein*, the *Shimwell* vein, *Spitskop*, *Hartebeestfontein*, *Edendale* mine, *Nooitgedacht*, *Kindergoed*, *Vergelegen*,

Baviaansdraai, etc., in the Transvaal. Galena occurs in Cape Province in the *Maitland Mines*, at *Uitenhage*, *Balloch*, near Hay, in *Namaqualand*, and in the *Mfongosi* and *Ngobevu* valleys in Natal.

A subconcordant quartz vein is included in the quartzites of the Ngwavuma hill side, Hlatikulu district, *Swaziland*. The vein extends more than 2,000 ft. along the strike, attaining a thickness of 2 ft. Pyrite, chalcopyrite and sphalerite are associated with galena in up to 1½ inch blebs. Grades attain 0.22% Pb, 21.2 dwt./ton Ag and 0.9 dwt./ton Au.

SILVER

In Africa two-thirds of the silver found occurs in sulphides, the other third occurring as a gold alloy, mainly in South Africa. Although silver is frequently included in galena, lead is not the dominant metal of large deposits. Galena is, indeed, a relatively rare mineral at Kipushi (Katanga), the copper–zinc deposit that supplied 40% of the continent's silver. The lead/silver ratio equals one there. Silver occurs in copper deposits in Southern Rhodesia and in the zinc deposit of Broken Hill, central Zambia. In Africa, argentiferous galenas contain 0.05–0.1% silver. Apparent ratios of recovered lead and silver average 1,500/1 at Tsumeb, Southwest Africa, 1,400/1 in Morocco, and 1000/1 at Broken Hill. Algeria, Tunisia and Kenya also supply silver.

Primary gold contains 10–30% Ag. However, the silver content of alluvial, partially reprecipitated, gold decreases to as little as 7%. South African gold contains 10% Ag. The behaviour of silver in the banket does not resemble alluvial conditions. Surprisingly, only 2.5% Ag is recovered from the primary gold of Ghana, which is considerably less than in the Congo or Tanganyika. Some silver shows in Ethiopia, Gabon, Upper Volta, Bechuanaland, Uganda, Mozambique, Swaziland and other countries.

CONGO

Silver is associated whith gold in the deposits of Kilo–Moto and Kivu–Maniema (e.g., Kamituga and Namoya). The principal silver deposit, however, is *Kipushi*, 20 miles south of Elisabethville (Fig. 88; see 'Zinc' chapter). Dolomite has been replaced in a ramifying funnel, not far from the sub-vertical contact of calcschists. While impregnated calcschists carry copper sulphides almost exclusively, lenses of sphalerite lie in the upper levels farther from the contact. Argentiferous galena crystallizes as the last primary sulphide, replacing sphalerite and overlapping the second stage of copper–zinc mineralization, characterized by bornite. Chalcocite is largely a secondary mineral, containing an increasing quantity of inclusions in depth. Chalcocite stringers frequently penetrate into bornite. Interestingly enough, minute films of silver have been identified in chalcocite. Stromeyerite and covellite belong to the same paragenesis.

MOROCCO

A majority of the Kingdom's silver is found in the lead–zinc ores of Bou Beker and Mibladen (Table CXXVIII).

TABLE CXXVIII
IMPORTANT MOROCCAN SILVER DEPOSITS

	Silver content of galena (%)	*Approximate share of supply (%)*
Bou Beker	0.025—0.041	50
Touissit	0.025—0.041	15
Aouli	0.030	7
Mibladen	0.025	5

At *Bou Beker* and *Touissit*, two types of mineralization have been distinguished. Most of the silver is contained in coarse grained galena, impregnating in parallel east-southeast and south-southeast striking runs the upper 3–10 ft. of Liassic (Lower Jurassic) limestones. Less silver and lead occur in the replacement bodies of lower layers. Notwithstanding the oxidization of the replacement bodies of Mibladen, the veins of Aouli contain more silver (Fig. 90).

At *Mefis*, a brachy-anticline of Devonian schists and limestones (BOULADON and JOURAVSKY, 1952) has been injected by diorites. Reefs fill open fissures. The zone of oxidization extends to a depth of 150 ft. It includes hematite and 0.2% Ag, i.e., six times more than the sulphides. Concentrates of the fissure filling breccia of Djebel Aouam contain 0.15% Ag.

Erdouz and Assif el Mal are located in the High Atlas, south of Marrakech. The Erdouz swarm has filled in several phases fissures, mainly in Cambrian dolomites, but also in limestones and schists. The paragenetic sequence includes large grained sphalerite, crushed galena, some chalcopyrite, and pyrite. The ramifying Assif el Mal vein occurs in vertical, Silurian schists. The paragenesis also comprises baryte. At both localities silver values of concentrates may attain 0.2%.

At *Goundafa*, to the south, fissure filling breccia follows the contact of early Cambrian (Georgian) limestones. Here irregular veins contain the same coarse grained minerals and 0.035% Ag. The Koudiat el Hamra vein follows the foliation of Paleozoic strata in the vicinity of Marrakech. The mineral association is given in Table CXXIX.

TABLE CXXIX
KOUDIAT EL HAMRA VEIN

	Silver (%)
galena	0.15—0.4
sphalerite inclusions	0.05—0.2
pyrrhotite	0.6 —0.7
tetrahedrite	12 —15

South of Bou Beker the Sidi Lahcen veins and impregnations affect Paleozoic schists and arkoses. Arsenopyrite is also present in this paragenesis. Many other galenas, like those of Rehamna and Beni Tadjit, contain less than 0.03% Ag.

The *Bou Azzer* cobalt concentrates normally contain only 0.001–0.003 % Ag. However, the ores of Veins 5 and 7 include 0.05–0.07 % Ag. On the border of, and as far as 150 ft. in the diorite wall-rock of Vein 5, skutterudite, chalcocite and bornite have been replaced by native silver.

At *Tiouit*, on the northern edge of the Anti-Atlas, a swarm of ramifying fissures traverses a granitic window and its country rocks. The total width of the principal fissure attains 30 ft. The ore contains 0.022% Ag and 15 dwt. gold per ton.

SOUTH AFRICA

The silver content of gold consistently averages 8.7–9.2%, varying, however, in the gold fields and in the stratigraphic column. Silver content of the bullion tends to decrease with depth. It is believed that silver has been reconcentrated during the reprecipitation of gold, either within one or between several horizons of banket. Contrary to what happens during the transfer of gold from primary to secondary deposits, the silver content increases in the successive paleo-alluvial cycles. The relatively young Ventersdorp Contact Reef and the Black Reef surprisingly contain more silver than the underlying Witwatersrand bankets from which they are believed to have been derived (HARGRAVES, 1963).

Silver occurs *(1)* at Argent near Pretoria, in galena; *(2)* at Roodepoortjie, in bornite; *(3)* in the Willows in tetrahedrite. Silver also shows in the nickel horizons of the Bushveld.

THE REST OF AFRICA

Tunisia. At Sidi Amor Ben Salem a fissure dividing Cretaceous and Miocene limestone has been impregnated by galena containing 0.038% Ag. Other lead deposits also include silver.

Algeria. At El Abed, Oued Zounder silver shows in the galenas accompanying sphalerite. At Mesloula, galena, containing 0.03% Ag impregnates Lower Cretaceous limestones between marls of the same epoch and of the Triassic. At Kef Oum Theboul silver is included in pyrite.

Kenya. In Kenya, electrum occurs in the Macalder gold–copper paragenesis, east of Lake Victoria.

Tanganyika. Silver content is variable in the Tanganyika gold fields as is indicated in Table CXXX (in the order of their importance as silver resources).

At Geita, gold and silver impregnate metavolcanics and ironstones.

The remaining argentiferous sulphide ores of the Mpanda shear are of low grade.

TABLE CXXX

SILVER CONTENT OF TANGANYIKA GOLD FIELDS

	Nature of accumulation	*Silver (%)*
Geita	primary	25—40
Musoma	primary	7—11
Saza, Lúpa	primary	40—50
others, Lúpa	primary	18—22
Lúpa	alluvial	8—12

Zambia. The Broken Hill deposit has been strongly oxidized. Native silver is associated to willemite at Excelsior Zinc. The copper ores of Mangula contain silver.

Southern Rhodesia. The lead–zinc–copper ores of the Copper Queen and -King contain silver (1 ounce of silver per ton). Silver also reports in gold bullion.

Southwest Africa. At Tsumeb in *Southwest Africa,* silver is included in galena and in tetrahedrite–tennantite.

Angola. Silver shows in silicified and crush zones at Quiongua, near Pungo Andongo, Quissol, near Malanga and near Duque de Bragança.

CADMIUM

Traces of cadmium show in a majority of sphalerites. Avoiding veins, cadmium concentrates in zinc deposits of the replacement type in Africa. Greenockite, the cadmium sulphide, occurs at Kipushi in the Congo. The cadmium content of zinc ores, however, is higher at Tsumeb, in Southwest Africa. Both deposits are late Precambrian.

The Prince Leopold Mine is situated near *Kipushi,* 20 miles south of Elisabethville, on the Zambian border (see 'Zinc' chapter). In this funnel of hydrothermal replacement in carbonates, sphalerite is second in abundance after chalcopyrite, and is rather coarse grained and dark coloured. The mineral has crystallized after pyrite, with the copper sulphides. Greenockite precipitates on sphalerite crystals and in concentrates the zinc/cadmium ratio averages 77/1. Cadmium also shows at Broken Hill, in central Zambia.

At *Tsumeb,* zinc is one of the less important constituents of the copper–lead ores. In the funnel of carbonate replacement, sphalerite occurs between 1,000 and 2,600 ft. of depth, copper sulphides continuing deeper. The zinc/cadmium ratio of concentrates is exceptionally favorable, oscillating between 20/1 and 14/1.

Cadmium is a trace element of the replacement deposit of Bou Beker–Touissit in northeastern Morocco. In fine grained, irregular ore shoots *(a)* zinc and cadmium are more abundant than in bedded runs of galena *(b)* (Table CXXXI).

TABLE CXXXI

GRADES OF IRREGULAR ORE SHOOTS AND BEDDED RUNS

	Irregular ore shoots (%)	*Bedded runs (%)*
Lead	8	72
Zinc	56	3
Cadmium	0.4	0.001

GERMANIUM

Germanium, a rare metal, occurs at Kipushi (Fig. 88, Congo) and Tsumeb (Fig. 92, Southwest Africa) in reniérite and germanite. Traces of germanium generally show in zinc minerals, and in coal as well. Some germanium reports in the bedded copper ores of the Aïr Mountains, Niger Republic, and in the zinc replacement deposit of Broken Hill. African coal, including much ash, is believed to contain little germanium. Reniérite is included in chalcopyrite, sphalerite and galena at *Kipushi*. Reniérite also occurs in independent idio- or hypidiomorphic grains or spindles. Its period of crystallization overlaps that of early tennantite. The mineral (Cu, Fe, Ge, Zn, As) S, contains 6.4–7.8% Ge, but no gallium. Iron is more abundant than in germanite.

Germanite, the mineral showing the highest germanium and gallium concentrations in the world (10%, 2%), is only known to occur at *Tsumeb*. Reniérite, first discovered at Kipushi, Katanga, crystallizes as a replacement product of germanite and of the host rock. 'Orange bornite' is probably identical to reniérite (SCLAR and GEIER, 1957). Both enargite and tennantite contain germanium. In the calcic environment of the zone of oxidization germanium is released and reprecipitated in a basic lead sulphate, containing 8% GeO, the arsenate bayldonite (500–5,000 p.p.m. Ge), mimetite, olivenite, and anglesite and cerussite (50–500 p.p.m. Ge). Most of the germanium is in the quadrivalent state. Finally, the oxidized germanium is releached by supergene solutions in alkali germanates.

GALLIUM

The world's principal source of gallium is included in the germanites of *Tsumeb* (1.8%). Sphalerite contains 0.001–0.02% Ga. With bauxites including up to 0.01% of the metal, the Guinean deposits probably contain much of the world resources.

SELENIUM

In Africa, especially in the Zambian Copperbelt, selenium is associated with copper ores, but this metal also shows in pyrites and uranium minerals of Katanga and northern Zambia.

INDIUM

Sphalerite contains 0.0024% In, which also shows in chalcopyrite, galena, cassiterite and coal.

RHENIUM

The molybdenum/rhenium ratio average 15,000/l. Molybdenum being scarce in Africa, rhenium can be found in copper, titanium, niobium and rare earth ores.

VANADIUM

Titaniferous magnetite seams of layered ultrabasics contain 0.1–2% vanadium. The principal occurrence of this type is the Bushveld. In addition, vanadium shows as an impurity in the lateritic iron ores derived from ultrabasics and in the ilmenite of beach sands. Vanadium also precipitates in the supergene alteration zone of lead veins and impregnations. Finally the metal is believed to have been leached from shales. However, deposits of this type acquire importance only in hydrothermal lead–zinc districts like Tsumeb. Vanadium also combines with uranium in hydrothermal deposits and phosphates.

SOUTH AFRICA

Titaniferous magnetites of the upper layers of the Bushveld ultrabasics contain up to 2.4% V_2O_5. The bands dip 10–25° inwards, but the vanadium content of the upper bands adjacent to the granites is low. The schematic sequence of injection and layering of Bushveld ores is: chromium, platinum, vanadium, and iron, followed by the tin paragenesis. Intimately intergrown magnetite and subsidiary ilmenite occur both in the eastern and western branches of the complex. The eastern branch extends 100 miles from Languitsig through Stoffberg, Roossenekal and Magnet Heights to the Olifants River. At Stoffberg, one of the bands contains more than 1.6% V_2O_5 with others including 1% V_2O_5. However, there is a gap in the outcrops 20 miles north of Languitsig. On the other hand, short seams of titaniferous magnetite are oriented parallel to the main bands between Lydenburg and Groblersdal. The western branch strikes west, extending 80 miles from the Pyramids north of Pretoria, beyond Rustenburg. This branch, interrupted by the Pilanesberg, extends west and northeast of this disturbance towards the Crocodile River (Fig. 46, 50, 99).

Pipes penetrating into the ultrabasic layers also contain vanadium, and the chromitite seams, lying considerably lower, include 0.05–0.81% V_2O_5, averaging 0.2%.

In addition, the titaniferous magnetites of Tugela, north of Durban and Rooiwater, and near the Antimony Bar, also carry vanadium. Vanadium is associated with lead and zinc in the quartz veins of Kindergoed, east of Languitsig.

In the *Marico* district, east of Mafeking, supergene telethermal vanadium is associated with lead–zinc parageneses impregnating dolomites of the Transvaal System. At Kafferskraal, nodules of galena are surrounded by a crust of cerussite, followed by pyromorphite and vanadinite. Here vanadinite crystallizes in irregular layers of wad, nests and in joints. The adjacent Doornhoek occurrence is similar. Finally, at Nooitgedacht, vanadium accompanies limonite and pyromorphite in a quartz vein, intersecting lavas and quartzites.

SOUTHWEST AFRICA

The secondary vanadium mineralization covers larger areas of the *Otavi Highlands* than the concentrated lead–copper deposits. *(1)* Tsumeb occurs in the northern alignment of occurrences. Scattered deposits have been found *(2)* along another east-northeast trending belt, 20 miles south, and *(3)* further south on the edge of lower Otavi sediments. The Tsumeb alignment consists of *Uris* and *Karawatu* near *Bobos*, *Friesenberg*, *Tsumeb West*, *Mine*, *South*, and of *Bahnhofsberg*. Zone 2 consists of *Gaus*, *Uitsab*, *Gaguas*, and *Olifantsfontein*, and Zone 3 of *Guchab*, *Schlangenthal*, *Uitkomst*, *Gross Otavi*, *Nagaib*, and *Aukas Mountain*. Abenab lies 30 miles east of Tsumeb (SCHNEIDERHÖHN, 1931; Fig. 91).

At Tsumeb West lead–copper vanadate stringers penetrate sandy infillings and silicified dolomite along *(a)* the obliterated schistosity, *(b)* the borders of the dolina and also *(c)* the country rock. Mottramite occurs at Tsumeb.

In the *Otavi* highlands descloizite and cuprodescloizite cluster with silicified dolomite concretions in sand and neogene limestone, forming a pre-breccia. Lead–zinc sulphides do exist under such occurrences. Vanadium content is low, however, and biogene vanadium appears to have originally infiltrated the shales (SCHNEIDERHÖHN, 1931).

In the clays of the principal deposit, *Abenab-West*, (VERWOERD, 1957) the lead content of the original zoned galena–cerussite assemblage increases, while the vanadium content decreases. Willemite is marginal and chalcopyrite scarce. The original hoodzone of the reef, located in a tectonized area of carbonates, also contained pyrite, some sphalerite, bournonite, tennantite and chalcopyrite. Secondary minerals include covellite, anglesite, mimetite, greenockite, smithsonite, malachite and wulfenite, accompanied by limonite, goethite, calcite, quartz, and clay silicates. At some distance secondary willemite occurs in a peripheral reef. The vanadium is believed to have been leached out from Otavi shales during a subarid period of oxidization and then precipitated in two vanadinite, mimetite cycles. The second vanadium cycle is distinguished by the zones of earlier vanadinite decreasing, descloizite substituting and increasing outwards.

THE REST OF AFRICA

Vanadium occurs in the *Congo* in the bedded copper–gold–platinum–palladium paragenesis of Ruwe, Kolwezi. Vanadium occurs with the same metals in the hydrothermal Shinkolobwe uranium deposits. Cuprodescloizite accompanies the complex copper–zinc ores of Kipushi. The second, oxidized ore body of Broken Hill, central Zambia, contains 0.75% V. However, vanadium generally occurs at some distance from the zinc–lead shoots. Vanadinite and descloizite are believed to have been derived from shales and phyllites containing 0.012% V. Vanadium is, indeed, soluble in sulphates under acid conditions in the presence of carbonates. Vanadium also shows in the contemporary mineralization of the Lower Congo geosyncline, in the zinc paragenesis of the Niari basin, *Congo Republic*, and in the copper paragenesis of Mavoio and Bembe, *Angola*. The faulted, eastern, contact of the geosyncline strikes north-northeast. Gossans have developed from sulphides at the contact of limestones at Lueca and Quinzo and descloizite–montramite crystallized in collapse breccias and stringers. Furthermore, uranovanadates precipitated in the hoodzone of Mounana, Gabon.

The vanadium mica, roscoelite, crystallizes in patches of the graphite gneisses of Angonia, Mozambique. Concentrations contain 0.11–0.29% V_2O_5. The titaniferous iron ores of Liganga, southwestern Tanganyika, contain 0.61–0.88% V_2O_5. Vanadium shows in the auriferous placers of Bilu, Ethiopia, and in the iron–copper–gold paragenesis of Akjoujt, Mauritania. Vanadinite is associated with hematite, limonite, goethite, calcite, cerussite, anglesite, malachite, quartz and gypsum in the hoodzone of the Jebel Tadaout veins, near Taouz in the *Moroccan* High Atlas. Here vanadinite concentrates along a line of druses on the footwall and in another band. Vanadium also occurs in the lead deposit of Keb Cheboul, western Tunisia.

Phosphates contain traces of the uranovanadates tyuyamunite and carnotite.

MERCURY

Mercury is the scarcest metal in Africa. Telethermal mercury precipitated in fissures and joints in Tunisia but it is also associated with baryte and galena (Fig. 56). Impregnations are known to occur in other parts of the continent.

TUNISIA

The lead–zinc–mercury impregnations of *Oued Maden* occur in Upper Cretaceous (Senonian) sediments of a Triassic–Cretaceous monocline in the Atlas Flysch (GOTTIS and SAINFELD, 1952). Oued Maden is located on the Algerian frontier, 30 miles from the coast. The As–Sb–Cu–Pb–Zn paragenesis is related to a durcicrust of limonite and baryte, covering a sub-vertical, north-northeast striking, open fault. In the northern section, north-northwest and east striking joints and shears, traversing limestones, form a sub-perpendicular couple. The permeable spaces are filled in by black clays accompanied by scattered cinnabar and metacinnabarite, representing an overall mercury grade of 0.5%. In the underlying marls and marly limestones fine-grained galena, accompanied by sphalerite and some cinnabar, impregnates east-westerly fissures and horizontal joints.

THE REST OF AFRICA

In the stockwork of thin fissure veins of Bir-Beni Salah, eastern *Algeria*, sphalerite and cinnabar are accompanied by galena. The sandstones, marls and brecciated limestones of Ras-el-Ma contain 0.4–0.8% Hg. At Taght, in the Aurès Mountains, fine sub-vertical veins intersect a similar Lower Cretaceous (Neocomian) column. Cinnabar, closely associated with galena, mercury and sphalerite, fills in the veins and impregnates the country rocks.

Southeast of Vitengeni and north of Mombasa, *Kenya*, thin veins and impregnations of cinnabar are associated with galena–baryte veins.

The Harrington Kop and Monarch Kop strike parallel to the Antimony Bar in the Murchison Range, *South Africa*. Lenticles of cinnabar, accompanied by eglestonite, outcrop along 400 yards in the carbonaceous schists. The *Harington Kop* veinlets contain 0.35% Hg. At Kaalrug and Singerton, north of Swaziland, cinnabar precipitated on joints and fissures. At some distance mylonitized cherts and quartzites strike north-northeast, 3 miles from the Havelock asbestos mine in the Pigg's Peak district of northwestern *Swaziland*. Quartz stringers penetrate into the softened, chlorite–sericite schists separating the quartzites, both obliquely and parallel to the shear. While traces of cinnabar have been widely disseminated in the sheared beds in an area of at least 800 by 750 ft., concentrations are limited to stringers and blebs. Samples contain 0.01–1.63% Hg.

FLUORINE

Fluorite crystallizes in veins from the early to the late hydrothermal phase, and frequently in the same environment as lead and zinc. In the principal deposits irregular ore bodies partially replace carbonates of various ages. The largest occur in the dolomites of the 2,200 million years old Transvaal System. However, North African deposits are Jurassic and younger. Since fluorine is particularly mobile, impermeable inliers in carbonatic rocks act as barriers, controlling its precipitation.

Impure fluorite also occurs with hematite in veins in granitic aureoles and carbonatites. Topaz is a typomorphic mineral of greisens, with fluocarbonates and fluophosphates characterizing carbonatite–alkaline complexes.

SOUTH AFRICA

The largest deposits of South Africa are the fluorite bodies replacing dolomites of the Transvaal System, east of Ottoshoop, a village situated east of the southeastern corner of Bechuanaland. Other deposits include: *(1)* ramifying, lenticular veins and quartz–hematite veins related to the megacrystalline Bushveld granite; *(2)* alkaline complexes; *(3)* post-Karroo fissure fillings in the Transvaal; and veins of *(4)* Zululand and *(5)* the northwestern Cape.

Ottoshoop

Fluorspar occurs in a 20 mile long east trending zone (Fig. 46, 50). Irregular fluorite bodies replace dolomite and its contact with chert at the top of the Dolomite Series. Frequent veinlets are of minor importance. Large deposits developed beneath relatively impermeable layers of chert, shale or quartzite, which barred the ascension of fluor. While dolomite has been enriched in talc at the contact of fluorite, tremolite tends to occur adjacent to sphalerite pipes and veins, e.g., at Buffelshoek. Tremolite has been displaced by talc in the deposits distinguished by predominant fluorite. On the other hand, sulphides have been disseminated at the base of fluorite deposits and as far as a few ft., in the wall rock. Brucite impregnates altered dolomite.

The *Oog van Malmanie* 'pipe' developed as an elongated mushroom. Its stem consists of dolomite breccia in a groundmass of fluorite, accompanied by some pyrite, pyrrhotite, chalcopyrite and galena. Pure, colourless and pale blue fluorite of optical quality has been extracted from the central top sector of the 'mushroom'. At *Buffelshoek* occur a *(1)* sphalerite–fluorite pipe, *(2)* alternating laminae of fluorite and wad extending to Witkop and *(3)* a tabular body beneath shale. One of the *Witkop* occurrences resembles the latter, continuing in depth, however, in ramifying veins. In other sectors of Witkop, dolomite, overlain by chert, has been partially replaced by fluorite and leached out. Other occurrences include Karroebosch, Naawpoort, Rhenosterfontein, Strydfontein, Wintershoek and Leeuwbosch.

Bushveld

Fluorite occurs in five areas:

(1) Slipfontein, near the bend of the western Bushveld, 40 miles north of Rustenburg.

(2) In a 30 mile long, northeast trending zone, west of Warmbad (60 miles north of Pretoria). This zone includes Tooyskraal, Cyferfontein and Ruygtepoort (quartz-hematite).

(3) Buffelsfontein and Vischgat, 30 miles northeast of Warmbad.

(4) Brobbelaars Hoek, 100 miles north of Warmbad and 40 miles from the Bechuanaland border.

(5) The belt connecting Pretoria with Marble Hall in the northeast.

At Slipfontein a fluorite ring was found in a quartz pipe occurring in granite. In addition, at Tooyskraal fluorite stringers replace a quartz vein. In the *Gillspar Mine* at Ruigtepoort, coarse fluorite and remnants of chloritized granite alternate in a large, flat lying lens. Its rim of quartz–chlorite greisen is surrounded by granite. Specularite stringers and nests occur both in fluorite and quartz. A quartz–hematite–fluorite lode appears to be related to the lens. At *Buffelsfontein* fluorite veins form a stockwork, and at Grobbelaarshoek ramifying fluorite veins, traversed by neogene quartz, intersect granophyre.

At *Kromdraai*, 56 miles northeast of Pretoria, green and colorless fluorite is traversed and locally accompanied by hematite. The ore body, surrounded by felsite, measures 120 by 70 yards. At Roodeplaat, near Pretoria, pure white fluorite occurs in an agglomerate of felsite and trachyte. At Wydhoek, in the Pilanesberg complex of the western Bushveld, a 14 ft. wide, banded fluorite lode cuts foyaite. Some of the syenitic bands include apatite and feldspar. Other occurrences include Leeuwfontein and Walmansthal.

At Hartebeestvlei, Hartebeestpoort, Knoppieskraal, Morgenzon and Rietfontein, near the Rooiberg tin field, fluorite impregnates post-Karroo faults.

Western Cape Province and Zululand

At Dyasonsklip, 60 miles east of the southeastern corner of Southwest Africa, 1–3 ft. thick fluorite veins crystallized in a larger vein of quartz-breccia. In the same area, the fluorite of the smaller Blauwskop deposit is purer. Fluorite shows in numerous pegmatites, e.g., at Onseepkans and Styerkraal.

At *Hlabisa*, 40 miles from the coast and 70 miles from Swaziland, extensive stockworks of fluorite and quartz have developed in a 12 mile long and 6 mile wide belt of fissures. The fissures intersect both early Precambrian granites and post-Carboniferous sediments. The veins average a length of 800 yards and have a maximum length of several miles. The Antoinette and Tainton's deposits are located in this belt.

SWAZILAND

The deposits of the *Hlatikulu* district are related to the same young rhyolites as those of Zululand. At concession 31, a 2 mile long fissure, traversing porphyritic granite, has been filled in by at least two fluorite veins. The 2–5 ft. thick main vein strikes N 20°W, dipping 65–85°E. Little fluorspar accompanies chalcedony, opal and geode quartz in the upper portion of the vein, but below a shear, the vein contains up to 90% CaF_2. In its best portion, its thickness varies between 8 and 12 inches with a maximum of 2 ft. Fluorite frequently concentrates on the footwall. A basic dyke forms the hangingwall locally and may have controlled the ascension of fluorine. The smaller vein strikes N 15°E.

A similar 1 ft. wide vein occurs on concession 31. Between Hluti and Mhlosheni other veins have been found in silicified shears.

MOROCCO

Deposits are situated in two areas. The Jebel Tirremi is located in the northeast, west of the Touissit–Bou Beker lead–zinc deposit. Another zone extends from Achemèche, 30 miles southwest of Meknès, to the Jebel Zrahina in the eastern part of the Paleozoic window of Central Morocco, distinguished by lead, antimony and tin deposits.

At Achemèche, fluorite occurs in a narrow, 7 mile long belt of Lower Carboniferous, Viséan limestone lenses, in the aureoles of a deep seated stock. At *El Hammam*, the most favourably located deposit of the Achemèche area, veins and lenses outcrop on both sides of a ridge. On the side of the Oued Beth, 800 ft. long outcrops trend east, striking north-northeast on the flank of the Oued Bout Nokret. The ore, averaging 90% CaF_2, is accompanied by quartz, calcite and iron oxides. Late quartz, galena, pyrite and pyrrhotite also occur in the adjacent fluorite veins of Bergamou, Gouaiada, and Moufrès South.

The veins and lenses of the *Jebel Tirremi* occur at the faulted contact of a small, Middle Liassic (Lower Jurassic) dome of limestone and dolomite. They are related to apatite bearing aiounite and fasinite of the Alpine orogeny. The veins strike north and east. They contain 85–90% CaF_2, replacement ore bodies showing higher grades. Remnants of the country rocks have been partially replaced. The paragenetic sequence includes calcite, fluorite and baryte.

TUNISIA

Replacement deposits of fluorite occur in the Lower Jurassic, Liassic Djebel Staa, and Djebilet el Kohol, in the Upper Jurassic, Tithonian, Hammam Zriba and in veins of the Djebel Oust, Hammam Jedidi, all in the Zaghouan area. The *Staa* ore bodies, extending for 60–200 ft. and attaining 3–6 ft. of thickness, are located along east striking fissures. At Zriba, the upper limestone bed has been replaced along a distance of 1,700 ft. and a thickness of 3 ft. This ore body contains 50% CaF_2.

THE REST OF AFRICA

In the south-central Sudan, violet and colorless coarse fluorine occurs at Jebel Semeih, south of El Obeid. At Bihendula, in northern Somalia, fluorite is found in a quartz vein. Other occurrences include Dananjiekh, Gol Ebed, Boqda and Daimoleh Wein.

In Nigeria, fluorite accompanies sphalerite and galena in the Akwana and Arufu deposits. Fluorite also shows in the tin fields of Jos. In the *Congo* two generations of violet and green fluorite veins, traversing granite, have been distinguished in the Mokama tungsten–tin deposit. Fluorite also occurs in the limestones of Wanie Rukula in the northeast, and in the Upper Katangian Kakontwe Stage. In Uganda, fluorite shows in the Singo pegmatites and in the Buswale and Mubende granite. In the Nyanza district of southwestern Kenya, fluorite veins have been found in carbonatites and gold fields, and at Rata.

In the Chilwa Island carbonatite, Malawi up to 200 ft. long and several ft. wide bands of fluorite have been found. In western Mozambique, fluorite occurs south of Mt. Zungi, near Chemba, and, according to unconfirmed reports, also near Macossa. In the Wankie district of *Southern Rhodesia*, two types of fluorite have been distinguished: Euhedral fluorite crystals fill in veins, but anhedral fluorite is associated with quartz and chalcedony. At Okasu, near Otjiwarongo northwest of Windhoek, in *Southwest Africa*, fluorite occurring in greywacke is related to carbonatites and syenite. Other occurrences include the Omaruru district, Bethanie, Garub and Warmbad in the south.

BARIUM

Baryte, a sulphate, crystallizes at low temperatures at the end of the hydrothermal sequence. The purest and whitest baryte occurs in lead (–zinc) districts or veins located in limestones. The principal deposits of this type are situated in the Algerian and Moroccan Atlas. Veins occasionally develop in chert, sandstone, quartzite and lava, e.g., in South Africa and Swaziland. Unfortunately, when baryte accompanies siderite and, rarely, iron minerals crystallizing at higher temperatures, it is generally iron stained. Baryte related to evaporites, mainly gypsum, may even precipitate from vadose water.

The largest deposits of baryte, however, are carbonatites, especially sövite, apatite sövite and apatite–magnetite rock. Strontiferous baryte is believed to crystallize at the end of the sequence. Baryte is abundant in the carbonatite soils of Uganda (Fig. 3).

ALGERIA

Baryte occurs along the whole length of the Tell Atlas and in the galena veins. The principal deposits, however, are (Fig. 93) concentrated in Grand Kabylia, east of Algiers and in the vicinity of the city. The first area includes the workings of Palestro and Bou-Mahni, the

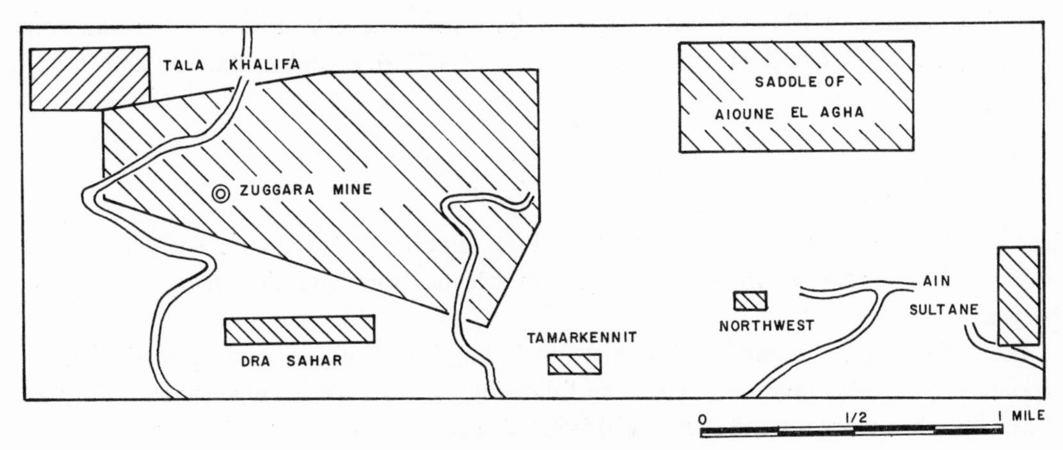

Fig. 93. The baryte deposits of Keddara.

latter now discontinued. Baryte is exploited at Bouzega, Rivet, St. Pierre and St. Paul near Algiers (SCHNELL, 1961b). Baryte veins occur in three rocks (ROCHE, 1962): *(1)* white, lamellar (high grade) baryte in limestone, e.g., at Keddara and Palestro; *(2)* impure baryte forming the gangue of siderite in shales and marly limestones of Gouraya and Mouzaïa; *(3)* hard sugary baryte in the metamorphics of Bou Mahni and Affensou.

Keddara

These deposits include 60% of the country's reserves. All are comprised in a rectangle of 3 by $^2/_3$ mile. The stratigraphic column consists of:

550 ft. Lower Jurassic (Lower Lias) grey limestone
270 ft. Lower Eocene (Lutetian) Foraminifera limestones
330 ft. latest Lutetian alternating sandy limestone and marl; unconformity
1,700 ft. Oligocene sandstone and conglomerate

Three orogenic phases have been distinguished:
(1) Post-Oligocene, east striking anticlinal warps; *(2)* late Lower Miocene thrusts and shears; *(3)* younger, northeast striking bloc faults.

Baryte, accompanied by rare lead and copper sulphides, crystallized in the youngest fissures, traversing limestone at the contact of unconformous layers.

The fissure filling swarms of veins 1–6, Tala Roumi, and des Chênes (Oaks), occur at Zouggara in faults dipping 45–70 °E and intersecting Lutetian–Liassic limestones. The swarms are known to outcrop for 1,700–3,000 ft. The thickness of the veins varies between 1 and 2 ft. Vein 3 includes the largest reserves, followed by No. 4, Tala Roumi and No. 5–6. Baryte of veins 5–6 and of the limestone quarry contains respectively 97.3 and 96.9% $BaSO_4$, 0.9 and 0.1% SiO_2, 1 and 2% CaO and 0.5 and 2% $Fe_2O_3+Al_2O_3$. The s.g. reaches the mark of 4.39.

Other occurrences have been localized in an extension of Zouggara, at the saddle of plot 4935, Ain Sultane, on the northwestern limb, Sidi Yous and also at Tamarkennit, Tala Khalifa and Bou Zegza.

At the Narrows of *Palestro*, baryte replaces the east striking contact of Lutetian limestone and Oligocene Flysch, also favoring a perpendicular line. Occurrences are restricted to the northern limb of *écailles*. Thirty ore bodies average 30 ft. in length and 3 ft. in thickness, attaining maxima of 100 and 10 ft. respectively. Calcite has crystallized in the wall zone. The baryte is of an excellent grade, having a s.g. of 4.41–4.47 and containing almost 98% $BaSO_4$.

Bou Mahni

The Bou Korai baryte body measures 90 by 40 ft., dipping 65° in gneiss. The Chabert vein is 100 ft. long and 10 ft. thick, dipping 20° in schists. Both contain sugary, ferruginous baryte.

The *Affensou* veins occur in micaschists, including lenses of pyrite and graphite bearing cipolinos. The veins favor a crush zone following the contact of augengneiss. They are generally subconcordant and strike N 20°–50°W. The principal sub-vertical lode is up to 40 ft. thick. The lode pinches to a width of 3 ft. along the strike, disappearing at the contact of gneiss. Other veins are 2–25 ft. thick. Baryte exhibits banding in this large deposit, being purer in the centre of the main lode than at the walls. The ore averages 97% $BaSO_4$, 1.6% Fe_2O_3, 1% Al_2O_3, 0.4% CaO.

Near *Gouraya*, fifteen siderite–baryte veins traversing Cretaceous calcschists have been exploited. The subvertical ramifying veins strike N 50°E, attaining up to 7 ft. of width. Banded baryte tends to concentrate at the walls, the end of the veins, and in their upper reaches. Since baryte is lamellar, it unfortunately includes iron oxides ($Fe_2O_3+Al_2O_3$ 2.3%), only 94% $BaSO_4$. The density reaches 4.33.

The *Mouzaï les Minesa* veins are similar. The longest vein of the Nemours sector extends 2,000 ft. along the strike, with widths apparently averaging 3 ft. The ore contains 84.7% $BaSO_4$ and 9% Fe_2O_3 (s.g. 4.30).

At *Koudiat el Madène*, 11 miles from Rivet, baryte, accompanied by hematite and galena, partially replaces fractured Liassic and Lutetian limestones. The limestones are underlain by marl and Permian sandstones, and overlain by sandstone and clay. They occur in the core of écailles, rising from Miocene folds. Hematite bodies strike north along the western contact of the chain. Coarse grained baryte tends to crystallize along the wall of fissures, also impregnating the veins proper. Other veins are also known to exist in the area.

At the dam of *Hamiz* a vein, outcropping along a stretch of 3,000 ft., intersects sandstones. Its width attains 16 inches. Lamellar baryte contains 95.15% $BaSO_4$, its density being 4.33.

At Eldough, concordant veins of baryte occur in gneisses, also accompanying sulphides at Collo and Bouzaréa (DALLONI, 1939). Thick veins of baryte have been found in the copper district of Tenia (Mouzaïa). Abundant baryte is associated with iron at Sidi Safi (hematite), Miliana, Tenès–Cherchel (siderite) and in the Lower Cretaceous, Aptian deposits of Medina and Ouenza. At Tessala, baryte accompanies Triassic anhydrite. Finally, veins of baryte intersect ancient gneisses at Stora and Lower Jurassic, Liassic strata of the Ouarsenis.

MOROCCO

Baryte is widespread in the country, but only easily accessible deposits are exploited. The *Jebel Ighoud*, the continent's principal individual producer, is located in the western Jebilet, 45 miles southeast of Safi on the road to Mogador and 25 miles from Souk es Sebt township. Cambrian limestones rise from a 3 mile long, and up to $^2/_3$ mile wide, band of schists and quartzites. Parallel veins of baryte strike east-northeast in the limestone cores. In the west, the four principal veins extend into the schists. The largest lode, dipping 45°S, forms the southern border of the limestone outcrop. Along a distance of 1,700 ft., the lode bulges to a thickness of (locally) more than 170 ft., its outcrop narrowing to approximately 10 ft. beyond this stretch. Baryte is coarse grained, pinkish and of good grade, containing 96% $BaSO_4$ and less than 2.5% SiO_2. The breccia like filling of the fissure also includes calcite and schist relics. Galena has been sparsely disseminated in the hangingwall. Other sulphides are rare, but other veins of the area carry zinc, copper, and lead containing 0.5% Ag (see 'Silver' chapter, Koudiat el Hamra). A brecciated body of white baryte occurs 600 ft. north of the main dyke and is believed to be its extension.

Tizi n'Tichka is located 70 miles southeast of Marrakech on the Ouarzazate road, at an altitude of 6,500 ft. High grade, white baryte contains 98% $BaSO_4$. The *Tessaout* baryte vein is located on the northern side of the High Atlas. At Mibladen, abundant baryte and galena replace marly (argillaceous) carbonates.

Glib en Nam is located in the manganese area of northeastern Morocco. In other veins such as those of the Tazzeka Mountains, Sidi Lahcen, Djebel Aouam, Imini and Taouz, baryte is included in the lead–zinc–silver paragenesis. Baryte also occurs at Tessaout.

NORTH, WEST AND CENTRAL AFRICA

Tunisia. At Hammam Sedidi, baryte is associated with fluorite in two veins 3–10 ft. thick and several hundred yards long. The mineral also occurs in the Djebel Staa and at Bou Jaber. Baryte accompanies calcite in the lead veins of Rhar ed Deba, Zeflana, Sidi Amor ben Salem and Djebel Agab (GOTTIS and SAINFELD, 1952).

In northern Ghana, low grade baryte has been found in the limestones of the White Volta, near Daboya. In *Nigeria*, baryte is abundant in the lead–zinc belt of Abakaliki-Zurak (see 'Zinc' chapter). Up to 6 ft. thick veins occur at Lefin, in Ogoja Province, Aba-Gbandi,

Akire and Keana in Benue Province and Dumgel in Adamawa. In the lower *Congo*, veins of baryte outcrop at Madimba and on the banks of the Bangu River. Baryte also occurs in the Lueshe carbonatite and in the Copperbelt. In eastern Zambia, in the Luangwa valley and south of Mporokoso, northwest of Lake Nyasa, a 6 inch bed of baryte is underlain by 200 ft. of plateau shales. Baryte rubble has been found south of the Kafwe River at the contact of Precambrian granite and schist. In the Zambezi valley, south of Kafwe, baryte float is related to gypsum.

EASTERN AFRICA

Somalia. In the Bihendula Range, south of Berbera, baryte accompanied by flurotite fills in cracks in gneisses. The Ardahh vein outcrops over 300 yards, averaging a thickness of 2 ft. A sample contains 90% $BaSO_4$, 1.5% SiO_2 and only 0.3% FeO. Other occurrences include Dananjiehh, Gol Ebed and Lau, and also Unkah and Kul in the Horn. Baryte replaces Jurassic–Cretaceous limestones at Arargob, 12 miles southeast of Candala. Baryte and iron oxides impregnate a vertical crush zone traversing the Eocene limestones of Galan Galo, 12 miles west of Candala.

East Africa. At Mugabuzi Hill, east of the Ruwenzori, baryte is associated with hematite lenses in granitic gneiss. However, the soils of the Sukulu carbonatite, northeast of Lake Victoria, constitute one of the continent's largest deposits (see 'Phosphate' and 'Niobium' chapters). In North Valley, highest grades reach the mark of 3.7%. At the head of West Valley better grades appear to exist, maxima being attained at the entrance to South Valley (Uganda).

In the Vitengeni lode, southeastern Kenya, iron-stained baryte is accompanied by lead and zinc. Other occurrences include the Sabaki Valley in the same area, Didimtu, and other districts. In Tanganyika, stringers of baryte traverse the dolomites of Kigoma and Handeni. Baryte also occurs in the sulphide deposit of Mpanda and in the Lupa gold field. In Malawi, baryte is associated with quartz and apatite at Nathace, near Mlanje. The mineral also occurs in the Kangankunde carbonatite.

Madagascar. The veins of Andavakoera, in the north, are ½–7 ft. thick, and extend several hundred yards. Baryte constitutes half of the veins, but the veins of neighbouring Ankitokaza–Andafia are more favorably located. The width of the Bereziky veins attains 3–7 ft. Samples from the Boabasatrana and Andrafilava veins of Andavakoera contain respectively 94 and 97% $BaSO_4$, 4 and 1% SiO_2 and 1% Fe_2O_3 and Al_2O_3.

SOUTHERN AFRICA

At Gamsberg, south of the Southwest African frontier, pink and white baryte outcrops over $1^2/_3$ mile, with its width varying between 20 and 40 ft. The ore body is conformable to manganiferous iron ore in an anticline of quartzite. Baryte contains 91–98% $BaSO_4$, 1–8% SiO_2 and 0.6–2% Fe_2O_3. At Blauwboschkuil, south of Postmasburg, baryte veins traverse lavas. At Blaauberg, in the northern Transvaal, baryte is the last of the vein minerals. High grade baryte of adjacent Goudmyn is of post-Karroo age. At Schoonoord and

Heemstede, on the road connecting Barberton with the Havelock asbestos deposit in Swaziland, ramifying baryte veins replace chert. Greyish white baryte contains 94% $BaSO_4$, 3.5% SiO_2 and 0.75% Fe_2O_3. Other occurrences include the Richtersveld, the Letaba district, Goudmyn and Springfontein.

The veins of the Londosi valley (Swaziland) are 5–6 ft. thick. Up to 4% Zn is associated with the baryte. At Droxford, baryte is related to slickensided quartzites located in early Precambrian schists.

In Southwest Africa, baryte occurs on the Khomas plateau and, with galena, at Grootfontein.

STRONTIUM

Strontium, a twin element of barium, is fairly abundant in the earth's crust, but its concentrations are rare. Epithermal baryte veins consistently contain strontium (e.g., 0.3–0.7% in Madagascar). However, strontianite and celestite impregnating limestones are more frequent. The continent's only former active deposit, Tirrhist in Morocco, belongs to this type. Largest concentrations are found in carbonatites, of which strontium is a typomorphic element. The Ba/Sr ratio rises to 4 in the East African complexes. Strontium bearing calcite, pyrochlore and other minerals are known in several localities. Their distribution shows that strontium has a longer span of crystallization in these deposits than in veins.

Tunisia. At Bazina and Mezzouna, and also at Bechater, Fedj el Adoum, Kebbouch, Dogra Zag and Tir, strontianite and celestite are associated with the sphalerite-galena impregnations bordering Triassic diapyrs. Strontium minerals also impregnate Upper Cretaceous layers in the Gafsa phosphate basin.

Morocco. At Tirrhist in the Jebel Fazzaz, celestite forms replacement bodies in massive Middle Jurassic limestones of Dogger age. The deposit is located southeast of Casablanca and south of Meknès. Fissures of the sub-vertical beds have been impregnated by baryte and finely disseminated pyrite, sphalerite and galena. Pyrite has been redistributed in even finer spherules after the injection of gabbro sills, which also caused the recrystallization of limestones and marl inliers.

Carbonatites. In the Lueshe carbonatite and its vicinity in the eastern Congo, the Ba/Sr ratio reaches 4/1. Strontium is consistently present in the complexes northeast of Lake Victoria. Pandaite, the strontium pyrochlore, has been identified at Mbeya, southwestern Tanganyika. While sövite of Chilwa Island, in Malawi, exhibits high Ba/Sr ratios, neighbouring limestones contain little strontium.

SECTION V

Ultrabasic group

With important platinum, chromium and asbestos deposits, this group is particularly well represented in Africa (Fig. 5). The world's principal deposits of these metals are related to the 2,000 million year old ultrabasic 'dyke' complexes of South Africa and Southern Rhodesia, but other occurrences of these metals, as well as asbestos, talc and magnesite, appear earlier. On the other hand, vermiculite mineralization also culminated 2,000 million years ago, but other carbonatic–ultrabasic occurrences of this mineral are much younger. This leads us to diamonds which appear in pipes and dykes of the same Cretaseous phase.

CHROMIUM

Chromium, a typically African metal, is largely restricted to two or three ultrabasic cycles and facies of southern Africa. The largest high grade deposits of Southern Rhodesia are believed to be 3,300 million years old. Similar disseminated chromite occurs in 2,800 million year old serpentinized dunites of that country. This type of occurrence is also known to exist in Sierra Leone, Madagascar, Togo and Egypt. While the Sierra Leone occurrences may belong to the first cycle, those of Madagascar may be attributed to the second cycle. Limited to South Africa and Southern Rhodesia, the third cycle of low grade deposits is, however, the most important. Before 2,000 million years, the pyroxenites, olivinites and anorthosites of the Bushveld and the Great Dyke were injected successively along rectilinear and arcuate, 50–300 mile long, rifts. Within these ultrabasics chromite seams have been rhythmically interlayered.

SOUTHERN RHODESIA

Near the western edge of the central Great Dyke, *Selukwe*, one of the world's principal chrome deposits, is situated 80 miles east of Bulawayo. Near Fort Victoria, *Mashaba* (Fig. 36, 50) is situated 40 miles east of the Dyke on the latitude of Bulawayo. In the Gwanda region, the *Belingwe* district lies 40 miles east, and *Aer* 40 miles south, of the southern extremity of the Dyke. Chromite layers are included in the 350 mile long *Great Dyke* crossing the country at an angle of N 17°E. The principal deposits tend to occur in its northern complex near Salisbury. From these seams eluvia have also been eroded.

Selukwe

Intruded into Precambrian rocks, a larger dunite body has undergone metamorphism and folding with its host-rock (Fig. 94). The post-mineralization Côte d'Or granite stock constitutes the southwestern limit of a 4 by 2 mile area enclosing the chromitite lenses. Auriferous quartz veins traverse both the granite and the metabasites, further west. The ultrabasics have been completely altered to serpentine, siliceous serpentine, chert talc, carbona-

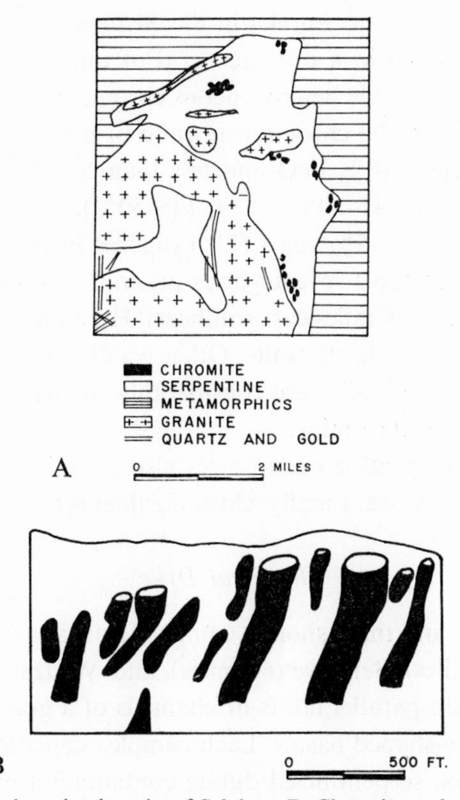

Fig. 94. A. The chromite deposits of Selukwe. B. Chromite pods at Selukwe Peak.

ceous talc and chlorite schists. The silica bearing serpentines and the talc schists, believed to have been metasomatized from pyroxenite, contain all the chromitite, with common serpentines and chloritoschists derived from dunite and gabbro remaining barren. In the talc and talc-carbonate schists, chromite occurs in crystals, clusters and massive bodies. Fat sub-vertical lenses and funnels attain 500 ft. of length, 50 ft. of width, and 100–400 ft. of depth. However, narrow veins do not extend to such depth. Although ramifications are rare, inclusions are frequent. The major chromite bodies are scattered in tight groups of three to twelve *(a)* $^2/_3$ mile northwest, *(b)* 2 miles north, *(c)* $^2/_3$–2 miles east-southeast, *(d)* $^2/_3$ mile south, and *(e)* 2½ miles south-southeast of Selukwe, mainly on Selukwe and Selukwe Peak, and also on Sebanga Poort and Adare. In the mines, lenses strike subparallel within their group. The large lenses contain 47–48% Cr_2O_3, 12% iron oxides, 13% Al_2O_3, 17% MgO, 5% SiO_2, i.e., metallurgical grade ore. Rare heterogeneous bodies of refractory chromite consisting of 38–40% Cr_2O_3, 15% FeO, 14% Al_2O_3, 16% MgO and 9% SiO_2, include a greater quantity of gangue.

High grade metallurgical, metallurgical and refractory chromite respectively attain Cr/Fe ratios of 3–3.3, 3–3.2 and 2.1–2.4. Although the composition of individual lenses varies, the ore is rather homogeneous within each body. While most of the chromite is believed to have formed by segregation, some of the amorphous chromite found in silicified serpentine might be a secondary product.

Other Disseminations

At Mlimo, Mapanzuri, Inyala and Eureka in the *Belingwe* district, isolated irregular and sub-vertical lenses are arranged in a distorted grid of ultrabasic schlieren surrounded by granite. The lenses include variable heterogeneous chrome-picotite. Metallurgical chromite contains 52–53% Cr_2O_3, with the chrome/iron ratio averaging only 2.8/1. On the other hand, refractory ore contains 16% FeO and less than 5% SiO_2. High grade hard lump metallurgical chromite carries 47–50% Cr_2O_3 (45–50%), 15% FeO (20%), 12% Al_2O_3 (15%), 14% MgO (12%) and 5.5% silica (2%) (figures in parentheses refer to chemical grade ore). Cr/Fe ratios for the two categories attain 2.7–3 and 2–2.3 respectively. The chromite trend extends to *Aer* 40 miles east-southeast. Here small schistose lenses suspended in granite contain refractory grade chromite. Other occurrences are located 25 miles south.

At *Mashaba* high grade ore bodies are considerably smaller but similar. In this district the principal deposit is the Prince mine.

Minor talc schist and serpentine occurrences also contain some chromite, e.g., at the Wanderer Mine and near Surprise. Finally, chromite float is found in the Umtebekwe basin.

The Great Dyke

In several cycles one long and three shorter composite ultrabasic complexes — Musengezi (30 miles), Hartley (200 miles), Selukwe (60 miles), and Wedza (50 miles) — were injected along feeders in vertical, sub-parallel faults or channels of a graben. Near the surface they spread in sheets in syncline-shaped basins. Each complex consists of rhythmically repeated groups of magmatic layers: serpentinized dunite containing a maximum of 5% orthopyroxene, overlain by harzburgite (5–50% pyroxene), olivine pyroxenite (5–50% olivine), pyroxenite (maximum 5% olivine) and gabbro. Serpentines exclusively derived from dunite attain a thickness of 3,800 and 2,800 ft. respectively in the Hartley and Wedza complexes. The Hartley complex pitches southward. Consequently, serpentine layers outcrop farther from the synclinal axis. At a depth of 600 ft. 25% olivine has been preserved, corresponding to a Fo_{94} composition.

After consolidation the major cycle of serpentinization took place in subsurface conditions with meteoric water flowing along zones of permeability to a depth not exceeding 1,000 ft. in the absence of faults. Following faulting and the injection of post-Karroo dolerite dykes into the Great Dyke (WORST, 1960) serpentinization proceeded.

The metadunite is overlain by poïkilitic harzburgite, granular harzburgite grading into olivine (Fo_{91}) pyroxenite, and by picrite–troctolite containing olivine (Fo_{86}), biotite, augite and plagioclase. Furthermore, serpentinized olivine embays enstatite altering to antigorite. The harzburgite layers attain a thickness of 1,000 ft. Cumulating 2,400 ft. of thickness in the Hartley complex and 800 ft. in the Wedza complex, serpentine and harzburgite layers separate seven pyroxenite bands from each other. At the top of the 'column' clinopyroxene (augite) appears along with platinum, with pyroxene (enstatite) grains measuring 1–6 mm in diameter.

With their thickness attaining 3,000 ft. at Hartley and 1,200 ft. at Wedza, gabbro and norite sheets have been partly eroded. In these, combined pyroxenes exceed the proportion of feldspars (An_{65-80}). The clinopyroxene is augite. At Wedza heterogeneous quartz-gabbro terminates the sequence.

The outcrop of the Dyke bulges at larger gabbro ellipses, diminishing in width where eroded to lower levels. The magmatic layers of constant thickness form flat sectors of concentric circles. Their dips vary between 10° and 30°, but the granitic country rocks dip at a greater angle. Both dip and longitudinal pitch increase towards the gabbro masses. In its northern 10 miles the Hartley complex is less than 3 miles wide, consisting of a narrow outer pyroxenite band, the rest being serpentinite and harzburgite. Along the following 70 miles pyroxenite covers the axis of the Dyke and more than half of its total outcrop. Gabbros appear 10 miles to the south. Quartz-gabbros occupy considerable areas of the Wedza complex, but here serpentine outcrops are limited in extent.

The Dyke has been broken by transverse faults and slightly tilted in its sections, also affecting its western satellite, the Umvimeela Dyke, but not extending into the country rocks. On the other hand, longitudinal faults are less prominent.

Genesis of the Great Dyke

Geologists believe that magma was injected in several cycles from chambers arranged along the present alignment of the complexes through feeders, spreading then in balconies or sills along sub-horizontal planes of parting. Layers consolidated before the injection of the next cycle, except for the later phases of acceleration when standard primary destratification also took place. The outflow of heavy magma caused saucer type subsidence followed by the development of a graben, especially in the vicinity of the heavily weighted feeders. In fact, the layers tend to pitch towards the feeder areas now represented by four gabbroic zones. At Hartley, the longest and widest of the complexes, even the thickness of layers is greater than in the other complexes. The number of layers is also variable. Trough faulting caused sagging of the horizontal sills, producing the present synclinal structure which extends to a depth of 7,000 ft.

The Great Dyke has several satellites, namely the 20 mile long Zambezi complex in the north, which includes chromite seams, and four minor complexes in the south.

Chromite in the Great Dyke

The inward drop (dip) of lower seams increases towards the contact of the Dyke, e.g., from

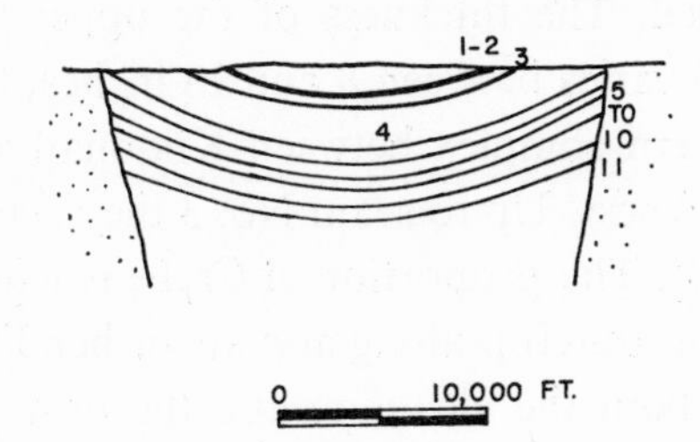

Fig. 95. Chromite seams (1–11) of the Great Dyke.

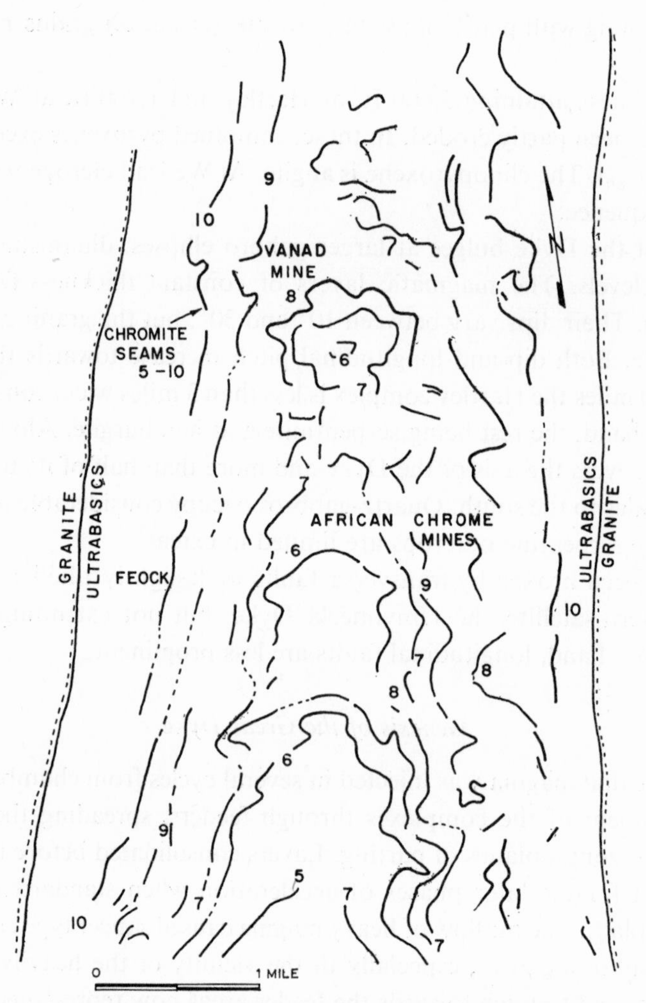

Fig. 96. The northern part of the Great Dyke.

5° to 25°. With their thickness varying only on a local scale, the seams are known to extend more than 200 ft. inwards. In the Wedza complex, chromitite of seam No. 2 grades into the overlying harzburgite (Fig. 95, 96), disappearing rapidly, however, in the underlying serpentine. Prior to the downwarping of the complexes, chromite was possibly settled from a separate injection. Nevertheless, chromite crystals are enclosed in pyroxene and olivine of the harzburgite. At the end of the sequence, the segregation of chromitite forming the hangingwall seams overlapped the consolidation of the dunite.

Chromite is sparsely disseminated in the complexes. At regular intervals, up to eleven stratiform chromite seams settled. The thickness of the upper layers of normally chemical grade chromite (49% Cr_2O_3) varies between 8 and 18 inches; their Cr/Fe ratio attains 2.3. Well developed in the northern complex between Kildonan and Vanad, the lower seams attain only 2–6 inches of thickness. Up to seam No. 3 they achieve metallurgical grade and a Cr_2O_3 grade of at least 48%. The proportion of Cr/Fe reaches 2.8. In the present erosion pattern of the Dyke, the seams outcrop along arcuate or bending contour lines.

The magnesia content of both the chromite and the host dunite increase downwards. Moreover, the Cr/Fe ratio increases with chromic oxide content. In the Hartley complex the composition of seams No. 5–6 and No. 8–10 is similar. However, late hangingwall chromite contains less chromium and magnesium, and more iron than the following seam No. 1. Up to 40% of the ore is friable. Typical compositions are given in Table CXXXII.

TABLE CXXXII

COMPOSITION OF CHROMITE

Grade of ore	*Cr_2O_3 (%)*	*FeO (%)*	*SiO_2 (%)*	*MgO (%)*	*Al_2O_3 (%)*	*Cr/Fe ratio*
friable metallurgical	48–50	14	5	17	10	3–3.1
friable chemical	44–47	18	6	14	12	2–2.1
lump chemical	42–46	18	6	14	12	2–2.1

Since the host layers contain as much as 3% chromite, experts believe that large eluvial deposits were derived from disseminated chromite. In transversal valleys and streams, the thickness of soil containing 3–35% fine grained (±3 mm) chromite varies from 2 inches to 2 ft. along the contacts of the Dyke, and averages 18 inches. The composition of disseminated chromite differs from the layered chromitite: Chromium values are very high, but silica content is particularly low; e.g., 53–55% Cr_2O_3, 19% FeO, 2% SiO_2, 8% MgO, 13% Al_2O_3 and Cr/Fe: 2.3–2.5.

SOUTH AFRICA

Although chromite is also disseminated in peridotite at Isitilo and Tugela Rand (Natal), and in serpentinite in the Pietersburg district, the principal deposits are situated in the Bushveld (Fig. 46, 50).

The Bushveld

As stated in the 'platinum' chapter, chromitic pyroxenite, anorthosite, norite, porphyritic pyroxenite, pyroxenite pegmatite, gabbro, ferro-gabbro, dunite and magnetite were injected subsequently and rather frequently in discordance as separate intrusions along arcuate channels (COERTZE, 1958). Near the surface the ultrabasics spread around the central Bushveld in slightly basined sills or synclinal structures, similar to the Rhodesian Great Dyke. Possibly reacting to the emplacement of the heavy mafic layers, the centre was updomed and later occupied by the granitic intrusions (COUSINS, 1959a, b).

We can distinguish four troughs and sills: *(1)* bending in the southwest in a half circle from Pretoria to Thabazimbi, with *(2)* its western basin extending in the Marico district toward Bechuanaland, *(3)* the slightly curved arc extending in the east from Lydenburg towards Potgietersrus, and *(4)* the separate cross-cutting outlier north of the Bushveld (KUPFERBÜRGER and LOMBAARD, 1937).

In the 3 mile wide *Eastern* belt commercial chromite deposits extend for 70 miles from Jagdlust to Kafferskraal and in the western belt for 100 miles from Elandsfontein to Elandskuil. From west to north the *Western* belt consists of the *(a)* Marikana–Rustenburg, *(b)* Rustenburg–Boshoek (Fig. 99), *(c)* Pilanesberg and *(d)* Bier Spruit sectors. From south to north the eastern zone encompasses the *(a)* Dwars River, *(b)* Steelpoort–Mooihoek, *(c)*

Groothoek–Mecklenburg and *(d)* Chromite Hill sectors. In the Marico structure four chromite sills are interlayered. However, we find gaps of mineralization *(1)* north and east of Pretoria, *(2)* west and north of Potgietersrus, *(3)* for 50 miles north of Wonderfontein, and *(4)* west of the Olifants River.

Chromitite

Geologists have found up to 25 chromite seams in the mafic intrusions from the Merensky Reef downwards to a depth of 4,000 ft. (FOURIE, 1959). Beyond this sequence chromite is rarely disseminated. We can divide the seams into three groups, as is indicated in Table CXXXIII.

TABLE CXXXIII

DEPTH OF CHROMITE SEAMS BELOW MERENSKY REEF

Locality/group	*Upper*	*Middle*	*Lower*
Eastern belt, north	?	(1,400) ft.	3,000 ft.
Marikana–Rustenburg	600–800 ft.	1,000–1,800 ft.	1,100–2,200 ft.
Rustenburg–Boshoek	500–900 ft.	1,200–2,500 ft.	2,100–3,100 ft.
Bier Spruit	200 ft.	1,300–1,600 ft.	2,200–2,800 ft.

In fact, the Lower and Middle Groups lie near each other, generally in pyroxenite at the base of the pyroxenite–norite (critical) zone. On the other hand, the Upper Group is more frequently interlayered with anorthosite and norite than with pyroxenite. An anorthosite layer underlies the seams rather often, with pyroxenite overlying it. Moreover, chromite tends to be disseminated in the contact layer of the host-rock if *over*lain by pyroxenite or if *under*lain by anorthosite (Fig. 97). In the hangingwall values generally continue beyond protruding domes. Chromitites which are interlayered with anorthosite rocks have suffered more dislocation than in pyroxenite. Indeed, a chrome-pyroxenite, following the injection of the first norite–anorthosite, appears to be the host-rock of most of the chromite (cf. 'Platinum' chapter). Normally dips vary between 4 and 12°, but at Rustenburg dips attain 10–25°. Basin and secondary synclinal folds also contribute exceptions.

South African ore generally contains less chromium than Rhodesian ore; it is also more friable. As in the Great Dyke, the quality of the ore generally decreases upwards in the sequence. Silica content, however, is low. Except for Ruighoek, Buffelsfontein, Kroonendal, Elandskraal, Vogelstruisnek, Mecklenburg and Swartkop, ore of chemical grade is rare. In the Eastern Belt, Cr_2O_3 content of the Lower Group varies between 42 and 46%, and in the Western Belt between 44 and 50%. On the other hand, the Cr/Fe ratio ranges between 1.8 and 2. Middle and Upper Group chromites – in fact, chrome-picotites – contain 32–46%

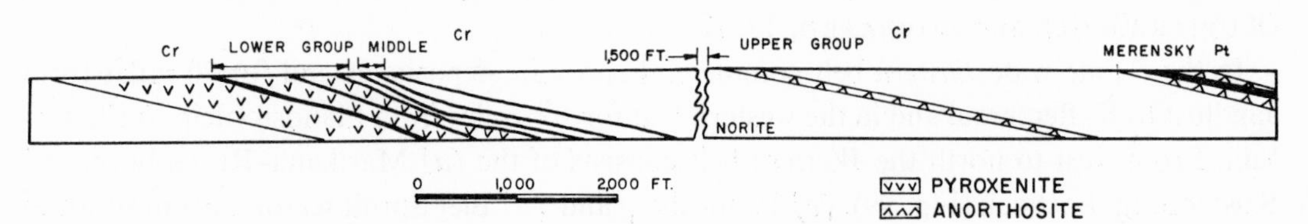

Fig. 97. Schematic south (left) – north(right) section through the Bushveld ultrabasics.

Cr_2O_3, with the proportion of chromium and iron varying between 1.5 and 1.75. But the average grade of Bushveld chromite is even lower: 60% of the exploited ore contains less than 44% Cr_2O_3, with 30% attaining a grade of 44–48%, and less than 10% a grade of over 48%.

The Eastern Belt

In this belt only the Lower Group is well developed, consisting of two to six seams in the 'old' chromitic pyroxenite. The Main Seam probably extends for 40 miles, but at its southern end the Group is found in a variable environment. At Tweefontein (Dwars Sector) the principal seam attains 2 ft. of thickness. Higher in the sequence an 18 inch thick seam is of lower grade. Possibly because of a change in the dip, the principal seam of Winterveld (Steelpoort Sector) dipping 12°, thickens to 41 inches. Later basic dykes and sills were injected. Attaining a width of 15 inches, the upper, more siliceous seam is underlain by 2 ft. of pyroxenite or gabbro. Pegmatite and a dunite pipe traverse the 3 ft. thick Onverwacht Seam. At Twyfelaar (Groothoek Sector) several tectonized horizons carry chromite.

At Waterkop the belt bends northwestward. On the northeastern slopes of Chromite Hill the following sequence outcrops: a 4–6 ft. thick chromite seam, a pyroxenite seam 2½–5 ft. thick, and a 1 ft. thick chromite seam. Both above and below them thinner seams are interlayered. However, seams attributed to the Middle and Upper Groups are indistinctly developed. At the Dwarfs Bridge chromite is entirely interlayered in anorthosite. At Onverwacht the outcrop of the two groups, disturbed by titanomagnetite inclusions, is at a distance of a mile, and it widens further north of Maandagshoek. Finally, seams attributed to the Middle Group emerge on the southwestern flank of Chromite Hill and further northwest in pyroxenite.

Rustenburg

The *Western Belt* is particularly well developed in the 60 mile long southern Rustenburg sectors, and in its northernmost 15 miles. The Main Horizon of the Lower Group has been divided. At Bultfontein, Boekenhoutfontein, Kroonendal and Rietfontein near *Rustenburg*, we can divide the Lower Group into a Main Horizon and Leader Seam averaging respectively 3 ft. and 1–1½ ft. in thickness. They are separated by 8–18 inches of pyroxenite. A couple of ft. beneath the Main Horizon, three to six thin seams of the Roamer Horizon extend over considerable distances. At Kroonendal we can distinguish fifteen seams in the Lower Group. At Boekenhoutfontein the Main Seam, containing 39% Cr_2O_3, thickens to 6 ft. along a strike length of 20 ft. Chrome-pyroxenite and pyroxenite separate the Main Seam from the Leader Seam (Fig. 99).

The Lower Group concludes the sequence of the chrome-pyroxenite: in twin horizons of variable anorthosite, indistinct elongate lenses and stringers of chromitite are located. At Kroondal, rather steeply dipping layers include 1 ft. of chromitite, chromiferous anorthosite (11 inches), chromitite (8 inches), an anorthosite horizon (10 inches), and 4 ft. of chromitite and norite. Geologists have correlated the four Waterkloof seams with those of Kroondal. However, at Kookfontein thinner seams of chromitite have been trenched. Furthermore, at Haakdoorndrift magnetite penetrates from a dunite pipe into the chromitite along the layering and in stringers. Although chromitite occurs as high as the platinum bearing porphyritic pyroxenite forming the footwall of the Merensky Reef, most of the chromite of the

Upper Group extends in seams and disseminations in and near a varied band of anorthosite. Underlying this indicator five thin seams can be correlated at Pritchard's Workings, Kroondal, and east of Rustenburg Station.

The Northwestern Belt

The *Pilanesberg* sector extends for 15 miles. At Boschhoek a 30 inch thick seam is interlayered in pyroxenite. At Palmietfontein 7 ft. of pyroxenite divide two 18 inch thick seams. Resembling the Roamers, experts believe that the Magazine Seam can be correlated to the Rustenburg Main Seam. The Vogelstruisnek seams average 1 ft. of thickness, and contain 46% Cr_2O_3. On the other hand, at Groenfontein the lowest of three seams, attaining a thickness of 3 ft., contains 44% Cr_2O_3. Here chromiferous titanomagnetite forms the footwall of the third seam. Nearby at Ruighoek chromitite is overlain by pyroxenite. Finally, at the northern end of this sector numerous seams occur in pyroxenite between Witleifontein and Cyferkuil.

Emerging after a chrome gap the *Bier Spruit* sector extends 20 miles. At Varkvlei and Nooitgedacht a rather thin seam of the Lower Group outcrops. Similarly, the Middle and Upper Group seams are also thin. At neighbouring Zwartklip, 4 ft. thick seams of the Upper Group are located 100 ft. below the Merensky Reef. After a gap of 10 miles six seams of the Lower Group are divided by 60–200 ft. of pyroxenite between Schildpadnest and Swartkop. Here too the Magazine Seam can be correlated with the Rustenburg Main Seam, as were the Leaders and Roamers. However the seams do not attain a thickness of 1 ft. At the contact of the pyroxenite and anorthosite, 6 miles northeast the number of Middle Seams increases between Schildpadnest and Haakdoorndrift from 1 to 5 and from a dissemination to a cumulative thickness of more than 10 ft. Only 250 ft. beneath the Merensky Reef a 6 inch thick seam represents the Upper Group. Indeed, at Haakdoorndrift six Upper Group seams outcrop 250 ft. below the Merensky Reef.

MALAGASY REPUBLIC

In the northern part of the Malagasy Republic moderately large chromite deposits occur, principally in talc schists. While the chromiferous metabasites are surrounded by rocks of the 1,800 million year old Vohibory System, nickel tends to occur in ultrabasics emplaced in the 2,600 million year old Graphite System.

Andriamena, the principal Malagasy chrome deposit (Fig. 98), is located on the High Plateaus 110 miles north of Tananarivo and 80 miles west of the Lake Alaotra railway. Between pyroxenites and amphibolites in the east, and migmatites and gneisses in the west, ultrabasics have been emplaced in a 30 mile long and 15 mile wide belt. We can distinguish three facies (GIRAUD, 1960):

(1) In a 250 by 150 mile area, bodies of diorite, gabbro and ultrabasics, affected by dynamometamorphism and recrystallization, represent the host-rocks.

(2) Occupying large areas, more metamorphic but pre-migmatic derivatives of these: pyroxenites, amphibolites and plagioclase gneisses.

(3) Containing actinolite, tremolite, chlorites, talc and chromium, soapstone lenses of up to 600 ft. favour sheared and altered zones.

These facies have evolved between more than 2,600 and 1,800 million years, thus covering the period extending between the formation of the Mashaba–Belingwe type of deposit in Rhodesia and the Great Dyke–Bushveld phase. The ultrabasics occupy part of a synclinorium extending further south among early Precambrian polymetamorphic rocks (see 'Graphite' chapter). Synorogenic faults strike N 20°E, followed by later parallel and transverse faults.

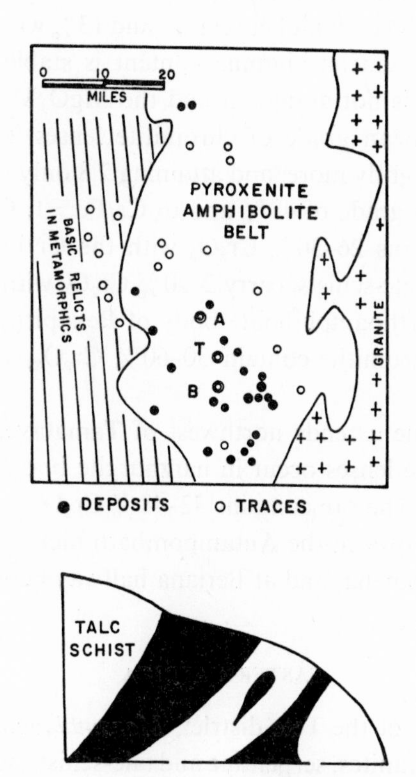

Fig. 98. The chromite deposits of Andriamena (*A*), Bemanevika (*B*) and Telomita (*T*). (After GIRAUD, 1960).

We can distinguish four stages of evolution or hydrothermal metasomatism of the soapstones: *(1)* partly tremolitized amphibolite, *(2)* actinolite–tremolitites, *(3)* clinochlore–pennine schists, and *(4)* talc-schists derived from tremolite and/or pyroxenes, also containing zoïsite, hornblende and pyrite. These metabasites contain only 0.1–0.5% Cr_2O_3, 0.01–0.06% NiO_2, and 0.1–1.5% TiO_2. However, during the main metamorphic cycle, hydroxyl affecting the alteration of the ultrabasics was probably derived from the country rocks, migrating into sheared or low pressure zones. Their products, interlayered lenses of chromiferous talc-schists (and chlorite-schists), attain a maximum length of 300 ft. and a thickness of 30 ft. Experts distinguish six types of mineralization: *(1)* random and *(2)* oriented or layered disseminations, *(3)* undulating aggregates or stringers, all in talc-schists, *(4)* massive bands of chromite in the talc-schist lenses or *(5)* at their contact with pyroxenites and gneisses, and *(6)* lenticular subconcordant chromitite in pyroxenite.

South of Telomita, the *Bemanevika* lens of high grade chromite extends more than 1,000 ft. About 3 miles south of Ambodiketsa two chromitite lenses have the following dimensions: 7 by 2 ft. and 13 by 7 ft. Furthermore, joints filled by vermiculite and tremolite cut a sheet of 25 by 5 ft. The talc-schists enclosing the chromitite are surrounded by biotite-amphibole migmatites. Northeast of Manakana, and south of Ambodiketsa, similar bands of chromitite are interlayered with talc tremolitites. While colluvial boulders of chromitite are scattered on lateritic surfaces, the ore also concentrates in placers. The composition of the highly magnesian chromite appears to vary with the degree of metamorphism.

Whereas in the west, in the poor Maevatanàna occurrences, tremolite is typomorphic, at Andriamena the indicator parent minerals are orthopyroxenes. Here Cr_2O_3 content varies between 49 and 53%, and MgO content between 11 and 13% with the Cr/Fe ratio oscillating from 1.7 to 2.7. But at Andriamena alumina content is stable and the Cr/Cr+Al ratio attains only 0.8. Ferric iron is not abundant and the MgO/MgO+FeO ratio varies only between 0.4 and 0.5. The mean grade of chromitite lenses is 30–35% Cr_2O_3, with the Cr/Fe ratio averaging 2 or slightly more, and attaining 2.8 only rarely near the surface. The ore can be concentrated to a grade of 48% and to Cr/Fe>1. On the other hand, *Maevatanàna* chrome-picotite contains 26–30% Cr_2O_3 with the Cr/Fe ratio averaging 0.5! The chromitite lenses, located in talc-schists, carry 2–20% Cr_2O_3 with the Cr/Fe ratio reaching a maximum of 1.2/1. In the orthoamphibolite body of Bekapirijy (south of Maevatanàna), several lenses of magnetic chromitite contain 30–60% Cr_2O_3, with the Cr/Fe ratio averaging only 0.7/1.

Another chromiferous zone extends northwest of Tamatave. At *Ranomena*, 20 miles from the harbour, chromitite lenses occur in harzburgite bodies interlayered with middle Precambrian metamorphics. The ore contains 42–48% Cr_2O_3, with the Cr/Fe ratio attaining 1.1–2. Chromite also shows in the Antampombato nickel ores, west of Tamatave, at Ambodiriana north of Ranomena, and at Beriana halfway between Andriamena and the northern tip of Madagascar.

EASTERN AFRICA

In the Nhantreze Mountains of the Tete district, *Mozambique*, chrome-picotite (23% Cr) occurs in peridotites among granites, serpentine and talc-schists, with gold and platinum concentrating in alluvia. At the confluence of the Murropaci and Namirrue Rivers, the Munuca occurrence is related to northeast striking north-Mozambiquian ultrabasics. At Nkenza, in southwestern *Tanganyika*, small masses of chromite are disseminated in serpentinite. At Twamba, near Karema east of Lake Tanganyika, chromite forms aggregates of nodules in talc. In Kenya chromite (containing 54.6% Cr_2O_3) shows on Mount Sekerr at Baragoi, Southern Embu, and Dobel in the north. North of the Moroto Mountains in northwestern Uganda sheared and boudinaged chromitite veins now form pods in serpentinites. These serpentinized ultrabasics are emplaced in amphibolites and gneisses, but the serpentinites include chromitite only in a 4 mile long belt. The ore generally contains antigorite, talc and 2–5 dwt. platinum per ton.

Chromite and nickel are disseminated in the ultrabasic complexes of the Hargeisa district in northwestern Somalia, and in a 15 mile long north-northeast striking serpentine belt following the eastern bank of the Barca River as far as the Anseba in northern Erythrea, *Ethiopia*. We can distinguish eight chromiferous patches here. At some distance, in the *Sudan*, chromite occurs at Qala en Nahl.

In the 10 mile long, serpentine, talc–carbonate schist range of Baramiya–Ras Shait-Salatit of southeastern *Egypt*, chromite forms lenses attaining 30 ft. of length, chromite being also disseminated in the metabasites (AMIN and AFIA, 1954). Similar segregations are found at Mudargag, Muelith and Abu Dahr. Peridotites and dunites were first emplaced in the metamorphic basement, and then were altered. In the alteration rim of peridotites, fat oriented chrome-picotite lenses occur, mainly associated with talc and carbonates. Among a few hundred, the longest lenses attained 75 ft. This spinel contains 30–45% Cr_2O_3. Similarly, at Hafafit chromiferous serpentine lenses are embayed by quartz–feldspar gneisses. In the same district we can see anthophyllite asbestos and vermiculite.

WESTERN AFRICA

In a 250 mile long north-northeast trending alignment the earliest Precambrian range of Mount *Togo* cuts across southeastern Ghana, Togo and northern Dahomey. In the east norites, gabbros and serpentinized peridotites border this chain (ARNAUD, 1945). In *Togo* the principal ultrabasic bodies are located between (Fig. 22, 54) Palimé and Atakpamé at Djabataouré and Tchitchao, and between Láma–Kara and the Dahomeyan frontier. Similar intrusives occur in the Atacorian Akwapimian, at Anum in *Ghana*, at Ayagbé north of Palimé and Atakpamé in Togo, and in the Buemian at Bontomo in northwestern *Dahomey*. Similarly, the serpentinized ultrabasics of Anum (Ghana), Moliendo, Ahito, Patégan, Djéti and Evon (Togo) contain chromium.

In Togo, 80 miles from Lomé, *Mount Ahito* is bordered on the east by gneisses, amphibolites and quartzites. Here a siliceous serpentinite occurring in metamorphics is interlayered with chlorite schists. Magnetite and chrome-picotite are disseminated in serpentinite, with the latter also forming 3–10 ft. thick lenses. Eluvia contain 41–43% Cr_2O_3, primary ores 39–42% (Cr/Fe = 3). The Mount Djéti serpentinites, which contain picotite pockets, form a wedge in amphibolites and some actinolite–dolomite schists. At Tiélé and Bontomo, in Natitingou County of *Dahomey*, large elongated serpentine lenses, bordering upper Precambrian jaspylites and quartzitic sandstones, contain some chromite.

Chromite has also been found at Nzérékoré on both sides of the Guinea/Liberia border, and in the laterites of the Kaloum peninsula, Guinea, where the ore is derived from dunite.

At Hangha, 180 miles by rail from Freetown, the capital of *Sierra Leone*, the sub-vertical Kambui schist belt trends north-northeast. A 200–300 ft. thick band of gneisses is surrounded here by 6,000 ft. thick hornblendites. Injected along a shear zone, lenticular serpentinized dunite bodies, attaining a thickness of up to 250 ft, include scattered chrome-picotite and normally sub-vertical chromite bodies (DUNHAM et al., 1958; Fig. 17, 54). In a chain structure, three disseminated chromite bands are arranged with chromitite bodies exhibiting indistinct primary stratification. The 3–20 ft. thick ore bodies are believed to have formed in several cycles before a magmatic front; the post-metamorphic, co-shear ultrabasics did contain chrome clinochlore, serpentine and brucite. During a subsequent sodic phase of pegmatitization chromiferous tremolite and muscovite were formed through the leaching of chlorite, along with anthophyllite, talc, phlogopite, some quartz and feldspar. Whereas phlogopite altered to vermiculite, and other micas and feldspars to clays, chromite and anthophyllite have resisted supergene alteration. Nevertheless, we can still find some interstitial talc in chromite containing 43–48% Cr_2O_3.

At *Bender–Yawie*, in the Gori Hills of Sierra Leone, an independent chromite inclusion occurs in granite. At some distance the Jaluahun chromite–mariposite lens outcrops in gneissose granite.

The ultrabasic belts of the Hoggar, southern Algeria, probably carry chromium. Moreover, chrome-picotite is disseminated in the serpentinites of the Graara–Bou Azzer, Morocco, and further east in the 150 ft. long and 3 ft. thick lens of the Jebel Inguijem.

NICKEL

Interest in African nickel deposits has been revived lately. Although the metallogeny of nickel resembles the genesis of the older type of chromites in Madagascar and Africa proper, concentrations of nickel are apparently more related to processes of serpentinization. Furthermore, platiniferous layers of ultrabasics contain some nickel in southern Africa, with concentrations increasing in late diathremes. Ultrabasic pegmatites coexist with these, acid pegmatites occurring close to Malagasy (and New Caledonian) deposits. Nickel mineralization culminated between 2,600 and 1,800 million years.

MADAGASCAR

Nickeliferous ultrabasics are intrusive into two major areas belonging to two epochs. The first category includes the 150 mile long and several miles wide band of rocks of the more than 1,700 million year old Vohibory System extending from Lake Alaotra, south of Ambatovy. The second, or Valozoro, group is located 150 miles south of Tananarivo in a zone of graphitic gneisses attributed to the 2,600 million years old Graphite System. The chromiferous ultrabasic lenses extending north of Tamatave also include harzburgite (pyroxene-peridotite) bodies containing 0.2–0.4% nickel. West of Manakara nickel bearing laterites cover the Manama peridotite. Finally, west of Ambatovy nickel also shows at Ampangabe on the Javo River.

The large gabbro-syenite stock of Amtampombato is situated in the Moramanga district, sixty miles southwest of Tamatave Harbour. Post-tectonic peridotitic differentiates of the intrusive are superficially altered and covered by durcicrusts at *Ambatovy* and Anamalay (DELBOS and RANTOANINA, 1961). Fissure filling veins of garnierite and stringers of chromite traverse peridotite outcrops with some cobalt showing in heterogenite. The lateritized peridotite contains 1.16% NiO. The durcicrusts are surrounded by 1.8 square miles of thick lateritic clays. The upper 3 ft. of the clays average 0.97% Ni, 1.53% Cr, 0.045% Co, 0.99% Ti, 42.34% Fe and 4.77% SiO_2. Since it increases with depth, nickel content attains 1–1.7% NiO at 20 ft., and even 2.3% lower. The durcicrusts of Ambatovy average 6½ ft. of thickness, containing 0.97% (0.1–3%) Ni, 0.61% Cr, 1.96% Ti, 47.07% Fe and 11.52% Al_2O_3. Southeast of Ambatondrazaka, Ambalotarabe or Nickelville is located between Ambatovy and Lake Alaotra. This alteration zone of a small dome of peridotite includes garnierite and nickeliferous talc.

Situated 20 miles south of Ambositra, the *Valozoro* harzburgite stock, occupying 120 acres, is intrusive into pyroxene gneisses, mica-schists and quartzites. In this area sub-vertical peridotite outcrops trend north-northeast. Enstatite and olivine have been altered to serpentine, and subsequently weathered into lateritic clay. The nickel content of harzburgite attains 0.45%. Water has penetrated and serpentinization has progressed along fissures and joints, with the porous rock averaging 2.5% Ni. Geologists attribute this concentration to the leaching of silicates and to the replacement of dissolved magnesia by nickel (MASCLANIS, 1956). Indeed, green and yellow serpentine contains 2–3% Ni.

Serpentine affected by argillization consists of red clay and relics of green serpentine. The nickel content of this product oscillates from 1.5% to 2.5%. This alteration zone has two facies. A layer of variable thickness (0–30 ft.), generally in sharp contact with the low grade harzburgite, consists of homogeneous red or brown argillaceous serpentine. In this layer nickel values tend to be high because relics of the ultrabasics are scarce. The second facies, attaining a thickness of 3–15 ft. consists of lateritic clay, embaying ½–5 ft. wide boulders of nickeliferous serpentine or low grade peridotite, which are surrounded by some serpentine. Boulders are abundant, their number decreasing locally. This facies generally overlies the preceding one. Finally, the transition zone is covered by red talc and chlorite bearing clays distinguished by leaching and variable grades of nickel (0.2–2.5%), averaging 1.2%.North and east of Valozoro nickel also shows at Fandriana and Ampasary.

SOUTH AFRICA

The principal deposits include two or more horizons of the Bushveld complex, comprising the Merensky Reef and Vlakfontein, and the Insizwa district in eastern Griqualand (Cape Province). Trevorite (nickel magnetite), partly altered to the nickel–magnesium silicate neponite, impregnates the contact zone of the Jamestown complex, concentrating particularly in green schists at *Sheba* Bridge 15 miles north of the Barberton gold centre. Nickel also occurs in the Witwatersrand banket and in veins on the Ngwekwene River in Zululand, 100 miles north of Durban, and in many other places.

At Kroonendal, Klipfontein and nearby Rustenburg, nickel–copper sulphides are by-products of Merensky Reef platinum (see 'Platinum' chapter). Although the nickel content has been estimated to be less than 0.3%, this is the major source of nickel in Africa. At Middellaagte pentlandite, pyrrhotite and chalcopyrite are also disseminated *beneath* the Merensky horizon with values ranging from 0.4 to 1.7%.

Between *Vlakfontein* and *Bierkraal* nickel occurs in pipes traversing the layered ultrabasics of the western outlier of the Bushveld. There are similar structures in the Pretoria, Marico, Lydenburg, and Groblersdal districts, and also in a hortonolite dunite pipe at Klipfontein. At Vlakfontein a pipe extends to a depth of more than 500 ft. This ore body includes pyrrhotite with 3% Ni, (plus 1% Cu) but platinoids, gold and silver are scarcer. The accumulated residual sulphides and the pegmatites have been injected, in clusters of diathremes, into the consolidated gabbro, harzburgite, bronzitite and other ultrabasics along zones of weakness. Moreover, the capping of the pipes has also been altered to opaline gossan at Bakhoutrandje, Tweelaagte, Vogelstruisnek, Davidskuil, Liliput and Groenfontein.

Around Mount Ayliff of eastern Cape Province, 100 miles southwest of Durban, an undulating, 1,000–3,000 ft. thick, basin-like layered sheet has been injected into Lower Beaufort (Middle Karroo) sediments. Remnants of olivine hyperite, picrite, troctolite and alternating olivine and quartz-gabbros rise in the *Insizwa*, Ingeli, Tabankulu and Tonti ridges. Finally, granitic xenoliths and contact rocks close the sequence. At the bottom of the olivinites disseminated pyrrhotite and chalcopyrite include a maximum of 2% Ni, 1% Cu and 1 dwt. platinoids per ton (1.6 g/ton). In the same basal horizon pyrrhotite–sulphide–

silicate veins, sheets and masses contain 3–6% Ni, 1.5–3% Cu, 4 dwt. platinoids per ton with chalcopyrite–cubanite bodies carrying 4–5% Ni, 15–20% Cu, 0.3–0.4% Co, 6–30 dwt./ton of platinoids, and 0.2–5 dwt./ton gold.

THE REST OF AFRICA

In Swaziland, copper and nickel are associated in two horizons of the intrusive layered gabbro of the asymmetric Usushwana complex. At Magogaphate, between the Macloutsie and Shashi Rivers, 65 miles southeast of Francistown in Bamangwato country of northwestern Bechuanaland, ultrabasics have formed in early Precambrian amphibolites and schists. A 2,000 ft. long belt of nickeliferous limonitic and earthy gossan covers the serpentines here. At a depth of 100 ft. the nickel content still attains 0.42–0.71%. At *Noel*, Gwelo district in the south of *Southern Rhodesia*, the arsenides chloantite and niccolite impregnate a west-southwest trending shear zone at its intersection with transversal fissures. Furthermore, nickel sulphides occur in serpentine, e.g., in the Trojan deposit south of *Bindura*. In the Hartley district considerable segregations of nickeliferous pyrrhotite outcrop in the Empress gabbro at *Ngondomo*, west of Gatooma.

Nickel appears with cobalt and molybdenum in the hydrothermal post-uranium parageneses of the *Congo* Copperbelt. At Shinkolobwe nickel sulphides are abundant in the low levels, though not in the deepest ones. In the zone of oxidization nickel is associated with uranium, cobalt, gold and palladium. Geologists believe that in this exceptionally young (600 million years old) deposit nickel has been remobilized. In addition to these, chromium and nickel occur in the 2,000 million year old ultrabasic bodies of the southern Congo, especially southeast of Luluabourg in the serpentines of the 10 mile long and 1 mile wide lenticular dyke of the Lutshasha River.

On the other hand, magnetites, disseminated in norite of the Blantyre district (Malawi), contains 40% Ni, 1% Cu, 0.3% Co and traces of platinum.

In east-central *Tanganyika* narrow veins of magnesite intersect the serpentinized ultrabasic body of Mwahanza, 60 miles north-northeast of Dodoma. Here nickeliferous magnetite contains 0.6% Ni and 0.1% V_2O_5. Relics of magnetic hornblendite contain 2.4% Ni, but serpentine includes garnierite and 0.2%–0.5% Ni. Occurring in a 40 mile long alignment of ultrabasics east of Lake Victoria, serpentines of Ngasamo and Wamangolo average 0.2% Ni (maximum 1.3%). East of the southern basin of Lake Tanganyika the serpentine and chlorite–magnetite schist zone of the Kabulwanyele ultrabasics contains less than 0.7% Ni to a depth of 30 ft. However, values in the soil cover are heterogeneous. Farther north pyrite and pyrrhotite of the Kilembe copper–cobalt deposits in western Uganda contain 1–2% Ni.

The laterites covering the dunite of Kakoulima near Conakry, the capital of Guinea, contain traces of nickel, as do laterites east of Kambui, Sierra Leone, and the serpentine belts of the Hoggar in southern Algeria. In another area the copper–iron paragenesis of Akjoujt (Mauritania) includes nickel. In the ultrabasics of the *Graara*, east of the Azegour cobalt deposit (130 miles southeast of Marrakech) in *Morocco*, nickel is associated with cobalt. Elongated bodies of young Precambrian serpentinites are subconcordant in amphibolites and quartzdiorites. The principal nickel parageneses of Ambed 1 and Aït Aman 51 are located at the western and eastern ends of a 10 mile long band of serpentinite, the first of these at the contact of a detrital-volcanic series, the other of quartzdiorites. Both of these occurrences are distinguished by accompanying magnetite and hematite. A typical paragenesis is that of Ambed 1 and Irhtem, including skutterudite, rammelsbergite, nickelite and chalcopyrite. On the other hand, vein 51 does not contain cobalt minerals, but nickelite, gersdorffite and rammelsbergite. The mineral association of vein 52 is intermediate in composition. Along part of vein 5, nickelite, skutterudite, copper oxides, calcite and talc impregnate altered diorite. This paragenesis carries silver. In the zone of supergene alteration nickel appears in black earthy cobalt oxides, e.g., at Ambed 2 (compare with the black ore of Kolwezi).

At Abu Swayel in southeastern *Egypt*, garnetiferous quartz–biotite schists surround a 600 ft. long and 120 ft. wide ellipse of amphibolite and hornblendite. Linnaeite, violarite, chalcopyrite and pyrrhotite are disseminated in a ramifying ore shoot in what is probably a shear zone. On Saint John's Island in the Red Sea, garnierite leached from ultrabasics occurs with peridot. The ore contains 5–9% Ni. East of the Barca River in northern Erythrea (Ethiopia) garnierite is disseminated, mainly along the western endocontact of a 10 mile long serpentine belt. We can distinguish nine narrow ore bodies (USONI, 1952). Finally, nickel occurs in the Kebre Mengist gold field of southern Ethiopia, and in the ultrabasics of north-western Somalia.

PLATINUM, PALLADIUM AND RHODIUM

The Bushveld Complex in South Africa is the world's principal deposit of platinoids. In addition, minor occurrences have been located in Southern Rhodesia, Sierra Leone, Algeria, Ethiopia and the Congo. The platinum group is related to the ultrabasic cycle of 2,000 million years. However, platinum that has been disseminated in the older serpentinites, and hydrothermal and supergene alteration deposits is of minor importance.

THE ULTRABASICS OF THE BUSHVELD

The Bushveld Complex occupies 15,000 square miles of the 300 by 100 mile oval basin of Transvaal System sediments (Fig. 46, 50; COUSINS, 1959a, b). The outer zones of the complex consist of basic rocks while the inner zones are granites. Geologists believed that the complex formed a lopolith until recent geophysical surveys failed to indicate ultrabasic intrusives underlying the central or lower part of the Bushveld. Like the Great Dyke of Southern Rhodesia, and not unlike carbonatite and granite ring complexes, the ultrabasic zones of the complex appear to have been injected in several stages as incomplete ring dykes. Near the surface their sub-vertical dip appears to have changed into sub-horizontal dips, and consequently the eastern and western wings are independent (NOAKES, 1961). While the western arc is a thin concordant sheet, the eastern branch cuts across the Transvaal System, starting at Potgietersrus.

The complex consists of:

(1) The chill or contact zone of 500–1,000 ft.

(2) The differentiated or so-called Critical Zone of 4,000–5,000 ft. with platinoid and chromium mineralization (Fig. 97).

(3) The unlayered Main Zone of 8,000–10,000 ft. At the top of the Main Zone layered white anorthosite, 1–6 ft. thick, carries vanadium bearing titanomagnetite and shows the imprint of catathermal metasomatism (compare with pneumatolytic replacement processes in various ultrabasic massives of the Kola peninsula).

(4) The quartz content increases and the pyroxene content decreases in the upper 1,000 ft. of red granite, which is distinguished by tin occurrences.

The Merensky Reef (discovered in 1926 by Dr. Hans Merensky) is near the top of the differentiated zone, above the chromite deposits. It is underlain by anorthosite and norite. A chromite band of less than 1 inch represents a sharp contact. The Merensky Reef has an average width of 1–18 ft. at Rustenburg (maximum 30 ft.). Above the reef occurs another thin band of chromitite and anorthosite, followed by 3 ft. of pyroxenite and 35 ft. of anorthosite and norite. Some chromite occurs at the base of the Bastard Reef, a low grade platinum deposit. The column above the Bastard Reef includes pyroxenite, and then alternating norite and anorthosite.

The Merensky Reef consists of chromitite, porphyritic pyroxenite (bronzite–diallagite) and diallage pegmatite, in that order. Accessory minerals of the diallagite include biotite, quartz, sulphides and chromite. On the other hand, the chromitite consists of feldspars, orthopyroxene and diallage with chromite inclusions.

The genetic history of the Merensky Reef consists of the successive intrusion of the following sheets: *(1)* a xenolithic chromitite layer embedded in anorthosite, *(2)* porphyritic pyroxenite, and *(3)* pegmatitic pyroxenite (COERTZE, 1958). A similar sequence was injected beneath the Merensky Reef in the Western Bushveld, the rhythmical emplacement of pyroxenite pegmatite tripling the extent of the platiniferous sheet. Apparently the pressure which ultimately pushed the intrusive sequence of the so-called Critical Zone into its present position caused some warping in the underlying ultrabasics (FERINGA, 1959). The first consolidated and elevated portions floated on and into the magma, like the rock islands of lava lakes, finally crystallizing too. The resulting xenolithic structure hindered the development of the Merensky Reef; for example, between Turfbult and Zwartklip in the northeastern Bushveld, a post-intrusive transverse fault complicates the tectonics of the latter locality. Cross fault zones also intersect Turfbult. Furthermore, the ferro-gabbro, intruded in complete discordance through the entire sequence, has eliminated the platiniferous and chromiferous horizons over 17 miles, doubling the length of the titanomagnetite sills.

THE BUSHVELD DEPOSITS

The Reef is more regular in the area of Rustenburg (Fig. 99) than in the Lydenburg–Pietersburg district. Intralayer contacts, however, are indistinct. The dip of the Reef is generally 10° to the north, but it may attain as much as 25°. The western arc of the Merensky Reef is known to extend over 125 miles, while the north–south trending eastern branch is 80 miles long. Nonetheless, a length of only 12 miles is being mined. The reef is normally 6,000 ft. above the contact. The surface of the Transvaal sediments however, is irregular, and at Brits the Reef is only 3,000 ft. above the floor (BEATH et al., 1961).

In the northeastern Potgietersrus branch the platinum horizon includes porphyritic, feldspathic, bronzitite lenses. In the eastern Lydenburg–Pietersburg arc, lower values are evenly distributed. The mean Pd/Pt ratio is 1, the palladium content of the precious metals varying between 44 and 62%. In the principal deposit at Rustenburg platinoids are more concentrated. Layers of destratification and values remain homogeneous for several miles and along several ft. of dip. The proportion of palladium to platinum varies here from 12–24%, to 66–78%. The platinoids occur near both the upper and lower contacts of the Reef. The basal 2 inches of platinoids and chromite are more important, the principal mineral being ferroplatinum. In the upper 1–2 inches the platinoids occur mainly as sulphides and sulpho-arsenides, sperrylite and cooperite, accompanied by a few per cent of iron, nickel and cobalt oxides.

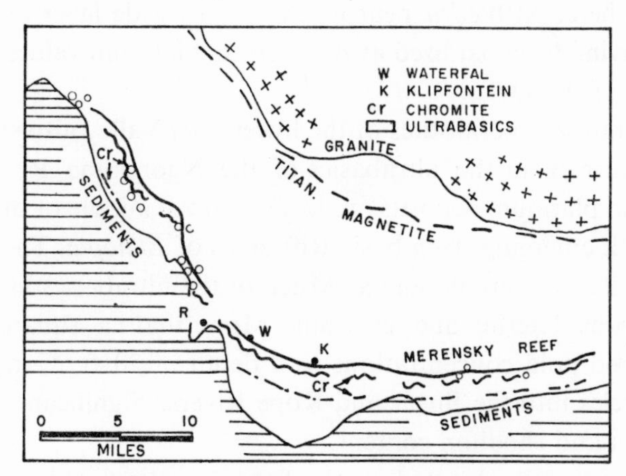

Fig. 99. Chromite deposits of the Rustenburg area.

Two interesting features of the footwall are domes (or koppies) and potholes. Circular–conical koppies project into the Reef and thin it out. Their diameter is less than 100 ft. The layering of rocks remains unchanged. Platinoid values diminish, however, some platinum being found in the chromitite.

Potholes are larger, folded, and spiral-like and used to be interpreted as eddies. Their diameter attains between 50 and 500 ft. in the Rustenburg deposit, their depth varying between a few and 100 ft. In the Union district north of Rustenburg they are 200–700 ft. wide. Faults and fissures are rare in the Reef.

Two other types of deposits are of only minor importance: in the upper reaches of the differentiated zone, sub-vertical pipelike, hortonolite–dunite dykes have a diameter of 40 –70 ft. Both the diameter and ferroplatinum values diminish with depth. Only 3 pipes of the Lydenburg district have been worked. The outer zone of the Onverwacht 330 pipe consists of chrysotile dunite, and the inner zone of hortonolite–hyalosiderite dunite. In the Willemskop pipe on the Driekop, platinum is found in iron-olivine segregations and veins of the dunite. Mooihoek 147 is a smaller deposit.

Impregnation deposits cover a large area of altered dolomite and crush zones, underlying the Bushveld's northern extension in the Potgietersrus district. In the bedded ironstones of Tweefontein, platinum arsenides, stibiopaladinite, copper, nickel and platinoid sulphides occur in shear zones and pegmatites. At Vaalkop 126 and Zwartfontein platinum impregnates contact-metamorphic dolomite. Post Karroo fissure veins contain platinum near Naboomspruit in the northern Transvaal.

THE REST OF AFRICA

As the platinum pyroxenites of the Great Dyke of *Southern Rhodesia* rather closely underlie the norite, they do not correspond to the Merensky Reef, being considerably higher in the sequence of layered intrusions.

East of Selukwe two or three platinum horizons outcrop for 30 miles in thickened pyroxenites. The richer lower horizon has been injected into a band of pyroxenites containing platinum amygdaloids. The next horizon occurs 100 ft. higher, extending north to Makwiro near Belingwe. The long platiniferous horizon, distinguished by traces of nickel and copper, underlies the norite here. At *Wedza*, nearby, the 3–6 ft. wide layer outcrops for 15 miles, the ore sulphides having been oxidized at the surface. Platinum values are known to occur at the northern end of the Dyke.

The ancient serpentinized peridotites of the lower *Sabi* Valley also show platinum traces with palladium occurring in the ultrabasics of the Ngondoma River at Mafungabusi.

The Yubdo–Soddo platinum deposits are located in the middle of the Wallega gold zone in a Precambrian belt emerging from basic Rift lavas of *Ethiopia*. The dunite stock is surrounded by pyroxenites and amphibolites. Much of the ultrabasics have been altered to a siliceous, chromiferous laterite and eluvium. More than a square mile of this crust contains platinum and gold concentrations to a depth of 20 ft. Concentrates have been eroded in several cycles into the Birbir and Kope Rivers. Significantly, palladium content is lower than iridium and rhodium content.

The gold placers of Golella, located north of the Tawalata gold area 80 airmiles north-northeast of Monrovia, Liberia, invariably contain platinum, e.g., Gborbor's creek, Freeman's creek and tributaries of the so-called Platinum Creek. Platinum was supposed to

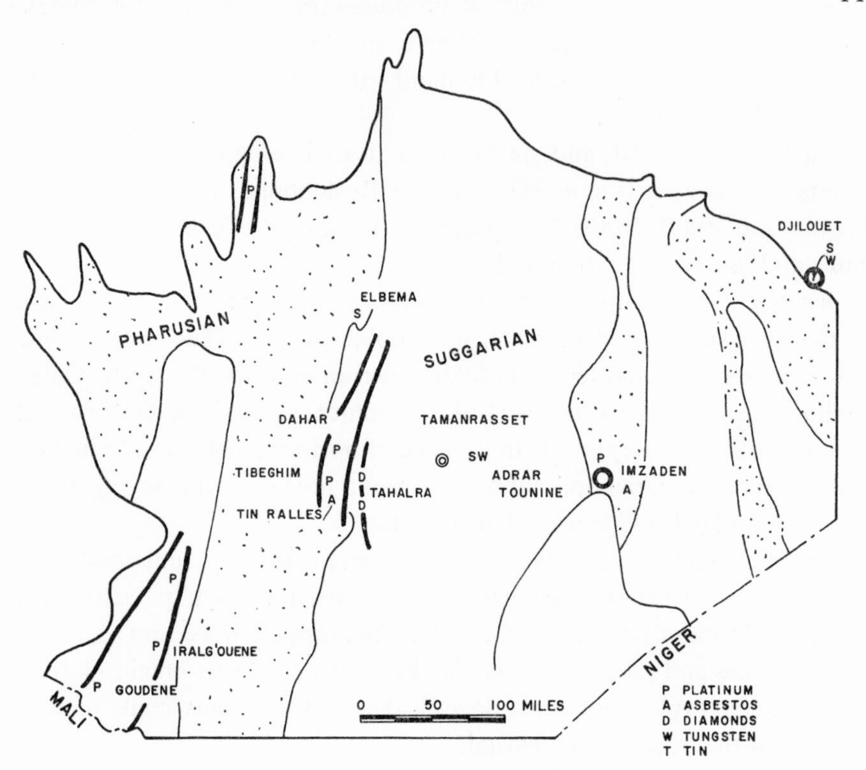

Fig. 100. The Hoggar.

constitute 50% of the precious metals. The prevailing country rock in the area is granite (SRIVASTAVU, 1961). Platinum has also been disseminated in the ultrabasic complex of the Colony peninsula, near Freetown, *Sierra Leone*. Platinum is widespread in pseudo-stratified bodies of basic intrusives and serpentinites of the Saharan *Hoggar* (Algeria; Fig. 100). Swarms of elongated serpentinite bodies, probably derived from dunite and peridotite, are parallel to the regional trend. Some of the serpentinites have been carbonatized. Platinum values form distinctive zones, intersecting lithologic contacts. In the Tibeghim stock and at G'oudène, platinum and ferroplatinum are associated with magnetite, chromite, some ilmenite and zircon.

More platinum has been recovered from the gold placers of the Lenda, west of Lake Edward in the *Congo*. Finally, palladium is an accompanying metal of the Shinkolobwe uranium–nickel ores, occurring with platinum in the copper–cobalt–gold paragenesis of the Kolwezi district.

CHRYSOTILE ASBESTOS

In the principal Southern Rhodesian deposits, chrysotile genesis can be summarized as follows: In the normal magmatic sequence, the ultrabasics preceded the intrusion of granite, which has invaded both the metamorphic rocks and the dunite. Some of the ultrabasics were assimilated, but minor lenses have resisted granitization; subsequently vapours and solutions liberated by the consolidation process invaded the roof pendants, serpentinizing their rim and penetrating the interior along fissures. Volumes increased considerably when olivine altered to serpentine, exerting pressure on the contacts of the ultrabasic lenses. The resulting shear, tangential and vectorial movements affected a heterogeneous assemblage, consisting of soft serpentine and hard ultrabasics, bursting open in fissures. During this phase of decompression and tension, chrysotile crystallized into fibres from overheated antigorite. Shear and faulting have produced similar tensional conditions in the Barberton–Swaziland district, where granitic intrusives are distant. Finally, carbonatic, magnesian and siliceous solutions liberated at the end of the alteration process invaded the outer zones of weakness, precipitating in talc and cherty serpentine.

In South Africa, pyrometasomatic asbestos developed in dolomites serpentinized by underlying diabase sills (Fig. 5).

SOUTHERN RHODESIA

The chrysotile deposits are arranged along a 100 mile long, broad east-northeast trending belt which extends from Filabusi, 50 miles from Bulawayo, across the southern end of the Great Dyke, to the greatest concentration at Shabani, and thence to Mashaba. At Lanninhurst, Silver Oak, Geta and Thorrwood, minor occurrences are aligned along a line striking east between West Nicholson and Gwanda, south of the Dyke. The major Ethel deposit is located along a transverse fault, intersecting the Great Dyke, 50 miles north-northwest of Salisbury (Fig. 36, 50).

Filabusi

The synclinal Filabusi roof pendant covers 230 square miles, its central width attaining

10 miles (FERGUSON, 1934). Three series have been distinguished: *(1)* sericitic quartzites and schists, *(2)* metalavas altered to hornblende, chlorite schists, and *(3)* phyllites including limestone, banded ironstone and conglomerate. The major syncline pitches northwest, bordered by complicated folds. In the metalavas, near the granitic contact, serpentinite and talc-schist follow a 20 mile long arc. At the core of the synclinal structure, outcrop two or more bands of serpentinite bending sharply southwards at Filabusi across the regional trend, and including the principal asbestos deposits, Pangani and Norma. Ormonde, Recompense and Wynne are located in the eastern belt, farther (½–2 miles) from the granite than at Shabani. At Filabusi proper, where several serpentine belts outcrop, asbestos occurs in those nearer to the granite. In the serpentinized rocks, the $^5/_8$ inch wide seams tend to strike in two directions, but they also branch out in many other directions; they concentrate in indistinctly defined 'lodes' or swarms, 100 ft. wide or more, generally exhibiting a concordant dip. Magnesite and opal veinlets cut the lenses, which are finally silicified by supergene solutions.

The Pangani lens consists of indistinctly oriented seams; fibres are generally shorter than an inch. The Norma lode, also at nearby Panasequa, strikes northeast. The Wynne deposit at Drumbulchan is located in the exocontact in serpentine, a mile from the granite; slip fibre has developed there along faults traversing the seams. The Croft 'lode', at Wallingford, is concordant with serpentinite, interbedded between metalavas and sericitoschists. Arcuate chrysotile seams, sometimes merging, swarm in the 'lode' in 5–20 inch wide bands and seams of short 'ribbon fibre' alternating in tight rhythmical bands.

The Filabusi field ends in a nose which continues in a narrow peridotite–serpentine roof screen towards Belingwe and Shabani in the northeast; this trend has been faulted and displaced by the Great Dyke. The principal deposit of *Gurumba Tumba* is situated halfway between the Dyke and Belingwe, wedged between harzburgite and a gold and copper bearing quartz vein, bordering granite. The main slip fibre body is 50 ft. thick, narrowing to half that thickness in depth; numerous seams dip south. Magnesite, containing some fibre, is abundant along the hangingwall, but tenors decrease progressively towards the footwall. In the *White* deposit on the Dowe stream near Belingwe the serpentine, including cross fibre seams, outcrops between harzburgite and carbonate rocks.

Shabani

On the eastern side of the Great Dyke a 10 mile wide roof screen of greenstones extends 20 miles, bending from the southeast towards the south, where it bulges. A narrow rim of banded ironstones borders this roof pendant. Ultrabasic bodies float on the granite ½–1 mile east of the greenstones, and smaller bodies lie north of Shabani (Fig. 101). The Ad Valorem Hills spread in the east, and small lenses outcrop halfway before the large remnant of the Vukwe Mountains in the south.

A 10 mile long and 1–2 mile wide mega-xenolith of dunite is concordant with the earliest Precambrian rocks. The southeast striking schist belt has been intruded by younger granite. In the south, the batholitic apophyse forms a ½–1 mile wide ledge between the dunite and its host-rocks. In addition, along the median axis dunite cores and lenses have been preserved from subsequent serpentinization, which expands from three fissure systems forming large angles. The endocontact of the ultrabasics has been altered along tectonized

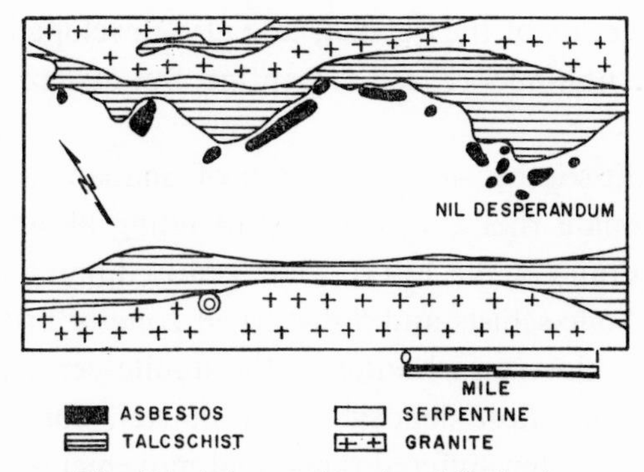

Fig. 101. Asbestos deposits of Shabanie.

shear zones to cherty serpentine, actinolite, chlorite, talc, talc–carbonate and similar schists. This alteration halo or zone is only 1,000 ft. wide in the southwest, but in the northeast the alteration zone pinches and swells from 500 to 3,000 ft. Since the granite includes elongated but unassimilated schlieren of this facies of alteration, the latter is believed to envelop the whole ultrabasic body. The world's principal chrysotile deposits extend for 2½ miles in serpentine and cherty serpentine, just inside the talcified shear zones. Normally only talcified tongues divide the various bodies which are believed to be remnants of a long 'lode'. Had talcification progressed, the deposit would have been destroyed. The seams strike parallel to the shear and contact zones for considerable distances, but they also follow other directions. The mode of occurrence can be best visualized thus: lenses of 'brittle fibre' or talcose asbestos lie sub-concordant in the talc schist zone and in the outer rim of the serpentine, partly altered to talc; the principal seams constitute an arborescent, leaf-like pattern with ramifications of serpentine surrounding lenses of less serpentinized dunite. Asbestos develops along the axes of the branches, which coalesce near the rim of the talc zone through the complete alteration of dunite.

There have been distinguished three types of chrysotile: Abundant *cross fibre* lies perpendicular to the walls, or sub-perpendicular if slanted by movements. Fibres may also be bent locally or fractured in part of their length, especially at the ends, thus adopting an integral-like shape. The fibres grow from the walls to a length of up to 2 inches, meeting at the axial plane, with seams attaining a maximum width of 4–6 inches.

Slip fibre, lying flat on planes of displacement or in bunches, is the product of intensive shear and fatigue of the asbestos, affected by movements of varied orientation. As slip chrysotile is frequently associated with fibrous brucite, this aggregate contains less asbestos than supposed. *Brittle fibre*, of low value, is the product of similar tectonic effects affecting the zone of talcose alteration. The chrysotile has suffered partial talcification, resulting in the midst of the talc zone in pyrophyllite pseudomorphs after chrysotile.

While areas containing only 1% fibre can also be worked, important zones include 3%. Seams and stockworks of similar values build up big lenses of indistinct contours in the serpentine, attaining a strike length at Shabanie mine of 2,000 ft. and a width of 200–300 ft. The Orphan's Luck body is isolated 1,500 ft. west of the principal lodes of Shabanie, and the westernmost 'lode' is all but surrounded by talc-schists. The following ore body is located in superficially serpentinized dunite. The main 'lode', situated in the centre of the belt, is terminated by talc-schists in the east. Magnesite seams related to the latter are localized in shears; in other sectors of the deposit, talc also developed along shears. Shrinkage cracks are frequent in the non-mineralized serpentine, but quartz veinlets are not restricted to any specific area.

At *Nil Desperandum*, the southeastern extension of Shabani, several 'lodes' of asbestos remain in the talc–carbonate rocks, with talc protruding along apophyses into the ore bodies. However, the contours of the 150 ft. wide, gently dipping sheet are ill defined. The largest 'lode' lies between talc-schists and a disturbed zone of brittle fibre in partially serpentinized dunite. The southwestern border of the dunite–serpentine body contains only scattered asbestos. The *White Rose* deposit, on its shorter, northern edge, carries narrow ($^1/_4$ inch) seams which have often suffered tangential movements.

Southern Shabani (Belingwe)

Serpentinites, intrusive into pre-Bulawayan pyroclastics, extend for more than 10 miles south of Belingwe, with their width diminishing towards the south. Talc–carbonate schists prolong this trend south of High Peak Farm, also bordering the belt in the West, along a sheared contact. The crest of the ridge has been silicified, asbestos outcropping for a considerable distance. In the Vukwe Mountains, an embryo-shaped, 3 mile long serpentine band borders a larger belt of early Precambrian ultrabasics underlying and probably invading earlier metasediments. The serpentinite body is largest in the north, measuring a mile. Chloritization in serpentinized olivine produces an oriented texture, and carbonate rock inclusions form narrow bands in the serpentine.

The Askari, Kloof and Slip deposits are also located in this area. At *Askari* a dolerite body divides serpentine from the granite; high grade slip fibre follows parallel, also east-striking fault planes there. At *Kloof* the fibre bands trend north-northwest, forming an acute angle with the orientation of the serpentine body, which dips west. The quantity of asbestos does not appear to be a function of the abundance of magnesite here. In the *Slip* (Rhomonte) deposit, chrysotile lenses occur in fault planes near the chilled contact zone of granite and a system of pairs of serpentine; magnesite seams developed later in them. Magnesite appears to be pene-contemporaneous with chrysotile.

Magnesite is also abundant in the Biltong extension. At *Biltong* the asbestos is located 500 –1,000 ft. from the contact, and high grade chrysotile occurs on both sides of the fault plane of an epidiorite dyke; other dolerites are parallel to ore bands. An ore zone is 30 ft. wide, containing 1 inch seams. Brittle fibre is found both with the principal seams and in separate stringers.

The Bend–Vanguard belt, at Belingwe, follows a largely north-northeasterly trend. At *Bend*, the proto-serpentine ultrabasics have been intruded in a sheet into the Upper Bulawayan; they later participated in the folding. Cross fibres, $^1/_{16}$–$^3/_4$ inch long, have developed in seams of closely packed lenses. The main ore bands, trending east-northeast, lie 100 ft. from each other; they are sub-parallel to the metadolerite–serpentine contact.

In the *Vanguard* deposit, 21 miles south of Belingwe, serpentine is intrusive in earliest Precambrian greenstones, which form a syncline. The deposits are located at the widening of the serpentine belt. They strike north, and a dolerite dyke borders them in the east. Swarms of seams trend north-northwest and grade into each other; they include, from east to west:

(1) The lenticular main 'lode', 500 ft. long, greatest width 150 ft.
(2) The west 'lode'
(3) The central slip, each 1,000 ft. long, 150 and 80 ft. wide, respectively, separated by:
(4) Hard serpentinite, containing ribbon fibre and narrow barren contact zones.
(5) A 500 ft. wide poor zone including isolated cross fibre patches, followed by:
(6) The arcuate west slip, converging with its twin (slip fibre tenors are high).

Hard serpentine concludes the sequence in the west. Seams of long cross fibre are oriented randomly; magnetite borders them. Slip seamlets exhibit zones of magnesite, magnetite and brittle slip fibre.

In the *Mashaba* district serpentine also surrounds the dunite; however, the development of the outer zone of talc metasomatism is less important. Fibre, attaining grades of only 0.7–1.2%, is more prilly and harder than at Shabani.

Great Dyke

Serpentine and brucite form from enstatite dunite, by the auto-hydration (?) of forsterite from solutions accumulating under the sub-horizontal layers and liberated by faulting. Pyroxenites have not been affected. Serpentinization of the fault walls is essential for asbestos development. The transverse *Ethel* fault, near Umvukwe, displaces half a mile east the northern part of the Dyke. Vertical, chrysotile cross fibre seams are grouped in parallel lenses, forming a U-shaped zone of 2 by 800 ft.; they extend towards the east in a median, lenticular zone and two converging bands. Low grade or barren serpentine divides the 5–10 ft. thick zones from each other. High grade chrysotile fibres of 4–25 mm constitute 1% of the rock, accompanied, however, by brucite. Minor transverse seams traverse the serpentine between the asbestos zones, which are situated at a certain distance from the fault. Granite outcrops 600 ft. north, and a post-serpentine dolerite dyke runs parallel to the zone with its apophyses destroying asbestos.

The zone extends to Brotherton, 2 miles in the east. Here a vertical chrysotile seam is changed to fibrous magnetite while intersecting chromitite. Asbestos also occurs in the serpentine, along the longitudinal fracture of the Dyke.

The *Ass* deposit is located southwest of Belingwe along an east-southeast striking fault displacing the Great Dyke towards the west, a few miles north of its southern edge. In the lower serpentine zone of the Dyke, 4 inch long, sub-cross fibres fill in a 3 ft. thick band parallel to the fault plane.

SOUTH AFRICA

The principal deposits occur in earliest Precambrian serpentines of the Barberton district, eastern Transvaal, northwest of the Swaziland border. At Carolina, 40 miles southwest of Barberton, a peculiar type of chrysotile deposit was found in the Dolomite Series in contact with dolerite sills (Fig. 46). In fact, minor occurrences of this type are spread over the Pietersburg–Lydenburg hornblende asbestos field, appearing also at Graskop, north of Barberton and around Pretoria. Minor earliest Precambrian occurrences include: *(1)* Driefontein and Honingsklip, 30 miles northwest of Johannesburg; *(2)* Doornkraal, in the northern Transvaal on the Bechuanaland border; *(3)* Maryland, near Messina on the

Rhodesian frontier, and *(4)* the Steynsdorp–Josefsdaal field, expanding into northwestern Swaziland.

A somewhat younger belt extends east of Barberton to Magnesite Siding. The deposits of the Eshowe N'kandh'la areas, Sitilo, Tugela Randt, and others in Zululand 60 miles north of Durban, are interbedded with serpentine, adjacent to an elongated body of granite.

Dolomite

Long, high quality, cross fibre has grown in the bedding planes of serpentinized dolomite, on and above the contact of intrusive basic sills which extend along the strike. The grade of the chrysotile tends to decrease along the dip as a function of the distance from the surface. The principal deposits include:
Carolina district: *(1)* the Congo–Vaal deposit at Rietfontein; *(2)* the Carolina deposit, at Diepgezet and Zilverkop; *(3)* Badplaas at Goedverwacht; and, in addition: *(4)* Elandshoek near Montrose Falls; *(5)* Olifantsgeraamte, Graskop, Normandale, Kalkkloof, Appeldoring, Engelschedraai, Uitkomst and Rietfontein.

At Goedverwacht a typical column shows: dolomite, basic sill (25 ft.), asbestos, dolomite (50 ft.), basic sill (20 ft.), shale. At Diepgezet, groups of seams tend to concentrate above and near the contact of the upper sill; in no case are they further away than 5 ft. Unaltered dolomite does not contain seams. Olivine has crystallized in the sills in layers, generally along bedding planes which have subsequently been serpentinized. Serpentine continues along the dip, but chrysotile disappears (similar asbestos is known to occur in the Asbestos Cañon, Arizona, U.S.A.).

Near *Sabie*, on the Crocodile River, similar deposits of long fibre have developed *above* diabase sills. Pyro-metasomatic effects can be witnessed in dolomite for several ft. from the sills; not *all* sills, however, have serpentinized the dolomite. Elandshoek is the only deposit where asbestos grew on a dyke and not on a sill.

Serpentine

The ultrabasic-serpentine zone, overlain by younger sediments, extends 50 miles from Kaapsehoop to Hectorspruit. Although resting on granite in a gently warped and basined structure, the distribution of serpentine is not a function of the granitic relief. Joubertsdaal is situated 30 miles northwest of Barberton and Diepgezet, 15 miles south. The belt includes (VAN BILJOEN, 1959): *(1)* the Havelock deposit (Swaziland), Msauli at Diepgezet, near Barberton, *(2)* New Amianthus, Munnik Myburgh and Sunnyside near Kaapsehoop, *(3)* Stolzburg, Sterkspruit and Doyershoek, *(4)* the Barberton deposit, at Koedoe, near Malelane, *(5)* Kalkkloof near Badplaas.

The standard, twice-repeated sequence above the intrusive appears to consist of dull-green fibrous serpentine, and then dark blue-black serpentine (probably incompletely altered ultrabasics). Each layer may measure 100 ft. in thickness. The paler fine or medium grained rock consists, in fact, of microscopic, randomly oriented, fibrous serpentine including disseminated and uniform stichtite (a hydrated chrome magnesite) and ilmenite stringers. However, the darker, barren serpentine carries none of these accessories. Nonetheless more silica probably originates in a pyroxenite or amphibolite (A. L. HALL, 1930). Contact phenomena appear to control the localization of the two asbestos alignments. The east striking Griffin Line of tightly spaced, thin, concordant seams, follows the limit of the two varieties of serpentine (or olivinite and pyroxenite?); it extends 10–15 ft. into green serpentinite. The more important Ribbon Line strikes north at the top of the serpentinite, overlain by quartzites, shales or a dolerite sill, i.e., unconnected with ultrabasics. The upper zone of the 'Ribbon' is 5–7 ft. thick, including 15–30 thin sub-concordant seams which are equidistant below the cover. In the 2–3 ft. thick lower lenses, thick seams are wide spaced. Conditions are rather similar in the eastern sector.

At *Kalkkloof*, in the Carolina district, at least four favourable zones have been distinguished in steeply dipping, pale green serpentinite, alternating with the barren, blue variety.

SWAZILAND

The *Havelock* deposit is located at the foot of Emlembe Mountain, near the northwestern border and 12 miles south of Barberton in South Africa. Serpentinite is included in the early Precambrian sequence along a 1½ mile long stretch (VAN BILJOEN, 1959; Fig. 102). The layers consist of: *(1)* shale; *(2)* 150–200 ft. of lower chert; *(3)* 200 ft. of schist, its total width intensely sheared and oxidized at the surface, and in a width of 20 ft. underground; *(4)* 60–350 ft. of light green granular chrysotile serpentinite, in gradational contact; with *(5)* 100 ft. of dense, darker green, barren serpentinite; *(6)* 150 ft. of hangingwall sill of diabase; *(7)* shale and schist with bands of chert.

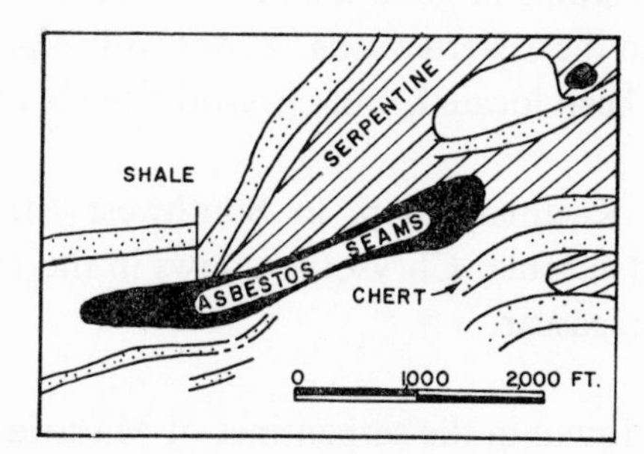

Fig. 102. Asbestos seams of Havelock. (After VAN BILJOEN, 1959).

At some distance from the ore body alternating bands of serpentinite and black and white chert are located in the sericite–chlorite–talc shales. In the western section, especially at the footwall of the ore body, serpentinite has been replaced by talc (-calcite) bearing magnesite. This rock borders the oblique strike fault which traverses the middle of the serpentinite sequence. In the western sector, this fault has sheared and partly destroyed chrysotile. Minor faults form a large angle with this feature. In the east their wall rocks have been ferruginized. Long fissures are filled in by magnesite and calcite or slip fibre. The diabase, amphibole and subsidiary epidote and chlorite, sill splits into several bands; thin diabase dykes also intersect the ore body. At the contact of these dykes, hematite and some magnetite have been disseminated in the fine grained chocolate coloured serpentinite.

The sequence dips 55–65° south and the ore body is located at an undulation of the strike from N80°E–N60°E. The ore body extends ³/₄ mile, and averages a width of 110 ft. At some distance from Havelock Mine, light green serpentinite carries only traces of asbestos. In the ferruginous serpentinite of the valley seams have been replaced by what is believed to be ferro-chrysotile. The mineralized light green serpentinite consists of colourless cores of chrysotile disposed in a grid, and surrounded by thin seams of pale yellow serpentine; it contains grains and veinlets of brucite, while the dark green variety does not. Locally, isotropic grains are intersected by stringers of serpentine exhibiting higher bi-refringence.

Stockworks of cross fibre seams represent 3.5% of the pale serpentine, and seams dipping 70°S and 15°N appear to be frequent. Workings have reached a depth of more than 600 ft. Fibre length averages $^1/_2$–$^3/_4$ inch, with extremes reaching to less than $^1/_8$ and more than 2 inches. Zoned grains and microscopic stringers of asbestos also occur. The core of these grains may be a (chrome?)-spinel or serpentine replacing the latter.

THE REST OF AFRICA

Silicified chrysotile was found south of Serowe. At Moshaneng, 8 miles west of Kanye, dolomites of the Transvaal System have been altered and serpentinized near the upper contact of a dolomite sheet; the ore contains 2% chrysotile, the bulk being of grade three.

In the Gara Jabbe and Gara Garad Hills, near the Harar Jijiga road in eastern *Ethiopia*, serpentine and amphibolite are interlayered between gneisses and micaschists. Chrysotile, developed in fissures in amphibolite, is longer but less pure. Asbestos is also believed to show at Logare and Got in southern Ethiopia, and in the Maat Nusc Valley and at Shamege, in Northern Erythrea.

In northern Somalia, thin seams of good quality asbestos occur near Lafarug. At Umm Sagata, near Qala en Nahl in the eastern *Sudan*, a zone with $^1/_2$–2 inches of $^1/_8$–3 inch fibre per foot of serpentinite has been located, but chrysotile tends to be replaced by silica and titano-magnetite.

Near Swakopmund and Keetmanshoop, in Southwest Africa, brittle chrysotile has crystallized in serpentinized marble. Chrysotile shows in the serpentinites of Vohibory–Sakoa in southwestern Madagascar.

Mozambique. Asbestos is found in the serpentines of Marávia and 20 miles from Fíngoè in the Tete district; the Mount Capapa pyroxenites also contain low grade asbestos. Traces of chrysotile show in the Usagara Mountains, Tanganyika, and long fibres occur at Mahagi on Lake Albert in the Congo, and in the Lower Congo, e.g. at Matadi.

Asbestos has been identified in serpentine at Anum, Akuse, Zanidaw, and Cape Three Points, in Ghana.

Morocco. Chrysotile seams occur in the 100 mile long serpentinized fissure extending from Tissoufra to Nkob, Bou Offroh, the Bou Azzer cobalt deposit and beyond Inguijem in the east-southeast. Seams may form tight packs as at Bou Offroh, the principal deposit, or strike randomly. The width of asbestiferous serpentine (or carbonate) zones varies between 2 and 50 ft. The pinching and swelling Bou Offroh zone extends more than 1,500 ft. along a fissure adjacent to amphibolites and garnetites. The ore contains 7–10% chrysotile, but the Bouznournak and N'kob deposits are poorer.

TALC

Impure soapstones, pure talc and massive steatite occur in the hydrothermally altered serpentine rim of generally sheared ultrabasics. Talcification and carbonatization follow the development of chrysotile or anthophyllite. Other deposits are related to amphibolite and dolomites. Stains and grains of magnetite, limonite, quartz and calcite determine the grade of the material.

NORTH AND EAST AFRICA

At Bou Azzer in Morocco talc follows the contact of serpentinites, but at Siroua and N'kob the mineral is related to dolomites (cf. 'Nickel' chapter).

At *Wadi Gulan el-Atshan* in Egypt, metamorphosed peridotites, dolerites and andesites have been intruded by granite forming a hornfelsic rim. Steatite occurs at Atshan and Mikbi (Fig. 12), but talc is found at Darhib. Peridotites have been transformed into tremolite–chlorite rock, dolerites into amphibolite, and andesite into tremolitic amphibolite. At Reidi and Atshan metasedimentary granulites and amphibolites, accompanying the shear zone of talc and talc–tremolite–chlorite carbonate, strike west-northwest. Slickensided talc, coloured green by malachite, attains a width of 150 ft. Tremolite fibres are replaced by talc and carbonate, the latter also forming a brittle, granular chlorite–talc–carbonate–tremolite rock. Finally, we also find altered metavolcanic dykes and remnants of granulite and epidosite.

In the Qala en Nahl (or Nahal) district of the eastern *Sudan* (northwest of Lake Tana) large deposits of talc and magnesite are known to occur. We can distinguish two generations of basic and ultrabasic rocks. The older group is intrusive into the basal schists. A younger group of basic rocks followed the Paleozoic extrusion of acid and intermediate volcanics. Massive granites conclude the sequence, affected by faulting, shearing and a hydrothermal phase. Moreover, the older ultrabasics have been affected by serpentinization, carbonatization and steatitization. They strike north-northeast, that is, parallel to the gneissose granites.

Talc lenses occur at Farar, Dankalia, northeastern Ethiopia, and at Nkongora and Ruimi Falls in western Uganda. On the other hand, soapstone has been located at Zeu Hill in northwestern Uganda. In southern *Kenya*, e.g., at Machakos and West Suk, actinolite–antophyllite–talc schists and talc developed at the contact of early Precambrian dolomitic limestones and ultrabasics. In northeastern *Tanganyika*, between Mount Kilimanjaro and the coast, the talc of Hedaru contains some limonite. Far away in central Tanganyika the largest deposit of the Kikombo area, near Dodoma, is Kikaza Hill. These altered ultrabasics measure 1 mile by $^1/_4$ mile. On the Kapeni Stream and at Ntonda in the Kirk Mountains of northern Malawi, talc-schists and actinolite schists develop by hydrothermal alteration of ultrabasics.

THE REST OF AFRICA

The principal deposit of the Manica district in western Mozambique is situated at Rotanda in the Muza Valley. Abundant talc surrounds the serpentinized ultrabasics of Southern Rhodesia.

South Africa. The Transvaal occurrences are similar to those described in the 'Southern Rhodesian' section of the 'Asbestos' chapter. In the Soutpansberg district south of the Limpopo, talcose rocks form a band between the serpentine aureole of ultrabasic or amphibolite and Precambrian gneisses. At Brindisi and Lintie the band extends 2½ miles and attains a width of 1 ft., but three larger bodies associated with crush zones measure 400 by 100 ft. This low grade talc contains chromite grains and serpentine relics. At neighbouring *Rotterdam* high grade talc seams occur, adjacent to pegmatites. At Barneveld, Gemsvale, Kaapmuiden, Elandsfontein, Noordkaap, Strathmore and Uitkyk, north of Swaziland, massive talc seams, 1 inch to 1 ft. thick, follow shears in talc–carbonate schists associated with amphibole- and/or chlorite schists. At Mount Vernon and Toggekry in the Ngutu district 130 miles north of Durban, tremolite accompanies talc in the aureole of serpentine intruded by granite. Finally, at Inanda low grade steatite has been altered from serpentinized hornblendite and limestone intruded by granite.

Near Gege in the Mankaiana district of *Swaziland*, a more than 6 ft. wide pyrophyllite body is interbedded between two horizons of massive quartzites near the base of the Middle Precambrian Pongola Series. The series dips 12° and the material contains 30.16% Al_2O_3 and 1.33% MgO. Its colour varies from white to greyish white. However, some of the pyrophyllite has been stained pink by ferro-manganesian solutions percolating from the overlying shales into which the talc grades.

Soapstones and talc-schists are abundant in the Vohémar, Tamatave, Tsaratanàna and Antsirabe districts of Madagascar, with high grade talc occurring at Sahatany. In Nigeria geologists have located low grade talc-schists in Ilorin, Oyo, Zaria and Niger (Abuchi) province. While in Ghana experts have identified talc-schists in the Awudome Hills at Anum and Pudo, talc occurs at Ajumako in the south.

MAGNESITE

Low grade magnesite is abundant in dolomites and rauhaugite (carbonatite). Furthermore, magnesium continues to precipitate from the waters spilling over from Lake Kivu between Bukavu and Shangugu, mainly in Rwanda. Magnesia precipitates in salts and emanates from volcanoes. However, the principal high grade deposits occur in the supergene or hydrothermal aureole of alteration of serpentinites. As ultrabasics are more abundant in southern Africa than in the rest of the continent, magnesite is also more frequent in those areas.

SOUTH AFRICA AND TANGANYIKA

At Magnesite Siding near Kaapmuiden, north of Swaziland, magnesite occurs in altered meta-ultrabasics. In addition, in *South Africa* at Aapiesboomen, Aapiesdoorndraai and Mooifontein near Burgersdorf, magnesite precipitated from leached magnesian pyroxenites in a 50 mile long discontinuous arc following the northeastern contact of the Bushveld complex. North of the Antimony Bar in the Southerland Range, lenses of amorphous magnesite outcrop in serpentinite, sandwiched between ultrabasics and granite. At *Tovey*, west of Messina (see 'Copper' chapter) on the Rhodesian border, serpentinite occupies a 3,500 by 900 ft. basin in mafic gneisses. Weathered serpentinite alters to magnesite along joints and partially in zones of shattering, but fresh serpentinite is not transformed into magnesite. At *Grasplaats*, 15 miles south of Messina, a magnesitic dolomite vein extends for 600 ft. attaining a width of 2–5 ft. In the Letaba gold district south of Messina thin veins of hard magnesite traverse locally silicified carbonatite serpentine.

Tanganyika. At Merkerstein, west of the Kilimanjaro, magnesite accompanies sheared serpentinites, also crystallizing in joints and fractures traversing unsheared serpentine. West of Merkerstein (in the Rift Valley) at the foot of Mount Gelai, a durcicrust covers massive magnesite underlain by soft magnesite. Geologists believe that magnesian gases of the volcano have precipitated with carbonatic solutions of adjacent Lake Natron. At Chambongo, south of the Kilimanjaro, white magnesite seamlets constitute 34% of the serpentinite they traverse. The ore body, containing 46% MgO, is believed to be 200 ft. thick and to dip 25°. In the area of Haneti and Itiso northwest of Dar es Salaam, e.g., at Iyobo, Mnakura and Mwahanze Hill, several swarms of magnesite veins intersect serpentinites. The width of veins varies from 1 inch to 2 ft. at Mnakura, and from 8 inches to 2 ft. at Iyibo where their length attains up to 300 ft. Other occurrences include Mount Vuju near Handeni, the Lugalla Hills, the Nsamya Hills and the Pangani Valley.

THE REST OF AFRICA

At Baramiya in southeastern *Egypt* regional carbon dioxide metasomatism has produced talc magnesite schists. At Umm Salatit outcropping magnesite lenses, formed by supergene solutions, attain 4 ft. of thickness. We have discussed the magnesite deposits of Qala en Nahl in the eastern Sudan in the 'Talc' chapter. In the serpentinites of Shamege, Ethiopia, nodular magnesite fills in veins and reticulated stringers. At Lolung in northern Uganda, and at Lolukai on the Sudanese border, magnesite occurs in altered serpentine.

Magnesite veins intersect dunite bodies of Kinyiki Hill at Kipiponi, of Magongo Hill near Kitui in the Embu and Baragoi districts of Kenya, and also at Merti in the north. The ramified Kinyiki lenses rarely attain a thickness of 2 ft., but magnesite forms 30% of this rock.

Magnesite is abundant in *Southern Rhodesia*. Magnesite veinlets intersect the asbestiferous serpentine lenses: e.g., at Filabusi chrysotile tends to resist carbonation more than serpentine. Because of exposure to alteration magnesite is, however, more abundant in the asbestos. Stringers of opal also traverse the upper reaches of the lodes. But in the zone of alteration, irregular magnesite gash veins attain only a few inches of thickness. At Croft, Pangani and Norma magnesite is most abundant in the upper 100 ft., becoming rarer in unaltered serpentine, and disappearing in unserpentinized dunite. In Mozambique on the Rhodesian border, magnesite is related to the serpentinites of the Manica schist-belt, and north of the Zambezi to the ultrabasics of Mount Atchiza.

Veinlets of magnesite invade the altered peridotite of Ampangabe Jabo 15 miles north of Tananarivo, of Antanetibe and the serpentinites of Vohibory. Finally, magnesite veinlets traverse serpentinite at Cape Three Points, Ghana and at Bou Azzer–Graara, Morocco.

ICELAND SPAR

Calcite of optical (Nicol) quality is a hydrothermal mineral crystallizing in basic rocks. At

Rehoboth, Gibeon and Swakopmund in the southwestern part of *Southwest Africa*, iceland spar occurs in vugs of Jurassic, Karroo basalts. In *South Africa* optical spar is known in the areas of Kenhardt, Calvinia and the Orange River in the southwest, and near Aliwal North, Lady Grey, Rosemead and Aberdeen. However, the principal finds – Dagab, De Paarden, Klerkshoop, Korannakolken, Verdorsputs, Vleyen and Zooafskolk – are located in the Kenhardt division near the marshes of the Sak River, 100 miles south of the southeastern corner of Southwest Africa. In lenses, veins and irregular aggregates occurring in decomposed Karroo dolerite, optical spar accompanies common calcite. At Jan Gora and Kransgat, 100 miles southwest of the preceding group, calcite accompanied by prehnite forms veins, also impregnating the dolerites.

Conversely, at Blauwskop close to the southeastern corner of Southwest Africa, iceland spar has crystallized in decomposed mylonitized gneisses. At Kristalkop, south of the Orange River, veins of optical spar occur in shales.

VERMICULITE

Through alteration at and below the surface vermiculite has been hydrated from phlogopite pockets associated with magnetite–apatite, pyroxenite and serpentinite zones of the ultrabasic rings of carbonatites. The principal occurrences are in the copper–phosphate deposit of Phalaborwa in northeastern South Africa. However, vermiculite bodies developed in ultrabasics traversed by felspathic pegmatites are minor, with nests disseminated in acid pegmatites being even smaller. With a few exceptions this type of deposit is restricted to eastern Africa.

THE PHALABORWA CARBONATITE

One of the world's principal vermiculite deposits is (Fig. 46) located near the Kruger Park in the pyroxenite rim of the *Phalaborwa* carbonatite of the northeastern Transvaal. This complex extends 4 miles in a north-northwesterly direction with its width attaining 1 ½ miles. The (dolomite) rauhaugite core *(a)* occupies 1 mile by ½ mile surrounded by a thin inner ring *(b)* of altered phoscorite (serpentine–apatite–magnetite) and an outer ring *(c)* of diopside pyroxenite. Near the outer edge of this zone feldspars are disseminated. However, the following 1 ½ mile wide ring of fenitized gneisses, *(d)* probably including rheomorphic syenite bodies, is discontinuous. Numerous alvikite and orthoclasite veins have been injected into the pyroxenite ring. Serpentine, apatite and magnetite occur in both inner zones, and a pegmatitic pyroxenite zone occupies the centre of the northern ultrabasic ring. In addition to these, post-carbonatite dolerite dykes cut the complex. Finally, secondary limestones leached from the carbonatite cover much of the complex.

In the inner rim of the pyroxenite, injected by phoscorite bodies and stringers, fine mica is most abundant near the surface. Vermiculite occurs only on the ultrabasics, especially close to serpentinized (probably olivine pyroxenite) patches. Furthermore, vermiculite has also been disseminated with apatite (C. M. SCHWELLNUS, 1938). The principal vermiculite areas are located: *(1)* in the north-central pegmatitic pyroxenite ½ mile from the limit of the carbonatite plug, *(2)* ¾ mile south of the edge of the core, and *(3)* in a mile long zone near the southeastern rim of the pyroxenite.

We find vermiculite hydrated from phlogopite in pockets and irregular bodies of apatitic pyroxenite pegmatite. At a depth of 150 ft. rich and poor streaks of vermiculite alternate. The Phalaborwa complex contains considerably less olivine than other carbonatites. Indeed, ferromagnesian fractions appear to have been fixed late in phlogopite crystallizing with apatite. Another characteristic of the complex is its cover of calcrete, which may have influenced the hydration of phlogopite. We can distinguish two types of vermiculite. The honey brown variety (2V = 0°) of the north is distinguished by an exfoliation coefficient of 26. Since it contains 5% combined water, 12% Al_2O_3 and 6.5% Fe-oxides, against 2.5% H_2O, 7% Al_2O_3, 8.5% Fe for the southern black variety (2V = 15°), which expands only elevenfold, this variety is believed to be a product of more intensive alteration.

THE PHALABORWA DEPOSITS

Old workings illustrate the mode of occurrence and paragenesis of the ore: Along a 2,000 ft. long, north-northeast oriented alignment, trending parallel to the axis of the complex, more than ten ft. wide aggregates of vermiculite plates are included in altered red serpentine. The best honey-yellow flakes have been found at a depth of 10–12 ft. Certain low grade flakes exhibit zoning of a yellow rim and dark brown centre. Similarly, the surficial drab mica is of inferior quality. At Loole No. 199 farm vermiculite flakes have penetrated into the calcrete.

A mile southeast of the Loolekop the series of apatite–vermiculite patches contains mixed mica or incompletely hydrated phlogopite, but good vermiculite has been found under the cover. Subvertical vermiculite–apatite bodies are arranged in pyroxenite 1 mile south of the Loolekop along a north–south trending axis. Apatitic pyroxenite borders a 12 ft. wide body consisting of *(a)* 4 ft. of vermiculite including streaks of apatite, *(b)* 3 ft. of interlayered and lenticular vermiculite, vermiculite–apatite–pyroxenite, apatitic pyroxenite, and *(c)* 5 ft. of apatite–diopside rock including vermiculite lenses. Vermiculite crystals bordering pyroxenite inclusions are particularly large. In another occurrence of this zone, a 1–2 inch thick wall zone of vermiculite has developed along one side of ramified apatite bodies. These minor occurrences are included in pyroxenites devoid of serpentine.

At 'Valentine's Beryl Pit' a vermiculite–apatite–pyroxene body appears to be related to nearby serpentinites. In the former main working the proportion of these three components has been 65%, 30% and 5%. However, the ratio differs in neighbouring pockets. Good grade vermiculite crystals have diameters of 9 inches, measuring a maximum of 6 inches in thickness. On the other hand, the diopside crystals may attain 1 ft. in diameter. At Wegsteek, 1 mile south-southwest of the Loolekop, apatite has been found with subordinate pyroxene in vermiculite. The Mamba pit measures 25 by 15 ft., extending to 15 ft. of depth. As in other occurrences, apatite has been disseminated only in the core of the granular vermiculite body. Near the southern edge of the complex, at one of the Land A Quarry apatite veins, stringers of vermiculite, also projecting into it, form the wall of the 1 ft. wide vein. The apatite contains 5% feldspar and some diopside. In this area isolated vermiculite flakes and grains measuring up to 4 ft. in diameter have been found.

Vermiculite also occurs in other parts of the northeastern Transvaal, e.g., at Nooitgedacht, Redhill and Natkruit.

NORTHEASTERN AFRICA

At *Hafafit* in the ultrabasic belt of southeastern *Egypt*, large and small xenoliths of serpentinized olivine–chlorite tremolitites measuring up to 400 ft. are preserved in granitic gneisses. Hornblende crystallized in the gneisses rimming the xenoliths. The serpentines have been cut by feldspar veins and carbonatized at their contact. The feldspar lenses are surrounded by a thin rim of oriented anthophyllite with 1–3 ft. of random anthophyllite occurring between the feldspar and the serpentine. Vermiculite forms pockets in the feldspar and between the latter and anthophyllite (AMIN and AFIA, 1954). However, in the gneiss the feldspar dykes change into barren pegmatites.

South of Port *Sudan* vermiculite occurs in an area located 16 miles south of Tohamiyam in serpentinite bodies surrounded by gneissose granite (ABDULLA, 1958).

Northeast of Lake Victoria, *Uganda*, vermiculite is abundant in the Sukulu and Bukusu carbonatites. Although the diameter of Bukusu attains 5 ½ miles, the carbonatite core is only 3 miles wide. Foyaites are the dominant alkaline rocks. Accessory minerals include sphene, zircon and pyrochlore. Vermiculite, probably derived from phlogopite, is a component of the magnetite–apatite (phoscorite) ring which extends 6 miles in the southwestern sector. Highest values include Namekara, Nakhupa and Surumbusa (Taylor in BARNES, 1961). The large Namekara deposit is covered by 14 ft. of loose magnetite cover, extending to a depth of more than 50 ft. Highest grade black varieties of vermiculite are scarce. Moreover, medium grades (–22 lb./cubic ft. density) are less abundant than lower grades. In *Kenya*, northeast of the preceding deposits and of Mount Elgon, vermiculitic micas occur in the West Suk district. Other Kenyan deposits are located east of Nairobi and north of the Kilimanjaro at Sultan Hamud, Machakos and in the Taita Hills in lenses surrounded by ore in fissures cutting Precambrian folds. The Kinyiki and the particularly high grade Kipiponi occurrences follow the contact of dunite bodies.

SOUTHEASTERN AFRICA

At Kwekivu, south of Mount Kilimanjaro in *Tanganyika*, large books of high grade vermiculite form small pockets in a pegmatite traversing ultrabasics. The Mziha occurrence in the Uluguru field of eastern Tanganyika is similar, with vermiculite also occurring in the Mikese pegmatites of this field. At Lufusi in central Tanganyika the mineral is located in the alteration rim of ultrabasics intersected by pegmatites.

The Kapirikamodzi deposit is located 3 miles north of Mpatamanga Bridge in the Blantyre district of southern Malawi. Vermiculite zones and pockets average a weight of 100 tons. Although of high grade, pegmatitic vermiculite is fine grained and only 5% are commercial. Ore bodies extend to a depth of more than 50 ft. West of Kapirikamodzi vermiculite occurs in the anorthosite–granodiorite zone of Tete in *Mozambique* on the lower Mavudzi River at Tundumula and near Matema.

Vermiculite is known to occur in several localities in *Southern Rhodesia*, especially in the Mtoko field west of Tete. Furthermore, the large Shawa carbonatite consists of a plug, magnetite–apatite sövite, serpentinous rauhaugite, an intermediate zone of serpentinite and dunite in the south, and the fenite ring including rheomorphic syenite and late apatite ijolite. The width of the intermediate zone expands from 0 to $^3/_4$ mile. Vermiculite occurs in the eastern and northeastern sectors of this ring. Its veins, derived from phlogopite, traverse melteigite and ijolite. On the other hand, in the northwest vermiculite has been altered from biotite (TYNDALE–BISCOE, 1959).

Vermiculitic micas occur in *Madagascar* in the Ampasimpaona and Ambatofotsy pegmatites of the Itasy field, at Maromby, Andreba East, Ambohimanakana, Anony and Ambakaka, in the Alaotra field and at Antsampandrano 20 miles north of Anjozorobe.

Finally, vermiculite is an alteration product of mica in the Mondangal pegmatite of southwestern Angola, of the serpentinites of Hangha in eastern Sierra Leone, and also in the Moroccan Anti-Atlas.

DIAMONDS

During the Cretaceous the world's principal kimberlites were emplaced along fissures that were formed in zones of weakness near and on the edges of shields or basins. Furthermore, diamonds also occur in the 2,200 million years old breccias of Ghana, and sparsely in the Witwatersrand banket. We distinguish four major belts:

(1) The West African belt, extending from Sierra Leone through Guinea and the central Ivory Coast to southern Ghana.

(2) The belt surrounding the Congo basin, from the Central African Republic through the orogenies bordering the Great Lakes in the west, to Kasai/Angola and the Kwango.

(3) The South African belt following the southwestern edge of the Transvaal orogenies along the Karroo transgression.

(4) The Southwest African belt which follows the Atlantic coast for 500 miles.

Carbonatite alignments of the same cycle are generally nearer to the middle of the shields. Smaller diamond zones occur under similar conditions southeast of Lake Victoria and in western Mali (Fig. 5).

THE GENESIS OF DIAMONDS

Both kimberlite veins and pipes were extruded under great pressures in several cycles, as a highly plastic mush of larger and smaller xenoliths, saturated by CO_2, H_2O, and other fluids (MEYER DE STADELHOFEN, 1961). Like carbonatite, protokimberlite appears to have developed at great depth as a solid-fluid system by partial refusion. Kimberlitic tuff and dust distinguish the first explosive phase. Second generation kimberlite re-cements brecciated agglomerates. Basic dyke rocks, frequently found in association with kimberlite, constitute a genetically related, less mobile facies. Since it causes expansion, autohydration of the kimberlites enhances their intermingling with host rocks.

Artificial diamonds crystallize at 100,000 atm and 2,500° C. Their study tends to prove that the natural stones are formed through a rather long *pt* range with opaque cubes characterizing the lowest, and transparent octahedra the highest, temperatures. Dodecahedra and complex forms are believed to be intermediate. The Sierra Leone kimberlite veins are thought to have formed in the highest *pt* conditions, followed by the Premier mine and the first Bakwanga phase. The second Bakwanga phase, followed by the Kimberley group and Mwadui, belong to lower *pt* conditions. Moreover, re-cementation of both opaque and transparent diamonds indicate fracturation and regressive cycles. Swift chilling, i.e., rapid decompression, appears to favour the development of high quality diamonds. However, diamonds are scarce in eclogites which crystallize at high but constant temperatures.

Let us enumerate a number of interesting features: kimberlites frequently contain phlogopite; veins are probably more abundant than pipes; several pipes have been extruded from dykes. Diamond grades decrease rapidly with depth. Thus, a majority of the stones has not been found in kimberlite but in the altered carbonatized serpentinites overlying them. Indeed, 85% of the world's diamonds come from the mixed breccias of Bakwanga (Congo). Diamonds have been eroded from the top of the pipes and veins and redeposited in alluvia. Stones were frequently recycled and often occur at the base of a sedimentary series. Such commercial concentrations extend more than 200 miles from the supposed original deposit.

SIERRA LEONE

Sefadu, the diamond centre, is located 40 miles (Fig. 17) northwest of the triangular border of Sierra Leone, Guinea and Liberia. The majority of the diamonds has been found here in the hydrographical system of the *Bafi*. Stones have been carried for 100 miles downstream by its collector, the *Sewa*, as far as Sumbuya where the tide reaches. The principal diamond placers include the Moinde, the Meya, and the Woyie, which locally contains 250 carats per cubic yard, stones of 50–100 carats, and the world's largest alluvial gem of 770 carats. Terraces rise several miles from the present river bed. The Kenja stream, 80 miles south of Sefadu, having cut through the Moa terraces, has redeposited diamonds.

Roof-screens of magnetic and amphibole-quartzites and talc-schists remain in the granitic Basement of Sefadu, which was injected by dolerite swarms. Geologists have recognized two major northerly faults here. From the confluence of the Bandafayi and Gborora, a swarm of vertical carbonatic kimberlite veins extends en échelon 12 miles N60 °E through Yengema and Koidu (Fig. 103). Beyond the Moinde River the width of the swarms is 1–3 miles. The thickness of each lens varies from a few mm to a few ft., but they also widen to blowouts and to a pipelike body. The kimberlite was injected or extruded in several phases, carbonatizing the country rock. Geologists have identified another swarm zone 30 miles south (GRANTHAM and ALLEN, 1960).

The surface veins have been altered to serpentinite, and principally form pseudomorphs after olivine and contain carbonates, chlorite, talc, phlogopite, some ilmenite and garnet.

Occurring also in the groundmass, serpentine and primary carbonate veinlets traverse the rock. A sample includes 58% serpentine, 24% phlogopite, 9% calcite, 5% metallics,

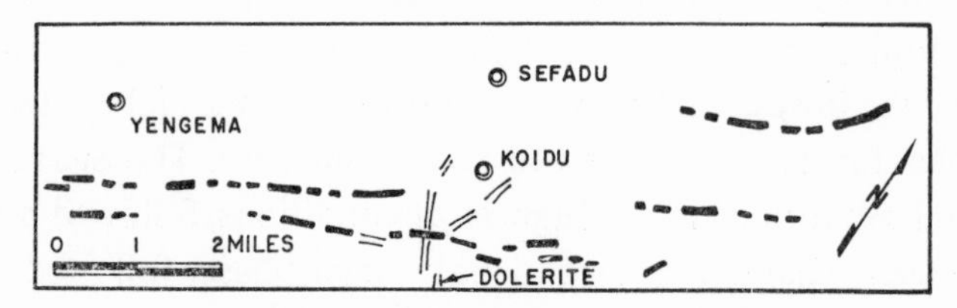

Fig. 103. Occurrence of kimberlite dykes.

2% perovskite and 1% fluorapatite. Apatite, accompanied by perovskite and magnetite, forms an assemblage similar to phoscorite, a carbonatite rock. Magnetite occurs both in the serpentine and phlogopite, where it occasionally exhibits inverse pleochroism. On the other hand, ilmenite and magnetite bearing agglomerate surrounds perovskite margins. The perovskite contains 3% rare earth oxides, 1% pentoxide of columbium and approximately 0.2% thorium oxide. Furthermore, 0.01% ZrO_2 reports in kimberlite analyses. Garnet, with the pyrope molecule predominating, is scarce, but chrome diopside is absent. Although it contains phlogopite, the kimberlite groundmass more resembles the composition of basaltic kimberlite than that of mica peridotite. The rock contains variable amounts of xenoliths, especially biotite granodiorite (±amphibole), which frequently constitute a greater part of the rock than the kimberlitic matrix. The kimberlite has reacted with the granodioritic inclusions, incorporating fractured crystals in its groundmass. Occasionally a younger phase invades older kimberlite and surrounds inclusions. Whereas a deep-lying serpentine contains garnet, carbonate and iron ore, eclogite consists of euhedral diopside–jadeite, anhedral pyrope–almandite, phlogopite, perovskite and ilmenite, rimmed by perovskite.

In the placers diamonds are associated with framesite (black stones ± 10–20% impurity), ilmenite, magnetite, chromite, garnet, gorceixite, corundum, zircon and rutile. Alluvials may be green-opaque, grey, coated or fractured. Frequently high grade gems form glassy octahedra. The average size is ½ carat. Pure and especially opaque diamonds found in the kimberlite veins are also rounded, while some of the abraded stones have acquired a later rim. Geologists believe that the regrowth has taken place in depth. Furthermore, a black substance, possibly graphite, forms inclusions.

Larger gems are more frequently twinned according to the spinel law than smaller ones. Opaque stones also twin according to other laws. Frequently asymmetric or elongated, transparent octahedra and their derivatives exhibit growth layers and shields. However, plates or crinkles cover only part of the crystal faces. Rare green speckled stones are believed to have been coloured by radioactive exposure. As at Bakwanga, symmetrical or spherical opaque stones contain a transparent core. The coating may also be banded or spotty and it is rough and brittle. The colour of the coating varies from green to grey. Cubes and dodecahedra are common among opaque stones, cubes being particularly abundant among the smaller crystals. Some of the fractured opaque stones have been recemented by a transparent diamond substance.

GUINEA, LIBERIA AND IVORY COAST

According to reports, diamonds would show in the country. However, these traces appear to be confined to the Sierra Leone (whence most of the diamonds are imported) and Guinea borders (SRIVASTAVU, 1961).

At a distance of 35 miles the Lofa basin follows the Sierra Leone border. From south of Bomi Hills the field extends 65 miles to beyond *Wea-Sua* in the north, in the Suehn Bopulu district and Grand Cape Mount county (SRIVASTAVU, 1961). In addition, diamonds reportedly occur in the Todee District and around Tapeta.

Beyond Wea-Sua in the north diamonds are also said to occur in and around Zoi Town. This, however, does not preclude the occurrence of diamonds south of Bomi Hills and between Wea-Sua and Zoi and beyond. In widths of from 4 to 6 miles this diamond belt extends on either side of the Lofa River, including the Butulu, Wea-Gay, Wea-Farma, Wea-Day, Wea-Mata, Wea-Jua-Gba and Bonmabor Creeks. In the northern extension of this alignment, diamonds have been found near Zoi (Zue) at the headwaters of the Mano River, 100 miles from the coast and 15 miles from the Sierra Leone border.

The Sanekole field extends from Bahn in the south to Zulowoi, 7 miles northwest of Sanekole in the north. It covers a belt of 35 miles in the triangle of the Guinea and Ivory Coast borders south of Nimba. Creeks which are reported as showing concentrations include the Yee-Hoh, Tene-Yee, Kolayah, Kori-Yee and Yen-Yee.

In gravels, placers of creeks, and streams of these fields, diamonds appear to be accompanied by rutile, ilmenite, corundum, gold and magnetite.

The Lofa River belt diamonds differ greatly in quality, appearance and size from the diamonds of the Sanekole area. The former are mostly of gem quality and smaller in size, while the latter are industrial and larger in size. Finally, diamonds have also been reported to occur in the Todee district around Tapeta (Tappita), 100 miles from the coast and 30 miles from the Ivory Coast border, and 30 miles northeast of Monrovia.

The diamond district of *Upper Guinea* extends from Kankan, in the north, to Beyla and Nzérekoré near the Liberian border in the east, and thence through Macenta and Gueckédou to the northwestern corner of Liberia. Rising 50 miles north of Macenta, tributaries of the Moa, Milo and other streams shed the waters of the same mountains. The Beyla–Macenta field contains the greatest number of stones, followed by the areas of Férédou, Banankoro, Bonro and Feniara. Furthermore, diamonds are found at Bounoudou west of Beyla, Bassam, Koulountou, Tominé and in the High Maceni. As in Sierra Leone diamonds have been eroded from sub-vertical, 1–3 ft. thick, kimberlite breccia veins. However, the Banankoro placers are eluvial–alluvial.

In the *Ivory Coast* the diamond province, underlain mostly by gneisses and granites, covers most of the northwestern half of the country. Deposits occur at Ferkessedougou, Korhogo, Seguela, Daloa, Bouaflé and Bouaké. Diamonds also occur in the Cavally at Nuon at the border with Liberia and Guinea. In the west the Sassandra, one of the big rivers, contains some diamonds 25 miles upstream from the Piehon Bridge. One stone has also been found in the Niéza tributary. The principal field is located at Tortiya on the left bank of the Bou River, a tributary of the Bandama. North of Seguela and at Toubabouko (Camp Bohy) is a minor district. The placers occur both in terraces and flats under a cover of a few inches or ft. The Tortiya gravels, supplying a high proportion of gems, have been derived from the underlying schists (Fig. 54).

GHANA

The Birim field is located 70 miles northwest of Accra, but the smaller Bonsa area is situated in southwestern Ghana (Fig. 21, 54) 30 miles from Axim and Takoradi. Isolated diamonds have also been found in the Obosum River and downstream in the Volta, in the southwest near the border, and along the Upper Volta frontier (JUNNER, 1943).

The Birim field covers 400 square miles, extending further downstream. However, the principal concentrations occur in the triangle of Birrimian graywackes, metalavas and amphibolites. The hypotenuse of this triangle extends from Aboabo 17 miles to the northeast beyond Akwatia. About 4 miles of the Atiankama basin are mineralized, but the other diamond carrying tributaries of the Birim extend scarcely farther than 2½ miles. The left hand affluents of Beduawara, Sandy Creek and Atiankama–Esuabena drain the waters of lenses of ultrabasic metalavas and pyroclastics. This triangular area is 5 miles long and extends to a maximum width of 3 miles in the south. In addition, disseminated diamonds have been found northwest of Birim, but no diamonds are found in streams flowing on granite. The line of maximal concentration trends parallel to a series including ultrabasics, tuffs, tuffaceous sandstones and phyllite. The ultrabasics are, in fact, porphyroblastic actinolite–tremolite schists which contain talc with or without chlorite, disseminated rutile, and fragments of the fine-grained rock and phyllites.

We can distinguish at least three *alluvial* diamond cycles: lateritic hill top gravels, terraces 90–100 ft. above the Birim, and the present hydrographic system. Greater concentrations are limited to small streams of low gradients (especially to hollows and potholes of the bedrock) and to their confluences. As velocities increase, grades decrease. Frequently the sandy gravel 100–150 ft. from the stream bed carries the greatest concentrations. However, good stones are also washed into the outer argillaceous gravels. Downstream in the Birim, Supong, Amaw and other streams, the size of the diamonds diminishes. Similarly, in the semi-solid terraces diamonds tend to be deposited in irregular concentrations near the bedrock, e.g., at Aboab, Jyedem and Takrowasi on the Birim, and between the confluence of the Supong and Atiankama. Terrace diamonds have been recycled by recent tributaries of the Birim. Only along the mineralized axis of Atiankama–Bawdua – especially at the contact of actinolite schists at Beduawara, Eduboui and Atiankama – do ancient hill gravels contain some encrusted low grade diamonds. The principal placers include the systems of the Esuboni, Abansa–Beduawara, Supong–Tuasenda and Atiankama–Esuabena, and others northwest of the Birim River, including the Subin and Subinsa.

About two-thirds of these diamonds are industrial. They are small and usually grey in colour, but sometimes they are white. Green diamonds (cf. Witwatersrand) are found at the sources of the Beduawara. Concentrates carry staurolite, ilmenite, rutile, tourmaline, chrysoberyl, gold and gorceixite. Furthermore, ilmenite and rutile also show as inclusions in diamonds.

The *Bonsa* field also forms a triangle, with its 15 mile long hypotenuse trending northeast. Near the contact of quartzite and conglomerate lenses of the Banket series, the placers flow on Lower Precambrian, Tarkwaian Kawere conglomerates. At the head of the streams diamonds have been recovered from altered Kawere detritus. Moreover, at Ntronang diamonds have actually been found in the auriferous banket. Averaging only 0.03 carats in size, the stones resemble those of Birim: two-thirds are boart or industrials. However, although considerably less staurolite is associated with the concentrates, limonite, ilmenite, tourmaline, magnetite, rutile and gorceixite do appear. All these placers are pre-Recent. Mineralized tributaries of the Nfutu include the upper Supong, Anakwabu, Aden and Dubirim, with the Tisini and Enikawkaw draining the deposits further east. Finally, north of the Bonsa the Pintotoum and Aboum carry diamonds.

CENTRAL AFRICAN REPUBLIC

We can distinguish two groups of deposits: the Berberati district in western Ubangi (near the southeastern corner of the Cameroon), and the Kotto–Ouadda district near the Sudanese border.

In the *Berberati*–Carnot district, conglomerates and arkoses of the lower Kwango and sandstones of the upper Kwango series surround a large branch of the basement (Fig. 53). However, earlier lower Karroo tillites outcrop infrequently. Finally, lateritic sands cover the series. The lower Kwango sea has flooded the area from the south. Diamonds occur downstream from the point of crossing of the rivers with these 200 ft. thick Lower Cretaceous beds. Indeed, in a 40 by 20 mile area, pre-Recent placers of the right hand tributaries of the Mambéré River — the Goudjembé, Sama, Sangema, Ngoélé and Ouabembé — rest on Kwango sandstones, especially at changes of flow velocity such as canyons and potholes. On the other hand, terraces exhibiting variable values lie as high as 120 ft. above the Ngoélé, Goudjembé, Batouri and Bolé. The size of diamonds shows a sympathetic relationship with the size of the gravel, as do values occasionally. The concentrates contain tourmaline, rutile, staurolite and minor quantities of gorceixite and kyanite, but chromium is practically absent. Colourless diamonds crystallize in cubes and convex octahedra.

Diamond carrying placers have been derived from the contact of the basement and sandstone in the Guilingala (a left bank tributary of the Lobaye), and from the basement in the Lopo (an affluent of the Mambéré). In addition, diamonds occur at Sapoua and Dongo, south of Berberati, and at Baroudo and Banzia. The deposits occur downstream from the point of intersection of rivers with the 200 ft. thick lower Kwango series (Lower Cretaceous) which has flooded the area from the south.

At Mouka–Ouadda, in the Bria district of eastern Ubangi (Fig. 13; DELANY, 1959), the basement is overlain by arkosic and fine grained sandstones, argillites and surface deposits. Although not limited to the sandstone zones, diamond carrying placers are believed to have been derived mainly from this contact. Their morphology is similar to Congolese stones. The diamonds are transparent, non-coloured or green-brown.

ANGOLA–CONGO

The Dundu (Lunda, Angola)–Tshikapa (Kasai, Congo) district occupies the hydrographic basin of a north–south oriented, 200 mile long and 60 mile wide, area. (N.B. Portuguese

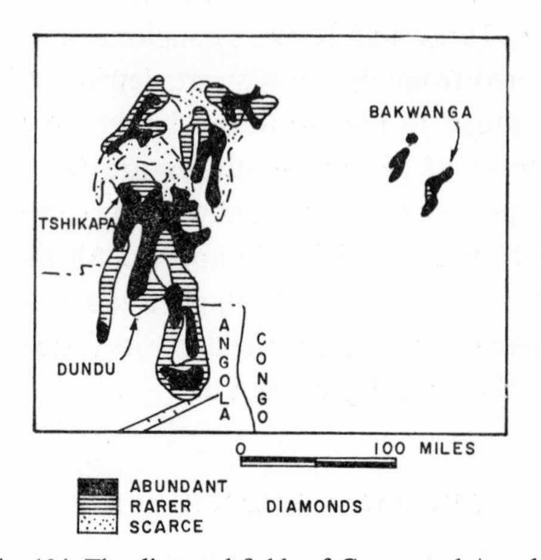

Fig. 104. The diamond fields of Congo and Angola.

spelling follows the Congolese names in parentheses.) Placers of the Kasai basin include the lower and middle sections of the left hand tributaries of Tshikapa (Chicapa), Longatshimo (Luachimo) and Tshiumbe (Chiumbe)–Lubembe (Luembe). In addition, they include discontinuous parts of the Kasai River, which extends for 130 miles, minor right hand affluents, and, further northeast, the lower Luebo and Lulua. The southern, Angolan, quarter of the district contains most of the remaining reserves (Fig. 104).

Geology of the district

We can distinguish two Mesozoic diamondiferous–sedimentary cycles (FIEREMANS, 1960): *(1)* the widespread Lungundi fluviate and delta phase, and *(2)* the subsequent Kalambo phase, partially eroding *1*.

Trending N60°E, the Lucapa graben limits the field in the south. This Rift Valley is at the northeastern end of the proto-Karroo alignment of ultrabasic–carbonatite complexes extending from Mossamedes through Nova Lisboa towards the Congo. The principal detrital deposits are located between Maludi and Cassanguidi in the drainage of the Luembe, along a 40 mile section downstream from the southern rim of the Lucapa graben. We find other concentrations in the Luana drainage near to its confluence with the Chiumbe, and in the northern rim of the Rift Valley around Mussolegi. Around Sanfuto and Fucauma diamonds concentrate in the Chiumbe 50 miles downstream (i.e., north) near the Congolese border. However, kimberlites outcrop at the end of the graben 60 miles southwest of Mussolegi.

The Basement complex consists of biotite and amphibolite gneisess, granites, micaschists, amphibolites and quartzites. Quartzites, schists and granites of the early Precambrian Kibara system outcrop along the diamond bearing areas of the Chiumbe and Luembe. Only the lower and middle Karroo are known, and they consist respectively of arenaceous and arenaceous–argillaceous sediments. In the Kwango (Calonda) Series we find four layers: *(1)* diamond bearing conglomerates, *(2)* cross-bedded violet sandstones, *(3)* sandstones with interbedded conglomerates and argillites, and *(4)* silicified sandstones. The Kalahari System consists of polymorphous sandstones overlain by a loose cover. We can distinguish five late tectonic cycles:

(1) the north-northwest trending fractures (Nyasa alignment) affecting the lower Karroo,

(2) post middle-Karroo east-northeast striking faults, which include the 70 mile long and 10 mile wide Lucapa Rift Valley,

(3) a kimberlite injection at the base of the Kwango,

(4) a late Cretaceous peneplane formation, and

(5) the Middle Cretaceous erosion.

Kimberlites

At the western end of the Rift the largest kimberlite body, of ½ by ½ miles, is sub-intrusive into lowest Karroo strata at the confluence of the Camazambo and the Chicapa (Tshikapa) between N27°W trending faults. As at Bakwanga the kimberlite is partly intrusive in and partly concordant with the sediments. A well developed breccia contains 50% kimberlite proper, plus xenoliths of granite, gneiss, amphibolite, quartzite and vein quartz, as well as inclusions of lower and middle Karroo tillite, sandstones and argillites. However, on the Camafuca kimberlite is more basaltic. This part is believed to be nearer the centre of the body. In addition, kimberlite, almost devoid of xenoliths, outcrops in the north along the Cassapa. Nodules attaining a length of 4 inches consist of hornblende–garnet pyroxenites carrying some ilmenite. Rarer nodules consist of pyroxene hornblendite.

Representing 34% of the harder rock, both of these varieties have been carbonatized. The kimberlites have been altered to 'blue ground'. 'Yellow ground', which extends to a depth of 25–30 ft., even contains lateritic fragments. Olivine has been completely replaced by antigorite and chrysotile. The surface of kimberlite has a characteristic chestnut 'escuro' colour. Ilmenite and diopside (2%) are abundant, but zoned pyrops, chromiferous almandite–spessartite, magnetite and titano-magnetite are rarer. Garnet represents 27% of the unaltered minerals. Enstatite, kyanite, andalusite, phlogopite, calcite (2%), octahedra and dodecahedra of industrial and gem diamonds are scarcer.

The Camuzenze phlogopite-kimberlite breccia and tuff, 4 miles south, is intersected by numerous calcite veins. The sub-stratified kimberlite breccia of Caindjamba and the tuff of Caquele outcrop 1,500 and 2,500 ft. south of the Camazambo body. Between the lower and middle Karroo the diamond bearing breccia of Cangoa has been emplaced 4 miles north-northwest of Camazambo. This mixed rock contains abundant calcite, enclaves of kaolinized feldspar, and heavy minerals of the acid suit. Furthermore, kimberlite tuff outcrops 2 miles northeast on the Caxixima. About 10 miles south of Camazambo the 500 by 1,200 ft. oval of carbonatized serpentine of the Camuanzana kimberlite contains 1 ½ inch granules of ilmenite, chrome pyrope, diopside and magnetite. On the other bank of the Chicapa tuffs outcrop on the Camaumbo. In the midst of a rich alluvial diamond zone the Canzololo tuff, which extends 90 ft., is situated 5 miles southeast of the Camazambo. Although this tuff grades into sediments, its contacts with the surrounding gneisses are sharp. Near the southern rim of the graben and the Luachimo River the Carivé body is situated 20 miles east-northeast of the main kimberlite area. Strongly resembling Camazambo, at 50 ft. of depth the kimberlite contains a great deal of ilmenite and magnetite as well. However, another facies has been intensely serpentinized.

The diamonds are not fractured. Although 40% of the stones are coloured, they are rarely opaque. We can summarize their morphology as shown in Table CXXXIV (REÁL, 1959).

TABLE CXXXIV
DIAMOND FORMS

	Camafuco		*Mussolegi*	*Iondi*
	kimberlite %	*alluvials* %	*kimberlite* %	*kimberlite* %
Rhombic dodecahedron	30	50	20	27
Dodecahedron octahedron	19	13	13	13
Octahedron dodecahedron	5	10	21	11
Octahedron	13	8	25	26
Other habit	5	2	3	1
Twins	4	2	1	3
Aggregates	3	1	1	1
Rounded stones	0	1	1	2
Imperfectly crystallized	19	10	12	15

Placers

The diamond carrying lower Kwango conglomerate of desert origin consists of pebbles of gneiss, quartzite, quartz, agate and argillite. It attains a thickness of 3–6 ft. At Nachitango and Cachimbungo the formation contains 6 ft. boulders, but the composition of the pebbles infrequently depends upon the Basement. Along the Cononda River the formation directly overlies the Basement, and in the Mussolégi district the conglomerates cover large basins. Sandstones of the Kalahari erosion surface rise at 2,500–2,700 ft. of altitude and cover large areas of central and southern Africa.

The principal placers of Angola are located between north-northwest trending faults spaced at 5 miles which extend from Fucauma and Sanfuto on the Chiumbe and the Congolese border, to Cassanguidi near the Luembe and the northern rim of the Lucapa Rift. Many placers of the main area cover the vicinity of faults belonging to the same trends as the kimberlites further west. Deposits of this type include, for example, the Cabuaqece and Saga Rivers, the Calembe–Mondji and Lussaca zones, the Nachitango–Calonda confluence, and the Cassanguidi, Furiz and Luachimo Rivers. North of Andrada placers carry diamonds downstream (east) in the tributaries of the Luembe. In the Graben the group of alluvia continues west of the fault to Maludi. At Mussolégi, on the Luana, another placer zone is located at the intersection of a parallel fault with the northern rim of the Graben.

Tertiary placers include several cycles of eluvia and alluvia. In addition, gravel terraces, which attain a thickness of 30 ft., contain sand lenses. They are distinguished by a sizing and rounding of alluvials. Apparently most of the deposits are derived from successive erosions of the preceding sediment. In the main area of Mussolégi, Luana, Iondi and Luembe, 20–35% of the alluvials are industrial. On the other hand, in the Tshikapa district opaque diamonds are rare, as are zoned stones or polychrome twins.

CONGO

Diamonds are disseminated in gravels along a vast but discontinuous arc surrounding the Congo basin. It starts in the north in the Central African Republic and turns southeast of Stanleyville through Kivu and Katanga. The arc then bends west towards the world's principal deposits at Bakwanga and continues to the Dundu (Angola)–Tshikapa field and the Lovua–Kwango district south of Léopoldville. Gem diamonds of the Uéle gold belt occur along the contact of Middle Precambrian schists and granite 150 miles northeast of Stanleyville. Diamonds have been recovered in numerous gold and tin placers of Kivu–Maniema, e.g., in the Omate in the Kima district, at Kaseka in the Kailo district, at Messaraba near Kalima, at Kibila near Moga, and at many other places. Near Shabunda diamonds are even more abundant in the Mutandulwe placers, as there are dykes of amphibolite and dolerite in most of these diamond areas. West of the Congo–Lualaba the middle Lomami carries rare diamonds of the Bakwanga type. East of Bakwanga a group of pipes which contains several diamonds pierces the Katangian sediments of the Kundelungu plateau. However, their altered kimberlite differs from the Bakwanga breccia (Fig. 105). Diamonds are also disseminated in areas of Kasai–Lunda other than the Bakwanga and Dundu districts. However, between the Kwango and Lovua the hydrographical system is only poorly mineralized, for these rivers eroded less of the Kwango Series.

Bakwanga is situated 130 miles northeast of the northeastern corner of Angola. Here 35 miles of the Bushimaie River are mineralized, including a concentration at Bakwanga and another upstreamtoo. With the Luilu River flowing northward, we find the placers of the Lubi and Lukula 30 miles to the northwest.

Rocks of the Basement outcrop only in the valleys. Gneissose granites, including amphibolites and gabbros, are overlain by 700 ft. of earliest Precambrian (?, cf. Katangian)

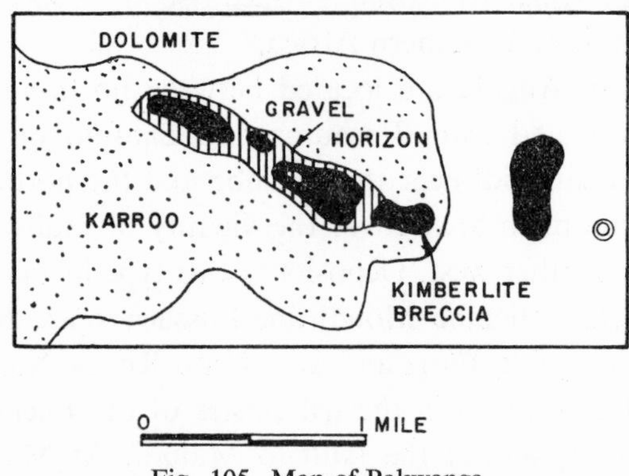

Fig. 105. Map of Bakwanga.

strata consisting of *(a)* 120 ft. of shaly sandstone, *(b)* 100 ft. of shales, calcschists and black silicified limestones, and *(c)* 450 ft. of calcareous breccia, shales and dolomite. In the Karroo we can distinguish three series of sandstones and argillites resting on basal conglomerates. While in the west the latest of these, the Kwango Series, covers a considerable area, only the middle or Lualaba Series outcrops in the east. Following them, Kalahari sands are underlain by their silicified basal breccias and limestone. Finally, the Plio-Pleistocene cycle of erosion developed the present peneplain and its red limonitic sands.

The Bakwanga breccia

At *Bakwanga* proper, early Precambrian (?) dolomitic limestones, Karroo sandstones, gravels and sands outcrop. The dull, warped, undulating limestones, dipping north, are tightly or broadly bedded. An upper horizon of silicification includes chert and neogene quartz. Karstic alteration, circulating solutions and streams have cut dolinas, caves and deep valleys into the red surface.

In a mile wide band the limestones are covered by typical lower Karroo sandstones, including flat lenses of dark shales. They follow the axis of the Kanshi–Bushimaie spur and start a mile west of Bakwanga. The thickness of these strata is in excess of 150 ft. Attaining a maximum of 10 ft. of thickness, the old limonitic gravel horizon is mainly conserved along the sides of the hills. Diamonds are disseminated in the ill-sorted, angular to rounded gravel. This gravel contains boulders particularly on the hillsides downslope from the kimberlite breccia. Several miles from the pipes the old gravel also serves as a source of secondary concentration. The sequence is covered by 70 ft. of brown Kalahari sand.

South of Bakwanga kimberlitic breccia, while also piercing the lower Karroo Lualaba sandstones, has filled in the Katsha pipe (Fieremans, 1955). Between the Bushimaie and the lower Kanshi we can distinguish at least five similar centres in and west of Bakwanga. They extend along a zone of weakness for 1 ½ miles in an east–west trending arc. Measuring 2,000 by 1,000 ft., the Bakwanga oval is oriented north–south. Further west the bodies of the Disele spur are arranged in the Karroo in the direction of the presumed zone of weakness, with their dimensions attaining respectively 1,000 by 700 ft., 1,800 by 1,000 ft., 400 by 200 ft., and 1,800 by 700 ft. These breccias appear to be interlayered or mingled with fragments of Karroo sandstones to a depth of 300 ft., and partly cemented by dolomitic relics. The breccia proper consists of *(1)* more or less silicified angular boulders of the pierced limestones and dolomites, attaining a diameter of 3–10 ft., *(2)* smaller sub-angular xenoliths of red schists, originally accompanied by sandstones, underlying the early Precambrian limestones, and *(3)* numerous rounded inclusions of granites and, less frequently, biotitic plagioclasite and eclogite. Rounded magnesian ilmenite grains occasionally measure ½–¾ inch in diameter, with industrial diamonds reaching a record of 460 carats. Chrome diopside, garnet and zircon also belong to the paragenesis. The bluish colour of the groundmass changes to green-gray and finally to red at the surface and in lateral digitations. Carbonatization has progressed along veinlets and in granules of 1–4 mm. Because of largely supergene silicification, late stringer calcite disappears at the periphery of the bodies. Chlorite is scarce.

The Placers of Bakwanga

Detrital deposits may be found in the (late Karroo) Kwango sandstones and in the late Cretaceous–mid Tertiary sediments. Near the Bushimaie River downstream of Bakwanga, sandstones overlain by sandstone boulders of the Kalahari (Tertiary) durcicrust contain diamonds. Diamonds and other eroded particles of the breccia have been rearranged on the late Tertiary surface *(1)* in the gravel sheet, *(2)* in the overlying red sand layer of 1 ft. or so and *(3)* in the fragments of soft sandstone embedded in sand which follows in the sequence. They were later redeposited in *(4)* dry karstic valleys filled in by gravel of chert, silicified limestone and breccia fragments, *(5)* alluvial and slope accumulations, *(6)* eluvia proper and *(7)* various alluvial placers north (i.e., downstream) of Bakwanga. The blue clayey bedrock of the river beds, which is partly derived from the pipes, also carries diamonds, with great concentrations accumulating in the valley of the Bushimaie and Kanshi.

We distinguish three groups of deposits: Bakwanga, Lukelenge and Tshimanga. The *Bakwanga* placers extend downstream in the Kanshi–Bushimaie basin, from the station 3 miles upstream. The *Lukelenge* placers, 7 miles north, represent a partial reconcentration of the Bakwanga diamonds and continue sporadically downstream in the Sankuru (the Bushimaie's collector) beyond Pania Mutombo. South of Bakwanga the *Tshimanga* placers, located on both sides of the Bushimaie at the confluence of the Katsha, have been carried 20 miles down the Tshimanga River. This area is underlain by the contact of the Basement and early Precambrian (?) strata.

Opaque diamonds tend to crystallize in cubes, octahedra, dodecahedra and their double or triple combinations, with transparent stones developing hexatetrahedra, octotrihedra and dodecatetrahedra (but not cubes). While the low temperature variety exhibits both twins, transparent stones favour only the spinel twin. Large opaque diamonds have an euhedral transparent core suggesting that the latter crystallized first. Frequently both varieties are joined together.

TANGANYIKA

Southeast of Lake Victoria kimberlite pipes are aligned along three arcs. Extending from Usongo through Kisumba and Uduhe 50 miles to Mwadui, the first of these trends north-northeast. The second connects Kisumba with Ilole in the east-northeast. Thirdly, the Mabuki pipe (Fig. 33, 106) is located between Mwadui and Lake Victoria. At Mabuki, Kisumba and Uduhe the surface deposits which overlie kimberlite contain diamonds.

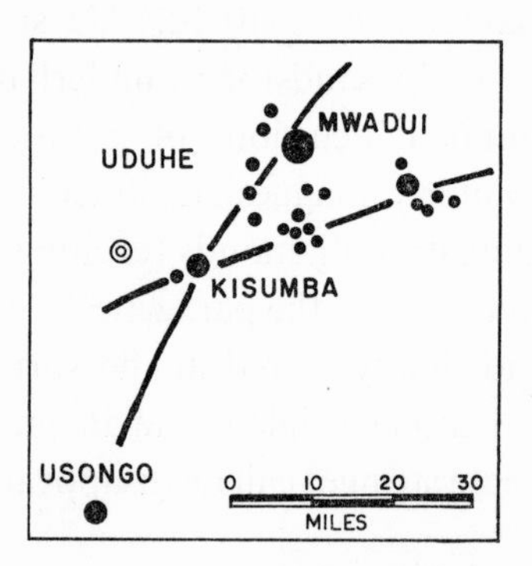

Fig. 106. The diamond pipes of Shinyanga.

Southeast of these arcs the third lineament is located on the Iramba Plateau. This arc connects Mtawira with Kisiriri in the north. Other pipes occur in the Ruhuhu Valley east of Lake Nyasa.

Covered by pre-Recent sediments the *Mwadui* crater (Fig. 107), located south-southeast of Lake Victoria, has a diameter of 4,000 ft. (TREMBLAY, 1957). The pipe was mainly extruded through Precambrian granites. The column of crater rocks has undergone intense alteration and rearrangement (Table CXXXV; HARRIS, 1961).

TABLE CXXXV
COLUMN OF MWADUI CRATER ROCKS

	ft.
Soil, gravel, silcrete	10–30
Shale, bedded lapilli tuffs	150
Grey sediments, green bedded lapilli tuffs, granite agglomerate, some kimberlite	500
Kimberlite breccia	?

The pipe, largely filled by breccias, is believed to contract to a fifth of the diameter of the crater. The extrusives were to important depths saturated by CO_2 and H_2O, with intense carbonatization at the surface. Volcanic sediments form a synclinal warp in the crater. We can distinguish the following units: *(1)* in the centre, 350 by 100 ft. of cream tuffaceous sediments; *(2)* surrounded by a 200–500 ft. wide ring of yellow bedded serpentine tuffs; *(3)* a lapilli and serpentine tuff zone attaining a width of up to 1,000 ft. in the northwest and from 0 to 200 ft. in the east; and *(4)* in the north a 3,000 by 800 ft. area of pink lapilli tuffs underlain by breccia.

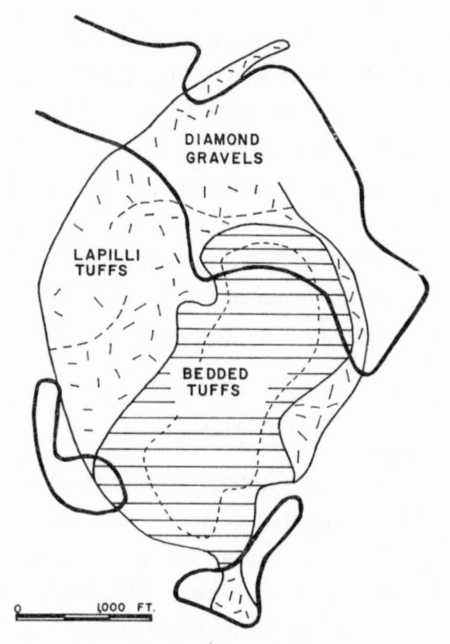

Fig. 107. The Mwadui Crater.

The beds dip steeply at the edge of the crater, levelling out inwards. They consist of alternating layers of *(1)* dark grey sediments and *(2)* olive green tuffs and muds. Each new tuff fall starts with granitic lapilli which separate layers. Moreover, the grey tuffaceous fluvio-lacustrine mudstones and siltstones include beds of granitic material. At the top of these sediments we can distinguish a garnetiferous horizon. The magnesia content of the tuffs attains only 1–18%, but the yellow chrysotile–montmorillonite tuffs containing 4–15% magnesia have been reworked by water. Light, coarse and dark fine ashes alternate in the cream coloured sediments forming the centre of the crater lake. These sediments consist essentially of silica.

Geologists believe that the *lapilli tuffs* were derived from the tuff ring. Above the 150–200 ft. level the upper pink lapilli tuffs surround the crater. These consist of a brecciated agglomerate of kimberlite and granite lapilli, and serpentine fragments of 2–12 inches with accompanying iddingsite, jasper and garnet. In addition, late calcite veinlets criss-cross the rock. At the surface this material is weathered to an iron montmorillonite. Below the pink lapilli, green contact or intrusion breccia, consisting of fractured quartz and feldspar filled in by chlorite and chalcedony, surrounds the crater. The degree of silicification and fragmentation decreases from a chloritized zone conveniently accepted as the edge of the crater.

On the perimeter of the crater several inches to 6 ft. of grey *conglomerate* lie on the tuffs, overlain by up to 30 ft. of silcrete. The silcrete may have been cemented partly by hot springs. The quantity of diamonds is a function of the thickness and depth of the conglomerate, and an inverse function of the distance from the silcrete and the periphery of the crater. Beyond the crater diamonds appear to concentrate in channels below the silcrete. The red gravels attain a maximum thickness of 10 ft. Diamonds eroded from the sub-silcrete conglomerate have reconcentrated in the gravel, particularly in the north. Finally, at Alamasi, bordering the Mwadui crater in the east, up to 20 ft. thick gravels carry diamonds.

Ilmenite, baryte, limonites and manganese nodules fill in sinkholes. All kimberlitic ilmenites contain columbium and more than 6% Mg. Furthermore, we can distinguish two types of garnet exhibiting reflective indices of 1.745 and 1.751 respectively. Both contain chromium. Between 20 and 130 ft. of depth the grade attains 10–14 carats per 100 tons of rock. Exhibiting various colours and sizes, the stones average a weight of 0.24 carats. Twinning is fairly frequent.

The host rock of the diamonds is considered to be the early tuff ring. Indeed, grades diminish rapidly in tuffs and explosive material of later cycles. Finally, in and around the crater the diamonds have been reconcentrated by alluvial activity.

ATLANTIC COAST

Diamond occurrences extend from the Cunene in Angola (Fig. 53, 108) to the Dooru Bay northwest of Cape Town, with the majority concentrating north of the mouth of the Orange River (De Villiers and Söhnge, 1959). Diamonds show in Conception Bay 300 miles north-northwest of the Orange, and at Kolmanskop near Elizabeth Bay 130 miles from the river. However, stones found between Lüderitz Bucht and Swakopmund exhibit different characteristics. Far in the south the Buffels River has eroded its stones from the poorly mineralized pipes of Gamvep. In addition, in the Schaap River west of Springbok possibly continental diamonds are found 600 ft. above sea level.

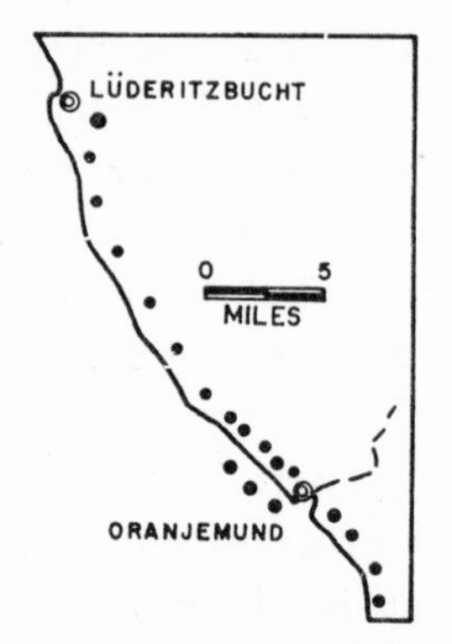

Fig. 108. The Diamonds Coast.

The main belt extends from Pomona in the Lüderitz Bucht, 100 miles north of the Oranjemund, 50 miles south to Port Nolloth. In the sub-arid climate of the Namib, the residual *Pomona* deposits have been concentrated from sands at considerable altitudes by torrential rain in depressions of the desert or 'Märchental'. The wind has blown the sands out, thereby achieving a pre-concentration (Schneiderhöhn, 1931). A fossile hydrographic system was last active in the Middle Miocene, and probably achieved its greatest importance between the Tertiary and Cretaceous. Climatic variations also appear to have played a role in the process of concentration. The Lüderitz Bay stones, which average only 0.1–0.25 carats, contain more fragments than those of Alexander Bay. The greatest diamond weighed 52 carats. The Bogenfels terraces, 25 miles south of Pomona, are genuinely marine. Other placers occur further south at Charmais.

The principal deposits now extend in a 30 mile long layer from Mittag through Uubvley to *Oranjemund*, which includes Area G and State Diggings south of the river. Diamonds occur in the interstices of large boulders and at the base of the beds of beach-walls, probably influenced by the delta of the proto-Orange which has cut violently through the Upington Gorge. The pebbles and boulders are rudimentarily stratified. The diamonds are of better quality and higher grade than south of the river; the biggest stone weighed almost 250 carats.

In *Alexander Bay* we can distinguish four marine terraces above the bedrock level, at 145 ft. (Oyster level) and at the upper, middle and lower terrace at Dinkel's surface. All three terraces outcrop between Cape Voltas and the Cliffs 33 miles south. Below the high cliffs, diamond bearing gravels attain 15 ft. of thickness. Approximately 110 ft. above the bedrock gypsum and anhydrite cement the base of the rich 2.5 ft. thick upper terrace. The higher beds form lenses, but south of the Orange River the Upper Terrace rises to 200 ft., distinguished by intermittent pay streaks. The middle level (100–35 ft. above the sea), including the 'Operculum Terrace', extends several miles south to the Holgat River. This bed also carries *Ostrea prismatica*. At a maximum height of 25 ft. above the bedrock, the Dinkel's surface gravels outcrop for $^{3}/_{4}$ mile, grading into beaches. Their fauna shows that the two top terraces have been deposited in warm water. After the erosion of the 168 ft. surface, diamonds were deposited in the 145 ft. terrace. Following this the middle terrace rose to 45 ft. and was eroded. Following warping and further elevation, the Orange River migrated north and the lower terrace was deposited. Finally, diamonds were also eroded into surface and sea sands. Potholes of differential alteration are characteristic of the bedrock. Concentrations in submarine sands are extensive.

In fact, the diamond bearing gravels are storm beaches. Since each bed is transgressive, the thickness of each level diminishes inwards. The terraces have been eroded by waves. Consequently, the placers form separate strips extending sub-parallel to the coast $1^{1}/_{2}$–3 miles inland. The thickness of the beds varies between a few inches and 20 ft.; they are covered by inland dune sand and calcrete. The size of the placer material varies from sand to boulders, with pebbles consisting of jaspers, red polychrome or banded ironstones various intrusives and sediments. The average size of diamonds attains 5 carats and a value of $25/carat at Oranjemund, but this diminishes considerably between the Buchuberg Hills and the Cliffs. In Alexander Bay stones of 70 and 100 carats are common, attaining a maximum size of 220 carats. Common forms include octahedra and hexoctahedra, but diamonds which have suffered long transportation are rare.

At Oubeep, 12 miles east-southeast of Port Nolloth (150 miles due south-southeast of the border and 3 miles inland), gravels rising at an altitude of 220 ft. contain 5 ft. quartzite boulders, diamonds, staurolite and hematite. The diamond bearing shingles of Kleinzee and its extension on the Buffels River 80 miles south-southeast of Oranjemund (west of the O'okiep copper deposits) have risen to altitudes of 24, 44, 115 and 210 ft. About 3 miles upstream late wash and fluviatile placers also carry stones. Finally, in the Vanrhynsdorp district, halfway between the Orange River and the Cape, the Klipvley Karoo Kop and Graauw Duinen diamondiferous terraces emerge 50 ft. above the sea.

SOUTH AFRICA

Diamonds are restricted to two major belts and (Fig. 46, 50) their outliers: *(1)* the deposits of the Atlantic Coast, discussed in the preceding chapter, extend from Alexander to Lambert's Bay in the south-southeast; *(2)* on the Jurassic rim of the Karroo the principal belt extends for more than 300 miles just north of northeast along the Vaal River. The northernmost area is the alluvial Lichtenburg field, which connects with the Vaal belt proper between Johannesburg and Klerksdorp. This field bulges into the placers of Schweizer-Reneke, and descends thence along the Orange River through Kimberley to Prieska in Griqualand. Pipes concentrate around Kimberley, fanning northwest to Barkly-West, northeast beyond Boshof (40 miles), and southwest to Koffiefontein and Jagersfontein (80 miles). Furthermore, minor pipes occur in the belt around Postmasburg (cf. 'Manganese' Chapter), in the Virginia gold field of the Orange Free State, in the Lichtenburg field and at Swartruggens. Among more than 250 pipes, only 25 contain significant quantities of diamonds, seven of which contribute the great majority of the output. On Elandfontein, 25 miles east of Pretoria, the Premier pipe provides half of the country's diamonds. The Dutoitspan, Bultfontein and Wesselton pipes 4 miles southeast of Kimberley, along with Jagersfontein, provide almost 40%. Excluding the coast, only 8% are derived from kimberlite fissure lenses and 2% from the various alluvial fields.

The arrangement of pipes on alignments suggests that, like carbonatites, they originated in deep seated rifts or fissures. Furthermore, vents intruded into the same pipe are rooted in distinct dykes. Indeed, one of the three Kimberley vents erupted from a 3–7 ft. thick dyke, located at a depth of 2,000 ft. and beneath an eroded column. Other veins have been injected between the vents. Like many minor carbonatite–alkaline vents, ultrabasic funnels form groups. We even find several in the same diathreme; thus at Kimberley one of the adventory vents coalesces at 850 ft. of present burial with the principal funnel rooted beneath the dyke.

Pipes

Eruptive breccias and tuffs of autohydrated olivinites, incorporating xenolithic fragments of pierced strata, plugged the pipes in several cycles. We can distinguish two varieties of olivinite: *(1)* kimberlite proper, originally a basalt of olivine (forsterite) in a monticellite matrix, favouring pipes rather than dykes, and *(2)* phlogopite peridotite lamprophyres. In the altered matrix of perovskite (containing 1% Nb_2O_5), apatite, magnetite and chromite, some phlogopite and ilmenite as well as pyrope, enstatite and diopside, accompany the olivine. Phlogopite acquires importance among the phenocrysts of the mica peridotite. The matrix of this rock also contains serpentine and calcite. As carbonatites the brecciated kimberlite has often been remobilized and reinjected into rocks of previous cycles. Similarly, sövite veins which prove late incipient carbonatization intersect kimberlites. However, the composition of kimberlite varies in different pipes and in the plugs themselves.

The size of the xenoliths varies from millimeters to hundreds of feet. E.g., the quartzite bloc in the midst of the Premier pipe measures 600 by 1,500 ft. and extends to more than 900 ft. of depth. Hypabyssal xenoliths include eclogite, harzburgite, lherzolite, pyroxenite and bronzitite down to the phenocrysts of the kimberlite matrix. Fragments of the country rock include Karroo sediments and Basement gneisses, zonally carbonatized and zeolitized.

Few diamonds occur in the kimberlite proper which alters successively to 'hardebank', 'blue ground' and 'yellow ground', a supergene alteration product. 'Blue ground' is a serpentine including disseminated ilmenite, chrome bearing diopside, enstatite, pyrope and serpentinized pseudomorphs after the phenocrysts. Serpentinization, extending to a depth of 3,000 ft., is a hypogene autometasomatic phenomenon (as in dunite bodies, cf. 'Asbestos' chapter). During a later hydrothermal phase chrysotile, pyrite and marcasite precipitated.

The *Premier pipe* is largely cylindrical and measures 2,700 by 1,500 ft. in diameter. We can distinguish several extrusive cycles in the indistinctly serpentinized kimberlite tuff and breccia. Apart from the mega-xenolith deep inclusions are frequent, as are thick and narrow carbonatite veins and dykes which also contain serpentine and magnetite. The funnel of the old Kimberley mine has an upper diameter of 1,500 ft. which diminishes to 1,200 ft. at 300 ft., of depth. This pipe has been economically exploited to a depth of 3,000 ft. The kimberlite of Bultfontein is completely brecciated. At De Beers carbonate plates have been found in partly serpentinized kimberlite at a depth of 600 ft. Anhedral perovskite contains 2–4% rare earths, thorium and 0.6% columbium oxides. Other pipes have elliptical cross sections.

Morphology of diamonds

Diamonds are cogenetic rare constituents of kimberlite, each facies carrying its distinctive suit. The ratio of economical serpentinized kimberlite/diamonds varies between 4,000,000/1 –700,000/1. Since eclogites contain a few diamonds it has been suggested that the stones originated in great depth ($>$ 50 miles). However, with values decreasing rapidly in depth great concentrations appear to be stable in altitudes of extrusive decompression. Indeed, at Kimberley diamond values decrease from 0.2 carat per cubic yard at the present erosion surface to 0.17 at a depth of 3,000 ft.; at De Beers from 0.7 carat per cubic yard to 0.25 at 2,000 ft., and in the Premier Mine from 0.6 carat per cubic yard to 0.1 at a depth of 450 ft.! However, because of inclusions, sub-surface values are lower at Dutoitspan and Bultfontein.

The unit value of diamonds differs in each pipe. Sizes and colours are variable, with the major mines averaging roughly one carat per diamond. The Cullinan, a fragment of a larger crystal, weighed 3,300 (or 3,020) carats, i.e., 1½ lb. The Premier Mine is also famous for the best blue-whites and also contains dark diamonds, fragments and boart. Yellow ('Cape' or 'Silver Cape') diamonds occur frequently in Dutoitspan in sizes as great as 200 carats, but Wesselton supplies many 'white' octahedra. Bultfontein stones exhibit a rough and etched surface, covering inclusions. Although smaller stones frequently crystallize in dodecahedra, larger diamonds favour octahedra.

Dykes

Deposits in carbonatized *dyke systems* are situated in three areas: *(1)* Nooitgedacht and Winkelhaak near Swartruggens 100 miles west of Pretoria, *(2)* Bellsbank Estate, Doornkloof, Sover and Mitchemanskraal near Barkly West, 30 miles northwest of Kimberley, and *(3)* Wynandsfontein south of the Orange Free State gold field. The Swartruggens dykes form a vertical chain structure of contorted and ramified lenses which intersects Middle Precambrian lavas and an earlier system of parallel quartz veins. The pinching and swelling, intensely carbonatized lenses consist of a series kimberlite veins. Their width varies from several inches to more than 5 ft. Moreover, carbonatization extends beyond the dykes into the lavas. Unlike the pipes, the dykes tend to contain homogeneous values in depth, although they mainly consist of industrial diamonds (0.7 carat per cubic yard). In the Bellsbank Estate secondary limestones have been redeposited from carbonatized kimberlite lenses, as on the slopes of East African sövites. This 2.5 mile long arcuate dyke zone consists of sub-vertical, 1½–4 ft. thick veins. In its middle a mile long branch widens to 100 ft. On the main zone 3 other blows or incipient pipes, measuring 600 by 90 ft. and 130 ft. in diameter, have developed. However, the mineralization is patchy and attains only 0.07 –2 carats per cubic yard.

Placers

The erosion of diamond deposits started early. The peculiar green diamonds of the *Witwatersrand*, which show a sympathetic relationship to the distribution of iridosmine, have been eroded from earlier cycles of disseminated diamonds recognized in many parts of Africa. Although scarce crystals have been found in all the districts of the system, most of the production comes from Modder B in the East Rand.

Since values increase upwards in the diathremes, many diamonds have been redeposited from up to 3,000 ft. of rocks eroded from the pipes. According to their morphology, geologists have determined the origin of part of the diamonds, but others come from unknown sources. The principal paleo-alluvial fields include those of Lichtenburg–Ventersdorp, situated between Johannesburg and the Bechuanaland border and the Harts–Vaal system, especially in the arc of Schweizer–Reneke Bloemhof, and the areas southwest of Klerksdorp and Barkly West. Forming a 90 mile long and 12 mile wide belt, the *Lichtenburg–Ventersdorp* runs, lenses and solution holes constitute partially eroded remnants of a Plio–Pleistocene river system flowing on the karst of the Dolomite Series. The principal placers of this field include Grasfontein, Goedvooruitzicht, Klipkuil, Ruitgelaagte, Uitgevonden and Welverdiend. Out of four beds pierced in deeper holes, two contain diamonds. The grade of half carat diamonds attains 0.03 carat per cubic yard.

The first Harts–Vaal–Orange hydrographic system developed before the Karroo. In the *Schweizer–Reneke* field, scattered Tertiary gravel contains diamonds 400 ft. above and 20 miles from the recent rivers. Similarly, Lower Pleistocene terraces of the 40, 30 and 20 ft. levels of the Vaal River (e.g., at Christiania, Windsorton and Longlands) and paleo-alluvial deep leads or channels (e.g., at Droogeveld southwest of Barkly-West) also carry stones. We can consider heavy (s.g. 3.3) banded pebbles of quartz, iron oxides and garnets as characteristic indicators. Finally, along the Vaal River eddy holes containing diamonds are pre-Recent.

THE REST OF AFRICA

In *Basutoland* a few diamonds have been found in the (Fig. 46) Kao River at Sekameng, Malibamatsa, Lampahane and in a vein at Butha Buthe. In Bechuanaland the stones of Foley and the Macloutsie River are believed to have been derived from Karroo conglomerates, while the diamonds of Somabula in Southern Rhodesia are attributed to the Stormberg Stage. Northeast of Bulawayo diamonds show in the drainages of the Shangani and

Insizwa, where pipes are also found. A few stones have been recovered at Ndarugu and in the Kakamega gold field of Kenya east of Lake Victoria, as well as in the Buhweju gold field of Uganda, east of Lake Albert. Other occurrences include an island of the Congo River near Brazzaville (?), Lomecha on the Upper Lukenie in the Congo Republic, Makongonio in southern Gabon, the Archambault district of southern Chad, the Adamawas of western Cameroon, and Birnin Gware and Maradu in northern Nigeria.

In southwestern *Mali* geologists have found large diamonds (98 carats) in the Doundé River at Sansanto, 2.5 miles north of *Kénieba*, and at Guindissou. Accompanied by dolerite dykes, the pipes cross the Basement and Paleozoic sandstones. At Sekonomala the kimberlite breccia has been intensely carbonatized by veins and matrix calcite even at a depth of 300 ft. (MARVIER, 1959). Olivine has been altered to chrysotile and antigorite. Although chromite occurs, there is no diopside. Similarly, diamonds are scarce, but typical kimberlite minerals show in a 25 mile long area trending north–south in Birrimian schists. For instance, a 300 ft. wide halo of large grained kimberlite ilmenite surrounds the pipes of Kobato, Sekonomata and Toromaya (Fig. 54).

SECTION VI

Tin–rare metal group

Minerals of this paragenesis are polygenetic. They are found in the aureole of granitic intrusives, in pegmatites and carbonatites and in various placers. Typical paragenetic associations are given in Table CXXXVI.

TABLE CXXXVI

THE TIN PARAGENESES

Niobium–zirconium–thorium–cerium	carbonatite, albitized granite, alluvia
Tin–niobium–tantalum	pegmatites, alluvia
Mica	acid and ultrabasic pegmatites
Thorium–uranium	ultrabasic pegmatites
Beryllium–lithium–tantalum	albitized pegmatites
Bismuth $\pm$ tungsten	altered pegmatites, veins
Tin $\pm$ tungsten	hydrothermally altered acid intrusives, greisen, quartz veins, eluvia
Tungsten–molybdenum	microgranite, veins
Thorium–cerium	veins, beach sands
Titanium $\pm$ zirconium $\pm$ cerium	(ultrabasics), beach sands

The most important deposits of the principal metal, tin, are in fact alluvial placers. Some of the metals crystallize as a late phase in granites and carbonatites. However, several of them achieve their maximum concentration during the last stages of development of pegmatites. Nonetheless, the hydrothermal phase proper is not of primary importance.

TIN

Cassiterite impregnates hydrothermally altered or sheared granites, granite prophyries, felsites and the country rocks of the Bushveld. In pegmatites tin does not favour albitized zones to the same extent as niobium (columbium), but tends to accumulate in micaceous border zones and quartz–mica formations. Furthermore, cassiterite has been eroded into placers from quartz veins and greisens. Since tungsten is relatively rare in Africa, the association of tin and niobium is more distinct than on other continents. The principal deposits include the placers of the eastern Congo, Nigeria and Rwanda. Tin is less abundant in Southwest and South Africa (Fig. 6).

The matrix of the Older Granites consists of potash feldspar, quartz, biotite with some hornblende, sphene and oligoclase and orthoclase phenocrysts. Some crystals, and especially biotites, are oriented, and xenoliths of Basement rocks are frequent. Older granites generally contain more lime and less soda and potash than the younger variety. The trend of foliation is north–south with frequent local variations. Although older granites contain few rare metals their pegmatitic aureole is the main source of tantalum in Nigeria (Fig. 24, 53).

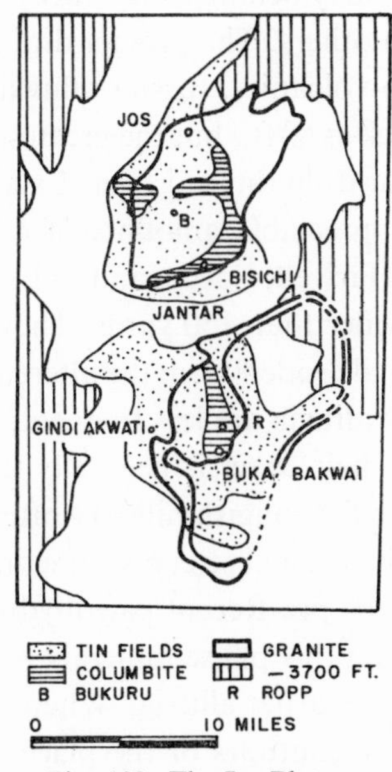

Fig. 109. The Jos Plateau.

The Younger Granites intrude the Basement and the older granites. Their isolated complexes of elliptical shape are aligned along a 250 mile long and 70 mile wide belt. Trending north-northeast it extends from Afo (on the contact of the Cretaceous transgression of the Benue–Niger) to Zinder (in the Niger Republic), and reemerges in the Central-Saharan Air. More than a hundred complexes outcrop; the distance between them varies from 10 to 30 miles off the plateau. While the smallest have a diameter of two miles, large Liruei–Kano, spreads over 40 miles. Whereas individual eroded complexes rise as 'inselbergen', the Jos Plateau emerges from the plain of the tributaries of the Benue to a height of 3,000–4,000 ft. because more than a dozen complexes were intruded near each other there. In fact, Younger Granites cover more than one-third of the outcrop area. They resist erosion better than the Older Granites and the Plateau follows their north–southerly trend. The southern Afo complex was partly protected from erosion by a dyke, and off the Plateau it remains the only structure of commercial interest.

The Jos Plateau is located in north-central Nigeria (Fig. 109) 150 miles south of the big northern city of Kano. Altogether it covers an area of 80 by 30 miles, but the central mineralized area covers only 50 by 15 miles. A barren strip of seven miles separates the northern areas, which lie in the zone of influence of the Jos–Bukuru complex, from the southern areas, which are a product of the Ropp–Tenti structure. Although smaller complexes also influenced the pattern of ore distribution, the greatest concentrations are definitely located in the southwestern part of the Jos and Ropp complexes.

The Plateau shows a mature morphology of flat valleys alternating with low hills which often indicate the old land surface. Outcrops of harder granitic phases form blocks or hills with dykes bending in arcuate ridges. Erosion of the country rocks undoubtedly started after the emergence of the complexes. The first continental cycle appears to belong to the Mid-Tertiary when the Fluvio-volcanic Series was formed. Geologists believe that the sub-basalt alluvia are Quaternary and that the mineralized paleo-hydrographic system is pre-Recent. Of course, the old alluvia may belong to several cycles. Indeed, concentrates eroded in one cycle are often recycled in others. The present drainage, however, is barren.

Beneath and within the fluvio-volcanics, fossile alluvia contain large boulders as well as wash. The cover may be as thick as 60 ft., but the greatest depth of the Gona–Sabon Gida lead attains 180 ft. We find similar deep tin leads near Exlands Camp in the Southern Area. The old placers only contain appreciable amounts of columbite where they intersected niobian phases (Rayfield Gona, Harwell, Forum and Buka–Bakwai). Fluvio-volcanic leads may be longer than a mile and wider than 150 yards. They carry abundant heavy minerals because they have received material eroded from the altered granites which weathered in the same cycle. The newer sub-basalt alluvia are less frequent, but they may contain important deposits, e.g., at Buka–Bakwai. Their cover is generally thin but quite compact. Laterites easily form on basaltic rocks, with flat longish hillocks often indicating old leads. However, they may also form basins, and the present streams sometimes flow parallel to them.

With its very flat long valleys, the pre-Recent paleohydrographic system is undoubtedly the most important. Indeed, during this phase minerals were redeposited directly from the granites and greisens and also from earlier alluvia. Whereas the two basaltic alluvia hardly form complete drainages, bedrock contours of the placers now reveal a complete hydrographic system. In the eastern part of the Rayfield area and north and south of Ropp the principal leads are more than 4 miles long. The width of the valleys varies 100 to 500 ft.; borders are generally quite sharp. The thickness of overburden attains 10–20 ft. while the thickness of fine grained wash attains only 3–6 ft. The slope of these valleys is not significant and they often appear to belong to the middle or lower parts of streams. When they cut older alluvia there is also captation of ore minerals, e.g., at Sabon Gida. Exploited grades average 0.4–1 lb./cubic yard (0.015–0.03%). For further details we refer the reader to the 'Niobium' chapter.

CONGO

From the equator the tin belts extend 750 miles west of Lake Edward as far as the Copperbelt. The principal deposits are located in two districts, Maniema and northern Katanga. The Maniema district covers a 200 mile long and 130 mile wide area between *(a)* the Congo–Lualaba and *(b)* Lake Kivu and the northern end of Lake Tanganyika (Fig. 110). For 200 miles the North Katanga belt follows the Upper Lualaba River from Manono (located 300 miles north of Elisabethville) south-southwest through Mwanza to the Busanga eluvia and greisens near Kolwezi. South of Manono the *Mitwaba* placers and pegmatites are the only significant deposits of another long and parallel belt (Fig. 51).

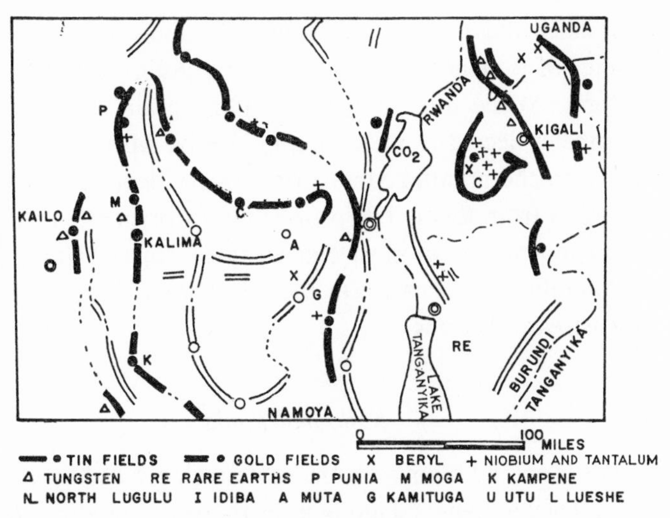

Fig. 110. The mineral deposits of the eastern Congo.

Maniema Province

Intruded into Precambrian schists, the Kasese batholith occupies the northeastern quarter of Maniema and Central Kivu provinces. The Aïssa, Kalima and Kihembwe granitic stocks have been intruded along a north–south oriented alignment nearer to the Lualaba River. In addition, metabasites are interbedded with the schists, and dolerites occur along the periphery of many stocks. Low plateaus of unfolded latest Precambrian (Katangian) red sandstones outcrop on the edges of this orogeny. The lower Karroo transgression of the Congo basin has flooded the system of valleys of the Ulindi–Lugulu, Elila and other rivers penetrating eastwards to the mountain ranges of Kivu, bordering the Graben. The tin belts follow the contacts of the stocks, but the mineralization is generally asymmetric. The westernmost belt rises in the Kailo district, plunging below the Karroo. From Punia in the north, the Western Tin belt extends 200 miles to south of Kampene. This mineralized axis continues towards northern Katanga. The Eastern Tin belt forms an arc occupying the southwestern sector of a circle. It follows the western contour of the Kasese batholith from Tshamaka beyond Kasilu for 230 miles.

The contact of granitic intrusives controls the mineralization which, west of Kasese, originally consisted of quartz veins, cassiterite stringers and greisens. Pegmatites acquire an important role at the contact of the batholith, and they predominate east of the Rift Valley in the Rwanda tin belts. Most of the deposits are related to the muscovite granite rim of the stocks, although the central core of biotite only shows traces. In the primary deposits economic concentrations are scarce. Altered greisenic or granitic bedrocks, including stringers and disseminated cassiterite, contain good grades, especially in the Kailo and Kalima districts. However, values tend to diminish with depth. The importance of eluvial and alluvial placers and terraces decreases eastwards, i.e., towards the higher mountain ranges distinguished by more intensive erosion. We distinguish several alluvial cycles. The major hydrographic systems acquired their primitive shapes in the Permian; the present placers, however, appear to be largely Holocene. The width of smaller valleys attains 20–50 ft., while larger rivers contain 200 ft. wide placers. Within these values form distinct runs.

The thickness of smaller alluvia does not exceed 3 ft. and widens to 20 ft. in major rivers. Attaining a diameter of 2 inches in terraces, quartz grains and pebbles are considerably larger than in Nigeria. At the foot of the eluvia the grain size of cassiterite diminishes rapidly, but the diameter decreases under 1 mm slowly. Maximum concentrations are located at 0–1 mile from the granitic contact or primary deposit. Furthermore, commercial values are rare 4 miles from them. The proportion of prismatic and pyramidal cassiterite is variable. We find tourmaline in most concentrates, topaz being more abundant in the west, garnet and columbite in the east. On the other hand, ilmenite, magnetite and martite prevail in the downstream sections.

Kailo and Punia

At *Kailo* three deposits spread mainly on the eastern slopes of Mount Kenye, the Mokama and Mususa stocks. They are aligned along a north–south trending transverse faulted anticline. The eluvial and alluvial Metsera placers have been eroded from a quartz vein and stockworks. The Lonioma placers divide this deposit from the largely eluvial Mokama deposit (to be discussed in the 'Tungsten' chapter). The terraces and eluvia of Mususa have also been derived from the cupola of greisenized metasediments. The abundance of topaz distinguishes both of these deposits. In the Kailo and Kitsha Rivers cassiterite is carried 3 miles downstream.

The small Nianga deposit is located on the northeastern edge of *Punia*, a satellite of the Aïssa stock. Along a distance of 35 miles the eastern edge of the stock is mineralized following the Saulia drainage from Amabidi to Saulia, Hamandele, Kenge and Ndonga. The principal deposits are alluvial placers, with less important eluvia and terraces. Along a granitic spur the deposits extend further south from Kubu to Kubitaka, Kakuku and Ulu. These include both eluvial and alluvial placers.

Kalima–Moga

The *Kalima* stock is 20 miles long and 15 miles wide (Fig. 111). The mineralized zone forms

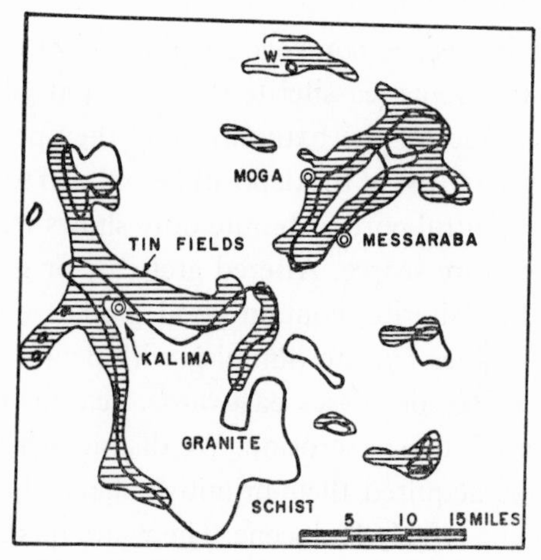

Fig. 111. Tin fields of the Kalima district.

an irregular cross extending from the satellite massif of Tuparaka 30 miles south along the western contact to Amekupi. From Kalima a branch extends eastwards to Nzilu and southwest towards Bunza. Each of the branches is 8 miles long, and they average a width of 2 miles. Deposits include, from north to south: Balendelende, Baselela, Abuki near Kalima, Bisamengo, Kakaleka, Kakula, Yubuli, Kabiala, Isongo, Avuanga and Amekupi. The eastern branch consists of Mukwale, Nakenge, Salukwango, Muswangilo and Nzilo, and the southwestern one includes Kakula, Kikambe, Bengobiri and Bunza. Minor placers extend along the southern extension of the Mesaraba trend, from Mikenzi to Makambo and Muzimba, along a gulf of schists in the granitic relief. However, the southeastern side of the main stock is barren. As far as 8 miles from the Kalima stock, the small satellite stocks of Atondo, Lubile and Moka–Alisele have produced important deposits.

The cassiterite has been eroded from dykes, veins and stockworks into eluvial and alluvial placers. However, the grade of only a few veins warrants exploitation. On the other hand, greisenized upper reaches of the granite contain numerous stringers, for example at Lubile. Indeed, the limit of eluvia and bedrock cannot always be exactly distinguished. Mineralized terraces of large rivers such as the Amicupi are 700 ft. wide, while the thickness of the gravel attains 15 ft. Furthermore, recent smaller valleys are also mineralized.

The 13 mile long and 3 mile wide *Messaraba*–Kibila stock or spur extends north-northeast towards the Katangian Lubiadja Plateau. The Permian transgression of the Ulindi divides it from its parent intrusive, the Kalima stock. Unlike most major intrusives, but like smaller ones, the Messaraba stock is mineralized along its whole periphery. The western rim consists of the deposits of Basumbu, Ilambo, Kalombo, Bimpombe, Nkongomeka, Tubaraka, Ngunde, Kisubili and Kibila. The eastern rim includes Biratwane, Musala, Madjakala, Munkuku and Kakange. At Tubaraka eluvial placers contain cassiterite near a system of parallel dykes. At Munkuku and Messaraba the upper reaches of granitic sills contain cassiterite. At Kibila a mile long belt of eluvia and quartz lenses rests on the endocontact of the stock, and includes remnants of roof pendants. Here topaz crystals measuring up to 2 inches occur. East of Musala the Kankwese deposit is located on the eastern edge of the similarly named satellite stock. Two anticlines which project seven miles west-northwest from the Messaraba stock are distinguished by particularly intense mineralization at their extremity. Indeed, at *Misoke* and Tundu, at the end of the northern anticline, small granitic stocks and sills are intrusive into metamorphic schists. However, at Kabobo and Kamikuba granite does not outcrop. Similarly at Makundju, at the end of the southern anticline, only the tourmalinization indicates the rise of the granitic relief.

Surrounded by abundant doleritic intrusions, the Upper Kihembwe stock rises south of the Karroo transgression of the Elila River. At Kamilanga a dolerite body located a mile from the granite contact has been sericitized and tourmalinized. Cassiterite is disseminated in the altered rocks, with elongated low-temperature prisms crystallizing in a series of parallel quartz veins and lenses.

The Eastern Tin belt

At Tshamaka 20 ft. thick terraces of the Lowa River include important deposits. The cassiterite is believed to have been eroded from underlying eluvia and from the northwestern edge of the Kasese batholith. The Kasese district covers a two-pronged 12 mile deep embayment in the granitic relief, and includes Bilu, Kamabea, Sinsibi, Nduma, Mya, Munu, Iliba, Milama, Balumbu, Kabimbi and Kadjekelela. We find large tin bearing quartz lenses, dykes and stockworks in several localities, e.g., at Bilu and Iliba. Balumbu is exceptionally located several miles inside the granite. Mubano is situated further south on the edge of another gulf of schists. In addition, the satellite stock of Momi is surrounded by the Bakwame and Mubilina placers. Along the southern rim of the batholith, the North Lugulu district proper extends 70 miles from Lokolia to Kibugiri, Tshonka, Niabesi, Wameri, Ezeze and Bionga to Swiza. The belt spreads 1 mile from the contact, and widens to 3 miles near smaller satellite stocks. These bulges are distinguished by great concentrations. The Lubilokwa, Lubilu and Nietubu placers cover a plateau between the satellites and the parent batholith. In this district pegmatites, cassiterite stringers, stanniferous quartz veins and tourmaline greisens coexist. The Kasilu, Nzovu and Nzibi deposits, 40 miles southeast of Bionga, also of mixed origin, follow the western rim of a satellite stock (Fig. 114).

The eastern rim of the batholith is poorly mineralized. Sukumakanga, Lowa Nord and Umate are near its northeastern corner. Satellite stocks contributed the concentrates of Muhulu, Obaye, Nyamasa, Tulakwa, Iseke, Utu, Tshamaka, Basse Luka and Kabunga. At Mumba and Numbi west of Lake Kivu cassiterite, derived largely from pegmatites, is partly covered by tuff. Placers have also been found on Idjwi Island in the lake. In addition, a 50 mile long belt extends from Luntunkulu through Mudubwe to Mwana and Nzombe, and thence to Ngussa, Kibi and Kiandjo (see 'Gold' chapter). The *Nzombe* shear zone extends for a mile in granite near its contact. This 10–30 ft. wide ramified zone is filled by lenses and sub-parallel bands of schists and quartz. The paragenesis includes cassiterite, pyrite, arsenopyrite, pyrrhotite and siderite.

Manono

At Manono–Kitotolo the world's largest ore-pegmatite (Fig. 112) has been intruded into banded schists and quartzites of the Middle Precambrian Kibara system. Each lens is 3 miles long, averaging a width of 1,300 ft. Both the pegmatites and the country rock trend

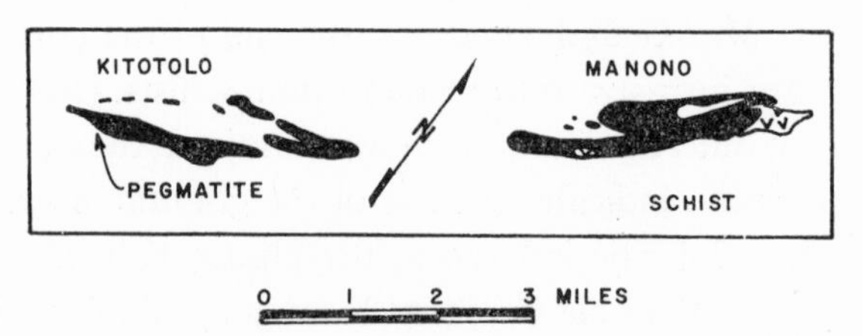

Fig. 112. The Manono pegmatite.

N 50 °E. Both dykes are sub-vertical and sub-concordant. North of these pegmatites, several zoned lenses protrude. Attaining a width of 30 ft. or more, the eluvial and lateritic cover was mined first. Indeed, the pegmatite is weathered to an average depth of 200 ft. Grades are about 0.08 % cassiterite and 0.006 % columbite. Below this the hard rock is also mineralized. Although the pegmatite is zoned the picture is complicated by a number of larger or smaller xenoliths. Parallel to their borders run concordant pegmatite zones, much in the same way as at the wall of the country rock.

We can distinguish four zones: *(1)* potash feldspar, quartz, some mica; *(2)* potash feldspar, quartz, some plagioclase (albite–andesite); *(3)* plagioclase (mainly albite), quartz, large spodumene, some mica; *(4)* potash feldspar, plagioclase, spodumene.

We can perhaps explain the presence of andesite by the intrusion of one or two series of post-Kibarian dolerite dykes.

Muscovites carry traces of lithium. Spodumene, of which Manono is one of the major deposits, contains 6% of lithia. Furthermore, beryl, tourmaline and apatite occur with traces of lithic mica and fluorite.

While tourmalinization characterizes the outer contact in the mica-schists, dolerites–amphibolites are biotitized. Quartz bands also occur at several contacts, and tourmaline bearing outer pegmatite zones exhibit relic schistosity. We find greisen patches in both the thin contact zone and the pegmatite. Autunite (uranocircite) traces can be found in the mica-schist.

Although patches of albite are richer, the ore is fairly evenly distributed. About 50% of the concentrates are smaller than 0.5 mm. Cassiterite grades vary from 0.01 to 0.4% and probably average 0.05%.

RWANDA

The Rwandese tin districts are centered at *Katumba* (Fig. 31) (Gatumba) and the capital *Kigali.* Extending into Tanganyika, Uganda and Burundi, they are located 40 and 70 miles east of Lake Kivu. The principal alluvial deposits belong to the drainage of the Nyavarongo River, a source of the Victoria Nile. Important primary deposits are situated at Katumba, Kigali, Rwinkwavu and Rutongo.

The *Katumba* stock is surrounded by pegmatites and quartz veins, and forms a north–south trending, 15 mile long and 3 mile wide belt. A majority of the deposits is on the west bank of the Nyavarongo River. However, the *Rugendabare* vein, surrounded by a biotite zone, is situated on the east bank. Dykes occurring in the vicinity of the granite tend to be oriented parallel to its contact, but the original or regional trends prevail farther away. Moreover, only dykes located farther than $^2/_3$ mile away carry commercial concentrations. At Katumba the longest dykes extend for 3 miles, with small lenses also forming series or swarms. The elongated ramified Gatsibu pegmatite has developed at the contact of schists and metabasites. At some distance the porphyric *Kirengo* pegmatite, which includes quartz–mica lenses, intersects asbestine schists. Most mineralized pegmatites contain only muscovite, although biotite occurs at Lugaragata on the Nyavarongo River. Tourmalinization affects both sides of the pegmatitic contacts, e.g., at Katumba South. The pegmatites generally contain schörl and beryl, and apatite develops in pockets.

The Kigali zone is largely hydrothermal. However, pegmatites are also known to occur at Bugalula, Mamfu, Budjumu and Kuluti. They are located 1–3 miles from the nearest concordant cupolas of granite and aplite. The Budjumu lens is, in fact, a quartz–mica formation. In the Musha–Ntunga district southeast of Kigali, pegmatites and quartz–mica dykes are also associated south of Lake Mohazi. This area has been affected by regional tourmalinization. North of Kigali the Rutongo vein intersects the planes of schistosity of sandstones. At Mamfu stringers cut through bedding planes with chains of quartz lenses favouring anticlinal folds. At Musha stringers of quartz traverse a 3 ft. wide discordant pegmatite lens. A concordant quartz vein projects from the pegmatite into the schists.

Conversely, the Kisanzi quartz vein includes fragments of schists. Lenticular stringers of mica cut the *Rwinkwavu* vein, but no pegmatite is found in this area east of Lake Mohazi.

THE REST OF EAST AFRICA

The Rwandese tin zones extend to the northwestern corner of *Tanganyika* and to southwestern *Uganda*. We distinguish six belts (Fig. 51):

(1) Bitale–Rwanyinyira in the Kigezi tungsten belt (southwest Uganda), also including Karambo, Ruhangiro and Kulitara.

(2) Further north the bismuth belt, where tin is an accessory metal, including Kayonza and Kabuga.

(3) Northern Kigezi–Kihanda, Kanungu.

(4) The Ihunga–Kitofa pegmatite belt corresponding to the east Rwanda tin zone, and also including Lyasa, Rwabuchacha, Kaina, Nyakahoko, Nyabubale, Rugaga, Burama Ridge and Ruhuma.

(5) The Mwirasandu belt, including Kyamugasha, Migvera, Kashozo, Kitembe, Kazomo, Swata, Namaherere, Twemengo and Lwamuire in Uganda and Kagaga and Bihanga in Tanganyika.

(6) The Karagwe tin field, consisting of Kichwamba, Naniankoko, Ruzinga, Ntundu and Kikagati in Uganda, and Murongo, Rugasha, Katera, Lwamosi, Kafalu, Rwabushoga and Kaborishoke (Kyerwa) in Tanganyika.

Tanganyikan deposits are essentially hydrothermal–pneumatolytic, but in Uganda pegmatites also occur. In pegmatites cassiterite is associated with columbite–tantalite, but not with beryl. In the external Uganda quartz veins cassiterite tends to concentrate with muscovite and sericite at changes of strike or ramification. Most of the valuable veins are a few ft. thick, but at Rwanyinya we can see stockworks. Tourmalinization is intense. On the other hand, cassiterite covers the walls of the Bihanga veins.

Veins traverse the phyllites and quartzites of Murongo ridge. Murongo proper is the only internal (granite) occurrence of the belt. The Rugasha swarm lies in the southern extension of the belt. The Katera veins are sericitic. In this area we also find placers. Nevertheless, the principal tin deposit of Tanganyika is Kyerwa. The wall rock affects the composition of the Kigarama veins of the deposit, but at adjacent Rwabushoga tin is located in quartz–mica veins. The main working, Kaborishoke, consists of a bedrock of stockworks, residual breccia and eluvia. Similarly, a considerable part of Tanganyikan cassiterite has been reconcentrated in arenas.

Cassiterite concentrates east of the Congo–Nile divide in the Mumirwa area of Burundi.

SOUTHERN RHODESIA

The *Kamativi* swarm of pegmatites was formed in a 25 mile long and 1–2 mile wide east–west trending belt of garnet–mica schists bordered by gneissose granite. This district is situated in the Gwaai drainage in the northwestern corner of the country. At its western end the belt turns southwest. The sub-vertical tin pegmatites either are aligned parallel to the belt or form large sub-horizontal sills. Smaller lenses include Kapata, Sam and Gwaai, but older tourmaline pegmatites are barren.

Tin pegmatites of the Salisbury district include Enterprise, Patronage, Pope, Meadows and Jack, with other lenses occurring at Poorti in the Shamva district, at Umkaradzi near Darwin, and in the Odzi and Mtoko belts. In the Bikita lithium–beryl pegmatites, cassiterite crystallizes in the margins of quartz–mica bodies, and also accompanies quartz lepidolite zones. Cassiterite forms pockets in the Nigel lens and in the Rurgwe extension (10 miles southwest).

SOUTH AFRICA

Within and around the northern part of the granitic phase of the Bushveld Complex, we distinguish three tin fields: *(1)* 15 miles northwest of Potgietersrus, 120 miles north-northeast of Pretoria, *(2)* at Stavoren 100 miles northeast of Pretoria, and *(3)* at Rooiberg 70 miles north of the capital (Fig. 46, 50).

In the narrow 14 mile long Potgietersrus tin belt lie three successive and intrusive granites covered by pegmatite, micrograpite and granophyre. This district includes (Fig. 113) the

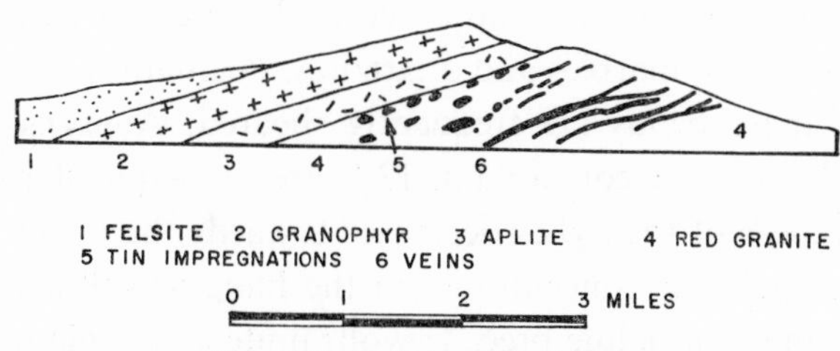

Fig. 113. Section of Zaaiplaats.

Groenfontein, Groenvley, Roodeport, Salomon's Temple and Zaaiplaats deposits. Here 20–3,000 ft. long, and 4 inch–40 ft. wide, irregular ramified zoned funnels cut through the latest microgranites. The core of the replacement pipes consists of sericite or a sericite–orthoclase which contains calcite and 8–25% cassiterite. It is surrounded by tourmalinite and white quartz. Moreover, in the last granite (preceding the microgranite) a red orthoclase inner zone and a sericitized wall zone develop. In the sericitic satellite lenses developing at the base of the microgranite, coarse grained cassiterite is associated with sulphides, tungsten and fluorspar. Syngenetic cassiterite is disseminated in the late granite, e.g., at Zaaiplaats, also impregnating the microgranite (Fig. 113).

At *Mutua Fides* cassiterite is associated with sericite near or beneath pegmatites crystallizing in late granite. At *Stavoren* cassiterite accompanied by sulphides, calcite and tungsten forms irregular concentrations in zoned and brecciated replacement bodies or stringers. A breccia zone is mineralized further south at Zusterhoek. Cassiterite has also impregnated agglomerates in felsite covered by tuff at Naboomspruit, Welgevonden and Rietfontein. Conversely, the tuff is impregnated at Dornhoers.

At *Warmbad* west of Rooiberg, cassiterite is found in pegmatites, quartz and breccia bodies of the late granite which intersects felspathic quartzite roof pendants. This swarm forms a triangle extending 15 miles from Weynek to Kwarriehoek and southward to Leeuwpoort. Thrust breccias, developed by the injection of the granites, occasionally contain cassiterite and sulphides. The *Rooiberg* veins and impregnations contain tourmaline, cassiterite, ankerite, sericite and sulphides. At Leeuwpoort sub-horizontal thrust faults, reverse faults and bedding planes exhibit a variable paragenesis including cassiterite, tourmaline, orthoclase, ankerite, magnetite, hematite and pyrite. In quartzite covered by shale, the mineral association of the *Vellefontein* fissures is similar. On the other hand, in the Nieuwpoort stockworks cassiterite is accompanied by sericite and some pyrite.

SOUTHWEST AFRICA

In the Damara highlands four east-northeast striking tin belts are distinguishable: the Brandberg–Gontagab, Northern (or Uis), Central, and southern (or Kranzberg–Omaruru) tin belts. The latter, which is the longest, attains a length of 70 miles. The mineralization is predominantly pegmatitic. Important swarms include: Kudugabis north, and Uis–Lewatei, west of Brandberg, Kudubis, Davib, Chatpütz and Ameib at the southern contact of the Erongo stock, Neineis, Aubinhonis, Paukwab and Kohero on the Omaruru northwest of Erongo, Sandemap and Ebony west of Usakos, Otjimbojo and Nooitgedag northeast of Karibib (Fig. 38, 53, 117).

Pegmatites may be internal in the granitic core of anticlinal batholiths, or external in biotite schists, hornblendites, marbles and quartzites. Pegmatites concentrate where strikes change, and indeed, these bends are frequently sheared. Analysis shows that south of Erongo 80% of 1,700 dykes are concordant, 13% are perpendicular and 7% are diagonal to foliation. The thickness of the dykes decreases towards the contact. Particularly at the intersection of dykes, cassiterite concentrates on the hangingwall side of albitized pegmatites. As a rule garnet and tourmaline precede wolframite and lithia minerals. North of the *Brandberg*, pegmatites develop in sub-vertical fractures which traverse lower Damara schists parallel to the zoning of two small granitic windows. The outer aplitic–pegmatitic zone of the granite includes xenoliths. Silicification, sericitization and greisenization control the distribution of tin here. In addition, some cassiterite is scattered between quartz veins. Wolframite characteristically accompanies cassiterite, and eluvia have developed on these zones.

At *Uis* and Arandis (where arandisite has been discovered) cassiterite occurs in quartz veins. At Paukuab (Paukwab) 7 ft. thick quartz–muscovite formations contain 2% cassiterite, some biotite, garnet, tourmaline and topaz. Internal veins which cut granite are narrower but richer. Fairly late eluvia cap these formations. In the poorer pegmatites of *Ameib* sporadic pockets or replacement bodies of up to 500 pounds occur. These low-lying dykes undulate, and cassiterite concentrates near the bulges. Swarms attain a grade of up to 0.3%. Turning to the Otjimbojo pegmatite we see several large pockets and numerous stringers with light coloured cassiterite.

Finally, some tin shows in the Orange River belt.

THE REST OF AFRICA

Alluvia carry some tin in the Mayombe and at Mayoko–Mossendjo, Congo Republic. West of Mayo Darlé (Adamawa), in the western Cameroon, eluvial placers contain some cassiterite, as do the alluvia of Mayo Taparé.

We have seen that young ring complexes of the Sahara contain tin. Near Agadès in the

Aïr (Niger Republic) cassiterite has been eroded from the biotite granite of Tarrouadji and El Mecki (see 'Tungsten' chapter; Fig. 49). Tin also shows in the Movinio stock in the Adrar of the Iforas, Mali. In the Hoggar of the Algerian Sahara cassiterite is related to the small, youngest 'Traouirt' ring complexes intruded into the Pharusian (later Precambrian) belts (Fig. 100). Quartz veins and greisens occur both in and above alkaline, biotite and microcline granites. In the Adrar In Tounine stock, near Tamanrasset, swarms of quartz lenses bordered by mica occur on both sides of the contact. Some of them contain tin and tungsten (e.g., Assaren). In the Adrar Elbema, 90 miles northwest of Tamanrasset, stanniferous quartz veins bordered by mica favour the exocontact. They form a belt 1 mile long and 1,000 ft. wide. However, in the northeastern Hoggar the differentiated Djilouet stock belongs to another type. Here quartz veins crystallizing in the alkaline granites of the endocontact contain cassiterite, wolframite and scheelite.

Cassiterite occurs at Tawalata on the Liberia–Sierra Leone border, in the Falémé pegmatite field of southwestern Mali, and in the Bir Oumgreine district of Mauritania.

In Morocco, 30 miles east of Casablanca the post-Hercynian Oulmès stock has been intruded into Upper Llanvirn rocks along an anticline (see 'Molybdenum' chapter). El Karit, the only tin bearing quartz veins of any importance, traverse both the tourmaline–muscovite granite and the schists. Minor occurrences of the stock include Tafer el Haj, Aklaye and Tarmilet (where tungsten and molybdenum also show) (AGARD et al., 1952).

Finally at Igla, Nuweibeh Abu, Dabbab el Mueilha and other localities in east-central Egypt, small stanniferous alluvial placers have been eroded from quartz veins containing topaz, fluorite and feldspar, with or without wolframite.

TUNGSTEN

Unlike other minerals of the tin group, tungsten is largely restricted to Rwanda and the eastern Congo. Indeed, Kifurwe and Mokama contain a majority of the continent's resources. In addition, minor deposits are located in Uganda in the extension of the Rwandese tungsten belt. We also find small occurrences in Southern Rhodesia, Southwest Africa and South Africa. The restricted development of wolfram bearing stockworks and veins contrasts with the wide distribution of mineralized Precambrian pegmatites! We distinguish two types of deposits: *(1)* epigenetic deposits in the aureole of granitic stocks, and *(2)* polygenetic occurrences in schist belts.

CONGO

Tungsten is irregularly distributed in the tin provinces of Maniema–Kivu (see 'Tin' chapter). West of Lake Albert there is a minor occurrence at Butsha, but the Ndesa, Ngawe, Etaetu and Lundjulu deposits are located nearer to the Albert–Edward Rift than to the Congo–Lualaba. At *Etaetu*, wolframite occurs in quartz veins and eluvia. Again, in the mountains bordering the Western Rift but further south, a belt of minor occurrences extends from Gombo through Mudubwe and the Kadubu anticline to Kibu, southwest of Lake Kivu.

If we return to the River we see abundant tungsten in the Kailo district, particularly at Mokama. However, in the adjacent Mususa greisens the tin/tungsten ratio is 20/1. Conse-

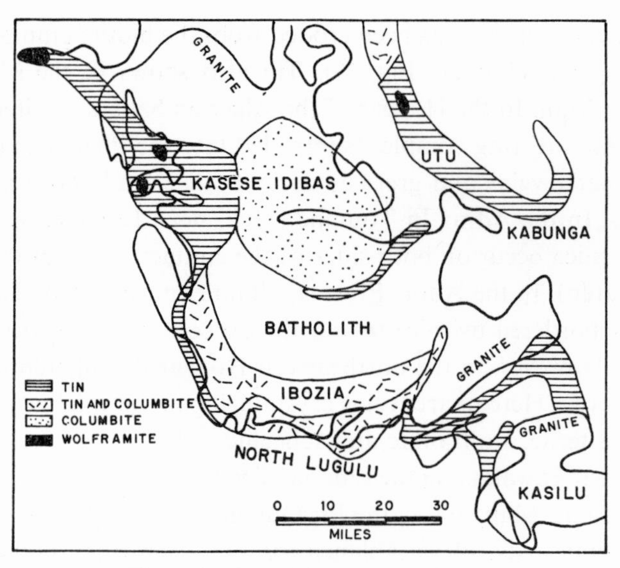

Fig. 114. Plan of the Kasese batholith.

quently columbite–tantalite is rare. The Mutuvutuvu ferberite deposit is situated 12 miles east of Mokama. The Kalulu occurrence 80 miles to the south probably belongs to the same belt. West of Kalima we find traces of wolframite, e.g., at Bunza, Bengobiri, Misobo, Kibongobongo and Yubuli. Nearby, in the Moga–Messaraba district, coarse wolframite occurs in quartz veins and lenses of *Misoke* in the vicinity of an irregular ramified granite body. Accompanying minerals include sericite and tourmaline here, but tin bearing kaolinized feldspars carry hardly any columbite. In the eluvia of Misoke, Kabobo and Tundu the tin/tungsten ratio varies between 5/1 and 30/1. Reinite was found at Kamilanga (Kampene, south of Kalima). Containing wolframite but little cassiterite, the *Kasowe* eluvia cover a quartzite ridge traversed by veins. These deposits are located 1 ½ miles from the northwestern contact of the Kasese batholith. In the Kasese district the principal eluvial and vein deposits of wolframite include Sinsibi, Kisima and Bakwame (Fig. 110, 114).

MOKAMA

The *Mokama* stock, which occupies 1 square mile, has been intruded into folded, interbedded sandstones and schists. Underlain by sugary quartz, alternating layers of topaz and quartz cover its summit. Along the southeastern contact of the stock greisenization is intense in an arc 1 mile long and 100–500 ft. wide. Prior to the intrusion of the granite a triangular body of amphibolite was emplaced southeast of the stock. Around the amphibolite, amphibole quartzite lenses have developed at the contact of sandstones. Near the stock the amphibolite has also been saussuritized. The quartz–oligoclase–muscovite granites exhibit incipient microclinization. Similarly, greisenization is a progressive phenomenon. Although topaz is reduced to traces 200 ft. below the surface, quartz and sericite extend considerably farther.

East of the stock more than 100 sub-vertical veins fill fissures in a 1 ½ by 1 mile triangle of greisens and metamorphics. However, only 12 of the veins acquire importance. In chronological sequence we distinguish three sets of fissures: *(1)* N30 °E (veins No. 4, 7, 8, 10 and CX), *(2)* N 60°W (veins No. 3, 5 and X), *(3)* N 70°E (veins No. 1, 6, 8, 9). Faults of type *2* cut type *1* veins. Vein No. 1 cuts veins No. 3 and 5, with a fault of this type limiting the field in the north. Systems *1* and *2* form a pair, but on the other hand, *3* is perpendicular to the regional anticlinal axis. The thickness of the major veins varies between 1½ and 3 ft. No. 6 is, in fact, a composite 25 ft. wide fault zone containing several veins divided by quartzite. Reef 1, which extends for 1½ mile, represents half of the reserves. Here the tungsten/tin ratio attains 1.5/1, but in the other veins this proportion reaches 9/1 (except for the stanniferous quartz vein No. 9). However, in surface greisens the ratio equals 0.1/1. Values of 1% are common. While cassiterite is fine grained and evenly distributed, wolframite is largely coarse grained and even occurs in 1 ton pockets. Wolframite, accompanied by pyrite, chalcopyrite, ferriferous sphalerite and green muscovite, forms perpendicular lenses in the midst of the veins or stringers. A rim of muscovite flakes develops at the walls of several veins. Sulphide values increase with depth, but 400 ft. below the surface cassiterite grades tend to decrease. Contemporaneous scheelite is finely disseminated, but late galena veinlets are parallel to some of the quartz veins. Greenish fluorite is disseminated, and violet fluorite also occurs in long thin veinlets.

Important eluvia mainly cover the eastern and southern slopes of Mount Mokama as well as the piedmont as far as 1 mile from the contact. The tungsten/tin ratio increases and inverts rapidly as a function of the distance from the veins. Indeed, in the alluvia wolframite does not concentrate farther than 1,000 ft. from the stock.

EAST AFRICA

In the late Precambrian orogenies which extend east of Lake Kivu we distinguish eight tungsten belts:

(1) Kabera, Lutsiro, Muregeya and Rwoza are located in Rwanda on the Congo/Nile divide.

(2) Tsharwa and Kintaruli are situated between the Nyanza–Gatumba batholith and the lava flows of the border.

(3) Along an anticlinal axis which connects Bumbogo (Yanza and Nyakabingu), near Kigali the capital, with Kifurwe, Kitwa and Bugarama in Rwanda and Kirwa in Uganda, the principal tungsten belt extends 70 miles. The Murore deposit is located on the same line further south in *Burundi.*

(4) Including Nyhanga, Bahati, Rwaminyinya and Mutolere, an 8 mile long series of deposits is located in Uganda close to *3.*

(5) The Rusongo (Rushunga), Nyamulilo, Ruhezaminda, Mpororo belt (Uganda) strikes parallel to and north of *4.*

(6) While Nyaigulu and Ruhiza (Luhizha), the southern members of the following belt, contain tungsten, the intermediate deposits, Ntendule and Kabuga, include bismuth and tin, however, the northern members only carry bismuth.

(7) Comprising Bugambira, Rwanguba and Mutshero, the wolfram belt of eastern Rwanda extends south-southeast of *5* and *6.* Rwinkwavu is situated east of this axis.

(8) Finally, some tungsten also occurs in the 30 mile long and 5 mile wide Karagwe tin field in the northwestern corner of Tanganyika.

RWANDA

The axis of the mineralized folds strikes N 20°W, turning towards N 50°W in Uganda. In the *Bumbogo* anticline, strike veins and a perpendicular system of fissure fillings cut pyrite bearing mica-schists and graphitic schists, overlain by barren quartzites. Sub-vertical veins appear to contain more ferberite and anthoinite. The schists include 0.02–0.1% tungsten, but the ferberite contains only 1% MnO.

One of the continent's largest deposits, *Kifurwe* (Fig. 31), is located 40 miles from the Congolese and 12 miles from the Ugandese border. The mountains rimming Lake Luhondo rise above an altitude of 7,700 ft. Here an anticlinal fold of graphitic schists strikes N 40°W. We can distinguish three ranges of accordion type folding: 100–1,000 ft., 3–10 ft., and 1 inch–1 ft. Two types of veins traverse these complex folds. *(1)* In the axial plane of the folds we find 4–15 inch thick micaceous quartz veins, which locally overlap into the upper reaches of limbs. Covered only by a film, some of the joints remain open. The distance between fissures is 20–30 ft. *(2)* Thin parallel veinlets traverse secondary folds. High grade stockworks and dense grids of stringers are connected with concordant strike veins. Thick strike veins are barren, but the altered graphitic schists are mineralized. The ferberite contains 0.9% MnO. Finally, late dolerite dykes cut the folds, and gold is disseminated in the alluvia.

On Kidwa Island in Lake Bulera tungsten bearing quartz veins occur in graphitic schists, covered by barren quartzites.

At *Bugarama* on the frontier, because of the different resistance of quartzite, ferberite bearing quartz veins outcrop only in permeable graphitic schists. A system of fissures is parallel, and another is perpendicular, to the bedding plane. Sericite accompanies ferberite in brecciated patches.

Bugambira is located in Rwanda between Kigali and the Tanganyikan frontier. In this tin deposit micaceous quartzites occupy the core of an anticline covered by tourmalinized schists. The secondary folds have been sheared, faulted, brecciated and injected by ferberite bearing quartz veins. Conversely, the cassiterite bearing quartz veins are perpendicular to this system.

In the quartzites of the northern cupola of the Katumba stock, *Lutare* is situated west of Kifurwe. Here wolframite stringers crisscross a megacrystalline quartz–feldspar pegmatite, but tin does not show in the graphitic horizon. At *Nyankuba*, on the Congo/Nile divide, alluvial placers of the upper Satinsky torrents include wolframite flakes. West of the Nyanza batholith the *Lutsiro* deposits, Mujebeshi, Kabera, Muregeya and the Rwoza placers are located in a band of graphitic schists adjacent to tin occurrences. In addition, Tsharwa and Kintaruli are situated halfway between Kifurwe and them.

THE REST OF EAST AFRICA

The Murore ferberite–tungstite veins outcrop in northeastern *Burundi* in the intensely tourmalinized aureole of a small stock. In the quartz–mica veins of Chamunyanya (Karagwe field, Tanganyika) wolframite and hematite are associated, injected along the bedding planes of graphitic schists. Grade is an inverse function of the thickness of the quartz–hematite veins of Karugu. Other Tanganyikan occurrences include Kibanda and Kazumeru.

At Kirwa in *Uganda* the mineralized graphitic horizon attains a thickness of 600 ft. Ferberite and reinite occur in stockworks here. At *Bahati* and Nyhanga, quartz and mica filled in a 20 ft. thick brecciated tension fracture which forms a large angle with the strike of the graphitic schists. Some mica, tourmaline, feldspar and beryl accompany wolframite. Furthermore, silicification and tourmalinization have affected the wall rock. At Rushunga tungsten concentrated in the brecciated sandstones of an anticlinal dragfold rather than in the graphitic horizon. At Nyamulilo veins fill joints, cutting secondary folds. Representing grades of 0.06%, ferberite nodules of less than 50 μ are embayed by black micaceous schists here. The nodules also contain some sericite. In the Mpororo veins schistose relics remain, scheelite structures being preserved in reinite. At *Ruhizha* ferberite concentrated in shears, traversing tightly folded schists.

West of Kampala a swarm of parallel quartz veins, bordered by mica, cuts the kaolinized Singo granite. These veins contain wolframite and tourmaline.

SOUTHERN AFRICA

Scheelite is an accessory mineral of the *Southern Rhodesian* gold–quartz veins. In addition, this mineral accompanies copper sulphides in veins traversing the lower Sabi basalts (Limpopo basin), e.g., in the Hippo deposit. Scheelite nests and disseminations occur in large quartz lenses included in granites and schists of the Gatooma, Gwelo, Filabusi and Bulawayo belts (see 'Gold' chapter). Occurring as far as the pegmatites of Inchope, Manica field (Mozambique), scheelite crystallizes in the outer zone of the Fort Victoria–Bikita schist belt. The Mazoe district also includes scheelite. Furthermore, we can find wolframite in the Sequel deposit in northwestern Southern Rhodesia, at Essexvale in Karoi, and in the alluvia of Machinga and Catoa, western *Mozambique*. The Ribao pegmatite is distinguished by a paragenesis of wolframite, bismuth and sphalerite.

The tungsten deposits of the O'okiep district in Namaqualand, *South Africa*, tend to concentrate in the 350 ft. thick 'wolfram schists', which consist of acid intrusives and metamorphics. This series is overlain by the Concordia gneiss in the north (see 'Copper' chapter). In the Nababeep, Kliphoog, Narrap and Tweedam veins and lenses, scheelite, bismuth, molybdenite and chalcopyrite accompany ferberite. However, the Biegies deposit is pegmatitic. On the Orange River, 50 miles to the north, both quartz veins and pegmatites carry scheelite in distinct horizons which are frequently in contact with metabasites. Such occurrences include Goodhouse, Vioolsdrif, Kaalbeen, Henkries, Koubankoppie, Isis, Nous and Xochasib. The wolfram bearing quartz veins of the Collinskop and Gordonia lie in the extension of the *Southwest African* Warmbad district. Scheelite also occurs in the pegmatites of Warmbad.

Northwest of Windhoek the Kranzberg is the principal tungsten deposit of the Erongo Mountains in the Omaruru tin belt. Rising isolated southwest of Windhoek, the concordant pegmatites of the Natas stock contain megacrystalline scheelite, apatite, tourmaline, hematite and zoned copper sulphides. With pockets attaining a weight of ½ ton, scheelite represents as much as 30% of the mass of adjacent pegmatites. Scheelite tends to occur with late calcite in the quartz core in zoned lenses. Scheelite includes gold. When carbonates are abundant, molybdenum tends to replace scheelite. Finally, north of Walvis Bay in the Brandberg, Gontagab belt, tungsten accompanies tin in quartz veins and eluvia (cf., Congolese deposits).

THE REST OF AFRICA

At Goutchoumi in the Cameroon wolframite is associated with sphalerite and galena. Farther north large crystals also occur in the aureole of the late Yédri granite, Tibesti Mountains, northern Chad. Wolframite is a typical rare accompanying mineral of the stanniferous quartz veins of 'young', i.e., Cretaceous, ring complexes of the Nigerian Arc, especially on the Jos Plateau. In its extension the eluvia of the Adrars Elmeki and Guissat, 75 miles northeast of Agadès in the *Niger Republic*, include cassiterite and wolframite. Similar late granites of Laouni, In Tounine and Djilouet in the Hoggar of southern Algeria are distinguished by the same parageneses.

In the Atlas Mountains tungsten occurs at Azegour (see 'Molybdenum' chapter), in the Tichka and Oulmès stocks (Morocco), and in the Belelieta veins of eastern Algeria. In the southeastern desert of the United Arab Republic wolframite shows in the Qash Amer quartz veins bordered by mica at Abu Kharif, Um Bissila and Zarget Naam, with some scheelite appearing at El Muilha.

MOLYBDENUM

This metal is scarce in Africa. Indeed, the only deposits of some importance are located in Morocco and South Africa. Molybdenite impregnates shear or contact zones in or adjacent to granites, but pegmatite and vein occurrences remain insignificant.

Morocco

Azegour is situated in the High Atlas between Marrakech and Agadir. Here a Hercynian granite is intrusive in earliest Cambrian (Georgian) limestones, transforming its aureole into tactites, cipolinos and siliceous marbles. While chalcopyrite is most abundant in para-amphibolite, molybdenite and scheelite form different series of patches or columns of concentration in the tactite. Molybdenite is frequently disseminated farther from the granite than scheelite (Adrouss, Entifa). The paragenesis also includes pyrite, sphalerite, hematite, magnetite and fluorite, rarely galena, tetrahedrite, linnaeite, bismuth, pyrrhotite, idocrase. However, little scheelite persists in the Tisgui vein near Azegour.

Near Azegour the *Tichka* stock intrudes Cambrian–Silurian schists and limestones. The western aureole of this stock, and of granitic apophyses, is mineralized. Although low grades are found in hornfelses, most of the molybdenite and scheelite concentrates in garnetite. The principal deposits include Isk Imoula, Figri, Djebel Afadad, Agadir n'Maïne and Amsiouni, all in garnetite (and pyroxenites).

The Sidi bou Othmane deposit is contact-metasomatic. Finally, some molybdenite is associated with cassiterite and wolframite in the Tarmilet vein of the Oulmès district in the Ment Mountains.

South Africa

North of the Zaaiplaats tin field at *Groenvley* and Appingendam molybdenite accompanies arsenopyrite, sphalerite, occasionally cassiterite, tourmaline, quartz, carbonates and fluorspar. Mineralized lenses and pipes follow fissures of the Bushveld granite. The Groenvley deposit averages 3% molybdenite. Furthermore, molybdenite and molybdite show in the luxulianites of the Stavoren tin field. Molybdenite is also disseminated in pipes which developed in the microgranite of Klipplaatdrift, and in a fissure zone of Houtenbek in the Groblersdal area (eastern Bushveld). At some distance gold, uranium, thorium and arsenic are also included in the paragenesis in the joints of Klipdrift. At Rietfontein (northern Bushveld) a quartz vein also contains arsenopyrite, bornite, fluorite and monazite. On the other hand, the mineral association of the Hartebeestfontein granite dykes, which traverse felsite, consists of Ni–Co–Au–Ag and arsenopyrite. Molybdenite also occurs in pegmatites at Zwartkloof and Witfontein in the Warmbad tin field (northwestern Bushveld), with emeralds at Gravelotte (eastern Transvaal), and at Subeni, Dumisa and on the Umgeni Stream, Buffalo Valley (Natal).

South of the Orange River molybdenite shows in the copper district of O'okiep. South of O'okiep coarse molybdenite is disseminated in the quartz core of the Garies pegmatite. Near the Cape molybdenite is associated with arsenopyrite, chalcopyrite, cassiterite and wolframite in the quartz veins of Helderberg, Langverwacht and Hazendal.

In western Natal on the Hlatimbe River the basal portion of a thick Karroo sandstone bed is impregnated and contains a maximum of 1.75% MoO_2. The rare mineral ilsemannite, molybdic ochre and marcasite occur there.

The rest of Africa

In the Sula Mountains of central *Sierra Leone* late-kynematic granite, which includes roof pendants, fills in the core of an early Precambrian anticline (MARMO, 1962b). Near the Qankatana Stream molybdenite is disseminated in slightly sheared granite. In the bed of the Mapoko Stream (Kambui Hills), pegmatites and quartz veins invade tremolitite, amphibolite, serpentinite and quartzite at the contact of granite. In a biotitized shear zone molybdenite is disseminated on the border of the veins (MARMO, 1962a; POLLETT, 1952). Molybdenite also shows in the Nyandehun and Petema aplites, in the Wungi Stream granite and in placers, and with wulfenite in the Sande River (Dalakuru).

Molybdenite shows *(a)* at Mankwadzi and north of Kanyangbo, southwestern and northwestern Ghana, *(b)* in a 150 ft. long and 15 ft. wide granite dyke in the Kigon Hills, Nigeria, *(c)* with uranium in the Ekomedion pegmatite in the Cameroon, *(d)* in the uranium–cobalt–nickel paragenesis of Shinkolobwe, Congo, and *(e)* in the gold fields of Southern Rhodesia, e.g., at Seigneury and Hartley.

In the Mbabane district of *Swaziland* granites and granitic biotite gneisses strike N 30°E. On the Komati River up to 2½ ft. thick stringers, external pegmatites, and up to 1 ft. thick quartz veins averaging 0.20% MoS_2 contain fine grained molybdenite and the yellow ochre molybdite. In addition to the northern ore zones, south of the Komati a zone attains a length of 1,200 ft. and a width of 150 ft.

In the pyroxenite pegmatites of southern Madagascar traces of molybdenite accompany phlogopite, e.g., at Mafilefy, and it also shows in the pyrite vein of Manantenina. Similarly, molybdenite occurs in southern Malawi. In Tanganyika molybdenite occurs in the quartz veins of Sango and Butiama 25 miles southeast of Lake Victoria, at Saza in the Lupa gold field, at Kyerwa in the Karagwe tin field, and in the granites of Nzega and Lion Rock near

Dodoma. While molybdenite shows in the Toro district of western Uganda, in Kenya its traces appear in a granite of Kavirondo Bay of Lake Victoria, in syenite in western Maragoli, in the Maralal and Taita Hill pegmatites and in the Mrima carbonatite on the coast. In the Shangalla plain of Wallega province, western Ethiopia, pinching and swelling pegmatites invade north-northeast striking gneisses, mica-schists and amphibolites of the Kilta Stream. In the wall zone of these less than 20 inch thick lenses some molybdenite is disseminated. Finally, in the Burama district of northern Somalia molybdenite is associated with uraniferous pyrochlore and zircon in nepheline–cancrinite syenites of a carbonatite.

NIOBIUM (COLUMBIUM)

Niobium occurs in the east African carbonatites. The Lueshe complex (Congo) contains two-thirds of the continent's resources. Similarly, the granitic ring-complexes of the Nigerian–Saharan arc contain niobium, as do late granites of Uganda and Katanga. In addition we know of a few pegmatites and placers in which columbium is more important than tin: Madondonet in Sierra Leone, Mayoko–Mossendjo in the Congo Republic, Warran Weis and the Theb Valleys in Somalia, Odzi Siding and Miami in Southern Rhodesia, the Namaqua field in South Africa, and Sandemap, Pankaub, Tsomstaub and Stoeckle's Claim in Southwest Africa. However, in most deposits of this type tantalum is more valuable than niobium. Consequently, we will describe them in the next chapter.

Including more than one hundred complexes, carbonatite belts extend from Somalia to Glenover in South Africa. The important complexes are located in an area extending east of the ranges bordering Lakes Albert, Victoria and Nyasa and north of the Mid-Zambezi. However, other complexes may contain important deposits of other substances, or concentrations of columbium which, compared to placers, are still considerable. Unfortunately, in the Mrima carbonatite south of Mombasa in Kenya a majority of the reserves have been reprecipitated in phyllosilicates. In another context the Angolan rare earth carbonatites lie in the extension of the kimberlite belts. Here a contact zone of Longonjo carries 0.24% Nb_2O_5.

The belt of the Great Lakes extends from Bingu west of Lake Albert, through Lueshe west of Lake Edward to Kawezi and Ngualla, Mbulu, Musensi and Mbeya in the Rukwa–Nyasa Rift. But among numerous carbonatites located south and east of Lake Nyasa only Chilwa is of interest. A double belt of a dozen complexes extends from the Ethiopian border to Ruri Bay on Lake Victoria. This belt includes Sukulu. While Nkumbwa lies near the intersection of the Luangwa and Rukwa Rifts (i.e., south of Mbeya), Kaluwa, Uma, Mwamboto, Nachombo and Chasweta rise at the intersection of the Zambezi and Rufunsa Grabens.

LUESHE

The Lake Edward Rift is 250 miles wide, and has a thrust of 6,000 ft. Including the *Lueshe* complex, the Kasali (Fig. 51) horst rises between the Main and Ruindi Graben. This complex forms a 2 by $1^1/_4$ mile oval with a 600–2,000 ft. wide rim of acmite sövite surrounding the core of cancrinite syenite (De Kun, 1961). However, the southeastern part of the carbonatite is dolomitic. In the east rings of acmite–amphibole albitite, adinole and albitized quartz–

garnet schists form a narrow fenite band with a larger ring of fenitized sheridanite schist surrounding the rest of the complex. Within the fenites, rheomorphic syenites appear in the south. Conversely, cancrinite veins traverse the core, and alvikites (calcite dykes) are particularly abundant in its rim. While veins of calcite, calcite–tourmaline and siderite also traverse the sövite, aplitic quartz dykes intersect the fenitized schists. From 24% at the contact, the magnesium content of the sheridanite schists decreases to 1%. Concomitantly, potash replaces magnesia.

Pyrochlore is disseminated in sövite and alvikite, in the cancrinite syenite and even in andalusite schists of the outer fenite ring. However, rauhaugite (dolomite) contains only traces of columbium. Between the syenite core and the sövite ring the deposits themselves are located in limonitic alteration products of calcitic aegyrinite or acmite sövite. Lower values connect a larger and smaller high grade (0.2% Nb_2O_5) area. Olive- and grey-green pyrochlore is particularly coarse, averaging more than 2 mm in diameter and even attaining half an inch. Although zircon is the principal accompanying mineral, pyrrhotite, pyrite and strontium bearing barytes also occur. Finally, geologists have discovered in a biotitite black cubes of lueshite, the soda niobate. Columbite replaces pyrochlore.

OTHER CARBONATITES

Sukulu is located near Victoria Nyanza (Fig. 33, 52) in eastern Uganda (see 'Phosphate' chapter). In this complex pyrochlore and zircon mainly crystallize in schlieren. Baddeleyite, 'knopite', pyrite, pyrrhotite, chalcopyrite and gold also occur in rauhaugite. We distinguish three processes of weathering: differential erosion, leaching (depth 80 ft.), and alluvial concentration. Partly in gullies, transported soils form the principal deposits. They contain 20–25% magnetite and 0.2% Nb_2O_5. Soil pyrochlore is lighter than the primary mineral.

In the 2½ mile wide sövite core of *Ngualla*, southwestern Tanganyika, bands of rauhaugitic schlieren crystallize. Surrounding the core, the width of the carbonatized contact breccia and fenitized breccia of granite exceeds 1 mile. Pyrochlore has been altered to columbite in the vicinity of late kimberlite.

Agglomerates and tuffs cover more than half of adjacent *Mbeya*. The sövite–rauhaugite core spans $^2/_3$ mile here. Carbonatization of the inner ring is followed by the desilicification of orthoclasites, silica being expelled into the bordering garnet schists. Veins and dykes of alnöite, alvikite and beforsite are abundant. Pyrochlore is mainly disseminated in sövite, and concentrates in schlieren, apatite sövites and ring dykes. Accompanying minerals include hematite, pyrite, pyrrhotite, zircon, cassiterite, sphalerite and galena. Columbite crystallites tend to replace pyrochlore.

In southern Malawi the Chilwa carbonatite, or Çisi, occupies an area of 1½ by $^3/_4$ mile. This island consists of *(1)* a core of manganiferous siderite, *(2)* an inner ring of ankeritic sövite, and *(3)* a large sövite belt. Agglomerate, contact and partially rheomorphic feldspar breccia outcrop mainly between the carbonatite and the ring of orthoclase, oligoclase and acmite fenites. Furthermore, alkaline and basic rocks form smaller bodies and dykes. Pyrochlore concentrates in schlieren and injection bands, forming a discontinuous ring in the inner third of the sövite belt. We can distinguish two facies: *(1)* pyrochlore–acmite sövite, replacing orthoclase, and *(2)* pyrochlore–apatite–tremolite–phlogopite–ankerite in shear zones.

NIGERIA

Following the law of asymmetry, deposits are concentrated at or near the southern edge of the 350 square miles of the Jos complex (Fig. 109), whence 80% of the columbium prior to 1965 has been extracted. Other deposits favour the Ropp–Tenti complex south of Jos and Afo, near the Benue, off the plateau. Most of their volume consists of very low grade rings or phases. The mean value of other complexes may indeed be less than 15 g/ton. However, differentiation favoured by ring complex injection has built up interesting concentrations.

We can only consider weathered granites as deposits. Probably more altered granite deposits existed in the Pliocene. However, they were later eroded and redeposited in the various placers. Containing 100 dwt. of columbium, the rich Rayfield–Gona phase covers less than 10 square miles. Distinguished by relics of roof pendants, Harwell, an even richer phase, hardly covers 1 square mile. Whereas granites contain quartz, potash feldspar or perthite in the poor phases, they include albite (plus biotite) in the niobian phases. Altered albite granites are plastic.

The centre of the Ropp–Tenti complex is situated 20 miles south of Rayfield. This complex is surrounded by dyke-like or elongated masses of microgranite, granite porphyry, quartz diorite and riebeckite granite. The long axis of this polyhedron extends 25 miles, while the short one measures 15 miles; its phases show some similarity with the northern one.

The late Buka–Bakwai phase has a large outcrop of 6 by 4 miles, but its grade is only a fourth of the Rayfield phase. The feldspars of this sugary granite cupola form aggregates. Equidimensional quartz and 'mud' zircon are associated with the central units of this complex. Similarly, brown opaque, equidimensional and somewhat dull hafnium bearing 'mud' zircon is typical for the three rich phases of the Jos complex: Rayfield–Gona, Forum and Harwell. Values of zircon may exceptionally reach 1%. Thorite and xenotime are typical for the Harwell phase. In addition, thorite is abundant at Afo. Thorite and orangeite have a tabular, short pyramidal or prismatic habit, but xenotime is pale and prismatic. Resembling columbite, its distribution is streaky. The latter, corroded and pitted and sometimes coke-like, is the most typical mineral of this late phase. Attaining 12% Ta_2O_5, its tantalum content is higher than in other phases where it averages 7%.

Fergusonite commonly occurs in porphyries (Federi). On the other hand, pyrochlore appears with riebeckite at Dorowa in the Ropp–Tenti complex, at Kaffo, etc. Topaz is abundant in the greisens, but the older granites contain garnet and tourmaline. The most probable grain size of most Harwell minerals is between 60 and 100 mesh.

The distributive provenance of cassiterite and columbite is not the same. In the placers their ratio depends on the rocks from which they were eroded. It is probably a fair estimate to say that at present the tin/columbium ratio of placers is approximately 10/1. Moreover, the ratio of tin reserves to niobium reserves (including granites) appears to be 2/1. In the north, the world's most important columbite areas extend from Bukuru and the Forum Hills eastwards and northwards towards the Harwell phase and then beyond Rayfield. In this old drainage the ratio varies between 0.5/1 and 1.5/1. In the northwest, in the neighbourhood of N'gell, another columbite rich area of 12 miles belongs to several alluvial epochs, with placers at depths of 30, 80, 100 and 180 ft.! These deposits follow the incomplete ring of late albite granites, and carry their columbite west and northwards in the old drainage. Columbite grade varies from 7 to 220 dwt./cubic yard.

In addition, some columbite has also been mined in other provinces at Rishi, Ningi, Oro, Akwa, Okpossum and other locations. These small deposits are either alluvia associated with Younger Granites or with older pegmatites, or eluvia covering the latter. In the Liruei Hills in southern Kano Province the tin/niobium ratio is 3/1. However, this proportion is lower near the Liruei–Kano lode. In Bauchi Province, east of the plateau, the deposit with the highest ratio is that of Rishi and Pengal. South of the plateau in Benue Province columbite occurs at Ningi and in the Fagam and Abu Hills. In Ilorin some columbite is found at Oke–Onigbin and at Oro. In coastal Calabar Province ilmenite, monazite, zircon, garnet, cassiterite and columbite are reported at Akwa, Ibame, Vyana, Okpossu, Ikpofia and Ekporam.

HAFNIUM

Hafnium, the hidden element, is only found in zirconium minerals. Although pegmatitic zircon may contain 10% hafnium or more, this type of concentrate is limited. However, in the Nigerian ring-complexes abundant zircon contains up to 3.5% Hf quite consistently. Indeed, a considerable part of the hafnium now used in the world comes from this source.

TANTALUM

Carbonatitic pyrochlore containing 0.1–0.4% Ta_2O_5 represents the principal, though still unrecoverable, resource of tantalum. On the other hand, ion exchange can facilitate the separation of tantalum from abundant Nigerian columbites containing 5–8% Ta_2O_5. Columbite–tantalite included in placers and pegmatites meanwhile remains the source of the metal. Because of the greater value of tantalum, concentrates averaging a Nb/Ta ratio of less than 2/1 will be discussed in this chapter. Placers of the Idiba and North-Lugulu (Congo) districts represent a majority of the world's current resources of the metal. Moreover, important pegmatitic deposits are located in the Congo, Mozambique, Southern Rhodesia and the rest of southern Africa.

CONGO

Minor placers and pegmatites include: *(a)* Teturi, Gawe, Etembo, Enehe, Elota, Liha, northwest of Lake Edward; *(b)* Mabuka, Samuda and Bongo, further west; *(c)* Lundjulu, Mutiko, Tshamaka, Masingu, Kanzoro, Utu, Matemba and Idambo, north of the Kasese batholith; *(d)* Mumba, Numbi, Tshiganda, Kobokobo, Mwana and Makalapongo west of Lake Kivu; *(e)* the placers of the central tin belt in the Matelemona (Aissa stock), at Mususa (Kailo district), Lutshurukulu (Kalima), Nyama and Pene Koka (Kihembwe stock); *(f)* Luizi, Kibumba, Maiko, Lukasi and Mitwaba, in northern Katanga.

Columbite–tantalite is associated with cassiterite in most of these occurrences.

The Idiba and North Lugulu districts are located 100 miles west of Lake Kivu (Fig. 114). The *Idibas* cover a 2,000 square mile area of the roof pendants of the Kasese batholith which extends from Kampulu east to Kalukangala, Nkumua and Kashinda and north to Ito, Nyamilenge and Nkenge. Deposits extend southeast of Nkumua to the upper Lokalia and Kasilu.

Mica schists, gneisses, migmatites and alternating schistose and porphyritic–pegmatitic layers have been invaded by two mica granites. Even in areas excluding roof pendants, dolerite dykes and sills cut the granite. Although pegmatites of the area are of no commercial importance, torrential valleys attain 3–10 ft. of width, depths of 1–3 ft., and lengths of 200–1,500 ft. In these thin gravel beds, and on the bedrock below large boulders, columbite–tantalite, tourmaline and locally xenotime concentrate. Of course, greater rivers like the Kanzuzu, Kalukangala and Ona include larger placers, but grades are lower in them. Columbite–tantalite grains of 2 mm–1 inch average 30% Ta_2O_5 and 0.3% U_3O_8. The world's only major tapiolite deposit is located on the northwestern rim of the batholith at Mubilina. Here the diameter of the largest tapiolite grains attains an inch.

In the *North Lugulu* placers the cassiterite/columbite–tantalite ratio averages 13/1. Although smaller valleys are also mineralized, larger rivers like the Makulumabili, Matumpa, Mulongondima, Tshonka and their tributaries include the principal resources. The deposits extend in a 100 mile long and 3 mile wide arc along the southern contact of the batholith (DE KUN, 1959). Furthermore, columbite–tantalite is also disseminated in altered pegmatites, mainly in their inner zone. The mineral contains 20–45% Ta_2O_5. As in the Idibas, its indicator mineral is pink garnet.

In the Manono pegmatite (see 'Tin' chapter) the Sn/Nb–Ta ratio averages 20/1, increasing to 100 in micaceous zones. Columbite–tantalite disseminated in the quartz–microcline phase contains 45% Ta_2O_5. However, the later variety, which concentrates in albitic phases, averages only 30% Ta_2O_5 (Mn/Fe = 1.5). In fact both are fine intergrowths of various members of this isomorphic group. Finally, thoreaulite occurs in the quartz lenses of the northern wall zone.

RWANDA AND BURUNDI

A pegmatite province covers southwestern Uganda, central Rwanda and northwestern Burundi. In *Burundi* pegmatites have been injected into the folds of the Congo/Nile divide extending south-southeast of Lake Kivu. The Ndora lens occurs in the southern part of this gold belt, while Mogare outcrops further east. However, the principal deposits are located in the Republic of *Rwanda* in (Fig. 31, 110): *(1)* a 70 mile long belt extending northward from the Katumba–Nyanza batholith, *(2)* a 20 mile long northern aureole of the batholith extending from Katumba and Tshubi eastward to Kigali the capital, *(3)* a 30 mile long zone following orogenic axes between Kigali and Mbuye in the southeast on the Burundi border, *(4)* a north-northwest and north trending zone located between Kigali and the Tanganyika border in the east, and *(5)* scattered in the Congo/Nile divide between Lake Kivu and the Nyanza batholith.

A majority of deposits occurs in zone *1* which includes: *(a)* in the north Lukagarata, Shaki, Kadaba and Ndiza, *(b)* the principal occurrences in the centre of Gatumba (Katumba), Bijojo, Kirengo, Luhanga, Kavumu, Burenga and Shori, and *(c)* in the south Kinyoni. The Kigali tin area divides zones *2* and *3*, but the Mayaga pegmatite is isolated south of this area. Zone *3* extends from Rukoma to Ntunga and Mbuye. Zone *4* includes the Bugalula–Bibale and Gikaya–Sinda trends.

In Rwanda and Uganda we can distinguish two types of tantalite–columbite: a widely disseminated, fine grained, homogeneously composed variety, and coarse crystals of variable composition restricted to specific particularly albitized zones. Most tantalum bearing members of the isomorphous group are included in the second type. Moreover, most pegmatites are muscovitic, marginal or external. Eluvia and alteration zones which cap pegmatites are particularly interesting. Furthermore, placers are abundant, e.g., at Bijojo, Mutaho and Nyungwe.

EASTERN AFRICA

The irregular pegmatites of Ankole, southwestern *Uganda*, are aligned along tectonized northwest trending (kyanite) schist folds. Columbite–tantalite and beryl typically occur in kaolinized feldspar on the shattered altered limit of the quartz core or bodies. The Nyabakwere lens consists of microcline, but albite and perthite accompany columbite at Bulema in the southwestern corner of the country. Microlite occurs in the same area, but fergusonite is only a trace element of columbite concentrates. Euxenite is found at Nanseke.

The principal deposits include Rwentali, Rwenkanga, Kashozo, Lyasa and Rwabuchacha, plus many smaller occurrences. Niobo-tantalates also occur in a belt extending west of Kampala the capital, which includes Nampeyo Hill, the type locality of bismuto-tantalite.

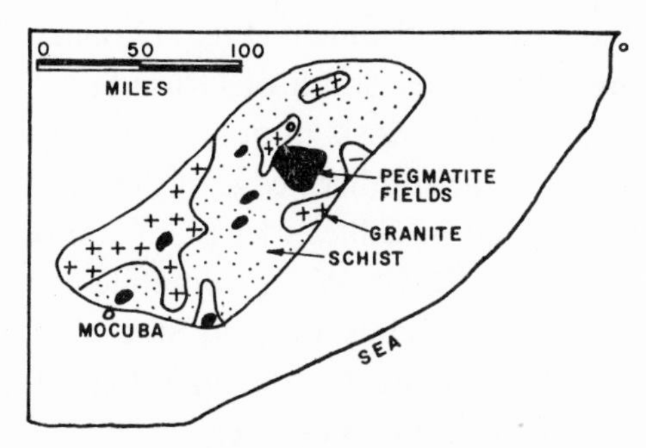

Fig. 115. The Alto Ligonha field.

Although some tantalite is disseminated in northeastern Uganda, the Kenya and Tanganyika pegmatites (e.g., Machakos and Uluguru Mountains) have no commercial significance, nor have the lenses of the Humbeleh Range, Warran Weis and Theb Valleys in Somalia. Similarly, betafite of the alkaline syenites of Tambani (Malawi) is not commercial.

Conversely, in northern *Mozambique* columbite–tantalites of the Alto Ligonha and Alto Molucuè districts (see 'Beryllium' chapter) contain 30–45% Ta_2O_5, with a specific gravity of 5.9–6.2. While columbite crystals of the perthite zone are large, pinacoidal grains originating in the albite–lithia zones are fan-like and tabular. At Macoteia euxenite–samarskite is associated with columbite–tantalite. Stibiobismuto-tantalite has been discovered at Muiâne, and it accompanies euxenite, samarskite, allanite and davidite in the Mavudzi Valley. Moreover, euxenite and allanite have also formed commercial concentrations. Other pegmatites include Muiâne, Boa Esperança, Cavale and Nahia. Let us note that Alto Ligonha samarskites are 408 and 465 million years old (Fig. 115).

Niobotantalates are widespread in eleven *Malagasy* pegmatite fields, e.g., in *Berere* (Antsakoa, Befilaho, Beampy, Fig. 48, 52), Andriamena, Antsirabe (Antsirabe, Sama) and Ampandramaika (I, Ambatofotsy, Marivolanitra). Generally columbite–tantalite occurs in regional zones farther from the intrusive than niobo-tantalates of the rare earths. However, in each pegmatite columbite–tantalite is confined to one or two zones, mainly those of intermediate rock. The limit of the core is particularly favourable to concentrations, e.g., at Ambatoharanana–Ambony (Berere field). Finally, samarskite occurs at Mbora near Mouneyres.

SOUTHERN AFRICA

In *Southern Rhodesia* microlite is disseminated in the Benson beryl pegmatite of the Mtoko field in the Odzi belt. On the other hand, at Kamativi tantalite accompanies tin (see 'Tin' and 'Beryl' chapters). Beneath the core of the Mdara lens tabular tantalite is found in purple lepidolite aggregates in the albitized layer. In the *Bikita* field tantalite accompanies lithium minerals (see 'Lithium' chapter). In the Nigel sheet tantalite occurs in the rhythmical mica–quartz formations, with microlite favouring albite or lepidolite. Nearby the Al Hayat petalite body intersects tantalum bearing albite (Fig. 116). In addition, some microlite occurs at Ranger, and it accompanies tantalite at Patronage.

Tantalite–columbite of variable composition occurs in the Omaruru field of central *Southwest Africa*, in the Goantagab belt 150 miles northwest, and in the Orange River belt bordering South Africa. In the Erongo Schlucht (Fig. 117, Omaruru) tantalite is associated with the intermediate zinnwaldite zone. South of Grosz Spitzkopje, Gossow's pegmatite attains a length of 1,000 ft. The core consists of quartz and muscovite, with tantalite associated with the latter and albite at the southern edge of the zone. At Okongava the mineral is part of a similar paragenesis, but in Cillier's lens tantalite occurs in albitized feldspar. Similarly, at Sandemap, Davib and Davib West, columbite–tantalite is found in an intermediate feldspar zone. At Pankaub, Tsomstaub and Stoeckle's claim columbite shows in quartz bodies related to the Salem granite. On the other hand the mineral occurs in albitized and greisenized zones at Kohero and Kranzberg.

In the Orange River belt at Umeis columbite–tantalite is disseminated in an intermediate zone, but north of this lens a quartz core contains some tantalite and beryl. On the *South African* (i.e., southern) bank of the Orange River columbite–tantalite is disseminated in

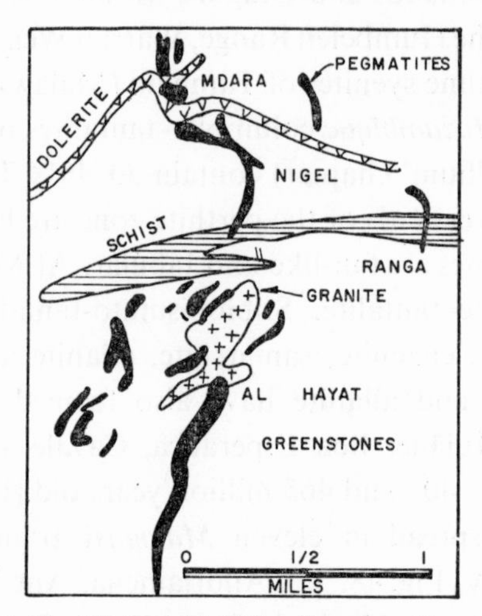

Fig. 116. The Bikita field. (After TYNDALE–BISCOE, 1951).

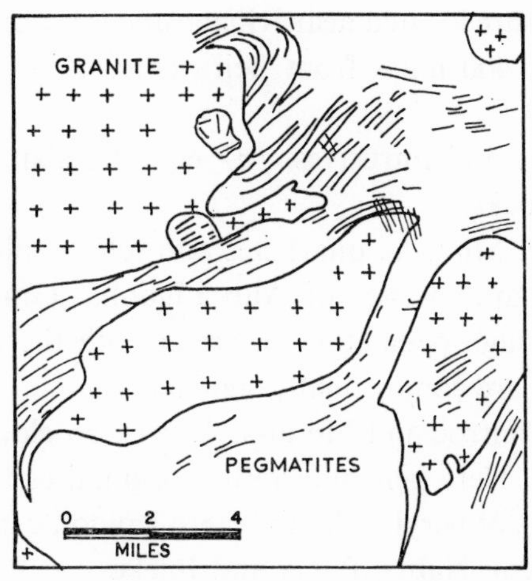

Fig. 117. The Erongo area. (After GEVERS and FROMMURZE, 1929).

pegmatites and placers of the Jackalswater–Uranoop–Spodumenkop area. In the north-eastern Transvaal the placers of the Palakop near Phalaborwa and adjacent Leeuwspruit nearby contain columbite–tantalite eroded, respectively, from muscovite and beryl pegmatites.

Finally, in Swaziland yttro-tantalite is associated with cassiterite in placers and pegmatites of the Kelly mine.

THE REST OF AFRICA

In central *Nigeria* the zoned pegmatite sill of Old Jemaa mainly carries columbite–tantalite in undulations of the walls. Other deposits include Wamba, Abuja, Tarkwashara and Gudi as well as the Egbe field south of the Niger. The whole isomorphic group is known to occur in the district, related mainly to albitization. In the southwestern Cameroon some columbite–tantalite is associated with cassiterite in placers with gold at Betare Oya. Pegmatites of the eastern Chad contain traces. Finally, columbite–tantalite is included in the placers of Bouaké in the central Ivory Coast, in the pegmatites of the Falémé district of southwestern Mali, and at Sidi bou Othman and Jebilet in western Morocco.

BERYLLIUM

Beryllium is more evenly spread than the earth acids in the pegmatite fields of Middle and Southern Africa. *Mozambique*, Madagascar, Rwanda–Uganda, the Congo, Southern Rhodesia, Southwest and South Africa remain, indeed, the world's principal sources of this metal (Fig. 6). In them beryl tends to occur in indistinctly zoned pegmatites affected by incipient replacement processes. Dissemination type deposits undoubtedly exist in Africa, but no data are available on the beryllium content of African coals.

MOZAMBIQUE

Deposits of beryl and niobo-tantalates occur on both sides of the Mozambique Straits.

In Mozambique pegmatites are located near Ribaue and Alto Ligonha in the Alto Molucue district of Niasa Province, 400 miles from Beira (Fig. 115). They are injected into the following country rocks:

Precambrian III: granites and intrusives, *(a)* equigranular, mineralized, *(b)* sheared, *(c)* porphyritic, *(d)* basic rocks.

Precambrian II: younger: Mirrucui quartzites and schists, older: in a geosyncline polymetamorphic ortho- and para-gneisses of Alto Ligonha. Granitic cupolas deform these southwesterly trends. Although pegmatites occur in both the gneisses and the metasediments, only external lenses are commercially significant.

In the triangle of the Namirroe and Metuisse Rivers, an elongated granitic dome is intrusive in Mirrucui schists. Here generally radial pegmatites traverse the schists. In this zone the principal producers of beryl, columbite and lepidolite include Nanro, Nacuissupa, Murrapane, Nihira, Mirrucui, Nahia, Piteia and Injela.

In the Nahora area pegmatites occur in a north–south trending tectonic belt of Mirrucui schists and quartzites which is invaded by the Namuiie granite. These pegmatites are parallel to the tectonic belt. The most important producers of columbite, tantalite, bismuthite and beryl include: Nahora, Namicaia, Namitaca, Alata, Mocachaia, Macula, Muacumuano and Nahavarra.

In roughly zoned pegmatites, beryl and columbite occur in two zones: border zone with plagioclase, quartz muscovite, schörl, garnet; wall zone, containing plagioclase, quartz, muscovite, biotite, perthite, *columbite*; intermediate zones: *(1)* idem as wall zone with book mica; *(2)* perthite zone: megacrystalline perthite, some quartz, *beryl*, *columbite*, bismuthite, cassiterite; *(3)* quartz – plumose muscovite zone; *(4)* megacrystalline plagioclase, spodumene, quartz, amblygonite, *lepidolite* (lithia); *(5)* lepidolite–albite in variable proportions; core: lenticular mass of white or rose quartz.

The principal mines include Muiane south of Alto Ligonha, which contains beryl, columbite–tantalite and red tourmaline, and Boa Esperança north of Alto Ligonha, which carries beryl, samarskite, bismuth, bismuthinite and topaz. Poorer pegmatites of the Alto Ligonha district include Cavale and Nahia, while others occur in the Ribaue district. At Macoteia tantalo-niobates, tourmalines, davidite, garnet, apatite, fluorite, rutile, hematite, fluorite and cassiterite accompany beryl.

MALAGASY REPUBLIC

Along the axis of *Madagascar* spread 14 pegmatite fields (Fig. 48, 52). The Antsakoa and Analila pegmatites of the Berere field, and A4 and F3 of Malakialina and Ambondro of the Analalava field contribute more than half of Malagasy beryl. The only other economically significant field is Ampandramaika (BESAIRIE, 1961a). External beryl pegmatites almost exclusively traverse mica schists, gneisses and amphibolite (but not granite). Beryl concentrates on the margin of the core.

The 25 mile long, important *Berere* field is located 150 miles north of Tananarivo. Here a brachy-anticline of leptites transformed into myrmekitic migmatites is overlain by pyroxenites. Pegmatites are unrelated to intrusives, and their zoneographic pattern is inverted. We distinguish three pegmatite shapes: dykes, lenses and ellipsoids. Within these distinctly zoned bodies, beryl occurs on the margin of the core and in the intermediate, partly albitized, megacrystalline perthite zone. Half of the beryl production of the field, as well as columbite, originates from the narrow quartz–perthite–albite–muscovite zone that surrounds the inner perthite of the *Antsakoa* lens, a 650 ft. long and 250 ft. wide body. Spurs of the quartz core penetrate into the large perthite zone, but do not reach the mineralized zone. The 20–50 ft. thick wall zone consists of quartz, muscovite, plagioclase, muscovite and garnet.

Nearby lies the small Antsakoa II lens. At some distance the distinctly zoned Analila pegmatite attains a length of 800 ft. and a width of 200 ft. Here a 10–40 ft. thick perthite core with accessory quartz is surrounded by the intermediate zone. The latter zone contains beryl and columbite, especially in the southern part of the hangingwall. In addition, the Ambararata and Imagoaka I pegmatites also contain beryl.

The Tsaratanana field contains little beryl. Its southern extension, the Vohombohitra field, can be subdivided into several swarms. However, gem beryl occurs in the quartz–perthite–muscovite zone of the asymmetrically zoned Anariafia and Ambalanirana swarms, and also in the indistinctly zoned Ankerana and Manakana swarms. Similarly, we find some beryl in the Ambohipihaonana swarm and in the Ampasangoaika lens of the Betsiriry field. South of Tananarivo the principal deposit of the Vavavato field is the Miandrarivo pegmatite. South of Vavavato there are three small fields. The Sahatany field is of minor significance, but some beryl is disseminated in the indistinctly zoned pegmatites of the Analalava field, e.g., at Manampa and Ambondrona, and also in the Vorondolo field (Ambohimanitra and Farihitsara). South of these sister fields, gem and industrial beryl occurs in the Ambolo, Soavina and Bebaranga pegmatites of the Ambatofinandrahana field.

Near the western edge of the Vohibory System rocks, the *Ampandramaika–Malakialina* field is situated 200 miles southwest of Tananarivo. This elliptical field attains a length of 40 miles and a width of up to 15 miles. Geologists have found a small stock here, but no granitic 'herd', although schists have been migmatized along an anticlinal axis. Finally, sheet-like bodies of granite have developed in the migmatites. The pegmatites occur in mica, kyanite, staurolite and garnet schists, surrounded by gneisses. In the 300 ft. long Ampandramaika I lens beryl is associated with columbite. Nearby the 660 ft. long Marivolanitra dyke bulges at the surface. Similarly, the Malakialina F3 pegmatite pinches out in depth. In the Mania Valley the long 90 ft. wide *Malakialina A4* pegmatite outcrops. In this dyke prismatic beryl forms 5–50 ft. long pockets between the large core and the intermediate albite–muscovite (–perthite) zone, but some beryl also occurs in other zones. The largest prisms (up to 270 tons) tend to grow vertically. The beryl averages 12% BeO. While its tint is blue in the core and white or light green on its margin, in the intermediate zone beryl is yellow. Perthite partly surrounds the zone, followed by the wall zone.

Swarms include Iandanoa, Andaboroa, Amboloano, Beseva and Bamokoty. Out of a total of 1,000 explored pegmatites, 100 contain beryl.

CONGO

Because of the sympathetic relationship between tantalum and beryllium we find beryl in several Congolese pegmatite fields (e.g., North Lugulu).

The *Kobokobo* swarm of pegmatites is located near Kamituga, 90 miles southwest of

Lake Kivu. The 700 ft. wide Lusungu pod, the principal beryl deposit of the Congo, is included in this group. An albite–oligoclase–muscovite zone occupies two-thirds of its area. Since albitization has been particularly intense in the poorly mineralized east, more microcline has been preserved in the west. The 250 ft. long and 70–20 ft. wide core consists of perthite, but a somewhat shorter body of pale green lithia muscovite rises in the albite zone. A 90–130 ft. wide perthite zone follows in the east. This zone is bordered by an 8 ft. wide wall zone of albite, and a similarly thin border zone of muscovite and quartz (compare to other pegmatitic 'greisen'). Finally, late quartz bodies and veins traverse several zones.

In the outer rim of otherwise normal perthite, beryl bodies are ditributed along a sinusoid band 250 ft. long and 20 ft. wide. An albite lens is intercalated between part of this 'critical zone' and the perthite. Non-oriented, rarely columnar bodies of beryl have a diameter of ½–2 ft., but at the northern end of the critical zone three bodies have expanded to a total length of 60 ft. Other nests have been spaced 20–50 ft. from each other. All but one of them occur approximately 20 ft. from the wall. Beryl includes columbite–tantalite, cyrtolite, loellingite and scorodite, but triphyllite has been altered to dufrénite, huréaulite and heterosite. Uraninite, autunite, apatite and fine grained columbite–tantalite overlap towards the centre, the 'critical zone'. On the other hand, amblygonite is mainly disseminated in the lens of lithic mica. In addition, some beryl and bismuth show in the eastern external zone.

The muscovite is 900 million years old, while uraninite attains an age of 830 million years. However, cyrtolite ages range between 660 and 860 million years.

RWANDA AND UGANDA

Among the pegmatite fields of *Rwanda*, beryl is abundant at Katumba and Bijejo, and also occurs near Kibuy on Lake Kivu. A pegmatite vein containing beryl belongs to the swarm of altered kaolinic pegmatites of the Nyavarongo (see 'Tin' chapter). On the other side of the River, beryl accompanied by tourmaline is part of one of the upper zones of the *Kirengo* pegmatite. The interior of beryl crystals has often been altered to a powdery pegmatitic derivate, sometimes leaving only a prismatic, 1 mm thick film of beryl.

In *Uganda* beryl occurs: *(a)* in the pegmatite fields of Ankole in the southwest near the Rwanda border, e.g., at Shumba, Nyabukweri, Nyaruwanya and Rwabuchacha; *(b)* at Bulema close to the Congo border; *(c)* at Mbale and Nampeyo Hill west of Kampala; *(d)* in the Lunyo pegmatite and Karonge granite east of the capital, and *(e)* near Kaabong in the northeastern corner of Uganda.

The pegmatites are indistinctly zoned and frequently kaolinized. Although some beryl is widely disseminated in the pegmatites (e.g., in the wall zone at Kazumu), the majority of coarse beryl is restricted to the pockets frequently located at the border of the core (e.g., Bulema). In addition, intermediate albite–muscovite zones contain ore at Ishasha (BARNES, 1961). Conversely, stanniferous quartz–muscovite formations include beryl at Namahehere (with euclase) and Mwirasandu.

SOUTHERN RHODESIA

Beryl is widespread in the pegmatites of the Rhodesian shield. Both large and fine grained beryl crystallizes in the *Bikita* field (see 'Lithium' chapter). In the Mdara (Mauve Kop) pegmatite, pockets of columnar beryl which occur in the central quartz bodies are related to albitization, but anhedral beryl aggregates have crystallized earlier. Lepidolite tends to replace beryl, which turns yellow in the eluvia. Beryl is an accessory mineral of the lithia zones of the Al Hayat pegmatite. Nearby it forms the wall zones of the *Bikita* lens where beryl is in contact with the country rock in the footwall, a micaceous border zone ('greisen') intercalating at the hangingwall. A considerable quantity of beryl has originated in the eluvia covering the southern end of Bikita (Fig. 116).

In the Salisbury district, beryl is found with mica at Mazoe and Mauve. Massive beryl occurs in the Augustus claims 18 miles north-northeast of Salisbury and at Hotspur, Mops and Leander. In the Pope pegmatite, twelve miles east, white, massive, and pink hexagonal beryl coexist. The principal beryl pegmatites of the Mtoko and Mrewa fields include Benson and Matake. At Rabbit Warren a beryl mass, 18 by 12 ft. wide and 15 ft. high, is included in a massive white quartz body. Beryl also occurs in the zoned mica pegmatites of the northern Urungwe field, notably Cajol, Ngesi Valley, Squirl and Chaminoka.

The Rhodesian beryl pegmatites frequently consist of: *(1)* a core of quartz stringers $\pm$ lepidolite *(2)* an albite-lithia silicate intermediate zone, and *(3)* an up to 1 ft. thick mica wall zone. Beryl concentrated during the albitization of microcline. Its size increases downwards towards the limit of the core and attains its maximum beneath it.

THE REST OF AFRICA

The principal beryl district of *Southwest Africa* is Karibib. In the southern part of Van der Malde's pegmatite, in the Erongo Schlucht (Fig. 117), beryl occurs in an intermediate zone. At Davib Ost beryl is associated with perthite in indistinctly zoned pegmatites. At Rössing, a pegmatite concordant to biotite quartzites, beryl is associated with microcline in the intermediate quartz–tourmaline zone. However, at the Kleine Spitzkopje beryl accompanies topaz. Moreover, beryl, tourmaline and quartz intensively impregnate the country rocks. Similarly, in the Donkerhoek pegmatites west of Windhoek, beryl, columbite, apatite and sulphides occur in an intermediate zone.

In Tantalite Valley in the Orange River belt, beryl occurs in the intermediate zone of the Umeis pegmatite. In addition, beryl is associated with quartz and muscovite in the dyke located on the bend of the Krom River.

At Middel Post the largest known beryl pegmatite of Namaqualand, in *South Africa*, extends 100 ft. and bulges to 30 ft. In the phase of albitization beryl is related to late quartz, accompanied by schörl, garnet, phosphates and titanates. Similarly, the Koegas lens traverses a gabbro sill. In the northeastern Transvaal we know of several beryl pegmatites: at Leeuwspruit, west of the Phalaborwa complex, tourmaline and columbite–tantalite accompany beryl.

Beryl pegmatites outcrop west of Tchatchou in Dahomey, at Falémé in Mali at Laouni and at the Adrar Elbema in the Hoggar (Algeria). In Egypt gem beryl is associated with schörl and garnet at Zabara, Sikait, Nugrus and Um Kabu, and it occurs in pegmatitic quartz which traverses micaceous schists in the southeast. In *Morocco* aggregates of coarse beryl are found in zoned pegmatites of Zenaga, 40 miles southwest of Ouarzazate. In the Angarf lens the grade attains 6% beryl, containing 12% B_2O_3.

LITHIUM

In the Congo the Manono–Kitotolo pegmatite represents the world's largest accumulation of lithium. However, the high grade Bikita pegmatites in Southern Rhodesia provide an overwhelming majority of the world's supply. Other deposits are located in Southwest Africa and Rwanda. Lepidolite represents two-thirds of the output with petalite, spodumene and amblygonite contributing 20, 5 and 5%, respectively. Spodumene tends to develop during the later phases of unzoned pegmatites, characterized by albitization. At the same time, the other lithium minerals begin to crystallize and reach maxima at the end of the replacement process.

MIDDLE AFRICA

Reaching to unknown depths, the Manono–Kitotolo pegmatite (cf., 'Tin' chapter) occupies almost 1 ½ square miles. Spodumene is a component of two of the four zones or repeated bands, accompanied by albite and quartz *(3)* and kaolinized microcline and albite *(4)*. With numerous plates that extend for more than 3 ft., spodumene of band *(3)* is megacrystalline. After microcline, spodumene which contains 6% LiO_2 has crystallized with the first albite. Typical samples of pegmatite rock contain more than 1.5% LiO_2. At the surface lithia has been leached from two alteration products of spodumene, killinite containing 2–4% LiO_2, and a kaolinic end-product containing only 1% LiO_2 (Fig. 112).

Amblygonite occurs in the albitized pegmatites of Kivu Province in the Congo, in Rwanda–Uganda, and on the Congo/Nile divide in Burundi: e.g., at Kobokobo (Kivu), Gatumba (central Rwanda), in the Ankole beryl district (Uganda), and with petalite at Ruhuma, Kigezi district (Uganda). Furthermore, the quartz–mica formations of Mwirasandu and Lwamuire contain some amblygonite. The principal deposit of the Kampala pegmatite belt is the Mbale Estate.

SOUTHERN RHODESIA

The Fort Victoria schist belt forms part of the early Precambrian east-northeast trending arc, which continues further east in the Odzi–Umtali–Vila de Manica belt. Surrounded by granite, granodiorite and gneisses, this 70 mile long arc bulges in its middle to a width of 14 miles. While near Mashaba, at the western tip of the belt, asbestos and chromite concentrate, tin appears at Rurgwe and Bikita at the eastern end. Here calc schist, granulite and banded ironstone are interbedded with sub-vertical greenstones which vary in texture from massive epidiorites to foliated hornblende schists. Skarn veins located nearer to the contact than the pegmatites contain scheelite, epidote and garnet (Beardmore).

However, in the *Bikita* field arcuate lenticular pegmatite sheets, which dip from several to 60° eastward, intersect the greenstones. The principal alignment comprises, from south to north: Bikita, Al Hayat, Nigel and Mdara (Mauve Kop). A ramified lens of metasediments outcrops in the middle of the area, but late dolerites have been injected in the north. South of the schists a ½ mile long body of granite has been intruded. The Nigel Valley lens lies east of Nigel, while Ramga cuts the schists. Numerous lenses occur west of the microgranite, e.g., at Ledingham Hill (Fig. 116).

The *Mdara* lens (TYNDALE–BISCOE, 1951) consists of fine grained albite–oligoclase and a large internal zone of quartz bodies containing beryl. Coarse quartz–muscovite remains barren. Mauve lepidolite covers the lens, and rhythmical quartz–mica formations develop south of it. Further south columnar lepidolite aggregates are embayed in the quartz–albite zone. A small cassiterite bearing vein follows the strike. The total length of the Mdara–Nigel complex attains $^{3}/_{4}$ mile. The altered quartz–albite–muscovite–garnet microgranite contains lenses, schlieren and concentrations of pegmatite and lepidolite. The *Al Hayat* pegmatite borders this stock on the east. Petalite, accompanied by spodumene, eucryptite and beryl, developed south of the microgranite later than tantalum bearing albite. It forms a dyke 1,600 ft. long and 150 ft. wide. This body represents 350,000 tons to a depth of 20 ft. We can distinguish petalite, which contains 2.1–3.6% LiO_2, from quartz by its cleavage (001). While white spodumene and quartz form lenses in the petalite, amblygonite–montébrasite also occurs in quartz.

The Bikita pegmatite comprises *(1)* a footwall zone of beryl, *(2)* a cobbled zone of feldspar partly replaced by nodular aggregates of lepidolite, *(3)* a lepidolite core, *(4)* a spodumene–petalite–amblygonite zone, *(5)* a beryl zone, and *(6)* a mica border zone. (Zone *4* corresponds to the Al Hayat petalite body.)

The quartz–microcline–oligoclase pegmatites have only been albitized in the mineralized zone. We can recognize both sugary albite and cleavelandite, but quartz–muscovite formations are barren. Biotite, zinnwaldite, altered micas, tourmaline and rutile impregnate the country rock. Micaceous zones (previously termed greisens) frequently border at least one side of the lithium zone, and stringers also penetrate into it. Albitization is particularly intense in these zones which average 4% LiO_2. Widespread and sometimes coarse topaz proves the abundance of volatiles and their affinity to greisens.

The lithium minerals are 2,600 million years old. *Lepidolite* is the most important lithium ore, while petalite and spodumene play a minor role. Amblygonite and eucryptite are less significant. We can distinguish three lepidolite tints: *(1)* deep mauve lepidolite covers, e.g., the Mauve Kop; *(2)* colourless twinned lepidolite, related to zinnwaldite, associated with topaz, quartz and two facies of zircon, whose pockets replace albite, muscovite granite or beryl; *(3)* a pale lilac lepidolite which carries cassiterite at Bikita.

Amblygonite in four lenses contains 9% lithia, but the remainder characteristically has been replaced by sodium. In addition, hydroxyl prevails over fluor. This paragenesis includes pollucite, the only cesium mineral, and cassiterite. Fine grained lepidolite is disseminated in these lenses, with books and crystals occurring mainly in the core and on its margin. Occasionally amblygonite occurs in pockets. Other typomorphic accompanying minerals include rubbellite, zinnwaldite, cookeite, paragonite (hallerite, an alteration product of feldspars), cassiterite and disseminated quartz grains. Muscovite, beryl, tourmalines, baddeleyite (a generally carbonatitic zircon mineral), and garnets (almandite, melanite, etc.) tend to occur in non-lithia zones where topaz also crystallizes.

Lepidolite occurs in several other pegmatites, e.g., in the Pope Claims (Salisbury district) where it embays tantalite.

SOUTHWEST AFRICA

Lithium minerals are particularly abundant in the Karibib field which surrounds Erongo between Walvis Bay and Windhoek. Other pegmatite fields are located 100 miles north of this district and along the Orange River. The *Karibib* field covers an area of 60 by 50 miles, with important deposits extending from Usakos to Otjokatjongo in the east, and concentrating near Kohero north of Karibib (west of Omaruru). Most lithium pegmatites are external and occur in marbles, dolomites and schists. They are frequently concordant, but country rocks do not control their crystallization. The pegmatites develop in three tension fissure systems. Vertical faults and thrusts of the Karlsbrunn and Albrechtshöhe fields are cogenetic with the latest Precambrian post-Damara orogeny. On the other hand, the late joints of the Tsaobismund area prolong anticlinal folds.

The internal *Erongo Schlucht* (Fig. 117) pegmatite outcrops intermittently along a mile. This irregular distinctly zoned body dips steeply west. The first intermediate zone carries lithia, the second muscovite. This is followed by a large body of albitized feldspar and quartz, and by the quartz core.

The lithia-muscovite is grey-green, but zinnwaldite is pink-red. Muscovite is sometimes chloritized. Albitized feldspar has a sugary texture, and contains some topaz and apatite. Zoning may be schematized as given in Table CXXXVII.

A pegmatite of the western part of the Erongo Schlucht attains a length of 1,500 ft., while lepidolite bodies form on the margin of the northern core. Further south amblygonite, topaz, quartz and albite occur in a similar position. Adjacent lenses also occasionally carry lepidolite. A megacrystalline pegmatite has been injected into marbles and schists of the northern part of *Okongava Ost* 15 miles from Karibib. Here a micaceous zone divides the lepidolite body from the quartz core. This body measures 800 ft. by ½–15 ft., averaging 5 ft. In the southern part of this locality xenoliths of silicified amphibolite remain. Quartz stringers traverse the intermediate lepidolite zone, but the wall is micaceous. At Kaliombe near Albrechtshöhe, lepidolite and topaz form the eastern margin of a 250 ft. long and 25 ft. wide body which bulges into 60,000 tons of lepidolite. A banded lepidolite/feldspar zone follows this. Conversely, spodumene accompanies quartz in an inner intermediate zone of the Donkerhoek pegmatite.

In the Orange River belt lepidolite is somewhat scarce. The inner quartz lenses of a pegmatite which outcrops on the border of Umeis and Kinderzitt are partially enveloped by lepidolite. The lithia zone, surrounded by perthite, also contains albite and quartz.

TABLE CXXXVII

PEGMATITE ZONES

1	*2*	*3*
granite	granite	granite
feldspar–quartz	coarse tourmaline	quartz–muscovite, partly chloritized
zinnwaldite	coarse tourmaline	
feldspar–quartz		lithia–muscovite
	lithia–muscovite	
coarse muscovite–quartz–tourmaline		
quartz–feldspar–much tourmaline	coarse quartz–feldspar	quartz–muscovite
fine granite–pegmatite	much tourmaline	
granite	granite	granite

THE REST OF AFRICA

On the *South Africa* side of the Orange River spodumene, containing only 4–7% lithia, predominates in the 3 ft. thick Noumaas pegmatite. Furthermore, the mineral constitutes 40% of the mass of other lenses. Let us mention a spodumene plate which attains a length of 5 ft. Similarly, lepidolite forms large bodies at Sleight's and Jackalswater. The lithium belt also includes Spodumene Kop, Narrabees, Kokkerboomrand, Henkrieswater and Mount Stoffel.

In *Madagascar* lepidolite, amblygonite and triplite occur in the zoned Somoaly beryl pegmatite, and also in the columbite pegmatites of the Ampandramaika–Malakialina field. The principal amblygonite deposit, Ambandrano (Ikalamavony field), also contains lepidolite, lithia tourmaline, albite and cleavelandite at the margin of the core. In the Anjahamiary pegmatite (southeastern Madagascar) pockets of lepidolite, albite, some beryl and monazite form four alignments. The same minerals along with spodumene occur in the albitized pegmatites of the Sahatany field south of Antsirabe and at Ambalahamatsara east of Ambatofinandrahana.

In the *Alto Ligonha* pegmatite field, *Mozambique*, lithium minerals are widespread. The principal mineralized pegmatites include Murrapane, Mirrucui, Nacuissupa, Nahia and others (see 'Beryllium' chapter). Microcline perthite is the principal mineral of the intermediate zones of these pegmatites. Lepidolite and amblygonite occur mainly in the albitized inner intermediate zone, frequently intergrown with plates of cleavelandite. However, spodumene has been altered or kaolinized. In addition, some lepidolite occurs in the Hombolo pegmatite in eastern Tanganyika.

In *Mali*, south of Bamako, four pegmatites of Bougouni include 160,000 tons of spodumene containing 6.5–7% LiO_2. The largest of these is Sinsinkourou. Finally, lithium silicates also occur in the Falémé field of southwestern Mali.

RUBIDIUM AND CAESIUM

Both of these rare metals occur in albitized (sodolithic) pegmatites. The principal source of *rubidium* is lepidolite. This mica contains up to 3.5% of the metal, while pollucite contains a maximum of 3.7%. Consequently, the Bikita and Karibib pegmatites (Southern Rhodesia and Southwest Africa) represent the world's principal high grade resources (viz., preceding chapters). Widespread microcline and common micas constitute the low grade resources.

Lepidolite also contains *caesium* (or cesium), another heavy alkali metal. Transparent pollucite, the only caesium mineral, strongly resembles quartz. However, cubic pollucite contains 25–35% Cs_2O, and the maximum attains 42.5%. Large deposits are located at Bikita and in the Karibib field. In the *Bikita* pegmatite, pollucite occurs on the margin of and in the core. A 600 by 60 ft. mass of pollucite has crystallized in the lepidolite body. Stringers of lepidolite penetrate into the pollucite, which includes petalite aggregates of several feet, disseminated microcline, lepidolite and quartz. In addition, pollucite occurs in other Rhodesian pegmatite fields. In the Karibib pegmatites pollucite lenses border the core, followed by quartz–potash, feldspar–muscovite–beryl zones. Probably pollucite could also be located in other African pegmatite districts.

SCANDIUM

Scandium is related to the rare earth metals, and also shows as a trace in uranium slimes. One of the world's major deposits of the high grade scandium mineral, thortvéitite (or befanamite), is the Befanamo pegmatite located 100 miles north of Tananarivo in the upper Betsiboka drainage, Madagascar. Thortvéitite has been disseminated in the flat-lying lenticular core. Until 1955, 85 lb. were produced there. North of Befanamo befanamite also occurs in the Berere beryl–columbite field.

BISMUTH

Unlike the majority of deposits elsewhere, in Africa bismuth occurs as an accessory mineral of pegmatites. In Uganda and Mozambique only minor deposits are active. Other occurrences are located in Middle and Southern Africa.

In southwestern *Uganda* the bismuth district of *Kigezi* is located near the Congo border. Here nine deposits cluster in an area of 5 by 2 miles. Bismutite, sometimes surrounding a core of bismuth, is disseminated in limonite bodies. Whereas in the northeast Muramba, Kyambeya and Lwenkuba are pegmatitic, Kitwa, Kitawulira and Rwanzu are hypothermal deposits. The diameter of the 2 ft. thick lenses of Muramba attains 150 ft. Bismutite forms smaller masses in limonitized siderite. These bodies are believed to have been emplaced in shear zones with veins following farther from the granite in reverse faults. Pyrite has been limonitized in them. Wolframite and gold occur in both varieties, but tin appears only in the megacrystalline bodies. Furthermore, west of Kampala geologists have discovered bismuto-tantalite or ugandite in the Nampeyo Hill pegmatite.

Bismuth also occurs in the Katumba pegmatite field, at Buranga, Rwanda, and on the Congo/Nile divide in Burundi. In the eastern *Congo* native bismuth crystallizes at Ngussa and Mayamoto with gold, cassiterite and pyrite. It occurs in the shear and crush zones of the long fault dividing the younger Precambrian Itombwe syncline from the surrounding Middle Precambrian folds. Weighing up to 10 pounds, sub-angular lumps of native bismuth occur in the placers covering the pegmatite fields of the North Lugulu tin–columbium district (cf. 'Tin' chapter), especially at the upper Lubilu River, and also in the Nietubu and Lubilokwa drainages. Frequently a yellow patina, which shares the diffraction pattern of synthetic bismutite or more rarely a black patina of bismuthinite, covers the bismuth (Hutton in DE KUN, 1959). Moreover, fine grained bismuth occurs frequently as a trace element in other parts of the Kivu–Maniema tin province. In the complex tin and gold bearing eluvia of Pizon, east of Kalima, white altered grains and pebbles of bismuthinite are abundant.

East of Lake *Tanganyika* the lead–copper ores of Mukwamba contain 0.025%. Similarly, bismuth shows in the Uluguru pegmatites and the auriferous quartz veins of Musoma and Lupa.

In the Alto Ligonha pegmatite field of northern *Mozambique* native bismuth, some bismutite and bismuthinite are associated with the beryllium–tantalum paragenesis. Bismuth tends to occur in the pegmatites in the inner intermediate lithia zone, but the alluvia and eluvia of *Muiane* are the principal deposits. In this area the Boa Esperança pegmatite carries both bismuth and bismuthinite, topaz, beryl and samarskite. In addition, geologists have discovered rare stibio-bismuto-tantalite–columbite in this district. In the *Malagasy* pegmatites some bismutite occurs, e.g., in the Ambopihaonana swarm southwest of Tananarivo, at Betalala in the Betsiriry field, and at Andranagoaka and Ambohimena in the Vorondolo field.

In *South Africa* bismuthinite is included in a copper–gold paragenesis in the Dientje dyke east of the Bushveld, and occasionally occurs in the auriferous quartz veins of the Lydenburg field (e.g., Finsbury). In the tin districts of the central Bushveld bismuth minerals appear in the Stavoren and Mutue Fides fields, while bismuthinite accompanies arsenopyrite in the lenses of the Geweerfontein felsite and the impregnations of Rhenosterhoekspruit near Rooiberg. The principal occurrences, however, are located in the Orange River pegmatite belt, where bismuth accompanies lithium minerals, beryl and columbite–tantalite at Keimoes at the eastern end of the zone. At the western end of the belt, at Witkop, Noumas and Uranoopkloof, copper is also included in the paragenesis. Similarly, silver bearing bismuthinite occurs in the Erongo pegmatite field, *Southwest Africa.*

In Morocco chemists have determined traces of bismuth in cerussite and galena at Aouli (0.09%), Azegour and Missour in the east. Bismuth crystallizes in calcite veins occurring in the tactites of the Azegour molybdenum–tungsten deposit, and in stringers traversing dolomite and quartz further east at El Khemis.

MICA

In potash pegmatites, muscovite mica occurs in one or both sides of the wall zone or at the border of the core. Unlike pegmatites containing metallic minerals, mica deposits are scarcely affected by replacement processes. Conversely, mica pegmatites are frequently controlled by the wall rocks, but metamorphism and shearing affects them unfavourably. The principal scrap mica deposits are in South Africa, while Angola and Tanganyika provide sheet mica. Pegmatitic phlogopite bodies are included in narrow pyroxenite bands in Madagascar. In addition phlogopite occurs in carbonatites (Fig. 6).

SOUTH AFRICA

In the eastern Transvaal the mica belt extends from the confluence of the Macloutsie and Stoman streams, 50 miles (Fig. 46, 50) eastward beyond the Malelane. The width of the belt is 5–8 miles. Interlayered relic-screens of hornblende, talc, magnesite and serpentine schists remain in the gneissic granite. In addition to the foliated granites, massive quartz–orthoclase–biotite granites outcrop. However, ferro-magnesian minerals are better represented in the gneisses. Quartzo-felspathic and biotite bands alternate in them. The number of pegmatites increases towards the Phalaborwa carbonatite which, in fact, divides the belt into northern and southern branches. Megacrystalline irregular lenses of the southern branch contain commercial concentrations of muscovite. Mainly aligned along the regional east-northeast trend, pegmatite swarms are abundant at Mica Siding, Islington, Inyoka, Excellence, and Melalane east of Phalaborwa. Concordant and discordant lenses extend to a length of up to 1,000 ft. and attain a width of 40 ft. However, most of the ramified lenses undulate.

Muscovite books, accompanied by quartz, are particularly abundant in megacrystalline phases between albite and albite–oligoclase, particularly at the limit of quartz–felspar zones. However, in less coarse grained zones mica is disseminated in pockets. In the eastern sector pockets are more scattered. Muscovite is also found infrequently in quartz or feldspar. At Islington and Macloutsie small mica books concentrate in the wall zone. The proportion of pegmatite to cut mica varies between 1,600 and 500/1. Moreover, good grade flakes generally represent 2–10% of the mica pockets. Whereas biotite is found mainly in the western sector, apatite occurs in the east. Spinel and tourmaline are scarcer.

Pegmatites differ east and west of Selati. East of Selati zoning is more distinct; pegmatites and mica are frequently megacrystalline. Pale green muscovite is of moderate quality, flakes may buckle and edges fray, and 1–2% is of commercial quality. West of Selati pegmatites and mica are not as coarse, biotite appears in an apical position, and zoning is less distinct. However, the quality of pale brown muscovite is better: 5–10% is saleable.

At Islington xenoliths of gneiss are suspended in the pegmatite, while muscovite tends to concentrate in the wall zone near feldspar. Locally, small lenses are included in an envelope of quartz and an outer band of biotite. Biotite and muscovite alternate at both Islington and Macloutsie, and occur singly in megacrystalline zones. At Macloutsie the quantity of biotite increases towards the walls, but downwards towards the core the size and quantity of muscovite increases. Muscovite is also found in the wall zone at other localities. Biotite is abundant in another asymmetric dyke. Muscovite is adjacent to the footwall microgranite contact, perpendicular and near to the hangingwall of schists. However, near Zandspruit muscovite avoids the quartz bodies. At Birdcage muscovite pockets rimmed by fine grained material develop between quartz and albite. Furthermore, at the Olifants River 12 ft. thick mica books occur in the wall zone of a quartz–feldspar dyke.

Finally, mica pegmatites outcrop in the Piet Retief district west of and at Melmoth, south of Swaziland, in the Umtwaluni area south of Durban, and at Steinkopf south of O'okiep.

SOUTHERN RHODESIA

The principal mica pegmatites are located in the northwest in the Miami field of Urungwe district, at Darwin and at Ubique near Wankie (Fig. 36, 50). The *Miami* field proper covers 17 by up to 10 miles. Its satellite, the Rkwesa field, is half this size. At the contact of a narrow band of sillimanite schists, small stocks of biotite granite pierce sillimanite gneisses. Since mica pegmatites are distinguished by metamorphic control, few pegmatites are found in the neighbouring staurolite schists (WILES, 1961). Occurring most frequently 3–4 miles from the nearest stock, commercial deposits are external, but some pegmatites are marginal (beryl is more widely disseminated). Porphyroblastic quartz and sillimanite and quartz–mica formations develop in the country rock. Commercial pegmatites are zoned and partly replaced. They belong to the younger phase of microclinization and exhibit sharp contacts. The age of the pegmatites is 480–530 million years.

Mica develops in the wall zones which occasionally attain their maximum thickness at their juncture on the edges of the pegmatite lenses (e.g., Hendren V). Such shoots extend for up to 600 ft. Although favouring the hangingwall, mica may occur on both walls. The basal plane of books is generally perpendicular to the wallrock, but this rule does not hold for columnar mica (e.g., Locust). The wall zone consists of albite–oligoclase and some quartz with mica. Along the stretch distinguished by hangingwall mica, the Last Hope pegmatite attains a thickness of 5 ft. At Catkin most of the mica concentrates in the hangingwall, while the footwall is mineralized at lower levels. However, some of the lenses contain mica all over, e.g., Grand Parade South, Karnie and Hari. Mica crystallizes largely in sheets before plagioclase and quartz. Since its precipitation is retarded if accompanied by microcline or massive quartz, mica which occurs in such phases is of low grade (wedge or fishbone type).

Ruby mica occurs in two-thirds of the field. Spotted mica is found in *(1)* an east-northeast trending zone in the north-central part of the field, *(2)* in a smaller zone of the east, and *(3)* at the southern tip. Brown-green mica occurs around Last Hope in the south and at Devonia, Hari and Dog 2 in the east, while green mica appears at neighbouring Durham. The 2,000 ft. long and 10 ft. wide Grand Parade pegmatite has been emplaced in a north striking fracture. Books which attain a diameter of 5 ft. occur here. However, at Grand Parade South biotite is in excess of muscovite. The Locust pegmatite attains a length of 300 ft., a width of 8 ft. and a depth of at least 220 ft. Since mica concentrates at xenoliths of gneiss and tourmaline, blue beryl and microcline accompany it; much of the mica is broken. Finally, in the unzoned Lynx pegmatite biotite and muscovite have been disseminated.

TANGANYIKA

The principal mica field, Uluguru, and its northeastern extension, Mikese, form a 70 mile long and 30 mile wide belt, 130 miles west of Dar es Salaam (Fig. 52). The *Uluguru* field proper strikes north, with the principal pegmatites following the Mbakana River between Mgeta and Kikeo. The nearest granite outcrop is at 80 miles! Most of the tabular dykes and rare lenses develop in fissures or faults (HARRIS, 1961) which traverse meta-anorthosite.

TABLE CXXXVIII

ZONED DYKES

Zones	*Mineral associations*
Border	Fine grained quartz, oligoclase (microcline), muscovite
Wall	Feldspar, subordinate quartz and medium grained muscovite, coarse muscovite shoot
Intermediate	Plagioclase–perthite–quartz, mica
Inner intermediate	Perthite–quartz, mica
Core	Quartz

Others intersect limestones, while at Msongozi they cut garnet–biotite gneisses. About two-thirds of the dykes are oriented within 10° of North. The country rocks are consistently rich in mica or feldspar, and wall rock alteration is intense. Unzoned barren pegmatites are believed to be older than the zoned dykes, in which we can distinguish five zones (see Table CXXXVIII).

The feldspar may be either microcline perthite or oligoclase, but some of the zones may be absent or discontinuous. The dykes average 13.3 ft. in width. Randomly oriented compact books of mica form shoots or zones of several ft. in width, especially on the hanging-wall side, which penetrate into the neighbouring zones. Furthermore, books of mica also follow the margin of the core, penetrating into the quartz. In the intermediate zones books or clusters of mica and quartz are rarely scattered, but disseminated fine grained muscovite is more abundant. The colour of the mica is green to brown, with ruby tints occasionally rimming it. Generally the mica is slightly wavy and less frequently, cross grained. Large sheets are found at Lugala. Biotite and vermiculite also occur in the shoots. Younger quartz–ilmenite veins may form the extension of the 610 million year old pegmatites. The order of crystallization of the paragenesis is: monazite, allanite, magnetite, niobo-tantalates, ilmenite, garnet, zircon, apatite, tourmaline and sulphides. Finally, uranium minerals crystallized during a hydrothermal phase.

In the north-northeast trending *Mikese* field short, narrow and irregular lenses tend to intersect the biotite gneiss and the axis of the field. In this field mica is of homogeneous colour.

The Jumbadimwe swarm is located northwest of Morogoro, the centre of the Uluguru Mountains. The dimensions of the lenses vary here from 100 to 200 ft. by 5–15 ft. However, only northeast striking pegmatites which contain white feldspar of more than 3 inches and no garnet are of commercial value. West of Uluguru the Mangalisa pegmatites strike south-southeast in biotite gneisses and migmatites. They contain ruby, brown and green mica. Other swarms of this district include Itende, Mafwemero and Irondo. South of the Kilimanjaro, sheets of the Usambara Mountains measure up to 4 by 2½ ft.

The pegmatites of Nkungwi–Sibwesa favour the quartz–garnet–mica gneisses of the horst connecting Lakes Rukwa and Tanganyika. In the northwest striking host rocks up to 100 ft. thick dykes trend north-northwest. Although abundant, muscovite has unfortunately been affected by shearing. Consequently the large mica is wavy. Accompanying beryl and orthite are typomorphic. Finally, mica pegmatites occur at Tungwa south of the western tip of Lake Rukwa, and at Bundali west of the northern end of Lake Nyasa.

ANGOLA

The pegmatite belt of northwestern Angola extends from Matadi, near Ambriz Harbour, 130 miles southeast through Quinzundo, Quizambilo, Cassalengues, Mabubas and Icau to Mussaca–Saca, east of Luanda. Generally individual pinching and swelling lenses, intrusive into Precambrian mica and garnetschists, strike north. Their width varies between a few and 50 ft. (T. C. F. HALL, 1949). The megacrystalline pegmatites contain microcline, some perthite and quartz, but quartz–mica formations and graphic granite are also frequent. Mica is disseminated. In the wall zone high grade books, which attain a maximum width of 3–4 inch, form bands and pockets. However, larger books are scarce. Lepidolite, beryl, garnet (up to 2 inch in diameter) and tourmaline are included among the accessory minerals.

The *Quizambilo* pegmatites contain more feldspar than quartz, and garnet also occurs. The dykes pinch and swell, and the surrounding micaschists are locally pegmatized by lit par lit injection. These ramifications occasionally contain graphite. Although the maximum width of the dykes is 60 ft., mica concentration is not a function of the thickness. Indeed, one of the richest lenses is only 15 ft. thick. The largest books of dark or light inclusion free mica attain a width of $2^2/_3$ ft. The length of one of the dykes of the Cassalengues swarm is in excess of 300 ft. Here books of pink mica attain a width of 2–4 inch.

On the southern coast the Mondangel pegmatites, situated 70 miles southwest of Benguela, include white mica and some vermiculite.

SUDAN

From G. Nakharu the *Rubatab* mica belt extends north-northeast to Wadi El Koro west of the Atbara bend of the Nile, but some pegmatites are also located east and north of the Nile. The district attains a length of 80 miles and a maximum width of 30 miles. Numerous lenses concentrate in the centre of the belt, between the Jebels Quarn, Damm El Tor and Kabsol, in a field of 20 by 7 miles. Other fields are situated (Fig. 13, 58): *(1)* further north at the Wadi El Koro close to the bend of the Nile, *(2)* at the Jebel Razam and Razam el Rawyan southwest of El Koro, *(3)* at Keheili and El Kab north of the Nile, *(4)* on the Khor El Ithnein southwest of the Jebel Kabsol, *(5)* at Esh Shereik east of the Nile, and *(6)* at Jebel Um Arafibia and Nakharu south of Damm El Tor.

The anticline of the central field pitches south-southwest. This fold consists of quartzites, gneisses, schists, sandstones, shales, calcareous shales and marble, as well as epidiorite and amphibolite, but there are hardly any outcrops of granite. Pegmatites have been affected by regional tectonics (KABESH, 1960), and appear to mainly invade psammopelitic rocks. They average 6–12 ft. in width, attain a maximum of 30 ft., and a length of up to 1,200 ft. Although distinctly zoned pegmatites are small and relatively rare, quartz cores or central quartz shoots may occur, representing a favourable indicator. In the absence of quartz inner perthite zones develop. Since the albitization of microcline has started, white feldspar constitutes another indicator mineral. Other favourable conditions include: *(a)* abundant biotite in the surrounding gneisses and schists, *(b)* pegmatite width of 7–15 ft., *(c)* a well defined strike, and *(d)* a steep dip. The pegmatites include tourmaline and garnet when they also occur in the country rocks. Concomitantly, tourmalinization of the wall rocks is frequent.

Muscovite is disseminated in a majority of the lenses, concentrating rather in the hangingwall side than on the footwall side adjacent to feldspar. Generally quartz, rather than feldspar, surrounds muscovite. Similarly, mica clusters are associated with quartz. The diameter of books averages from 4 to 8 inch, and the thickness attains 1–2 inch, with a maximum of 4 inch. Most of the mica is of ruby, good stained or stained quality. Muscovite may embay schörl, rarer spessartite and apatite. Finally, beryl is sparsely disseminated.

MALAGASY REPUBLIC

Muscovite occurs only sparingly in the pegmatite fields. However, Madagascar is distinguished by the world's principal deposits of phlogopite. Numerous dykes and smaller masses of the mica develop in Androyan pyroxenites, with large but isolated crystals occasionally occurring in related cipolinos (see 'Uranium' chapter). Leptites, ultrabasics and lime magnesia facies characterize the Androyan rocks. These rocks cover a 150 mile wide and 300 mile long triangle in southern Madagascar. Pegmatitization affected this old System only 500 million years ago. Phlogopite is abundant in *(1)* a 20–30 mile wide zone extending from Imanda 130 miles north to Benato, and *(2)* in a 70 mile long zone extending further east, between the sea and the bend of the Mandrare River in the north. This zone is bordered by *(3)* urano-thorite bearing phlogopite pyroxenites in the west and charnockites in the east. Mica is less abundant north of the Mandrare and south-southeast of Benato where the two zones join. From Benato a similar zone *(4)* extends 80 miles north to Morarano. West of zone *1*, at Vohitramboa *(5)*, and in its narrow pedicle *(6)* extending from Imanda 40 miles south to Sakamasy, deposits are also less frequent.

Zone *1* includes the two largest deposits, Benato in the north and Ampandandrava in the centre. In addition it includes (Fig. 48, 52), from south to north: Mafilefy, Mikoboka, Miary, Ambararata, Benato 2, Ambatomena, Marovala, Aritatabe and smaller occurrences. Zone *2* consists of Ambatoabo, Marotoka, Anara and smaller occurrences. Finally, Kalambatritra, Ambatomainty and Sakasoa are situated in the northern part of zone *3*.

In the thick pyroxenite band of *Benato* high quality, cleavable (STO) and softer silvery BSA mica occur, concentrated in several bodies. In deposit JS7 pockets continue to occur beneath 400 ft. Deposit JS9 consists of a north striking swarm of funnel, sac and drop shaped pockets connected by short channels, veins or areas of less abundant mica. The great 'dyke' attains a width of 40 ft., but decreases to 20 ft. at a depth of 150 ft. This feature contains several bodies of massive phlogopite, while the longest column develops at higher levels. In the south other ellipsoidal pockets, measuring 20 by 20 by 10–40 ft. (height), tend to occur at greater depths. The Fontana pocket has a diameter of 70 ft., including two large non-mineralized bodies (BESAIRIE, 1961b).

In the area of *Ampandrandava* numerous pyroxenite bands occur, with the principal swarm extending for 1½ miles. Several bands attain 30–200 ft. in thickness, while 10–50 ton pockets of high grade splittings (VSP and DSP) and FP mica are scattered in the ultrabasics. The Marovala pyroxenite lens attains a length of 2,150 ft. and a maximum width of 600 ft. In this body phlogopite pockets are aligned perpendicularly to the strike. In like manner columns, nappes and pockets of mica occur in the 1,500 ft. long and 60 ft. thick Vohitramboa dyke. Forming a transversal swarm, the sub-vertical phlogopite bodies of Morarano attain widths of 2–9 ft. Similar structures have developed in the core of the mile long Sakamasy pyroxenite band. At some distance, the ramified Miari Ambony band extends for $^3/_4$ mile. Here phlogopite forms veins in two 30–50 ft. thick strips, and also penetrates into the country rock.

Furthermore, phlogopite seams cut the leptites surrounding the 5–15 ft. thick pyroxenite band of Mafilefy. At Anara hypersthene charnockites and biotite gneisses surround the diopside bands, but elongated sub-vertical phlogopite–scapolite shoots are transversal and diagonal. The principal ore body forms a 180 ft. long and 15 ft. wide spiral or helix (BRENON, 1956).

THE REST OF AFRICA

Along the backbone of the Moroccan Anti-Atlas, Precambrian windows extend east-northeast of Ifni. Mica and beryl occur in pegmatites related to small granitic stocks located west of the Azegour cobalt deposit and south of Marrakech. The Timgharghine pegmatite attains a length of 300 ft. and a width of 40 ft. The maximum width of the wall zone attains 10 ft. Books of muscovite tend to be perpendicular to the tourmalinized micaschists.

Partly albitized microcline and quartz form the intermediate zone, but a long sericite–quartz vein traverses the quartz core.

Some mica occurs in the Falémé field of southwestern Mali, at Etuakrom, Salt Pond and the Ampia Ajumako area of south-central Ghana, and near Yipala in northern Ghana. Some muscovite is also found in the Niakado–Saria swarm, Upper Volta, and the Egbe field of Nigeria.

In Ethiopia two mica fields are known west of Massawa (Erythrea). Near Mount Arasoli the belt strikes north-northeast but, with a few exceptions, lenses are small. Near Kheru and Mount Lataiman 10 inch books of mica occur in the pegmatites. East of Harar the Jijiga field extends from Shebeli to Carara, Mount Bishuman, and Shaveli 40 miles in the north. The dykes strike northeast. In addition, some mica occurs in the Jobir and Gundru Valleys in Wallega Province. Ruby mica shows near Matthew's Range and in the Embu district of *Kenya*, while sheet mica is scattered in the pegmatite fields of Tsavo, Taita, Sultan Hamud and Kierra. In Mozambique the Alto Ligonha pegmatites include some mica, as do the pegmatites of northern Malawi which are adjacent to the Bundali field of Tanganyika. Other Malawi occurrences include the Kirk Mountains, Dowa, Dedza and Mzimba. Books of up to 5 inches occur near Mulungushi in the Broken Hill area of Central Zambia, and also in the pegmatite fields of Southwest Africa. Finally, in the pegmatites of the Mankaiana district of Swaziland, sheets and flakes of up to 3 inches are disseminated.

FELDSPAR

Used in the enamel, ceramic, detergent and electrode industries, fusible feldspar is widespread in the acid pegmatites of most African fields. Its workability is conditioned by *(1)* a high potash content and less than 10% soda, *(2)* the presence of less than 0.5% iron oxide, *(3)* the occurrence of valuable accompanying minerals, *(4)* transport facilities, and *(5)* the degree of development of the country of origin.

South Africa

In the Orange River belt the principal occurrences of potash feldspar include Sleight's Mine, Jackalswater as far as the Uranoop, Witkop, Noumaas and Henkries. Conversely, sodium feldspar shows with mica at Thankerton, Hannan, De Elbe, Ham, Dentz and Selati southeast of the Gravelotte antimony deposit. Unfortunately, the Na_2O content reaches 10–12% there. At Mica Station, plagioclase feldspar is intergrown with quartz. On the other hand, in the north in the Soutpansberg district feldspar is found at Faure and Kitchener.

The rest of Africa

In Ethiopia feldspars of the Hamasien Plateau (Erythrea) alter to kaolin. In Ghana feldspar occurs in the pegmatites of Princes Town, Winneba, Salt Pond, Cape Coast, Etuakrom and Mangoase. In the Lower Congo geologists have identified high grade feldspar at Cul-de-Boma, Monolithe, Kungu, Busin and other localities. Similarly, the mineral occurs in the pegmatite districts of the eastern Congo (e.g., the Idiba field) and in the Katumba field of Rwanda. In neighbouring Uganda microcline is found at Bulema and Nyabakweri, and also at Wabiyinja and Nakabale. In addition experts have located feldspar in the Kinyiki pegmatite in southeastern Kenya. In Tanganyika feldspar is found in the Uluguru Mountains. A dyke near Dodoma (central Tanganyika), and bostonites of the Lupa gold field (north of Lake Nyasa), consist of feldspar. In Madagascar the mineral is abundant in the Malakialina field, and also occurs 7 miles from Tananarivo on the Tamatave road.

PIEZO-ELECTRIC QUARTZ

Quartz of piezo-electric quality crystallizes from pegmatitic solutions in vugs at the intersection of fissures and joints, especially in quartzites. Furthermore, eluvia and alluvia have been eroded from quartz veins. The principal deposits are in the *Malagasy Republic*, where ornamental and foundry quartz also occur.

In *Madagascar* industrial quartz is spread in districts (Fig. 48, 52) occupying 500–4,000 square miles. These areas form a broad arc extending from the Bay of Antongil north of Tamatave, to the axis of the Island and bending to its southeastern corner. The principal deposits of Ramartina and Antambolehibe, however, lie isolated in the late Precambrian Quartzite Series 100 miles southwest and south of Tananarivo.

At Beombiaty or Cristallina, the principal deposit of the *Ramartina* area, geodes are filled in by kaolin, altered from feldspars and quartz at the intersection of bedding planes and joints. Although the quartz is of good quality, some crystals contain impurities. At *Antambolehibe* geodes form non-oriented pockets in the Itremo quartzites between Betaimboraka and Ambatofinandrahana. On the Horombe River at Takodara, a quartz vein and eluvia contain the crystals. East of Horombe other deposits are situated at Farafangana. Prisms of the highest grade occur in the southeast between Andriavory and Pehasy at Tsivory.

Workers have recovered some quartz at Ankazobe near Tananarivo, at Ambatondrazaka and at Vohémar in the northeast. Crystals occur in bands of quartzites at Sahasinaka, in quartz veins at Vondrozo, and in eluvia eroded from veins of milky quartz in the area of Antongil.

In the area of Rubeho of the Kilosa district, *Tanganyika*, workers have collected prisms in eluvia which are believed to have been eroded from geodes of a sheared quartz vein. Nearby at Idibo the crystals form aggregates in a veinlet traversing a weathered basic rock. Similarly, at Kinyiki Hill, *Kenya*, quartz has been transferred from veins into eluvia, but at Tseikuru (North Kitui) quartz occurs in pegmatites. To pass to western Uganda, crystals have been recovered from the swamps of the Buhweju gold field. In Southern Rhodesia, the quartz of Broadside near Gwelo and of Tweeny is used in metallurgy.

THORIUM

Thorianite occurs in ultrabasic pegmatites of Madagascar, and also in carbonatites, particularly in Southern Africa. Thorite, thoriferous alteration products and limonites are found in carbonatites, while thorite also appears in granitic ring complexes. A high degree of differentiation appears to be the only link between this variety of deposits. In larger deposits, however, thorium combines with the cerium earths, especially in monazite and bastnaesite. While bastnaesite and fluocarbonates occur in sövite and rauhaugite of carbonatites, monazite is one of the minor accessory minerals of beach sands, e.g., in the delta of the Nile. Finally, high grade monazite deposits of South Africa are hydrothermal, but the extraction of thorium–rare earth minerals from pegmatites is uneconomical.

SOUTH AFRICA

Western Cape Province. The continent's principal deposit is located at *Steenkampskraal*, halfway between Cape Town (Fig. 53) and the Orange River, where an irregular and subvertical lode has developed in a shear zone which traverses ancient granite-gneisses. The outcrop measures 900 by up to 13 ft. Accompanied by apatite, zircon, copper and iron minerals, 3–6% fine grained monazite partly fill in the lode to a depth of at least 400 ft. The monazite averages 8% thoria and more than 40% rare earth oxides. At Roodewal monazite and its accompanying minerals crystallize in a 250 ft. long and up to 18 ft. wide lode in granite-gneiss affected by hydrothermal alteration. At Uilkip, the paragenesis is related to pods and stringers, but at Karragas monazite and xenotime show in a quartz vein. While monazite accompanies magnetite in the quartz veins of Townlands, Banke and Speelmanshoek, the mineral is associated with hematite at Gordonia and with copper at Concordia. Indeed low grade, rare earth–thorium minerals are abundant in the Orange River pegmatite belt.

Large grained monazite is disseminated in quartz veins and pegmatites filling in joint planes of the *Bushveld* granite porphyry. The monazite contains 1–4% thorium and 23–65% combined rare earth oxides. Fluorite and sulphides are included in the paragenesis. We have already described the carbonatite deposits in the uranium chapter, to which we refer the reader.

In the Klerksdorp gold field, rounded monazite is disseminated in the coarse granitic conglomerate of the ancient Dominion Reef System. These grains are more abundant in the reworked upper reef than in the auriferous lower reef. Monazite shows in the Middle Ecca (Karroo) sandstones of eastern Transvaal, e.g., between Rustig and Zooifontein. While on the east coast the ilmenite sands of Morgan Bay, Port Saint Johns, Umkomaas, Umgababa, Isipingo, Umhlali, Illovo and Brighton Beach contain 0.3% monazite, on the west coast monazite reports in the beach sands of Cape Voltas, Zout River and Strandfontein.

WEST AND SOUTHWEST AFRICA

Urano-thorianite and thorianite occur in large pegmatites traversing early Precambrian amphibole–pyroxene schists in the Adrar of the Iforas, Mali. In the Ennedi, northern Chad, basal Ordovician conglomeratic sandstones carry urano-thorianite.

Monazite occurs in placers of Odienné, flowing on Birrimian granites in the northwestern Ivory Coast. Furthermore, the monazites of the Mount Dô and Peko areas contain 9% ThO_2. Monazite is also found in the Mkushi placers, as is thorite on the Menencha and Mamantombwa Hills. Monazite shows in the beach sands of Ghana and at Bétéro in Dahomey. Several rings of the younger granite complexes of the *Niger Republic* and *Nigeria* consistently contain monazite, thorite and one of its varieties, orangeite. The latter appears to concentrate in the late albitic phases (see 'Columbium') chapter). Whereas euxenite has been identified in the Oued Fomballera (Niger Republic), monazite also occurs in the alluvia flowing on the Precambrian Basement.

On the Nova Lisboa Plateau of *Angola* twelve alkaline–carbonatite complexes form three belts. These belts fan out in a northeasterly direction. The carbonatite core of several complexes is small; the largest, Longonjo, has a diameter of 4½ miles. The bedrock and soil are believed to average, respectively, 0.83 and 1.03% ThO_2, 0.62 and 0.87% Ce_2O_3, and 0.09 and 0.11% La_2O_3. At Bailundo and Capuia, inferred grades of combined thorium–rare earth oxides oscillate from 0.1 to 5.8%, but values are not as high in the complexes of Coola, Canata and Chinga.

Radioactive minerals are scattered in the Kaokoveld and in the Umeis granite of Southwest Africa.

EASTERN AFRICA

In *Egypt*, in the delta of the Nile the black sands of Rosetta and Damietta include some monazite containing 5.4% ThO_2. Recovered concentrates consist of 67% ilmenite, 24% magnetite, almost 6% zircon, 3% garnet and 0.2% monazite.

Monazite occurs in the Dacata and Errer Streams in the area of Harrar, Ethiopia, and in a pegmatite located near Daarbuduq, northern Somalia. The latter contains 10% equivalent ThO_2. In addition, monazite is included in the black sands of the Juba delta (see 'Titanium' chapter), and also in the estuaries of the *Kenya* coast. Nearby, monazite is the principal ore mineral of the Mrima carbonatite south of Mombasa. The weathered zone of this complex extends to a depth of more than 100 ft. Similarly, the Buru Hill complex near Muhoroni is also radioactive. At Kalapata, in the northeastern corner of *Uganda*, quartz–ilmenite–rutile nodules of biotite gneiss contain monazite (11% ThO_2). Conversely, in southeastern Uganda a majority of the thorite of the Lunyo granite is included in magnetite. Columbite and cassiterite also belong to this paragenesis. Although monazite is widely scattered in the placers of the pegmatite fields of Rwanda and the eastern *Congo*, in the Mudubwe veins (Central Kivu Province) monazite is associated with wolframite.

In *Tanganyika*, west of Dar es Salaam, monazite and bastnaesite bearing bands are included in the rauhaugites (dolomite) of the Wigu Hill carbonatite, but green monazite occurs in dykes. In addition, sheared gneisses of North Mara contain 3–5% thorite. In Eastern Zambia the carbonatites of Nkumbwa Hill and of the Feia district include monazite and thorium minerals. We find orthite, gadolinite and euxenite in the pegmatites of Alto Ligonha, Mozambique, e.g., at Macoteia, in the Mavudzi Valley, and near Cape Delgado. Finally, monazite float shows in many parts of Southern Rhodesia.

MALAWI AND MALAGASY

At Kasupe in southern *Malawi* thorite, monazite and xenotime are associated with pyrochlore and zircon in albite–granite dykes and pegmatites of the Zomba syenite complex. Indeed, in the eluvia of the Mlanje Plateau fine grained monazite contains 17% ThO_2.

While thorite and monazite are associated in the pegmatites of Tambani, carbonatite zones of Chilwa and Tundulu carry 0.8–4.5% ThO_2. On the southern shore of Lake Nyasa the monazite (17% ThO_2) bearing raised beaches of Monkey Bay have been partially derived from aplites. Inland from the southeastern shore, monazite, ilmenite and zircon are abundant, but in the area of Chikulo material deposited at higher levels of the Lake has been rearranged. Similarly, thoriferous monazites occur at the northern end of the Lake, at Fort Johnston, and on the Palombe Plain.

On the southeastern coast of *Madagascar*, heavy minerals have been eroded from the Basement. Along favourable reefs, wind and currents concentrated them between Manakara and Fort Dauphin, especially at Ialanamainty, Vohibarika and Antete. Recovered concentrates of the upper reaches of the dunes consist of 89% ilmenite, 6% zircon and 5% monazite. Tests show that the mineralization of the dunes of Antete is homogeneous. Moreover, thorium and uranium show in betafite, euxenite, monazite and biotite in the pegmatites of Itasy and Vakinarkaratra, and in columbite–tantalite, beryl, uraninite and muscovite at Ambositra and Fianarantsoa. For a description of the principal deposits of uranothorianite included in the phlogopite pyroxenites of southern Madagascar, we refer the reader to the 'Uranium' chapter.

RARE EARTH METALS

A description of thorium bearing rare earth minerals can be found in the preceding chapter. Usually the rare earth metals are divided into two groups. The cerium earths, or light group, are considerably more abundant than the heavier yttrium earths. Whereas the latter concentrate mainly in alluvial xenotime, the cerium group occurs both in monazites of beach sands and carbonatites, and in bastnaesites and fluocerian minerals of carbonatites.

In Ethiopia yttro-tantalite collected at Qushersher, four miles northeast of Harar, contains 3% cerium and 1.5% yttrium. Similarly, the monazites of the Buhweju gold field and Ankole pegmatites of southwestern Uganda contain little thorium. Xenotime is abundant in parts of the Idiba and North Lugulu fields of the eastern *Congo* (see 'Tantalum' chapter). In these concentrates xenotime accompanies almandite.

At Karonge in *Burundi*, 20 miles southeast of Bujumbura and east of the northern tip of Lake Tanganyika, bastnaesite is accompanied by baryte, galena and pyrite. The bastnaesite contains 31.1% Ce_2O_3, 39.4% $(La_2, Di_2)\,O_3$, 0.25% Y_2O_3, and 0.10% ThO_2. Rauhaugite dykes have been injected into fenitized gneisses at Wigu Hill in Tanganyika. These ore bodies contain minerals of the bastnaesite group and monazite in concentrations of up to 20%. They measure up to 500 ft. in length and 4 ft. in width. On the other hand, pegmatitic orthites of Milembule contain 6.8% cerium oxide and 7.4% of lanthanide oxides.

In the Kangankunde carbonatite of southern *Malawi* we can distinguish three vents. Monazite, florencite–synchysite–bastnaesite and pyrite (accompanied by sphalerite and galena) have been introduced, during the final stage of carbonatization, into a small core of mobilized ankerite orthoclasite and siderite carbonatite. To a depth of 165 ft. the inner half of the core averages 4.3% of low thoria monazite, which mainly contains cerium, lanthanum and neodymium. Moreover, another area contains 5.8% monazite to a depth of 290 ft.

In Swaziland yttro-tantalite, monazite and cassiterite are associated in the placers of the Mbabane River, with some yttro-tantalite showing in concordant pegmatites. A monazite sample contains 22% Ce_2O_3 and 25% other rare earth oxides.

In the Orange River pegmatite belt, orthite and rare earth tantalates co-exist. Bastnaesite of the Mutua Fides tin field, *South Africa*, contains 30% cerium group oxides and 37% yttrium group oxides, but a calcium silicate related to cenosite contains 45% yttrium group oxides. The same minerals also occur at Groenvley and Appingendam in the Potgietersrus field.

TITANIUM

Titanium is an abundant and widespread light metal. We distinguish four types of concentrations: submarine sediments, beach sands and dunes, alluvia, and ultrabasic complexes. With a few exceptions, in Africa only coastal resources are in the commercial category. As a rule, several generations of beaches succeed each other. On such shores ilmenite, rutile and zircon are cyclically deposited in proportions ranging from 60/1/30–10/1/3. Monazite is a rare accompanying mineral. We can only partially trace the origin of the concentrates to rocks of the hinterland, which therefore raises the problem of neogenesis. Currents, bars and lagoons control the location of South African, Senegalese and Mozambique deposits (Fig. 6).

Ilmenite, martite and magnetite are the principal components of continental alluvia which include the valuable tin and gold placers. Important rutile placers are located in the southern Cameroon. Moreover, titano-magnetite is disseminated in ultrabasic complexes, and also forms seams. These resources are large but of low grade.

SOUTH AFRICA

Black sands concentrate in the beaches *(a)* of the east coast, e.g., near Saint Lucia Bay from Umhlali to Umgababa (Fig. 50) on both sides of Durban, near Port Saint Johns and East London, and *(b)* on the west coast, e.g., in the Vanrhynsdorp district 150 miles north of Capetown. The principal exploitations were located at *Umgababa* and Isipingo, south of Durban. The origin of the ilmenite is not known with certainty. Indeed, in the sands the ilmenite/magnetite ratio attains 20/1. However, the opposite is generally true in inland rocks, except for sediments such as Ecca (Permian) sandstones. We distinguish three types of deposits: old red sands, recent sands and lagoonal alluvia.

The 40 yard–4 mile wide band of older, cross-bedded Holocene sands is separated from the coast, probably because of erosion. From the beaches the dunes emerge 30–100 ft., and rarely to almost 500 ft. Averaging 7–8%, the black sands constitute 2–25% of the sand. They consist of 70–80% ilmenite, 2–5% rutile, 2–5% leucoxene, 9% zircon, 2–8% magnetite, and traces of monazite reaching 0.3%. Consequently, the ilmenite/rutile/zircon ratios are in the range of 10/0.7/1. In the sands values of ilmenite containing 50% TiO_2 attain 9%. Furthermore, a capping of the dunes, locally 30 ft. thick, carries values of 25%. The quartz grains of 0.7–0.05 millimetre form most of the sand, and are accompanied by feldspar, pyroxene and garnet. Solutions derived largely through the oxidization of ilmenite consoliate and enrich the iron rich part of the dunes. The colour of the sands changes rhythmically, probably because of isostatic movements.

The recent sands and dunes have derived some black sand from the older beaches. Such a recent dune is exploited at Isipingo. The average grain size of the younger sands is three times as large as the preceding one! Black sands have also been washed into estuaries and lagoons. With TiO_2 values ranging from 42 to 50%, Cr_2O_3 reaching 0–0.2%, and V_2O_5 0.1–2.8%, South African ilmenite is of low grade. At Umgababa a linear relationship exists between rutile, zircon and ilmenite values of all grain sizes (Table CXXXIX).

TABLE CXXXIX

GRAIN SIZE OF MINERALS

Size (mm)	*Ilmenite (%)*	*Rutile (%)*	*Zircon (%)*
0 –0.10	9	4	8
0.10–0.12	24	13	24
0.12–0.14	31	29	32
0.14–0.16	18	26	18
0.16–0.18	10	14	10
0.18–0.20	5	9	6
0.20–0.30	3	5	2

The principal *western* mineralized beaches and dunes are situated between the Zout River and Strandfontein on Rietfontein extension, at Graauw Duinen (grey dunes), and at Gellwal Karoo (COERTZE, 1958). Two dunes constitute more than 90% of the 350,000 ton ilmenite reserve, evaluated only to a depth of 5 ft. Concentrations extend for 5 miles. As at Umgababa magnetite is scarce, although it is abundant in the hinterland. Remarkably frequent garnet has been eroded from the Stoutrivier gneisses. Ilmenite values range from 7 to 30% (average 18%), with 0–2% rutile (mean 1%). This ilmenite contains 41% TiO_2, 35% Fe_2O_3, 19% FeO and 0.7–1% V_2O_5 (double the standard specifications). Zircon also shows a large dispersion: 2–10%, averaging 6%. Other components include 30% garnet, 30% quartz, 13% pyroxene, and 1% magnetite. Moreover, some diamonds are also supposed to be disseminated here.

OTHER BEACHES

Mauritanian coastal occurrences include Arguin Island, Cape Timiris (with garnet concentrations), El Mahara 25 miles south of the cape, and Lemsid and Gendert 45 and 25 miles from Nouakchott, the capital. The Senegalese titanium sands will be discussed in the 'Zircon' chapter. Unlike the Senegal River, the Gambia carries titanium minerals almost exclusively, with the ratio of recovered ilmenite and rutile varying between 50 and 15/1.

In the western Ivory Coast, black sands of the Sassandra delta even contain some cassiterite. In Sierra Leone the 35 mile long Sherbo rutile deposit contains 30 million tons.

At Rosetta and Damietta in *Egypt*, black sands carried by the Nile have been sorted and rearranged in the delta by tidal waters and distributed in seams on the beaches. Between 1932 and 1955 recovered ratios of the components averaged 64% ilmenite, 30% magnetite,

5% zircon, 0.3% garnet and 0.02% monazite. Rutile, pyroxene, amphiboles and epidote are also present (EL SHAZLY, 1959).

In *Somalia* black sands carried by the Juba (Giuba) have been deposited at its last meander at Giumbo, and also on both sides of a sand bar obstructing its delta. Thin, heavy (black) and light layers alternate. Concentrates contain 15% TiO_2, but hardly any monazite and zircon. However, in Kenya the black sands of Formosa Bay and Malindi include less than 1% TiO_2, while ilmenite contains 0.12% V_2O_5 and 0.015% Cr_2O_3. At Bagamoyo in Tanganyika, ilmenite grades attain a maximum of 55%, while rutile grades at Kunduchi and Mbweni, near Dar es Salaam, reach 5–15%. Between Pebane, Matirre, Almandia and the Moebaze River north of the Zambezi, black sands cover the Mozambique beaches, bars and lagoons. The ratio of components is fairly stable. Grades vary between 33 and 62% ilmenite, 0.2–0.5% rutile, 1.5–5.5% zircon and 0.6–0.8% monazite. In the alluvia of Incomáti, near Marracuene north of Lourenço Marques, the ilmenite/zircon ratio attains 5/1. North of Tamatave in northeastern Madagascar, opposite Sainte Marie Island, we find the greatest concentrations in the embayment extending from Foulpointe to Pointe Larrée. The protected position of these beaches appears to have preserved the black sands. However, among the rocks of the hinterland only dolerite dykes, injected into migmatites and mica schists, include some ilmenite. The ilmenite contains only 37–42% TiO_2, up to 1% MnO, and traces of chromium and phosphorus. In addition, ilmenite concentrates in the southeastern Fort Dauphin beaches (see 'Rare Earth' chapter).

OTHER DEPOSITS

Inland placers are known to occur at Chikulo on Lake Nyasa, at Gangantan in Mali, in the Aïr (Niger Republic) and in Nigeria, while ilmeno-rutile occurs at Natitingou and ilmenite at Couffo in Dahomey. Ilmeno-rutile is also included in the alluvia of Tonkolili, Sierra Leone. In the *Cameroon,* east of Yaoundé rutile concentrates in the alluvia of the Nyong drainage. Rutile impregnates the albitized Giftkuppe granite in the Omaruru pegmatite field of Southwest Africa, crystallizing in fissures.

Titanomagnetite occurs in the ultrabasic complexes of the Bushveld (South Africa), the Great Dyke (Southern Rhodesia), Liganga (Tanganyika), Colony (Sierra Leone) and many others. Also forming stringers, ilmenite is disseminated in various labradorite–hypersthene anorthosites, e.g., at Jangoa near Miandrivazo, Madagascar (BESAIRIE, 1959).

ZIRCONIUM

Zircon is a common accessory mineral of granites, carbonatites, alkaline and syenitic rocks, and concentrates especially in differentiated ring complexes. The mineral has been eroded and transferred into placers and then into coastal sands, which represent the principal commercial source of the metal. The Senegal beaches contain proportionately five times as much zircon as the South African deposits (see 'Titanium' chapter). Some zircon has also been recovered from the coasts of Gambia, Egypt, Mozambique and Madagascar.

Zircon is an accessory mineral of placers in granitic areas. The concentration of zirconium attains its peak in the late albitic phases of granitic complexes, e.g., on the Jos Plateau in Nigeria. Zircon may attain as much as 5% in these complexes, and their value is increased by the *hafnium* content. Alkaline syenites contain zircon, e.g., in Malawi, and zircon pegmatites also occur in Madagascar. Zircon is an accessory mineral of carbonatites, including the Lueshe (Congo), Ugandan and Angolan complexes. Rarer baddeleyite occurs at Sukulu in Uganda.

SENEGAL

From the Mauritanian border through the Gambian coast (Fig. 15), the coastal deposits extend to Guinea. The beaches extending from Saint Louis du Sénégal to Cayar, northeast of Dakar, contain several million tons of black sands in concentrations of 1–2%. Moreover, the old dunes spreading 70 miles between Nianing, Bargny, Siendou near Rufisque (east of Dakar), Joal, Pointe de Fata, the Saloum delta and Sangomar Point include one million tons of concentrates in grades of 10–12%. The centre is situated at *Djifère* on the Sangomar Peninsula. Conversely, deposits of the Gambia delta have been only irregularly worked for titanium. Consequently little zircon has been recovered. South of Gambia the sands of the Casamance delta near Niafourang and Diogue include 600,000 tons of heavy minerals in concentrations of 15%.

Black sands are deposited in recent beaches, especially during the hibernal tides or high tides preceding and following the equinox. These black sands form seams of an inch or two. Being heavier than quartz they infiltrate or descend 2 ft. into the sands. Although overburden sands attain 5 ft. of thickness, old dunes also contain important deposits. Grades depend naturally on selection. Thus, for years sands containing 40–60% ilmenite, 15–40% zircon and 20–25% quartz were worked. The Pointe Fata sands, 8 miles north of Djifère, contain 40–80% of heavy minerals which average 66% ilmenite, 10% zircon, 2% rutile, 21% quartz and 1% other components. Unlike Umgababa, the grain dimensions of components differ: ilmenite ranges between 0.12 and 0.28 mm, zircon between 0.10 and 0.13 mm, and quartz between 0.15 and 0.45 mm, but a majority of the quartz is larger than 0.2 mm. The 'ilmenite' is, in fact, an oxidized rutile containing 55–60% TiO_2 (averaging 57–58%). Zircon includes 66% Zr_2O.

Although believed to originate in Basement rocks, part of the heavy minerals are thought to have been deposited earlier in Eocene sands. Some were also deposited in late continental strata which reach the coast between Cape Rouge and Cape de Naze and north of Joal. Winds have blown the Eocene sands away from intermediate zones. Possibly the Pan'tior Somone, Tarare and Sine rivers carried the minerals, and then rearranged them by coastal currents which flow mainly in a southerly direction.

SECTION VII

Carbon-fuel group

Malagasy and East African graphite are believed to be of sedimentary origin. Although graphitization reached its peak more than 2,400 million years ago, we know little about the early Precambrian organisms from which graphite may have been derived. Carbonaceous materials reappear in other Precambrian cycles, e.g., 2,200, 2,000, 1,400 and 600 million years ago. Hydrocarbons first show in the Witwatersrand gold field, but the first economic concentrations appear only in the Cambrian in the Northern Sahara. After various Paleozoic phases, the accumulation of gas reaches its peak in the Triassic reservoirs of Algeria, and oil in the Eocene sediments of Libya. While small reserves of anthracite are found in the Carboniferous of Morocco and in the Permian of Natal and Swaziland, the principal coal resources are restricted to the Permian early Karroo, mainly in Southern Africa. Later post or late Karroo bituminous sediments of the Triassic cover the bend of the Congo River. Extending from the Eocene to the Pleistocene, lignite resources are moderate. Finally, peat continues to evolve in coastal swamps and small basins inland (Fig. 7).

GRAPHITE

Graphite is the first form of carbon to appear in Africa! The world's largest graphite deposits are located in Madagascar. Significantly another Gondwanian island, Ceylon, is also an important source of the mineral. Some graphite occurs in the Mozambique belt, the structural equivalent of the Malagasy Graphite System, as far north as Kenya. Although this System is older than 2,400 million years, experts believe that the deposits are sedimentary. Indeed, graphitic layers are interbedded in concordant bands of gneissic rocks. Graphite is also disseminated in Zambia, Malawi, Rhodesia, Uganda, the Congo, and South and Southwest Africa. Some of these deposits may belong to the earlier Precambrian, but in the Ivory Coast graphite is included in more than 2,200 million year old schists.

MALAGASY REPUBLIC

Although some graphite shows in granite, and considerably more in pegmatites and quartz veins (east of Mount Vohibantaza, Ampanihy area) traversing graphitic schists, all major deposits are included in migmatites and paragneisses. Within them graphite belongs to certain stratigraphic horizons transformed by various degrees of metamorphism, e.g., *(a)* mica, talc, garnet, sillimanite and kyanite schists, *(b)* quartzites, *(c)* cipolinos, *(d)* leptites,

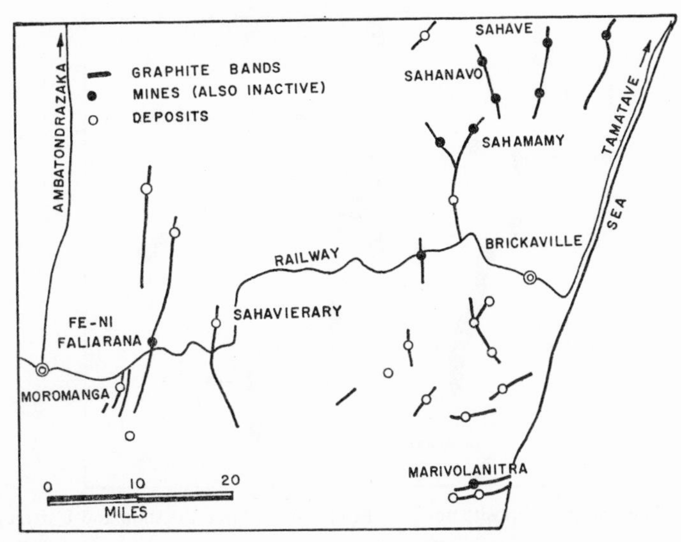

Fig. 118. The eastern part of Malagasy Republic. (After BESAIRIE and LAUTEL, 1951).

gneisses and migmatites, and *(e)* also to ultrabasics, pyroxenites and wernerites. Generally striking south-southwest, graphite belts extend from the north to the southern tip of the island. We can distinguish three groups: east, centre and south.

In the *south* highly metamorphic leptites of the Ampanihy group are typical host rocks of flake graphite. Among them the richest are amygdaloidal, orbicular, granoblastic or intensely microfolded. Moreover, coarse-grained leptites derived from paragneiss also constitute a favourable medium. In the *eastern* (Fig. 118) part of the island (Tamatave) graphite is irregularly disseminated in granoblastic, less schistose rocks and orthogneisses. Here graphite has concentrated in definite bands of migmatites, in alternating gneisses and micaschists, and in various schists of the Manamopotsy Group. Although belonging to a similar stratigraphic horizon, this formation is less metamorphic than the Ampanihy Group. Both quartz and sillimanite are frequent. The quartz-monzonitic–garnet–amphibole migmatites of the Brickaville facies are good ore indicators. In the *centre* of the island we find graphite in sillimanite gneiss roof pendants surrounded by migmatites. Graphite may be interstitial or included in quartz or feldspar with which it is syngenetic. Although their mode of deposition differs, the host-rocks of Ceylonese and Malagasy graphite are most similar. The more than 2,400 million years old Graphite System forms the backbone of Madagascar.

TAMATAVE BELT

In the mining area between *Tamatave* and Brickaville (Fig. 119) we can distinguish two Precambrian groups of rocks on sheets Moramanga–Brickaville and Ambodilazana–Tamatave. All occurrences of graphite are included in the Manamopotsy paragneisses, mica- and amphibole schists. Sillimanite, corundum and garnet are typomorphic, but scapolite parapyroxenites and quartzites are rare. Similarly, migmatites and amphibole garnet gneisses of the Brickaville Facies are the only significant granitoids. Graphitic paragneisses, banded migmatites or, rarely, quartzites form alignments traversed by barren pegmatites. These trends generally follow the schistosity, although they also cut

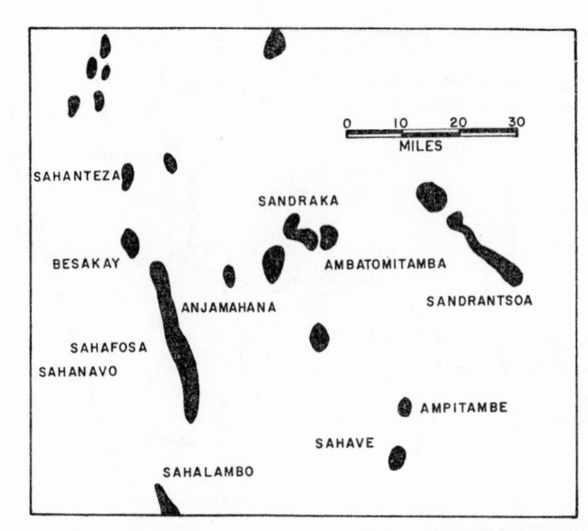

Fig. 119. The graphite deposits north of Brickaville. (After BESAIRIE and LAUTEL, 1951).

it in a sub–perpendicular tectonic line. Mineralized pegmatites and quartz veins are rare. However, enclaves or compartments of granitic Brickaville orthomigmatites also include graphite. Faults dividing them from the graphitic schists are nontronitized. Furthermore, graphite is included in the garnet crystals in garnetites, embayed in quartz and sillimanite granoblasts.

Located on the left bank of the Ranofotsy River 60 miles south of Tamatave at *Sahamany*, a large but discordant xenolith of graphitic gneiss is included in migmatites, orthogneisses and granodiorites. Dipping 35° S, this band strikes N 75° E. Its thickness is 8–15 ft.

The *Sahanavo* deposits are located nearby, 15 miles from the coast at Ambalarondra. The northern sector of this alignment consists of Marovato, Ambalojatsy, and Mankarana 2 miles to the south. This enclave of graphite–sillimanite paragneiss is included in granite and Brickaville orthogneisses. In the Sahalambo sector we see this column:

(1) a cover of lateritic clays and boulders of granite;
(2) 12 ft. of a grey graphitic layer;
(3) and *(4)* a barren and a poor blue layer including a bed of pyroxene leptite;
(5) a 1 ft. thick limonite–hematite sill containing graphite;
(6) an up to 6 ft. thick graphitic nontronite layer;
(7) a thin kaolin seam;
(8) sillimanite–garnet gneisses, graphite–garnet rock;
(9) altered metamorphics;
(10) laterites derived from granite.

Both dipping 35° W, beds strike N 35° W in the north and N 10° W in the south. Thin tourmaline pegmatite veins traverse the southern Mankarana layers.

North of Sahanavo the *Antsirakambo* deposit, which dips 45°, is oriented N 30° W. In the unaltered quartz-flake graphite layer, which contains some kaolinized feldspar, the nontronitic beds are distinguished by values of up to 95% C. In the lowest bed of the layer nests of nontronite and pyrite are disseminated. In addition, a grey graphite gneiss contains sillimanite.

The *Ampangadiatany* graphite zone, 2 miles north, dips 40° and outcrops at a width of 50 ft. The grade attains 10%.

The *Ambatomitamba* deposit is situated on the Fanandrana River 1 mile north of Andronobolaha. There have been distinguished three layers, which dip only 20°. From 50 ft. of graphite gneiss in the west, a fault divides the felspathic footwall gneisses, injected by quartz veins and traversed by a diabase dyke. Diabase and nontronite partly fill this fault. Feldspars underline ptygmatic folds. Similarly, small pegmatites carry feldspar, quartz and muscovite. Exhibiting a more distinct banding of quartz, feldspar and muscovite, 60 ft. of overlying lateritized gneisses contain less graphite.

At some distance the 3 mile long *Marovintsy* graphite zone, dipping 45°, bulges to a maximum width of 1,300 ft.

Faliarano is located 200 miles south of Tamatave, 3 miles from Perinet station. Here gneisses, quartz mica schists and quartzites form the country rock, strike N 10°–20° E, and dip 30° W. Measuring 1 mile from east to west, the graphite zone extends along the strike. Brickaville migmatites form the eastern border, but piano key type compartments complicate the structure. Lateritized basalt dykes traverse quartz–graphite gneisses. The ore includes 4–8% graphite which contains 92–95% carbon. The proportion of fine graphite is higher than the average in the fine-grained and bedded gneisses. Farther south, at *Andasifatatelo*, stringers of quartz or graphite cut the altered graphite gneiss, which is also intersected by a basic dyke.

Near Fanandrana in the small *Antsirabe* deposit, roof pendants, migmatites and a shear zone contain graphite. Other deposits of this belt are:

South (*Brickaville*): Concession 105, Ambalarondro, Vatoharana, Mankarano, Ambodiriana, Marofody–Andranofianjavatra, Ampangadiatana, Faliarano, Masse, Ambatovy, Sahave, Fanovana, Ampositabe, Marivolanitra, Ankatsakafobe, Ambinaninony, Alavakorana, Ambalavero, Ambodilalona–Ambodimanga, Ambodipeso, Ampasipotsy, Ampotombia, Andoanavolobe, Andranomangatsiaka, Anevorano, Betainsira, Raboana, Sahabe, Tsaraloaka.

North *(Tamatave)*: Sahamaloto, Ampisaka, Ampitabe-Sahamitrony, Sandraka, Sandratsoa, Mahalina, Besakay, Sahafosa, Ambalafotaka.

OTHER MALAGASY BELTS

A majority of the deposits of the extreme *south* is (Fig. 52) included in leptites. Micaceous gneisses are rarer host rocks, and pegmatites, cipolinos and pyroxenites are even scarcer. South of Benitra glandular, migmatic leptites are distinguished by the greatest concentrations. Between Ampanihy and Bekily graphite alignments extend for 90 miles. In the *south-southeast* occurrences are rare, but their number increases again in the leptynites of the area of *Ranohira* and in the gneisses and pyroxenites of *northern Ihosy*. Since they are lateritized, gneisses of the *Fianarantsoa* degree sheet, e.g., of Benenitra and Isaha–Vorondolo are soft and economical to work. In the area of Ambositra (Manarinony, Itandroka), schists, micaschists and some of the Ivato Valley limestones carry graphite. On sheet *Ampasinambo* but a few deposits occur, namely Ambodiraidoka, Ambatoharanana, Anjamana and Ankeratrana. However, on the *Mandoto* sheet graphite is abundant in the Betafo–Miandrarivo–Mandoto paragneisses. At the base of a syncline they form a graphite concentration near a migmatic front. Nevertheless, intensive lateritization of the micaschists and gneisses has obliterated surface ore. Beneath them we can distinguish seven graphitic alignments. Most of the graphite is flake, but powdery and aggregated ore also occurs. Deposits include Hiarambao, Ankaramalaza, Vorondolo, Ampizarandrano, Ankaramahery and Miandrarivo. Several of them were worked formerly.

On the *Antsirabe* sheet reports quote only four occurrences. The thickness of these graphitic layers varies between 10 and 15 ft. In the *Ambatolampy* degree sheet south of the capital, Tananarivo, geologists distinguish the following graphitic zones: Morarano–Ampangabe–Ambohikambana, Bemasoandro–Faravato–Ambatovaventy–Sarobaratra, west and north of Tsinjoarivo comprising the migmatic Andavabato deposit and Ambatoafo. Other deposits include: Andrangaranga, Antsambavy, Ambatomitely, Amparihimena, Ambalanarivo, Ampotsapetsa, Bemasoandro, Vavavato, Antanifotsikely, Andohanilaka, Faravohitra, Ambohimandroso, Antsampandrano, Antsofimbato, Antanisoa, Ambatolampy, Antsiriribe, Tamiana, Antananarivokely, Andravoravo, Analamarina, Lohavohitra, Ambatomainty, Ifandra, Befotaka, Antsahalava, Antsahabelona, Anosibe, Anevokambo, Ambohidrazana, Andrianavalona.

In the *Anosibe–Vatomandry* sheets geologists have mapped numerous deposits east of a line which runs parallel to the coast 6 miles east of Anosibe. Dipping west, swarms of flake graphitic zones strike north. The host rocks are lateritized. Graphite is associated with sillimanite in half of these occurrences. At Analatsara, graphite is included both in the granitic (quartz–feldspar) and metamorphic constituents of migmatites. On the other hand, at Tetezambato graphite is limited to thin stripes of the banded migmatite. However, the ore also embays quartz blows in the graphite layers of Marovintsy. Other deposits include Andranotsara, Itomampy, Mahasoa, Ambinanindrano, Sahanimorona, Ampanaofanana and Ambalafandra.

Interestingly enough, the Ngilomby, Ambohijafy and Ambohitsivalana deposits of the Soavinandriana sheet occur in Androyan paragneisses, believed to be somewhat younger than the Graphite System. However, in the centre of the island little is known about a number of occurrences. These are the following:

In the Miarinarivo area: Amperifery, Tsarazafy, Zoma Bealoka, Ambatolampy, Ambohimiangara, Fiantsonana, Vodivohitra, Soamananety, Amparihimena, Amberobe, Ambatomilika, Apano, Antanimievitra.

In the Tananarivo area: Ambodinonoka, Beampombo, Ambatofotsy (Behenjy), Tafaina, Antanetibe, Anosibe, Ambike, Ambohidrabiby, Ambatofahavalo.

We have already described the favourably located deposits of the Brickaville and Tamatave districts. Although outcrops are rarer north of this area, a few have been located on sheets Maevatanàna (in the Ambongabe Chain and the upper Kamorokely Valley), Andilamena, Marovato, Marotandrano (the Beanana Range including Mangeny, Tampoketsa and Sahavalanina) and Mandritsara.

EAST AFRICA

The Malagasy graphite zones continue in the Mozambiquian belt. In felspathic megacrystalline biotite gneisses of Metochéria (northern Mozambique), graphite stringers follow a pair of planes and a horizontal line. Vertical beds strike N 40° E, with larger layers including gneissose relics. The thickness of stringers attains ½ inch. At Itotone, Netia, abundant graphite is disseminated in gneisses, traversed by diabase and pegmatite dykes. In the extension of this belt graphitic quartzites form a 1,000 ft. wide zone in the Chilungula Hills of southern *Tanganyika*. Other occurrences of the same district include Nanganga and Ndanda. West of Dar es Salaam, Mtombozi is the richest graphite band of the Matombo district, Uluguru Mountains. Indeed, its grades average from 12 to 25% (HARRIS, 1961). Graphite also occurs at Idete and Tambi in central Tanganyika, and at Daluni in the north.

In the extension of this zone we find numerous deposits in southeastern *Kenya*. In the Tsavo–Mtito Andei district, parallel and thick early Precambrian graphitic bands extend for several miles. Carbon content exceeds 6%. These occurrences extend south of Voi. Near Nairobi the schists of Machakos contain a maximum of 25% graphite. In the Loldaika Hills, Namanga, and in southern Kitui grades of 6–10%, and 2–23%, have been recovered respectively. Furthermore, graphite also occurs in northern Kitui, near Loperot and the Merti Plateau.

Graphite occurs in gneisses, limestones and charnockites of the northern frontiers of Uganda. East of the Ruwenzori a 6 inch thick band of graphite was found in the Moboku drainage. Carbonaceous schists belong to the stratigraphic column of the late–Middle Precambrian rocks of southwestern Uganda, Rwanda and the eastern Congo. In the Bahr el Gasal district of the southwestern Sudan, graphite occurs in the Bongo drainage and on the Yambio–Meridi road.

In central *Malawi* the principal concentrations of flake graphite are at Dowa. Here a 30 mile long and 1 mile wide band of quartz feldspar gneiss contains up to 10–20% graphite. The diameter of the flakes attains 1–2 mm. Interlaminated biotite contaminations remain peripheral. Graphite follows planes of anisotropy, also disseminating beyond them. On the other hand, the Rivi–Rizi River deposit near Geya is heterogeneous. Amorphous graphite occurs at Mposadala and on the Wamkurumadzi River north of Neno. South of Dowa, the Angonia deposit is situated in western *Mozambique* where a mile long belt of early Precambrian gneisses strikes N 35° W. The contact of gneisses and limestones concordant to granitic intrusives is favourable for graphitization. Furthermore, randomly oriented high grade graphite stringers occur in the Chiziro gneisses. Dipping 24° E, this band strikes north-northwest.

THE REST OF AFRICA

Graphite flakes are disseminated in Katanga and in the Petauke–Sasare district of *Eastern Zambia*. At Mvuvye, Mkonda and Sasare, graphitic gneisses and granulites form tight horizons in migmatized biotite gneisses, pyroxene granulites, lime-granulites and limestones injected by granite (DRYSDALL, 1960). The graphite-bearing sheets are small but frequent. Graphite also occurs in Southern Rhodesia (Graphite King) and in the Hlatikulu district, Swaziland.

Along the northern frontier of *South Africa* flake graphite is associated with marble, interbedded with early Precambrian gneisses, which occur between the Magalekwane and Pafuri rivers. Deposits include Gumbu, Khonongakop, Grafietmyn, Steamboat and Waterpoort. Massive graphite has developed from carbonaceous Transvaal Shales west of Carolina, and from Karroo coal, injected by dolerite at Malonga, in the northeastern corner of the country. At Maandagshoek a xenolith of the Bushveld gabbro has been graphitized.

In southern *Southwest Africa* graphite occurs in the Keetmanshoop gneisses at Bethanie and Aukam, east of Lüderitz Bay. Carbonaceous schists are widespread, e.g., at Otjivarongo. Graphite occurs at Maio Butabe, Hossare Nuwa, Gayam and Jauro Jalo in Cameroon, incorporated in Nigeria, at Birnin Gwari, northern Nigeria. On the Nzérékoré–Man road in the northwestern Ivory Coast, the altered micaschists of Lola contain up to 20% graphite flakes and average 9%. The band attains a length of 300 ft. and a width of 200 ft. The mineral also occurs at Toulepleu, Sierra Leone, and in the Kédougou district of southeastern Senegal.

At Sidi bou Othmane north of Marrakech, Morocco, probably early Carboniferous (Visean) limestone lenses, included in schists, have been affected by regional metamorphism and consequently graphitized. The exocontact of cipolinos and tactites contains the highest grades.

ANTHRACITE

Small basins of anthracite have developed in northeastern Morocco and in the area of southern Swaziland, northern Natal, South Africa. The former is Carboniferous, but in the latter, Permian coal was upgraded into anthracite through the injection of dolerite (Fig. 7).

MOROCCO

The principal field, *Djerada,* is located south of Oujda. This south-southwest trending alignment extends to the Christian field southeast of Rabat, and in the east to Oran in Algeria (Fig. 10, 59).

The stratigraphic column of Djerada consists of Lower Silurian, (Llandoverian) phtanites (AGARD et al., 1952), of Lower Carboniferous (Visean) shales, lava flows and limestones, and of Lower–Middle Carboniferous (Namurian) shales, including inliers of sandstones and conglomerates. While, the thin (Namurian) strip south of the syncline consists mainly of the upper, Ardennian, *Reticuloceras, Anthracoceras* and Homoceratoid carrying bed, in the wider northern part the Lower Namurian, *Gravenoceras* and *Homoceras* carrying layer is prevalent. The thickness of this stage varies from 300 to 600 ft. Upper Carboniferous, Lower Westphalian sandstones and conglomerates are then divided by *Goniatites* shales. The first continental layer is lower Sidi Brahim, but only one of the four coal seams of this layer is commercial; it is located south of Sidi Messaoud Mohammed. A marine transgression is followed by the upper Sidi Brahim layer of calcareous and *Productus rimberti* carrying shales. The Djebel Chekhar Series includes several beds of carbonaceous shales, high ash coal seams and a thin anthracite seam, but in the west this series thins out.

The Upper Westphalian starts with the Big Conglomerate beds of sub-angular Silurian pebbles, and a graptolite fauna. The conglomerate delta is the thickest in the northwest, and the Djerada layer is 1,400 ft. thick. While brown Moscovian, *Spirifer mosquensis* and *S. myatchkovensis* carrying limestones do not occur above the estuary, they are found 90 ft. above the conglomerate in other parts of the basin. This bed is overlain by five anthracite seams, each with thicknesses varying between 1 and 2½ ft., and a total thickness of 7 ft. Type III anthracite is amorphous bright coal. The two overlying, thin coal seams show an

Anthracomya phillipsi fauna; their top layer is attributed to the Houve Phase of the Westphalian. *Estherieulla reumauxi, Estheria simoni, Phyllopodes* and *Anthracomya phillipsi* have been collected at the roof of the three lower anthracite seams, along with rare plants which belong to the Bruay Stage of the Westphalian. On the other hand sandstones cover two anthracite seams and conglomerate in the northwest. The sea invaded the basin four times, and the last, Moscovian transgression — between the accumulation of the two upper seams — is distinguished by *Spirifer strangwaisi, Homoceratoides kitchini, Marginifera pusilla* and *Anthracoceras aegiranum.*

Whilst the strike of major Asturian Phase (Westphalian) faults is parallel to the west-southwest trending, Hercynian anticline, faults of the (Permian) Saalian Phase are perpendicular. Thrusts may attain as much as 900 ft. Since the northern limb of the fold is flat and its southern flank dips steeply, the anticline pinches in the east.

The coal alignment extends for a few miles towards the east, and far to the west under the Hammada (stone desert) of Guir in the Tafilalet and under the Cretaceous Hammada of Tizimi. However, the Carboniferous does not outcrop.

The *Christian* basin is located 70 miles east of Casablanca. The faulted Silurian–Devonian syncline trends northeast. Indeed, it has been refolded by the Sudetan Stage of the Upper Westphalian. In this southern basin conglomerates and sandstones include several layers of thin Bruay Stage coal seams. Late Carboniferous, (Stephanian) anthracite lenses and thin seams also occur in the Ida ou Zal basin of the High Atlas, as do carbonaceous shales at Oued Zat and Ourika.

We now pass to the other end of Africa.

SWAZILAND

While the Lower (Permian, Middle Ecca) Coal Zone (Fig. 47) located along the Stegi–Manzini Road is sub-bituminous, the Upper (Upper Ecca) Coal Zone contains bright, ash rich anthracite. This fuel of conchoidal fracture is parted by dull coal and shale. The Upper Zone of the southern sector appears to be the equivalent of the Lower Zone of the north!

The Upper Zone lies 2,300 ft. above the base of the Karroo. The averages are given in Table CXL.

Pyrite occurs in the bright coal of the lowest horizon. Bands of carbonaceous shale,

TABLE CXL

THE UPPER ZONE COAL MEASURES OF SWAZILAND

	Sediments (ft.)	*Coal (ft., in.)*		*Minable seam (ft., in.)*	
Coal and shale	65	9	9	4	9
Sandstone	32				
Coal and shale	58	18	0	3	10
Sandstone	31				
Main Seam and shale	90	25	9	20	6
Sandstone	19				
Coal and shale	97	15	3	3	4
Sandstone	43				

with a thin stripe of limestone, are interbedded with the remarkably constant Main Seam. The bright coal carries pyrite, siderite and calcite with mixed dull and bright pyritic coal in its lower beds. The Zone averages 71.0% fixed carbon, 21.0% ash, 6.0% volatiles, 1.0% moisture, 1% sulphur, and a calorific value of 12,000 B.T.U./lb. The third zone consists of bright coal, with shaly partings and alternating bands of dull, shaly and bright coal at the base. Similar oscillations have been observed at the top of the fourth zone (DAVIES, 1961).

In Mineral Concession 2 the zone is intensely faulted. Nearby in Crown Mineral area 9 and at Nsalitshe semi-anthracite seams average, respectively, 4 and 3 ft. of thickness. Finally in Concession 27, extending from the Stegi road to the Mtendekwa River, dolerite sills, injected at several levels, have burnt one seam.

THE REST OF AFRICA

Similarly in Natal, *South Africa*, Karroo coals (Fig. 45) have been injected by dolerite. Some anthracite has developed in the Vryheid and Klip River fields. Vryheid anthracite averages 79.7% fixed carbon, 9.4% ash, and a calorific value of 13,500 B.T.U./lb. In the principal anthracite collieries, at Ngwibi Mountain, the thickness of seams (affected by dolerite injections) has been measured to be, respectively, 4, 3, 6, 2, 2 and 4 ft. Anthracite is also found at Elandsberg, Alpha, Natal and Impati, between Utrecht and Ladysmith. Klip River field anthracite contains 2% sulphur, and has an ash fusion temperature that reaches 1,350° C.

Sub-anthracite coal occurs in Southern Rhodesia. In the Ruhuhu basin of southwestern Tanganyika anthracite is found in the upper horizon of the Mchuchuma field and in the lower horizon of the Ngaka field.

Far in the north, anthracite lenses have been identified in the Oran and Constantine area of Algeria. This concludes the description of the meagre anthracite resources, and leads us to:

BITUMINOUS COAL

An overwhelming majority of African, and of other Gondwana, coal is of Permian age. Indeed, Karroo covers large areas of southern and middle Africa. The carbon-bearing Middle–Upper Karroo Ecca Stage, however, is limited to: *(1)* the eastern part of South Africa and adjacent countries; *(2)* the Limpopo valley and its extensions; *(3)* the Zambezi valley; *(4)* the basins surrounding Lake Nyasa; and *(5)* the upper Lualaba and other valleys in the southeastern Congo.

A majority of the coal accumulated in South Africa. Unfortunately most of it is non-coking; however, Wankie, in Southern Rhodesia, is a notable exception. Triassic coal occurs in southern Africa. Cretaceous coal measures appear in Nigeria, and the only major (Fig. 7) Carboniferous basin is located in the Algerian Sahara. Coal has not been found, however, in most African countries.

SOUTH AFRICA

In addition to the Triassic Molteno–Indwe field, north of East London, the coal fields spread 400 miles in the Permian, Ecca Series of the Karroo from near Kimberley to the Klip River in Natal. The shape of the Karroo coal basin is between that of a triangle and half ellipse, its northern point being situated halfway between Pretoria and Swaziland (Fig. 46, 50). The western limit partly follows the Vaal River, but the bulging eastern limit is oriented north–south. The main fields follow the eastern limit of the Karroo along a 150 mile long and 70 mile wide oval, but the western limb is less developed and the centre of the basin lies too low for recovery. Other fields are located at Springbok Flats north of Pretoria, and at Waterberg on the Limpopo extending into Bechuanaland in the west, to Soutpansberg in the east, and thence to the Sabi field of Southern Rhodesia.

From *Natal* the Karroo basin extends 80 miles north (Fig. 45) to Ermelo, in the eastern Transvaal. Between Paulpietersburg and Newcastle, seams of coking coal have been partially transformed into anthracite through the injection of dolerite, and locally even into graphite, and in a number of cases destroyed. In the Paulpietersburg–Vryheid fields, an eastward bulging Karroo gulf, five seams out of a total of ten are of commercial significance. Their soft, coking and sub-coking coal contains 62.8% fixed carbon and 16.7% ash, and their calorific value averages 12,500 B.T.U./lb. At some distance, in the Utrecht–Newcastle field, the principal collieries include Elandsberg and Belelasberg. In the Klip River field between Ladysmith, Newcastle and the Drakensberg the top and bottom seams are of commercial importance. This coal averages 17.4% ash and has a calorific value of 12,200 B.T.U./lb. In the eastern areas the coking coal of the top seam attains 6–10 ft. of thickness, but north of Ladysmith the top seam contains 20% ash.

In the Heatonville–Somkele area of Zululand, north of Durban, two Triassic (i.e., young) but (nevertheless) mature seams are significant.

The *Carolina–Ermelo* field, which includes Breyten, Ermelo and Estantia, is situated west of Swaziland in the northeastern corner of the coal basin. Several inconsistent but thick seams and measures exist, as at neighbouring Waterberg (Fig. 44). In the area of Breyten, seam A attains 1–2 ft. of thickness; however the coal is brighter and more homogeneous towards Ermelo. The low grade seam B is 7–12 ft. thick, and appears more than 100 ft. below A. At Estantia the position and thickness of seam C is also most variable; at Naudelsbank, south of Carolina, eight seams have been proved. Calorific value attains 11,500 B.T.U./lb. and ash content 15.4%. South of this, dolerites have pervaded five variable seams of the *Ermelo–Wakkerstroom* field. Between the Pongola and Assagai Rivers coal of the 4–5 ft. thick second seam is coking. Further west, in the *Bethal* field, the 10 ft. thick middle seam, amongst three, is of commercial value.

Witbank–Middelburg, east of Johannesburg, is the continent's principal producer. The total thickness varies between 50 and 400 ft. and averages 200 ft., but only one or two out of a maximum of ten seams are worked. Rather complete columns at the Navigation and Clydesdale collieries include (Fig. 44, 120):

(a) Dwyka tillite; *(b)* seam No. 1 approximately 5 ft.; *(c)* 5–10 ft. of sandstone; *(d)* No. 2, the principal seam, 20 ft.; *(e)* 10–20 ft. of shale; *(f)* 35 ft. of sandstone, including thin seam No. 3; *(g)* seam No. 4, 20 ft.; *(h)* 25–35 ft. of shale and sandstone; *(j)* seam No. 5, approximately 6 ft.; *(k)* shale, sandstone and soil.

Seam No. 2 has a calorific value of 12,100 B.T.U./lb. and an ash content of from 10 to 15%. Approximately half of its total thickness of 20 ft. is of commercial value; the lower 3 ft. of the seam are bright, with similar bands occurring higher in the column. In the south-

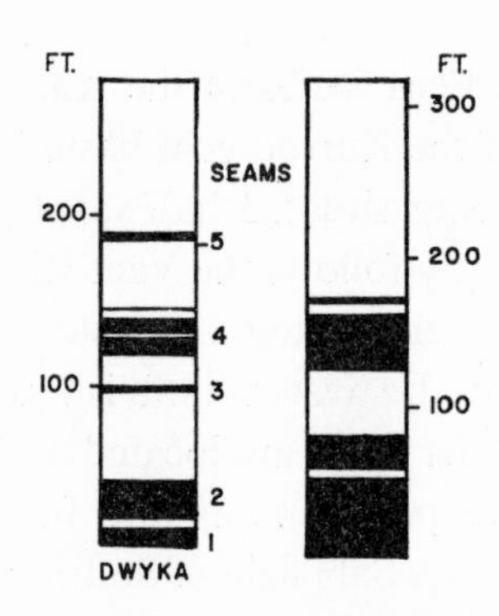

Fig. 120. Coal seams of Navigation Colliery (left) and Springfield Colliery (right).

western part of the field the upper layer of seam No. 4 is of good grade. Unfortunately the bright No. 5 seam has been eroded at many localities. The coal averages 9–20% ash and has a calorific value of 12,600 B.T.U./lb.

The *Springs* field, the westernmost outlier of the great Karroo coal basin, is situated 40 miles east of Johannesburg. There Ecca coal measures have been deposited on and then have sunk into the karstic surface of the Dolomite Series. Consequently the lenses are most variable in thickness and extent.

The *Vereeniging* field (Fig. 43) is conveniently located south of Johannesburg. In a typical column, 10–13 ft. of basal tillite of the Karroo lie on the Precambrian Dolomite Series. The coal measures include:

more than 300 ft. of arenaceous cover;
30 ft., seam No. 3;
70 ft., shale and sandstone;
20 ft., seam No. 2 with shaly partings;
7 ft., conglomerate, rearranged from tillite and shale;
14 ft., seam No. 1, lying near the floor.

The calorific value of the seams ranges around 9,700 B.T.U./lb.; the ash content of No. 1 and No. 3 is 20–30%, while the middle seam attains 15%. As at Vereeniging, in the *Vischkuil–Delmaas* field, the Bottom Seam frequently lies on the Dolomite Series. In addition, coal lenses occur in the goldfield near *Nigel.*

In the central depression of the Bushveld, the Springbok Flats synclines strike northeast, connecting with the main coal basin of the Transvaal through Pyramid. Parallel to this trend faults strike northeast and east. The principal, Upper Ecca (Permian) zone attains 35 ft. of thickness, including several seams of bright, sub-coking coal, but the lower seam, located at the base of the Ecca Stage, is of low grade.

The *Waterberg* field contains the continent's largest reserves. It covers 730 square miles in the northwestern Transvaal and connects with the Malapye field in Bechuanaland. More than seven seams and coal bearing horizons have been distinguished which represent a maximum aggregate thickness of 430 ft.! Seam No. 1 attains 5 ft. of thickness and 11,600 B.T.U./lb. of calorific value, and seam No. 2 averages 10 ft. of thickness containing dull coal and occasional shaly partings. However, the thickness of the principal seam, No. 3, is variable, and its lower portion includes mixed coal. Zone 6 consists of several seams and shaly partings containing only 6–9% ash. Finally, the ash content of seams No. 3–7, of zone 7, increases to 20–25%.

Farther north, the *Soutpansberg* field connects with the Sabi–Limpopo basin. The Karroo beds, which strike east and dip slightly north, are arranged in long, fault-bordered belts there. Seams and shales alternate, as in Rhodesia. The most interesting part of the field extends from Waterpoort to Nzheble Dam. The coal contains up to 30% of volatiles and much ash, but, further east towards the Kruger Park, the coal has been upgraded through the injection of dolerite.

BECHUANALAND

Bechuanaland is covered by Kalahari sands, but the Karroo outcrops on its eastern edge. The Mamabule basin (Fig. 46, 50) covers 17 square miles in southern Bamangwatoland and thus extends through the Tuli bloc the Waterberg coal field of South Africa. East of the railway, the converging Mabuane and Zoetfontein faults form the limits of exploration. Non-coking steam coal in found in the (Permian), arenaceous Middle Ecca Stage, which consists of:

1– 7 ft. thin Upper Coal, locally found in mudstone;
30–63 ft. upper carbonaceous sediments;
16–23 ft. Upper Coal, predominantly dull, massive;
46–70 ft. sandstone;
5–11 ft. Lower Coal, semi-bright, banded.

Lying under a shallow cover, the three seams average 4, 18 and 8 ft. of thickness, respectively. The last and lowest seam is the most persistent, and is also of high grade (Table CXLI).

TABLE CXLI

THE COAL SEAMS OF BECHUANALAND

Coal	*Fixed Carbon (%)*	*Volatiles (%)*	*Ash (%)*	*Moisture (%)*	*B.T.U./lb.*
Thin Upper	49.5	30.0	14.7	5.8	10,800
Upper	51.0	24.2	18.7	6.1	10,000
Lower	54.0	25.6	14.7	5.7	10,500

The *Morapule* basin is located in central Bamangwatoland, 6 miles west of Palapye and 100 miles north of Mamabule. Coal measures occur at the base of the Ecca Shales which overlie sandstones. Unfortunately, their quality and thickness varies laterally. Towards the eastern sub-outcrop the seams thin out; in the west, however, best grades lie at a depth of 350–400 ft. In this area the useful thickness of the seam varies between 25 and 30 ft., and the calorific value lies between 11,000 and 11,500 B.T.U./lb., while the ash content oscillates between 12,6 and 14%.

SWAZILAND

In southern Swaziland, the 650 ft. thick Middle Ecca Coal Zone of the Stegi–Manzini Road (Fig. 47) is interbedded between sandstones. Four seams have been correlated to those of the Vryheid–Paulpietersburg field in Natal. The 4–28 ft. thick Main Seam, which averages 16 ft. in thickness, lies 60–120 ft. above the banded and impure Coking Seam. A shaly parting locally divides the lower, bright coal of this seam from the upper dull coal. The Main Seam averages: 72.9% fixed carbon, 13.2% volatiles, 12.1% ash, 1.4% moisture, and 0.4% sulphur.

The first Upper Seam, averaging 3 ft. 8 inches in thickness, lies 55–90 ft. above the Main Seam; it is distinguished by constant thickness, a grade of 12,300 B.T.U./lb., and an ash content of 18%. On the other hand, the thickness of the second Upper Seam varies from 8 inches to 4½ ft., with a calorific value of 12,400 B.T.U./lb., 65.9% fixed carbon, 18.4% ash and 15% volatiles. The Upper Coal Zone is anthracitic. The seams of Nsalitshe, St. Phillips and Nhloya respectively average 3 ft. 5 inches, 2 ft. 10 inches and 4 ft. in thickness. The contents are given in Table CXLII.

TABLE CXLII

COMPOSITION OF UPPER COAL ZONE IN SWAZILAND

	Fixed carbon (%)	*Ash (%)*	*Moisture (%)*	*Sulphur (%)*	*B.T.U./lb.*
Nsalitshe	59.5	16.3	8.0	0.5	9,500
St. Phillips	64.9	14.6	5.1	0.8	10,400
Nhloya	53.9	12.6	10.6		9,000

SOUTHERN RHODESIA–ZAMBIA

The Limpopo and Zambezi Valleys limit the Rhodesian shield in both the south and the north. Both valleys were inundated during the Permian and coal developed.

The Sabi and Mkwasini basins of the Limpopo Valley, *Southern Rhodesia*, contain high ash coal seams which are dissected by dolerite dykes. In the *Lower Sabi*, the Coal Group (Fig. 50) is of Permo-Triassic, Ecca to Molteno age. The formation of 0–900 ft. lies 9–270 ft. above the base of the Karroo. At Mkushwe, the lowest seam of non-coking, sub-anthracitic coal attains a thickness of 46 ft., with ash averaging 36%! The upper basin is located further east than the lower seam B, which bulges to 7 ft. (32% ash), and seam A, which measures 6 ft. (37% ash).

Near the confluence of the Bubye and the Limpopo an 8 ft. thick seam of coking coal is found. Good quality coal also occurs in the Tuli field, northwest of Beitbridge.

In the Sabi–Limpopo basin, the sequence consists of up to 230 ft. of conglomerates, grits and sandstones, sandy shales and subordinate limestones. The main coal seam opens the Upper Series, which is similar in composition to the lower beds. Finally, another coal seam occurs high in the column.

Karroo beds cover the *Mid-Zambezi* Valley along a 150 mile long stretch. Along the southern rim of this trough, coal fields extend from Wankie (50 miles east-southeast of Victoria Falls) to neighbouring Entuba and Lubimbi, to as far east as Sessami and Mafungabusi, and bend northward to Marowa and Nebiri. Gwembe and Kandabwe are located on the northern Zambian edge of the basin, and Sebungu, Lubu and Sengwe lie on the perimeter of a pre-Karroo window in the midst of the sediments. Coal has been deposited on the irregular, old land surface in discontinuous basins (Fig. 121).

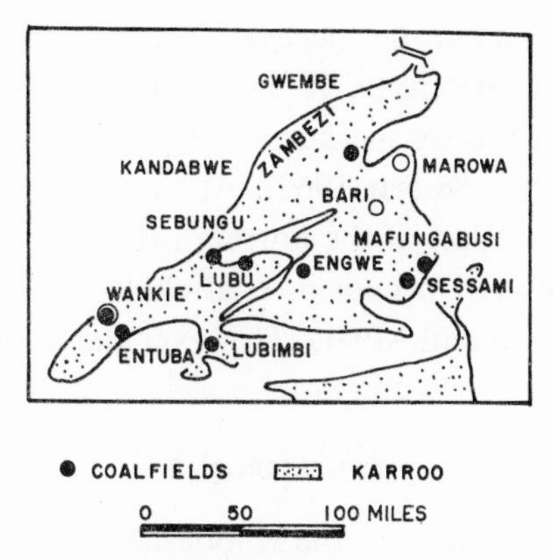

Fig. 121. The Mid-Zambezi coal fields.

At *Wankie*, the Permian (Ecca) column consists of:

approximately 750 ft. of Madumabisi Shale with coal seams;
75–120 ft. of red, micaceous upper sandstone;
0– 80 ft. of fireclay;
10–110 ft. of black shale and coal;
6– 42 ft. of main coal seam;
10– 20 ft. of sandy micaceous shale;
130 ft. of lower, coarse and fine-grained sandstone.

Karroo sediments have been deposited on a slightly undulating, subsiding shore, which is steeper on its northern edge. Seams average a thickness of 16 ft. at Wankie, and consist of 6 ft. of lower grade coal that contains 6% ash. Where developed, the Main Seam rests on the Footwall Series. The maximum thickness of this seam attains 40 ft., but it grades out laterally and vertically into carbonaceous shales, and from Wankie eastwards, as the floor steepens, its quality deteriorates. Wankie coal forms excellent coke, averaging 65.7% fixed carbon, 23.8% volatiles, 9.8% ash, 0.8% moisture, and a calorific value of 13,500 B.T.U./lb. The ash content of 5% increases upwards in the column. There is no cyclical recurrence of coal.

The *Madumabisa* series consist of *(a)* up to 400 ft. of medium grey shales interbedded with white mudstone and a few coal seams, *(b)* 200 ft. of grey shales and limestones which exhibit cone in cone structures, *(c)* 600 ft. of khaki brown shales including concretions, and *(d)* 300 ft. of pale shales and septarian limestones. The spotted Madumabisa coal seams are low grade and attain a thickness of from 2 inches to 2 ft.; they bulge to low grade, 10 ft. thick seams where carbonaceous shale is present. Siderite forms streaks and spherules in the coal.

Further east, at *Sebungwe*, 100 ft. of basal tillite and fluvio-glacial sediments cover the Basement. Numerous coal seams, which are especially thick at the base, are interbedded in a thicker column of black shale upon them. East of Sebungwe, the Mafungabusi field covers 17 square miles, with coal measures attaining 6 ft. of thickness and an ash content of 13.29%. High grade coal seams, which average 10 ft. of thickness, extend to Lubimbi on the Shangani River. The Sengwe seams, 60 miles northeast of Lubimbi, are thinner, with shale covering partings.

We now cross the border into *Southern Zambia*. A thin coal seam is interbedded in the lower Karroo of Kandabwe and in the Sjegega stream in the Gwembe district (Zambezi Valley). These basins cover, respectively, 50 by 5 and 90 by 10 miles. The seams have been correlated with the K2 and K5 lower measures of Wankie, with Umi–Kaonga representing a connecting stage (TAVERNER–SMITH, 1962). At Luana, southeast of Broken Hill, there is a small isolated coal basin. West of the northern tip of Lake Nyasa, at the head of the Luangwa Valley, low grade coal occurs in coal shales at Pangala, Songa, Luwuwushi, Luhoka, Lusangani, Luwumbu and Mwila. However the thickest seam is only 4 ft. wide.

MOZAMBIQUE

Extending in an easterly direction from the Rhodesias to Lake Nyasa, the Karroo of the Lower Zambezi Valley sedimented in a pre-existing trough. Coal measures outcrop between Caringe and Chicoa, and for 50 miles east of Tete. Aligned along the northern edge of the valley, three tectonic cycles have divided them into five fragments, or the Moatize (Fig. 34), Revugo and Morongodzi basins. Towards this northern edge, the basal shale laterally changes into conglomerate. In addition, as we move east the thickness of the coal measures expands, but sandstones intercalate between them. With their thickness diminishing upwards, seams tend to form groups. At *Moatize* the thickness of the seams varies from 1 to 21 ft. and totals 40 ft. At Revugo it varies from ½ to 7 ft., with an additional seam of 13 ft. and numerous seamlets, but at Morongodzi only three seams are left. Carbonaceous shales and sandstones are intercalated between the seams, in which bright coal forms only thin laminae, in these fields. The ash content varies between 12 and 25%. *Glossopteris indica*, *G. browniana*, *G. boancai* and *Gangamopteris cyclopteroides* indicate Upper Ecca age. In addition, coal shales are also known to occur on the eastern shore of Lake Nyasa and in the Lugenda basin on the Rovuma River (which forms the border of Tanganyika).

MALAWI

From Mozambique, the Zambezi–Tete basin extends to the Sumbu–Nkombedzi field southwest of Zomba, the capital of Malawi. At Nkombedzi coal shales, injected by dolerite dykes, lie on the Basement. Northwest and elsewhere striking faults dissect the lenses, which were formed through the carbonation of shales. Further east, at Chiromo, the seams even become sub-anthracitic. The flora, nevertheless, indicates Permian–Upper Ecca–Lower Beaufort age. The three northern coal fields are part of the extension of the Tanganyika Kiwira field. They are situated at the southern rim of a structural basin which extends from the Sengwe River east of Fort Hill to Nkana, south of Karonga, and thence to Waller Mountain, opposite the Ruhuhu basin. The *Nkana* field (Fig. 34, 52) is located in a triangular Karroo fragment which attains a maximum width of 3 miles. In the Kapembe Hills the measures extend for 2½ miles eastwards from Mwansalano beyond the Makulumu stream, but a transverse fault intersects in the middle. On the western boundary of the down fault, along the Ndamutakwa stream, 5 mile long thinner coal seams outcrop. The coal measures frequently dip 15–25° south. The Karroo sequence consists of 350–1,500 ft. of lower conglomerates and sandstones, 50–300 ft. of shale and coal measures (corresponding to the lower coal measures of Ruhuhu), and the cover of mudstone, limestone and sandstone. In the west two 7 ft. seams have been distinguished along with three similar seams in the east. The coal contains 30% ash (BLOOMFIELD, 1957).

The Mount Waller (Chombe) field is situated in the faulted bloc of Livingstonia, 3,000 ft. above Lake Nyasa. There are two to three southward dipping seams which include non-coking coal that contains as much as 12–40% ash.

TANGANYIKA

Small and large coal basins are arranged in the Rift Valleys which connect Lakes Tanganyika and Nyasa. In the southwestern corner of Tanganyika in the Ruvuma depression, which includes Mhukuru and the smaller Njuga basin, two coal fields are located. Another field, Lugenda, is situated in Mozambique in the lower Ruvuma drainage, and a small field lies in Mbamba Bay on Lake Nyasa, north of the Maniamba field of Mozambique. However, the country's principal coal basin, Ruhuhu (Fig. 52), which consists of the Ketewaka–Mchuchuma and Ngaka fields, is situated 80 miles southeast of the northern tip of Lake Nyasa, opposite the Malawi Waller Mountain field. West of the northern tip of Lake Nyasa and north of the Malawi Nkana basin, is located the Kiwira–Songwe field, including the small Ilima colliery. The small Galula field is situated in the trough south of Lake Rukwa. The Muze field appears north of the Lake, and the Namwele Mkomolo field rises above the trough. Finally coal reappears at Albertville, west of Lake Tanganyika.

The axis of the Mhukuru syncline has been faulted. The main seam, interrupted by shaly partings, averages 22% ash and has a calorific value of 10,400 B.T.U./lb. The calorific value of the thin Njuga seams is similar. In the Karroo segment of Mbamba Bay, which dips 5–10° southwest, coal seams totalling 4 ft. of thickness average 10,900 B.T.U./lb. and 13.6% ash.

In the *Ruhuhu* basin, Karroo fragments lie on the Basement of a tectonic trough, subperpendicular to the Nyasa Rift. The column is given in Table CXLIII.

The trough has jigsaw-shaped limits which narrow to a width of 15 miles towards the

TABLE CXLIII

THE COLUMN OF THE RUHUHU BASIN

Ft.	*Sediment*
440	Pink sandstones and marls
1,200	Sandstones
300	Lower bone beds
700–1,000	Calcareous, sandstones, siltstones and mudstones
330	Upper Coal Measures, including grey argillaceous shales with coal seams, ironstones and sandstones
450	Marls and sandstones
450	Lower Coal Measures
1,700	Basal conglomerate and sandstone, including gritty sandstone with coals seams, carbonaceous shale and limestone

Lake. The principal faults and mosaics strike northeast, but others are sub-perpendicular. The coal basins form minor blocs in the major mosaics. The Wanda field occurs near the Lake; the Ketewaka–Mchuchuma field is located in the northeastern corner of the western mosaic, and the Ngaka (Mbuyura) and Mbalawala fields lie in the southwestern part of the eastern mosaic. The seams of the Ngaka field total a thickness of 10–14 ft. At Mchuchuma, 10% of the 100–150 ft. thick upper measures and 20% of the lower 100–200 ft. thick measures consist of coal, but only the 5 ft. thick seam 4 is coking. Ash content decreases rapidly with the thickness of seams.

TABLE CXLIV

SOME TANGANYIKA COAL SEAMS

	B.T.U./lb.	*Ash (%)*
Mchuchuma	12,800	15
Mbalawala	12,000	16
Ngaka	12,400	15

The Karroo dips 25° east at *Kiwira*, on the Songwe River. Traversed by southeast striking faults, thin impure coal seams grade into mudstones and shales. Consequently this coal averages only 11,500 B.T.U./lb., but does contain 15% ash. At Galula similar seams occur in a down faulted block, and at Muze a total of 15 ft. of coal has accumulated in a Karroo segment of the Rift. The elongated, tilted blocks of Namwele and Mkomolo include 3 ½ ft. of low grade coal (8,100–9,400 B.T.U./lb.), containing 25–31 % ash.

CONGO

The Karroo covers considerable areas on the eastern edge of the Congo basin. However, the coal measures of its early Lualaba stage were only deposited in small lakes. The principal deposits are located near the upper Lualaba Valley and the Upemba Graben, and also at Luena north of Jadotville in Katanga, Greinerville in the Lukuga Valley (Fig. 51), Albertville west of Lake Tanganyika, and Walikale west of Lake Kivu. The latter seams, which are flexured and jointed, mainly carry *Gangamopteris* which indicates Dwyka, Upper Carboniferous age.

West of *Luena*, Permian (lower Karroo) coal has sedimented from deltas in four basins of Basement schists and quartzites along a 12 mile long, north-northeast trending arc. Erosion of the Kisulu and Luena basins has started in the north, while Kaluku and Kalule lie in the south. At *Luena*, a 2 mile long and ½ mile wide oval is oriented in a northeasterly direction, and bulges towards the southeast. The column is given in Table CXLV.

A total of four seams may be present, attaining 35 ft. of thickness. The coal contains pyrite and *Glossopteris*. The warped sediments dip 25° to the southeast, and sub-horizontally to the northwest, but open post-Karroo fissures cut the strata. At *Kisulu* coal seams dip slightly south, and the lowest pinches out northwards; two seams average 3 ft. of thickness. There have been found two seams at Kaluku, and three at Kalule North, which respectively attain 10 and 8 ft. of thickness.

Hard coal is abundant and contains 5.5% moisture, 20.5% ash, 34% volatiles and 40.0%

fixed carbon. The calorific value reaches only 8,300–9,200 B.T.U./lb. for humid coal, with 25% ash, and 10,600–11,200 B.T.U./lb. when reduced to 15% ash.

The *Lukuga* coal is resinous, containing pollen and relics of wood; five seams and seven seamlets attain a total thickness of 15–20 ft. Since the measures carry *Glossopteris*, *Phyllo-teca* and *Cyclodendron*, they are of Permian (Ecca) age. The measures have been strongly dissected south of Albertville and also at Greinerville. The coal contains 5.6–5.9% moisture, 15.3–19.4% ash, 31% volatiles and 43.4–46.8% fixed carbon.

TABLE CXLV

THE COLUMN OF THE LUENA BASIN

Ft.	*Sediment*
3	Soil and eluvia
35	Argillite
4	Coal
20	Dark shales with interbedded thin seams of coal
10	Coal
18	Grey shale
6	Low grade coal
	Black shale, pyritic sandstone

MADAGASCAR

In the southwestern part of the island, irregular, faulted seams of non-coking coal are aligned in the 60 mile long, middle Onilahy Valley. Coal basins include Imalato, Ianapera, Vohibory, Sakoa and Sakamena Fig. 48, 52; (BESAIRIE, 1961a). Between the Bevalaha field and the intensely faulted Vohipotsy field, the *Sakoa* basin extends for 11 miles, while the Sakoa monocline (dipping 20°) outcrops along a stretch of 7 miles. The coal measures are underlain by tillite and black shale and covered by the Red Series. There are five Permian, Karroo coal seams, I and II lying lowest. Table CXLVI refers to seams III–V.

TABLE CXLVI

SOME PERMIAN COAL SEAMS OF SAKOA

No.	*Thickness (ft.)*	*Ash (%)*	*Volatiles (%)*
III	4– 6	32	24
IV	11–22	17	26
V	16–30	22	31

As regards ash content, only seam IV is of commercial significance: this seam, including shaly partings, thins out from 30 ft. of thickness at Mahasora, in the south, to 10 ft. in the north. Seam IV contains 10 ft. of usable coal; affected by up and down faulted 'piano key' faults in the central sector. The calorific value of seam IV attains only 11,900 B.T.U./lb., and the cutinous and spore bearing coal is of hard, dominantly dull quality. Some calcite also aggregates.

The Sakamena field, 10 miles southwest of Sakoa, extends for 7 miles. Coal outcrops dip 20°, and the seams are even less regular than at Sakoa. In the centre of the basin at Analahiva, the intermediate seams attain respectively 6 and 4 ft. of thickness and contain 14 and 24% ash, but the ash content of other seams is even greater.

NIGERIA

The *Cretaceous* coal belt extends for more than 100 miles on the eastern bank of the Niger, south of its tributary, the Benue. Reserves increase from Inyi towards Enugu (Fig. 122) in the north-northwest, thence 40 miles to Ezimo and Orukpa, and 50 miles to Okapa, Ogboyaga and Odokpono, where the coal measures turn towards Dekina with the Benue escarpment. Minor coal occurrences include Owo near Benin, Lafia and Auchi in Benin province, Agbaja in Kabba province, and Lamja in the east.

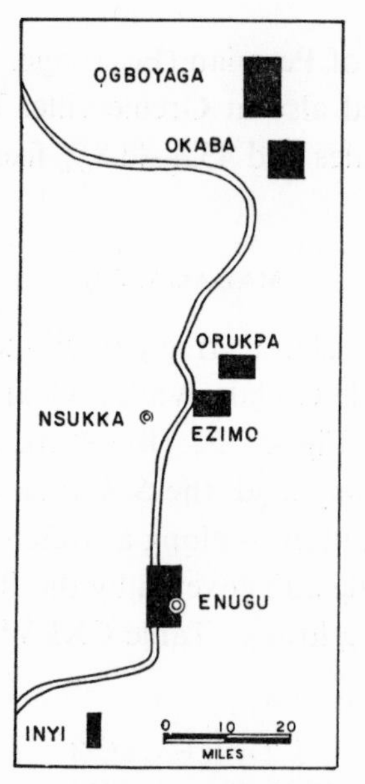

Fig. 122. The Nigerian coal fields.

The basin of the low grade Upper Coal Measures extends from the Mamu River to Enugu–Ezike. The seams are of variable thickness, although generally thin. The hydrated Inyi seams, overlain by sandstone, attain 3½–4 ft. of thickness.

All the other basins are located in the non-coking Lower Coal Measures. At *Enugu* only the third of five seams is commercially exploitable, but along the upper Ekulu seam 4 attains 3–4 inches of thickness. Seam 3 attains 2½–6½ ft. of thickness. In its middle it includes 5 inches of a shaly parting, the width of which increases south of the Hayes Mine. North of the Iva colliery, the lower section of seam 3 attains $2^1/_4$–6 ft. of thickness, with the upper bed thinning out. The seam outcrops for 7 miles; the principal reserves, however, are located west and northwest of the present workings. Along the *Ivoku* River near Ukana, and 3½ miles north along the Egodo near Okpatu, $3^3/_4$ ft. thick seams are known to exist.

The *Ezimo* seam, near the university town Nsukka, outcrops with a thickness of 4–5 ft., which increases to 6 ft. in depth. Reserves of the adjacent Ibagwa seam, near Orukpa, are considerably larger. The thickness of the horizon varies from 4½ to 12 ft., but only attains 3–7 ft. in the 2 mile long outcrop. The seam dips west.

The important *Okaba* seams unite to an 8 ft. thick measure in the northwest. On the other hand the Odokpono seam outcrops for 3 miles, and attains $2^1/_4$–7 ft. of thickness. The country's largest basin is located in the extension of the Okaba at *Ogboyaga*. The 5 ft. thick seam, extending for 3 miles to the north, bends 4 miles west with the Benue Valley.

ALGERIA

The Colomb Béchar–Kenadsa basin is located in the Sahara 300 air miles south-southwest of Oran. The coal measures extend west beyond the Oued Messouar into Morocco. The Rouzaud and '18 inch' seams extend for 15 miles in a slightly curved arc south of the Barga el M'Hamed and Barga el Gada to Béchar Djedid, on the Oued Bechar. However, the '14 inch' seam is known only as far as 1½ miles west of Djedid. Further south, the Bonouvrier horizon extends for 3 miles, and turns towards the northeast beyond Béchar Djedid. Since Carboniferous beds were folded into basins and brachy-synclines, by the Hercynian orogeny and divided by domes and short anticlines, the coal measures now form three secondary synclines. Upper Carboniferous marls and shales, interbedded with sandstones and thin layers of limestone, were transgressed by Middle Cretaceous (Cenomanian) strata of the Ksour Chain. The underlying Middle Carboniferous, Westphalian coal measures lie on lower (Moscovian) sandstones and limestones, and on the marine fossiliferous, Lower Carboniferous (Dinantian) limestones of the Djebels Béchar, Mzarif and Antar. Up to 17 coal seams are interbedded in sandstones and marls, but only two seams, which average 16 inches of thickness (locally pinching to only 8 or 12 inches), are of commercial importance. Continuing under the Cretaceous cover, the Main Seam has a northerly dip of 23°. This coal cokes readily, and contains 22–25% volatiles, 15–25% ashes, and 0.7% sulphur. However, infiltration of water is frequent. The lower seams contain 76.7% carbon, and grades diminish rapidly towards the surface.

Further south, near Sfaïa, the bituminous coals of the Abadla–Ksiksou basin contain 35% volatiles.

THE REST OF AFRICA

Thin seams of coal are interbedded with the Upper Cretaceous, Nubian sandstones of the northern Sudan. On the coast of *Somalia*, Cretaceous coal shows in the Hedhed Tug, 12 miles from Onkhor, and averages a calorific value of 5,660 and an ash content of 12%. Other occurrences include Biyo Gora and Subera. West of the Kenya coast thin seams have been drilled in Permo-Triassic sediments, but the sediments have been downfaulted at the MacKinnon Road. In neighbouring Uganda thin seams occur at Entebbe, Dagusi Island, and Bugiri. Finally in western *Basutoland*, thin seams of coal outcrop over a distance of 80 miles, e.g., at Butha–Buthe, Lechesa, Penepena and Mohale's Hoek, where half foot seams average a calorific value of 11,300 B.T.U./lb.

LIGNITE

Small deposits of lignite are known to occur in many countries of Africa, notably in those that do not contain coal. The Tertiary deposits of Nigeria appear to be extensive. It should be noted that peat is also abundant in the moors of High Africa and in the coastal swamps.

NORTH AFRICA

In *Libya*, at Shekshuk, Jaran e Nalut, Nalut, Gesrel-Hag, Josh and Tiggi Cabao, seams and lenses of low grade lignite are found, generally interbedded with marl, carbonaceous clay and limestone. The Shek–Shuk seams, interbedded between clays and dunes, attain $1^2/_3$ ft. of thickness.

In *Algeria*, lignite accumulated at Marceau, Smendon, Boukhanéfis, Djebel Amour and Bou Maïz, and in early Cretaceous sediments of the northern Sahara.

In *Morocco* thin lenses of lignite are found in (Autunian) beds of Khenifra, in Triassic layers at Skoura, in the Lias of Sefrou and Tamda, in the Dogger of Boulemane, and in the Miocene sequence of Guercif and Rhafsaï.

WEST AFRICA

In the tidal creeks of Sierra Leone, north of Newton (near Freetown) hydrous, unfortunately non-cokable, and sulphurous lignite outcrops between clays of the Pleistocene Bullom Series. Its calorific value is not in excess of 7,900 B.T.U./lb. In Upper Volta, near Ouagadougou (the capital), 20 inches of lignite lie under a cover of 15 ft. On the coast of Dahomey near Porto Novo (the former capital), 3 ft. of lignite are found at a depth of 200 ft. and again at a depth of 600 ft., covered by vegetal clays. We now pass to the Niger Republic: 4–8 inch thick seams are located at Tanout, 1–3 ft. of lignite lie at a depth of 130 ft. at Tahoua, and at Sarou vegetal clays 20 inches thick are interbedded in common clays.

Nigeria. The 60 mile long lignite belt follows a northeasterly trend from Orlu, Umu, Ezeala and Umuahia to Oba an Nnewi, and thence across the Niger to the principal field, (Fig. 24) Ogwashi–Asaba, and to Obomkpa. The early Tertiary lignite measures of *Ogwashi-Asaba–Mgbiligba* consist of a 17 ft. thick main seam, 12 ft. of argillaceous shale and an upper seam averaging 8 ft., covered by 250 ft. of sands and clays. Along the Nnemagadi River, near *Obomkpa*, the lower of two or three lignite horizons achieves 8 ft. of thickness and a fairly good quality. Numerous other seams occur in this field. Up to 5 ft. thick seams outcrop at Okpanam, Uku Nzu, and the Atakpo and Oboshi Rivers, near Ibusa. Other occurrences include Ogwashi–Uku, Illah, Ubiaja and Ogbefu, further north. Finally, the seam of Agbos town attains 7 ft. of thickness.

Lignite occurs at Thysville in the Lower Congo. At Cabo Ledo, south of Luanda in *Angola*, limestone and lignite alternate. The warped sediments strike east and dip 20°. The thickness of numerous seams varies between 1 and 3 ft. The lignite contains 69.6% ash, 6.2% moisture, 3.3% sulphur and only 0.6% fixed carbon, but the quality of the material improves in the upper seams.

EASTERN AFRICA

Carbonaceous material occurs in the Nubian Series of the northwestern Sudan.

Ethiopia. At Adi Ugri, in Erythrea, almost 3 ft. thick lignite lenses are found in 30 ft. of Miocene marls. In the 6 square miles of the *Chelga* basin near Amar in Gondar, five 1½–3 ft. thick seams average 5.2% ash, 0.6% sulphur, and a calorific value of 8,100 B.T.U./lb. A 130 mile long belt extends along the escarpment of the Rift Valley from Waldia south to Ankober (Fig. 13, 58). The calorific value of the seamlets of the Wahi Titu, near Ukchiali, varies between 7,100 and 10,300 B.T.U./lb., but the lignites of Guonto, near Waldia, contain less sulphur. On the other hand, the Debra Brehan lignite has been covered by trap. At Mush, the calorific value of a 14 ft. seam, and of an 8 ft. thick seam, varies from 6,900 to 7,200 B.T.U./lb. Furthermore, lignite seams outcrop at Tiannu and Gherba, near Ankober, 80 miles northeast of Addis Ababa.

North of the capital, the principal outcrop of the Debra Libanos area is the Zega Wadeb. The seams of Sululta lie between trap, but the Soddu seams, 130 miles south of the capital, occur in clays. In Wallega province (western Ethiopia) a 130 mile long belt follows the valley of the Alaltu and that of the Didessa River in the southeast. Three basins have been distinguished: Jarso, Alaltu–Nejo and Badessa. The Nejo seams are separated by a clay parting. The lignites of Gute Seddo contain little sulphur and their calorific value varies between 5,600 and 9,900 B.T.U./lb., with Didessa lignite averaging 8,800 B.T.U./lb. Resting on granite, the seams and other sediments are covered by basalt.

In northern *Somalia*, Eocene lignite shows at Daban, near Berbera, and Cretaceous material occurs near Onkhor. In addition thin lignite seams are found at Durbo. As in Ethiopia, lignite may occur in the Kenya Rift Valley, near Lake Stéphanie; lignite is also interbedded with the Pleistocene clays of Mui in Kitui. In Uganda, lenses of lignite are located in the Kisegi Valley, Toro.

SOUTHERN AFRICA

Madagascar. Bituminous shale and lignite occur in the area of Sambaina, southwest of Tananarivo. In the Soamandrariny sector three seams lie under a shallow cover. A 2 ft. thick seam dips slightly. The lignite contains 25% moisture, but when dried the material consists of 33% fixed carbon, 35% volatiles, 30% ash and 2% sulphur, and achieves a calorific value of 7,200–9,000 B.T.U./lb.

In southern Cape Province, *South Africa*, 13% tar can be extracted from the lignite of Knysna. Other occurrences are located at Albertinia, Phillipi, Strandfontein, in the Uitenhaage Series of Cape Province, and on the coast of Zululand.

BITUMINOUS SEDIMENTS

In the eastern Congo, Jurassic, bituminous shales cover vast areas. Indeed, the Stanleyville deposits are the only major occurrence of hydrocarbons in interior basins. Deposits of the Angolan coast are Cretaceous, but the impregnated sandstones of Madagascar and the carbonaceous shales of South Africa are of Karroo age.

NORTH AND WEST AFRICA

In *Morocco*, Middle Cretaceous bituminous limestones average 6% oil at: *(1)* Dar caïd Meknafi and Dar cadi Zelteni, southeast of Mogador; *(2)* northeast of Taroudant, south of the High Atlas; and *(3)* at Tourtit n'Aït Argi, northeast of Tinerhir.

The marls covering the phosphates of Youssoufia also carry bitumen.

On the *Ivory Coast* the thickness of the sand horizons, which rest on older Cretaceous sediments, decreases towards the west. Therefore it is believed that they were affected by pre-Tertiary erosions, followed by the deposition of Paleocene limestones and younger sands. At Eboïnda and Ebocco lenses of bituminous sands occur in the upper part of the sequence, and the belt extends into Apollonia, southwestern *Ghana*, where bitumen seeps. On the other hand, oil shales have been found at Poasi, 4 miles southwest of Sekondi. In Nigeria, bituminous sands show east of Lagos; Lower Eocene sandstones are impregnated by bitumen at Idakur and on the Awbiaumada River in the east.

CENTRAL AFRICA

In the *Congo* bituminous shales, argillites and, less frequently, sandstones are interbedded with the sediments of the *Stanleyville* Stage, and sometimes with the Loia Stage of the Lualaba Series (Karroo). The deposits are therefore considered to be of Upper Jurassic and, possibly, of Lower Cretaceous age. Their sub-horizontal layers, which average 30 ft. of thickness, cover vast areas in the bend of the Congo River. They are mainly located between Stanleyville and Ponthierville, but also south of Ponthierville, southwest of Stanleyville (Fig. 51) on the Lomami River, and downstream from Stanleyville as far as Basoko. The shales average a density of 1,82, but only 14–16% organic matter. They contain 40.5 gallons of oil and 19.5 lb. of ammonium sulphate per ton, 2.7% fixed carbon and 64.8% mineral residues. The most extensive layer yields 6% tar, that is, 23 gallons/ton. The results of the fractional distillation of tar and oil are given in Table CXLVII.

TABLE CXLVII

OIL AND TAR CONTENTS OF CONGOLESE OIL-DISTILLATES

Degrees (°C)	*Oil (%)*		*Degrees (°C)*	*Tar (%)*
0–120	12.6	gasoline	0–200	19.6
120–240	29.5	lamp oil	200–250	15.6
240–280	24.3	oil	250–300	16.0
280–360	22.3	paraffin oil	300–370	24.0
	11	residue		24.8

The Minjaro–Mekombi layer, which outcrops 25 miles south of Stanleyville, contains oil similar to Diesel oil.

Sediments have also been impregnated by bitumen at Banningville and Mpo (Kwango). On the coast bitumen is carried in Cretaceous (Albian) sediments at Mavuma, in Senonian rocks at Vonzo, and in others at Makungu Lengi. Various limestones respectively contain 11, 16 and 37% asphalt, while sands are impregnated by 12–18% bitumen. Along the

Angolan and Congolese coasts, from Maiombe in the Cabinda enclave to Mossamedes, Cretaceous bituminous limestones and sandstones frequently outcrop. The column of up to 150 ft. rests on a basal conglomerate and Precambrian rocks. Thin beds of bituminous shale are interbedded with the basal conglomerate of the Cretaceous System. The limestones were quarried at Caxito, near Dondo 70 miles east of Luanda.

Finally at Waki, in the Albertine Rift of Uganda, oil shales are found at a depth of 2,500–3,900 ft.

MADAGASCAR

At *Bemolanga*, in the northwestern bulge of Madagascar, extensive reserves of bitumen occur in Karroo sandstones of the Isalo stage. The sandstones outcrop in a 660 ft. thick structure which rises above underlying Precambrian rocks (Fig. 48, 52). Of the sandstone column 420 ft., i.e., 62% is permeable and averages 430 millidarcys and 5.3% bitumen (1–8%). Highest grades occur at the top; the characteristics of the bitumen and the ample water content, however, remain the same in depth. The sandstones of the surface contain 5.7–15% bitumen. The composition of the derivates is given in Table CXLVIII.

TABLE CXLVIII

THE COMPOSITION OF OIL-DERIVATES AT BEMOLANGA

Substances	%
Lamp oil	25.1
Light oil	24.3
Heavy oil	21.1
Gasoline	7.2
Coke	7.8
Gas and loss	4.5

At Sambaina, 70 miles south-southwest of Tananarivo, bituminous shales are included in the Pliocene lacustrine sediments. At *Antanifotsy* lignite reserves equal those of pyroshales. The column is given in Table CXLIX.

TABLE CXLIX

THE COLUMN OF ANTANIFOTSY

ft.	*Soft rock*
3	locally, lignite
23	sediments
4–5	pyroshale
33	sediments
3-4	pyroshale
5–7	lignite
83	sediments
2	lignite
100	sediments
	shales

The oil content varies from 9.4 to 15% in the layers and sectors of the deposit. On pyrogenation the extracted oil is light and contains little paraffin. While density increases from 0.78 g/cm^3 at 150 °C to 0.89 g/cm^3 at 270 °C, the sulphur content decreases from 0.67 to 0.55%. The calorific value remains 19,000 B.T.U./lb. at these temperatures.

At Mandrosohasina, a small adjacent deposit, only a layer of 4 ft. is considered to be usable. These pyropissite shales are distinguished by a high content of montanite wax; they contain 11% oil and 48% water, but the dried product averages 41.5% volatiles, 37.6% ash, 17% moisture, and a calorific value of 4,700 B.T.U./lb. Their pyrogenation provides 52.4% semi-coke, 18% oil, 8.9% gas and 20.7% water.

SOUTH AFRICA

On carbonization torbanite, a carbonaceous shale, yields 20–100 gallons of oil and 8–50 lb. of ammonium sulphate per ton. The torbanite seams of Mooifontein, at Ermelo, underlie low grade coal. However, in the Kromhoek area the material is interbedded with sandstone, and extends to Spruitfontein and Winter Plaats where it yields 35 gallons of oil and 10 pounds of sulphur per ton. At Driehoek, near Ermelo, torbanite and coal alternate, and at Winkelhaak, near Bethal, the width of lenses varies from 1 inch to 2 ft. Other occurrences include the surroundings of Blaawkop, Schimmelhoek, Sheepmoor, Kortlaagte and Winkelhaak in the Transvaal, and the Molteno beds of Natal.

BITUMEN

The Cretaceous asphaltite of Angola is the end product of hydrocarbon diagenesis in sediments, but the South African pyrobitumens fill in fissures which traverse Karroo rocks.

Angola. Lenticular bodies of asphaltite (libollite) are mainly interbedded in the crests of sandstone warps which generally pitch 30 °W. The asphaltite laterally passes into semi-solid bitumen, and thence into bituminous sandstones of the Quilungo beds. Although asphaltite occurs between Cabinda and Mossamedes, the principal deposits are localized north of the Cuanza and Luanda in a 60 mile long and 15 mile wide (Fig. 26, 53) belt. At Calucala, bitumen outcrops over $^2/_3$ of a mile. The upper layer, which is up to 65 ft. thick, is separated from the lower level of 3 ft. by 30–50 ft. of bituminous sandstone. On the other hand, the upper layer of asphaltite includes interbeds of bituminous sandstone. The main

TABLE CL

THE COMPOSITION OF ASPHALTITES IN ANGOLA

	Calucala (%)	*Quilungo* (%)
Fixed carbon	21.4	18.7
Volatiles	51.9	43.7
Ash	25.1	35.9
Moisture	1.6	1.7

horizon of Quilungo attains a thickness of 5 ft. The black, brittle asphaltite has a density of 1.2 and a calorific value of 6,000–7,000; the composition is given in Table CL.

Finally, pitch lakes of semi-solid bitumen are located at Sassa–Zau in the Cabinda enclave.

South African pseudocoal, a substance resembling asphaltic pyrobitumen, contains 47–92% fixed carbon, 3–30% volatiles, 1–30% ash, 0.6–17% moisture; and 0.6–1.8% sulphur. When heated to 900°C one can obtain 52–88% coke, 1–6% tar, 2–24% liquor, 8–23% hydrogen, gaseous hydrocarbon, carbon dioxide and other gases. The substance fills vertical and subconcordant fissures which are 2 inches –7 ft. wide. With calcite it also fills in a sheared bed. These branching veins occur in three stages of the Middle Karroo. They are located mainly around Merweville, 200 miles east-northeast of Capetown, and also at Oude Bosch, Maclear, Uithoek and Witkop.

OIL

In the *Sahara* hydrocarbons accumulated from the seas that flooded the area between the Cambrian and the Eocene. From the Precambrian window of the Hoggar towards the Mediterranean, successively younger sediments accumulated in epicontinental zones. Consequently, successively younger oil fields appear from the southwest towards the northeast. However, irrespective of age, the traps were frequently controlled by Precambrian–Paleozoic structures. While oil accumulated between the Cretaceous and the Miocene in the narrow basins of the Atlantic coast and of Suez, very little oil seeps from the Rift Valleys in the centre of Africa.

The *Sahara* is divided into two unequal parts by the chain of the Eglab, Tanezrouft, Hoggar, Aïr and Tibesti Mountains (Fig. 124). Little is known about oil south of this line! In the Niger Republic the *Djado* basin, before the Tibesti, is Paleozoic, but the Cretaceous overlaps at *Talak*, south of the Hoggar. Tests have also been carried out at Bamako, Mali, in the Chad, and in northern Cameroon.

The northern limit of the desert is the Saharan Atlas and, further east, the Mediterranean. The Miocene Gulf of Suez oil fields are, in fact, small coastal basins and the Egyptian Sahara has yielded little oil until now. The rest of this area can be divided into three unequal

TABLE CLI

IMPORTANT OIL DISTRICTS IN NORTH AFRICA

West	*Centre*	*East*
Algeria		*Libya*
Triassic		*Eocene*
Hassi er Rmel (gas) Hassi Messaoud Oued Chebbi		Syrte
Paleozoic		*Cretaceous–Tertiary*
In Salah (gas)		Cyrenaica Coast
Mouy dir-Azzel Matti Fort Polignac Hamada el Homra		
Reggan–Tindouf Djado (Mali)	Murzuk	Kufra–Erdi (Chad)

structural units, the East, the Centre and the West, and into three stratigraphic units, the Eocene, Triassic and Paleozoic basins. This can be seen in the pattern, given in Table CLI.

Between the West and the Centre, the volcanic Tibesti alignment trends due north at El Haroudj, Djebel Soda, and in the Garian Arch where it bends northwest through Djeffara. Separating the East from the Centre, the uplift of El Biod extends a northerly outlier of the Hoggar to the Tilrhemt Arch, south of the Atlas. Finally (not shown in the table), the MacMahon basin lies west of In Salah and south of Colomb Béchar.

EASTERN LIBYA

The Paleozoic Sea covered most of Libya, but we know little about the shales and sandstones of this period since many of these beds were eroded, in eastern Tripolitania and Cyrenaica, during the emergence which lasted from the Permian to the Lower Cretaceous. Farther west continental red beds alternate with lagoonal and marine sediments of this period. Nevertheless, until the Lower Cretaceous the Atlas trend extended from Tunisia to the Cyrenaica coast, but in the shield proper the emergence of lands continued in the Mesozoic. To compensate for this movement, what became the Gulf of Syrte commenced its subsidence, possibly enhanced by the faulting of the margins of this trough (COLLEY, 1963). On the eroded Cretaceous relief clastic sediments, with thick shales and sands, were first deposited inland. However, along the coasts carbonatic shales, silts, heterogeneous sands, salt, anhydrite and lagoonal limestones alternated. Fluctuations of the coastline intermingled shales and sands with them.

During the Cenomanian downwarping gathered impetus, and in the furrows of this relief basal sandstones, evaporites and dark shales sedimented. Finally a uniform mantle of carbonates covered the entire basin.

Carbonates continued to predominate from the late Cretaceous into the Tertiary. Danian shales were followed by post-Maastrichtian limestones, with which shales were again associated in the Paleocene. This epeirogenetic subsidence was accelerated by the development of a perpendicular trough during the Paleocene–Eocene, which developed between Southern Italy and the Bay of Syrte. Reefs and lagoons bordered this transversal feature and both vertical and lateral faults structured the basin. At the same time sedimentation was controlled and even locally interrupted by these movements.

In eastern Libya the Paleocene is particularly saline, but thick interbeds of dolomite and anhydrite were deposited during the Lower Eocene. In the Middle Eocene the subsidence continued, as shown by limestones and thin seams of interbedded lignite, and even by coal. In the Upper Eocene another emergent phase was initiated, as indicated by conglomerates and sands. This emergence reached its climax in the Oligocene. Further north the coastal chain of Cyrenaica was refolded and refaulted in the Eocene, and re-emerged twice during the Oligocene and the Pliocene–Pleistocene.

In the Middle Miocene eastern Libya continued its interrupted subsidence, followed by emergence. However, in the southern Syrte sedimentation continued locally until the Quaternary. Nevertheless, sub-horizontal Eocene and Oligocene beds outcrop in flat-lying (2–3°) domes and brachoïd anticlines.

As in Algeria, the pre-basin relief strongly affects the accumulation of hydrocarbons. Sediments were first deposited in the troughs, and then over the whole pre-Cretaceous land surface. The differential compaction of these along with tectonics, further impressed the imprint of the Paleozoic–Precambrian topography. Upper Cretaceous carbonates form more favourable traps than Lower Cretaceous sandstones, although basal sandstones, and partly dolomitized calcarenites of the coast, constitute the principal reservoirs. However, recifal facies do not appear to have developed into commercial fields yet. As in the northern Sahara, many of these reservoirs are located on uplifts, structural heights, or their edges. The Samah and Waha fields are good examples of Upper Cretaceous accumulations of oil.

The shallow waters of the Paleocene, which covered the fine-grained Danian sediments, were particularly favourable to the development of hydrocarbons. Conditions of life, the sedimentation of fractured clastic limestones and their subsequent diagenesis, controlled such reservoirs as Defa and Dahra. However, less oil shows in the carbonates, anhydrites and salt of the Lower Eocene. With Nummulites and other large Foraminifera, the thick carbonates of the Middle Eocene form better traps, particularly when overlain unconformably by shales, as at Gialo (COLLEY, 1963). Finally, at depths of less than 3,000 ft. Upper Eocene and Oligocene pays are lenticular within carbonates and sands with inliers of shale.

The oil fields lie along a 300 mile long east-southeast trending arc, roughly parallel to the Bay of Syrte (Fig. 123).

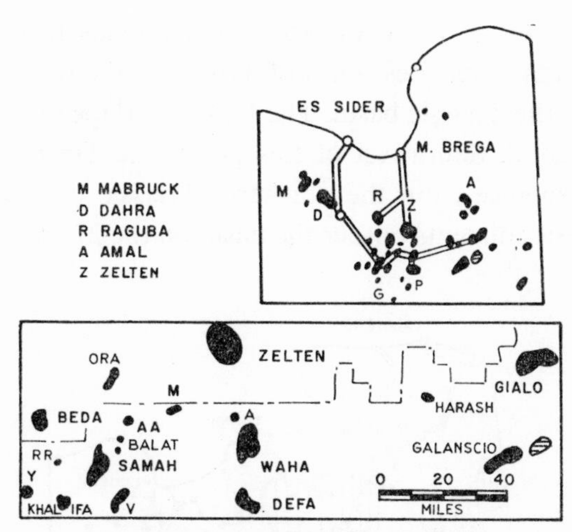

Fig. 123. The east Libyan oil fields.

At *Amal*, 120 miles east-southeast of the gulf, oil accumulated at a depth of 8,900 ft. in post-Tassilian or Cambro-Ordovician limestones. However, according to other sources the reservoir would be in Paleozoic quartzites and sand, surrounding a Precambrian dome or knob, 30 miles in diameter. The thickness of the reservoir beds attains 400 ft., with the output reaching 1,200 barrels per day of (34° API) oil. South of Amal, oil shows in fissured granites of El Rakb (A.1.12) at a depth of almost 10,000 ft. In the *Zelten* field, 120 miles south of Marsa el Brega on the Syrte coast, 300 ft. thick, sub-reef, Paleocene limestones yield 190,000 barrels per day from a depth of 5,500 ft. However, a Cretaceous sandstone reservoir lies lower. There is hardly any sulphur in the medium oil. West of Zelten at Raguba, the Raguba section consists of 150 ft. thick, detrital porous limestones. However, the Gargaf section consists of 200 ft. thick, fractured quartzitic sandstones.

While the Defa field is 60 miles south of Zelten, Beda is at the same distance to the south-southwest. At Beda the Paleocene reservoir is 90 ft. thick, and yields 5,000 barrels per day from a depth of 5,100 ft. Similarly, at Defa Paleocene limestones and dolomites yield oil. Between Beda and Defa, the Samah fields are block-faulted. A broad north striking horst yields oil, mainly from fractured quartzites, from a depth of 6,300 ft. This zone measures 9 by 4 miles. In the southeastern, or V segment of Samah oil flows from 7,200 ft. from a 2 by 2 ½ mile area. Similarly, at Waha (north of Defa) and further north at Jebel Pl-6 (15 miles south of Zelten), the productive level is Upper Cretaceous. At Gialo, east of Zelten, Eocene limestones of the 2,700 ft. level, and even Oligocene sediments, constitute reservoirs. These trends continue 100 miles south-southeast to the C 6 and C 7-65 fields. These are considered to be one of the largest fields of Libya. A production of 600.000 barrels is to be exported through Tobruk. The northwesterly Hofra-Mabruck (Mabruk) trend is parallel to the coast, 110 miles inland. Hofra yields 2,000 barrels per day of 43° API oil. On the other hand, *Dahra* includes a Paleocene limestone and a minor Upper Cretaceous reservoir. The pre-Paleocene *Mabruck* reservoir is at a depth of 5,700 ft., and has a yield that attains 3,500 barrels per day. Finally, the *Bahi* structure is Upper Cretaceous.

THE PALEOZOIC PROVINCES OF ALGERIA AND LIBYA

The east–west trending belt of Paleozoic sediments extends from Egypt to the meridian of Greenwich. The Garian Arch rises south of Tripoli in the triangle of the Eocene basin, the Paleozoic basin and the Triassic basin. The latter overlaps them on the triple border of Libya, Tunisia and Algeria. Southwest of this point, the Tinrhert and Tademaït arches divide the Paleozoic province from the northern, Triassic basin. North of the Hoggar, Upper Precambrian metasediments border the basin on the 27° parallel (Fig. 124).

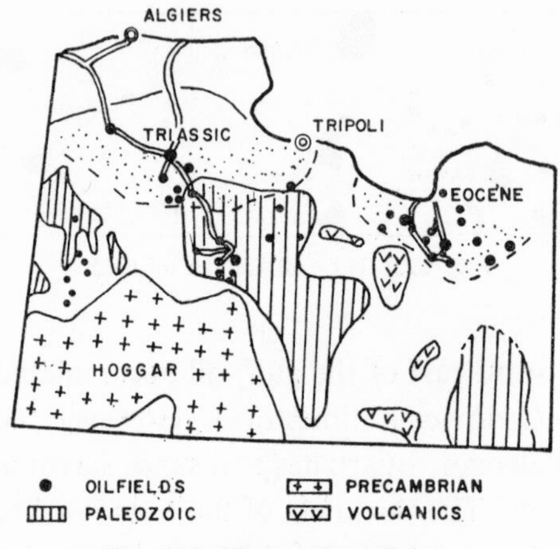

Fig. 124. The northern Sahara.

Geologic history

The thickness of *Cambrian*–Ordovician strata increases from 100 ft. in the west to 120–1,500 ft. in the Fort Polignac basin, but at Edjeleh–Zarzaïtine the thickness of the column is reduced to 300 ft., and it diminishes further along the Tiguentourine–Reculée arch to 150 ft. As the sedimentation of these formations was most regular in Libya, the thickness of the column attains 2,000 ft. in the Jebel Fezzan. Since unconformities and fissures are prevalent in the Cambrian–Ordovician of the Fort Polignac basin, Cambrian sandstones generally constitute better reservoirs than the overlying, more compact strata.

In the *Silurian*, a sequence of up to 1,500 ft., interbedded sandstone lenses form better reservoirs than do radioactive Graptholite shales and inliers of marl. However, part of the Silurian was eroded during the *Devonian*. In the Fort Polignac basin, the thickness of the Lower Devonian increases from 150 ft. in the east to 1,400 ft. in the west, where depressions surround and control reservoirs. Reflecting isostatic movements, the Devonian seas overlap the Paleozoic of western Libya. Favourable reservoirs are found in the calcareous and argillaceous Middle Devonian coral reefs, and also in sandstone inliers of the Tanezrouft. In the southeastern part of the Fort Polignac basin, the Upper Devonian (Khrenig) sandstones thin out; however, the thickness of the terminal reservoir level remains at 100–150 ft. The Djebel Fezzan–Tihemboka arches emerged during the late Devonian Brittany Phase. The Tournaisian overlaps the Paleozoic; this unconformity disappears further north however, where Hercynian movements prevail.

The 3,500 ft. thick *Carboniferous* shales and dolomites of southern Libya include inliers of anhydrite. The Visean lies in conformity on the Devonian here and marls and the Moscovian Stage terminate the sequence in the north.

The Carboniferous of the Fort Polignac basin consists of a thick column of shales which includes sandy inliers and rare carbonate lenses. The Issendjel Series, including two horizons of sandstone reservoirs (the lower or 'D' and the later Issaouane), generally attains a thickness of 1,500 ft. The Oued Asse–Kaifaf Series (including reservoirs of medium value) commences with a *Collenia* horizon, followed by *Archaediscus*. This series lies as low as 700 ft. above the Edjeleh field, but its thickness increases towards the west. Conversely, the thickness of the El Adeb Larache Series of *Profusinella* limestones, marls and anhydrite augments in the opposite direction. In Algeria, the Tiguentourine Series of red shales and anhydrite is contemporaneous with the latest Carboniferous (continental) sandstones of Libya, which are products of the Hercynian epeirogeny of Tripolitania. Intensity of late Carboniferous erosion diminishes from the northeast towards the southwest.

Various sediments represent the *Jurassic*–early Cretaceous continental climate, followed by the shallow seas of the Hamada el Homra in Libya. Finally, Tertiary sandstones and limestones occasionally outcrop.

The west Libyan oil fields

The basins of Tripolitania and northwestern Fezzan lie along a south-southwest trending, 400 mile long broad arc, which includes Bir Tlacsin, Bir el Rhezeil, Emgayet (east of the arc), Wadi Tahara, Atshan and El Haghe. However, the Oued Chebbi gas field, the northernmost part of the alignment, belongs to the northern Triassic basin.

At Bir Tlacsin, 120 air miles south-southwest of Tripoli, light petroleum (s.g. 0.8) has accumulated at a depth of 8,200 ft. in a 150 ft. thick sandstone lens of Upper Silurian shales. The output is 1,200 barrels per day. At Bir el Rhezeil, 120 miles southwest of Tlacsin, 800–1,500 barrels of light oil and gas flow per day from a supposedly Strunian (latest Devonian), 40 ft. thick and 5,000 ft. deep sandstone bed.

The reservoir of the Emgayet field, 70 miles east-southeast of Rhezeil, appears to be at least partly Paleozoic, possibly Upper Silurian, *Acacus* sandstone. The daily flow attains 1,500 barrels from a depth of 4,000 ft. At Wadi Tahara, halfway between the Fort Polignac and Bir el Rhezeil basins and similar to them, oil shows in either the Carboniferous or the Strunian sandstone. The Strunian lies at a depth of 4,200 ft. At Atshan, almost 500 miles south-southwest of Tripoli and 60 miles east-southeast of Tiguentourine, a lens of Lower Devonian sandstones yields 900 barrels per day of light oil (s.g. 0.8) from a depth of 2,000 ft. The lens is wedged and thins out at the top. At El Haghe, 40 miles south of Atshan, 400 barrels of oil per day flow from Ordovician beds, which lie at a depth of 2,000 ft.

Fort Polignac Basin

At *Tam Emellel*, on the Libyan border, oil shows in the Lower Devonian (at 3100 ft.) and in the Ordovician. The northeastern part of the Polignac basin has been intensely faulted by the Mesozoic and post-Jurassic rejuvenation of Paleozoic movements. In this sector the faulted *Ohanet* anticline trends north–south on the Tinrhert Plateau. At Ohanet, several periods of erosion have been identified in the Devonian and Cambrian, and oil accumulates in Lower Devonian sandstones at a depth of 7,500 ft. On the surface this structure measures 13 by 2 miles.

The structure of the southern fields is very different from the flat-lying basins of Hassi Messaoud and Hassi er Rmel. Both areas, however, share the subsequent imprint of Paleozoic tectonics. The *Polignac* monocline pitches towards the north. The faulted, Devonian, Djebel Assaoui Mellène anticlinorium borders the basin in the west, and steep dipping faulted anticlines constitute its eastern limit, which coincides with the Libyan frontier. Lower Paleozoic, epeirogenetic movements were followed by north–south trending Devonian warping there. The Hercynian orogeny completed the structuring of the basin (Fig. 125).

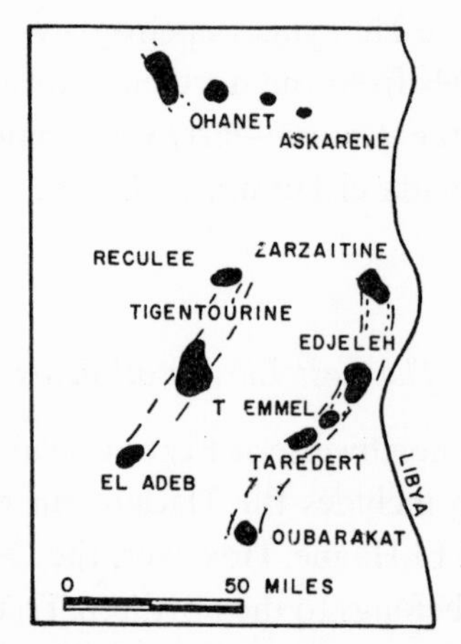

Fig. 125. The Fort Polignac oil fields.

The Upper Devonian is the most favourable reservoir in the Edjeleh field, but the early stages of the Issendjel and Oued Assékaïfaf Series are also productive. Larger fields are in the eastern fifth of the basin. They can be divided into two zones: *(1)* the north-northeast trending Edjeleh zone, on the Libyan border, includes Zarzaïtine, Arène, Edjeleh, Ouan Taredert, Tin Essameid, Oued Tissit, Oubarakat Daia and Ibekrane; *(2)* the northeast striking Tiguentourine belt includes Taouratine, Reculée, El Adeb Larache and Assekaifaf.

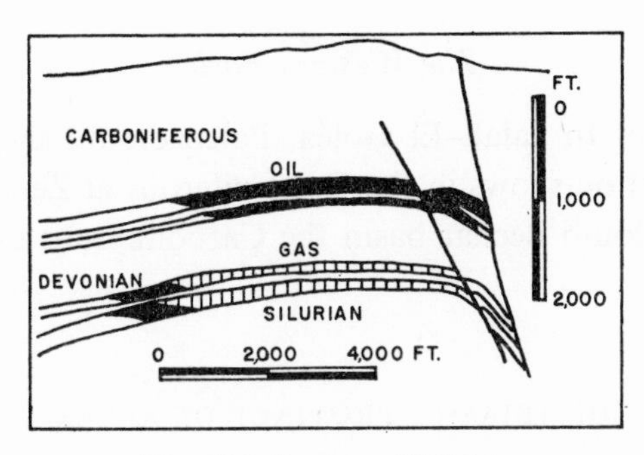

Fig. 126. Section of Edjeleh.

A northeast trending fault and twin northwest striking faults constitute the triangle shaped trap of the *Zarzaïtine* Series. The principal oil level is Silurian. The Devonian is estimated to contain an oil reserve of 1,600–1,900 million barrels between 4,000 and 5,000 ft. The surface dimensions of Zarzaïtine attain 9 by 5 miles. The Ouan Taredjeli extension-nose is trapped similarly by a fault.

The *Edjeleh* anticline is oriented north–south. The axes of the El Adeb Larache Series ellipse are 20 miles and 4 miles long. A north–south trending fault thrust down two Carboniferous oil levels and, further east, a sub-vertical fault constitutes the trap. The pitch is generally gentle. Although gas accumulates in two Devonian horizons, and oil accumulates in the west and in the Ordovician, the principal reservoir is Carboniferous. It includes a reserve of 1,000 million barrels at a depth of 1,000 ft. (Fig. 126). Oil accumulates as far as 2,600 ft. of depth.

The open, northwest trending Tin Essameid anticline, in the lower Erg Tan Othman series (Devonian), is 4 by 2 miles long. The axis of the northwest striking Djebel Ouan Taradert has been displaced towards the north. The gas bearing anticline is 6 miles long and 1½ miles wide; Ordovician oil lies at a depth of 3,500 ft. In the 5 by 1 mile Oued Oubarakat structure, in the lower Erg Tan Othman Series, Devonian oil is found at a depth of 2,400 ft. A west-northwest trending fault thrust down the southern part of this field. In a similar environment, Devonian oil rises from a depth of 1,800 ft. from the 3 by $1^1/_4$ mile Ikebrane structure of the Assékaïfaf Series.

The 10 by 6 mile *Reculée* (Receded) structure underlies the Zarzaïtine pericline in an embayment of this Series. In this reservoir Devonian oil was tapped at 4,900 ft., above Ordovician gas. The 5 by 3 mile *Tiguentourine* polyhedron is bordered *(a)* on three sides by anhydrite of the synonymous series, *(b)* in the south partly by marls, and *(c)* roofed by the top of the El Adeb Larache Series. Oil was found in two Devonian levels, but the main reservoir is Carboniferous; their depths are 1,640 and 2,600–4,100 ft.

Whilst Devonian oil of the 6 by 2 mile flat-lying *El Adeb Larache* anticline is hardly 50 ft. under the surface, gas accumulated at a depth of 1,600 ft., with oil accumulating as deep as 4,100 ft.

At Askarene the reservoir rocks belong to the Lower Devonian at a depth of 7,500 ft., but at Guelta, at similar depths, the oil level is Silurian. Finally, the Tin Fouye field is isolated west of the principal fields of the Polignac basin. The surface dimensions of this structure attain 6 by 3 miles. Lower Devonian sediments yield oil at a depth of 4,200 ft.

The Western Basins

In the long gas district of In Salah–El Goléa, Paleozoic oil shows locally. While in the Reggan basin (southwest) oil shows in the Carboniferous at Zegmir, and in the Devonian at Tanezrouft, in the Colomb Béchar basin the Carboniferous extends from Zousfana to Ben Zireg.

THE TRIASSIC PROVINCE OF ALGERIA

Dividing the Northern Sahara into an eastern and western basin, Paleozoic strata form a north–south trending ridge along the longitude of Algiers. A 3,000–6,000 ft. thick column of early Paleozoic sandy shales, siltstones and quartzites, including inliers of microconglomerate, has sedimented in most of the area (ORTYNSKI et al., 1959), but the southeastern part of both basins is covered by Erg dunes (Fig. 127).

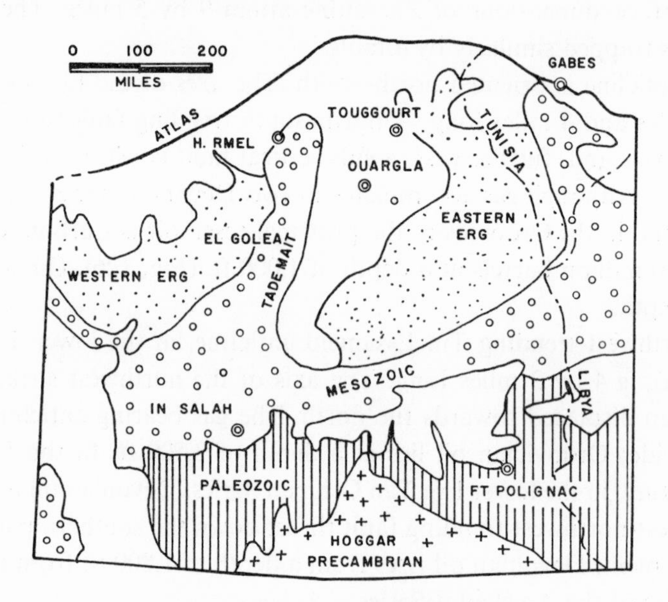

Fig. 127. The Algerian Sahara.

Geologic history

Immersion, transgression and, in other areas, erosion mirror the Taconian and Caledonian epeirogenies. Fine-grained Silurian sediments consist of black carbonaceous, calcareous and bituminous shales; they are radioactive at their base, as in the Paleozoic belt. Graptolites and orthoceres distinguish the Upper and Lower Silurian. Silurian sandstones are overlain by Lower *Devonian* (Siegenian) calcareous sandstones, siltstones and Bryozoa, Brachyopoda carrying lumachelles. Emsian (Middle Devonian) and Frasnian sediments include transgressive shales and siltstones. During the Famennian (Upper Devonian), southern and western Tripolitania having emerged, the Brittany phase of the Hercynian epeirogeny left its imprint on the whole region. Late *Carboniferous* sediments lie unconformably on the early Paleozoic here. Tournaisian transgression continued in certain areas until the Viséan. However, during the Moscovian Stage (Middle Carboniferous), following the last Carboniferous epeirogeny of southern Tunisia, the east–westerly direction of transgression was reversed. Whereas this area formed a Permian trough (like the Sicilian oilfields) the Tilrhemt Arch, south of the Atlas, emerged in the northeast as a complex Cambrian–Silurian arc and in the west as a Devonian coastline.

Expanding from the Permian synorogeny during the Triassic, sediments transgressed the northeastern Sahara and northern Tripolitania. However, the stratigraphy of the east and the west differ. The column includes transgressive conglomerates, sandstones and shales, distinguished by lateral pinching and changes of facies. They are overlain by a thicker column of dolomite shales, rock salt and anhydrite, constituting an ideal reservoir at Hassi Messaoud. During the subsidence of the eastern areas, tabular dolerite flows covered the western basin. Lower Jurassic, marine Lias sediments are overlain by Middle Jurassic sandstones. The continental, early Cretaceous is distinguished by shallow lignite lenses. While Middle Cretaceous (Cenomanian and Turonian) carbonates were overlain by lagoonal sediments in the east, the west suffered erosion.

The *western basin*, connecting with the Tindouf syncline (see 'Iron' chapter), is bordered by the Saharan Atlas, the central 'backbone' and the Eglab. Carboniferous sedimentation produced the Colomb–Béchar coal field. Several warps traverse this province but only its northern, Triassic sub-basin, bordered by the El Biod and Tilrhemt Arches, the Atlas, and the Great Western Erg, will be considered here. At Hassi er Rmel, considerable quantities of oil are mixed with gas and at Tilrhemt, further north, oil shows in Carboniferous beds. Moreover, oil also shows at Belkateief, southwest of Hassi er Rmel.

The *eastern basin* is surrounded by the central 'backbone' of the Chebka el Mzab, the Atlas bordering the African continent, the southern Tunisian shield, the Garian Arch and the Eastern Great Erg. The major subsidence of this province only started in the Mesozoic, when 10,000–12,000 ft. of sediments accumulated in it. In the middle of this province, a north-northeast trending structure (extending one of the Precambrian outliers of the Hoggar) includes the Hassi Messaoud field. Near Hassi Messaoud, the trend bends towards the east, i.e., in the direction of the Saharan Atlas. Various warps underlie the basin's eastern corner.

The genesis of both the Hassi er Rmel gas field, one of the largest in the world, and the oil field of Hassi Messaoud show the imprint of: *(1)* the unconformity, dividing the Paleozoic from the Mesozoic, *(2)* recurrent Mesozoic peneplanation, facilitating the concentration of hydrocarbons, and *(3)* compact covers of salt and anhydrite.

Hassi Messaoud district

The flat-lying ellipse of Cambrian–Ordovician sandstones occupies 700 square miles. A northeast trending fault borders the north-northeast striking oval structure in the west, and a parallel thrust cuts its eastern third. The northern part of the oval has a harmonious contour, but its southern border undulates dissymetrically. The top of the roof is situated east of the centre, and a secondary maximum or step lies west of the roof (Fig. 128).

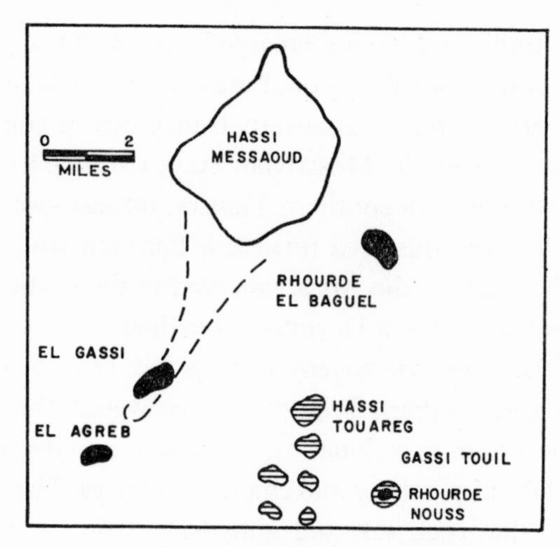

Fig. 128. The Hassi Messaoud district.

Towards the north, the Paleozoic Amguid arch has been eroded to the depth of the Cambrian. Indeed, these highlands remained emerged until the Upper Cretaceous. The sandstones are overlain by discordant shales, sandstones and Triassic salt. The borders of this very flat ellipse subsided during the Mesozoic, and consequently the Paleozoic pitches of as much as *one* degree diminished to a few tens of seconds. The 400 ft. thick Triassic sandstone bed yields light oil at a depth of 11,000 ft., but Cambrian sandstones also yield oil. The productive area extends over 250,000 acres (400 sq. miles).

At El Gassi, 50 miles south of Hassi Messaoud, Cambrian sandstones show oil. At *El Agreb*, 20 miles southwest, the same beds yield oil at a depth of 10,500 ft. However, north of Hassi Messaoud watei invades the Carboniferous Bordj Nili reservoir. A major field, *Rhourde el Baguel*, is situated east of Hassi Messaoud. Oil flows here from fractured Cambrian sandstones found at a depth of 9,100 and 12,000 ft. Conversely, south of these two fields, at Gassi Touil, two Triassic sandstone levels at 4,300–4,700 and 6,400–6,700 ft. form the reservoir.

NORTHERN ALGERIA AND MOROCCO

The *Oued Guétérini* reservoir is located in the Atlas (Fig. 56), 70 miles southeast of Algiers, in a Triassic–Lutetian (Eocene) allochtonous écaille of the South Tell nappe. The complexly folded stratigraphic column includes Upper Cretaceous (Maestrichtian) and Paleocene marls, overlain by the indicator horizon of Eocene limestone and silex. This sequence is transgressed by Miocene sandstones, glauconites and marls. Oil mainly accumulates in fissures of the limestone. The principal (Lower Eocene) Ypresian oil and gas reservoir covers 5 square miles, but other reservoirs of the same epoch remain insignificant. On the other hand, three productive Lutetian (Middle Eocene) lenses are interbedded in the marl/lumachelle interval. At Tliaouanet, in the Chélif basin, a small quantity of oil accumulated in Miocene sands overlying folded Cretaceous–Nummulitic anticlines. Finally, oil is also found at a depth of 3,600 ft. in Coniacian sediments, in the vicinity of the *Djebel Onk* phosphate–iron deposit.

The Pre-Rif Rharb depression of Morocco lies before the front of the Moroccan Atlas and Tableland, and the *Oued Beth* group of fields is located 60 air miles east of Rabat (Fig. 59). The Tselfat anticline has been affected by the diapyrism of its anhydrite core, and the Bou Draa anticline has been intensely faulted. In both fields most of the oil accumulated in jointed, Lower Jurassic (Toarcian) limestones, but at Tselfat some oil has been extracted from somewhat older Domerian sandy limestones, and at Bou Draa also from younger Aalenian limestones. The fissures of the block faulted monoclines of Aïn Hamra have been invaded by anhydrite, and oil has concentrated in Tortonian (Middle Miocene) sands and marls. This group of reservoirs is only 300–1,000 ft. below the surface.

The reservoir rocks of the Paleozoic basin are at a depth of 4,000–7,000 ft. They consist of chloritic or siliceous shaly quartzose and fissured and cavernous sandstones, partly filled by granitic marl. Block faults and unconformities of molassic marls are the traps of the Sidi Fili reservoir.

The Paleozoic horst of the Bâton field is bordered by Permo-Triassic and covered by Miocene sediments. The Oued Mellah field corresponds to a Paleozoic warp; however, some oil also migrated into calcareous Miocene sandstones (AGARD et al., 1952). Oil has also been trapped by faults in the Mers el Kharez and Tisserand reservoirs.

The most important reservoirs, the *Bleds Khatara* and *Eddoum*, are Triassic arkose sands. While Bled Eddoum also has a large Lower Paleozoic reservoir, the lower strata yield little oil at the Bled Khatara. At Sidi Fili oil has accumulated in both layers, but at Zrar it is only found in the Trias, and at *Bled el Defaa* it is found exclusively in the Paleozoic. The latest discovery is *Haricha*, but Aïn Hamra south of Tangier has been exhausted.

Southern Morocco contains other suitable basins at *Sous*, Essaouia–Mogador and Tarfaya. Sidi Rhalem, a major field, is located 16 miles southwest of Kechoula. A 100 ft. thick layer of faulted Argovian (Jurassic) limestone forms a reservoir here, at a depth of 5,600 ft.

THE WEST AFRICAN COAST

The width of the coastal basin varies considerably. It is 120 miles on the Moroccan border, 10 miles at Villa Cisneros, 20 miles at Nouakchott and 150 miles at Dakar. Precambrian rocks extend to the vicinity of the coast between the Portuguese Guinea/Guinea Republic border and Dahomey. The Paleozoic sedimentary column of the northern part of the *Spanish Sahara*, southeast of Cape Juby (and the Canary Islands), is believed to be of considerable thickness (RIOS, as quoted by HEDBERG, 1961), but not very favourable to oil concentration. Lying unconformably on pre-Triassic rocks in the south, Cretaceous, Eocene and Holocene sediments of the Hamada formations appear to be 6,000–15,000 ft. thick.

The Mauritanian coast links Rio, de Oro with the Senegalese basin, an open monocline. Precambrian tectonics and morphology appear to control the western pitch of the structure as well as subsequent sedimentation. After the Paleozoic erosion Jurassic marine transgression, Cretaceous subsidence and oil concentration followed successively. The thickness and the age of the strata diminish inland.

In *Senegal*, the sequence commences with Upper Jurassic, Kimmeridgian and Portlandian, *Iberina lusitanica* and *Pseudocyclammina jaccardi* carrying limestones. They are overlain by Lower Cretaceous (Valanginian–Aptian) clastic sediments, characterized by *Choffatella decipiens* near the coast and by *Pseudocyclammina hedbergi* inland. While sands are coarse-grained in the east, fine-grained sands of the west have been reworked. *Orbito-*

lina texana carrying Lower Cretaceous (Albian) siltstones, micaceous shales and sandstones, including carbonatic lenses, are roughly 5,000 ft. thick. In the overlying 6,000–11,000 ft. thick Cretaceous, Cenomanian–Maestrichtian green and black shales, which are accompanied by sand and siltstones, oil and gas appear (TENAILLE et al., 1960). The Paleo–Eocene cover of coastal shales, inland marls and limestones is 1,500 ft. thick. On the other hand, southeast of Kaolack the Basement is found at 10,800 ft. and at Koungheul at only 2,900 ft. A favourable alignment appears to be Rufisque–Diam–Nidé: Upper Cretaceous sands of the productive Diam–Nidé 4 lens, 25 miles northeast of Dakar, are at a depth of 3,340 ft. (Fig. 15).

The Senegalese basin extends into Gambia, Casamance and Portuguese Guinea, i.e., to Brikama 1 southwest of Bathurst, Sarakunda, and northwest of Bissau, where the Paleozoic lies at a depth of 3,200–5,700 ft. Following the narrow coastal strip of western Guinea, the basin expands in the Ivory Coast and Ghana, stretching 280 miles from Fresco to Axim, and approximately 30 miles inland, as far as a major fault. Oil seeps on both sides of the border. The Basement may lie as deep as 18,000 ft.; Jurassic–Cretaceous clastic sedimentation was followed by intense subsidence, as shown by up to 8,000 ft. of marine (Lower Cretaceous) Albian–Aptian shales and arenaceous sediments. Subsequently, in the west coarser Middle Cretaceous sediments were covered by Upper Cretaceous shales, a rock which also composes most of the Paleo–Eocene. Plio–Miocene sands and clays close the sequence.

In *Nigeria*, the alignment of oil fields trends east–west from Oloibiri, east of the Niger delta, towards the triangle of Ebubu, Afam, Korokoru and Bomu, east of Port Harcourt. The fields are 25 miles inland. The most important is *Bomu*, followed by *Imo River*. Other fields include Afam, Oloibiri, Ebubu, Apara, Elelenwa, Oza, Korokoro, Umuechem 6 (or Igrita), Ughelli, Kokori, Krakama, Nun River, Eriemu and others.

GABON

Some oil shows near Douala, in Cameroon, but little is known about the potential of the thin sedimentary strip of Rio Muni. The Basement recedes south of the border, expanding somewhat in the Libreville basin. A Precambrian horst separates it from the lozenge shaped Gabon basin which extends for 180 miles, with its greatest width attaining 70 miles. The oil fields are located above the series of salt ridges which cover the southwestern part of the basin, i.e., Mandji Island (250 square miles), between branches of the Ogoué River and the sea. The structures trend south-southeast for 25 miles from the Cap Lopez field through Pointe Clairette and the Tchengué gas field, to M'béga, Ozouri and Animba, with the small Aléwana field lying east of Tchengué (PEGAND and REYRE, 1959; Fig. 25, 53, 129).

Jurassic (cf. Karroo) sediments overlie the Precambrian. Post-Neocomian (Lower Cretaceous) tectonic movements were followed by largely clastic sedimentation, and later by marine transgression and the precipitation of a 2,500 ft. thick salt layer. Salient features of post-Aptian stratigraphy are: *(1)* Albian marls, *(2)* 1,000–4,000 ft. thick Cenomano-Senonian clays divided by a Turonian (Middle Cretaceous) unconformity, and *(3)* 2,000–5,000 ft. of Paleo–Miocene clays and terminal sands.

Tectonic movements have concentrated the original salt layers into slabs or ridges ½–2 miles long and almost as high. Almost 100 faults and folds are known, with curved ridges and minor salt bodies lying offshore.

In the *Pointe Clairette* and *Cap Lopez* fields, the salt ridges started their upward drift during the Upper Albian, and this movement was accelerated in Cenomanian times. Marine sands were heaped before the salt ridges, thus forming the reservoirs of the four oil horizons of Clairette. They are covered by clays, dolomites and silts. Oil of the lower Clairette

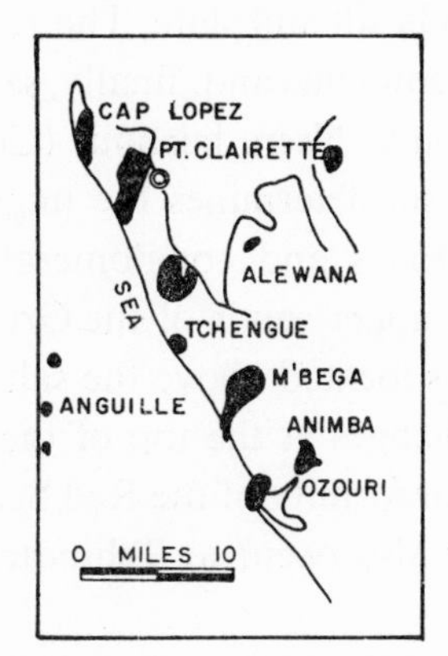

Fig. 129. The Gabon oil fields.

horizon is similar to Cap Lopez oil, but the upper Clairette horizon was impregnated during the middle of the Port Gentil phase, and generally affected by later tectonic movements. Some oil migrated into the Ikando dolomites and finally, after the Animba discordance, the fissured Eocene clays and siltstones were saturated. Most of the Cap Lopez oil permeated fractured silicified Eocene shales. On the other hand, the Tchengué Océan horizon is Lower Senonian.

Part of the M'béga crest of clays was uplifted, and later cut by a Miocene fault. Oil accumulated in several Eocene–Miocene phases, but at *Aléwana* oil was trapped in a small Oligocene graben. The two southern fields of the peninsula, Animba and Ozouri, share the same salt ridge. Both have suffered several phases of erosion and dislocation. Whilst the Ozouri concentration is post-Miocene, the age of Animba remains ambiguous. The older, inland fields overlie the broad Fernan–Vaz horst and much of the red, fluviatile Cenomanian (Lower Cretaceous) Series. At Rembo Kotto, four oil horizons correspond to argillaceous silts, divided by Coquina beds and overlain by Turonian limestones. At some distance, the Upper Cretaceous *Gongo* field is smaller.

At Illigoué, at a depth of 7,900 ft., 23 ft. of Pointe Clairette sands form the reservoir. The sands lie in a block faulted structure which overlies a salt slab.

Offshore, 650 ft. of Lower Senonian detrital sediments constitute the Anguille reservoir at a depth of 7,900 ft. Further south, at Espadon and Dorade, some oil also shows above salt slabs. Similar strata continue for as far as 100 miles south-southeast.

CONGO REPUBLIC AND ANGOLA

The sedimentary strip narrows at the Gabon–Congo Republic border, widens to a maximum of 50 miles at the Congo estuary, contracts completely between Ambrizete and Ambriz in the Congo province of Angola, and finally expands into the oval-shaped Cuanza basin. This basin is 150 miles long and 60 miles wide.

The Precambrian is overlain by 900 ft. of sandstone between the Kouilou and the Congo. Above a layer of black slates follow 1,000 ft. of clastic, biotite and chlorite bearing sediments. Overlying them, 3,000 ft. of dark coloured shales include inliers of continental sediments, which correspond to the Gabon Cocobeach formation. The alternating stratified potash and halite salt series, accompanied by some anhydrite, at the upper levels commences above 300 ft. of green, Ostracoda silt and slate. The 1,500–1,800 ft. thick red cover consists of carbonatic rocks, marine sediments and, finally, sands (PEGAND and REYRE, 1959).

At *Pointe-Indienne*, north of Pointe Noire harbour (Congo Republic), the relief of the Basement rather than salt diapyrism determines the migration of gas and oil. The 15 ft. thick, coarse lower Chela sandstones and conglomerates contain some hydreous gas; however, the limestones, silts and upper sands of the Green pre-salt, silt-shale Series form the main reservoir. The reservoir is located above the salt and best developed in the upper reaches. It disappears by facies changes at the top of the Green Series. A fault limits the field in the northeast, and the unconformity of the Red Series defines the trap. At some distance, the lower oil bearing sands also occur at Tshibote (Fig. 53; MOODY and PARSONS, 1963).

In the *Cabinda* enclave (i.e., between the two Congos), oil mainly shows in the pre-salt Chela sands. Similar formations are believed to extend to the *Congo* estuary (Congo), where Cretaceous oil-bearing sediments outcrop east of Matadi. Oil shows here at Lindu.

The *Angolan* oil fields of Cacuaco, Luanda, Benfica, Bom Jesus and Catete lie between the bend of the Cuanza River and the sea (Fig. 130). However, Galinda is located south of

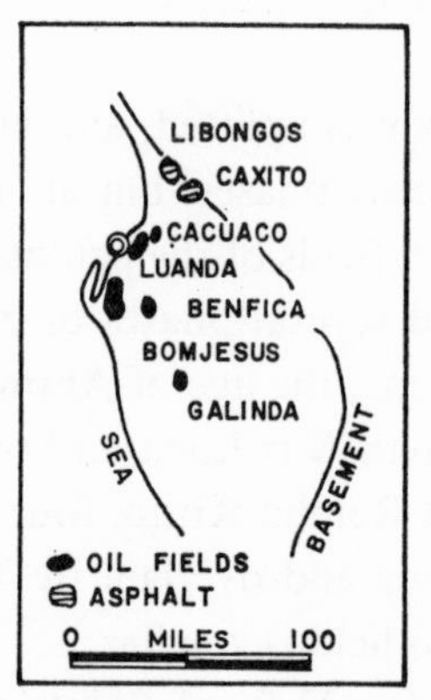

Fig. 130. The oil fields and asphalt deposits of Angola.

the river. In the central part of the Cuanza basin sediments lie on a folded and faulted Basement, which affected their folding patterns. Lower, or pre-Aptian clastic sediments, including carbonatic cement, lie along the coast. Inland the marine facies of the Cuvo reservoir attains a thickness of 1,000 ft. consisting of rock and potash salts, with clay inliers. At Benfica, oil permeated probably Upper Aptian (Lower Cretaceous), fissured, dolomitized limestones, which were covered by 8,000 ft. of alternating lagoonal anhydrite and dolomite during the Lower Albian. Finally, fissured Upper Albian Laguna limestones, with the Cabo Ledo reservoir, constitute the lowest layer of the Oligocene–Albian column of marls. They are interrupted by thin sandstone and limestone beds. The largest reservoir, Tobias is at a depth of 2,500 ft.; it yields 14,000 barrels of 37.1 API oil per day.

We will now consider the tectonics of the field: west of the Basement, a monoclinal border zone is overlain by tightly folded structures with parallel north-northwest trending axes. Sediments have been faulted in the whole Cuanza basin; throws are as deep as 3,000 ft.! Since sub-salt, i.e., lower sediments, have been less affected than upper layers, geologists believe that salt diapyrism transferred small scale but deep seated movements, magnifying them in the large disturbances of the upper strata. Conversely, in the Galinda anticline of the southern, Quissama part of the basin, only folding took place and salt movements were moderate. Since Angolan oil has accumulated in the sands (cf., Chela of Congo) *beneath* the salt, the best reservoirs are related to diastrophic diapyrism. Indeed, under prolonged compression a diapyric ridge pierced the dolomites of the Cabo Ledo anticline. Salt of some of the elongated diapyrs outcrops, but where salt diapyrism is intensive or dynamic the upper lying strata crumble, e.g., in the Morro de Tuenza horst.

Further south, the Walvis Bay basin of Southwest Africa is believed to be favourable to oil accumulation.

EGYPT

The Egyptian oil fields are located on both sides of the Gulf of Suez, and the principal fields face each other halfway along the gulf. In the Sinai, which is out of the scope of this work, the Sudr and *Asl* fields of Eocene–Miocene age cluster 30 miles south of Suez; Abu Rudeis, Sidri, Wadi Feiran and Belayim, all of Miocene age, and Abu Durba (in Nubian sandstones) trend roughly north–south. The new Belayim finds of Middle Miocene age are the most important oil field of Egypt. At coastal Belayim, the top of the reservoir is at a depth of 5,150 ft., at marine Belayim, its bottom is at 9,400 ft.

On the African coast the Rahmi, Ras Bakr, Ras Gharib and K'reim fields trend north-northwest. The *Ras Gharib* field has reservoirs both in the Carboniferous and in the Miocene, and it has produced more than half of Egypt's crude oil. On the other hand, the new finds of Ras Bakr and K'reim, on both sides of Gharib, has both Cretaceous and basal Miocene reservoirs. Finally, the distant Gemsa (cf., 'Sulphur') and Hurghada fields of Middle Miocene age are at the entrance of the Gulf (Fig. 12, 58).

The Suez Gulf coast can be divided into five units: *(1)* the Great Plain extending from the Red Sea chain to *(2)* the Sufr limestone range, *(3)* the granitic, volcanic, Esh Mellaha Hills, *(4)* the Coastal Plain including oil basins, and *(5)* the Jebel Zeit and its insular extension.

Overlying 3,000–6,000 ft. of Miocene gypsum, Plio–Miocene sandstone, limestones and oyster beds are followed by Pliocene–Pleistocene beds of *Pecten*, and covered by late coral reefs of the Red Sea fauna. The two seas were apparently connected during part of the Miocene anhydrite precipitation. Similarly, in the Pliocene the Red Sea inundated the area as did the Mediterranean later. At Jebel Abu Gerfan, a nose of Nubian (Cretaceous) sandstone appears to form the core of an essentially Miocene warp of *Pecten* limestones and shales. Overlying coral reefs, with slightly dipping anhydrite, form the trap. Dolomite concludes the sequence.

EASTERN AFRICA

In *Sudan*, areas believed to be favourable include Port Sudan, the Suakin Islands and Muhammas Qol. In addition, there are the areas of the Erythrean Red Sea coast, the generalized area of the Somali coast, the Kenya coast and Lamu embayment near Mombasa, the Ruvuma on the Tanganyika/Mozambique border, and the coast further south. Triassic seas have flooded the regressive Paleozoic–Cretaceous shelf sea of Ogaden in Ethiopia, and the Gumbura limestones mark the hiatus.

In the Albertine (i.e., Western) Rift Valley of Uganda and the Congo, oil seeps from the Pleistocene Kaiso lacustrine clay, sand and bone beds. Oil also seeps at Bujumbura, Burundi, on the northern tip of Lake Tanganyika.

At the Cape of Africa the Karroo has been drilled at Zitzi Kamma, on Table Mountain.

To pass to Madagascar, post-Cambrian sediments, consisting of the northwestern Majunga and the larger western Morondava basin, cover the western third of the island. Unfortunately fresh water invades oil of the Isalo formation in southern Madagascar, 40 miles southeast of Ankazoaba along the Tsimiroro and Maroboaly horsts, east and southeast of Maintirano.

GAS

Hydrocarbon gas is closely associated with oil. Consequently we have discussed the geologic history of gas fields in the oil chapter, with the exception of the Paleozoic In Salah belt, where gas predominates. A majority of African gas reserves, however, are concentrated in the Triassic sandstones of Hassi er Rmel (also in Algeria). Furthermore, gas is known to occur in the Paleozoic Polignac basin, the Eocene basin of Libya, and on the Atlantic coast.

ALGERIA

The Triassic Province

Hassi er Rmel, the continent's largest gas field (Fig. 57, 124, 131), is located south of Algiers and west of Hassi Messaoud. A north-northeast trending, uplifted flat dome of Triassic sandstones, divided by argillaceous beds and bordered by low lying strata, constitutes three gas reservoirs. In the south, the most important part of the field, salt overlaps the sandstones. The total thickness of the sandstone reservoirs varies between 50 and 150 ft. While permeabilities oscillate between 400 and 700 millidarcys, porosities range from 16 to 20%. Out of a total area of 900 square miles, the productive area occupies 700 square miles.

With over 3,000 million cubic ft. of gas produced each day Hassi Messaoud is a large gas field. *Hassi Touareg–Nezla* is situated 40 miles southeast of El Gassi in the Hassi Messaoud district. The long anticline forms the northern nose of the El Biod uplift, between a late Paleozoic–Triassic zone of subsidence in the east and a stable area in the west. The (Lower Cretaceous) Albian is transgressive on the Middle Jurassic and unconformities are apparent between the Ordovician or Silurian and Triassic. Indeed, gas concentrates at the top of Lower Triassic and Ordovician sandstone structures. The Gassi Touil field is believed to be of the same magnitude as Hassi er Rmel. It is located southeast of Hassi Touareg. The

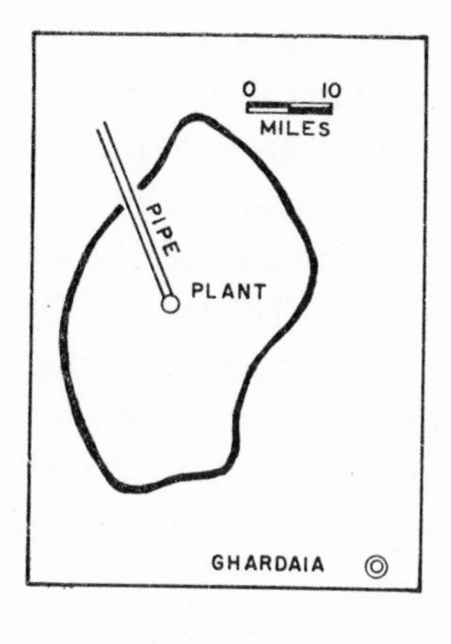

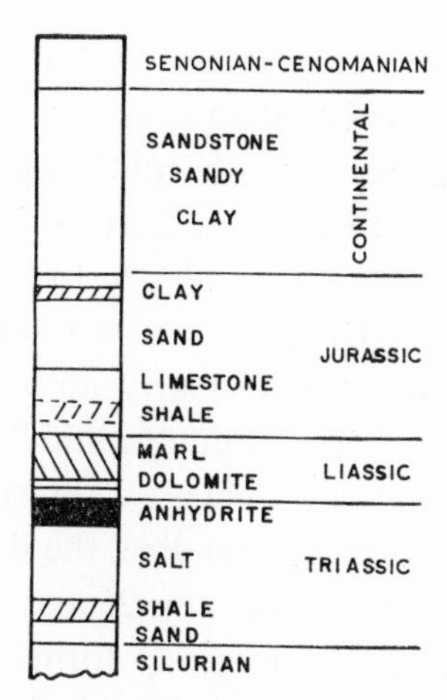

Fig. 131. Plan and column of the Hassi er Rmel gas field.

gas condensate pay of the upper Triassic horizon is more than 200 ft. thick. South of Gassi Touil and also on the El Biod uplift, gas accumulates at Rhourde Nouss in the Devonian and Triassic as deep as 9,100 ft.

The In Salah District

Northwest of the Hoggar, localized lenses of microconglomeratic Cambrian sandstones overlap sandstones and shales, which are followed by Ordovician (Arenigian) Graptolite sandstones and quartzites. In the western prong of the Hoggar Silurian sediments attain a thickness of up to 3,000 ft. Middle Silurian limestone inliers in the north, and sandstone lenses in the south, reflect the Caledonian epeirogeny. Whereas *Anarcestes plebejus*, *Uncinulus kayseri* and *Paraspirifer cultrijugatus* are found in the Emsian–Eifelian Stage, *Chonetes sarcinulata*, *Leiorhynchus megacostalis* and *Tropidoleptus carinatus* distinguish the Givetian and Frasnian Stages.

Between Mouydir and Ahnet the best reservoir sandstones are aligned along a northwest trending belt of shallow sediments. The Carboniferous begins with Tournaisian shales, followed by large-grained sandstones. During the Visean 700–1,200 ft. thick shales were overlain by up to 1,200 ft. of Garet Debh Series sandstones associated with some shale. In the Tanezrouft 120 ft. thick reservoirs of this series are covered by 450 ft. of anhydrite and shale. The Namurian Stage then begins with the Djebel Berga Series of limestones and shales which include sandstone lenses. In the late Carboniferous Azzal Matti Series, shales alternate with thin sandstone limestone lenses. Finally, conglomerate relics bear witness to the Hercynian orogeny. In this sequence, 700 miles from the Oran coast and roughly at the same distance from Agadir, Lower Devonian gas is found with traces of Paleozoic oil. The principal structures are located at *In Salah* in the Tidikelt and in the northern Ahnet. Whilst four major groups form a triangle with its northern tip at In Salah, the isolated Ordovician Zini field, 40 miles southeast of In Salah, contains 110,000 million cubic ft. of gas. The Ordovician Meredoua lens, 150 miles south of In Salah (west of the Hoggar), contains approximately 50,000 million cubic ft. of gas. The Gara Azzal Matti gas field is situated 140 air miles southwest of In Salah, that is 90 miles south of the nuclear centre of Reggan (Fig. 57, 124).

At *In Salah* gas accumulates in Lower Devonian, 40 ft. thick sandstones, which cover 100 square miles and contain a reserve of 450,000 million cubic ft. In the *Djebel Berga–Oued Djaret* group, 50 miles south–southwest of In Salah, the thickness of the Lower Devonian reservoir at Djebel Berga l is 75 ft. This field covers 120 square miles and the order of magnitude of reserves is similar to In Salah. The Oued Djaret 1 and 101 fields, located in Lower and Upper Devonian strata respectively, are similar. The field of Djaret 1 contain 140,000 cubic ft. of gas. Thirty air miles south of this group, the alignment of the Bahar el Hammar, Tirechoumine and Tirebadine fields is oriented towards the northwest. Bahar el Hammar represents a Cambrian–Ordovician reserve of 10,000 million cubic ft. At Tirechoumine and Tirebadine, the thickness of the gaseous Lower Devonian and (Carboniferous) Tournaisian attains 25 and 18 ft., respectively.

The Mahbes Guenatir group forms a triangle 70 miles south-southwest of In Salah. At *En Bazzane* 10,000 million cubic ft. of gas has accumulated at the Devonian/Silurian contact, while the Cambrian–Ordovician Mahbes Guenatir lens contains 7,500 million cubic ft. of gas. The thickness of the Silurian *Mouhadrine* reservoir reaches almost 30 ft., but the lower Devonian Thara 201 layer is only 20 ft. thick. In the same strata, *Thara* 101 contains 1,000 million cubic ft. of gas.

Fort Polignac basin

Gas tends to accumulate in the Ordovician, Devonian and Carboniferous. At Zarzaïtine and in the Djebel Ouan Taradert, gas is found in the Devonian and Middle Carboniferous. However, at Edjeleh gas was only trapped in two Devonian horizons. On the other hand, at Tiguentourine and Reculée, gas accumulated earlier than oil in Ordovician sediments at a depth of 5,400 and 6,700 ft. respectively. Similarly, at El Adeb Larache gas is found below oil in a 1,600 ft. deep Carboniferous reservoir. Gas is trapped in the Carboniferous at Ouan, in three Devonian levels at *Taouratine*, Arena, Daia and Assekeifaf, and in the Ordovician at Ihansatene, Oued Tissit and Assekeifaf.

LIBYA

At *Oued Chebbi*, 70 miles south of the coast near the Tunisian border, Triassic sandstones outcrop unconformably upon a large, northwest trending, Ordovician anticline on the edge of the Paleozoic basin. The gas reservoir is at a depth of 3,000 ft.

At Bir el Rhezeil in western Tripolitania, 240 miles southwest of Tripoli, 1.3 million cubic ft. of gas per day flow with oil from late Devonian sandstones. This field is located between the Tlacsin and Emgayet oil fields.

In east Libya, the Hofra-Mabruck basin of the Syrte includes several gas fields: Zahaligh is located 20 miles south of Mabruck, but its reservoir is shallower than the latter; it yields 2.2 million cubic ft. of gas per day. The GI 32 field is of Eocene–Paleocene age. About 30 miles south-southwest of Gialo, gas accumulates in Cretaceous sands at more than 9,800 ft. of depth. The EE field is located southwest of this BB field.

THE REST OF AFRICA

The Cap Bon peninsula, near *Tunis*, was formed around the north-northeast trending Djebel Sidi Abd–er–Rahmane anticline. Surrounding a core of Eocene, *Hanthemna* marls, Lower Miocene, Burdigalian limestones appear in the north. Middle Eocene marls and Oligocene *Fortuna* sandstones enclose the 350 ft. thick inner zone or series. It is underlain by 900 ft. of Triassic–Upper Jurassic limestones, Middle Cretaceous marls, and an alternating sequence of younger Cretaceous marls and limestones. Gas indices are found at lower levels. Significant accumulations occur only in Barremian–Neocomian (Lower Cretaceous) light-grey sandstone horizons, interbedded in olive-green pyrite-bearing shales at a depth of 5,100 ft. The somewhat calcareous, arkosic sandstone beds are 10–25 ft. thick with porosity attaining 15%.

In the Senegalese basin northeast of Dakar, gas accompanies oil in the 6,000–11,000 ft. thick column of Middle–Upper Cretaceous shales. In addition, gas sands surround the Diam Niadé oil lens as far as Cayar 1. In Nigeria gas occurs, e.g., at Ibotio. The Tchengué gas field is situated in the middle of the north-northwest trending Ogoué–Port Gentil basin of *Gabon*. It is believed that gas migrated there from an Eocene deposit, and was then trapped in a trough of argillaceous sands between a salt ridge and a Miocene discordance. Gas is also abundant in other fields of the Ogoué Basin. Indeed, the gas/oil ratio attains 1,150 at M'Bega (3,300 ft. of depth) and 1.010 at Batanga (6,200–6,300 ft. of depth). The ratios of other fields (expressed in cubic ft. per barrel) ranged in 1962 from 850 at Illigoue (7,900 ft. of depth) to 200 at Alewana (2,900 ft. of depth). Further south, in the Pointe Indienne field of the Congo Republic, wet gas shows in arenaceous sediments of the Chela Stage.

In addition to Pande, gas and condensate accumulate below 4,700 ft. of depth at Buzi, near Beira in Mozambique.

Finally, to the gas reserves of the Rharb oil fields we can add the Kechoula gas field of the Eassaouria basin of southern Morocco. Undoubtedly, numerous other gas fields will be found in Africa during campaigns of exploration.

CARBON DIOXIDE AND METHANE

Carbon dioxide frequently escapes from the volcanic Rift Valleys of eastern Africa. Between the Congo and Rwanda this gas, along with methane, is absorbed in the peculiarly layered waters of Lake Kivu.

CONGO–RWANDA

The Kivu basin has been drained through the Ruindi Valley towards Lake Edward and the Albertine Nile. However, after the eruption of the volcanic chain of the Virunga Mountains had barred this pass, Lake Kivu was created at an altitude of 4,670 ft. The flow of the Kivu drainage was then inverted, the Ruzizi River spilling over between Bukavu in the Congo and Shangugu in Rwanda, and entering Lake Tanganyika and the Congo drainage. Since Lake Kivu occupies a system of valleys, it has an irregular shore line and numerous islands, the largest of which is Idjwi (Fig. 51, 110). The maximum dimensions of the Lake are 70 by 20 miles. Methane and carbon dioxide are absorbed by its waters below 210 ft. of depth and, between 600 ft. and its maximum depth of 1,620 ft., gas accumulation is such that gas separates from this water when pumped to the surface. Experts estimate a reserve of 10.7 cubic miles of gas, or 12.6 cubic miles expressed in methane, which equals the calorific power of 250 million barrels of gas-oil. The deposit is dynamic and continues to form. Indeed, its tapping would enhance the development of gas in the residual waters. Since the relative nitrate and phosphate content of the gas free waters returned to the lake would increase, lacustrine vegetation would develop even faster. At present, life has only been observed above a depth of 200 ft. Since decomposed fossils undergo anaerobic fermentation, their water sinks, becomes heavier, and thus produces the deposit. Although the proportion of methane and carbon dioxide is variable, the gas content and the layering of the lake is not. The risks of a turnover are not great. Sulphuric acid is abundant in the lower layers.

THE REST OF AFRICA

In *Kenya*, carbon dioxide escapes from several bore holes of the Western Rift Valley. On its western wall, the gas of Esegeri consists of 97.2% CO_2 and 2.2% O_2. As gas escapes under a pressure of 80 lb./sq. inch, it carries water with it. However, the pressure of the gas escaping at Kerita, from the eastern flank of the Rift, probably reaches only 30 lb./sq. inch. A gas-composition is given in Table CLII.

TABLE CLII

COMPOSITION OF GAS FROM KERITA

CO_2	97.8%
CH_4	1.1%
N	0.8%
O_2	0.2%
Ar	0.07%

Carbon dioxide also escapes from Mount Margaret near Kedong, from the soils of Lake Magadi (Kenya), and from springs of the Songwe Valley near Mbeya, in southwestern Tanganyika.

In Natal, *South Africa*, carbon dioxide escapes from a Cretaceous fault zone, traversing Karroo tillite, at Bongwana and around Harding. It is accompanied by 1% nitrogen, oxygen and argon. The gas is believed to be liberated from carbonates by sulphuric acid dissolved in vadose waters. Finally, the gases of the gold fields mainly consist of methane.

HELIUM

The only useful inert gas, helium, escapes with other gases from the faulted areas of eastern Africa. Helium is also included in the alteration products of uranium.

Tanganyika. At Maji Moto ('hot water'), 40 miles east of Musoma, on Lake Victoria, the gas which escapes from brine-filled pools and springs averages 13% helium, while the rest is nitrogen. The gas is endogene, ascending through a shear zone, and yields approximately 6 cubic ft. of helium per hour. At Manyeghi Springs in the Rift Valley yield is estimated at 7 cubic ft./hour. Gas containing 5 % helium escapes from vents, heated to 100 ° F, and thus yields 7 cubic ft. of helium per hour. In the same sector, the Mponde, Tarkwa, Conga and Hika springs contain 7–10 % helium. Similarly, helium shows in the gases escaping from the Rift at Golai and on the shore of Lake Eyasi.

Gases of the Orange Free State goldfield, *South Africa*, contain 77 % methane, 14 % nitrogen, 8 % helium and 0.5 % argon. The helium is believed to be a decay product of tucholite and uraninite.

SECTION VIII

Sediments

We can distinguish four categories of sedimentary rocks: *(1)* biogenic sediments, including phosphates and diatomite, *(2)* evaporites, including salt, soda, potash, nitrates, sulphur and gypsum; *(3)* chemical sediments: limestone; *(4)* residual sediments: bentonite, sepiolite, kaolin and other clays.

Post volcanic processes play a role in the development of phosphate and lime deposits of carbonatites, sulphur, alum and montmorillonite. Not only does supergene alteration control the accumulation of residual deposits, but this phenomenon also affects the other categories. While the largest phosphate deposits are restricted to latitudes 33° and 34° N, evaporites accumulate in the northern Sahara and on the eastern coasts of Africa (Fig. 8).

PHOSPHATES

THE GENESIS OF PHOSPHATE DEPOSITS

We distinguish three types of phosphate deposits: carbonatites, bedded sediments and guano. The accumulation of late Tertiary–Quaternary guano in the islands, along the coasts, and in Tanganyika proves the importance of biogene factors in the cycle of phosphorus. Although Precambrian carbonatites do contain phosphates (Phalaborwa), the first important carbonatite cycle, the Cretaceous, precedes and parallels the sudden appearance of phosphorus in the sea. Apatite (and gorceixite) crystallize during the later carbonatite phases in sövite cores, dykes and ultrabasics. It also forms phoscorite or magnetite apatite. Finally apatite weathers to staffelite, a mineral occurring frequently in marine phosphates.

The most favourable environment of the marine phosphates is the sea of synclinal warps or shallow shelves surrounded by flat shores and slowly eroded in fine-grained limestone, marl and clay. Indeed, its transparent waters facilitate photosynthesis and some oxygen transfer by agitation, which is so important for Algae (cryptogames) and marine organisms. Sub-tropical temperatures, currents supplying salts, the presence of carbon dioxide and lime, and an alkaline pH (approximately 8) appear to be essential for the precipitation of bicalcic phosphate, hydroxyl-apatite and fluoapatite.

It is appropriate to recall here the divisions of the Cretaceous–Eocene stratigraphic column (Table CLIII) distinguished by the sedimentation of phosphates along latitude 33° N in Morocco, 34° N in Algeria–Tunisia, and also in Egypt, Senegal and Togo.

The abundance of sharks, reptiles, mollusks and micro-organisms in North African

TABLE CLIII

THE UPPER CRETACEOUS–EOCENE STRATIGRAPHIC COLUMN

	Middle	Lutetian
		Ypresian
	Lower	Thanetian
Eocene		
	initial	Montian
	terminal	Danian
		Maestrichtian
		Campanian
Upper	Senonian	Santonian
Cretaceous		Coniacian
	Turonian	
	Cenomanian	

phosphates indicates shallow but agitated plankton rich seas covering epicontinental shelves communicating with the high sea. Calcareous algae and the rather thin (250–300 ft.) Cenomanian–Eocene column also indicate strong currents in shallow seas, as does the criblometry of free quartz. While quartz is included in phosphate grit, bone fragments never are; nevertheless shallow coprolithic phosphate beds are recurrent features. Phosphates accumulated in Morocco in a non-transgressive sea during long periods. However, each time the beginning of phosphate concentration is abrupt, and it reaches its climax in the Ypresian in the eastern gulf of the Moroccan basin. In Algeria–Tunisia the climax of phosphatogenesis is older, Thanetian. Phosphate-bearing waters are believed to have penetrated into the Djebel Onk–Gafsa basin along a southeast trending channel. The small Egyptian deposits appear to be Cretaceous, viz. Campanian–Maastrichtian, and much of the phosphate is believed to have been derived from Mesozoic bone fragments.

The original phosphorus of the high sea may have been derived from submarine or coastal volcanism of the Moroccan shelf and of the straits of Malta–Pantellaria in Tunisia–Algeria. Low phosphorus sediments and microorganisms are alternative sources. Phosphorus is believed to have been fixed by ascending carbon dioxide rich currents, below the subsurface phytoplankton layer, and flocculated by organogene ammonia. It was deposited on the continental shelves, characterized by an abundant ichtyan and saurian fauna, along with continental detriti. These seas also absorbed phosphorus redeposited in coproliths and reconcentrated from bones by carbon dioxide leaching. Some of the phosphate beds have been reworked by submarine erosion. In Algeria and Tunisia, at least, repetitive phosphate deposition is believed to have been triggered by the opening of straits admitting phosphate rich water of the high sea. In fact, distinctive features of this province include abundant clastic material and pebble horizons.

MOROCCO

All deposits are Cretaceous and early Eocene, and most are located in a broad east-northeast trending 60 mile long belt of the Moroccan Tableland. However we also find smaller deposits west and east of Marrakech (AGARD et al., 1952) (Fig. 10, 59).

The western tip of the principal Oulad Abdoun district is situated 20 miles south of Casablanca. Continuing eastward in a 30 mile long branch, the Phosphate Plateau forms a triangle consisting of the principal or *Khouribga–Oued Zem* zone, the El Borouj zone in the southwest, and the Kasba Tadla arc. Beginning 15 miles from Safi Harbour, the Ganntour proper extends for 10 miles. While the principal deposit of Youssoufia is at its western end, north of El Kelaa des Srarhna phosphate layers extend beyond its eastern tip in the direction of Oulad Abdoun.

Between Mogador Harbour and Marrakech other occurrences include the Meskala, Chichaoua and Imi n'Tanout–Oued Zat. In addition, east of Agadir phosphates outcrop at the Oued Erguita, south of the High Atlas at Imini, and at Khela d'Ouarzazate–Skoura.

We will see that paleo-oceanography and stratigraphy determine the distribution of phosphate beds. We can summarize the common features of Moroccan 'phosphate paleontology' as follows:

(1) An abundance of microfauna and flora in the phosphate beds contrasts with the scarcity of invertebrates (especially mollusks), and also with the extreme frequency of gastropodes in both roof and wall limestones of many beds.

(2) The rather homogeneous column of Lower Maastrichtian but barren ichtyan sediments, and the Upper Maastrichtian phosphate complex of Youssoufia, contrast with the composite stratification of Neocomian marls, basal bone beds and Maastrichtian phosphates of Khouribga. However, a common feature of both basins is shallow water and shelf algae. Between the two major deposits, El Borouj stratigraphy is considered as transitional.

(3) Shallow water Spuellaria are embayed in the core of phosphate nodules. Radiolaria are most abundant at Khouribga,

(4) Foraminifera occur both in the ooliths and in the groundmass, *Frondicularia* being most abundant in the roof of Montian phosphates of Youssoufia,

(5) Even in the poorer Montian and Maastrichtian phosphate beds, mollusks are rare, but Cardites, Ostreae and Vulsellidea are more common.

Khouribga

The Oulad Abdoun district is roughly triangular; for 70 miles its longer or northern side trends east–west. While Senonian sediments cover the northern corners of the triangle and most of its borders, a Lutetian limestone layer occupies its core. Except for a thin band of

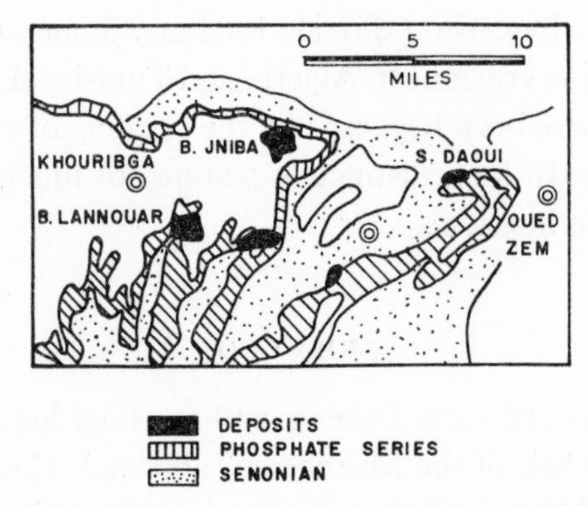

Fig. 132. Occurrence of phosphate deposits at Khouribga.

Turonian sediments bordering the Senonian between Khouribga and Oued–Zem, hardly any other rocks outcrop in the vicinity of the phosphates. The phosphate 'shore line' is most irregular. A continuous undulating band extends for 70 miles as far as Oued–Zem, with its width varying between 5 and 30 miles. At Bled el Hasba, outliers of the main phosphate band embay a large Lutetian bank. Between Oued–Zem and El Boruj, several ramifying bands of phosphate undulate for 50 miles. The total width of the branches may attain 12 miles. Important workings are centred in these branches between Khouribga and Oued–Zem (Fig. 132) at *Bou Jniba*, *Bou Lannuar*, *R. Hatane*, *Foum Tizi* and *Sidi Daoui-Kerkour er Rieh*. Northwest of El Boruj lies a 20 mile long and 3–12 mile wide ramifying phosphate lens.

Stratigraphy

The slightly dipping, 300 ft. thick sedimentary series of the Khouribga Plateau covers tightly folded Silurian violaceous schists and quartzites. Middle Cretaceous marls, sandstones and limestones underlie Cenomanian marls and limestones. They are overlain by a Turonian indicator horizon of limestone. We find another indicator layer among the upper marls of the Lower Senonian which covers much of the eastern and western areas of the Plateau. Indeed, the basin was extremely shallow at the end of the Senonian.

The Maastrichtian–Danian stage consists of odorous, yellow marly phosphate limestone, with sandy phosphates overlying a bone bed at André Delpit. Tricalcic phosphate values of bed III range between 55 and 60%. The characteristically abundant fauna includes *Globidens*, *Lecidon*, *Mosasaurus* and *Lamna biauriculata*, *Rhombodus*, *Enchodus*, *Corax*, *Orchosaurus* and *Sclerorynchus*. At Oued–Zem lenses of phosphatic limestone, accompanied by sandy phosphates, represent the *Montian*. Carried over from the Maastrichtian, *Lamna appendiculata* and *Odontaspis tingitana* are associated with *Odontaspis macrota*, *Myliobatis* and small *Lamna obliqua*, all Cretaceous–Eocene fossils.

Generally 3–6 ft. thick, the *Thanetian* is most developed in the east at Sidi Daoui. In this sector it attains a thickness of 30 ft. at *Kerkour er Rieh*. In the same area, at Bou Jniba tricalcic phosphate values of bed III increase from 70 to 74% until they reach 80%. Whereas sandy phosphate marls, limestones and silex are hardly differentiated in the east, in the western areas calcite levels interrupt the phosphates. From the Montian, the Eocene fauna carries over *Odontaspis macrota* and *Lamna obliqua*, accompanied by *Odontaspis cuspidata*, *Trichiurus*, *Squalus crenatidens*, *Ginglymostoma*, *Dyrosaurus*, *Myliobatis* and others. Moreover, the ichtyan fauna is carried over into the Ypresian phosphates, which we will discuss later.

Consisting of marls and limestones, the barren Lower *Lutetian* cover concludes the sequence. This series is 15–30 ft. thick at Oued–Zem and 60–75 ft. thick at Bou Jniba. Between three *Thersitea* limestone layers, marls and interbedded silex report including three marly inliers. The lowest of these beds is characterized by *Hemithersitea marocana*. Other silicified invertebrates include various *Hemithersitea* and *Thersitea*, and also *Pseudoliva* and *Nautica crassatina*. Although well developed at Oued–Zem in the east, the intermediate limestone bank of compact detrital and silicified Lamellibranchiae is only 10 ft. thick in the west. Nevertheless, few Lamellibranchiae are left in the top layer of crystalline or sandy limestone, but both silex and phosphate traces reappear. Finally, the flexure of the Souk Tleta of the Beni Oukil drives the Eocene under the Quaternary plane of the Beni Amir.

Ypresian phosphates

Most of the exploited phosphate beds are Ypresian (Fig. 133). The majority of bed No. 1 which contains 76–78% tricalcic phosphate, is powdery. However, the thin beds 0 of Bou Jniba and A–B of Bou Lannouar, associated with silexes and marly limestones, are less interesting. In the phosphate bed four new fishes appear, namely *Xiphiorynchus*, *Cylindracanthus*, *Brychaetus* and *Pristis*. While the vertebrate fauna is very abundant, invertebrates remain rare.

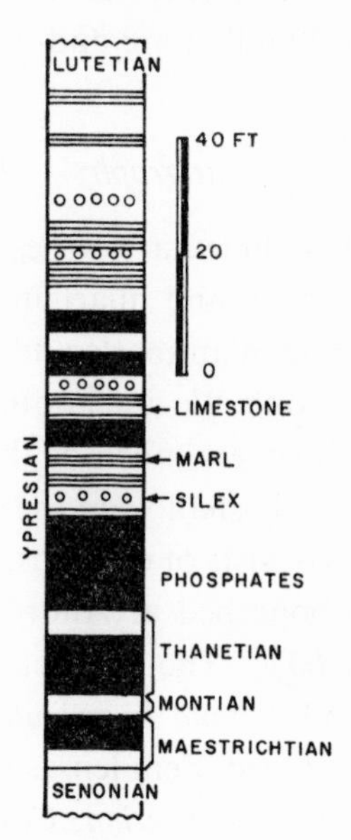

Fig. 133. The column of Khouribga.

At *Bou Lannouar* adjacent to Khouribga, bed No. 1 is well-developed as a 5–6 ft. thick layer of powdery coprolithic phosphate. The bed is crossed by two or three alignments of mounds of silex and a 1 ft. layer of marl is interbedded near its top. Conversely, bed 0 is thin and irregular, comprising, in fact, two alignments (A and B) of powdery phosphate. Thick silex marl and limestone cover this syngenetic sequence.

At *Bou Jniba*, halfway between Khouribga and Oued–Zem, the silex of bed No. 1 of Bou Lannouar is missing. However the interbedded marls attain a thickness of 3 ft. The sequence includes interbedded layers of *(a)* silex and phosphate, *(b)* red argillaceous phosphate, *(c)* phosphatic limestone, subsequently silicified into phosphate silex, marls and limestone, *(d)* the well-developed 2.5 ft. thick bed 0 of powdery high grade phosphate, and *(e)* a thin series of marl and limestone which terminates the Ypresian.

Overlying the thick Thanetian complex at Kerkour er Rieh near Oued–Zem, bed No. 1 further thins out to 3 ft. Within this irregular bed the ratio of alternating phosphate, marl and limestone is most variable.

The mineralogy of phosphates

Its colour may vary from grey yellow to greenish or red, but phosphate rock whitens as it dries. Impure marly Maastrichtian phosphates are yellow. Phosphate grains (measuring 0.1–0.2 mm), coproliths and rare quartz are embedded in a calcic or argillaceous matrix.

Grain diameters show a characteristic maximum. Tricalcic phosphate content rises with lime from 65% in the Maastrichtian to 75% in the Ypresian. Grades are also functions of local factors.

Many phosphate grains consist of fluocollophanite accompanied by occasional calcite lamellae, and sometimes surrounded by staffelite, a mineral also reporting as an alteration product of apatite in deeper carbonatite soils. In high grade beds clay embays the grains. On the other hand, in some grains phosphate precipitates around a quartz or zircon core. However grains of free quartz constitute only 1.5% of the volume of the phosphate rock. In other grains we can find Radiolaria, Foraminifera and rare Diathomea relics. Coproliths as big as half an inch may attain a grade of 76% of tricalcic phosphate! Reptile-bone fragments occur frequently. even constituting true bone beds beneath Maastrichtian bed III between Sidi Daoui and Faoum Tizi.

The chalky groundmass or gangue includes Foraminifera relics, with big nodules of late calcite in bed No. II. Larger calcite grains may embay phosphate grains when diagenetic recrystallization prevails. With the calcite ratio increasing on both contacts of phosphate beds and at intermediate levels, the matrix changes to phosphatic limestone. Moreover, the ground mass is transformed by opalization into silex, which includes phosphate grains at the contact of calcareous phosphates. Such silex horizons are interbedded with other sediments at various levels and localities (e.g., El Borouj). Taking the place of phosphates during facies changes, silex also precipitates in lenses or thin beds.

Youssoufia

The Main branch of the *Ganntour* deposit extends for 25 miles east of Youssoufia. East of this locality it bulges to a width of 7 miles and narrows progressively to 2 miles. A north trending branch of 8 by 3 miles is attached to the main body at Youssoufia. An Upper Lutetian bank of *Thersitea* limestone lies south of the Maastrichtian–Lutetian phosphates. Relics, Lower Maastrichtian marls and limestones form much of the northern contact line of the phosphate beds. Quaternary sediments surround this complex. While lacustrine Pliocene limestones outcrop south of the *Thersitea* bank, we find smaller remnants of Neocomian marls, red sandstone and a Paleozoic stock of granite in the north.

Maastrichtian–Montian

Following a period of emersion, earlier Maastrichtian fish marls were overlain by 60–70 ft. thick *latest Maastrichtian* phosphate beds consisting of *(1)* phosphate grit, *(2)* phosphate marls, *(3)* slightly phosphatic limestones, and *(4)* powdery phosphate. Grit *1* contains only 55–60% tricalcic phosphate, while bed C, or III in local terminology, is 7–9 ft. thick. While the fish and reptile fauna of *1* and *4* includes *Odontaspis*, *Enchodus* and *Corax pistodontus*, *Plesiosaurus* and *Leiodon*, frequently *Alectryonia* and *Heligmopsis*, characterize the marls. Furthermore, in bed *3*, or the Baculite bank, various *Cardita* and *Corbula* have also lived.

Following white marls interbedded with silex, the principal phosphate layer (A or I) sedimented during the *Montian*. This layer exhibits values of 73% tricalcic phosphate with a fauna of *Ostrea canaliculata* or *eversa*, accompanied by Lamellibranchiae and Gastropodes. Comprising *Odontaspis tinguitana* carried over from the Maastrichtian, as well as *O. macrota*,

Lamna obliqua and *L. appendiculata*, *Myliobatis dixoni* and *Cinglymostoma africanum*, the ichtyan fauna is transitional. On the other hand, the *Cardita coquandi* limestone bank is distinguished by a Paleocene fauna of several geni of *Nautilus* associated with *Lucina moevusi* and *Turritella delettrei*. This bank is a common feature of all Moroccan phosphate deposits including Oued–Zem, but excluding Khouribga.

Thanetian–Lutetian

The 75–90 ft. thick *Thanetian–Middle Lutetian* consists of *(1)* phosphatic marls, *(2)* white marls overlying *(3)* a thick layer of phosphatic marls, interrupted in its upper reaches by bituminous and white *Frondicularia* marls, and *(4)* alternating beds of marl and silex.

We can easily subdivide the *Upper Lutetian*, so-called *Thersitea* Bank, into at least three units: *(1)* siliceous marine limestones in which *Carolia*, *Cardita*, *Meretrix*, *Mactra*, *Melongena*, *Pleurotoma*, various *Ostrea* and other Lutetian–Bartonian molluscs are preserved, *(2)* a transitional level in which marls alternate with limestone, and *(3)* large and round-grained sandstones which prove the emergence of the Ganntour at the foot of the Jebilet Highlands, as does warping and moderate faulting. Although incipient or early phosphatization lacks intensity, the maximum of phosphate sedimentation advances from the Thanetian to the Montian here. Although grades are lower because of tectonic disturbances, the mineralization is strikingly similar to Khouribga. However, organic relics are more abundant at Youssoufia, as are phosphatic silices at El Borouj.

ALGERIA

Between the two Atlases the Suessonian 'Phosphate Sea' inundated western and central Algeria, embaying the northeastern corner of the Algerian Sahara and the northern half of Tunisia. Disseminated phosphorites also occur in the narrow channel of the western Oranian area, e.g., at Bab-Tounaï and Beni Ouarsous. Between Oran and Algiers, guano has covered the floor of the Retaïma caves near Inkermann. However, commercial phosphate deposits are restricted, to two distinctive belts. The first of them extends in a wedge between the two Atlases, from 100 miles south of Algiers, through the Sétif Tableland south of the Bougie oil harbour, to the Kouif near Tébessa, and thence to the Djebel Kechaib in Tunisia. In the south the second belt covers an 80 by 30 mile area between Djebel Onk and Négrine in Algeria (Fig. 11, 134) and Gafsa in Tunisia. The phosphatic waters flowed into this basin from the northwest through the straits of Zoui. In this channel the upper reaches of Danian–

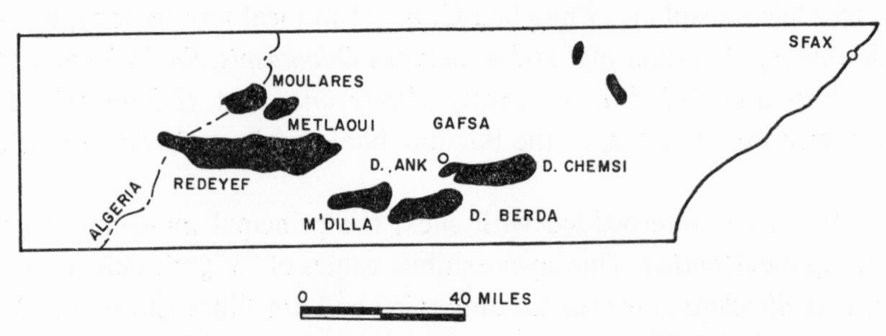

Fig. 134. The phosphates of Algeria and Tunisia.

Thanetian marls have frequently been affected by rubefaction (reddening), corresponding to emergent epeirogenetic movements. Gypsum and red layers are interbedded further west in Algeria in the El Kantara area. At the end of this emergence appear detrital spheres of *Venericardia beaumonti*, followed by a 1 ft. thick phosphate bed in the Straits of Zoui. Although reworking accumulates phosphate beds, erosion has not been followed by phosphatization at Fedj-Reddima in the area of the Chéria domes.

Indistinct phosphatization started during the Middle Cretaceous Albian, continued in the Senonian at Tébessa, and in the Maastrichtian in the Djebel Mzaïta. The westernmost Eocene layers occur above the silex between Titteri and Boghari in the Dra el Abiod syncline. This 6 ft. thick layer contains 33% tricalcic phosphate. Traces found in the Sidi Aïssa anticlines link this area with Sétif. Moreover, bituminous and frequently siliceous phosphates, underlying silex limestones impregnated by oil, show grades of 63% at Mzaïta, 63–70% at Bordj-Redir, and 58% at Tocqueville. On the other hand, the siliceous beds which extend in the south to Maâdid attain 1½–2½ ft. of thickness. Between the Sétif Plateau and the Tébessa–Djebel Kouif–Roumela deposits lie the Souk-Ahras occurrences. Horizontal Eocene strata which cover the centre of an Upper Cretaceous syncline consist of *(1)* dark argillaceous marls containing anhydrite, *(2)* bituminous marls and limestones including granules of silex and phosphatic horizons, and *(3)* the Nummulitic silex limestone covering the Dyr Highland of 900 square miles. However, the limestone cover is thinner at Kouif than in the Dyr. In the Kouif out of five phosphate beds only the upper three have been worked. From them workers concentrated dried and decarbonated ore which contains 65.5% tricalcic phosphate.

The stratigraphy of Djebel Onng

In the Tunisian Gafsa basin, its southern M'dilla–Redeyeff arc extends for 20 miles into Algeria. The Djebel Onng deposit, 30 miles north of this branch, boasts reserves of 120 million tons containing 61% tricalcic phosphate, and 420 million tons including 53–58% tricalcic phosphates.

Largely Senonian marls, measuring 900 ft. in thickness, are overlain by 60 ft. of *Maestrichtian*, slightly phosphatic, limestones and 250 ft. of silex and Inocerame limestone. Above these 130 ft. of *Ostrea overwegi* limestone and a ferrugineous layer conclude this series. Some 80 ft. of marls contain in the Négrine Chain, Danian *Ostrea overwegi*, *O. bomilcaris* and *O. tissoti*; they include two thin phosphatic layers. Nevertheless, at Metlaoui these phosphates carry Maastrichtian *Corax pristodomtus*, *Scapanorhyncus raphiodon* and *S. subulatus*. Later 165 ft. of Danian marls are followed by 200 ft. of Thanetian lumachelle limestones. Phosphatized relics of Lamellibranchae and Gastropodes lie at the base of 65 ft. of marls, but a limestone layer is interbedded at the top of this sequence. In the upper layers of the Jbel Djemidjma, 100 ft. of pseudo-oolithic and coprolithic *Thanetian* phosphate are distinguished by *Ostrea multicostata* D. ESH., var. *phosphatica* D. OUV., *Lucina*, *Porchati*, *Cardita* and *Turritella*. About 6 ft. of phosphatic limestone, 6 ft. of coprolithic pebble phosphate, and 50 ft. of *Ypresian* alternating limestones, marls, phosphates and silex follow. They are overlain by 100 ft. of silex limestone, including phosphatic inliers, 65 ft. of Lower Lutetian marly limestones, 500 ft. of gypsum limestone (at Aïn Fouris), and 100 ft. of green gypsum marls. During the Burdigalian–Pontian Period, 2,000 ft. of alternating layers of clays and sand sedimented. The sequence is concluded by 500 ft. of Pliocene conglomerate (VISSE, 1951).

In the eastern Djebel Onng the Miocene transgression has cut the phosphate bed into a wedge-like shape. Averaging $^{3}/_{4}$ mile in width, the commercial deposit extends for 5 miles and bulges in the east to 1 ½ miles. The anticline strikes east-northeast. Although its northern limb slopes only 15–20°, the southern flank dips 50–90°. *Inoceramus* limestones halted the erosion of the dissymmetrical ridge. A similar polyclinally faulted brachy-anticline lies 1 mile south in the Djebel Djemidjma. Phosphates have been well conserved along its southern limb. On the other hand, the northern phosphate beds extend for 5 miles from the Djebel Fouris in the west, through the Kef es Sennoun and the Djebel Tarfaya, to Rass Mergueb and Tir in the east.

The phosphates of Djebel Onng

In the *Djebel Djerissa* Thanetian phosphates attain 180 ft. of thickness. They are covered by 11 ft. of phosphatic limestone and 7 ft. of coprolithic phosphates. A major arcuate fault cuts the anticlinal axis at an acute angle. Consequently, the northern limb or plateau remains sub-horizontal, but the faulted southern limb dips 45–50° south. Located on the plateau, the main bed consists of five types of phosphates: 60 ft. pseudo-oolithic, 20 ft. coprolithic, 12 ft. pseudo-oolithic, 4 ft. including marly seamlets, and 9 ft. large-grained phosphate. Because of accompanying exocalcite, the bed averages only 55% tricalcic phosphate. Towards the northwest calcium and silica contents derived from fossils increase, generally affecting either the low or upper levels of the phosphate bed. Reserves attain 110 million tons. More than half of this quantity is concentrated in the plateau, where the ratio of overburden to ore is also particularly favourable (0.2/1).

The *Bled Djemidma* syncline slopes towards the west-southwest between the Onng and Djemidma limbs. Lower Miocene sands cover the phosphatic beds. In three pits they attain 70–120 ft. of thickness, at a depth of 230–300 ft. In the western sector the base of the phosphate has been calcified, and this is particularly intensive in the southwest. Between the flexures of the trough the commercial deposit occupies 1.8 square miles. Probable reserves attain 300 million tons. At three points overburden/ore ratios vary between 2 and 4.

Situated at the eastern extremity of the Djebel Onng anticline, the *Rass Mergueb* and *Tir* sectors are covered by Quaternary sands. Coprolithic beds, attaining 110 ft. of thickness, contain only 53% tricalcic phosphate, 25% $CaCO_3$ and a reserve of 7 million tons. The *Djebel Tarfaya* rises north of the central and western bled Djemidma. In the eastern part of Tarfaya a thrust has doubled the thickness of the ore beds to almost 200 ft. However, satellite faults facilitated rubefaction, the development of silex, and the penetration of sand. Conversely, transgressive Burdigalian sediments have decalcified part of the upper bed. Therefore grades vary between 60 and 63% tricalcic phosphate. In the eastern sector the thickness of coprolithic (high grade) ore increases. Concomitantly, the thickness of pseudo-oolithic phosphates and intercalations diminish. Grades vary between 58 and 61% here. The reserves of Tarfaya attain 12 million tons.

The wide Kef es Sennoun is covered by slightly warped Ypresian and Lutetian limestones. At Garet ech Chela, and at other parts of the plateau, we can distinguish minor flexures and fractures. The phosphatic series of the southern slopes include marls overlain by

Gastropode limestones, *(a)* 11 ft. of fine-grained pseudo-oolithic phosphates, *(b)* 53 ft. of fine-grained coprolithic but partly crossbedded phosphate, *(c)* 45 ft. of coprolithic phosphate, *(d)* 15 ft. of white coprolithic phosphate including pebbles of silex, limestone and bone and *(e)* phosphatic silex limestone (15 ft.). In this large, ½ mile wide tabular area, the greatest thickness attains 120 ft. under a cover of 220 ft. The coprolithic levels exhibit the best values, while grades diminish along the borders towards the Djebel Fouris in the southwest. The upper 45 ft. of coproliths contain 60–63% tricalcic phosphate, but this diminishes to 55% in the peripheral beds. Reserves reach 110 million tons (61% grade); the overburden ore ratio is also favourable.

TUNISIA

The Algero–Tunisian phosphate belt trends west for 200 miles north of parallel 38° N and the Chott (lake) el Djerid west of the Gulf of Gabès. The principal deposits of Tunisia include the Redeyef–Metlaoui basin (50 by 10 miles), the Aïn Moulares lenses, the twin beds of M'dilla and Djebel Berda (15 by 7 miles each) further north along the frontier, and the Djebel Ank–Djebel Chemsi basin (20 by 7 miles). Minor occurrences halfway between Sfax and the frontier include *Mezzouna, Maknassy, Meheri Zebbeus* and the *Djebel Mechaib* (Fig. 134). South of Constantine and southwest of Tunis, at *Kalaa–Djerda* and in the Kuif, small deposits of the Tébessa overlap the border.

Stratigraphy

Cretaceous. Alternating Lower Maastrichtian chalk and limestones, distinguished by a *Botrychoceras–Orbitoides* fauna overlie 400–900 ft. of Campanian marls. The later part of 200–300 ft. of marls and limestone inliers carries *Hemiaster, Inoceramus regularis, Ostrea villei* and *O. dichotoma*, and thus belongs to the Upper Maastrichtian. It is overlain by 150–250 ft. of *Inoceramus* limestones. The Cretaceous is apparently characterized by a neritic subsidence. Later a rubefied bank signals the late Maestrichtian emergence.

The fauna of Danian marls includes *Roudaireia, Scapanorhyncus* and various *Ostrea.* A *Venericardia beaumonti* limestone bed concludes the Danian. The depth of the Danian sea was variable, and the present Chotts (salt-lakes) emerged as shelves.

Eocene. A fauna of *Ostrea eversa, Echinantus* and *Cardita coquandi* distinguishes the overlying transitional beds, while *Thersitea verrucosa, Lucina moevusi, Pseudoliva fissurata, Venus grenieri* and *Venericardia pectucularis* characterize both the silt and lumachelle facies of the Thanetian. Lumachelle lenses, interbedded with marls and limestones, are most abundant in the southeastern Tunisian part of the basin between Gafsa and Négrine. A gypsum bed divides this complex from the 60 ft. thick (marine) marls in which the '*Thersitea* bank' is interbedded. While most of the phosphate series is Thanetian, its top might be Lower Ypresian. The phosphate fauna includes *Ostrea multicostata phosphatica* and *O. multicostata, Dinosaurus phosphaticus, Cardita gracilis, Terebratula kiski, Thersitea verrucosa, Lucina porchati* and *Aporrhais decoratus* accompanied by *Elphidium, Cibicides vulgaris* and *Anomalina acuta.* In the Gafsa basin a *Nautilus* conglomerate concludes this peak of phosphatization.

Ypresian limestone facies vary between *Ostrea multicostata* lumachelles in the Gafsa basin, marly silex limestones at the Djebel Onng, and coral reefs in the northwest.

Epeirogenetic movements of the Lutetian are reflected by marine marls, followed by lagoonal gypsum limestones, and then again by marls. The Pliocene orogeny concludes the Burdigalian–Pliocene emergence of southern Tunisia (see 'Petroleum' chapter).

Geologic history

Southern Tunisia emerged during the early Eocene, the movement progressing from the south towards the north, as indicated by two *Ostrea overwegi* lumachelles. The Straits of Zoui in eastern Algeria acted as a sub-surface shelf several times, admitting or excluding marine currents, facilitating the concentration of faunistic placers of bone fragments, coproliths and limonite–phosphate nodules. The Danian sea was not deeper than 30 ft., and phosphate sedimentation is related to the opening of the Straits' shelf.

There can be distinguished three epeiro–orogenetic cycles: Maastrichtian sub-emergence and warping, Burdigalian warping of the Djebels Onng, Mrata and Zréga around the Fériana dome, and finally Quaternary folding.

During the Thanetian and Lutetian, lagoonal and later glauconite facies preceded phosphate sedimentation. During the Danian and Ypresian, phosphatization was also followed by marine ingression. The latest concentrations correspond to phosphatic interbeds of late Thanetian–Danian marls. However, phosphatogenesis started in the Maastrichtian at the Djebel Berda and Kef el Aouidja, and terminated only during the Upper Lutetian in the Négrine chain and the Djebel Abiod.

VISSE (1952) has divided nine Thanetian phosphate beds into two series that are separated by a horizon of silicified phosphates. Beds No. 8 (the earliest), 6, 2 and 1 cover all workings, namely the Oueds el Dakla and Zireg in the Djebel Mrata, the Djebel Sif el Leham, Stah es Souda, the Tables Brachim, Zimra and Chouabine, Tseldja north and south, Jaat Cha and Kef el Dour. Bed No. 7 has only been found in the four latter localities, and Beds No. 3–5 are missing in the Djebels Mrata and Sif el Leham. Bed No. 4 has also disappeared from Tseldja South, Jaat Cha and Kef el Dour. The thickness of the more important beds attains 30 ft. at M'dilla in beds 1 and 3, and 6 ft. in beds 5 and 6. The combined thickness of beds 1 and 2 is more than 20 ft. in the five last localities, almost 20 ft. in the Djebel Mrata, and 10–12 ft. at other outcrops. While beds 6, 7 and 8 may attain a thickness of more than 3 ft. each at the Oued Jaat Cha, Kef el Dour and M'dilla, beds below level 3 are rarely more than 3 ft. thick (bed 8 at Sif el Leham, Stah es Souda, Table Chouabin and Tseldja North, bed 4 at Table Brachim and Tseldja South).

Phosphatization

Phosphate layers are *(a)* either thick and rare or *(b)* thin, conformous and repetitive. Whereas case *(a)* applies to emergent straits of the eastern Moularès–Oued el Zireg subbasin and outlying Table Brachim at Redeyef, case *(b)* prevails in more distant areas of the basin, e.g., Djebel Onk and most of Moularès. The correlated phenomena of erosion, reworking and phosphatization spread from the straits. While both the sixth phosphate layer and its roof are missing along the Oued el Zireg–North Tseldja axis, layer 2b anomalously overlies layer 4 along the northern slopes of Djebel Alima at Oued el Zireg. Gravel and smaller boulders, as well as smaller size material, has been eroded from layer 3 and its interbedded *Aporrhais decoratus–Cardita gracilis* limestone horizon. These detriti have been reconcentrated in three pebble horizons in the lower part of layer 2. Erosion was particularly intense before the accumulation of layer 2 along the main marine channel at North Tseldja, though it diminished east of Moularès at Sif el Leham. Following periods of erosion and reworking, these straits provided space for thicker phosphate accumulation in layer 8 of Sif el Leham and in layer 6 of Metlaoui–Kef ed Dour.

Between Table Brachim and Djebel Chouabine in the Redeyef basin, *Cardita–Aporrhais* limestone beds have been partly eroded from layers 3 and 4, and reconcentrated in a pebble horizon. Similarly, at Stah es Souda the phosphate limestones of layer 8 of Djebel Mrata have been laterally reworked into pebbles. Indeed, thin and hard interbeds have been selectively eroded by strong channel currents from elastic phosphates.

Deposits

In the *Djebel M'dilla* (south of Gafsa) layers 5, 6, 7 and 8 have been best developed. Here limestones and marls are overlain by 4 ft. of phosphate and an interstratified limestone horizon. Lumachelle limestones and marls include the main 12 ft. thick phosphate layer and an interbed of marls. A series of *(1)* marls, *(2)* a thick layer of alternating gritty and siliceous phosphate marls and pebbles, *(3)* marls and *(4)* the phosphate-bearing limestone roof conclude the sequence. At this southern border of the basin both sand and lumachelle are abundant.

Near *Metlaoui* at 'Coup de Sable', 20 miles west of Gafsa, 18 phosphate layers occur. The two lower horizons of 3 and 2 ft. mainly contain coproliths. In a column of marls with subsidiary horizons of limestone and rare pebbles, the thickness of the phosphate beds oscillates between 2 and 5 inches. The thickness of the first major phosphate layer attains 6 ft., but two horizons of large and small pebbles are also bedded here. A horizon of bone fragments and perforated pebbles divides this layer from 2½ ft. of *Ostrea multicostata* phosphates, overlain by a lumachelle of various *Ostrea*. The next phosphate layer of 8 ft. is distinguished by an interbed of large limestone pebbles. In the overlying repetitive column marls alternate with a conglomerate consisting of limestone and gravel in a phosphate groundmass. *Ostrea multicostata* limestones conclude the sequence. Thus the lower phosphate beds and part of the intermediate layer beds have been thinned out, but layers 1 and 2 remain unchanged.

Near *Redeyef* (20 miles east of Gafsa and north of Chott Gharsa) at the Fort Espagnol in the Djebel Chouabine, a 3 ft. thick bone horizon is overlain by a gypsoïd and coprolithic phosphate bed including pebbles. We find the next large-grained phosphate bed of 2 ft. at the top of a column of marl, limestone, pebble and gypsum. However, most of the following phosphate beds only attain 3–8 inches of thickness. Among them six are compact rock phosphates; one of them changes into a *Thersitea* limestone reef, though laterally. In fact, the layer consists of marls alternating with some phosphate and limestone. Above the hard phosphate horizon most of the beds are coprolithic or interfoliated with marl. However, we find little limestone beneath the pebble horizon and bone fragment bed. We can then see a 3 ft. thick, somewhat coprolithic, phosphate layer overlain by a 2 ft. thick, similarly repetitive, series of more compact phosphates. Thinner beds of foliated phosphates interbedded with marls complete this facies. The thickest phosphate bed (7 ft.) is oolithic, but it includes several pebble horizons. Almost 2 ft. of conglomerate embaying *Ostrea*, Gastropoda and shell fragments in a phosphatic groundmass divide this bed from the next, a 4½ ft. thick layer of oolithic phosphate. Marls and phosphate breccia overlain by Ypresian limestones conclude the sequence.

Thus the lower phosphate beds No. 5–8 are hardly developed. In fact, the only well-developed phosphate layer combines beds 1 and 2 in a 13 ft. thick column, interrupted by marls and conglomerates and characterized by *Cardita gracilis*, *Aporrhea* and *Ostrea multicostata phosphatica.*

At *Eastern Moularès* phosphate sedimentation is different: The sequence commences with 6 ft. of coprolithic phosphate of bed 8 showing interstratified gypsum and pebble alignments. Gypsum occurs in several overlying phosphate beds and among the country rock marls. Although most of them are thin, the last is 2 ft. thick. In the overlying column, seven phosphate beds alternate with marls. *Cardita gracilis* is abundant in the 2½ ft. of coprolithic phosphate of bed 4. Until 5 ft. of oolithic phosphate are reached in beds 2 or 1, marls, coprolithic phosphate beds, pebble horizons and occasional phosphatic limestones alternate. Finally, the top of the column consists of 3 ft. of marls, 2½ ft. of phosphate and 15 ft. of phosphatic conglomerate, covered by a limestone roof.

EGYPT AND LIBYA

Discontinuous Upper Cretaceous phosphate lenses are found at Sofeggin, *Libya*, which attain a maximum of 8 inches of thickness. Phosphatic limestones occur above at Santitiggo. The Lower Eocene stratigraphic discontinuity eliminates the possibility of finding phosphates in western Syrte. However, coprolithic phosphates and bone relics show in the Upper Eocene at Zella.

Deposits occur on latitude 25° N in Egypt. The most conveniently located deposits of the *Safaga* and *Quseir* districts (Hume, 1927) are situated along the sedimentary strip of the Red Sea coast, approximately 100 air miles south of the entrance of the Gulf of Suez. Minor deposits of the Nile Valley are mainly located on the east bank between Idfu and Qena, 100 miles north of Aswan; they include Sibaiya and Mahamid. Phosphates also outcrop in the *Kharga* and *Dakhla* oases of the Sahara, 100 and 250 miles west of the preceding occurrences (El Shazly, 1956).

Campanian–Maestrichtian beds of oolithic collophanite grit more than 6 ft. thick carry shells, bones, rock fragments and quartz grains in an argillaceous ground mass. Carbonate or hydroxyl apatite occurs in the bones and teeth of reptiles, saurians and fish. Detrital material has apparently sedimented in Egyptian phosphates. Uranium content attains 0.008% at Sibaiya and 0.004% U_3O_8 at Quseir.

SENEGAL

Northeast of Dakar the phosphate district is underlain by Cretaceous sediments in which Maestrichtian sands and sandstones represent the lower hydrostatic level. Following the Paleocene phosphate nodules were disseminated in two marine, Lower Eocene horizons, topped by Upper Ypresian phosphatic clays and Middle Eocene Lutetian clays with hard nodules and masses containing up to 85% tricalcic phosphate (ANCELLE, 1959). Laminated Lower Lutetian marls were then transgressed by beds of large phosphate lenses, quartz and sand with thin inliers of clay and Foraminifera (*Nummulites*) silexes. This fine-grained (μ–mm size) and high grade (80–85% tricalcic) phosphate forms most of the Taïba and Lam–Lam deposits. A 5–10 ft. thick residual piedmont of hard rock-phosphate, argillaceous sand, boulders of sandstone and basalt covers vast areas of Cayor (Kellé Mékhé, Pire Goureye), Sine Saloum and Djoloff, but the iron aluminium rock contains only 75% tricalcic phosphate. Pire Goureye and Se Méké are minor deposits. Lutetian sediments cover 15 by 7 miles of the phosphate zone which extends for 50 miles between Tivaouane and Louga, while sub-phosphatic laterites cover large areas of the Thiès Plateau and concentrate at Pallo as aluminium phosphate. Late eolian sands terminate the sequence (Fig. 15).

Calcium phosphates

In the 10 square miles of the *Taïba* deposit we can distinguish three areas: *(a)* The western, 4 square mile area constitutes the deposit proper, *(b)* in a 3 square mile area the proportion of interbedded limestone ($CaO/P_2O_5 > 1.6$) and overburden increases, *(c)* the depth of the cover augments further in the northeast.

In addition, we can divide the northern part of Taïba proper into three zones:

(1) The Keur Mor Fall beds occupy 1,7 square miles. Covered by 55 ft. of overburden, they attain 18 ft. of thickness. Concomitantly, the grade increases from 58 to 82% tricalcic phosphate in the concentrates.

(2) The Ndomour Diop zone covers 1 square mile, with the thickness of the phosphate and overburden attaining respectively 25 and 85 ft., and the grade 61.5%.

(3) The Intermediate zone covers $^2/_3$ square mile. The 25 ft. thick deposit lies under a 90 ft. thick cover. The standard column consists of *(a)* a bedrock of montmorillonite and subsidiary palygorskite surrounding calcite, *(b)* a phosphate bed including silex nodules, clay, and *(c)* a discontinuous lateritic crust and sand.

At Thiès we can distinguish three types of calcium phosphate: *(1)* surrounded by staffelite, apatitic brecciated phosphate grains of $^2/_3$ inch sedimented from the sea: *(2)* resembling marls, bedded phosphates formed by the phosphatization of lime are contained in marls; *(3)* powdery phosphates or lenses with abundant montmorillonite surround detrital quartz and silexes which embay phosphatized fossils. This variety has precipitated in the sea, and phosphatic acid of the aluminium phosphates has been derived from it (LATRILHE, 1959).

South of Taïba at *Lam–Lam*, 20 ft. thick lenses contain 78% of powdery tricalcic phosphate. They occur in Upper Lutetian sands overlying foliated clays and 30 ft. of laterite. This sequence is covered by 15 ft. of sand. South of Thiès at *Keur Mamour* 30 ft. of greenish yellow unbedded phosphates are underlain by Ypresian calcareous marls, but the clays contain only 20–40% tricalcic phosphate, although scattered lenses average 70–90%.

Aluminium phosphates

The *Pallo Yungo* phosphates, which contain 28.5% P_2O_5, attain 20–30 ft. of thickness, outcropping or overlain by not more than 15 ft. of sand. Alumina grades in the area oscillate between 20 and 30%, but iron and silica content is more variable. Alternating marly limestones and clays underlie the aluminium phosphates (CAPDECOMME, 1953). Most of the fossils have been calcitized; however, they contain only 1–2% tricalcic phosphate in nodules or in relics of bony tissues. Beneath the aluminium phosphates at Pallo, e.g., montmorillonites embay detrital quartz and fine-grained phosphate.

The lime phosphates of *Lam–Lam* are underlain by foliated palygorskite. Recemented by staffelite, the lime phosphate breccia of Thiès may have been derived from sub-marine shelves. In porous bedded phosphates, lime phosphate lenticules are enveloped by layers of clay. Apparently collophanite has been replaced by calcite in the carbonatic layers of marls. Moreover, disseminated nodules, phosphatized Foraminifera and tissues are particularly widespread in montmorillonite that has concentrated in marine basins.

Most of the aluminium phosphates have been derived from lenses of pure powdery phosphate in the following sequence: The kaolinized and rubefied argillaceous border layer, containing phosphate patches or 'dolls', is overlain by a breccia which grades inwards into a porous ground mass embaying nodules. The nodular rock is covered by a powdery surface zone of augellite, which frequently surrounds pallite (LATRILHE, 1959). At Pallo, augellite also covers fissures in the underlying horizon. With crandallite filled cavities, early pallite (which contains more iron but less aluminium and sodium than millisite) constitutes most of the nodular rocks and breccia. During the continental leaching of calcite, paillite gel coagulated in an acid environment and crumbling produced the breccia. As it is unstable in the present climate we surmise that paillite has been derived from montmorillonite and powdery phosphate during alternating rainy and dry seasons. Then hyperaluminous augellite developed by surface leaching from pallite during the dry seasons. Concomitantly crandallite crystallized from descending solutions by hydrolysis. Paillite includes disseminated anatase and uranium. Similarly, uranium forms inclusions in tricalcic phosphates of the apatite type, partially entering the lattice of recrystallized minerals such as crandallite and staffelite.

TOGO–NIGERIA

Between the southeastern corner of Ghana and southwestern Nigeria the early Eocene coastline strikes N 60° E, traversing Togo and Dahomey (Fig. 54). However, only in *Togo* have the phosphate beds been well preserved from erosion. Mesozoic and Tertiary sediments filled in a triangle of lower Togo, which is only 1 mile wide, west of Lomé, the capital, although it expands to a width of 15 miles on the eastern frontier. Dipping a few seconds southeast, the monocline of Ypresian phosphates strikes N 50° E. The thickness of the phosphates varies between 5 and 20 ft., averaging 13 ft., and the main bed is locally underlain by a thinner layer of phosphates. The phosphates contain 25–40% clay and traces of uranium. A 12–30 ft. thick cover consists of soil formed on the reefs and argillaceous sands; in the southeast, however, its thickness increases to 80 ft. The overburden/phosphate ratio thus reaches the mark of 6/1. The phosphate belt extends for 18 miles and consists of five deposits: Adeta and Kpomé northeast of Lomé and west of Lake Togo, and Hahotoé, Dagbati and Momé northeast of the Lake. Hahotoé covers a 7 by 1 mile area, but the combined area of each of the other deposits only attains about 1 square mile (Fig. 135).

The Togolese phosphate belt extends through Lama and (Fig. 24) Toffo *(Dahomey)* to

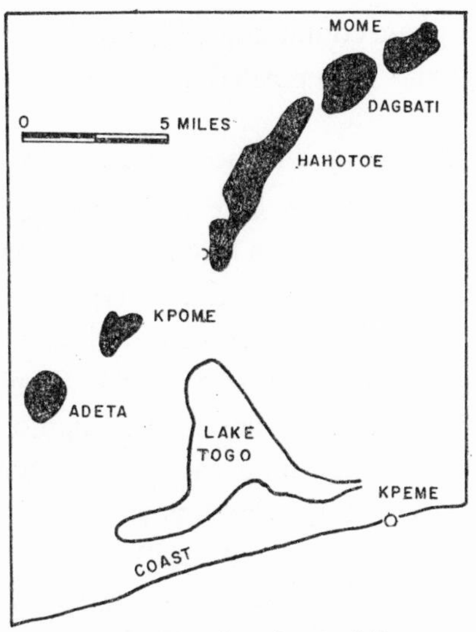

Fig. 135. The phosphates of Togo.

Ososhun, 20 miles from Lagos (in the southwestern corner of Nigeria). In Dahomey Lower Eocene beds are very thin, lying 200 ft. above the horizon of Nummulite limestone. Between Ifo Junction and Ajegunle at Ososhun, Middle Eocene Lutetian clay, marl and sandstone include lenses and several thin horizons of phosphate. Phosphate also occurs at Akinsinde, as does amorphous aluminium phosphate and disseminated wavellite further south.

THE REST OF AFRICA

Extending the phosphate zone of the Spanish Sahara, sandstones of the Bay of the Lévrier near Port Etienne in the northwestern corner of *Mauritania* contain 14% P_2O_5 and 38% SiO_2 (Ikiaone Island). Between Matam and Kaédi the Civé deposit is situated 200 miles east of Saint Louis on the northern bank of the Senegal River. Separated by thin beds of argillaceous shale, four lenticular layers, which contain 55–60% of tricalcic phosphate, attain a total thickness of 3 ft. In *Mali* the Tamaguilet deposit is interbedded in a north trending Eocene ridge of clays north of Gao on the Niger. Attaining 6–8 ft. of thickness, these powdery beds contain 60–70% tricalcic phosphate.

In the Congo Republic we find phosphates at Hollé, notably at Tchivoula; the ore contains 20–25% P_2O_5. In the *Congo*, Upper Cretaceous, Senonian, yellowish phosphatic limestones form a 7–15 mile wide, north-northwest striking, band. It extends from Zambi on the Congo estuary to Luali near the confluence of the Lukula and Shiloango Rivers, and then penetrates into the Cabinda enclave and south into the Lucunga Valley of northwest Angola. The phosphates are radioactive. (Glauconite sandstones of Inhoca near Lourenço Marques, Mozambique contain 2.7% phosphoric acid.)

GUANO

Mainland and coast

On Corail Island in the Los Archipelago of *Guinea*, water percolating from guano is believed to have developed phosphate containing 33% P_2O_5, 25% Al_2O_3, 6% Fe_2O_3, 2% CaO and 5% SiO_2. In *Somalia* guano is deposited on Mait Island near Cape Humbeis; the material contains 17% P_2O_5.

Tanganyika. On Minjingu Hill, 50 miles west-southwest of Mount Kilimanjaro, 10 million tons of phosphate have accumulated in lake beds. When the level of Lake Manyara was 250 ft. higher than now, phosphate was leached from guano and then precipitated on the periphery of the hill from carbonatic water. This type of phosphate has been cemented by silica; it contains 21.4% P_2O_5. Farther from the Hill, however, phosphates and clay alternate; this powdery phosphate averages 18.5% P_2O_5. Similar phosphatic lake beds rise in the Challi Hills in the centre of Tanganyika and at Chamoto Hill in the Southwest. Bat guano has been deposited in the limestone caves of Amboni in the northeast, and at Sukamawera in the southwest. The latter contains 10–20% P_2O_5. Rising 10 ft. above the ocean, Lantham Island occupies 1,000 by 500 ft. 40 miles southeast of Dar es Salaam. A thin layer of guano covers its coral limestone and contains 8.5% P_2O_5.

Guano covers small islands of Lake Nyasa, and also occurs south of Lusaka.

South Africa. At Zoetendalesvley in the northwestern Transvaal, iron–aluminium phosphate (probably derived from guano) occurs in breccia and veins underlain by dolerite. North of Port Elisabeth phosphate nodules are disseminated in lenses of early Karroo (Permian) shales of the Sundays River. In the Hoedjes Bay Peninsula, 70 miles north of Cape Town, phosphatic solutions percolating from guano into granite and quartz-porphyry produce a mixed rock which contains iron aluminium phosphate; phosphorite develops in lumachelles.

Southwest Africa. Guano is deposited on the Southwest African coast and on islands. While bat or cave guano contains little phosphorus, sea birds drop high grade guano on platforms erected on the coast.

Islands

Two islets known as *Kal Farun* rise 280 ft. from the sea off Abdelkuri between Socotra and the Horn of Africa.

In the Seychelles, Bird Island is a flat calcareous sandbank occupied by colonies of *Sterna fuscata.* The guano of adjacent Dennis Island contains 27.7% P_2O_5. On the Amirante Islands south of the Seychelles, some guano has been deposited on Saint Pierre and its neighbours. *Astove* Island, the principal deposit situated east of Saint Pierre, is a 2 miles long raised coral atoll. Near the deposits travellers reported seeing bones, possibly of giant tortoises. Northwest of Madagascar the guano of Assumption Island contains 61–73% CaP_2O_6 or 28–36% P_2O_5. In addition, the principal deposits in the ring of Aldabra are located on Picard or West Island. On Cargados or Carajos Island northeast of Mauritius, *Sterna fuscata* colonies deposit guano, and some guano may also come from the cays of Rodriguez Island, the easternmost of the Mascarenes.

In the Mozambique Channel near *Madagascar* guano develops on coral limestones between the Iles Glorieuses and Europa Island. On Juan de Nova Island the principal deposit consists of collophanite concretions and powder averaging 29.1% P_2O_5, i.e., 63.6% tricalcic phosphate.

In the south Atlantic explorers report guano soils, but add that no 'discrete deposition takes place' on Tristan da Cunha (HUTCHINSON, 1950). According to other reports there is guano on Egg Island off the Saint Helena coast, and on Saint Helena in Prosperous Bay a thin cake of phosphate contains 25% P_2O_5, with powder including 18% P_2O_5. Boatswain Bird Island is the main nesting ground of Ascension Island, where guano is also believed to have been dug from clinker. Finally, on the Ilheu das Cabras near Sao Thomé, trachyte alters to aluminium phosphate.

CARBONATITES

Middle Africa

Southwest of Lake Edward apatite occurs in the Lueshe carbonatite of the *Congo* (see 'Niobium' chapter).

Uganda. Bukusu and Sukulu, the principal deposits of the East Uganda carbonatite belt, are located northeast of Lake Victoria (Fig. 33). The Busumbu–Namekara–Nakhupa ridge (one of the outer rings of Bukusu) consists of apatite, magnetite and vermiculite. The Busumbu deposit measures 3,000 by 400 yards, with 30–200 ft. thick apatite soils which contain 3.2–44.6% P_2O_5. The deposit continues under the lateritic cover and 130 million tons of soil which average 13.1% P_2O_5. With depth values increase, and apatite locally recrystallizes as staffelite. We find apatite-bearing soils in three valleys of the Sukulu complex (Table CLIV).

The hard rock contains 20.4–33.9% P_2O_5.

TABLE CLIV
THE PHOSPHATE VALLEYS OF THE SUKULU COMPLEX

	Width (miles)	*Depth (ft.)*
North Valley	$^1/_2$	70
South Valley	$^2/_3$	220
West Valley	1	80

In *Kenya*, northeast of Lake Victoria up to 30% apatite occurs in the Ruri, Homa and Rangwa complexes, and south of Mombasa gorceixite shows in the zone of alteration of the Mrima complex. Apatite is found in the Mbeya complex, Tanganyika, at Isoka, Zambia and in the carbonatites of the Shire Valley, Malawi (Fig. 52).

In *eastern Zambia* four carbonatites have been emplaced on the Mozambique border in a downfaulted Karroo block of the Rufunsa Valley near the Zambezi: consisting of alternating layers of carbonatized apatite agglomerate and sövite, the concordant sill of Kaluwe is 6 miles long and 800 ft. wide. The intermediate phase of Kaluwe is particularly rich in phosphate. Indeed, soil values of more than 10% P_2O_5 appear to be restricted to this zone. The other carbonatites exhibit more conventional structures which include bands of fine-grained apatite at Nachomba and Mwambuto. Moreover, the soils of Mwambuto and Chasweta contain 3–3.5% P_2O_5.

Southern Africa

Southern Rhodesia. Along with its sister complexes Shawa and Chishanya, the *Dorowa* complex has been intruded, into Precambrian granodiorites 100 miles south of Salisbury. Irregular bodies and dykes of sövite, magnetite veins and apatite–vermiculite–magnetite stringers penetrate into the feldspar–augite fenites, occupying an ellipse of $1^1/_2$ by $^4/_5$ miles. The inner fenites contain only sugary apatite. Conversely, in altered yellowish ground decomposed from alnöite that replaces the syenitic fenites of the northern rim (TYNDALE-BISCOE, 1959), abundant phosphate is accompanied by magnetite. Other phosphates outcrop in the south. Reserves averaging 8% P_2O_5 attain 37 million tons. Pyroxenite (tveitasite) has been located in depth. Furthermore, in the Shawa complex sugary apatite is associated with magnetite in sövitic zones of the inner rauhaugite ring, in pyroxenites, and in late ijolite bodies of the outer ring.

South Africa. Apatite concentrations are greatest (10% P_2O_5) in the 400 ft. wide serpentine–olivine–magnetite or inner brecciated phoscorite ring of the *Phalaborwa* carbonatite (see 'Vermiculite' and 'Copper' chapters). Apatite crystallizes in the ground mass and in blebs of several ft. Containing 5% P_2O_5, spotted rock penetrates into the outer ring of pyroxenite, while late alvikite (calcite) veins traverse both formations. Resources of the western Loolekop attain 40 million tons, and they continue in depth. Apatite is also disseminated in the pyroxenite, and particularly in feldspar schlieren, but values are erratic.

DIATOMITE

The diatoms living on shallow shelves and in lakes extracted their silica from late Tertiary lavas. The distribution of these volcanoes predetermines the location of diatomite deposits. The largest occur in the Middle Miocene of western Algeria. The occurrences of the Rift Valley, Nigeria and southern Africa are of Pleistocene age.

Mascareignite resembles diatomite, but it is heavier, containing few diatoms. In the Kedong and Sagana Valleys and near Thomson's Fall in Kenya, mascareignite has been derived from residues of reeds and grasses by silicification. Spongiolite has been found on the shores of Lake Kivu, in the Congo.

ALGERIA

Diatomaceous earths occur in Miocene–Pliocene sediments in the synclinal warp that connects the plains of Macta (Fig. 11, 59), Mina and the lower Chélif River. Outcrops have been located along a 110 mile long stretch extending from St. Denis-du-Sig to (Fig. 136) Flatters in the east and northeast. The width of the syncline varies between 30 and 40 miles; its northern limb only reaches the coast in the west, between the oil harbour of Arzew and Mostaganem, while the chain of Dahra separates it from the shore further east. The southern limb of the fold rests upon the Beni Chougrane in the west and upon the Ouarsenis Mountains in the east. The sequence is given in Table CLV (TENAILLE et al., 1951).

Along the southern limb of the syncline, diatomite outcrops west and southwest of St. Denis-du-Sig. These deposits are being exploited. A 30 mile long and ½–1 mile wide band

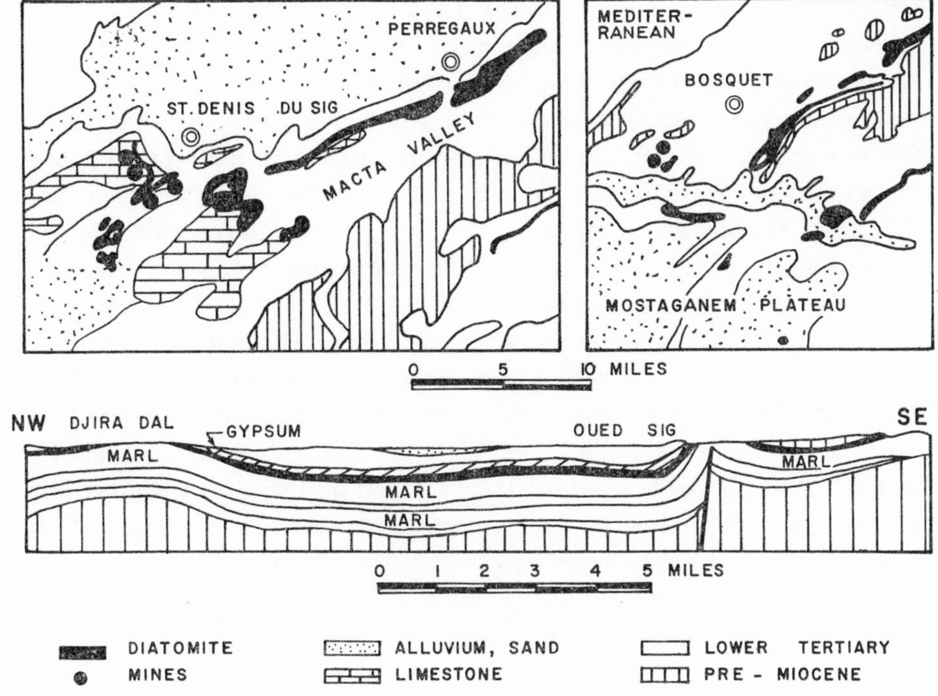

Fig. 136. The lower Chélif Valley. (After PERRODON, 1960).

TABLE CLV

STRATIGRAPHIC SEQUENCE OF CHÉLIF. ALGERIA

Holocene		Alluvium, locally sands, crusts and a lacustrine facies
Upper Pliocene	Villefranchian	red continental sediments
	Calabrian	lumachelle
		sand and continental silt with *Helix*
	Astian	marine sand
Lower Pliocene		blue marls, interbedded sandstone
(max. 2,200 ft.)		Lithothamnia limestone
	Plaisancian	sandstone
		GYPSUM AND GYPSEOUS MARL
		Lithothamnia limestone and/or
		DIATOMITE AND DIATOMACEOUS MARL
		sands of El Bordj
	Vindobonian	blue marl
Upper Miocene	(1,000–3,000 ft.)	basal sandstone
		continental sediments of Bou Hanifia
Lower Miocene	Burdigalian	marl
	(2,000–4,000 ft.)	sandstone and conglomerate

of diatomite extends from St. Denis beyong Perrégaux; diatomite also outcrops near L'Hillil and St. Aimé. Northeast of Bou-Guirate, a narrow band of tripoli, related to an undulation of the synclinal fold, is found in the centre of the basin.

On the northern limb, windows of diatomite rise between Bosquet and Mostaganem, and related outcrops continue for 20 miles east-northeast. The principal diatomite band of the northern limb extends for 80 miles from beyond the confluence of the Chélif and Mina rivers to Ain Zeft, Renault and Fromentin, where it turns east, thus terminating the basin. The width of this outcrop varies between 1,500 ft. and 1 mile. South of Fromentin, diatomites emerge along an axial undulation.

The thickness of the diatomaceous marls averages 600 ft. The diatomite consists of frustules and fragments of the fossil, with a silica content that generally oscillates between 80 and 90%. Part of the silica is believed to have been derived from submarine andesite flows and cinerite. Grades attain 74 and 86% at St. Denis. The Pelot quarry is situated 2 miles from that town. This diatomite layer is 50 ft. thick and dips 65°. At Cheurfas three beds have been distinguished; the thickness of each varies between 3 and 6 ft., and they dip 7–10°. Two beds of Douar El-Mehdi are respectively 10 and 2½ ft. thick and dip 15°.

The diatomites of Ourllis, near Bosquet, contain 88.2% SiO_2, 1% Fe_2O_3, 1% Al_2O_3, 0.3% CaO and 9% water. Here seven sub-horizontal, 3–5 ft. thick beds have been distinguished. Of adjacent Douar Mehalif, six beds are also sub-horizontal, and the thickness of each varies between 1½ and 3 ft.

In neighbouring *Morocco*, diatoms lived in small lakes, near Oulmès and Bou Gafer, which were dammed up by recent volcanoes.

NIGERIA

Laminated, high grade diatomite covers a 200–300 yard wide basin near *Bulabara*, Bornu province, northeastern Nigeria (VARLEY et al., 1956). The diatomite contains 78% SiO_2, 2% Fe_2O_3, 6% Al_2O_3, 1% MgO, 0.4% TiO_2 and 0.4% CaO. At *Abakire* (Fig. 24), Bornu province, 8 ft. of less laminated but more massive diatomite is interbedded in the gritty, compact clays, and a thin layer of sandstone underlies the base. Diatomite also outcrops between Abakire and Maiduguri, at Badice, Garin Maina and Bugzu, and lies at a depth of 150–200 ft. at Gujba, Bolokolo, and Sabon Kasasa. *Melosira granulata* RALFS probably sedimented in shallow Holocene lakes at Balaraba and Abakire. The fine fractions contain some insect relics. Other occurrences are believed to be of Tertiary, or Plio–Pleistocene age.

Diatomite lenses are found in Borkou, *Chad.*

EASTERN AFRICA

Diatomite occurs in the Rift Valley, east and south of Addis Ababa, *Ethiopia*, near the sources of the Webi Shebeli. In the Laga Tabo Valley, an argillaceous diatomite layer is covered by tuff.

Diatomite is found in northwestern *Uganda* at Panyango, Alui and Atar, and further north on the Ambosi River. At Panyango, along the Pleistocene scarp of the Nile, six diatomite horizons are interbedded in a 120 ft. thick column of clay, and overlain by 80 ft. of sand. Of the upper 55 ft. of clay 22 ft. consist of diatomite; the best quality is found in a thick bed, the lowest among three. Of this bed 4 ft. are of particularly high grade; the rest is of moderate to good quality. Only 8½ ft. of diatomite are found in the lower clays.

At Alui, a 5 ft. and a 3 ft. thick bed of diatomite lie under a few ft. of cover, and on the Atar, more than 3½ ft. of diatomaceous earths lie beneath an overburden of 8 ft.

Kenya. Pleistocene lake beds of the Rift Valley contain extensive deposits. The Kariandus occurrence is situated near Gilgil, 80 miles northwest of Nairobi. The thickest layer of diatomite is more than 100 ft. thick and it includes bands of tuff and pumice. The principal diatome is *Melosira;* it is dazzling white and averages 84.2% SiO_2 and 5.5% H_2O. It has a block density of 28 lb./cubic ft., and a powder density of 7½–9 lb./cubic ft., following screening through a 120 mesh B.S. sieve. After ignition, the material absorbs three to three and a half times its volume in water.

Other deposits include those of Soysambu, near Elmenteita, near Eburru, Subukia and Nderit on the Koora plains, and Gicheru in the Kedong valley.

West of Lake Victoria, in *Tanganyika*, 9 inch–3 ft. thick beds of diatomite are overlain by 9–13 ft. of overburden *(1)* between Nyakanyasi and Kikagati, and on both sides of the Kagera River, *(2)* between Nyakanyasi and Kyaka and *(3)* west of the last locality; the material contains 75–77% silica. Diatomite has been drilled in depth at Makutapora, in west-central Tanganyika, and diatomaceous earths outcrop in the Rift Valley.

SOUTHERN AFRICA

In *Malawi*, diatomaceous sediments cover the bed and shore of Lake Malombe, and they also occur in the Karonga district. Irregular beds of low grade diatomite outcrop in the 60 mile long belt that follows the *Mozambique* coast between Manjacaze, Bilene and Manhiça, north of Lourenço Marques, and also south of that city, at Nacala and at mile 17 of the Goba railway. The contents of two of them are given in Table CLVI.

At Sambaina, in central *Madagascar*, 4–5 inch thick beds of diatomaceous earths are interbedded with lignite, lacustrine clay and bituminous shale.

A belt of small Quaternary deposits of diatomaceous earths crosses *South Africa*, southern Bechuanaland and Southwest Africa, connecting Louis Trichardt, in the northern Transvaal, with Port Nolloth on the west coast. The thickness of beds generally varies from 3–6 ft. Occurrences include Kannikwa and Witdraai in the Western Cape, and Amersfoort, Tweeputkoppies, Souting and others in the Transvaal. In the Bankplaas swamp, near Ermelo, diatomite and phytolith are disseminated in 15 ft. of peat.

TABLE CLVI

COMPOSITION OF SOME DIATOMITES IN MOZAMBIQUE

	Bilene (%)	*Nacala* (%)
SiO_2	86.9	87.3
Al_2O_3	3.9	3.5
Fe_2O_3	0.7	0.7
CaO MgO TiO_2	0.1	1.4
K_2O Na_2O	—	0.2
H_2O	8.3	6.9

SALT

Sodium salt or halite is the most abundant salt and sodium mineral. In lagoons, halite starts precipitating in a dry, hot climate after hematite, limestone and gypsum, when the original volume of water is reduced to 9.5% and a density of 1.21 g/cm^3 can be measured. During this process, first $CaSO_4$, $MgSO_4$ and $MgCl_2$, then NaBr, and finally complex potash salts precipitate with NaCl. Since salt is plastic and light, it tends to form diapyric and slab-like structures. The principal cycles of salt deposition are Triassic in the northern Sahara, and Jurassic and Cretaceous in Tanganyika and Gabon. Salt is also actively evaporated in coastal ponds which imitate lagoonal conditions. In desert depressions, crusts of salt an inch to a ft. thick develop during the dry season; salt is also one of the few eolian sediments. Saline brines drain sediments, but many springs of the Rift Valley are of fumarolic origin.

NORTH AFRICA

Libya. Sebkhet, or salt pans, have been found along the coast at: *(1)* Zuara west of Tripoli, *(2)* south of Misurata, *(3)* south of Bengasi, *(4)* in the Syrte, northwest of the Zelten oil field, *(5)* at Jarabub on the Egyptian border, and *(6)* in a broad belt parallel to the Libyan line (west-northwest), extending from the Cufra oasis to north of Murzuk, roughly 400 miles from the coast.

The last belt can be divided into a 150 mile long eastern sector from Zerg to Tazerbo, and a 350 mile long western zone connecting Uau-en-Namus to Edri. The Syrte basin includes the Agheila, Marada and Sahabi districts; the Marada basin extends 90 miles. The Jarabub and Marada column consists of salt lenses in limestones, the karst surface of which received the infiltrating waters, from which salt concentrates by evaporation (Desio, 1943). Magnesium is abundant at Jarabub.

Salt is abundant on the *Tunisian* coast and in the chott, or salt lake, district of the south.

In *Algeria,* four types of salt deposits can be distinguished: rock salt layers, chotts (salt lakes), sebkhet (salt pans) and littoral ponds. Thick layers of salt are underlain by conglomerate and sandstone in the Triassic oil province of the northeastern Sahara, e.g., at Hassi Messaoud; dolomite and shale occur in the same sequence. Salt layers overlap Triassic sandstone in the Hassi er Rmel gas field and at Colomb Béchar, in the northwestern Sahara near the Kénadza coal field, the thickness of youngest Cretaceous salt layers reaches the mark of 1,600 ft.

More than 300 tons of salt are precipitated each season in the sebkhet of Amadror, in the northern Hoggar.

In the *Constantine* area of northeastern Algeria (Fig. 11), the salt ponds of Sétif, the lake of El Bahira, and the Djebel Kasbah basins are believed to drain Triassic salt. Salt lakes form swarms between Batna and Aïn Beïda. South of them chotts and rock salt diapyrs, covered by some gypsum, are aligned along the axis of the Saharan Atlas. One of these diapyrs was worked in the eleventh century at El Outaïa, according to El Bekri's description of Africa. The salt dome of Metlili covers only 1,000 by 2,000 ft. The playa of Télemine in the Miocene St. Cloud basin, which extends for 3 miles, acquires a 2 inch cover of salt each summer. The Arzew basin, which occupies 8 miles near Oran, is being salified by waters percolating through the sandstone.

In the sebkhet of *Oran*, which covers 1 sq. mile, more than 500 million tons of salt have accumulated from waters traversing the Triassic beds. The sebket Bou Ziane, in the Chélif Valley rejuvenates its salt supply each season. Each summer the Zahrez Chergui, spreading over 1 sq. mile, is covered by more than 1 ft. of salt. Reserves have been estimated at more than 300 million tons, but neighbouring Zahrez Rharbi contains only 200 million tons.

In *Morocco*, salt evaporates in sebkhet, or salt pans of haline soils, from meteoric water. The principal salt lake, *Zima*, is situated 15 miles inland from Safi harbour, near the Youssufia phosphate deposits, i.e., roughly 15 miles north of Marrakech. The hydrostatic table drains salt from Quaternary sediments derived from underlying Permian–Triassic salt beds, and accumulates it in the basin. The Baumé grade of these waters reaches the mark of 6.5–17 (sp.gr. 1.05–1.14).

The Triassic salt beds of the Pre-Rif (northwestern Morocco) include Dessenon, Taza, Touaba, M'soun and Nsala en Oudaïa. At *Taza*, blue marls are overlain by folded green and red marls, comprising a 10 ft. and an 80 ft. thick salt bed covered by Cretaceous Ammonites and pyrite marls. The Tissa salt mountain lies on the Wadi Lebène. South of this belt, 6,000 ft. thick andydrite–salt beds have been drilled.

Salt is being evaporated from the ocean between Casablanca and Safi at Sidi Brahim and Sidi Moussa, and also at Oued Moussa between Agadir and Ifni. Many other occurrences of salt have been known for centuries.

WEST AFRICA

Mauritania. Salt and clay alternate in the sebkhet of Idjil, which is uniquely underlain by Precambrian rocks. The salt is believed to be of eolian or pluvial origin, with the combined salt content of rain, fog and dew attaining at least 0.002‰.

In the area of Trarza, a series of sebkhet surround the Aftout between Toumbos and the capital, Nouakchott. Compact salt is being recovered at Twidermi and N'Terert 50 miles north of the Senegal river, the frontier, from January to July. At N'Terert eight layers of salt and clay alternate with mud; the five lower levels are called: Silchat el Mitkia (7 inches) Sikhat el Fahl, El Fahl (12 inches), Zrewila (2 inches) and Lehreicha.

In *Senegal*, salt is evaporated at Kaolack. The sebkha of Teghazza is located in the northwestern, empty corner of *Mali*, at the foot of a Devonian cuesta. Salt and clay alternate at the bottom of the sebkhet of Taoudéni, 400 miles north of Tombouctou.

Niger. Three areas of sebkhet can be distinguished in the Kaouar district in the northeast, between the Aïr and Tibesti Mountains: Seguedine in the north, Blima in the south, and Fachi on the edges of the Ténéré. Salt is recovered between April and October. Other salt sebkhet are located in the Teguidda N'Tesemt.

In *Chad*, salt is recovered in the Ennedi, Borku and Tibesti Mountains. Brine springs are abundant in the Cretaceous fold belt of the Benue in Nigeria, and near Mamfe in the Cameroon.

NORTHEASTERN AFRICA

In Egypt, salt is evaporated on the littoral and recovered from salt pans. On the coast of the northern *Sudan*, salt is recovered from the lagoon of Ras Roweiya. Thick beds of salt occur in the Selima Oasis, but desert gravels are also saline in large areas of the Sahara, e.g., *(1)* in the Butana between Khartoum and Rufâa, east of the Blue Nile, *(2)* west of the Nile, north of Elai, *(3)* in the Atbara bend and *(4)* southwest of Wadi Halfa.

Ethiopia. Salt is evaporated between Abd el Kader (Fig. 13, 58) and Gerat at Massawa, Wakiro and Assab, on the Erythrea coast. The salt plain of Dankali is distinguished by its potash deposits. East of the Webi Gestro, in southern Ethiopia, salt crusts are deposited from waters circulating in the Lug series. At El Dere, an area of ½ sq. mile is covered by salt and bordered by a 120 ft. high wall of clay. Another deposit, Dole, lies nearby in the Wadi Cullalo. The contents of the water, in g/l, is given in Table CLVII.

TABLE CLVII

SALT CONTENTS OF WATER IN ETHIOPIA IN g/l

	El Medo	*Dole*
NaCl	262	178
$MgCl_2$	9	5
$CaCl_2$	7	4
$CaSO_4$	2	2

The El Soda deposit developed by evaporation in the Mega crater, east of Lake Rudolf.

Somalia. The belt of Dole extends to Ajgerar. Rock salt is also believed to occur 25 miles east of Berbera, at Daga Der and in the Darraboh Hills; a sample contains 96% NaCl, 1.30% $CaSO_4$ and 0.45% $MgCl_2$. Large salt ponds are located on Hordio Peninsula, in the Bay of Hafun (Dante) south of Cape Gardafui; they have been established between the dunes and the coastline. Salt is also recovered at Gesira, south of Mogadishu and near Zeila.

CENTRAL AFRICA

In *Gabon* thick layers of salt have developed in the 250 sq. miles of the lower Ogowé oil field. The thickness of post-Neocomian (Lower Cretaceous) salt attains 2,500 ft. Following the Aptian, the layer was transformed into ridges ½–2 miles long and high. Such slabs of salt are found at Tchengué and Animba–Ozouri. At Pointe Clairette and Cap Lopez, salt has been domed up in the Upper Albian and Cenomanian stages of the Lower Cretaceous and covered by clay.

In the coastal belt of the Congo Republic, 1,000 ft. of halite and potash salts alternate, underlain by silt and slate. Some anhydrite is interlayered in the upper reaches.

In the *Congo* salt is recovered at Kalamoto, Mwashia and N'Ganza in Katanga. The salt of Guba is of excellent grade. Springs and brines deposit salt along the recent fractures of Kivu and northern Katanga. In eastern Zambia, salt is found at Mpika, east of Lake Bangwelu, and on the northeastern shore of Lake Mweru. In the lower Cuanza basin of *Angola* near Luanda, halite, potash salts and argillaceous inliers alternate in a 1,000 ft. thick column. Salt also occurs in the Cabo Ledo diapyr and the Morro de Tuenza horst.

EASTERN AFRICA

Salt is deposited from brines, related to Rift faulting, in Burundi and Rwanda. Lake

Katwe flooded a crater of tuffs, 250 yards from the northeastern shore of Lake Edward in *Uganda*. Pleistocene, fumarolic vapours and younger hot springs are believed to have contributed the salt. Table CLVIII gives the composition of the salts (BARNES, 1961).

TABLE CLVIII

COMPOSITION OF SALTS OF KATWE (UGANDA)

	Wind blown (%)	*Brine* (%)	*Compact* (%)
NaCl	55–95	51–85	52–77
Na_2SO_4	0.2–29	3–36	7–22
Na_2CO_3	0.3–13	1–11	

Saline Lake Bunyampaka (or Kasenyi) is also located in western Uganda. At Kibiro hot springs deposit sodium chloride, accompanied by some sulphate and potassium chloride.

On the *Kenya* coast, salt is evaporated north of Malindi. Halite is also recovered from the sodium carbonate of Lake Magadi.

Tanganyika. There have been distinguished four types of concentrations: *(1)* salt layers on the southern coast, *(2)* littoral evaporation, *(3)* pans and springs of the Eastern and Rukwa Rift Valleys, and *(4)* the brines of Uvinza.

At Mandawa Mahokondo, on the coast 100 miles north of the Mozambique border, several layers of salt have been deposited in an elongated dome. At Mbare, in the northern part of the dome, several 300 ft. thick, and one 700 ft. thick, layer of salt lie beneath 200–600 ft. of gypsum, anhydrite, shale (and salt). Salt even occurs below 10,000 ft. of depth. The rock salts essentially consist of sodium chloride and calcium sulphate, with up to 0.14% K. At Pindiro, in the adjacent Makangaga–Ruawa anticline, salt layers extend to a depth of more than 3,000 ft. Salt is being evaporated on the littoral at: *(1)* Kivindiani, south of the Kenya border, *(2)* Bagamoyo, Sadani, Changohela and Mkadini, north of Dar es Salaam, *(3)* Msasini, Kunduchi and Mjimwema, south of the city, and *(4)* Machole, Mtope and Mtwara, north of the Mozambique border.

Lake Natron is located in the Eastern Rift Valley (Fig. 32, 52) on the Kenya border. Fumarolic vapours of the carbonatite volcano Oldonyo Lengai and hot springs contribute the brine. They also form 80 sq. miles of crust which contains more than 8% sodium chloride, accompanied by sodium carbonate and some sodium sulphate, fluoride and phosphate. Further south, 80% of the salt of Lake Eyasi is halite. Partly eolian sodium chloride, and an equal amount of combined sodium carbonate and sulphate, accumulate at Lake Balangida. The brines of Ikasi and Manyoni, near the geographic centre of the country, contain varied sodium salts. The hot brines of Ivuna, southeast of Lake Rukwa, contain 0.42% solids; the salt contains 20% Na_2SO_4 and 5% KCl.

Thirty saline springs and seepages are located at Uvinza, 70 miles east of Kigoma on Lake Tanganyika. Near the Malagarasi River, brine seeps from fissures of Precambrian rocks. The salts of the Nyanza spring consist of 91–94% NaCl, 2–3% KCl, and sulphates and chlorides of calcium and magnesium. The yield of the springs varies seasonally, but they appear to be endogene and related to rift tectonics.

SOUTHERN AFRICA

Salt is widely evaporated near *Mozambique* city. Most of the remaining 50% of the country's supply is provided by salt ponds of Zambezia, the coast extending from Lourenço Marques 200 miles north and the Cape Delgado area, near the Tanganyika border. At Chibuto, saline soils extend in the Limpopo drainage between the Sangutane (a tributary of the Changane River) and Uatsotschane rivers. Brine contains 73 g of solids per litre there (Table CLIX).

TABLE CLIX
COMPOSITION OF MOZAMBIQUE SALTS

	Lourenço Marques district (%)	*Chibuto solids (%)*
NaCl	80.5–88.6	89.9
$CaCl_2$		3.5
$MgCl_2$	0.7– 1.9	—
$CaSO_4$	0.4– 0.9	2.0
$MgSO_4$	0.2– 0.9	4.0
KCl	0.1– 0.4	—
$CaCO_3$		0.5
H_2O	7.8–12.2	

Salt is deposited on the shores of Lake Cavela, in the Sofala district near the Southern Rhodesia border.

In arid *Bechuanaland*, salt is over-abundant, occurring e.g., in the Makarikari and Ngami pans, and extending to the headwaters of the Nossob in Southwest Africa. Gypsum, anhydrite and halite are interbedded on the coast; salt is evaporated from brine pans at Swakopmund and Cape Cross. Salt is much scarcer in *South Africa*: a broad belt of inland pans crosses Cape Province, extending from Vryburg, south of Mafeking, bulging around Kimberley and turning west towards Calvinia. Salt pans are also found at Soutpansberg, in the far north. At Hammanskraal, north of Pretoria, anhydrite is associated with halite, while sodium sulphate occurs in other pans. Glauberite tends to precipitate in winter (the summer of the northern hemisphere), and nitrates appear at Matsap. Sea salt concentrates in coastal pans.

SODA

Trona, the most common sodium mineral after salt, crystallizes when carbon dioxide is bubbled through waters containing 9% sodium carbonate. Nahcolite, the bicarbonate, precipitates from less concentrated solutions. Such conditions still prevail in the arid Rift Valley, e.g., at Magadi, in Kenya, and in Lakes Chad and Makarikari, in Bechuanaland (Fig. 8). Soluble sodium sulphates can only be preserved in arid lakes or chotts (playas).

EASTERN AFRICA

In the Wadi Natrun 400 miles west of Khartoum, *Sudan* ½–2 ich thick beds of pure natron occur more than 1 mile west of Jebel Kashaf. Trona is found in Ethiopia in the Dankali salt plane and at El Soda, east of Lake Rudolf.

Kenya. Lake *Magadi*, the continent's principal producer of soda ash, covers the floor of the Rift Valley. The lake is 11 miles long and 2 miles wide, and has a 4 by 1½ mile gulf that extends to the northwest. The Holocene trona beds are bordered in the west by narrow but younger alkaline lagoons, and they also appear on parts of other shore lines and in Little Magadi, north of Amagad proper (Fig. 13, 52). Middle Pleistocene chert beds outcrop mainly at the southern end of the lake, and Lower Pleistocene alkali trachytes surround the depression. Along the axis of the lake, a column has been drilled of which Table CLX gives the contents.

TABLE CLX
OCCURRENCE OF TRONA IN THE COLUMN OF MAGADI

	North (ft.)	*Intermediate (ft.)*	*Centre (ft.)*
Trona	70	20	20
Clay with some trona	10	10	3
Trona	15	—	60?
Clay with some trona	15	—	(5)
Clay and silt	30	25	—
Clay with some trona	5	(15)	(5)
Trona	—	(15)	>30
Clay with some trona		30	?
Chert			

Trona crystals that include some villaumite or sodium fluoride grow upwards. The surface is warped when the interstitial liquor is at some depth, which averages several inches to a few ft. Nahcolite or sodium bicarbonate is found at depth in core B. New, purer trona crystallizes at a rate of 3 inches per year in dredged out areas. The deposition of evaporites continues in the form of sodium bicarbonate. The contents of trona are given in Table CLXI.

The accumulation of trona is believed to have been controlled by (BAKER, 1955): *(1)* the large surface of recent volcanics, leached by *(2)* ground water concentrating at a low

TABLE CLXI
COMPOSITION OF TRONA IN KENYA

	Surface trona (%)	*Lake liquor (%)*	*B bore 90 ft. (%)*	*Soda ash (%)*
Na_2CO_3	45	16–18	66[1]	97.5
$NaHCO_3$	36	1–16	29	0.05
NaCl	1.7	5– 7	0.3	0.5
NaF	0.9	0.3	0.05	1.3
Na_2SO_4	0.06			0.4
H_2O	15			0.2
Alkali		17–19		

[1] Trona.

point of the Rift Valley, *(3)* supplied by alkaline springs, and *(4)* numerous cycles of evaporation in subsequent lakes.

Tanganyika. The largest deposits are the crusts and brines of Lake Natron, south of Magadi. They contain equal amounts of sodium carbonate and chloride (salt), unfortunately with the sulphate, fluoride and phosphate of sodium. Lake Balangida, further south, is covered by a crust of 1 inch that contains as much common salt as sodium carbonate and sulphide combined. Sodium also shows in the Bahi depression.

NORTH AND WEST AFRICA

Algeria. Sodium sulphate is particularly abundant in the salt lakes of the east: the salt crusts of the Chott Chergui contain 91% Na_2SO_4 and 3% $MgSO_4$, i.e., mirabilite, which turns into thénardite in the dry season. Chotts generally contain more magnesia than soda, but the composition of the crust changes seasonally. Epsomite accompanies sodium salts in the basins of Nador and Medea, and magnesium chloride occurs at Zahrez Rharbi. Sodium carbonate is believed to also precipitate in the crusts of the Djebel Zouabi.

Morocco. In a sub-Saharan sebkhet of the Hamada of Daoura, 40 miles east of Zegdou on the Algerian border on the latitude of Agadir (southwest of Colomb Béchar), Middle Cretaceous, Cenomanian–Turonian limestones are overlain by Hamada sediments. A 10 ft. thick bed of gypsum fills in the sebkhet, and includes two 1 ft. thick inliers of compact thénardite (Na_2SO_4).

Lake *Chad* receives the waters of the Shari and Logone rivers. Since yearly evaporation and seepage equal 80 and 8 inches respectively, the lake's waters are saline. Salinity increases from 45 g/l at the estuaries, in the south, to 90 in the middle of the lake, and 300 at its northern end. The saturation point is reached at 205 g/l (0.02%) because the water of the lake also seeps north, flowing below beds of wadis (dry valleys) and depressions. High grade, black and compact soda, and white light soda is quarried at a depth of less than 2 ft., e.g., at Kalia (Fig. 24). The ratio of components of the carbonate and bicarbonate is 950 sodium, 9 magnesium, 4 calcium, 36 chlore and some potash.

SOUTHERN AFRICA

Bechuanaland. Considerable amounts of brines and evaporates have accumulated in the delta of the Nata River (Fig. 50), and south of its estuary, in the *Sua* pan, the northeastern section of Makarikari. These pans are located in the northern Kalahari, 120 miles northwest of Francistown. Silcretes, clays and sands, partly covered by aeolian sand and alluvium, average a pH of 9.1; 356 g solids per l contain 85% NaCl, 5.7% carbonate, 5.1% bicarbonate and 1.3% sulphate. On a sand bar projecting into the northern Sua pan, brines rise up to a depth of 13–20 ft., with a water rest level of 7 ft. The yield of a bore hole was in excess of 3,500 gallons/h. The floor of the Makarikari pan itself consists of jelly-like clay, saturated by brine and covered by a 1 ft. thick crust during dry periods.

Soda is known to occur northwest of the Etosha pan in the northern part of Southwest Africa. In *South Africa* 26 miles north of Pretoria, brines containing 10% NaCl, 5% Na_2CO_3 and some bicarbonate rise from 200 ft. of mud which fills a sunken caldera. In the centre of the pan, saline clay and marl are overlain and interbedded with coarse gaylussite accompanied by silica and covered by 18 ft. of alternating trona and mud.

POTASH

Potash salt precipitates after sodium and magnesium salts when the mother liquor reaches a density of 1.32 kg/l, that is, 1.66% of the original volume of water. Such lagoonal conditions prevailed during the Quaternary in a former gulf of the Red Sea in Ethiopia, and during the Cretaceous on the Congo coast. Fumarolic activity facilitates the precipitation of potash, and Rift lavas contain up to 8% K, i.e., more than three times the average of the earth's crust.

ETHIOPIA

The Salt Plane of the Dankali extends 150 miles south of Mersa Fatma Heri on the Red Sea. Alternating sediments and tuffs form the walls of this depression, even below the level of the sea. The trough periodically communicated with the sea, from which it was separated by movements related to the late cycle of volcanic eruptions. The last lake appears to have developed during the Pluvial, but between January and March, in the rainy season, part of the depression is still flooded. The *Dallol* deposits (Fig. 13, 58) occur 40 miles from the sea, within an oval of 3.5 by 2 miles. The Black Mountain protudes into the western part of this oval; in its northern part salt fills in a crater-like feature, and fumarolic and solphataric gases erupt from the northwest. In the cold waters of a geyser ½–1% KCl is dissolved. Saline solutions seep into a lake in the centre of Dallol; their temperature varies between 80° and 100° C. The composition of salts recovered from the lake differs (Table CLXII).

TABLE CLXII

COMPOSITION OF SALTS IN DALLOL LAKE

	Solution (g/l)	*Salt (%)*
KCl	54	70–80
$MgCl_2$	478	3–10
Fe_2Cl_6	5	
$MgSO_4$	3	
Br	2	0– 0.3
NaCl		7– 8
$CaCl_2$		0.1– 0.2
$CaSO_4$		traces– 0.1

In the 150 ft. thick column of Dallol 1–3 inch thick layers of salt and clay alternate. Sylvite is deposited at the bottom of the lake, accompanied by some carnallite. Nevertheless, potash grades appear to increase from 0.5% in the lower layers to 6% in a 4 inch thick bed that lies 3 ft. below the surface; potash also tends to precipitate near the springs. There have been distinguished four types of salts: *(1)* the southeastern sector of the oval contains more than 40% carnallite and magnesium chloride; *(2)* carnallite prevails in the northwestern corner, and since 1926 it has been deposited along the long axis of the oval; *(3)* in the southwestern sector rock salt, consisting of halite, 27% sylvite, kieserite and anhydrite has been deposited; *(4)* in a smaller area southeast of the Black Mountain, a 16 inch thick crust of salt occurs which contains more than 80% KCl.

THE REST OF AFRICA

At the floor of the sebkhet (salt pans) of Jarabub and Marada, in eastern Libya, potash salts are found underlying halite. Potash is found in the Zaizis sebkhet in Tunisia.

Potash salts overlie the horsts of Lake Azingo, in the Gabon oil field. In the *Congo Republic* at Hollé and north of the Kouilou river, in the area of Pointe Noire, a 1,500–2,000 ft. thick column of salts is overlain by 830–2,000 ft. of sediments. Numerous carnallite ($KCl.MgCl_2.6H_2O$) beds, up to 50 ft. thick, and rarer but thinner layers of non-hydrous sylvinite containing 30% K_2O have been distinguished in this column between 700 and 1,200 ft. of depth. The thickness of three layers ranges from 5 to 15 ft. each.

The Uvinza brines, east of *Lake Tanganyika*, contain 2–3% potassium chloride. Indeed, potash is generally more abundant in the springs of the Western Rift Valley than in those of the Eastern Rift Valley. For instance, the Ivuna springs on Lake Rukwa contain 5% KCl; the salt crust of Lake Balangida, however, includes only 2.7%.

In the Analavelona Hills and in the Menarandroy Valley of southwestern Madagascar, abundant glauconite is found in the green sands of the earliest Cretaceous Neocomian.

The leucite lavas of the Virunga chain north of Lake Kivu, in the Congo and partly in Rwanda, contain 8% K_2O. Adjacent leucite basanite and banakite flows contain 5.7% and 5.2% K_2O, respectively, in Uganda.

NITROGEN

Since soda and potash nitrates are highly soluble, they can only be preserved in arid climates, such as in the Sahara, where they seasonally precipitate in soils.

Soils of Ziban and the Oued Rhir, in *Algeria*, contain 12% nitrate, as do the oases of Sebâ and Guerrara. In the northern Sahara, nitrated earths have been worked at Gourara, Touat and Tidikelt. In the Touat Oasis, salpeter rises, especially after a humid season. The Gourara nitrate basins are aligned along the sub-meridian trough of the sebkhet Oulad Mahmoud, Kabersten and Meraguen; they are distinguished by a high sodium/nitrate ratio. The 2 inch crusts of *Sba*, near Touat, contain 34 % KNO_3 and 54% $NaNO_3$, but the grade of nitric acid does not exceed 4%. Waters of the eastern Sahara contain 20 mg nitrate per litre.

The soda nitre of Magadi, south of Nairobi, *Kenya*, is believed to have been derived from guano. At Kibaya and Tabora–Nzega, in Tanganyika, some potassium nitrate is produced from potash silicates by the urine of rock rabbits (HARRIS, 1956). Some nitrate has been found in the Prieska and Hay districts of northern Cape Province, South Africa. The Matsap salt pan contains abnormally high amounts of nitrate. At Mosikolelo, in Basutoland, salpeter is reported to have been formed on the floor of a cave through the action of animal refuse.

SULPHUR

Sulphur has precipitated in a reducing environment of limestone and anhydrite in East Africa, in areas distinguished by Rift volcanics, and in fumaroles. In the northern Sahara, sulphur occurs in the sedimentary column that also includes hydrocarbons.

Egypt. At Ranga, on the Red Sea, bands of sulphur (Fig. 12, 58) alternate with gypsum and shales, and it also occurs in lenses and pockets. In the Miocene evaporites of Gemsa, beds of clay and dolomitic limestone are intercalated into anhydrite. Sulphur, bitumen and fine-grained pyrite fill in bed joints and cavities in the limestone; they also replace it. The sulphur is believed to have precipitated from hydrocarbons ascending through fault planes.

Ethiopia. At Zariga, near Massawa, Erythrea, anhydrite layers contain 16–25% S (averaging 16–20%). In the Aush Valley of northeastern Ethiopia, 1 inch thick layers of sulphur are deposited in the crater of the fumarolic volcano of Dofan(e). The Kebrit Ale volcano is solfataric. About 2 miles west of Berbera, *Somalia*, impure sulphur impregnates the coral rag and sand of the '24 ft.' raised beach. The sulphur is believed to be related to an underlying fault which belongs to a young volcanic cycle.

Sulphur is reported to occur on Central Island in Lake Rudolf, Kenya. At Naivasha, the tuffs of an extinct fumarole average 12% of this element. Sulphur occurs in the limestone of the coastal belt of the Congo, and in the volcanoes of the Virunga Chain, on the frontier of the Congo, Uganda and Rwanda. Fumarolic sulphur fills in the inner crater of Kibo, on Mount Kilimanjaro, Tanganyika. Furthermore, the sandstones of Wingayongo are impregnated by 0.2% S deposited from springs, and the sands of Ras Matunda Island contain up to 0.8% S. Sulphuric acid accumulates in Lake Kivu (Congo–Rwanda).

Algeria. Native sulphur is associated with anhydrite and oil in most of North Africa: sulphur nodules are disseminated in Tertiary marls in the Constantine district between Héliopolis, Millesimo and Zaouria. The blue, bituminous marls of Seybousse are interbedded with five layers of sulphur, with each attaining 8–10 inches of thickness, and with Sahelian, gypsum-bearing clays; traces show as far as Hammam Meskoutine. At Djebel bou Kebch, sulphur accompanies anhydrite. Similar occurrences have been found at Tébessa, on the Tunisian border.

BROMINE, BORON AND IODIUM

Bromine is extracted from brines of salt lakes and from sea water, and generally converted into ethylene dibromide, an additive of gasoline. Bromine salts tend to precipitate between the sodium and potash stage. Bromine is extracted from salt at Fedala in Morocco. While lakes Katwe and Bunyampaka, in Uganda, are distinguished by a relatively high proportion of bromine, they contain only traces of iodium, an element which occurs in a similar environment. Boron is found in salts and brines of arid or sub-arid areas.

ALUM

Alumina, leached from rocks by vadose waters or gases containing sulphuric acid, combines with the latter to form aluminium sulphate. Readily soluble alum can only be preserved in an arid climate.

A thick layer of alum covers diorite porphyries in the Oued Maaza in the Rif, northern Morocco (see 'Kaolin' chapter). In Hamasien, Ethiopian Erythrea, alum is associated with kaolin. In *Kenya*, alum occurs in the vicinity of fumaroles of the Rift Valley, while pickeringite, the magnesia alum, is found at Voi. An aggregate of aluminium sulphate has been located in the Machakos district.

At Ampanihy, at the southern tip of *Madagascar*, lenses of alum are included in two beds of continental, early Tertiary sediments; the upper bed has a strike length of 2 miles. The alum contains 34.5% Al_2O_3 and 35.20% SO_3. At Vioolsdrif, south of the Orange River in *South Africa*, up to 2 inch wide veins of alum cross fibre and gypsum traverse altered shales. The alum is believed to have been formed by sulphuric acid, derived from the oxydation of pyrite, in the aluminous shale.

GYPSUM

Anhydrite and gypsum precipitate with some carbonate from lagoons and epicontinental shelves in a hot climate when the original volume of the water is reduced to 9–19%. Some gypsum continues to evaporate with salt later. Although anhydrite has been preserved in the late Precambrian Copperbelt and in even earlier sediments, its principal cycles are Mesozoic and younger. Several hundred ft. thick columns of the sulphates, dolomite and limestone precipitated during the Miocene along the Gulf of Aden and the Red Sea, but deposits of the East African coast are Jurassic. While the Lower Cretaceous deposits of Angola are included in a 2 mile thick column, in the northern Sahara, anhydrite has cyclically evaporated from the Devonian to the Tertiary (Fig. 8).

NORTHEASTERN AFRICA

In the Gulf of Suez, in *Egypt*, 3,000–6,000 ft. of Miocene gypsum is overlain by Plio–Miocene sandstone, limestone and lumachelle. At Jebel Abu Gerfan, Miocene anhydrite is underlain by Nubian (Upper Cretaceous) sandstone, while the Middle Miocene Gypsum Series extends further south along the coast. Gypsum and shale alternate at Ranga. At Gemsa, however, clay and dolomitic limestone are intercalated in the column of anhydrite. In the northern *Sudan*, several beds up to 30 ft. thick, extend along the Red Sea from 40 miles north of Port Sudan, beyond Khor Donganab. Ranges of hills (Fig. 13, 58) of the Gypsum Series rise from the coastal plain, and beds of gypsum also cover the entire island of Makawa. The purest and whitest gypsum was found in relatively thin beds of 2–12 ft. Gypsum also occurs in the Jebel Abiad, 100 miles west of Kandak on the Nile.

Ethiopia. The gypsum of the Erythrean coast is partly powdery. Such dry deposits occur, e.g., near Massawa. Up to 300 ft. thick beds of gypsum evaporated when the Dankali depression was cut off from the Red Sea. The cement works of Dire Dawa use the gypsum of Adele and Demale, near Aisha, on the border of the French Somali Coast. In Shoa, gypsum is believed to be interbedded with Mesozoic sediments. The mineral also occurs at Muger on the Fikshe plateau, and on the shore of Lake Rudolf.

SOMALIA

Between the Lower and Middle Eocene, a series of at least 300 ft. of gypsum, anhydrite and carbonates was deposited on the Horn of Africa. It covers a quadrangle or trapeze, the sides of which extend for 300 miles south and west of Cape Gardafui. The series is transgressive over lower Eocene limestones that cover the eastern corner of Ethiopia and southern Somalia; their contact follows the border of the former British Somaliland, a line extending south of Djibouti, the western coastline of the Red Sea. Layers of gypsum are thickest along the east-southeast trending axis of the platform. The Precambrian remnants which rise along the Gulf of Aden separate the Gypsum Series from the coast in northern Somalia (Fig. 14, 58). However, the Gypsum Series reappears on the coast between east of Berbera and Djibouti, along a 150 mile long and up to 15 mile wide strip. This small segment of the Series is more favourably located than the 14,000 sq. mile area it covers between Al Hills, Bulhar and Bohotleh, and which are believed to include 500 cubic miles of gypsum and anhydrite. Since the lagoonal series contains little salt, it is believed to have evaporated at temperatures ranging from 86°F to more than 150°F.

The *Suriah Malableh Ridge*, extending 10 miles west of Berbera Harbour, rises 350 ft. above the valley. Although most of the bands are lenticular, there is little oscillation of gypsum/anhydrite and sulphate/carbonate ratios along the strike, variations of carbonate ratios being largely limited to the top and bottom of the outcrop. Excluding an 11 ft. thick

TABLE CLXIII

OCCURRENCES OF GYPSUM IN SOME COLUMNS OF SOMALIA

Outcrop averages				*Three drill cores*					
ft.	*rock*	*grade (%)*	*carbonate (%)*	*ft.*	*rock*	*ft.*	*rock*	*ft.*	*rock*
1	impure gypsum			9	gypsum	10	limestone	24	gypsum
9	sulphates	94.2	2.6	4	limestone	4	shale	8	marl
7	mixed			24	gypsum	72	gypsum	2	limestone
8	dolomite			9	limestone	8	marl	24	gypsum
23	gypsum	91.7	2.0	11	gypsum	9	gypsum	13	limestone
10	impure gypsum			30	anhydrite	26	anhydrite	4	gypsum
42	gypsum	94.7	3.6	21	gypsum	6	sulphates	28	anhydrite
?	(no data available)			10	marl	18	gypsum	21	gypsum
7	anhydrite	96.7	0.9	8	gypsum	7	gypsum	3	shale
19	sulphates		1.5	6	limestone	5	mudstone	6	gypsum
42	gypsum	86.7	8.6	10	gypsum	9	gypsum	4	mudstone
3	carbonate			17	sulphates	8	marl	7	sulphates
32	gypsum	92.8	3.9	22	anhydrite	8	gypsum	6	limestone
30	gypsum	96.1	0.9	31	gypsum	9	anhydrite	11	sulphates
5	sulphates		3.2	4	dolomite	5	sulphates	41	anhydrite
27	gypsum	96.0	1.8	12	anhydrite	33	anhydrite	15	gypsum
3	impure gypsum			4	gypsum	4	sulphates	3	shale
30	gypsum	93.5	4.4	8	shale	21	gypsum	4	gypsum
				5	gypsum	13	sulphates	3	anhydrite
				8	anhydrite	14	mudstone	3	shale
				8	gypsum	8	sulphates	13	gypsum
				45	sediments	3	shale		

carbonatic bed, which includes 2 ½ ft. of sulphates at the top, grades of the 20–60 ft. thick bands consistently reach the mark of 85% (PALLISTER, 1958; PALLISTER and WARDEN, 1959; HUNT, 1954).

The study of sub-vertical cores drilled on the slopes above the outcrop show considerably more anhydrite than channel samples. Indeed, the anhydrite zone of the lower part of the holes thins out towards the southeast, i.e., the outcrop. There is also more lithologic variation in the sulphate beds; thus the 10 ft. thick 'marbled' zones which appear around 200 and 270 ft. of depth in the second drill hole, and around 140 ft. and 150 ft. of depth in the third drill hole, exhibit variations of sulphate content. Streaks, partings and thin lenses of ochreous silt or dolomitic limestone, and thin, partly bituminous shales are frequently interbedded. In Table CLXIII the column has been simplified by the elimination of thinner intercalations and alternating beds.

The deposits extend with dolomitic limestone to Ferfer, Shushuban, Jredami and Gardo in southern Somalia, and thin beds of gypsum-bearing limestone outcrop at El Wagit, Juba, Bardera (Juba), Erdeb Delbile and Gosha.

EAST AFRICA

In Uganda gypsum occurs on the shores of Lake George and at Kibuko. Occurrences of gypsum extend in *Kenya* from Mandera in the northwestern corner of the country, to El Wak and Wajir near the Somali border, thence to the Tana Valley between Garissa and Bura, and to North Kitui and the southern coast. Gypsum is included in the column of Jurassic sediments that also cover southern Somalia. At Mombasa, thin stripes of gypsum are interbedded with Jurassic shales. Gypsum has also precipitated in Tertiary and Pleistocene lakes and swamps. Such deposits have been located in the Tula Valley, near Garissa; in them gypsum either forms a continuous white band in clays or else pinkish but purer aggregates. Thin seams of gypsum occur on Mount Homa, and gypsite underlies the soils of Konza. Gypsum occurs in recent sediments at Mida Creek, near Malindi Harbour, and nearby it precipitates by solar evaporation in the Gongoni plant.

Tanganyika. Between Mandawa and Pindiro, 100 miles north of the Mozambique border, gypsum is interbedded with Jurassic limestone and shale. Gypsum occurs in a sub-diapyric position in anticlines which strike north-northwest, parallel to the coast. Gypsum outcrops on the Mbaru stream, near Mandawa, occurring on the banks under a cover of 50–100 ft. The 200 ft. thick layer, which contains 80–90% gypsum, is underlain by 400 ft. of anhydrite and salt, and possibly rests on a salt diapyr. The deposit is located 17 miles from the jetty of Rushungi. At *Mkomore*, 8 miles northwest of Mbaru, the width of gypsum and anhydrite appears to be from 1 to several hundred ft. Along the Pindiro scarp, 30 miles from Rushungi, gypsum outcrops over more than 1 mile. Banded beds of gypsum, interbedded with shale and limestone, attain widths of from several inches to 20 ft. The grade reaches the mark of 86%, but impure beds also occur. Gypsum also shows in the same area at Mtegu (Fig. 52).

At Msagali, 200 miles west of Dar es Salaam, gypsum impregnates layers of clay 5–50 ft. thick; consequently the grade is not in excess of 70%. At Itigi, approximately in the centre of Tanganyika, similar material forms 3–5 ft. thick layers. At Mkomazi, southeast of the Kilimanjaro, gypsite is interbedded with lacustrine clay, silt and sand along a distance of 6 miles. Sand and calcite fill in the open spaces between aggregates of selenite, and the grade averages 78%.

NORTH AFRICA

Libya. The late Cretaceous and Eocene of Cyrenaica is overlain by anhydrite and limestone. In the area of Atshan, in the southwest, inliers of anhydrite are interbedded with carbonaceous shale and dolomite.

Algeria. In the Edjeleh field in the southeast, Carboniferous *Profusinella* limestone and marl is interlayered with anhydrite; the thickness of the series increases in the east. The Tigentourine structure of 5 by 3 miles is bordered by anhydrite on three sides. In the Triassic fields of the northeastern Sahara, anhydrite is included in a sequence of dolomite, shale and salt. At Hassi er Rmel south of Algiers, lagoonal sedimentation started after the Middle Cretaceous, Turonian.

Triassic gypsum also outcrops in more accessible localities, e.g., east of Oran, in the Djebel Chegga, in the Mostaganem and Arzew Highlands, at Sidi bel Abbès and many other localities, and in the vast dome of Haboucha, in the Mina Mountains. Windows of gypsum can be followed through the hinterland of Algiers to the Babors in Kabylia. Bodies of gypsum emerge in the crests dividing the Chotts (saline lakes). The cyclical sedimentation of gypsum continues until the Miocene: At Dahra and in the Chélif plane, east of Oran, undulated layers of this epoch attain a width of 150 ft. (see 'Diatomite' chapter). Miocene gypsum is generally purer than Triassic gypsum. Finally, Plio–Holocene layers of gypsum reprecipitated from earlier evaporites at Laghouat, Djelfa, Biskra and many other localities.

Morocco. The principal deposits are located 5 miles inland from Safi, at *Sidi Ahmed Tiji*, and 17 miles inland and further east near *Youssoufia.* Upper Jurassic gypsum beds and lenses, 150–300 ft. thick, are included in a column of Cretaceous– Jurassic, black or light coloured, gypsum-bearing limestones. Gypsum beds dip slightly towards the west; their fluid structures fracture and pierce the overlying limestones, like diapyrs. At Zaouïa Sidi Ahmed Tiji two gypsum horizons can be distinguished; light green marl sandstones divide the thin lower bed from the thick upper lens. A whole mountain consists of gypsum at Sidi Ahmed Tiji; pure, sugary gypsum alters to regular beds of fibrous material there. Many other occurrences are known in Morocco (AGARD, et al., 1952).

WEST AND CENTRAL AFRICA

North of Nouakchott, in Mauritania, extensive layers of high grade gypsum of the coastal sebkhet (salt pan) are 2–8 ft. thick. Satin Spa type gypsum occurs at Pokoasi in Ghana. In Nigeria, gypsum shows in the Chad formation of Bornu, in the northeast, in Cretaceous sediments at Sokoto and the Benue Valley, and in Tertiary clays of Onitsha.

In the Congo Republic some anhydrite appears in the upper salt layers of the coast.

Angola. At Cuvo, near Luanda, anhydrite and dolomite alternate in an 8,000 ft. thick Lower Albian (Lower Cretaceous) column. North of Cabo Ledo, south of Luanda, seams less than ½ inch thick and lenticles of fibrous gypsum traverse Tertiary sandstones. At Ouche, near Benguela, undulating sub-horizontal layers of gypsum, which attain a thickness of 3–5 ft., outcrop or rest under a thin cover. Finally at Dombe Grande, 50 miles south of Benguela, up to 1 inch long crystals of transparent selenite are associated with sulphur in patches and seamlets of porous chalk; they are believed to have precipitated from percolating solutions (T. C. F. HALL, 1949).

Stalactites of gypsum occur in the Lovo Caves near Bango, in the Lower Congo, and anhydrite characterizes the lagoonal, Katangian sediments of the Copperbelt.

SOUTHERN AFRICA

At Chiromo, in Malawi, marls include seams of gypsum between 230 and 280 ft. of depth.

Madagascar. Seams of gypsum are included in clays and marls of the Upper Jurassic and earliest Cretaceous belt of western Madagascar at Ankay in the northwest, Belaza in the Tongobory area, Mahaboboka, Mampikony and Ambondromamy. At Ankay, the combined length of three lenses of sugary gypsum attains 500 ft., while their width averages 9 ft. At Belaza, 1 inch seams of fibrous gypsum distinguish the contact of Middle (Turonian) and early Cretaceous (Coniacian) sands and sandstones.

In the sub-arid but only moderately drained basins of *South Africa*, gypsite has developed from carbonatic rocks affected by sulphuric acid. At Bosvark, Riverton, Windsorton Road and Melkvlei, in the vicinity of Kimberley, up to 10 ft. of clays containing 8 inch lumps of gypsum rest on Karroo shales. The gypsiferous clays are covered by 3 ft. of barren clay and/or 1½ ft. of somewhat gypsiferous limestone. At Zoutpansfontein wind blows gypsum from a pan floored by lavas. In the Vanrhynsdorp division of the western Cape, gypsiferous clay lies on shale and limestone (Droë River, Onder Aties, Wolwenest and others). Gypsum also shows at Kalk Gat Vlakte, Goedverblyf and in the pans of the belt extending from Port Nolloth to Prieska and Strydenburg. Thick layers of anhydrite and gypsum occur on the west coast. At Woodlands, near Port Elizabeth, powdery gypsum has developed from leached shales. Finally at Ngobevu, in Natal, gypsiferous clays cover a basined dolerite sheet.

Gypsum outcrops at Swakopmund, in Southwest Africa. Anhydrite and gypsum are frequently interbedded with limestone in the diamond belt near Oranjemund.

LIMESTONE AND CEMENT

Since a majority of Africa is occupied by Precambrian rocks, high grade limestone is rarer than on other continents. Indeed, Precambrian limestones are heterogeneous and often magnesian, as are the carbonatites of eastern and southern Africa. The principal, carbonate rich, Precambrian cycles are the 2,200 million year old Transvaal System and the 650 million year old Katangian. Karroo limestone beds are thin, but Jurassic and Cretaceous limestones are abundant on the East African coast and in Algeria. Quaternary limestones are found on the coast, and calcretes and travertines are found inland. Since the utilization of limestone, mainly as cement rock, depends on communications, only few deposits are worked and the description will be limited to a few examples (Fig. 8).

NORTH AFRICA

Algeria. The Lower Jurassic, Liassic limestones of Filfila, near Philippeville, are of excellent quality; banded marble forms their lower layers. While Cretaceous limestones are exploited at Constantine, Tébessa and Bou Saâda, early Tertiary, Nummulitic rock was already used by the Romans. The limestone occurrences of northern Tunisia are similar to those of Algeria.

Morocco. The principal sites of marble include the Oueds Akrech (black), Cherrat (white), Sidi Lachemi, Imi Moqqorn, Yquem (white and rose), Cherrat (white), Bou Acila and Boured (green). The principal quarries of limestone are located at the Devonian Oueds Akrech and Yquem mentioned above. Lime is extracted from limestone at Oued Cherrat, from Plio–Holocene limestones between Fès and Meknès and from calcareous sandstones between Casablanca and el Qenitra. The following rocks are mixed in cement plants: *(1)* Middle Cambrian shales and Quaternary, calcareous limestones at Casablanca, *(2)* Miocene clays and lacustrine limestones of Saïs at Meknès, *(3)* Middle Cretaceous marls and Pliocene shells at Agadir.

WEST AFRICA

Limestone is scarcer in the Precambrian, but some limestone shows in the Cambrian secondary deposits and in the coral reefs of the coast. The argillaceous limestones of Bargny, e.g., provide excellent Portland cement in Senegal.

In *Ghana*, beds of oyster shells and soil form the banks of the Volta River between Amedua and Tefle; the pure shells can be used for the burning of white lime. Limestone of good grade, however, is not abundant. The Nauli horizon of Appolonia connects Kangan and Twendani, on the coast, with the area of Edu on the Ivory Coast border. This material is suitable for the manufacture of cement. A thick bed of dolomitic limestone has been located east of Akubi, in Ashanti; a 3 ft. thick horizontal bed of limestone has been discovered at the foot of Longoro Juju Hill, and a thick bed has been drilled at Wenchi. In the North, a sub-horizontal bed of dolomitic limestone (of cement quality) connects the Black Volta, west of Buipe Ferry, with the Baka area. Impure limestones outcrop at Daboya and between Nyambo and Jimam.

In Togo clay separates limestone beds from the phosphates. In Dahomey, at Arlhan, extensive limestones lie between 1 ½ and 7 ft. of depth, with a clayey parting separating two beds. Thick limestones extend into Abeokuta and Ijebu Province in *Nigeria.* The cement factory of Nkalagu, east of Enugu, uses local limestones and marls; the belt extends to Igumale in the north. Extensive deposits of marble occur at Jakura, northwest of the confluence of the Niger and Benue rivers, and at Ukpilla north of Auchi (Fig. 24). Limestone outcrops near Gombe and Numan along the Benue and Gongola rivers, and at Kwakuti near the Niger. Large limestone deposits are known to occur in Sokoto province in the north, but beds are thinner at Siluko in the west, and at the Owam, Osse and Awle Rivers in Benin Province. Dolomitic limestones are found at Igbetti, Elebu, Itobe and Burum.

NORTHEASTERN AFRICA

Egypt. Middle Carboniferous carbonates outcrop at Um Bogma and in the Oweinat Oasis, and Middle Cretaceous limestones occur at Arif el Naga, in Sinai. Limestones are included in the Bajocian, Oxfordian and Lusitanian stages of the Jurassic (Khashm el Galala, Maghara), but thick beds of Cretaceous and Eocene limestone are more widely distributed at Abu Rawash, in the Sinai, in middle Egypt and in the deserts. The type locality of Pliocene limestone is Kom el Shellul.

There is an abundant supply of limestone in the *Sudan*, e.g., at Shereik, south of Abu Hamed, east of the Atbara bend of the Nile and southwest of Suakin, on the coast. Marble is found, e.g., at Dirbat Well. Extensive deposits of limestone have been located in the Jebelein area, 230 miles south of Khartoum. The limestones of Nyefr Regaig contain 52.4–55.6% CaO, and 0.3–3.3% MgO, but no sulphur or phosphorus. The Nyefr Abu Dakheira beds unfortunately contain 18% MgO. The smaller Bagger el Haddari deposit, located between the two Niles further northeast, contains 49.6–53.4% CaO and 0.5–3.0% MgO. Minor occurrences of the area include the Jebel Dud, Mashata and Sagadi.

Ethiopia. In Erythrea, limestone is abundant at Keshiret and Semaït, travertines occurring e.g., at Mai Edem. The limestones of Enda Eish contain 98% $CaCO_3$. Marble outcrops in Adigrat at Antalo, Teramini and Dahalac Island. In Ethiopia proper limestone is abundant in Tigrai, Harar and Ogaden; the latter is also interbedded with sediments along the Blue Nile in Shoa and Amhara. Travertines have been located at Burca Gadaburba. The cement works of Gurgussum, near Massawa, use Madrepore limestone, while those of Dire Dawa utilize adjacent materials. Marble outcrops at Guder and Zega Wadeb. Limestone that can be used in construction has been located at Adi Rassi and near Harar.

Limestone is abundant in *Somalia* at Ras Antara in the north, at El Wegit near Mogadishu, the capital, and in the south near the Juba River. Limestones of Brava, Mogadishu, Shebeli and the Bur region are also used for building stone.

CENTRAL AFRICA

In Gabon the limestones of Achouka, on the Ogoué River, can be utilized for the manufacture of cement. In the Central African Republic the Bobassa deposit is conveniently located near the capital, Bangui. The limestone–dolomite occurrences of Dolisie and of the Mayombe are also relatively near to Pointe Noire on the coast of the Congo Republic.

Limestone is relatively scarce in the *Congo*. The principal deposits belong to the Katangian, to the contemporaneous Lindi System of the north, and to the Shale–Limestone System of the Lower Congo. Between Leopoldville and the Congo estuary, several carbonatic horizons have been distinguished in the overlying Shale–Sandstone System: dolomites, 1,100 ft. of plaquette limestones (Bulu), 1,000 ft. of oolithic crystalline limestone (Luanza), 1,000 ft. of siliceous and sandy limestone (Lukunga), and 1,000 ft. of oolithic limestone. In northeastern and eastern Congo limestones are included at the base of the probably contemporaneous Lindi System; they outcrop, e.g., *(1)* at the confluence of the Lindi River, *(2)* at Wenya Rukula and Kewe, between Stanleyville and Ponthierville, and *(3)* in a long belt extending in the north halfway between the frontier and the Lindi River (Fig. 29, 51).

Carbonatic sediments are abundant in Katanga; the purest ones are generally the pink Kakontwe limestones. The cement works of Lubudi use Katangian limestones. The crystalline limestone of Lukela, Lower Congo, and the *Collenia* limestones of Katanga can be used as building stones. Limestone horizons are included in the Kwango sediments which overlie the Karroo. Younger limestones and travertines are related to the Rift faulting, but Middle Precambrian, carbonatic sediments are relatively scarce in Kivu.

The first limestones of southern Zambia are of early Precambrian age; these rocks outcrop east of Lusaka and in the lower Luangwa Valley north of Petauke. Katangian sediments include several carbonatic horizons, such as *(1)* the lower Kundelungu Kakontwe limestones of the Copperbelt; *(2)* the Kundelungu limestones bordering the Luapula River, 70 miles east of Elisabethville, near Fort Rosebery; and *(3)* the middle stage of the Broken Hill and Lusaka Series. Limestone beds outcrop near Maseki, south of Lusaka, and east and west of Mazabuka, further southwest. However, limestones interbedded in the arenaceous Karroo are thin.

EAST AFRICA

Limestone is scarce in Burundi, Rwanda and Uganda. Travertines are found in the Ruzizi Valley. Calcretes and sinters are found at Muhokya and Dura 100 ft. above Lake George and bigger deposits of lake limestone occur at Hima. However, the most accessible marble quarry is at Moroto. The carbonatites of Sukulu and Tororo can be used; however, they contain phosphorus in apatite (magnesium content rises to 8% at Napak). Precambrian marble is more abundant in *Kenya* than in Uganda; unfortunately the sites are located far from the lines of communication. However, the lime-burning plant of Turoka, near Kajiado, uses this type of rock. The Athi River cement plant is supplied by low-magnesia crystalline limestone from Sultan Hamud, local kunkar and decomposed volcanic ash. Jurassic and Pleistocene limestones and corals are available in the east; they are used for lime burning and the manufacture of cement at Mombasa. The Bambury factory uses Pleistocene coral, Jurassic shale and the gypsum of Roka. The late Tertiary carbonatites of Lake Victoria and the kunkar of the Rift Valley can also be used. Miocene, Pleistocene and Holocene lake limestones outcrop respectively at Koru, Mount Homa and Makindu–Kibwezi (Fig. 33).

Jurassic–Holocene limestones are abundant on the *Tanganyika* coast. Widespread crystalline, Precambrian and thin, Karroo limestones and sövite are generally less pure. Late Precambrian limestones tend to be dolomitic. At Uvinza, a Pleistocene limestone has been quarried and thick travertines are used for lime-burning at Mbeya; durcicrusts are utilized at Moshi and elsewhere. Jurassic limestones, 70 ft. thick, outcrop at Tanga. The upper 20 ft. of the coral limestone of Mjimwena is suitable for the manufacture of cement, and at Wazo Hill, also near Dar es Salaam, the lower 20 ft. of limestone and the top soils could be mixed. Both limestone and green marls are abundant on the southern coast, e.g., at Mandawa. Suitable occurrences of the hinterland of Dar es Salaam include Lugoba, Tarawanda, and the area between Matari and Chalinze.

SOUTHEASTERN AFRICA

The crystalline, Precambrian limestones of *Malawi* are dolomitic. The 7,000 ft. long and 150 ft. thick bed of Lirangwa appears to average 20% magnesia. The limestone of Matope and the Karroo carbonates of the middle Shire are similar. However, the limestone of Changalumi, near Zomba, which is used in the manufacture of cement, contains only 1% magnesia. Malawi sövites also contain little magnesium. A 10 mile long stretch of the bed of Lake Malombe, now flooded, consists of cement clay. The clays of Mpata, in the north, have similar qualities.

In *Mozambique*, limestones of Precambrian, Cretaceous and post-Cretaceous age have been distinguished. Precambrian limestones have been distinguished at Chiúre and Montepuez in the district of Cape Delgado, at the Tanganyika border. Calc schists, striking northeast and dipping 18 °N, are burnt at Unango, in the Jeci Mountains of the Niasa district. Mozambique and Malawi share the Mauze carbonatite. The sövite of Munetaca, near Itoculo, is of high grade. Secondary limestones developed on Lake Chirua and Amaramba, on Mozambique Island and Lumbo. At Chirombe, near the Zambezi, Precambrian limestones border granite; similar rocks are widely used at Matema and Marávia, in the Tete district of western Mozambique. At Mossurize, in the Manica and Sofala district on the border of Southern Rhodesia, high grade sedimentary limestones extend for 50 miles, with a width of 1–10 miles. While they contain only 0.5% MgO and 0.9% SiO_2, those of Cheringoma average 0.9% MgO and 5.3% SiO_2. At Vila Machado heterogeneous crystalline limestones have been found at the eastern edge of the Precambrian. The outcrops of Báruè are of similar age, but limestones are Tertiary–Quaternary south of the Save River. A band of Pleistocene deposits extends 20 miles from the coast from Maxixe, near Mucoque, 140 miles south of Homoíne, near Inhambane. The marine, Eocene limestones of Maputo and Mangulane, south of Lourenço Marques, are dolomitic.

Limestones are abundant in the sedimentary belt of *Madagascar*, but they are frequently unsuitable for lime burning because of their magnesium content. Precambrian cipolinos tend to be dolomitic (Diavonaomby) or siliceous (Fianarantsoa). Upper Cretaceous limestone and clay are mixed in the cement plant of Amboanio, near Majunga. At distant Soalara, near Tuléar in the southwest, 150 ft. thick beds of limestones and the clays of Ambatobe can be used; coal occurs conveniently nearby at Sakoa. At Antsirabe, 84% of the Ibity cipolinos plus 16% of lacustrine clay can be mixed in cement. Northeast of the capital cipolinos have been located at Ambatondrazaka, and clays have sedimented near Moramanga.

SOUTHERN AFRICA

In *Southern Rhodesia*, five types of limestone have been distinguished: Precambrian or crystalline, Karroo, calcite lodes and veins, travertine and calcrete. As elsewhere, the composition of Precambrian limestone is variable. The quarries of Colleen Bawn, near Gwanda, supply this type of rock for the manufacture of cement; the Sternblich deposit near Salisbury and those of Shamva are similar. Lime is burnt at Snow White, near Salisbury, and on a smaller scale at Lalapanzi. The Moosgwe lime works, at Melsetter, operate in 70 ft. high travertine cliffs. Limestone is also quarried at Redcliff and Lambourne, in the Hartley district, and dolomitic limestone is won in the Sinoia district and at the Houghton mine, near Fort Victoria. High grade, flux limestone is exploited in the Early Worm deposit. Around Sambawisa, near Wankie, calcite lodes are actively worked for metallurgical purposes (Fig. 36).

The late Precambrian limestones of the Karibib district, Southwest Africa, have been transformed into attactive, banded, undulating or folded and finely brecciated marbles.

Limestone is abundant in *South Africa*, but all too often it is distant from centres of industrial activity. Early Precambrian limestone is quarried at Steinkopf, Holrivier, Boplaas and Loerie, in Cape Province, and is used for the manufacture of cement at De Hoek. Limestone lenses occur in the Postmasburg syncline in the Dolomite Series of Griqualand, and also in the 150 mile long belt which extends from Griquatown to Bromley; this bed is quarried at Bowden. Dolomite of the same age is also abundant in the Transvaal, and marble occurs at Marble Hall. High grade travertine and secondary limestone have developed from the former in a belt which extends from west of Mafeking and Taung to Buxton, Thoming, Boetsap, the Ulco cement plant and further west to Douglas. Similar limestones are quarried at Ottoshoop and Kapsteel, and used in the manufacture of cement at Benadesplaats and Lychtenburg. Low grade limestones derived from Karroo rocks are used in the northern Free State around Henneman and notably at Whites. At Kalkheuvel, Doornkuil and Graspan, near Warmbad, and at Zebedia and Rietfontein, near Groblersdal, limestone has been derived from Karroo basalt. Deposits of sandy shells extend from East London to Saldanha Bay, and large bodies outcrop between Bushmans River and Uitenhage and between Mossel Bay and Cape Agulhas; cement is manufactured from this material at Philippi. Other cement plants include Port Elizabeth, in Cape Province, and Hercules, Industria, Jupiter, Orkney and Roodepoort in the Transvaal (Fig. 46).

While the Phalaborwa carbonatite of the eastern Transvaal contains 3–10% magnesia, the rauhaugite core of Spitskop, in the north, is surrounded by pure sövite.

In *Basutoland* limestone nodules and interbeds rarely show in Karroo rocks, but on Nsalitshe Hill, in Swaziland, a vein of calcite of sub-Nicol quality traverses Karroo sediments.

BENTONITE

While alkaline members of the montmorillonite–attapulgite clays are distinguished by a high coefficient of swelling, the calcic members of the group are characterized by a bleaching and detergent effect. These clays alter in a lagoonal–volcanic environment from ash and lavas, and from the argillaceous sediments they are interbedded with or from magnesian rocks. The principal deposits are related to the late Tertiary, Alpine volcanism of the Moroccan Atlas and to the Quaternary.

MOROCCO

At *Providencia*, 20 miles from Melilla, a fluid rhyolite is covered by breccia and partly altered to bentonite. The white bentonite mass extends for 1,200 ft., attaining 60–90 ft. of width and a thickness of at least 180 ft. The relic structure of the rhyolite ground mass can be seen in the transitional zone; alteration progresses along brecciated fractures there. Opal, fluorite and gypsum accompany montmorillonite. Similar occurrences, like the Jbel *Tidinit* (Fig. 10, 59) and the Oued Aguilmane, have also been derived from submarine rhyolite flows by hydrothermal alteration. At the Oueds Bohoua, Sgangane and Ras Tarl, homogeneous and agglomeratic pumice matrices have been partly replaced by montmorillonite.

The Miocene *Rio Masin* syncline is filled in by green marls interbedded with sanidine–chalcedony, pumice and pyroclastics. Montmorillonite has precipitated in the ground mass of sandstones, volcanic gravels and pumices. At Tamouat Arroumai, montmorillonite–kaolinite lenses are interlayered in a 15 ft. thick current bedded horizon. However, the high grade, 40 ft. thick layer of bentonite of Taouriat Hamet consists of hydrobiotite and vitreous particles. Similar occurrences include Bled Ghassoul and Maaza.

Fuller's earth. The principal deposit, at *Camp Berteaux* (Melg-el-Ouidane), is located 60 air miles south of Melilla on the Mediterranean; indices of the same area include *Beni Koulal* on the Wadi Za, Moul el Bacha, Dar Slimane and Tikkerdadine. In the Camp Berteaux basin, a 350 ft. thick late Tortonian (Middle Miocene) series of sandstones, gypsum marls, limestones and smectite lies unconformably on Middle Jurassic, Dogger sandstones; Pontian (Upper Miocene), continental sediments overlie the sequence. The smectites represent *(1)* a lacustrine or lagoonal stage of sedimentation or *(2)* the diagenesis of marls affected by interbedded Miocene cinerites of the Djebel Guilliz volcano. At Camp Berteaux the smectite column consists of marl, clay and cinerites; cinerite lenses also occur in the Fuller's earth beds which contain 50% silica, 16% alumina and 5% magnesia. At *Beni Koulal*, a six ft. thick smectite layer is interbedded with sandstone and gypsiferous marls (Fig. 10, 59; AGARD et al., 1952).

Detergent clay. The Djebel Rhassoul is located in the Middle Moulouya, 70 miles south-southeast of Fès. The Tertiary, sedimentary column of the syncline consists of marl, gypsum, marl, limestone and silex, laterally changing into lacustrine Oligocene limestone. The detergent clay or 'Rhassoul' beds are included in the upper marls, comprising *(a)* a bed of silex and *(b)* partly schistose, laminated montmorillonite, green marls or limestones. This cycle is repeated several times, the thickness of the detergent clay beds diminishing and their lime content increasing upwards in the sequence. The montmorillonite contains 57% silica and 27% magnesia. The clays, possibly eroded from talcschists, sedimented under lacustrine–lagoonal conditions in a sub-desertic environment.

THE REST OF AFRICA

Algeria. Bentonite is known to exist in the Oligocene at Maillot, in the Middle Miocene Helvetian at Tsighaout, and in the Holocene of Constantine at Mchounèche, Tighanimine and Sillègue.

In Kenya, bentonite-bearing nontronite–illite clays occur in the volcanic sequence near Nairobi and in Pleistocene beds north of Mount Kenya. Montmorillonite shows in the Pleistocene–Holocene beds of Lake Amboseli on the Tanganyika border. Southeast of Lake Natron, in northern *Tanganyika*, magnesium montmorillonite embays patches of magnesite; the intensity of gelling of this hectorite (?) amounts to 70% of that of Wyoming bentonites. Bentonite, altered from volcanic ash, is believed to exist in the Songwe Valley, in the southwest and also near Dar es Salaam and Morogoro. A variety of sandy white clay ('mbuga') is also distinguished by swelling, e.g., at Wembere. In Mozambique, montmorillonite outcrops at Impamuto, 35 miles from Lourenço Marques.

An extensive, 8 ft. thick layer of bentonite rests on the Eocene limestone of Beomby, near Ejeda, in *Madagascar*. The clay consists of 40% attapulgite and 60% illite, and has a Stormer viscosity that attains 48 when it is mixed with ten times its volume of water.

Bentonite-bearing montmorillonite occurs at Barry, near Salisbury in Southern Rhodesia. At Arcadia, in the Orange Free State (*South Africa)*, 60% montmorillonite and 40% quartz have been derived from Karroo mudstone. The attapulgite–montmorillonite deposits of Springbok Flats, in the centre of the Bushveld, are alteration products of 20 ft. of Karroo basalt.

MEERSCHAUM

Bands and lenses of sepiolite, a hydrated silicate of magnesium, are interlayered with sedimentary limestones in Somalia and on the Tanganyika/Kenya border. While montmorillonite appears at the latter, gypsum is associated with the first of these occurrences.

Central *Somalia* is covered by Eocene limestones. In the area of *El Bur*, on the southern edge of the Mudugh Plain 200 miles north of Mogadishu, extensive deposits of meerschaum are underlain by limestone, gypsum and anhydrite. The sepiolite also appears to be overlain by limestone. The mineral outcrops in a radius of more than 10 miles from El Bur, and especially 7 miles northwest and 1 mile west of the Dusa Mareb road. If the meerschaum is interbedded with the sediments the deposit may extend to vast areas; on the other hand, sepiolite may only form lenses or replacement bodies that 8 ft. thick pits have not pierced. The material contains detrital quartz, some feldspar, disseminated carbonate, iron oxides, and stringers of gypsum which traverse it. The fibrous substance is hard and white at the surface, but brownish in depth; the composition is given in Table CLXIV.

TABLE CLXIV
COMPOSITION OF SEPIOLITE

Component	*Percentage*
MgO	15.8%
SiO_2	42.2%
CaO	12.0%
CO_2	7.9%
Fe_2O_3	2.7%
Al_2O_3	1.0%
H_2O	17.8%

The border of Kenya and Tanganyika crosses *Lake Amboseli* northwest of Kilimanjaro. A substance similar to Turkish meerschaum occurs as concordant, banded masses, slips and cavity fillings in secondary, dolomitic limestone of Quaternary age. In Kenya it is interbedded with montmorillonite and in Tanganyika it is covered by clay. The sediments are believed to have been warped by the expansion of bentonitic clays. In Tanganyika sepiolite bodies with a diameter of several ft. are found; the most attractive, 1,600 ft. long area is parallel to the shore (Fig. 33).

KAOLIN

The clay minerals of the kaolinite–halloysite group are alteration products of the feldspars of generally Precambrian pegmatites, granites and occasionally other intrusives and gneisses. Impure kaolin is also found in residual clays of river beds, and in arkosic and other sediments. Only a limited number of explored deposits is described here.

Morocco. At *Souk-es-Sebt* of *Sgangane*, south of Melilla, kaolinite and alum are found resting on diorite porphyry beneath alternating layers of trachyandesite and pumice. On Mount *Maaza*, seams of halloysite have developed: *(1)* at the contact of delta limestones and dune sands; *(2)* in travertines and at their contact with montmorillonite cinerites, impregnating reed; and *(3)* in stringers intersecting limestone.

The kaolin of Ropp on the Jos plateau of northern *Nigeria* is of excellent quality. It was derived from the Younger Granites; feldspar of the Abuja pegmatite also alters to kaolin.

In three localities of the Khor Odrus–Khor Eishaff area, southwest of Port Sudan, medium quality kaolin was altered from diorite, but the material is stained by ferromagnesian minerals. Kaolin frequently develops on the Erythrean plateau, in *Ethiopia*, especially in Hamasien. At Zolot, Ad Zanaf, Adi Goorbati and Bet Meka, near Asmara, feldspars derived from schists have been transformed into kaolin. In the Burs of southern Somalia granite weathered to kaolin.

Congo. In the granitic and pegmatitic areas of Maniema, Kivu and North Katanga provinces, kaolin is abundant and used locally (pembe). Kaolin has also precipitated in river beds of the western provinces, and from carbonaceous shales of the Lower and Middle Congo.

Kaolin is abundant in the pegmatite provinces of central Rwanda, e.g., in the area of Gatumba–Kigali, in southwestern Uganda and near Kampala (the quartz of this kaolin can be removed). At Kisai, near Masaka, kaolin has altered from shale. Few deposits have been explored in *Kenya*: in the Ndi Hills near Voi, in the southeast in the Machakos district east of Nairobi and in the Rift Valley, e.g., at Fort Hall, kaolin has been derived from Precambrian pegmatites and gneisses. At Eburru, kaolin is a product of solfataric alterations of lavas. At Pugu Hill, near Dar es Salaam, Tanganyika, kaolin constitutes 30% of a 600 ft. thick bed of Neogene sandstones; at Magoye, the zone of weathering is 20–50 ft. thick. In the area of Matambe, east of Lake Nyasa, kaolin of low grade china quality has altered from gabbro and granite. Impure kaolin has been located at Malangali. In Malawi subceramic kaolin has been identified near the sources of the Rivi-Rivi, south of Chinteche, near Lake Nyasa and on Kumbanchenga Hill. These last two are gritty and coarse; the first was derived from anorthosite granulite. At Boa Esperança, northeast of Alto Ligonha, *Mozambique*, pegmatites have been kaolinized, and the Moriangane deposit near Vila Manica contains 98.5% kaolin.

In *Madagascar*, pure kaolin of the kandite type has been identified at the contact of Precambrian and sedimentary rocks at Ampanihy; the material contains 41.7% Al_2O_3, 39.0% SiO_2 and 3% TiO_2. Deposits are found near Kokomby and on the Androka Road. While high grade kaolin developed from the Andilona pegmatite, near Alaotra, the pegmatites of the Tananarive area have not produced good material.

Kaolin is exploited northwest of Salisbury in Southern Rhodesia. In *South Africa*, kaolin was altered from granite, pegmatites and gneiss at *(1)* Potgietersrus and Witrivier, in the Transvaal, *(2)* at Inanda, in Natal, *(3)* at Fish Hoek and Kommetjie near Cape Town and *(4)* in the Vanrhynsdorp district of western Cape Province. Kaolin derived from shale, however, is impure. In the Mankaiana district of Swaziland kaolin occurs in concession No. 50, and halloysite has altered from the feldspars of concession No. 29.

CLAY

Refractory ceramic and brick clays are mixed layer sheet silicates, generally including illite and hydrated micas. The following genetic categories can be distinguished: residual and river clays, sedimentary layers, alteration products, mainly of shales and basalts. Argillites are abundant in the Karroo and in later sediments, and lavas alter to clay in eastern Africa. Fluviatile clays are widespread, but only a few examples, mainly industrially exploited deposits, are described here.

NORTH AND WEST AFRICA

Algeria. Cretaceous marls are used for pottery in the Tell Atlas, and Miliana, upper Cretaceous and Senonian argillite is used in the Babors. Lower Cretaceous, Neocomian and Barremian brick clays are used at Sétif, but Suessonian and Medjanian marls are utilized at Bône and Calle. Miocene, Helvetian and Pliocene gray clays are favoured for local pottery.

Morocco. The most frequently used clays are of *(1)* Permian–Triassic age at Fedala, *(2)* Upper Jurassic (Callovian–Oxfordian) at Oujda, *(3)* Cretaceous at Safi and Agadir, *(4)* Miocene at Rabat, Salé, Fès, Meknès and Taourirt, and *(5)* Quaternary substances at Fedala, Casablanca, el Qenitra, Marrakech and Agadir. Many of them are impure and marly; they mainly consists of illite.

Kaolinite–attapulgite clay is burned at Pout, in Senegal, but at Kaolack and Taky other clays are utilized. In *Ghana*, brick clays are known to occur in numerous localities, e.g., 5 miles from Accra near the Kumasi railway, in the Subin Valley, near Dunkwa (alluvial), in the lagoons of Winneba near Sekondi, at Inchaban (gypsiferous), Lake Bosumtwi (lacustrine), and in the area of Tamale and Bui in the north. Pottery clay has been located at Kwahu, between Kommenda and the estuary of the Pra, at Eddubura, and at Dukludja in the east. High grade ceramic clay occurs along the coastal strip of Togo.

Clay is available in most of *Nigeria*, but not always suitable for the manufacture of bricks. Thick beds of white clay located in Abeokuta Province are used in the pottery industry of Ikorodu. Clay of suitable quality also occurs in Benin, Onitsha and Owerri Provinces; fireclay intercalates in the Lower Coal Measures of Enugu, where shale and other clays are also found. Pottery clays are exploited near Abuchi, near Abuja, Ikorodu and Okigwi, and residual clay is exploited at Kwali, near Abuja. A light coloured clay is derived from the Older Basalts of the Jos plateau, and white clays also occur at Arkilla, Dakingari, Garkidda, Kutigi and other localities.

CENTRAL AND EASTERN AFRICA

Deposits of ceramic clay are known to occur near Bangui, the capital of the Central African Republic. Various clays are frequently used for the building of houses in the Congo.

Residual and Nile clays are widely used in Egypt. Ethiopian clays belong to *(1)* Cainozoic interbeds of the sedimentary column, *(2)* the base of traps, and *(3)* to inliers of basalt. High grade clay is rare in Ethiopia, although it occurs at Zaga. While fluviatile clays are abundant in *Somalia*, for example 15 miles from Mogadishu, on the Webi Shebeli, lagoonal clays have been located on the coast at Danane, south of Mogadishu and at Brava.

In Uganda swamp clays can be used for bricks, and pottery clays are won north of Lake Victoria at Buku, Mukono and Nansane. Brick clays are used on a larger scale in Kenya at Mombasa, Muthaiga, and Gatharaini near Nairobi. In Tanganyika, hill-slope loams, sandy clays and river valley silts are used for brick burning or, in their absence, the calcareous 'mbuga' soil.

In Malawi residual clays altered from gneiss or syenite are used with success; in the Blantyre area, the supply is adequate. The Karroo shales of *Madagascar* are siliceous but clays and marls are abundant in the Upper Jurassic and Cretaceous. In the Betsimitatatra basin, surrounding Tananarive, 3–5 ft. of sandy clay lie at shallow depth. The clays of the Ambohimandroso plain and the Matsiatra Valley are also used for brick burning. Marly clays have been analysed at Tuléar, the Table Mountain consisting of oyster marls.

SOUTHERN AFRICA

Clay is abundant in the Zambezi basin of *Southern Rhodesia;* refractory and brick clay are recovered at Wankie, as well as clay utilized in the manufacture of pipes nearby at Fridon. The clay of Tipton, near Lalapanzi, is used as a refractory.

At Burgersdorp in the Cape, *South Africa*, mixed layer montmorillonite–hydromica, a ceramic clay, has developed at the contact of Karroo shale and dolerite. The foundry clay of Addo, near Port Elizabeth, consists of 85% illite and 15% montmorillonite. At Albertinia on the south coast, 5 ft. of Tertiary, mixed clays are quarried, and the hydromuscovite of Pretoria is used in tile burning. Mixed refactory clays have generally been derived from Karroo sediments; occurrences include Vereeniging and the Witwatersrand, notably Olifantsfontein, Marievale, Boksburg, Brakpan, Heronmore, Witbank, Hammanskraal and Gezubuza. Residual, burning clays have been derived from Transvaal System shales, from Karroo shales on the south coast and from marine clays at Port Elizabeth.

PIGMENT

Although graphite and carbonaceous rocks are used as a black pigment and kaolins as a white one, the largest supply of red and yellow ochre is derived from altered, ferruginous rocks, and particularly from shales and clays.

South Africa. Lenses and pockets of excellent yellow ochre have developed on folded Paleozoic shales in an 18 by 7 mile area which extends between Riverdale and Albertinia along the south coast. The depth of alteration is locally in excess of 70 ft. ,and the ratio of recovery of high and low grade ochre averages 3/1. Quarries include Zout Vley, Zout Pan, Tartouwa, Drooge, Snymanskraal and others. Red and yellow ochre is also known to occur in the Cape at Humansdorp, in the Swellendam and Bredasdorp districts, and at Ga-Mopedi in Griqualand.

At Hezeldene, Novembersdrift and Sub-Palmyra, between Newcastle and the Buffalo River in northern Natal, a siderite bed of the Karroo has been altered to hematitic, dark red oxide. Other pigment quarries of Natal include Pinetown, Groenkloof and Hopevale.

At Uitgedacht, near Bethal, Transvaal, Karroo shales have altered to red and yellow oxide; but the Weltevreden and Elandsfontein pigments are derived from Precambrian Pretoria shales. At Waterval iron–manganese umber forms lenses in leached dolomite, which also covers it. Larger deposits of umber are found at Leeuwkloof and Hartebeesthoek.

In Basutoland pyrite has weathered to brown oxide, and ochre has developed in the Sephonghong River where a volcanic dyke traverses sandstone.

Southwest of the Kilimanjaro, in *Tanganyika*, the ferrugineous soils of volcanic Mount Monduli are used by the Masai. Yellow ochre occurs with phyllites at Tandalo and Lupalilo, near Lake Nyasa, and at Kagongo on Lake Tanganyika in an 8 ft. thick bed of altered shale. In Kenya red, brown and yellow colours are won from oxidized clays. Green clays and chlorite schists are used in Kitui, Lodwar and Turkana, in the north, graphite being utilized for black colouring. The Congolese use various ferrugineous clays as a pigment. In Ghana, pigment clays are derived from phyllite near Jamise between Humzibre and Sefwi Bekwai, at Huni and Foso, and at Intabrawso.

SECTION IX

Miscellaneous

There are six categories of substances in this group: *(1)* volcanic materials like pumice, pozzuolana and perlite; *(2)* nepheline in intrusives; *(3)* metamorphic minerals like corundum, kyanite, sillimanite, andalusite, wollastonite, and at least partially garnet; and the functional but polygenetic categories of *(4)* semi-precious stone; *(5)* building, and road building stone; and *(6)* sand and silica. The latter are of only minor mineralogical interest, but they continue to acquire importance with each phase of economic development.

PUMICE

Pumice or volcanic ash is used as an abrasive and as a light weight pozzuolonic building material. Pumices were deposited in the Atlantic islands, the coastal Atlas and the Rift Valley during the Neogene and Tertiary.

The pumices of Tamechkout, near Sgangane, northeastern Morocco, have sub-pozzuolanic characteristics; they are interlayered with kaolinite and related to effusives.

The *Canary Islands* lie along a line extending west-northwest from the fault that divides the Atlases from the Sahara. Pumice is included in the volcanics of the Canary Islands as it is in the Azores, further north (Fig. 55).

The *Cape Verde* Islands are located west of Dakar; the archipelago consists of the Santo Antão–San Nicolau line, which strikes east-southeast, and of the arc connecting Brava, San Tiago, Sal and other islands. The probably Precambrian core of the islands is sur-

TABLE CLXV

COMPOSITION OF PUMICES

	Gamboesas (%)	*Ribeira Fria* (%)	*Fundao* (%)	*Brejo* (%)
SiO_2	46.2	47.0	47.3	46.4
Al_2O_3	18.4	18.9	19.5	18.1
Fe_2O_3	1.9	1.8	1.5	2.4
CaO	2.4	1.4	1.3	1.3
MgO	1.7	1.4	1.6	1.9
alkalis	8.8	10.0	9.5	10.7
water	19.4	18.7	19.2	18.4

rounded by trachyte, phonolite, andesite and other effusives and limestone. Pumice occurs on Santo Antão and San Tiago. Analyses of four deposits of Santo Antão are given in Table CLXV.

Inliers of pumice are found in the diatomites Kariandus in southwestern Kenya. Large reserves are also available in the Rift Valley at Longonot, Lake Naivasha, Menengai and Rongai. Between Tukuyu and Mbeya, in southwestern *Tanganyika*, considerable areas are covered by pumice and volcanic ash of the Rungwe and Ngozi volcanoes. The slopes of Ngozi, 6 miles from Mbeya town, are easily accessible. Pumice occurs at Mount Meru and Arusha; tuff and ash is also abundant in the north. In Malawi, pumice is reported to occur at Karonga, and similar materials have erupted in central and northern Madagascar and in the Comoro Islands.

POZZUOLANA

Pozzuolanas are volcanic rocks, ashes and sediments, affected by volcanics, that expand when mixed in cement. Pozzuolanic materials occur in the East African Rift and in the Malagasy volcanics.

In *Ethiopia*, pozzuolanic sands cover the sides of Mount Dula, a volcano situated in the Gulf of Zula (Erythrea); at Dogali and Saati basalt flows have transformed soils into pozzuolana. Large deposits of Las Addas and Bishoftu, near Mount Boseti, are related to recent volcanism, and cinerites have been deposited in a lake near Addis Ababa. Other deposits occur at Sirie. Several volcanic rocks of the Kenya Highlands and the volcanic ashes of Arusha, northern Tanganyika, are pozzuolanic, as is the red ochre of Monduli Moutain. The pumices and ashes of the Rukwa trough, when mixed with one fifth of their weight in lime and with water, also produce good pozzuolana.

In the volcanic areas of Diégo Suarez, Antsirabe–Ankaratra and Itasy pumice-materials are abundant. Analyses show that the Antsirabe material, which contains 41% SiO_2, 29% Al_2O_3, 10% Fe_2O_3 and 2% TiO_2, can be considered as pozzuolana; however, the Joffreville and Roussettes rocks are only sub-pozzuolanic.

PERLITE

Perlite or vitreous rock distinguished by an expansive effect is found among rapidly solidified lavas, particularly in the Triassic volcanics of Mozambique.

A narrow band of Triassic, Drakensberg tuffs and basalts extends from the Zambezi, along the border of *Mozambique* with Southern Rhodesia and South Africa, through Swaziland to Komatipoort in the northeastern corner of Zululand. Large deposits of perlite are found at Libombos (the Lebombo monocline) in the Muguene Mountains of the Boane district, southwest of Lourenço Marques. The cracked acid lava consists of green volcanic glass and blebs with streaks of altered feldspar. The blackish volcanic glass of Upper Karroo, Stormberg age found in concessions 49 and 74 in the Stegi district, and of mineral area 11 of the Hlatikulu district, in *Swaziland*, resembles the perlite of the Libombos range.

Obsidian, some of it suitable for expansion as perlite, occurs in Kenya, e.g., at Magadi and Naivasha.

NEPHELINE

Fusible nepheline is relatively rare in Africa, although it occurs in the alkaline-carbonatite areas of the southeast, e.g., in Malawi.

Fairly pure nepheline syenite occurs at Bou Agrao, south of Midelt in Morocco, and at Kpong in eastern Ghana. In Kenya Tertiary lavas and urtite, ijolite of the Homa Bay carbonatite, contain fine grains of the mineral. Nepheline is abundant in the alkaline province of southern *Malawi*: the zoned syenites of Mauze Hill, straddling the Mozambique border, include 29.5% nepheline. North of Zomba Mountain foyaites and pulaskites of five ring complexes, which occupy 50 sq. miles, contain respectively 50% and 10% nepheline. At Nathace Hill, an up to 1,800 ft. wide syenite ring-dyke contains 24% nepheline. Other nepheline occurrences of the alkaline-carbonatite suit include Nkalonje, Chilwa Island (Çisi), and Salambidwe. The nepheline gneisses of Tambani and Port Herald outcrop in areas of 30 and 7 sq. miles, respectively. The Ilomba complex of northern Malawi also includes nepheline syenite.

In Madagascar, nepheline syenites are known to outcrop at Nosy Komba, on the Ampasindava peninsula, and at Tsaratanàna. Numerous other African carbonatite complexes include nepheline-bearing bodies and dykes.

CORUNDUM

Corundum is a typical metamorphic mineral of schists, gneisses, ironstones and basic rocks, invaded by pegmatites, granite or syenite. Feldspar pegmatite lenses may contain up to 90% corundum, accompanied by sillimanite, in which case they become corundite or corundum rock. Since corundum is hard and resistant, the mineral remains on the weathered surface of the rocks (and in gravels) from where it is recovered. The principal corundum-producing areas are the northeastern parts of Southern Rhodesia and South Africa and the adjacent districts.

SOUTHERN AFRICA

In *Southern Rhodesia* five corundum areas can be distinguished: *(1)* the Mazoe drainage in the northeastern corner of the country, extending into the Tete district of Mozambique; *(2)* near Amandas, north of Salisbury; *(3)* between Marandella, Mangwendi and Rutape, southeast of Salisbury; *(4)* the Upper Sabi, west of Umtali and *(5)* between Singwesi and Beitbridge, connecting with the northern Transvaal district.

In the Mazoe district, between Ndiri and Talland (Fig. 36), fine-grained corundum granulite forms a long, sub-vertical lens in an itabirite horizon, intersected by a dolerite dyke. On *O'Brien's* prospect, purplish-green corundum, accompanied by andalusite and chrome-bearing fuchsite, represents 60% of the rock in which corundum has been reconcentrated in boulder float. At Somerset, a body resembling the preceding occurrence is veined by quartz stringers and fuchsite rock. Corundum has also been located in the Rufunsa drainage of eastern Zambia, near the Mozambique border.

In western *Mozambique* corundum is abundant at Zobuè, 50 miles west of Blantyre (in Malawi), and at Moatize near Canchoeira, 15 miles east of Tete, where corundum feldspar pegmatites strike northwest in basic rocks intruding gneisses, amphibole schists and quartzites. Muscovite pegmatites do not contain corundum. In the smaller, Zobuè deposit the metamorphics have been intruded by granite and syenite. It is included in pegmatites and gravels at Cachissene and Chissindo, in the Niasa district, and at Revuè near Vila Pery, on the border of Southern Rhodesia, pegmatites traversing biotite gneisses and garnet aplites contain red corundum. At Tambani, Port Herald and Malawi Mountain, in southern Malawi, corundum and zircon have been eroded from pegmatites invading nepheline gneisses.

South Africa. The deposits of the northern and northeastern Transvaal are extensive. The largest fields (Fig. 46) include *(1)* Louis Trichardt, Pietersburg, the Magabeen Mountains and Blouberg; *(2)* Lilliput and Tatchankop; and *(3)* the Leydsdorp–Malelane belt. Early Precambrian, isoclinal folds of basic perknites, invaded by pegmatite and granite, comprise 7,000 ton lenses of corundum rock. In the northern Transvaal, common pegmatites grade into plumasite or coarse corundum plagioclasite. In the rarer marundites of the eastern Transvaal, and Kranskop in Natal, feldspar is hydrothermally replaced by magarite, corundum being affected by incipient gibbsitization. Accessory garnet, tourmaline and apatite are included in this paragenesis.

While garnet and sillimanite accompany corundum in the hornblende gneiss of Cassel, feldspar surrounds the mineral in the biotite gneisses and schists of Pella Mission (see 'Sillimanite' chapter). At Kabis and Narries in Namaqualand, and Kranskop in Natal, corundum has crystallized in pegmatites which intrude basic rocks and gneisses.

The bulk of the corundum supply, however, comes, from boulders and eluvial pebbles covering the weathered corundum rocks and their slopes in the northern and eastern Transvaal. The Piet Retief and Steinkopf deposits in the Transvaal and Namaqualand supply similar material.

THE REST OF AFRICA

In *Madagascar*, corundum is recovered from weathered syenites, pegmatites, sakenites and micaschists, and from eluvial and alluvial placers derived from them. At Ambohitranefitra, near Beforona on the Tananarive–Tamatave road, reddish corundum and pink corundite have been eroded from corundum syenites and their contact zone. They were reconcentrated at the foot of the hills and in the beds of the Sahamaloto and Marafody rivers (Iaroka drainage) and in the Tsarafosa stream. The corundum zone extends south of Beforona to the contact of the Lakato stock, as far as Masse. Southeast of Antanifotsy and northeast of Fandriano, in the Ambatolampy–Ambositra area (south of the capital), corundum has crystallized in several zones of mica schists affected by granite. In the gravels of Mount Ampahanivorona and Ambatomaladia, near Vatomandry, crystals of up to several pounds have been found. Northeast of Ejeda, e.g., at Anavoha and Vohitany as well as near Sakeny, northwest of Ihosy, corundum has been disseminated in hard, felspathic pyroxenites, sakenites (plagioclasites) and granites.

At Mleha, in west-central *Tanganyika*, pockets of corundum occur in a hornblende pegmatite traversing serpentines, but at Mlali corundum is included in biotite schists and on Ilende Hill it is found in mica schists sandwiched between pegmatites. At Kinyiki Hill, southern *Kenya*, up to 4 ft. long, but cracked, crystals have been found in eluvia at the contact of dunite intrusive in hornblende schists. Other occurrences include *(1)* the Taita Hills, Machakos, Loldaika, Baragoi and the kyanitite of Taveta, in Kenya; *(2)* the Karamoja district of Uganda; *(3)* the placers of Kamabea, in the Kasese tin district of the eastern Congo; and *(4)* the alluvia north of Mayoko in the Congo Republic. In the ultrabasic bodies of Hargeisa, northern Somalia, corundum spinel and corundum iron ore coexist.

In Ghana, corundum is found in the gneisses of Neibwale and in the diamond field of Birim. The mineral similarly accompanies diamonds in Grand Cape Mount County in Liberia, and also occurs near Golella. In Sierra Leone grains and fragments concentrate in placers near granite–schist contacts.

KYANITE

Kyanite, andalusite and sillimanite, occasionally accompanied by corundum, typically crystallize in metamorphic schists and gneisses. Kyanite is recovered from weathered rock or from adjacent eluvia and alluvia.

Kyanite is abundant in the placers of Jang and Gurupe, in northern *Ghana*, and in a tributary of the Kulapawn, the Ginawa and Guliguli, near Wa. Extensive placers of kyanite are located in the upper Niyba River between Edea and Puma in southwestern *Cameroon*.

Southwest of Port Sudan, a kyanite–quartz gneiss contains 58.4% Al_2O_3, 37.7% SiO_2, 1.2% Fe_2O_3 and 0.9% CaO. On Erusi Hill in northwestern *Uganda* kyanite constitutes up to 10% of a muscovite–garnet schist. At Azi, a 300 yard long and 3 ft. thick lens of schists contains 80% kyanite and 5% rutile. Mica soils the 6 mile long band of kyanite schists of Kamera, in the southwest.

In southeastern *Kenya*, a band of kyanite schists extends between Taveta and Tsavo from Murka to Kevas and Loosoito. At Murka one of the bands, which lies between biotite gneiss and quartz schist, includes lenses of kyanitite surrounded by sillimanite and some corundum; the schist averages 75% kyanite. Boulders of kyanite cover the slopes of the hill. Schists of Sultan Hamud, south of Nairobi, contain 25–31% kyanite. A similar occurrence has been located 28 miles to the southwest at Isiolo, south of Voi and near Loita, while at Machakos kyanite has been found in a pegmatite (Fig. 33).

Kyanite schists and gneisses frequently occur in the Pare Mountains of northern *Tanganyika*. At Chankuku, a more than 200 ft. long and 10 ft. wide lens has been located, while the Uguruwa band possibly extends for more than 3 miles. The Ugweno, Chaborati and Longwana occurrences, however, are of low grade. Eluvial kyanite has been found at Kitwai, in Masailand, in the Mpwapwa district, and on Chala Hill in the west.

The largest deposit of Malawi is situated in the Kapiridimba Mountains: lenses of schists averaging 15% kyanite include pegmatitic bodies of kyanitite. Abundant kyanite is disseminated in the gneisses of the Kirk Range and Neno, in central Malawi, e.g., near Chingoma stream and in schists at Chakumbela, further south. In Mozambique kyanite is found in the gravels of Muaguide in the north, in the Mevuzi River in the west, and between Bárue and Zóbuè, near the border of Southern Rhodesia.

Kyanite is frequently included in the Precambrian Vohibory schists of eastern *Madagascar;* the principal deposits are situated in the Mananjary area, halfway between Tananarive and Fort Dauphin. About 2 ½ miles northeast of Ampasimazava, two parallel and approximately 15 ft. wide bands of kyanite–mica schists are 300 ft. distant from each other. Sillimanite has been eroded from the band, located 1 mile southeast of Vohilava and transferred into the placers of Saka. Similar deposits are known to occur near Fénérive Soanierana–Ivongo, north of Tananarive and Ampandramaika, in south-central Madagascar.

In Southern Rhodesia kyanite schists are abundant in the northern mica fields of Mtoko, Miami and west of Sabi. Kyanite also occurs at Half Way Kop, near Tati, in northeastern Bechuanaland. At Rehoboth, south of Windhoek, in the Precambrian remnants of central Southwest Africa, kyanite schists rise frequently; kyanite is recovered at Kyanitkop and 50 miles south, at Uisib. The mineral occurs at Blaauwvlei, in South Africa, halfway between the Cape and the Orange River.

SILLIMANITE

Sillimanite, accompanied by corundum and less frequently by kyanite, occurs in schists and quartzites, in which it also forms lenses. Sillimanite is restricted to the metamorphic areas of southern and eastern Africa.

In the Congo sillimanite is found, e.g., in the quartzites of Albertville. Sillimanite occurs in Kenya at Kitui, Embu, Machakos, in the Loita Hills, south of Voi and at Maralal. Up to 30% sillimanite and kyanite appear in the metamorphic schists of Chankuku in northeastern Tanganyika, while possibly less occurs at Uguruwa. Sillimanite and even sillimanitite are abundant in *Madagascar*: The quartzites of Mananjary, on the southeastern coast are of particularly high grade. Lenses of sillimanitite have been located near Andavabato on the Tsinjoarivo road and at Vondrozo, and fibrolite has been identified in the gneisses of Androba and Tsibondroina.

Sillimanite accompanies kyanite in the schists of Miami and Mtoko, in Southern Rhodesia, and in those of Rehoboth in Southwest Africa. At Pella Mission, in the Orange River belt of *South Africa*, granulites and gneisses have been granitized. At Swartkoppies a range of block jointed corundum–sillimanitite rises from sands, and at Pella West similar boulders rest on granulite. The rock averages 53% sillimanite, 41% corundum, 3% ilmenite and 1% rutile. At Hotson, Koenabib, and Gamsberg up to 200 ft. long lenses of sillimanite, with considerably less corundum, are located at a constant distance from a metaquartzite indicator horizon. Sillimanite also reports at Mogane on the Southern Rhodesian border, in the Bushveld xenoliths, and in Insuzi and Mozaan rocks at Prieska and the Barberton district.

ANDALUSITE

Accumulations of andalusite are the least abundant in the kyanite group because andalusite is more distinctly restricted to the metamorphic aureole of granitic intrusives and placers, rather than to boulder deposits on weathered rock.

In *Morocco* the mineral occurs near granitic cupolas in the Precambrian schists of Tazeroualt and Iguerda, and in the Hercynian cycle at Zaër and Sidi bou Othmane. In *Ghana* extensive deposits of andalusite have been located at Abodum, 4 miles east of Bekwai. Andalusite occurs in the fenitized aureole of the Lweshe carbonatite in the Congo and in the diamond placers of Kasai.

In the Marico district, near Mafeking *(South Africa)*, andalusite has crystallized in the outer zone of metamorphism of the Bushveld. Andalusite has reconcentrated in the sands of the Little Marico River between Kromellenboog and Veeplaas, and in the Doorn River, the Wilgeboomspruit and the Vaalkop stream. Typical concentrates contain 52–57% Al_2O_3, 35–44% SiO_2 and 2–4.5% Fe_2O_3. Andalusite also reports at Burgersfort, Pietersburg, and in the Insuzi and Mozaan formations.

WOLLASTONITE

This lime silicate is a skarn mineral occurring at the contact of ultrabasic or basic rocks and carbonates, especially in Madagascar and the Sudan.

In *Morocco* wollastonite occurs in lime rich schists adjacent to granite stocks. At Azegour and Sibara beds attain up to 30 ft. of thickness. Smaller occurrences include Jebel Aouam, Ment, Tichka and Sidi bou Othmane.

Sudan. At Dirbat Well (Fig. 13), on both sides of the Khor Haiet 70 miles west-northwest of Port Sudan, (titaniferous) gabbro and possibly granite have marmorized limestones transforming calco-magnesian sediments into skarns. Irregular skarn bands and aggregates consist of variable proportions of grossularite, schorlomite, diopside and idocrase (vesuvianite). Wollastonite occurs with them, with marble, and in monomineral, keel-shaped lenses up to 100 ft. long and 20 ft. wide. Diopside inclusions, quartz stringers and laminae have been observed in the wollastonitite, distinguished by a buff-coloured and rough elephant-hide like surface. There have been distinguished seven bodies in a mile long band in the northeast, and 25 others outcrop in an area of $^1/_3$ sq. mile in the southwest. Sorted samples include 70% wollastonite and 25% diopside; the wollastonite contains 43–48% CaO, 51% SiO_2, and 1–6% Mg, Al and Fe oxides.

Wollastonite covers ½ acre in *Kenya*, at Epiya Chapeyu on the Yatta plateau. The mineral occurs in Precambrian crystalline limestones in the Homa Bay carbonatite district, e.g., at Ali Gollo and in ijolite at Usaki. In *Madagascar*, wollastonite pegmatites and wollastonitites are most abundant in the calcic-ultrabasic rocks of the Androyan System, at the southern tip of the Island and north of Behara (see 'Mica' and 'Uranium' chapters.

GARNET

Garnets are polygenetic. Almandite–spessartite occurs in granitic cupolas, gneisses and pegmatites, while pyrope favours kimberlite, melanite carbonatites and andradite skarns. Since garnets are resistant and equidimensional they tend to concentrate in placers and, less frequently, in beach sands. Although garnet is a typomorphic mineral of granitic–pegmatitic districts, important deposits are mainly restricted to Madagascar and eastern Africa.

The largest garnet deposits are located in southwestern *Madagascar*. In the Ampanihy district garnet has been disseminated in an elongated band of metamorphic schists of the early Precambrian Graphite System. The principal occurrences include Ianara, Ianatratra and Besosa, and the latter two also contain precious garnet. The bands of garnet-bearing schists of Pinjo 20 miles southeast of Ampanihy, and of Ambalavato 3 miles northeast of Besosa, have not yet been explored. While the chlorite schists of Ampandramaika, in central Madagascar, contain coarse garnet, the silicate is also disseminated in the Bekily and Betroka districts in garnetites and beach sands (Fig. 48).

In *Morocco*, garnet occurs in the aureole of limestone and schist of Precambrian stocks at Azegour, Tichka, Sidi bou Othmane, Jebel Aouam, Sibara and Ment. Garnet is also abundant at Bou Agrao in the Lower Jurassic limestones intruded by nepheline syenite. Garnet–cummingtonite schists and banded ironstones are in contact in the Sula Mountains of Sierra Leone. In Ghana, garnet occurs in the diamond fields, the alluvia of Osudoku Mountain and Adaklu–Ho, the Mankwadzi schists and the Ho gneisses.

In the Sudan, almandite–spessartite crystallizes in the pegmatites and gneisses of the Rubatab belt, and grossularite and uwarovite are included in the skarns of Dirbat Well.

In the eastern Congo fine-grained garnet is abundant in the columbite concentrates of the granitic areas such as the Idiba district (see 'Tantalum' chapter). Garnet occurs in schists at Boma, and Ituri in Katanga. The garnet pegmatites of central Rwanda, and the garnet schists and gneisses of northwestern Uganda, are of Precambrian age. Melanite occurs in the carbonatites of western Kenya, and other garnets concentrate from Precambrian rocks in streams. At Namaputa, in *Tanganyika* on the Mozambique border, abrasive garnet is recovered from altered, basic gneiss. At Nambunju, 100 miles to the north, large garnet is extracted from gneiss. In the Pare Mountains, east of the Kilimanjaro, garnetite, garnet schists, gneisses and gravels are found at Sangaruma, Buiko, Kichaa and Lolobukoi. In central Tanganyika, a garnet skarn and eclogite are located, respectively, near Tambi and at Nyarumba.

The ilmenite–rutile sands of Port Herald, *Malawi*, average 11% garnet. Lenses of almandite–grossularite garnetite have been discovered in the gneisses of Fort Johnston in the north. Melanite sövite forms band in the Chilwa carbonatite. The drainages of Malibu, Mingonha and Muaguide, in northernmost *Mozambique* carry garnet. Near the Duembe River, west of the Malawi border, andradite has crystallized in limestones and schists intruded by granite. In the Nhamitombo Mountains, near the Southern Rhodesia border, garnet has been disseminated in schists.

In Southwest Africa, extensive deposits of grossularite are known at Swakopmund, and garnet also occurs in the Karibib pegmatite field.

SEMI-PRECIOUS STONES

Semi-precious and ornamental stones are of varied origin. Numerous minerals include varieties that can be considered as gem stones, but at least half of them are actually forms of silica. Lapidaries are most active in Madagascar and South African Transvaal. They operate less frequently in Southwest Africa, Southern Rhodesia and eastern Africa. Among many other gem stones Malagasy beryls and Griqualand jasper are famous. The rarer precious stones of Madagascar include yellow danburite, rhodizite, scapolite and green kornerupine.

Mainly albitized pegmatites contain beryl, feldspar, garnets, amethyst, rose quartz, spodumene, topaz and tourmaline; ultrabasics include garnet, serpentine and peridot; the latter, and agate–chalcedony, also occur in basic lavas. Corundum and cordierite stones are metamorphic, while jasper and crocidolite distinguish metasediments. Opal is sedimentary, but malachite is diagenetic. Stones are frequently recovered from weathered surfaces and gravels.

In western South Africa *amber* has been located in the Hay district, and blue and yellow *apatite* occur in the Gordonia district. Lime alabaster or *aragonite* of ornamental quality is deposited by the sources of Mahatsinjo, near Analavory, Itasy in Madagascar.

BERYL

Pale blue *aquamarine* is found *(1)* in Madagascar at Marijao (Betsiboka), and Tongafeno (Betafo); *(2)* in Southern Rhodesia at Karoi; *(3)* at Selaty and in Namaqualand, in South Africa; *(4)* at Rössling and Spitzkopje in Southwest Africa. Non-coloured, transparent *beryl* occurs at Mahazoma, in Madagascar, and golden yellow *heliodor* appears at Itremo and similar Southwest African pegmatites.

Green *emeralds* show in the beryl pegmatite of Androfia, in Madagascar, and emerald is produced at the Zeus Mine in Southern Rhodesia. Around Letaba, the principal South-African deposit in the eastern Transvaal, up to 10 inch long emeralds are found in ancient biotite schists, near beryl pegmatites. In larger crystals, a beryl core is surrounded by light to dark green emerald and biotite flakes. Other occurrences include Baviaanskop and Uitvalskop. In Madagascar, pink *morganite* of the albite-lithia pegmatites of Sahatany, south of Antsirabe, and of Anjanabonoina is noteworthy. Morganite also shows in the northeastern Transvaal and Namaqualand in South Africa, and in the pegmatites of Rössling and Spitzkopje, in central Southwest Africa.

CALCITE, CHRYSOBERYL, CORDIERITE AND CORUNDUM

Onyx marble forms up to 20 ft. long bodies north of the Isser, in Algeria; this form of calcite is scarcer at Sidi Hamza and Oued Chouly, south of Tlemcen, than at Tekbalet. The red onyx of Rhumel, near Constantine, is more attractive than that of Mansoura.

Chrysoberyl occurs in the diamond fields of Birim and Bonsa in Ghana, and also in the Nantia and Beshea Su streams in the east. Similarly, chrysoberyl accompanies diamonds in the Congo. Some chrysoberyl has been found in the alluvia of Chania, near Tika, Kenya. Golden yellow cymophane of 1–3 carats occurs in Madagascar. Minuscule crystals have developed in the Lake Alaotra pegmatites.

Cordierite or inexpensive 'water sapphire' is distinguished by blue and yellow pleochroism. The mineral is abundant in the Androyan leptites and gneisses of southern Madagascar, e.g., at Ihosy, Ankaditany and Fort Dauphin; it is recovered from weathered gneisses and from the gravels of Bekily, and ornamental cordierite is extracted at Tsilaizina, While shales of the Pretoria Series have been altered to cordierite in South Africa, only the blue cordierite of the Hout Bay and Cape Town granite is recovered as a gem stone.

Corundum. Red *ruby* occurs in the placers of Chania and Thika, in Kenya. A ruby-bearing zoisite amphibolite of Longido, in northern Tanganyika, is marketed as 'anyolite', an ornamental stone. Pink *sapphire* has been identified in the Bonsa diamond field in Ghana; a few sapphires have also been collected in Kivu Province in the Congo. At Kinyiki Hill, Kenya, sapphire has altered from corundum. Sapphires have been infrequently found in the lapilli of the Plio–Pleistocene Ankaratra volcano, in Madagascar. Tabular, white and rose corundum of the pyroxenites and sakenites of Ejeda, Madagascar, have been cut as gems. Wine-coloured, pink and blue stones of the corundum fields of the northern Transvaal are cut en cabochon, others occurring in Namaqualand.

DIOPSIDE, EPIDOTE, FELDSPAR, FLUORSPAR

Green *diopside* of gem stone quality has been won from Itrongay, in Madagascar.

The *epidote* of Mahazoma, Madagascar, can be cut en cabochon as can the ancient, epidotized granites of the eastern Transvaal and the multicoloured okkolite of Keimoes, in northern Cape Province.

Feldspar. Green microcline *amazonite* occurs in the pegmatites of Machakos and Embu in Kenya, and east of Luguruni, in northern Tanganyika. Ornamental amazonite is abundant in the albitized pegmatites of Madagascar; large crystals are found at Mahabe (Tsaratanana), others occuring at Ambohibelona and Andina, near Ambositra. The amazonite of Ambatofinandrahana, Madagascar, and honeydew of the central Transvaal and of the Upington area, in northern Cape Province, can be cut en cabochon. *Labradorite* of Ankafotia and Saririaka in southern Madagascar, and of the eastern Transvaal, is distinguished by its play of colour. *Moonstone* and *sunstone* occur, respectively, in the Vaal River and the Letaba district, Transvaal. Beautiful yellow, but semi-soft, *ferrugineous orthoclase* is recovered at Itrongay and Ampisopiso, north of Ampandrandava, in Madagascar.

Fluorspar is abundant and multicoloured but unfortunately soft, e.g., in the Ottoshoop district of the western Transvaal (Table CLXVI).

TABLE CLXVI

TRANSVAAL FLUORSPARS

Colour	*Locality*
Purple, blue, green, amber	Oog van Malmanie
Purple	Leeuwfontein
Purple, green	Potgietersrus tin field
Green	Valsfontein (Groblersdaal)
Mauve, green, banded	Hlabisa (Zululand)

GARNET TO SERPENTINE

Garnet. Red or lilac *almandite* pyrope occurs at Ampanihy, and in the gneisses and amphibole pyroxenites of Ankaditany, Madagscar. In the Soutpansberg district of the northern Transvaal, red and brownish almandite has crystallized in schists and pegmatites. At Buffelsfontein, Turffontein and other localities, generally green *grossularite* and some uvarovite or 'South African jade' accompany chromite in more than 1 ft. thick inclusions of the Bushveld gabbro. Chromiferous red *pyrope*, or 'Cape Ruby', is associated with diamonds, and pyrope also occurs in the alluvia of the Vaal and Orange drainages. The precious yellow *spessartite* of the albitized Mahazoma and Tsilazaina pegmatites, in Madagascar, is unfortunately fine-grained.

Nephrite *jade* is exploited in Southern Rhodesia.

Ornamental *malachite* and other hydroxides and silicates of copper are obtained from the Katanga Copperbelt, and especially from Kolwezi.

Peridot. A stockwork of stringers and nests of olivine traverses the serpentinite of St. John Island, which is situated off the Egyptian coast in the Red Sea. Peridot is believed to have crystallized from solutions derived from semiconsolidated or leached ultrabasics. In Ethiopia the peridots of Kod Ali Island, east of Edd, and of an islet of the Bay of Bahar Assoli, are related to basaltic dykes; other occurrences include the Sagan Valley and the area south of Adaito Ela, northeastern Ethiopia.

Green *prehnite* crystallizes in lavas and especially in Cape dolerite.

Valuable rare minerals or specimens for mineralogical collections are obtained from the complex copper–uranium ores of Katanga, their type locality, and from the pegmatite fields of Madagascar (thortveitite, ampangabeite).

Abundant *rhodonite* of Mahafaly, Madagscar, is occasionally distinguished by an ornamental peach flower colour.

The *serpentine* of Vohibory, Madagascar, is yellow. The verdite of Nordkaap, in the eastern Transvaal, is dark green and occasionally mixed with corundum-bearing rock. Green serpentine of Collo and Oued Madar, Algeria, is used as an ornamental stone.

SILICA

Agate and chalcedony occur in the placers of Kwango, Kasai and the Leopoldville district of the Congo. Agates have also been found in the amygdales of a volcanic rock at Tshala. At Kigoma and Kisulu, east of Lake Tanganyika, generally cracked agate covers amygdales of late Precambrian volcanics, but agate is less abundant on the Chafukwa in the southwest. In Mozambique agates are found in the Corumana Mountains, near Lourenço Marques and on Mount Zungi, in the Manica district. In South Africa, agate has reconcentrated from amygdaloidal lavas of various ages in the placers and soils of western Griqualand, western, central and southeastern Transvaal (Springbok, Lebombo), and Zululand. A stone similar to cairngorm is found near Piet Retief, and common agate occurs in the Drakensberg lavas of Basutoland.

Amethyst veins are frequent in the Lower Congo, Kasaï, and in the pegmatite fields of Maniema and Kivu, and most abundant in the Astrida–Nyanza district of Rwanda. Amethyst and rose quartz are associated in the Dodoma, Mpwapwa, Handeni and Kilosa districts of Tanganyika. Amethyst is also abundant in the pegmatite fields of Madagascar and Southwest Africa, occurring in the quartz veins of the Jukskei and Hennops, near Pretoria, and at Keimoes. Mauve quartz has been located at Goodhouse and Postmasburg in Cape Province.

The red carnelian of the Hay and Prieska gravels (South Africa) is derived from amygdaloidal lavas. *Chalcedony* is associated with agate in the Congo and in South Africa; at Mont aux Sources, in the Drakensberg, blue chalcedony occurs. Multicoloured chalcedony and jade can be recovered from the surface of the basalt flows of western Madagascar and from adjacent streams; Analalava, Marovoay and Maintirano can be reached rather easily.

Citrine is abundant in Madagascar, translucent, green chrysoprase being obtained from Itiso, central Tanganyika. The multicoloured, ornamental crocidolite of Griqualand, variously known as bull's, cat's, hawk's or tiger's eye, is the silicified variety of riebeckite cross fibre (see 'Asbestos' chapter). Green quartz, quartzite and quartz schist occur at Sandor, in the Soutpansberg district, and near Gravelotte and Leydsdorp in the Transvaal.

Jasper is associated with chalcedony at Analalava and in the surficially silicified pegmatites of Ampanihy and Bekily, in southern Madagascar, jasper is found in the banded ironstones of Southern Rhodesia and frequently so in the iron–asbestos–manganese belt of Griqualand. While banded jasper is abundant in several Lower Griquatown layers, red and greenish jasper is less abundant in the later stages of the sequence, e.g., at Honingkrans, Lucasdam and Eureka. Jasper also occurs in the gravels of the west coast of South Africa.

Opaline quartz has been located on the Cataracte Plateau, west of Leopoldville, and in southern Katanga, Congo. Opaline chert is found in Tertiary lake beds at Thomson's Fall, Kenya, and veins of opal traverse the trachyte of Faratsino, in Madagascar.

Rock crystal is associated with piezoelectric quartz in the Malagasy deposits, and also occurs frequently in South Africa.

Rose quartz is abundant in the Malagasy pegmatite fields, e.g., at Ampandramaika. This variety has been recovered from *(1)* Andrianampy and Samiresy; *(2)* in South Africa from Namaqualand, Kenhardt and Gordonia, in the Orange River belt, and from Barend in the northern Transvaal; and *(3)* in Southwest Africa. Smoky quartz is found at Zaaiplaats, Barberton and in Namaqualand. Quartz with Venus hair is recovered at Mananara–Maroantsetra with the piezoelectric variety.

SPINEL TO TOURMALINE

Spinels of various colours are obtained in southern Madagascar; a dark blue, ten carat crystal was found at Betroka. Black iron picotite of Nosy Mitsio Island, off the coast of Ambilobe, can be used as a jewel of mourning. Yellowish *spodumene* occurs in the albitized pegmatites of Madagascar; pink kunzite is scarcer.

Topaz of poor quality is abundant in the greisens of the Maniema tin fields, the largest crystals occurring at Kibila in the Moga district. Other deposits include adjacent Kisubili, and topasolites occur at Mokama in the Kailo district. Large alluvial concentrations of topaz are located north of Albertville. Massive topaz veins the Lunyo pegmatite in Uganda, and some topaz occurs in the gravels of Chania in Kenya. Colourless topaz has been recovered from the Analamisaka pegmatite, Ifempina and Mahabe, in Madagascar.

Tourmalines of various colours are abundant in the pegmatite districts of Maniema and Kivu in the Congo, in central Rwanda, and in the Alto Ligonha pegmatite field in Mozambique. Red rubellite and rarer blue indigolite occur in the albitized pegmatites of Madagascar, e.g., at Sahatany, Anjanabonoina and Ambalahifotsy.

STONE

Quarries of building and road building stone are of value when they are located near communication centres. As Precambrian rocks are valuable but hard, softer sandstone and carbonatic rocks are widely used. The level of development of each country determines the utilization of these materials, as shown in the following examples:

SOUTHERN AFRICA

South Africa. Sandstones are the most widely used building stones. Mainly Table Mountain and Cretaceous sandstones are quarried on the Cape, but Karroo sandstones are used in the Eastern Province. Sandstone exceptionally outcrops in Natal at Durban and Pietermaritzburg. However, Karroo sandstones are abundant in the eastern Orange Free State and the southwestern Transvaal, as are Witwatersrand rocks at Johannesburg. Early Precambrian granite is exploited between Johannesburg and Pretoria, and in the Barberton district, where ornamental stone and serpentine is also won. The Bushveld granite and gabbro are also attractive.

Early Precambrian marble occurs in the northern Transvaal (Kairo) and at Port Shepstone. Limestones intruded by the Bushveld have frequently been marmorized, e.g., at Marble Hall and Buffelspoort. The marbles of the Vanrhynsdorp district, in western Cape Province, are most attractive as are the travertines of Insinuka, near Port St. John, in Natal. Varied 'kopje' stones, the slates of Pretoria and Ottosdal wonderstones, a metamorphosed ash, are also quarried.

Crushed dolerite mixed with sand is used as a *road building* gravel in the Orange Free State, the eastern Transvaal and Transkei, and weathered but not micaceous granite is successfully utilized on the east coast. The grit, marl and shale used in other areas are but second choices. Among Transvaal laterites, a conglomeratic sandy variety is favoured. Dolomitic road-metal of Pretoria is abrasive, but adhesive to tars. Witwatersrand quartzites have been more extensively used as aggregate than acid intrusives, basic rocks, some tillite and even sandstones of foundation quality.

The best building sandstones of Basutoland are the Red and Molteno beds. Cave sandstone, when bedded, is used as paving material, while dolerite serves for surfacing roads.

EASTERN AFRICA

In Burundi and Rwanda granite and metamorphic schists are used most frequently, but in Uganda quartzite is utilized where granite does not outcrop. Amphibolite is exploited at Jinja.

Building stone is not abundant near the major towns of *Kenya.* A thin band of tuff has been quarried near Nairobi, Kedowa and Nyeri. Coral blocks are utilized along the coastal strip, and lateritic ironstone outcrops near Kisumu. The Precambrian marble of Turoka, near the Rift Valley, and granites of the west provide good ornamental stone. Ballast and crushed stone for concrete are generally available. Tertiary lavas are used in the Highlands and Triassic sandstones are employed on the coast. Various rocks, mainly Precambrian, are quarried in Tanganyika.

With a variety of crystalline rocks, *Madagascar* offers an ample choice of building stone. Granite dykes are quarried near the capital, e.g., at Ambatomaro. While granitic migmatites and green charnockites and pyroxene gneisses are also used in the Precambrian areas, beautiful blue, red and green banded leptites are limited to the southern part of the island. The rose granite of Ibity and Andringitra, and the cipolino marbles of Ibity and Sahatany can be polished. Dolerite is quarried in the coastal area of Tamatave, e.g., at Mahatsinjo, but limestones are used for building in the sedimentary belt of the west. In road building, soft but resistant extrusives are favoured, such as basalts, trachytes, dellenites and phonolites; they are found in the Ambatolampy, Antsirabe, Arivonimamo and Diego Suarez districts. The red soils of the south and west, as well as sandy laterites, can be used for the surfacing of roads.

THE REST OF AFRICA

Algeria. Paleozoic sediments are harder than more recent rocks of the coast, but they are not as well located. Windows of crystalline limestone are quarried at Bouzaréa, Bône and Philippeville, and limestones of various epochs are also crushed. While Nummulitic sandstones can easily be exploited, limestones and diatomites are used in Oran, and slate is used at Chiffa, Djebel Doui and Trara.

Morocco. The Precambrian and Hercynian granites and rhyolites are only utilized at Oulmès and Tiflet, but the volcanic rocks of the Oujda district, Middle Atlas and Central Morocco are also suitable for construction. Light basalts occur at Outigui, near El Hajeb, and at Koudiat, near Ifrane.

In *Ghana*, sandstone of good quality is located at Ajena, Elmina and Komenda in the south, and at Bolgatanga in the north. The Precambrian granites and gneisses covering central and northern *Nigeria* can be conveniently crushed. Basalts can be extracted on the Jos, Biu and Longonda plateaux and in southern Cameroon, where soft white trachyte also outcrops.

Sandstone and limestone are widely used in Egypt. In *Ethiopia*, west of Massawa, an augite teschenite (basalt) has been tested with granite outcropping at Nafasit and Kulluku in Erythrea, and at Kombolcha and in Harar Province in Ethiopia proper. Tuffs are extensively used at Entotto and Akaki, near Addis Ababa, and in Gondar. Trachyte and basalt are abundant in Shoa. Limestones are generally quarried in *Somalia*, and a quartz breccia of the Webi Shebeli is also attractive.

Congo. Limestones, sandstones and granites are most frequently used. In the western Congo, granite is quarried near Boma, quartzite at Matadi, limestone between Songololo and Inkisi, and red sandstone between Inkisi and Leopoldville. Polymorphous (impure) sandstones are exploited in the same area and near Lisala in the north. Limestones, sandstones, metamorphic schists and granites are quarried in the eastern provinces, as is red sandstone near Kindu. Unfortunately the volcanic rocks of Kivu weather rapidly.

SAND AND SILICA

While low grade sand for building and road building is widely available, the distribution of glass sand is frequently restricted to lacustrine deposits. Abrasive quartz is also won from sandstones and quartzite.

Sands of several epochs are used in *Algeria*: Jurassic sands are used at Saïd-Tlemcen, Eocene–Oligocene sands at Sidi bel Abbès, and Holocene and Sahelian sands at Marceau and Rovigo.

In *Morocco*, sands containing 95% silica are exploited at Sidi Larbi, near Casablanca. Impure, siliceous sands are found in the Cretaceous of the Phosphate Plateau, at Azegour, in the Oligocene at Dabes, in the Pliocene at Saïs, and in the Quaternary of Mamora. Large quantities of sand are processed in the titanium deposits of the Senegal coast.

In southwestern Ghana, glass sands are located at Petepong, Tarkwa, Aboso, Eduapriem and Abontiakon. Sand is available in most of Nigeria, and glass sand is abundant in the false-bedded sandstones of the Enugu escarpment.

Sand is readily available in the Nile valley and delta in Egypt and Sudan, as it is in the Gulf of Suez. Extensive deposits of Ethiopia include Guri and Ad Temarian in Erythrea, 300 ft. thick beds in Dankali country, the shores of Lakes Tana and Rudolf, and the probably Mesozoic sands of Finche in Shoa. Sand is abundant on the Somali coast, and calcareous sand accumulates at Ras Antara, Bender Zinda near Mogadishu, and in the Bur area.

Siliceous sands cover the coast near Dolisie in the Congo Republic, and the banks of the Ubangi at Bangui, in the Central African Republic. Some of the sands of the Leopoldville–Thysville area, in the Congo, are pure. In *Uganda* good sands are available near the Lakes and in the Rift, but the best glass sands have been eroded from quartzites on the shore of Lake Victoria. Deposits are located at Entebbe, Kabagoga, Bukakata and in the bays of Nalumuli, Nyima and Nyoba; 60–95% of the sand grains are of 36–100 mesh B.S. Sands of the highest grade, those of Kome Island off Entebbe, contain 99.95% silica. Similarly, the best glass sands of Tanganyika have been eroded from late Precambrian sandstones and deposited at Bukoba, on the western shore of Lake Victoria. The coastal sands of Tanganyika are less suitable for this purpose but in Kenya, e.g., at Mombasa and Sokoke, sands occur which are suitable for the manufacture of colourless glass. In Kenya, river sands are mixed into concrete, e.g., at Athi River, near Nairobi.

The sands of the Ikopa and Sisaony streams, near Tananarive, *Madagascar*, are micaceous. The glass sands of the Anjozorobe road do not contain lime, and those of Masse and Moramanga contain 98–99% silica and 0.1–0.2% iron oxides. While argillaceous, Pliocene sands are located at Amborovy and Marohogo, near Majunga, dune and alluvial sands are used at Tamatave.

Building sand is available in most of *South Africa*, e.g., north of the Magaliesberg, in the Transvaal. Foundry sands are recovered in the Witkop Valley of the Witwatersrand, at Barkley Bridge near Kimberley, and at Nahoon near East London. Clayey sand is obtained at Pretoria and Clairwood, near Durban. Extensive deposits of glass and moulding sand occur between Zeekehoek and Silverton in the 50 mile long Moot Valley, near Pretoria. At Uitzicht and Rietfontein, the belt measures several hundred yards in width; the sand averages 98.8% silica. Other occurrences include Zandfontein and Zeekoehoek, and poorly graded silica sands are found at Pienaarspoort.

The biggest deposits developed by largely fluviatile action affecting the Cape Flats, near Cape Town. While the glass sand contains 99.7% SiO_2, 0.2% Al_2O_3, 0.04% TiO_2 and 0.01%

Fe_2O_3, calcareous, aeolian sand concentrates in pans. At Philippi, the sands include: 0–6 ft. glass sand; 6–7 ft. laterite; 7–11 ft. glass sand; 11 ft. gravel.

Similar deposits are found at Kuilsrivier and Kleinmond.

Compact rocks, mainly sandstones, can be crushed to obtain sand and silica; quartzites can also be processed to manufacture grinding stones or abrasives.

In Morocco, powdery silica occurs in the gossans of the Kettara hematite lode. Precambrian and Ordovician quartzites occur in the Anti-Atlas (e.g., Jebel Bani), Middle Cambrian occur at El Hank, near Casablanca, and those of central Morocco are Devonian. Siliceous, Miocene sandstones are exploited at Marchand. Quartzites are quarried at Nyafomang, in southeastern Ghana, and pure quartz forms veins at Zuarungu. Colourless glass is manufactured from the pegmatitic quartz of Sultan Hamud, Kenya. At Dar es Salaam, Tanganyika, excellent sand is extracted from the Pugu sandstone.

On Special Grant 191 at Wankie, in Southern Rhodesia, a close-grained ganister type sandstone is won for refractory purposes. Metallurgical quartz is mined at Broadside, 8 miles northeast of Gwelo, and at Tweeny. At Weltevreden, in the central Transvaal, South Africa, a quartz body occurs at the contact of granite and schists. The silcrete of Mossel Bay and Riverdale, 200 miles east of Cape Town, contains 98.4% SiO_2. Transvaal quartzites are less favoured for the manufacture of silica bricks, but quartzites of Mound's Concession, near Barberton, could be ground to silica. Fairly pure, altered quartzites are located in Swaziland in the Pigg's Peak district, and in the Insuzi Series of the Mankaiana district.

SECTION X

Water and soil

While hydrocarbons (and even gases) have long been described as mineral resources, another liquid – water – has not always been recognized as such. Similarly, biogenic sediments (like phosphates) and surface covers (like lateritic bauxites and iron ores) have long been accepted in the mineral realm, but soils are not always considered as belonging to the latter. However, water and soil are the most important geologic resources. It is unfortunate that, because of past and present climates, Africa should be so poorly endowed in these substances. Happily, the same factors and its particular structure have provided this continent with the world's largest hydropower potential. Water is thus scarce, but water power is plentiful.

GROUND-WATER

The exploration and supply of adequate water resources remains the principal problem of African geology. Unfavourable conditions are caused by the combination of three circumstances: *(1)* the extensive Precambrian areas provide only secondary aquifers; *(2)* the sedimentary areas of the Sahara, and of the interior basins extending from the Congo to the Kalahari, are affected by high evapotranspiration; and *(3)* the coastal strip of sediments tends to be narrow. Examples illustrating these conditions follow.

NORTH AFRICA

While ground-water conditions of northern Tunisia recall those of eastern Algeria, southern Tunisia is part of the Chott region of the Sahara. Water is generally scarce in northern *Algeria.* Triassic aquifers tend to be contaminated by salts. Water circulates in the Liassic (Lower Jurassic) limestones. Favourable sandstones appear in the Lower Cretaceous. A good aquifer has developed in the area of Tébessa at the contact of marls and Eocene limestone. The sandstones of the east Algerian and Tunisian Flysch and the Miocene coastal belts provide good aquifers. In sub-arid Morocco, coastal sediments, Jurassic limestones and Paleozoic sediments yield ground-water, but the southern part of the kingdom can be described as sub-Saharan.

THE SAHARA

Notwithstanding the investment in hydrologic exploration, water remains scarce in the desert, which covers almost 4,000,000 sq. miles of Africa. The principal free aquifer is situated in the Lower Cretaceous, intercalated continental sediments which attain a thickness of up to 1,800 ft. This basin covers 250,000 sq. miles in the eastern part of *Algeria* and in the Hoggar, and has a theoretical capacity of 470,000,000 million cubic ft. of water. The aquifer lies at a depth of 250 ft. at El Goléa, but at Touggourt it descends to 5,000 ft.; this formation is the principal water resource of the Mzab area. In the valley of the Saouras tunnels, or 'foggaras', belonging to a chain of oases drain the continental, Hamada limestones and Quaternary sediments. The yield of the 'foggaras' varies locally, but not seasonally, from 5 to 500 gallons per minute. The ground-water of the eastern Erg is unfortunately saline, but the sediments of the east Algerian Sahara provide a favourable artesian basin. The Wadi Rhir, including the richest date growing oases between Ouargla, Chott Melrhir and Touggourt, drains Quaternary paleo-alluvia, but the Paleozoic sediments surrounding the Hoggar only occasionally yield water.

In desert-covered *Mauritania* three hydrological areas can be distinguished, namely the coastal sediments, the metamorphic and granitic rocks of the Reguibat, and the Taoudéni basin which also covers part of Mali. Senegal is situated in the coastal belt, and the Senegal River provides secondary aquifers. The Niger Valley supplies a majority of the water resources of Mali and the Niger Republic. In addition, the underflow of the wadis Tesellaman–Assakaraï, Azaoud and Tilemsi, the Pliocene–Quaternary source of the Middle Niger, is also exploited. The eastern part of the Niger Republic is arid, and water is particularly scarce in the northern Cameroon in the district of Ngaundéré. In neighbouring *Chad*, Cretaceous sediments are believed to underlie the valley of the Logone and to hold water. East of Fort Lamy the Precambrian basement lies at a depth of only 530 ft., but the contact zone yields little water. In the Precambrian areas of eastern Chad which connect with the Sudan, paleo-alluvia covered by younger sediments are believed to hold water, as do the underflows of wadis.

Water is scarce in the *Sudan*. In the Kerma Basin in the north, the Cretaceous Nubian Series, which includes sandstones, is 800 ft. thick. In the northwest, the seepage of the Nile raises the water table for 60 miles west of the river. In the area of Wadi Hassuna, near Khartoum, water is restricted to faults traversing the granite and the Nubian Series.

The ground-water divide of Meidob–Tageru separates the Libyan and central-Sudan hydrographic provinces: between Khartoum and El Fasher, continuous water tables exist in the Nubian and Umm Ruwada Series. While water seeps from the Main and Blue Nile for as far as 60 miles, the clay sealed beds of the White Nile do not admit it. The water supply is particularly inadequate in Kordofan and Darfur in the west, especially in the Precambrian areas. Water is more abundant in the Umm Rawda Series where these sediments are thick, but only isolated bodies of water occur in the Qoz sand and clay. In the Nuba Mountains of southern Kordofan, clay overlies granite and gneisses and, some ground-water has been tapped from old channels or khors, springs and fissures. At Qala en Nahl, near the Ethiopian border, fractured serpentinite imbibes water, while talc and chlorite schists remain impermeable.

The Nubian sandstones occasionally provide water in Libya. A majority of the ground-water of Egypt, however, is derived from the alluvia of the Nile.

WEST AND CENTRAL AFRICA

The area of West Africa which is not covered by the Sahara, i.e., the belt extending from Guinea to the Cameroon, is underlain by Precambrian granite gneisses and schists which only provide secondary aquifers. However, the situation is more favourable in the areas of Guinea, Mali and Dahomey which are covered by Paleozoic sediments. The coastal strip, expanding into a broader zone in Nigeria, is more favoured.

Water supply is inadequate in the arid, but heavily populated areas of northern and central *Nigeria*, and especially in Bornu, Bauchi, Katsina, Plateau, Kano, Zaria, Sokoto, Niger, Kabba and Benue provinces. Indeed, the old granites and gneisses are capable of storing only little water in their shallow zone of alteration. Some water is imbibed by arenaceous interbeds of the Chadian clays which lie at a depth of 300–750 ft. At Maiduguri in Bornu, the hypothetical water level, which favours the contact of sediments and granite, is at a depth of more than 3,200 ft. Water is scarce even in such areas of heavy rainfall as the Adamawas, Benin, Owerri and Onitsha provinces, where the water table lies at a depth of 600 ft. An artesian flow has been achieved at Oji River in Onitsha, but unfortunately, impermeable sediments cover large areas of the south. Water saturated sandstones are known to occur, however, between Kontagora, Mokwa and Bida.

Ground-water conditions of the southern Cameroon, the Fort Archambault district of Chad, the Gabon, the Congo Republic and part of the Central African Republic are similar to those of West Africa. Indeed, these tropical areas are underlain by Precambrian rocks which hinder the development of free, primary aquifers. Part of the Central African and Congo Republics, however, is occupied by Mesozoic sediments of the Congo basin which provide more ground-water.

The *Congo* can be divided into eight ground-water provinces. The largest of these is the Congo basin proper, surrounded on three sides by the river, and by the Kasai Plateau in the south. Evaporation is particularly intense in the centre of the basin, covered by Kalahari sands. Nevertheless, aquifers have been drilled in alternating layers of arenaceous and pelitic sediments, particularly in areas underlain by Mesozoic rocks. Free aquifers have been located in Katanga, in the Kundelungu and Roan Series, but the Precambrian rocks of western Katanga are less favoured. Only the northeastern part of Kasai appears to provide abundant ground-water; further west in Kwango, Tertiary sands which extend into Angola are considered to be good aquifers. West of Leopoldville, only alluvia retain the waters of the geosyncline of late Precambrian sediments which connects the Gabon with Angola. Unfortunately the ground-water of the coastal strip has been polluted by the sea.

Since the northern borderland of the Congo basin lies on the equator, the seasonal variation of the piesometric level is slight: water can be found near the wide alluvial flats. Limestones hold water north and east of Stanleyville, but the Precambrian areas of Maniema show only secondary aquifers. The Rift Valley of the Great Lakes contains abundant water resources in the young lavas and arenaceous sediments.

Except for the coastal strip of sediments and the western areas, which are occupied by

Precambrian rocks, Angola is covered by low yielding Kalahari sands as is Barotseland, the western part of Zambia. However, the Katangian sediments, which extend as far as Broken Hill and south of Lake Tanganyika, provide good aquifers.

EASTERN AFRICA

The altered lavas of Ethiopia, alternating with sediments, provide water. However, water is scarce in the sedimentary areas of Somalia and eastern Kenya because of excessive evaporation. Since Precambrian rocks cover most of *Uganda*, and western Kenya, only the upper 150–250 ft. of altered material generally yields 100–250 gallons of water per hour. The water table is found at a depth of 200–400 ft. in sediments yielding more than double this flow, but borehole water is hard. The situation is similar in tropical Sudan. The Graben of the Great Lakes extends to Western Rwanda and Burundi. The juvenile relief and intensive erosion of most of Rwanda and large areas of Burundi hinder the development of adequate aquifers (and of soils). Since valleys are narrow, little water seeps into alluvial gravel. The situation is somewhat more favourable in southeastern Burundi.

In the volcanic Kenya Highlands, and in northern Tanganyika, the walls of the Eastern Rift Valley facilitate the tapping of water. As a majority of Tanganyika is covered by Precambrian rocks, only secondary aquifers are available there.

Precambrian rocks cover two thirds of *Mozambique*; they retain water only in their zone of weathering. As sandstones and shales alternate in the Karroo of the Zambezi Valley, this area boasts a certain potential of ground-water. Conditions are similar in Malawi, where the sedimentary Shire Valley provides large resources. A continuous, free aquifer exists in the 90,000 sq. miles of southern end east-central Mozambique occupied by Cretaceous and post-Cretaceous sediments (De Freitas, 1959). In the dry season streams dry up and the quality of the water of the rivers deteriorates. Even at a depth of 100–130 ft. the ground-water is frequently saline, and in the coastal belt the lower limit of salinity has occasionally not been pierced at a depth of 500 ft. Fortunately, south of the Save River the horizons of alternating diatomaceous and argillaceous sediments and the Tertiary limestone belt of Morrumbane are distinguished by a greater potential of ground-water.

The southern part of *Madagascar* is sub-arid. At Sambava, Antalaha and Ambanja water has been drilled in alluvial sands, while south of Antsohihy an aquifer has been pierced at a depth of 400 ft. in Karroo sandstones. Tuléar and Bezaha are supplied by artesian water, and in the Bobaomby peninsula and the Vohémar district, in the north, drilling has been carried out for the stock breeding industry. In the southwest, water rises from a depth of 330 ft. at Ankatrafay, from 370 ft. at Erada and from only 100 ft. at Ambatolahy. At Manera Jurassic (Argovian) sandstones, lying at a depth of 3,400 ft., are imbibed by water. South of Befandriana, Cretaceous (Santonian) sandstones lie below basalt flows at 3,000 ft. of depth. The Cretaceous sandstones of Marovoay–Boboka, in the northwest, also contain an abundant water supply.

SOUTHERN AFRICA

The southern sub-continent can be divided into three ground-water areas: the Kalahari Desert, the Karroo, and the Precambrian areas of South Africa, Southern Rhodesia and

Southwest Africa. The western part of Southern Rhodesia is occupied by sand and Karroo sediments.

Since most of *Bechuanaland* is covered by the Kalahari, water is scarce: Kalahari Beds constitute the aquifers in the Northern Crown Lands; in the vicinity of Lake Ngami (Batawana land), water tends to be saline, and in the central Kalahari potable water is only found in Middle Ecca (Karroo sandstones). While 7,190 p.p.m. of solids are dissolved in water at a depth of 980 ft., at Boritse pan, only 470 p.p.m. of solids are found in a quartzite aquifer northwest of Sekhuma pan on the Ghanzi cattle route. Again, the ground-water of the south-central Kalahari (Bakwena and Bangwaketse country) unfortunately contains 10,000 p.p.m. of solids. The fossile waters of the inert Cave Sandstones do not become saline. The yield of the early Precambrian rocks of Barolong country is low. In Bamalete country, Middle and late Precambrian Transvaal and Ventersdorp rocks are favoured for drill sites; in Bamangwato country the success ratio of bores is 80%. As the yearly water requirements of Lobatsi town are now in excess of 85,000,000 gallons, the gravels of a flood plain are tapped.

The eastern 60,000 sq. miles of *Southwest Africa* are covered by Kalahari sand. Water is stored in an undulating, south trending band, representing a thickness of 1–4 ft. of sand. When the sand cover reaches a thickness of 20 ft. it absorbs all the rain water, which is later lost by evaporation. The Precambrian areas of Southwest Africa are also arid.

Karroo sediments cover 40% of the area of *South Africa;* while most of them have low permeabilities, those narrow zones which have been baked or altered by dolerite dykes yield considerable quantities of water. In areas of igneous rocks such as the eastern Transvaal and western Cape Province, water is drawn from the fractured zone dividing the weathered and hard rock; only isolated basins or troughs hold ground-water there. Fault-planes have a high yield if they are connected with a storage aquifer.

DEUTERIUM

Deuterium develops in undisturbed columns of water. Such conditions prevail in the lakes which fill the bottoms of the craters of the Congo, East Africa and Ethiopia. Samples of the volcanic lakes of Kichwamba, in southwestern Uganda, contain 8–12% heavy water.

HOT SPRINGS

Thermomineral springs may be of vadose or endogene origin. The endogene springs are particularly frequent in the volcanic and faulted areas of Ethiopia, and in the periphery of the Eastern and Western Rift Valley in Kenya, Tanganyika, western Uganda, Rwanda, Burundi, the eastern Congo, Malawi, Mozambique and Madagascar. Hot springs tend to be located on faulted alignments. The waters and brines carry variable amounts of salts, sulphur and carbonates, depending on the composition of the original gases and the rocks through which the water ascends. Three countries have been selected to illustrate the distribution of springs.

The most important springs of *Ethiopia* are located in Dankali country, in the east *(1)* between Ada Ela and Harsah, south of the Salt Plane; *(2)* in the Dobbi Valley, near the French Somali Coast; and *(3)* in the Auash Valley. In Erythrea, the Eletta Mareb and Ali Hasa springs are used commercially, as is the soda-bearing Filoha spring near Addis Ababa. Springs cluster southeast of Lake Tana; others are aligned in the Janjero belt, 100 miles southwest of Addis Ababa, and in southwestern Ethiopia near the western limit of the traps and Lake Zuai of the Rift Valley.

In *Mozambique*, five areas of springs are distributed along the edge of the Precambrian, extending from the area of Pebane Harbour (150 miles southwest of Mozambique City) 450 miles southwest, towards the bend of the Limpopo River on the border of Southern Rhodesia. Concentrations include those west of Pebane and Quelimane, and 11 clusters in the south between the Beira railway and the Save River. Other springs are located along the southeastern border of Malawi, i.e., at the faulted contact of the Precambrian and the Shire Valley and the upper Zambezi. Their distribution shows the connection between rising water and Rift tectonics. Thermonimeral springs of the second area include those of Chipice, Nhamassero, Rupice, the Zónuè and Rupízi rivers and the Báruè area. Waters of the Morrambula spring in Zambezia contain 530 g/l of solids, mainly soda and sulphur, and 130 g/l of carbonic acid. Springs of cold hyposaline water are frequently found in the area of Libombos, the volcanic hinterland of Lourenço Marques. Other cold, endogene springs include Moamba, Metololo and Inquingire.

Thermomineral springs concentrate in *Madagascar*, *(1)* along the axis of the island between Tananarive–Itasy and Namorona; *(2)* in a belt extending from the latter towards the west coast; *(3)* in the south, near the west coast (Tsarimpioky and Ranomay); *(4)* at Ranomofana Tanosy, in the southeast; *(5)* along the east coast, south of Tamatave; and *(6)* in the graben of Antongil Bay extending between Doany and Andavokoera towards the north-northwest. Only the Antsirabe and Namorona sources have been developed.

STEAM

Geothermal steam is found in the same geologic environment as hot springs. Steam is believed to escape from the Rift Valley of Ethiopia, Tanganyika and the Great Lakes, but only in Kenya has it been utilized. In the Toro district of western Uganda steam accumulates at a depth of more than 5,000 ft. In Kenya steam jets escape from the Rift volcanics in the Longonot and Menengai, south of Lake Naivasha. Steam is utilized at Eburru.

HYDROPOWER

Africa is distinguished by a capacity of 204,000,000 kW, that is 40% of the potential of the world, or 3.5 times the potential of North America or Europe. These power resources are most remarkable, since one-third of the continent is covered by desert. But in the remaining areas, tropical rains and waterfalls, i.e., structural denivellations combine to produce 130,000,000 kW in the Congo basin (Fig. 7) and approximately 60,000,000 kW in the Nile and Zambezi drainages, plus relatively minor resources elsewhere. The principal power sites are located where water falls *(1)* from the Precambrian hinterland to the narrow coastal strip (e.g., Congo, Kouilou), *(2)* to the Karroo surface (e.g., Stanley Falls, Victoria Falls), or *(3)* to other sedimentary levels (e.g., Niger, Aswan) and *(4)* at the overflow of the Great Lakes (e.g., the Victoria Nile, Shire).

NORTH AFRICA

The principal hydropower resources of *Tunisia* are largely restricted to the Medjerda basin in Small Kroumiria. The falls of the Medjerda, near Trajan's Bridge, can produce 7,500 kW and 9,000 kW at El Lil, but the Sidi–Salem site has a greater capacity of 13,000 kW. Other plants include Lower Medjerda (5,300 kW) and Nebeur. The Ben Metir dam and the Fernana Falls are conveniently located near Tunis. *Algeria* is handicapped by its semi-arid climate but its structure of mountain ranges which rise from the coastal strip favours the development of waterfalls. Indeed, the country's hydropower sites are aligned in one band, 15–40 miles from the coast, frequently above irrigation schemes. The principal resources of Small Kabylia include, respectively, the Oued Agrioun (21,000 kW), the upper and lower Oued Djendjen (10,000 and 28,000 kW) and the Oued Bou Sellan with 14,000 kW. Furthermore, the waters of the Chott Chergui, south of Oran, can be utilized on the canals connecting this lake with the plain of Relizane. Other sites are given in Table CLXVII.

TABLE CLXVII

ALGERIAN POWER SITES

Site	*Power (kW)*
Ghrib	8,500
Bakhada	3,800
Beni Bahdel	3,800
Aïn Temouchent	4,600
Foum el Gherza	700
Hamiz	2,700
Beni Bahdel	3,400
Bou Hanifa	5,300
Perrégaux	1,000
Baukada	3,500
Zardizao	3,800
Oued el Ksob	1,500
Boghni, lower	3,000
Michelet	7,600
Lower Maillot	11,500
Chabet Saïad	5,300
Oued Fodda	15,000

Morocco. A majority of the hydropower resources are concentrated in the Oued el Abid, with those of Fez being much smaller. The Oued el Abid flows from the Atlas towards the coast, west of Casablanca, with dams rising at Bin el Ouidane and Afourer near the head waters, at Kasba Zidania, near Beni Mellal on its tributary the Oum er Rebia, and above the coastal plain between Im'fout and Machou (Table CLXVIII).

TABLE CLXVIII

HYDROPOWER IN MOROCCO

River	*Plant*	*Power (kW)*
Oued el Abid	Afourer	69,000
	Bin el Ouidane	105,000
Oum er Rebia	Im'Fout	31,000
	Daourat	17,000
	Sidi Saïd Machou	15,000
Oued N'fio	Lalla Takerkoust	11,000
Oued Beth	El Kansera	10,000

WEST AND WEST CENTRAL AFRICA

The Félou plant, on the Senegal near Kayes in *Mali*, only generates 450 kW, but the river could be regulated and harnessed to produce more power. Similarly the Sotuba dam near Bamako is very small, but the Kénié Falls of the Niger and the waters of the Sansanding dam could also by utilized. When, several thousand miles downstream, the Niger descends at Kainji to the Cretaceous plains of northwestern *Nigeria*, 860,000 kW could be generated (Fig. 24).

The Konkouré River drains the waters in *Guinea* of the Fouta Djalon Mountains towards the sea. Water falls from Precambrian Upper Guinea to the plain of tabular Ordovician–Silurian sediments. The Konkouré is dammed at Souapiti, providing a minimum of 350,000 kW. The Kaleta, Grandes Chutes (or Big Falls) of the Sarnou near Conakry and Forecariah sites can supply 116,000 kW. At Foumion, on the Niandan (a tributary of the Upper Niger), 500,000 cubic ft. of water can be impounded above a head of 80 ft.

About 80 miles northeast of Abidjan, the capital of the Ivory Coast, the Ayamé River falls 60 ft. to the level of the coastal strip, thus providing a capacity of 19,200 kW with an additional 30,000 kW downstream.

Ajena is located in *Ghana* 70 miles upstream from the estuary of the Volta River, near the Togo border. The dam rests on the folded Buem formation, which includes *(1)* quartzites, thinner dolomites, shales and greywacke in the west; *(2)* shale below the beds of the river; and *(3)* quartzites in the east. At Misikrom, the Volta, is 1,000 ft. wide, 300 ft. of this width resting on a rocky bed on the western bank. Initial capacity attains 617,000 kW.

The Sanaga River drains the central *Cameroon*; its waters fall from the level of the Precambrian highlands to the coastal strip near Edea. The capacity of the Edea plant east of Douala is being increased to 145,000 kW.

While the Poubara Falls of the Upper Ogoué in southern *Gabon* can supply 6,000–9,000 kW, in western Gabon the lower Nyanga River and the Kinguélé Falls, on the M'Bei, can respectively provide 15,000 and 5,000–16,000 kW. The Kouilou, a relatively small river, flows in the *Congo Republic* from the area of Mossendjo towards the Atlantic, north of Pointe Noire. The river traverses the late Precambrian geosyncline of the Lower Congo and the Mayombe Chain, and its valley contracts at Sounda, thus creating a head of water above the coastal plain. Plans call for the building of an 800,000 kW plant at this site, and studies prove that infiltrations from the artificial lake into the carbonatic rocks of the

Shale–Limestone System would be negligible. For the time being the Djoué plant, near Brazzaville, supplies 30,000 kW.

The Cuanza River drains extensive areas of the Precambrian plateaus of western *Angola* and drops to the coastal plain southeast of Luanda, the capital. In its initial phase the Cambambe dam provides 90,000 kW, but the capacity of the plant can be increased to 570,000 kW. The Mabubas plant, near Luanda, the Catumbela dam on the Biopio River, near Lobito and the turbines of Matala, on the Cunene near Mossamedes, respectively supply 3,000, 5,000 and 10,000 kW.

CONGO

Hydropower reserves of the Congo basin have been estimated at 125,000,000–135,000,000 kW, with a margin of error of approximately 26,000,000 kW. Among the big rivers of Africa, the Congo is particularly favoured because its catchment area of 1,600,000 sq. miles extends on both sides of the equator. Rainfall and seasons alternate north and south of this line, but the flow of the Congo is affected only slightly by these variations. The flow increases from 470,000 cubic ft. per second at Stanley Falls, near Stanleyville, to an average of 1,150,000 cubic ft. per second (five times the Niagara) 1,000 miles downstream in the Stanley Pool, near Leopoldville; even in the driest days the flow has been measured to attain 1,050,000 cubic ft./sec. there. While the sources of the Congo–Lualaba spring from an altitude of 5,350 ft., Stanleyville is 1,200 ft. and Leopoldville 1,000 ft. above the sea. Between the latter and Matadi Harbour, the river traverses the late Precambrian geosyncline of the Lower Congo and the Mayombe Range, losing 900 ft. of altitude before it reaches Matadi Harbour and the coastal strip; thus it accumulates a potential of 85,000,000 kW ($\pm$ 7%), that is, 17% of the world total or 70 times as much as the Hoover Dam. The Great River cuts through bands of volcanic rocks, dolerites, quartzites and schists at the Cataracts of Inga. According to the Van Deuren Plan, part of the flow of the Congo could be diverted into the Matamba Valley there, providing a head of 320 ft., but a more modest programme calls for the utilization of the adjacent Sikila Falls, which only provides a head of 133 ft. The potential of these sites respectively attains 25,000,000 and 3,000,000 kW of power.

The *Inkisi River*, a tributary of the Lower Congo, is distinguished by a hydropower potential of 300,000 kW, of which 75,000 are particularly easy to harness. The Zongo and Sanga plants, located respectively at 90 and 40 miles from Leopoldville, generate 31,000 and 25,000 kW, but the small turbines of Kwilu only provide 5,000 kW. In addition, a small plant operates at Boali near Bangui, the capital of the Central African Republic.

The *Stanley Falls* connect the middle and upper (Lualaba) basin of the Congo between Stanleyville and Ponthierville (in the Congo). Over 32 rapids flow 470,000 cubic ft. of water, with the sites boasting a potential of about 1,250,000 $\pm$ 300,000 kW. The Tshopo River, an affluent of the Congo that descends from the late Precambrian plateau into the Congo basin at Stanleyville, was easier to harness; generators already supply 17,000 kW there, although the capacity of the site is greater. The Kibali–Ituri drainage is tributary to the Congo; its headwaters have been utilized in the Kilo–Moto gold field in the northeastern corner of the country, with the Shari, an affluent of the Ituri, supplying 15,000 kW and the Arebi, a tributary of the Kibali, 3,000 kW. The Budana plant is situated only 20 miles from

Lake Albert, that is, the Nile drainage. However, the Dubele plant is at a distance of 150 miles.

Numerous plants have been built to exploit water falling from higher levels to the Karroo in the *Maniema* tin and gold districts. The combined capacity of these sites is in excess of 20,000 kW. At Zalya near Kamituga, Magembe near Namoya, Kama near Kampene, Lubilu near Lulingu, Lutshurukulu near Kalima, and Belia near Punia, waterfalls are aligned at the contact of granite and the Karroo. The Lubiadja near Moga and the Ambwe near Kailo descend from the late Precambrian (Lindi) level to that of the Karroo. The extensive capacity of the Gorges de l'Enfer, the falls of the Congo–Lualaba, near Kongolo, is not needed yet. The sources of the Kiyimbi are situated near Lake Tanganyika, but this river flows towards the graben of the Luama, a tributary of the Lualaba; the site can supply 50,000 kW to Albertville. The Lukuga River regulates the water level of Lake Tanganyika, seasonally transferring its overflow to the Lualaba (Fig. 29).

The Piana–Mwanga plant, on the Lovua River 50 miles from Manono, in North Katanga, generates 35,000 kW. A favourable site exists on the Luvua at its overflow from Lake Mweru, and on the Luapula River at the Giraud Falls (of 50 ft.) where it leaves Lake Bangwelu, with flows varying from 400–2,600 cubic ft./sec. The waters of the late Precambrian, Kundelungu and Biano plateau, and of the older Kibara Mountains, are drained by the Lualaba flowing 1,000 ft. below. However, the hydropower resources of the Katanga Copperbelt have been better utilized: above its confluence with the Lufupa at *Zilo*, near Kolwezi, the Lualaba loses more than 1,200 ft. of altitude, and a flow of 1,600 cubic ft./sec generates 120,000–276,000 kW. At the *Cornet Falls*, east of Jadotville, the Lufira River falls 380 ft. thus providing an energy of 59,000 kW, with the Koni Falls supplying 21,000 kW downstream.

The Kalule plant generates 7,000 kW near Lubudi, and near Bakwanga turbines of the Lubilash River supply 12,000 kW.

Lake Kivu was separated from Lake Edward and the Nile drainage by the outpouring of Rift lavas. The volcanic ranges of northwestern *Rwanda* and the mountain lakes provide the setting of the Taruka plant. As the other end of the lake, the Ruzizi River carries its overflow to Lake Tanganyika. At the debouchure between Bukavu, in Kivu Province of the Congo, and Shangugu, in southwestern Rwanda, the river falls towards the level of the plain; the Ruzizi dam connects the two countries there, and generates 18,000 kilowatts (maximum 42,000 kW), also providing power for the adjacent Kingdom of *Burundi*. The Malagarasi River, an affluent of Lake Tanganyika, divides Burundi from Tanganyika; its capacity has been estimated at 40,000 kW.

THE NILE

Egypt. Cyclical and seasonal floods have made the Nile famous. At Aswan the flow is attributed to these sources: January–December, White Nile, 25,000 cubic ft./sec; August–November, Blue Nile, 115,000 cubic ft./sec; August–September, Atbara, 40,000 cubic ft./sec.

The crest of the flood travels 10 days from Roseires in the Sudan, near the Ethiopian border, to Aswan; it averages 0.14 (maximum 0.2) lb. of silt per cubic ft. of water, which poses considerable problems. The original dam was built in 1902 and raised in 1912 and

1933. Only 100,000 kW of the installed 344,000 kW capacity are used during floods, but the reservoir of the High Dam will permit the permanent use of this energy and also provide an additional 2,100,000 kW. Indeed, high Sadd-el-Aali will hold the record for artificial lakes, damming up the waters of the Nile for 300 miles as far as the Sudan.

North of latitude 25 °N, dams have been built at Isna, Nag Hammadi and Asyút. The Fayum Oasis could be used as a regulator of the flow, if the river were diverted to it. (The second cataract at Wadi Halfa has a capacity of 520,000 kW, but only 420,000 kW are used.)

A different project concerns the Quattar depression in northwestern Egypt: a mile long channel and a 40 mile long tunnel could divert sea water into this basin, the lowest point of which is 370 ft. below the sea. At the 165 ft. level 23,000 cubic ft./sec. could be evaporated from an area of 5,000 sq. miles. This would reduce the water level at a rate of 4.2 mm/day and provide a capacity of 300,000 kW, of which 230,000 kW could be regularly used. However, salt would sediment at the bottom of the depression after 160 years.

Sudan. The capacity of four waterfalls of the Nile, which lie between Khartoum and Aswan, has been estimated (Table CLXIX).

TABLE CLXIX
POWER POTENTIAL IN THE SUDAN

Cataract	*Full year (1,000 kW)*	*Seasonal (1,000 kW)*
Fourth	500	450
Semna	500	380
Fifth	200	190
Sixth (Sebaloka)	100	80
Total	1,300	1,100

The Sennar dam is situated on the Blue Nile south of Khartoum; the fifth cataract is downstream from Berber and the confluence of the Atbara; parallel 20 °N crosses between the fourth (Merowe) and third cataracts, and the sites of Dal and Semna are located above Wadi Halfa and the second cataract, i.e., the Egyptian frontier (Fig. 13).

The Blue Nile leaves Lake Tana (at an altitude of 6,000 ft.) at Bandar Giorgis in *Ethiopia*, and falls spectacularly downstream at Tis Essat. The Awash River falls at Coka from the high plateaus to the Danakil plain, where it supplies 14,000 kW. The Nile flows out of Lake Victoria at Jinja, east of Kampala, in *Uganda*. The Owen Falls plant at Jinja can supply 150,000 kW, and the Bujagali Falls further north have a capacity of 100,000 kW. Other sites are known to exist on the Victoria Nile and on the Albertine Nile or Bahr el Gazal.

Power is exported from Owen Falls to Kenya, but at Seven Forks, 70 miles from Nairobi, the falls of the Tana River can be harnessed to provide 100,000 kW. The total capacity of smaller plants has been measured at 26,000 kW in Kenya, and 20,000 kW in Tanganyika. An energy of 68,500 kW can be obtained from the rapids of the Pangani River, and other power sources await development on the Rufiji and Ruvu Rivers.

ZAMBEZI

Although considerably shorter than the Nile, the Zambezi drainage does not have a much smaller hydropower potential (Table CLXX).

TABLE CLXX
POWER SITES OF THE ZAMBEZI

Power (kW)	*Locality*	*River*
750,000	Victoria Falls	Zambezi
750,000	Kariba	Zambezi
300,000	Kafue	Kafue
230,000	Malawi	Shire

The Zambezi descends at Livingstone from the Precambrian Angola plateau to the Permian Valley which divides Zambia and Southern Rhodesia. Farther downstream from the Victoria Falls, the Mid-Zambezi Valley contracts considerably, thus facilitating the construction of the Kariba dam. The 250–400 ft. high dam spans a 1,600 ft. wide gorge, 150 miles upstream from the Mozambique border. The eastern end of the dam is anchored in late Precambrian quartzites, but the structure generally rests on early Precambrian, biotite gneisses. Between these banks, the Zambezi follows a fault which strikes northeast. The artificial Kariba Lake (Fig. 36) will permit, it is hoped, a flow of 37,000 cubic ft./sec, thus generating 750,000 kW. The power potential of the Zambezi in Mozambique is even greater.

The Kafue River flows from the Northern Zambian Copperbelt to Kafue Flats, where it starts its descent to the Mid-Zambezi. A 1,530 ft. long and 400 ft. high dam would be capable of providing 300,000 kW of power there. Meanwhile the Mulungushi and Lunsemf-wa plants, near Broken Hill in central Zambia, and the installations of Zomba (Fig. 34) in Malawi, generate 40,000 and 600 kW respectively. The Shire River, the overflow of Lake Nyasa, is capable of producing 300,000 kW; however, this tributary of the lower Zambezi unfortunately loses altitude through rapids and small waterfalls, and its capacity is therefore expensive to utilize. Small turbines supply 1,000 kW in Southern Rhodesia.

SOUTHERN AFRICA

The sedimentary plain divides the Precambrian areas of northern and western *Mozambique* from the coast, thus providing a head of water. Since the Chicamba dam impounds 55,000,000 cubic ft. of water, its reservoir facilitates the increase of the capacity of the Revuè plant, near Beira Harbour and Vila Pery, above the rate of 17,000 kW. In addition on the Movene, near Lourenço Marques, a head of water of 140 ft. supplies 10,000 kW, along with the additional flow of the Incomati, 83,000 kW. On the Lúrio, in the north, 90,000 kW could be developed. Other plans call for dams on the Luo and Lugela rivers.

The complex Orange River project in *South Africa* (Fig. 46) will serve irrigation and also generate power. Part of the water of the Orange will be diverted through the Ruitge–Steynsburg Canal into the Fish River. The capacity of the plants (in kW) is given in Table CLXXI. Finally, the Edelwini plant on the Usutu river, in Swaziland, will generate 10,000 kW.

TABLE CLXXI

THE ORANGE RIVER PROJECT

	Orange			Fish River	
	initial (kW)	*ultimate (kW)*		*initial (kW)*	*ultimate (kW)*
De Aar	11,500	45,000	Steynsburg	4,000	10,000
Petrusville	12,000	18,000	Middelburg		6,000
Douglas	6,500	13,000			14,500
Boegoeberg	1,100	4,800	Cradock		9,000
Augabries	4,200	7,200	Hofmeyr		10,000
Kakamas	900	1,500			8,000
		1,100	Bedford	1,000	9,000
		4,500			9,000
Vioolsdrif		800	Sondags River		9,000
		900	Graaff Reinet		4,500

MADAGASCAR AND OTHER ISLANDS

The Ikopa flows from Tananarive north to Majunga (Fig. 48). The river falls from an earlier to a younger Precambrian level 100 miles north of Tananarivo, and thence to the Cretaceous plain near Maevatanana and its confluence with the Betsiboka, a tributary on which other dams can be built (APERTET, 1961; Table CLXXII).

TABLE CLXXII

THE MALAGASY POWER POTENTIAL

River	*Waterfall*	*Power (1,000 kW)*	*Height of dam (ft.)*	*Surrounding rocks*
Ikopa	Mahavola	400	270	oriented granite
Ikopa	Antafofo	100		migmatite, laterite
Ikopa	Vohitsera	320	130	granite, laterite
Ikopa	Isandrano	144	20	granite
Firingilava	Firingalava	540		quartzite, laterite
Andriamena	Marokoloy	270		gneiss
Ikopa	Antanandava	390	50	coarse granite
Betsiboka	Vohombohitra	120	340	leucocratic granite
	Ambondiroka	210	80	gneiss, augen gneiss
Betoafo	Manara	100		
Tsinjorano	Mania	100		

Installed capacity is unfortunately lagging at 12,000 kW at Mandraka, 10,000 kW at Antelomita (Ikopa), 5,000 kW at Volobé, near Tamatave, and 2,000 kW each at Namorona and Manandona.

Since the Langevin River loses 430 ft. of altitude at Passerelle, in the southeastern corner of Réunion Island, a capacity of 36,000 kW can be generated there in addition to numerous other existing, but small, plants. Finally, we should mention that even a tropical island as small as Saõ Thomé, in the Bight of Guinea, boasts hydropower plants.

SOIL

Soils are the alteration products of rocks. While climate influences soil formation in the deserts and in the tropics more than in temperate areas, in Africa the dynamics of soil evolution are such that even minor variations of the bedrock determine major pedologic changes. Indeed, high temperatures and heavy rains facilitate the leaching of bases, the supply of which is particularly short among the acid granites, gneisses and schists of Precambrian Africa, as well as in the quartz sands of the Sahara and the Kalahari. Thus, relatively good soils are restricted to the basic volcanics of the East African and Ethiopian Highlands and the Cameroon, and to the alluvia of the major arteries — the Nile, Niger, Congo and Zambezi — and to smaller rivers.

There can be schematically distinguished twelve groups of soil provinces:

(1) In North Africa, a 20–200 mile wide band of largely *chestnut soils* covers the Atlases.

(2) In the Sahara, a 1,000 mile wide band of *desert* extends between latitudes 32°N and 17°N, and is bordered in the northwest and south by 150 mile wide belts of *light-brown steppe* soils.

(3) The Sahel, a 200 mile wide belt undulating between latitudes 10°N and 15°N, consists of a northern, *chestnut* and a southern, *black soil* zone.

(4) In West Africa, a 400 mile wide band of *red-brown soils* extends between latitudes 12°N and 6°N as far as Ethiopia.

(5) In eastern Ethiopia, an irregular zone of *chestnut soils* extends to Mombasa (Kenya).

(6) In the Horn of Africa, *light-brown steppe* soils cover Somalia and adjacent areas.

(7) The equatorial area of *red tropical soil* occupies the coastal strip of the Gulf of Guinea, the Gabon and the Congo between latitudes 4°N and 3°S.

(8) The Plateaus, covered by *red-brown soils*, connect Angola, the southern part of the Congo basin, and eastern Tanganyika as far as latitude 12°S and eastern Mozambique.

(9) The Zambezi Basin, a 500 mile wide band of *slightly leached soils*, extends from southern Angola to western Mozambique and thence southward to Durban.

(10) The Southern Steppes of *light-brown soils* cover southwestern Angola, Southwest Africa and Bechuanaland.

(11) A belt of *chestnut soils* surrounds *(10)* in a 250 mile wide ring which lies mostly in South Africa, cf. *1*.

(12) In Madagascar, *laterite* plateaus are surrounded by various smaller zones.

A comparison of soil and climate maps thus shows a distinct zoneography, with soils distributed along east–west trending belts of latitudes, with the exception of the Horn and southern Africa, which are influenced by other structural and meteorological patterns. A glance at the geological map of Africa will reveal a considerable control of the formation of soils. Indeed, areas *1* and *2* (north) correspond to the folded sediments of the Atlas; *2* (south) and *3* correspond to the young tabular sediments of the Sahara, with light-brown soils covering the Precambrian window of the Hoggar. Again, area *4* coincides with the Precambrian Guinea Shield and the orogenies surrounding the Congo Basin in the north, and areas *5* and *6* are, respectively, controlled by the Rift Valleys and Highlands and by the sedimentary basin of Somalia. However, south of the equator groups of soil provinces tend to include various geological units, but even there a closer examination shows that *(a)* the laterites of Madagascar mainly cover the Precambrian Highlands, that *(b)* Precambrian orogenies covered by solonchak soil rise from the Southern Steppes of sediments and sands *(10)*, and that *(c)* the strip of brown forest soils of the south coast covers the Paleozoic Cape orogeny.

While most of Morocco and northern Algeria are covered by chestnut soils similar to those of Europe, strips of brown forest soils and podzolized soils as well as saline solonyets are also included. The red 'Hamri' or sandy soils have absorbed more bases than the black 'Tirs', and terra rossa has developed on limestones of the Atlas where iron oxides are abundant.

The Sahara covers most of Tunisia, Algeria, Mauritania, Mali, Niger, Libya, Chad, the Sudan and Egypt with mountains and plain deserts of stone and sand. However, the alluvia of the Nile, and cultivation of several millenia (e.g., the use of lime, nitrates and phosphates), have developed a strip of light-brown soil and sandy, highly permeable soils in the delta. The +2 mm fraction of the Badob soils of the Gesira in the Sudan consists of lime, and aeolian and alluvial sedimentation control the soil pattern. In the desert, ground-water controls the formation of scarce soils and the development of oases.

Red-brown soils cover Upper Guinea, the northern Ivory Coast and Ghana, southern Mali, Upper Volta, Togo, Dahomey, northern and central Nigeria and Cameroon, the Central African Republic and southern Sudan. However, laterites (occasionally bauxitic) extend from Guinea to Abidjan, in Gabon, Cameroon and elsewhere. Yellow soils cover granites and gneisses of the shields; black and red soils, however, occupy the Ethiopian plateau. Tropical rains rapidly degrade the equatorial soils, and indeed, exchangeable bases of Congo soils appear to average only 1 mval. In East Africa three types of red soils have been distinguished: lateritic, non-lateritic (covering acid rocks) and podzolized, the latter surrounding the red soils of the Highlands.

Light coloured sandy soils cover the Plateau inland from the coastal ridge of Angola, but black soils make their appearance in Southern Rhodesia, and their prairie variety reaches even greater expanses in the High Veld of the eastern Transvaal. Finally, humus develops on the south coast of Africa.

APPENDIX I

CONVERSION FACTORS

1 long ton equals 2,240 pounds equal 1.016 metric tons.
1 short ton equals 2,000 pounds equal 0.907 metric tons.
1 metric ton equals 1,000 kilograms (kg) equal 1,000,000 grams (g).
1 metric ton equals 5.86 barrels of cement.
1 metric ton equals 7.64 barrels of Algerian oil.
1 metric ton equals 7.7 barrels of Libyan oil.
1 metric ton equals 28.60 flasks of mercury.
1 metric ton equals 32,151 troy ounces.
1 pound (lb.) equals 16 ounces avoirdupois equal 454 grams.
1 ounce (oz.) avoirdupois equals 28.35 grams.
1 ounce troy equals 31.10 grams.
1 pennyweight (dwt.) equals 1.555 grams.
1 metric carat equals 0.2 grams.
1 cubic foot equals 0.028 cubic meters (m^3).
1 square mile equals 2.59 square kilometers (km^2).
1 mile equals 1.61 kilometers (km).
1 yard equals 91.5 centimeters (cm).
1 foot (ft.) equals 30.5 centimeters.
1 inch equals 2.54 centimeters.
1 pound per cubic yard equals 593 grams per cubic meter.
1 pound per short ton equals 500 grams per metric ton.
1 pound per long ton equals 446 grams per metric ton.
1 troy ounce per short ton equals 31.2 grams per metric ton.
1 troy ounce per long ton equals 30.6 grams per metric ton.
1 pennyweight per short ton equals 1.71 grams per metric ton.
1 pennyweight per long ton equals 1.50 grams per metric ton.
1 pennyweight gold per any ton equals 1.75$ per any ton.
1 litre per second (l/sec) equals 3.6 cubic meters per hour (m^3/h) equal 15.8 gallons per minute.
1 kilowatt (kW) equals 1.34 horsepower (U.S.).
1 billion used in the American sense equals 1,000 million.

APPENDIX II

LIST OF COMPANIES, INDIVIDUALS AND ORGANIZATIONS ENGAGED IN OR HAVING INTERESTS IN THE CEMENT, MINING, PETROLEUM AND POWER INDUSTRIES OF AFRICA

ADDI OU MOHA OU ZAID, Gourrama; Aïn Ber (Morocco).

ADILIA SANTA RITA DE LIMA (HERDEIROS), Erati (Mozambique).

AER CHROME MINES (PVT.) LTD., P.O. Box 946 Bulawayo (Southern Rhodesia).

AERO SERVICE CORP., 210 East Courtland Street, Philadelphia 20, Pa.; 375 Park Avenue, New York, N.Y. (U.S.A.).

AFRICAN AND EUROPEAN INVESTMENT CO., P.O. Box 2567 Johannesburg (South Africa); 40 Holborn Viaduct, London E.C.1 (Great Britain).

AFRICAN ASBESTOS-CEMENT CORP., 102 Locarno House, Locarno Street, P.O. Box 7374 Johannesburg; P. Bag, Carolina, E.Tvl. (South Africa).

AFRICAN ASSOCIATED MINES (PVT.) LTD., P.O. Box 1100 Bulawayo (Southern Rhodesia).

AFRICAN CHROME MINES LTD., 103 Mount Street, London W.1 (Great Britain); P.O. Box 124 Selukwe (Southern Rhodesia).

AFRICA-CITIES SERVICE PETROLEUM CORP., 60 Wall Street, New York 5, N.Y. (U.S.A.).

AFRICAN ASBESTOS MINING CO., Southern Rhodesia; 77–79 Fountain Street, Manchester 2, London (Great Britain).

AFRICAN CONSOLIDATED MINES LTD., P.O. Box 1778 Salisbury (Southern Rhodesia).

AFRICAN EXPLOSIVES AND CHEMICAL INDUSTRIES LTD., Modderfontein, Johannesburg (South Africa).

AFRICAN GEOPHYSICAL CO., Pilgrim Street, P.O. Box 118 Barberton; P.O. Louw's Creek, E. Tvl. (South Africa).

AFRICAN GOLDEN OCHRE CO., P.O. Box 18 Albertinia (South Africa).

AFRICAN GRANITE AND INDUSTRIAL CORP., 211 Holland Place, 71 Fox Street, Johannesburg (South Africa).

AFRICAN INVESTMENT TRUST LTD., 1 Cornhill, London E.C. 3 (Great Britain).

AFRICAN MANGANESE CO., 103 Mount Street, London W. 1 (Great Britain); Dagwin, Nsuta (Ghana).

AFRICAN METALS CORP. (AMCOR), Vereeniging, Tvl. (South Africa).

AFRICAN MINING AND TRUST CO., Anglovaal House, 56 Main Street, P.O. Box 1054 Johannesburg (South Africa).

AFRICAN PETROLEUM TERMINALS LTD., 135 East 42nd Street, New York 17, N.Y. (U.S.A.); Dakar (Senegal); Monrovia (Liberia); Abidjan (Ivory Coast); Accra (Ghana); Lagos (Nigeria).

AFRICAN RESEARCH AND DEVELOPMENT CO., 75 East 55th Street, New York 22, N.Y. (U.S.A.).

AFRICAN SALT WORKS (PTY.) LTD., 64 Eloff Street, P.O. Box 3317 Johannesburg (South Africa).

AFRIKANDER LEASE LTD., 28 Harrison Street, Johannesburg; P. Bag, Klerksdorp, W. Tvl. (South Africa).

AFRIKANDER PROPRIETARY GOLD MINES LTD., Mutual Buildings, Harrison Street, P.O. Box 1370 Johannesburg (South Africa).

AFROPEC, 26 Rue Laffitte, Paris 9e (France).

AGIP BRAZZAVILLE, S.A., B.P. 209 Bacongo, Brazzaville (Congo Republic).

AGIP CAMEROUN S.A., B.P. 4015 Douala (Cameroons).

AGIP CASABLANCA, Casablanca (Morocco).

AGIP CONGO S.A.R.L., B.P. 8065 Léopoldville (Congo).

AGIP (CÔTE D'IVOIRE) S.A., B.P. 1874 Abidjan (Ivory Coast).

AGIP (DAHOMEY) S.A., Carré 211, B.P. 420 Cotonou (Dahomey).

AGIP DJIBOUTI S.A., B.P. 128 Djibouti (French Somali Coast).

AGIP ETHIOPIA LTD., Addis Ababa (Ethiopia).

AGIP GABON S.A., Gros Bouquet, B.P. 3045 Libreville (Gabon).

AGIP LIBERIA CORP., P.O. Box 423 Monrovia (Liberia).

AGIP LTD., Nairobi (Kenya).

AGIP MADAGASCAR S.A., Tananarive (Malagasy Republic).

AGIP MINERARIA SPA., San Donato Milanese, Milan (Italy).

AGIP MINERARIA (SUDAN) LTD., Khartoum (Sudan).

AGIP (NIGERIA) LTD., Lagos (Nigeria).

AGIP SIERRA LEONE LTD., P.O. Box 752 Freetown (Sierra Leone).

AGIP SOMALIA SPA., P.O. Box 168 Mogadishu (Somali Republic).

AGIP (SUDAN) LTD., P.O. Box 1155 Khartoum (Sudan).

AGIP (TOGO) S.A., B.P. 974 Lomé (Togo).

AGMI, see: Bureau d'Assistance Géologique et Minière.

AIR CARRIER SERVICE CORP., 1744 G Street N.W., Washington 6, D.C. (U.S.A.).

AKIM CONCESSIONS LTD., 15–16 Basinghall Street, London E.C. 2; 107 Cheapside, London E.C. 2 (Great Britain).

ALADDINS FLUORSPAR MINE., 128 London House, Johannesburg; P.O. Darnall (South Africa).
ALAJI, M. IARAH AND GOLDBERG DIAMOND CO., Monrovia (Liberia).
ALAMASI LTD., Mwadui (Tanganyika).
ALETTA IRON ORE QUARRIES, P.O. Box 78 Dundee, Ntl. (South Africa).
ALEXANDRIA PORTLAND CEMENT CO., P.O. Box 75 Cairo (U.A.R.).
ALFSTROM GOLD MINE (PTY.) LTD., P.O. Box 19 Barberton, E.Tvl. (South Africa).
ALGEMENE BANK NEDERLAND N.V., 32 Vijzelstraat, Amsterdam (The Netherlands).
ALHERI MINING CO., Jos (Nigeria).
ALI BEN MOHAMMED, Tazzarine, par Missour; Tazzarine (Morocco).
ALLIED MINERAL DEVELOPMENT CORP., P.O. Box 4711 Capetown; P.O. Box 7093 Johannesburg (South Africa).
ALLUVIALE SANDWERKE, P.O. Box 125 Groot Marico, Tvl. (South Africa).
ALMA SOUTWERKE, P.O. Florisbad (South Africa).
AL NASR OILFIELDS CO., Shell House, 6 Sharia Orabi, P.O. Box 1634 Cairo (U.A.R.).
ALPHA ANTHRACITE CO., P.O. Box 4280 Johannesburg; P.O. Hlobane, Ntl (South Africa).
ALPHA CONSOLIDATED WITBANK COLLIERIES LTD., 63 Commissioner Street, Johannesburg; P.O. Box 7 Kendal, Tvl. (South Africa).
ALPHA FREE STATE HOLDINGS LTD., 42 Marshall Street, Johannesburg (South Africa).
ALPINE (BARBERTON) GOLD MINES LTD., 2–3 Salisbury Court, Fleet Street, London E.C. 4 (Great Britain).
ALUMINIUM CO. OF CANADA (ALCAN), 1 Place Ville Marie, Montreal, Quebec (Canada).
ALUMINIUM A. G. MENZIKEN, Menziken, Aargau (Switzerland).
ALUMINIUM (CANADA) LTD., Berkeley Square House, Berkeley Square, London W. 1 (Great Britain).
ALUMINIUM LTD., Head Offices: 1155 Metcalfe Street, Montreal 2 (Canada).
ALUMINIUM LTD. (CANADA) S.A., 13 Quai d'Ile, Geneva (Switzerland).
ALUMINIUM LTD. INC., 620 Fifth Avenue, New York, N.Y. (U.S.A.).
ALZI, see: Société Algérienne du Zinc.
AMALGAMATED BANKET AREAS LTD., P.O. Box 26 Tarkwa (Ghana).
AMALGAMATED COLLIERIES OF SOUTH AFRICA LTD., 44 Main Street, P.O. Box 2567 Johannesburg; P.O. Box 97 Vereeniging; P. Bag Meerlus, Tvl (South Africa); 40 Holborn Viaduct, London E.C. 1 (Great Britain).
AMALGAMATED GRANITE INDUSTRIES (PTY.) LTD., 203 Slegtkamp Street, Hercules, Pretoria (South Africa).
AMALGAMATED TIN MINES OF NIGERIA LTD. (ATMN), 55–61 Moorgate, London E.C. 2 (Great Britain); Rayfield, Jos; Buka Bakwai (Nigeria).
AMAX, see: American Metal Climax.
AMCOMETALL GMBH., Blittersdorfplatz 29 Frankfurt/Main (W. Germany).
AMCOR, see: African Metals Corp.
AMERADA PETROLEUM CORPORATION, 120 Broadway, New York 5, N.Y.; 218 West 6th Street, Tulsa, Okla.; 550 South Flower Street, Los Angeles (U.S.A.).
AMERICAN ABRASIVE CO., P.O. 245 Pietersburg, N. Tvl. (South Africa).
AMERICAN INTERNATIONAL OIL CO., 555 Fifth Avenue, New York 17, N.Y. (U.S.A.).
AMERICAN LIBYAN OIL CO., Tripoli (Libya).
AMERICAN METAL CLIMAX LTD., (AMAX), 1270 Avenue of the Americas, New York 20, N.Y. (U.S.A.).
AMERICAN OVERSEAS PETROLEUM LTD. (AMOSEAS), 485 Lexington Avenue, New York, N.Y. (U.S.A.); Tripoli (Libya); Port Harcourt (Nigeria); Algiers (Algeria).
AMERICAN POTASH AND CHEMICAL CORP., 99 Park Avenue, New York 16, N.Y. (U.S.A.).
AMERICAN OVERSEAS PETROLEUM (SPAIN) LTD., Vila Cisneros (Spanish Sahara).
AMERICAN POTASH AND CHEMICAL CORP., 3000 West 6th Street, Los Angeles 54, Calif. (U.S.A.).
AMERICAN–SOUTH AFRICAN INVESTMENT CO., 54 Marshall Street, Johannesburg (South Africa).
AMETAL S.A., 1 Rue du Commerce, Geneva (Switzerland).
AMI, see: Ausonia Mineraria.
AMIF, see: Société Ausonia Minière Française.
AMOCO INTERNATIONAL S.A., 41 Rue Versonnex, Geneva (Switzerland).
AMOSA (PTY.) LTD., P.O. Boxes 8644 and 2533, Skefko House, 1 Rissik Street, Johannesburg (South Africa).
AMOSEAS, see: American Overseas Petroleum Ltd.
ANDALUSITE REFRACTORIES (PTY.) LTD., 25 Main Reef Road, P.O. Box 131 Roodepoort, Tvl. (South Africa).
ANDERSEN, W. P., GRANITE STONE WORKS, 41 Zendelings Street, Rustenburg, W. Tvl. (South Africa).
ANGLO-AFRICAN TRUST LTD., 205/212 His Majesty's Building, Joubert Street, P.O. Box 3910 Johannesburg (South Africa).
ANGLO-ALPHA CEMENT LTD., P.O. Box 6810 Johannesburg (South Africa).
ANGLO-AMERICAN CORP. (CENTRAL AFRICA) LTD.: new name for Rhodesian Anglo-Americain Ltd.

ANGLO-AMERICAN CORPORATION OF SOUTH AFRICA LTD., 44 Main Street, P.O. Box 4587 Johannesburg; Consolidated Share Registrars Ltd., 76 Main Street, Johannesburg (South Africa); 70 Jameson Avenue Central, P.O. Box 1108 Salisbury C. 4; Charter House, Seborne Avenue, Bulawayo (Southern Rhodesia); Leslie Pollak House, Kitwe (Zambia); 1 Tulbagh Street, Welkom, O.F.S. (South Africa); 40 Holborn Viaduct, London E.C. 1 (Great Britain).

ANGLO AMERICAN INVESTMENT TRUST LTD., 44 Main Street, P.O. Box 4587 Johannesburg (South Africa); 40 Holborn Viaduct, London E.C. 1 (Great Britain).

ANGLO AMERICAN PROSPECTING CO. (AFRICA) LTD., 44 Main Street, P.O. Box 4587 Johannesburg (South Africa).

ANGLO-AMERICAN VULCANIZED FIBRE CO., P.O. Box 247 Mbeya (Tanganyika).

ANGLO BASE METAL MINING CO., P.O. Box 54 Postmasburg (South Africa).

ANGLO-FRENCH CONSOLIDATED INVESTMENT CORP., 94 Commissioner Street, Johannesburg (South Africa).

ANGLO FRENCH EXPLORATION CO. LTD., 208–224 Salisbury House, London Wall, London E.C. 2 (Great Britain); Gold Fields of South Africa Ltd., 75 Fox Street, Johannesburg (South Africa).

ANGLO METAL CO. LTD., Garrard House, 31–45 Gresham Street, London E.C. 2 (Great Britain).

ANGLO RAND MINING AND FINANCE CORP. LTD., Gold Fields Bldg., 75 Fox Street, P.O. Box 1167 Johannesburg (South Africa); 49 Moorgate, London E.C. 2 (Great Britain).

ANGLO-SWEDISH MINERALS LTD., 39 Hillroad, Wimbledon, London S.W. 19 (Great Britain).

ANGLO-TRANSVAAL COLLIERIES LTD., Anglovaal House, 56 Main Street, P.O. Box 7727 Johannesburg: P.O. Van Dyksdrif, Tvl.; P.O. Box 6 Estantia; P.O. Breyton (South Africa).

ANGLO-TRANSVAAL CONSOLIDATED INVESTMENT, Anglovaal House, 56 Main Street, P.O. Box 7727 Johannesburg (South Africa); Bilbao House, 36 New Broad Street, London E.C. 2 (Great Britain).

ANGLO-TRANSVAAL INDUSTRIES LTD., P.O. Box 7727 Johannesburg (South Africa).

ANGLOVAAL RHODESIAN EXPLORATION CO. (PVT.) LTD., Exploration House, corner Fifth Street and Manica Road, P.O. Box 2378 Salisbury (Southern Rhodesia).

ANIC SPA., San Donato Milanese, Casella Postale 3587 Milan (Italy).

ANTAR PÉTROLES DE L'ATLANTIQUE, Crédit Commercial de France; Crédit Lyonnais Paris (France).

ANTOINETTE FLUORSPAR MINE (PTY.) LTD., 119 Volkskas Building, 76 Market Street, Johannesburg; P.O. Box 52 Nongoma (South Africa).

APEX MINES LTD., 75 Fox Street, P.O. Box 1083 Johannesburg (South Africa).

ARCADIA ASBESTOS MINES (PTY.) LTD., 36 Pritchard Street, P.O. Box 9111 Johannesburg; P.O. Box 227 Krugersdorp (South Africa).

ARCTURUS MINES LTD., Lonrho House, P.O. Box 80 Salisbury (Southern Rhodesia).

ARGENT LEAD AND ZINC CO., 75 Fox Street, Johannesburg (South Africa).

ARISTON GOLD MINES (1929) LTD., P.O. Box 1 Prestea (Ghana).

ARNHOLD TRADING CO., Creechurch House, 37–45 Creechurch Lane, London E.C. 3 (Great Britain).

ARNHOLD, WILHELMI AND CO., South Africa Centre, 253 Bree Street, P.O. Box 4307 Johannesburg (South Africa); P.O. Box 2511 Salisbury (Southern Rhodesia).

ARNOLD, G. P., P.O. Box 8 Aughrabies, C.P. (South Africa).

ARUNA, E., Kisor (Uganda).

ASBESTOS INVESTMENTS (PTY.) LTD., P.O. Box 2167 Bulawayo (Southern Rhodesia); P.O. Box 1424 Johannesburg (South Africa).

ASBESTOS REFINING CO., P.O. Box 891 Bulawayo (Southern Rhodesia).

ASHANTI GOLDFIELDS CORP. LTD., 10 Old Jewry, London E.C. 2 (Great Britain); Obuasi (Ghana).

ASSEIL., Tripoli (Libya).

ASSOCIATED ASBESTOS LTD., Anglovaal House, 56 Main Street, P.O. Box 1054 Johannesburg; P.O. Box 72 Pietersburg, N. Tvl. (South Africa).

ASSOCIATED MANGANESE MINES OF SOUTH AFRICA LTD., P.O. Box 7727 Johannesburg; P.O. Mancorp Mine, Postmasburg (South Africa).

ASSOCIATED MINES CO., 3 Manshaat El Kataba Street, Cairo (U.A.R.).

ASSOCIATED PORTLAND CEMENT MANUFACTURERS LTD., Takoradi (Ghana).

ASSOCIATED TUNGSTEN MINES LTD., P.O. Box 204 Upington (South Africa).

ASSOCIATED ORE AND METAL CORP., 56 Main Street, P.O. Box 1054 Johannesburg (South Africa).

ATBARA CEMENT FACTORY, Atbara (Sudan).

ATHOL JONES MINES LTD., Jos (Nigeria).

ATLANTIC EXPLORATION CO., Vila Cisneros (Spanish Sahara).

ATLANTIC REFINING CO., 1111 Delancy Street, Newark, N.J. (U.S.A.); Vila Cisneros (Spanish Sahara).

ATLANTIC SALT CO., P.O. Box 129 Swakopmund (Southwest Africa).

ATMN, see: Amalgamated Tin Mines of Nigeria Ltd.

ATONIC CHROME MINE, P.O. Mabieskraal, Rustenburg, W. Tvl. (South Africa).
AUSONIA MINERARIA (AMI), Viale Liszt, Eur-Roma (Italy); Via San Antonio, Maria Zaccaria I, Milan (Italy); 41 Boulevard Latour–Maubourg, Paris 7e (France).
AUXILACS, see: Société Auxiliaire Industrielle et Financière des Grands Lacs Africains.
AUXILIAIRE MAROCAINE DES MINES, 79 Avenue Hassan II, Casablanca (Morocco).
BACH, ERNEST, Ambia (Behara) (Malagasy Republic).
BACHRAN, H. G., P.O. Box 47 Okahandja (Southwest Africa).
BADPLAATS ASBESTOS CO., 46 Marshall Street, P.O. Box 957 Johannesburg; Goedverwacht, Carolina, E. Tvl. (South Africa).
BAHATI MINES LTD., Kabale (Uganda).
BALMORAL GOLD MINING CO., City Trust House, 106 Fox Street, P.O. Box 742 Johannesburg; P.O. Box 5 Knights (South Africa).
BALUBA MINES LTD., P.O. Box 1479 Salisbury (Southern Rhodesia); Lusaka (Zambia).
BAMALETE MANGANESE CO. LTD., Lobatsi (Bechuanaland).
BANCO BURNAY, Rua dos Fanqueiros 10, Lisbon (Portugal).
BANCROFT MINES LTD., Lusaka; Leslie Pollak House, Nkana, Zambia.
BANQUE DE L'UNION PARISIENNE, 8 Bld. Hausmann, Paris 9e (France).
BANQUE DE PARIS ET DES PAYS BAS, 3 Rue d'Antin, Paris 2e (France).
BANQUE D'INDOCHINE, 96 Bld. Hausmann, Paris 8e (France).
BANQUE INTERNATIONALE DE LUXEMBOURG, 18 Place de la Liberté, Luxembourg (Luxembourg).
BANQUE LAMBERT, 2 Rue d'Egmont, Brussels (Belgium).
BANQUE NAGELMACKERS FILS ET COMPAGNIE, 12 Place de Louvain, Brussels (Belgium).
BANQUE NATIONALE POUR LE COMMERCE ET L'INDUSTRIE (BNCI), 16 Bld. des Italiens, Paris (France).
BANZET, ARTHUR, Rue de l'Aisne, B.P. 29 Fort-Dauphin (Malagasy Republic).
BARBERTON CHRYSOTILE ASBESTOS LTD., P.O. Box 957 Johannesburg; Barberton, E. Tvl. (South Africa).
BARBERTON MINES LTD., 63 Commissioner Street, P.O. Box 5958 Johannesburg; Fairview, P.O. Caledonian, Barberton, E. Tvl.; Three Sisters, P.O. Louwscreek, E. Tvl. (South Africa).
BARBOSCO MINES (PTY.) LTD., 6 Hollard Street, P.O. Box 11 Johannesburg; Klerksdorp, W. Tvl. (South Africa).
BARKER, W. R. (PTY.) LTD., 423–436 Church Street West, P.O. Box 847 Pretoria (South Africa).
BARKER, W. R. (PVT.) LTD., 64 Abercorn Street, P.O. Box 640 Bulawayo (Southern Rhodesia).
BARTA AND PARTNERS (PTY.) LTD., P.O. Box 11016 Johannesburg (South Africa).
BARTON MAYHEW RYDER AND CO., ABC Chambers, 27 Simmonds Street, P.O. Box 1639 Johannesburg (South Africa).
BARYTES MINING CO., 10 Milne Street, Vulcania; P.O. Box 240 Brakpan, Tvl.; P. Bag Schoonoord, Barberton, E. Tvl. (South Africa).
BATAAFSE INTERNATIONALE PETROLEUM MAATSCHAPPIJ, 30 Carel van Bylandtlaan, The Hague (The Netherlands).
BAUXICONGO, see: Société de Recherches et d'Exploitation des Bauxites du Congo.
BEATRICE GOLD MINING CO., P.O. Box 1167 Johannesburg (South Africa).
BÉCÉKA-MANGANÈSE, 46 Rue Royale, Brussels (Belgium); Kisenge (Congo).
BECHUANALAND EXPLORATION CO., 11 Haleo House, Fife Street, Bulawayo (Southern Rhodesia); 19 St. Swithin's Lane, London E.C. 4 (Great Britain).
BECKER, E. M., P.O. Box 2 Karibib (Southwest Africa).
BELGIKAMINES S. A., 27 Rue du Trône, Brussels; 42 Rue Royale, Brussels (Belgium); c/o COBELMIN, Bukavu (Congo); Kima via Punia (Congo); Kampene (Congo).
BELGO-AMERICAN DEVELOPMENT CORP., 511 Fifth Avenue, New York 17, N.Y. (U.S.A.).
BELINGWE MINING AND INVESTMENT (PVT.) LTD., P.O. Box 3112 Salisbury (Southern Rhodesia).
BELINGWE INVESTMENT AND DEVELOPMENT CO., Belingwe (Southern Rhodesia).
BELLEVUE COLLIERY CO., corner Commissioner and Kruis Streets, P.O. Box 989 Johannesburg; P.O. Box 10 Ermelo, Tvl. (South Africa).
BEND ASBESTOS LTD., P.O. Box 1426 Bulawayo (Southern Rhodesia).
BENSIMON, J., Ksar es Souk; Taltfraout (Morocco).
BEPI, see: Bureau d'Études et Participations Industrielles.
BERGER, J., P.O. Box 35 Karibib (Southwest Africa).
BERGMANN, W. O., P.O. Box 1727 Düsseldorf 1 (W. Germany).
BETHLEHEM STEEL CORP., 25 Broadway, New York, N.Y. (U.S.A.).
BETTY AND DICKSON LTD., 511-6 City Trust House, 106 Fox Street, P.O. Box 742 Johannesburg (South Africa).
BIA, see: Bureau Industriel d'Algérie, Algiers (Algeria).
BIKITA MINERALS (PVT.) LTD., P.O. Box 128 Fort Victoria (Southern Rhodesia).

BILLITON MAATSCHAPPIJ, N.V., 19 Louis Couperusplein, P.O. Box 190 The Hague (The Netherlands); 1270 Avenue of the Americas, Room 2601, New York 20, N.Y. (Montanore Inc.) (U.S.A.); 80 Richmond Street W., Room 1401, Toronto (Canada).

BISICHI TIN CO. (NIGERIA) LTD., 52 Leadenhall Street, London E.C. 3 (Great Britain); Jos (Nigeria).

BITTER, H., P. O. Box 29 Swakopmund (Southwest Africa).

BJORDAL MINES, P. O. Box 50 Kabale (Uganda).

BLATT, P., P.O. Box 103 Swakopmund (Southwest Africa).

BLESBOK COLLIERY LTD., P.O. Box 2567 Johannesburg; P. Bag Meerlus (South Africa).

BLINKPAN KOOLMYNE BEPERK, P.O. Box 9598 Johannesburg; P.O. Box Van Dyksdrift (South Africa).

BLINKPOORT GOLD SYNDICATE LTD., General Mining Building, 6 Hollard Street, Johannesburg (South Africa); Winchester House, Old Broad Street, London E.C. 2 (Great Britain).

BLOEMFONTEIN CONSOLIDATED INVESTMENT CORP. LTD., Cape House, 54 Fox Street, Johannesburg (South Africa); 3 London Wall Buildings, London E.C. 2 (Great Britain).

BLYVOORUITZICHT GOLD MINING CO. LTD., The Corner House, 77 Commissioner Street, P.O. Box 1056 Johannesburg; P.O. Box 1 Blyvooruitsig (South Africa); 4 London Wall Buildings, London E.C. 2 (Great Britain).

BNCI, see: Banque Nationale pour le Commerce et l'Industrie.

BOBTSCHI, W. (ETS.), B.P. 17 Fort-Dauphin (Malagasy Republic).

BONGWAN GAS SPRINGS CO., Port Shepstone, Ntl. (South Africa).

BORAX CHEMICALS LTD., 35 Piccadilly, London W. 1 (Great Britain).

BOSCHOEK PROPRIETARY CO., 1 Broad Street Place, London E.C. 2 (Great Britain); 17 Harrison Street, Johannesburg (South Africa).

BOSS ASBESTOS MINES (PVT.) LTD., P.O. Box 93 Mashaba (Southern Rhodesia).

BOTRON MIJNBOU MAATSKAPPY, P.O. Box 4711 Cape Town (South Africa).

BOUGROFF, ALEXANDRE, Antanimora-Sud (Malagasy Republic).

BOUVIN, SVEN (PVT.) LTD., P.O. Box 2112 Salisbury (Southern Rhodesia).

BRITISH PETROLEUM CO., Britannic House, Finsbury Circus, London E.C. 2 (Great Britain).

BP-HUNT CO., Tripoli (Libya).

BP-SHELL PETROLEUM DEVELOPMENT CO. OF KENYA, Prince Charles Street, Mombasa (Kenya).

BP-SHELL PETROLEUM DEVELOPMENT CO. OF TANGANYIKA, Gerazani Creek, Dar es Salaam (Tanganyika).

BP-SHELL PETROLEUM DEVELOPMENT CO. OF ZANZIBAR, Zanzibar.

BP-SHELL PETROLEUM REFINING CO. OF NIGERIA, Alese (Nigeria).

BP SOUTHERN AFRICA (PTY.) LTD., Mutual Building, 21 Parliament Street, P.O. Box 664 Cape Town (South Africa).

BP (SUDAN), see: Shell Red Sea.

BRACKEN MINES LTD., 74–78 Marshall Street, Johannesburg; P.O. Box 73 Evander (South Africa).

BRAKPAN MINES LTD., 44 Main Street, Johannesburg (South Africa).

BRANCO, Diogo Maria L. S. C., Cabinda (Angola).

BRANDHURST CO., Vitry House, Queen Street Place, London E.C. 4 (Great Britain).

BREACHER, W. KG., Hauptstrasse 179, Idar–Oberstein 2 (W. Germany).

BREMANG GOLD DREDGING CO., P.O. Box 128 Dunkwa (Ghana).

BRITISH ALUMINIUM CO., Norfolk House, St. James's Square, London S.W. 1 (Great Britain); P.O. Box 1 Awaso (Ghana).

BRITISH GAS BOARD, Woodall house, Lordship Lane, London N. 22 (Great Britain).

BRITISH GRANITE QUARRIES LTD., 79b Rosettenville Road, P.O. Box 2262 Johannesburg (South Africa).

BRITISH INSULATED CALLENDER'S CABLES LTD., 21 Bloomsbury Street, London W.C. 1 (Great Britain).

BRITISH IRON AND STEEL CORP. (ORE) LTD., (BISC), 7 Old Park Lane, Piccadilly, London W. 1 (Great Britain).

BRITISH METAL CORP., State Building, 43 Gresham Street, London E.C. 2 (Great Britain); Corner Commissioner and Kruis Streets, P.O. Box 9527 Johannesburg (South Africa); Eagle House, Abercorn Street, P.O. Box 1544 Bulawayo; Bennett House, Moffat Street, P.O. Box 2366 Salisbury (Southern Rhodesia).

BRITISH ORE INVESTMENT CORP. LTD., 7 Old Park Lane, Piccadilly, London W. 1 (Great Britain).

BRITISH SOUTH AFRICA CO. (BSA), 11 Old Jewry, London E.C. 2 (Great Britain); Charter House, Salisbury (Southern Rhodesia); P.O. Box 2000 Lusaka; P.O. Box 1508 Ndola (Zambia).

BRITISH STEEL CORP. LTD., 7 Old Park Lane, Piccadilly, London W. 1 (Great Britain).

BRITISH STANDARD PORTLAND CEMENT CO., Bamburi, Mombasa (Kenya).

BRITISH TITAN PRODUCTS CO., 10 Stratton Street, London W. 1 (Great Britain).

BRITSTOWN SALT WORKS (PTY.) LTD., P.O. Box 1016 Cape Town; P.O. Box 2 Sodium (South Africa).

BRGM, see: Bureau de Recherches Géologiques et Minières.

BRMA, see: Bureau de Recherches Minière Algérie.

BROCKMANN, M. H. C., P.O. Box 4 Karibib (Southwest Africa).

BROUDE TUNGSTEN SYNDICATE, P.O. Box 28 Upington (South Africa).

BRUFINA, see: Société de Bruxelles pour la Finance et l'Industrie.

BRUSIUS, A., P.O. Box 97 Usakos (Southwest Africa).

BSA, see: British South Africa Company.

BUFFALO FLUORSPAR MINE, P.O. Box 51 Naboomspruit (South Africa).

BUFFALO OXIDES (PTY.) LTD., P.O. Box 81 Newcastle, Ntl. (South Africa).

BUFFELS CHROME MINE (PTY.) LTD., 411 Charter House, P.O. Box 7269 Johannesburg (South Africa).

BUFFELSFONTEIN CHROME MINE, P.O. Box 82 Marikana (South Africa).

BUFFELSFONTEIN GOLD MINING CO., P.O. Box 1173 Johannesburg; P. Bag, Stilfontein (South Africa).

BUHEMBA MINES LTD., P.O. Box 36 Musoma (Tanganyika).

BUKURU LTD., P.O. Box 12 Bukuru (Nigeria).

BULTFONTEIN CHROME MINE (PTY.) LTD., P.O. Box 9 Boshoek (South Africa).

BUREAU D'ASSISTANCE GÉOLOGIQUE ET MINIÈRE (AGMI), Tunis (Tunisia).

BUREAU DE RECHERCHE DE PÉTROLE, 7 Rue Nélaton, Paris 15e (France).

BUREAU DE RECHERCHES ET DE PARTICIPATION MINIÈRES, 27 Avenue Urbain Blanc, Rabat; Hassian ed Diab (Morocco).

BUREAU DE RECHERCHES GÉOLOGIQUES ET MINIÈRES (BRGM), 8 Rue Léonard de Vinci, Paris 16e (France); 74 Rue de la Fédération, Paris 15e (France); B.P. 24 Birmandreis (Algeria); B.P. 268 Dakar (Senegal); B.P. 1335 Abidjan (Ivory Coast); B.P. 431 Brazzaville (Congo Republic); B.P. 458 Tananarive (Malagasy Republic); B.P. 343 Yaoundé (Cameroons); B.P. 386 Bobo-Dioulasso (Upper Volta); B.P. 175 Libreville (Gabon).

BUREAU DE RECHERCHES MINIÈRES D'ALGÉRIE (BRMA), Birmandreis 24, Algiers (Algeria).

BUREAU D'ÉTUDES ET DE DOCUMENTATION DE LA CHAMBRE SYNDICALE DU ZINC ET DU CADMIUM, 39 Rue St. Dominique, Paris 7e (France).

BUREAU D'ÉTUDES DE RECHERCHES ET D'INTERVENTIONS INDUSTRIELLES ET MINIÈRES, Palais du Gouvernement, Algiers (Algeria).

BUREAU D'ÉTUDES ET PARTICIPATIONS INDUSTRIELLES (BEPI), Rabat (Morocco).

BUREAU D'INVESTISSEMENT D'ALGÉRIE, Algiers (Algeria).

BUREAU INDUSTRIEL D'ALGÉRIE (BIA), Algiers (Algeria).

BURGER, E. P., Burgershof, P.B. Platveld, Otavi (Southwest Africa).

BUSHENYI MINES, P.O. Box 1645 Kampala (Uganda).

BUSINESS BUYING AND INVESTMENT CO., 804 Maritime House, Loveday Street, P.O. Box 7193 Johannesburg (South Africa).

BUSUMBU MINING CO., P.O. Magodes Station (Uganda).

CABINDA GULF OIL CO., Luanda (Angola).

CABOL ENTERPRISES LTD., Toronto (Canada); Cotonou (Dahomey).

CADO, see: Ciments Artificiels d'Oranie.

CAIRO ELECTRICITY AND GAS ADMINISTRATION, 53, 26th July Street, Cairo (U.A.R.).

CAISSE D'ÉQUIPEMENT D'ALGÉRIE, Algiers (Algeria).

CALCINED PRODUCTS LTD., Chamber of Mines Building, Johannesburg (South Africa).

CALIFORNIA TEXAS OIL CORP., 380 Madison Avenue, New York 17, N.Y. (U.S.A.).

CALTEX (AFRICA) LTD., Sanlam Centre, Heerengracht, P.O. Box 714 Cape Town (South Africa).

CALTEX (EGYPT) S.A.E., Cairo (U.A.R.).

CALTEX OIL (KENYA) LTD., Nairobi (Kenya).

CALTEX OIL (TANGANYIKA) LTD., Dar es Salaam (Tanganyika).

CALTEX OIL (UGANDA) LTD., Kampala (Uganda).

CAM AND MOTOR GOLD MINING CO., Doncaster House, Salisbury (Southern Rhodesia); 59 Gresham Street, London E.C. 2 (Great Britain).

CAMARA MUNICIPAL DA BEIRA, P.O. Box 95 Beira (Mozambique).

CAMEL, see: Compagnie Algérienne de Méthane Liquide.

CANADIAN DELHI OIL CO., Rabat (Morocco); 505 Fifth Avenue, Calgary, Alberta (Canada).

CANADIAN EXPLORATION LTD., Isaka (Tanganyika).

CANADIAN-LIBERIAN IRON ORE CO. (LTD.), Vancouver (Canada).

CAPE ASBESTOS CO., 114–116 Park Street, London W. 1 (Great Britain).

CAPE BLUE MINES (PTY.) LTD., Skefko House, 1 Rissik Street, P.O. Boxes 2533, 8644 Johannesburg; P.O. Pompom via Vryburg (South Africa).

CAPE LIME CO., P.O. Box 4 Worcester, C.P. (South Africa).

CAPE METALS (PTY.) LTD., P.O. Box 2 Springbok, Namaqualand, C.P. (South Africa).

CAPE PORTLAND CEMENT CO., Argus Building, St. George Street, P.O. Box 1067 Capetown (South Africa).

CAPITAL GOLD AND EXPLORATION CO., 54–62 Riebeeck Street, Cape Town (South Africa).

CARBONEZ, A., Buco Zan (Maiombe); Cabinda (Angola).
CARDOSO E CASTRO, ANTONIO AUGUSTO, Tete (Mozambique).
CAREP, see: Compagnie Algérienne de Recherches et d'Exploitations Pétrolières.
CARNEIRO, ARMANDO MARTINS R., Cabinda (Angola).
CARRIG DIAMONDS LTD., P.O. Box 957 Johannesburg (South Africa).
CASAL, BERNARDINO LOURENÇO, Vila Gamito (Mozambique).
CAST, see: Consolidated African Selection Trust Ltd.
CATG, see: Compagnie pour l'Application des Techniques Géophysiques.
CAYCO LTD., Anyinatsu (Ghana).
CCCI, see: Compagnie du Congo pour le Commerce et l'Industrie.
CDDC, Boungou (Central African Republic).
CEFEM, see: Compagnie Française du Méthane.
CELB, see: Companhia Eléctrica do Lobito et Benguela.
CEMENTIR, Garoua (Cameroons).
CEMENTOS CANARIAS S.A., Consejo de Ciento 290, Barcelona (Spain); Santa Cruz de Tenerife (Canary Islands).
CEMENTOS ESPECIALES S.A., Secretario Artiles no. 65, Las Palmas (Canary Islands).
CEMENTOS MARROQUIES C.M.A., Plaza de Ben Azus, Tetuan (Morocco).
CEMENTOS TANGER S.A., Rembrandt no. 17, Tanger (Morocco).
CEMLIM (PTY.) LTD., P.O. Box 12 Port Shepstone, Ntl. (South Africa).
CENTRAL AFRICAN PETROLEUM REFINERIES LTD., Umtali (Southern Rhodesia).
CENTRAL AGENCIES AND IMPORT CO., 205 Main Street, P.O. Box 2660 Johannesburg (South Africa).
CENTRAL ASBESTOS CO., 5 Fenchurch Street, London E.C. 2 (Great Britain); P.O. Box 10213 Johannesburg (South Africa).
CENTRAL DESERT MINING CO., P.O. Box 20 Port Sudan (Sudan).
CENTRAL ELECTRICITY AND WATER ADMINISTRATION, P.O. Box 1380 Khartoum (Sudan).
CENTRAL ELECTRICITY BOARD, Royal Road, Curepipe (Mauritius).
CENTRAL ELECTRICITY CORP., P.O. Box 40 Lusaka (Zambia).
CENTRAL MINING AND FINANCE CORP. (GOLD MINES) see: Central Mining Finance Co.
CENTRAL MINING AND INVESTMENT CORP., 1 London Wall Buildings, London E.C. 2 (Great Britain).
CENTRAL MINING FINANCE LTD., 1 London Wall Buildings, London E.C. 2 (Great Britain).
CENTRAL SELLING ORGANISATION, 40 Holborn Viaduct, London E.C. 1 (Great Britain).
CENTRAL SOUTH AFRICAN LAND AND MINES LTD., P.O. Box 2105 Johannesburg (South Africa).
CENTRE D'ÉTUDES GÉOLOGIQUES ET MINIÈRES, 74 Rue de la Fédération, Paris 15e (France).
CENTRE D'ÉTUDES ET DE RECHERCHES DES PHOSPHATES MINÉRAUX (CERPHOS), 47 Rue de Liège, Paris (France).
CENTRE D'INFORMATION DE PLOMB, 10 Place Vendôme, Paris 1er (France).
CENTRE TECHNIQUE DE L'ALUMINIUM, 87 Boulevard de Grenelle, Paris 15e (France).
CEP, see: Compagnie d'Exploration Pétrolière.
CEREBOS LTD., 67 Paterson Road, P.O. Box 3251 Port Elizabeth; P.O. Teviot (South Africa).
CERPHOS, see: Centre d'Études et de Recherches des Phosphates Minéraux.
CFA, see: Compagnie Financière Africaine.
CFL, see: Compagnie des Chemins de Fer du Congo Supérieur aux grands Lacs.
CFP, see: Compagnie Navale des Pétroles.
CFP (A), see: Compagnie Française des Pétroles (Algérie).
CFPS, see: Compagnie Française de Prospection Sismique.
CGG, see: Compagnie Générale de Géophysique.
CHAMBISHI MINES LTD., Lusaka (Zambia).
CHARBONNAGES DE LA LUENA, 10 Rue Brédérode, Brussels (Belgium); Elisabethville; Luena (Congo).
CHARBONNAGES NORD-AFRICAINS, 27 Avenue Urbain-Blanc, Rabat (Morocco); Djirada, (Morocco); M. Fegueux, 47 Rue de Liège, Paris (France).
CHARTERED EXPLORATION LTD., Charter House P.O. Box 364 Salisbury (Southern Rhodesia); Exploration House, Lusaka (Zambia).
CHARTER HOLDINGS LTD., 220 Jeppe Street, P.O. Box 7788 Johannesburg (South Africa); Robinson House, 51 Union Avenue, Salisbury (Southern Rhodesia).
CHARTERLAND AND GENERAL LTD., 19 St. Swithin's Lane, London E.C. 4 (Great Britain).
CHASE INTERNATIONAL INVESTMENT CORP., 1 Chase Manhattan Plaza, New York, N.Y. (U.S.A.).
CHASE MANHATTAN BANK, 1 Chase Manhattan Plaza, New York, N.Y. (U.S.A.).
CHAUX ET CIMENTS DU MAROC, 23 Rue Emile Menier, Paris 16e (France); 237 Boulevard Moulay Ismael, Casablanca (Morocco).

CHEVRIER, H., Boulhaut; Aïn Kseub (Morocco).
CHIBULUMA MINES LTD., Lusaka; Chibuluma (Zambia).
CHICAGO–GAIKA DEVELOPMENT CO. LTD., The Spring House, Spring Street, Ewell, Surrey (Great Britain).
CHILANGA CEMENT LTD., P.O. Box 99 Chilanga (Zambia).
CHINGACHURU EXPLORATION CO., 107 Selrho House, corner Rhodes and Selborne Avenues, Bulawayo (Southern Rhodesia).
CHISANGWA MINES LTD., Leslie Pollak House, Nkana (Zambia).
CHRISTIEN F. F. AND CO., Blackhorse Lane, Walthamstow, London E. 17 (Great Britain); P. Bag 577 Sinoia (Southern Rhodesia); P.O. Box 66 Morogoro (Tanganyika).
CHROME CO., 103 Mount Street, London W. 1 (Great Britain).
CHROME CORP., Main House, 96 Màin Street, P.O. Box 6612 Johannesburg (South Africa).
CHROME INTERESTS (PVT.) LTD., P. Bag 187-H Salisbury (Southern Rhodesia).
CHROME MINES OF SOUTH AFRICA LTD., P.O. Box 1125 Johannesburg (South Africa).
CHULLIAT, A., 38 Boulevard Danton (Casablanca); Azguemerzi (Morocco).
CIMAR, see: Compagnie Industrielle des Pétroles du Maroc.
CIMENTAL, see: Cimenteries d'Albertville.
CIMENTERIES D'ALBERTVILLE (CIMENTAL), 112 Rue du Commerce, Brussels (Belgium); Albertville (Congo).
CIMENTERIES DU RWANDA ET DU BURUNDI, 112 Rue du Commerce, Brussels (Belgium); Kisenyi (Rwanda).
CIMENTS ARTIFICIELS D'ORANIE (CADO), 30 Boulevard Clémenceau, Algiers (Algeria).
CIMENTS ARTIFICIELS TUNISIENS, 47 Rue du Portugal, Tunis (Tunisia).
CIMENTS DU KATANGA, Lubudi (Congo).
CIMENTS DU KIVU, Bukavu (Congo).
CIMENTS MÉTALLURGIQUES, Jadotville (Congo); 6 Rue Montagne du Parc, Brussels (Belgium).
CIPAO, see: Compagnie Industrielle des Pétroles de l'Afrique Occidentale.
CITIES SERVICE CO., 60 Wall Street, New York 3, N.Y. (U.S.A.); Dakar (Senegal).
CITY DEEP LTD., P.O. Boxes 1056, 1411 Johannesburg (South Africa); 4 London Wall Buildings, London E.C. 2 (Great Britain).
CLAY INDUSTRY LTD., Ikeja (Nigeria).
CLAY PRODUCTS LTD., P.O. Box 1997 Bulawayo (Southern Rhodesia).
CLIMAX MOLYBDENUM CO., Tolstrasse 70, Zürich (Switzerland).
CLIMAX MOLYBDENUM CO. OF EUROPE LTD., 2 Cavendish Place, London W. 1 (Great Britain).
CLIMAX MOLYBDENUM CO. OF MICHIGAN, 1 Avenue de l'Observatoire, Paris 6e (France).
CLYDESDALE (TRANSVAAL) COLLIERIES LTD., Trust Buildings, corner Fox and Loveday Streets, Johannesburg; P.O. Coalbrook; P.O. Van Dyksdrift (South Africa); 10 Upper Grosvenor Street, London W. 1 (Great Britain).
CMC, see: Compagnie Minière du Congo Français.
CMOO, see: Compagnie Minière de l'Oubangi Oriental.
COBALT DEVELOPMENT INSTITUTE, 35 Rue des Colonies, Brussels (Belgium).
COBELMIN, see: Compagnie Belge d'Entreprises Minières.
COBRA EMERALD MINE, P.O. Box 7 Gravelotte, E. Tvl. (South Africa).
COFIMER, see: Compagnie Financière pour l'Outremer.
COFININDUS, see: Compagnie Financière et Industrielle.
COFIREP, see: Compagnie Financière de Recherches Pétrolières.
COFOR, see: Compagnie Générale de Forage.
COFRAMET, see: Compagnie Franco-Américaine des Métaux et des Minérals.
COLLINS, T. K. A. and J. E. A., Mbarara (Uganda).
COLONIAL DEVELOPMENT CORP., 33 Hill Street, London W. 2 (Great Britain).
COMILOG, see: Compagnie Minière de l'Ogoué.
COMINIÈRE, see: Société Commerciale et Minière du Congo.
COMIPHOS, see: Compagnie Minière et Phosphatière.
COMITÉ SPÉCIAL DU KATANGA (CSK), 51 Rue des Petits Carmes, Brussels (Belgium); Élisabethville (Congo).
COMMERCIALE MINERARIA CONTINENTALE, Via dei Bardi 49, Florence (Italy).
COMMISSARIAT À L'ÉNERGIE ATOMIQUE, 69 Rue de Varenne, Paris 7e (France); B.P. 282 Tananarive (Malagasy Republic).
COMMISSARIAT TUNISIEN À L'ÉNERGIE ATOMIQUE, Tunis (Tunisia).
COMPAGNIA IMPRESE ELETTRICHE DELL' ERITREA (CONIEL), Via deg. Hailu Chebbedè, 36, Asmara (Ethiopia).
COMPAGNIA RICERCHE IDROCARBURI (CORI), 50 Via Tevere, Rome (Italy); Tripoli (Libya).
COMPAGNIE ALGÉRIENNE DE RECHERCHES ET D'EXPLOITATIONS PÉTROLIÈRES (CAREP), 105 Avenue Raymond-Poincaré, Paris 16e (France); 2 Place Centrale, Hydra, Algiers 8e (Algeria); Oued Guétérini (Algeria).

COMPAGNIE ALGÉRIENNE DU MÉTHANE LIQUIDE (CAMEL), 22 Place Vendôme, Paris 1er; 29 Place du Marché Saint Honoré, Paris 1er (France).

COMPAGNIE ALGÉRIENNE SCHIAFFINO, Algiers (Algeria).

COMPAGNIE AUXILIAIRE DE NAVIGATION, 9 Rue de la Bruyère, Paris 9e (France).

COMPAGNIE BELGE D'ENTREPRISES MINIÈRES (COBELMIN), Bukavu; Lulingu; Kima; Kailo; Moga; Kampene (Congo); 42 Rue Royale, Brussels (Belgium).

COMPAGNIE CAMÉROUNAISE DE DÉPÔTS PÉTROLIERS, 26 Rue de la Pépinière, Paris 8e (France); B.P. 299 Douala (Cameroons).

COMPAGNIE CAMÉROUNAISE DE L'ALUMINIUM PÉCHINEY-UGINE (ALUCAM), B.P. 54 Edéa (Cameroons); 23 Rue Balzac, Paris 8e (France); B.P. 1090 Douala (Cameroons).

COMPAGNIE CENTRAFRICAINE DES MINES, 6 Avenue de Messine, Paris 8e (France); Berbérati (Central African Republic).

COMPAGNIE CENTRALE D'ÉCLAIRAGE ET DE CHAUFFAGE PAR LE GAZ, P.O. Box 241 Alexandria (U.A.R.).

COMPAGNIE CENTRALE DE DISTRIBUTION D'ÉNERGIE ÉLECTRIQUE, 42 Avenue de la Grande Armée, Paris 17e (France); Cotonou (Dahomey).

COMPAGNIE COLONIALE DE MADAGASCAR, Rue Lacoste, Tananarive (Malagasy Republic).

COMPAGNIE CONTINENTALE DES MINÉRAIS, 15 Rue Vézélay, Paris 8e (France).

COMPAGNIE D'ANVERS, Antwerp (Belgium).

COMPAGNIE DE MOKTA EL HADID, 60 Rue de la Victoire, Paris 9e (France); Behisaf, Département de Tlemcen (Algeria); 5 Avenue de l'Armée Royale, B.P. 241 Casablanca (Morocco); B.P. 1848 Abidjan (Ivory Coast).

COMPAGNIE DE PARTICIPATIONS, DE RECHERCHES ET D'EXPLORATION PÉTROLIÈRES (COPAREX), 1 Rue d'Astorg, Paris 8e (France); 7 Place Centrale, Hydra; El Gass–El Agreb (Algeria).

COMPAGNIE DE PRODUITS CHIMIQUES ET ÉLECTROMÉTALLURGIQUES ALAIS, FROGES ET CAMARGUE (PÉCHINEY), 1 Rue Capitaine Broussole, Marrakech; Iguerda (Morocco).

COMPAGNIE DE PRODUITS CHIMIQUES ET ÉLECTROMÉTALLURGIQUES PÉCHINEY, 23 Rue Balzac, B.P. 87-07 Paris 8e (France); 9 Cours de Verdun, Lyon (France); B.P. 1487 Tananarive (Malagasy Republic); Fria (Guinea).

COMPAGNIE DE RECHERCHES ET D'EXPLOITATION DE PÉTROLE AU SAHARA (CREPS), 12–16 Rue Jean Nicot, Paris 7e (France); 5 Rue Daguerre, Algiers; In Amenas; Zarzaïtine; Edjeleh; Tiguentourine; El Adeb Larache (Algeria).

COMPAGNIE DES CHEMINS DE FER DU BAS CONGO AU KATANGA, 44 Rue Royale, Brussels (Belgium).

COMPAGNIE DES CHEMINS DE FER DU CONGO SUPÉRIEUR AUX GRANDS LACS (CFL), 24 Avenue de l'Astronomie, Brussels (Belgium); Service Minier, Kindu (Congo).

COMPAGNIE DES CIMENTS MALGACHES, P.O. Box 302 Majunga (Malagasy Republic).

COMPAGNIE DES EAUX ET D'ÉLECTRICITÉ DE L'OUEST AFRICAIN, Dakar (Senegal).

COMPAGNIE DES MINES D'OR DU GABON (ORGABON), Étéké par Mouila (Gabon).

COMPAGNIE DES MINES DU DJEBEL TOUILA, 1 Rue de Constantine, Tunis (Tunisia).

COMPAGNIE DES MINES D'URANIUM DE FRANCEVILLE, B.P. 578 Libreville (Gabon); Mounana par Moanda (Gabon); 16 Rue Le Pelletier, Paris 9e (France).

COMPAGNIE DES PÉTROLES D'AFRIQUE OCCIDENTALE (COPETAO), Dakar (Senegal).

COMPAGNIE DES PÉTROLES D'ALGÉRIE (CPA), 7 Rue Daguerre (Algiers); Ouargla (Algeria).

COMPAGNIE DES PÉTROLES FRANCE–AFRIQUE (COPEFA), 7 Rue Nélaton, Paris 15e; 34–42 Avenue Raymond-Poincaré, Paris 6e (France).

COMPAGNIE DES PÉTROLES TOTAL (LIBYE) (CPTL), 5 Rue Michel-Ange, Paris 16e (France); P.O. Boxes 531, 984, 230 Sciara Istiklal, Tripoli (Libya).

COMPAGNIE DES PHOSPHATES DE CONSTANTINE, 12 Avenue Marceau, Paris 8e (France); Le Kouif (Bône); 8 Rue de Frandre, Bône (Algeria).

COMPAGNIE DES PHOSPHATES DE TAÏBA, Taïba, Dakar (Senegal).

COMPAGNIE DES PHOSPHATES DU DJEBEL ONK, 12 Avenue Marceau, Paris 8e (France).

COMPAGNIE DES PHOSPHATES ET DU CHEMIN DE FER DE GAFSA, 67 Rue Marcescheau (Tunis); 9 Rue Mozagran, Tunis; Metlaoui (Tunisia); 60 Rue de la Victoire, Paris 9e; 41 Avenue Hoche, Paris 8e (France).

COMPAGNIE DES POTASSES DU CONGO, Hollé; Brazzaville (Congo Republic).

COMPAGNIE DES PRODUITS CHIMIQUES ET RAFFINERIES DE BERRE, 55 Rue d'Amsterdam, Paris 8e (France).

COMPAGNIE DES TRANSPORTS PAR PIPE-LINES AU SAHARA (TRAPSA), 12–16 Rue Jean Nicot, Paris (France).

COMPAGNIE DE TIFNOUT TIRANIMINE, 52 Avenue Hassan II, B.P. 'ONA 657' Casablanca; Tifnout; Tanourat; El Borj; Bachkoun; Finnt (Morocco).

COMPAGNIE D'EXPLOITATIONS ET DE CHIMIE APPLIQUÉE, 29 Rue Général d'Amade, Oujda; Gara Ziad (Morocco).

COMPAGNIE D'EXPLORATION PÉTROLIÈRE (CEP), 7 Rue Nelaton, Paris 15e; 12 Rue Jean Nicot, Paris 7e; 76bis Grande Rue, Chambourcy, S. et O. (France); 6 Boulevard Saint Saëns, Algiers; Tamadenet; Askarene (Algeria).

COMPAGNIE DIAMANTIFÈRE DU DAR-CHALLA, Onadda (Central African Republic); 6 Avenue de Messine, Paris 8e (France).

COMPAGNIE DU CHEMIN DE FER DU NORD, 26 Rue Laffitte, Paris 9e (France).

COMPAGNIE DU CONGO POUR LE COMMERCE ET L'INDUSTRIE (CCCI), 13 Rue Brédérode, Brussels (Belgium).

COMPAGNIE DU KATANGA, 13 Rue Brédérode, Brussels (Belgium).

COMPAGNIE ÉQUATORIALE DE MINES, 4 Rue de Pentièvre, Paris 8e (France); El Akhouat (Tunisia); Teboursouk (Tunisia).

COMPAGNIE FINANCIÈRE AFRICAINE (CFA), 112 Rue du Commerce, Brussels (Belgium).

COMPAGNIE FINANCIÈRE BELGE DES PÉTROLES (PETROFINA), 31–33 Rue de la Loi, Brussels (Belgium); Luanda (Angola).

COMPAGNIE FINANCIÈRE DE RECHERCHES PÉTROLIÈRES (COFIREP), 20 Rue Washington, Paris 8e (France).

COMPAGNIE FINANCIÈRE DE SUEZ, Paris (France).

COMPAGNIE FINANCIÈRE ET INDUSTRIELLE (COFININDUS), 71 Rue Royale, Brussels (Belgium).

COMPAGNIE FINANCIÈRE POUR LE DEVELOPPEMENT ÉCONOMIQUE DE L'ALGÉRIE, Algiers (Algeria).

COMPAGNIE FINANCIÈRE POUR L'OUTREMER (COFIMER), 13 Rue Paul-Valéry, Paris 2e (France).

COMPAGNIE FRANÇAISE DE DISTRIBUTION DES PÉTROLES EN AFRIQUE, 26 Rue de la Pépinière, Paris 8e, 5 Rue Michel-Ange, Paris 16e (France).

COMPAGNIE FRANÇAISE DE PROSPECTION SISMIQUE (CFPS), 15bis Rue Ballu, Paris 9e (France).

COMPAGNIE FRANÇAISE DE RAFFINAGE, 11 Rue du Docteur Lancereaux, Paris 8e; 22 Rue Boileau, Paris 16e (France).

COMPAGNIE FRANÇAISE DES PÉTROLES (CFP), 5 Rue Michel-Ange, Paris 16e (France).

COMPAGNIE FRANÇAISE DES PÉTROLES (ALGÉRIE) CFP(A), 5 Rue Michel-Ange, Paris 16e (France); Hassi Messaoud (Algeria).

COMPAGNIE FRANÇAISE DES PÉTROLES (GESTION ET RECHERCHES), 5 Rue Michel-Ange, Paris 16e (France); Kent House, Market Place, London W. 1 (Great Britain).

COMPAGNIE FRANÇAISE DU MÉTHANE (CEFEM), Algiers (Algeria).

COMPAGNIE FRANÇAISE POUR LE FINANCEMENT DE LA RECHERCHE ET DE L'EXPLOITATION DU PÉTROLE (REPFRANCE), 75 Avenue des Champs Elysées, Paris 8e (France).

COMPAGNIE FRANCO-AFRICAINE DE RECHERCHES PÉTROLIÈRES (FRANCAREP), 7 Rue Nelaton, Paris 15e; 36 Avenue Raymond-Poincaré, Paris 16e (France); El Gassi; Ohanet; Groupe Vauban, Bloc IV, Rue Louis Barthou, Houssein–Dey, Algiers (Algeria).

COMPAGNIE FRANCO-AMÉRICAINE DES MÉTAUX ET DES MINÉRALS (COFRAMET), 22 Rue du Général Roy, Paris 8e (France).

COMPAGNIE GAZIÈRE D'AFRIQUE, Dakar (Senegal).

COMPAGNIE GÉNÉRALE DE FORAGE (COFOR), 39bis Rue de Chateaudun, Paris 9e (France); 116 Boulevard du Telemly, Algiers (Algeria).

COMPAGNIE GÉNÉRALE DE GÉOPHYSIQUE (CGG), 50 Rue Fabert, Paris 7e (France); 'Dans les Arbres', Chemin Shakespeare, La Redoute, Algiers (Algeria); 6 Rue des Fabres, Marseille; 96 Avenue Verdier, Montrouge, Seine (France); 30 Avenida del Generali Stimo 9, Madrid 16 (Spain); 7 Sciara Enasser, B.P. 97 Tripoli (Libya); 29 Rue Blanchot, Dakar (Senegal); B.P. 6031 Dakar–Etoile; B.P. 435 Yaounde (Cameroons); P.O. Box 12–49 Lagos (Nigeria); 31 Boulevard du Général Moinier, Rabat (Morocco).

COMPAGNIE GÉNÉRALE DE MADAGASCAR, 13 Rue Clémenceau, B.P. 432 Tananarive (Malagasy Republic).

COMPAGNIE GÉOLOGIQUE ET MINIÈRE DES INGÉNIEURS ET INDUSTRIELS BELGES, 4 Rue de la Science, Brussels (Belgium); Manono (Congo).

COMPAGNIE INDUSTRIELLE DES PÉTROLES DE L'AFRIQUE OCCIDENTALE (CIPAO), Dakar (Senegal).

COMPAGNIE INDUSTRIELLE DES PÉTROLES DU MAROC (CIMAR), Rabat (Morocco).

COMPAGNIE INTERNATIONALE DE L'ÉTAIN, 112 Rue du Commerce, Brussels (Belgium).

COMPAGNIE INTERNATIONALE POUR LA PRODUCTION D'ALUMINE FRIA, 9e Avenue, Conakry (Guinea); 40 Avenue Hoche, Paris 8e (France).

COMPAGNIE LYONNAISE DE MADAGASCAR, B.P. 188 Tananarive; Ambatomitamba, Tamatave province (Malagasy Republic).

COMPAGNIE MÉTALLURGIQUE ET MINIÈRE, 8 Rue Bellini, Paris 16e (France).

COMPAGNIE MINIÈRE D'AGADIR, 57 Avenue Hassan II, B.P. 71 Casablanca; Idikel (Morocco).

COMPAGNIE MINIÈRE DE CONAKRY, B.P. 610 Conakry (Guinea); 77 Rue la Boëtie, Paris 8e (France).

COMPAGNIE MINIÈRE CONGOLAISE DES GRANDS LACS AFRICAINS (MGL-CONGO), Kamituga (Congo).

COMPAGNIE MINIÈRE DE L'OGOUÉ (COMILOG), B.P. 759 Pointe-Noir (Congo Republic); 62bis Avenue d'Iéna, Paris 16e (France); Moanda (Gabon).

COMPAGNIE MINIÈRE DE L'OUBANGUI ORIENTAL (CMOO), 4 Rue de Penthièvre, Paris 8e (France); Berberati (Central African Republic).

COMPAGNIE MINIÈRE DE L'UREGA (MINERGA), 24 Avenue de l'Astronomie, Brussels (Belgium); c/o COBELMIN, Bukavu (Congo); Lulingu (Congo).

COMPAGNIE MINIÈRE DES GRANDS LACS AFRICAINS (MGL), 24 Avenue de l'Astronomie, Brussels (Belgium); Bukavu; Kamituga (Congo).

COMPAGNIE MINIÈRE DU CAMÉROUN, Betare Oya (Cameroons).

COMPAGNIE MINIÈRE DU CONGO FRANÇAIS (CMC), Brazzaville (Congo Republic).

COMPAGNIE MINIÈRE DU DJEBEL GUSTAR, 154–156 Rue de l'Université, Paris 7e (France); Syndicat Minier Africain (Symaf), 112 Rue du Commerce, Brussels (Belgium); Djebel Gustar par Behagle, Département de Sétif (Algeria).

COMPAGNIE MINIÈRE DU DJEBEL MANSOUR, 198 Rue de l'Aviation française, Casablanca; Tiouit (Morocco).

COMPAGNIE MINIÈRE DU LUALABA, see: Société Minière du Lualaba.

COMPAGNIE MINIÈRE D'URANIUM DE FRANCEVILLE, Franceville (Gabon); 69 Rue de Varenne, Paris 7e (France).

COMPAGNIE MINIÈRE DU RUANDA–URUNDI (MIRUDI), 24 Avenue de l'Astronomie, Brussels (Belgium).

COMPAGNIE MINIÈRES DU SOUSS, 21 Rue Descartes, Meknès; Sidi M'Bark (Morocco).

COMPAGNIE MINIÈRE ET INDUSTRIELLE DU MAROC, 22 Rue du Languedoc, Rabat; Jebel Ighoud (Morocco).

COMPAGNIE MINIÈRE ET MÉTALLURGIQUE, 3 Boulevard Mohammed V, Casablanca (Morocco); 8 Rue Bellini, Paris 8e (France); B.P. 135 Marrakech (Morocco).

COMPAGNIE MINIÈRE ET PHOSPHATIÈRE (COMIPHOS), 12 Avenue Marceau, Paris 8e (France); Le Kouif (Algeria).

COMPAGNIE MINIÈRE TUNISIENNE, 91 Avenue de Carthage, Tunis (Tunisia).

COMPAGNIE MINIÈRE ZAMBÉZIENNE, 13 Rue Brédérode, Brussels (Belgium).

COMPAGNIE NAVALE DES PÉTROLES (CNP), 162 Rue Faubourg Saint Honoré, Paris (France).

COMPAGNIE NORD-AFRICAINE DE L'HYPERPHOSPHATE RENO, Sfax (Tunisia).

COMPAGNIE NOUVELLE DE FORAGES PÉTROLIERS (FORENCO), 162 Rue Faubourg Saint Honoré, Paris 8e (France).

COMPAGNIE ORIENTALE DES PÉTROLES D'ÉGYPTE (COPE), Caïro (U.A.R.).

COMPAGNIE POUR L'APPLICATION DES TECHNIQUES GÉOPHYSIQUES (CATG), 65 Avenue des Champs Elysées, Paris 8e (France).

COMPAGNIE REYNOLDS DE GÉOPHYSIQUE (CRG), 18–20 Place de la Madeleine, Paris 8e (France); Rue Auguste Cahours, Côte Rouge, Houssein Dey, Algiers (Algeria).

COMPAGNIE ROYALE ASTURIENNE DES MINES, Touissit; Jebel Aouam (Morocco); 15 Rue Danvilliers, Casablanca; B.P. 406 Oujda (Morocco); 8 Plaza de España, Madrid 13 (Spain); 42 Avenue Gabriel, Paris 8e (France); 12 Place de la Liberté, Brussels (Belgium).

COMPAGNIE RWANDAISE D'EXPLOITATION MINIÈRE (COREM), Kigali (Rwanda).

COMPAGNIE SALINIÈRE DU MAROC (CSM), Rue du Lieutenant Ferand, Casablanca (Morocco).

COMPAGNIE SÉNÉGALAISE DES PHOSPHATES DE TAÏBA, B.P. 1713 Dakar (Senegal); 88 Avenue Kleber, Paris 16e (France).

COMPAGNIE SHELL DE RECHERCHES ET D'EXPLOITATION AU GABON (COSREG), Port Gentil (Gabon).

COMPAGNIE TOGOLAISE DES MINES DE BENIN, B.P. 362 Lomé (Togo); 12 Avenue Marceau, Paris 8e (France).

COMPAGNIE TUNISIENNE DES PHOSPHATES DU DJEBEL M'DILLA, 26 Rue d'Angleterre, Tunis (Tunisia).

COMPANHIA CARBONIFERA DE MOÇAMBIQUE, Moatíza (Mozambique).

COMPANHIA DE CIMENTOS DE MOÇAMBIQUE, Beira; Lourenço Marques (Mozambique).

COMPANHIA DE CIMENTOS SECIL DO ULTRAMAR, Avenue Paulo Dias de Novaís 82-1°, Luanda (Angola).

COMPANHIA DE COMBUSTIVEÍS DO LOBITO PURFINA, Lobito (Angola).

COMPANHIA DE DIAMANTES DE ANGOLA (DIAMANG), 12, 2nd Floor Rua dos Fanqueiros, Lisbon (Portugal); Dundu; Cassanguidi; Andrada; Maludi; Camissombo (Angola); 42 Rue Royale, Brussels (Belgium).

COMPANHIA DE PESQUISAS MINEIRAS DE ANGOLA, Pema, Buco Zan; Maiombe; Cabinda (Angola).

COMPANHIA DE PETROLEOS DE ANGOLA (PETRANGOL), Luanda; Tobias (Angola).

COMPANHIA DE PIPELINE MOÇAMBIQUE, Beira (Mozambique).

COMPANHIA DO MANGANÉS DE ANGOLA, Avenue Antônio Augusto de Aguiar 25, 20 Esq., Lisbon (Portugal); Avenue Alvaro Ferreira 22, Luanda; Quitota (Angola).

COMPANHIA DOS ASFALTOS DE ANGOLA, Libongos (Angola).

COMPANHIA DOS BETOMINESOS DE ANGOLA, Husso Norte, Luanda (Angola).

COMPANHIA DO ZAMBEZIA, Missao, Tete; Angonia (Mozambique).

COMPANHIA DOS CIMENTOS DE ANGOLA, Caixa Postal 157 Lobito (Angola).

COMPANHIA ELÉCTRICA DO LOBITO ET BENGUELA (S.A.R.L.), (CELB), P.O. Box 98 Lobito (Angola).

COMPANHIA MINEIRA DE LOBITO SARL, Rua Diogo Cão 26, P.O. Box 228 Nova Lisboa (Angola); Avenue Sidónio Pais 2–40, Lisbon (Portugal).

COMPANHIA MINEIRA DE LOMBIGE, Cassinga; Luanda (Angola).

COMPANHIA MINEIRA DE MOMBASSA, Luanda; M'Bassa (Angola); Avenue Antonio A. Aguiar 25–20, Lisbon (Portugal).

COMPANHIA MINEIRA DO LOBITO, Lobito; Cassalengues; Cuima (Angola).

COMPANHIA MINEIRA MONARCH LTDA., Vila Manica (Mozambique).

COMPAÑIA ESPAÑOLA DE MINAS DE RIF, Ouichan (Uixán) (Morocco).

COMPAÑIA DE MINERALES, Avenida José Antonio, 88 Edificio España, Madrid (Spain).

CONCESSION CORUNDUM MINE, P.O. Box 8 Concession (Southern Rhodesia).

CONCH INTERNATIONAL METHANE LTD., Villiers House, Strand, London W.C. 2 (Great Britain).

CONGOVIELMONT, see: Société Congolaise de la Vieille Montagne.

CONIEL, see: Compagnie Imprese Elettriche dell'Eritrea.

CONNEMARA MINE, Hunter's Road (Southern Rhodesia).

CONORADA PETROLEUM CORP., 630 Fifth Avenue, New York 20, N.Y. (U.S.A.); Tunis (Tunisia).

CONSAFRIQUE, see: Consortium Européen pour le Développement des Resources Naturelles de l'Afrique.

CONSOLIDATED AFRICAN MINES LTD., S.A. Centre, 253 Bree Street, P.O. Boxes 4307, 8543 Johannesburg (South Africa).

CONSOLIDATED AFRICAN SELECTION TRUST LTD. (CAST), Selections Trust Buildings, Mason's Avenue, London E.C. 2 (Great Britain).

CONSOLIDATED BLUE ASBESTOS CORP., P.O. Box 9 Limeacres, Kuruman (South Africa).

CONSOLIDATED CHROME CORP., Annan House, 86 Commissioner Street, P.O. Box 10905 Johannesburg (South Africa).

CONSOLIDATED COLLIERIES LTD., P.O. Coalville (South Africa).

CONSOLIDATED COMPANY BULTFONTEIN MINE LTD., 36 Stockdale Street, Kimberley, C.P. (South Africa).

CONSOLIDATED DIAMOND MINES OF SOUTHWEST AFRICA LTD., 36 Stockdale Street, P.O. Box 616 Kimberley, C.P. (South Africa); Oranjemund (Southwest Africa).

CONSOLIDATED FLUORSPAR MINES (PTY.) LTD., Volkskas Building, 76 Market Street, Johannesburg; 11 Jean Street, Zeerust, W. Tvl. (South Africa).

CONSOLIDATED FREE STATE GOLD FIELDS LTD., 201 Swiss House, 86 Main Street, Johannesburg (South Africa).

CONSOLIDATED GLASS WORKS LTD., P.O. Box 7727, Johannesburg; P. Bag 544 Pretoria (South Africa).

CONSOLIDATED GOLD FIELDS OF SOUTH AFRICA LTD., 49 Moorgate, London E.C. 2 (Great Britain); P.O. Box 2552 Salisbury; P.O. Box 27P Que Que (Southern Rhodesia).

CONSOLIDATED MAIN REEF MINES AND ESTATE LTD., 77 Commissioner Street, P.O. Box 1056 Johannesburg; P.O. Box 2 Maraisburg, Tvl. (South Africa).

CONSOLIDATED MINES SELECTION CO. LTD., 40 Holborn Viaduct, London E.C. 1 (Great Britain).

CONSOLIDATED MURCHISON (TRANSVAAL), Anglovaal House, 56 Main Street, Johannesburg; P. Bag Gravelotte, E. Tvl. (South Africa).

CONSOLIDATED PETROLEUM CO., Shell Centre, London S.E. 1 (Great Britain).

CONSOLIDATED REFRACTORIES (PTY.) LTD., P.O. Box 4711 Cape Town (South Africa).

CONSOLIDATED SECRETARIAL, FINANCIAL AND TRUST CO., St. Barbara House, Moffat Street, P.O. Box 2311 Salisbury (Southern Rhodesia).

CONSOLIDATED TIN MINES, P. O. Box 93 Omaruru (Southwest Africa).

CONSORTIUM EUROPÉEN POUR LE DEVELOPPEMENT DES RESSOURCES NATURELLES DE L'AFRIQUE (CONSAFRIQUE), Paris (France).

CONSTANTINE PHOSPHATE CO., see: Compagnie des Phosphates de Constantine.

CONSTANTINOU, O. CH., P.O. Box 145 Mbarara (Uganda).

CONTINENTALE ERZ GESELLSCHAFT, 29 Berliner Allee, Düsseldorf (W. Germany).

CONTINENTAL GEOPHYSICAL CO., 602 Continental Life Building, Fort Worth, Texas (U.S.A.).

CONTINENTAL OIL COMPANY (DELAWARE), 1300 Main Street, Houston 2, Texas; 30 Rockefeller Plaza, New York 20, N.Y.; 1000 South Pine Street, Ponca City, Okla. (U.S.A.).

CONTINENTAL OIL CO. OF LIBYA, 1300 Main Street, Houston 2, Texas (U.S.A.); Es Sider (Libya).

CONTINENTAL ORE (AFRICA) (PVT.) LTD., P.O. Box 3411 Salisbury (Southern Rhodesia).

CONTINENTAL ORE CORP., 500 Fifth Avenue, New York 36, N.Y. (U.S.A.); 1 Hay Hill, Berkeley Square, London W. 1 (Great Britain); 215 St. James' Street West, Montreal (Canada).

COOP, see: Société Coopérative de Pétrole.

COOPÉRATIVE D'ACHAT ET DE DÉVELOPPEMENT DE LA RÉGION MINIÈRE DE TAFILALET, Tafilalet (Morocco).

COOPÉRATIVE DU DJEBEL HAMRA, Foussana (Tunisia).

COPAREX, see: Compagnie de Participations de Recherches et d'Exploration Pétrolières.

COPE, see: Compagnie Orientale des Pétroles d'Égypte.

COPEFA, see: Compagnie des Pétroles France–Afrique.

COPETAO, see: Compagnie des Pétroles d'Afrique Occidentale.

COREM, see: Compagnie Rwandaise d'Exploitation Minière.

CORI, see: Compagnie Ricerche Idrocarburi.

CORK ASBESTOS MINES (PTY.) LTD., 18A Vorster Street, P. Bag 1304 Pietersburg, N. Tvl. (South Africa).

CORNER HOUSE INVESTMENT CO. LTD. ,The Corner House, 77 Commissioner Street, P.O. Box 1056 Johannesburg (South Africa).

CORONATION COLLIERIES LTD., 44 Main Street, P.O. Box 2567 Johannesburg; P.O. Box 2 Witbank, Tvl. (South Africa).

CORONATION SYNDICATE LTD., 9th Floor, Hollard Place, 71 Fox Street, Johannesburg (South Africa); 1 Cornhill, London E.C. 3 (Great Britain).

CORY, WM. AND SON LTD., Fenchurch Street, London E.C. 3 (Great Britain); Tripoli (Libya).
COSREG, see: Compagnie Shell de Recherches et d'Exploitation au Gabon.
COSTA, JOSE AMIEL DA, Vila Manica (Mozambique).
CPA, see: Compagnie des Pétroles d'Algérie.
CPTL, see: Compagnie des Pétroles Total (Libya).
CRAIG STANTON AND CO., 271 Route de Médiouna, Casablanca; Tisguililane (Morocco).
CRÉDIT FONCIER AFRICAIN, 112 Rue du Commerce, Brussels (Belgium).
CREPS, see: Compagnie de Recherches et d'Exploitation de Pétrole au Sahara.
CRG, see: Compagnie Reynolds de Géophysique.
CROESUS GOLD MINING CO., 801 Loveday House, Marshall Street, Johannesburg (South Africa).
CROWN MINES LTD., The Corner House, 77 Commissioner Street, P.O. Box 1056 Johannesburg; P.O. Box 102 Crown Mines (South Africa).
CRYSTAL SALT CO., Property Centre, 12 New Street, P.O. Box 11315 Johannesburg; P.O. Box 17 Waterpoort (South Africa).
CSK, see: Comité Spécial du Katanga.
CSM, see: Compagnie Salinière du Maroc.
CULLINAN REFRACTORIES LTD., Olifantsfontein (South Africa).
DAGGAFONTEIN MINES LTD., 44 Main Street, Johannesburg; P.O. Box 1 Daggafontein (South Africa); 40 Holborn Viaduct, London E.C. 1 (Great Britain).
DAHO-AMERICAN OIL CO., Cotonou (Dahomey).
DALTON ASBESTOS MINING CO., Zambesi House, 44 Von Wielligh Street, P.O. Box 7788 Johannesburg (South Africa).
DAWN GOLD MINING CO., Queens (Southern Rhodesia).
DEA, see: Deutsche Erdöl AG.
DE BEERS CONSOLIDATED MINES LTD., 36 Stockdale Street, Kimberley, Griqualand West (South Africa).
DE BEERS HOLDINGS LTD., 44 Main Street, Johannesburg (South Africa).
DE BEERS INVESTMENT TRUST LTD., P.O. Box 616 Kimberley (South Africa).
DE BEERS PROSPECTING (RHODESIAN AREAS) LTD., 70 Jameson Avenue Central, P.O. Box 1108 Salisbury (Southern Rhodesia).
DELAREYVILLE SALT CO., P.O. Box 1377 Johannesburg; P.O. Delareyville (South Africa).
DELIMCO, see: German–Liberian Mining Co.
DEMIERRE AND COMPAGNIE S.A., 5 Corraterie, Geneva (Switzerland).
DENAIN-ANZIN, S.A., 12 Rue d'Athènes, Paris 9e (France).
DERBY AND CO., 42 Marshall Street, P.O. Box 4829 Johannesburg (South Africa).
DE RYCK, O., 96 Avenue Lyautey, Meknès; Adrar (Morocco).
DEUTSCHE BANK, Königsallee 43–47, Düsseldorf (W. Germany).
DEUTSCHE ERDÖL A.G. (DEA), Mittelweg 180, Hamburg 13 (W. Germany); Tripoli (Libya).
DEUTSCHE SCHACHTBAU UND TIEFBOHRGESELLSCHAFT, Waldstrasse 39, Lingen/Ems (W. Germany).
DEVELOPMENT AND RESOURCES CORP., 50 Broadway, New York, N.Y. (U.S.A.).
DIAMANG, see: Companhia de Diamantes de Angola.
DIAMIMPEX DIAMOND CO., Monrovia (Liberia).
DIAMOND CORP. LTD. (DE BEERS), Consolidated Building, Stockdale Street, Kimberley (South Africa); 2 Charterhouse Street, London E.C. 1 (Great Britain); Freetown (Sierra Leone).
DIAMOND CORP. OF SIERRA LEONE, Freetown (Sierra Leone).
DIAMOND DISTRIBUTORS INC., Monrovia (Liberia).
DIAMOND IMPORT EXPORT CO., Monrovia (Liberia).
DIAMOND MINING AND UTILITY CO. (SWA) LTD., Indosa Buildings, Lüderitz (Southwest Africa); 415–416 Bristol Building, corner MacLaren and Marshall Streets, Johannesburg (South Africa).
DIAMONDS PRODUCER'S ASSOCIATION, P.O. Box 616 Kimberley (South Africa).
DIAMOND TRADING AND PURCHASING CO., P.O. Box 616 Kimberley; P.O. Box 7727 Johannesburg (South Africa).
DIAMSTONE DIAMOND CO., Monrovia (Liberia).
DIMATIT, see: Société Nord-Africaine de l'Amiante-Ciment 'Dimatit'.
DIVIDE CHROME MINES LTD., St. Barbara House, Sixth Floor, Moffat Street, P.O. Box 2311 Salisbury (Southern Rhodesia).
DODWELL AND CO., 18 Finsbury Circus, London E.C. 2 (Great Britain); 120 Wall Street, New York 5, N.Y. (U.S.A.); P.O. Box Central 297 Tokyo (Japan); P.O. Box 3110 Salisbury (Southern Rhodesia).
DOISHAL GOLD MINING CO., P.O. Box 843 Khartoum (Sudan).
DOMINION BASE METALS (PVT.) LTD., P. Bag 5 Filabusi (Southern Rhodesia).

DOMINION REEFS (KLERKSDORP) LTD., Gold Fields Building, 75 Fox Street, P.O. Box 1167 Johannesburg; P.O. Box 107 Klerksdorp, W. Tvl. (South Africa).
DOORNFONTEIN GOLD MINING CO. LTD., Gold Fields Building, 75 Fox Street, P.O. Box 1167 Johannesburg; P. Bag Welverdiend (South Africa).
DOS SANTOS, CARLOS MARQUES, Cabinda (Angola).
DOS SANTOS, JOSÉ, Cabinda (Angola).
DOUGLAS COLLIERY LTD., P.O. Box 550 Johannesburg; P.O. Van Dyksdrif (South Africa).
DOWSON AND DOBSON LTD., P.O. Box 7764 Johannesburg (South Africa).
DRIEFONTEIN GOLD MINING CO. LTD., 511-6 City Trust House, 106 Fox Street, P.O. Box 742 Johannesburg; P.O. Box 5 Knights (South Africa).
DRILLING SPECIALTIES, 129 South State Street, Dover, Delaware (U.S.A.); Algiers (Algeria).
DUBLIN CONSOLIDATED ASBESTOS MINES (PTY.) LTD., P.O. Box 1100 Bulawayo (Southern Rhodesia).
DUBOIS, A., Taouirirt par Oujda; Jebel Tirremi (Morocco).
DUNDEE ANTHRACITE LTD., New Marlborough House, corner Eloff and Commissioner Streets, P.O. Box 8543 Johannesburg (South Africa).
DUNDEE COAL CO., 89 National Mutual Buildings, 275 Smith Street, P.O. Box 372 Durban (South Africa).
DUPONSEL, P. ET R., Avenue de l'Indépendance, Tananarive (Malagasy Republic).
DURABEES SPODUMENE MINE, P.O. Box 19 Nababeep (South Africa).
DURA LIMESTONE CONCESSION, P.O. Mbarara (Uganda).
DURAN, R., 129 Avenue Mers Sultan, Casablanca; Aït Labbès (Morocco).
DURBAN NAVIGATION COLLIERIES LTD., SASOL, P.O. Box 450 Pretoria (South Africa).
DURBAN ROODEPOORT DEEP LTD., The Corner House, P.O. Box 1056 Johannesburg; P.O. Box 193 Roodepoort, Tvl. (South Africa).
DUSSOL, GILBERT, Rue Colbert, Tananarive (Malagasy Republic).
DWARSBERG LOODMYN BEPERK., P.O. Box 1415 Pretoria (South Africa).
EAST AFRICAN CONCESSIONS LTD., Geita (Tanganyika).
EAST AFRICAN MARKETING CO., P.O. Box 10939 Nairobi (Kenya).
EAST AFRICAN MINING AND DEVELOPMENT CO., P.O. Box 410 Dar es Salaam (Tanganyika).
EAST AFRICAN OIL REFINERIES LTD., P.O. Box 5342 Nairobi (Kenya); Mombasa (Kenya).
EAST AFRICAN PORTLAND CEMENT CO., Nairobi, (Kenya).
EAST AFRICAN POWER AND LIGHTING CO., P.O. Box 30099 Nairobi (Kenya).
EAST AFRICAN SALT MINES, P.O. Box 413 Dar es Salaam (Tanganyika).
EAST CHAMP D'OR GOLD MINING CO. LTD., Consolidated Building, corner Fox and Harrison Streets, P.O. Box 590 Johannesburg; P.O. Box 11 Luipaards Vlei, Tvl. (South Africa); 10–11 Austin Friars, London E.C. 2 (Great Britain).
EAST COAST MINING AND DEVELOPMENT CO., Cape Town (South Africa).
EAST DAGGAFONTEIN MINES LTD., 44 Main Street, P.O. Box 4587 Johannesburg; P.O. Box 6 Daggafontein (South Africa).
EASTERN NIGERIA DEVELOPMENT CORP., Enugu (Nigeria).
EAST GEDULD MINES LTD., Union Corporation Building, 74–78 Marshall Street, P.O. Box 1125 Johannesburg; P.O. Box 222 Springs, Tvl. (South Africa).
EAST RAND CONSOLIDATED LTD., 120 Moorgate, London E.C. 2 (Great Britain).
EAST RAND PROPRIETARY MINES LTD., Corner House, Johannesburg; P.O. Box 57 East Rand (South Africa).
EASTERN PROVINCE CEMENT CO., P.O. Box 2016 Fort Elizabeth (South Africa).
EASTERN RAND EXTENSIONS LTD., General Mining Building, 6 Hollard Street, Johannesburg (South Africa).
EASTERN TRANSVAAL CONSOLIDATED MINES LTD., Anglovaal House, 56 Main Street, P.O. Box 7727 Johannesburg; P.O. Box 182 Agnes, Barberton; Mamre, P.O. Slaaihoek; New Consort P.O. Noord Kaap; P.O. Sheba (South Africa).
EAUX, GAZ, ÉLECTRICITÉ DE TUNIS, 38 Rue de Besançon, Tunis (Tunisia).
ECHO SALT WORKS, P.O. Florisbad (South Africa).
EDINBURGH DEVELOPMENT (PVT.) LTD., P.O. Box 3615 Salisbury (Southern Rhodesia).
EGNEP (PTY.) LTD., Skefko House, 1 Rissik Street, P.O. Boxes 8644 and 2533 Johannesburg; P.O. Penge (South Africa).
EGYPTIAN BLACK SANDS CO., El Houria Road, Alexandria (U.A.R.).
EGYPTIAN GENERAL MINING ORGANIZATION, 5 Goheini Street, Dokki, Cairo (U.A.R.).
EGYPTIAN GENERAL PETROLEUM ORGANIZATION, 44 El Mesaha Street, Dokki, Cairo (U.A.R.).
EGYPTIAN INDEPENDENT OIL CO. S.A.E. (EIOC), 15 Chérif Pasha Street, Cairo; 9 Chérif Pasha Street, Alexandria (U.A.R.).
EGYPTIAN IRON AND STEEL CO., Aswan (U.A.R.).
EGYPTIAN PHOSPHATE PRODUCTION AND TRADING CO., 23 Tarat Harb Street, Cairo (U.A.R.).

EGYPTIAN QUARRIES AND MARBLE CO., 62 Mostafa El Maraghi Street, Helwan (U.A.R.).
EIOC, see: Egyptian Independent Oil Co.
ELANDSDRIFT CHROME MINE, P.O. Box 82 Marikana (South Africa).
ÉLECTRICITÉ DE FRANCE, 23 Rue de Vienne, Paris 8e (France).
ÉLECTRICITÉ ET GAZ D'ALGÉRIE, 2 Boulevard du Télémly, Algiers (Algeria).
ELECTRICITY CORP. OF NIGERIA, Lagos (Nigeria).
ELECTRICITY DEPARTMENT, New England, Freetown (Sierra Leone).
ELECTRICITY DIVISION, Ministry of Construction and Communication, P.O. Box 521 Accra (Ghana).
ELECTRICITY SUPPLY COMMISSION, 204 Smit Street, P.O. Box 1091 Johannesburg (South Africa).
ELEKTROSOND, Belgrade (Yugoslavia); Tunis; Kef; Kelaa Edjenam (Tunisia).
ELLATON GOLD MINING CO. LTD., General Mining Building, 6 Hollard Street, Johannesburg; P. Bag P.O. Klerksdorp, W. Tvl. (South Africa).
EL NASR PHOSPHATE CO., 1, 26th July Street, Cairo (U.A.R.).
EL NASR SALINES CO., El Houria Road, Alexandria (U.A.R.).
EL PASO FRANCE, Algiers (Algeria).
EL PASO NATURAL GAS CO., 1 Chase Manhattan Plaza, New York, N.Y. (U.S.A.).
ELWERATH CO., see: Gewerkschaft Elwerath.
EMPRESA DO COBRE DE ANGOLA, Uige, Bembe (Angola).
EMPRESA MINEIRA DO ALTO LIGONHA, Predio Ecsal, Caixa Postal 1152 Lourenço Marques (Mozambique).
ÉNERGIE DU MALI, B.P. 69 Bamako (Mali).
ÉNERGIE ÉLECTRIQUE D'AFRIQUE ÉQUATORIALE FRANÇAISE, B.P. 295 Brazzaville (Congo Republic).
ÉNERGIE ÉLECTRIQUE DU MAROC, Rabat (Morocco).
ENGELHARD HANOVIA, see: Engelhard Industries.
ENGELHARD INDUSTRIES INC., 113 Astor Street, Newark 2, N.J. (U.S.A.).
ENGELHARD INDUSTRIES LTD., Hanovia Ceramic Division, 154 Vauxhall Street, London E.C. 2 (Great Britain).
ENI, see: Ente Nazionale Idrocarburi.
ENTE NAZIONALE IDROCARBURI (ENI), 50 Via Tevere, Rome (Italy).
ENTORES LTD., 14–24 Finsbury Street, London E.C. 2 (Great Britain).
ENTREPOSTO COMERCIAL DE MOÇAMBIQUE, Beira (Mozambique).
ENYATI COLLIERY LTD., P.O. Box 1381, 24–26 Beach Grove, Durban (South Africa).
ERFRUST ASBESTOS MINE, Erfrust, Prieska (South Africa).
ESSEXVALE LIME WORKS (PVT.) LTD., P.O. Box 1497 Salisbury (Southern Rhodesia).
ESSO INTERNATIONAL INC., 15 West 51st Street, New York 19, N.Y. (U.S.A.).
ESSO LIBYA INC., P.O. Box 385 Tripoli (Libya); Marsa el Brega (Libya); 30 Rockefeller Plaza, New York 20, N.Y. (U.S.A.).
ESSO MEDITERRANEAN INC., 81 Rue de l'Aire, Geneva (Switzerland).
ESSO OVERSEAS TRADING LTD., 50 Stratton Street, London W. 1 (Great Britain).
ESSO PETROLEUM CO., 36 Queen Anne's Gate, Westminster, London S.W. 1 (Great Britain).
ESSO RESEARCH LTD., Abingdon, Berkshire (Great Britain).
ESSO SAHARA INC., 30 Rockefeller Plaza, New York 20, N.Y. (U.S.A.); 41 Avenue Georges V, Paris 7e (France).
ESSO SIRTE INC., 30 Rockefeller Center, New York 20, N.Y. (U.S.A.); P.O. Box 565 Tripoli; Marsa El Brega; Zelten; Raguba, (Libya).
ESSO STANDARD ALGÉRIE S.A., 11 Boulevard Victor Hugo, Algiers (Algeria).
ESSO STANDARD (EAST AFRICA) LTD., Esso House, Queensway P.O. Box 30200 Nairobi (Kenya); Kampala (Uganda); Dar es Salaam (Tanganyika); Zanzibar (Zanzibar); Saint Denis (Réunion); Port Louis (Mauritius); Victoria (Seychelles Islands); B.P. 24 Tananarive (Malagasy Republic).
ESSO STANDARD EASTERN INC., 15 West 51st Street, New York 19, N.Y. (U.S.A.).
ESSO STANDARD LIBYA INC., P.O. Box 385 Tripoli; Marsa el Brega (Libya).
ESSO STANDARD (MADAGASCAR) S.A., 15 Rue Clémenceau, P.O. Box 54 Tananarive (Malagasy Republic); Saint Denis (Réunion); Moroni (Comoro Islands).
ESSO STANDARD (NEAR EAST) INC., 81 Route de l'Aire, Geneva (Switzerland), Cairo Airport (U.A.R.); Tripoli (Libya); 30 Rockefeller Plaza, New York 20, N.Y. (U.S.A.).
ESSO STANDARD S.A. FRANÇAISE, 82 Avenue des Champs Elysées, Paris 8e (France).
ESSO STANDARD (SOUTH AFRICA) (PTY.) LTD., 30 Simmonds Street, Johannesburg (South Africa).
ESSO STANDARD TUNISIE S.A., 12 Avenue de Paris, Tunis (Tunisia).
ESSO WEST AFRICA INC., Investment House, 21/25 Broad Street, P.O. Box 176 Lagos (Nigeria); 81 Route de l'Aire, Geneva (Switzerland); 30 Rockefeller Plaza, New York 20, N.Y. (U.S.A.).

ÉTABLISSEMENTS CH. DE TAILLAC ET CIE., Route de l'Abattoir, Isotry, Tananarive (Malagasy Republic).
ÉTABLISSEMENTS GALLOIS, B.P. 159 Tananarive (Malagasy Republic).
ETHEL ASBESTOS MINES LTD., P. Bag 3 Mtoroshanga; P.O. Box 2112 Salisbury (Southern Rhodesia).
ETHIOPIAN CEMENT CORP., see: Imperial Cement Corp.
ETHIOPIAN ELECTRIC LIGHT AND POWER AUTHORITY, P.O. Box 1233 Addis Ababa (Ethiopia).
EURAFREP, see: Société de Recherches et d'Exploitation de Pétrole.
EURAFRIC DIAMOND CO., Monrovia (Liberia).
EX-LANDS NIGERIA LTD., Jos (Nigeria); 120 Moorgate, London E.C. 2 (Great Britain).
EXPLORATION GÉOPHYSIQUE ROGERS, 34 Avenue des Champs Elysées, Paris 8e (France); 88 Boulevard du Télemly, Algiers (Algeria).
EXTRAIMINE, see: Société d'Exploitation et de Traitement des Minerais.
FAIRCHILD AERIAL SURVEYS, 224 East 11th Street, Los Angeles 15, Calif. (U.S.A.).
FALCONBRIDGE AFRICA LTD., Kilembe (Uganda).
FALCONBRIDGE NICKEL MINE LTD., Bank of Nova Scotia Building, 44 King Street West, Toronto (Canada).
FALCON MINES LTD., 120 Moorgate, London E.C. 2 (Great Britain); Dalny Mine, P.O. Chakari (Southern Rhodesia).
FARREN SLATE QUARRIES (PTY.) LTD., 406 Chequer House, Queen Street, P.O. Box 76 Pretoria (South Africa).
FAUNCE TAINTON LTD., Winchester House, 91b Main Street, P.O. Box 3444 Johannesburg (South Africa).
FEDERALE MYNBOUW BEPERK, Sanlam Building, 63 Commissioner Street, P.O. Box 5958 Johannesburg (South Africa).
FEDERAL POWER BOARD, Salisbury (Southern Rhodesia).
FELIXBURG MINES (PVT.) LTD., Felixburg (Southern Rhodesia).
FELLI, FRANÇOIS, B.P. 11 Fort-Dauphin (Malagasy Republic).
FELS MINERALS AND EXPORT CORP., 818–830 Pan Africa House, Troye Street, P.O. Box 3198 Johannesburg (South Africa).
FERREIRA ESTATE CO., The Corner House, 77 Commissioner Street, P.O. Box 1056 Johannesburg (South Africa); 4 London Wall Buildings, London E.C. 2 (Great Britain).
FIAT, Turin (Italy).
FINAREP, see: Société Financière des Pétroles.
FIRST NATIONAL CITY BANK OF NEW YORK, 55 Wall Street, New York 15, N.Y. (U.S.A.); Monrovia (Liberia).
FISONS LTD., Sasolburg; Modderfontein (South Africa).
FLEISCHMAN, BURD AND CO., 1271 Avenue of the Americas, New York 20, N.Y. (U.S.A.).
FLEVET, B., Fianarantsoa (Malagasy Republic).
FLUITJIESKRAAL SALT WORKS, P.O. Orange River (South Africa).
FLUORSPAR AND MINERALS, P.O. Box 21 Germex (South Africa).
FLUORSPAR EXPORT (PTY.) LTD., His Majesty's Building, Joubert Street, P.O. Box 3518 Johannesburg (South Africa).
FONDERIES PEÑARROYA–ZELLIDJA, Oued el Heimer par Oujda (Morocco).
FORAFRANCE, see: Société Française de Forage.
FORAGES ET EXPLOITATIONS PÉTROLIÈRES, 262 Rue Faubourg St. Honoré, Paris 1er (France).
FORBES AND THOMSON (BYO.) (PVT.) LTD., Gwanda (Southern Rhodesia).
FORCES, see: Société des Forces Hydroélectriques du Bas-Congo.
FORENCO, see: Compagnie Nouvelle de Forages Pétroliers.
FORMINIÈRE, see: Société Internationale Forestière et Minière.
FORUM EXTENDED LTD., 66 Gresham Street, London E.C. 2 (Great Britain).
FORUM MINES LTD., P.O. Box 26 Bukuru (Nigeria).
FOSFAAT-ONTGINNINGSKORPORASIE (EIENDOMS) BEPERK (FOSKOR), P.O. Box 1 Wegsteek Phalaborwa (South Africa).
FOSKOR, see: Fosfaat-Ontginningskorporasie (Eds) Beperk.
FOWLIE REID AND WILLS LTD., 25–35 City Road, London E.C. 1 (Great Britain).
FRANCAREP, see: Compagnie Franco-Africaine de Recherches Pétrolières.
FRASER'S QUARRIES (PTY.) LTD., Burt Drive, P.O. Box 7124 Port Elizabeth, (South Africa).
FREDDIES CONSOLIDATED MINES LTD., Consolidated Building, corner Fox and Harrison Streets, P.O. Box 590 Johannesburg (South Africa).
FREE STATE DEVELOPMENT AND INVESTMENT CORP. LTD., Consolidated Building, corner Fox and Harrison Streets, P.O. Box 590 Johannesburg (South Africa); 10–11 Austin Friars, London E.C. 2 (Great Britain).
FREE STATE GEDULD MINES LTD., 44 Main Street, P.O. Box 4586 Johannesburg (South Africa); 40 Holborn Viaduct, London E.C. 1 (Great Britain).
FREE STATE GOLD AREAS LTD., Cape House, 54 Fox Street, Johannesburg (South Africa).
FREE STATE SAAIPLAAS GOLD MINING CO., Gold Fields Building, 75 Fox Street, P.O. Box 1167 Johannesburg (South Africa); 49 Moorgate, London E.C. 2 (Great Britain).

FREE STATES SYNDICATES LTD., Cape House, 54 Fox Street, P.O. Box 4280 Johannesburg (South Africa).
FRIA CO., see: Compagnie Internationale pour la Production d'Alumine Fria.
FROBEX LTD., (SUCCESSOR OF FROBISHER), 85 Richmond Street West, Toronto (Canada).
GAFRODIAM DIAMOND CO., Monrovia (Liberia).
GAIGHER FRAUENDORF MANGAAN, P.O. Box 164 Zeerust (South Africa).
G. AND W. BASE AND INDUSTRIAL MINERALS (PTY.) LTD., Beresford House, 86 Main Street, P.O. Box 6211 Johannesburg (South Africa).
GAUCHÉ AND NORTJE, P.O. Box 58 Vredendal (South Africa).
GAZIELLO CO., Dakar (Senegal).
GEDULD PROPRIETARY MINES LTD., Union Corporation Building, 74–78 Marshall Street, P.O. Box 1125 Johannesburg; P.O. Dersley (South Africa).
GEITA GOLD MINING CO., Geita (Tanganyika).
GELSENBERG BENZIN A.G., Johannstrasse 8, Gelsenkirchen–Horst (W. Germany).
GEMINI ASBESTOS LTD., P. Bag Pietersburg, N. Tvl. (South Africa).
GENERAL EXPLORATION ORANGE FREE STATE LTD., General Mining Building, 6 Hollard Street, P.O. Box 1242 Johannesburg (South Africa).
GENERAL GEOPHYSICAL CO. OF FRANCE, 4 Square Rapp, Paris 7e (France).
GENERAL MINING AND FINANCE CORP., General Mining Building, 6 Hollard Street, P.O. Box 1242 Johannesburg (South Africa).
GENERAL OVERSEAS TRADERS (PTY.) LTD., Smal Street, P.O. Box 10059 Johannesburg (South Africa).
GENERAL REFRACTORIES LTD., Ncheu (Malawi); 25 Whitehall, London S.W. 1 (Great Britain).
GENOUX, GEORGES, (Établissements), 3 Rue Amiral-Pierre, Tananarive (Malagasy Republic).
GÉOMINES, see: Société Géologique et Minière des Ingénieurs et Industriels Belges.
GEOPHOTO EXPLORATIONS LTD., 305 Ernest and Crammer Building, Denver 2, Colo. (U.S.A.); Tripoli (Libya).
GEOPHYSICAL SERVICE INC., Exchange Bank Building, Dallas 35, Texas (U.S.A.); Tripoli (Libya); Lagos (Nigeria).
GÉORUANDA, 4 Rue de la Science, Brussels (Belgium).
GERMAN–LIBERIAN MINING CO., (DELIMCO), Bong (Liberia); Düsseldorf (W. Germany).
GETTY OIL CO., Pennsylvania Building, Wilmington, Delaware (U.S.A.).
GEWERKSCHAFT ELWERATH, Hindenburgstrasse 28, Hannover (W. Germany).
GEWERKSCHAFT EXPLORATION, Düsseldorf (W. Germany).
GHAIR, see: Ghanian Italian Petroleum Co.
GHANA COOPERATIVE DIAMOND UNION LTD., P.O. Box 1139 Accra (Ghana).
GHANA DIAMOND MARKETING BOARD, Accra (Ghana).
GHANAIAN ITALIAN PETROLEUM CO. (GHAIP), Asylum Down, P.O. Box 3183 Accra (Ghana).
GHANAIAN ITALIAN PETROLEUM CO. (GHANAIP), Tema (Ghana).
GHANAIP, see: Ghanian Italian Petroleum Co.
GHANA STATE MINING CORP., P.O. Box M39 Accra (Ghana); P.O. Box 26 Tarkwa (Ghana).
GIANT REEFS GOLD MINING CO., 43 Market Street, Johannesburg; P. Bag New Union Louis Trichardt, N. Tvl. (South Africa).
GIBELINO, FRANCESCO, Tete (Mozambique).
GLEN ALLEN MINES (PTY.) LTD., via Prieska (South Africa).
GLOBE AND PHOENIX GOLD MINING CO., 35 Old Jewry, London E.C. 2 (Great Britain); Sebakwe; Que Que (Southern Rhodesia).
GOLD AND BASE METAL MINES OF NIGERIA LTD., 120 Moorgate, London, E.C. 2 (Great Britain); Jos (Nigeria).
GOLDEN WEST MANGANESE CORP., P.O. Box 3198 Johannesburg (South Africa).
GOLD FIELDS FINANCE CO. (S.A.) LTD., Gold Fields Building, 75 Fox Street, Johannesburg (South Africa).
GOLD FIELDS MINING AND INDUSTRIAL LTD., 49 Moorgate, London E.C. 2 (Great Britain); 123 William Street, New York 38, N.Y. (U.S.A.).
GOLD FIELDS MINING CORP. LTD., 1855 Scarth Street, Regina, Saskatchewan (Canada); Executive Offices: 11 Adelaide Street West, Toronto (Canada).
GOLD FIELDS OF SOUTH AFRICA LTD., Gold Fields Building, 75 Fox Street, P.O. Box 1167 Johannesburg (South Africa); P.O. Box 176 Dar es Salaam (Tanganyika).
GOSSOUW, R., Swakopmund (Southwest Africa).
GOVERNMENT GOLD MINING AREAS CONSOLIDATED LTD., Consolidated Building, corner Fox and Harrison Streets, P.O. Box 590 Johannesburg; P.O. Box 14 State Mines (South Africa).
GOVERNMENT PETROLEUM REFINERY, Suez (U.A.R.).
GRACE, W.R. AND CO., 7 Hanover Square, New York (U.S.A.). (Operator: TEXAS GULF PRODUCING CO.).
GRAFITES DE MOÇAMBIQUE, Natía (Mozambique).

GREENWOOD POULTON AND CO., Union House, Queen Victoria Street, Cape Town (South Africa).
GRIQUALAND ASBES (PTY.) LTD., P.O. Box 89 Kuruman (South Africa).
GRIQUALAND EXPLORATION AND FINANCE CO., Fenchurch House, 5 Fenchurch Street, London E.C. 3 (Great Britain); P.O. Box 69 Kuruman (South Africa).
GRIQUALAND IRON ORE (PTY.) LTD., P.O. Box 10, 898 Johannesburg (South Africa).
GRIQUALAND WEST DIAMOND MINING CO., DUTOITSPAN MINE LTD., 36 Stockdale Street, Kimberley, C.P. (South Africa).
GROBLER, P.J.L., P.O. Box 391 Fort Victoria (Southern Rhodesia).
GROENFONTEIN TIN MINE, P.O. Box 11 Potgietersrus, N. Tvl. (South Africa).
GROOTVLEI PROPRIETARY MINES LTD., Union Corporation Building, 74–78 Marshall Street, P.O. Box 1125 Johannesburg; P.O. Box 445 Springs, Tvl. (South Africa).
GUGGENHEIM BROTHERS, 120 Broadway, New York City, N.Y. (U.S.A.).
GULF EASTERN CO., 100 West 10th Street, Wilmington, Del. (U.S.A.); Gulf House, 2 Portman Street, London W. 1 (Great Britain).
GULF OIL CORP., Gulf Building, P.O. Box 1166 Pittsburg 30, Pa. (U.S.A.).
GULF OIL OF LIBYA, Tripoli (Libya).
GWANDA NICKEL SYNDICATE, P.O. Box 839 Bulawayo (Southern Rhodesia).
GYPSUM INDUSTRIES OF SOUTH AFRICA LTD., 16 Frederick Street, P.O. Box 9656 Johannesburg; P.O. Box 80 Germiston, Tvl.; P.O. Box Paarden Eiland, C.P. (South Africa).
GYPSUM PRODUCTS LTD., P.O. Box 1192 Dar es Salaam (Tanganyika).
HALLIBURTON CO., P.O. Box 1431 Duncan, Okla. (U.S.A.); Tripoli (Lybia); Lagos (Nigeria).
HAMBRO'S BANK LTD., 41 Bishopsgate, London E.C. 2 (Great Britain).
H. AND B. ASBESTOS CO., 212 Van Riebeeck Buildings, Grobler Street, Pietersburg, N. Tvl. (South Africa).
HANS MERENSKY MANAGEMENT SERVICES LTD., 1409 African Life Building, Rissik Street, P.O. Box 7651 Johannesburg (South Africa).
HARMONY GOLD MINING CO., Corner House, 77 Commissioner Street, P.O. Box 1056 Johannesburg; P.O. Virginia (South Africa).
HARPER, S.E., P.O. Dett (Southern Rhodesia).
HARTEBEESTFONTEIN GOLD MINING CO. LTD., Anglovaal House, 56 Main Street, Johannesburg; P. Bag 800 Stilfontein (South Africa); Bilbao House, 36 New Broad Street, London E.C. 2 (Great Britain).
HARTESKLOOF ASBESTOS MINE (PTY.) LTD., 814 Saker's Corner, 34 Eloff Street, Johannesburg; P. Bag 1341 Pietersburg, N. Tvl. (South Africa).
HARVEY ALUMINUM CO., 19200 Southwestern Avenue Torrance, Los Angeles, Calif.; The Dalles, Ore., St. Croix, Virgin Islands (U.S.A.).
HAWKER HART MINERALS (PVT.) LTD., Zutphen Farm, Hunter's Road (Southern Rhodesia).
HAYFIELD SALT WORKS, P. Bag Hayfield, via Belmont (South Africa).
HECHE ET GIRARD, 3 Rue Amiral-Pierre, Tananarive (Malagasy Republic).
HENCHERT, B.V., P.O. Box 82 Karabib (Southwest Africa).
HENDERSON CONSOLIDATED CORP., 9th Floor Hollard Place, 71 Fox Street, P.O. Box 1146 Johannesburg (South Africa).
HENDERSON'S TRANSVAAL ESTATES LTD., 1 Cornhill, London E.C. 3 (Great Britain).
HENRY GOULD LTD., Elandsdrift, Tvl. (South Africa).
H.E. PROPRIETARY LTD., 49 Moorgate, London E.C. 2 (Great Britain).
HERDEIROS, see: Adilia Santa Rita de Lima.
HERING, M., P.O. Box 33 Karibib (Southwest Africa).
HIBON, RAPHAËL, Andranondambo (Behara) (Malagasy Republic).
HILTON QUARRIES (PTY.) LTD., P.O. Hilton Road, Natal (South Africa).
HIPOLITO, VIRGILIO, Tete (Mozambique).
HOLDRYHILL NICKEL MINING AND EXPLORATION CO., Shava (Southern Rhodesia).
HOLLANDSCHE METALLURGISCHE BEDRIJVEN, Arnhem (The Netherlands).
HOLLAND SYNDICATE, Apryensua (Ghana).
HOMESTAKE GOLD MINING CO. LTD., Lonrho House, P.O. Box 80 Salisbury (Southern Rhodesia).
HOMESTAKE MINING CO., 100 Bush street, San Francisco, Calif. (U.S.A.).
HORNGATE MINE, P. Bag. 1352 Pietersburg, N. Tvl. (South Africa).
HOUILLÈRES DU SUD-ORANAIS, 101 Avenue Raymond Poincaré, Paris 16e (France); Kenadsa, Colomb Béchar (Algeria).
HÜHLE, J.E.E., P.O. Box 13 Karibib (Southwest Africa).
HUMAN, P.J., Derust, Omaruon (Southwest Africa).
HUMBLE OIL AND REFINING CO,. 1216 Main Street, P.O. Box 2180 Houston 1, Texas (U.S.A.).
HUME PIPE CO., P.O. Box. 204 Germiston, (South Africa).

HUNTING SURVEYS LTD., 6 Elstree Way, Boreham Wood, London (Great Britain); 1450 O'Connor Drive, Toronto (Canada); 10 Rockefeller Plaza, New York City, N.Y. (U.S.A.).

HUNT OIL CO., 700 Mercantile Bank Boulevard, Dallas 1, Texas (U.S.A.); Tripoli (Libya).

HUSKY OIL CO., Tunis (Tunisia).

ILIMA COLLIERY AND LIME WORKS, P.O. Box 24 Tukuyu (Tanganyika).

IMPALA MINES, 65 Magor House, Fox Street, Johannesburg; P.O. Box 77 Gravelotte, E. Tvl. (South Africa).

IMPATI ANTHRACITE LTD., Hattingspruit (South Africa).

IMPERIAL CEMENT FACTORY CORP., Dire Dawa (Ethiopia).

IMPERIAL CHEMICAL INDUSTRIES LTD. (ICI), Millbank, London S.W. 1 (Great Britain).

IMPERIAL LIMESTONE QUARRIES, P.O. Box 962 Dar es Salaam (Tanganyika).

IMPERIAL SMELTING CO., 6 St. James Square, London S.W. 1 (Great Britain).

INDEPENDEX S.A.F., 12 Rue Chabanais, Paris 2e (France); 1 Place du Maréchal Lyautey, Algiers (Algeria).

INDUSTRIAL AND AGRICULTURAL DEVELOPMENT CORP., Accra (Ghana).

INDUSTRIAL DIAMONDS OF SOUTH AFRICA (1945) LTD., 415–416 Bristol Building, Corner MacLaren and Marshall Streets, Johannesburg (South Africa).

INDUSTRIAL MINERALS LTD., P.O. Box 45 Dar es Salaam (Tanganyika).

INDUSTRIA RAFFINAZIONE OLI MINERALI (IROM), 15 Piazza di Spagna, Rome; 4, Via dei Petroli, Porto Marghera, Venice (Italy).

INSIZWA MINING CO., Umbogintwini, Durban (South Africa).

INTERCONTINENTAL TRADING CO., 90 West Street, New York 6, N.Y. (U.S.A.).

INTERFOR, see: Société Internationale Commerciale et Financière de la Forminière.

INTERMINE, see: Société Internationale d'Exploitation Minière au Marve.

INTERNACIONAL IMPORTADORA E EXPORTADORA LTDA., Rua do Almada 443, Porto (Portugal).

INTERNATIONAL AFRICAN AMERICAN CORP., 1030 West Georgia Street, Vancouver B.C. (Canada); Monrovia (Liberia).

INTERNATIONAL DRILLING CO., Houston, Texas (U.S.A.); Tripoli (Libya).

INTERNATIONAL EGYPTIAN OIL CO., C.P. 4174 San Donato Milanese, Milan (Italy).

INTERNATIONAL EXPLORATION INC., San Antonio 5, Texas (U.S.A.); Tripoli (Libya).

INTERNATIONAL MANAGEMENT AND ENGINEERING GROUP, London (Great Britain).

INTERNATIONAL MINERALS AND CHEMICALS, Skokie, Ill. (U.S.A.).

INTERNATIONAL MINERALS AND CHEMICALS FRANCE, S.A., 20 Place Vendôme, Paris 1e (France).

INTERNATIONAL NICKEL AND MINERALS SEPARATION CO. (MOND), Thames House, Milbank, London S.W. 1 (Great Britain).

INTERNATIONAL SELLING CORP., 220 East 42nd Street, New York 17, N.Y. (U.S.A.).

INTER-RECOIN, 14 Avenue Marcel-Olivier, Tananarive (Malagasy Republic).

INYATI COPPER MINE, P.O. Box 94 Headlands (Southern Rhodesia).

IRENE MINING CORP., J.B.S. Building, Fox and Commissioner Streets, P.O. Box 9688 Johannesburg (South Africa).

IRI, see: Istituto per la Riconstruzione Industriale.

IROM, see: Industria Raffinazione Oli Minerali.

IRON DUKE MINING CO., P.O. Glendale (Southern Rhodesia).

ISCOR, see: South African Iron and Steel Industrial Corp.

ISTITUTO PER LA RICONSTRUZIONE INDUSTRIALE (IRI), Via Vittoria Veneto 89, Rome (Italy).

ITAMA MINES LTD., P.O. Box 50 Kabale (Uganda).

ITO, C. AND CO., 36, 2-chome, Honmachi, Higashi-ku, Osaka (Japan).

IZOUARD, RENÉ, Rue du Père-Camboué, Tananarive (Malagasy Republic).

JAMES CRANSHAW (PTY.) LTD., P.O. Box 40 Newcastle, Ntl. (South Africa).

JANTAR NIGERIA CO. LTD., 38 Finsbury Square, London E.C. 2 (Great Britain); Jos (Nigeria).

JEANNETTE GOLD MINES LTD., 44 Main Street, P.O. Box 4587 Johannesburg; P.O. Box 161 Allanridge (South Africa); 40 Holborn Viaduct, London E.C. 1 (Great Britain).

JENNY, LES FILS DE, B.P. 9 Fort-Dauphin (Malagasy Republic).

JCI, see: Johannesburg Consolidated Investment Co.

JOHANNESBURG CONSOLIDATED INVESTMENT CO. LTD (JCI), 10–11 Austin Friars, London E.C. 2 (Great Britain); Consolidated Building, corner Fox and Harrison Streets, P.O. Box 590 Johannesburg (South Africa).

JOHNSON A. AND CO., 21 West Street, New York, N.Y. (U.S.A.).

JOHNSON LILLY LINA, Vila Manica (Mozambique).

JOINT MINERAL SALES LTD., Montana; P.O. Box 399 Pietersburg, N. Tvl. (South Africa).

JOS HOLDINGS LTD., 7 Warwick Court, Holborn, London W.C. 1 (Great Britain).

JOS TIN AREA LTD., Jos (Nigeria).

JUMPERS CONSOLIDATED GOLD MINING CO., LTD., 511–6 City Trust House, 106 Fox Street, P.O. Box 742 Johannesburg; P.O. Box 15 Cleveland (South Africa).
KADOLA MINES LTD., Livingstone House, 48 Jameson Avenue Central, P.O. Box 1479 Salisbury (Southern Rhodesia).
KADUNA PROSPECTORS LTD., 78 Highlands Heath, Putney, London S.W. 15 (Great Britain); Jos (Nigeria).
KADUNA SYNDICATE LTD., 78 Highlands Heath, Putney, London S.W. 15 (Great Britain).
KAFFIR CREEK TALC MINES, P.O. Box 197 Germiston, Tvl. (South Africa).
KAISER ALUMINUM AND CHEMICAL CORP., Kaiser Center, 300 Lakeside Drive, Oakland 12, Calif. (U.S.A.).
KAISER AND REYNOLD ALUMINIUM CO., Richmond, Va. (U.S.A.).
KALKKLOOF ASBESTOS MINE, P. Bag Machadodorp, E. Tvl. (South Africa).
KAMATIVI TIN MINES LTD., Bradlow Buildings, Abercorn Street, Bulawayo; P.O. Wankie (Southern Rhodesia).
KANSANSHI COPPER MINING LTD., Lusaka (Zambia).
KANYEMBA GOLD MINES LTD., African Exploration Company Ltd., 9th Floor Hollard Place, 71 Fox Street, P.O. Box 1146 Johannesburg (South Africa); Umsweswe (Southern Rhodesia).
KAPLAN, B., Mtoko (Southern Rhodesia).
KARUGU AND CHAMUNYANA MINES CO. LTD., P.O. Box 66 Bukoba (Tanganyika).
KATUNDA, G.W., Kikagati (Uganda).
KEFFI TIN CO., Jos (Nigeria).
KEIR AND CAWDER LTD., 58 Marshall Street, P.O. Box 9893 Johannesburg (South Africa).
KELLY AND CO., P.O. Box 21 Mbabane (Swaziland).
KENDALL COLLIERY CO., P.O. Box 163 Witbank, Tvl. (South Africa).
KENNECOTT COPPER CORP., 161 East 42nd Street, New York 17, N.Y. (U.S.A.).
KENTAN GOLD AREAS LTD., Princes House, 95 Gresham Street, London E.C. 2 (Great Britain); Geita (Tanganyika).
KENYA SHELL CO., Mombasa (Kenya).
KEWANEE OIL CO., 40 Morris Avenue, P.O. Box 591 Bryn Mawr, Pa. (U.S.A.).
KEWANEE OVERSEAS OIL, see: Kewanee Oil Co.
KHAN, M.K. AND C.S. PATEL, P.O. Box 39 Kabale (Uganda).
KIKEO MICA MINING COOPERATIVE SOCIETY, P.O. Box 66 Morogoro (Tanganyika).
KILEMBE COPPER COBALT LTD., Room 2200, 25 King Street West, Toronto 1, Ontario (Canada).
KILEMBE MINES LTD., P.O. Box 1 Kilembe (Uganda).
KILMARNOCK ASBESTOS MINES LTD., P.O. Box 18 Mashaba (Southern Rhodesia).
KIMBERLEY GYPSUM SUPPLIES (PTY.) LTD., 16 Frederick Street, Johannesburg; Rietpan, P.O. Riverton (South Africa).
KIMBERLEY SALT CO., P.O. Box 362 Kimberley (South Africa).
KINORÉTAIN, see: Mines d'Or et d'Étain du Kindu, S.A.
KINOSHITA AND CO., 5, 2-chome, Takaracho, Chuo-ku, Tokyo (Japan); Conakry (Guinea).
KINSHO–MATAICHI CO., P.O. Box Nibonbashi no. 186 Tokyo (Japan); Lagos (Nigeria); 80 Wall Street, New York 5, N.Y. (U.S.A.).
KIRWA WOLFRAM MINES LTD., Kabale (Uganda).
KIVINDANI SALT WORKS, P.O. Box 157 Tanga (Tanganyika).
KIVUMINES, Bukavu; Kigulube; Kasese; Utu (Congo).
KLERKSDORP CONSOLIDATED GOLD FIELDS LTD., 1 Broad Street Place, London E.C. 2 (Great Britain); 27 Simmonds Street, Johannesburg (South Africa).
KLIPPOORTJIE COLLIERIES LTD., 63 Commissioner Street, P.O. Box 5958 Johannesburg; P.O. Ogies (South Africa).
KOBE MINING CO., Kobe (Japan); Que Que (Southern Rhodesia).
KONINKLIJKE NEDERLANDSCHE PETROLEUM MAATSCHAPPIJ N.V., Carel van Bylandtlaan 30, The Hague (The Netherlands).
KONONGO GOLD MINES LTD., 49 Moorgate, London E.C. 2 (Great Britain).
KOORNFONTEIN KOOLMYNE BEPERK, 63 Commissioner Street, P.O. Box 5958 Johannesburg; P.O. Vandyksdrif (South Africa).
KRANTZBERG MINES LTD., P.O. Box 18 Omaruru (Southwest Africa).
KRUPP A.G., Rheinhausen; Essen (W. Germany).
KÜLCHE, L. RUD. (PTY.) LTD., P.O. Box 9960, Johannesburg (South Africa).
KURUMAN CAPE BLUE ASBESTOS (PTY.) LTD., 34 Eloff Street, Johannesburg; P.O. Box 105 Kuruman (South Africa).
KUWAIT DEVELOPMENT FUND, Kuwait (Kuwait).
KWAGGAFONTEIN COLLIERY, P.O. Box 7 Carolina, E. Tvl. (South Africa).
KYERWA SYNDICATE LTD., P.O. Box 1023 Mbarara (Uganda).
LAMCO, see: Liberian American Swedish Minerals Co.
LANCASTER OUTCROPS LTD., P.O. Box 742 Johannesburg; P.O. Box 295 Krugersdorp, Tvl. (South Africa).
LANGEBAAN PHOSPHATE QUARRIES, Iscar House, Johannesburg (South Africa).

LANGUE, CAMILLE, Behara (Ambovombe) (Malagasy Republic).

LANNINHURST ASBESTOS LTD., P.O. Box 1722 Bulawayo (Southern Rhodesia).

LAVINO (S.A.) (PTY.) LTD., P.O. Box 7419 Johannesburg; P.O. Steelpoort (South Africa).

LAZARD FRÈRES ET COMPAGNIE, 5 Rue Pillet-Will, Paris 9e (France).

LDC, see: Liberia Development Co.

LEITÃO, ALVARO MARTINS, Cabinda (Angola).

LESLIE GOLD MINES LTD., Union Corporation Building, 74–78 Marshall Street, Johannesburg; P.O. Box 74 Evander (South Africa).

LES TRAVAUX SOUTERRAINS (LTS), 36bis Avenue de l'Opéra, Paris 2e (France).

LIBANON GOLD MINING CO. LTD., Gold Fields Building, 75 Fox Street, P.O. Box 1167 Johannesburg; P.O. Box 5 Venterspost (South Africa).

LIBERIA DEVELOPMENT CO., (LDC), Monrovia (Liberia).

LIBERIA MINING CO., 70 Pine Street, New York, N.Y. (U.S.A.).

LIBERIAN AMERICAN SWEDISH MINERALS CO. (LAMCO), Nimba (Liberia).

LIBERIAN EUROPEAN DIAMOND TRADING CO., Monrovia (Liberia).

LIBYA CEMENT CO., Tripoli (Libya).

LIBYAN AMERICAN OIL CO., Tripoli (Libya).

LIBYAN ATLANTIC CO., Tripoli (Libya).

LIBYA SHELL INC., Tripoli (Libya).

LIBYA SHELL N.V., 30 Carel van Bylandtlaan, The Hague (The Netherlands); Tripoli (Libya).

LING, H. A., 20 Kew Drive, Highlands, Salisbury (Southern Rhodesia).

LOIZEAU, L., Karuhinda, Kigezi (Uganda).

LOMAGUNDI WATER BORING AND MINING (PVT.) LTD., P.O. Box 101 Sinoia (Southern Rhodesia).

LONDON AND ASSOCIATED INVESTMENT TRUST LTD., 120 Moorgate, London E.C. 2 (Great Britain).

LONDON AND RHODESIAN MINING AND LAND CO. LTD., 1 Cornhill, London E.C. 3 (Great Britain); Lonrho House, P.O. Box 80 Salisbury (Southern Rhodesia).

LONDON AND SCANDINAVIAN METALLURGICAL CO., 39 Wimbledon Hill Road, London S.W. 19 (Great Britain).

LONDON NIGERIAN MINES LTD., Jos (Nigeria).

LONDON TIN CORP. LTD., 55–61 Moorgate, London E.C. 2 (Great Britain).

LONGWOOD ASBESTOS MINING CO., P. Bag 5 Filabusi (Southern Rhodesia).

LONRHO LTD., Salisbury; Umtali (Southern Rhodesia).

LOPYS, A. ET CIE., Antanimena, Route de Majunga, Tananarive (Malagasy Republic).

LORAINE GOLD MINES LTD., Anglovaal House, 56 Main Street, Johannesburg; P.O. Box 167 Odendaalsrus (South Africa); Bilbao House, 36 New Broad Street, London E.C. 2 (Great Britain).

LORELEI COPPER MINES LTD., Indosa Buildings, P.O. Box 161 Lüderitz (Southwest Africa).

LOUIS, ARSÈNE ET CIE., Antanimena, Tananarive (Malagasy Republic).

LOUISA MICA AND BERYLLIUM MINE, P.O. Box 135 Duiwelskloof, E. Tvl. (South Africa).

LTS, see: Les Travaux Souterrains.

LUAPULA MINES LTD., Livingstone House, Jameson Avenue Central, P.O. Box 1479 Salisbury C. 4 (Southern Rhodesia).

LUBIMBI COAL AREAS LTD., Kirrie Building 4th Floor, 94 Abercorn Street, Bulawayo (Southern Rhodesia).

LUCINADA UMBELUZI MINA LTDA., Impamputo (Mozambique).

LUIPAARDSVLEI ESTATE AND GOLD MINING CO. LTD., Gold Fields Building, 75 Fox Street, P.O. Box 1167 Johannesburg; P.O. Box 53 Krugersdorp (South Africa); 49 Moorgate, London E.C. 2 (Great Britain).

LUVEVE STONE CRUSHING AND DEVELOPMENT CO., Cowdray Paek, Luveve, Bulawayo (Southern Rhodesia).

LUX ORE AND CHEMICAL LTD., 419 Nikkatsu Building, Yurako Cho, Chiyoda-ku, Tokyo (Japan).

LYDENBURG ESTATES LTD., 40 Holborn Viaduct, London E.C. 1 (Great Britain); 44 Main Street, Johannesburg (South Africa).

LYDENBURG GOLD FARMS CO. LTD., Gold Fields Building, 75 Fox Street, P.O. Box 17679 Johannesburg (South Africa); 49 Moorgate, London E.C. 2 (Great Britain).

LYDENBURG PLATINIM LTD., General Mining Building, 6 Hollard Street, Johannesburg (South Africa); Winchester House, Old Broad Street, London E.C. 2 (Great Britain).

MACALDER–NYANZA MINES LTD., P.O. Box 3233 Nairobi (Kenya); 33 Hill Street, London W. 1 (Great Britain).

MACHOLE SALT WORKS LTD., P.O. Box 21 Lindi (Tanganyika).

MACK, JOHN AND CO., Golden Valley (Southern Rhodesia).

MAGADI SODA CO., Magadi (Kenya).

MAGELLAN PETROLEUM CORP., 1 Exchange Place, Jersey City 3, N.J. (U.S.A.).

MAINSTRAAT BELEGGINGS (EDMS.) BEPERK, 63 Commissioner Street, Johannesburg (South Africa).

MAKERI SMELTING CO., Jos (Nigeria).

MALAWI ELECTRICITY SUPPLY COMMISSION, P.O. Box 186 Blantyre (Malawi).

MALLIG DIAMOND LTD., Standard Bank Chambers, Marshall Street, Johannesburg; P.O. Box 43 Zwartruggens (South Africa).

MALMANI MINING AND DEVELOPMENT CO., P.O. Box 2 Zeerust, Tvl. (South Africa).

MANFROY, H., El Karit, par Oulmès; Oulmès-El Karit (Morocco).

MANGANESE CORP., 94 Commissioner Street, Johannesburg (South Africa); 2 Queen Anne's Gate, Westminster, London S.W. 1 (Great Britain).

MANGANESE IRON MINING CO. LTD., P.O. Box 8186 Johannesburg; Manganore, Postmasburg (South Africa).

MANGANESE SYNDICATE, P.O. Box 115 Krugersdorp, Tvl. (South Africa).

MAPANZURI CHROME MINES (PVT.) LTD., P.O. Box 963 Bulawayo (Southern Rhodesia).

MARATHON INTERNATIONAL OIL CO., 539 South Main Street, Findlay, Ohio (U.S.A.).

MARATHON INTERNATIONAL PETROLEUM (GB) LTD., 36–38 Berkeley Square, London W. 1 (Great Britain).

MARATHON OIL CO., 539 South Main Street, Findlay, Ohio (U.S.A.).

MARBLE LIME AND ASSOCIATED INDUSTRIES LTD., 8 Wright Boag Road, P.C. Box 7711 Johannesburg; Noordkaap Station; P.O. Box 24 Kroondal (South Africa).

MARICO MINERAL CO. (PTY.) LTD., 27 Smuts Avenue, Vereeniging, Tvl. (South Africa).

MARIEVALE CONSOLIDATED MINES LTD., Union Corporation Building, 74–78 Marshall Street, P.O. Box 1125 Johannesburg; P.O. Marieshaft (South Africa); Princes House, 95 Gresham Street, London E.C. 2 (Great Britain).

MARINE DIAMOND CORP., Oranjemund (Southwest Africa).

MARLIME CHRYSOTILE ASBESTOS CORP., 8/32 Wright Boag Road, Johannesburg (South Africa); Moshaneng via Lobatsi (Bechuanaland).

MARLIN TRUST (PTY.) LTD., 505–515 Stock Exchange Buildings, Hollard Street, P.O. Box 5782 Johannesburg (South Africa).

MARINE DIAMONDS LTD., Radio City, Tulbagh Square, Cape Town (South Africa).

MARTIN, FRANK AND CO., Smit Street, Industries West, P.O. Box 197 Germiston; P.O. Gravelotte, E. Tvl. (South Africa).

MAULAK DIAMOND CO., Monrovia (Liberia).

MASHABA GOLD MINES (PVT.) LTD., Mashaba (Southern Rhodesia).

MASHABA RHODESIAN ASBESTOS COMPANY LTD., Rock Secretariat Ltd., 24 Broad Street Avenue, London E.C. 2 (Great Britain); Kirrie Buildings, P.O. Box 77 Bulawayo (Southern Rhodesia).

MATTE SMELTERS (PTY.) LTD., P.O. Box 143 Rustenburg, W. Tvl. (South Africa).

MAURETANIA MINES (PVT.) LTD., P.O. Box A 112 Avondale, Salisbury (Southern Rhodesia).

MAYFAIR GOLD MINING CO. LTD., 511-6 City Trust House, 106 Fox Street, P.O. Box 745 Johannesburg; P.O. Box 36 Crown Mines (South Africa).

MAZISTA SLATE QUARRIES LTD., P.O. Box 353 Roodepoort, Tvl. (South Africa).

MAZOE CONSOLIDATED MINES LTD., Lonrho House, P.O. Box 80 Salisbury (Southern Rhodesia).

MCCREA, R. T., P.O. Box 1329 Bulawayo (Southern Rhodesia).

MCCREEDY TIN MINE, Mbabane (Swaziland).

MERRALLI, S., P.O. Box 81 Mbarara (Uganda).

MERRIESPRUIT (ORANGE FREE STATE) GOLD MINING CO. LTD., Corner House, 77 Commissioner Street, P.O. Box 1056 Johannesburg; P. Bag 60 P.O. Virginia (South Africa); 4 London Wall Buildings, London E.C. 2 (Great Britain).

MESSIAS, AURÉLIO VICENTE, Luanda (Angola).

MESSINA RHODESIA SMELTING AND REFINING CO., see: Messina(Transvaal) Development Co.

MESSINA (TRANSVAAL) DEVELOPMENT CO., Messina, N. Tvl. (South Africa); Cape House, 54 Fox Street, Johannesburg; P. Bag Messina, N. Tvl. (South Africa); Central Africa House, First Street, Salisbury; P.O. Bag 53 Fort Victoria (Southern Rhodesia); 29–30 St. James's Street, London S. W. 1 (Great Britain).

METALKAT, see: (a) Société Métallurgique du Katanga. (b) Société Métallurgique Katangaise.

METALLURG INC., 99 Park Avenue, New York 16, N.Y. (U.S.A.).

MGL-CONGO, see: Compagnie Minière Congolaise des Grands Lacs Africains.

MIAMI BERYL CO., P.O. Box 1972 Salisbury (Southern Rhodesia).

MICOUIN, E., Chez MM. Micouin et Pochard, 9 Place du Général Leclerc, Tananarive (Malagasy Republic).

MICUMA, see: Société Anonyme des Mines de Cuivre de Mauritanie.

MIDDLE WITWATERSRAND (WESTERN AREAS) LTD., Anglovaal House, 56 Main Street, P.O. Box 7727 Johannesburg (South Africa); Bilbao House, 36 New Broad Street, London E.C. 2 (Great Britain).

MIFERMA, see: Société Anonyme des Mines de Fer de Mauritanie.

MILE CEMENT CO., Rank (Sudan).

MILUBA, see: Société Minière du Lualaba.

MINAS CATIPO LTDA., Tete (Mozambique).
MINAS DE LICE LTDA., Alto Molocué (Mozambique).
MINE DE NEBEUR, Contrôle Civil du Kef (Tunisia).
MINE MANAGEMENT ASSOCIATES (MMA), Bomi Hill (Liberia).
MINERAIS PASICOS LTDA., Pebane (Mozambique).
MINERAIS ET MÉTAUX S.A., 61 Avenue Hoche, Paris 8e; 28 Rue Arthur Rozier, Paris 19e (France).
MINERALIA MINING CO., P.O. Box 1303 Johannesburg; P.O. Box 100 Postmasburg (South Africa).
MINERALI E METALLI, Via Durini 9, Milano (Italy).
MINERALS AND CHEMICALS PHILIPP CORP., 350 Park Avenue, New York, N.Y. (U.S.A.).
MINERALS RESEARCH SYNDICATE LTD., Jos (Nigeria).
MINERAL TRADING CORP. OF SOUTHWEST AFRICA, City Centre, Kaiser Street, P.O. Box 762 Windhoek (Southwest Africa).
MINERAL WEALTH GENERAL CO., 1, 26th July Street, Cairo (U.A.R.).
MINERARIA SOMALA SPA., San Donato Milanese, Milan (Italy); Mogadishu (Somali Republic).
MINERGA, see: Compagnie Minière de l'Urega.
MINES D'AOULI, 1 Rond-point Saint-Exupéry, Casablanca; Mibladen par Midelt (Morocco).
MINES DE FER DE MILIANA, 13 Rue Clauzel, Algiers; Quais Philippeville, Philippeville, Département Constantine (Algeria).
MINES DE L'ATLAS CENTRAL, 9 Rue de Foucauld, Casablanca; Tizi Mizar (Morocco).
MINES DE MIDKANE, 38 Boulevard de la Résistance, Casablanca; Midkane (Morocco).
MINES DES ZENAGA, 10 Rue Bendahan, Casablanca; Angarf (Morocco).
MINES DE TIMEZRIT, 21 Avenue Montaigne, Paris 8e (France); Timezrit par El Maden (Algeria).
MINES DEVELOPMENT SYNDICATE (WEST AFRICA) LTD., 120 Moorgate, London E.C. 2 (Great Britain).
MINES D'OR ET D'ÉTAIN DU KINDU, S.A. (KINORÉTAIN), 42 Rue de Royale, Brussels (Belgium).
MINES DU DJEBEL HAMRA (M. DELGA), 76 Rue du Portugal, Tunis (Tunisia).
MINES DU DJEBEL ZEBBEUS (M. DELGA), 76 Rue du Portugal, Tunis (Tunisia).
MINES DU KEF AGAB (M. DELGA), 76 Rue du Portugal, Tunis (Tunisia).
MINES ET GRAPHITE DU MAROC, 9 Rue de Toul, Casablanca; Frag el Ma (Morocco).
MINÉTAIN, see: Société des Mines d'Étain du Ruanda et de l'Urundi.
MINEX (PTY.) LTD., 808 Volkskas Building, 76 Market Street, Johannesburg (South Africa).
MINIÈRES ET CARRIÈRES DE RIBET-EL-MADEN, 49 Rue Sadi-Carnot, Algiers; Mine d'El Maden par Ribet (Algeria).
MINING AND TECHNICAL SERVICES LTD., 120 Moorgate, London E.C. 2 (Great Britain).
MINISTRY OF POWER STATIONS CONSTRUCTION, Moscow (U.S.S.R.).
MINING AND METALLURGICAL AGENCY LTD., Trafalgar House, 11 Waterloo Place, London S. W. 1 (Great Britain).
MIRUDI, see: Compagnie Minière du Ruanda-Urundi.
MITSUBISHI METAL MINING CO., 6 Otemachi 1-chome, Chiyoda-ku, Tokyo (Japan).
MITSUBISHI SHOJI KAISHA, 20, 2-chome Marunouchi, Chiyoda-ku, Tokyo (Japan).
MITSUI AND CO., 2-1 Shiba Tamuracho Minato-ku, Tokyo (Japan); 530 Fifth Avenue, New York, N.Y. (U.S.A.).
MLIMO CHROME MINES (PVT.) LTD., P.O. Box 26 Bulawayo (Southern Rhodesia).
MLOTA MINE (PVT.) LTD., P.O. Box 10 Eiffel Flats (Southern Rhodesia).
MMA, see: Mine Management Associates.
MOBIL EXPLORATION ÉQUATORIAL AFRICA INC., P.O. Box 564 Port Gentil (Gabon).
MOBIL EXPLORATION NIGERIA INC., P.O. Box 31 Port Harcourt (Nigeria).
MOBIL OIL CAMEROUN, Rue Joffre B.P. 4058 Douala (Cameroons).
MOBIL OIL CO. LTD., Caxton House, Tothill Street, London S.W. 1 (Great Britain).
MOBIL OIL CONGO S.C.R.L., 2 Avenue Van Gele, P.O. 2400 Léopoldville (Congo).
MOBIL OIL CO. OF CANADA, Tripoli (Libya).
MOBIL OIL DE CANARIAS S.A., Leon y Castillo 309, Apartado de Correo 34, Las Palmas (Canary Islands); Ceutà; Melilla; Fernando Póo; Bata (Rio Muni).
MOBIL OIL DE L'AFRIQUE ÉQUATORIALE, Avenue Maréchal-Foch, B.P. 134 Brazzaville (Congo Republic); Bangui (Central African Republic); Libreville (Gabon); Fort Lamy (Chad).
MOBIL OIL DE L'AFRIQUE OCCIDENTALE, B.P. 3120, 4 Rue Salva, Dakar (Senegal); Lomé (Togo); Nouakchott (Mauritania); Cotonou (Dahomey); Conakry (Guinea); Abidjan (Ivory Coast); Niamey (Niger); Ouagadougou (Upper Volta); Bamako (Mali); Khartoum (Sudan).
MOBIL OIL EAST AFRICA LTD., Hughes Building, Delamere Avenue, Nairobi (Kenya); 37–43 Sackville Street, London W. 1 (Great Britain).
MOBIL OIL EGYPT S.A., 1097 Corniche El Nil, Garden City, Cairo; Asl, Ras Sudr (U.A.R.).
MOBIL OIL FRANÇAISE S.A., 46 Rue de Courcelles, Paris 8e (France).

MOBIL OIL GHANA LTD., Ghana House, P.O. Box 450 Accra (Ghana).
MOBIL OIL GUINÉE S.A.R.L., Lapaternelle Building, 1st Boulevard and 8th Avenue, P.O. Box 305 Conakry (Guinea).
MOBIL OIL LIBERIA INC., Bushrod Island, P.O. Box 342 Monrovia (Liberia).
MOBIL OIL LIBYA LTD., Bedri Building, Sciara Ahmed Scarif, Tripoli (Libya).
MOBIL OIL MAROC, 53 Rue Allal Ben Abdallah, B.P. 485 Casablanca (Morocco).
MOBIL OIL NIGERIA LTD., Broad and Kakawa Streets, P. Bag 2054 Lagos (Nigeria).
MOBIL OIL NORD-AFRICAINE, 29 Rue Didouche Mourad, Algiers (Algeria); Tunis (Tunisia).
MOBIL OIL NORTHERN RHODESIA LTD., Lusaka (Zambia).
MOBIL OIL NYASALAND (PVT.) LTD., Blantyre (Malawi).
MOBIL OIL PORTUGUESA, Rua de Horta Seca 15, Lisbon (Portugal); Rua Pereira Forjaz 110-4°, Caixa Postal 330 Luanda (Angola); Bissau (Portuguese Guinea); São Tomé.
MOBIL OIL RWANDA BURUNDI S.A.R.L., B.P. 1482 Usumbura (Burundi); Kigali (Rwanda).
MOBIL OIL SIERRA LEONE LTD., Bank of West Africa Building, Oxford and Wilberforce Streets, P.O. Box 548 Freetown (Sierra Leone); Bathurst (Gambia).
MOBIL OIL SOUTHERN AFRICA (PTY.) LTD., Boston House, 44/46 Strand Street, Cape Town (South Africa); Windhoek (Southwest Africa); Lourenço Marques (Mozambique); St. Helena.
MOBIL OIL SOUTHERN RHODESIA (PVT.) LTD., Salisbury (Southern Rhodesia).
MOBIL OIL SUDAN LTD., Barlaman Avenue, P.O. Box 283 Khartoum (Sudan).
MOBIL PETROLEUM ETHIOPIA INC., 13 Dejatch Afework Street, Asmara (Ethiopia).
MOBIL PRODUCING ALGERIA INC., 54 Rue de Londres, Paris 8e (France).
MOBIL PRODUCING SAHARA INC., 54 Rue de Londres, Paris 8e (France); Ohanet; Askarene (Algeria).
MOBIL PRODUCING TUNISIA INC., 219 Avenue de Paris, Tunis (Tunisia).
MOBIL REFINING CO. SOUTHERN AFRICA (PTY.) LTD., P.O. Box 956 Durban (South Africa).
MOBILREX, P.O. Box 564 Port Gentil (Gabon).
MOBIL SAHARA, 54 Rue de Londres, Paris 8e (France).
MODDERFONTEIN B. GOLD MINES LTD., Corner House, P.O. Box 1056 Johannesburg; P.O. Modder East, Tvl. (South Africa).
MODDERFONTEIN EAST LTD., Corner House, 77 Commissioner Street, P.O. Box 1056 Johannesburg; P.O. Modder East, Tvl. (South Africa).
MOINHOS, JOAQUIM FERNANDES, Hotel Club, Lourenço Marques (Mozambique).
MONAZITE AND MINERAL VENTURES (PTY.) LTD., 44 Main Street, Johannesburg (South Africa).
MONTROSE EXPLORATION CO., Coventry House, 3 South Place, London E.C. 2 (Great Britain); 6 Hollard Street, P.O. Box 2283 Johannesburg; P.O. Box 245 Pietersburg, N. Tvl. (South Africa).
MOUNTAIN MINERALS (PTY.) LTD., Volkskas Building, 76 Market Street, Johannesburg; P. Bag Baumann, Warmbaths, Tvl. (South Africa).
MOZAMBIQUE GULF OIL CO. (MOZGOC), Beira (Mozambique).
MOZGOC, see: Mozambique Gulf Oil Co.
MSAULI ASBESTOS MINING AND EXPLORATION CO., 818/821 Annuity House, P.O. Box 3384 Johannesburg; P. Bag Barberton, E. Tvl. (South Africa).
MTD (MANGULA) LTD., Central Africa House, First Street, Salisbury; Mangula (Southern Rhodesia); 29–30 St. James's Street, London S.W. 1 (Great Britain).
MTWARA SALT WORKS LTD., P.O. Box 1646 Dar es Salaam (Tanganyika).
MUFULIRA COPPER MINES LTD., Mufulira; Lusaka (Zambia).
MULLER, PAUL, Vila Manica (Mozambique).
MUNNIK MYBURGH CHRYSOTILE ASBESTOS LTD., 44 Main Street, P.O. Box 2567 Johannesburg; P.O. Kaapsche Hoop, E. Tvl. (South Africa).
MURPHY CORP., Murphy Building, El Dorado, Ark. (U.S.A.).
MUTUE FIDES TIN LTD., P.O. Box 8653 Johannesburg (South Africa); Kapata Tin Mine; Tshontanda Tungsten Mine (Southern Rhodesia).
MWINILUNGA MINES LTD., Livingstone House, 48 Jameson Avenue Central, Salisbury C.4 (Southern Rhodesia).
MYLROIE, A., P.O. Box 99 Bindura (Southern Rhodesia).
NAKAIMA, M. N., P.O. Box 1042 Kikagati (Uganda).
NANAK CHAND LTD., P.O. Box 40 Musoma (Tanganyika).
NANTWICH SALT WORKS LTD., THE, P.O. Box 625 Kimberley; P.O. Riverton (South Africa).
NAP, see: Société Nord-Africaine du Plomb.
NARAGUTA EXTENDED AREAS LTD., Jos (Nigeria).
NARAGUTA KARAMA AREAS LTD., Exploration House, Fishmonger's Hall Street, London E.C. 4 (Great Britain); Jos (Nigeria).

NARAGUTA TIN MINES LTD., Jos (Nigeria).

NATAL AMMONIUM COLLIERIES (1946) LTD., 27 Simmonds Street, P.O. Box 1639 Johannesburg; P.O. Mount Ngwibi, Ntl. (South Africa).

NATAL ANTHRACITE COLLIERY LTD., 24 Beach Grove, P.O. Box 1381 Durban, Natal; P.O. Langkrans (South Africa).

NATAL CAMBRIAN COLLIERIES LTD., Consolidated Building, corner Fox and Harrison Streets, P.O. Box 590 Johannesburg; P.O. Dannhauser, Ntl.; P.O. Ballengeich (South Africa).

NATAL COAL EXPLORATION CO. LTD., 44 Main Street, P.O. Box 2567 Johannesburg; P.O. Box 160 Newcastle, Ntl. (South Africa); 40 Holborn Viaduct, London E.C. 1 (Great Britain).

NATAL NAVIGATION COLLIERIES AND ESTATE CO., 809 Barclay's Bank Building, Field and Smith Street, P.O. Box 993 Durban, Ntl.; P.O. Glencoe; P. Bag 657 Newcastle, Ntl. (South Africa).

NATAL PORTLAND CEMENT CO., 56 Main Street, Johannesburg; P.O. Box 22 Port Shepstone; 138 National Mutual Building, Smith Street, Durban; P.O. Bellair (South Africa).

NATAL STEAM COAL CO., 708-9 Payne's Buildings, West Street, Durban, Ntl.; P.O. Wessel's Nek (South Africa).

NATIONAL IRON ORE CO., Monrovia (Liberia).

NATIONAL MANGANESE MINES LTD., A.B.C. Chambers, Simmonds Street, P.O. Box 6839 Johannesburg; P.O. Box 33 Sishen (South Africa).

NATIONAL MINING AND EXPLORATION CO., P.O. Box 8543 Johannesburg; P.O. Box 100 Postmasburg (South Africa).

NATIONAL OIL CO. OF LIBYA (NOCOL), Tripoli (Libya).

NATIONAL PORTLAND CEMENT CO., P.O. Box 21 Claremont; P.O. Ratelfontein (South Africa).

NATIONAL QUARRIES (PTY.) LTD., P.O. Box 1091 Cape Town (South Africa).

NATIONAL SALT CO., Varsch Vlei, 137 Cullinan Building, Johannesburg (South Africa).

NCHANGA CONSOLIDATED COPPER MINES LTD., Lusaka; Nchanga; Leslie Pollak House, Nkana (Zambia).

NDOLA COPPER REFINERIES LTD., Woodgate House, Cairo Road, P.O. Box 851 Lusaka; Ndola (Zambia).

NELSON'S QUARRIES (PTY.) LTD., 38 Essex Terrace, Westville, P.O. Box 2268 Durban; P.O. Box 30 Westville, Durban (South Africa).

NETHERBURN SYNDICATE, P.O. Lalapanzi (Southern Rhodesia).

NEW AFRICAN MICA CO., Morogoro (Tanganyika).

NEW AFRICAN MINING CO., P.O. Box 169, Morogoro (Tanganyika).

NEW AMIANTHUS MINES (PTY.) LTD., P.O. Box 1100 Bulawayo (Southern Rhodesia); P.O. Box 171 Kuruman (South Africa); Emlembe (Swaziland).

NEWCASTLE PLATBERG COLLIERY LTD., 24/26 Beach Grove, Durban; P.O. Elandslaagte (South Africa).

NEW CENTRAL WITWATERSRAND AREAS LTD., 44 Main Street, P.O. Box 2567 Johannesburg (South Africa).

NEW CONSOLIDATED GOLD FIELDS, P.O. Box 1167 Johannesburg (South Africa).

NEW JAGERSFONTEIN MINING AND EXPLORATION CO., 36 Stockdale Street, Kimberley, C.P. (South Africa).

NEW JERSEY ZINC CO., 35 Main Street, Franklin, N.J. (U.S.A.).

NEW KLEINFONTEIN CO., Northern Trust Building, 28 Harrison Street, P.O. Box 2105 Johannesburg; P.O. Box 2 Benoni (South Africa); 62 London Wall, London E.C. 2 (Great Britain).

NEW KLERKSDORP GOLD ESTATES LTD., P.O. Box 7727 Johannesburg; P.O. Box 97 Klerksdorp (South Africa).

NEW LARGO COLLIERY LTD., 44 Main Street, P.O. Box 2567 Johannesburg; P.O. Box 63 Kendal (South Africa).

NEWMONT MINING CORP., 300 Park Avenue, New York 22, N.Y. (U.S.A.).

NEWMONT OIL CO., 300 Park Avenue, New York 22, N.Y. (U.S.A.); Algiers (Algeria).

NEW PIONEER CENTRAL GOLD MINING CO., General Mining Building, 6 Hollard Street, Johannesburg (South Africa).

NEW STATE AREAS LTD., Consolidated Building, corner Fox and Harrison Streets, P.O. Box 590 Johannesburg (South Africa); 10–11 Austin Friars, London E.C. 2 (Great Britain).

NEW VAN RYN GOLD MINING CO., P.O. Box 9688 Johannesburg; P.O. Van Ryn (South Africa).

NEW WITWATERSRAND GOLD EXPLORATION CO., Gold Fields Building, 75 Fox Street, P.O. Box 1167 Johannesburg (South Africa).

NICKEL CORP. OF AFRICA LTD., P.O. Box 7597 Johannesburg; P.O. Box 2 Mount Ayliff (South Africa).

NIGEL GOLD MINING CO., Mutual Buildings, 4th Floor, Harrison Street, Johannesburg (South Africa).

NIGERIAN AGIP OIL CO., Port Harcourt (Nigeria).

NIGERIAN CEMENT CO., THE, P.O. Box 331 Enugu, Kalagu (Nigeria).

NIGERIAN CONSOLIDATED MINES LTD., Jos (Nigeria).

NIGERIAN GAS CORP. LTD., Lagos (Nigeria).

NIGERIAN LEAD-ZINC MINING CO., P.O. Box 51 Abakaliki, via Enugu (Nigeria).

NIGERIAN TIN AND EXPLORATION CO., Jos (Nigeria).

NILE CEMENT CO., THE, Khartoum (Sudan).

NISHO CO., 30 ,3-chome Imabashi, Higashi-ku, Osaka (Japan).

NOCOL, see: National Oil Co. of Libya.

NORTHERN AFRICAN MINING AND FINANCE LTD., Bon Accord House, 19 Harrison Street, Johannesburg (South Africa).

NORTHERN ASBESTOS MINES (PTY.) LTD., 19/29 Roeland Street, Cape Town; Donkerhoek, Pietersburg (South Africa).

NORTHERN LIME CO., Lewis and Marks Building, 65 President Street, P.O. Box 4610 Johannesburg (South Africa).

NORTHERN MERCANTILE AND INVESTMENT CORP., City Wall House, Chitwell Street, London E.C. 1 (Great Britain).

NORTHERN NATAL NAVIGATION COLLIERIES LTD., 28 Natal Bank Chambers, P.O. Box 993 Durban (South Africa).

NORTHERN RHODESIA CO., 19 St. Swithin's Lane, London E.C. 4 (Great Britain).

NORTHERN RHODESIA LIME CO., Chilanga; Lusaka (Zambia).

NORTHMEAD GOLD MINING CO., P.O. Box 9688 Johannesburg (South Africa).

NOUVELLE SOCIÉTÉ MINIÈRE MALGACHE 'L'ESPÉRANCE', chez M. Krafft, Avenue de la Réunion, Tananarive (Malagasy Republic).

NTJ MINING CO. AND TRIBUTORS (PTY.) LTD., P. Bag 1352 Pietersburg, N. Tvl. (South Africa).

NTUMBI REEFS LTD., P.O. Box 21 Chunya (Tanganyika).

NYANZA SALT MINES LTD., P.O. Box 45 Dar es Salaam (Tanganyika).

NYASALAND PORTLAND CEMENT CO., Heavy Industrial Area, P.O. Box 523 Blantyre (Malawi).

OASIS OIL CO. OF LIBYA INC., 1270 Avenue of the Americas, New York 20, N.Y. (U.S.A.); Oasis Oil Building, Sciara Giama el Magarbad, P.O. Box 395 Tripoli; Samah; Gialo; Defa (Libya).

OCEAN SCIENCE AND ENGINEERING INC., Washington 25, D.C. (U.S.A.).

OCP, see: Office Chérifien des Phosphates.

OFFICE CHÉRIFIEN DES PHOSPHATES (OCP), Avenue Urbain Blanc, Rabat; Khouribga; Youssoufia (Morocco).

OFFICE NATIONAL DES MINES (ONM), 19 Rue Al-Djazira, Tunis (Tunisia).

OFFICINE MECCHANICHE (OM), Algiers (Algeria); Via Bissolati 57, Rome (Italy).

OFFSHORE CO., 169 Lafayette Street, P.O. Box 1268 Baton Rouge 1, La. (U.S.A.).

OFFSHORE PETROLEUM EXPLORATIONS LTD., Port Harcourt (Nigeria); Britannic House, Finsbury Circus, London E.C. 2 (Great Britain).

OLDHAM, H. AND SONS (PVT.) LTD., Kirrie Buildings, P.O. Box 77 Bulawayo (Southern Rhodesia).

OLD HERRIOT MINE LTD., P.O. Box 742 Johannesburg; P.O. Box 42 Cleveland (South Africa).

OLIN–MATHIESON CHEMICAL CORP., 460 Park Avenue, New York 22, N.Y. (U.S.A.).

OM, see: Officine Mecchaniche.

OMNIREX, see: Omnium de Recherches et Exploitations Pétrolières.

OMNIUM DE GÉRANCE INDUSTRIELLE ET MINIÈRE, 3 Rue Pégoud, Casablanca; M'Koussa; Tafgout; Tourtit; Enta; Smala; Masser Amane (Morocco).

OMNIUM DE RECHERCHES ET D'EXPLOITATIONS PÉTROLIÈRES (OMNIREX), 280 Boulevard Saint Germain, Paris 7e (France); Hassi Touareg (Algeria).

OMNIUM FRANÇAIS DE PÉTROLES S.A., 280 Boulevard Saint Germain, Paris 7e (France).

ONM, see: Office National des Mines.

O'OKIEP COPPER CO. LTD., 300 Park Avenue, 19th floor, New York 22, N.Y. (U.S.A.); P.O. Box 17 Nababeep, C.P. (South Africa).

ORANGE FREE STATE INVESTMENT TRUST LTD., 44 Main Street, P.O. Box 4587 Johannesburg (South Africa).

ORANGE RIVER SALTWORKS LTD., P.O. Box 1125 Johannesburg (South Africa).

ORE SALES AND SERVICES LTD., 103 Mount Street, London W. 1 (Great Britain).

ORE TRADING CORP., 135 Broadway, New York 6, N.Y. (U.S.A.).

ORIENTAL TRUST CO., 43 Market Street, Johannesburg (South Africa).

ORIENT MANGANESE CO., P.O. Box 7193 Johannesburg (South Africa).

ORNASTONE QUARRIES (PTY.) LTD., P.O. Box 7953 Johannesburg (South Africa).

OTAVI MINING CO., P.O. Box 9876 Johannesburg (South Africa).

OTHON, JULIEN, Avenue Marcel-Olivier, B.P. 120 Tananarive (Malagasy Republic).

OVERSEAS DEVELOPMENT CORP., P.O. Box 3233 Nairobi (Kenya); Njombe, Songea (Tanganyika).

OVERSEAS MINERAL RESOURCES DEVELOPMENT CORP. ASSOCIATION, 24, 3-chome Kanda Nishiki-cho, Chiyoda-ku, Tokyo (Japan).

PADRÃO, ERNESTO TEIXEIRA, Cabinda; Quibaxi; Dembos; Uige (Angola).

PALABORA HOLDINGS LTD., Corner Market and Fraser Streets, Johannesburg (South Africa).

PALABORA MINING CO., Atlantis House, corner Market and Fraser Streets, Johannesburg (South Africa).

PALA KOP MINERALS, P.O. Box 23 Mooketsi (South Africa).

PALILUX, Luxembourg (Luxembourg).

PALMIET CHROME CORP., Belmore House, corner Main and Harrison Streets, P.O. Box 1331 Johannesburg; Rustenberg, W. Tvl. (South Africa).

PAN AMERICAN INTERNATIONAL OIL CO., 630 Fifth Avenue, New York, N.Y. (U.S.A.).

PAN AMERICAN LIBYA OIL CO., Tripoli (Libya); 630 Fifth Avenue, New York, N.Y. (U.S.A.).
PAN AMERICAN SAHARA OIL CO., Algiers (Algeria); 630 Fifth Avenue, New York, N.Y. (U.S.A.).
PAN AMERICAN UNITED ARAB REPUBLIC OIL CO., Cairo (U.A.R.).
P. AND M. CORP., P.O. Box 1717 Bulawayo (Southern Rhodesia).
PANGANI ASBESTOS MINES LTD., P.O. Box 8 Filabusi (Southern Rhodesia).
PARSONS, R. M. CO., see: Ralph M. Parsons Co.
PAULSEN, H, P.O. Box 1042 Kikagati (Uganda).
PEAK MINE (PVT.) LTD., P.O. Box 16 Mashaba (Southern Rhodesia).
PEAK QUARRY CO. (PTY.) LTD., 37 Durham Avenue, Salt River (South Africa).
PEAT, MARWICK, MITCHELL AND CO., Shell House, Fort Street, Bulawayo (Southern Rhodesia).
PÉCHINEY, see: Compagnie de Produits Chimiques et Électrométallurgiques Péchiney.
PELLA REFRACTORY ORES S.A. (PTY.) LTD., P.O. Box 140 Karasburg, (Southwest Africa); Pella Mission, Namaqualand (South Africa).
PEÑARROYA, see: Société Peñarroya–Maroc.
PERA TRADING CO., Livorno (Italy).
PET MANGAAN, Netherlands Bank Building, P.O. Box 340 Germiston, Tvl. (South Africa).
PETRANGOL, see: Companhia de Petroleos de Angola.
PETRANGOL, see: Sociedade dos Petroleos de Angola.
PETROCONGO, 52 Rue de l'Industrie, Brussels (Belgium).
PETROFINA, S.A., see: Compagnie Financière Belge des Pétroles.
PÉTROLES DE DJIBOUTI, Djibouti (French Somali Coast).
PETROLIBIA, Tripoli (Libya).
PETROPAR, see: Société de Participations Pétrolières.
PETROSAREP, 42 Avenue Raymond-Poincaré, Paris 16e (France).
PETROSUD SPA., 8 Via Cernaia, Milan (Italy).
PHILIPP BROTHERS, 350 Park Ave, New York, N.Y. (U.S.A.).
PHILLIPS PETROLEUM CO., 129 South State Street, Dover, Del. (U.S.A.); Tripoli (Libya).
PHILLIPS PETROLEUM CO. OF LIBYA, Tripoli (Libya).
PHILLIPS PETROLEUM FRANCE S.A., 37 Avenue d'Iéna, Paris 16e (France).
PHOENIX COLLIERY LTD., Consolidated Building, corner Fox and Harrison Streets, P.O. Box 590 Johannesburg (South Africa); 10–11 Austin Friars, London E.C. 2 (Great Britain).
PHOENIX PRINCE GOLD MINING CO. LTD., 35 Old Jewry, London E.C. 2 (Great Britain); Bindura (Southern Rhodesia).
PIERREFITTE, see: Société Générale d'Engrais et Produits Chimiques Pierrefitte.
PINHEIRO, João da Costa, Caixa Postal 60 Mocuba (Mozambique).
PLATRIERES DE BALLAH CO., 37 Talaat Harb Street, Cairo (U.A.R.).
POFADDER TRANSPORT (PTY.) LTD., Pofadder; P.O. Box 4711 Cape Town (South Africa).
PONS CHROME MINES LTD., P.O. Box 1125 Johannesburg (South Africa); P.O. Box 74 Selukwe (Southern Rhodesia).
PORTLAND CEMENT CO. OF TANGANYIKA, Wazo Hills, Dar es Salaam (Tanganyika).
POSTMASBURG ORES (MANGANESE) (PTY.) LTD., P.O. Box 8186 Johannesburg (South Africa).
POTGIETERSRUS PLATINUM LTD., Consolidated Building, P.O. Box 590 Johannesburg (South Africa); 10–11 Austin Friars, London E.C. 2 (Great Britain).
PREMIER PORTLAND CEMENT CO. (RHODESIA) LTD., Charter House, Selborne Avenue, P.O. Box 699 Bulawayo (Southern Rhodesia).
PREMIER (TRANSVAAL) DIAMOND MINING CO., 36 Stockdale Street, P.O. Box 616 Kimberley, C.P.; P.O. Box 44 Cullinan, (South Africa); 40 Holborn Viaduct, London E.C. 1 (Great Britain).
PREPA, see: Société de Prospection et Exploitations Pétrolières en Alsace.
PRESIDENT BRAND GOLD MINING CO., 44 Main Street, P.O. Box 4587 Johannesburg; P.O. Box 64 Welkom, O.F.S. (South Africa).
PRESIDENT STEYN GOLD MINING CO. LTD., 44 Main Street, P.O. Box 4587 Johannesburg; P.O. Box 2 Welkom, O.F.S. (South Africa).
PRETORIA NORTH DEVELOPMENT CO., P.O. Chloorkop, Johannesburg (South Africa).
PRETORIA PORTLAND CEMENT CO., P.O. Box 3811 Johannesburg; P.O. Box 7 Slurry; P.O. Pienaars River; P.O. Kingswood (South Africa).
PREUSSAG, see: Preussische Bergwerke und Hütten A.G.
PREUSSISCHE BERGWERKE UND HÜTTEN A.G. (PREUSSAG), Leibnizufer 9, Hannover (W. Germany).
PREXMIN S.A., 18 Rue Delord, B.P. 832 Tananarive (Malagasy Republic).
PRIMROSE GOLD MINING CO. (1934) LTD. 511-6 City Trust House, 106 Fox Street, P.O. Box 742 Johannesburg; P.O. Box 193 Germiston (South Africa).

PRINCESS GOLD MINING CO., P.O. Box 742 Johannesburg; P.O. Box 190 Roodepoort, Tvl. (South Africa).
PRODUITS CHIMIQUES PÉCHINEY–SAINT GOBAIN, 2 Place de la Libération, Levallois-Perret, Seine (France); 38 Avenue Hoche, Paris 8e (France); B.P. 36 Thiès (Senegal).
PROVINCE MINES LTD., 1000 Winchester House, Loveday Street, P.O. Box 7193 Johannesburg (South Africa).
PUREFINA, S.A., 33 Rue de la Loi, Brussels (Belgium).
PURE OIL CO., 620 East Broad Street, Columbus, Ohio (U.S.A.).
PURFINA TUNISIENNE, Biserta (Tunisia).
PYROLO MINING CO., 62 Piet Joubert Avenue, Monument, P.O. Box 375 Krugersdorp, Tvl. (South Africa).
QUE QUE MINES LTD., P.O. Box 123 Que Que (Southern Rhodesia).
RABEROJO, PHILIPPE, 6 Rue Guillain, Tananarive (Malagasy Republic).
RAFFINERIES CHÉRIFIENNES D'HUILES DE PÉTROLE, Route de Camp Boulhaut, Casablanca; Bled Zerga (Morocco).
RAHAMEFY, JEAN, 13 Rue Nicolas Mayeur, Tananarive (Malagasy Republic).
RALPH M. PARSONS CO., Los Angeles, Calif. (U.S.A.).
RAND AMERICAN INVESTMENTS (PTY.) LTD., 54 Marshall Street, P.O. Box 9123 Johannesburg (South Africa).
RANDFONTEIN ESTATES GOLD MINING CO., Consolidated Building, corner Fox and Harrison Streets, P.O. Box 590 Johannesburg; P.O. Boxes 2 and 11 Randfontein, W. Tvl. (South Africa).
RAND LEASES (VOGELSTRUISFONTEIN) GOLD MINING CO., Anglovaal House, 56 Main Street, Johannesburg; P.O. Box 1 Florida, Tvl. (South Africa); Bilbao House, 36 New Broad Street, London E.C. 2 (Great Britain).
RAND MINES LTD., Corner House, P.O. Box 1056 Johannesburg (South Africa); 4 London Wall Buildings, London E.C. 2 (Great Britain).
RAND SELECTION TRUST CORP., 44 Main Street, P.O. Box 4587 Johannesburg (South Africa).
R.A.N. MINES LTD., P.O. Box 37 Bindura (Southern Rhodesia).
R.A.P., see: Régie Autonome des Pétroles.
RAW ASBESTOS DISTRIBUTORS LTD., Asbestos House, 77/79 Fountain Street, Manchester (Great Britain).
RAYMOND INTERNATIONAL INC., 140 Cedar Street, New York 6, N.Y. (U.S.A.).
RAZAFY–CALLIXTE, Avenue du Maréchal-Joffre, Tananarive (Malagasy Republic).
REEF NIGEL EXPLORATION GOLD MINING CO., 75 Fox Street, P.O. Box 8653 Johannesburg (South Africa).
REGIDESO, see: Régie de Distribution d'Eau et d'Électricité du Congo.
RÉGIE AUTONOME DES DISTRIBUTIONS D'EAU ET D'ÉLECTRICITÉ, 48 Rue Mohammed, Diouri, Casablanca (Morocco).
RÉGIE AUTONOME DES PÉTROLES (R.A.P.), 12–16 Rue Jean Nicot, Paris 7e (France).
RÉGIE DE DISTRIBUTION D'EAU ET D'ÉLECTRICITÉ DU CONGO (REGIDESO), 30 Rue Marie de Bourgogne, Brussels (Belgium); Léopoldville (Congo).
REMNA, see: Société de Recherches Minières en Afrique.
RENÉ, G., Karahinda (Uganda).
REPFRANCE, see: Compagnie Française pour le Financement de la Recherche et de l'Exploitation du Pétrole.
REPUBLIC STEEL CORP., 405 Lexington Avenue, New York, N.Y. (U.S.A.).
RES ASBESTOS MINES LTD., P.O. Box 11 Salisbury (Southern Rhodesia).
RESOURCES AND DEVELOPMENT CORP., see: Development and Resources Corp.
REYNOLDS GEOPHYSICAL CO., Houston, Texas (U.S.A.).
REXCOLLAH, J. AND SONS LTD., Jos (Nigeria).
RHENOSTERFONTEIN FLUORSPAR MINES (PTY.) LTD., P.O. Atlanta, Brits, Tvl. (South Africa).
RHOANGLO, see: Rhodesian Anglo American Ltd.
RHODESIA BROKEN HILL DEVELOPMENT CO., Lusaka; Broken Hill (Zambia).
RHODESIA CEMENT LTD., Rhocem House, Selborne Avenue and Main Street, P.O. Box 1515 Bulawayo; Colleen Bawn (Southern Rhodesia); 33 King William Street, London E.C. 4 (Great Britain).
RHODESIA CHROME MINES LTD., 103 Mount Street, London W. 1 (Great Britain); P.O. Box 124 Selukwe (Southern Rhodesia).
RHODESIA CONGO BORDER POWER CORP., Afcom House, P.O. Box 819 Kitwe (Zambia).
RHODESIA COPPER REFINERIES LTD., Lusaka (Zambia).
RHODESIA–KATANGA CO. LTD., Princes House, 95 Gresham Street, London E.C. 2 (Great Britain).
RHODESIA MICA MINING CO., P. Bag 577 Sinoia (Southern Rhodesia).
RHODESIA MONTELO ASBESTOS LTD., Robinson House, 51 Union Avenue, Salisbury (Southern Rhodesia).
RHODESIAN ALLOYS LTD., Gwelo (Southern Rhodesia).
RHODESIAN ALUMINA CO., O'Brien's Prospect, Mazoe (Southern Rhodesia).
RHODESIAN AND GENERAL ASBESTOS CORP., P.O. Box 1100 Bulawayo (Southern Rhodesia).
RHODESIAN ANGLO AMERICAN LTD. (RHOANGLO), Lusaka; Leslie Pollak House, Nkana (Zambia).
RHODESIAN CAMBRAI MINES (PVT.) LTD., P.O. Lalapanzi; P.O. Box 155 Gwelo (Southern Rhodesia).
RHODESIAN CEMENT LTD., Gwanda (Southern Rhodesia).

RHODESIAN CHROME MINES LTD., P.O. Box 124 Selukwe (Southern Rhodesia).

RHODESIAN CORP. LTD., York House, Eighth Avenue, Jameson Street, Bulawayo (Southern Rhodesia); 120 Moorgate, London E.C. 2 (Great Britain).

RHODESIAN EMERALD MINES (PVT.) LTD., P.O. Box 3520 Salisbury (Southern Rhodesia).

RHODESIAN INDUSTRIAL MINERALS AND AGENCIES (PVT.) LTD., Belmont, P.O. Box 8132 Bulawayo (Southern Rhodesia).

RHODESIAN IRON AND STEEL CO. (RISCO), Throgmorton House, Jameson Avenue Central, P.O. Box 3491 Salisbury; Redcliff; Bulawayo (Southern Rhodesia).

RHODESIAN METALLURGICAL AND MINING CO., P. Bag 2 Gatooma (Southern Rhodesia).

RHODESIAN MINING ENTERPRISES (PVT.) LTD., P.O. Box 6 Salisbury (Southern Rhodesia).

RHODESIAN SELECTION TRUST EXPLORATION LTD., Livingstone House, 48 Jameson Avenue Central, P.O. Box 1479 Salisbury C. 4 (Southern Rhodesia); Lusaka (Zambia).

RHODESIAN SELECTION TRUST LTD., Woodgate House, Cairo Road, P.O. Box 851 Lusaka (Zambia).

RHODESIAN VANADIUM CORP., 420 Lexington Avenue, New York 17, N.Y. (U.S.A.); Eagle Star House, Gordon Avenue, P.O. Box 2729 Salisbury; Vanad Mine; Birkdale Mine; Sutton Mine (Southern Rhodesia); Bahati Mines, P.O. Box 27 Fort Rosebery (Zambia).

RHOKANA CORP. LTD., Lusaka; Leslie Pollak House, Nkana (Zambia).

RHONDA CHROME MINES (PVT.) LTD., P.O. Box 1544 Salisbury (Southern Rhodesia).

RIBEIRO, VICENTE E CASTRO, Tete (Mozambique).

RIBON VALLEY TINFIELDS LTD., Jos (Nigeria).

RICHFIELD OIL CORP., 708 Third Avenue, New York City, N.Y. (U.S.A.); Vila Cisneros (Spanish Sahara).

RICHMOND SECRETARIAT LTD., Crosskey House, 56 Moorgate, London E.C. 2 (Great Britain).

RIETFONTEIN CONSOLIDATED MINES LTD., Gold Fields Building, 75 Fox Street, P.O. Box 1167 Johannesburg; P. Bag Elandsfontein (South Africa); 49 Moorgate, London E.C. 2 (Great Britain).

RIMROCK TIDELANDS INC., Tunis (Tunisia).

RIO RITA MINES (PTY.) LTD., P.O. Box 315 Randfontein, Tvl. (South Africa).

RIO TINTO CO. LTD., Barrington House, 59 Gresham Street, London E.C. 2 (Great Britain).

RIO TINTO FINANCE AND EXPLORATION LTD., Barrington House, 59 Gresham Street, London E.C. 2 (Great Britain).

RIO TINTO MANAGEMENT SERVICES (CENTRAL AFRICA) (PVT.) LTD., Angwa Street and Speke Avenue, Salisbury (Southern Rhodesia).

RIO TINTO MINING CO. OF SOUTH AFRICA LTD., Atlantic House, corner Market and Fraser Streets, Johannesburg (South Africa).

RIO TINTO (RHODESIA) LTD., Doncaster House, corner Angwa Street and Speke Avenue, Salisbury (Southern Rhodesia).

RIO TINTO RHODESIAN HOLDING CO., Doncaster House, corner Angwa Street and Speke Avenue, Salisbury (Southern Rhodesia).

RIO TINTO RHODESIAN MINING LTD., Doncaster House, corner Angwa Street and Speke Avenue, Salisbury; Cam and Motor, Gatooma (Southern Rhodesia).

RISCO, see: Rhodesian Iron and Steel Co.

RIVERSDALE ANTHRACITE COLLIERY (PTY.) LTD., P.O. Box 209 Vryheid, W. Tvl. (South Africa).

R.M.B. ALLOYS LTD., Witbank, Tvl. (South Africa).

ROAN ANTELOPE COPPER MINES LTD., Lusaka; Ndola (Zambia).

ROBERTS VICTOR DIAMONDS LTD., 27 Simmonds Street, P.O. Box 7400 Johannesburg; P.O. Victor, O.F.S. (South Africa).

ROBINSON DEEP LTD., Gold Fields Building, 75 Fox Street, P.O. Box 1167 Johannesburg (South Africa).

ROGERS GEOPHYSICAL CO., 3616 West Alabama Street, Houston 6, Texas (U.S.A.).

ROGOFF, E.I., LTD., Zambesi House, 44 Von Wielligh Street, P.O. Box 7296 Johannesburg (South Africa).

ROLAND, RENÉ, Betroka (Malagasy Republic).

ROLLET (SOCIÉTÉ D'EXPLOITATION DES MICAS–SEMIR), 2 Rue Raysbaud, Tananarive (Malagasy Republic).

ROODEPAN MINERALS (PTY.) LTD., P.O. Box 47 Krugersdorp, Tvl. (South Africa).

ROODEPOORT GOLD MINING CO. (1957) LTD., 511-6 City Trust House, 106 Fox Street, Johannesburg; P.O. Box 190 Roodepoort, Tvl. (South Africa).

ROOIBERG MINERALS DEVELOPMENT CO., Gold Fields Building, 75 Fox Street, P.O. Box 1167 Johannesburg; P.O. Box Rooiberg, Warmbaths, Tvl. (South Africa).

ROSE CHROME MINES (PVT.) LTD., P. Bag 43 Gwelo (Southern Rhodesia).

ROSE DEEP LTD., The Corner House, P.O. Box 1056 Johannesburg; P.O. Box 6 Germiston, Tvl. (South Africa); 4 London Wall Buildings, London E.C. 2 (Great Britain).

ROSS MCINTYRE AND PARTNERS (PVT.) LTD., P.O. Box 140 Shabani (Southern Rhodesia).

ROSTAING (ETS), 22 Rue Théodore-Villette, B.P. 99 Tananarive (Malagasy Republic).

ROTSCHILD FRÈRES AND CO., 21 Rue Lafitte, Paris 9e (France).
ROYAL MCBEE CORPORATION (A. RYAN), 850 Third Avenue, New York, N.Y. (U.S.A.).
R.S.T. MINE SERVICES LTD., Woodgate House, P.O. Box 851 Lusaka (Zambia).
RUIGHOEK CHROME MINES (PTY.) LTD., Main House, 96 Main Street, P.O. Box 6612 Johannesburg (South Africa).
RUSTENBURG CHROME MINES (PTY.) LTD., 28 Thorpe Street, Selby, P.O. Box 1366 Johannesburg; P.O. Box 68 Kroondal, Tvl. (South Africa).
RUSTENBURG PLATINUM MINES LTD., Consolidated Building, P.O. Box 590 Johannesburg, Transvaal; P.O. Box 143 Rustenburg, Tvl.; P.O. Box 58 Northam (South Africa).
RUTALA MINES, P.O. Box 2276 Salisbury (Southern Rhodesia).
RYAN NIGEL INVESTMENT CO., Trust Buildings, corner Fox and Loveday Streets, P.O. Box 155 Johannesburg (South Africa).
SABA, see: Société Anonyme des Barytes Algériennes.
SACEM, see: Société Anonyme Chérifienne d'Études Minières.
SADD-EL-AALI EXECUTIVE ORGANIZATION, Cairo; Asswan (U.A.R.).
SAE, see: Société Égyptienne pour le Raffinage et le Commerce du Pétrole.
SAFAGA PHOSPHATE CO., 1 26 July Street, Cairo (U.A.R.).
SAFOR, see: Société Anonyme de Forage.
SAFREP, see: Société Anonyme Française de Recherches et d'Exploitation de Pétrole.
SAICI, see: Società Agricola Industriale per la Cellulosa Italiana.
SAINT ANNE'S MINING CO., P.O. Box 274 Salisbury (Southern Rhodesia).
SAINT BARBARA CHROME MINE, 505 Permanent Building, Commissioner Street, P.O. Box 2413 Johannesburg (South Africa).
SAINT BEZART, Marcel, Betroka (Malagasy Republic).
SAINT HELENA GOLD MINES LTD., 74 Marshall Street, P.O. Box 1125 Johannesburg (South Africa).
SAINT JOSEPH LEAD CO., 230 Park Avenue, New York, N.Y. (U.S.A.).
SALINES DU MAROC, Immeuble du Parc, Fédala; Taza (Morocco).
SALISBURY PORTLAND CEMENT CO., P.O. Box 3898 Salisbury (Southern Rhodesia).
SALNOVA, Coegas (South Africa).
SAM, see: Société Auxiliaire de Mines.
SAMIR, see: Société Anonyme Marocaine-Italienne de Raffinage.
SAND AND CO. LTD., 95–107 Bree Street, Johannesburg (South Africa).
SANDRAMINE, B.P. 1370 Abidjan (Ivory Coast).
SANDWANA MINES LTD., Angwa and Speke Avenues, Salisbury; Belingwe (Southern Rhodesia).
SANGHA MINE, N'Dem (Central African Republic).
SANTA FE DRILLING CO., P.O. Box 310 Whittier, Calif. (U.S.A.).
SAP, see: Société Africaine des Pétroles.
SAPPHIRE PETROLEUM CO., Niamey (Niger Republic).
SARATOGA SALT WORKS LTD., P.O. Box 301 Kimberley, Salt Lake (South Africa).
SAREMCI, see: Société Anonyme de Recherche et d'Exploitation Minière en Côte d'Ivoire.
SAREMCO, see: Société Anonyme de Recherches et Exploitation Minière Centre Oubangui.
SASOL, see: The Durban Navigation Collieries Ltd.
SASOL, see: South Africain Coal, Oil and Gas Corp.
SCHANDL, L., Kabale (Uganda).
SCHINAZI, J., 171 Rue Blaise Pascal, Casablanca; Jebel Rheris; Mouhajibat (Morocco).
SCHWEIZERISCHE ALUMINIUM A.G., Buckhauserstrasse 5, Zurich 9-48 (Switzerland).
SEA DIAMONDS LTD., Radio City, Tulbagh Square, Cape Town (South Africa).
SEDAO, see: Società Elettrica Dell' Africa Orientale.
SEGANS, see: Société d'Étude du Transport et de la Valorisation des Gaz Naturels du Sahara.
SEHR, see: Société d'Exploitation des Hydrocarbures d'Hassi er Rmel.
SEISMOGRAPH SERVICE CORP., 6200 East 41 Street, Tulsa, Okla. (U.S.A.).
SEKUKUNI CHROOM (EIENDOMS) BEPERK, 110 Pieter Neethlinggebou, Centraal St., Pretoria; Mecklenburg Mine, P.O. Driekop (South Africa).
SELATI MINE, P.O. Box 2413 Johannesburg; P.O. Mica, N. Tvl. (South Africa).
SELATI RIVER SILICA AND FELDSPAR MINES, P.O. Box 197 Germiston, Tvl.; P.O. Gravelotte, E. Tvl. (South Africa).
SELECTION TRUST LTD., Selection Trust Building, Mason's Avenue, London E.C. 2 (Great Britain).
SELTRUST INVESTMENTS LTD., Mason's Avenue, London E.C. 2 (Great Britain).
SELUKWE GOLD MINING AND FINANCE CO., 19 St. Swithin's Lane, London E.C. 4 (Great Britain).
SEMA, see: Société d'Exploitation Minière de l'Androy.

SENDI, A., P.O. Box 1012 Kikagati (Uganda).
SERAM, Algiers (Algeria); 50 Via Tevere, Rome (Italy).
SEREM, see: Société d'Études, de Recherches et d'Exploitations Minières.
SEREPT, see: Société de Recherches et d'Exploitation des Pétroles en Tunisie.
SERMAN, see: Société d'Exploitation et de Recherches Minières dans l'Afrique du Nord.
SERVIÇOS MUNICIPALIZADOS DE AGUA ET ELECTRICIDADE (SMAE), P.O. Box 43 Luanda (Angola).
SHELL AND BP SOUTH AFRICAN PETROLEUM REFINERIES (PTY.) LTD., P.O. Box 45 Isipingo Beach, Ntl. (South Africa).
SHELL-BP PETROLEUM DEVELOPMENT CO. OF NIGERIA LTD., 40 Marina, Lagos; Port Harcourt; Bomu (Nigeria).
SHELL CHEMICAL EASTERN AFRICA CO., Nairobi (Kenya).
SHELL CO. OF GHANA LTD., Shell Centre, London S.E. 1 (Great Britain).
SHELL CO. OF LYBIA LTD., Sciara el Magharba, P.O. Box 402 Tripoli (Libya).
SHELL CO. OF SIERRA LEONE LTD., Shell Centre, London S.E. 1 (Great Britain).
SHELL D'AFRIQUE OCCIDENTALE S.A., Dakar (Senegal).
SHELL DE TUNISIE S.A., Tunis (Tunisia).
SHELL DU MAROC S.A., Rabat (Morocco).
SHELL EGYPT, Cairo (U.A.R.).
SHELL INTERNATIONAL PETROLEUM CO., Shell Centre, London S.E. 1 (Great Britain).
SHELL OIL CO., 50 West 50th Street, New York 20, N.Y. (U.S.A.).
SHELL OVERSEAS EXPLORATION CO., Shell Centre, London S.E. 1 (Great Britain).
SHELL RED SEA, Khartoum (Sudan).
SHELL WEST AFRICA LTD., Lagos (Nigeria).
SHER, see: Sociedade Hidro-Eléctrica do Revuè.
SHERWOOD STAR GOLDMINING CO., York House, Eighth Avenue, Bulawayo (Southern Rhodesia).
SIBÉKA, see: Société d'Entreprise et d'Investissement du Bécéka.
SIERRA LEONE DEVELOPMENT CO., City House, Finsbury Square, London E.C. 2 (Great Britain).
SIERRA LEONE ORE AND METALS LTD., Freetown (Sierra Leone).
SIERRA LEONE SELECTION TRUST LTD. (SLST), Selection Trust Building, Mason's Avenue, London E.C. 2 (Great Britain); Freetown; Sefadu (Sierra Leone).
SIERRA LEONE STATE DEVELOPMENT CORP. (SLOC), Freetown (Sierra Leone).
SILVERDALE INVESTMENTS LTD., P.O. Box 1096 Pretoria (South Africa).
SIMÃO, Maria Alzina, Mocuba (Mozambique).
SIMMER AND JACK MINES LTD., Gold Fields Building, 75 Fox Street, P.O. Box 1167 Johannesburg; P.O. Box 192 Germiston, Tvl. (South Africa); 49 Moorgate, London E.C. 2 (Great Britain).
SIMON, E.E., Karibib (Southwest Africa).
SINAI MANGANESE CO. S.A.E., 1 Sharia El Bustan, Cairo (U.A.R.).
SINAREX, see: Société Nationale de Recherche et d'Exploitation Minières.
SINCLAIR AND CO., Bon Accord, P.O. Box 954 Pretoria (South Africa).
SINCLAIR MEDITERRANEAN PETROLEUM CO., 600 Fifth Avenue, New York 20, N.Y. (U.S.A.); Algiers; Rhourde el Baguel (Algeria).
SINCLAIR OIL CORP., 600 Fifth Avenue, New York 20, N.Y. (U.S.A.).
SINCLAIR SAHARA S.A., 52 Avenue des Champs Elysées, Paris 8e (France); 4 Boulevard Camille Saint-Saëns, Algiers (Algeria).
SINCLAIR SOMAL CORP., Mogadishu (Somalia); 600 Fifth Avenue, New York 20, N.Y. (U.S.A.).
SLAMI, see: Société Lyonnaise Agricole Minière et Industrielle.
SLASTO CO. OF AFRICA LTD., International House, corner Loveday and Kerk Streets, Johannesburg (South Africa).
SLDC, see: Sierra Leone State Development Corp.
SLST, see: Sierra Leone Selection Trust Ltd.
SMAE, see: Serviços Municipalizados de Agua et Electricidade.
SMALL MINES AND GENERAL INVESTMENTS LTD., 511-6 City Trust House, 106 Fox Street, P.O. Box 742 Johannesburg (South Africa).
SMBA, see: Société Anonyme des Mines de Bou Arfa.
SMEO, see: Société Minière Est Oubangui.
SMGI, see: Société des Minerais de la Grande-Ile.
SMI, see: Société Minière Intercoloniale.
SMYTH, J. H., P.O. Box 2 O'okiep, C.P. (South Africa).
SMZ, see: Société Minière du Zamza.
S.N. MAREP, see: Société Nationale de Matériel pour la Recherche et l'Exploitation du Pétrole.
S.N. REPAL, see: Société Nationale de Recherche et d'Exploitation des Pétroles en Algérie.

SOBAKI, see: Société Belgo-Africaine du Kivu.
SOBIASCO, see: Société Bitume et de l'Asphalte du Congo.
SOCIBEMA, see: Société des Ciments et Bétons Manufacturés.
SOCIEDADE ANGOLANA DE MINAS, Caixa Postal 5637 Luanda; Muxexe (Angola).
SOCIEDADE COMERCIAL E MINEIRA DO DANDE, Luanda (Angola).
SOCIEDADE DOS PETROLEOS DE ANGOLA (PETRANGOL), Luanda (Angola).
SOCIEDADE HIDRO-ELÉCTRICA DO REVUÈ (SHER), P.O. Box 39 Vila Pery (Mozambique).
SOCIEDADE MINEIRA DA HUÍLA, Luanda (Angola).
SOCIEDADE MINEIRA DE MAIOMBE, Buco Zan; Cabinda (Angola).
SOCIEDADE MINEIRA DE MALANJE, Luanda; Quiluco (Angola).
SOCIEDADE MINEIRA DE MARROPINO, Pebane (Mozambique).
SOCIEDADE MINEIRA DE MAVITA, Mavita (Mozambique).
SOCIEDADE MINEIRA DE MELELA, Caixa Postal 47 Quelimane (Mozambique).
SOCIEDADE MINEIRA DE MOCUBELA, Pebana (Mozambique).
SOCIEDADE MINEIRA DE ZAMBEZIA, Caixa Postal 63 Mocuba (Mozambique).
SOCIEDADE MINIERA DO INCHOPE, Vila Machado (Mozambique).
SOCIEDADE MINEIRA DO ITOTONE, Caixa Postal 94 Moçambique (Mozambique).
SOCIEDADE MINEIRA DO LOMBIGE, Vila Salazar; Cazengo; Cuanza Norte (Angola).
SOCIEDADE MINEIRA DO LUEZA, Buco Zan; Cabinda (Angola).
SOCIEDADE MINEIRA E INDUSTRIAL DE MOÇAMBIQUE, Caixa Postal 157 Nampula (Mozambique).
SOCIEDADE MINEIRA LUSO-FRANCESA DE ANGOLA, Caixa Postal 153 Luanda (Angola).
SOCIEDADE NACIONAL DE PETROLEOS DE MOÇAMBIQUE, Caixa Postal 417 Lourenço Marques (Mozambique).
SOCIEDADE NACIONAL DE REFINAÇÃO DE PETROLEOS (SONAREP), Caixa Postal 1866, Lourenço Marques (Mozambique).
SOCIETÀ AGRICOLA INDUSTRIALE PER LA CELLULOSA ITALIANA (SAICI), Viale Cernaia 8, Milan (Italy); Algiers (Algeria).
SOCIETÀ ANONIMA DE CIMENTI AFRICA ORIENTALE, Gurgussum, Massawa (Ethiopia).
SOCIETÀ CIMENTIERE DEL TIRRENO, Viale Gorizia 24, Rome (Italy); Garoua (Cameroons).
SOCIETÀ FINANZIARA SIDERURGICA, Viale Castro Pretorio 122, Rome (Italy).
SOCIETÀ ELETTRICA DELL'AFRICA ORIENTALE (SEDAO), Avenue Imperatrice Menen 87, Asmara (Ethiopia).
SOCIETÀ MINERARIA SIDERURGICA, Via San Giacomo di Carignano 13, Genoa (Italy).
SOCIETÀ SALINE DI MASSAWA, Massawa (Ethiopia).
SOCIÉTÉ AFRICAINE D'ÉLECTRICITÉ, B.P. 2020 Dakar (Senegal); Pt. Étienne (Mauritania); Niamey (Niger); Ouagadougou (Upper Volta).
SOCIÉTÉ AFRICAINE DE RAFFINAGE, Dakar (Senegal).
SOCIÉTÉ AFRICAINE DES MINES, 22 Rue du Languedoc, Rabat; Imarira (Morocco).
SOCIÉTÉ AFRICAINE DES PÉTROLES (SAP), Dakar (Senegal); Abidjan (Ivory Coast).
SOCIÉTÉ ALGÉRIENNE DE PRODUITS CHIMIQUES ET D'ENGRAIS, 18 Avenue de Carthage, Tunis (Tunisia).
SOCIÉTÉ ALGÉRIENNE DU ZINC (ALZI), 26 Rue Geoffroy Lasnier, Paris 4e (France); B.P. 20, El Abed près Bou Beker, Oujda (Morocco).
SOCIÉTÉ ANONYME CHÉRIFIENNE D'ÉTUDES MINIÈRES (SACEM), 5 Avenue de l'Armée-Royale, B.P. 241 Casablanca; Marrakech-Principal Morocco; B.P. 8113 Casablanca-Oasis; Imini (Morocco).
SOCIÉTÉ ANONYME DE FORAGE (SAFOR), 36 Boulevard Victor Hugo, Limoges, Haute Vienne; 18bis Rue d'Anjou, Paris 8e (France).
SOCIÉTÉ ANONYME DE RECHERCHES ET EXPLOITATION MINIÈRE CENTRE OUBANGUI (SAREMCO), Berberati (Central African Republic).
SOCIÉTÉ ANONYME DE RECHERCHE ET D'EXPLOITATION MINIÈRE EN CÔTE D'IVOIRE (SAREMCI), 8 Rue Lafayette, Paris 9e; 13 Rue Lafayette, Paris 9e (France); B.P. 1368 Abidjan (Ivory Coast); Tortiya par Katiola (Ivory Coast).
SOCIÉTÉ ANONYME DES BARYTES ALGÉRIENNES (SABA), Route de Rivet, Oued Smar, Maison Carrée (Algeria).
SOCIÉTÉ ANONYME DES MINERAIS, 11bis Boulevard Prince Henri, Luxembourg (Luxembourg).
SOCIÉTÉ ANONYME DES MINES DE BOU ARFA (SMBA), 101 Rue Saint Lazare, Paris 8e (France); Bou Arfa par Oujda; Bou Arfa (Morocco).
SOCIÉTÉ ANONYME DES MINES DE CUIVRE DE MAURITANIE (MICUMA), 10 Place Vendôme, Paris 1er (France); Akjoujt (Mauritania).
SOCIÉTÉ ANONYME DES MINES DE FER DE MAURITANIE (MIFERMA), Fort-Gourad (Mauritania); B.P. 42 Port Étienne (Mauritania); 87 Rue La Boëtie, Paris 8e (France).
SOCIÉTÉ ANONYME DES MINES DE L'ADRAR, 1 Rond-point Saint-Exupéry, Casablanca; P.O. 865 Casablanca (Morocco).
SOCIÉTÉ ANONYME DES MINES DU ZACCAR, 21 Avenue Montaigne, Paris 8e (France); Zaccar (Algeria).
SOCIÉTÉ ANONYME DES MINES ET FONDERIES DE ZINC DE LA VIEILLE MONTAGNE, Angleur près Liège (Belgium); 19 Rue

Richer, Paris 9e (France); Bou Caïd par Orléansville (Algeria).

SOCIÉTÉ ANONYME DU DJEBEL CHIKER, Taza; Aïn el Aouda (Morocco).

SOCIÉTÉ ANONYME DU DJEBEL HALLOUF, TUNISIE, 26 Rue d'Angleterre, Tunis; Souk-el-Khemis (Tunisia).

SOCIÉTÉ ANONYME FRANÇAISE DE RECHERCHES ET D'EXPLOITATION DE PÉTROLE (SAFREP), 7 Rue Nelaton, Paris 15e; 12 Rue Jean Nicot, Paris 7e (France); Vila Cisneros (Spanish Sahara); Groupe Vauban, Rue Louis Barthou, Hussein-Dey, Algiers (Algeria).

SOCIÉTÉ ANONYME MAROCAINE ITALIENNE DES PÉTROLES (SOMIP), B.P. 410 Casablanca; Rabat; Safi (Morocco).

SOCIÉTÉ ANONYME MAROCAINE ITALIENNE DE RAFFINAGE (SAMIR), Mohammedia (Morocco).

SOCIÉTÉ ANONYME MINIÈRES ET CARRIÈRES DE RIVET EL MADEN, 13 Rue de Turin, Paris 9e (France); 49 Rue Hassiba Ben Bouali, Algiers (Algeria).

SOCIÉTÉ AUSONIA MINIÈRE FRANÇAISE (AMIF), 41 Boulevard de la Tour-Maubourg, Paris 8e (France).

SOCIÉTÉ AUXILIAIRE DE FRANCEVILLE, Franceville (Gabon); 69 Rue de Varenne, Paris 7e (France).

SOCIÉTÉ AUXILIAIRE DE MINES (SAM), Ouandjia (Central African Republic).

SOCIÉTÉ AUXILIAIRE INDUSTRIELLE ET FINANCIÈRE DES GRANDS LACS AFRICAINS (AUXILACS), 24 Avenue de l'Astronomie, Brussels (Belgium).

SOCIÉTÉ AUXILIAIRE MINIAIRE COLONIALE, 17 Rue de la Chancellerie, Brussels (Belgium).

SOCIÉTÉ BELGO-AFRICAINE DU KIVU (SOBAKI), 16 Rue d'Egmont, Brussels (Belgium); Bukavu (Congo).

SOCIÉTÉ CHARBONNIÈRE DE L'OCÉAN INDIEN, Rue Rabearivelo, Antsahavola, Tananarive (Malagasy Republic).

SOCIÉTÉ CHÉRIFIENNE D'ENGRAIS ET DE PRODUITS CHIMIQUES, see: Office Chérifien des Phosphates.

SOCIÉTÉ CHÉRIFIENNE DES ENGRAIS PULVÉRISÉS, Berrechia, Mohammedia (Morocco).

SOCIÉTÉ CHÉRIFIENNE DES MINES, 2 Rue de Sfax, Rabat; Glib en Nam (Morocco).

SOCIÉTÉ CHÉRIFIENNE DES PÉTROLES, 27 Avenue Urbain Blanc, Rabat; Petitjean (Morocco).

SOCIÉTÉ CHÉRIFIENNE DES SELS, 5 Rue Martinière, Rabat; Lac Zima; El Ayasna (Morocco).

SOCIÉTÉ COMMERCIALE ET MINIÈRE DU CONGO (COMINIÈRE), 5 Rue de la Science, Brussels (Belgium).

SOCIÉTÉ CONGOLAISE DE LA VIEILLE MONTAGNE (CONGOVIELMONT), 23 Rue Belliard, Brussels (Belgium).

SOCIÉTÉ CONGOLAISE D'ÉLECTRICITÉ, 299 Avenue Lt. Valcke, B.P. 499 Léopoldville (Congo).

SOCIÉTÉ CONGOLAISE D'ENTREPOSAGE DE PRODUITS PÉTROLIERS (SOCOPETROL), Léopoldville (Congo).

SOCIÉTÉ CONGOLAISE DES PÉTROLES SHELL, Léopoldville (Congo).

SOCIÉTÉ CONGOLAISE–ITALIENNE DE RAFFINAGE (SOCIR), Léopoldville, Banana (Congo); 50, Via Tevere, Rome (Italy).

SOCIÉTÉ COOPÉRATIVE DE PÉTROLE (COOP), Cairo (U.A.R.).

SOCIÉTÉ DAHOMÉENNE D'ENTREPOSAGE DE PRODUITS PÉTROLIERS, Cotonou (Dahomey).

SOCIÉTÉ DE BANQUE ET DE PARTICIPATIONS, 10 Rue Volney, Paris 2e (France).

SOCIÉTÉ DE BRUXELLES POUR LA FINANCE ET L'INDUSTRIE (BRUFINA), Rue de la Régence, Brussels (Belgium).

SOCIÉTÉ DE FIBRE ET MÉCANIQUE (SOFIMEC), Douala (Cameroons).

SOCIÉTÉ DE GESTION DU DÉPÔT D'HYDROCARBURES DE TAMATAVE, B.P. 54 Tananarive; Tamatave (Malagasy Republic).

SOCIÉTÉ DE GESTION ET D'EXPLOITATION INDUSTRIES CHIMIQUES, Rue Montagne du Parc, Brussels (Belgium).

SOCIÉTÉ DE LA RAFFINERIE D'ALGER, 126bis Rue Michelet, Algiers (Algeria).

SOCIÉTÉ D'ÉLECTRO-CHIMIE, D'ÉLECTRO-METALLURGIE ET DES ACIÉRIES ÉLECTRIQUES D'UGINE, 10 Rue du Général-Foy, B.P. 722-08 Paris 8e (France); Ranomena par Tamatave; Andriamena (Malagasy Republic).

SOCIÉTÉ DE L'OUENZA, 78 Avenue d'Iéna, Paris 16e (France); Ouenza, Bône (Algeria).

SOCIÉTÉ D'ÉNERGIE ÉLECTRIQUE DE LA CÔTE D'IVOIRE, P.O. 1345 Abidjan (Ivory Coast).

SOCIÉTÉ D'ÉNERGIE ÉLECTRIQUE DE PORT GENTIL, Port Gentil (Gabon).

SOCIÉTÉ D'ENTREPOSAGE DE PRODUITS PÉTROLIERS, Bangui (Central African Republic).

SOCIÉTÉ D'ENTREPRISE ET D'INVESTISSEMENT DU BÉCÉKA (SIBÉKA), 46 Rue Royale, Brussels (Belgium).

SOCIÉTÉ DE PARTICIPATIONS PÉTROLIÈRES (PETROPAR), 7 Rue Nelaton, Paris 15e; 31 Rue Marbeuf, Paris 8e (France).

SOCIÉTÉ DE PROSPECTION ET EXPLOITATIONS PÉTROLIÈRES EN ALSACE (PREPA), 12 Rue Jean-Nicot, Paris 7e (France); 12 Avenue du General Yusuf, Algiers (Algeria).

SOCIÉTÉ DE RECHERCHES AU KATANGA (SOREKAT), c/o 42 Rue Royale, Brussels (Belgium).

SOCIÉTÉ DE RECHERCHES ET D'EXPLOITATION DE PÉTROLE (EURAFREP), Algiers (Algeria); Rourde el Baguel (Algeria).

SOCIÉTÉ DE RECHERCHES ET D'EXPLOITATION DES BAUXITES DU CONGO (BAUXICONGO), 46 Rue Royale, Brussels (Belgium); Sumbi par Tshela (Congo); P.O. 2399 Léopoldville (Congo).

SOCIÉTÉ DE RECHERCHES ET D'EXPLOITATION DES PÉTROLES DU CAMÉROUN (SOREPCA), Douala (Cameroons).

SOCIÉTÉ DE RECHERCHES ET D'EXPLOITATION DES PÉTROLES EN TUNISIE (SEREPT), 6 Rue René Caille, Tunis (Tunisia); 34–42 Avenue Raymond-Poincaré, Paris 16e (France).

SOCIÉTÉ DE RECHERCHES ET EXPLOITATIONS DIAMANTIFÈRES, chez la Compagnie de l'Afrique Noire, Brazzaville (Congo Republic); chez la Compagnie Centrafricaine des Mines, 6 Avenue de Messine, Paris 8e (France).

SOCIÉTÉ DE RECHERCHES MINIÈRES DU SUD-KATANGA (SUD-KAT), 6 Rue Montagne du Parc, Brussels (Belgium); Élisabethville; Kasekelesa (Congo).

SOCIÉTÉ DE RECHERCHES MINIÈRES EN AFRIQUE (REMNA), 112 Rue du Commerce, Brussels (Belgium); Kalima (Congo).
SOCIÉTÉ DES ARGILES DE BOU ADRA, 18 Avenue du Père de Foucauld, Rabat; Jebel Hariga; Tamdafelt (Morocco).
SOCIÉTÉ DES BAUXITES DU MOYEN-CONGO (BAMOCO), 42 Rue Royale, Brussels (Belgium).
SOCIÉTÉ DES CHARBONNAGES DE LA SAKOA, Bureau de Recherches Géologiques et Minières, Ampandrianomby, Tananarive (Malagasy Republic).
SOCIÉTÉ DES CHAUX HYDRAULIQUES ET CIMENTS D'ALGÉRIE, 19 Rue de la République, Lyon (France); Bougie (Algeria).
SOCIÉTÉ DES CIMENTS ARTIFICIELS DE MEKNÈS, Route de Fès, Meknès (Morocco).
SOCIÉTÉ DES CIMENTS D'AGADIR, Route de Mogador, Agadir (Morocco).
SOCIÉTÉ DES CIMENTS DU CONGO, 13 Rue Brédérode, Brussels (Belgium); Lukala; Stanleyville (Congo).
SOCIÉTÉ DES CIMENTS ET BETONS MANUFACTURES (SOCIBEMA), B.P. 600 Douala (Cameroons).
SOCIÉTÉ DES CIMENTS PORTLAND DE BIZERTE, Baie de Sebra, Bizerte (Tunisia).
SOCIÉTÉ DES CIMENTS PORTLAND DE HÉLOUAN, Hélouan, Cairo (U.A.R.).
SOCIÉTÉ DES FORCES HYDROÉLECTRIQUES DE L'EST DU CONGO (FORCES), 22 Rue de Livourne, Brussels (Belgium); Bukavu (Congo); Usumbura (Burundi).
SOCIÉTÉ DES FORCES HYDROÉLECTRIQUES DU BAS-CONGO (FORCES), 22 Rue de Livourne, Brussels (Belgium); Léopoldville (Congo).
SOCIÉTÉ DES GEMMES DE MADAGASCAR, 7 Rue de Liège, B.P. 117 Tananarive (Malagasy Republic).
SOCIÉTÉ DES GRAPHITES DE LA SAHANAVO, 21 Rue Marc-Rabibison, B.P. 164 Tananarive (Malagasy Republic).
SOCIÉTÉ DES HYPERPHOSPHATES RENO, Safi (Morocco).
SOCIÉTÉ DES MINERAIS DE LA GRANDE-ÎLE (SMGI), 23 Rue de l'Amiral d'Estaing, Paris 16e (France); B.P. 3 Fort-Dauphin (Malagasy Republic).
SOCIÉTÉ DES MINERAIS RARES DE MADAGASCAR (SOMIRAMAD), 28bis Avenue de l'Indépendance, B.P. 230 Tananarive (Malagasy Republic).
SOCIÉTÉ DES MINES D'AÏN-KERMA, 4 Rue de Rome, Paris 8e (France); 10 Rue Negrier, Bône (Algeria).
SOCIÉTÉ DES MINES D'AMBATOBE, chez le Syndicat Lyonnais de Madagascar, 33 Rue de Liège, B.P. 62 Tananarive (Malagasy Republic).
SOCIÉTÉ DES MINES D'ANTIMOINE DE L'ICH OU MELLAL, 34 Boulevard de la Gare, Casablanca; Ich ou Mellal; Aïn Koheul (Morocco).
SOCIÉTÉ DES MINES D'AOULI, 1 Rond-Point Saint-Exupéry, Casablanca; Aouli; Mibladen; Midelt (Morocco).
SOCIÉTÉ DES MINES DE BOU SKOUR, 52 Avenue Hassan II, Casablanca (Morocco).
SOCIÉTÉ DES MINES DE FER DE KHANGUET, 60 Rue de la Victoire, Paris 9e (France); Le Kouif, Bône (Algeria).
SOCIÉTÉ DES MINES DE FER DE MÉKAMBO (SOMIFER), 8 Place Vendôme, Paris 1er (France); Libreville (Gabon); Bethlehem, Pa. (U.S.A.).
SOCIÉTÉ DES MINES DE FER DE MILIANA, 5 Avenue Sainte-Foy, Neuilly sur Seine, Seine (France); 13 Rue Claudel, Algiers; El Halia; Philippeville Quai, Philippeville (Algeria).
SOCIÉTÉ DES MINES DE L'ADRAR, 1 Rond-Point Saint-Exupéry, Casablanca; Taouz; Mefis; Chib er Ras; Tizi n'Ressas (Morocco).
SOCIÉTÉ DES MINES DE L'ASSIF EL MAL, 69 Rue Alexandre 1er, Marrakech; Assif el Mal (Morocco).
SOCIÉTÉ DES MINES DE BOU SKOUR, 26 Rue Michel de l'Hospital, Casablanca; Bou Skour (Morocco).
SOCIÉTÉ DES MINES DE DOUARIA, Place Amiral Darrieu, Bizerte (Tunisia).
SOCIÉTÉ DES MINES DE L'OUED MADEN, 47 Rue de Portugal, Tunis (Tunisia).
SOCIÉTÉ DES MINES DE PALESTRO, Parc Gatlif, Mustapha Supérieur, Algiers (Algeria).
SOCIÉTÉ DES MINES DE POURA, Poura par Boromo (Upper Volta); 2 Rue Lord-Byron, Paris 8e (France).
SOCIÉTÉ DES MINES DE SAHANAVO, Sahanavo, Tamatave Province, (Malagasy Republic).
SOCIÉTÉ DES MINES DE SAINTE MARIE, 52 Avenue d'Amade, Casablanca; El Hammam; Bergamou (Morocco).
SOCIÉTÉ DES MINES DE SAKASOA, Betroka (Malagasy Republic).
SOCIÉTÉ DES MINES DE SEL DE MOGADOR, Mogador; Ida ou Iazza (Morocco).
SOCIÉTÉ DES MINES DE SIDI KAMBER, 3 Rue Théophile-Bressy, Algiers; Mine de Sidi Kamber par Sidi Mesrich, Département de Constantine; Mine d'Aïn Barbar par Bugeaud, Département de Bône (Algeria); 12 Place Vendôme, Paris 1er (France).
SOCIÉTÉ DES MINES D'ÉTAIN DU RUANDA ET DE L'URUNDI (MINÉTAIN), 33 Rue des Colonies, Brussels (Belgium).
SOCIÉTÉ DES MINES DE TIMEZRIT, 21 Avenue Montaigne, Paris 8e (France); Timezrit, Bougie (Algeria).
SOCIÉTÉ DES MINES DE ZELLIDJA, 1 Place Mirabeau, Casablanca; Bou Beker par Oujda; Bou Beker (Morocco); 26 Rue Geoffroy L'Asnier, Paris 4e (France).
SOCIÉTÉ DES MINES DE ZINC DU GUERGOUR, 4 Rue de Rome, Paris 8e (France); Aïn Sedjera par Lafayette, Département de Sétif (Algeria).
SOCIÉTÉ DES MINES D'OR DE KILO–MOTO, 1 Place du Luxembourg, Brussels (Belgium); Kilo (Congo); Watsa (Congo); 6 Avenue Renkin, Léopoldville (Congo).

SOCIÉTÉ DES MINES DU CONGO SEPTENTRIONAL (SOMINOR), 41 Rue Jean Staes, Brussels (Belgium).

SOCIÉTÉ DES MINES DU DJEBEL AZERED (SOMIDA), 8 Rue d'Avignon, Tunis (Tunisia).

SOCIÉTÉ DES MINES DU DJEBEL SALRHEF, 129 Rue Verlet Hanus, Marrakech; Jebel Salrhef (Morocco).

SOCIÉTÉ DES MINES ET FONDERIES DE ZINC DE LA VIEILLE MONTAGNE, 23 Rue Belliard, Brussels (Belgium); 19 Rue Richer, Paris 9e (France); Mines de l'Ouarsenis, Bou-Caïd, Orléansville (Algeria).

SOCIÉTÉ DES PÉTROLES BP D'AFRIQUE OCCIDENTALE, 2 Avenue Albert Sarraut, Dakar (Senegal).

SOCIÉTÉ DES PÉTROLES BP D'ALGÉRIE, 76 Avenue Général Yusuf, Algiers (Algeria).

SOCIÉTÉ DES PÉTROLES BP DU CONGO, B.P. 15 Usumbura (Burundi).

SOCIÉTÉ DES PÉTROLES BP DU MAROC, Place Zallaga, Casablanca (Morocco).

SOCIÉTÉ DES PÉTROLES BP DU TUNISIE, 86 Avenue Hedi Chaker, Tunis (Tunisia).

SOCIÉTÉ DES PÉTROLES D'AFRIQUE ÉQUATORIALE (SPAFE), B.P. 524 Port-Gentil (Gabon); 7 Rue Nélaton, Paris 15e (France); B.P. 761 Pointe-Noire (Congo Republic).

SOCIÉTÉ DES PÉTROLES DE MADAGASCAR, Tananarive (Malagasy Republic).

SOCIÉTÉ DES PÉTROLES DU CONGO, Léopoldville (Congo).

SOCIÉTÉ DES PÉTROLES DU SÉNÉGAL (SPS), 77 Boulevard Malesherbes, Paris 8e (France); 2 Avenue Albert Sarraut, Dakar (Senegal).

SOCIÉTÉ DES PHOSPHATES TUNISIENS ET DES ENGRAIS DE PHOSPHATES ET PRODUITS CHIMIQUES, 40 Rue Marcescheau, Tunis (Tunisia).

SOCIÉTÉ DES PRODUITS BARYTIQUES NORD-AFRICAINS, 280 Boulevard Saint Germain, Paris 7e (France).

SOCIÉTÉ DES SALINES D'ASSAB, Assab (Ethiopia).

SOCIÉTÉ DES TERRES RARES DU SUD DE MADAGASCAR (SOTRASUM), B.P. 1487 Tananarive (Malagasy Republic); B.P. 87–07 Paris (France).

SOCIÉTÉ DE TRANSPORT DE GAZ NATUREL D'HASSI ER R'MEL À ARZEW (SOTHRA), 9 Rue de l'Aspirante Denise-Ferrier, Hydra, Algiers (Algeria); 37 Avenue Pierre 1er de Serbie, Paris 8e (France).

SOCIÉTÉ DE TRANSPORTS DE PÉTROLES À L'EST SAHARIEN (TRAPES) (CEP), Haoud-el-Hamra; Bougie (Algeria).

SOCIÉTÉ D'ÉTUDE DE LA CIMENTERIE DU NORD-CAMEROUN, Paris (France); Garoua (Cameroons).

SOCIÉTÉ D'ÉTUDE DU TRANSPORT DU GAZ D'HASSI ER R'MEL PAR CANALISATIONS TRANSMEDITERRANÉENNES, 93 Avenue de Neuilly, Neuilly sur Seine (France).

SOCIÉTÉ D'ÉTUDE DU TRANSPORT ET DE LA VALORISATION DES GAZ NATURELS DU SAHARA (SEGANS), 7 Rue Nélaton, Paris 15e (France).

SOCIÉTÉ D'ÉTUDES DE RECHERCHES ET D'EXPLOITATION MINIÈRES (SEREM), 12 Rue Bergé, Tananarive (Malagasy Republic).

SOCIÉTÉ D'ÉTUDES D'EXPLOITATIONS MINIÈRES DE L'ATLAS, Rue du Professeur Roux, Agadir; Tachdamt; Tamegra; El Borj (Morocco).

SOCIÉTÉ D'ÉTUDES ET D'APPLICATION DES MINERAIS DE THIÈS, Rue Faubourg Saint Honoré, Paris (France).

SOCIÉTÉ D'EXPLOITATION DES HYDROCARBURES D'HASSI ER RMEL (SEHR), 1 Rue Campocosso, Parc d'Hydra, Algiers 8e (Algeria).

SOCIÉTÉ D'EXPLOITATION DES POTASSES DE HOLLÉ, Brazzaville; Pointe Noire (Congo Republic).

SOCIÉTÉ D'EXPLOITATION DE TOURTIT ET D'ÉTUDES MINIÈRES, Rue du Sous-Lieutenant Préjean, Casablanca; Tirrhist (Morocco).

SOCIÉTÉ D'EXPLOITATION ET DE RECHERCHES MINIÈRES DANS L'AFRIQUE DU NORD (SERMAN), 1 Rue de Constantine, Tunis (Tunisia).

SOCIÉTÉ D'EXPLOITATION ET DE TRAITEMENT DES MINERAIS (EXTRAIMINE), 63 Boulevard de la Gare, Casablanca; Mejma Sline (Morocco).

SOCIÉTÉ D'EXPLOITATION MINIÈRE DE L'ANDROY (SEMA), B.P. 42 Fort-Dauphin (Malagasy Republic).

SOCIÉTÉ DIAMANTIFÈRE DE LA CÔTE D'IVOIRE (SODIAMCI), B.P. 1368 Abidjan; Tortiya (Côte d'Ivoire).

SOCIÉTÉ DU BITUME ET DE L'ASPHALTE DU CONGO (SOBIASCO), 42, Rue Royale, Brussels (Belgium).

SOCIÉTÉ DU DJEBEL DJERISSA, 60 Rue de la Victoire, Paris 9e (France); Djerissa (Tunisia).

SOCIÉTÉ DU DJEBEL-ONK, 12 Avenue Marceau, Paris 8e (France); 8 Rue de Flandre, Bône (Algeria).

SOCIÉTÉ DU QUARTZ DE MADAGASCAR, Rue Nicolas-Mayeur, Tsaralalàna; B.P. 320 Tananarive (Malagasy Republic).

SOCIÉTÉ ÉGYPTIENNE POUR LE RAFFINAGE ET LE COMMERCE DU PÉTROLE S.A.E., Cairo (U.A.R.).

SOCIÉTÉ EUROPÉENNE DES DÉRIVÉS DU MANGANÈSE, Terture (Belgium).

SOCIÉTÉ FARCAHAL, 81 Rue Alexandre 1er, Marrakech; Agouni; Tazigzaout; Tifrit el Madène (Morocco).

SOCIÉTÉ FINANCIÈRE DES PÉTROLES (FINAREP), 41 Avenue de l'Opéra, Paris 2e (France).

SOCIÉTÉ FRANÇAISE DE FORAGE (FORAFRANCE), 7 Rue Pillet-Will, Paris 9e (France); 21 Rue no. 3 Lotissement Dar Naama El Biar, Algiers (Algeria).

SOCIÉTÉ FRANÇAISE DES PÉTROLES BP, 19–21 Rue de la Bienfaisance, Paris 8e (France).

SOCIÉTÉ FRANÇAISE D'EXPLORATION BP, 21 Rue de la Bienfaisance, Paris 8e (France).

SOCIÉTÉ FRANÇAISE POUR LE COMMERCE EN EXTRÊME-ORIENT (SOFICOMEX), Central P.O. Box 403 Tokyo (Japan).
SOCIÉTÉ GÉNÉRALE AFRICAINE D'ÉLECTRICITÉ (SOGELEL), 31 Rue de la Science, Brussels (Belgium); Élisabethville (Congo).
SOCIÉTÉ GÉNÉRALE DE BELGIQUE, 3 Rue Montagne du Parc, Brussels (Belgium).
SOCIÉTÉ GÉNÉRALE D'ENGRAIS ET PRODUITS CHIMIQUES PIERREFITTE, 4 Avenue Vélasquez, Paris 8e (France).
SOCIÉTÉ GÉNÉRALE DES FORCES HYDROÉLECTRIQUES DU KATANGA (SOGEFOR), 6 Rue Montagne du Parc, Brussels (Belgium); Élisabethville (Congo).
SOCIÉTÉ GÉNÉRALE DES GRAPHITES, B.P. 117 Tananarive (Malagasy Republic).
SOCIÉTÉ GÉNÉRALE DES MINERAIS, 31 Rue du Marais, Brussels (Belgium).
SOCIÉTÉ GÉNÉRALE INDUSTRIELLE ET CHIMIQUE DE JADOTVILLE (SOGECHIM), Jadotville (Congo).
SOCIÉTÉ GÉNÉRALE INDUSTRIELLE ET CHIMIQUE DU KATANGA (SOGECHIM), 6 Rue Montagne du Parc, Brussels (Belgium).
SOCIÉTÉ GÉNÉRALE MÉTALLURGIQUE DE HOBOKEN, Hoboken (Belgium).
SOCIÉTÉ GÉNÉRALE POUR L'INDUSTRIE, Rue Bovy Lysderg 17, Geneva (Switzerland).
SOCIÉTÉ GÉOLOGIQUE ET MINIÈRE DES INGÉNIEURS ET INDUSTRIELS BELGES (GÉOMINES), 4 Rue de la Science, Brussels (Belgium).
SOCIÉTÉ GÉOLOGIQUE ET MINIÈRE DU ZAMBÈZE, Tete (Mozambique).
SOCIÉTÉ GUINÉENNE D'ENTREPOSAGE, Conakry (Guinea).
SOCIÉTÉ H.A. DE HEAULME, B.P. 37 Fort-Dauphin (Malagasy Republic).
SOCIÉTÉ HELLÉNIQUE DES MINERAIS ET MÉTAUX, 7 Rue Merlin, Athens (Greece).
SOCIÉTÉ INTERNATIONALE COMMERCIALE ET FINANCIÈRE DE LA FORMINIÈRE (INTERFOR), 46 Rue Royale, Brussels (Belgium).
SOCIÉTÉ INTERNATIONALE D'EXPLOITATION MINIÈRE AU MAROC (INTERMINE), 179 Route des Ouled Ziane, Casablanca; Narguechoum; Tanourat (Morocco).
SOCIÉTÉ INTERNATIONALE FORESTIÈRE ET MINIÈRE (FORMINIÈRE), Tshikapa (Congo); 46 Rue Royale, Brussels (Belgium).
SOCIÉTÉ ITALO-TUNISIENNE D'EXPLOITATION PÉTROLIÈRE, Tunis (Tunisia).
SOCIÉTÉ IVOIRIENNE DE RAFFINAGE, Abidjan (Ivory Coast).
SOCIÉTÉ 'LE MOLYBDÈNE', 81 Rue Colbert, Casablanca; Azegour (Morocco).
SOCIÉTÉ 'LES MINES RÉUNIES', Mine de Sidi-Amor-Ben Salem (Tunisia).
SOCIÉTÉ LYONNAISE AGRICOLE MINIÈRE ET INDUSTRIELLE (SLAMI), c/o Compagnie Lyonnaise de Madagascar, B.P. 188 Tananarive (Malagasy Republic); 12 Rue Faubourg Saint Honoré, Paris 8e (France).
SOCIÉTÉ MALGACHE DES MINERAIS ET GEMMES (MM. KRAFFT ET CO.), Avenue de la Réunion, Tananarive (Malagasy Republic).
SOCIÉTÉ MAROCAINE DES MINES ET PRODUITS CHIMIQUES, 1 Place Mirabeau, B.P. 269 Casablanca; B.P. 1 Oued Zem; Djebel Irhoud; Safi; Tessaout; Marrakech (Morocco).
SOCIÉTÉ MAROCAINE D'EXPLOITATIONS MINIÈRES, Bou Arfa; Jebel Klakh; Djahifat; Foum Defla; Mechkakour (Morocco); 101 Rue Saint Lazare, Paris 9e (France).
SOCIÉTÉ MAROCAINE ITALIENNE DE RAFFINAGE (SAMIR), Mohammedia (Morocco).
SOCIÉTÉ MAROCAINE-ITALIENNE DES PÉTROLES (SOMIP), 17 Rue Bullet, Casablanca; Safi (Morocco).
SOCIÉTÉ MÉTALLURGIQUE DU KATANGA (METALKAT), 7 Rue de la Chancellerie, Brussels (Belgium); Kolwezi (Congo Republic).
SOCIÉTÉ MÉTALLURGIQUE KATANGAISE (METALKAT), Kolwezi; Jadotville (Congo).
SOCIÉTÉ MINIÈRE DE BAKWANGA, Bakwanga (Congo).
SOCIÉTÉ MINIÈRE DE BOU AZZER ET DU GRAARA, 52 Avenue Hassan II, Casablanca; B.P. 'Ona 657' Casablanca; Bou Azzer; Bou Offroh (Morocco).
SOCIÉTÉ MINIÈRE DE CARNOT (SOMICA), Carnot (Central African Republic).
SOCIÉTÉ MINIÈRE DE KISENGE, Kisenge, par Élisabethville (Congo).
SOCIÉTÉ MINIÈRE DE KSIBA, 34 Boulevard de la Gare, Casablanca; Bou Ouchèn (Morocco).
SOCIÉTÉ MINIÈRE DE LA LUETA, 46 Rue Royale, Brussels (Belgium).
SOCIÉTÉ MINIÈRE DE LA TÉLÉ, 46 Rue Royale, Brussels (Belgium).
SOCIÉTÉ MINIÈRE DE L'ATLAS MAROCAIN, 1 Rond-Point Saint-Exupéry, B.P. 865 Casablanca; Ksar-es-Souk; Keba; Megta Sfa; Taklimt (Morocco).
SOCIÉTÉ MINIÈRE DE L'EST-OUBANGHI, 4 Rue de Penthièvre, Paris 8e (France); Ambilo par Bria (Central African Republic).
SOCIÉTÉ MINIÈRE DE LUEBO, 46 Rue Royale, Brussels (Belgium).
SOCIÉTÉ MINIÈRE DE LUESHE, Lueshe, Kivu (Congo); 270 Park Avenue, New York, N.Y. (U.S.A.).
SOCIÉTÉ MINIÈRE DE MUHINGA ET DE KIGALI (SOMUKI), Rutongo lez Kigali (Rwanda).
SOCIÉTÉ MINIÈRE DES ABDA-AHMAR, Safi; Chemaïa; Sidi Rahmoun (Morocco).
SOCIÉTÉ MINIÈRE DES GUNDAFA, 81 Avenue Moinier, Casablanca; Toundout; Kheneg el Brak; Ouicheddène; Ounein (Morocco).

SOCIÉTÉ MINIÈRE DES REHAMNA, 1 Rond-Point Saint-Exupéry, Casablanca; Ouled Hassine (Morocco).
SOCIÉTÉ MINIÈRE DES TORBA, 18 Avenue du Père de Foucauld, Rabat; Camp Berteaux (Morocco).
SOCIÉTÉ MINIÈRE DE TIRZA, 12 Avenue Dar el Maghzen, Rabat; Mguedh; Tirza (Morocco).
SOCIÉTÉ MINIÈRE DU BÉCÉKA, 46 Rue Royale, Brussels (Belgium).
SOCIÉTÉ MINIÈRE DU DAHOMEY–NIGER, Agadès (Niger Republic); 1 Rue de Stockholm, Paris 8e (France).
SOCIÉTÉ MINIÈRE DU DJEBEL AOUM, 15 Rue de Danvilliers, Casablanca; M'Rirt (Morocco); 42 Avenue Gabriel, Paris 8e (France).
SOCIÉTÉ MINIÈRE DU DJEBEL TAZZEKA, 24 Rue Marcel Chapon, Casablanca; Dar Izid; Boujada (Morocco).
SOCIÉTÉ MINIÈRE DU HAUT-GUIR, 1 Rond-Point Saint-Exupéry, Casablanca; Beni Tadjit (Morocco).
SOCIÉTÉ MINIÈRE DU KASAÏ, 46 Rue Royale, Brussels; 2 Rue de Brédérode, Brussels (Belgium).
SOCIÉTÉ MINIÈRE DU KHANGUET, Le Khanguet Kef près Béja (Tunisia).
SOCIÉTÉ MINIÈRE DU LUALABA (MILUBA), 24 Avenue de l'Astronomie, Brussels (Belgium); c/o COBELMIN, Bukavu (Congo); Kima par Punia (Congo).
SOCIÉTÉ MINIÈRE DU MICOUNZON, Étéké par Mouila (Gabon).
SOCIÉTÉ MINIÈRE DU NORD-AFRICAIN, 12bis Rue Raspail, Montfleury (Tunisia).
SOCIÉTÉ MINIÈRE DU SIROUA, 58 Rue Chevandier de Valdrôme, Casablanca; N'Kob (Morocco).
SOCIÉTÉ MINIÈRE DU TIZI N'RECHOU, 41 Rue Général Margueritte, Casablanca; Tizi n'Rechou (Morocco).
SOCIÉTÉ MINIÈRE DU ZAMZA (SMZ), Bria; Berberati (Central African Republic).
SOCIÉTÉ MINIÈRE EST OUBANGUI (SMEO), Yalinga (Central African Republic).
SOCIÉTÉ MINIÈRE ET AGRICOLE DE GARN-ALFAYA, Garn Alfaya; Majembia par Tadjerouine (Tunisia).
SOCIÉTÉ MINIÈRE ET FORESTIÈRE, B.P. 37 Fort-Dauphin (Malagasy Republic).
SOCIÉTÉ MINIÈRE ET MÉTALLURGIQUE D'AOULOUZ, 52 Avenue d'Amade, Casablanca; Tasdremt (Morocco).
SOCIÉTÉ MINIÈRE ET MÉTALLURGIQUE DE PEÑARROYA, 12 Place Vendôme, Paris 1er (France); Fonderie de Mégrine S.M.M.P., Mégrine; 47 Rue de Portugal, Tunis (Tunisia).
SOCIÉTÉ MINIÈRE GAZIELLO ET CO., B.P. 311 Dakar (Senegal); Thann, Haut-Rhin (France).
SOCIÉTÉ MINIÈRE INTERCOLONIALE (SMI), Ouadda (Central African Republic).
SOCIÉTÉ MINIÈRE INTERTROPICALE, Berbérati (Central African Republic); chez la Compagnie Centrafricaine des Mines, 6 Avenue de Messine, Paris 8e (France).
SOCIÉTÉ MINIÈRE MAROCAINE D'OUJJIT, 4 Rue d'Algérie, Casablanca; Jebel Oujjit (Morocco).
SOCIÉTÉ MINIÈRE OGOUÉ LOBAYE, Berbérati (Central African Republic); Kellé (Congo Republic); Société Immobilière Roll Berthier, 14 Rue Alfred Roll, Paris 17e (France).
SOCIÉTÉ NATIONALE ALGÉRIENNE DE TRANSPORTS ET COMMERCIALISATION DES HYDROCARBURES, Algiers; Arzew (Algeria).
SOCIÉTÉ NATIONALE DE MATÉRIEL POUR LA RECHERCHE ET L'EXPLOITATION DU PÉTROLE (S.N. MAREP), 12–16 Rue Jean Nicot, Paris 7e (France); 26 Broadway, New York 4, N.Y. (U.S.A.); Algiers; Hassi Messaoud (Algeria).
SOCIÉTÉ NATIONALE DE RECHERCHE ET D'EXPLOITATION DES PÉTROLES EN ALGÉRIE (S.N. REPAL), 105 Avenue Raymond Poincaré, Paris 16e (France); Chemin du Réservoir, Hydra, Algiers; Hassi Messaoud (Algeria).
SOCIÉTÉ NATIONALE DE RECHERCHE ET D'EXPLOITATION MINIÈRES (SINAREX), Bangui (Central African Republic).
SOCIÉTÉ NATIONALE DES MINES DE BENI SAF, BERIM, Palais du Gouvernement, Algiers; Beni Saf (Algeria).
SOCIÉTÉ NATIONALE DES MINES DE ROUINA, BERIM, Palais du Gouvernement, Algiers; Miliana (Algeria).
SOCIÉTÉ NATIONALE DES MINES DE ZACCAR, MILIANA, BERIM, Palais du Gouvernement, Algiers (Algeria).
SOCIÉTÉ NATIONALE DES PÉTROLES D'ALGÉRIE, HYDRA, Algiers (Algeria).
SOCIÉTÉ NATIONALE DES PÉTROLES D'AQUITAINE, 16 Cours Albert 1er, 2 Rue Bayard, Paris 8e (France); 2 Boulevard Saint-Saëns, Algiers (Algeria).
SOCIÉTÉ NATIONALE POUR LA PRODUCTION DU CIMENT, 1 Soliman Pasha Street, Cairo (U.A.R.).
SOCIÉTÉ NORD-AFRICAINE DE L'AMIANTE-CIMENT 'DIMATIT', 81 Rue Lapérouse, Casablanca; Tif Dra (Morocco).
SOCIÉTÉ NORD-AFRICAINE DES CIMENTS LAFARGE, 88 Rue Michelet, Algiers (Algeria).
SOCIÉTÉ NORD-AFRICAINE DU PLOMB (NAP), Zellidja (Morocco).
SOCIÉTÉ NOUVELLE DES MINES D'AÏN-ARKO, B.P. 20 Zellidja Bou Beker par Oujda; El Abed; Tlemcen (Morocco); 26 Rue Geoffroy l'Asnier, Paris 4e (France).
SOCIÉTÉ NOUVELLE DES MINES DE SIDI-BOU-AOUANE, 40 Rue Mareschau, Tunis; Mines de Sidi-Bou-Aouane par Souk-El-Khemis (Tunisia); 23 Rue de l'Amiral d'Estaing, Paris 16e (France).
SOCIÉTÉ NOUVELLE D'EXPLOITATIONS MINIÈRES, 69 Boulevard de la Résistance, Casablanca; Khaloua-Satour (Morocco).
SOCIÉTÉ OUEST-AFRICAINE DES CIMENTS (SOCOCIM), B.P. 29 Rufisque (Senegal).
SOCIÉTÉ PEÑARROYA-MAROC, 1 Rond-Point Saint-Exupéry, Casablanca (Morocco).
SOCIÉTÉ PÉTROLIÈRE DE GÉRANCE (SOPEG), 64 Rue Charron, Paris 8e (France).
SOCIÉTÉ POUR LE DÉVELOPPEMENT MINIER DE LA CÔTE D'IVOIRE (SODEMI), Abidjan (Ivory Coast); 50 Broadway, New York, N.Y. (U.S.A.).

SOCIÉTÉ POUR LE TRANSPORT DES HYDROCARBURES SAHARIENS AU LITTORAL ALGÉRIEN (TRAPAL), Algiers (Algeria)
SOCIÉTÉ SACHRO-OUGMAR, 72 Rue Lamoricière, Casablanca; Tikirt; Tifernine (Morocco).
SOCIÉTÉ SAHARIENNE DE RECHERCHES PÉTROLIÈRES (SSRP), 21 Rue de la Bienfaisance, Paris 8e (France); 126bis Rue Michelet, Plateau Saulière, B.P. 46 Algiers (Algeria).
SOCIÉTÉ SHELL D'ALGÉRIE, 46 Boulevard Saint-Saëns, Algiers (Algeria).
SOCIÉTÉ SHELL DE L'AFRIQUE ÉQUATORIALE, Brazzaville (Congo Republic).
SOCIÉTÉ SHELL DE TUNISIE, Tunis (Tunisia).
SOCIÉTÉ TOGOLAISE D'ENTREPOSAGE, Lomé (Togo).
SOCIÉTÉ TUNISIENNE D'ENGRAIS SUPERPHOSPHATÉS, Sfax (Tunisia).
SOCIÉTÉ TUNISIENNE DES PHOSPHATES D'AÏN-KERMA, 70 Rue de Corse, Tunis (Tunisia).
SOCIÉTÉ TUNISIENNE D'ÉTUDES, DE COOPÉRATION ET DE DÉFENSE DE L'INDUSTRIE PHOSPHATIÈRE (STECDIP), 26 Rue d'Angleterre, Tunis (Tunisia).
SOCIÉTÉ TUNISIENNE D'EXPLOITATION PHOSPHATIÈRES (STEPHOS), 9 Rue Mazagran, B.P. 144 Tunis; Kalaa Djerola (Tunisia); 4 Avenue Velasquez, Paris 8e (France).
SOCIÉTÉ TUNISIENNE MINIÈRE ET MÉTALLURGIQUE, 47 Rue de Portugal, Tunis (Tunisia).
SOCIÉTÉ TUNISO-ITALIENNE DE RAFFINAGE (STIR), Tunis (Tunisia).
SOCIÉTÉ 'WOLFRAM DU ZGUIT', 10 Avenue de Champagne, Rabat; Oulmès-Zguit (Morocco).
SOCIR, see: Société Congolaise Italienne de Raffinage.
SOCOCIM, see: Société Ouest-Africaine des Ciments.
SOCONY MOBIL OIL CO., 150 East 42nd Street, New York 17, N.Y. (U.S.A.).
SOCONY OVERSEAS OIL CO., 37–43 Sackville Street, London W. 1 (Great Britain).
SOCONY SOUTHERN AFRICA (PTY.) LTD., Corner Isando Road and Brewery Street, Islando, Kempton Park, Tvl. (South Africa); Mbabane (Swaziland); Lobatsi (Bechuanaland); Windhoek (Southwest Africa).
SOCOPETROL, see: Société Congolaise d'Entreposage de Produits Pétroliers.
SODEMI, see: Société pour le Développement Minier de la Côte d'Ivoire.
SODIAMCI, see: Société Diamantifère de la Côte d'Ivoire.
SOFICOMEX, see: Société Française pour le Commerce en Extrème-Orient.
SOFIMEC, see: Société de Fibre et Mécanique.
SOGECHIM, see: (a) Société Générale Industrielle et Chimique de Jadotville.
(b) Société Générale Industrielle et Chimique du Katanga.
SOGEFOR, see: Société Générale des Forces Hydroélectriques du Katanga.
SOGELEL, see: Société Générale Africaine d'Électricité.
SOGUINEX, Macenta (Guinea); Mason's Avenue, London E.C. 2 (Great Britain).
SOMALI GULF OIL CO., Mogadishu (Somali Republic).
SOMICA, see Société Minière de Carnot.
SOMIDA, see: Société des Mines du Djebel Azared.
SOMIFER, see: Société des Mines de Fer de Mékambo.
SOMINOR, see: Société des Mines du Congo Septentrional.
SOMIP, see: Société Anonyme Marocaine Italienne des Pétroles.
SOMIRAMAD, see: Société des Minerais Rares de Madagascar.
SOMUKI, see: Société Minière de Muhinga et de Kigali.
SONAREP, see: Sociedade Nacional de Refinação de Petroleos.
SOPEG, see: Société Pétrolière de Gérance.
SOREKAT, see: Société de Recherches au Katanga.
SOREPCA, see: Société de Recherches et d'Exploitation des Pétroles du Cameroun.
SOTHRA, see: Société de Transport de Gaz Naturel d'Hassi er R'Mel à Arzew.
SOTRASUM, see: Société des Terres rares du Sud de Madagascar.
SOUAREZ, A., 39 Rue de la Marne, Meknès; Jebel Mirsan (Morocco).
SOUTH AFRICAN ASBESTOS TRADING (PTY.) LTD., P.O. Box 8613 Johannesburg (South Africa).
SOUTH AFRICAN COAL ESTATES (WITBANK) LTD., 44 Main Street, P.O. Box 2567 Johannesburg (South Africa).
SOUTH AFRICAN COAL, OIL AND GAS CORP. (SASOL), P.O. Boxes 1 and 32 Sasolburg (South Africa).
SOUTH AFRICAN GENERAL INVESTMENTS TRUST, P.O. Box 4587 Johannesburg (South Africa).
SOUTH AFRICAN IRON AND STEEL INDUSTRIAL CORP. (ISCOR), P.O. Box 450 Pretoria; P.O. Thabazimbi; P.O. Sishen (South Africa).
SOUTH AFRICAN LAND AND EXPLORATION CO. LTD., 44 Main Street, Johannesburg; P.O. Box 200 Witbank, Tvl.; P.O. Box 202 Brakpan, Tvl. (South Africa).
SOUTH AFRICAN MUTUAL LIFE ASSURANCE SOCIETY, P.O. Box 66 Cape Town (South Africa).
SOUTH AFRICAN MANGANESE LTD., Albatross House, Johannesburg; Lohathla, Griqualand (South Africa).

SOUTH AFRICAN MINERALS CORP., Frederick and Sauer Streets, P.O. Box 1494 Johannesburg; Fox and Harrison Streets, P.O. Box 590 Johannesburg; P.O. Eerstemyn Henneman (South Africa); P.O. Otjosundu (Southwest Africa).

SOUTH AFRICAN NATIONAL LIFE INSURANCE CO., Sanlam Building, corner Andries and Pretorius Streets, Pretoria (South Africa).

SOUTH AFRICAN OXIDES (PTY.) LTD., P.O. Box 240 Brakpan, Tvl.; Hazeldene, P. Bag 602 Longueval; P.O. Dundee (South Africa).

SOUTH AFRICAN PETROLEUM REFINERIES (PTY.) LTD., Durban (South Africa).

SOUTH AFRICAN TORBANITE MINING AND REFINING CO., 56 Main Street, P.O. Box 7727 Johannesburg (South Africa).

SOUTH AFRICAN TOWNSHIPS, MINING AND FINANCE CORP., 44 Main Street, P.O. Box 4587 Johannesburg (South Africa).

SOUTH AMERICAN GOLD AND PLATINUM CO., 535 Fifth Avenue, New York 17, N.Y. (U.S.A.).

SOUTHERN CALIFORNIA PETROLEUM CORP., 4250 Wilshire Boulevard, Los Angeles 5, Calif. (U.S.A.).

SOUTHERN RHODESIAN CHRYSOTILE CORP., Olifantsfontein (South Africa); Gurumba, Tumbwa, Belingwe (Southern Rhodesia).

SOUTH ROODEPOORT MAIN REEF AREAS LTD., General Mining Building, 6 Hollard Street, P.O. Box 1173 Johannesburg P.O. Box 11 Roodepoort, Tvl. (South Africa).

SOUTHWEST AFRICA CO. LTD., 49 Moorgate, London E.C. 2 (Great Britain); Grootfontein (Southwest Africa).

SOUTHWEST AFRICA GEMS (PTY.) LTD., P.O. Box 42 Swakopmund (Southwest Africa).

SOUTHWEST AFRICA LITHIUM MINES, P.O. Box 1517 Windhoek (Southwest Africa).

SOUTHWEST AFRICAN SALT CO., P.O. Box Swakopmund (Southwest Africa).

SOUTHWEST FINANCE CORP., THE, P.O. Box 616 Kimberley (South Africa).

SOUTH WITBANK COAL MINES LTD., 205/212 His Majesty's Building, Joubert Street, P.O. Box 3910 Johannesburg; P.O. Oogies (South Africa).

SOUTHERN DIAMOND CORP., Oranjemund (Southwest Africa).

SOVINSKY, J., P.O. Steinkopf (South Africa).

SPAARWATER GOLD MINING CO. LTD., 120 Moorgate, London E.C. 2 (Great Britain); P.O. Box 1167 Johannesburg; P.O. Box 12 Dunnottar (South Africa).

SPAFE, see: Société des Pétroles d'Afrique Équatoriale.

SPITZKOP COLLIERY (PTY). LTD., P.O. Box 171 Springs, Tvl.; P.O. Box 448 Ermelo, Tvl. (South Africa).

SPRINGBOK COLLIERY LTD., 44 Main Street, P.O. Box 2567 Johannesburg; P.O. Vandyksdrif (South Africa).

SPRINGFIELD COLLIERIES LTD., 44 Main Street, P.O. Box 2567 Johannesburg, Tvl.; P.O. Grootvlei (South Africa).

SPRINGS MINES LTD., 44 Main Street, P.O. Box 4587 Johannesburg; P.O. Box 54 Springs, Tvl. (South Africa).

SPS, see: Société des Pétroles du Sénégal.

SPYROPOULOS, T.S., P.O. Box 44 Kabale (Uganda).

SSRP, see: Société Saharienne de Recherches Pétrolières.

STAFFORD MAYER CO., 24/26 Beach Grove, P.O. Box 1381 Durban (South Africa).

STANDARD OIL CO.(OHIO), Midland Building, Cleveland 15, Ohio (U.S.A.); Vila Cisneros (Spanish Sahara).

STANDARD OIL CO. OF CALIFORNIA, 225 Bush Street, San Francisco 20, Calif.; 605 West Olympic Boulevard, Los Angeles (U.S.A.).

STANDARD OIL CO. OF INDIANA, 910 South Michigan Avenue, Chicago 80, Ill. (U.S.A.).

STANDARD OIL CO. OF NEW JERSEY, 30 Rockefeller Plaza, New York 20, N.Y. (U.S.A.).

STANDARD ORE AND ALLOYS CORP., 120 Wall Street, New York, N.Y. (U.S.A.).

STANHOPE GOLD MINING CO., 511-6 City Trust House, 106 Fox Street, P.O. Box 742 Johannesburg; P.O. Box 46 Cleveland (South Africa).

STAR ASBESTOS CO. (PTY.) LTD., 203 Jubilee House, 158 Simmonds Street, Johannesburg; Kaapsche Hoop, P.O. Box 9 Nelspruit, E. Tvl. (South Africa).

STAR COLLIERY, P.O. Box 77 Newcastle, Ntl. (South Africa).

STAR DIAMONDS (PTY.) LTD., 47 Main Street, P.O. Box 9 Johannesburg; Theron Station (South Africa).

STATE ALLUVIAL DIAMOND DIGGINGS, Alexander Bay, Namaqualand, C.P. (South Africa).

STECDIP, see: Société Tunisienne d'Études, de Coopération et de Défense de l'Industrie Phosphatière.

STEPHOS, see: Société Tunisienne d'Exploitation Phosphatières.

STERKSPRUIT CHRYSOTILE ASBESTOS (PTY.) LTD., Ingram's Corner, Twist Street Hillbrow, P.O. Box 8967 Johannesburg; P.O. Badplaats via Carolina, E. Tvl. (South Africa).

STILFONTEIN GOLD MINING CO., General Mining Building, 6 Hollard Street, P.O. Box 1173 Johannesburg; P.O. Box 1 Stilfontein (South Africa).

STINKHOUTBOOM MANGAAN, P.O. Box 163 Zeerust (South Africa).

STIR, see: Société Tuniso-Italienne de Raffinage.

STRICKLAND, A., Bundali Hills (Tanganyika).

STRONG AND MOORE (PTY.) LTD., P.O. Simonstown, C.P. (South Africa).

SUB NIGEL LTD., Gold Fields Building, 75 Fox Street, P.O. Box 1167, Johannesburg; P.O. Box 14 Dunnottar (South Africa).

SUDAN PORTLAND CEMENT CO., P.O. Box 96 Atbara (Sudan).

SUD-KAT, see: Société de Recherches Minières du Sud-Katanga.

SUKULU MINES LTD., Ralli House, Grant Street, P.O. Box 442 Kampala; P.O. Box 115 Toronto (Uganda).

SUMITOMO SHOJI KAISHA LTD., 15 5-chome, Kitahama, Higashi-ku, Osaka; 8 1-chome, Marunouchi, Chiyoda-ku, Tokyo (Japan).

SUN OIL CO., 1608 Walnut Street, Philadelphia 3, Pa. (U.S.A.); Vila Cisneros (Spanish Sahara).

SUNRISE OCHRE MINES, Matlock (Great Britain); Albertinia (South Africa).

SWARTKOPS SEESOUT BEPERK, P.O. Box 5958 Johannesburg; P. Bag Swartkops (South Africa).

SWAZILAND BARYTES LTD., P.O. Box 457 Johannesburg (South Africa).

SWAZILAND IRON ORE DEVELOPMENT CO., 44 Main Street, Johannesburg (South Africa); Darkton (Swaziland).

SWAZILAND MINES LTD., P.O. Box 8727 Johannesburg (South Africa); Mbabane (Swaziland).

SYMAF, see: Compagnie Minière du Djebel Gustar.

SYMAF, see: Syndicat Minier Africain.

SYMETAIN, see: Syndicat Minier de l'Étain.

SYNDICAT DE RECHERCHES DE FER ET DE MANGANÈSE DE CAMEROUN, c/o BRGM, Yaoundé (Cameroons); 74 Rue de la Féderation, Paris 15e (France).

SYNDICAT DES BAUXITES DU CAMEROUN, c/o BRGM, Yaoundé (Cameroons); 74 Rue de la Féderation, Paris 15e (France).

SYNDICAT DES PROSPECTEURS ET EXPLOITANTS MINIERS DE MADAGASCAR, 15 Rue Gourbeyre, B.P. 1379 Tananarive (Malagasy Republic).

SYNDICAT LYONNAIS DE MADAGASCAR, 3 Rue du Président-Carnot, Lyon; 59 Rue de Provence, Paris 9e (France); 9 Rue Béréni, B.P. 716 Tananarive (Malagasy Republic).

SYNDICAT MINIER AFRICAIN (SYMAF), 112 Rue du Commerce, Brussels (Belgium).

SYNDICAT MINIER DE L'ÉTAIN (SYMETAIN), 112 Rue du Commerce, Brussels (Belgium); Kalima; Punia (Congo).

TANESCO, Dar es Salaam (Tanganyika).

TANGANICA-ITALIAN REFINING CO. (TIPER), Dar es Salaam (Tanganyika).

TANGANYIKA CONCESSIONS LTD. (TCL OR TANKS), Tanganyika House, Salisbury (Southern Rhodesia); Princes House, 95 Gresham Street, London E.C. 2 (Great Britain).

TANGANYIKA CORUNDUM CORP., P.O. Box 45 Dar es Salaam (Tanganyika).

TANGANYIKA CRYSTALS LTD., Dar es Salaam (Tanganyika).

TANGANYIKA DIAMOND AND GOLD DEVELOPMENT CO., Standard Bank Chambers, 46 Marshall Street, P.O. Box 957 Johannesburg (Southern Africa); 62 London Wall, London E.C. 2 (Great Britain).

TANGANYIKA HOLDINGS LTD., Princes House, 95 Gresham Street, London E.C. 2 (Great Britain).

TANGANYIKA ELECTRIC SUPPLY CO., P.O. Box 9024 Dar es Salaam (Tanganyika); Messrs. Balfour, Beatty and Co., Bow Bells House, Bread Street, London E.C. 4 (Great Britain).

TANGANYIKAN ITALIAN PETROLEUM REFINING CO., Dar es Salaam (Tanganyika).

TANGANYIKA MEERSCHAUM CORP., P.O. Box 45 Dar es Salaam (Tanganyika).

TANGANYIKA SHELL CO., Dar es Salaam (Tanganyika).

TANGOLD MINING CO., Kiabakari, Musoma (Tanganyika).

TANKS, see: Tanganyika Concessions Ltd.

TANTALITE VALLEY MINERALS (PTY.) LTD., P.O. Box 140 Karasberg (Southwest Africa).

TAVISTOCK AND UITSPAN COLLIERIES LTD., 505-515 Stock Exchange Buildings, Hollard Street, P.O. Box 5782 Johannesburg; P.O. Box 78 Oogies (South Africa).

TAYLOR AND HOAR LTD., Standard Bank Chambers, Marshall Street, P.O. Box 957 Johannesburg (South Africa).

TCL, see: Tanganyika Concessions Ltd.

TENNANT, C., SONS AND CO., 100 Park Avenue, New York 17, N.Y. (U.S.A.).

TENNECO OIL CO., 220 South Harrison Street, East Orange, New Jersey (U.S.A.).; Freetown (Sierra Leone).

TENNESSEE SIERRA LEONE INC., Freetown (Sierra Leone).

TEXACO AFRICA LTD., 135 East 42nd Street, New York 17, N.Y. (U.S.A.); Freetown (Sierra Leone); Monrovia (Liberia); Dakar (Senegal); Abidjan (Ivory Coast); Cotonou (Dahomey); Lomé (Togo); Accra (Ghana).

TEXACO CANARIAS S.A., Las Palmas; Santa Cruz de Teneriffe (Canary Islands).

TEXACO GUINÉE S.A.R.L., Conakry (Guinea).

TEXACO INC., 135 East 42nd Street, New York 17, N.Y. (U.S.A.).

TEXACO OVERSEAS PETROLEUM CO., 135 East 42nd Street, New York 17, N.Y. (U.S.A.).

TEXACO PETROLEUM CO., Luanda (Angola).

TEXACO SMPP, Rabat (Morocco).

TEXAS GULF PRODUCING CO., 200 Park Avenue, New York, N.Y. (U.S.A.); Tripoli (Libya).
TEXAS PETROLEUM CO., 135 East 42nd Street, New York 17, N.Y. (U.S.A.); Luanda (Angola).
THOMAS MEIKLE TRUST AND INVESTMENT CO., 25–35 City Road, London E.C. 1 (Great Britain); P.O. Turk Mine (Southern Rhodesia).
THORNWOOD ASBESTOS MINES (PVT.) LTD., P.O. Gwanda (Southern Rhodesia).
THYSSEN HÜTTE, A., Essen (W. Germany).
TIDEWATER OIL CO., 4201 Wilshire Boulevard, Los Angeles 5, Calif.; 660 Madison Avenue, New York 21, N.Y.; 815 Walker Avenue, Houston 1, Texas (U.S.A.); Vila Cisneros (Spanish Sahara).
TIN AND ASSOCIATED MINERALS LTD., Dogon Dutse, P.O. Box 462 Jos; P.O. Box 162 Jos (Nigeria).
TIPER, see: Tanganica-Italian Refining Co.
TIRA MINES LTD., Busia (Uganda).
TOGO-AMERICAN OIL CO., Lomé (Togo); Dallas, Texas (U.S.A.).
TOLWE MINING CO., P.O. Box 3198 Johannesburg; Potgietersrus, N. Tvl. (South Africa).
TORDJMANN, E., Erfoud; Hassi Hisbiha (Morocco).
TORO LIME CO., Muhokya via Fort Portal (Uganda).
'TOTAL' COMPAGNIE FRANÇAISE DE DISTRIBUTION, 157 Avenue de Neuilly, Neuilly, Seine (France).
TOTAL OIL PRODUCTS (NIGERIA) LTD., 26 Rue de la Pépinière, Paris 8e (France); P.M.B. 2143 Lagos (Nigeria).
TOTAL OIL PRODUCTS (PTY.) LTD., 15 Anderson Street, P.O. Box 11277 Johannesburg (South Africa).
TOTAL OIL PRODUCTS RHODESIA (PVT.) LTD., Livingstone House, Jameson Avenue, Salisbury (Southern Rhodesia).
TRAFIK AB GRÄNGESBERG-OXELÖSUND, Stockholm (Sweden).
TRANSVAAL AND DELAGOA BAY INVESTMENT CO., 'Unitas', 42 Marshall Street, P.O. Box 550 Johannesburg (South Africa); Fenchurch House, 5 Fenchurch Street, London E.C. 3 (Great Britain).
TRANSVAAL ASBESTOS LTD., 46 Marshall Street, P.O. Boxes 957 and 1739 Johannesburg; P.O. Box 80 Pietersburg, N. Tvl. (South Africa).
TRANSVAAL COAL CORP., 44 Main Street, P.O. Box 2567 Johannesburg (South Africa).
TRANSVAAL CONSOLIDATED LAND AND EXPLORATION CO., Corner House, 77 Commissioner Street, P.O. Box 1056 Johannesburg; P.O. Vandyksdrif (South Africa); 4 London Wall Buildings, London E.C. 2 (Great Britain).
TRANSVAAL CORUNDUM CO., 56 Main Street, P.O. Box 1054 Johannesburg; P.O. Box 72 Pietersburg, N. Tvl. (South Africa).
TRANSVAAL GOLD MINING ESTATES LTD., Corner House, 77 Commissioner Street, P.O. Box 1056 Johannesburg; P.O. Box 2 Pilgrim's Rest, E. Tvl. (South Africa); 4 London Wall Buildings, London E.C. 2 (Great Britain).
TRANSVAAL LAND AND DEVELOPMENT CO., 131 Annan House, Commissioner Street, Johannesburg (South Africa).
TRANSVAAL MANGANESE (PTY.) LTD., P.O. Box 4587 Johannesburg (South Africa).
TRANSVAAL NAVIGATION COLLIERIES AND ESTATE CO., 98 Market Street, P.O. Box 4220 Johannesburg; P.O. Vandyksdrif (South Africa).
TRANSVAAL-ORANGIA EXPLORATION LTD., 6 Hollard Street, P.O. Box 1007 Johannesburg (South Africa).
TRANSVAAL ORE CO., P.O. Box 7651 Johannesburg; P.O. Box 4 Phalaborwa (South Africa).
TRANSVAAL VANADIUM CO., 44 Main Street, P.O. Box 697 Johannesburg; Ferrobank Siding, Witbank, Tvl. (South Africa).
TRANSWORLD PETROLEUM S.A. FRANÇAISE, Algiers (Algeria).
TRAPAL, see: Société pour le Transport des Hydrocarbures Sahariens au Littoral Algérien.
TRAPES, see: Société de Transports de Pétroles à l'Est Saharien.
TRAPSA, see: Compagnie des Transports par Pipe-lines au Sahara.
TREASURE TROVE DIAMONDS LTD., Barclay's Bank Building, corner Commissioner and Kruis Streets, Johannesburg; Treasure Trove, 27K, Kimberley (South Africa).
TRUST AND MINING CO., P.O. Box 47 Omaruru (Southwest Africa).
TSHOBA COAL SYNDICATE LTD., 10 Leslie Street, P.O. Box 2161 Durban; P. Bag 1305 Vryheid (South Africa).
TSUMEB MINING CORP. LTD., Tsumeb (Southwest Africa); 300 Park Avenue, New York 22, N.Y. (U.S.A.).
TUNGI LTD., P.O. Box 11 Morogoro (Tanganyika).
TURNER AND NEWALL LTD., Asbestos House, 77–79 Fountain Street, Manchester 2 (Great Britain).
TURNER ASBESTOS LTD., Emene (Nigeria).
TWEEFONTEIN INVESTMENTS LTD., 1 Cornhill, London E.C. 3 (Great Britain).
TWEEFONTEIN UNITED COLLIERIES LTD., Hollard Place, 71 Fox Street, P.O. Box 1146 Johannesburg; P.O. Coalville (South Africa).
TYGERSBERG QUARRIES (PTY.) LTD., P.O. Box 4151 Cape Town (South Africa).
UGANDA CEMENT INDUSTRY LTD., P.O. Box 74 Tororo (Uganda).
UGANDA DEVELOPMENT CORP., Kampala (Uganda).

UGANDA ELECTRICITY BOARD ,P.O. Box 559 Kampala (Uganda); Uganda House, Trafalgar Square, London W.C. 2 (Great Britain).
UGANDA LIME CO., Tororo (Uganda).
UGANDA SHELL CO., Kampala (Uganda).
UGF, see: United Geophysical Company of France.
UGP, see: Union Générale des Pétroles.
UIP, see: Union Industrielle des Pétroles.
UIPA, see: Union Industrielle des Pétroles (Algérie).
UIS TIN MINING CO., P.O. Uis via Windhoek (Southwest Africa).
UITKYK ASBESTOS (PTY.) LTD., Haenertsburg 246 N. Tvl.; P.O. Box 470 Pietersburg, N. Tvl. (South Africa).
ULTRA HIGH PRESSURE UNITS LTD., P.O. Box 7727 Johannesburg, (South Africa).
ULUGURU MICA MINING COOPERATIVE SOC., P.O. Box 40 Morogoro (Tanganyika).
UMGABABA MINERALS LTD., 44 Main Street, P.O. Box 4587 Johannesburg; Umgababa via Durban (South Africa); 40 Holborn Viaduct, London E.C. 1 (Great Britain).
UNIÃO COMERCIAL DE AUTOMÓVEIS LTDA., Luanda (Angola).
UNIÃO DE MICAS LTD., Quizambilo; Luanda (Angola).
UNION AMERICAN TRUST LTD., Annan House, 86 Commissioner Street, P.O. Box 10905 Johannesburg (South Africa).
UNION CARBIDE CORP., 270 Park Avenue, New York 19, N.Y. (U.S.A.).
UNION CARBIDE ORE CO., 270 Park Avenue, New York 19, N.Y. (U.S.A.).
UNION CHIMIQUE BELGE, 4 Chaussée de Charleroi, Brussels (Belgium).
UNION COLLIERIES LTD., P.O. Box 957 Johannesburg: P. Bag Breyten (South Africa).
UNION CORP. LTD., Corporation Building, 74–78 Marshall Street, P.O. Box 1156 Johannesburg (South Africa); Princes House, 95 Gresham Street, London E.C. 2 (Great Britain).
UNION DES MICAS, Ampandrandava par Beraketa, Province de Tuléar (Malagasy Republic).
UNIÓN ELECTRICA DE CANARIAS S.A., Viera y Clavijo 1, Las Palmas (Canary Islands).
UNION ÉLECTRIQUE D'OUTRE-MER, 54 Rue de Lisbonne, Paris 8e (France); Lomé (Togo).
UNION FINANCIÈRE INTERNATIONALE POUR LE DÉVELOPPEMENT DE L'AFRIQUE, 14 Rue d'Athènes, Paris 9e (France).
UNION FINANCIÈRE POUR L'ENTREPRISE GÉNÉRALE ET LES EXPLOITATIONS MINIÈRES, 1 Place Mirabeau, Casablanca; Oum Kiad (Morocco).
UNION FREE STATE MINING AND FINANCE CORP., Cape House, 54 Fox Street, P.O. Box 4280 Johannesburg (South Africa).
UNION GÉNÉRALE DES PÉTROLES (UGP), 25 Rue Jasmin, Paris 16e (France).
UNION INDUSTRIELLE DES PÉTROLES (UIP), 7 Place Vendôme, Paris 1er (France).
UNION INDUSTRIELLE DES PÉTROLES (ALGÉRIE) (UIPA), Algiers (Algeria).
UNION KYANITE, Rehoboth (Southwest Africa); Cape Town (South Africa).
UNION LIME CO., P.O. Box 7727 Johannesburg; Ulco via Kimberley (South Africa).
UNION MICA MINE, 62 Webber Street, Selby, P.O. Box 6996 Johannesburg (South Africa).
UNION MINERALS LTD., Annan House, Commissioner Street, P.O. Box 8947 Johannesburg; P.O. Box 399 Pietersburg, N. Tvl. (South Africa).
UNION MINIÈRE, see: Union Minière du Haut Katanga.
UNION MINIÈRE DE L'ATLAS OCCIDENTAL, 43 Avenue Victor Hugo, Tanger; 42 Avenue de l'Armée Royale, Casablanca; B.P. 716 Marrakech (Morocco).
UNION MINIÈRE DE TUNISIE, 95 Avenue Mohammed V, Tunis (Tunisia).
UNION MINIÈRE DU HAUT KATANGA, 6 Rue Montagne du Parc, Brussels (Belgium); Élisabethville; Jadotville; Kolwezi (Congo).
UNION OIL CO. OF CALIFORNIA, 610 Fifth Avenue, New York, N.Y. (U.S.A.).; Vila Cisneros (Spanish Sahara).
UNION PHOSPHATIÈRE AFRICAINE (UPHA), 41 Avenue Hoche, Paris 8e (France).
UNION PLATINUM MINING CO., Gold Field Building, 75 Fox Street, P.O. Box 1167 Johannesburg (South Africa); 49 Moorgate, London E.C. 2 (Great Britain).
UNION POUR LA RECHERCHE ET L'EXPLOITATION PÉTROLIÈRES SAHARIENNES S.A. (UNIPETROL), Postfach 52 Kassel (W. Germany).
UNION TIN MINES LTD., P.O. Box 8653 Johannesburg; P.O. Naboomspruit (South Africa).
UNIPETROL, see: Union pour la Recherche et l'Exploitation Pétrolières Sahariennes.
UNITED AFRICA CO., Lever House, London (Great Britain); Freetown (Sierra Leone).
UNITED AFRICAN EXPLORATIONS LTD., 28 Austin Friars, London E.C. 2 (Great Britain).
UNITED AFRICAN INVESTMENTS LTD., Anglovaal House, 56 Main Street, P.O. Box 7727 Johannesburg (South Africa).
UNITED GEOPHYSICAL CO. OF FRANCE (UGF), 194 Rue de Rivoli, Paris 1er (France).
UNITED GEOPHYSICAL CORP., 2650 East Foothill Boulevard, Pasadena, Calif. (U.S.A.); Tripoli (Libya).

UNITED HOLDINGS (PTY.) LTD., Mutual Buildings, Harrison Street, P.O. Box 1370 Johannesburg (South Africa).
UNITED MINING CO., P.O. Box 1364 Khartoum (Sudan).
UNITED STATES STEEL CORP., 71 Broadway, New York 6, N.Y. (U.S.A.); Moanda (Gabon).
UNITED TIN AREAS OF NIGERIA LTD., 120 Moorgate, London E.C. 2 (Great Britain).
UNITED VERMICULITE AND BASE MINERALS LTD., P.O. Box 2887 Johannesburg (South Africa).
UPHA, see: Union Phosphatière Africaine.
URUWIRA MINERALS LTD., Mpanda (Tanganyika).
UTRECHT COLLIERIES LTD., Corner House, P.O. Box 1056 Johannesburg (South Africa).
VAAL REEFS EXPLORATION AND MINING CO., 44 Main Street, P.O. Box 4587 Johannesburg (South Africa).
VAAL RIVER SALT WORKS LTD., P.O. Box 3600 Johannesburg (South Africa).
VALITON, MME., Rue Colbert, B.P. 706 Tananarive (Malagasy Republic).
VANADIUM CORP. OF AMERICA, Corporate Office: 100 West 10th Street, Wilmington, Del.; Executive Office: Graybar Building, 420 Lexington Avenue, New York 17, N.Y.; Oliver Building, Pittsburgh; Field Building, Chicago, Ill.; Book Tower, Detroit, Mich.; Union Commerce Building, Cleveland, Ohio (U.S.A.).
VAN DYK CONSOLIDATED MINES LTD., Union Corporation Building, 74–78 Marshall Street, P.O. Box 1125 Johannesburg; P.O. Van Dyk's Mine (South Africa).
VANGUARD ASBESTOS MINES, Belingwe (Southern Rhodesia).
VAN RYN SAND CO., P.O. Box 249 Benoni, Tvl. (South Africa).
VAN STRATEN, J. G. EN SEUNS, P.O. Box 13 Danielskuil (South Africa).
VAN WYK BROTHERS (PTY.) LTD., P. Bag 1334 Pietersburg, N. Tvl. (South Africa).
VEEDOL MINERALS SWA (PTY.) LTD., Lüderitzbucht (Southwest Africa).
VELLEFONTEIN TIN MINING CO., P.O. Box 8653 Johannesburg; P.O. Rooiberg (South Africa).
VENTERSPOST GOLD MINING CO. LTD., Gold Fields Building, 75 Fox Street, P.O. Box 1167 Johannesburg; P.O. Box 5 Westonaria (South Africa).
VENUS MANGANESE (PTY.) LTD., Ottoshoop (South Africa).
VEREENIGING BRICK AND TILE CO., 27 Smuts Avenue, P.O. Box 49 Vereeniging, Tvl.; P.O. Bankop (South Africa).
VEREENIGING ESTATES LTD., 27 Smuts Avenue, P.O. Box 49 Vereeniging, Tvl. (South Africa); 40 Holborn Viaduct, London E.C. 1 (Great Britain).
VEREINIGTE ALUMINIUM WERKE A.G., Am Nordbahnhof, Bonn (W.Germany).
VERGENOEG MINING CO., P.O. Box 13 Ottoshoop (South Africa).
VEROLITE ASBESTOS CO. LTD., 804 Maritime House, Loveday Street, P.O. Box 7193 Johannesburg; P.O. Box 399 Pietersburg, N. Tvl.; P.O. Kranskop, Ntl. (South Africa).
VICTORIA FALLS ELECTRICITY BOARD, P.O. Box 46 Livingstone (Zambia).
VIERFONTEIN COLLIERIES LTD., 44 Main Street, P.O. Box 2567 Johannesburg; P. Bag Bourketon (South Africa).
VILLAGE MAIN REEF GOLD MINING CO. (1934) LTD., Anglovaal House, 56 Main Street, P.O. Boxes 7727 and 5622 Johannesburg (South Africa).
VIRGINIA-MERRIESPRUIT INVESTMENT (PTY.) LTD., Corner House, 77 Commissioner Street, P.O. Box 1056 Johannesburg (South Africa).
VIRGINIA ORANGE FREE STATE GOLD MINING CO., Anglovaal House, 56 Main Street, P.O. Box 7727 Johannesburg; P. Bag 60 Virginia (South Africa).
VLAKFONTEIN GOLD MINING CO., Gold Fields Building, 75 Fox Street, P.O. Box 1167 Johannesburg; P.O. Box 22 Dunnottar (South Africa); 49 Moorgate, London E.C. 2 (Great Britain).
VOGELGARNETSMINE, 5 Fifth Avenue, Bergview East, Messina, N. Tvl. (South Africa).
VOGELSTRUISBULT GOLD MINING AREAS LTD., Gold Fields Building, 75 Fox Street, P.O. Box 1167 Johannesburg; P.O. Box 218 Springs, Tvl. (South Africa); 49 Moorgate, London E.C. 2 (Great Britain).
VOLTA ALUMINIUM CO. (VALCO), Accra (Ghana).
VOLTA RIVER AUTHORITY, Accra (Ghana).
VOORSPOED ASBES MAATSKAPPY BEPERK, P. Bag 1334 Pietersburg, N. Tvl. (South Africa).
VRYHEID CORONATION LTD., 44 Main Street, P.O. Box 2567 Johannesburg; P.O. Hlobane (South Africa).
VRYHEID (NATAL) RAILWAY, COAL AND IRON CO., Cotts House, Camomile Street, London E.C. 3 (Great Britain); Barclays Bank Building, Field and Smith Streets, Durban; P.O. Hlobane (South Africa).
VULCAN MINERALS (PVT.) LTD., P.O. Box Gwelo (Southern Rhodesia).
W. AND B. ASBESTOS CO (PTY.) LTD., P.O. Box 175 Pietersburg, N. Tvl. (South Africa).
WANDRAG ASBESTOS (PTY.) LTD., 704 Garlick House, 26 Harrison Street, Johannesburg; P.O. Box 114 Kuruman (South Africa).
WANDSCHNEIDER, THERESE C.M., Cabinda (Angola).
WANKIE COLLIERY CO., Charter House, Selborne Avenue, P.O. Box 1997 Bulawayo; P.O. Box 123 Wankie (Southern Rhodesia); 40 Holborn Viaduct, London E.C. 1 (Great Britain).

WATERVAL (RUSTENBURG) PLATINUM MINING CO., Gold Fields Building, 75 Fox Street, P.O. Box 1167 Johannesburg (South Africa); Watson Co., Abidjan (Ivory Coast).

WATSON CO., 3A Maiden Lane, New York, N.Y. (U.S.A.).

WAVERLEY GOLD MINES LTD., 75 Fox Street, P.O. Box 8653 Johannesburg (South Africa); 49 Moorgate, London E.C. 2 (Great Britain).

WEBSTER MARINE SALTS (PTY.) LTD., P.O. Box 73 Swakopmund (Southwest Africa).

WEIDNER, P., P.O. Box 12 Warmbad (Southwest Africa).

WELGEDACHT EXPLORATION CO., The Corner House, 77 Commissioner Street, P.O. Box 1056 Johannesburg (South Africa); 4 London Wall Buildings, London E.C. 2 (Great Britain).

WELKOM GOLD MINING CO., 44 Main Street, P.O. Box 4587 Johannesburg; P.O. Box 1 Welkom, O.F.S. (South Africa).

WENMAN, WILLIAMS MAGNESITE (PTY.) LTD., 2 Delvers Street, P.O. Box 2524 Johannesburg; P.O. Box 1 Malelane, E. Tvl. (South Africa).

WEST AFRICAN ALUMINIUM LTD., P.O. Box 242 Accra (Ghana).

WEST AFRICAN FINANCE CORP., 10 Old Jewry, London E.C. 2 (Great Britain).

WEST AFRICAN GOLD CORP., P.O. Box 1 Prestea (Ghana).

WEST AFRICAN PORTLAND CEMENT CO., P.O. Box 642 Lagos; P. Bag 1011 Ikeja (Nigeria).

WEST AND SOUTH AFRICAN GOLD MINES, P.O. Box 205 Yaba (Nigeria).

WEST AND SOUTH AFRICAN MINING SERVICES, P.M.B. 1030 Apapa (Nigeria).

WEST DRIEFONTEIN GOLD MINING CO., Gold Fields Building, 75 Fox Street, P.O. Box 1167 Johannesburg; P. Bag Carltonville, Tvl. (South Africa).

WESTERN AREAS GOLD MINING CO., Consolidated Building, corner Fox and Harrison Streets, Johannesburg (South Africa).

WESTERN DEEP LEVELS LTD., 44 Main Street, P.O. Box 4587 Johannesburg (South Africa).

WESTERN GEOPHYSICAL CO., 933 North LaBrea Avenue, Los Angeles 38, Calif. (U.S.A.).

WESTERN HOLDINGS LTD., 44 Main Street, P.O. Box 4587 Johannesburg; P.O. Box 3 Welkom, O.F.S. (South Africa).

WESTERN REEFS EXPLORATION AND DEVELOPMENT CO., P.O. Box 4587 Johannesburg; P.O. Box 92 Klerksdorp, W. Tvl. (South Africa); Chunya (Tanganyika).

WESTERN RIFT EXPLORATION CO., P.O. Box 35 Chunya (Tanganyika).

WESTERN SELECTION AND DEVELOPMENT CO., 120 Moorgate, London E.C. 2 (Great Britain).

WESTERN ULTRA DEEP LEVELS LTD., 44 Main Street, P.O. Box 4587 Johannesburg (South Africa).

WEST RAND CONSOLIDATED MINES LTD., General Mining Building, 6 Hollard Street, P.O. Box 1173 Johannesburg; P.O. Box 1 West Rand, Tvl. (South Africa); Winchester House, Old Broad Street, London E.C. 2 (Great Britain).

WEST RAND INVESTMENT TRUST LTD., 44 Main Street, P.O. Box 4587 Johannesburg (South Africa).

WEST SPAARWATER LTD., Hollard Place, 71 Fox Street, P.O. Box 2420 Johannesburg (South Africa); 120 Moorgate, London E.C. 2 (Great Britain).

WEST VLAKFONTEIN GOLD MINING CO., Hollard Place, 71 Fox Street, P.O. Box 2420 Johannesburg (South Africa); 120 Moorgate, London E.C. 2 (Great Britain).

WEST WITWATERSRAND AREAS LTD., Gold Fields Building, 75 Fox Street, P.O. Box 1167 Johannesburg (South Africa); 49 Moorgate, London E.C. 2 (Great Britain).

WHITE'S S.A. PORTLAND CEMENT CO., Edura Building, 40 Commissioner Street, P.O. Box 2484 Johannesburg; White's, O.F.S.; Lichtenburg, E. Tvl. (South Africa).

WILFORD GOLD MINING CO. LTD., 511-6 City Trust House, 106 Fox Street, Johannesburg; P.O. Box 190 Roodepoort, Tvl. (South Africa).

WILLIAM BAIRD AND CO., City Wall House, Chiswell Street, London, E.C. 1 (Great Britain).

WILLIAMSON DIAMONDS LTD., Mwadui (Tanganyika); 40 Holborn Viaduct, London E.C. 1 (Great Britain).

WILLIAMSON EXPLORATION LTD., Mwadui (Tanganyika).

WINDSOR FERROALLOYS LTD., Que Que (Southern Rhodesia).

WINKELHAAK MINES LTD., Union Corporation Building, 74–78 Marshall Street, Johannesburg; P.O. Box 1 Evander (South Africa).

WINTERFELD, T.C.L., CHROME MINES (PTY.) LTD., P.O. Box 5865 Johannesburg; P.O. Box 3 Steelpoort (South Africa).

WINTERSHALL A.G., August Roster Haus, Postfach 52 Kassel (W. Germany).

WINTERSHALL SAHARIENNE S.A. POUR LA RECHERCHE ET L'EXPLOITATION PÉTROLIÈRES (WISAREP), Postfach 52, Kassel (W. Germany).

WISAREP, see: Wintershall Saharienne S.A. pour la Recherche et l'Exploitation Pétrolières.

WITBANK COAL HOLDINGS LTD., 44 Main Street, P.O. Box 2567 Johannesburg (South Africa).

WITBANK COLLIERY LTD., Corner House, 77 Commissioner Street, P.O. Box 1056 Johannesburg; P.O. Box Witbank, Tvl. (South Africa); 4 London Wall Buildings, London E.C. 2 (Great Britain).

WITBANK CONSOLIDATED COAL MINES LTD., Hollard Place, 71 Fox Street, P.O. Box 1146 Johannesburg; P. Bag Oogies, Tvl. (South Africa).

WIT EXTENSIONS LTD., Gold Fields Building, 75 Fox Street, P.O. Box 1167 Johannesburg (South Africa).

WITTMAN INC., G.H., 111 Broadway, New York 6, N.Y. (U.S.A.).

WITWATERSRAND DEEP LTD., Gold Fields Building, 75 Fox Street, P.O. Box 1167 Johannesburg (South Africa).

WITWATERSRAND GOLD MINING CO., Consolidated Building, P.O. Box 590 Johannesburg; P.O. Box 206 Heidelberg, Tvl. (South Africa); 10–11 Austin Friars, London E.C. 2 (Great Britain).

WITWATERSRAND NIGEL LTD., Hollard Place, 71 Fox Street, P.O. Box 2420 Johannesburg (South Africa); 120 Moorgate, London E.C. 2 (Great Britain).

WONDERSTONE 1937 LTD., P.O. Box 1054 Johannesburg; P.O. Box 10 Ottosdal (South Africa).

YONDER BASE MINERALS (PTY.) LTD., Permanent Buildings, Commissioner Street, P.O. Box 2313 Johannesburg; P.O. Box 10 Mica, N. Tvl. (South Africa).

YOUNG, W., P.O. Box 1491 Bulawayo (Southern Rhodesia).

ZAAIPLAATS TIN MINING CO., United Building, corner Pretorius and Andries Streets, Pretoria; P.O. Box 179 Johannesburg; P.O. Box 77 Potgietersrus, N. Tvl. (South Africa); 10 Upper Grosvenor Street, London W. 1 (Great Britain).

ZAMBESIA EXPLORATION CO., Princes House, 95 Gresham Street, London, E.C. 2 (Great Britain).

ZANDPAN GOLD MINING CO., 56 Main Street, Johannesburg (South Africa).

ZANZIBAR ELECTRICITY BOARD, P.O. Box 235 Zanzibar (Zanzibar).

ZEVENFONTEIN SALT WORKS LTD., P.O. Box 99 Rustenburg, W. Tvl. (South Africa).

ZINC ET ALLIAGES S.A.R.L., 34 Rue Collange, Levallois–Perret, Seine (France).

ZUINGUIN NATAL COLLIERIES LTD., P.O. Box 101 Vryheid (South Africa).

ZULU COAL CO., 900 Mansion House, 12 Field Street, P.O. Box 541 Durban; P.O. Box 10 Mtubatuba, Ntl. (South Africa).

References

ABDULLA, M. A., 1958. Annual report for the period July 1955 to June 1957. *Sudan, Geol. Surv. Dept., Bull.*, 4: 1–21.

AFRIQUE OCCIDENTALE FRANÇAISE. DIRECTION GÉNÉRALE DES MINES ET DE LA GÉOLOGIE, 1956. L'industrie minière en Afrique Occidentale Française. *Encyclopédie d'Outre-Mer*, pp.1–22.

AFRIQUE OCCIDENTALE FRANÇAISE, 1958. Résultats récents obtenus en matière de prospection minière en A.O.F. *Chronique Mines Outre-Mer Rech. Minière*, 261: 75–79.

AGARD, J., BOULADON, J., DESTOMBES, J., HURON, O., JEANETTE, A., JOURAVSKY, G., LEVY, R.-G., MORIN, PH., MOUSSU, R., OWODENKO, B., PERMINGEAT, F., RAGUIN, E., SALVAN, H. et VAN LECKWIJK, W., 1952. Géologie des gîtes minéraux marocains. *Div. Mines Géol., Notes Mém.*, 87: 1–416.

AICARD, P., 1957. Le Précambrien au Togo et du nord-ouest du Dahomey. *Gouvt. Gén. Afrique Occidentale Franç., Bull. Direc. Mines Géol.*, 23: 1–228.

AMERADA PETROLEUM CORP., 1961, 1962, 1963. *Annual Reports.*

AMERICAN METAL CLIMAX INC., 1962, 1963. *Annual Reports.*

AMIN, M. S. and AFIA, M. S., 1954. Antophyllite–vermiculite deposits of Hafafit. *Econ. Geol.*, 49 (3): 77–87.

ANCELLE, H., 1959. *Le Phosphate de la Compagnie Sénégalaise des Phosphates de Taïba.* Roneoscript, pp.1–7.

ANDERSON, R. B., 1961. The mineral resources of Southern Rhodesia. *Commonwealth Mining Met. Congr., 7th, Johannesburg, 1961*, pp.25–91.

ANDREW, G., 1952. Iron ores in the Anglo-Egyptian Sudan. Symposium sur les gisements de fer du monde. *Congr. Géol. Intern., Compt. Rend., 19e, Algiers, 1952*, 1: 187–189.

ANGLO-AMERICAN CORP. OF SOUTH AFRICA, 1960–1963. *Annual Reports.*

ANONYMOUS, 1959. Report of the commission on refractory minerals. Chapter 10: Report on the platinum group metals. *Natl. Acad. Sci.–Natl. Res. Council*, 154 (2).

ANONYMOUS, 1961. Estatística mineira. *Angola, Serv. Geol. Minas, Bol.*, 3: 64–76.

ANONYMOUS, 1962a. Panorama minier de l'Afrique en 1961, Juillet–Août. *Ann. Mines*, 1962, pp.430–459.

ANONYMOUS, 1962b. L'industrie minière au Congo-Léopoldville et au Ruanda-Burundi. *Ind. Trav. Outre-Mer*, 107.

ANONYMOUS, 1962c. *Geological Literature on Southwest Africa, 1838–1961.* Southwest Africa Geological Services, Windhoek, 47 pp.

ANONYMOUS, 1963. Panorama de l'Afrique en 1962. Juillet–Août. *Ann. Mines*, 1963.

APERTET, J., 1961. L'électrification des pays d'outre-mer de la zone franc. *Construction*, 16(1): 13–17.

ARCHAMBAULT, T., 1960. *Les Eaux souterraines de l'Afrique occidentale.* Nancy.

ARNAUD, M. G., 1945. Les ressources minières de l'Afrique occidentale. *Gouvt. Gén. Afrique Occidentale Franç., Direc. Mines Géol.*, 8: 1–100.

ARNOULD, M., 1954. Le manganèse dans le Nord-Ouest de la Côte d'Ivoire. *Direc. Mines Géol.*

AUBAGUE, M., 1957. Les gisements de fer de la région de Makakou-Mékambo. *Afrique Équator., Haut Comm. Rep., Bull. Direc. Mines Géol.*, 8: 45–52.

AUTUN, P., 1960. Sur la genèse et les propriétés de stannines et de varlamoffites du Maniéma. *Bull. Serv. Géol., Congo Belge Ruanda-Urundi*, 9: 1–31.

BAKER, B. H., 1958. Geology of the Magadi area. *Kenya, Colony Protectorate, Mining Geol. Dept., Geol. Surv. Kenya, Rept.*, 42 (1958): 1–81.

BANCROFT, J. A., 1952. *Mining in Northern Rhodesia.* British South Africa Company, London, 74 pp.

BARBER, W. and JONES, D. G., 1960. The geology and hydrology of Maiduguri, Bornu Province. *Records Geol. Surv. Nigeria, 1958*, pp.5–21.

BARNES, J. W., 1961. The mineral resources of Uganda. *Geol. Surv. Uganda Bull.*, 4: 1–90.

BARNES, J. W. and BROWN, I. M., 1958. *The Geological Environment of Copper in Uganda.* Commission Coopération Technique en Afrique, Léopoldville, pp.219–222.

REFERENCES

BAUD, L., 1956. Les gisements et indices de manganèse de l'Afrique Equatoriale Française. *Intern. Geol. Congr., 20th, Mexico, 1956, Rept. Symp. Yacimentos Manganes, II (Africa)*, pp.9–38.

BEATH, C. B., COUSINS, C. A. and WESTWOOD, R. J., 1961. The exploitation of platiniferous ores of the Bushveld Igneous Complex. *Commonwealth Mining Met. Congr., 7th, Johannesburg, 1961, Trans.*, 1: 217–244.

BÉCÉKA–MANGANÈSE, 1960. *Rapports.*

BECHUANALAND PROTECTORATE, GEOL. SURV. DEPT., 1958–1962. *Annual Reports.*

BEERMAN'S ALL MINING YEARBOOK, 1961, 1962. Beerman, Cape Town, pp.1–480.

BELGO–AMERICAN DEVELOPMENT CORP., 1962. *Background Information concerning Union Minière du Haut-Katanga*, pp.1–11.

BENKIRANE, M., 1961. Situation et perspectives d'avenir de l'industrie minière marocaine. *Mines Géol. (Morocco)*, 15(3): 7–12.

BERNAZEAUD, J., 1959. Premières données sur le gisement d'uranium de Mounana. *Chronique Mines Outre-Mer Rech. Minière*, 27: 311–315.

BERTHOSSA, A., KANYAMAHANGA, CH. et LEPERSONNE, J., 1963. *Géologie, Mines, Volcans. La Géologie et les Minéralisations du Ruanda.* Service Géologique de la République Rwandaise, Bujumbura, pp.1–14.

BESAIRIE, H., 1959. Les gisements de titane de Madagascar. *Malgache, Rép., Ann. Géol. Madagascar*, 26: 139–150.

BESAIRIE, H., 1961a. Carte minière (Madagascar), 1/2,500,000 et 1/2,000,000. *Malgache, Rép., Bur. Géol.*

BESAIRIE, H., 1961b. Les resources minérales de Madagascar. *Malgache, Rép., Bur. Géol.*, 151: 1–116.

BESAIRIE, H., GUIGUES, J., LAPLAINE, L. et LAUTEL, L., 1951. Le graphite à Madagascar. *Malgache, Rép., Trav. Bur. Géol.*

BESSOLES, B., 1955. Notice explicative sur la feuille Yalinga ouest. *Afrique Équator. Franc., Haut Comm. Rép., Bull. Direc. Mines Géol.*, 1955: 1–22.

BETIER, G., 1946. Les mines et carrières. *Encyclopédie Coloniale et Maritime*, Vol. Algérie et Sahara, pp.1–22.

BETIER, G., 1952. Extractive industry in Algeria. *Encyclopédie Mensuelle d'Outre-Mer*, 9: 1–20.

BETIER, G., 1952. L'industrie extractive en Algérie. *Encyclopédie Mensuelle d'Outre-Mer*, 9: 1–20.

BETIER, G., 1959. L'industrie minière en Algérie. *Encyclopédie d'Outre-Mer*, pp.1–18.

BETTE, R., 1945. Puissance hydraulique existante dans le bassin du Congo. *Bull. Inst. Roy. Colonial Belge*, 16: 1.

BETTENCOURT DIAS, M., 1961. Os Pegmatitos do Alto Ligonha, *Moçambique, Provincia, Direc. Serv. Geol. Minas, Mem. Commun, Bol.*, 27: 17–36.

BLANCHOT, A., 1955. Le Précambrien de la Mauritanie occidentale. *Gouvt. Gén. Afrique Occidentale Franç., Bull Direc. Mines Géol.*, 17: 1–215.

BLANCHOT, A., 1958. *Le Gisement de Cuivre d'Akjoujt (Mauritanie, A.O.F.).* Commission Coopération Technique en Afrique, Léopoldville, pp.267–270.

BLONDEL, F., 1952. Les gisements de fer de l'Afrique Occidentale Française. Symposium sur les gisements de fer du monde. *Congr. Géol. Intern., Compt. Rend., 19e, Algiers, 1952*, 1: 5–9.

BLOOMFIELD, K., 1957. The geology of the Nkana coalfield. *Nyasaland Geol. Surv. Dept., Bull.*, 8: 5–18.

BOARDMAN, L. G., 1961a. Further geological data on the Postmasburg and Kuruman manganese ore deposits, Northern Cape Province. *Trans. Proc. Geol. Soc. S. Africa, Spec. Paper.*

BOARDMAN, L. G., 1961b. Manganese in the Union of South Africa. *Commonwealth Mining Met. Congr., 7th, Johannesburg, 1961, Trans.*, 1: 201–216.

BOINEAU, R., 1961. Études géologiques relatives aux grands projets d'équipement. *Inst. Équator. Rech. Études Géol. Minières*, 14: 15–24.

BOLGARSKY, M., 1950. Étude géologique et description pétrographique du Sud-ouest de la Côte d'Ivoire. *Gouvt. Gén. Afrique Occidentale Franç., Bull. Direc. Mines Géol.*, 9: 1–170.

BONHOMME, M., 1962. Contribution à l'étude géochronologique de la plate-forme de l'Ouest Africain. *Ann. Fac. Sci. Univ. Clermont, Géol. Minière*, 5: 1–62.

BONIFAS, M., 1959. Contribution à l'étude géochimique de l'altération latéritique. *Mém. Serv. Carte Géol. Alsace Lorraine*, 17: 1–159.

BOOCOCK, L., 1958. *The Geological Environment of Copper Deposits in the Bechuanaland Protectorate.* Commission Coopération Technique en Afrique, Léopoldville, pp.249–254.

BOULADON, J. et JOURAVSKY, G., 1952. Les gîtes de manganèse de Maroc, une description des gisements du Précambrien III. *Intern. Geol. Congr., 20th, Mexico, 1956, Rept. Symp. Yacimentos Manganes, II (Africa)*, pp.217–248.

BOULANGER, J., 1958. *Les Gîtes de Cuivre de la Région Vohibory (Sud-ouest de Madagascar) et leur Liaison Géologique.* Commission Coopération Technique en Afrique, Léopoldville, pp.255–262.

BREIL, P., 1959. Aperçu sur l'hydrogéologie du sud et de l'ouest Cameroun. *Intern. Geol. Congr., 20th, Mexico, 1956, Assoc. Serv. Geol. Africains, Rept.*, pp.417–422.

BRENON, P., 1956. Étude du gisement de mica phlogopite d'Anara. *Rappt. Ann. Serv. Géol., Direc. Mines Géol. Madagascar*, 1956: 511–530.

REFERENCES

BRENON, P., 1958. Contribution à la géologie des gisements de thorianite de Madagascar. *Bull. Soc. Géol. France*, 3.

BUCHANAN, M. S., 1958. The Enegu coals, UDI Division, Onitsha Province. *Geol. Surv. Nigeria, Bull.*, 1955: 5–11.

BUREAU MINIER DE LA FRANCE D'OUTRE-MER, 1957. Compte rendu de l'exercice. *Rapports Annuels*, 1956–1957: 4–31.

BURSILL, C., LUYT, J. F. M. and URIE, J. G., 1961. The Bomvu Ridge iron ore deposit. *Trans. Proc. Geol. Soc. S. Africa, Spec. Paper*.

BYRAMJEE, R. et MEINDRE, M., 1956. Le gisement de manganèse de Guettara. *Intern. Geol. Congr., 20th, Mexico, 1956, Rept. Symp. Yacimentos Manganes, II (Africa)*, pp.179–198.

CALEMBERT, L., 1952 (1954). Sur les conditions de gisement de certains gîtes métallifères nord-africains et leurs causes. *Congr. Géol. Intern., Compt. Rend., 19e, Algiers, 1952*, pp.129–142.

CAMERON, E. N. and EMERSON, M. E., 1958. Origin of certain chromite deposits of the eastern part of the Bushveld complex. *Geol. Soc. Am., Abstr. 1958*, p.1545.

CAPDECOMME, L., 1953. Étude minéralogique des gîtes de phosphates alumineux de la région de Thiès, Sénégal. *Congr. Géol. Intern., Compt. Rend., 19e, Algiers, 1952*, 11: 103–118.

CARTER, G. S., 1958. The Liganga titaniferous magnetite occurrences. *Records Geol. Surv. Tanganyika*, 8: 67–71.

CAUSSE, R., 1962a. L'activité minière en République Centrafricaine. *Ind. Trav. Outre-Mer*, 10 (106): 731–736.

CAUSSE, R., 1962b. *Les Mines en République Centrafricaine en 1961.* Bangui, pp. 1–22.

CEDAT, R., 1962. Note sur l'industrie minière du Cameroun en 1961. *Cameroun, Territ., Bull. Direc. Mines Géol.*, 1962, pp.1–9.

CENTRE DE RECHERCHES ET D'INFORMATION SOCIO-POLITIQUES, 1961. *Morphologie des Groupes Financiers.* CRISP, Brussels, 486 pp.

CHARBONNAGES NORD-AFRICAINS, 1962. Assemblée générale annuelle des Actionnaires. *Reports.*

CHEMERY, J., 1960. Histoire de la mise en valeur minière des territoires d'Afrique centrale. *Publ. Bur. Études Géol. Minières*, 21: 1–175.

CHERMETTE, A., 1938a. Le gisement de chromite de Bontomo. *Gouvt. Gén. Afrique Occidentale Franç., Bull. Direc. Mines Géol.*, 1.

CHERMETTE, A., 1938b. Le titane au Dahomey. *Gouvt. Gén. Afrique Occidentale Franç., Bull. Direc. Mines Géol.*, 1.

CHERMETTE, A., 1946. Guinée (Chaîne de Niandan Banié). *Rappt. Ann. Sect. Géol., Gouvt. Gén. Afrique Occidentale Franç.*, 1946: 32–36.

CHERMETTE, A., 1949. Esquisse physique et géologique du Togo. *Gouvt. Gén. Afrique Occidentale Franç., Bull. Direc. Mines Géol.*, 11: 5–12.

COETZEE, C. B., 1958. Ilmeniethoudende sand langs die Weskus in die distrik Vanrhynsdorp. *Geol. Surv. S. Africa, Bull.*, 25: 1–17.

COETZEE, C. B., 1960. The geology of the Orange Free State goldfield. *Geol. Surv. S. Africa, Mem.*, 49: 1–198.

COERTZE, F. J., Intrusive relationships and ore-deposits in the Western part of the Bushveld Igneous Complex. *Trans. Proc. Geol. Soc. S. Africa*, 61: 387–400.

COLLEY, B. B., 1963. Libya: Petroleum Geology and Development. *World Petrol. Congr., Proc., 6th, Frankfurt, 1963*, 6 pp.

COLONNA-CIMARA, J., 1961. Le démarrage de l'exploitation du gisement de phosphate du Bas-Togo. *Ind. Trav. Outre-Mer*, 94: 1–8.

COMPAGNIE MINIÈRE DES GRANDS LACS AFRICAINS, M. G. L., 1959–1962. *Rapports.*

COMPAGNIE ROYALE ASTURIENNE DES MINES, 1962. *Rapports.*

CONSOLIDATED DIAMOND MINES OF SOUTHWEST AFRICA, 1963. *Annual report.*

CONSOLIDATED GOLD FIELDS OF SOUTH AFRICA, 1960–1963. *Annual reports.*

CONTINENTAL OIL CY, 1961. *Annual Report*, pp.1–35.

COOPER, W. G. G., 1936. The Bauxite deposits of the Gold Coast. *Gold Coast Geol. Surv., Bull.*, 7: 1–35.

COOPER, W. G. G., 1957. The geology and mineral resources of Nyasaland. *Nyasaland Geol. Surv. Dept., Bull.*, 6: 1–49.

CORTESÃO, J. L., 1958. *Notes on some Copper Deposits in Northern Angola.* Commission Coopération Technique en Afrique, Léopoldville, pp.263–266.

COSSON, J., 1959. Essai de contrôle géologique de quelques minéralisations du Gabon Central. *Chronique Mines Outre-Mer, Rech. Minière*, 27: 203–213.

COUSINS, C. A., 1959a. The Bushveld Igneous Complex. *Platinum Metals Rev.*, 3: 1–3.

COUSINS, C. A., 1959b. The structure of the mafic portion of the Bushveld Igneous Complex. *Trans. Proc. Geol. Soc. S. Africa*, 62: 179–202.

C.R.E.P.S., 1960–1961. *Rapports Généraux Annuels.*

CUNHA GOUVEIA, J. A., 1960. Notas sobre os fosfatos sedimentares de Cabinda. *Angola, Serv. Geol. Minas., Bol.*, 1, p.49.

DALLONI, M., 1939. *Géologie Appliquée de l'Algerie.* Masson, Paris, 888 pp.

REFERENCES

Daniels, J. L., 1960. Minerals and rocks of Hargeisa and Borama districts, Somaliland. *British Somaliland, Min. Nat. Res., Mineral Res. Pamphlet*, 3: 1–24.

Davidson, C. F. and Atkin, D., 1952. On the occurrence of uranium in phosphate rock. *Congr. Géol. Intern., Compt. Rend., 19e, Algiers, 1952*, 11: 13–31.

Davies, D. N., 1961. The tin deposits in Swaziland. *Trans. Proc. Geol. Soc. S. Africa, Spec. Paper*.

Davies, D. N., 1962. The zones of the Ecca Series, Karroo System in Swaziland. *Geol. Surv. Mines Dept. Swaziland, Ann. Rept. 1961*, pp.14–22.

De Beers Consolidated Mines Ltd., 1963. *Annual Report*.

De Chetelat, E., 1938. Le modelé latéritique de l'ouest de la Guinée Française. *Rev. Géograph. Phys. Géol. Dyn.*, (1): 1–120.

De Dorlodot, L. et Mathieu, F. F., 1929. Esquisse géologique des environs de Duru. *Ann. Soc. Géol. Belg., Bull.*, 51 (1927–1928): 109–114.

De Freitas, A. J., 1959. *A Geologia e o Desenvolvimento Ecónomico e social de Moçambique*. Junta de Communicaçoes, Lourenço Marques, 396 pp.

Degallier, R., 1960. Questions actuelles d'hydrogéologie en Afrique Occidentale et méthodes d'étude. *Notes Bur. Rech. Géol. Minières*, 61.

De Jager, D. H. and Von Backström, J. W., 1961. The sillimanite deposits in Namaqualand near Pofadder. *Geol. Surv. S. Africa, Bull.*, 33: 1–49.

De Kock, W. P., 1932. Lepidolite deposits of South-West Africa. *Trans. Proc. Geol. Soc. S. Africa*, 35: 97–113.

De Kun, N., 1955. Les pegmatites du Nord Lugulu. *Ann. Soc. Géol. Belg., Bull.*, 73: 27–30.

De Kun, N., 1957. On the Central-African Metallogenetic Province. *Geol. Rundschau*, 46 (2): 494–505.

De Kun, N., 1959a. Les gisements de cassitérite et de columbotantalite du Nord Lugulu. *Ann. Soc. Géol. Belg., Mém.*, 82: 81–196.

De Kun, N., 1959b. Die Zinn–Niob-Tantal Lagerstätten des Bezirkes von Nord-Lugulu. *Neues Jahrb. Mineral., Abhandl.*, 95 (1): 106–140.

De Kun, N., 1960a. La vie et le voyage de Ladislas Magyar dans l'intérieur du Congo en 1850–1852. *Acad. Roy. Sci. Outre-Mer (Brussels), Classe Sci. Nat., Mém.*, 4: 605–636.

De Kun, N., 1960b. Die Zinn–Niob–Tantal–Zirkon und Lanthaniden Lagerstätten von Nigerien. *Neues Jahrb. Mineral., Monatsh.*, 6: 100–107.

De Kun, N., 1961. Die Niobkarbonatite von Afrika. *Neues Jahrb. Mineral., Monatsh.*, 46: 124–135.

De Kun, N., 1962. The economic geology of columbium (niobium) and of tantalum. *Econ. Geol.*, 57: 377–404.

De Kun, N., 1963a. *Abstracts. Engineering and Applied Science at Columbia University*, pp.73–74.

De Kun, N., 1963b. Carbonatites, comment. *Econ. Geol.*, 58 (3): 446–447.

De Kun, N., 1963c. The stratiform copper deposits in Africa, review. *Econ. Geol.*, 58: 632–633.

De Kun, N., 1963d. The mineralogenetic provinces of Africa, I. *Econ. Geol.*, 58: 774–790.

De Kun, N., Cavadias, G., Gegan, A. and Brewer, W., 1964. *Report on Industrial Development for the Malagasy Republic*. Laramore, Douglass and Popham, New York, N.Y., 1: 257 pp., 2: 175 pp., 3: 116 pp.

Delany, F., 1959. Étude des grès de Mouka-Ouadda et des gisements diamantifères de l'Oubangi oriental. *Gouvt. Gén. Afrique Occidentale Franç., Bull. Direc. Mines Géol.*, 12: 41–45.

Delavesne, Y., 1958. Les gisements d'hydrocarbures de l'Afrique du Nord. *Ann. Mines*, 1958 (12): 795–814.

Delbos, L. et Rantoanina, M., 1961. Les gisements fer–nickel des environs de Moramanga (Madagascar). *Malgache Rép., Trav. Bur. Géol.*, 106 (1961): 1–61.

Delclaud, C. et Martel, A., 1959. Résultats des travaux de reconnaissance sur le champ de gaz de Hassi er Rmel. *Bull. Inst. Franç. Pétroles*, 14: 4–57.

Deleau, P. et Thiery, P., 1953. Les gîtes d'antimoine du département de Constantine. Le Gîte d'Aïn-Kerma. *Publ. Serv. Carte Géol. Algérie, Bull.*, 1: (1953): 7–72.

De Magnée, I., 1960. Contribution à la connaissance du tungstenbelt ruandais. *Acad. Roy. Sci. Outre-Mer (Brussels), Classe Sci. Nat.*, 11 (7).

Derricks, J. J. and Vaes, J. F., 1956. The Shinkolobwe uranium deposit. *Proc. Intern. Conf. Peaceful Uses At. Energy, U.N., Geneva, 1955*, 2: 94–138.

De Saint Ours, J., 1960. Étude générale des pegmatites. *Rappt. Ann. Serv. Géol., Direc. Mines Géol. Madagascar*, pp.73–88.

Des Ligneris, X. et Bernazaud, J., 1960. Le gisement de Mounana, Gabon. *Comm. Énergie At. (France), Bull.*, 38: 4–17.

Desio, A., 1943. L'esplorazione mineraria della Libia. *Coll. Sci. Doc. Africa Italia, X, Ist. Studia Polytech. Intern., Milan*, 333 pp.

Development and Resources Corp., 1962. *A Program for Geologic Exploration and Minerals Development 1962–1964*, New York, N.Y., 1–33, A: 1–49; B: 1–35.

REFERENCES

De Villiers, J., 1956. The manganese deposits of the Union of South Africa. *Intern. Geol. Congr., 20th, Mexico, 1956, Rept. Symp. Yacimentos Manganes, II (Africa)* pp.39–72.

De Villiers, J. and Söhnge, P. G., 1959. The geology of the Richtersveld, S. Africa. *Geol. Surv. S. Africa, Mem.*, 48: 1–266.

Drysdall, A. R., 1960. Graphite of the Petauke District, Eastern Province. *Northern Rhodesia, Geol. Surv., Rept.*, 14: 1–28.

Duhoux, P. V., 1950a. La pétrogénèse et la métallogénèse du domaine minier de Kilo-Moto. *Ann. Soc. Géol. Belg., Mém.*, 73: 171–244.

Duhoux, P. V., 1950b. *Les Itabirites du Nordest de la Colonie.* Comité Spécial du Katanga (50e anniversaire), Élisabethville, 2: 486–492.

Dumon, E., 1953. Note sur l'anticlinal de Sidi Abd-er-Rhamane au Cap Bon. *Congr. Géol. Intern., Compt. Rend., 19e, Algiers, 1952*, 16: 113–118.

Dunham, K. C., Phillips, R., Chalmers, R. A. and Jones, D. A., 1958. The chromiferous ultrabasic rocks of Eastern Sierra Leone. *Overseas Geol. Mineral Resources (Gt. Brit.), Suppl. Ser., Bull. Suppl.*, 3: 1–44.

Dunn, S. C., 1911. Notes on the mineral deposits of the Anglo-Egyptian Sudan. *Sudan, Geol. Surv. Dept., Bull.* 1: 1–70.

El Shazly, E. M., 1956. Notes on the mining map of Egypt. *Intern. Geol. Congr., 20th, Mexico, 1956, Rept. Assoc. Serv. Geol. Africa*, 1: 139–148.

El Shazly, E. M., 1959. Controls of Tertiary ore deposition in Egypt. *Chronique Mines Outre-Mer Rech. Minière*, 27 (275): 139–148.

El Zoghby, M., 1955. Les ressources minérales de l'Égypte. *Rev. Ind. Minière*, 1955: 1072–1082.

Ente Nazionale Idrocarburi (E.N.I.), 1963. *Annual Report and Statement of Accounts.*

Esso, Standard Oil Company (New Jersey), 1961–1963. *Report*, pp.1–31.

Eyssautier, L., 1952. L'industrie minière du Maroc. *Div. Mines Géol., Notes Mém.*, 88: 1–184.

Fawley, A. P., 1958. *The Environment of some Copper Deposits near Mpanda, Tanganyika.* Commission Coopération Technique en Afrique, Léopoldville, pp.209–212.

Ferguson, J. C., 1934. The geology of the country around Filabusi. *Southern Rhodesia Geol. Surv., Bull.*, 27: 1–179.

Feringa, G., 1959. The geological succession in a portion of the north-western Bushveld. *Trans. Proc. Geol. Soc. S. Africa*, 62: 219–238.

Fick, L. J., 1960. The geology of the tine pegmatites of Kamativi, Southern Rhodesia. *Geol. Mijnbouw*, 39: 472–491.

Fieremans, C., 1955. Étude géologique préliminaire des conglomérats diamantifères d'âge mésozoïque au Kasaï. *Mém. Inst. Géol. Univ. Louvain*, 19 (2): 223–294.

Fieremans, C., 1960. Étude critique des classifications des formations diamantifères au Kasaï et dans le Lunda. *Mém. Inst. Géol. Univ. Louvain*, 21: 249–278.

Fockema, A. R. P. and Austin, A. L., 1956. Manganese deposits in Northern Rhodesia. *Intern. Geol. Congr., 20th, Mexico, 1956, Rept. Symp. Yacimentos Manganes, II (Africa)*, pp.273–292.

Fockema, A. R. P., Fockema, P. D. et Marais, J. A. H., 1961. Gypsum occurrences in South Africa. *Trans. Proc. Geol. Soc. S. Africa*, Spec. Paper.

Fourie, G. P., 1959. The chromite deposits in the Rustenburg area. *Geol. Surv. S. Africa, Bull.*, 27: 1–45.

Française, République, 1959. L'industrie minière dans les Territoires d'Outre-Mer en 1958. *Admin. Gén. France d'Outre-Mer, Inspection Gén. Mines Geol.*, pp.1–31.

Française, République, 1959. Note sur les Gisements de Fer d'Afrique. *Inspection Gén. Mines Géol.*, pp.1–15.

Française, République, 1960. L'industrie minière en 1959. *Inspection Gén. Mines Géol.*, pp.1–63; 1–15.

Française, République. L'exploitation des gisements de phosphates du Sénégal et du Togo. *Direc. Affaires Écon. Plan, 1er Bur.*, pp.1–7.

Française, République. La Compagnie Sénégalaise des Phosphates de Taïba. *Direc. Affaires Écon. Plan, 1er Bur.*, pp.1–3.

Frankel, J. J., 1958. Manganese ores from the Kuruman District, Cape Province, S. A. *Econ. Geol.*, 53: 577–597.

Fratschner, W. Th., 1960. Die Laterite der südöstlichen Boé (Portugiesisch Guinea). *Geol. Mijnbouw*, 39: 500–511.

Freire de Andrade, 1953. Diamond deposits in Lunda. Companhia Diamantes de Angola (Diamang), Serv. Culturaís, Publ. Cult. 17, Lisbon. *Reports.*

Frondel, C. and Ito, J., 1957. Geochemistry of germanium in the oxidized zone of the Tsumeb Mine. *Am. Mineralogist*, 42: 743–753.

Furon, R., 1957. *Le Sahara.* Payot, Paris, 302 pp.

Furon, R., 1961. *Les Ressources minérales de l'Afrique.* Payot, Paris, 284 pp.

Garlick, W. G., 1958. *Structures of the Northern Rhodesian copper deposits.* Commission Coopération Technique en Afrique, Léopoldville, pp.159–180.

REFERENCES

Garlick, W. G., 1961. Chambisi. In: F. Mendelsohn (Editor), *The Geology of the Northern Rhodesian Copperbelt.* Macmillan, London, pp. 281–296.

Geomines, 1963. *Rapports, Exercice 1961–1962.*

Ghana Geological Survey, 1962. References in the reports of the director of Geological Surveys to occurrences of economic minerals in Ghana. *Bull. Ghana Geol. Surv.*, 28: 1–63.

Ghana Information Services, 1962. *The Volta River project.* Press release, 173.

Giraud, P., 1960. Les roches basiques de la région d'Andriamena et leur minéralisation chromifère. *Malgache, Rép., Ann. Géol. Madagascar*, 27: 1–95.

Giri, J., 1961. L'industrie minière de la République Gabonaise en 1961. *Direc. Mines Gabon*, 1–21.

Giri, J., 1961. L'industrie minière au Gabon. *Mines Mét.*, 3553: 421–423.

Giri, J., 1961. L'industrie minière gabonaise. *Ind. Trav. Outre-Mer*, 9, 87: 151–158.

Glacon, J., 1958. Les minéraux du nickel, du cobalt et du bismuth dans les minéralisations du Nord de l'Algérie. *Bull. Soc. Franc. Minéral. Crist.*, 81: 7–9.

Glacon, J. et Mougin, D., 1951. Quelques observations sur le gîte plombo-zincifère de Boukdema et sur le talc qu'on y rencontre. *Compt. Rend.*, 234 (25): 2471–2473.

Glangeaud, L., 1952. Mode de formation des gisements épitéléthermaux. Symposium sur les gisements de fer du monde. *Congr. Géol. Intern., Compt. Rend., 19e, Algiers, 1952*, 1: 46–59.

Godfriaux, I., Lamotte, M. et Rougerie, G., 1957. La série stratigraphique du Simandou (Guinée française). *Compt. Rend.*, 245: 2343.

Golubinow, R., 1938. Les bauxites de Tougué. *Gouvt. Gén. Afrique Occidentale Franç., Bull. Direc. Mines Géol.*, 1: 78–88.

Gottis, Ch. et Sainfeld, P., 1952. Les gîtes métallifères tunisiens. *Congr. Géol. Intern., Compt. Rend., 19e, Algiers, 1952*, 2: 1–106.

Grantham, D. R. and Allen, J. B., 1960. Kimberlite in Sierra Leone. *Overseas Geol. Mineral Resources (Gt. Brit.)*, 8 (1): 5–25.

Groen, H., 1961. Luansobe. In: F. Mendelsohn (Editor), *The Geology of the Northern Rhodesian Copperbelt.* Macmillan, London, pp.406–410.

Groeneveld, D., 1961a. The Potgietersrus tin-mining area. *Trans. Proc. Geol. Soc. S. Africa, Spec. Paper.*

Groeneveld, D., 1961b. A composite description of the chrome deposits of the Bushveld Igneous Complex. *Trans. Proc. Geol. Soc. S. Africa, Spec. Paper.*

Groenewald, P., 1958. *The Geological Environment of the Copper Deposits of the Union of South Africa.* Commission Coopération Technique en Afrique, Léopoldville, pp. 223–230.

Grosemans, P., 1959. La bauxite dans le Bas-Congo. *Acad. Roy. Sci. Outre-Mer (Brussels)*, 1: 457–469.

Gross, W. H. and Strangway, D. W., 1961. Remanent magnetism and origin of hard hematites in Precambrian banded ironstone formation. *Econ. Geol.*, 56: 1345–1362.

Gübelin, E. J., 1958. Emeralds from Sandawana. *J. Gemmology*, 6: 8.

Guernsey, T. D., 1952. *Mineral Occurrences in Northern Rhodesia.* British South Africa Company, Salisbury, 89 pp.

Habgood, F., 1956. *Report on the Recent Investigation of the Sumbu-Nkombedzi Coal field.* Geol. Surv. Nyasaland. Roneoscript, pp.1–29.

Habgood, F., 1961. Report on the investigation of the Kapiridimba kyanite deposit, Ncheu District. *Nyasaland, Records Geol. Surv. Nyasaland*, 1: 63.

Hall, A. L., 1920. Mica in the Eastern Transvaal. *Geol. Surv. S. Africa, Mem.*, 13: 1–95.

Hall, A. L., 1930. Asbestos in the Union of South Africa. *Geol. Surv. S. Africa, Mem.*, 12: 1–152.

Hall, T. C. F., 1949. The mineral resources of Angola. *Angola, Serv. Geol. Minas*, 1949, pp.1–64.

Hamou, A., 1961. L'activité minière au Maroc en 1960. *Mines Géol. (Morocco)*, 15(4): 13–24.

Harder, E. C., 1952. Examples of bauxite deposits illustrating variations of origin. Problems of Clay and Laterite Genesis. *Am. Inst. Mining Met. Engrs., Symp.*

Harder, E. C. and Greig, E. W., 1960. Bauxite, the ore of aluminium, inindustrial minerals and rocks. *Am. Inst. Mining Met. Petrol. Engrs., Inst. Metals Div., Spec. Rept. Ser.*, 1960: 65–85.

Hargraves, R. B., 1963. Silver-gold ratios in some Witwatersrand conglomerates. *Econ. Geol.*, 58(6): 952–970.

Harris, J. F., 1960. Geological investigations, sampling and diamond-drilling at Manyeghi helium-bearing hot springs, Singida District. *Records Geol. Surv. Tanganyika*, 8: 86.

Harris, J. F., 1961. Summary of the geology of Tanganyika. 4, Economic Geology. *Tanganyika Geol. Surv. Dept., Mem.*, 4(1): 1–44.

Haughton, S. (Editor), 1958–1962. *Abstracts of Literature.* Commission Coopération Technique en Afrique, Léopoldville.

Hedberg, H. D., 1961. Petroleum developments in Africa in 1960. *Bull. Am. Assoc. Petrol. Geologists*, 45 (7): 1145–1185.

REFERENCES

HERRMANN, F., 1950. *Les Richesses minérales du Monde*. Payot, Paris, 247 pp.

HOLLOWAY, H. L., 1958. Alluvial gold field in Ethiopia. *Mineral. Mag.*, 98 (2): 73–79.

HOLMES, A., 1959. A revised geological time scale. *Trans. Edinburgh Geol. Soc.*, 17: 185–216.

HOLMES, A. et CAHEN, L., 1951. Géochronologie africaine, 1956. *Acad. Roy. Sci. Outre-Mer (Brussels), Classe Sci. Nat., Mém.*, 52.

HOLMES, R., 1954. *Mineral Deposits of Somalia*. World Mining Consultants Inc., New York. 83 pp.

HOLZ, H. R., 1960. Fluorspar in South Africa. *Pit Quarry*, 52: 172–175.

HOSE, H. R., 1962. Bauxite mineralogy. *Am. Inst. Mining Met. Engrs., Aluminum Symp.*

HOTTIN, G., 1961. Recherches de bauxites sur les Tampoketsa de la région centrale (Campagne, 1960). *Malgache, Rép., Trav. Bur. Géol.*, 104.

HOTTIN, G. et MOINE, B., 1963. État des recherches de bauxite à Madagascar. *Comité Natl. Malgache Géol.*, 31: 1–13.

HUGE, J. et EGOROFF, A., 1947. Ressources minérales du Congo. *Bull. Serv. Géol. Congo Belge Ruanda-Burundi*, 3: 21–35.

HUME, W. F., 1927. The phosphate deposits of Egypt. *United Arab. Rep., Geol. Surv. Mineral Res. Dept., Paper*, 41.

HUNT, J. A., 1954. Gypsum-anhydrite. *Somaliland, Protect., Geol. Surv. Mineral Res. Pamphlet*, 1.

HUNTER, D. R., 1962. The mineral resources of Swaziland. *Swaziland, Geol. Surv. Mines Dept., Bull.*, 2: 1–112.

HURST, H. E., 1952. *The Nile*. Constable, London, 326 pp.

HUTCHINSON, G. E., 1950. The biogeochemistry of vertebrate excretion. *Am. Mus. Novitates*, 96: 1–554.

JACOBSEN, J. B. E., 1961. The Geology of the Alaska mine. *Trans. Proc. Geol. Soc. S. Africa, Spec. Paper.*

JACOBSEN, W., 1961. The geology of the Mangula copper deposits, Southern Rhodesia. *Trans. Proc. Geol. Soc. S. Africa, Spec. Paper.*

JEANNETTE, A., 1958. Nouvelles observations sur les gîtes de pyrites et magnétite du Maroc nord-oriental (région de Mélilla). *Compt. Rend. Soc. Géol. France*, 1958 (7): 153.

JEANNETTE, A., 1961. Les ressources minérales du Rif nord-oriental. *Mines Geol. (Morocco)*, 14 (4).

JORDAAN, J., 1961. Nkana. In: F. MENDELSOHN (Editor), *The Geology of the Northern Rhodesian Copperbelt*. Macmillan, London, pp.297–327.

JUNNER, N. R., 1943. The diamond deposits of the Gold Coast. *Gold Coast, Geol. Surv., Bull.*, 12: 1–52.

KABESH, M. L., 1960. Mica deposits of Northern Sudan. *Sudan, Geol. Surv. Dept., Bull.*, 7: 1–55.

KABESH, M. L. and AFIA, M. S., 1959. The Wollastonite deposits of Dirbat Well. *Sudan, Geol. Surv. Dept., Bull.*, 5: 1–54.

KABESH, M. L. and AFIA, M. S., 1961. Manganese ore deposits of the Sudan. *Sudan, Geol. Surv. Dept., Bull.*, 9: 1–31.

KEEP, F. E., 1929. The geology of the Shabani mineral belt, Belingwe District. *Southern Rhodesia, Geol. Surv., Bull.*, 12: 1–193.

KEEP, F. E., 1961. Amphibole asbestos in the Union of South Africa. *Commonwealth Mining Met. Congr., 7th, Johannesburg, 1961*, 1: 90–121.

KIEKEN, M. et WINNOCK, E., 1956. Le champ de l'Oued Guetérini. *Intern. Geol. Congr., 20th, Mexico, 1956, Symp. Yacimentos Petrol. Gas*, pp.23–43.

KILIAN, C., 1957. Pétrole et cuivre dans le Sahara central. *Compt. Rend.*, 245 (15): 1250.

KIMBLE, G. H. T., 1960. *Tropical Africa*, 2. Twentieth Century Fund, New York, N.Y., 603 pp.

KRENKEL, E., 1957. *Geologie und Bodenschätze Afrikas*. Akademie Verlag, Leipzig, 597 pp.

KUPFERBURGER, W. and LOMBAARD, B. V., 1937. The chromite deposits of the Bushveld Igneous Complex. *Geol. Surv. S. Africa, Bull.*, 10: 1–48.

KUPFERBURGER, W., BOARDMAN, L. G. and BOSCH, P. R., 1956. New considerations concerning the manganese ore deposits in the Postmasburg and Kuruman areas. *Intern. Geol. Congr., 20th, Mexico, 1956, Rept. Symp. Yacimentos Manganes, II (Africa)*, pp.73–88.

LABORATORIO NACIONAL DE ENGENHERIA CIVIL (LISBON), 1952. Estudo das Pozzolanas de Cabo Verde. *Bull. Mens. Inform.*, 38. Roneoscript, 1: 1–13; 2: 1–3; 3: 1–7.

LACROIX, A., 1914. Les latérites de la Guinée. *Nouvelles Arch. Musée*, 5, V: 256–356.

LAMMING, C. K. G., 1956. World lithium resources. *Mineral Mag.*, 247: 334–335.

LATEGAN, P. N., 1961. The coal mining industry of South Africa. *Commonwealth Mining Met. Congr., 7th, Johannesburg, 1961*, 1: 39–48.

LATRILHE, ED., 1959. Contribution à l'étude des phosphates alumineux de Thiès. *Haut Comm. Rép. Afrique Occidentale Franç., Bull. Serv. Géol.*, 25: 1–89.

LECLERC, J. C., RICHARD-MOLARD, J., LAMOTTE, M., ROUGERIE, G. et PORÈRES, R., 1955. La chaine de Nimba. *Mém. Inst. Franç. Afrique Noire*, 43.

LETHBRIDGE, R. F. and PERCIVAL, F. G., 1954. Iron deposits at Fort Gouraud, Mauritania, French West Africa. *Trans. Inst. Mining Met.*, 63: 285–298.

LIEBENBERG, W. R., 1955. The occurrences and origin of gold and radioactive minerals in the Witwatersrand System, the Dominion Reef, the Ventersdorp Contact Reef and the Black Reef. *Trans. Proc. Geol. Soc. S. Africa*, 58: 101–254.

LOMBARD, J. et NICOLINI, P. (Rédacteurs), 1962. *Gisements Stratiformes de Cuivre en Afrique.* Assoc. Serv. Géol. Afrique, Paris, pp.1–211.

LOSSEL, P., 1953. Les grès à gaz du champ du Cap Bon, Tunisie. *Congr. Géol. Intern., Compt. Rend., 19e, Algiers, 1952*, 16: 119–124.

LUCAS, G. M. L., 1952 (1954). Relations de la structure géologique et de la minéralisation plombo-zincifère dans la région de Ghar-Rouban. *Congr. Géol. Intern., Compt. Rend., 19e, Algiers, 1952*, 12: 339–366.

LUCAS, G., 1956. Gisements et indices de manganèse en Algérie. *Intern. Geol. Congr., 20th, Mexico, 1956, Rept. Symp. Yacimentos Manganes, II (Africa)*, pp.123–130.

MANSOUR, A. O. and ISKANDER, W., 1960. An iron ore deposit at Jebel Abu Tulu. *Sudan, Geol. Surv. Dept., Bull.*, 6: 1–41.

MARCHANDISE, H., 1958. Le gisement et les minerais de manganèse de Kisenge. *Ann. Soc. Géol. Belg.*, 67: 187–210.

MAREE, S. C., 1958. *The Geology and Ore Deposits of Mufulira, Northern Rhodesia.* Commission Coopération Technique Afrique, Léopoldville, pp.147–158.

MARMO, V., 1961. Antophyllite asbestos in central Sierra Leone. *Schweiz. Mineral. Petrog. Mitt.*, 37 (31): 1–117.

MARMO, V., 1962a. Geology and mineral resources of the Kangari Hills schist belt. *Geol. Surv. Sierra Leone, Bull.*, 2: 1–116.

MARMO, V., 1962b. The molybdenum bearing granite of the Wankatana River, Sierra Leone. *Overseas Geol. Mineral Resources (Gt. Brit.)*, 8: 416–427.

MARTHOZ, A., 1955. L'industrie minière et métallurgique du Congo belge. *Acad. Roy. Sci. Outre-Mer (Brussels), Classe Sci. Tech., Mém. in 8°.*, 1: 1–60.

MARVIER, L., 1959. Sur la présence de kimberlites dans la partie occidentale du Soudan français. *Intern. Geol. Congr., 20th, Mexico, 1956, Assoc. Serv. Géol. Africains*, pp. 437–442.

MASCLANIS, P., 1956. Observations sur le gisement de nickel de Valozoro. *Rappt. Ann. Serv. Géol., Direc. Mines Géol. Madagascar*, pp.121–124.

MATHERON, G., 1955. Le gisement de fer de Gara Djebilet. *Bull. Sci. Écon. Bur. Rech. Minières Algérie*, 2: 53–63.

MATHERON, G., 1956. Utilisation de la géochimie au *Bur. Rech. Minières Algérie. Rev. Ind. Minière*, pp.275–280.

MCDERMOTT, E. K. and OXLEY-OXLAND, G. ST. J., 1961. Tin mining in South Africa. *Commonwealth Mining Met. Congr., 7th, Johannesburg, 1961*, 1: 145–158.

MCKINNON, D. and SMIT, N. J., 1961. Nchanga. In: F. MENDELSOHN (Editor), *The Geology of the Northern Rhodesian Copperbelt.* Macmillan, London, pp.234–275.

MCNAUGHTON, J. H. M., 1958. *The Environments of Copper Deposits in Nyasaland.* Commission Coopération Technique en Afrique, Léopoldville, pp. 197–198.

MEINDRE, M., 1959. Principales minéralisations du Hoggar. *Chronique Mines Outre-Mer Rech. Minière*, 27 (271): 9–22.

MENDELSOHN, F. (Editor), 1961. *The Geology of the Northern Rhodesian Copperbelt.* Macmillan, London, 533 pp.

MERO, J. L., 1962. Ocean-floor manganese nodules. *Econ. Geol.*, 57: 746–767.

MEYER DE STADELHOFEN, C., 1961. Les kimberlites de Bakwanga. *Chronique Mines, Outre-Mer Rech. Minière*, 29 (297): 17.

MICHEL, P., CLARACE, P., LAURIOL, E., VERRIEN, J. P., CONRAUD, A. et AUFRERE, G., 1959. Les problèmes pétroliers paléozoïques de la bordure nord du Hoggar. *World Petrol. Congr., Proc., 5th, N.Y., 1959*, pp.747–785.

MILLOT, G. et DARS, R., 1959. L'archipel des îles de Cos. *Serv. Géol. Prospection Minière, Dakar*, (2): 48–56.

MINES D'AOULI, 1962. *Compte Rendu de l'Assemblée Générale Ordinaire*; pp.1–17.

MOINE, J., GORICHON, A. et CANCE, E., 1963. *Valorisation des Minerais de Manganèse de l'Imini.* Roneoscript.

MOODY, D. J. and PARSONS, M. C., 1963. Petroleum developments in Africa in 1962. *Bull. Am. Assoc. Petrol. Geologists*, 47 (7): 1348–1396.

MOORE, L. R., 1960. Summary report on the Ketewaka-Mchuchuma coalfield. *Records Geol. Surv. Tanganyika*, 8: 54.

MOREAU, M., 1958. Les gisements d'uranothorianite du Sud-Est de Madagascar. *Comm. Énergie At. (France)*, Rappt.

MOREAU, M., 1959. Les gisements d'uranothorianite du Sud-Est de Madagascar. *Chronique Mines, Outre-Mer Rech. Minière*, 27 (315): 1–26.

MOREL, S. W., 1958. The geology of the Middle Shire Area. *Nyasaland, Geol. Surv. Dept., Bull.*, 10: 1–66.

MURDOCK, TH. G., 1954. The mica deposits and industry of Angola. *U.S. Bur. Mines, Mining Trade Notes*, 39 (3). *Spec. Suppl.*, 42: 1–38.

MURDOCK, TH. G., 1955. Bécéka's industrial diamond mining operations at Bakwanga. *U.S. Bur. Mines, Mining Trade Notes*, 46: 1–23.

NASSIM, G. L., 1949. The discovery of nickel in Egypt. *Econ. Geol.*, 44 (2): 143–150.

REFERENCES

Newmont Mining Corp., 1957–1963. *Annual Reports.*

Nicolini, P., 1959. Contribution à l'étude des minéralisations stratiformes du Moyen Congo. Le synclinal de la Nyanga. *Bull. Direc. Mines Géol.*, pp.1–150.

Niedermüller, W., 1955. Die Blei-Zinkgänge von Guerrouma und Nador-Chait im Tell Atlas. *Berg-Huettenmaenn. Monatsh. Montan. Hochschule Leoben*, 100: 180–181.

Nigeria, Federation of, 1957. *Minerals and Industry in Nigeria. Report.*

Noakes, L. C., 1961. Bushveld guide. *Commonwealth Mining Met. Congr., 7th, Johannesburg, 1961.*

Nyasaland, Ministry of Natural Resources, 1963. *Mineral Resources of Nyasaland*, pp.1–30.

Obermuller, A., 1941. Description pétrographique et étude géologique de la région de la Guinée française. *Gouvt. Gén. Afrique Occidentale Franç., Bull. Direc. Mines Géol.*, 5: 1–207.

Obermuller, A., 1959. Rapport sommaire sur la recherche géologique et la prospection minérale fin 1958. *Chronique Mines Outre-Mer Rech. Minière*, 27 (280): 337–405.

Obermuller, A., 1962. L'industrie minérale au Libéria. *Chronique Mines Outre-Mer Rech. Minière*, 30 (309): 174–180.

O'Brien, P. L. A., 1958. *Copper Deposits and their Environment in Northern Rhodesia.* Commission Coopération Technique en Afrique, Léopoldville, pp.133–146.

Office Chérifien des Phosphates. *Bull*, 12 pp.

Office Chérifien des Phosphates, 1961. *Morrocan Phosphates.* Roneoscript, 58 pp.

Office Chérifien des Phosphates, 1962. *Report.*

Ohio Oil Co., 1961. *74th Annual Report.*

O'okiep Copper Mining Co., 1963–1964. *Annual report.*

Oosterbosch, R., 1962. La minéralisation du système du Roan au Katanga. In: J. Lombard et P. Nicolini (Rédacteurs), *Gisements Stratiformes de Cuivre en Afrique.* Assoc. Serv. Géol. Afrique, Paris, pp.71–136.

Oosterbosch, R. and Schuiling, H., 1951. Copper mineralization in the Fungurume region, Katanga. *Econ. Geol.*, 46 (2): 121–148.

Ortynski, I., Perrodon, A. et De Lapparent, C., 1959. Esquisse géologique et structurale des bassins du Sahara Septentrional. *World Petrol. Congr., Proc., 5th, N.Y.*, pp.705–727.

Ostrander, F. T., 1963. The place of minerals in economic development. *Am. Inst. Mining Met. Petrol. Engrs., 92nd Ann. Meeting*, 24 pp.

Paine, L. A. and Wilson, P. F., 1954. Mineral resources of Southwest Africa. *Mineral Mag.*, 91 (6): 329–335.

Pallister, J. W., 1958. Suria Malableh sypsum-anhydrite deposit, Estimate of reserves. *Somaliland Protect. Geol. Surv.* Typescript, pp.1–10.

Pallister, J. W. and Warden, A. J., 1959. Core-drilling on Suria Malableh gypsum/anhydrite deposit. *Somaliland Protect. Geol. Surv.* Roneoscript, pp.1–13.

Pauly, E., 1962. Geology and mineralization of the Lueca region. *Angola, Serv. Geol. Minas, Bol.*, 5: 43–58.

Peccia-Galletto, J., 1960. Le gisement de manganèse du Grand-Lahou (Côte d'Ivoire). *Ann. Mines*, 1960 (49): 711–720.

Pegand, G. et Reyre, D., 1959. Les champs de pétrole de l'Afrique Occidentale Française. *World Petrol. Congr., Proc., 5th, N.Y.*, 1 (30): 551–573.

Perrodon, A., 1960. Aperçu des principaux résultats obtenus en Libye. *Bull. Assoc. Franc. Techniciens Petrole*, 142: 617–630.

Perry, J. D., 1961. The significance of mineralized breccia pipes. *Mining Eng.*, 120 (13): 367–376.

Phaup, A. E., 1961. The gold mines of Southern Rhodesia. *Trans. Proc. Geol. Soc. S. Africa, Spec. Paper.*

Phillips, K. A., 1958. *A regional Outline of certain metalliferous Zones and their Bearing upon some Problems of Granitisation in Northern Rhodesia.* Commission Coopération Technique en Afrique, Léopoldville, pp.119–132.

Pienaar, P. J., 1961. Bwana Mkubwa area. In: F. Mendelsohn (Editor), *The Geology of the Northern Rhodesian Copperbelt.* Macmillan, London, pp.467–486.

Polinard, E., 1946. Les gisements de manganèse à polianite et hollandite de la Haute Lulua. *Inst. Roy. Colonial Belge, Mém.*, 16 (1): 1–41.

Polinard, E., 1952. Les richesses minérales du Congo belge. *Encyclopédie du Congo Belge*, 2: 471–570.

Pollett, J. D., 1952. The geology and mineral resources of Sierra Leone. *Colonial Geol. Mineral Resources*, 2 (1): 3–28.

Pons, A. et Quezel, P., 1958. Premières remarques sur l'étude palynologique d'un guano fossile du Hoggar. *Compt. Rend.*, 246 (15): 2290–2292.

Prain, R. L., 1962a. *The Basic Factors of the Copper Industry Today.* Speech at Mining Club, New York. Roneoscript.

Prain, R. L., 1962b. *Rhodesian Selection Trust Group of Companies.* Statement by the Chairman.

Prirogine, A., 1956. Concentration des minerais de wolfram et de niobium-tantale au Congo belge et au Ruanda-Urundi. *Acad. Roy. Sci. Outre-Mer (Brussels), Classe Sci. Tech. Mém. in 8°, Nouv. Sér.*, 4 (1): 191 pp.

PULFREY, W., 1958. *The Geological Environment of Copper Deposits in Kenya.* Commission Coopération Technique en Afrique, Léopoldville, pp.213–218.

PULFREY, W., 1960. The geology and mineral resources of Kenya. *Kenya, Colony Protect., Mining Geol. Dept., Geol. Surv. Kenya, Bull.*, 2: 1–41.

PURSER, D. A. C., 1962. The trend of mineral production of Swaziland from 1907–1960. *Swaziland, Geol. Surv. Mines Dept., Bull.*, 1: 23–29.

QUEZEL, P. et MARTINEZ, C., 1958. Étude palynologique de deux diatomites du Borkou (Tchad). *Bull. Soc. Hist. Nat. Afrique Nord*, 49: 5–6.

QUINN, H., 1962. *Ethiopia, the Mineral Industry in 1961.* Commission Coopération Technique en Afrique, Léopoldville. Roneoscript, pp.1–2.

RAGUIN, E., 1961. *Géologie des Gîtes Minéraux.* Masson, Paris, 679 pp.

RAMDOHR, H., 1957. Recherches microscopiques sur les minerais du gisement du Guelb Moghrein, Akjoujt. *Gouvt. Gén. Afrique Occidentale Franç., Bull. Direc. Mines Géol.*, 20: 193–257.

RAMDOHR, P., 1958. New observations on the ores of the Witwatersrand and their genetic significance. *Trans. Proc. Geol. Soc. S. Africa*, 61 (1959): 1–50.

RAND SELECTION CORP., 1961–1963. *Annual Reports.*

RANOUX, J., 1956. Études métallogéniques sur les gîtes de Cavallo. *Bull. Sci. Écon. Bur. Rech. Minières Algérie*, 4: 67–81.

RATH, R. und PUCHELT, H., 1957. Indigolith von Usakos. *Neues Jahrb. Mineral., Monatsh.*, 9: 206–208.

REÁL, F., 1959. Intrusões kimberlitícas da Lunda. *Port., Min. Econ., Direc.-Geral Minas Serv. Geol., Serv. Geol. Port., Mem.*, 5 (1959): 1–116.

REED, F. R. C., 1949. *The Geology of the British Empire.* Arnold, London, 764 pp.

RESELMAN, P., 1962. Water for a 1,000 years. *S. African Panorama*, June, 1962.

RHODEN, N. H., 1961. Géologie du gisement de fer d'Ouichane. *Mines Géol. (Morocco)*, 14 (4).

RHODESIAN ANGLO-AMERICAN LTD., 1960–1963. *Annual Reports.*

RHODESIAN SELECTION TRUST LTD., 1962, 1963. *Annual Reports.*

ROBERT, M., 1931. An outline of the geology and ore deposits of Katanga. *Econ. Geol.*, 26: 531–539.

ROBERT, M., 1946. *Le Congo Physique.* Vaillant–Carmanne, Liège, 449 pp.

ROCHE, J., 1962. Inventaire et étude des gisements de barite du département d'Alger. *Bull. Sci. Écon. Bur. Rech. Minières Algérie.* Typescript, pp.1–16.

ROPER, H., 1956. The manganese deposits at Otjosundu, S.W. Africa. *Intern. Geol. Congr., 20th, Mexico, 1956, Rept. Symp. Yacimentos Manganes, II (Africa)*, pp.115–122.

SADD-EL-AALI, MINISTRY OF, U.A.R., 1962. *Sadd-el-Aali Project*, pp.1–22.

SADRAN, G., MILLOT, G. et BONIFAS, M., 1955. Sur l'origine des gisements de bentonite de Lalla Maghnia (Oran). *Publ. Serv. Carte Géol. Algérie, Bull.*, 5: 213–234.

SAGATZKY, N., 1954. La géologie et les ressources minières de la Haute-Volta méridionale. *Gouvt. Gén. Afrique Occidentale Franç, Bull. Direc. Mines Géol.*, 13: 11–225.

SAHLI, E. W., 1961. Antimony in the Murchison Range. *Commonwealth Mining Met. Congr., 7th, Johannesburg, 1961, Trans.*, 1: 181–201.

SCHNEIDERHÖHN, H., 1931. *Mineralische Bodenschätze im Südlichen Afrika.* NEM Verlag, Berlin, 111 pp.

SCHNELL, 1961a. L'industrie du gaz naturel. *Serv. Mines Algérie.* Roneoscript, pp.1–16.

SCHNELL, 1961b. Les industries minières en Algérie. *Serv. Mines Algérie.* Roneoscript, pp.1–21.

SCHNELL, 1961c. Recherches et exploitation de pétrole. *Serv. Mines Algérie.* Roneoscript, pp.1–10.

SCHOKALSKAJA, S. J., 1953. *Die Böden von Afrika.* Akademie Verlag, Berlin, 408 pp.

SCHUILING, H. et GROSEMANS, P., 1956. Les gisements de manganèse du Congo belge. *Intern. Geol. Congr., 20th, Mexico, 1956, Rept. Symp. Yacimentos Manganes, II (Africa)*, pp.131–142.

SCHWELLNUS, C. M., 1938. Vermiculite deposits in the Palaboroa area. *Union S. Africa, Dept. Mines, Geol. Surv. Div., Bull.*, 11: 1–27.

SCHWELLNUS, J. E. G., 1961. Bancroft. In: F. MENDELSOHN (Editor), *The Geology of the Northern Rhodesian Copperbelt*, pp.214–233.

SCHWOERER, P., 1959. Problèmes hydrogéologiques du Nord-Cameroun. *Intern. Geol. Congr., 20th, Mexico, 1956. Assoc. Serv. Géol. Africains*, pp.443–446.

SCLAR, CH. B. and GEIER, B. H., 1957. The paragenetic relationships of germanite and rieniérite from Tsumeb, Southwest Africa. *Econ. Geol.*, 52 (6): 612–631.

SÉNÉGAL, REPUBLIQUE DU — SERVICE DES MINES ET DE LA GÉOLOGIE, 1961. *Programme Quadriennal de Recherches Géologiques et Minières*, pp.1–13.

SÉNÉGAL, REPUBLIQUE DU — SERVICE DES MINES ET DE LA GÉOLOGIE, 1961. *Programme Quadriennal 1961–1965*, pp.1–29.

SERVANT, J., 1956. Les gisements et indices de manganèse de l'Afrique Occidentale Française. *Intern. Geol. Congr., 20th, Mexico, 1956, Rept. Symp. Yacimentos Manganes, II (Africa)*, pp.89–114.

SERVICE, H., 1943. The geology of the Nsuta manganese ore deposits. *Gold Coast, Geol. Surv., Mem.*, 5: 1–32.

SHARPE, J. W. N., 1961. The Empress nickel–copper deposit — Southern Rhodesia. *Trans. Proc. Geol. Soc. S. Africa Spec. Paper.*

SIERRA LEONE DEVELOPMENT CY., 1961. Iron ore from Marampa. *Report.*

SIERRA LEONE, MINES DEPT., 1960–1963. *Reports.*

SKINNER, W. E., 1962a. *Mining Yearbook.* Skinner, London, 790 pp.

SKINNER, W. E., 1962b. *Oil and Petroleum Yearbook.* Skinner, London, 782 pp.

SMIT, N. J., 1961. Mimbula. In: F. MENDELSOHN (Editor), *The Geology of the Northern Rhodesian Copperbelt.* Macmillan, London, pp.275–280.

SOCIÉTÉ BELGO-AFRICAINE DU KIVU (SOBAKI), 1960–1962. *Rapports.*

SOCIÉTÉ D'ENTREPRISE ET D'INVESTISSEMENT DU BÉCÉKA (SIBÉKA), 1962. *Rapports.*

SOCIÉTÉ DES MINES DE ZELLIDJA, 1962. *Rapport Annuel.*

SOCIÉTÉ DES MINES D'OR DE KILO-MOTO, 1959–1963. *Rapports du Conseil d'Administration.*

SOCIÉTÉ GÉNÉRALE DE BELGIQUE, 1962. *Rapports.*

SOCIÉTÉ NATIONALE REPAL, 1960–1962. *Rapports Annuels d'Activité.*

SOCIÉTÉ NATIONALE REPAL — C.F.P.(A), 1963. *Situation des Travaux de l'Association.*

SÖHNGE, P. G., 1952. The geology of Tsumeb. *S. African Inst. Mech. Engrs., J.*, 2: 1023–1025.

SÖHNGE, P. G., 1958. *The Environment of Copper Deposits in South-West Africa.* Commission Coopération Technique en Afrique, Léopoldville, pp.231–248.

SÖHNGE, P. G., 1961. The geology of the Tsumeb Mine. *Trans. Proc. Geol. Soc. S. Africa, Spec. Paper.*

SOREM, R. K. and CAMERON, E. N., 1958. Manganese ores of the Nsuta deposit, Ghana. *Econ. Geol.*, 53 (7): 927–928.

SOULÉ DE LAFONT, D., 1956. Le Précambrien moyen et supérieur de Bondokou, Côte d'Ivoire. *Gouvt. Gén. Afrique Occidentale Franç., Bull. Direc. Mines Géol.*, 22 (1956): 13–169.

SOUSA TORRES, A. i PIRES SOARES, J. M., 1946. Formações sedimentares de arquipélago de Cabo Verde. *Min. Colónias Junta Missões Geográf. Invest. Coloniais, Mem. Sér. Geol.*, 3: 1–396.

SOUTH AFRICA, DEPT. MIN. GEOL. SURV., 1959. *The Mineral Resources of the Union of South Africa.* Geol. Surv. S. Africa. Pretoria, 622 pp.

SOUTH AFRICAN MINING AND ENGINEERING YEARBOOK, 1961–1962. pp.1–813.

SOUTHERN RHODESIA, MINES DEPT., 1962. Chromium in Southern Rhodesia. *Rapport.*

SOUTHERN RHODESIA, MINING IN, 1962. *Annual Reports.*

SPALDING, J., 1932. The asbestos mine at Shabani, S. Rhodesia, 1935. *Trans. Inst. Mining Met.*, 41: 362–381.

SPENCE, W. I., 1961. The Messina copper mine. *Commonwealth Mining Met. Congr., 7th, Johannesburg, 1961, Trans.*, 1: 123–144.

SRIVASTAVU, S., 1961. Report on the mineral resources of Liberia. *Bur. Natl. Res. Surv. Liberia*, Typescript.

STAM, J. C., 1960. Some ore occurrences of the Mississippi Valley type in Equatorial Africa. *Econ. Geol.*, 55 (10): 1708–1715.

STANTON, W. I., KORPERSHOEK, H. R. e SCHERMERHORN, L. J. G., 1962. Carta geologica, notícia explicativa da Folha Sul B–33/0–1 (S. Salvador). *Angola, Serv. Geol. Minas, Bol.*, 1962: 1-20.

STAS, M., 1959. Contribution à l'étude géologique et minéralogique des bauxites du Nord-est du Mayumbe. *Acad. Roy. Sci. Coloniales (Brussels), Bull. Séances*, 1: 470–493.

STOCKLEY, G. M., 1935. Outline of the geology of the Musoma district. *Tanganyika Territ., Geol. Surv. Dept., Bull.*, 7: 1–64.

STOCKLEY, G. M., 1947. *Report on the Geology of Basutoland.* Comptroller of Stores. Maseru, pp.1–115.

STOCKLEY, G. M., 1948. Geology of the north, west and central Njombe district. *Tanganyika Territ., Geol. Surv. Dept., Bull.*, 18: 1–70.

STRAUSS, C. A., 1961a. The iron ore deposits in the Thabazimbi area, Transvaal. *Trans. Proc. Geol. Soc. S. Africa, Spec. Paper.*

STRAUSS, C. A., 1961b. The iron ore deposits in the Sishen area, Cape Province. *Trans. Proc. Geol. Soc. S. Africa, Spec. Paper.*

SWAZILAND, GEOLOGICAL SURVEY AND MINES DEPT., 1959–1963. *Annual Reports.*

SYMÉTAIN, 1959–1963. *Rapports et Comptes Annuels.*

SWIFT, W. H., 1961. An outline of the geology of Southern Rhodesia. *Southern Rhodesia Geol. Surv., Bull.*, 50: 1–73.

TAVERNER-SMITH, T., 1962. Recent exploration in the Kandabwe coal area. *Trans. Proc. Geol. Soc. S. Africa*, 61 (1958): 5–18.

TAYEB, G., 1956. Géologie et minéralisation du massif éruptif de Cavallo. *Bull. Sci. Écon. Bur. Rech. Minières Algérie*, 4: 1–7.

TAYLOR, J. H., 1954. The lead-zinc-vanadium deposits at Broken Hill. *Northern Rhodesia Geol. Surv. Rept.*, 4 (4): 335–365.

TEIXERA FAISCA, M. L., 1961. Sobre a exportaçao e produçao de minerios em Angola. *Angola, Serv. Geol. Minas Bol.*, 2: 59.

TENAILLE, M., 1951. *Carte Géologique du Bassin du Bas-Chélif.* SN REPAL, Algiers.

TENAILLE, M., NICOD, M. A. and DE SPENGLER, A., 1960. Petroleum exploration in the Senegal, Mauritanian and Ivory Coast coastal basins. *Am. Assoc. Petrol. Geologists, 45th Ann. meeting, Repts.*

THAYER, TH. P., 1953. The iron deposits of Western Liberia. *Congr. Géol. Intern., 19e, Compt. Rend., Algiers, 1952*, 10 (10): 47.

TOOMBS, R. B., 1962. A survey of the mineral industry of South Africa. *Can. Dept. Mines Tech. Surv., Mineral Resources Div., Mineral Inform. Bull.*, 58: 1–271.

TRANSVAAL AND ORANGE FREE STATE, CHAMBER OF MINES, 1962. Statistical suppl. Analysis of working results of gold and gold and uranium mining members. *Mining Survey*, pp.38–46.

TRANSVAAL AND ORANGE FREE STATE, CHAMBER OF MINES, 1963. *Annual Report.*

TRANSVAAL AND ORANGE FREE STATE, CHAMBER OF MINES, 1964. *Mining Survey, 1. Statistical Suppl.*, pp.29–36.

TREMBLAY, M., 1957. The Geology of the Williamson Diamond Mine. Ph. D. Thesis, McGill Univ., Montreal, Quebec.

TSUMEB CORP., STAFF, 1961. Geology, mining methods and metallurgical practice at Tsumeb. *Commonwealth Mining Met. Congr., 7th, Johannesburg, 1961, Trans.*, 1: 159–180.

TYNDALE-BISCOE, R., 1951. The geology of the Bikita tin-field. *Trans. Proc. Geol. Soc. S. Africa*, 54: 11–23.

TYNDALE-BISCOE, R., 1959. Alkali ring-complexes in Southern Rhodesia. *Intern. Geol. Congr., 20th, Mexico, 1956, Rept.*, pp.335–340.

TYNDALE-BISCOE, R. and STAGMAN, J. W., 1958. *Copper deposits in Southern Rhodesia.* Commission Coopération Technique en Afrique, Léopoldville, pp.181–196.

UGANDA MINES DEPT., 1960–1962. *Annual Report.*

UNION MINIÈRE DU HAUT-KATANGA, 1906–1956. *Reports.* Cuypers, Bruxelles, 265 pp.

UNION MINIÈRE DU HAUT-KATANGA, 1956. *Evolution des Techniques et des Activités sociales.* Cuypers, Bruxelles, 355 pp.

UNION MINIÈRE DU HAUT-KATANGA, 1960–1963. *Rapports Annuels.*

UNITED NATIONS DEPT. OF ECONOMIC AND SOCIAL AFFAIRS, 1959. *Economic Survey of Africa since 1950.* New York, N.Y., pp.1–248.

UNITED STATES BUREAU OF MINES, 1962. Mineral Facts and Problems. *U.S., Bur. Mines, Bull.*, 585 (1962): 1–1016.

UNITED STATES BUREAU OF MINES, 1962. *Minerals Yearbook*, Pittsburgh, Pa., 1: 531 pp., 2: 1206 pp., 3: 1410 pp.

UNITED STATES BUREAU OF MINES, 1964. *Commodity Data Summaries*, pp.1–168.

USONI, L., 1952. *Risorse minerarie dell'Africa orientale.* Ministerio Africa Italiana, Roma, 553 pp.

VAES, A., 1959. L'industrie minière du Congo belge et du Ruanda-Urundi en 1958. *Ann. Mines Belg.*, 11: 1092–1122.

VAN BILJOEN, W. J., 1959. *The Nature and Origin of the Chrysotile Asbestos Deposits in Swaziland and the Eastern Transvaal.* Ph. D. Thesis. Univ. Witwatersrand, pp.10–22; 123–124.

VAN BILJOEN, W. J., 1960. Summary from student's notes. *Havelock Asbestos Mines, Publ.*, pp.21–23.

VAN EEDEN, O. R., 1939. The mineral deposits of the Murchison Range, east of Leydsdorp. *Geol. Surv. S. Africa, Mem.*, 36: 1–152.

VAN STRATEN, O. J., 1960. A note on the chemical composition of some ground waters from the Bechuanaland Protectorate. *Records Geol. Surv. Bechuanaland, 1957–1958*, p.24.

VARLAMOFF, N., 1958. Les gisements de tungstène au Congo belge et au Ruanda-Burundi. *Acad. Roy. Sci. Coloniales (Brussels), Classe Sci. Nat.*, 82.

VARLAMOFF, N., 1960. Contribution à l'étude de la métallogénie des minéralisations wolframifères du Tibesti (République du Tchad). *Acad. Roy. Sci. Outre-Mer, Classe Sci. Nat., Bull.*, 12.

VARLEY, E. R., 1951. *Vermiculite.* H. M. Stationary Office, London, pp.1–70.

VARLEY, E. R., DU PREEZ, PH. D. J., PARKIN, E. A. and SCOTT, E. I. C., 1956. Nigerian diatomite. *Colonial Geol. Mineral Resources (Gt. Brit.), Suppl. Ser., Bull. Suppl.*, 2 (1956): 176–181.

VASSIVIÈRE, J. et FEILLÉ, G., 1961. L'aménagement hydroélectrique de la Chute d'Ayamé I. *Construction*, 16 (1): 28–35.

VERMAES, F. H., 1952. The amphibole asbestos of South Africa. *Trans. Proc. Geol. Soc. S. Africa*, 55: 199–230.

VERWOERD, W. J., 1957. The mineralogy and genesis of the lead–zinc–vanadium deposit of Abenab West in the Otavi Mts. *Ann. Univ. Stellenbosch, Ser. A*, 33 (1957): 235–319.

VIÉ, G., 1961. Aménagements hydroélectriques projetés à Madagascar. *Construction*, 16 (1): 25–27.

VINCIENNE, H., 1956. Observations géologiques sur quelques gîtes marocains de manganèse syngénétique. *Intern. Geol. Congr., 20th, Mexico, 1956, Rept. Symp. Yacimentos Manganes, II (Africa)*, pp.249–268.

VINCIENNE, H., 1957. Étude métallogénique du minéral du Guelb Moghrein à Akjoujt. *Gouvt. Gén. Afrique Occidentale Franç., Bull. Direc. Mines Géol.*, 20: 257–295.

VISSE, L., 1951. *Le Gisement de Phosphate de Chaux du Djebel Ongg.* Roneoscript, pp.1–158.

VISSE, L., 1952. Genèse des gîtes phosphatés du Sud-Est Algéro-Tunisien. *Congr. Géol. Intern., Compt. Rend., 19e, Algiers, Monogr.*, 1 (27): 1–58.

VISSER, D. J. L., 1958. The geology and mineral deposits of the Griquatown area. *Afr. Geol. Surv. S. Africa Explanatory Sheet*, pp.1–72.

VON KNORRING, O., 1960. Some geochemical aspects of a columbite-bearing granite from South-East Uganda. *Nature*, 188: 204.

WACRENIER, PH., 1961. Recherches de bauxite au Logone et au Mayo-Kébi. *Inst. Équatorial Rech. Études Géol. Minières*, 14: 37–42.

WAGNER, P. A., 1928. The iron deposits of the Union of South Africa. *Geol. Surv. S. Africa, Mem.*, 26: 1–268.

WAY, H. J. R., 1962. The results of eighteen years' organised geological work in Swaziland. *Swaziland, Geol. Surv. Mines Dept., Bull.*, 1: 4–13.

WELTER, C., 1960. Étude de la zone chromifère d'Ambodiriana. *Rappt. Ann. Serv. Géol., Direc. Mines Géol. Madagascar*, 1960: 119.

WHITTINGHAM, J. K., 1958. *The Geological Environment of Copper Deposits in Tanganyika.* Commission Coopération Technique en Afrique, Léopoldville, pp. 199–208.

WILES, J. W., 1957. The geology of the eastern portion of the Hartley gold belt. *Southern Rhodesia, Geol. Surv. Bull.*, 44 (2): 1–180.

WILES, J. W., 1961. Geology of the Miami mica field. *Southern Rhodesia, Geol. Surv. Bull.*, 51: 1–235.

WILLIAMS, F. A., 1956. The identification and valuation of decomposed columbite-bearing granites of the Jos-Bukuru younger granite complex. *Trans. Inst. Mining Met.*, 65 (11): 169–175, 563–564.

WILSON, N. W. and MARMO, V., 1958. Geology, geomorphology and mineral resources of the Sula Mts. *Geol. Surv. Sierra Leone, Bull.*, 1: 1–91.

WINFIELD, O., 1961. Chibuluma. In: F. MENDELSOHN (Editor), 1961. *The Geology of the Northern Rhodesian Copperbelt.* Macmillan, London, pp.328–342.

WOODTLI, R., 1959. Description de quelques sondages dans des itabirites du Nord-est du Congo belge. *Intern. Geol. Congr., 20th, Mexico, 1956, Assoc. Serv. Géol. Africains*, pp.469–476.

WOODTLI, R., 1961a. *L'Europe et l'Afrique.* Centre de Recherches européennes, Lausanne, pp.1–302.

WOODTLI, R., 1961b. Gold impregnation deposits in the Moto area (Central Africa). *Econ. Geol.*, 54: 603–607.

WOODTLI, R., 1961c. Relationship of general structure to gold mineralization in the Kilo area (Central Africa). *Econ. Geol.*, 56: 584–591.

WORLD OIL, 1962. *Africa.* 154 (8): 158–168.

WORST, B. G., 1956. The geology of the country between Belingwe and West Nicholson. *Southern Rhodesia, Geol. Surv. Bull.*, 43: 1–218.

WORST, B. G., 1960. The Great Dyke of Southern Rhodesia. *Southern Rhodesia, Geol. Surv. Bull.*, 47: 1–234.

WORST, B. G., 1961. Chromite in the Great Dyke of Southern Rhodesia. *Trans. Proc. Geol. Soc. S. Africa, Spec. Paper.*

WORST, B. G., 1962. The geology of the Buhwa iron ore deposits. *Southern Rhodesia, Geol. Surv. Bull.*, 53: 1–114.

REFERENCES

Index

Minerals are listed under the heading of the commodity derived from them (e.g., sphalerite under zinc, apatite under phosphates).
Lakes, Mounts, etc. are listed under the heading place name (e.g., Lake Victoria under Victoria, Lake).
Numbers given in parenthesis refer to illustrations.

B

[1] See in text under Madagascar.
[2] See in text under Egypt.

C

D

H

I

P

Q

R

T

U